OXFORD BROOKES UNIVERSITY
LIBRARY HEADINGTON

ONE WEEK LOAN

This book is issued for ONE WEEK only.
A fine will be levied for each day or part of a day overdue.
Books may be renewed by telephoning Oxford 483133
and quoting your borrower number.
You may not renew books that are reserved by another borrower.

11. OCT. 2002
12. NOV 2002
28. FEB 2003
09. MAY 2003
30. SEP 2003
10. OCT 2003
22 JAN 2004
23 FEB 2007
26 NOV 2007
23 FEB 2009

OXFORD BROOKES
UNIVERSITY
LIBRARY

00 697135 09

Petroleum Provinces
of the
Twenty-first Century

Edited by
Marlan W. Downey
Jack C. Threet
William A. Morgan

AAPG Memoir 74

Published by
The American Association of Petroleum Geologists
Tulsa, Oklahoma, U.S.A.
Printed in the U.S.A.

Copyright © 2001
The American Association of Petroleum Geologists
All Rights Reserved
Printed in the U.S.A.
Published December 2001

ISBN: 0-89181-355-1

AAPG grants permission for a single photocopy of an item from this publication for personal use. Authorization for additional copies of items from this publication for personal or internal use is granted by AAPG provided that the base fee of $3.50 per copy and $.50 per page is paid directly to the Copyright Clearance Center, 222 Rosewood Drive, Danvers, Massachusetts 01923 (phone: 978/750-8400). Fees are subject to change. Any form of electronic or digital scanning or other digital transformation of portions of this publication into computer-readable and/or transmittable form for personal or corporate use requires special permission from, and is subject to fee charges by, the AAPG.

Association Editor: Neil C. Hurley, 1997–2001; John C. Lorenz, 2001–2004
Technical Editor: William A. Morgan
Geoscience Director: Robert C. Millspaugh
Publications Manager: Kenneth M. Wolgemuth
Special Publications Editor: Hazel Rowena Mills
Copy Editor: Michael H. Blechner
Consulting Editor: Kathy A. Walker
Cover Design: Rusty Johnson
Production: ProType Inc., Tulsa, Oklahoma
Printing: The Covington Group, Kansas City, Missouri

ACC. NO. 69713509 FUND GEOL
LOC HW CATEGORY WEEK PRICE 2500
30 JUL 2002
CLASS No. 553.282 WAL
OXFORD BROOKES UNIVERSITY LIBRARY

The American Association of Petroleum Geologists (AAPG) does not endorse or recommend any products or services that may be cited, used or discussed in AAPG publications or in presentations at events associated with the AAPG.

This and other AAPG publications are available from:

The AAPG Bookstore
P.O. Box 979
Tulsa, OK 74101-0979
Telephone: 1-918-584-2555 or 1-800-364-AAPG (U.S.A.)
Fax: 1-918-560-2652 or 1-800-898-2274 (U.S.A.)
www.aapg.org

Canadian Society of Petroleum Geologists
No. 160, 540 Fifth Avenue S.W.
Calgary, Alberta T2P 0M2
Canada
Telephone: 1-403-264-5610
Fax: 1-403-264-5898
www.cspg.org

Geological Society Publishing House
Unit 7, Brassmill Enterprise Centre
Brassmill Lane, Bath BA13JN
U.K.
Telephone: +44-1225-445046
Fax: +44-1225-442836
www.geolsoc.org.uk

Affiliated East-West Press Private Ltd.
G-1/16 Ansari Road, Darya Ganj
New Delhi 110 002
India
Telephone: +91-11-3279113
Fax: +91-11-3260538
E-mail: affiliat@nda.vsnl.in

WALLACE EVERETTE PRATT

1885–1981

Wallace Everette Pratt, one of the most distinguished petroleum geologists of all time, was born in Phillipsburg, Kansas, on March 15, 1885, and died in Tucson, Arizona, on December 25, 1981.

After receiving three degrees from the University of Kansas, Pratt worked for the U.S. government as a geologist in the Philippines before returning to the United States to begin a career in petroleum geology. When Humble Oil and Refining Co. hired Pratt as its first geologist in 1918, it was a small, struggling company. When he retired for health reasons 27 years later, Pratt was vice president and a member of the executive committee of Standard Oil Company (N. J.), now ExxonMobil, which had been Humble's parent company since 1919. In those 27 years, Humble became the leading producer of oil in the United States and holder of the country's largest petroleum reserves. Humble's success was due largely to Wallace Pratt. Under his direction, the company pioneered many of the exploration methods and strategies that we now take for granted, including the use of micropaleontology, geophysical prospecting, and leasing of extensive blocks of undeveloped acreage.

Pratt's contributions to the geological community were at least as important as his role in his company. One of the founders in 1917 of the predecessor of AAPG, fourth president of AAPG (1920), and first recipient of AAPG's Sidney Powers Medal (1945) and Human Needs Award (1972), Pratt was untiring in his help and encouragement of geologists and oil and gas finders in general. Perhaps his most lasting contribution to his profession was his insistence on the importance of new ideas, faith, persistence, venturous spirit, and vision. Pratt's great 1952 paper, "Toward a philosophy of oil-finding" (*AAPG Bulletin*, v. 36, no. 12, p. 2231–2236 [reprinted in the *AAPG Bulletin*, v. 66, no. 9 (1982), p. 1417-1422]), summed it up best:

> Where oil is first found, in the final analysis, is in the minds of men. The undiscovered oil field exists only as an idea in the mind of some oil-finder. When no man any longer believes more oil is left to be found, no more oil fields will be discovered, but so long as a single oil-finder remains with a mental vision of a new oil field to cherish, along with freedom and incentive to explore, just so long new oil fields may continue to be discovered.

Dedication and Acknowledgments

As cochairmen of the Pratt II Conference, we and all who attended the conference are most pleased to dedicate AAPG Memoir 74, *Petroleum Provinces of the Twenty-first Century*, to Fred A. Dix Jr., retired executive director of AAPG and the AAPG Foundation. In his retirement, he has continued to serve the AAPG Foundation in many ways, including his key role in helping to plan, organize, and conduct the conference. For his specific contribution toward the success of the conference and his very accomplished leadership of AAPG for more than two decades, it is most appropriate to honor Fred Dix by dedicating this volume to him.

Marlan W. Downey
Jack C. Threet
Cochairmen
Wallace E. Pratt Memorial Conference on Petroleum Provinces of the Twenty-first Century

Fred A. Dix Jr.

AAPG wishes to thank the AAPG Foundation and its board of trustees for sponsorship of the Wallace E. Pratt Memorial Conference on Petroleum Provinces of the Twenty-first Century, held in San Diego, California, on January 12–15, 2000, and for publication of this book.

Members of the board of trustees at the time of the conference were Lawrence W. Funkhouser, chairman; John J. Amoruso; Paul H. Dudley Jr.; James A. Gibbs; Eugene F. Reid; and Jack C. Threet. On July 1, 2001, Threet succeeded to the chairmanship, Funkhouser was named chairman emeritus, and William Fisher became a new trustee.

The AAPG Foundation, established in 1967 by the AAPG leadership, carries out scientific, educational objectives through the use of charitable, tax-deductible gifts and contributions from AAPG members and corporate entities.

Contributions toward publication of this book are applied to production costs, thus directly reducing the book's purchase price and making the volume available to a larger readership.

About the Editors

Marlan W. Downey, a former president of the American Association of Petroleum Geologists, is senior fellow at the Institute for the Study of Earth and Man at Southern Methodist University in Dallas, Texas. He received a B.A. degree from Peru State College in Nebraska and B.S. and M.S. degrees in geology from the University of Nebraska. Downey worked for Shell Oil for 30 years, retiring in 1987 as president of Pecten International. He joined ARCO in 1987 and served successively as senior vice president of exploration, president, and special adviser to the board of ARCO International. Downey then moved to academia, becoming Bartell Professor of Geology at the University of Oklahoma and chief scientist of the university's Sarkeys Energy Center.

Downey has published dozens of articles on the subjects of seals for hydrocarbons, effective management, energy policy, and understanding risk. He was chairman of the first Hedberg Conference on Seals, the first Hedberg Conference on Understanding Exploration Risk, the Conference on Unconventional Methods of Exploration, and the AAPG Conference on a National Energy Policy. He was a cochairman of the Wallace E. Pratt Memorial Conference on Petroleum Provinces of the Twenty-first Century, held in 2000.

Jack C. Threet is president and CEO of Threet Energy, Inc., an independent oil and gas exploration company in Houston, Texas. A native of Illinois, Threet received a bachelor's degree in geology from the University of Illinois in 1951, after which he joined the exploration department of Shell Oil Company as a junior stratigrapher. In his 36-year career at Shell, he held numerous positions of increasing responsibility in exploration and production in the United States, Netherlands, Australia, and Canada. Threet culminated his Shell career in Houston, serving nine years as vice president and head of exploration. He retired from Shell in 1987, and in 1989, he joined the ranks of independent explorers.

Threet is an active member of the American Association of Petroleum Geologists and a trustee associate and chairman of the board of the AAPG Foundation. He is also a trustee of the American Geological Institute Foundation and past vice chairman of the Offshore Technology Conference. He was a cochairman of the Wallace E. Pratt Conference on Petroleum Provinces of the Twenty-first Century, held in 2000.

William A. Morgan is a senior explorationist in the Predictive Stratigraphy Group of Conoco Inc. in Houston, Texas. He received B.S. and M.S. degrees in geology from the University of Wisconsin, Madison. Since then, he has been employed for more than 20 years by Conoco, holding a variety of positions in research, exploration, and development. Morgan's primary technical interests lie in the stratigraphy, sedimentology, and diagenesis of hydrocarbon reservoirs, particularly those in carbonate successions. He has applied those interests to integrating core, well-log, and seismic data and developing sequence-stratigraphic/reservoir frameworks in many areas of the world.

Morgan has published several papers, principally on the stratigraphy and depositional facies of reservoirs in carbonate successions. He is a member of several geological societies and has been especially active in the American Association of Petroleum Geologists and the Society for Sedimentary Geology (SEPM), having held committee chairmanships in both organizations. He was chairman of AAPG's Grants-in-Aid Committee and received Certificates of Merit for cochairing the AAPG Summits on Committees. He is an associate editor of the *AAPG Bulletin* and is the SEPM technical program chairman for the 2002 AAPG/SEPM Annual Meeting.

Preface

In December 1984 in Phoenix, Arizona, the American Association of Petroleum Geologists Foundation sponsored the Wallace E. Pratt Memorial Conference on Future Petroleum Provinces of the World. This highly successful conference achieved its goals of (1) presenting expert opinions as to where and why large oil and gas fields would yet be found and (2) raising funds for the construction of the Pratt Tower building at AAPG Headquarters in Tulsa, Oklahoma. The papers from the conference were published as AAPG Memoir 40, *Future Petroleum Provinces of the World.* Sales of 3500 copies testify to its usefulness.

Nearly 15 years later, Fred Dix, longtime executive director of the AAPG and the AAPG Foundation and then executive director emeritus, properly sensed that it would be timely at the start of the new millennium to hold a second Pratt Conference. He suggested the title Petroleum Provinces of the Twenty-first Century, or Pratt II for short.

The AAPG Foundation trustees quickly endorsed the concept, asked us to be cochairmen, and selected January 12–15, 2000, and San Diego, California, as the date and place.

To make the conference both interesting and authoritative, we chose to invite from around the world highly regarded explorationists whom we believed had the credentials to tell us *where* and show us *why* major accumulations of oil and gas would be found in their particular provinces of interest.

A few provinces that we thought were important are not included because we lacked willing experts to share their views with a public audience. We also regret that several papers were withdrawn at the last moment because permission to publish was not granted by the countries involved.

Michel T. Halbouty, who was general chairman of the original Pratt Conference, in 1984, graciously agreed to deliver the keynote address at Pratt II. In typical Halbouty fashion, Mike, 90 years young, opened the conference with a stirring speech expressing his view that huge quantities of undiscovered hydrocarbons exist throughout the world. He chided prognosticators who contend that sunset is already upon our petroleum industry. His speech, presented as the first paper in this volume, set the tone for the conference. As speaker after speaker came to the podium to give expert assessments of remaining discovery potential in their areas of interest, it became obvious that they agreed that there are enormous quantities of oil and gas yet to be found and profitably produced throughout the world, especially with the advent of new technology. There will continue to be challenging and rewarding work for thousands of petroleum geologists, geophysicists, and engineers for a long, long time. We are convinced that all 300 registrants at Pratt II shared that view and returned home invigorated and full of exploration ideas useful in this new millennium.

We are proud to have cochaired the Pratt II Conference, and we are pleased to share with you the edited proceedings recorded in this Memoir.

We especially commend to your attention the paper (originally a luncheon address) by Peter Gaffney. He provides a stimulating, lighthearted, but painfully serious account of the realities of today's exploration world with budgets driven by analysts, costs driven by pseudotechnology, and corporate actions driven by investment bankers.

We sincerely thank all the authors whose excellent papers were instrumental in the success of this conference, as well as all the registrants who supported the conference with their attendance, their comments, and their financial contributions to the AAPG Foundation. We also thank the staff at AAPG and the AAPG Foundation for their key roles in conducting this conference. The volume benefited greatly from the wise editorial review of William A. Morgan. Finally, we extend special thanks to Fred Dix and Larry Funkhouser for their vision and tireless dedication to this conference.

Marlan W. Downey
Jack C. Threet
Cochairmen
Wallace E. Pratt Memorial Conference on Petroleum Provinces of the Twenty-first Century

Table of Contents

Dedication and Acknowledgments iv

About the Editors v

Preface vi

Petroleum Resources of the Twenty-first Century

Chapter 1
Exploration into the New Millennium 11
Michel T. Halbouty

Chapter 2
Twenty-first-century Energy: Decline of Fossil Fuel, Increase of Renewable Nonpolluting Energy Sources 21
John D. Edwards

Chapter 3
Discoveries of the 1990s: Were They Significant? 35
Robert Esser

Chapter 4
Oil: Are We Running Out? 45
David Deming

Chapter 5
Exploration: Tomorrow's Charge, Tomorrow's Challenge 57
Peter D. Gaffney

North America and the United Kingdom

Chapter 6
New Oil in Old Places: The Value of Mature-field Redevelopment 63
Robert M. Sneider and John S. Sneider

Chapter 7
Natural-gas Hydrates: Resource of the Twenty-first Century? 85
Timothy S. Collett

Chapter 8
Petroleum Resources of Canada in the Twenty-first Century 109
Keith Skipper

Chapter 9
Alaska: A Twenty-first-century Petroleum Province 137
Kenneth J. Bird

Chapter 10
Hydrocarbon Exploration Opportunities in the Twenty-first Century in the United Kingdom 167
J. R. V. Brooks, S. J. Stoker, and T. D. J. Cameron

Chapter 11
Exploration Opportunities in the Greater Rocky Mountain Region, U.S.A. 201
Fred F. Meissner and M. Ray Thomasson

Asia

Chapter 12
Future Hydrocarbon Potential of Kazakhstan 243
Igor Effimoff

Chapter 13
New Frontiers for Hydrocarbon Production in the Timan-Pechora Basin, Russia 259
Bret J. Fossum, William J. Schmidt, David A. Jenkins, Vladimir I. Bogatsky, and Boris I. Rappoport

Chapter 14
Future Petroleum Production from Indonesia and Papua New Guinea 281
J. V. C. Howes

Chapter 15
Australian Petroleum Provinces of the Twenty-first Century 287
Ian M. Longley, Marita T. Bradshaw, and John Hebberger

Chapter 16
Geologic Challenges of Exploration: Onshore China, with Special Focus on the Tarim and Junggar Basins 319
Barry J. Katz

Central and South America

Chapter 17
The Gulf of Mexico Basin South of the Border: *The* Petroleum Province of the Twenty-first Century 337
Alfredo E. Guzmán and Benjamín Márquez-Domínguez

Chapter 18
Future Petroliferous Provinces of Venezuela 353
F. E. Audemard and I. C. Serrano

Chapter 19
Present and Future Petroleum Provinces of Southern South America 373
Carlos M. Urien

Chapter 20
Foz do Amazonas Area: The Last Frontier for Elephant Hydrocarbon Accumulations in the South Atlantic Realm 403
M. R. Mello, R. Mosmann, S. R. P. Silva, R. R. Maciel, and F. P. Miranda

The Middle East and Africa

Chapter 21
Major Hydrocarbon Potential in Iran 417
Porter L. Versfelt Jr.

Chapter 22
Libya: Petroleum Potential of the Underexplored Basin Centers—A Twenty-first-century Challenge 429
Donald C. Rusk

Chapter 23
The Petroleum Potential of Egypt 453
John C. Dolson, Mark V. Shann, Sayed Matbouly, Colin Harwood, Rashed Rashed, and Hussein Hammouda

Chapter 24
Paleozoic Stratigraphy and Hydrocarbon Habitat of the Arabian Plate 483
G. Konert, A. M. Afifi, S. A. Al-Hajri, K. de Groot, A. A. Al Naim, and H. J. Droste

Chapter 25
Lower Congo Basin, Deep-water Exploration Province, Offshore West Africa 517
J. Leite Da Costa, T. W. Schirmer, and B. R. Laws

Chapter 26
Hydrocarbon Exploration in the Berkine Basin, Grand Erg Oriental, Algeria 531
Michael D. Cochran and Lee E. Petersen

Index 561

Petroleum Resources of the Twenty-first Century

Halbouty, M. T., 2001, Exploration into the new millennium, *in* M. W. Downey, J. C. Threet, and W. A. Morgan, eds., Petroleum provinces of the twenty-first century: AAPG Memoir 74, p. 11–19.

Chapter 1

Exploration into the New Millennium

Michel T. Halbouty
Michel T. Halbouty Energy Co., Houston, Texas, U.S.A.

I am entering my 71st year as a member of our profession and as an explorer for petroleum. During those 71 years, I have devoted much of my time and resources to activities pertinent to our profession and our Association.

As I venture forth in the last few golden years of my life, I find that the professions in the geosciences have undergone many changes, particularly in the last 20 years—some good, some very bad—but the professions have always looked to the future—and that is the reason this conference was convened at the beginning of this millennium.

I am indeed honored—and yes, very pleased—to have been asked to present the keynote address to this second Pratt Conference. The first was held in Phoenix, Arizona, December 2–5, 1984, and was called the Wallace E. Pratt Memorial Conference on Future Provinces of the World. I had the pleasure of presenting the opening address at that meeting on December 2, 1984 (Halbouty, 1984a).

That was the best conference of its kind ever held by our auspicious Association. I am honored to have had the privilege of chairing it and participating in it as a speaker. Fred Dix, who was Executive Director of AAPG, and Bill St. John, who was program chairman, contributed greatly to the success of the conference.

The papers from that conference were compiled in a volume as AAPG Memoir 40 (Halbouty, 1984b) and published less than one year after the conference. That book sold out quicker than any other AAPG publication. I tell you all of this with some braggadocio, because at 90½ years old, I am very proud of that conference, that publication, and what it achieved.

CONTRIBUTIONS OF WALLACE E. PRATT

Because this is the second Wallace E. Pratt Memorial Conference on Future Petroleum Provinces of the World, I think that it is proper to say something about the man whose name was attached to the first—and now to this second—conference.

There are not many of us left who knew Wallace Pratt. I look over the young faces in this room and I feel compelled to say something about Wallace that really sets the theme and the full essence of this conference named in his memory.

Pratt was born in Phillipsburg, Kansas, on March 15, 1885, and died on December 25, 1981, at the age of 96. He held multiple degrees from the University of Kansas.

Wallace Pratt was noted as one of the world's most eminent geologists, and he recognized early in his prolific career that the most valuable natural resource in the world is the human mind. He concluded one of his most memorable papers with the words, "Oil is found in the minds of men." Pratt firmly believed that without the proper use of the ultimate primary exploration tool—the mind—all other "tools," from whatever source, were inconsequential.

He was a founding member of AAPG and was its fourth president. In 1945, AAPG honored him with the first Sidney Powers Memorial Medal. Twenty-seven years later, in 1972, he was awarded the first AAPG Human Needs Award. Pratt and I were very close friends for over 40 years, and it was indeed a privilege and an honor for me when he asked me to be his citationist for the award.

His acceptance speech of the Human Needs Award was filled with the most beautiful of philosophic phraseology. One of the best was: "Earthly life is the only life in the universe known to man—so far! And Mother Earth is the only possible abode for life known to man—so far! There are within us the same chemical elements that make up the mountains, the pine trees, the seashore. We are indeed of the earth—brothers of the boulders, cousins of the clouds, and distant kin, by way of chemical tie-up, of the fossil plants and animals."

So it was that Pratt was a living symbol of the essence of geology. His creed was the science, and he lived as a student and teacher of it. His contributions to the science of geology will endure as long as the science of geology is practiced. We honor ourselves by naming this conference for Wallace E. Pratt—one of the all-time great geoscientists.

PETROLEUM INDUSTRY HISTORICAL PERSPECTIVE

The world's energy requirements are so interconnected today that the development and production of future petroleum supplies in all countries must be considered jointly.

In the industrialized, developed countries of the world, science and technology are rapidly advancing toward the best exploitation of those countries' remaining hydrocarbon reserves and potential. In the developing countries, exploration for commercial oil and gas deposits is intensifying.

Even though most of the world has begun a transition from conventional oil and gas energy toward alternative energy sources, it is evident that the major means of world energy survival will depend on oil and natural-gas supplies for most of the next century. Oil exploration has followed a very logical and economic pattern, which is to say that it has been carried on in those areas offering the greatest reward. This means the areas where there is the most favorable geology; where the chances are best of finding the most oil most cheaply; where it can be produced and marketed at the lowest cost; where the markets are the strongest, the prices the highest; and where stable political and economic conditions can be expected.

From the beginning of modern oil exploration until the 1930s, geomorphology and surface geology were the evidence sought for locating likely spots to drill—oil and gas seeps, areas where the sediments could be seen dipping away in all directions and forming a structural dome, suspicious mounds or hills in otherwise flat lands, and curvature of rivers and creeks around areas of higher elevation.

With the advent of geophysics and technology, exploration took a different course, and subsurface imaging controlled where wells would be drilled. Consequently, most of the easy-to-find accumulations of petroleum worldwide have been found (Figures 1, 2). This is not an original statement. You have heard it or read it on many occasions in recent years. As a reflection of this situation, statistics during the past decades show that worldwide discovery successes have been declining, not only in numbers, but also in quality or economic worth.

A Petroconsultants' report (1998) covering the decade of the 1990s noted that less than half the world's production has been replaced by new discoveries. It also reported that during the previous five years, the replacement ratio dropped to just over 40%. Meanwhile, gas discoveries are nearly keeping up with production.

Petroconsultants (1998) also noted that a poll of more than 100 internationally active companies rated the United Kingdom, followed by Australia, as the most attractive place on earth for future exploration and production investments. This is in contrast to the previous five years, in which Africa led the world in terms of both new oil discoveries and reserves replacement.

FUTURE DIRECTION

It is obvious that the thrust of future exploration will not be in the United States, because very little exploration has taken place in the last decade. Oil production in the United States has fallen at an average rate of 240,000 barrels per day per year. This accentuates the reason we are importing more oil each year. Our consumption increases and our production declines, necessitating greater imports. During the early years of the petroleum industry, the independents were its backbone, and they drilled more and found more oil and gas than all the majors combined—but this is no more. Recently, the independents have shown an uninspiring performance in exploration and development. In 1999, the accelerated drop in oil prices took a heavy toll among them, which drastically slowed exploration. The changes in the industry in just the last five years have had desultory effects on the independent. The "megamajors" rule the roost in every factor of the industry. Overall, drilling activity in the United States is near record low rates, whereas in foreign lands, exploration activities are increasing.

Worldwide, explorationists are still confining most of their exploratory efforts to searching for the obvious types of reservoirs, which are becoming scarce. To a large degree, this approach can be attributed to a generally lazy attitude in exploratory thinking which motivates us to

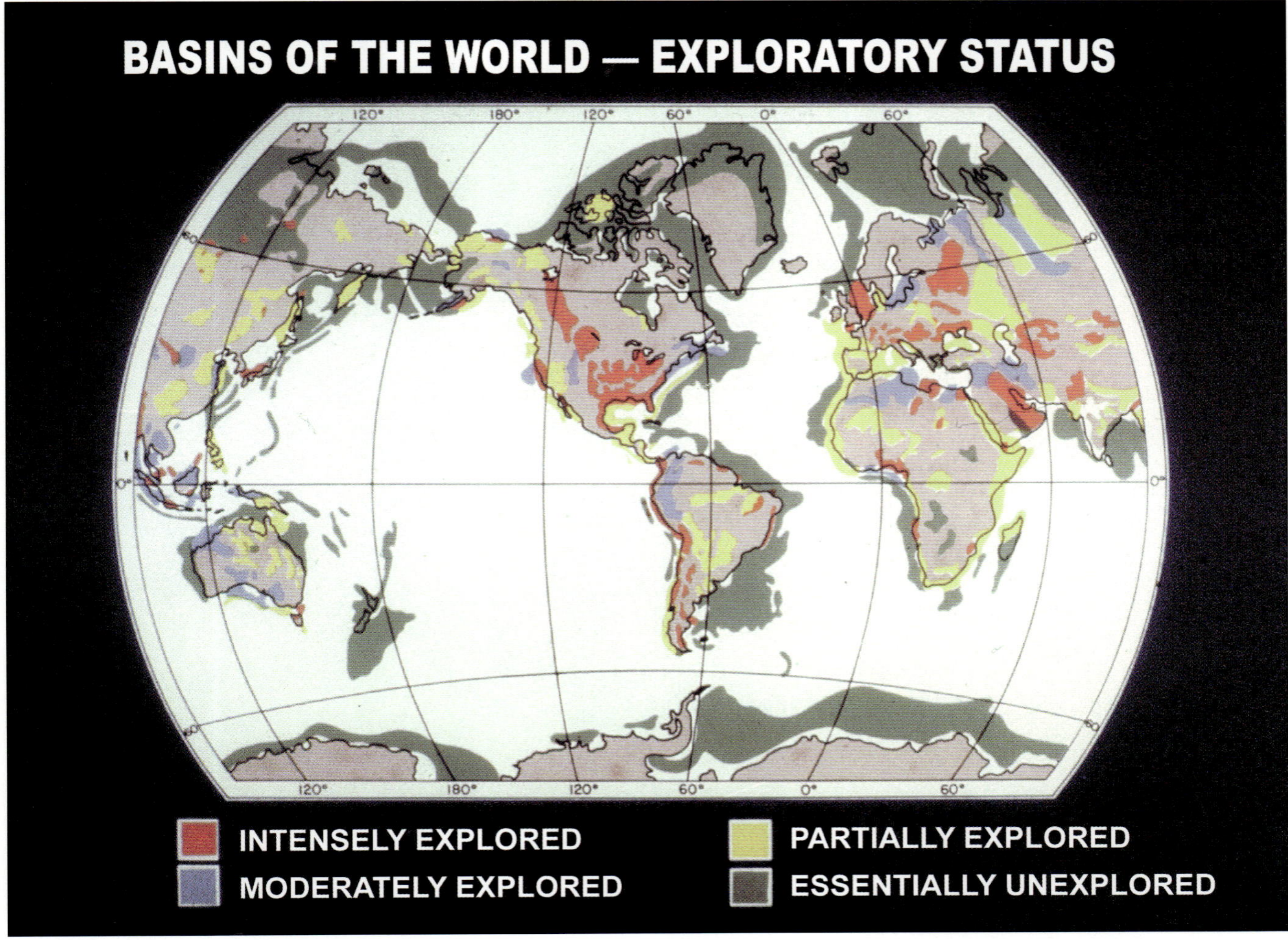

Figure 1. Petroleum exploratory status of the basins of the world (after St. John, 1980).

continue looking for structures which can be seismically imaged, thereby requiring very little cranium effort.

Psychologically, explorationists are pressured to look for anticlines, domes, and fault structures because this type of prospect is most likely to be acceptable and salable to management. In other words, we geologists and geophysicists have made little effort to *purposefully* search for the traps which are obscure, and we just continue to look for the salable anomaly which can be detected easily by seismology.

I certainly am not advocating that we should turn our backs on the obvious play, but I am strongly suggesting that we must diversify and concentrate a substantial part of the exploratory effort on looking for traps that are subtle and, consequently, are most difficult to find (Figure 3). This is discussed in more detail below.

Nevertheless, a vast petroleum reserve potential exists in the world (SEG, 1982; Halbouty, 1984c). As part of a unified successful exploration effort, each specialized geoscientific discipline, whether it is geology, geophysics, or petroleum engineering, must provide the bold and innovative thinking which will lead to future discoveries of oil and gas.

For many decades, there has been a continuing worldwide search for new reserves of petroleum. Yet there remain scores of geologically attractive areas which need to be extensively explored to meet the soaring demand for petroleum supplies (Figure 1).

Figure 2 shows the status of exploration in the sedimentary basins of the world. For brevity, many of the basins have been combined. Of the world's prospective sedimentary basins categorized as intensely explored, moderately explored, partially explored, and essentially unexplored, only 29% currently produce commercial hydrocarbons. Another 39% of the basins have been partially and moderately explored and tested, but have not yielded large quantities of commercial production. The remaining 32% of the world's basins have had little or essentially no exploration activity.

Onshore basins hold high potential in many parts of the world, especially in the more mature productive areas (Halbouty et al., 1970). Geologists and geophysicists have good opportunities to find additional petroleum reserves by using advanced technology. Three-dimensional seismic investigations, reprocessing

Figure 2. Sedimentary petroleum basins currently productive and potentially productive.

old seismic data, and the use of remote-sensing images from satellites and spacecraft are some of the methods they use to find clues for discovering "new" reserves in "old" areas.

The majority of the unexplored basins is considered to be the frontier areas of the world. These basins are located in harsh physical environments such as the Arctic, in deep water, or in remote interiors of continents. Many of these frontier regions are restricted because of disputes, political boundaries, or governmental regulations. Others have been bypassed because of a combination of remote location and low geologic potential.

Figure 2 shows the basins of the world, both those currently productive (red) and the potential areas for future exploration (purple). There is purple both onshore and offshore, but take a long look at the purple offshore. This is where the future high-cost exploration will take place and where large reserves of oil and gas will be discovered. Considering the vast offshore areas to be explored, it will take many decades for most of the areas to be evaluated. Purple covers a large part of the earth, and it is expected that the large megacorporations will be the explorers.

More and more attention will be paid in coming decades to the world's challenging offshore petroleum frontiers because, in today's offshore drilling, neither depth nor harsh environmental condition can be considered impossible (Hedberg, 1981). As explorationists look further to the oceans for major new reserves, especially to deeper waters and hostile environments such as the arctic regions, engineers are diligently working on designs of new drilling rigs and production platforms to meet the challenges thrust upon them (Harrison, 1979).

As long as there are inspiration, courage, innovative skills, and financial backing, there is no limit to what the global petroleum industry can accomplish—offshore and onshore. Forty years ago, it was felt that the entire North Sea was out of the question for exploration and production of petroleum because of the extremely hostile environment. Yet since 1969, when the first commercial discovery was made in the North Sea, the numerous oil and gas finds there have made that part of the world practically self-sufficient in petroleum energy.

As recently as 25 years ago, drilling in 200 to 300 feet of water was considered too hazardous and difficult. Production in over 1000 feet of water from a conventional platform has proven to be very common, whereas less than a decade ago, it was perceived to be quite dangerous. Today, production from 1000 feet of water is considered to be relatively shallow.

Drilling in water depths of 10,000 feet and beyond may be achievable in less than five years. Drilling water-depth records keep falling as exploration technology advances. The present record, set in the Gulf of Mexico, is 7 718 feet, drilled by Chevron with Global Marine's Glomar Explorer rig. The average depth for most offshore Gulf of Mexico wells is 475 feet.

Engineering and technological breakthroughs are fast opening vast areas heretofore considered almost impossible for exploration, such as the deeper portions of the Gulf of Mexico, the Mediterranean Sea, and the South China Sea, to name only a few.

SEARCHING FOR SUBTLE TRAPS

It is my opinion that there have been more new conceptual ideas in geoscience in the last 10 years than were promulgated in the preceding 50 years. The next few years will contribute even more breakthroughs to solving today's so-called impossible problems. In this regard, it is also interesting to note that explorationists pay too little attention to the deliberate search for those reserves which lie in subtle traps. We must turn more of our studies toward those not-so-obvious reservoirs of petroleum—those in stratigraphic traps, those lying below unconformities, those which are associated with buried geomorphological features, and those which may or may not be associated with structure (Figure 3).

The huge oil accumulation in the East Texas field is contained in a simple stratigraphic trap which occurs where the truncated edge of the Woodbine Sand crosses regional nosing on the west flank of the Sabine uplift (Halbouty and Halbouty, 1982) (Figure 4).

The size of the East Texas field is shown in Figure 5. Notice that the Woodbine Sand does not extend beyond the field (Figure 4). The pinch-out cuts off the sand so completely that no Woodbine oil escaped to the east.

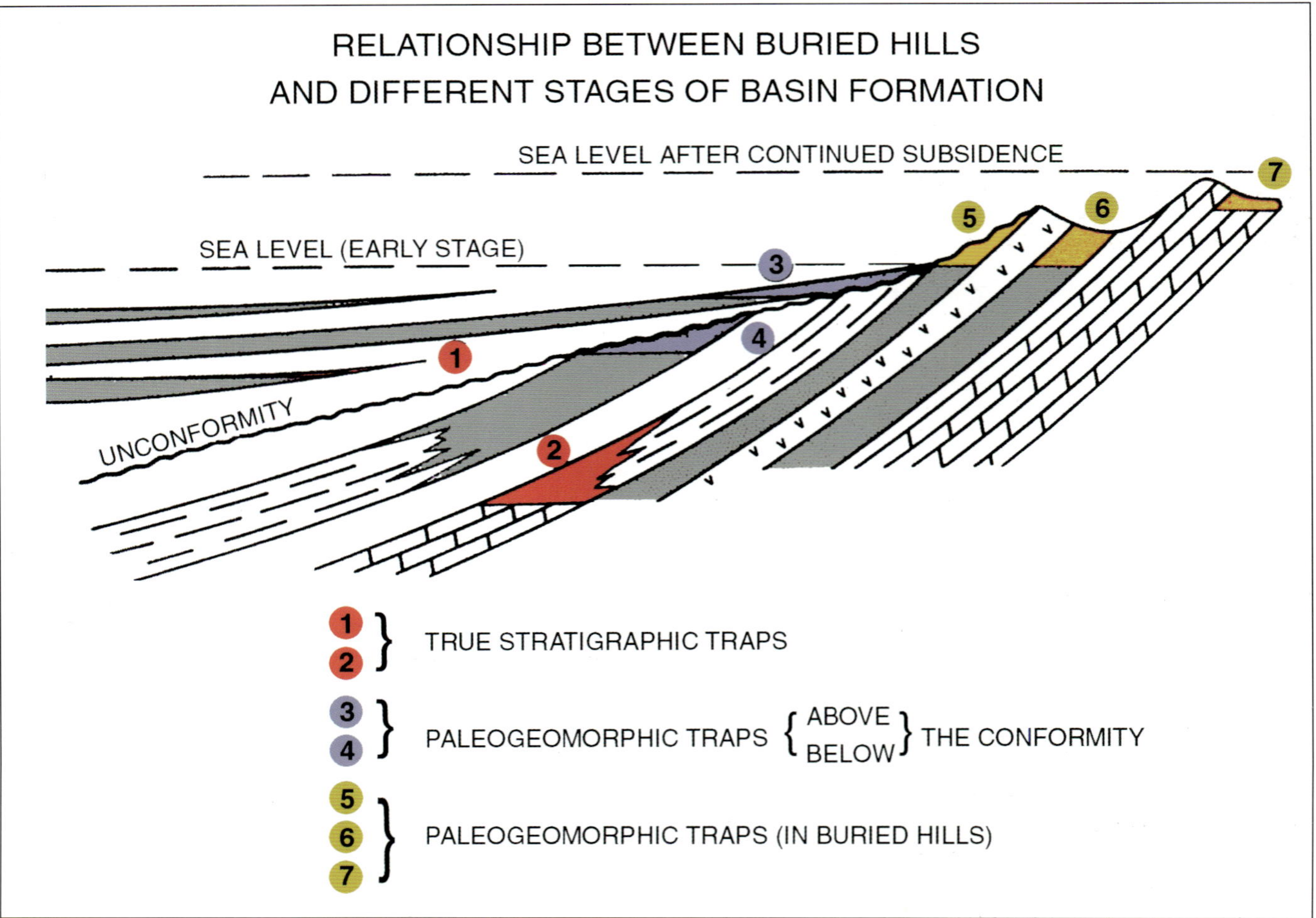

Figure 3. Subtle traps (after Martin, 1966).

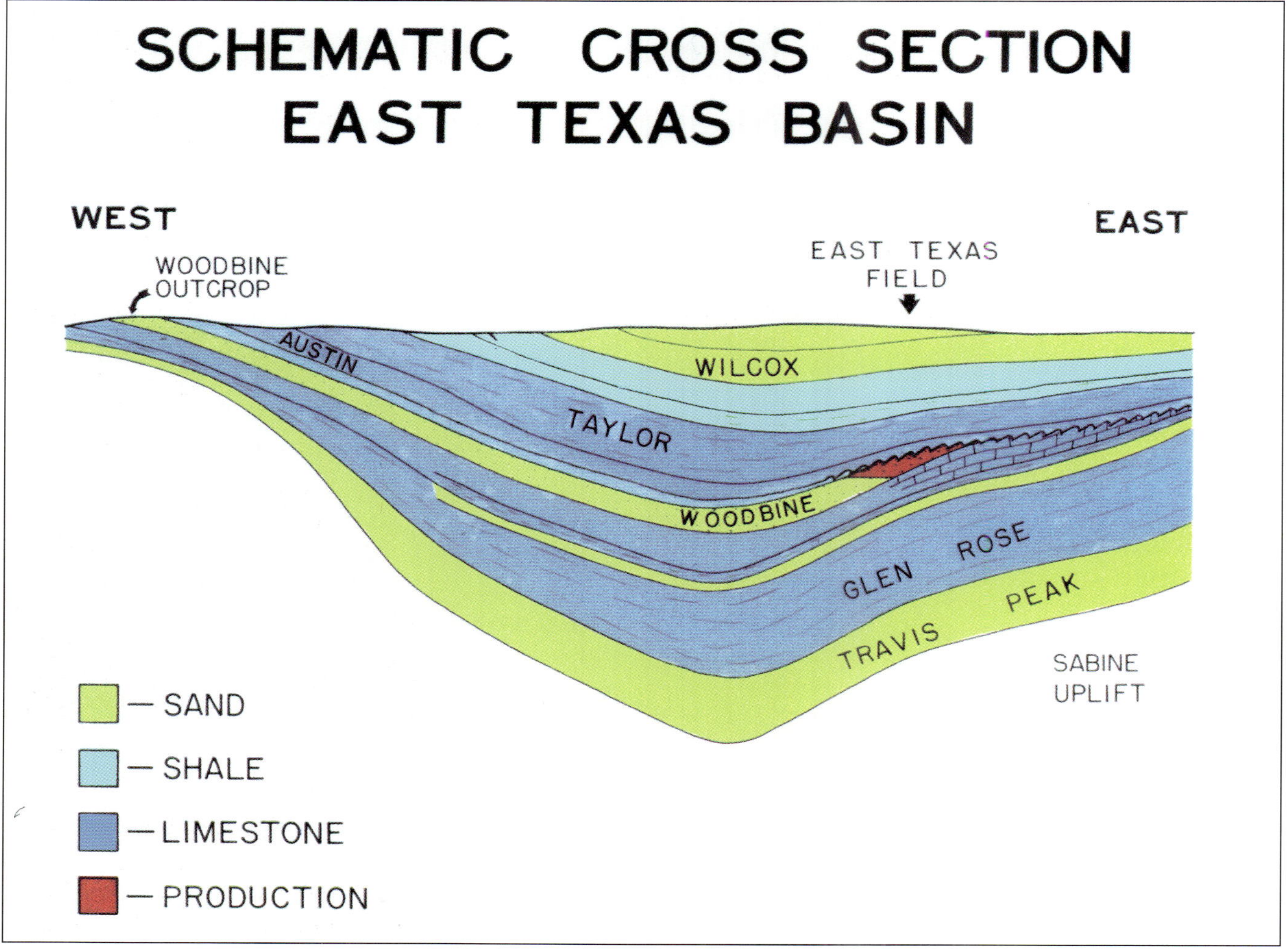

Figure 4. Generalized cross section, East Texas Basin, showing pinch-out of eroded Woodbine sandstone between Austin and Washita limestones, forming the trap for East Texas field on the western flank of the Sabine uplift. Note the outcrop of Woodbine beds west of the East Texas Basin.

This is the perfect stratigraphic trap, which accumulated 7 billion barrels of oil at shallow depths. Prior to its discovery, every major company and many independents, with their expertise and seismic black boxes, ran their equipment repeatedly over the area and agreed that it did not have the parameters for the accumulation of oil.

Isn't it possible that along the flanks of some monoclinal or uplift areas, which shows no elliptical closure, there might be another East Texas type of field waiting to be discovered by an ingenious geoscientist who thinks differently or by another Dad Joiner who happens to be a dreamer—but a real believer?

The geophysicists' computers and computer programs are tremendous aids in developing some of the scientific facts needed for finding subtle traps. Great advances have been stimulated by the use of seismic modeling, better data acquisition and data processing, and the extensive use of color seismic display formats. Mapping and discovering the elusive traps will require highly imaginative thinking combined with every applicable scientific discipline and tools—but the most important tool is the mind. It supersedes all the others.

Seismic control and interpretation can materially assist in locating the subtle trap, but geophysics is not the full answer. What is needed more than any other facet is the hard work of "pure geology" to ferret out the source and kind of sediments, the facies changes, paleogeologic history, tectonic effect on sedimentation, and the various kinds of environmental conditions which were present during deposition. That requires detailed study of stratigraphy, biostratigraphy, paleogeomorphology, paleogeography, paleostructure, and paleontology. That takes hard geological probing, but the resulting discoveries will warrant the effort.

I contend that the future large reserves will be found in subtle traps, and the explorationists must put their efforts toward them and not around them. These traps represent the last domain of petroleum exploration.

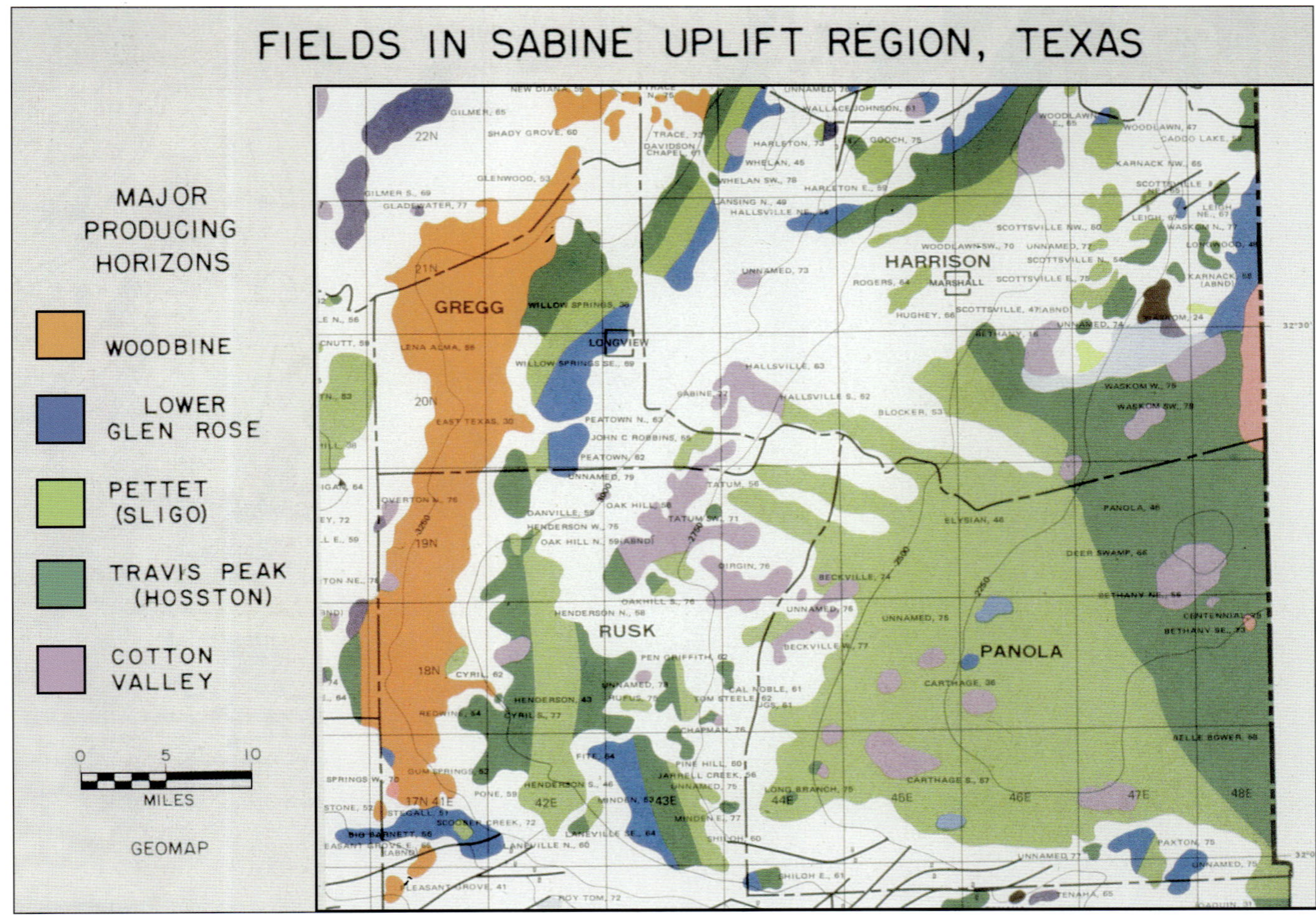

Figure 5. Productive areas east of East Texas field. The stratigraphic trap is considered to be one of the largest accumulations of oil in a single pinch-out.

FUTURE SUPPLIES OF OIL AND GAS

It has been predicted over and over for decades that the world would run out of oil and gas. Those pessimistic doomsayers have consistently been proven wrong. New oil and gas fields are being discovered almost daily, all over the world, in both known producing areas and in the frontier regions onshore and offshore.

Geoscientists are limited only by their imagination, innovation, and determination. In the coming decades, there will be tremendous strides in petroleum geology, geophysics, and petroleum engineering. The challenge for all of us—whether we are geologists, geophysicists, engineers, independent explorationists, or company or government explorationists—is to devise new concepts and skills to explore in areas long considered to be out of the question or impossible.

Look what industry has accomplished in the heretofore "impossible" arctic regions! Success in discovering, producing, and transporting oil in such a harsh environment has been phenomenal. The circumarctic region as a whole covers a vast territory. Many researchers feel that in the combined total arctic basins, there is the potential for accumulations of oil and gas that may equal those of the Middle East.

Exploration in arctic basins represents the most costly search ever, but the chances of discovering giant accumulations are excellent. Along the United States and Canadian continental shelves, there are seismic indications of thick layers of sedimentary rock with trapping features that could hold oil and gas accumulations equal to those already found in onshore fields. Prudhoe Bay and Kuparuk fields in Alaska and Urengoy field in Russia are prime examples of the giant field potential of the region (Carmalt and St. John, 1984).

There is another "impossible" area where reservoirs of petroleum eventually will be found if drilling is ever conducted on this remote continent—Antarctica, the last true frontier resource area on the earth. Its continental margins occupy an area roughly similar to the continental margins of North America and are composed of sediment wedges many kilometers thick.

Figure 6 shows a map of Antarctica with a map of Texas superimposed on it to illustrate the size of the con-

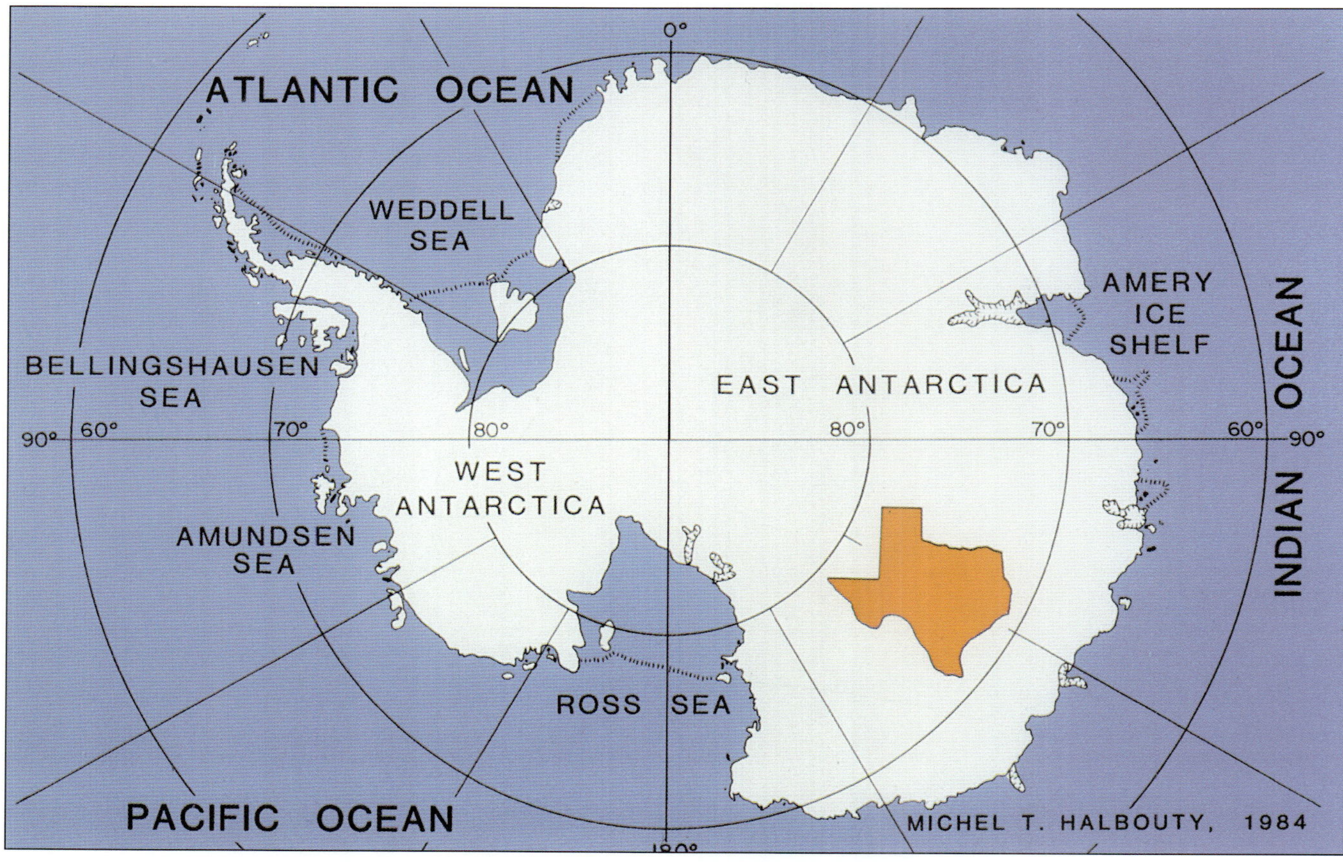

Figure 6. Size comparison of Antarctica and Texas.

tinent. Geoscientific studies and surveys have indicated that a sizable potential for petroleum exploration exists in the immediate offshore areas.

In his paper, "Antarctica—Geology and hydrocarbon potential" (St. John, 1984), Bill St. John wrote, "Based on limited data, geologists have identified 21 sedimentary basins for the Antarctic and immediately adjacent areas. These include 6 onshore, subglacial basins and 15 offshore basins. Excluding 11 basins considered to have little, or no potential, the other 10 basins contain an estimated 16.9 million cubic kilometers (4.05 million cubic miles) of sediment having a potential hydrocarbon yield of 203 billion barrels oil equivalent." St. John ends his excellent paper by stating, "Antarctica does have potential for large hydrocarbon reserves, and the technical expertise, ultimately, is available. The sole hindrance is jurisdiction—from whom do we get a drilling permit?"

When it becomes necessary to drill in harsh environments such as Antarctica, the cost to produce oil is very high, but if the world still demands the use of oil, those hostile places will be drilled and the high prices to use that oil will be paid accordingly. Antarctica is the epitome of the definition of a petroleum frontier. When we scan the petroleum potential areas of the world, it is obvious that as long as the desire exists to find new supplies of oil and gas (as it will for years and years to come), there will be an area in the world worth the risk of exploration.

In other words, the end of the oil era will be affected solely by economics. When the cost of finding oil and gas is absolutely prohibitive, exploration will stop and the beginning of the postpetroleum era will begin.

In my seven decades of petroleum exploration, I have observed that whenever substantial reserves of oil and gas have been discovered, they have always been brought to market from the hostile areas of the world. I am optimistic. I believe that as long as it is economical and as long as the geoscience professions, together with the petroleum engineers, increase their technology and methodology of searching and producing, discoveries will be made, even in the harshest of environments, such as the deep oceans, swamps, arctic ice, mountains, deserts, forests, jungles and—who knows—maybe many years from now, even in space.

As one familiar with the world's geologic potential, I firmly believe that in the future, at least as much oil and substantially more gas will be found than we have produced to date. Our giant fields of today were once the frontiers of yesteryear, and surely our continually advancing exploration concepts, methods, and technology will convert the frontier areas of today into the giants of tomorrow. The second Wallace E. Pratt Conference on Future Petroleum Provinces of the World and this resulting publication add immeasurably to the wealth of knowledge that men such as Wallace Pratt left as a legacy to us all.

REFERENCES CITED

Carmalt, S. W., and J. B. St. John, 1984, Giant oil and gas fields, 1984, *in* M. T. Halbouty, ed., Future petroleum provinces of the world: AAPG Memoir 40, p. 12–53.

Halbouty, M. T., 1984a, Basins and new frontiers: An overview, *in* M. T. Halbouty, ed., Future petroleum provinces of the world: AAPG Memoir 40, p. 1–10.

Halbouty, M. T., ed., 1984b, Future petroleum provinces of the world: AAPG Memoir 40, 708 p.

Halbouty, M. T., 1984c, Reserves of natural gas outside the Communist Block countries, *in* Proceedings of the 11th World Petroleum Congress: London, John Wiley & Sons, Ltd., v. 2, p. 1281–1292.

Halbouty, M. T., and J. J. Halbouty, 1982, Relationship between East Texas field region and Sabine uplift in Texas: AAPG Bulletin, v. 66, p. 1042–1054.

Halbouty, M. T., A. A. Meyerhoff, R. E. King, R. H. Dott Sr., H. D. Klemme, and T. Shabad, 1970, World's giant oil and gas fields, geologic factors affecting their formation, and basin classification, Part 1, *in* M. T. Halbouty, ed., Geology of giant petroleum fields: AAPG Memoir 14, p. 502–528.

Harrison, G. R., 1979, Exploratory drilling, the polar challenge, *in* Proceedings of the 10th World Petroleum Congress: London, Heyden & Sons, v. 2, p. 243.

Hedberg, H. D., 1981, Hydrocarbon resources beneath the oceans, *in* J. J. Mason, ed., Petroleum geology in China: Tulsa, PennWell Publishing Company, p. 249–253.

Martin, R., 1966, Paleogeomorphology and its application to exploration for oil and gas (with examples from western Canada): AAPG Bulletin, v. 50, p. 2277–2311.

Petroconsultants, 1998, Oil future less bright than gas: First Break, v. 16, p. 241–242.

Society of Exploration Geophysicists, 1982, Giants still exist in Canada: The Leading Edge, v. 1, n. 4, p. 26–47.

St. John, B., 1980, Sedimentary basins of the world and giant hydrocarbon accumulations: AAPG map and accompanying text.

St. John, B., 1984, Antarctica—Geology and hydrocarbon potential, *in* M. T. Halbouty, ed., Future petroleum provinces of the world: AAPG Memoir 40, p. 55–100.

Edwards, J. D., 2001, Twenty-first-century energy: Decline of fossil fuel, increase of renewable nonpolluting energy sources, *in* M. W. Downey, J. C. Threet, and W. A. Morgan, eds., Petroleum provinces of the twenty-first century: AAPG Memoir 74, p. 21–34.

Chapter 2

Twenty-first-century Energy: Decline of Fossil Fuel, Increase of Renewable Nonpolluting Energy Sources

John D. Edwards
Energy and Minerals Applied Research Center, University of Colorado, Boulder, Colorado, U.S.A.

ABSTRACT

The world must prepare for the transition to renewable nonpolluting energy sources to ensure the continuous flow of energy to the increasing population and expanding economies. World oil supply will meet demand until the peak plateau of world oil production is reached, which is estimated to be between the years 2010 and 2030. Ultimate oil recovery will range from a conservative 2750 billion barrels of oil (BBO) or an optimistic 3670 BBO. Declining production after peak oil production occurs will cause a global energy gap to develop because energy demand will continue to grow. This gap can be avoided by advance planning. Energy conservation, improved energy efficiency, expanded production of unconventional oil, and conversion of natural gas to liquids will help to extend the time of peak oil production. The long-term solution to energy supply is conversion to renewable, nonpolluting energy sources, which include solar, nuclear, hydroelectric, geothermal, wind, biomass, and hydrogen. Solar, nuclear, and hydrogen energy should become major power sources in the twenty-first century.

OBJECTIVES

The purpose of this paper (a revision of an earlier paper on twenty-first-century energy [Edwards, 1997]) is to inform the petroleum industry, governments, and the general public of the risks, challenges, and opportunities that will occur as peak world oil production is approached in the first quarter of the twenty-first century. The fact that earth resources inevitably control nations and individuals must be recognized (Youngquist, 1998). The transition to alternate energy sources is inevitable because fossil fuels, which now supply 85% of world energy, are exhaustible resources (Table 1).

Oil production, which now supplies 40% of the world's energy, will decrease after peak production occurs. The petroleum industry will then have opportunities to develop and supply renewable, nonpolluting energy instead of only fossil fuels. In anticipation of this transition, research and development of renewable energy sources should begin soon. World energy demand will continue to increase. Producing and marketing competitively priced, convenient, readily available renewable energy supplies will become profitable during the first quarter of the twenty-first century.

FORCES DRIVING INCREASING WORLD ENERGY DEMAND

Population growth, expanding industrialization, and improving lifestyles, principally in developing countries, are the causes of continued increase in world energy demand. World population passed six billion in October 1999 (Table 2) and is increasing by 80 million per year. During the twentieth century, global population tripled. Improvements in medicine, agriculture, education, and communication have caused decreases in mortality and

Table 1. United States and World Energy Consumption by Source (January 2000).

	United States		World	
Energy Source	BBOE	Percent	BBOE	Percent
Petroleum	6.8	41	27.1	40
Natural gas	4.2	24.2	14.4	21.2
Coal	3.9	23.2	16.7	24.7
Nuclear	1.3	7.9	4.5	6.7
Hydroelectric	0.73	3.7	1.6	2.4
Biomass	—	—	3.4	5
Total	**7.6**	**100**	**69.3**	**100**

Sources: World Resources Institute, 1992; BP, 1998; Petzet, 1999; Beck and Radler, 2000.

Table 2. Twentieth-century World Population Growth, Oil Production, and Oil Consumption.

	1930	1960	1975	1987	1999
World population (billions)	2	3	4	5	6
Time to add 1 billion people (years)	130	30	15	12	12
Annual world oil production (billion bbl)	1.5	7.1	20.4	20.4	24.3
Annual world oil consumption per capita (bbl)	0.75	2.37	5.1	4.08	4.05

Sources: Degolyer and MacNaughton, 1992, 1998; Population Reference Bureau, 1999.

fertility rates. However, even with decreasing fertility rates, the present population of two billion people younger than 20 (mainly in developing countries) guarantees continued population growth at least until the middle of the twenty-first century. The world's present annual population growth rate is 1.4%, which adds 80 million people each year. The population growth rate in developing countries is 1.7% (Table 3). World population will probably reach eight billion by 2025 and could reach nine billion by 2050 (United Nations, 1998; Population Reference Bureau, 1999). A reasonable possibility exists that world population will peak at or below 10 billion between 2100 and 2150.

Most industrial growth and 98% of population growth are expected to occur in developing countries. This growth will increase the demand and competition for natural resources, particularly energy. Table 4 shows world oil production versus demand, by region. Notice

Table 3. World Population, mid-1999.

	Population (millions)	Annual Population Growth (percent)
Developed Countries		
United States	272	0.6
Canada	31	0.4
Europe	728	–0.1
Japan	127	0.2
Australia	18	0.7
New Zealand	4	0.8
Total developed	1180	0.1
Developing Countries		
Central America and Caribbean	172	1.7
South America	339	1.7
Africa	771	2.5
Asia (excluding China and India)	1396	1.5
China	1254	1.0
India	987	1.9
Total developing	4802	1.7
Total world	5982	1.4

Source: Population Reference Bureau, World Population Data Sheet, mid-1999.

Table 4. Annual World Crude-oil Production and Demand.

	Production mid-1999 (billion bbl)	Demand 1998 (billion bbl)
Developed Countries		
United States and Canada	2.83	7.36
Western Europe	2.32	5.00
Eastern Europe	2.65	2.08
Japan	0.005	2.15
Australia and New Zealand	0.19	0.34
Developing Countries		
Africa	2.43	0.88
Latin America	3.43	2.09
Asia	2.38	4.34
Middle East	7.79	1.77
Total world	24.03	26.01
Developed countries	8.00 (33%)	16.93 (65%)
Developing countries	16.03 (67%)	9.08 (35%)
OPEC	10.14 (42%)	1.97 (7.6%)

Sources: Degolyer and MacNaughton, 1998; Oil & Gas Journal, 1998. The shortfall between production and demand is met by refinery processing gain and the use of other hydrocarbons, alcohol, stocks, etc.

that the developed countries produce one-third and the developing countries produce two-thirds of the world's oil. But the developed countries use two-thirds and the developing countries use only one-third of the world's oil. Per-capita oil consumption in developing countries is 1.9 barrels (bbl)/year; the average in developed countries is 14.3 bbl/year. The annual per-capita oil consumption in the United States is 25 bbl (Table 5). World population growth has outpaced world oil production since 1979 (Duncan and Youngquist, 1998). A major shift in demand from developed to developing countries can be expected in the next few decades. This change is driven by the predominant growth of population and industry in the developing world. Annual world oil production is estimated to continue to increase into the first quarter of the twenty-first century at a rate of about 1.5 %. If this rate persists, annual world demand for oil will increase from 24 billion bbl of oil (BBO) in 1999 to about 38 BBO in 2030, which is 104 million bbl/day.

WORLD PETROLEUM IN THE 1990s

Annual world crude-oil demand increased from 22 billion to 28 billion bbl from 1990 to 2000 and is expected to continue to rise. World conventional oil reserves have remained above 1000 BBO since 1993. They peaked at 1163 BBO in 1997 and were at 1016 BBO in January 2000. However, 300 BBO are suspect as "political reserves" booked by OPEC during 1987–1989 to support production quotas. These reserves may be real, but they may include probable and speculative oil from future discoveries and development. Cumulative world oil production reached 836 BBO in January 2000 (Degolyer and MacNaughton, 1998; Oil & Gas Journal, 1998; Petzet, 1999; Beck and Radler, 2000) (Figure 1).

In the 1990s, major oil and gas discoveries were made, most notably in deep, offshore waters of the U.S. Gulf Coast, Angola, Brazil, and the Caspian Sea. Significant additional oil and gas reserves were discovered on the northwest Australian shelf and in Indonesia, China, Iran, Iraq, Saudi Arabia, Egypt, Algeria, Nigeria, the North Sea, Alaska, Canada, Colombia, Venezuela, Peru, and Brazil.

New technologies were successfully applied, including rapid, computer-assisted data analysis, 3-D and 4-D seismic methods, extended-reach and underbalanced horizontal wells, new drilling tools, new logging and completion methods, seismic sequence stratigraphy, petroleum-systems analysis, and detailed reservoir-characterization studies. These techniques have improved success rates for wildcats and development wells and have added reserves in new and old fields.

Many new oil and gas discoveries will be made throughout the world in the next few decades. Deep-water discoveries in new areas, particularly associated with major deltas, undoubtedly will be made. Unfortunately, other than West Africa in the 1990s, no new giant oil provinces have been found since the North Sea and offshore Mexico in the 1960s. Peak world discovery volumes occurred 35 years ago, in the mid-1960s. Super-giant field discoveries—those with reserves greater than 5 BBO—have declined. However, recent discoveries, not yet recorded, in Saudi Arabia, Iran, Iraq, and the Caspian Sea—each one more than 10 BBO—have increased world reserves by more than 50 BBO.

In 1998 and 1999, the petroleum industry was forced to deal with excess oil supply and low prices resulting from reduced demand from shaky economies in Asia, the former Soviet Union, and Latin America. Companies downsized, restructured, merged, and cut budgets. Reserve replacement has declined for many companies because of declining production in old fields, sale of marginal production, plugging of stripper wells, record-low drilling rates, and low discovery rates.

Fortunately, oil prices recovered during 1999 because of production cuts by OPEC, Norway, and Mexico. Reduced production by OPEC and prices of more than $20/bbl have stimulated some renewed upstream activity. However, many companies apparently intend to maintain their economic yardsticks of profitable exploration and production operations at $10–$12/bbl. Crude-oil prices could remain above $20/bbl for the near future if excess OPEC productive capacity is kept off the market.

Table 5. Annual World Oil Consumption Per Capita, mid-1999.

	Population (millions)	Per-capita Oil Consumption (bbl)
Developed Countries		
United States	272	25
Canada	31	22
Western Europe	385	13
Eastern Europe	343	5
Japan	127	17
Australia and New Zealand	23	16
Total developed	**1180**	**14.3**
Developing Countries		
Africa	771	1.2
Latin America	512	3.7
Asia	3351	2.1
Middle East	167	8.2
OPEC	386	5.1
Total developing	**4802**	**1.9**
Total world	**5882**	**4**

Sources: Degolyer and MacNaughton, 1998; Population Reference Bureau, 1999.

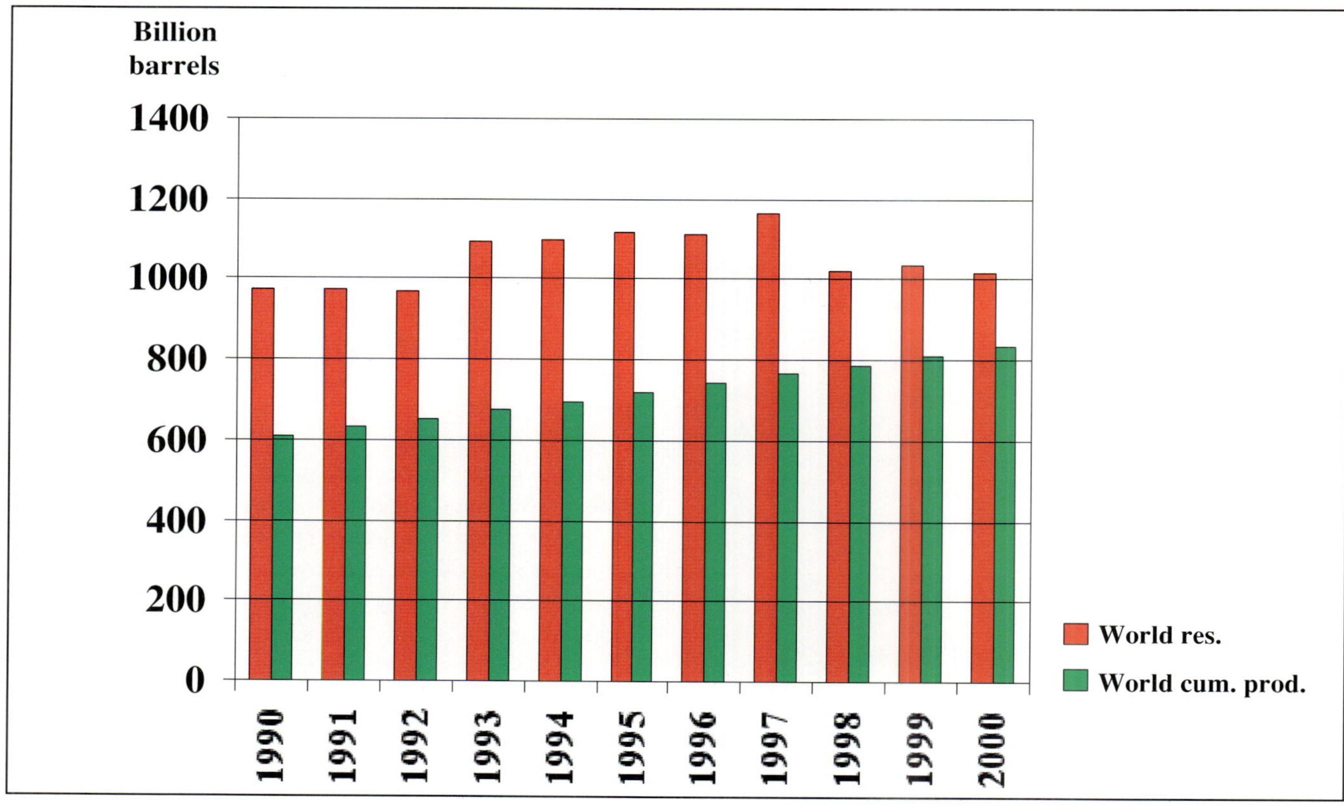

Figure 1. World crude-oil reserves and cumulative production, 1990–2000. Sources: Degolyer and MacNaughton, 1998; Petzet, 1999; Beck and Radler, 2000.

U.S. PETROLEUM IN THE 1990s

U.S. annual crude-oil production from 1990 to 2000 decreased from 2.7 to 2.1 BBO, while annual production of natural-gas liquids (NGL) remained flat, at about 0. 8 BBO. Total U.S. liquid production decreased from 3.43 BBO to 2.9 BBO from 1990 to 2000 (Degolyer and MacNaughton, 1998; Oil & Gas Journal, 1998; Petzet, 1999; Beck and Radler, 2000) (Figure 2). In contrast, imports of crude oil and products increased from 2.9 to 4.1 BBO/ year in the past decade. Total U.S. oil consumption increased from 6.35 to 7 BBO/year from 1990 to 2000 (Figure 3). U.S. reserves of crude oil and NGL have declined from 34 BBO in 1990 to 29 BBO in 2000. During this period, crude-oil reserves dropped from 26 BBO to 21 BBO, while NGL reserves remained at about 8 BBO (Figure 4). Reserve growth in existing fields has become the dominant component of reserve additions in the United States and is likely to remain so in the future (Klett et al., 1997).

ULTIMATE WORLD LIQUIDS PRODUCTION

Ultimate world liquids recovery is estimated to be 3670 BBO (case A, Table 6), but the ultimate could be only about 2750 BBO (according to Campbell, 1998, and Laherrère, 1999) (case B, Table 6). The ultimate liquids recovery of case A consists of 836 BBO cumulative crude-oil production, 100 BBO cumulative NGL production, 1016 BBO crude-oil reserves, 100 BBO NGL reserves, 583 BBO future oil discoveries and oil-field growth, 100 BBO NGL future discoveries and field growth, 570 BBO of unconventional oil, and 365 BBO equivalent (BBOE) from gas-to-liquid (GTL) conversion (Masters et al., 1991, 1997; Oil & Gas Journal, 1998) (Table 6). GTL conversion is estimated optimistically at 20% of world gas reserves and future discoveries (10,941 trillion cubic feet [tcf] of gas = 1823 BBOE [6 tcf gas = 1 BBOE]; 20% of 1823 BBOE = 365 BBOE). Mean values of undiscovered oil and gas are used. Unconventional oil resources include 270 BBO of eastern Venezuelan heavy oil and 300 BBO of western Canadian tar-sand oil (Meyer and DeWitt, 1990; Kahn, 1998). Neither liquids from oil shale nor liquids from coal-to-liquid conversion are counted.

If the projection of 3670 BBO ultimate production and the 1.5% demand growth rate are correct, the peak plateau of oil-production capacity will be reached from 2020 to 2030 at approximately 38 BBO/year, which is 104 million bbl/day (Figure 5). If the world ultimate production is only 2750 BBO, peak oil production will occur in 2010 at about 33 BBO/ year, which is 90 million bb/day (Campbell, 1998; Laherrère, 1999).

Table 6. Estimated Ultimate World Liquid Petroleum.

	Case A (this paper)	Case B (Campbell, 1999; Laherrère, 1999)
Conventional Oil		
Crude-oil cumulative production	836 BBO	800 BBO
NGL cumulative production	100 BBO	75 BBO
Crude-oil reserves	1016 BBO	800 BBO
NGL reserves	100 BBO	125 BBO
Future Discoveries and Field Growth		
Crude oil	583 BBO	200 BBO
NGL	100 BBO	50 BBO
Unconventional oil		700 BBO
Venezuelan heavy oil	270 BBO	
Canadian tar-sand oil	300 BBO	
Coal oil	???	
Shale oil	???	
Ultimate Liquid Oil Production	**3305 BBO**	**2750 BBO**
Natural Gas		
Cumulative production	2200 tcf (366 BBOE)	
Reserves	5150 tcf (858 BBOE)	
Future discoveries	5791 tcf (965 BBOE)	
Future total gas	10,941 tcf (1823 BBOE)	
Natural-gas-to-liquids Conversion[1]	**365 BBO**	
Ultimate Liquids Production[2]	**3670 BBO**	**2750 BBO**
Future Liquids Production[3]	**2734 BBO**	**1875 BBO**

Sources: Degolyer and MacNaughton, 1998; Masters et al., 1991; Oil & Gas Journal, 1998; Potential Gas Committee, 1998; Campbell, 1999; Laherrère, 1999.
[1]20% of future gas = 20% of 1823 BBOE. *[2]3305 BBO + 365 BBO.* *[3]3670 BBO – (836 + 100) BBO.*

Table 7. Estimated Ultimate United States Liquid Petroleum.

Conventional Oil	
Crude-oil cumulative production	180 BBO
Crude-oil reserves	21 BBO
NGL cumulative production	36 BBO
NGL reserves	8 BBO
Future Discoveries and Field Growth	
Crude oil	140 BBO
NGL	30 BBO
Ultimate oil production	415 BBO
Natural gas	
Cumulative production	840 tcf (140 BBOE)
Reserves	167 tcf (28 BBOE)
Future discoveries	1038 tcf (173 BBOE)
Ultimate recovery	2045 tcf (341 BBOE)
Future total gas	1205 tcf (200 BBOE)
Natural-gas-to-liquids Conversion[1]	**40 BBOE**
Ultimate Liquid Petroleum[2]	**455 BBO**
Future Liquid Petroleum[3]	**239 BBO**

Sources: Minerals Management Service, 1998; Degolyer and MacNaughton, 1998; Oil & Gas Journal, 1998; Potential Gas Committee, 1998.
[1]20% of future gas = 20% of 200 BBOE. *[2]415 BBO + 40 BBOE.* *[3]455 BBO – (180 BBO + 36 BBO).*

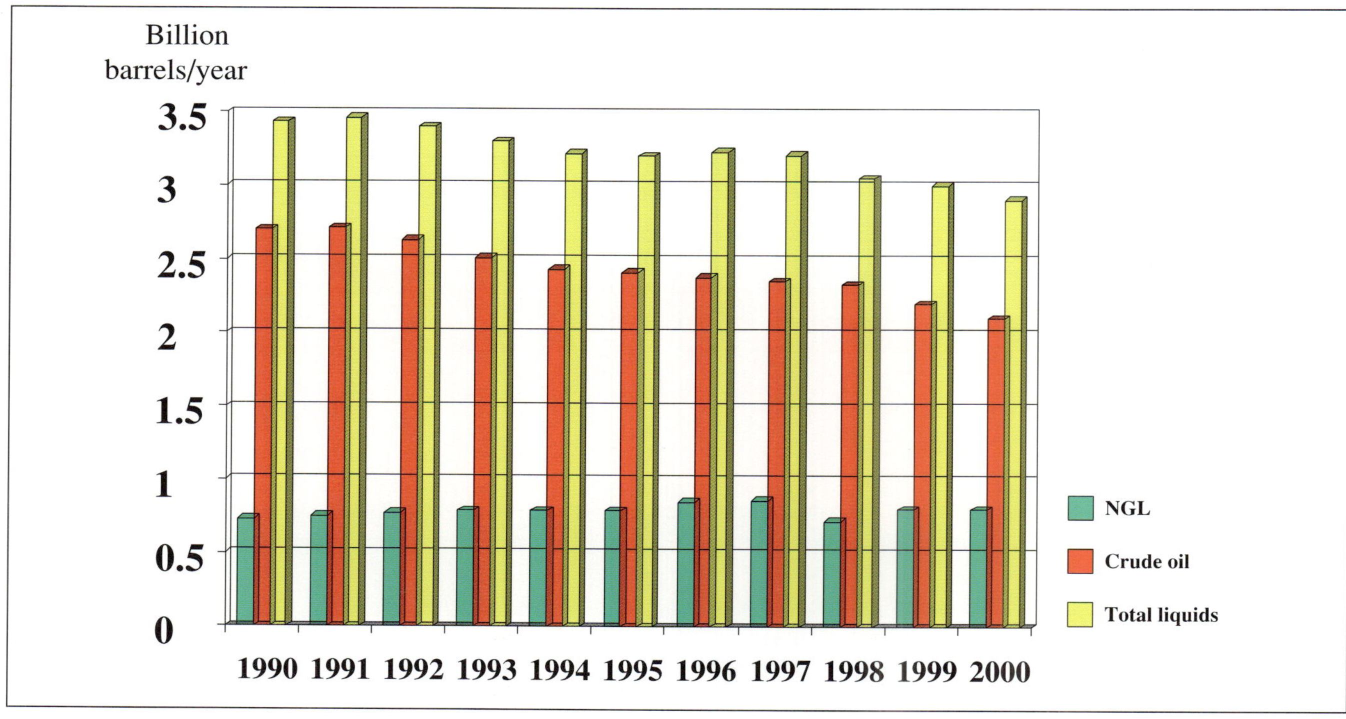

Figure 2. U.S. crude-oil, NGL, and total liquids production, 1990–2000. Sources: Degolyer and MacNaughton, 1998; Beck and Radler, 2000.

ULTIMATE U.S. LIQUIDS PRODUCTION

U.S. ultimate liquid-petroleum recovery is estimated to be 455 BBO (Table 7). This consists of 180 BBO cumulative crude-oil production, 36 BBO cumulative LNG production, 21 BBO crude-oil reserves, 8 BBO NGL reserves, 140 BBO future crude-oil discoveries and field growth, and 30 BBO future NGL discoveries and field growth (U.S. Geological Survey, 1995; Minerals Management Service, 1996; Masters et al., 1997; Degolyer and MacNaughton, 1998). An additional optimistic 40 BBO from GTL conversion is included. This is 20% of U.S. gas reserves and estimated future gas discoveries (1205 tcf gas = 200 BBOE; 20% of 200 BBOE = 40 BBOE) (Potential Gas Committee, 1998; Gas Research Institute, 1999) (Table 7).

Future U.S. oil discoveries are estimated to be 45.6 BBO offshore and 30.3 BBO onshore. Future field growth is estimated to be 60 BBO onshore and 4.1 BBO offshore (U.S. Geological Survey, 1995; Minerals Management Service, 1996; Masters et al., 1997). If United States oil consumption continues to grow at the rate of 1.5% per year, demand could rise to almost 9 BBO/year from 2020 to 2030, which is about 25 million bbl/day. Total U.S. liquid petroleum production in 2030 could still be about 3 BBO/year, which is 8 million bbl/day. To supply the total U.S. demand in 2030, imports of 6 BBO/year will be required, which is slightly more than 16 million bbl/day (Figure 6).

PEAK PLATEAU OF WORLD LIQUIDS PRODUCTION

Why should we be concerned about the near-term peak of world oil production? The problem is not if but when peak world oil-productive capacity will occur. There is a strong probability that peak world oil production will be reached in the first quarter of the twenty-first century. Peak oil production most likely will be reached between 2010 and 2030 (Table 8). The peak plateau of oil production would be suppressed and would occur earlier if the future discovery rate is low or if capital is not available to develop new production capacity. A gap in energy supply could develop at that time.

The 19 sources cited in Table 8 estimated ultimate economically recoverable oil and the year of peak oil production. These calculations have been made repeatedly in the past and will continue to be made in the future. Past projections have been too pessimistic. The scenarios presented in this paper may also be inaccurate, but they have credibility based on reasonable present estimates of oil reserves, future oil discoveries, and demand growth. Even with the addition of 1 trillion bbl of oil reserves above an optimistic 3670 BBO ultimate recovery, peak oil production would be extended for only about 10 years (Bartlett, 2000).

Oil and gas reserves will be generated from new discoveries, improved recovery technologies, and huge unconventional resources, the limits of which are unknown. These unconventional resources include bitumen, heavy oil, oil sands, coal-bed methane, deep-basin gas, biomass, and possibly oil shales and gas hydrates. Oil will be more expensive as supplies diminish. However, oil will be available throughout the twenty-first century and into the twenty-second century for high-value uses such as petrochemicals and air transportation.

Some authors believe there is no threat of an oil shortage in the foreseeable future. Price rises will generate new oil production. They believe market forces, new technologies, and adequate resources will ensure the growth of supply to meet rising demand (Fisher, 1991; Alelman and Lynch, 1997; Linden, 1998; McCabe, 1998;

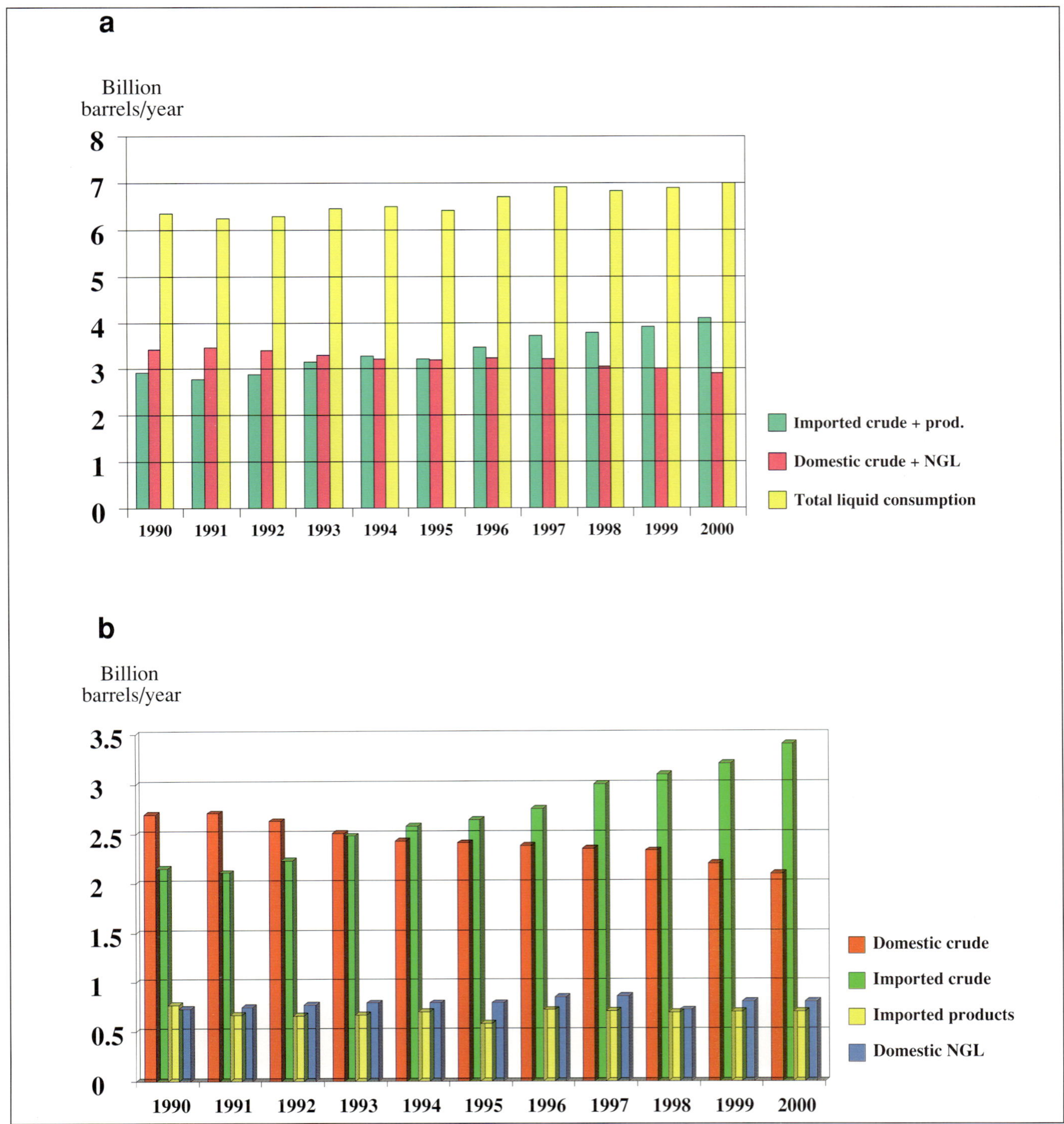

Figure 3. (a) U.S. petroleum consumption, 1990–2000. Sources: Degolyer and MacNaughton, 1998; *Oil & Gas Journal,* 1998; Beck and Radler, 2000. (b) U.S. petroleum consumption, 1990–2000. Sources: Degolyer and MacNaughton, 1998; *Oil & Gas Journal,* 1998; Beck and Radler, 2000.

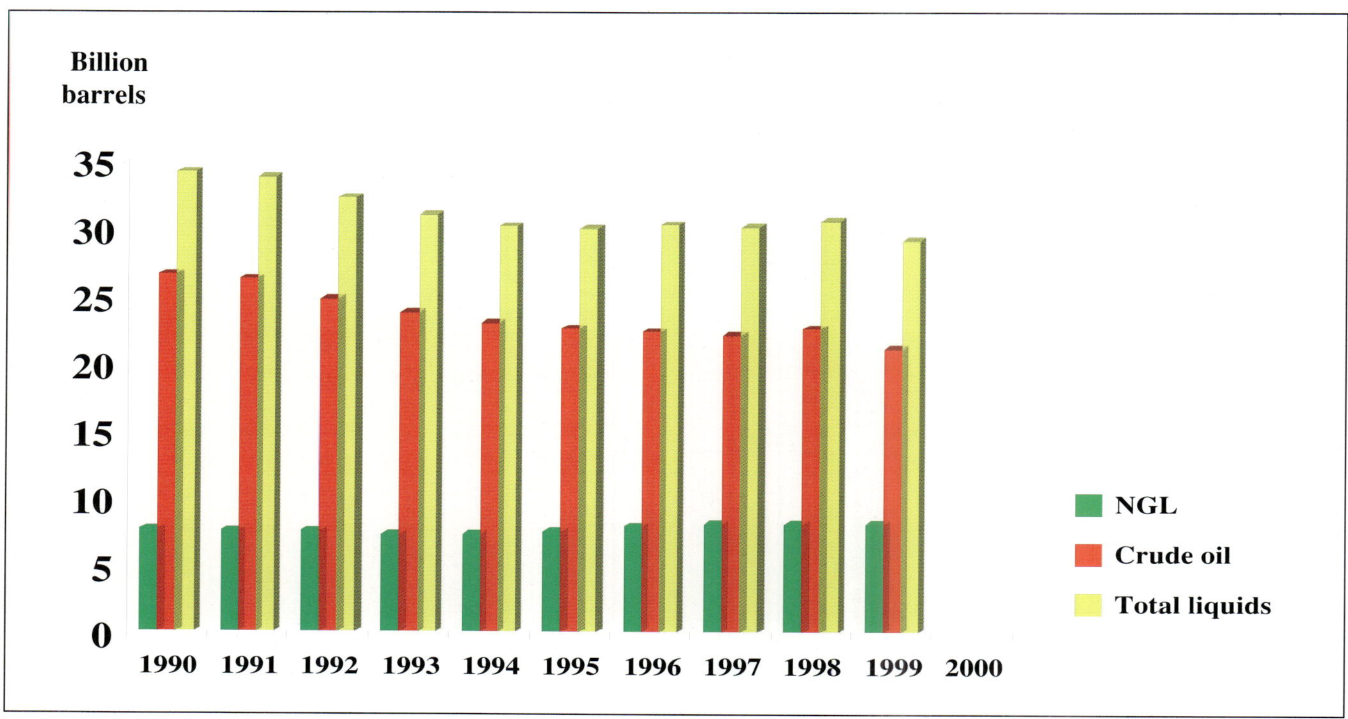

Figure 4. U.S. crude-oil, NGL, and total liquids reserves, 1990–2000. Sources: Degolyer and MacNaughton, 1998; *Oil & Gas Journal,* 1998; Petzet, 1999.

Table 8. Peak Year of World Crude-oil Production and Estimated Ultimate Recovery.

Author (Company)	Year	Estimated Ultimate Recovery (billion bbl)	Peak Year
Hubbert (Shell)	1969	2100	2000
Moody (consultant)	1978	3200	2004
Odell and Rosing (Delft University)	1983	3000	2025
Bookout (Shell)	1989	2000	2010
Townes (employed by AAPG)	1993	3000	2010
Campbell (consultant)	1994	1650	1997
Laherrère (consultant)	1994	1750	2000
MacKenzie (Western Research Institute)	1996	2600	2007–2019
Appleby (BP)	1996		2010
Ivanhoe (consultant)	1996		2010
van der Veer (Shell)	1997		2020
Edwards (University of Colorado)	1997	2836	2020
Barnabe (ENI)	1998		2000–2005
Schollnberger (Amoco)[1]	1998		2015–2035
Duncan and Youngquist (consultants)	1998		2006
International Energy Agency[2]	1998	2800	2010–2020
Energy Information Administration (U.S. Department of Energy)[2]	1998	4700	2030
Laherrère (consultant)[2]	1999	2700	2010
Edwards (University of Colorado)	this study[2]	3670	2020–2030

[1]*Oil and gas.*
[2]*Total liquids (crude and heavy oil, tar-sand oil, GTL).*

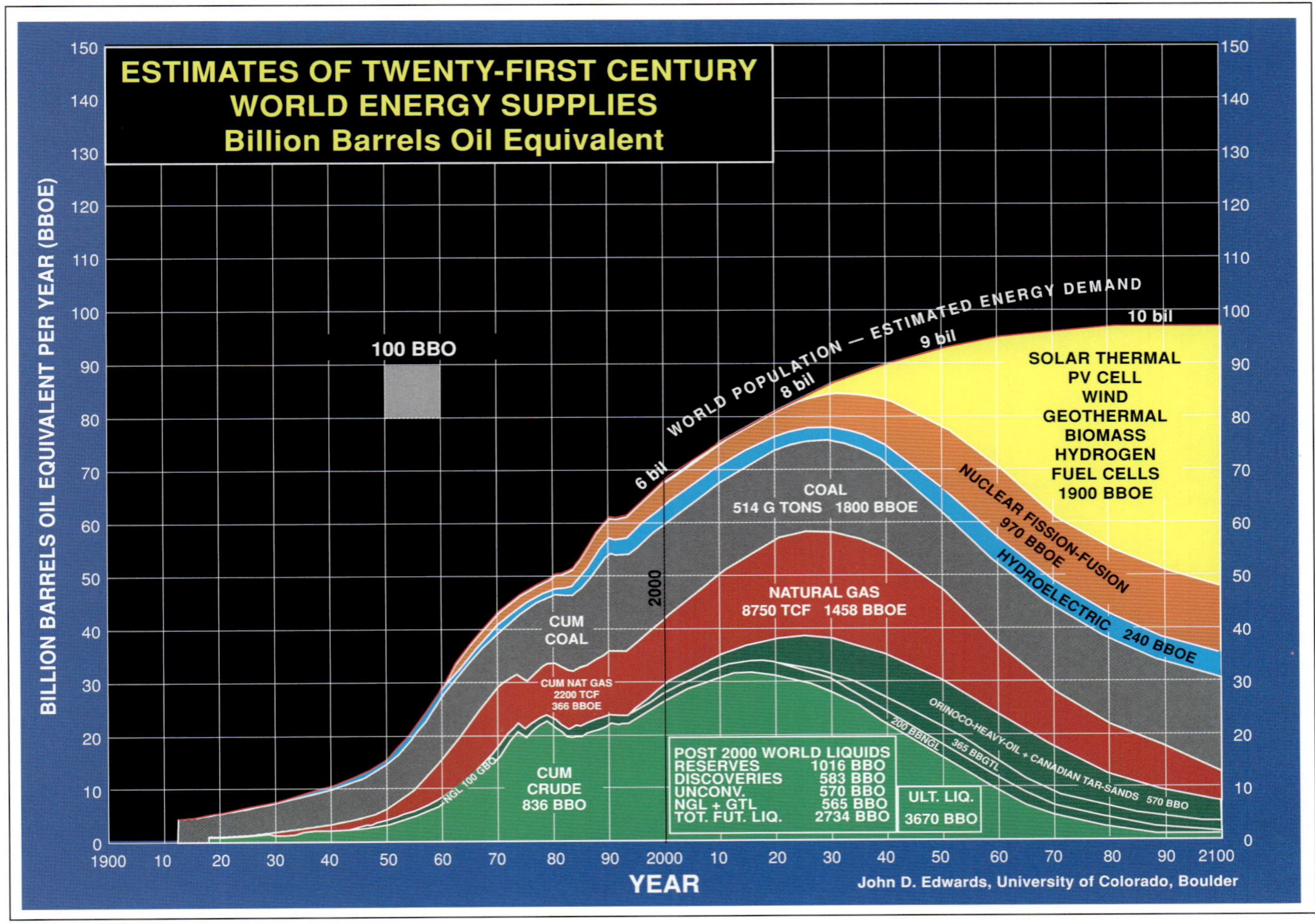

Figure 5. Estimates of twenty-first-century world energy supplies and population.

Lynch, 1999). A balanced discussion of the differing opinions of economists and geologists concerning ultimate oil reserves was presented by Riva (1999).

Fossil fuels are exhaustible resources. The amount of oil, gas, and coal that nature has formed in the past 500 million years is huge and unknown. All the oil and gas that exists in the earth will never be totally extracted, just as the last ton of coal in deep, thin seams will never be mined. U.S. Geological Survey estimates of world recoverable hydrocarbon resources have continued to increase as the perception of time-dependent limits recedes (Schmoker and Dylan, 1998). The total amount of oil ultimately produced will depend on the validity of present reserve estimates, recovery efficiency from existing fields, quantity and quality of future discoveries, new technologies, competition from alternate energy sources, and future energy demand and price changes.

In the second half of the twenty-first century, oil and gas production will decline. Oil and gas will be produced from lean, remote, deep, and expensive resources to extend production into the twenty-second century. These "bottom-of-the-barrel" reserves will be used for highest-value products, such as petrochemicals. None of the renewable energy sources, except biomass, can provide feedstocks for these unique products.

FOSSIL-FUEL CHALLENGES (OIL, GAS, COAL)

Fossil fuels produce 86% of world energy but are facing political and environmental constraints because of their polluting emissions. The energy industry must meet the clean-fuel challenge (Bowlin, 1999). Competitive renewable nonpolluting energy sources will ultimately replace fossil fuels. As this transition away from fossil fuels occurs, the potential problem of carbon-dioxide emissions causing global climate change will be mitigated.

Oil

Oil is the most valuable and most used fuel in the world today. It provides 40% of world and U.S. energy supply (Table 1). Oil, with its higher BTU density and ease of transport and storage, has no comprehensive substitute today. The world will remain critically dependent on oil well into the twenty-first century. It is impossible to imagine anything but a continued increase in the demand for oil (Robinson, 1998).

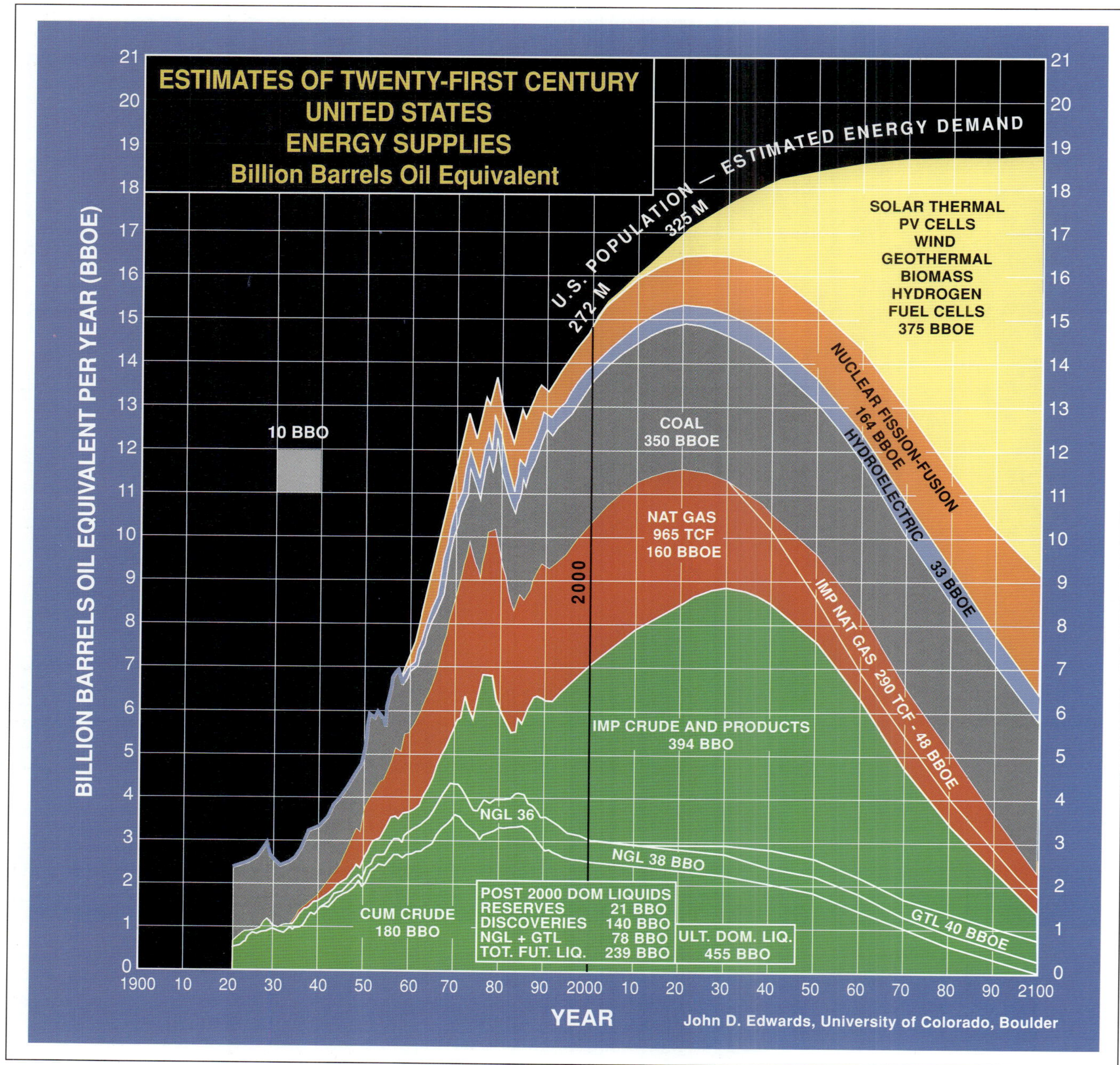

Figure 6. Estimates of twenty-first-century U.S. energy supplies and population.

The U.S. Geological Survey, in its ranking of world petroleum provinces, has identified 406 petroleum-bearing provinces among the world total of 954 geologic provinces. Exclusive of the United States, 76 provinces with the largest petroleum volumes contain 95% of the world's known petroleum. Future discoveries are likely to come from those provinces (Klett et al., 1997).

The price of oil will rise as the approach of peak oil production is recognized. Peak liquid-petroleum production can be extended by increased use of natural gas, GTL conversion, and expanded coal production. Increased oil production will come from eastern Venezuelan heavy-oil deposits, estimated to contain 270 BBO recoverable from an in-place resource of 1200 BBO, and from western Canadian tar sands, estimated to have 300 BBO recoverable from an in-place resource of 1686 BBO (Meyer, 1987; Meyer and Dewitt, 1990; Masters et al., 1991; Kahn, 1998). Oil production in the twenty-first century may also occur from tar sands and oil shales in the United States and elsewhere. However, even with the sources just listed and with new discoveries, increased recovery from old fields, conservation, and improved efficiency of use, demand for oil will increase beyond the capacity of petroleum supplies.

Natural Gas

World resources of natural gas are underrecognized and underused. The ratio of known oil fields to gas fields is about 2 to 1, which probably does not represent the natural endowment of gas to oil (Klett et al., 1997). World future gas resources are estimated to be only 10,941 tcf (Table 6). U.S. future gas resources are estimated to be only 1205 tcf (Table 7) (Masters et al., 1997; Potential Gas Committee, 1998). The U.S. estimate includes 167 tcf reserves, 896 tcf future discoveries, and 141 tcf coal-bed methane. Future discovery estimates for gas are very low. Natural gas will be aggressively searched for, discovered, and produced in the twenty-first century.

Enormous resources of natural gas are locked in methane hydrates in the deep oceans and Arctic tundra. As much as 100,000 tcf is estimated to be trapped in gas hydrates (Finley and Krason, 1989; Krason, 1994; Haq, 1998). If only a low percentage of this resource is ever commercially extractable, it could extend the life of natural-gas supplies for decades.

The infrastructure for worldwide distribution of natural gas and liquid natural gas (LNG) will expand and help fill the energy gap. Gas exploration and production are expected to expand rapidly as new distribution systems and markets are developed (True, 1999). Projects are planned or in progress to move stranded gas to available markets by pipeline or LNG tankers from northwestern Australia, Papua New Guinea, Indonesia, Malaysia, Vietnam, Kazakhstan, Oman, Abu Dhabi, Qatar, Iraq, Iran, Algeria, Nigeria, Trinidad, Colombia, Peru, Bolivia, and Argentina. Gas will reach markets as an economically competitive, preferred clean fuel.

Increasing environmental concerns will help natural gas become the most important fuel for electrical industries (Otto et al., 1999). Power plants with natural-gas-fired combined-cycle gas turbines and with gas-electric cogeneration are increasing the demand for gas throughout the world. Natural gas will also be the fuel used in fuel cells for automotive power. In the United States today, alternative fuels account for only 0.2% of total transportation fuel consumption (Chang, 1999). In the near future, fuel cells energized by natural-gas-to-hydrogen technology could power an increasing percentage of automotive transport and could supply selected commercial and residential electricity (Fouda, 1998; Bensabat, 1999; Chang, 1999). Hydrogen is the ultimate nonpolluting, renewable, and sustainable fuel of the future (Hefner, 1999). Hydrogen from natural gas and eventually from dissociation of water could become the perpetual energy source of the future.

GTL and possibly coal-to-liquid technologies will extend the life of the internal-combustion engine for transportation use and will help to continue to supply petrochemical feedstocks (Fouda, 1998). Arco-Syntroleum (Oil & Gas Journal, 1999b) and Chevron-Sasol (Matske, 1999) have begun GTL pilot projects. Chevron Overseas President Richard Matske said, "Gas-to-liquid technology is so promising that its development could create an entire paradigm shift throughout the petroleum industry" (Matske, 1999).

Coal

World coal production of more than 3 billion tons supplies 25% of the annual world energy demand and 37% of world electricity. U.S. coal production of more than 1 billion tons per year supplies 23% of domestic energy demand and 57% of domestic electricity. World demand for electricity is increasing by 5% per year (International Energy Agency, 1999). Coal, the lowest-cost fossil fuel, is abundant and widely distributed. It will continue to supply much of the world's expanding electricity demand in the twenty-first century.

However, as an energy source, coal is the most polluting of the fossil fuels. Combustion of coal is blamed for environmental damage, health problems, and elevation of the earth's surface temperature. China, the world's largest user of coal, obtains 73% of its energy from coal and is projected to become the largest emitter of greenhouse gases by 2015 (Ebel, 1998). China already has nine of the world's 10 most polluted cities (Dunn, 1999). Of its 88 major cities, 85 exceed World Health Organization air-quality guidelines for particulates, and half of those cities exceed sulfur-dioxide guidelines (Robinson 1998). In 1994, China became a net oil importer (Tempest, 1998). These circumstances caused China to expand coal production while trying to restrict combustion of high-sulfur coals.

Coal will continue to supply electricity in Asia-Pacific countries, particularly China and India (International Energy Agency, 1999). Near-term reduction of emissions from world coal-fired power plants is very unlikely, even though clean coal technology is starting to be used in developing countries. However, where natural gas is abundantly available, as in Europe, electric power plants are converting from coal to gas, thus reducing emissions.

RENEWABLE NONPOLLUTING ENERGY SOURCES

International consensus reached by the 1997 Kyoto Protocol supports development of renewable nonpolluting energy sources to replace combustion of fossil fuels. The goal is to reduce emissions of greenhouse gases below 1990 levels and thus avoid the possibility of human-induced climate change, global warming, and sea-level rise. Opportunities for reduction of carbon-dioxide emissions and sequestration are being pursued. However, significant reduction of carbon emissions by substitution of alternative fuels for petroleum-fueled vehicles and coal-fired electric power plants is very unlikely (Flannery, 1999). In fact, some climate scientists now recognize that meeting the Kyoto accord will not significantly reduce the potential increase of the earth's surface temperature in the twenty-first century. They are recommending that society find ways to adapt to the possibility of a warmer world (Science News, 1999).

Air and water pollution can be mitigated by switching to alternative energy sources that will replace fossil fuels. Conversion from fossil fuels to renewable, nonpolluting energy sources will result in significant reduction of carbon dioxide, sulfur dioxide, and nitrous oxides in the atmosphere.

Renewable nonpolluting energy in the next century will come from solar thermal-hydrogen generation and from solar thermal-electric, photovoltaic, hydroelectric, geothermal, wind, and biomass sources. Commercialization and implementation of technologies required for these renewable nonpolluting energy sources will take time, and time is one of our most limited resources. Renewable energy sources will fail if their enabling technologies do not improve, to ensure affordability and convenience of use (Crow, 1998).

Both BP Amoco and Royal Dutch Shell have purchased solar-power companies and are entering the photovoltaic (PV) cell market. BP, by its purchase of Solarex, expects to have a $1 billion PV business by 2010 with 20% of the world market (Oil & Gas Journal, 1999a). Shell International Renewables is investing $500 million in five years in solar and biomass projects and forestry. Shell International Gas and Enron are investing in gas-fired electric power plants. Both BP and Shell support sustainable development, which takes into account economic, environmental, and social considerations (van der Veer, 1997; BP Amoco, 1998). Many other oil companies have also adopted this credo. Some of them are listed in the "World Oil" category of the Dow-Jones Sustainability Group Index (see www.sustainability-index.com/description/components.html).

Nuclear power may plateau in the near future, but its expansion will be needed before 2050. Flavin and Lenseen (1999) present a negative view of the future of nuclear power. Modular, fail-safe, economically competitive nuclear electric power plants, with zero emissions, can be built to replace coal-fired power plants, but the nuclear-waste disposal problem must be solved. The WIP site salt mines in southeast New Mexico are an excellent long-term residence for nuclear waste. If nuclear power, the largest noncarbon energy source, continues to be discouraged in developed countries, then coal, natural gas, and renewable energy sources must increase to supply the expanding world demand for electrical energy.

SUSTAINABLE ENERGY SUPPLY

The incremental environmental impact of human activities will continue to concern governments and industries. Demand growth for all resources will exponentially expand the volume of waste that the earth, oceans, and atmosphere must absorb (Foster and Wise, 1999), affecting the quality of air, water, land, natural resources, and wildlife. The communication-technology "explosion" has increased the flow of information, has fostered the globalization of industry, and has increased the quality of life expectations in the developing world. It is also causing demand for more equitable distribution of resources. These events have placed greater demands on governments and businesses to improve performance and accountability.

Unrestrained growth of population and industry is occurring in the developing world. The pursuit by many governments and some international companies of competitive advantage to the exclusion of environmental and societal considerations, along with continued self-indulgent consumption in developed countries, is not conducive to establishment of a future sustainable world economy, environment, and society (Robinson, 1998). The bottom line of business in the twenty-first century must be to help society achieve three interlocked goals: economic prosperity, environmental protection, and societal equity (Elkington, 1998). Achieving these goals will require the collective international attention of governments and industries.

CONCLUSIONS

The energy scenario presented here is not the only possible outcome. However, it is a scenario for plausible, sustainable world-energy supply for the twenty-first century. In the next few years, three main themes appear to be obvious: (1) We should increase efficiency of energy use. (2) We should clean up the energy sources we are using. (3) We should continue to search for cleaner forms of energy (Robinson, 1998).

World oil production will increase to a peak plateau by 2020–2030 and then will decline during the rest of the twenty-first century. Natural gas and clean coal energy can help fill the energy gap created by declining oil production. Renewable, nonpolluting energy sources must expand to replace fossil fuels and supply the world's continuously increasing energy demand. The transition to renewable energy sources will create a sustainable energy supply that is more benign environmentally.

Research and development efforts should begin now to accelerate the transition to renewable, nonpolluting energy sources. A sustainable future world energy supply is a must. Oil and gas reserves must be preserved for future production of petroleum's highest-value products—petrochemicals. Future generations will question why we burned so much of our valuable hydrocarbon resource.

Future unknowns are daunting. The changes we can anticipate are awesome. In 428 B.C., Pericles said, "The key is not to predict the future, but to be prepared for it." Future changes are complex, risky, and threatening, but they offer challenges and opportunities. Sir Winston Churchill admonished that worrying should translate into advance planning. We must plan ahead for the transition to renewable, nonpolluting energy sources in the twenty-first century.

ACKNOWLEDGMENTS

I thank members of EMARC in the Department of Geological Sciences at the University of Colorado for assistance in preparing this paper. Paul Weimer's editing and suggested reorganization greatly improved the manuscript. Braden J. Van Matre prepared the computer graphics with significant assistance from Andrew Pulham and David Knapp. Helpful conversations with Albert Bartlett (professor emeritus of physics, University of Colorado) and Thomas Ahlbrandt (U.S. Geological Survey, Denver) added to the technical relevance of this paper. Donna Edwards' editing improved the text.

REFERENCES CITED

Alelman, M. A., and M. C. Lynch, 1997, Fixed view of resource limits creates undue pessimism: Oil & Gas Journal, April 7, p. 56–60.

Appleby, P., 1996, Reserves debate: Oil & Gas Journal, January 29, p. 40.

Bartlett, A. A., 2000, An analysis of United States and world oil production patterns using Hubbert-style curves: Mathematical Geology, v. 156.

Beck, R. J., and M. Radler, 2000, Liquids, gas demand hikes expected in U.S. this year: Oil & Gas Journal, v. 98, no. 5 (January 31), p. 44–61.

Bensabat, L. E.,1999, U.S. fuels mix to change in the next 2 decades: Oil & Gas Journal, July 12, p. 46–53.

Bernabe, F., 1998, Cheap oil: Enjoy it while it lasts: Forbes, June 15, p. 84–86.

Bookout, J. F., 1989, Two centuries of fossil fuel energy: Episodes, v. 12, no. 4, p. 257–262.

Bowlin, M. R., 1999, The last days of the age of oil: Calls on U.S. energy industry to meet clean fuel challenge: Cambridge Energy Research Associates 18th Annual Executive Conference, Boston.

BP, 1998, BP statistical review of world energy, June.

BP Amoco, 1998, Where BP AMOCO stands on climate change: www.bpamoco.com/about/policies/climate.htm.

Campbell, C. J., 1994, The imminent end of cheap oil-based energy, Part 1: Historical development of oil: Sunworld, v. 18, 4 p.

Campbell, C. J., 1998, The end of cheap oil: Scientific American, March, p. 78–83.

Chang, T., 1999, U.S. consumption of alternative road fuels growing: Oil & Gas Journal, July 12, p. 37–39.

Crow, P., 1998, Fossil fuels becoming more sustainable: Oil & Gas Journal, November 2, p. 38–39.

Degolyer, E. L., and L. W. MacNaughton, 1992, Petroleum statistics twentieth century, 126 p.

Degolyer, E. L., and L. W. MacNaughton, 1998, Twentieth century petroleum statistics, 130 p.

Dunn, S., 1999, King Coal's weakening grip on power, World Watch, v. 12, no. 5, p. 10–19.

Duncan, R. C., and W. Youngquist, 1998, The world petroleum life-cycle, Seattle, Washington; Institute of Energy and Man, 19 p.

Ebel, B., 1998, It's time to lift trade barriers with China: International Association for Energy Economics Newsletter, winter.

Edwards, J. D., 1997, Crude oil and alternate energy production forecasts for the twenty-first century: The end of the hydrocarbon era: AAPG Bulletin, v. 81, p. 1292–1304.

Elkington, J., 1999, Cannibals with forks: The triple bottom line of 21st century business. Personal communication, keynote address, Offshore Technology Conference, Houston, Texas, May 4.

Energy Information Administration, 1998, International energy outlook, April: U.S. Department of Energy, Energy Information Administration.

Finley, P., and J. Krason, 1989, Basin analysis, formation and stability: Summary report: Geological evolution and analysis of confirmed or suspected gas hydrate localities, v. 15: U.S. Department of Energy DOE/MC/ 21181-1950, 111 p.

Fisher, W. L., 1991, Future supply potential of United States oil and natural gas: The Leading Edge, December, p. 15–21.

Flannery, B. P., 1999, Global climate change: International Association for Energy Economics Newsletter, third quarter, p. 4–10.

Flavin, C., and N. Lenseen, 1999, Nuclear power nears its peak: World Watch Institute, July-August, p. 36–38.

Foster, G. D., and L. B. Wise, 1999,Transnational environmental threats: Tomorrow's hidden enemy: World Watch Institute, May-June, p. 7–9.

Fouda, S. S., 1998, Liquid fuels from natural gas: Scientific American, March, p. 92–95.

Gas Research Institute, 1999, Policy implications of the GRI baseline projection of U.S. energy supply and demand to 2015: Gas Research Institute, 36 p.

Haq, B. U., 1998, Gas hydrates: Greenhouse nightmare? energy panacea or pipe dream?: GSA Today, v. 8., no., 11, p. 1–6.

Hefner, R. A., 1999, The age of energy gases: 10th Repsol-Harvard Seminar on Energy Policy, Madrid, Spain, p. 7–26.

Hubbert, M. K., 1969, Energy resources, *in* P. Cloud, ed., Resources and man: San Francisco, Freeman, p. 157–242.

International Energy Agency, 1998, World energy prospects to 2020, prepared for the G8 Energy Ministers Meeting, Moscow, March 31: International Energy Agency.

International Energy Agency, 1999, The future role of coal markets, supply and the environment: Coal Industry Advisory Board, International Energy Agency, 150 p.

Ivanhoe, L. F., 1996, Updated Hubbert curves analyze world oil supply: World Oil, November, p. 91–94.

Kahn, S. A., 1998, The opening of Venezuelan petroleum sector: New investment opportunities: International Association for Energy Economics Newsletter, winter, p. 4–6.

Klett, T. R., T. S. Ahlbrandt, J. W. Schmoker, and G. L. Dolton, 1997, Ranking of the world's oil and gas provinces by known petroleum volumes: U.S. Geological Survey Open-File Report 97-463, 17 p.

Krason, J., 1994, Study of 21 marine basins indicates wide prevalence of hydrates: Offshore, August, 2 p.

Laherrère, J. H., 1994, Published figures and political reserves: World Oil, January, p. 33.

Laherrère, J. H., 1999, World oil supply, what goes up must come down, but when will it peak?: Oil & Gas Journal, February 1, p. 57–64.

Linden, H. R., 1998, Flaws seen in resource models behind crisis forecasts for oil supply, price: Oil & Gas Journal, v. 96, no. 52, p. 33–37.

Lynch, M. C., 1999, Oil scarcity, energy security and long term oil prices, lessons learned and unlearned: International Association for Energy Economics Newsletter, first quarter, p. 4–8.

MacKenzie, J. J., 1996, Oil is a finite resource, when is global production likely to peak?: Washington, World Resources Institute, 22 p.

Masters, C. D., D. H. Root, and E. D. Attanasi, 1991, World resources of crude oil and natural gas, *in* Proceedings of the Thirteenth World Petroleum Congress: West Sussex, England, John Wiley and Sons, p. 51–64.

Masters, C.D., D. H. Root, and R. M. Turner, 1997, World resource statistics geared for electronic access: Oil & Gas Journal, October 13, p. 98–103.

Matske, R., 1999, Sasol, Chevron form global GTL partnership: Oil & Gas Journal, June 14, p. 30.

McCabe, P. J., 1998, Energy resources—cornucopia or empty barrel?: AAPG Bulletin, v. 82, p. 2110–2134.

Meyer, R. F., 1987, ed., Exploration for heavy crude oil and natural bitumen: AAPG Studies in Geology 25, 731 p.

Meyer, R. F., and W. DeWitt Jr., 1990, Definition and world resources of natural bitumens: U.S. Geological Survey Bulletin 1944, 14 p.

Minerals Management Service, 1998, An assessment of the undiscovered hydrocarbon potential of the nation's outer continental shelf: U.S. Minerals Management Service, 53 p.

Moody, J. D., 1978, The world hydrocarbon resource base and related problems, *in* G. M. Philip and K. L. Williams, eds., Australia's mineral resources assessment and potential: University of Sydney Earth Resources Foundation Occasional Publication 1, p. 63–69.

Odell, P. R., and K. E. Rosing, 1983, The future of oil, world oil resources and use: London, Nichols Publishing, 224 p.

Oil & Gas Journal, 1998, Worldwide look at reserves and production: Oil & Gas Journal, v. 96, no. 52 (December 28), p. 38–39.

Oil & Gas Journal, 1999a, BP Amoco buys out Enron's stake in joint solar power firm: Oil & Gas Journal, v. 97, no. 15 (April 12), p. 25.

Oil & Gas Journal, 1999b, ARCO, Syntroleum start up pilot GTL plant: Oil & Gas Journal, v. 97, no. 32 (August 9), p. 24.

Otto, K., C. Whitley, R. Gist, and A. Chandra, 1999,World LPG trade patterns poised for rapid change: Oil & Gas Journal, v. 97, no. 24 (June 14), p. 47–53.

Petzet, A., 1999, Decline in world crude reserves is first since 1992: Oil & Gas Journal, v. 97, no. 51 (December 20), p. 91–93.

Population Reference Bureau, 1999, World population data sheet: Washington, Population Reference Bureau, 12 p.

Potential Gas Committee, 1998, Potential supply of natural gas in the United States: Potential Gas Agency, Colorado School of Mines.

Riva, J. P., 1999, Is the world's barrel half full or half empty? It depends upon whether you are an economist or a geologist: M. King Hubbert Center for Petroleum Studies, Colorado School of Mines, 6 p.

Robinson, C. P., 1998, Critical energy and environmental issues in the next century: Where can technology make a difference?: United States Association for Energy Economics, December, p. 4–7.

Schmoker, J. W., and T. S. Dyman, 1998, How perceptions have changed of world oil, gas resources: Oil & Gas Journal, v. 96, no. 8 (February 23), p. 77–79.

Schollnberger, W. E., 1998, Projections of the world's hydrocarbon resources and reserve depletion in the 21st century: Houston Geological Society Bulletin, November, p. 31–37.

Science News, 1999, Acclimating to a warmer world: Science News, v. 156, August 28, p. 136–138.

Tempest, P., 1998, China petroleum: A sense of history in the making: International Association for Energy Economics, winter, p. 8–10.

Townes, H. L., 1993, The hydrocarbon era, world population growth and oil use—A continuing geological challenge: AAPG Bulletin, v. 77, no. 5, p. 723–730.

True, W. R., 1999, Worldwide gas processing continues to expand, shift balance: Oil & Gas Journal, v. 97, no. 24 (June 14), p. 41–46.

United Nations, 1998, World population prospects: The 1998 revision: New York, United Nations.

U.S. Geological Survey, 1995, National assessment of United States oil and gas resources: U.S. Geological Survey Circular 1118, 20 p.

van der Veer, J., 1997, Royal Dutch Shell reinventing itself as total energy company: Oil & Gas Journal, November 24, p. 29–36.

World Resources Institute, 1992, World Resources, part IV, no. 21, Energy and Materials, Table 21.2, p. 316, 317: Oxford University Press.

Youngquist, W., 1998, Geodestinies: The inevitable control of earth resources over nations and individuals: Portland, Oregon, National Book Co., 500 p.

Esser, R., 2001, Discoveries of the 1990s: Were they significant?, *in* M. W. Downey, J. C. Threet, and W. A. Morgan, eds., Petroleum provinces of the twenty-first century: AAPG Memoir 74, p. 35–43.

Chapter 3

Discoveries of the 1990s: Were They Significant?

Robert Esser

Cambridge Energy Research Associates, Cambridge, Massachusetts, U.S.A.

Oil discoveries of the 1990s, ranging from 4 billion to 9 billion barrels (bbl) per year, were significant in location but less so in size. They were smaller in both size and number than in the past and are projected to supply only about 20% of consumption in 2005. To meet demand of 28 billion bbl of oil in 1999 and projected demand of 35 billion bbl of oil in 2010, new production must have originated not only from recent discoveries but also from other sources, including:

- pre-1990s discoveries made economic and accessible with new technology and with political rapprochement, including full development of existing reserves in the Middle East, especially in Iraq and Saudi Arabia
- the likelihood of a continuation of the recent trend toward increased annual discovery rates
- development of extra-heavy oil in Canada and Venezuela
- growth in gas-related liquids: condensate and natural gas liquids (NGLs), and gas converted to liquids
- continued technological advances to reduce costs and increase recovery rates in mature fields by identifying bypassed reserves

The outlook for long-term liquid supply shows strong growth from the mid-1990s to 2010, adding 25.4 million bbl per day (BOPD) to supply, of which 6.4 million BOPD was added between 1995 and 2000 (see Table 1).

Unless oil-demand growth slows because of replacement by natural gas and other energy sources, increased efficiencies, or some new technology, the other sources listed above will be called on to far exceed the production expectations for 2010 and to meet demand in 2020 that could exceed 40 billion bbl. This growth in supply is necessary to overcome significant declines in production, caused by resource constraint projected, for the United States and western Europe and increasing depletion rates in existing production in all areas.

Table 1. Forecast of World Petroleum Production Capacity (million bbl/day)*

1995	2000	2010
72.9	79.3	100.3

**Source: Cambridge Energy Research Associates*

ECONOMIC FRAMEWORK OF THE 1990s

The 1990s have been characterized by two complete "boom-and-bust"cycles resulting in five distinct price movements, each putting its stamp on exploration and production (E&P). After the invasion of Kuwait in 1990, companies and some Middle East countries moved to increase their productive capacity, resulting in a rapid increase in surplus capacity and low prices by late 1993 and early 1994. In a reaction to lower prices, companies reduced budgets, increased layoffs, and pulled back to focus on core areas. By the mid-1990s, surging demand reversed the decline in prices, and companies renewed activity by moving to prospective regions worldwide to build up supply. This expansion attained boom conditions into late 1997 with large budget increases and was accompanied by fears of rig and personnel shortages.

Suddenly, in mid-1997, the collapse of the Asian economy (eliminating more than 2 billion barrels of oil per day [BOPD] of previously anticipated demand for 1999) and a record warm winter led to the collapse in oil prices throughout 1998 and into early 1999. This brought about the largest cutback in E&P budgets since the mid-1980s as companies again focused on reducing spending levels on core areas and engaged in stepped-up merger activity. However, something was different this time because many companies maintained aggressive E&P activity in the highly prospective deeper-water areas, spurred by the increasing availability of deep-water drill ships and rigs and the increasing discovery success rate. In spite of reduced budgets, 1999 exceeded 1997 as the peak year for discoveries in the 1990s. Suddenly, in early 1999, oil prices doubled because of production constraint. This was followed only by a selective rebound in activity (largely geared to gas in North America) and a hesitant investment outlook for 2000, especially outside North America.

DISCOVERY CHARACTERISTICS OF THE 1990s

The dominant discovery pattern of the 1990s was the evolution of deep water as the primary new exploration target. This was evident in the Gulf of Mexico, West Africa (especially Angola and Nigeria), and Brazil, and was made possible by the combination of the surprising individual-well productive capability (a reflection of the high reservoir quality and completion techniques) and the large reduction in per-barrel costs resulting from new technology and the decrease in the discovery-to-production time period. Another discovery pattern was the shift in the early 1990s to increased gas exploration, especially in Southeast Asia, resulting in annual discovery rates of 40–60 trillion cubic feet (tcf) of gas and the construction of six grass-roots (new) liquefied-natural-gas (LNG) liquefaction facilities.

As a result of the shift of activity to high-potential frontier and newly opened, underexplored areas, the pace and size of the discoveries during the past few years increased. During the 1990s, 10 giant oil fields with recoverable reserves exceeding 1 billion bbl of oil each were discovered (Figure 1). These fields are located in Algeria (Oughroud), Brazil (Barracuda and Roncador), Colombia (Cusiana/Cupiagua), Angola (Girassol and Dalia), Nigeria (Agbami), Iran (Azadegan), Mexico (Sihil), and the U. S. Gulf of Mexico (Crazy Horse). In addition, there were 10 discoveries exceeding 0.5 billion barrels each. In the case of gas, 13 fields larger than 5 tcf each were discovered. These are located in Australia (Perseus, Evans Shoal), Azerbaijan (Shak Deniz), Indonesia (Peciko, Wiriagar), Oman (Saih Rawl), Papua New Guinea (Hides), Myanmar (Yadana), Russia (Shtok-

Figure 1. Giant discoveries of the 1990s: fields with reserves greater than 1 billion bbl of oil or 5 tcf of gas.

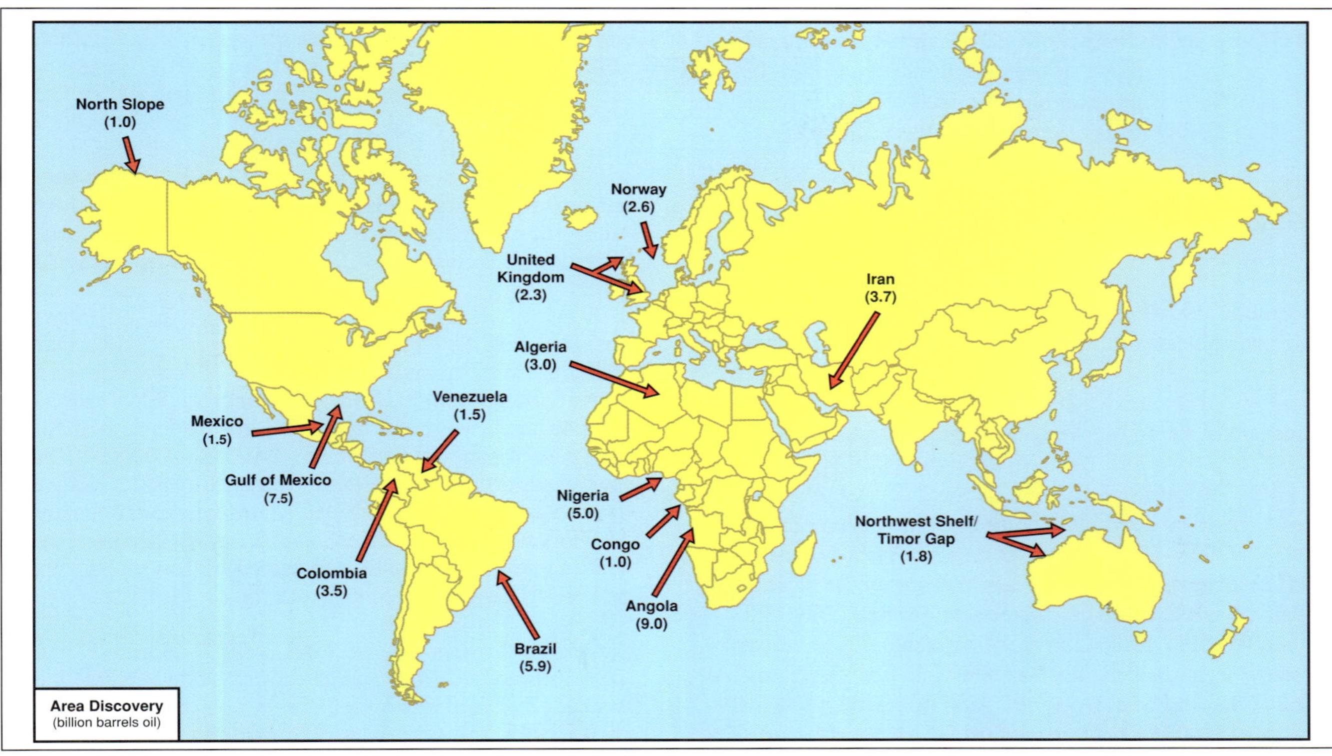

Figure 2. Areas with oil discoveries exceeding 1 billion bbl, 1990–1999. Numbers in parentheses are reserves in billions of barrels.

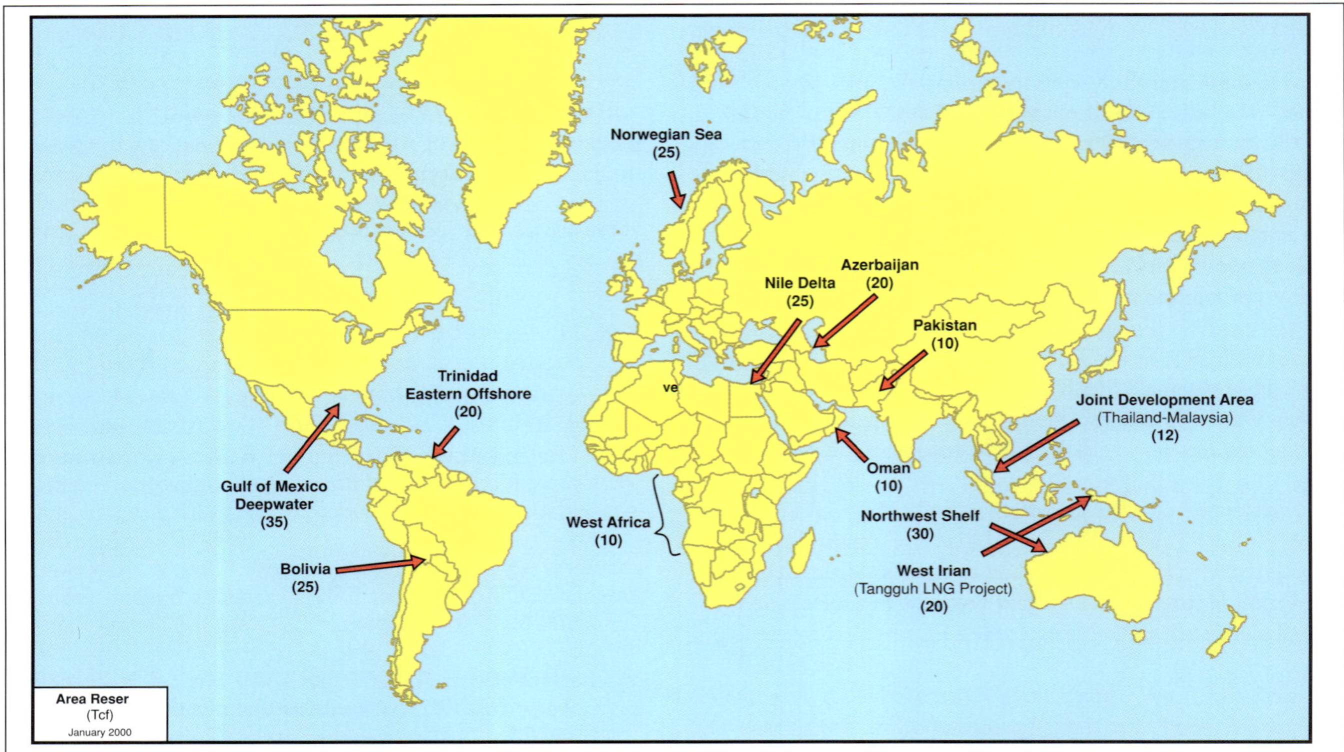

Figure 3. Areas with gas discoveries exceeding 10 tcf, 1990–1999. Numbers in parentheses are reserves in trillions of cubic feet.

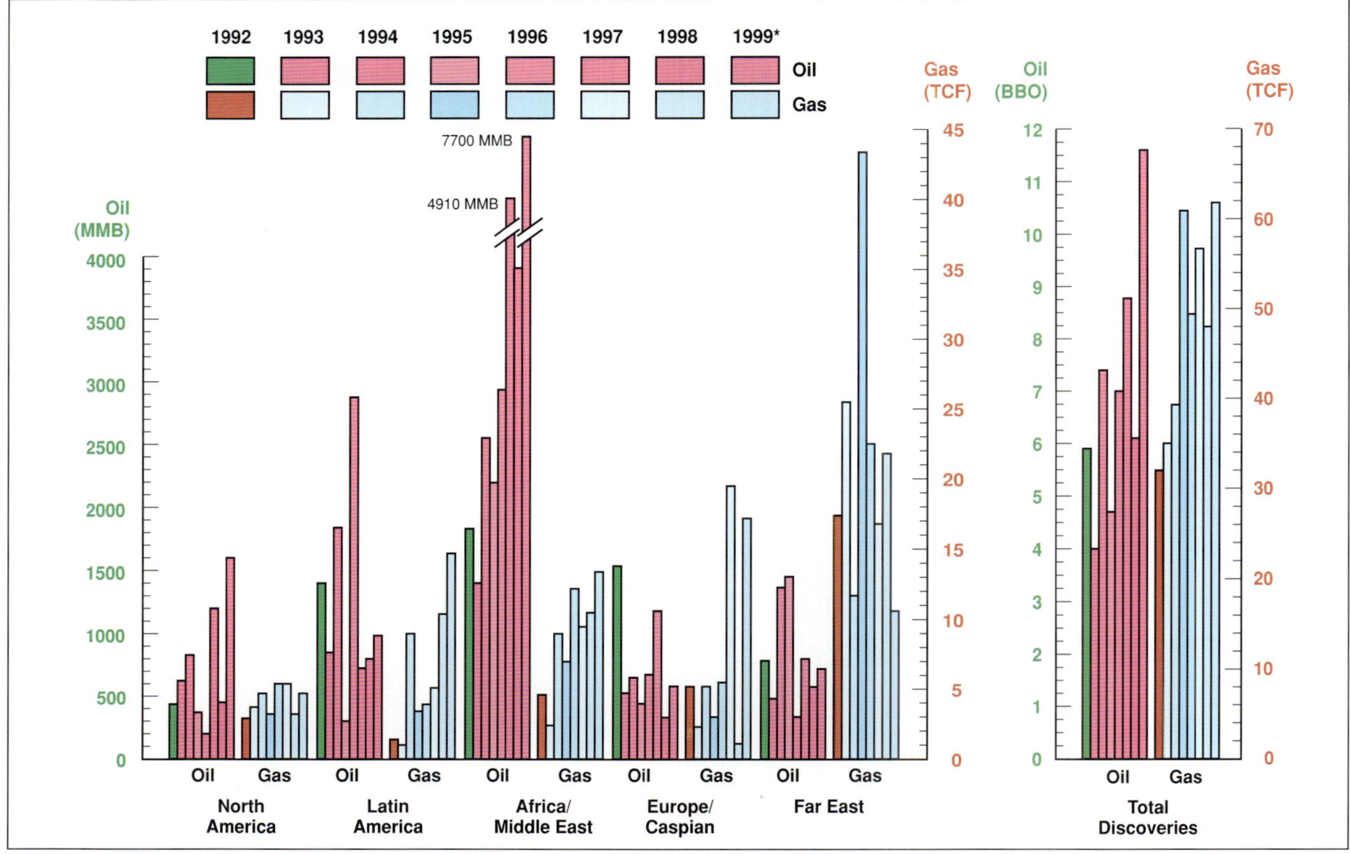

Figure 4. World oil and gas discoveries, 1992–1999.

manovskoye, Russanovskoye), Norway (Orman Lange), and Bolivia (Itau and San Alberto).

The most significant areas of discovery in the 1990s consisted of clusters of oil discoveries (Figure 2) led by Angola, the Gulf of Mexico, and Brazil. Gas discoveries (Figure 3) were led by the Gulf of Mexico, the Norwegian Sea, the North West Shelf of Australia, Indonesia (West Irian), Egypt, and Bolivia. It is these clusters of fields that resulted in large-area discoveries similar in size to some of the major discoveries of the 1970s.

As for discovery totals for the 1990s (Figure 4), Africa/Middle East exceeded by far all other areas for oil, with discoveries of 27 billion barrels of oil. The Far East led for gas, with discoveries of 175 tcf. However, major discoveries occurred throughout the decade in different areas. Examples include 1999 in North America with the discovery of Crazy Horse field in the Gulf of Mexico; 1996 in Latin America—Roncador field in Brazil; 1992 in Europe—Foinavon field, located west of the Shetlands, and Norne field in Norway; and 1994 and 1995 in the Far East—the Laminaria and Bayu Undan discoveries in the Timor Gap area.

INCREASED PACE AND SIZE OF NEW DISCOVERIES INTO THE NEW DECADE

Emphasis on the high potential E&P "hot spots," especially deep water, has resulted in an increase in the pace and size of new discoveries (Figure 5). Areas with the largest number of significant discoveries greater than 100 million bbl and 1.0 tcf between 1996 and 1999 include Angola, the Gulf of Mexico, Nigeria, Norway, Egypt, and the United Kingdom. The number of significant discoveries in recent years is increasing, with 50 discoveries in 1999, exceeding the 1998 level of 40 discoveries and up from 32 discoveries in 1996. The increased discovery level is a result of the following:

- There has been growing emphasis on highly prospective ultradeep-water areas characterized by large structures with high-quality reservoirs. These areas include the Gulf of Mexico, West Africa, and Brazil. The majority of the larger discoveries in recent years has been made in water deeper than 3000 ft.
- The late 1990s saw a rapid increase in the availability of drilling equipment capable of operating in waters deeper than 5000 ft.
- Countries previously closed to outside investment are opening as a result of political change or the privatization of national oil companies. Besides some of the Middle East countries, examples include the republics of the former Soviet Union, Algeria, Argentina, Venezuela and, most recently, Brazil.
- Discoveries have been made in areas undergoing initial exploration.

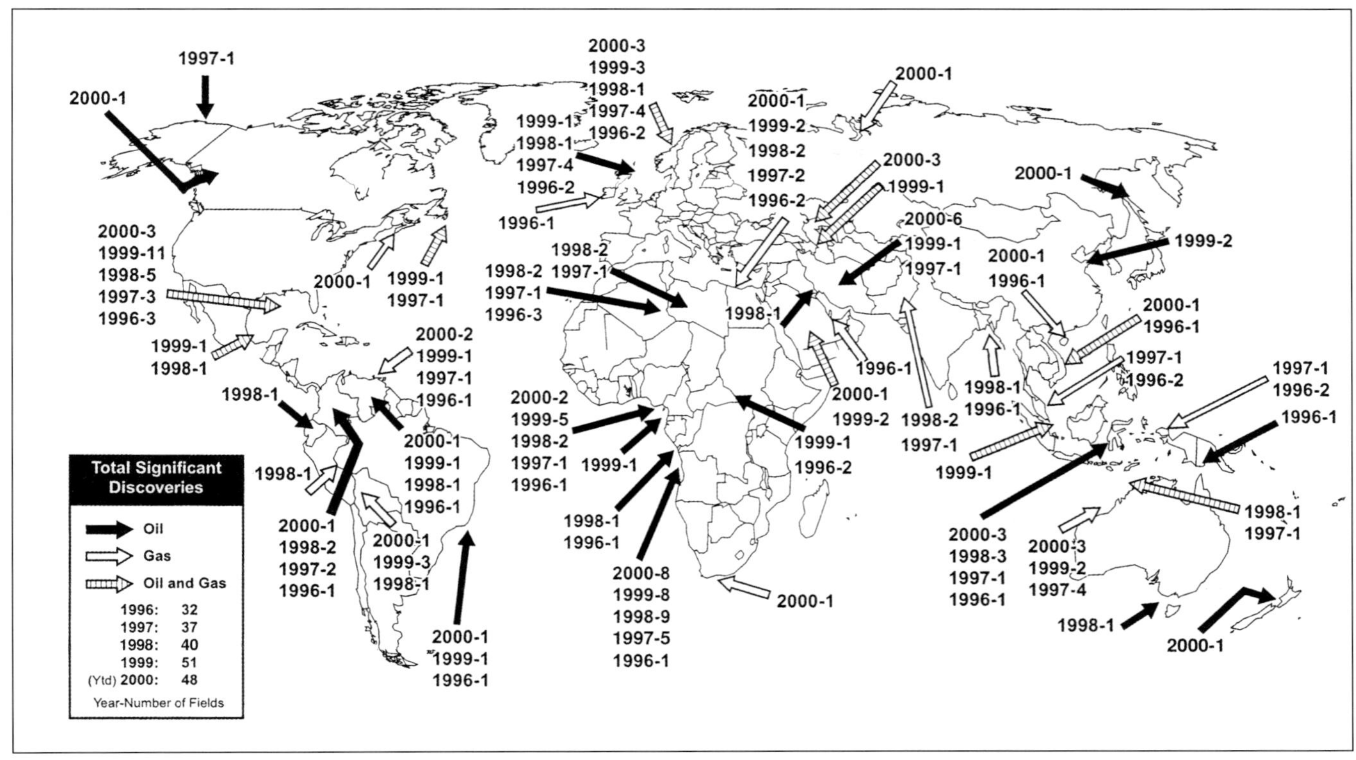

Figure 5. Significant discoveries (reserves larger than 100 million bbl of oil or 1 tcf of gas), 1996–2000.

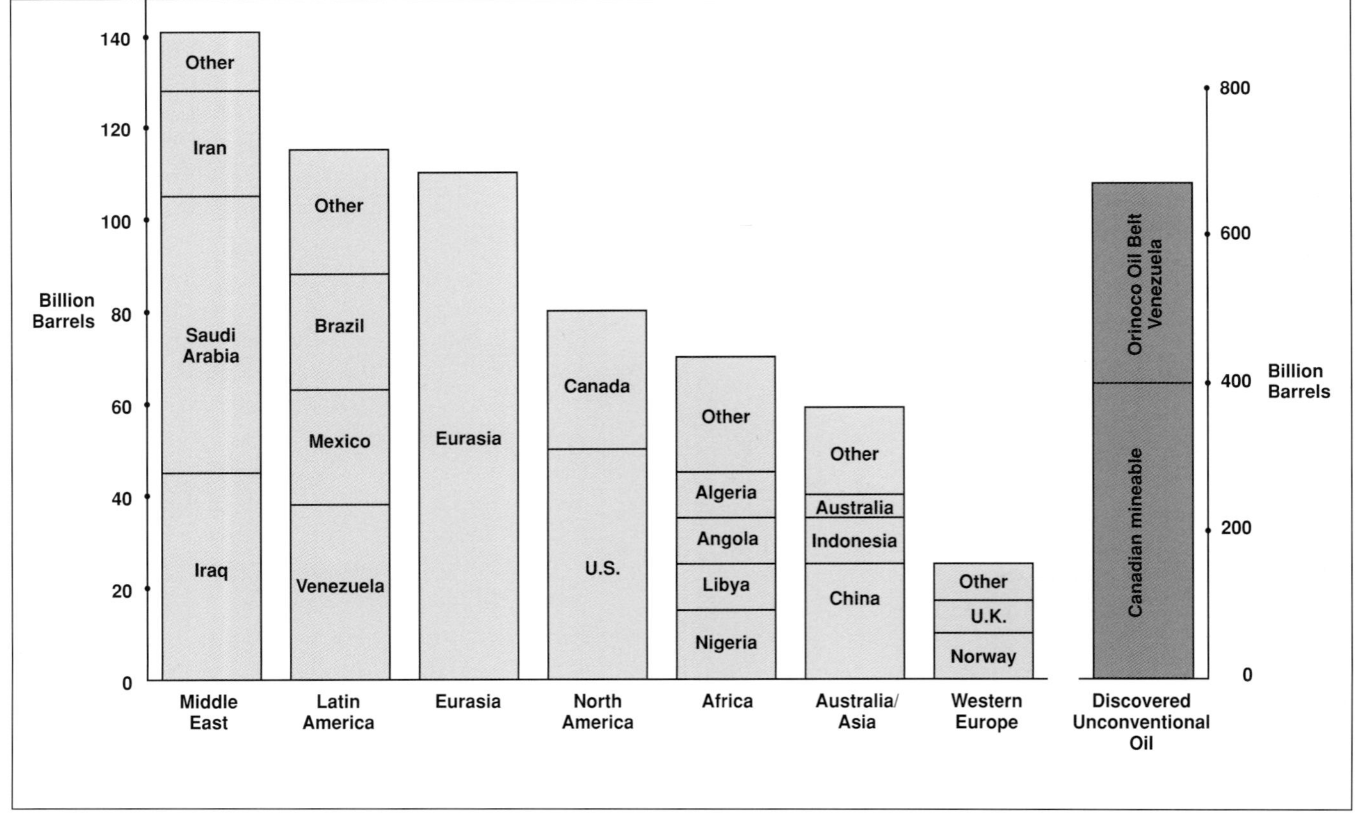

Figure 6. An estimate of the 600 billion bbl of world undiscovered oil-potential reserves (mean value).

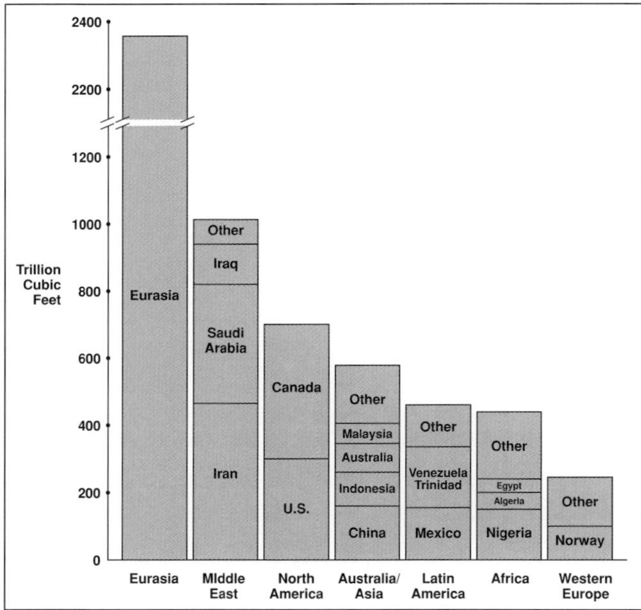

Figure 7. An estimate of the 5761 tcf of world undiscovered gas-potential reserves (mean value).

Because of the high prospectivity of these areas, it is probable that annual discovery rates will continue to increase in the new decade, eventually yielding more production than the discoveries of the 1990s. Thus, annual discovery rates exceeding 10 billion bbl are possible in the future, especially considering that in 1998–1999, the main focus of the newly minted supermajor oil companies was directed toward merger-related factors and away from exploration. Also, one very large major company was cutting fundamental investment levels. Thus, aggressive exploration and development activity from the largest companies has been constrained since 1997. The slow opening of the major countries of the Middle East to Western investment, leading to the revival of exploration, will eventually raise the discovery level in an area where exploration has been at an ebb for almost 20 years. Libya will also be the scene of increased exploration.

WORLD OIL AND GAS UNDISCOVERED POTENTIAL

The location of future discoveries is dependent on expectation of the potential for undiscovered oil and gas. Estimates for global oil and gas undiscovered potential are based on work by Charles D. Masters of the U.S. Geologic Survey (1994). In certain cases, such as Brazil, Colombia, the Gulf of Mexico, Angola, and Algeria, these estimates have been modified by Cambridge Energy Research Associates to accommodate recent discovery experience (Figure 6). Estimates of undiscovered conventional oil are similar in the Middle East, Latin America, and Eurasia. North America, Africa, and Australasia are in the second tier. It is within this tier (the Gulf of Mexico and West Africa) that the current large discoveries are being made. What is most important for the source of new production is the size of the extra-heavy unconventional oil resource of 675 billion bbl in Canada and Venezuela, which exceeds the potential undiscovered conventional oil estimate of 600 billion bbl.

The undiscovered gas potential of Eurasia is 41% of the world total of 5791 tcf (Figure 7). This is followed by the Middle East and North America. Interestingly, the leading area of the 1990s discoveries—the Far East—rates fourth in potential, behind North America.

E&P "HOT SPOTS"

The 1990s

During the 1990s, E&P activity was concentrated on individual company core areas that, collectively, are the global E&P "hot spots" (Figure 8). A few observations regarding the hot spots include:

- Deep-water areas of the Gulf of Mexico, Brazil, and West Africa lead in E&P activity.
- The Caspian Sea area, especially Azerbaijan and Kazakstan, continues to retain the high mid-1990s priority of many companies, even though the southern Caspian Sea has likely turned out to be gas-condensate prone, based on recent drilling. The Kashagan East test well in the northern Caspian Sea, with the objective of a Tengiz oil field look-alike, recently discovered a giant oil field.
- Other areas with strong E&P activity primarily related to oil include the eastern Canadian offshore, portions of the Atlantic Margin west of the Shetlands, the Berkine Basin of east-central Algeria, Central Africa (Sudan and Chad), and Iran.
- Gas-oriented hot spots, to supply increasing demand expectations as pipeline gas and as LNG, include eastern Canada, Bolivia, Nigeria, Qatar, Egypt, Libya, Oman, Pakistan, many countries in Southeast Asia, and the North West Shelf of Australia. Considerable condensate and NGLs will be associated with this gas.
- Unconventional extra-heavy oil operations are proceeding in the minable oil sands in Alberta, Canada, and the Orinoco oil belt in Venezuela. Production from each of these areas is projected to reach more than 1 million bbl/day by 2010.
- Hot spots of the 1970s and 1980s, which included Mexico, eastern Canada, the North Sea, Brazil, Qatar, Iraq, and the North West Shelf of Australia, remained as hot spots of the 1990s and will probably continue to see considerable activity. In some cases, drilling is being perpetuated by adding the domain of very deep water as an exploration objective.

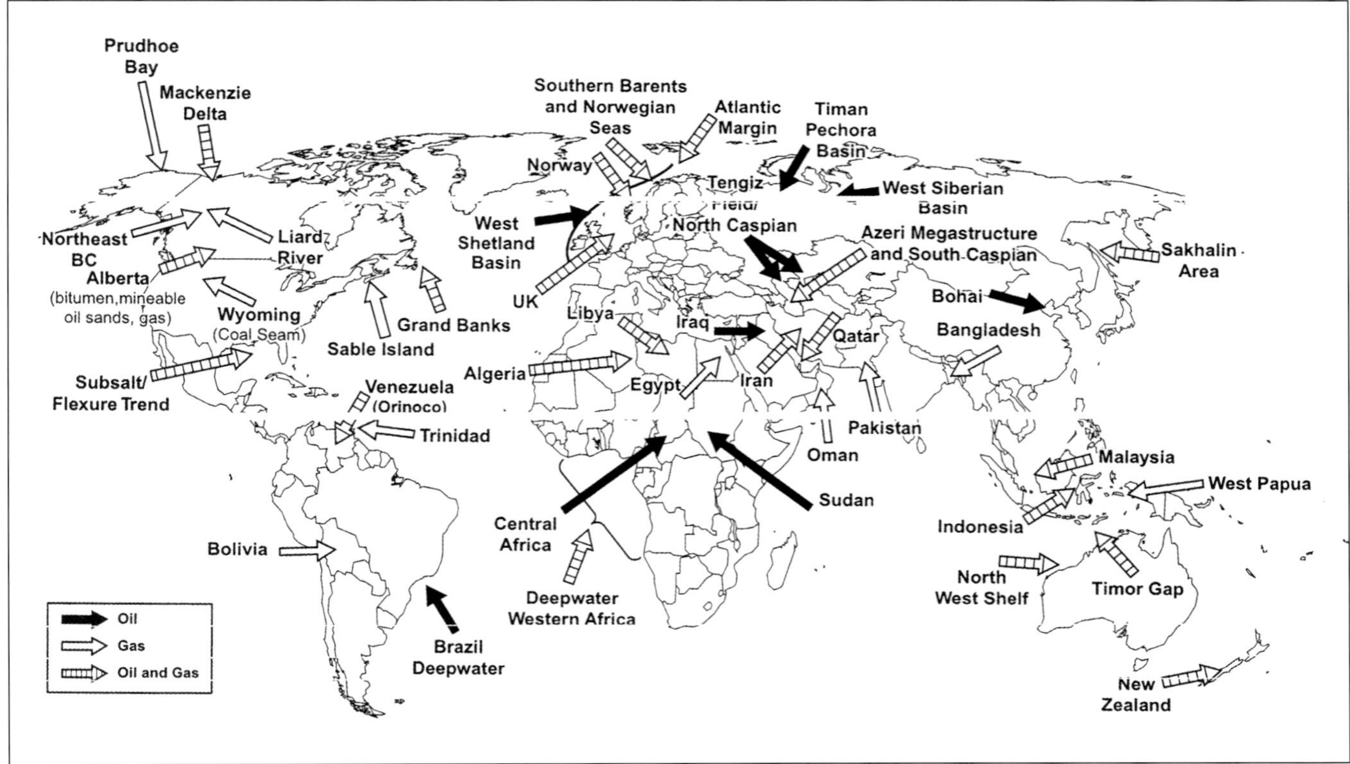

Figure 8. Major E&P hot spots of the 1990s.

Tomorrow's Hot Spots

In spite of the volatile crude-oil prices of 1998 and 1999, the industry has continued its quest for future hot spots, including the movement into deeper water, with leasing progressing beyond the 10,000-ft water depth (Figure 9). Two areas (the Falkland Islands and the Eritrean offshore) were unsuccessfully tested in 1999. In some cases, exploration drilling in future hot spots has occurred, resulting in some discovery and production. Examples include west Newfoundland, Mongolia, and Guatemala. In other areas, some larger discoveries have been made, but development efforts have stalled. Examples include the Norwegian portion of the Atlantic Margin, the East African gas play, the Russian Arctic, and east Siberia. In 2000, drilling was expected off western Greenland and the Atlantic Margin and, in 2001, in the Faeroe Islands. A similar map of future hot spots, as visualized in 1990, would have shown expectations for future E&P activity in areas with 1990s discoveries such as the Timor Gap/West Irian area, west of the Shetlands, the Egyptian offshore, the Berkine Basin in Algeria, the deep-water portions of West Africa, and the Philippines offshore.

In the United States, many highly prospective areas, such as the Arctic National Wildlife Refuge (ANWR) and the entire East Coast and West Coast offshore areas, remain off limits for exploration. However, these areas might well appear on a subsequent map of future hot spots.

FUTURE PRODUCTION LEVELS

Contribution of 1990s Discoveries

Development of the estimated 60 billion bbl of oil discovered in the 1990s is projected to yield more than 17 million bbl/day of production, 20% of projected 2005 demand of 85 million bbl/day. Of the 17 million bbl/day, 13 million bbl/day will be initial production in the late 1990s, peaking in 2005–2007. More than half of the total-capacity additions will emanate from four areas, including 3.2 million bbl/day from the Gulf of Mexico and 1.5 million bbl/day each from Angola, Nigeria, and Brazil. The timing of the appearance of much of the new productive capacity, after 2000, is related to the mid-1990s increase in discoveries in deeper water and the additional time required to appraise and develop those discoveries. The increase in production from deep water reflects the lowering of the water-depth "barrier" to production. In the 1980s, the barrier was 1800 ft of water, with the deepest oil production from 1760 ft. In the 1990s, oil production occurred from 6079-ft water depth in Roncador field in Brazil. By late 1999, water-depth production "barriers" were 6500 ft for oil and 5500 ft for gas. Recent developments show that the 7500-ft water-depth barrier will be surpassed by 2003, and both drilling and producing technology and the ability to deal with associated gas will allow production from 10,000-ft water by 2010.

Figure 9. Potential E&P hot spots of tomorrow.

Other Sources of Future Production

In addition to the 1990s discoveries, considerable new production to meet future demand will result from discoveries made before the 1990s, from unconventional extra-heavy oil and gas-related liquids, and from new technology and increased recovery rates.

Pre-1990s Discoveries Made Accessible and Economic

To supplement the production added by discoveries made in the 1990s, the development of fields discovered in the 1970s and 1980s (with some initiating production in the 1990s) will be required to meet demand. Fields in this category are projected to produce as much as 8.0 million bbl/day by 2007. They are projected to include undeveloped or partially developed fields in areas such as the Caspian (3.7 million bbl/day) and Brazil deep water (1.0 million bbl/day), plus 0.75 million bbl/day each from eastern Canada offshore, Venezuela, and western Canadian bitumen, with lesser amounts of heavy oil from California, the Gulf of Mexico deep water, the Beaufort Sea of Canada, the North Slope of Alaska, and eventually the Sakhalin area. The major pre-1990s discoveries under development and currently ramping up production include Tengiz field in Kazakstan and the Azeri megastructure in Azerbaijan in the Caspian Sea; Marlin and Albacora fields in Brazil; Hibernia and Terra Nova fields offshore eastern Canada; and Mars, Auger, and Brutus fields in the Gulf of Mexico.

Large existing undeveloped reserves in Saudi Arabia and the backlog of giant, undeveloped 1970s discoveries in Iraq will also be sources of major new production in the years after 2000. Development of new productive capacity in the Middle East will accelerate after 2005; more than 15 million bbl/day of new capacity is projected to be added beyond 2010. In addition, many of the fields in the Middle East have not yet benefited from modern technology, deep drilling below existing production, infill drilling, and secondary and enhanced oil-recovery techniques. An example of deeper drilling in known fields is in Ghawar field in Saudi Arabia, where major reserves have been discovered below the current production. The main producing countries of the Middle East will benefit from Western investment as major companies seek low-cost areas in which to invest.

Development of Extra-heavy Oil in Canada and Venezuela

New technology has reduced the cost of development of minable oil sands in Canada and production from the Orinoco oil belt in Venezuela. Volatile oil prices have resulted in hesitation in some Orinoco projects, but by 2010, production of at least 1.0 million bbl/day is projected. New horizontal drilling technology has increased individual well productivity in the Orinoco oil belt to more than 1000 bbl/day. Development of minable oil-sands expansions in Canada continued throughout the period of low oil prices. Expansions and new projects will result in production from this source of 0.9 million bbl/day by 2007 and well over 1.0 million bbl/day by 2010, compared with 0.35 million bbl/day in 1999. Expansion of these areas will continue to add large amounts of new capacity in both Canada and Venezuela beyond 2010.

Natural Gas, Gas-related Liquids/Condensates, and LNG

The shift in strategy to emphasizing gas exploration in the early 1990s resulted in the discovery of considerable gas, especially in Southeast Asia. The discovery of more than 450 tcf of gas in the 1990s is expected to be equaled in the next decade to supplement gas from the development of 1990s and pre-1990s discoveries, such as the North supergiant gas field in Qatar, to meet strong growth in global gas demand. Also, the flaring of large amounts of associated gas in a "post-Kyoto" world is likely to become untenable, leading to both initial local consumption and gas-to-liquids projects after initial use in reservoir pressure-maintenance operations in both stranded associated and nonassociated gas fields. Much of this gas will be delivered as LNG to both Asian markets and as spot supplies elsewhere.

Most of this gas is associated with condensate and will be processed to separate the NGLs. The outlook for these liquids is for strong growth to 2010, adding 2.7 million bbl/day from 1995 to 2000 and 6.0 million bbl/day between 2000 and 2010 (Table 2). In addition, the application of gas-to-liquids technology to exploit remote or "stranded" gas resources could eventually add 1–2 million bbl/day of products by 2020.

The future of LNG, considered lackluster in the mid-1990s, has brightened with the construction of six grass-roots liquefaction facilities, two of which have been greatly expanded during the initial construction phase. Because of the expansion of some of the existing producing facilities and new capacity from the North West Shelf of Australia and West Irian, LNG will add to supplies beyond 2005 as demand surges in China, India, and Pakistan.

Gas-demand growth is projected to continue to accelerate beyond 2000 as countries convert their energy base to natural gas and demand increases in those countries already using gas. In addition, the struggle to supply sufficient gas to the United States will require increasing imports from Canada and LNG from global sources.

New Technology and Increased Recovery Rates

During the 1990s, the application of new technology—especially 3-D seismic, geosteering (horizontal drilling), and new well-completion techniques—has contributed to the reduction of exploration risk and of finding and developing costs. Not only the discovery of hard-to-find fields (such as subsalt discoveries) but also the development of wedge-edge pay zones has been made economic with the advances in technology. As much as 3.0 million bbl/day of oil-capacity additions in the 1990s were made possible with horizontal and extended-reach drilling. In the future, the technological improvements will center around the second generation of geosteering, acquiring information ahead of the drill bit, thus providing an instantaneous correlation of subsurface well data to seismic data. The combination of expected technological improvements and the application of enhanced recovery techniques, including improved reservoir fracturing and stimulation to older fields throughout the world, is projected to continue to rise.

Table 2. Global Gas-associated Liquids Outlook (million bbl/day).*

	1995	2000	2010
Condensate	2.9	4.3	7.4
NGLs	5.2	6.5	9.4
Total	8.1	10.8	16.8

**Source: Cambridge Energy Research Associates*

CONCLUSION

The giants of the 1990s were significant in that they confirmed deep water as a primary source of new oil production. In addition, political rapprochement and privatization efforts have provided a second source of new production. A trend toward increased discovery rates in the late 1990s was projected to continue beyond 2000, but future production from this source will only partially meet the growing demand for oil. Continued development of fields discovered before the 1990s and the heavy-oil resources of Canada and Venezuela, combined with further technological breakthroughs that can raise overall recovery rates, will be required to supply the demand for oil.

The switch to an emphasis on gas in the early 1990s resulted in the discovery of large gas reserves. In spite of the temporary weakness in demand for LNG in the late 1990s, new liquefaction facilities are under construction and older facilities are being expanded, using gas from recent discoveries to meet the overall increasing growth in demand for gas.

REFERENCE CITED

Masters, C. D., E. D. Attanasi, and D. H. Root, 1994, presentation to 14th World Petroleum Congress, Stavanger, Norway.

Deming, D., 2001, Oil: Are we running out?, *in* M. W. Downey, J. C. Threet, and W. A. Morgan, eds., Petroleum provinces of the twenty-first century: AAPG Memoir 74, p. 45–55.

Chapter 4

Oil: Are We Running Out?

David Deming
School of Geology and Geophysics, University of Oklahoma, Norman, Oklahoma, U.S.A.

ABSTRACT

Predictions of imminent oil shortages have been made throughout the twentieth century. Although all previous predictions have been false, in recent years a new generation of predictions based on the Hubbert model has become ascendant and has attracted media attention. The Hubbert model assumes that a resource is limited and finite. Although conventional oil supplies are finite, it has proved difficult to estimate the size of the ultimate resource. In the last 50 years, estimates of the size of the world's conventional crude-oil resources have increased faster than cumulative production. The estimated size of the ultimate resource base will continue to increase in the future as unconventional fossil fuels come on-line. Oil production from Canadian tar sands has already begun. Unconventional oil resources such as tar sands and oil shales are likely to replace conventional oil and ensure a supply of petroleum for about 100 to 1000 years. The only uncertainty concerns the nature of the transition from conventional to unconventional oil resources. The transition may be slow and seamless with no economic disruptions, or it may be characterized by a difficult transition period.

In the long run, nuclear power has the potential to provide large amounts of power for very long periods of time if low-grade uranium is used in breeder reactors. The technology and resources to use nuclear power already exist. Limitations on the energy used by our technological civilization are not imposed by finite resources but by social and political attitudes.

INTRODUCTION

In the past few years, several predictions of imminent oil shortages have been made. In a widely publicized *Scientific American* article, Campbell and Laherrère (1998) wrote, "What our society does face, and soon, is the end of the abundant and cheap oil on which all industrial nations depend." Hatfield (1997) subtitled his short paper in *Nature* with "A permanent decline in global oil production rate is virtually certain to begin within 20 years." Ivanhoe (1995) warned of impending economic doom resulting from an imminent oil shortage:

> The global price of oil after 1999 should follow the simplest economic law of supply vs. demand—resulting in a major increase in crude and all other fuels' prices.... After the associated economic implosion, many of the world's developed societies may look more like today's Russia than the U.S.

Based on the work of Campbell and Laherrère (1998) and others, science reporter Richard Kerr (1998) authored an article in *Science* titled "The Next Oil Crisis Looms Large—and Perhaps Close."

The availability of crude oil is a critical issue. It is hardly an exaggeration to state that crude oil is the lifeblood of the world's industrial and technological civilization. Although coal and nuclear power can be used for the generation of electricity, there is no ready substitute for petroleum and natural gas in the transportation field. If a permanent decline in oil production is imminent, then some allocation of public and private resources toward the development of alternative energy sources may be a prudent investment. However, if such a decline is not imminent, then the development of alternative energy sources could be a wasteful diversion of valuable resources.

A HISTORICAL PERSPECTIVE

Warnings of oil shortages have been made for most of the twentieth century. In 1950, Leonard M. Fanning wrote "A Case History of Oil-Shortage Scares," in which he describes six such episodes until 1950. These are:

1. The Model T Scare, 1916
2. The Gasless Sunday Scare, 1918
3. The John Bull Scare, 1920–1923
4. The Ickes' Petroleum Reserves Scare, 1943–1944
5. The Cold War Scare, 1946–1947
6. The Cold Winter Scare, 1947–1948

To these, we may add two scares brought about by Middle Eastern politics in the 1970s:

7. The Arab Embargo Scare, 1973
8. The Iranian Revolution Scare, 1979

Doomsayers and their critics have a long history. Fanning (1950, p. 343) quoted Sir Edward Mackay Edgar as stating in 1922,

> Before long there will be a smash. An economic crisis is approaching. American demand for metals, cotton, and oil is so insatiable that a world-wide shortage in these commodities is inevitable. One hundred and fifteen million people are feverishly tearing from the Earth its irreplaceable wealth, using it to maintain a rate of growth unprecedented in all human history....

As long ago as 1918, A. J. Hazlett (quoted in Fanning, 1950, p. 322) wrote in the *Oil Trade Journal*:

> At regularly recurring intervals in the quarter of a century that I have been following the ins and outs of the oil business there has always arisen the bugaboo of an approaching oil famine, with plenty of individuals ready to prove that the commercial supply of crude oil would become exhausted within a given time—usually only by a few years distant.

Hazlett's words were echoed three years later, in 1921, by Thomas A. O'Donnell (quoted in Fanning, 1950, p. 342), president of the American Petroleum Institute, who said,

> We still have with us a rather overproduction of superscientists who are constantly measuring the petroleum resources of the world and point out various kinds of disaster following its exhaustion within a few years.

In 1920, the U.S. Geological Survey estimated that the world's supply of conventional recoverable petroleum was 60 billion barrels (bbl) (Fanning, 1950, p. 331). This proved to be slightly below the mark. By 1998, 800 billion bbl of oil had been produced, with 850 billion bbl in reserve (Campbell and Laherrère, 1998).

Written near the peak of the 1979 Iranian Revolution scare, the 1981 edition of a respected and widely used textbook on economic geology (Jensen and Bateman, 1981, p. 6) stated:

> The energy crisis has brought home ... that oil and gas reserves are finite. Such a situation was given little recognition by laymen even a few years ago, and the mineral economist feared overstating the case because of geological success in locating new petroleum deposits. They had cried "wolf" too many times in the past. Finally, however, the wolves are with us.

A figure from Jensen and Bateman's book (Figure 1) shows the United States entering an incipient 125-year "energy gap," projected to peak shortly after the year 2000. Less than 20 years later, in February 1999, U.S. gasoline prices were at a nearly all-time low. This

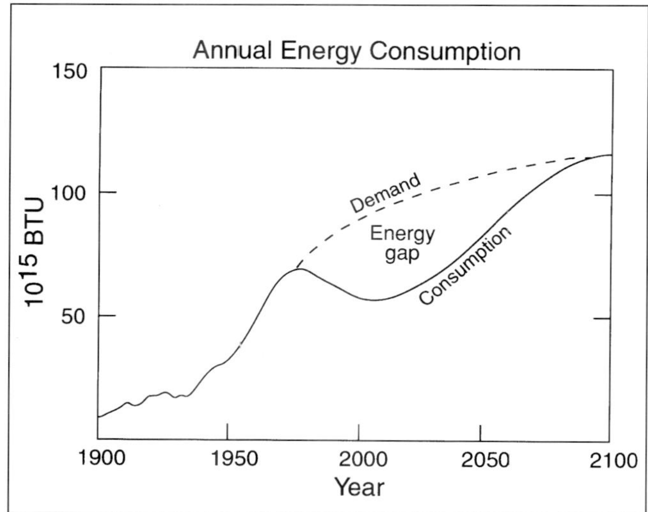

Figure 1. Past and predicted U.S. energy consumption, from the 1981 edition of a textbook on economic geology (after Jensen and Bateman, 1981, p. 10).

Figure 2. Cartoon illustrating that in February 1999, crude oil in the United States was cheaper than bottled water (copyright February 13, 1999, *The Daily Oklahoman*).

prompted one editorial cartoonist to picture oil in a race with bottled water to see which could attain the lowest price (Figure 2).

RESERVES, RESOURCES, AND THE ENERGY PYRAMID

Fossil fuels can be broadly categorized as either resources or reserves (Bird, 1989). Resources includes all fuels, both identified and unknown. Reserves are that portion of identified resources which can be economically extracted and exploited by using current technology. Reserves grow as technology changes, and new resources are discovered through ongoing exploration. It is therefore a fallacy to predict oil shortages by dividing reserves by current consumption rates. McCabe (1998) showed that in the 80 years from about 1915 through 1995, reserves of U.S. crude oil grew at the same rate as consumption (Figure 3). For 80 years, U.S. reserves of crude oil have ranged between 10 and 18 years of current consumption. McCabe (1998) concluded crude-oil reserves are comparable to food stocks held in a pantry or warehouse which are constantly replenished.

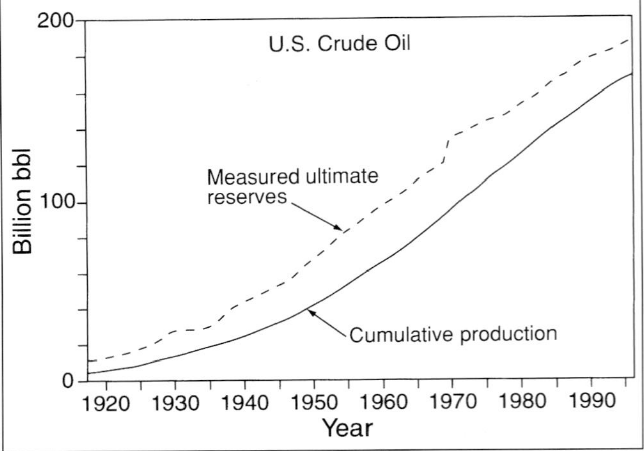

Figure 3. Reserves of U.S. oil versus cumulative production (after McCabe, 1998, p. 2114).

Surprisingly, estimates of both U.S. and world crude-oil resources (Figure 4) have also increased over time (McCabe, 1998). That increase of resources can be explained by the concept of the energy pyramid (Figure 5). A relatively small quantity of high-quality, low-cost hydrocarbons exists. Those resources constitute the top of the energy pyramid. However, there are larger quanti-

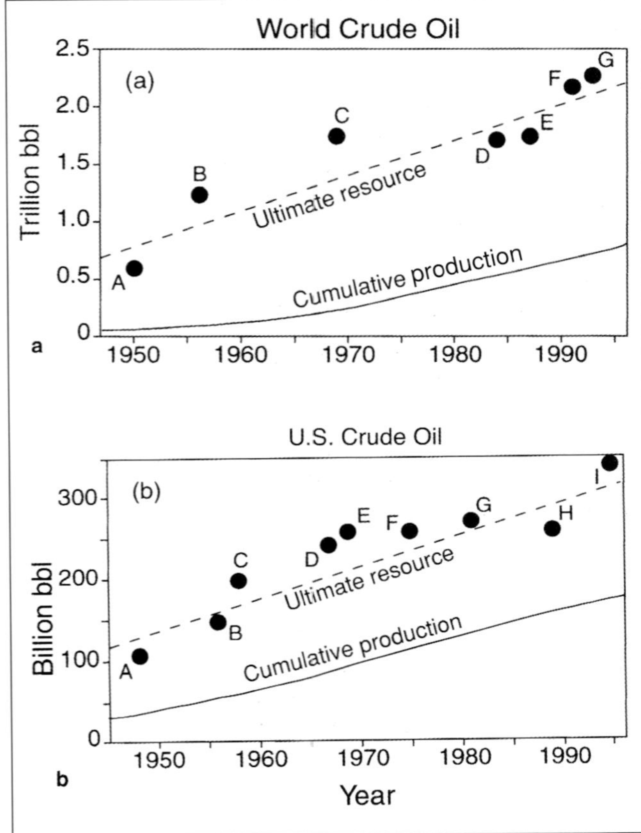

Figure 4. (a) Historical estimates of the total size of the world's crude-oil resource versus cumulative world production. Resource estimates are those selected by McCabe (1998) and are keyed as follows: A (Weeks, 1950), B (Hubbert, 1957), C (Hubbert, 1969), D (Masters et al., 1984), E (Masters et al., 1987), F (Masters et al., 1991), G (Masters et al., 1994). (b) Historical estimates of the total size of the U.S. crude-oil resource versus cumulative U.S. production. Resource estimates are those selected by McCabe (1998) and are keyed as follows: A (Weeks, 1948), B (Hubbert, 1957), C (Weeks, 1958), D (Hubbert, 1967), E (Hubbert, 1969), F (Miller et al., 1975), G (Dolton et al., 1981), H (Mast et al., 1989), I (U. S. Geological Survey National Oil and Gas Resource Assessment Team, 1995, and Minerals Management Service, 1996).

ties of lower-quality, higher-cost hydrocarbon resources. At present, the top of the energy pyramid contains conventional oil and gas resources. Simply stated, these are hydrocarbons which can be extracted by drilling holes in the ground and pumping them out. However, even if coal is excluded, conventional oil and gas constitutes less than 5% of sedimentary hydrocarbons (Dusseault, 1997). As time passes, man utilizes lower tiers of the energy pyramid, and resources which were formerly unknown or considered uneconomic are added to the resource base.

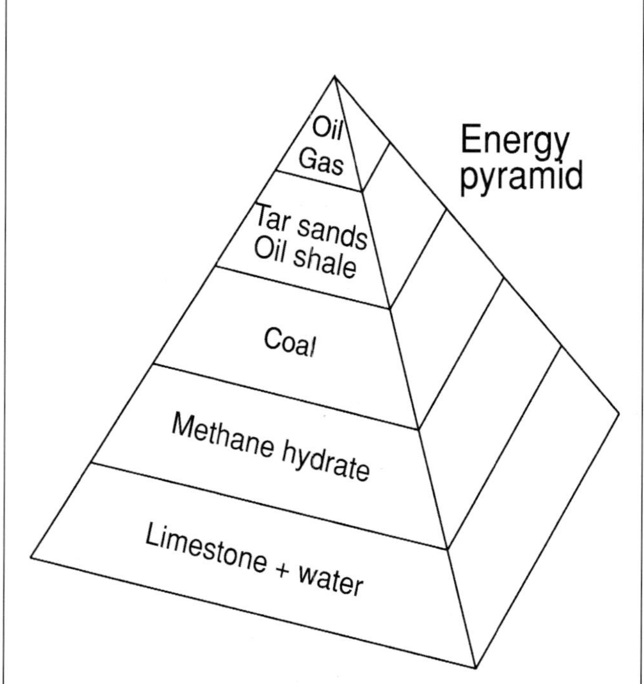

Figure 5. Hydrocarbon energy pyramid.

Unconventional hydrocarbon resources include tar sands, oil shale, coal, and methane hydrates. The inclusion of limestone and water at the base of the hydrocarbon pyramid (Figure 5) illustrates that as a last resort, synthetic hydrocarbons could be assembled from the carbon in limestones and the hydrogen and oxygen in water (Dusseault, 1997). Manufactured hydrocarbons would not be an energy source so much as a means of storing and transporting energy. With present technology, the process would be expensive and impractical. However, if a large-scale source of cheap energy such as nuclear power were available, it might eventually become economically feasible to engage in the mass production of synthetic hydrocarbons.

UNCONVENTIONAL HYDROCARBON RESOURCES

Tar Sands

In all probability, the next cut lower in the hydrocarbon pyramid will be the exploitation of tar sands. Large-scale mining of tar sands in Canada began in 1967. Extraction technology was updated in 1992 by the Canadian oil company Suncor (George, 1998). George (1998) reported that Suncor had been producing oil from Canadian tar-sand deposits profitably for five years. In 1998, the International Energy Agency (1998) estimated the cost of producing Canadian tar sands to be in the range of $12–15/bbl.

The size of the tar-sand resource is large (Table 1). George (1998) and the International Energy Agency (1998, p. 113) estimated that 300 billion bbl of oil could be recovered from Alberta tar sands using present technology. Although consumption rates are not constant

Table 1. Estimates of the size of unconventional oil resources.

Resource Type	Location	Estimate Size (billion bbl)	Years of U.S. Crude Oil Consumption*	Reference
All unconventional	North America	6000	1000	Bird (1989)
Heavy crude oil and bitumen	Worldwide	6200	1033	Meyer and Schenk (1985)
Heavy oil	Canada and Venezuela	600	100	International Energy Agency (1998)
Gas to liquids	Worldwide	150	25	International Energy Agency (1998)
Heavy oil	Canada and Venezuela	3460	577	Dusseault (1997)
Oil shale	Green River Formation, Colorado, Utah, Wyoming	1500	250	Smith (1981), McCabe (1998)
Oil shale	All U.S., including low grade	160,000	26,667	Duncan and Swanson (1965), McCabe (1998)

**Calculated using a 1997 consumption rate of 6 billion bbl/year.*

with time, it is sometimes more intuitive to express resource size in terms of years of supply.[1] The 1997 U.S. consumption of crude oil was about 6 billion bbl (British Petroleum, 1998). Based on this rate, the Alberta tar sands contain crude-oil equivalent to a 50-year supply. Another well-known tar-sand resource is the Orinoco deposit in Venezuela. The International Energy Agency (1998) estimated that 300 billion bbl of oil could be produced from the Orinoco deposits at a total extraction cost of \$15–17/bbl. Worldwide, Meyer and Schenk (1985) estimated world heavy crude-oil and bitumen resources to be 6200 billion bbl, with 890 billion bbl being considered "recoverable." These numbers are equivalent to 1033 and 148 years of 1997 U.S. crude-oil consumption, respectively.

Oil Shale

The extraction of crude oil from oil shales is somewhat more difficult and expensive than the processing of tar sands. However, the size of the potential resource is huge. Smith (1981) estimated that the Green River Formation in Colorado, Utah, and Wyoming contains 1500 billion bbl of oil, equivalent to 250 years of 1998 U.S. consumption. Duncan and Swanson (1965) estimated that if all oil shale in the United States is considered, the size of the potential resource is 160,000 billion bbl, or 26,667 years of U.S. consumption. Suncor Corporation is developing an oil-shale demonstration project in Australia. The Suncor technology involves baking oil out of crushed rock in a giant, drum-shaped kiln (George, 1998).

Gas to Liquids

Worldwide, there are large gas resources which cannot be exploited because of a lack of delivery systems (pipelines). The International Energy Agency (1998, p. 112) estimated that there are 1488 trillion cubic feet (tcf) of "stranded" gas which cannot reach markets. This 1488 tcf of gas is equivalent to 150 billion bbl of oil, or 25 years of U.S. supply at a 1997 consumption rate. Gas-to-liquids (GTL) technology allows these gas resources to be converted into a transportable form. The International Energy Agency (1998) described GTL technology as dating to 1923, but only recently "coming of age." Fouda (1998) reviewed several different GTL technologies under development and noted that even with existing technology, natural gas can be converted into liquid fuels at prices "only about 10 percent higher per barrel than crude oil."

Methane Hydrates

Methane hydrates are icelike mixtures of methane (CH_4) and water in which the gas molecules are trapped within a framework or clathrate of water molecules.

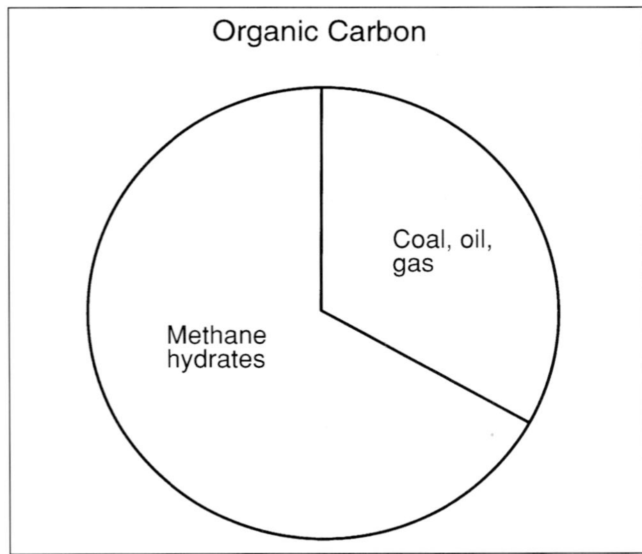

Figure 6. Relative amounts of organic carbon in methane hydrates compared with coal, oil, and gas (after Kvenvolden, 1993).

Methane hydrates are found in polar regions where temperatures are cold enough for permafrost to be present and offshore on the continental margins where cold water is found at depths greater than 300–500 m (Kvenvolden, 1993).

Methane hydrates are of interest for two reasons. Many estimates of the size of the hydrate resource are very large (Figure 6). Kvenvolden (1993) estimated the amount of methane in gas hydrates throughout the world to be about 700,000 tcf. As of 1997, the U.S. consumption rate for natural gas was 22.3 tcf/year (British Petroleum, 1998). Thus, the potential size of the methane-hydrate resource is equivalent to 31,390 years of 1997 U.S. consumption. Second, methane is a proven and relatively clean-burning fuel which can be used in motor vehicles as well as for electricity generation and home heating.

At the present time, it is uncertain if economic production of methane hydrates will occur at any time in the foreseeable future. Laherrère (1999a, b) pointed out that the size of the methane resource and its nature (concentrated versus dispersed) is very uncertain. Direct evidence of methane occurrences in thick, concentrated deposits is rare to nonexistent, and indirect indicators are "unreliable and highly speculative" (Laherrère, 1999a, b).

To produce natural gas from hydrates, the hydrate must be dissociated or broken down into gas plus water or gas plus ice. At the present time, there is no technology for the economic extraction of methane from hydrates. Methane hydrates are mostly viewed in the petroleum industry as a production nuisance which can clog pipelines and well bores. Three potential production

[1] Throughout this text, I convert raw resource estimates to years of supply. This is done to provide the reader with a more intuitive appreciation for the numbers quoted. I do not mean to imply that current usage rates will remain constant.

methods are recognized in the literature. These are (1) thermal stimulation, (2) pressure reduction, and (3) inhibitor injection. In brief, the only one of these three methods which can lead to large-scale economic production is thermal stimulation. This is because hydrate dissociation involves an endothermic phase change. It is thermodynamically feasible to supply the necessary energy if freed methane is burned, because the heat of dissociation is only about 10% of the heat released by methane combustion (Holder, 1984). The primary production challenge will be to devise a technology for delivering heat throughout hydrate reservoirs.

HUBBERT'S MODEL AND PREDICTIONS

M. K. Hubbert made two valuable contributions to the understanding of fossil-fuel production and depletion. First, he pointed out that production and consumption tend to increase exponentially during the initial phase of a resource's development. In 1950, it was common to calculate years of total supply by dividing reserves or resources by current production or consumption rates. Hubbert pointed out that such a method was likely to drastically underestimate the time during which adequate fuel supplies would be available, because consumption and production rates would increase in the future. Second, Hubbert pointed out that problems arise not when total supplies are exhausted, but when demand exceeds supply. If the production curve is symmetric, this point is reached when half of the ultimate production is reached.

Hubbert's model is based on two assumptions: first, that the supply of any resource is finite, and second, that the production rate of a resource will grow exponentially, peak, and then decrease exponentially as the resource is depleted. Hubbert applied these assumptions to make predictions of both U.S. and world oil production. The most famous of these predictions was made in a 1957 paper published in the American Petroleum Institute's *Drilling and Production Practice, 1956* titled "Nuclear Energy and the Fossil Fuels."[2] Based on ultimate oil-reserves estimates of 150 billion and 200 billion bbl, Hubbert (1957) predicted that oil production in the United States would peak as early as 1965, but no later than 1970. Oil production in the conterminous United States peaked in 1970 at 3.44 billion bbl.

In his earlier works, Hubbert was somewhat obscure regarding the mathematical formulas he used to produce his bell-shaped production curves. Although the "Hubbert curve" is bell shaped, it is not the Gaussian curve seen in probability theory (Laherrère, 1999c). A complete explanation of Hubbert's methods is found in a somewhat obscure symposium volume published by the National Bureau of Standards (Hubbert, 1982). In that work, Hubbert (1982) derives an expression for cumulative production Q (Hubbert, 1982, p. 49) and identifies it as the "logistic equation" originally derived by the Belgian demographer Verhulst. The more familiar plots of production rate *(dQ/dt)* follow immediately by taking the derivative with respect to time. Although Hubbert seems to have favored the logistic equation as the most logical approach, he (Hubbert, 1982, p. 139) claimed that his prediction made in 1956 was drawn by hand. In response to a question, he stated,

> In my figure of 1956 [published in 1957], showing two complete cycles for U.S. crude-oil production, these curves were not derived from any mathematical equation. They were simply tailored by hand subject to the constraints of a negative-exponential decline and a subtended area defined by the prior estimates for the ultimate production. Subject to these constraints, with the same data, I suggest that anyone interested should draw the curves himself. They cannot be very different from those I have shown.

Whatever the case, in reality there is no unique Hubbert curve. Hubbert himself (1949, p. 105) wrote,

> Thus we may announce with certainty that the production curve of any given species of fossil fuel will rise, pass through one or several maxima, and then decline asymptotically to zero. Hence, while there is an infinity of different shapes that such a curve may have, they all have this in common: that the area under each must be equal to or less than the amount initially present.

In fact, the production curves shown by Hubbert for coal (1949, p. 107) and for petroleum (1950, p. 104) are not bell shaped but are strongly asymmetrical, with post-peak production declining more slowly than prepeak production increased. In his 1957 paper and afterward, Hubbert seems to have exclusively used symmetrical production curves.

Despite the fame of Hubbert's 1956 prediction and the widespread acceptance of his model, some pointed criticisms of the Hubbert model can be made. McCabe (1998) demonstrated that there is no inherent reason that production of an energy resource should necessarily follow a Hubbert-type curve, and cited examples which did not. He also pointed out that the production history of some resources (e.g., Pennsylvania anthracite) has followed a Hubbert-type curve but is far from exhaustion.

[2] Hubbert made the prediction in a paper he presented at a meeting of the Southwest Division of the American Petroleum Institute, held in San Antonio, Texas, in March 1956.

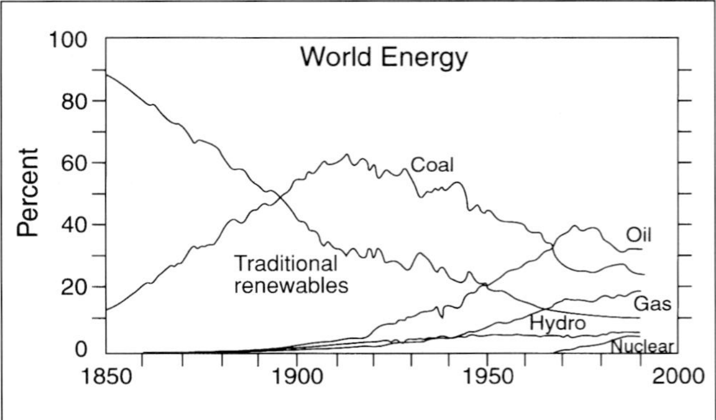

Figure 7. World energy consumption, 1850 through 1990 (after Nakicenovic et al., 1998).

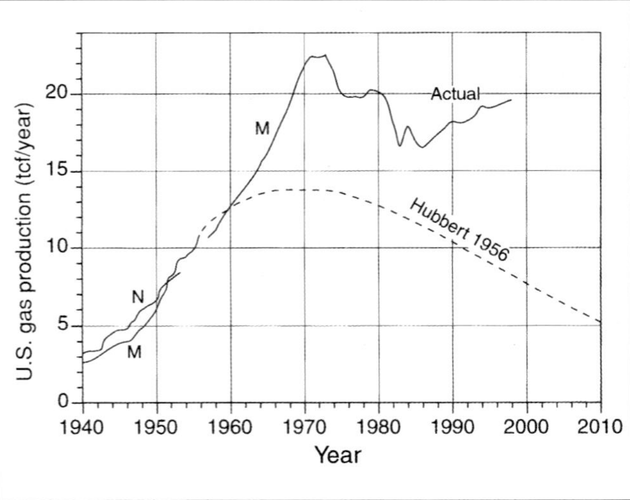

Figure 8. Production of natural gas (solid lines) in the United States (excluding Alaska) compared with Hubbert's 1956 prediction (dashed line) of future gas production. "M" indicates marketed gas, which is about 1% lower than net or total gas production ("N"). Data prior to 1957 are from Hubbert (1957, p. 18) and were obtained by digitizing his Figure 22. Data from 1957 through 1998 are from annual estimates of marketed gas made by the *Oil & Gas Journal.*

In some historical cases, production dropped because of substitution of alternative energy sources. For example, in the twentieth century, coal was largely replaced by oil and natural gas (Figure 7). McCabe (1998) showed that energy markets tend to be "open," as opposed to "closed." That is, the history of energy use is one of substitution. It is not clear, however, if this can be the case for petroleum. Coal and nuclear power can replace gas and oil for electricity generation, but at present there is no ready substitute for petroleum in the transportation sector.

Advocates of the Hubbert model (e.g., Ivanhoe, 1995; Campbell and Laherrère, 1998) often cite the accuracy of Hubbert's 1956 prediction of U.S. oil production as substantiation for the correctness of the Hubbert method. Ivanhoe (1995) characterized Hubbert's 1956 prediction as the "only truly valid scientific projection of future oil production," and claimed that since the 1970 production peak, oil production from the 48 conterminous U.S. states declined within 5% of Hubbert's 1956 prediction. Campbell and Laherrère (1998), in reference to Hubbert's 1956 prediction, said, "He was right: production peaked in 1970 and has continued to follow Hubbert curves with only minor deviations." However, McCabe (1998) objected that "the conclusion that Hubbert's methodology and assumptions must have been correct because his predictions were accurate is not logical." Furthermore, claims of predictive accuracy appear to be overstated. Hubbert's 1956 prediction of gas production proved to be grossly in error. As of 1998, U.S. gas production was more than twice as high as Hubbert had predicted in 1956 (Figure 8).

To correctly evaluate the accuracy of Hubbert's 1956 predictions, it is necessary to consider the context of his writing. In 1956, Hubbert made what he considered to be a "best" estimate of the total size of the U.S. crude-oil resource (given the symbol Q_∞ by Hubbert). This estimate, Q_∞= 150 billion bbl, included offshore areas of the conterminous United States but excluded Alaska. Hubbert (1957) then proceeded to consider an exaggerated estimate of the largest possible size of the U.S. crude-oil resource, to set an outer limit. Hubbert (1957, p. 18) set this at Q_∞ = 200 billion bbl, noting that the addition of 50 billion bbl to the resource base was "an amount equal to 8 East Texas oil fields."

Claims regarding the predictive accuracy of Hubbert's predictions appear to be largely based on the upper curve derived from Hubbert's upper limit of Q_∞ = 200 billion bbl (Figure 9). Hubbert presented this curve not as a best prediction but as an absolute upper limit. Even so, as early as 1970, actual production of U.S. crude oil had exceeded Hubbert's upper limit by 13%, and Hubbert's best estimate of future U.S. crude-oil production was badly in error. From about 1975 through 1995, Hubbert's upper curve was a fairly good match to actual U.S. production data (Figure 9). However, in recent years, U.S. crude-oil production has been consistently higher than what Hubbert (1957) considered to be the highest possible future levels. In light of this, it is strange that Hubbert's predictions have been characterized as a remarkable success. Ironically, Hubbert's subsequent revisions to his U.S. crude-oil predictions ultimately proved to be less accurate than the upper curve he drew in 1956, because he finally settled on a number for the size of the ultimate resource Q_∞ at about 170 billion bbl. Hubbert's last prediction of U.S. crude-oil production was made in 1980 and published in 1982 (Hubbert, 1982, p. 89). This prediction (Figure 10) is based on an estimate of Q_∞ = 170

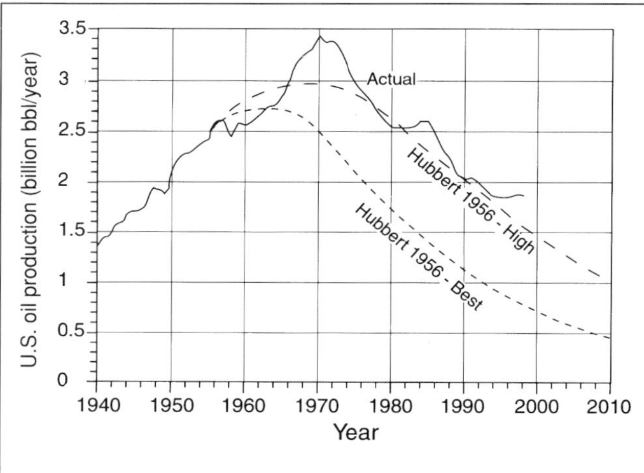

Figure 9. Crude-oil production (solid line) in the United States (excluding Alaska), compared with predictions (dashed lines) made by Hubbert in 1956. Production data prior to 1955 are from Hubbert (1957, p. 17) and were obtained by digitizing his Figure 21. Data from 1956 through 1998 are from annual estimates made by the *Oil & Gas Journal.*

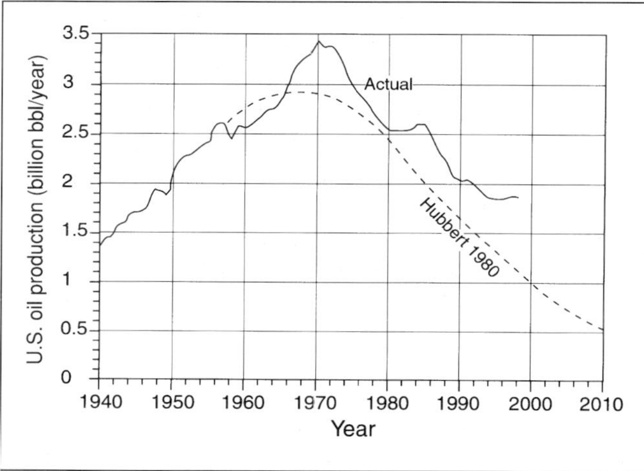

Figure 10. Crude-oil production (solid line) in the United States (excluding Alaska), compared with prediction (dashed line) made by Hubbert (1982) in 1980. Production data prior to 1955 are from Hubbert (1957, p. 18) and were obtained by digitizing his Figure 22. Data from 1956 through 1998 are from annual estimates made by the *Oil & Gas Journal.*

billion bbl, and thus is intermediate between the two curves published by Hubbert in 1957 (Figure 9). As of 1998, Hubbert's 1980 prediction was 39% too low.

Hubbert's primary success was his prediction that oil production in the United States would peak "at about 1965" (Hubbert, 1957, p. 18). Hubbert (1957) went on to claim that even if Q_∞ proved to be as large as 200 billion bbl, the peak would be "retarded only until about 1970." The actual peak occurred in 1970. Hubbert's best estimate (1957) of peak production in 1965 was in error by five years, with the actual production peak occurring at the outer limit of his uncertainty range. However, adherents to Hubbert's methods tend to present the peak-production prediction in the best possible light. For example, both Ivanhoe (1995) and Campbell and Laherrère (1998) claimed that Hubbert (1957) predicted that U.S. oil production would peak in 1969, plus or minus a year. This is true only if Hubbert's (1957) upper curve (Figure 9) is used and his prediction of peak oil production in 1965 is ignored.

The primary problem with Hubbert-type analyses is that they depend on an estimate of the total resource size. The greater the size of the resource, the longer the Hubbert peak is delayed. The history of resource assessments (Figure 4) shows that in the last 50 years, estimates of ultimate oil reserves have risen as fast or faster than cumulative oil production. For a Hubbert-type model to work, there must be some assurance that the estimated size of the resource base is accurate and will not increase in the future. McCabe (1998, p. 2122) points out that "all [resource] assessments have a strong subjective component and, at best, only identify those resources that appear to have some economic potential within the foreseeable future." McCabe (1998) goes on to state that "a finite resource cannot realistically be measured." As time proceeds, each succeeding assessment cuts down farther into the resource pyramid; thus, each succeeding assessment is larger than the previous. If this is true, then it is impossible to make an accurate or stable assessment of the size of the resource base, and all predictions based on a Hubbert model will fail. Lynch (1999) points out that "the Hubbert method has consistently produced bad forecasts," and cites predictions made by Campbell (1989, 1991).

Like other models, the Hubbert model is based on assumptions, and its validity should be judged by comparing predictions with facts. However, among not only its adherents but the larger world as well, the Hubbert model is dogmatically regarded as being absolutely correct, almost to the point of religious zealotry. In a posthumous tribute to Hubbert, the National Academy of Sciences (1990) referred to Hubbert's model—an empirical model based on assumptions—as a "mathematical proof." In reference to the Hubbert method, Lynch (1999) pointed out that all previous applications had systematically underestimated actual production, and he suggested,

> Since persistent errors in one direction are proof that a theory and/or parameters need revision (if not discarding), reviewing existing work is an important aspect of scientific discovery. Scientists are supposed to produce theories that are reproducible and verifiable. If not, they are practicing religion.

Hubbert apparently never considered that the size of the finite resource base might be a moving target, thus invalidating his entire approach. In an interview published in 1983 (Clark, 1983, p. 22), Hubbert proved himself to be both a doomsayer and a strict Malthusian:

> We are in a crisis in the evolution of human society. It's unique to both human and geologic history. It has never happened before and it can't possibly happen again. You can only use oil once. You can only use metals once. Soon all the oil is going to be burned and all the metals mined and scattered.... I think the world is seriously overpopulated right now. There can be no possible solutions to the world's problems that do not involve stabilization of the world's population.

Despite Hubbert's 1983 prediction of imminent shortages, the average price (in terms of wages) of nonferrous minerals (oil, coal, zinc, lead, platinum, tin, etc.) in the United States fell by more than 50% from 1980 to 1990 (Moore, 1995, p. 112).

NUCLEAR ENERGY

The history of energy use is one of substitution (Figure 7). Until about 1895, wood was the primary energy source for the world. Coal was the dominant source of the world's energy from 1895 through 1965. The age of oil began in 1965, and it remains the largest single energy source in the world today. Where will energy for the civilization of the future come from?

Wood is a largely renewable resource, but its energy density is inadequate for a technological civilization. The world contains large amounts of oil and gas, both conventional and unconventional. Barely a fraction of the world's coal resources has been depleted. However, the potential nuclear-energy resource is so large that it is virtually unlimited when measured in terms of today's energy demands.

Uranium consists of three natural isotopes in the following proportions: U-238 (99.3%), U-235 (0.7%), and U-234 (0.005%). The only uranium isotope which may be readily used in fission reactors is U-235. The much more abundant U-238 isotope is not fissionable, but it may be converted into fissionable plutonium under neutron bombardment by U-235 fission. Nuclear reactors may be burner or breeder reactors. Burner reactors fission the high-grade isotope U-235. Breeder reactors use U-235 to produce fissionable plutonium from the more abundant U-238 isotope. More fissionable material is produced than used in a breeder reactor. This does not violate the conservation of energy, because a breeder reactor simply unleashes the potential energy inherent in the relatively stable U-238 isotope.

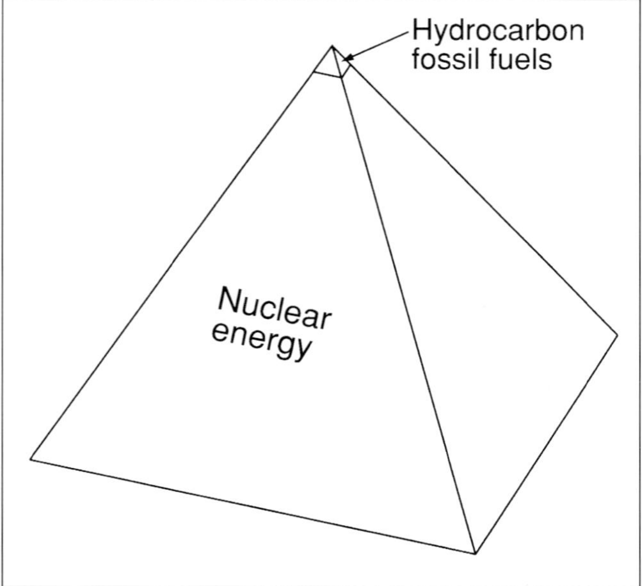

Figure 11. Energy pyramid. Potential nuclear-energy resources dwarf all hydrocarbon fossil fuels.

Finch (1997) estimated that at current use rates, the size of the world's U-235 resource was sufficient to supply 500 years of use. However, if U-238 is used in breeder reactors, the size of the resource becomes almost unimaginably large (Figure 11). Hubbert (1969) pointed out that the continental crust contains large amounts of low-grade uranium. For example, the Chattanooga Shale (found in Tennessee, Kentucky, Ohio, Indiana, and Illinois) contains a unit about 15 ft (4.6 m) thick which contains 0.006% uranium by weight. The energy density is such that 20 km^2 of this unit would provide the same energy as 200 billion bbl of oil (Hubbert, 1969, p. 227). Hubbert (1969) noted that the energy potentially obtainable from low-grade uranium extracted from crustal rocks is easily hundreds or thousands of times larger than all other fossil fuels combined. More recently, Cohen (1983) pointed out that uranium could be extracted from seawater at a cost of only $1000/lb. This cost, although high compared with a 1983 market price for uranium of $40/lb, would add only 0.03 cents/kW-hr to the cost of generated electricity. Uranium is also being constantly added to seawater through erosion of the continental crust. Cohen (1983) calculated that 16,000 metric tons of uranium/year could be drawn from seawater continuously for hundreds of millions of years without depleting the oceans. If breeder reactors are used, this amount of uranium would be sufficient to provide energy at a rate 25 times the world's electricity usage in 1983, or twice the world's total energy consumption in 1983.

Although nuclear power has great promise, it is being phased out in the United States. As of 1995, uranium accounted for 23% of electricity generated in the United

States (Finch, 1997). No new nuclear power plants have come on-line in the United States since 1993. Commercial nuclear power plants have operating lifetimes of 40 years. Starting in 2012, there will be a steady decrease in the amount of electricity generated through nuclear power if new plants are not built. If the present situation continues, nuclear power will cease to exist in the United States by 2033.

The use of uranium for nuclear power generation is controlled almost exclusively throughout the world by government policies and regulations. Government policies in turn are strongly influenced by social and political attitudes. Two well-publicized nuclear power-plant accidents have contributed to the formation of highly negative attitudes concerning nuclear power. In 1979, a nuclear power plant on Three Mile Island, about 10 miles south of Harrisburg, Pennsylvania, released some radioactivity into the surrounding environment. Although the magnitude of the release was inconsequential, public media played on public fears. In 1986, a catastrophic accident occurred at a nuclear power plant in Chernobyl, Ukraine. The Chernobyl disaster cemented opinion regarding the safety of nuclear power and essentially made further debate moot for perhaps a generation.

If the environmental problems and perceptions surrounding nuclear power plants cannot be overcome, there is no known resource or technology which can supply energy for the civilization of the future. In the long run, however, it may be best if today's generation of burner reactors is phased out. Although the amount of U-238 is very large, fissionable U-235 is scarcer. If we continue on the present course, supplies of U-235 may be depleted in burner reactors before they can be used in a new generation of self-sustaining breeder reactors. Hubbert (1969, p. 228) predicted that the failure to make the transition from burner to breeder reactors would constitute "one of the major disasters in human history."

CONCLUSIONS

1. The history of the twentieth century is one of repeated false predictions of imminent oil shortages.
2. Although conventional oil and gas deposits may be finite, estimates of their size have grown through the last 50 years as fast as cumulative production.
3. As production of unconventional oil resources in the form of tar sands has already begun, it becomes increasingly difficult to define the size of the world's oil and gas resource base. Unconventional oil resources such as tar sands and oil shales are sufficient in size to supply the world's petroleum needs for about 100 to 1000 years.
4. As conventional oil resources begin to become depleted, they will be replaced by unconventional oil resources. The nature of the transition is unknown at present. It may be seamless, with uninterrupted supplies and low prices, or there may be interruptions in oil supplies, high prices, and economic disruptions.
5. Although the Hubbert model was an important advance in the science of resource evaluation, its application in the last 40 years has resulted in depletion estimates that are systematically premature and inaccurate. Current and future application of the Hubbert model will remain problematic as long as resource estimates remain uncertain.
6. At present, the only known energy resource which can supply the world's needs for the distant future is nuclear power. Nuclear power is a long-term solution only if breeder reactors are used.

REFERENCES CITED

British Petroleum, 1998, BP statistical review of world energy, June 1998: London, CTD Printers, 40 p.

Bird, K. J., 1989, North American fossil fuels, *in* A. W. Bally and A. R. Palmer, eds., The geology of North America—An overview: The Geology of North America, Geological Society of America, v. A, p. 555–573.

Campbell, C. J., 1989, Oil price leap in the early nineties: Noroil, v. 17, no. 12, p. 35–38.

Campbell, C. J., 1991, The golden century of oil 1950–2050: The depletion of a resource: Dordrecht, Kluwer Academic Press, 345 p.

Campbell, C. J., and J. H. Laherrère, 1998, The end of cheap oil: Scientific American, v. 278, p. 78–83.

Clark, R. D., 1983, King Hubbert: The Leading Edge, v. 2, p. 16–24.

Cohen, B. L., 1983, Breeder reactors: A renewable energy source: American Journal of Physics, v. 51, p. 75–76.

Dolton, G. L., et al., 1981, Estimates of undiscovered recoverable conventional resources of oil and gas in the United States: U. S. Geological Survey Circular 860, 87 p.

Duncan, D. C., and V. E. Swanson, 1965, Organic-rich shale of the United States and world land areas: U.S. Geological Survey Circular 523, 30 p.

Dusseault, M. B., 1997, Flawed reasoning about oil and gas: Nature, v. 386, p. 12.

Fanning, L. M., 1950, A case history of oil-shortage scares, *in* L. M. Fanning, ed., Our oil resources (2nd ed.): New York, McGraw-Hill, p. 306–406.

Finch, W. I., 1997, Uranium, its impact on the national and global energy mix—and its history, distribution, production, nuclear fuel-cycle, future, and relation to the environment: U. S. Geological Survey Circular 1141, 24 p.

Fouda, S. A., 1998, Liquid fuels from natural gas: Scientific American, v. 278, p. 92–95.

George, R. L., 1998, Mining for oil: Scientific American, v. 278, p. 84–85.

Hatfield, C. B., 1997, Oil back on the global agenda: Nature, v. 387, p. 121.

Holder, G. D., V. A. Kamath, and S. P. Godbole, 1984, The potential of natural gas hydrates as an energy source: Annual Review of Energy, v. 9, p. 427–445.

Hubbert, M. K., 1949, Energy from fossil fuels: Science, v. 109, p. 103–109.

Hubbert, M. K., 1950, Remarks on fuels and energy, *in* Proceedings of the United Nations Scientific Conference on the Conservation and Utilization of Resources, Plenary Meetings: Lake Success, New York, United Nations Department of Economic Affairs, v. 1, p. 103–104.

Hubbert, M. K., 1957, Nuclear energy and the fossil fuels: Drilling and production practice, 1956: New York, American Petroleum Institute, p. 7–25.

Hubbert, M. K., 1967, Degree of advancement of petroleum exploration in the United States: AAPG Bulletin, v. 51, p. 2207–2227.

Hubbert, M. K., 1969, Energy resources, *in* Resources and man: San Francisco, W. H. Freeman, p. 157–242.

Hubbert, M. K., 1982, Techniques of prediction as applied to the production of oil and gas, *in* S. I. Gass, ed., Oil and gas supply modeling: National Bureau of Standards Special Publication 631, p. 16–141.

International Energy Agency, 1998, World energy outlook: Paris, International Energy Agency, 475 p.

Ivanhoe, L. F., 1995, Future world oil supplies: There is a finite limit: World Oil, October, v. 216, p. 77–88.

Jensen, M. L., and A. M. Bateman, 1981, Economic mineral deposits (3rd ed.): New York, John Wiley & Sons, Inc., 593 p.

Kerr, R. A., 1998, The next oil crisis looms large—and perhaps close: Science, v. 281, p. 1128–1131.

Kvenvolden, K. A., 1993, A primer on gas hydrates, *in* D. G. Howell, ed., The future of energy gases: U. S. Geological Survey Professional Paper 1570, p. 279–291.

Laherrère, J. H., 1999a, Uncertain resource size: Enigma of oceanic methane hydrates: Offshore, v. 59, p. 140–141, 160.

Laherrère, J. H., 1999b, Data show methane hydrate resource over-estimated: Offshore, v. 59, p. 156–158.

Laherrère, J. H., 1999c, World oil supply—What goes up must come down, but when will it peak? Oil & Gas Journal, February 1, v. 97, p. 57–64.

Lynch, M. C., 1999, The debate over oil supply: Science or religion?: Geopolitics of Energy, v. 21, no. 8, p. 8–16.

Mast, R. F., G. L. Dolton, R. A. Crovelli, D. H. Root, E. D. Attanasi, P. E. Martin, L. W. Cooke, G. B. Carpenter, W. C. Pecora, and M. B. Rose, 1989, Estimates of undiscovered conventional oil and gas resources in the United States—A part of the nation's energy endowment: U. S. Geological Survey and Minerals Management Service, 44 p.

Masters, C. D., D. H. Root, and W. D. Dietzman, 1984, Distribution and quantitative assessment of world crude oil reserves and resources: Proceedings of the 11th World Petroleum Congress, v. 2, p. 229–237.

Masters, C. D., E. D. Attanasi, W. D. Dietzman, R. F. Meyer, R. W. Mitchell, and D. H. Root, 1987, World resources of crude oil, natural gas, natural bitumen, and shale oil: Proceedings of the 12th World Petroleum Congress, v. 5, p. 3–27.

Masters, C. D., D. H. Root, and E. D. Attanasi, 1991, World resources of crude oil and natural gas: Proceedings of the 13th World Petroleum Congress, p. 51–64.

Masters, C. D., E. D. Attanasi, and D. H. Root, 1994, World petroleum assessment and analysis: Proceedings of the 14th World Petroleum Congress, v. 5, p. 529–541.

McCabe, P. J., 1998, Energy resources—Cornucopia or empty barrel?: AAPG Bulletin, v. 82, p. 2110–2134.

Meyer, R. F., and C. J. Schenk, 1985, An estimate of world resources of heavy crude oil and natural bitumen, *in* R. F. Meyer, ed., The Third UNITAR/UNDP International Conference on Heavy Crude and Tar Sands: Edmonton, Alberta, Canada, Alberta Oil Sands Technology and Research Authority, p. 73–83.

Miller, B. M., H. L. Thomsen, G. L. Dolton, A. B. Coury, T. A. Hendricks, F. E. Lennartz, R. B. Powers, E. G. Sable, and K. L. Varnes, 1975, Geological estimates of undiscovered recoverable oil and gas resources in the United States: U. S. Geological Survey Circular 725, 78 p.

Minerals Management Service, 1996, An assessment of the undiscovered hydrocarbon potential of the nation's outer continental shelf: Minerals Management Service OCS Report MMS 96-0034, 40 p.

Moore, S., 1995, The coming age of abundance, *in* R. Bailey, ed., The true state of the planet: New York, Free Press, 472 p.

Nakicenovic, N., A. Grübler, and A. McDonald, 1998, Global energy perspectives: Cambridge, England, Cambridge University Press, 299 p.

National Academy of Sciences, 1990, Tributes, M. King Hubbert: National Academy of Sciences Letter to Members, v. 19, no. 4, p. 26–27.

Smith, J. W., 1981, Oil shale resources of the United States: Mineral and Energy Resources, Colorado School of Mines, v. 23, no. 6, p. 1–20.

U. S. Geological Survey National Oil and Gas Resource Assessment Team, 1995, 1995 national assessment of United States oil and gas resource: U.S. Geological Survey Circular 1118, 20 p.

Weeks, L. G., 1948, Highlights on 1947 developments in foreign petroleum fields: AAPG Bulletin, v. 32, p. 1093–1160.

Weeks, L. G., 1950, Discussion of "Estimates of undiscovered petroleum reserves by A. I. Levorsen": Proceedings of the United Nations Scientific Conference on the Conservation and Utilization of Resources, 1949, v. 1, p. 107–110.

Weeks, L. G., 1958, Fuel reserves of the future: AAPG Bulletin, v. 42, p. 431–438.

Gaffney, P. D., 2001, Exploration: Tomorrow's charge, tomorrow's challenge, *in* M. W. Downey, J. C. Threet, and W. A. Morgan, eds., Petroleum provinces of the twenty-first century: AAPG Memoir 74, p. 57–60.

Chapter 5

Exploration: Tomorrow's Charge, Tomorrow's Challenge

Peter D. Gaffney
Gaffney, Cline & Associates, Alton, Hampshire, United Kingdom

As we enter this new century, we appear to be somewhat adrift in exploration, if not in geology. Indeed, we can ask, "Where are the great 'end-of-century' explorationists—the ones at the front end driving us forward today?" The worry is, of course, that many of us have already made our contribution—but where is that next generation of explorationists?

Alas, they probably haven't even come into our business, because our industry, like all the earth sciences and natural-resources businesses, has found it more and more difficult to attract the very best as our position in the public perception has continued to deteriorate. Indeed, the position of the explorationist and the geologist in oil companies has deteriorated in the past years. Even in the large national oil companies, with huge exploration possibilities, the role of the exploration manager or the senior geologist seems to have been reduced and perceptions of them diminished.

So in a sense, you can say "we lost the plot" somewhere in the period from the late 1970s perhaps to the middle 1980s—perhaps because of oil price, perhaps because we didn't find some of the big fields that we hoped to find, perhaps (in that terrible period from about 1987 to 1994) because we failed to recognize that 3-D seismic, good though it is, doesn't change the distribution of field sizes in nature and doesn't make those big billion-barrel fields easier to find in practice.

Whatever the reasons, if we face the facts, somehow in this period we lost the plot. So we do have some responsibility for our current situation. Let us not dwell on it any more right now. For the moment, let us look back on that marvelous century gone by—almost a fable—when there were real explorationists in this world.

TWENTIETH-CENTURY RESULTS

Just think what explorationists managed to accomplish in the last 100 years—most of it in the first 75 years! We found unbelievable quantities of oil, we delivered them to market cheaply, and we helped to develop the twentieth-century world of cheap fuel and, I would argue, relatively environmentally friendly fuel, compared with coal and firewood. We had our wars with the drillers, with the petroleum engineers, and with our managers. We had our technical difficulties with deep water, deep drilling, and cold and inhospitable situations. We had wars with all those nonbelievers in the possibilities of exploration success—but by and large, we won those wars.

We developed an industry that we can all look back on and be proud of—until along came a terrible virus that struck us explorationists dumb.

That virus, the AIB virus (the analyst and investment banker virus) was the virus that hit all industries toward the end of the twentieth century. There was little antidote, no antibiotics for it, and it was very long lasting. This was the virus that led to merger mania and management by head count in virtually all industries. In our industry, all but a few caught this dreadful disease.

We were caught up in an epidemic of mergers, downsizing, rightsizing, and capsizing, as we have continued to be these last 10 years.

What can we do except perhaps revert to basics and, like all geologists, start off where we came in? The present is indeed the key to the past, but can we make part of the understanding of the past a key to the future?

To answer this question, look at the world we are in, our industry, those problem areas, and the "Century 2000" challenges we all face.

CURRENT STATUS OF INDUSTRY

The Influence of Wall Street

We accept that the industry is run largely by Wall Street (or its equivalent in other parts of the world). There is no interest, little understanding, and no mind-set other than the making of short-term money. Wall Street, to a large extent, prays to an imaginary god—the shareholder—who may typically be a 29-year-old fund manager who will buy your shares in the morning and sell them in the afternoon. The direct shareholders are no longer that great cross section of individuals that they were, for example, in the United States of 50, 30, or even 20 years ago.

Our industry is confused at all levels, even today and even among the largest, and is running scared at most levels. It is an industry in which a billion-barrel discovery means very little to the largest companies in the land, companies that are largely subsumed by head count and the next merger.

Our companies have budgets, to a large extent driven by analysts, and costs, perhaps and regretfully driven by pseudotechnology, some of which all of us have encouraged. Tied to all these factors is this tremendous overfocus on cost.

We have been through earlier times when there was a lot of pressure on prices and costs. Yes, we knew we were going to have to reduce people. We knew the industry was going to have to restructure to some extent, but was the focus only on cost reduction? No, it was on throughput, improving production, reducing the cost per barrel or the cost per report.

Unfortunately, today our managers can see only one thing, and this comes from the very background that they have. You might say any of them can be stock-market heroes—all they need to do is stop exploration. Then they can cash-flow the company, and with that cash they can go out and buy or merge with some other company.

The questions are, "Will the company ultimately benefit? Will the state ultimately benefit? Will the shareholders ultimately benefit?" Yes, we know the investment bankers and company management will benefit, but it is not entirely clear yet what is really going to happen when the full cycle of 10 years or more takes effect.

Suffice it to say it is an old truth from many reports that, when we look back at them, the great majority of mergers in all industries did not meet the original expectations.

For those of you involved in raising funds in today's peculiar market, you know that eventual success virtually demands that you mention technology and appropriate buzzwords in every fund-raising document, preferably including *3-D seismic, horizontal drilling, deep water*, and *work-force reduction*. You must mention these items even if you do not intend to employ any people, run 3-D seismic, or drill horizontally; otherwise, you are unlikely to raise any money at all.

The pressure on "one size fits everybody" and on "one solution to all problems" is enormous. Our management schools have encouraged this approach, and the ideal situation for your project, your staff, your shareholders, and your assets rarely meets the analysis-and-solution culture that drives most high-level businesses.

Exploration and Production Balance

We used to say that the important thing was balance —because an oil and gas company needed to have exploration to feed its future, it needed to have exploration to get the advantage of new low-cost barrels to offset higher-cost declining barrels. That balance needed explorationists on the staff too, for they would often be the ones to challenge conventional wisdom.

Such balance seems to me to be missing in much of our industry at the moment. Of course, for the very large companies, where many of their deals now with national oil companies provide profits of 50 cents to a dollar net per barrel, even extremely large deals can have little impact on the bottom line or indeed on corporate futures. In this context, small though a hundred-million-barrel discovery might be, it may add more value than yet another foreign deal at a few cents per barrel.

When we look at Figure 1, we see that there are a large number of places where balance seems to be missing today, and I am sure you can think of a lot of companies where balance is totally absent.

Technology Promises Not Fully Realized

Then there is the matter of technology. Recent technology has not really found any big fields—what is finding the fields is still imagination and the technology that we had 30 or 40 years ago. We are in danger of being dumbed down by much pseudotechnology, particularly as our young people inevitably are taught about black boxes and their advantages.

The dry-hole ratios are, unfortunately, about the same. We continue to drill several thousand dry holes in the United States and internationally every year. There is little indication that we are really learning or using the learning we have achieved. Technology, in many respects, has not really impacted these numbers or ratios. Of course, we can assume that without some of the technology, the situation might have been worse, but that would just be an assumption. We know that in fact, many of our tools have increased the number of dry holes as we sought to materialize smaller and smaller potential discoveries.

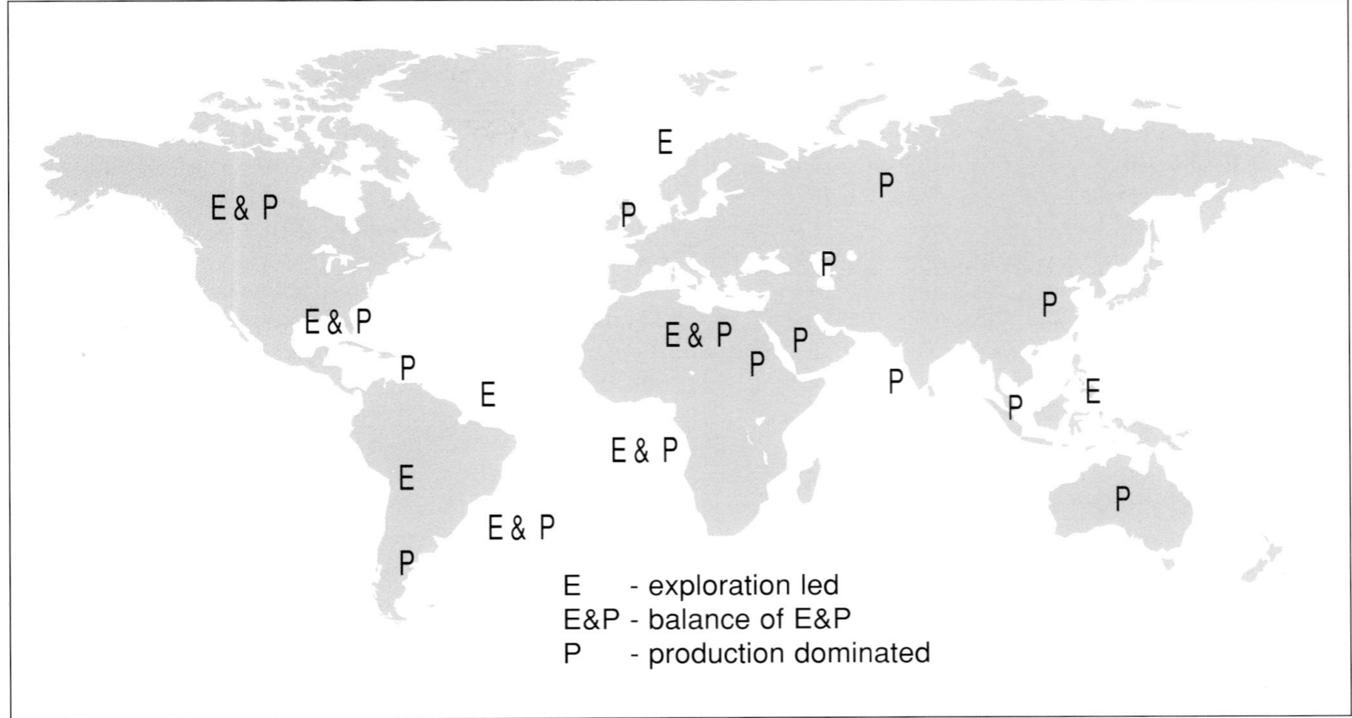

Figure 1. Exploration and production (E&P) hot spots. Balance is the exception.

Even our drilling methods, contrary to much belief, are little changed. If I think back to when my father joined the industry in the 1920s and I look for changes since then, I realize that most of our rigs around the world today possess absolutely marvelous "state-of-the-art" positioning devices, and we have put thousands more horsepower on the surface compared with 100 or so horsepower back then. But much of what we do now is still amazingly the same for the majority of rigs in our world. What we have done extremely well is learn to build our structures for drilling and operating in the harshest of climes and deepest of waters.

Of course, much as we did in our earlier lives many years ago when almost anything we couldn't understand was called a graywacke, in parts of our world today that is replaced by calling everything we do not understand a turbidite. Indeed, I think we can all be concerned now that virtually everything we stand on is about to slump if we are to believe the interpretations from many of our current explorationists.

TWENTY-FIRST-CENTURY CHALLENGES

This leads to the twenty-first-century challenges.

We need better risk analysis. It can be argued that other industries, particularly industries such as aerospace, have a much better handle on risk analysis than we have in our industry. Whether it is in exploration risk, reserves expectations, predicting plant performance, or estimating tanker-related oil-spill eventualities, we need a much greater effort across the board to provide decent risk analysis, leading to better evaluation of potential downsides.

From an exploration point of view, we want to look more at full-cycle economics, and such full-cycle economics needs to look at cumulative basin economics. Cumulative basin economics means that when we have already put $100 million into a play concept or basin arena, we look especially hard when the new budget suggests putting in another $100 million. The concept is just the same if we are playing with $5 million or $10 million.

We need to look at a more cost-effective approach to data acquisition and the whole wildcat drilling approach, acquiring the information we need as cheaply as we possibly can.

But let's be fair—we must not ignore some of the new tools which are either with us already or are on the horizon. Who knows what satellite radar, airborne fluorescence, or high-resolution magnetics will do for us in the next 10 years?

There is no doubt that we need to relook at 3-D seismic, as we are doing, and see how we can make more use of a truly excellent tool and perhaps do it much more cheaply and effectively.

There are many challenges out there—not least, identifying that stratigraphic trap, particularly the deep one, a problem that has been with us almost forever.

It is important that as earth scientists, we take a more proactive role in dealing with environmental issues in all our locations. We, rather than many others, have the tools and understanding of Mother Nature, and I believe we can defend and indeed encourage our fellow men and women to look to realistic environmental policies which are in the best interests of everyone. We should not abrogate this role.

We are in an environment where multidisciplinary teams and working together are supposed to be the approach of the day, following our overspecialization in the late 1970s and most of the 1980s. We need the geologists and the explorationists at the reservoir-engineering face just as explorationists need a better understanding of what the developer and the reservoir engineer can make work. The reservoir engineer and the petroleum engineer still deal, in effect, with 8.5-inch holes scattered around, and they base their entire estimates on relatively small samples. Reservoir characterization is still something that is a long way from being adequate, it costs far too much, and it takes far too long to carry out in any degree of detail. We need a lot of improvement here. A better relationship between the parties and their concepts will make this happen sooner rather than later.

Of course, we are in an E-commerce and Internet environment, and perhaps we would all be more comfortable being virtual geologists and working in a virtual explorationist's playground. Alas, while there may well be great opportunities for all of us in E-commerce, real oil and real gas will continue to be produced up real wells and will inevitably be discovered by real explorationists who look for clues in real rocks.

So the charge to you, the explorationist and the geologist, at the beginning of this new century first becomes an issue of education. We need a petroleum professional with a good dose of geology, engineering, geophysics, and economics. That education needs to be spiced with extensive common sense and a deep understanding of Mother Nature and her vagaries.

Can we get this just in the university? Possibly not! Perhaps we need to enhance the first few years of working in the industry. Perhaps we need to encourage our young men and women to spend more time in looking at the economics side, especially with a technical perspective.

Perhaps in that way there will be a greater opportunity for many of them to achieve commanding positions in our companies and industry. Without seeing a reasonable number of our technical people in the highest ranks of the industry, we will not see the right decisions for the future, and all of us in our profession, our industry, and our countries will suffer.

We need to remember that high tech won't save us. We also need to question whether resting most of our effort on the excellent performance of our geophysicist friends is the right way forward. The geophysicist sees the world quite correctly from his perspective—a geophysical one. I would argue that the geologist has a broader perspective, and it is a geologic view which encompasses an understanding of Mother Nature. In today's multidisciplined world, it is this understanding which needs to lead us forward.

We need to advise our current leaders of the benefits of an exploration barrel—that it can be substantially cheaper and a lot more profitable than a purchased barrel. The deal might be better (after all, which would you rather have, $8/barrel on 5000 barrels/day or $0.50/barrel service fee on 50,000 barrels/day?), the tax implications might be better, and you can use newer technology.

But how do we find that plot we lost? We need to look harshly at what we have done wrong and accept that we have contributed to the position we find ourselves in today. We recognize where we might do better. Let's not overfocus on any one technology. Let's look at the realities, let's do things cheaper, let's attract the better students and, most importantly, let's stand up and be counted again on policy and environmental questions.

We have the tools, we have the challenge, we have a much-needed commodity. The industry needs our assistance to move forward. I commend this challenge to you.

North America and the United Kingdom

Sneider, R. M., and J. S. Sneider, 2001, New oil in old places: The value of mature-field redevelopment, *in* M. W. Downey, J. C. Threet, and W. A. Morgan, eds., Petroleum provinces of the twenty-first century: AAPG Memoir 74, p. 63–84.

Chapter 6

New Oil in Old Places: The Value of Mature-field Redevelopment

Robert M. Sneider and John S. Sneider
Sneider Exploration Company Inc., Houston, Texas, U.S.A.

ABSTRACT

Large reserves remain and are economically recoverable in many mature oil fields. This study documents the successful search for such overlooked reserves in old fields. Small multidisciplinary teams studied several basins in North America and a few large fields in South America, searching for large volumes of low-risk reserves in poorly performing fields. The fields studied included those producing by primary recovery, with or without secondary recovery potential, as well as fields undergoing waterflooding. In the United States, more than 350 mature oil fields were evaluated for potential purchase from 1981 to 1997. The majority of the fields were in the Permian Basin of west Texas–New Mexico and in a coastal portion of the Gulf of Mexico Basin. In addition, some large fields were studied so property owners could be advised on rejuvenating or increasing production.

Finding large volumes of low-risk, presently nonproducing reserves within fields involves several steps. First, the teams search in reservoir systems that appear more massive and homogeneous than they really are. Second, the geoscience and engineering data are scanned to estimate original oil in place, percent recovery, and bypassed reserves. Third, the teams make an economic analysis, including improvement costs. Fields that are candidates for purchase have discovered low- to moderate-risk reserves amounting to at least 5% of the cumulative reserves already produced.

New reserves are discovered or exploited by applying one or more of the following: (1) improved drilling and completion technology; (2) identification of bypassed pay, especially very low resistivity pay; (3) new 2-D and 3-D seismic surveys; and (4) sequence-stratigraphic concepts. Additional reserves are found in both land-derived clastic and carbonate reservoirs in mature fields.

Forty-six mature fields were purchased in the Permian Basin of Texas–New Mexico and in the Gulf of Mexico Basin. In the fields purchased, 625 million barrels of oil equivalent (BOE) of proved and probable reservoirs were added at a cost of US $2.69/BOE. The average after-tax rate of return of the 46 fields is 21%. Based on these results, "exploring for new oil in old places" is an economically viable strategy for adding hydrocarbons to the world's reserves base.

INTRODUCTION

Exploration is essential to increase hydrocarbon reserves worldwide. Exploration is high risk and may be high cost, and international production is highly taxed in many areas. The time between discovery and first production can be several years, and the average size of fields discovered is decreasing in many basins. These factors can make exploration unattractive.

Another important source of increased reserves can be found within or adjacent to existing fields, especially in reservoirs producing by depletion or a weak water drive. Apparently recoverable reserves in such fields may be only 12% to 30% of the original oil in place. In United State basins, except in the deep-water offshore Gulf of Mexico Basin, most reserve additions in the recent past are being produced from improvements to mature fields. Nehring (1995) showed that in the United States from 1983 to 1992, about 85%, or 20 billion barrels (bbl), of proved oil-reserve additions were from old fields.

Is there good potential for adding reserves in old fields in mature basins? Our answer is yes. The application of new exploration and production technology, along with appropriate oil-field practice, is the driving force behind finding *new oil in old places.*

This paper presents four examples of reserve additions from mature fields. One example is a revitalized, originally failed waterflood in dolomite reservoirs. The second example shows reserves added in sandstones by applying good engineering practices, coupled with field-extension opportunities defined by new 2-D and 3-D seismic surveys. The third example, also in sandstones, shows reserve additions from field extensions defined by geological-geophysical-petrophysical engineering studies and supplemental recovery opportunities from an injection of water alternating with gas (WAG) recovery process. The fourth example, in sandstone and carbonate reservoirs, shows significant reserve additions by infill and horizontal wells and field extensions. Reserve additions in all four examples are a direct result of integrated, multidisciplinary teams of geoscientists and engineers focused on finding new reserves in and around mature fields.

BACKGROUND AND APPROACH

In the 1980s industry downturn, our small teams of geoscientists and engineers changed their search for hydrocarbons from exploration ventures to property acquisition. The teams searched for mature fields with large undeveloped oil and gas potential that could be produced profitably. They looked for fields that (1) could be waterflooded, (2) had waterfloods that were performing poorly, (3) had unrecognized pay in low-resistivity reservoir rocks, (4) had unrecognized reservoir compartments because of structural and/or stratigraphic changes, and (5) had unrecognized field extensions that might be defined by new 2-D or 3-D seismic surveys.

The teams worked mainly in the Permian Basin of west Texas–New Mexico and in the Texas-Louisiana Gulf of Mexico Basin from onshore to shallow water less than 200 ft deep (Figure 1). In those two basins, more than 350 fields were scanned for acquisition candidates. About 80 of those fields met the reserve/profit/technical criteria and became potential acquisition candidates. Forty-six potential candidates were purchased in competitive-bid sales from major oil companies.

Potential field-acquisition candidates were identified using geologic criteria and by production comparisons with analogous fields. They searched for economically marginal fields with reservoirs that (1) are interpreted as homogeneous and correlatable by the operator and (2) are composed of multiple stacked reservoir units that contain thin, continuous fluid-flow barriers, often below

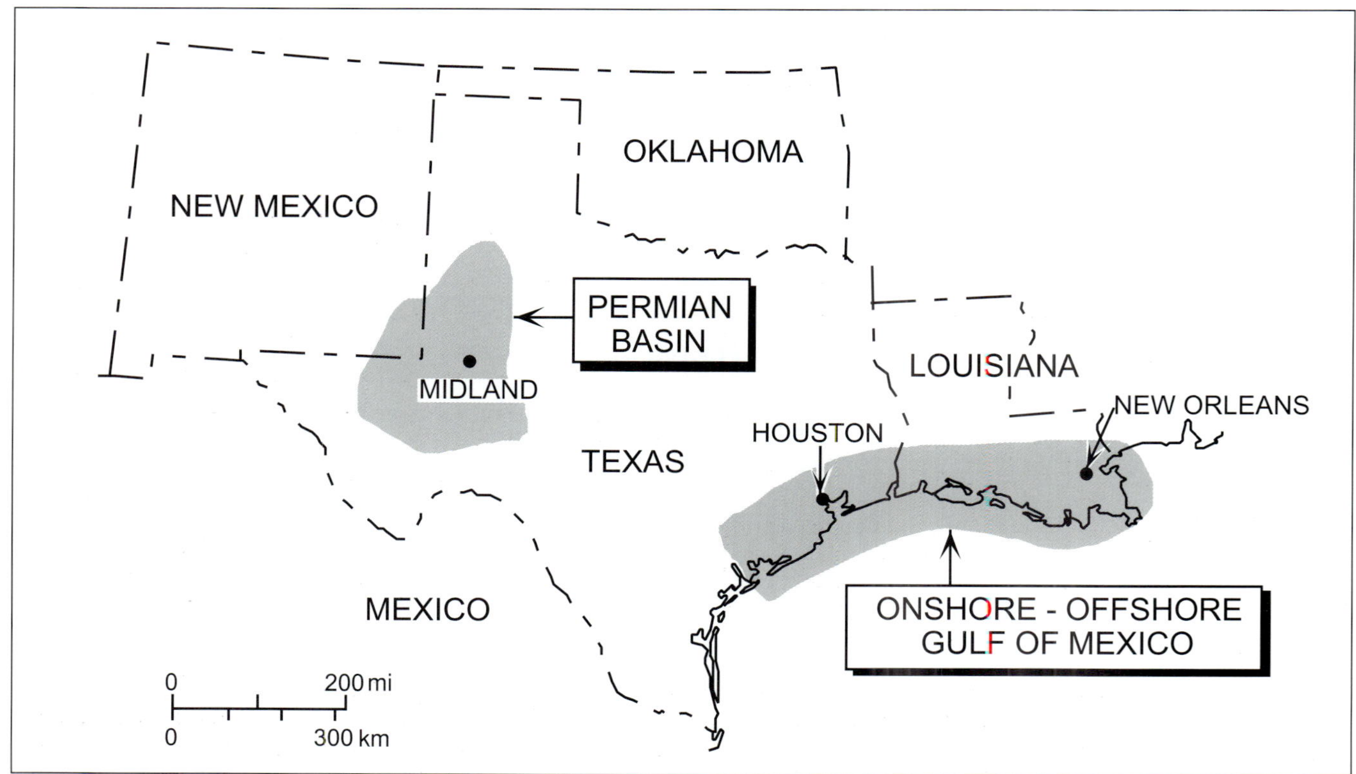

Figure 1. Location of basins where fields were acquired in west Texas, New Mexico, and onshore-offshore Texas and Louisiana.

well-log resolution. The depositional systems the teams concentrated on were (1) dolomitized tidal flat sequences; (2) oolite tidal bars; (3) mud-rich deltas; and (4) turbidites, especially shingled turbidites in front of deltas. In the Gulf of Mexico Basin, the teams concentrated on structurally complicated fields associated with growth faulting and intermediate to deep-seated salt. These structural styles often contain undrained reservoir compartments because of faults and facies pinch-outs that are difficult to identify with well control and older 2-D seismic surveys.

Production performance from analogous fields helps guide selection of acquisition candidates. Often, bids are based on extrapolation and interpretation of production-decline curves, as in the type "A" production curve (Figure 2). Most companies using this curve will estimate similar field reserves. In contrast to this approach, we look for acquisition candidates geologically similar to analogous fields that have production performance such as types "B," "C," and "D" (Figure 2). Production increases come from new wells in undrained reservoir compartments, infill wells, field extensions, workovers/recompletions, or installing a waterflood, or simply improving surface equipment, water quality, and injection profiles in waterflooded fields. The analogous fields serve as a guide for bidding on the acquisition candidates.

Sneider and Sneider (1998) detailed the steps to rapidly identify acquisition candidates, define potential opportunities, and arrive at a bid. The key steps are as follows. First, the team searches for anomalies in reservoir systems that appear to be more massive and homogeneous than they really are. This includes depositional environments that result in very thin vertical and horizontal flow barriers such as tidal flats, tidal bars, mud-rich deltas, and turbidites. Well-completion intervals are compared with the actual reservoir compartments to identify undrained or poorly drained pay intervals. Real or potential "thief zones" in supplemental recovery projects are identified, and remedies are considered to eliminate those zones.

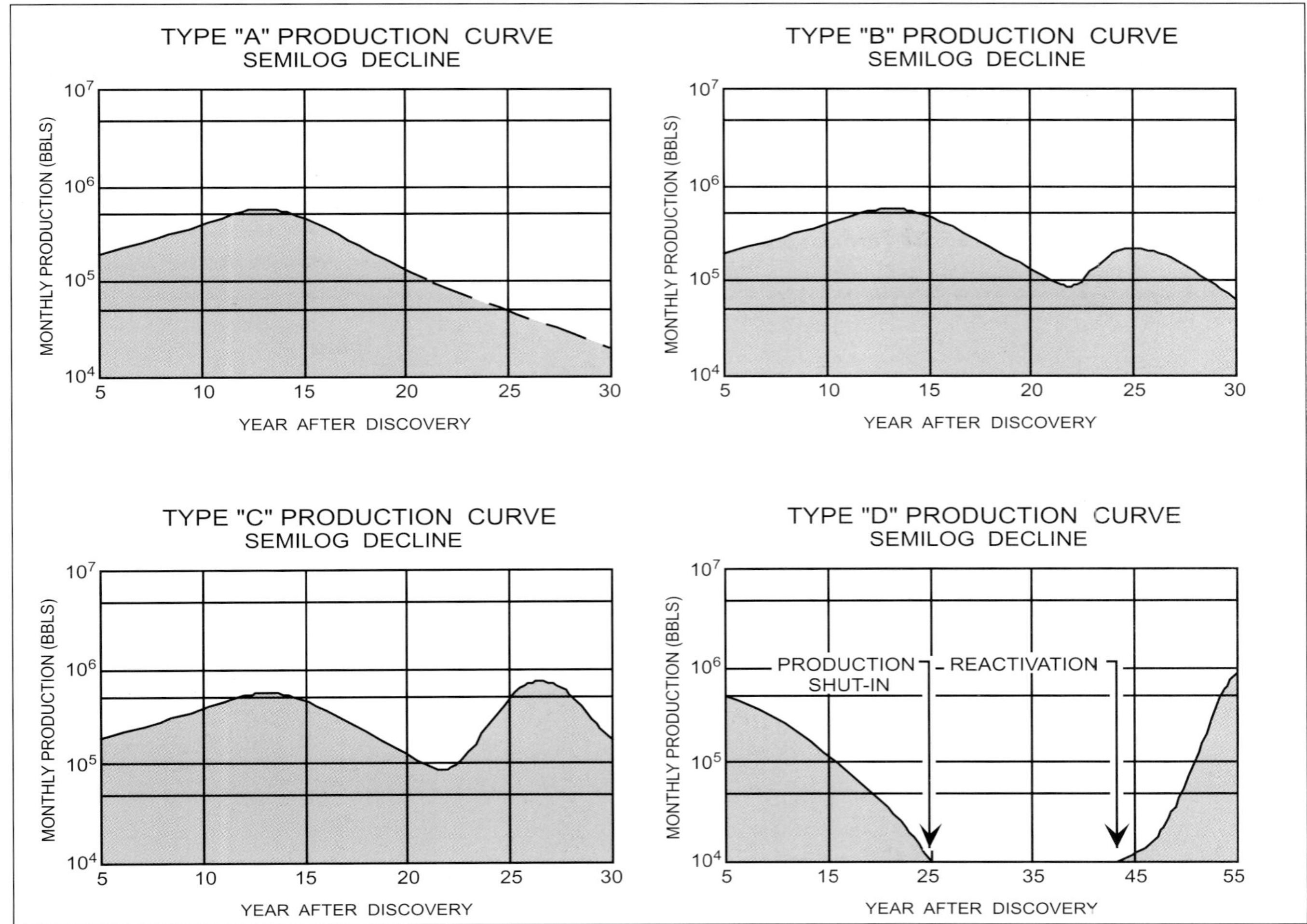

Figure 2. Type "A" graph illustrates a typical production-decline curve of a depletion reservoir. Type "B" production decline shows a secondary production increase resulting from infill drilling, recompletions, workovers, and/or flooding. Type "C" production curve shows that secondary production is higher than the initial production rate. Type "D" graph is typical of a rejuvenated marginal field. Production increases are the result of operations similar to types "B" and "C."

Second, the team organizes the geoscience and engineering data to quickly estimate original oil in place, percent recovery, and remaining reserves. Rapid scans of representative portions of a field identify undrained or poorly drained reserves. The team identifies workovers, recompletions, infill well locations, and "hidden" pays (e.g., low-resistivity, low-contrast pays that are difficult to evaluate). The value of reprocessing existing 2-D seismic data, acquiring new 2-D and 3-D seismic data, and applying sequence-stratigraphic concepts is ascertained. This second step identifies potential reserves.

The third step is an economic analysis that includes recovery-cost estimates for overlooked reserves. The certainty of these new reserves is estimated. This economic evaluation also includes an analysis of existing facilities such as platforms, active and inactive well bores, flow lines, facility consolidation and maintenance to improve operating efficiency, and known potential liabilities (e.g., environmental). Experience from analogous fields in the same or similar trends aids in selecting candidates to purchase. From the teams' experiences, fields that are candidates for purchase contain new low- to moderate-risk reserves amounting to at least 5% of the cumulative reserves already produced.

A forward-looking economic analysis of the candidate field is the final evaluation step before preparing and submitting a final purchase offer. This analysis is based on a cash-flow projection and calculated rate of return. The economic analysis includes cost estimates to recover the overlooked reserves, costs of upgrading existing facilities, needed maintenance to improve operating efficiency, and known and potential liabilities, including abandonment costs. The identification of acquisition candidates is an iterative process conducted by teams composed of geologists, geophysicists, petrophysicists, reservoir engineers, and operations and field engineers, assisted by lawyers and negotiators.

After purchasing a field, an implementation team executes the workover/recompletion program on previously identified opportunities to increase cash flow and reserves. Infill or replacement wells test the technical concepts for increasing reserves. Key intervals are cored and tested to evaluate the new reserve opportunities identified in the detailed field studies. If the new wells are unsuccessful, the concepts are wrong and the field is resold.

OPPORTUNITIES FOR PERFORMANCE IMPROVEMENTS AND RESERVE ADDITIONS

The scan of more than 350 fields in the Permian Basin and part of the Gulf of Mexico Basin (Figure 1) identified 82 fields that met reserve/profit/technical criteria and became acquisition candidates. Thirty-six candidate fields were producing by primary recovery without supplemental recovery potential, 14 fields were producing by primary recovery and had waterflood potential, and 32 fields were being waterflooded. The types of opportunities for reserve additions/performance improvements are shown in Figure 3.

In the fields producing by primary recovery with no supplemental recovery potential (Figure 3a), opportunities are primarily by field extensions (44%) identified by better reservoir characterization provided by geologic studies and new 2-D and 3-D seismic surveys. Infill wells to produce reserves in undrained reservoir compartments represent 29% of the opportunities. Bypassed pays, especially low-resistivity/high-irreducible water saturation pays and recompletions, are 21% of the opportunities. Workovers are 6% of the performance improvements.

In the 14 primary-recovery fields with supplemental recovery potential (Figure 3b), the major opportunity

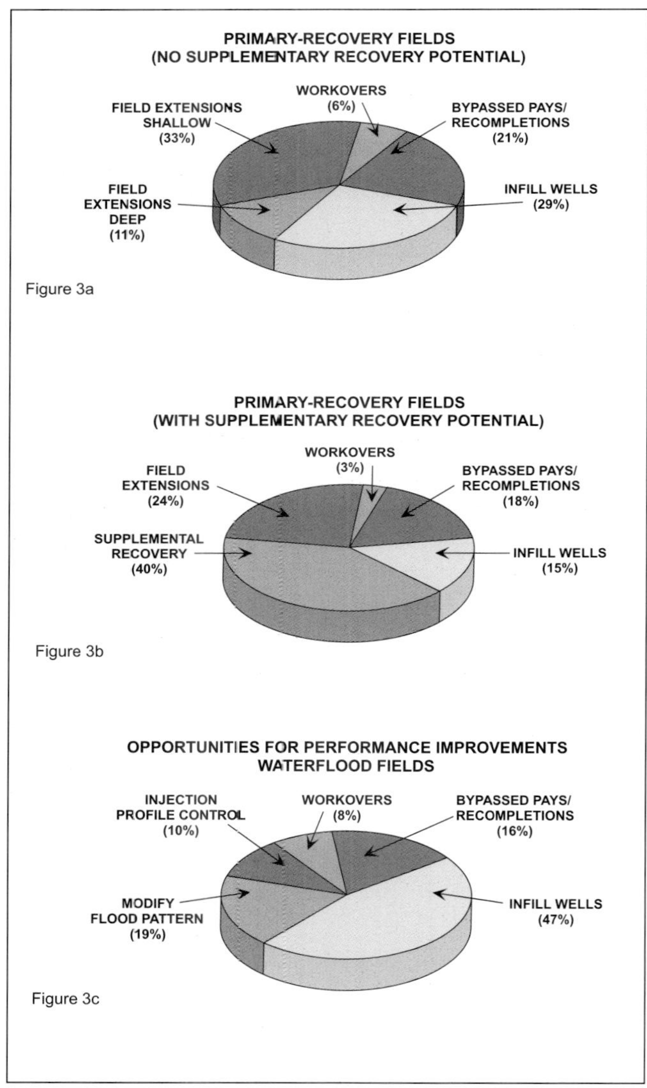

Figure 3. Opportunities for reserve additions and performance improvements: (a) primary-recovery fields with no supplemental recovery potential, (b) primary-recovery fields with supplemental recovery potential, (c) waterflood fields.

(40%) comes from installing a waterflood. Field extensions are 24% of the potential. Infill wells needed for the waterflood and bypassed pays/recompletions represent 15% and 18% of the opportunities, respectively. Workovers are 3%.

For fields under an existing waterflood (Figure 3c), the majority of the opportunities to improve performance and add reserves is by infill wells (45%) and flood-pattern modification (19%). These changes are to ensure injector-producer continuity and uniform sweep of the injected water. Use of production profiles to determine the rate of injected water ensures a uniform flood front and prevents premature water breakthrough. Profile control and bypassed pays/recompletions represent 26% of the opportunities for performance improvements and reserve additions.

EXAMPLES

Four examples illustrate the methodology, reserve additions, and performance improvement possible in mature fields. Two examples (the "C" waterflood unit in the Permian Basin and the "G" field in coastal Louisiana) were purchased as part of our property-acquisition program. The third example is based on a joint field study with PDVSA to increase production from a marginal field in Lake Maracaibo, Venezuela. The fourth example is based on Shell/Esso work in the Auk field area in the North Sea.

"C" Waterflood Unit, Permian Basin

The most effective waterfloods in the Permian Basin optimized production by using 20-acre well spacing, five- or nine-spot flood patterns, and new injector wells to control injection profiles and open up additional pays in lower resistivity, lower porosity, and higher water-saturation intervals. Avoiding or shutting off "thief" zones for injected water, using clean water, and ensuring that injected water connects with producing intervals by avoiding thin flow barriers increase waterflood efficiency.

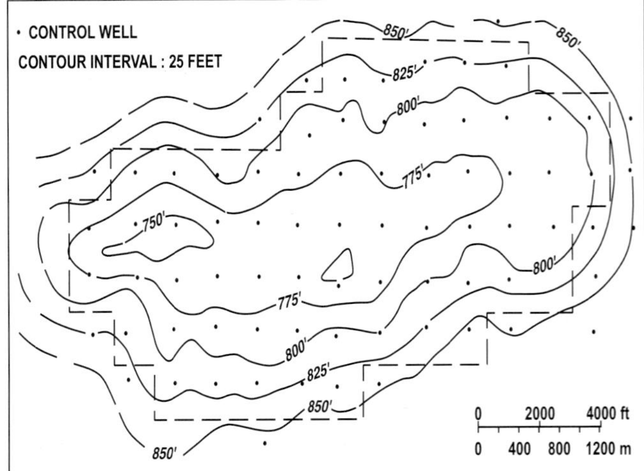

Figure 4. Structure map on top of the San Andres reservoir "C" waterflood unit, Permian Basin, Texas. Note the gentle anticline with four-way closure.

The criteria used to evaluate waterflood candidates to purchase are (1) 40-acre spacing, (2) peripheral or irregular water-injector patterns, (3) low secondary-to-primary (S/P) recovery ratio (less than 0.75), and (4) low-recovery efficiency (primary plus secondary production is less than 25% of the oil in place).

Target reservoirs are oolite bars and tidal flats. These environments commonly have thin, continuous impermeable flow barriers that affect connectivity between injectors and producers. The "C" field produces from dolomitized oolite bars on a gentle anticline at a depth of 5000 ft (Figure 4). The operator estimated 49 million bbl of oil in place in the 2500-acre flood unit. Primary development ended in 1960 with 58 producers on 40-acre spacing (Figure 5). Production decline started in 1960

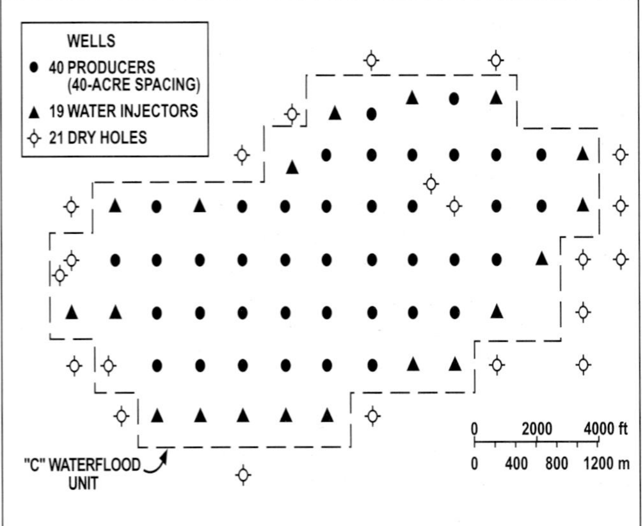

Figure 5. Map showing the location of injectors and producers at the start of the peripheral waterflood, San Andres "C" waterflood unit, Permian Basin, Texas.

and, by 1970 (time A, Figure 6), the operator had obtained approval to initiate a peripheral waterflood (time B). Water injection began in 1974; a modest increase in production began in 1975. For the next two years, production increased only slightly. The operator injected poorly filtered produced water. Most interior wells did not respond to the peripheral water injection. We estimated an S/P ratio of about 0.55 in 1974.

Our field study using logs, well cuttings, rock-log calibration, pressure data, and production response showed that the waterflood had significant secondary reserve potential in spite of the poor early response. The field was acquired at the end of 1979.

The "C" field produces from dolomitized oolite bars and tidal channels (Figures 7–9). The original operator correlated the reservoir unit as a continuous unit with no continuous flow barrier (Figure 7b). Study of well cut-

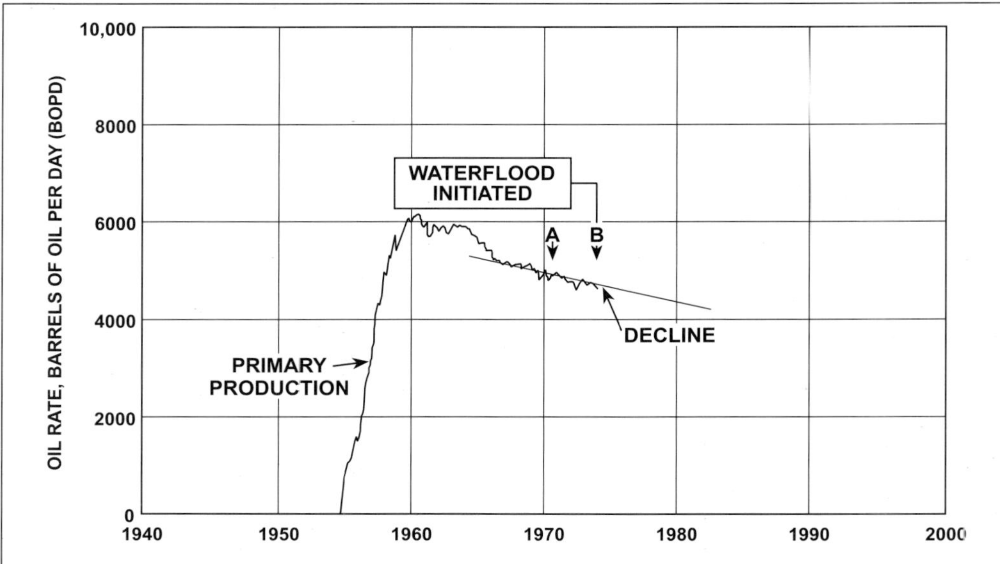

Figure 6. Graph showing primary production, San Andres "C" waterflood unit, Permian Basin, Texas. The peripheral waterflood was approved at time A and initiated at time B.

tings and two new cores taken after acquisition confirmed that the reservoir consisted of multiple, discrete oolite bars with tidal channels separated by thin, excellent-quality flow barriers of carbonate mud and clay (Figures 7c, d). Pressure tests and repeat formation tests (RFTs) confirmed the presence of unswept, low- to intermediate-pressured oil zones. Only producers located near injectors responded to the peripheral waterflood.

New wells with modern logs and two cores showed that the 8%-porosity pay cutoff used by the original operator was too high. Although porosity ranges between 3% and 12%, permeability depends on dolomite crystal size (Figure 8). Capillary pressure measurements, flood pot tests, and a field test showed that the lower limit of producible pay was 4% porosity and 0.1-md permeability. The new pay cutoffs significantly increased San Andres pay (compare Figures 9a and 9b).

The net pay in each subzone using the new cutoffs (Figures 10, 11) was mapped, as well as the flow barriers. Additional pay zones identified from the core-log studies and well tests were perforated; clean water was injected; and a small 20-acre, five-spot pilot was initiated (Figure 12). Appropriate volumes of water were injected selectively into the A1, A2, and A3 subzones. Six months after the initiation of the pilot, pressure increased and production increased slightly (time C, Figure 13).

With the encouragement from the pilot, the field was "downspaced" to 20 acres, and a five-spot pattern was developed fieldwide (Figure 14 and time D, Figure 15). Clean water (a mixture of new fresh water and clean produced water) was injected at a pressure of 75% of the formation fracture gradient. To improve sweep efficiency, zones with better permeability (i.e., "thief" zones) were plugged.

The 20-acre, five-spot pattern flood dramatically increased producible reserves. The S/P ratio increased from 0.55 to 1.75. The efficiency for primary and secondary recovery increased to more than 42%. An additional 4% to 6% of the oil in place may be recoverable from CO_2 flooding, based on analogy with CO_2 floods in nearby fields.

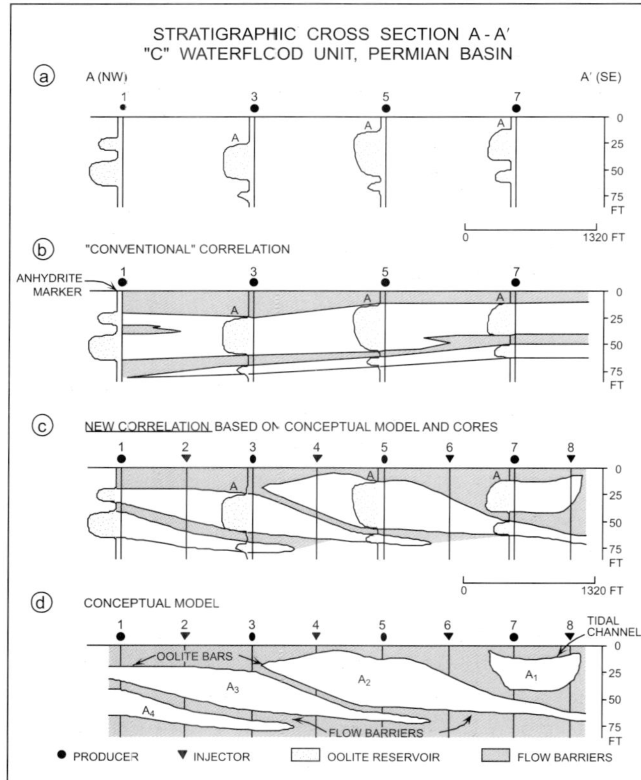

Figure 7. North-south stratigraphic cross section A-A′ showing (a) original gamma-ray log response, (b) original operator's correlation, and (c, d) new correlations of reservoirs and flow barriers based on oolite tidal bars and tidal channel models. Infill injector wells are required to effectively flood the reservoir flow units.

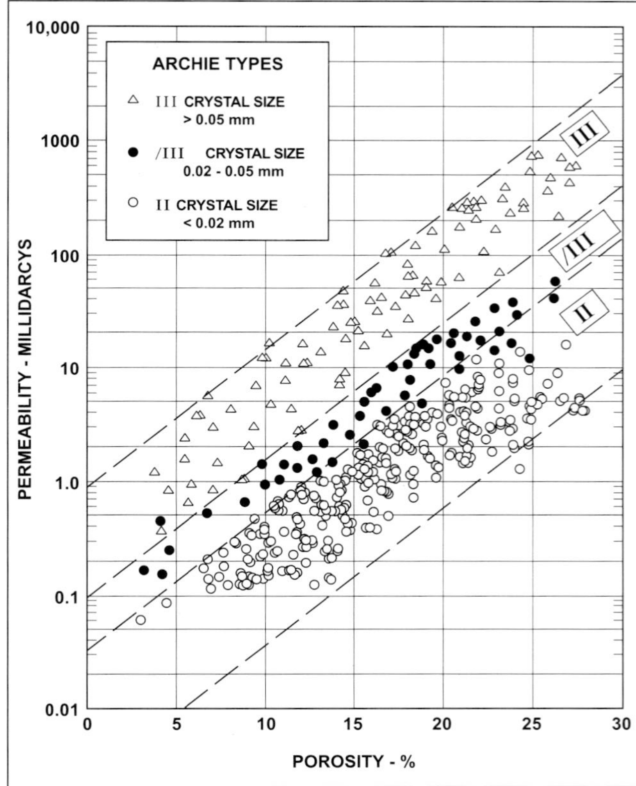

Figure 8. Crossplot of porosity versus permeability for different Archie rock types. Note that for the same porosity, permeability increases as the size of dolomite crystals increases. Dolomite crystal size reflects the size of the original limestone particles. The new lower limit of net pay is 4% porosity and 0.1-md permeability, based on field and laboratory tests.

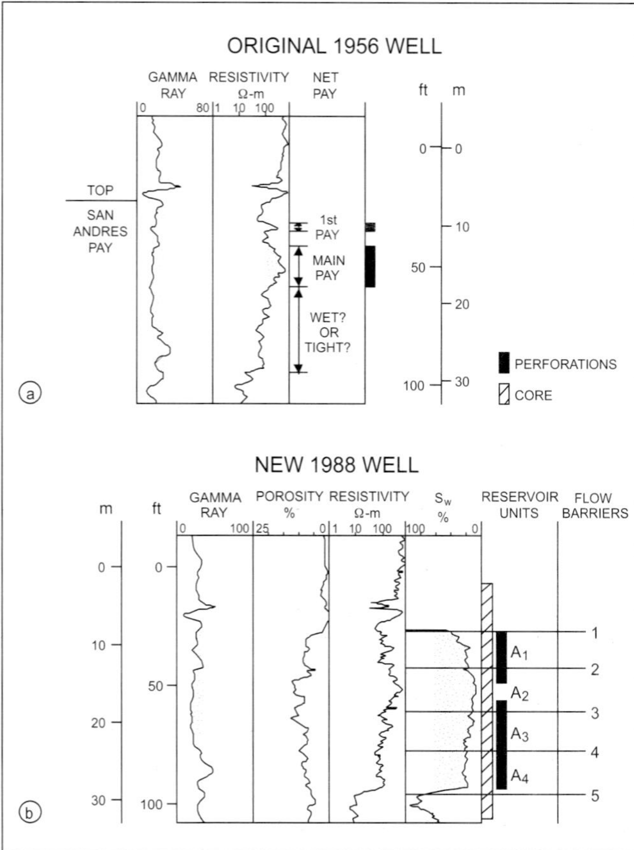

Figure 9. (a) Gamma-ray and resistivity logs of an original field well (1956), San Andres "C" waterflood unit, Permian Basin, Texas. Net pay is 7 m. (b) Gamma-ray neutron porosity and resistivity logs of a 1988 well drilled and cored in the same unit about 9 m from the 1956 well. Net pay in this well is about 20 m.

Reserve additions and performance improvements in the "C" waterflood unit (Figure 16) are mainly from infill wells (36%) and bypassed pays/recompletions (11%). Injection profile control and workovers add 7% and 3%, respectively.

"G" Field, Coastal Louisiana

A major oil company discovered the "G" field in the late 1930s. At the time of the acquisition, the cumulative production from the field was 170 million BOE. The structural trap for the field is a rollover anticline formed by a large growth fault. Large sealing faults subdivide the field into three blocks. Numerous small faults partially or completely offset reservoirs.

The field has a shallower hydropressured interval and a deeper geopressured interval (Figure 17). Before acquisition, all production came from the hydropressured reservoirs, which were deposited in a river-dominated delta complex. Seven wells in the 1940s drilled through the geopressured section; the wells found thick sands with abundant condensate and gas shows. Completion

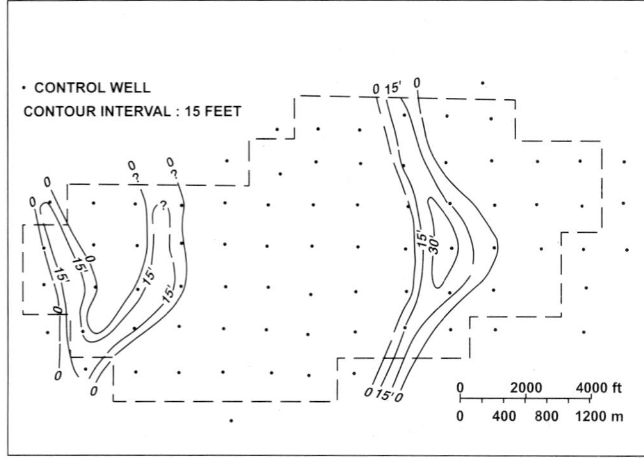

Figure 10. Net-pay isopach map of the "A_1" zone tidal channels, San Andres reservoir "C" waterflood unit, Permian Basin, Texas.

attempts in the geopressured reservoirs failed. Sands in the geopressured interval appear to be lenticular and difficult to correlate. Field studies based on a new 2-D seismic survey after acquisition suggest that the deeper sands are shingled turbidites with good continuity (Figure 18).

The "G" field has 27 reservoirs in the hydropressured interval between 5000 and 10,300 ft. Since acquisition, new wells found at least five reservoirs in the geopressured interval down to 17,000 ft.

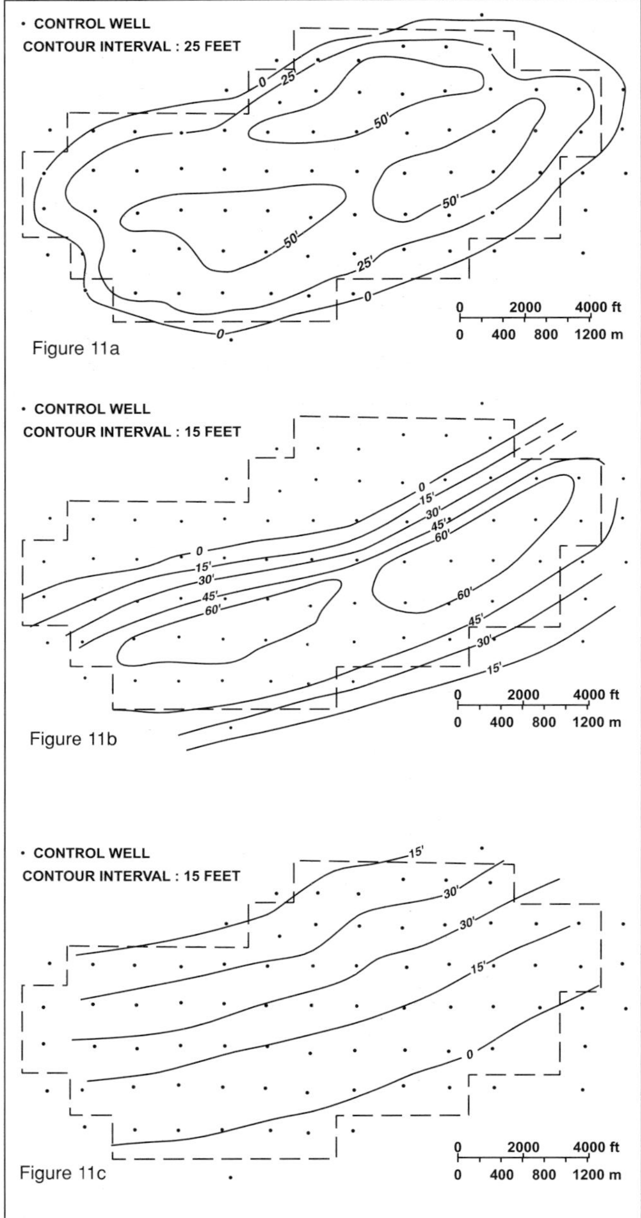

Figure 11. Net-pay isopach map of (a) combined oolite bar subzones "A2" and "A3," separated vertically by an effective flow barrier of shaly carbonate mudstone; (b) "A2" subzone—a shaly carbonate mudstone separates this subzone from the "A3" subzone; (c) "A3" subzone. "A2" and "A3" subzones are located in the San Andres reservoir "C" waterflood unit, Permian Basin, Texas.

From 1975 to 1989, the "G" field declined at 12% per year (Figure 19). The field has a solution gas drive with limited water support. At the time of acquisition, production was marginally economic at approximately 1100 bbl/day.

A multidisciplinary geoscience-engineering team quickly identified numerous opportunities before and after acquisition. Increased reserves came from infill wells in undrained or poorly drained reservoir compartments, workovers and recompletions (especially of lower-resistivity pay zones), and new field extensions. Lateral field extensions and the deeper geopressured reservoirs have significant upside potential.

A year after field acquisition, production increased from 1100 to more than 4000 BOE/day after recompletions and workovers of shut-in wells. During initial development, only the hydropressured portion of the reservoir system was completed. Thin shale beds within the reser-

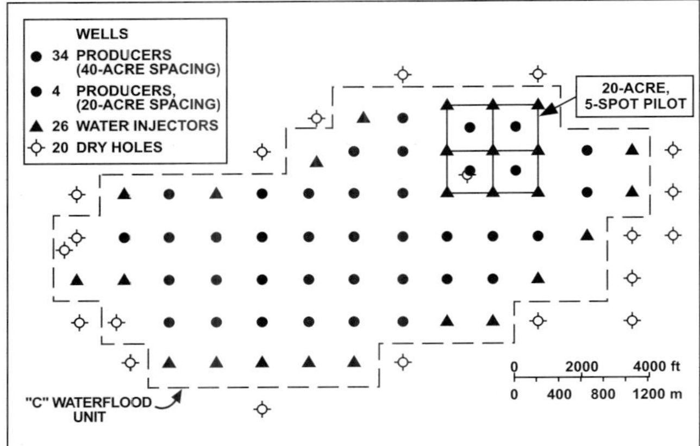

Figure 12. Map showing the location of 20-acre, five-spot pilot flood. The peripheral flood pattern of injectors and producers is the same as that in Figure 5. San Andres "C" waterflood unit, Permian Basin, Texas.

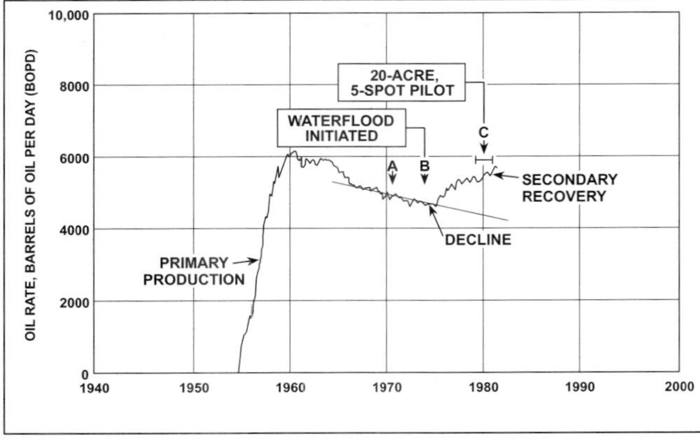

Figure 13. Graph showing oil production during primary recovery and secondary peripheral waterflood production and the 20-acre, five-spot pilot flood until December 1980. San Andres "C" waterflood unit, Permian Basin, Texas.

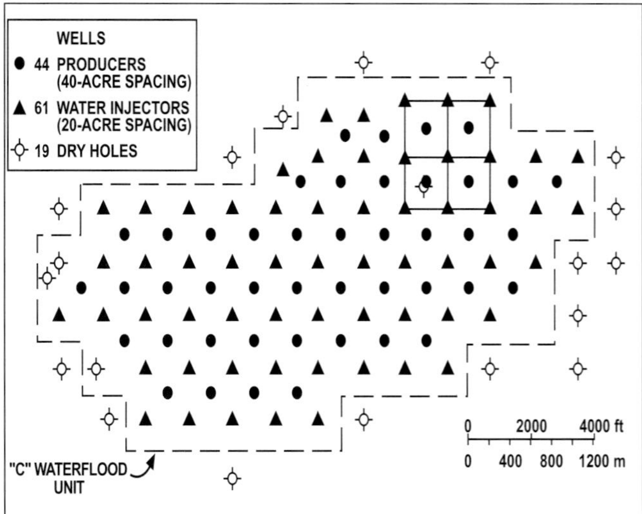

Figure 14. Map showing the expanded 20-acre, five-spot waterflood injection pattern beginning in late 1980 and the pilot flood area. San Andres "C" waterflood unit, Permian Basin, Texas.

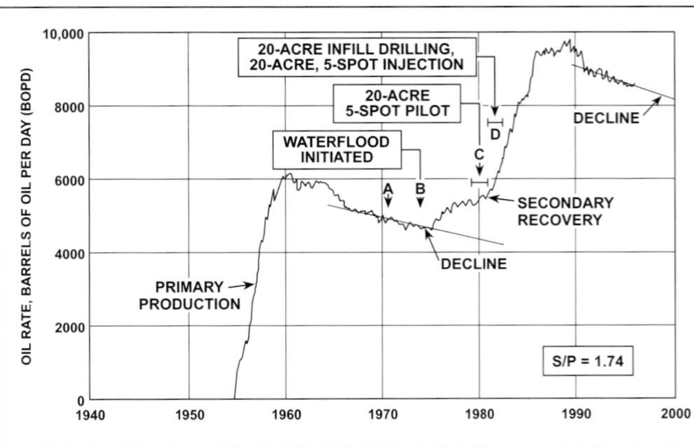

Figure 15. Graph illustrating primary and secondary waterflood production, San Andres "C" waterflood unit, Permian Basin, Texas. Note the significant increase in production starting at time D which resulted from a 20-acre, five-spot pattern waterflood and perforating of all pay zones.

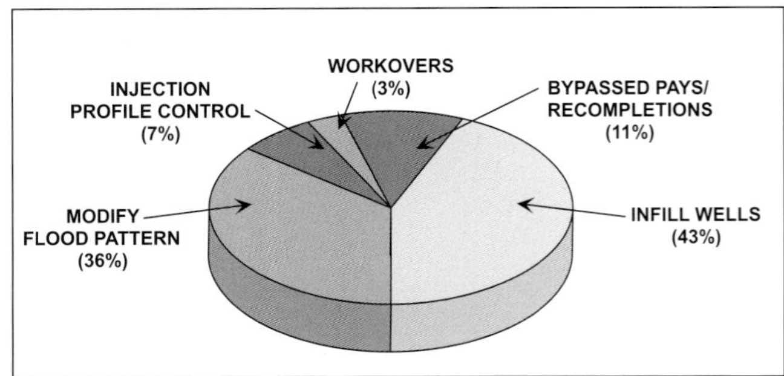

Figure 16. Opportunities for reserve additions and performance improvements, "C" waterflood unit, Permian Basin, Texas.

voir units effectively isolated the reservoir vertically. Wells were recompleted through the entire pay zone. Some of the recompleted wells produced at or near discovery pressure and rate.

Some "wet" zones were actually low-resistivity pays. The "X" sands (Figure 17) in one well alone added about 1000 bbl/day from a low-resistivity (<1.5 ohm-m) zone. The conventional pay in the 9500-ft "X" sands in this zone had gone to water 11 years earlier. A new well was drilled and cored (Figure 20). The high-resistivity zone that had gone to water 11 years earlier was now productive, suggesting that it had previously coned water.

Analysis of new core and production testing demonstrated that the zone below the high-resistivity pay was productive. To demonstrate the production potential, small zones were tested. The lower zones produced 990 bbl/day without water and at original pressure on 30-day tests. The original pay zone (i.e., the high-resistivity zone) produced 3112 bbl/day. The original pay zone produced approximately 300,000 bbl oil before watering out again; the low-resistivity pay zone continues to produce without significant water. Figure 21 shows the original hydrocarbon distribution mapped on the conventional pay only. Notice that pay occurs in one fault block. Figure 22

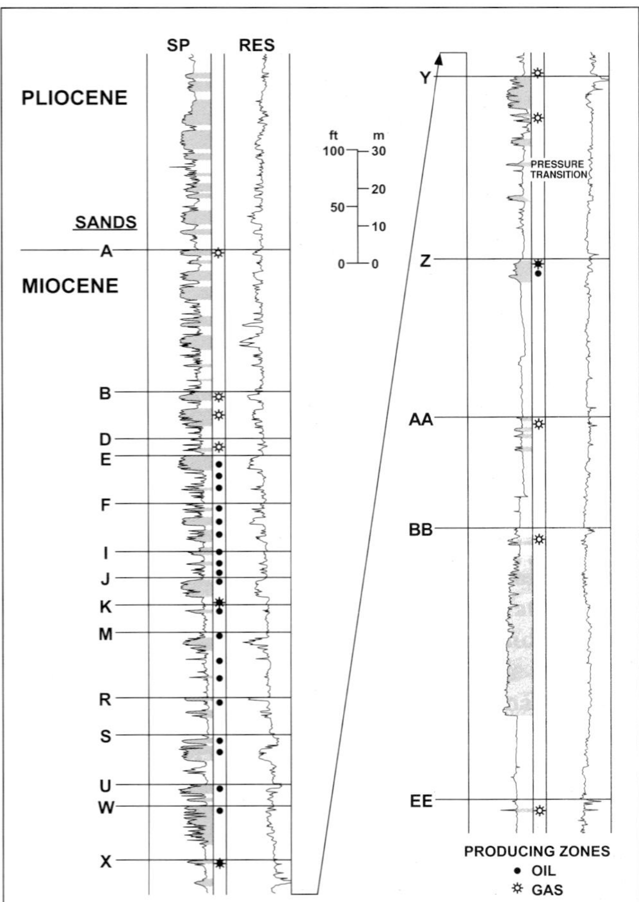

Figure 17. Composite-type log of the "G" field, coastal Gulf of Mexico Basin, Louisiana.

shows the distribution of hydrocarbons, including low-resistivity pay. Development of this low-resistivity pay added 14 million bbl of proved oil reserves. Newly identified low-resistivity pays in several zones doubled the proved reserves in the field.

The initial redevelopment study increased reserves and increased production to more than 5000 bbl/day. A new 3-D seismic survey run in 1994–1995 delineated numerous deep opportunities in the field. Three significant wells produce from the geopressured section now. One well has more than 125 net ft of pay.

We expect development during the next few years to increase production to more than 11,000 BOE/day through field extensions and in newly defined prospects in the deeper geopressured reservoirs (Figure 19).

Performance improvements and reserve additions are

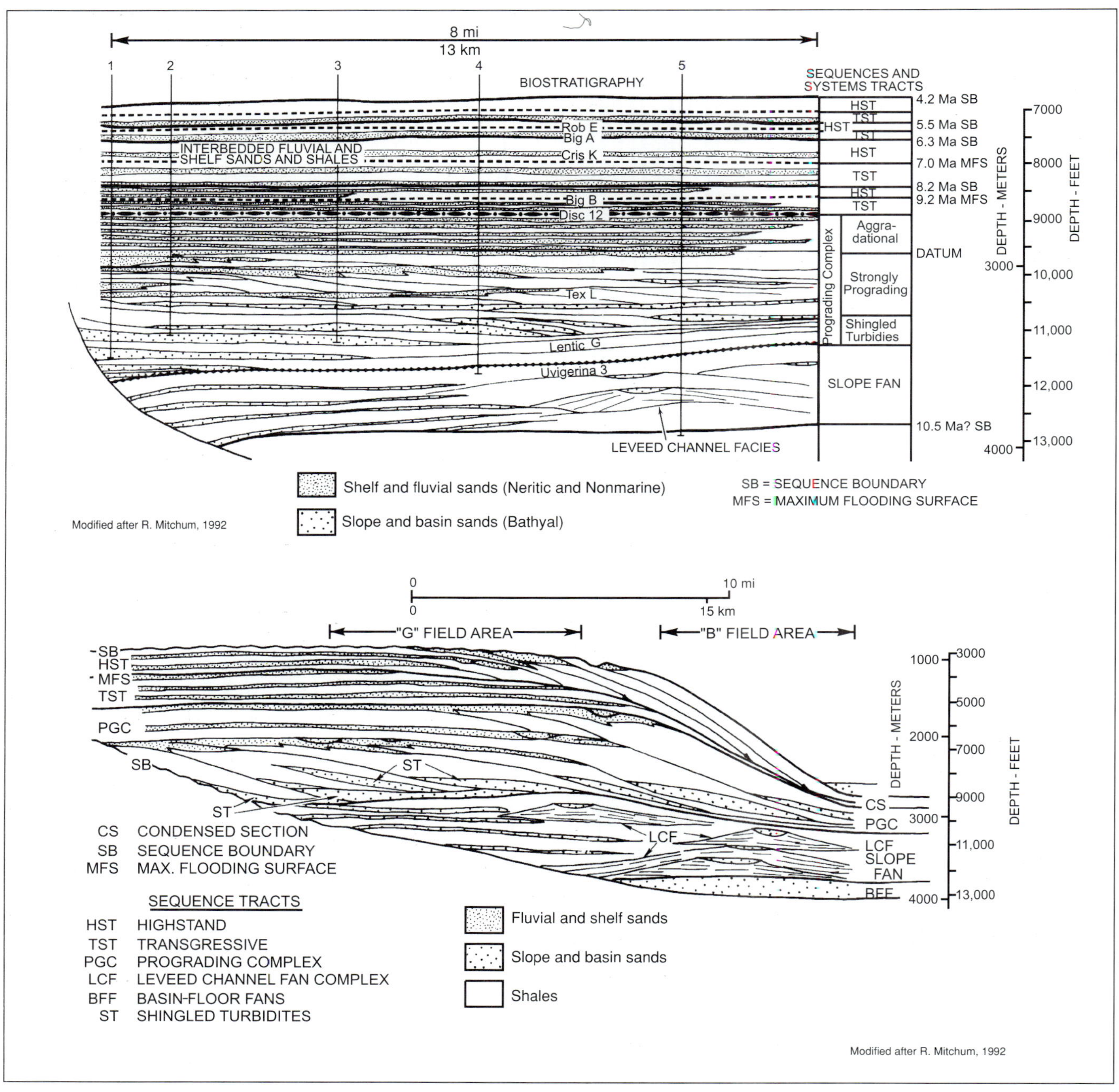

Figure 18. (a) Stratigraphic cross section of the "G" field based on six deep wells, paleodata, and 3-D seismic data. (b) Sequence-stratigraphic correlation of shallow and deep reservoir units based on well logs, 3-D seismic data, and some paleodata, "G" and "B" fields, coastal Gulf of Mexico Basin. Only two deep 1950s wells penetrated the sections below 3000 m. Both wells had gas-condensate shows and were recompleted in the normally pressured prograding complex reservoirs. Orientation from left to right is north-south.

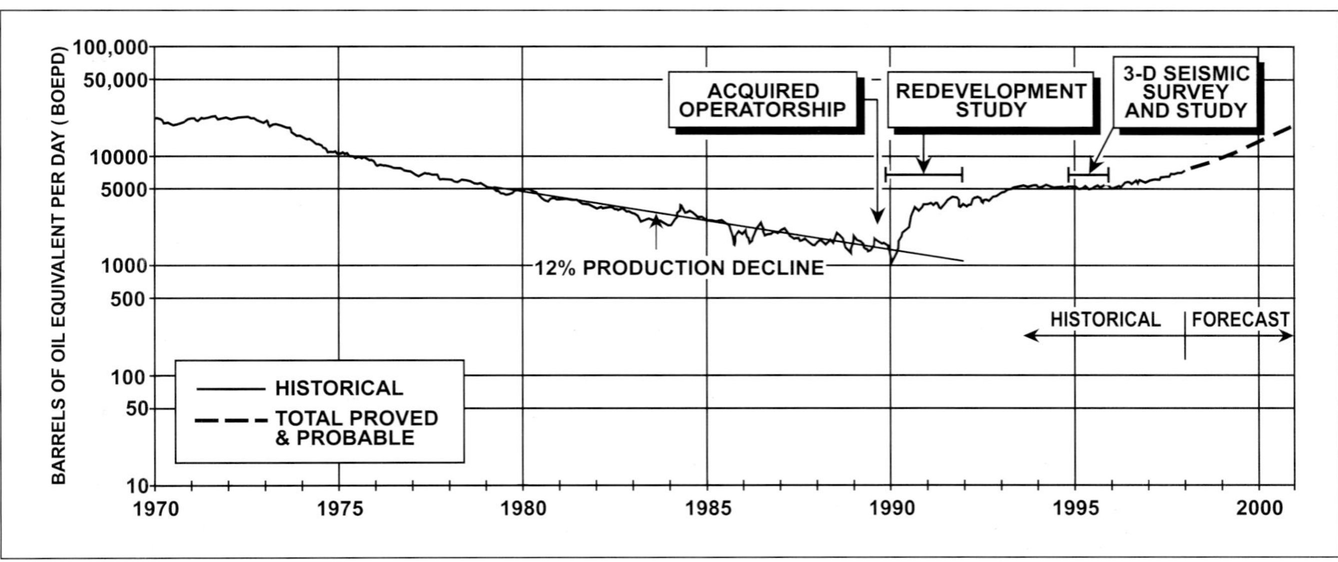

Figure 19. Graph of "G" field daily production in BOE/day (BOEPD). The field was acquired in 1989, and a field-redevelopment study was initiated. The production increase, which started in 1990, resulted from workovers and recompletions. A 1994–1995 3-D seismic survey documented additional shallow and deep new-well opportunities and field extensions.

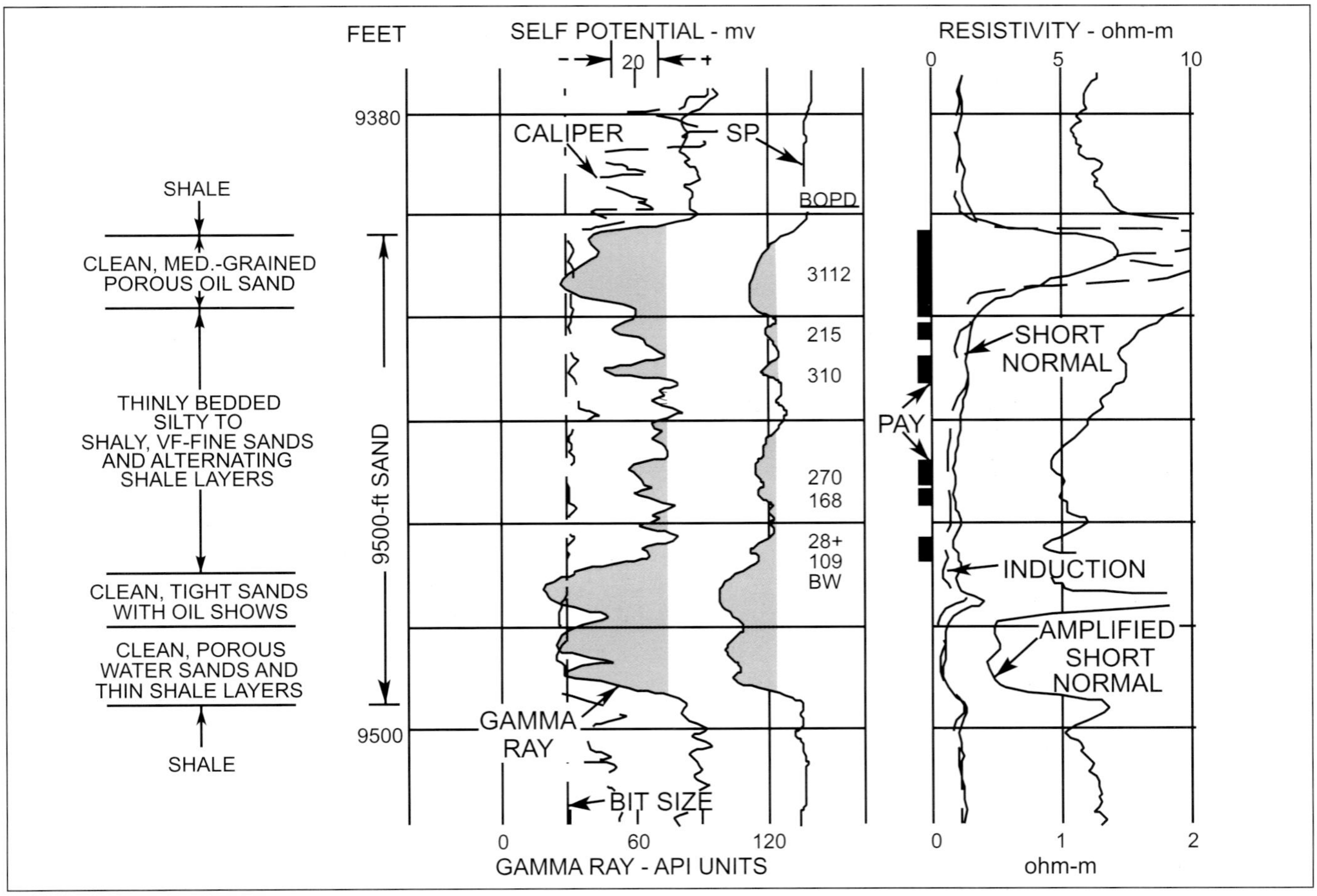

Figure 20. Well logs of the 9500-ft reservoir sand in a 1989 well, "G" field, coastal Gulf of Mexico. This new cored well is less than 10 m from a 1940s well that produced from the upper 20 ft of the 9500-ft sand. The low-resistivity zone below the high-resistivity zone was never tested because it was interpreted as wet. This low-resistivity zone production tested about 1000 BOPD. Each individual flow test was 30 days.

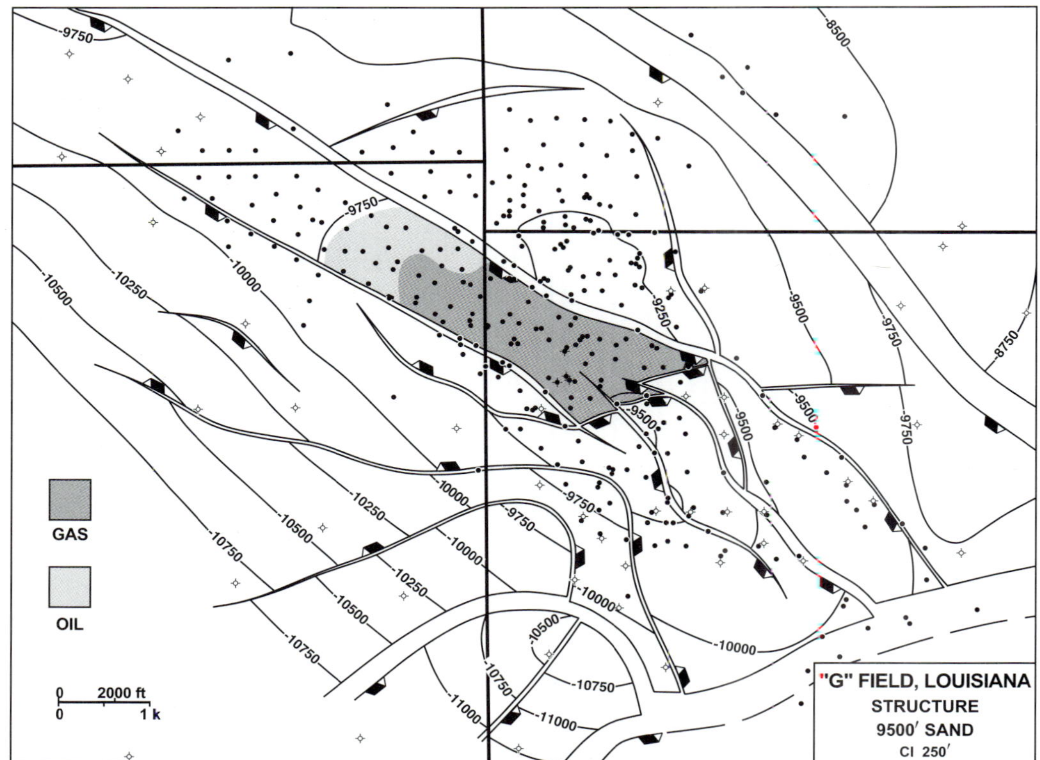

Figure 21. Structure map on top of the 9500-ft sand showing its original distribution of oil and gas, "G" field, coastal Gulf of Mexico, Louisiana.

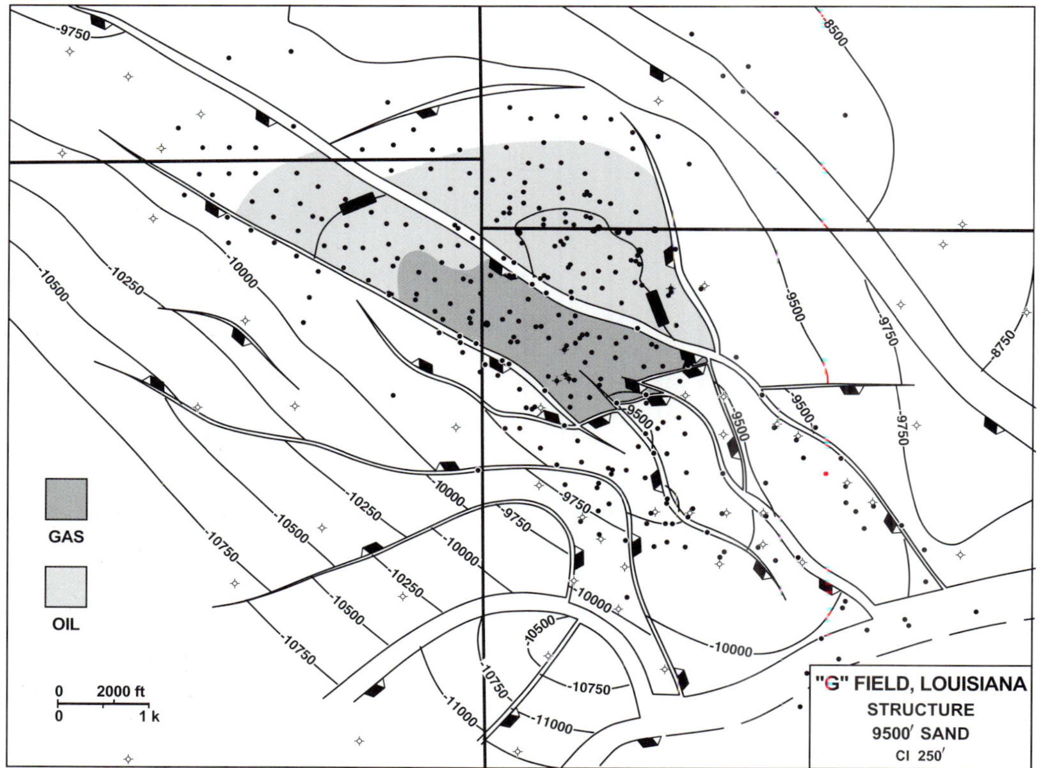

Figure 22. Structure map on top of the 9500-ft sand showing the new distribution of oil and gas resulting from recompleting the low-resistivity zones in the sand, "G" field, coastal Gulf of Mexico, Louisiana.

principally from field extensions (61%), bypassed pays/recompletions (18%), and infill wells (12%) (Figure 23).

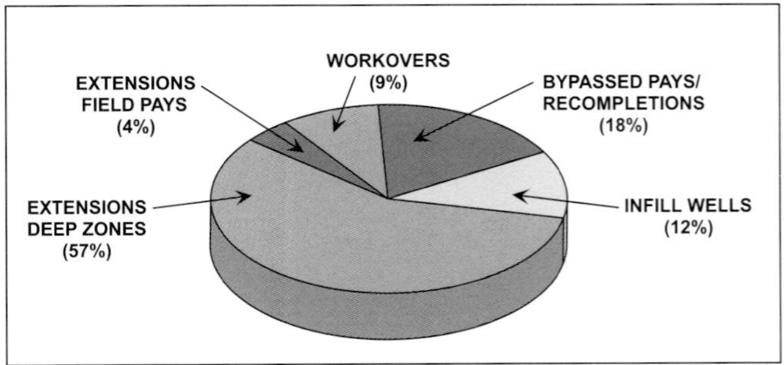

Figure 23. Opportunities for reserve additions and performance improvements, "G" field, Gulf of Mexico Basin, coastal Louisiana.

VLC-363, Block III Field, Lake Maracaibo, Venezuela

VLC-363, Block III field is a supergiant field located in eastern Lake Maracaibo that produces from the Eocene lower C reservoir from 63 wells (Figure 24). The field was found in the 1960s and had original oil in place (OOIP) of more than 1.7 billion bbl. It has produced about 200 million bbl of retrograde condensate and an indeterminate amount of gas, and it will recover 12% to 13% of the OOIP by primary recovery. The remaining potential is more than 200 million bbl.

A team of geologists, geophysicists, petrophysicists, and engineers from PDVSA and U.S. consultants started a rejuvenation project in 1994 (Sneider et al., 1999). The structure was mapped using a low-resolution 3-D seismic survey and well logs. The well logs were extremely important in identifying the faults because of the poor reso-

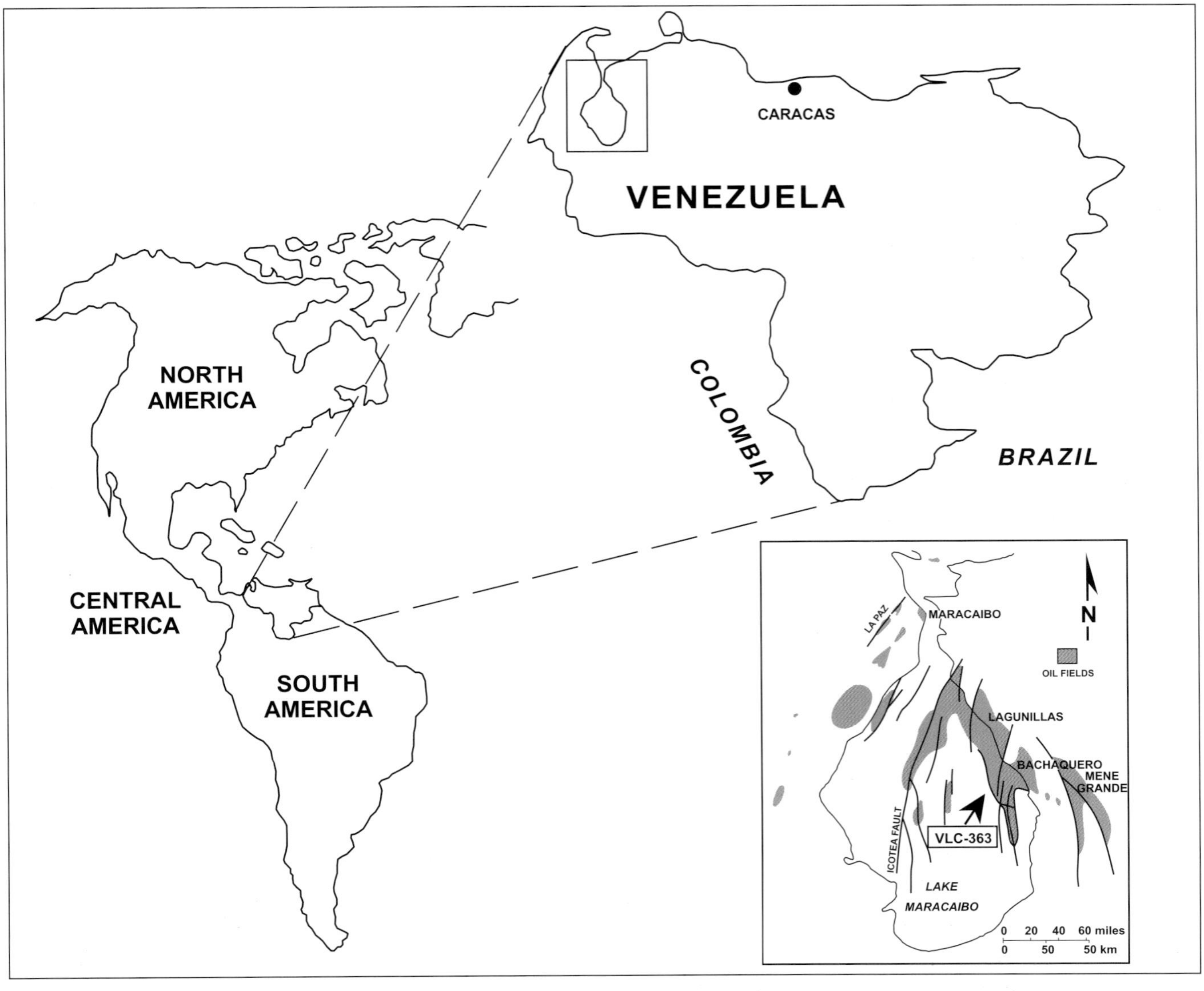

Figure 24. Location of VLC-363, Block III field, Lake Maracaibo, Venezuela.

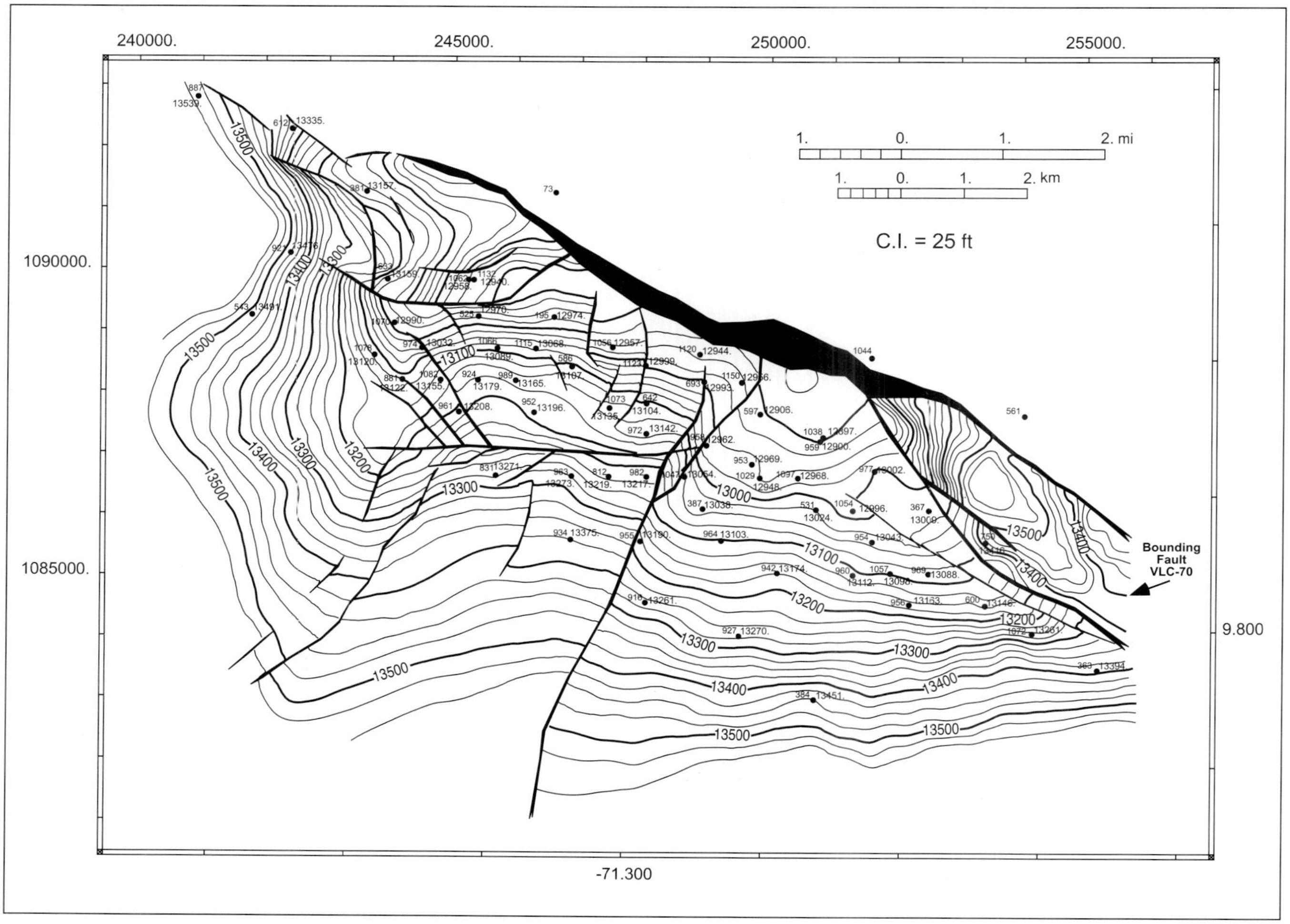

Figure 25. Structure map on top of the C455 reservoir, based on well logs and 3-D seismic data, VLC-363, Block III field.

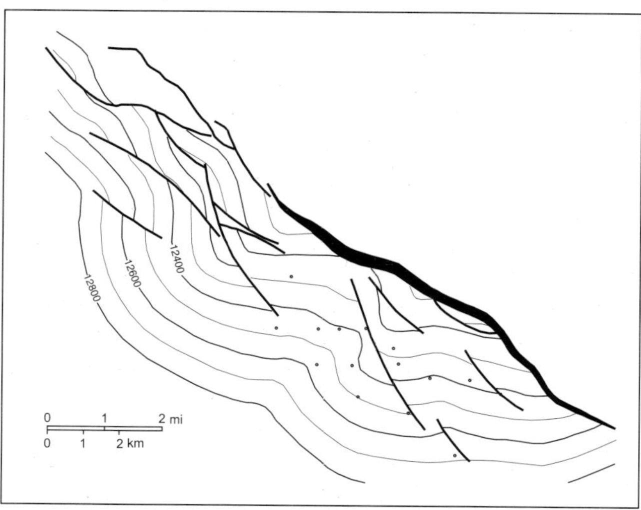

Figure 26. 1960s structure map on the top of the C4 reservoir based on well logs and 2-D seismic, VLC-363, Block III field.

lution of the seismic data, which could not fully resolve most of the faults.

The hydrocarbons are trapped in an upthrown faulted anticline on a down-to-the-northeast fault (Figure 25). The field lies between two major strike-slip faults. One bounds the eastern edge of the field, and the other is located about 15–20 km west of the field. The field has had at least three periods of structural evolution, including both extensional and strike-slip deformations.

There are more than 22 small faults with less than 150 ft of throw within the field. None of these small faults had been mapped previously (Figure 26). The faults subdivide the field into four distinct production regions (Figure 27).

The new structural interpretation explained several previously unexplained production anomalies, such as water producing from zones above oil pay and large differences in the producing water level across the field (Figure 28). The north-south fault in the center of the field separates a producing water level that is more than 350 ft lower in the east than in the west (13,025 ft in the east and 13,375 ft in the west). The throw along this

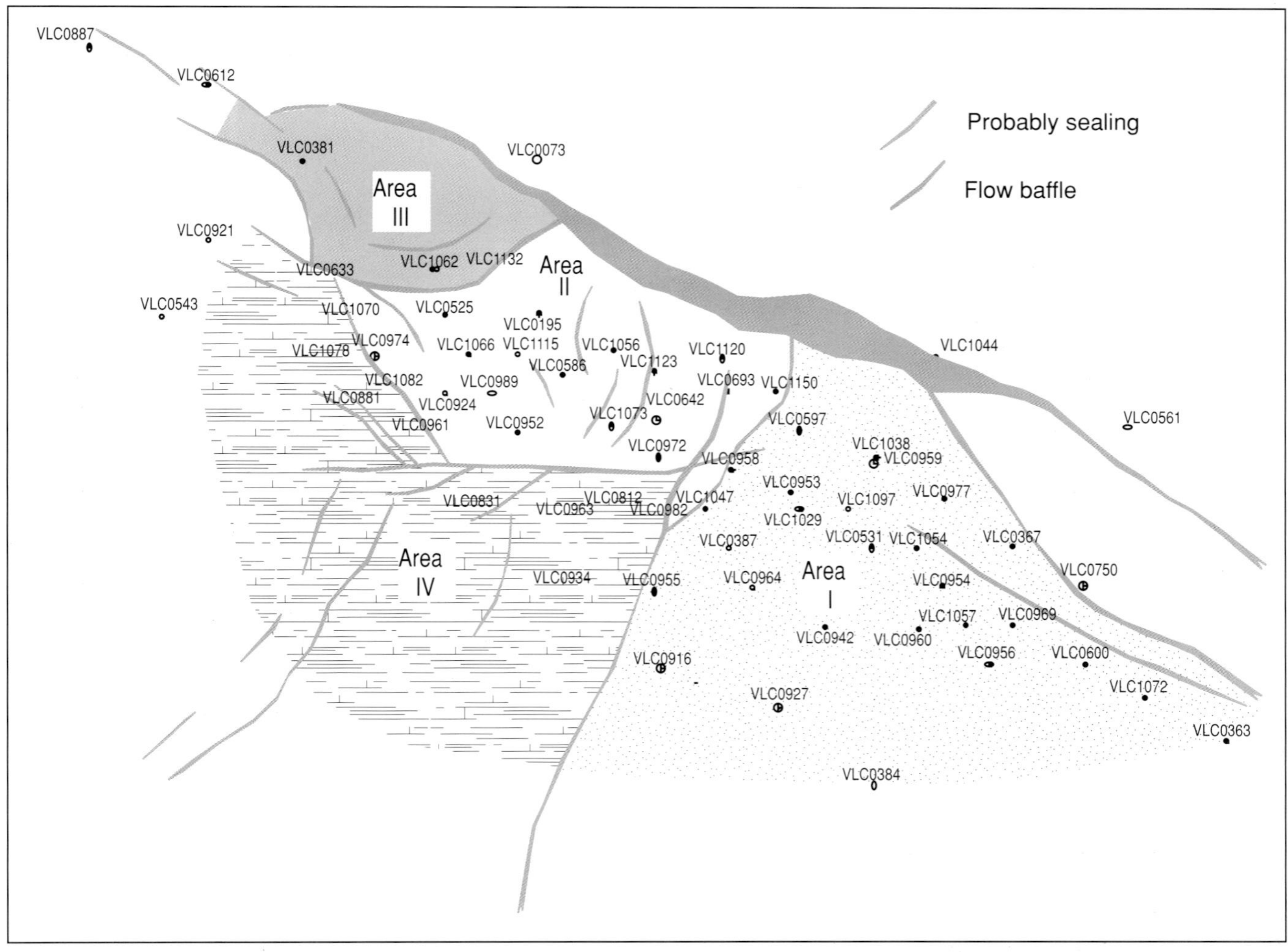

Figure 27. Sealing and partially sealing faults divide the field into four major production zones, VLC-363, Block III field.

fault ranges from approximately 125 ft in the south to less than 40 ft in the north.

The small east-west fault separating production area 2 from production area 4 has about 50 ft of throw, but it is associated with a production anomaly. It is clear from the pressure data that this fault leaks. Shortly after the map was made, the four wells to the south of the fault were worked over, and production increased to more than 1000 bbl/day in all four wells. The wells upstructure to the north of the fault produced at low rates with very high water cuts, and the wells were shut in.

Faults act as barriers and baffles to production (Figure 29). Production data indicate that faults with throws less than 100 ft leak. These small faults still affect production and have resulted in parts of the field being poorly drained. Additionally, these small faults will greatly affect any enhanced recovery project because individual reservoirs are usually less than 50 ft thick.

The Eocene C deposits consist of higher-frequency depositional cycles superimposed on an overall transgression (Figure 30). The lower part of the section from the C455 to the Guasare comprises predominantly fluvial-deltaic deposits, and rests unconformably on a major regional unconformity. The upper part of the section from the C4 to the C448 is predominantly marine.

The field contains 16 major flow units. Some of these were subdivided further for reservoir simulation. The major pay zones are the C448, C455, and C460.

The depositional environment has a major impact on reservoir and flow-barrier distribution. Thin flow barriers have a major impact on the primary and secondary recovery process and cause poor vertical communication in the field. The reservoirs contain numerous barriers and baffles to vertical flow. The shales between the flow units can support several thousand psi differences in pressure. The ratio of vertical to horizontal permeability (K_v/K_h) in the sandstones is approximately 0.6, but the effective K_v/K_h is less than 0.001 when considering the thin shale laminations within the sandstones. This means that there is poor vertical communication within the reservoirs.

The field went below the bubble point in the mid-1980s, resulting in numerous problems because the hydrocarbon is a retrograde condensate; therefore, the field has (1) significant oil shrinkage, (2) formation of

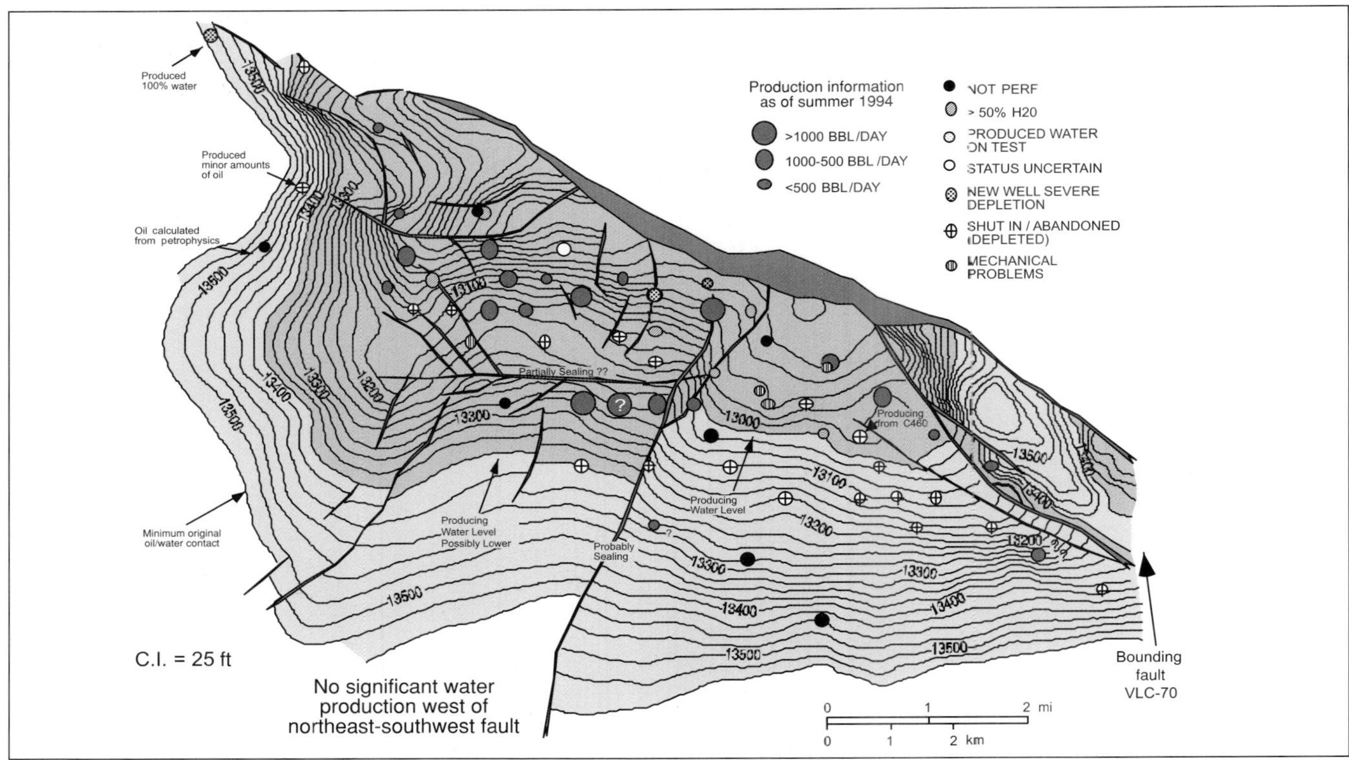

Figure 28. Daily production, hydrocarbon distribution, and well status in July 1995, superimposed on the C455 structure map, VLC-363, Block III field.

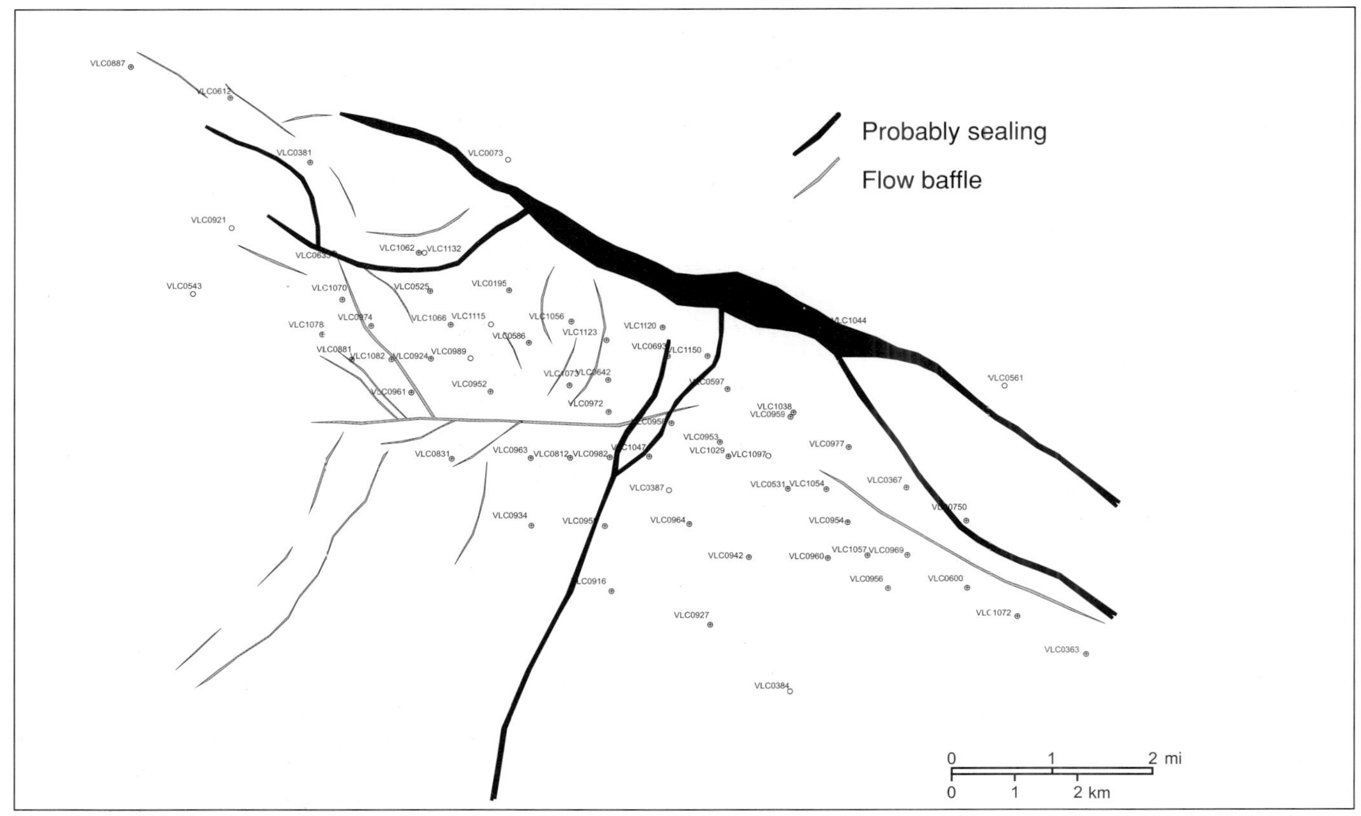

Figure 29. Map showing completely sealing faults that act as barriers to fluid flow, and faults that leak but act as partial barriers to fluid flow, VLC-363, Block III field.

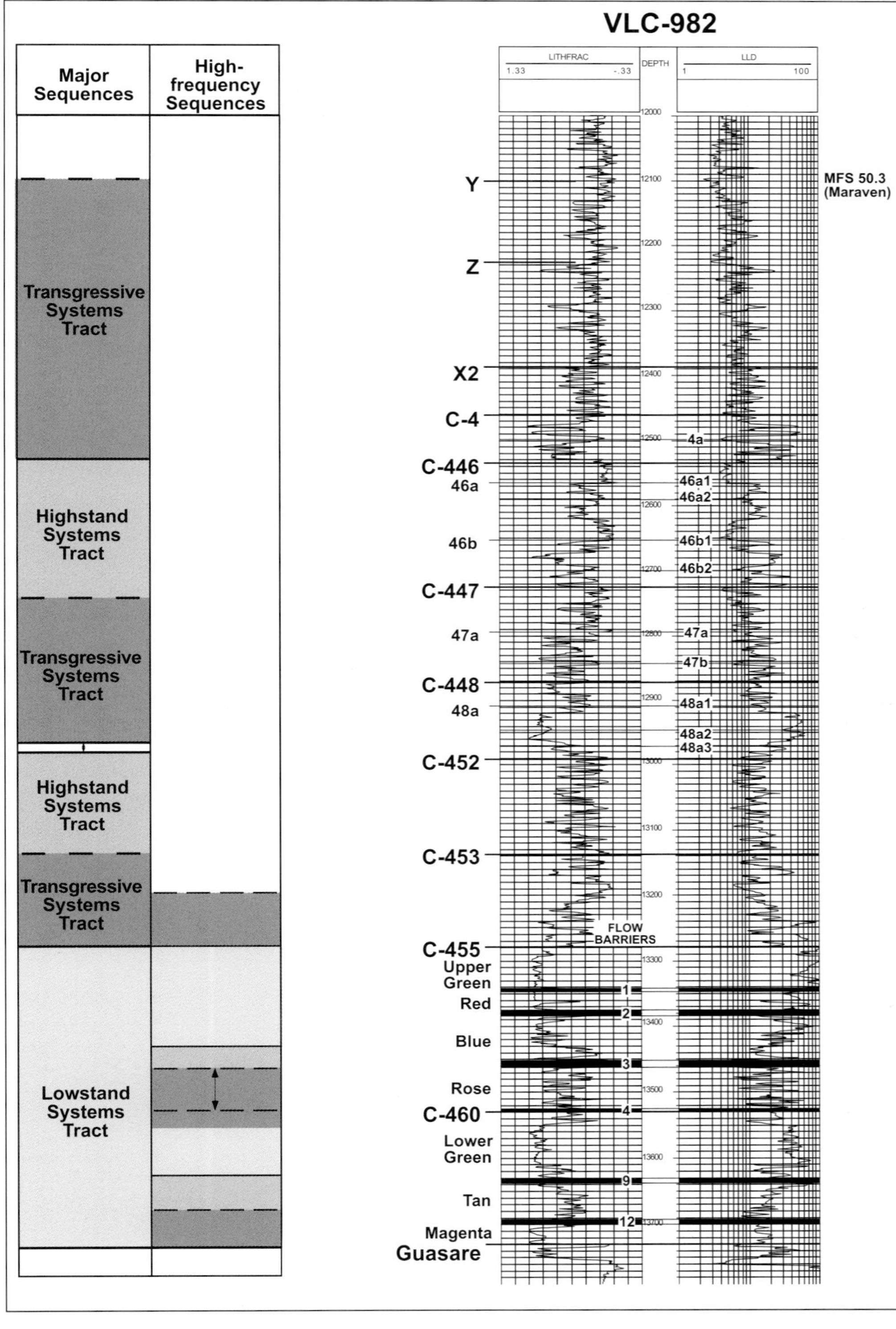

Figure 30. Type log for the VLC-363, Block III field, showing detailed correlation of flow units and barriers, and the sequence-stratigraphic interpretation of the succession.

numerous secondary gas caps underlying flow-unit boundaries, and (3) insufficient free gas available to significantly expand oil volume upon repressuring with water.

The extensive reservoir depletion resulted in very little oil movable by waterflood. Without repressuring, gas injection would have resulted in very little additional oil recovery, although with repressuring, the reservoir system could have become miscible. Without repressuring, injected gas will stream through existing high gas-saturated regions.

There are four main development opportunities with significant reserve potential: enhanced recovery, new wells in underdeveloped fault blocks, workovers, and recompletion in lower-resistivity zones. There are about 87 million bbl of potential reserve additions with primary recovery and 400 million bbl of potential reserve additions with enhanced recovery (Figure 31).

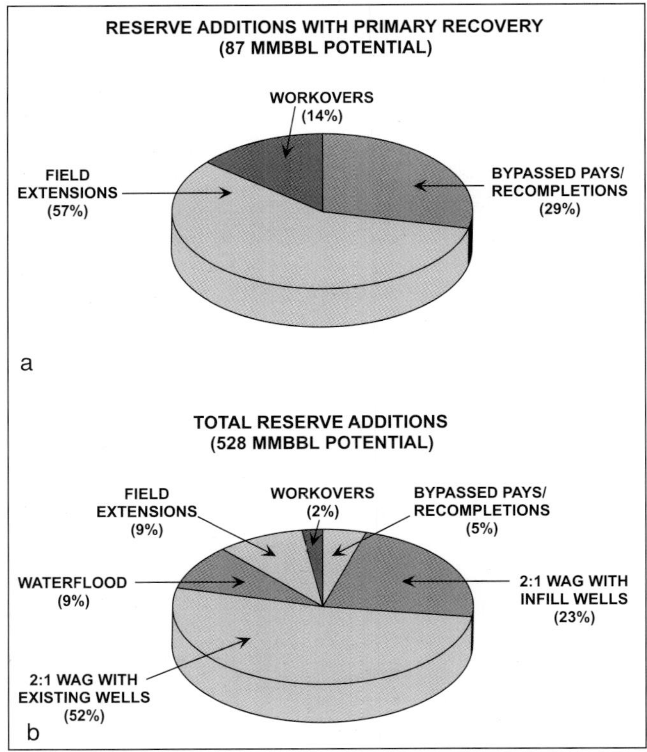

Figure 31. Diagrams showing (a) distribution of reserve additions and opportunities for production improvements with primary recovery, and (b) distribution of reserve additions and performance improvements, including enhanced recovery, VLC-363, Block III field. WAG refers to the injection of water alternating with gas.

A waterflood would recover less than 5% of the OOIP and is uneconomic (Figure 32). Gas injection would have similar results. If the field is filled with water and then gas is injected, there would be an increase in recovery of 11% of the OOIP, or 187 million bbl. A 2:1 injection of water alternating with gas (WAG) with the existing wells would recover an additional 16% of the OOIP, or 272 million bbl. If infill wells are drilled to produce a better pattern, 23% of the OOIP, or 391 million bbl of additional reserves, is possible.

There are at least three possible field extensions: to the northeast, to the southwest, and possibly on the southeastern flank of the field (Figure 33).

The northeast extension, in a separate fault block

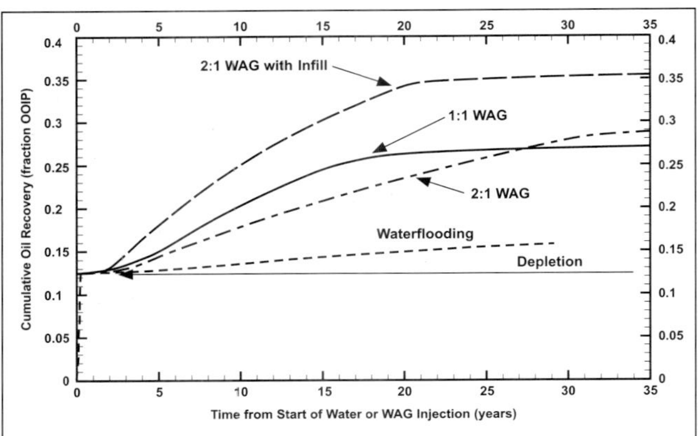

Figure 32. Plot showing recovery in time after water or WAG injection in years versus cumulative oil recovery as fractions of OOIP, VLC-363, Block III field.

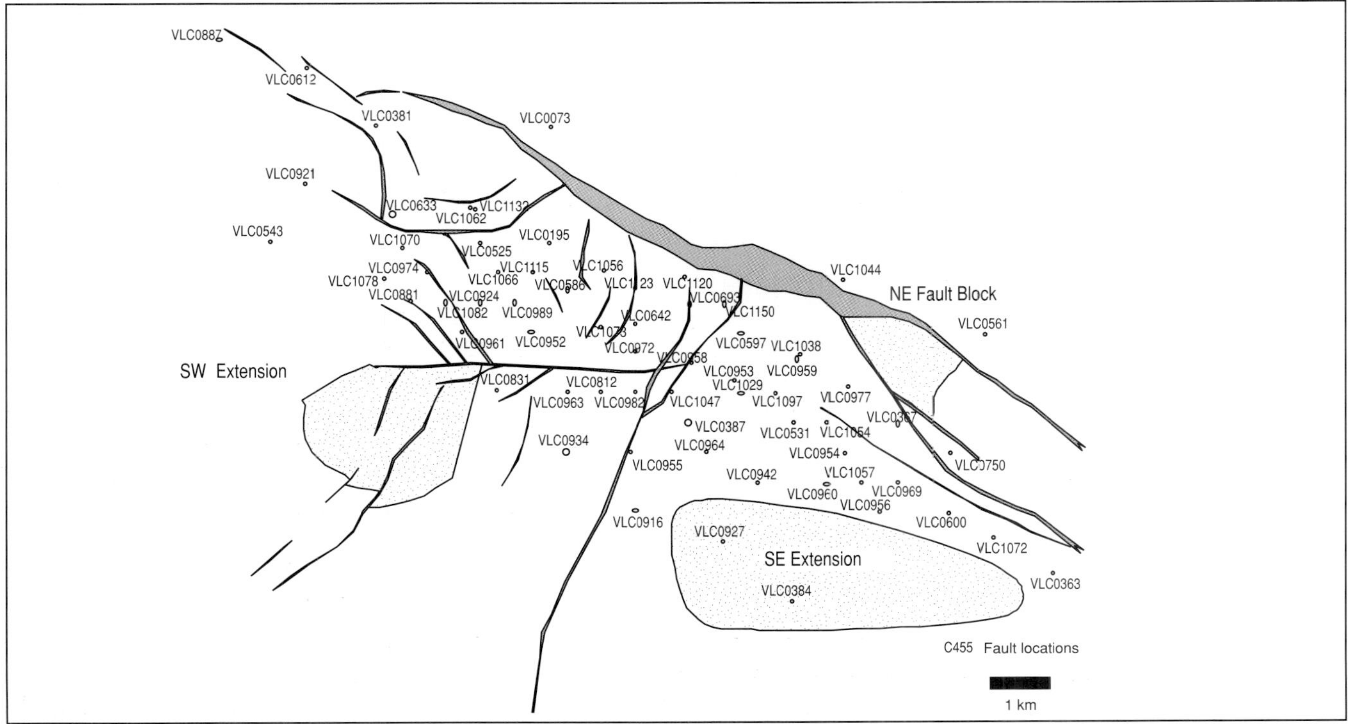

Figure 33. Location of potential field extension areas, VLC-363, Block III field.

from the main field, has one well (VLC-750) drilled in it (Figure 34). This well produced as much as 500 bbl/day and produced 200,000 bbl oil, which is only 3% of the OOIP in the small fault sliver (area 3). The entire reservoir interval was saturated with hydrocarbons, and the well had a 16% water cut at abandonment. Two additional structures are isolated from the VLC-750 well (Figure 25). The area northwest of the fault block is more than 200 ft above the VLC-750 well. The structure in area 1 contains about 7 million bbl of recoverable oil by depletion.

The hydrocarbon pore-volume map of the main reservoir shows that reservoir quality decreases to the east (Figure 35). The VLC-750 well is in the area of poorer reservoir quality. To the west in the fault block, the reservoir quality should improve. All the wells south of the northwest-southeast-trending fault south of the VLC-750 well produced more than 1 million bbl, with some producing more than 20 million bbl. Also, there are many recompletion opportunities in lower-resistivity zones that have not been completed.

Significant reserves and potential reserves were added to this field by (1) understanding the rock types and pay classification, (2) mapping horizontal flow barriers (mainly faults, but also facies changes), (3) recognizing vertical flow barriers and changing completion practices or recompleting zones accordingly, and (4) integrating the reservoir simulation with the reservoir performance and the geoscience in an iterative, not a linear, manner.

Auk Field, Central Graben, North Sea

In the North Sea, many of the fields found in the late 1960s and 1970s have reached maturity by primary recovery. Reserve additions and performance improvements in these mature fields have resulted from the application of new production technologies, including horizontal or multilateral wells. A good example is the Shell/Esso Auk field (Figure 36). The field went on production in 1975, had peak production in 1977–1979 of about 40,000 bbl oil/day (BOPD), and rapidly declined to less than 10,000 BOPD in 1988 (Frazer, 1998). The estimated ultimate recovery in 1988 was 93 million bbl, or about 13 million bbl remaining (Trewin and Bramwell, 1991). The estimated ultimate recovery in 1998 is about 180 million bbl, or about twice the 1988 estimate.

A major share of the reserve additions at Auk is from a large field extension (Auk North) found previously by seismic but considered uneconomic for many years. Previous work by Shell/Esso showed that Auk North production was not feasible to develop from a satellite platform or from extended-reach wells from existing Auk platforms. Improvement in economics that make Auk North development now possible for Shell/Esso comes from the use of (1) multilateral wells from existing Auk development wells and (2) electric submersible pumps deployed and retrieved on coil tubing without the need of a semisubmersible rig. The employment of those engineering advances, plus the drilling of new wells and infill wells, raised the recovery factor from 17% to 30% of the oil in place.

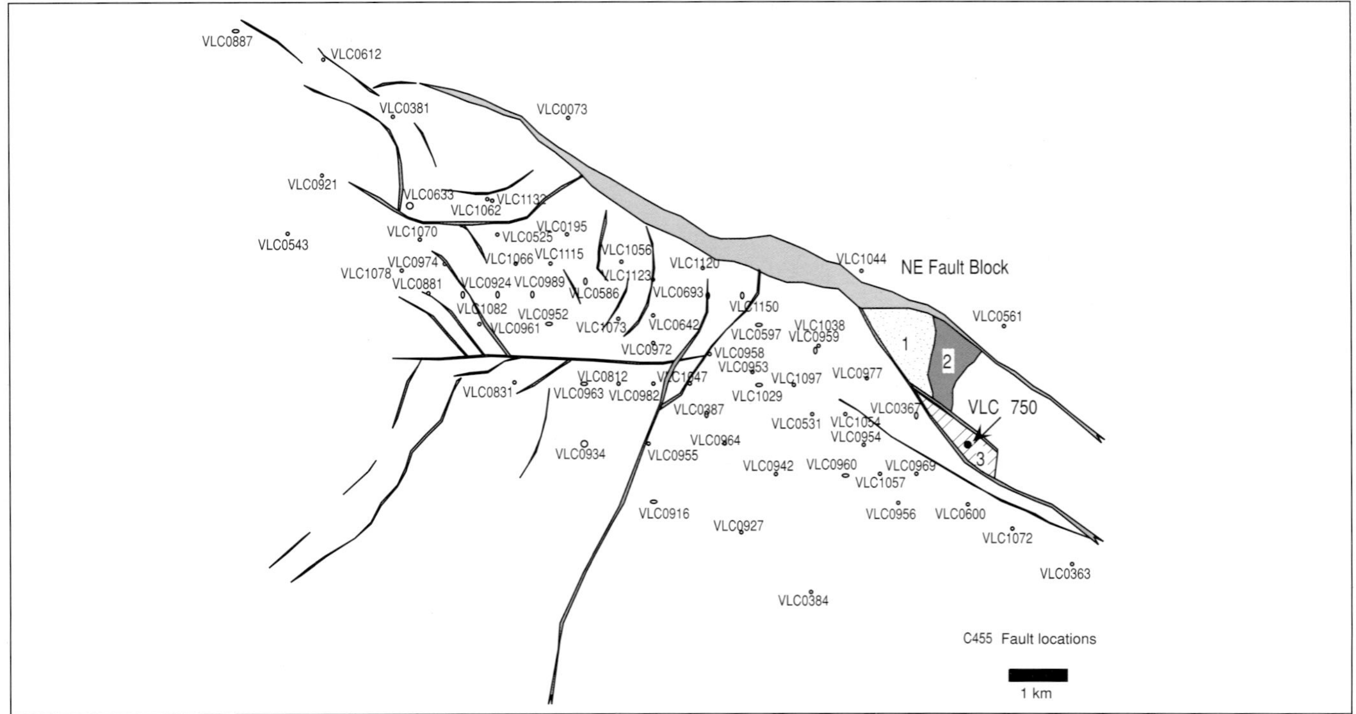

Figure 34. Location map of the northeast fault-block prospect, VLC-363, Block III field. Area 1 contains 456 acres and 34.6 million bbl OOIP, area 2 contains 278 acres and 22.6 million bbl OOIP, and area 3 contains 152 acres and 7.5 million bbl OOIP.

POTENTIAL FOR RESERVE ADDITIONS

What is the reserve-growth potential in and around aging marginal fields? Nehring (1995) showed that about 85% (20 billion bbl) of proved oil-reserve additions in the United States between 1982 and 1992 were from mature fields. Schollnberger (1998) estimated that future additional reserves of about 400 billion bbl of oil and 600 trillion cubic feet of natural gas are expected from mature fields. Schollnberger's estimates are based on Masters' 1994 U.S. Geological Survey report (see Masters et al., 1994).

Our estimate of additional reserves in and around the 82 mature fields we studied in the Permian Basin and part of the Gulf of Mexico Basin (Figure 1) is more than 6 billion BOE. This estimate is for fields that have produced at least 25 million BOE and have a potential for adding at least 5% of the cumulative production.

Hundreds of mature fields in the United States and Canada have undergone some type of redevelopment.

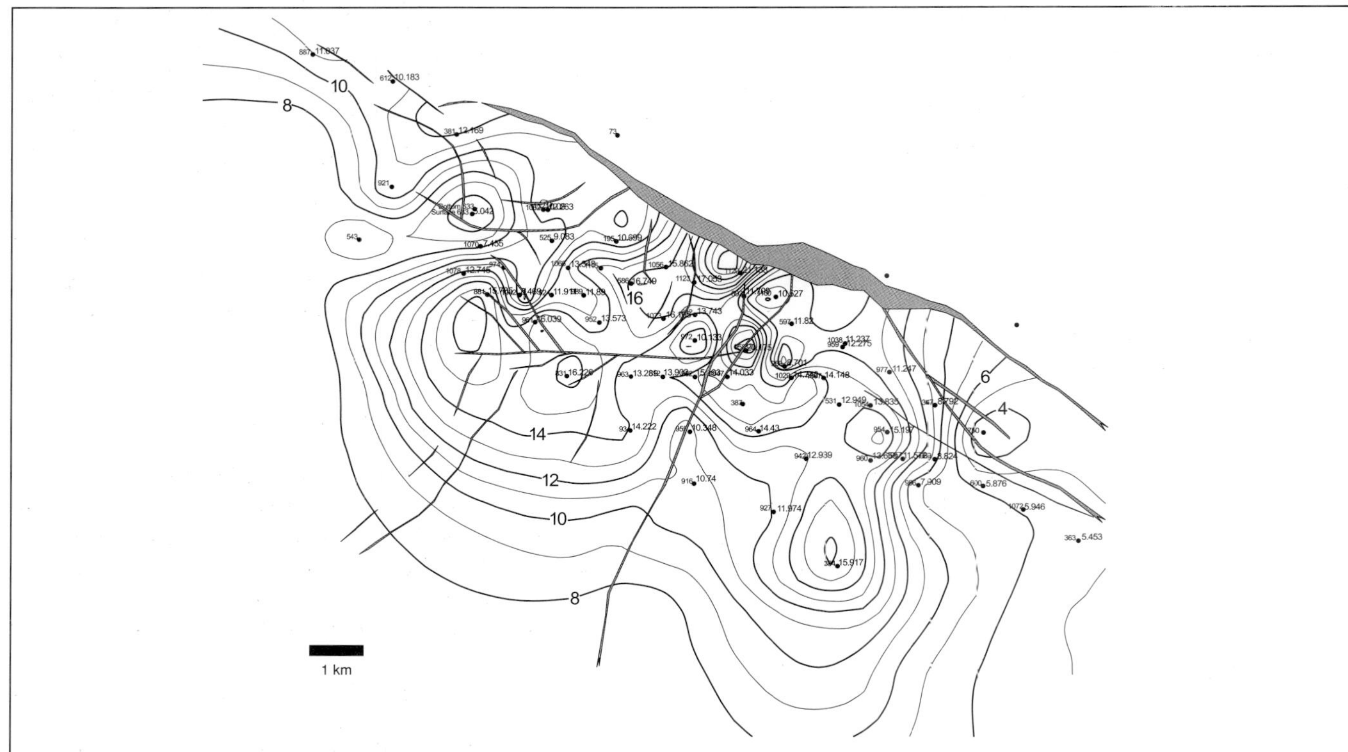

Figure 35. Hydrocarbon pore-volume (HPV) map for the C455 reservoir, VLC-363, Block III field. HPV = $(1 - S_w) \times$ (net feet of pay) × (porosity). Contour interval is 1 unit.

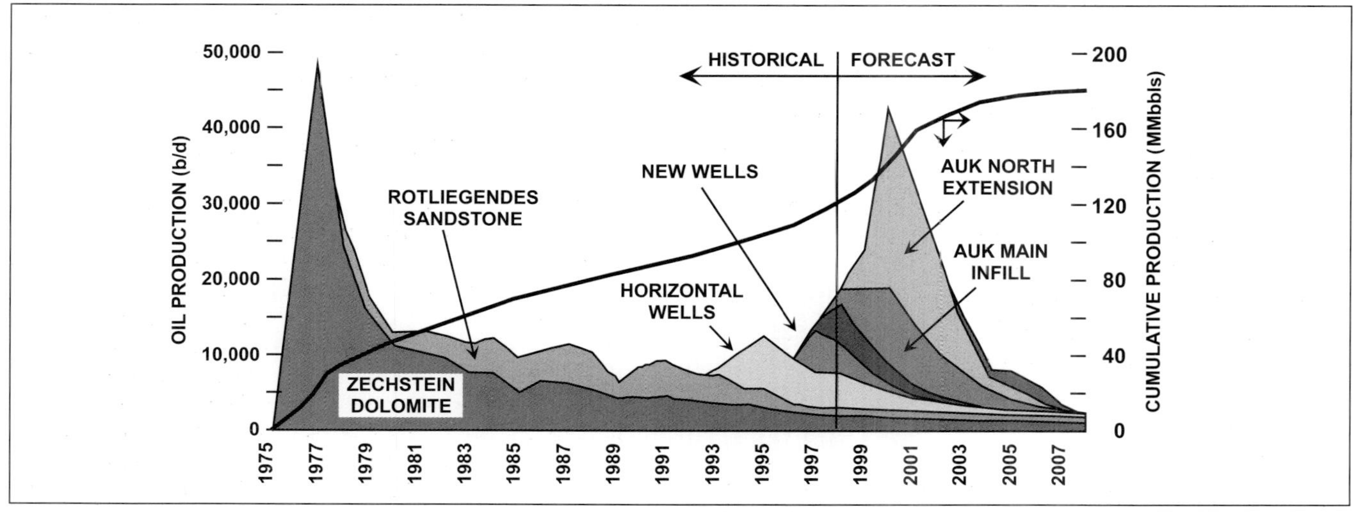

Figure 36. Production history and field development, Auk–Auk North field, Central Graben, North Sea. After Frazer (1998).

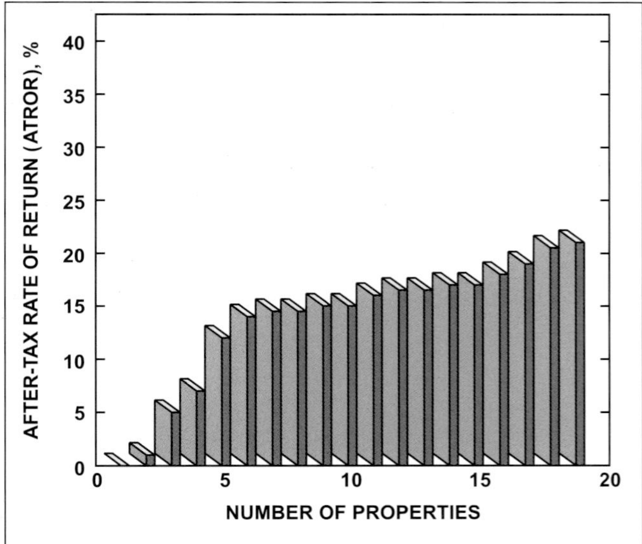

Figure 37. Value of 19 waterflood fields purchased in the Permian Basin, Texas and New Mexico. Value is expressed as the percent after-tax rate of return (ATROR). The two unprofitable fields were fractured, and water injection could not be controlled.

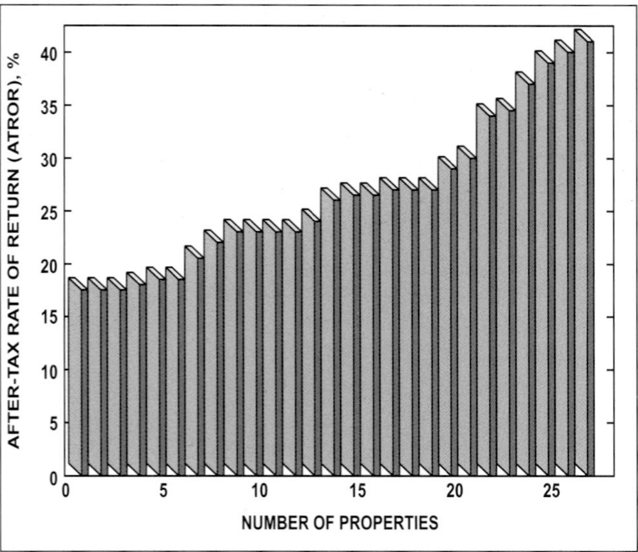

Figure 38. Value of 27 fields purchased in the Texas-Louisiana coastal Gulf of Mexico basin. Value is expressed as the ATROR%.

The potential for future reserve additions still appears excellent in many of those mature basins. Worldwide, adding reserves in and around mature fields is being recognized as an important source of low-risk and profitable reserve additions.

ECONOMICS OF RESERVE ADDITIONS

How profitable is adding reserves in and around mature fields? Our experience with the 46 fields we purchased gives some indication of profitability. Figures 37 and 38 show the after-tax rate of return (ATROR) for all 46 properties. The 27 properties producing by primary recovery are more profitable than the 19 waterflood fields. The average ATROR for all 46 properties is 21%. Forty-one properties have an ATROR of 12% to 41%. Three properties, all waterfloods, had 5% to 7% ATROR, which is acceptable economically. The two waterfloods with 0% to 1% ATROR are unprofitable. Those two projects failed because thief zones could not be plugged and a uniform flood front could not be maintained.

Payout (the time in years required to recover the after-tax total investment from the net cash flow) for the 46 properties is 1.0–3.1 years for the 27 primary production properties, 2.0–5.8 years for the 17 waterflood properties, and 8.0–12.8 years for the two uneconomic waterfloods.

Proved and probable reserves added in the 46 fields are 625 million BOE at an average cost of US $2.69/BOE. We expect that some additional reserves will be added in all the properties through improved recovery methods, CO_2 flooding, or field extensions.

CONCLUSIONS

Exploration is the key to finding large reserves. Redevelopment of many mature fields is an important additional source of profitable reserves. Not all mature fields can be redeveloped profitably. Environmental liabilities and the cost to fix and replace boreholes and producing facilities may exceed the value of proved and potential reserves.

We have been responsible for purchasing and initially redeveloping 46 properties for several oil companies. Forty-four field redevelopments are profitable and two are economic failures. The total proved and probable reserves for the fields are 625 million BOE at an average cost of US $2.69/BOE. The after-tax rate of return is 21% for all 46 properties. Based on our experience, redevelopment is a profitable and relatively low-risk means of increasing reserves in mature areas.

REFERENCES CITED

Frazer, F., 1998, Shell applying new research to marginal prospects: Offshore, August, p. 46, 50.

Masters, C. D., E. D. Attanasi, and D. H. Root, 1994, World petroleum assessment and analysis, *in* Proceedings, 14th World Petroleum Congress: John Wiley & Sons, p. 529–541.

Nehring, R., 1995, The impending crisis in U.S. oil production: What can be done about it: 1995 SPE Hydrocarbon Economics and Evaluation Symposium, SPE Paper 30059, 7 p.

Schollnberger, W. W., 1998, Projections of the world's hydrocarbon resources and reserve depletion in the 21st century: Houston Geological Society Bulletin 40, November, p. 31–37.

Sneider, J. S., R. M. Sneider, C. Coll, B. Cortiula, G. Gonzales, J. T. Kulha, J. D. Loren, D. Shaughnessy, and M. R. Todd, 1999, Rejuvenating a mature supergiant field VLC-363, Block III field, Lake Maracaibo, Venezuela—Reservoir characterization: AAPG International Conference, Birmingham, England, Extended Abstracts, p. 462–467.

Sneider, R. M., and J. S. Sneider, 1998, Rejuvenating marginal, aging oil fields: Is it profitable?: Petroleum Geoscience, v. 4, p. 303–315.

Trewin, N. H., and M. G. Bramwell, 1991, The Auk field, Block 30/16, UK North Sea, *in* I. L. Abbotts, ed., United Kingdom oil and gas fields, 25 years commemorative volume: Geological Society Memoir 14, p. 227–236.

ADDITIONAL REFERENCES from 1998 AAPG Annual Convention Poster Session

Farina, J. R., J. T. Kulha, and E. Arzola, 1998, Rejuvenating a mature supergiant field, VLC-363, Block III field, Lake Maracaibo, Venezuela: Reservoir performance and observations.

Loren, J .D., R. M. Sneider, and L. Rondón, 1998, Rejuvenating a mature supergiant field, VLC-363, Block III field, Lake Maracaibo, Venezuela: Introduction.

Loren, J. D., J. T. Kulha, and C. Coll, Rejuvenating a mature supergiant field, VLC-363, Block III field, Lake Maracaibo, Venezuela: Petrophysical evaluation.

Sneider, J. S., D. Shaughnessy, R. M. Sneider, G. González, and B. Cortiula, 1998, Rejuvenating a mature supergiant field, VLC-363, Block III field, Lake Maracaibo, Venezuela: Geologic framework.

Sneider, J. S., G. González, and B. Cortiula, 1998, Rejuvenating a mature supergiant field, VLC-363, Block III field, Lake Maracaibo, Venezuela: Reservoir characterization.

Sneider, J. S., G. González, J. T. Kulha, and D. Shaughnessy, 1998, Rejuvenating a mature supergiant field, VLC-363, Block III field, Lake Maracaibo, Venezuela: Development opportunities.

Todd, M. R., and E. Arzola, 1998, Rejuvenating a mature supergiant field, VLC-363, Block III field, Lake Maracaibo, Venezuela: Numerical simulation.

Collett, T. S., 2001, Natural-gas hydrates: Resource of the twenty-first century?, *in* M. W. Downey, J. C. Threet, and W. A. Morgan, eds., Petroleum provinces of the twenty-first century: AAPG Memoir 74, p. 85–108.

Chapter 7

Natural-gas Hydrates: Resource of the Twenty-first Century?

Timothy S. Collett
U.S. Geological Survey, Denver, Colorado, U.S.A.

ABSTRACT

Although considerable uncertainty and disagreement prevail concerning the world's gas-hydrate resources, the estimated amount of gas in those gas-hydrate accumulations greatly exceeds the volume of known conventional gas reserves. However, the role that gas hydrates will play in contributing to the world's energy requirements will ultimately depend less on the volume of gas-hydrate resources than on the cost to extract them.

Gas hydrates occur in sedimentary deposits under conditions of pressure and temperature present in permafrost regions and beneath the sea in outer continental margins. The combined information from arctic gas-hydrate studies shows that in permafrost regions, gas hydrates may exist at subsurface depths ranging from about 130 m to 2000 m. The presence of gas hydrates in offshore continental margins has been inferred mainly from anomalous seismic reflectors (known as bottom-simulating reflectors) that have been mapped at depths below the seafloor ranging from approximately 100 m to 1100 m. Current estimates of the amount of gas in the world's marine and permafrost gas-hydrate accumulations are in rough accord at about 20,000 trillion m^3.

Gas hydrate as an energy commodity is often grouped with other unconventional hydrocarbon resources. In most cases, the evolution of a nonproducible unconventional resource to a producible energy resource has relied on significant capital investment and technology development. To evaluate the energy-resource potential of gas hydrates will also require the support of sustained research and development programs.

Despite the fact that relatively little is known about the ultimate resource potential of gas hydrates, it is certain that they are a vast storehouse of natural gas, and significant technical challenges will need to be met before this enormous resource can be considered an economically producible reserve.

INTRODUCTION

The discovery of large gas-hydrate accumulations in terrestrial permafrost regions of the Arctic and beneath the sea along the outer continental margins of the world's oceans has heightened interest in gas hydrates as a possible energy resource. However, significant to potentially insurmountable technical issues need to be resolved before gas hydrates can be considered a viable option for affordable supplies of natural gas.

Disagreements over fundamental issues such as volume of gas stored within delineated gas-hydrate accumulations and the concentration of gas hydrates within hydrate-bearing reservoirs have demonstrated that we know very little about gas hydrates. Recently, however, several countries (including Japan, India, and the United States) have launched ambitious national projects to further examine the resource potential of gas hydrates.

These projects may help answer key questions dealing with the properties of gas-hydrate reservoirs, the design of production systems and, most importantly, the costs and economics of gas-hydrate production.

It is proposed in this paper that the evolution of gas hydrates as a viable source of natural gas, like any other unconventional energy resource (e.g., deep gas, shale gas, tight-gas sands, and coal-bed methane), will follow a predictable path from research and discovery to implementation (Figure 1). However, insurmountable barriers may exist along this pathway.

Today, most of the gas-hydrate research community is focused on three fundamental issues: *Where* do gas hydrates occur, *how* do gas hydrates occur in nature, and *why* do gas hydrates occur in a particular setting? However, relatively little has been done to integrate these distinct research topics or to evaluate how they affect collectively the ultimate resource potential of gas hydrates. Only after understanding the fundamental aspects of where, how, and why gas hydrates occur in nature will we be able to make accurate estimates of how much gas is trapped in the gas-hydrate accumulations of the world. Even with the confirmation that gas hydrates may exist in considerable volumes, significant technical, economic, and political issues need to be resolved before gas hydrates can be considered a viable energy resource.

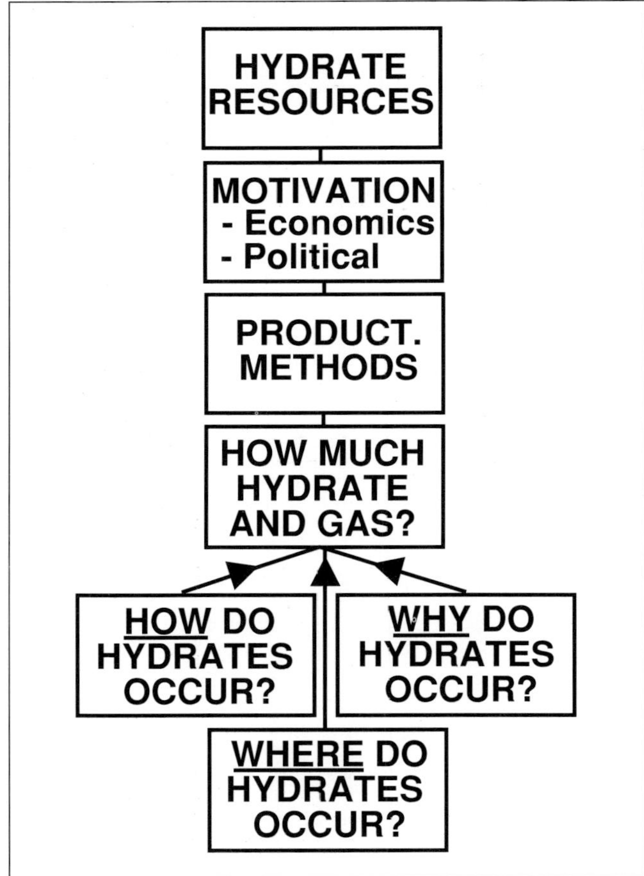

Figure 1. Flowchart depicting the evolution of gas-hydrate understanding from a nonproducible unconventional gas resource to a producible energy resource.

In this paper, I review the status of gas hydrates as a future energy resource, including the fundamental questions of where, how, and why; the technical and nontechnical factors controlling the ultimate resource potential of gas hydrates; published gas-hydrate volume assessments; and the production technology needed to extract the world's gas-hydrate resources. The paper concludes with a discussion of the economic and political motivations that may eventually lead to gas-hydrate production. However, before proceeding with assessment of future energy-resource potential of gas hydrates, I begin with a technical overview of gas-hydrate physical properties and a review of four relatively well characterized gas-hydrate accumulations.

GAS-HYDRATE TECHNICAL REVIEW

Under appropriate conditions of temperature and pressure (Figure 2), gas hydrates usually form one of two basic crystal structures, known as structure-I and structure-II (Figure 3). Each unit cell of structure-I gas hydrate consists of 46 water molecules that form two small dodecahedral voids and six large tetradecahedral voids. Structure-I gas hydrates can hold only small gas molecules such as methane and ethane, with molecular diameters not exceeding 5.2 Å. The chemical composition of a structure-I gas hydrate can be expressed as $8(Ar, CH_4, H_2S, CO_2)46H_2O$ or $(Ar, CH_4, H_2S, CO_2)\ 5.7H_2O$ (Makogon, 1981).

The unit cell of a structure-II gas hydrate consists of 16 small dodecahedral and eight large hexakaidecahedral voids formed by 136 water molecules. Structure-II gas hydrates may contain gases with molecular dimensions in the range 5.9–6.9 Å, such as propane and isobutane. The chemical composition of a structure-II gas hydrate can be expressed as $8(C_3H_8, C_4H_{10}, CH_2C_{12}, CHCL_3)\ 136H_2O$ or $(C_3H_8, C_4H_{10}, CH_2C_{12}, CHCL_3)17H_2O$ (Makogon, 1981). At conditions of standard temperature and pressure (STP), one volume of saturated methane hydrate (structure-I) will contain as much as 189 volumes of methane gas. Because of this large gas-storage capacity, gas hydrates are thought to represent an important source of natural gas.

An overview of gas-hydrate structures would not be complete without mentioning the newly discovered hydrate structure called structure H. The existence of this structure was determined by laboratory nuclear magnetic-resonance studies of Ripmeester et al. (1987) and is characterized by three types of cages. Structure H hydrates have been shown to be unique, with a number of large molecules able to fit into the largest cage. Structure H guest molecules include numerous naturally occurring substances, including adamantane, gasoline-range hydrocarbons, and naphthalene ingredients. For a complete description of the structure and properties of gas hydrates, see the summary by Sloan (1998).

Gas hydrates have been inferred to occur at about 50 locations throughout the world (Figure 4). However, only a limited number of gas-hydrate accumulations have been examined in any detail. In the following section of this paper, four of the best-known marine and onshore permafrost-associated gas accumulations are described.

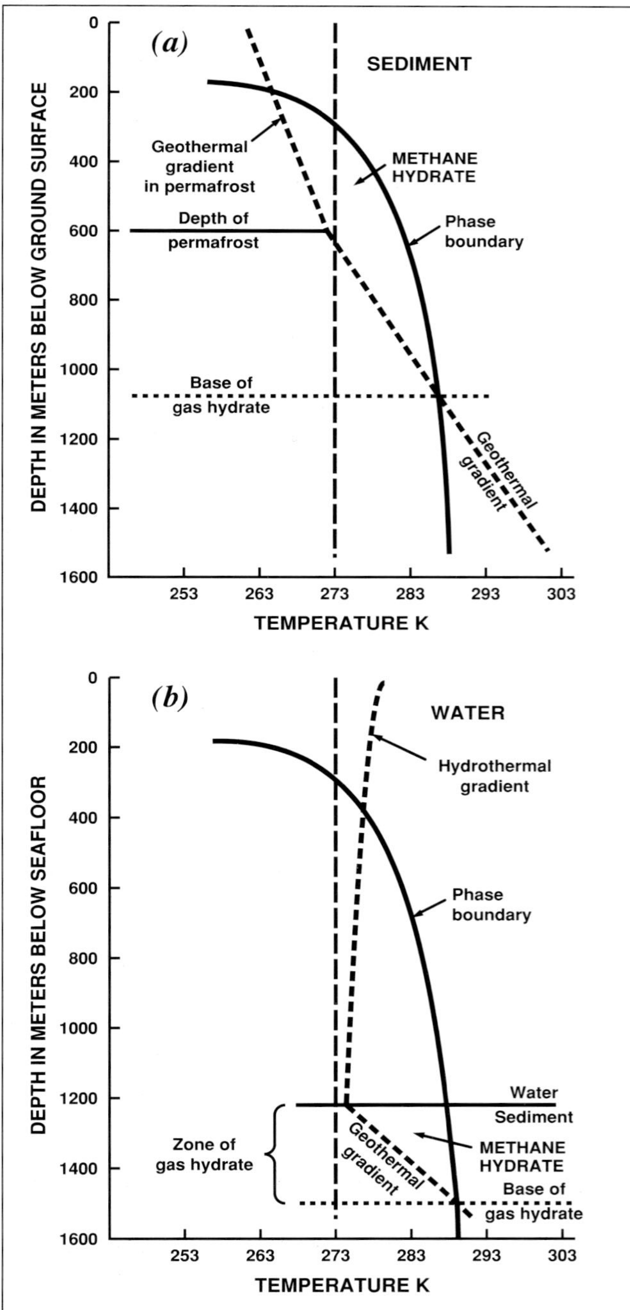

Figure 2. Graphs showing the depth-temperature zone in which methane hydrates are stable in (a) a permafrost region and (b) an outer continental-margin marine setting (modified from Collett, 1995).

Discussions pertaining to the volume of gas within each of the gas-hydrate accumulations described in the following section are included later in the "Energy-resource Potential" section of this paper. The four gas-hydrate accumulations considered (labeled in Figure 4) are (1) on the Blake Ridge along the southeastern continental margin of the United States, (2) along the Cascadia continental margin off the Pacific coast of the United States, (3) on the North Slope of Alaska, and (4) in the Mackenzie River Delta of northern Canada.

Blake Ridge

Seismic profiles along the Atlantic margin of the United States are often marked by large-amplitude bottom-simulating reflectors (BSRs), which in this region are believed to be caused by large acoustic-impedance con-

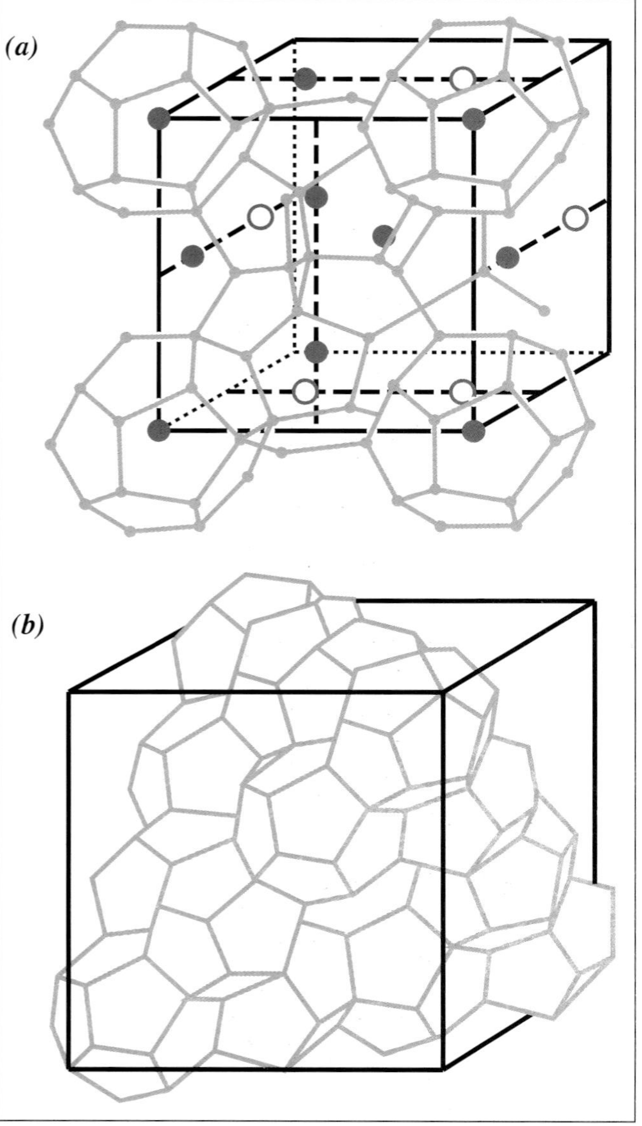

Figure 3. Two gas-hydrate crystal structures: (a) structure-I, (b) structure-II (modified from Sloan, 1998).

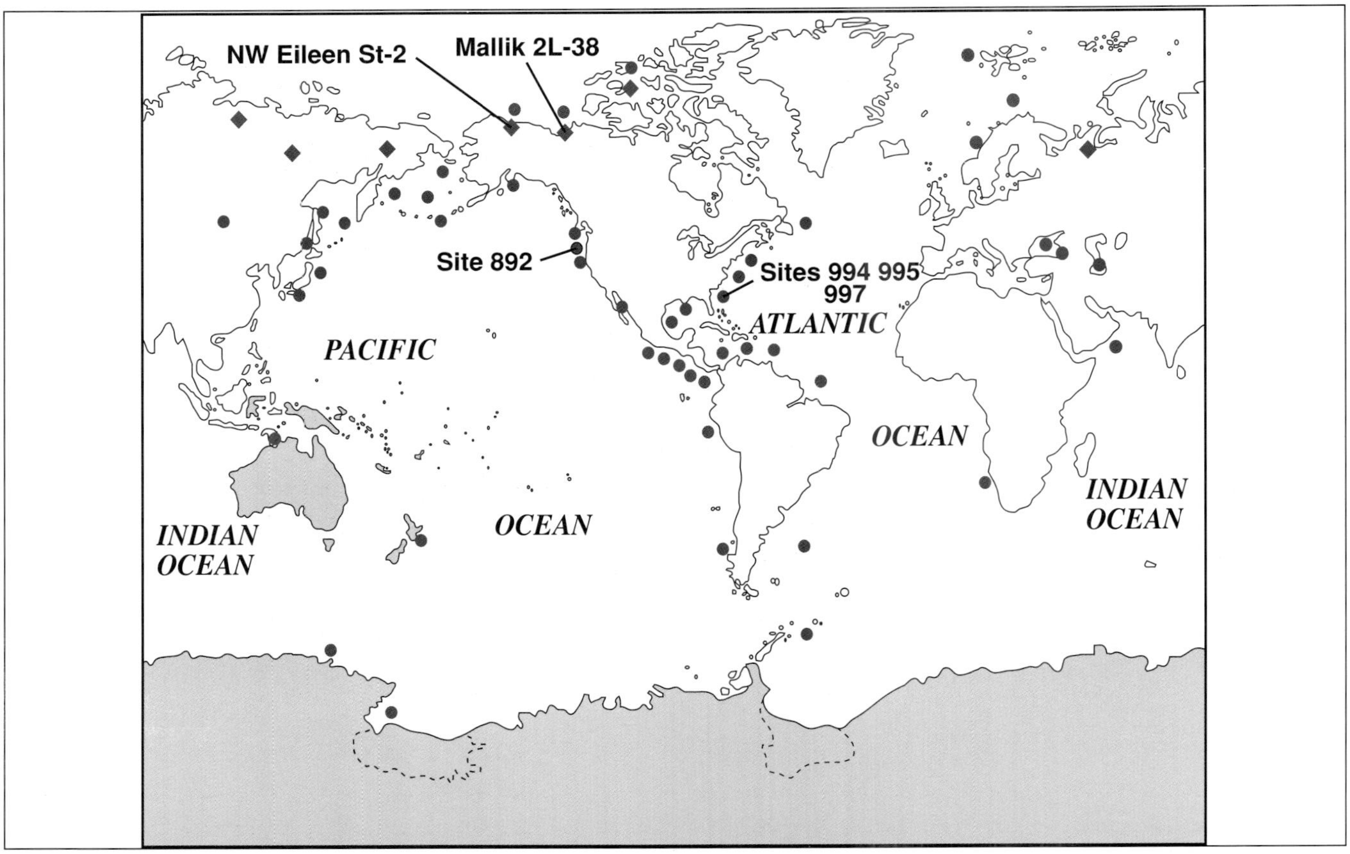

Figure 4. Location of known and inferred gas-hydrate occurrences (modified from Kvenvolden, 1993). Occurrences described in detail in this paper are labeled.

trasts at the base of the gas-hydrate stability zone that juxtaposes sediments containing gas hydrates with sediments containing free gas (Dillon et al., 1993; Lee et al., 1993). BSRs have been mapped extensively at two locations off the east coast of the United States: along the crest of the Blake Ridge and beneath the upper continental rise of New Jersey and Delaware (Tucholke et al., 1977; Dillon et al., 1993).

The Blake Ridge is a positive topographic sedimentary feature on the continental slope and rise of the United States (Figure 5). The crest of the ridge extends approximately perpendicular to the general trend of the continental rise for more than 500 km to the southwest from water depths of 2000 to 4800 m. The Blake Ridge is thought to be a large sediment drift that was built on transitional continental to oceanic crust by complex accretion of marine sediments deposited by longitudinal drift currents (Tucholke et al., 1977). The Blake Ridge consists of Tertiary to Quaternary sediments of hemipelagic muds and silty clay (Shipboard Scientific Party, 1996). The thickness of the methane-hydrate stability zone in this region ranges from zero along the northwestern edge of the continental shelf to a maximum thickness of about 700 m along the eastern edge of the Blake Ridge (Collett, 1995). The occurrence of gas hydrates on the Blake Ridge

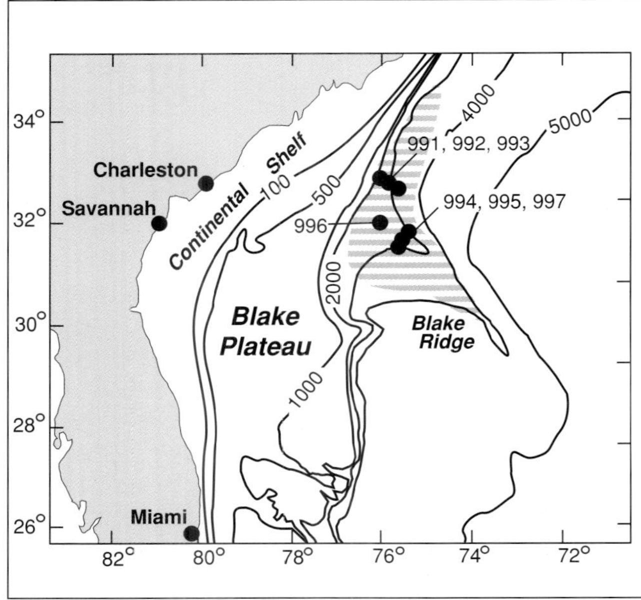

Figure 5. Physiographic map of the southeastern continental margin of North America. Location of ODP leg 164 drill sites are indicated. Also shown is the area (shaded) where gas-hydrate occurrence has been mapped on the basis of BSRs. Contours are in meters.

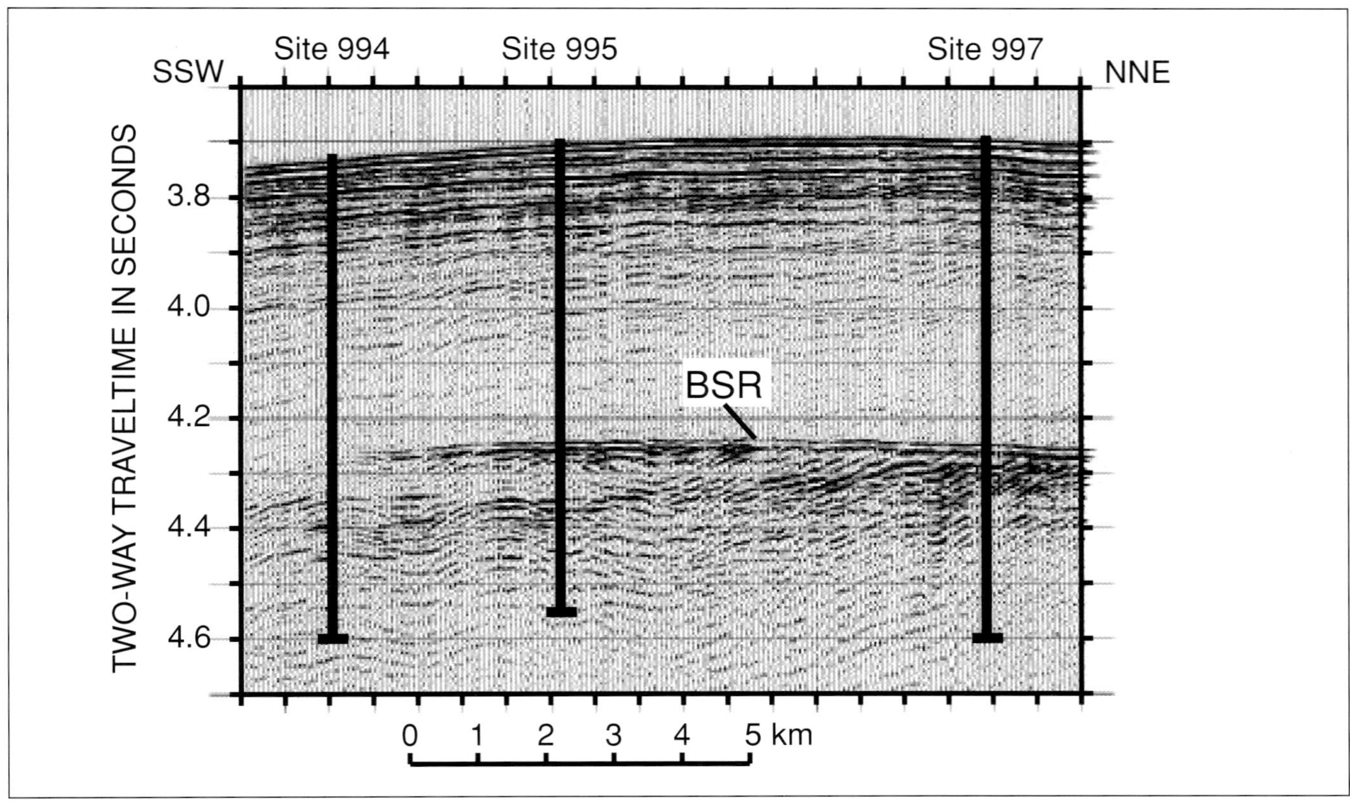

Figure 6. Seismic profile along which sites 994, 995, and 997 are located. Note that site 994 is not associated with a distinct BSR, although a very strong BSR occurs at sites 995 and 997.

was confirmed during leg 76 of the Deep Sea Drilling Project (DSDP) when a sample of gas hydrate was recovered from a subbottom depth of 238 m at site 533 (Shipboard Scientific Party, 1980).

Leg 164 of the Ocean Drilling Program (ODP) (Shipboard Scientific Party, 1996) was designed to investigate the occurrence of gas hydrate in the sedimentary section beneath the Blake Ridge (Figure 5). Sites 994, 995, and 997 comprise a transect of holes that penetrates below the base of gas-hydrate stability within the same stratigraphic interval over a relatively short distance (Figure 6). This transect of holes on the southern flank of the Blake Ridge extends from an area where a BSR is not detectable to an area where an extremely well developed and distinct BSR exists (Figure 6). The presence of gas hydrates at sites 994 and 997 was documented by direct sampling; however, no gas hydrates were conclusively identified at site 995 (Shipboard Scientific Party, 1996). Although a BSR does not occur in the seismic-reflection profiles that cross site 994, several pieces of gas hydrate were recovered from 259.90 m below seafloor (mbsf) in hole 994C, and disseminated gas hydrate was observed at almost the same depth in hole 994D. One large, solid piece (about 15 cm long) of gas hydrate was also recovered from about 331 mbsf at site 997 (hole 997A). Despite these limited occurrences of gas hydrates, it was inferred, based on geochemical core analyses and downhole logging data, that disseminated gas hydrates occur in the stratigraphic interval from about 190 to 450 mbsf in all the holes drilled on the Blake Ridge (Figure 7).

The depths to the top and base of the zone of gas-hydrate occurrence at sites 994, 995, and 997 were determined using interstitial-water chloride concentrations and downhole log data (Figure 7). Interstitial-water chloride concentrations were used to establish whether gas hydrate occurred within a given core sample, based on the observation that gas-hydrate decomposition during core recovery releases fresh water and methane into the interstitial pores, resulting in a freshening of the sediment pore waters. The observed chloride concentrations also enable an estimate to be made of the amount of gas hydrate that occurs on the Blake Ridge by calculating the amount of interstitial-water freshening that can be attributed to gas-hydrate dissociation. The estimated gas-hydrate saturations in the recovered cores had a skewed distribution, ranging from a maximum of about 7% and 8.4% at sites 994 and 995, respectively, to a maximum of about 13.6% at site 997. For a more complete discussion on the chlorinity-calculated gas-hydrate contents, see Shipboard Scientific Party (1996).

Natural gas-hydrate occurrences are generally characterized by an increase in downhole log-measured acoustic velocities and electrical resistivities and the release of unusually large amounts of methane during drilling. The well-log-inferred gas-hydrate-bearing stratigraphic interval on the Blake Ridge (190–450 mbsf; Fig-

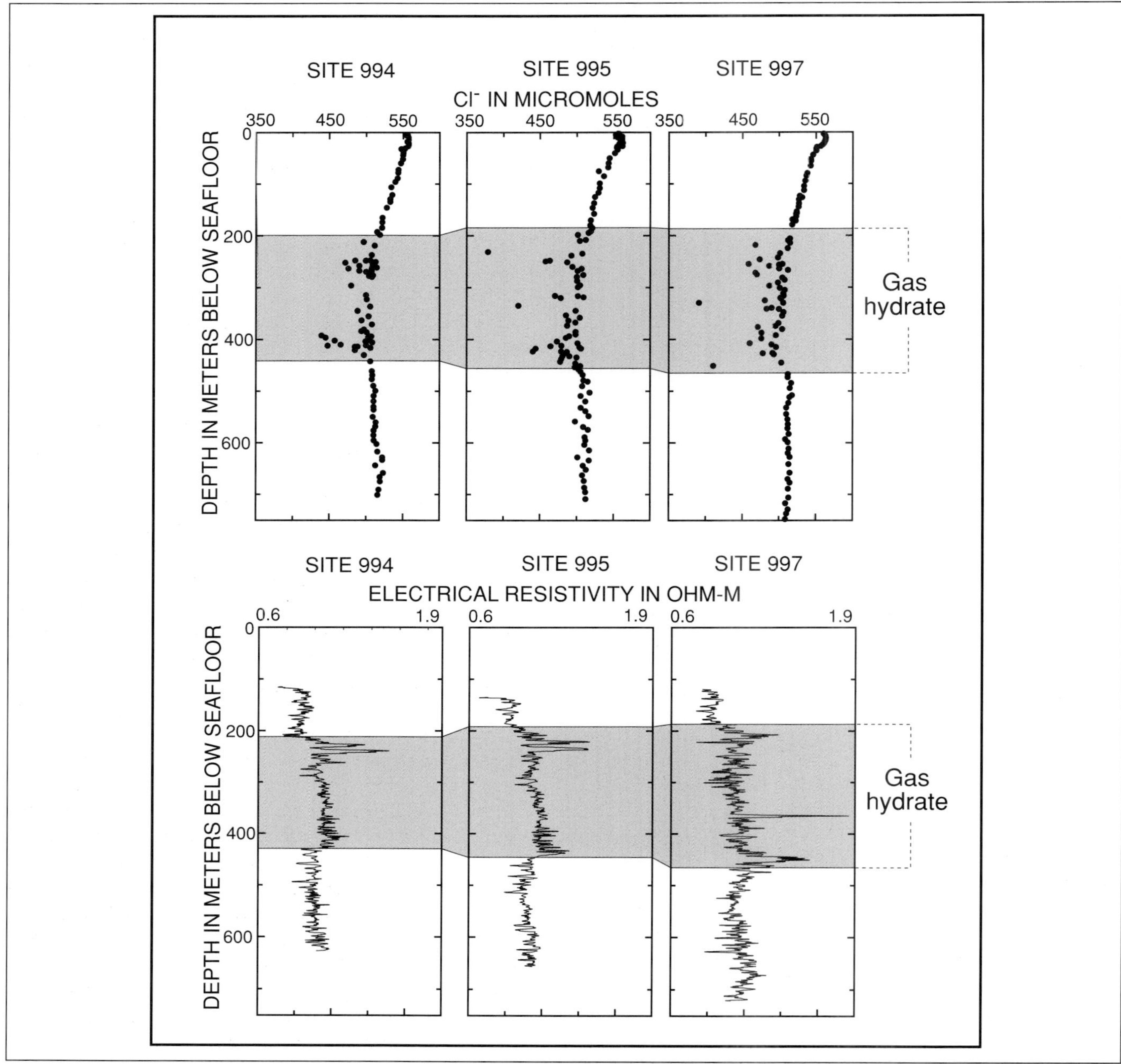

Figure 7. Chloride concentration profiles for interstitial waters collected from cores at sites 994, 995, and 997 and corresponding downhole electrical resistivity log data. Also shown (shaded) is the chloride-concentration and electrical-resistivity inferred gas-hydrate distribution (modified from Shipboard Scientific Party, 1996).

ure 7) is characterized by a distinct stepwise increase in both electrical resistivity (increase of about 0.1–0.3 ohm-m) and acoustic velocity (increase of about 0.1–0.2 km/s). The depth of the lower boundary of the log-inferred gas-hydrate-bearing interval on the Blake Ridge is in rough accord with the predicted base of the methane hydrate stability zone, and it is near the lowest depth of the observed interstitial-water chlorinity anomaly (Figure 7).

Cascadia Continental Margin

BSRs have been extensively mapped on the inner continental margin of northern California (Field and Kvenvolden, 1985). These constitute a single, inferred gas-hydrate accumulation that covers an area of at least 3000 km^2 on the Klamath Plateau and the upper continental slope at water depths ranging from 800 to 1200 m. Limited seismic data show that this regionally extensive inferred gas-hydrate occurrence extends northward

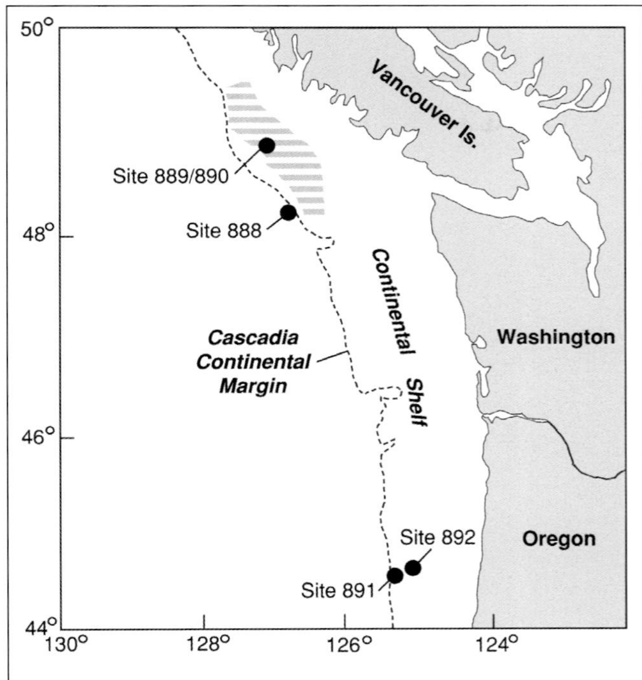

Figure 8. Physiographic map of the Cascadia continental margin of North America. Location of ODP leg 146 drill sites are indicated. Also shown is the area (shaded) where gas-hydrate occurrence has been mapped on the basis of BSRs (modified from Hyndman et al., 1996).

to offshore Canada (Hyndman et al., 1996) and seaward at least to the base of the slope (3000-m water depth).

The occurrence of gas hydrates on the Pacific margin of the United States was confirmed in 1989 when numerous gas-hydrate samples were obtained during seabed (0–6 mbsf) sediment coring operations (water depths ranging between 510 and 642 m) in the Eel River Basin (Brooks et al., 1991). Recovered gas-hydrate samples consisted of dispersed crystals, small nodules, and layered bands. The location of these gas hydrates coincides nearly, but not exactly, with the area of BSR-inferred gas hydrates described by Field and Kvenvolden (1985) along the northern California coast. Gas hydrates have also been recovered along the Cascadia margin from a relatively restricted zone within 17 m of the seafloor in three research core holes drilled during leg 146 of the ODP: holes 892A, 892D, and 892E (Figure 8) (Shipboard Scientific Party, 1994). All these core holes are located on the Oregon continental slope in about 675 m of water.

Leg 146 of the ODP (Shipboard Scientific Party, 1994) was designed to examine fluid movement in the Cascadia continental margin and to provide well-constrained estimates of the volume of fluid associated with accretionary sedimentary wedges. In addition, the presence of distinct BSRs on the Cascadia margin also provided an opportunity to examine the potential interrelation between the occurrence of natural-gas hydrates and BSRs. Four locations were drilled off the west coast of Vancouver Island and Oregon (Figure 8). In addition to the gas-hydrate crystals recovered in the near-surface (2–17 mbsf) sediments at site 892, downhole logs and a vertical seismic profile (VSP) at site 892 indicated free gas below about 71 mbsf and established at least a local correspondence to the BSR. However, the borehole surveys yielded relatively few useful gas-hydrate data.

Site 889, located off the west coast of Vancouver Island (Figure 8), yielded a wealth of data pertaining to the in-situ nature of gas hydrates on the Cascadia margin. Massive accumulations of gas hydrate were not encountered at site 889. Rather, indirect evidence from recovered cores and downhole geophysical surveys suggests that most of the gas hydrates at site 889 occur as finely disseminated pore-filling substances. Temperature measurements of the recovered cores and the dilution of pore-water salts suggest that about 10%–40% of pore space in the sediment is filled with gas hydrate at site 889 (Shipboard Scientific Party, 1994).

Gas hydrate was not conclusively identified at site 889 (Shipboard Scientific Party, 1994); however, its presence was inferred, based on geochemical analyses of cores and downhole geophysical surveys (VSPs), and borehole logging data in the depth interval from about 127.6 to 228.4 mbsf (Figure 9). Similar to the observations from the Blake Ridge boreholes, gas-rich cores, low interstitial-water chloride concentrations, and low temperature measurements in the recovered cores suggested the presence of gas hydrates at site 889 (Shipboard Scientific Party, 1994; Spence et al., 1995; Hyndman et al., 1996). In addition, sediment velocity data from downhole VSPs and ocean-bottom seismometer (OBS) surveys (Shipboard Scientific Party, 1994; Spence et al., 1995; Hyndman et al., 1996) indicate that gas hydrates occur in the 50- to 80-m-thick interval above the BSR (approximate depth of 230 mbsf) at site 889. Observed chloride anomalies were also used to estimate the amount of gas hydrate that occurs at site 889 by calculating the amount of interstitial-water freshening that can be attributed to gas-hydrate dissociation. The estimated volume of sediment porosity calculated to be occupied by gas hydrate in the recovered cores ranged from a minimum of about 5% immediately below the seafloor to a maximum of about 39% near the bottom of a well-log-inferred gas-hydrate occurrence at site 889 (Hyndman et al., 1996).

North Slope of Alaska

Previous North Slope studies (Collett, 1983, 1993; Collett et al., 1988) indicate that the Prudhoe Bay–Kuparuk River gas-hydrate accumulation (Figure 10) is restricted to Tertiary sediments of the Sagavanirktok Formation. The formation consists of shallow-marine shelf and delta-plain deposits composed of sandstone, shale, and conglomerate whose provenance was the Brooks Range, to the south. The Sagavanirktok Formation includes the informally named West Sak and Ugnu sands.

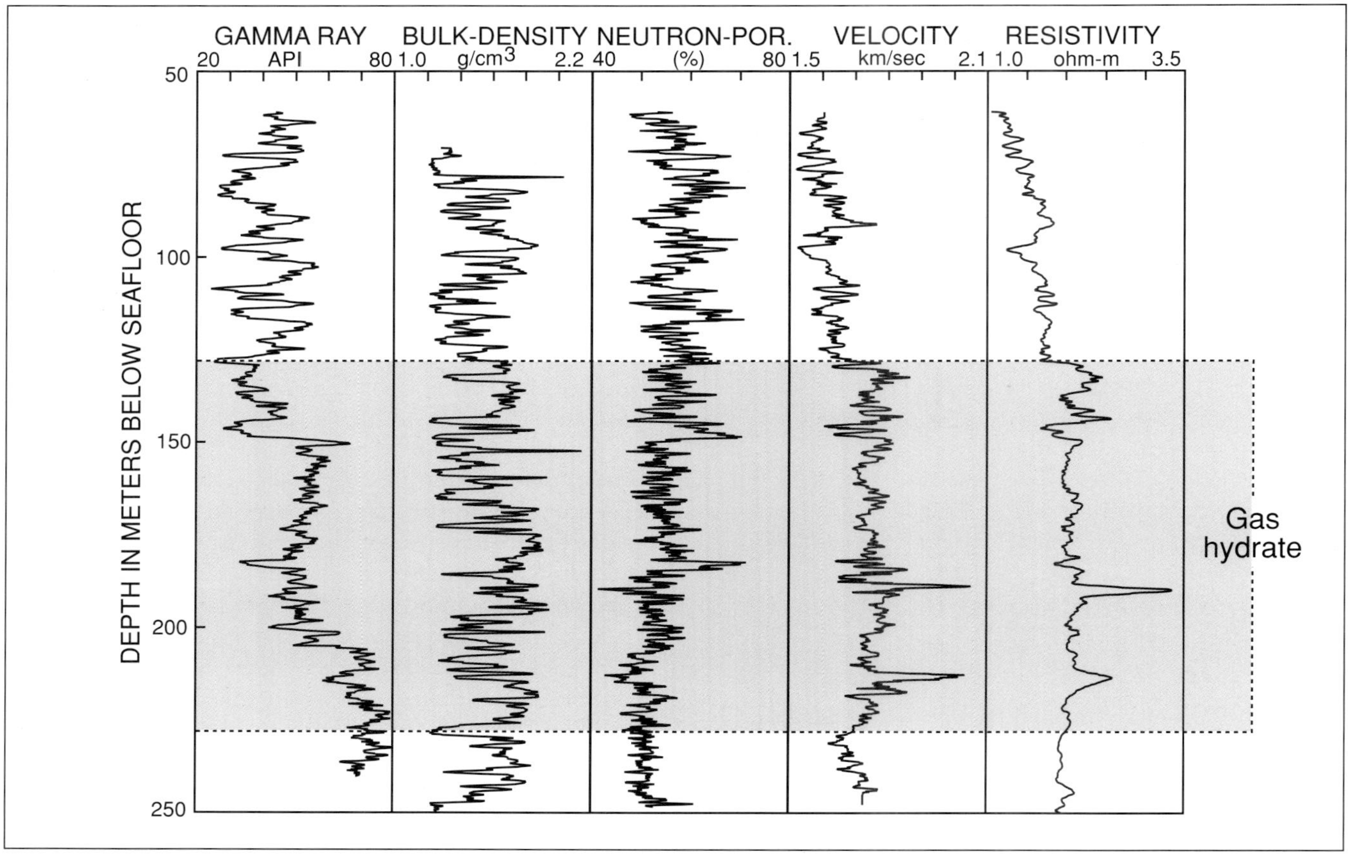

Figure 9. Downhole log data from site 889, Cascadia continental margin. Data shown include the natural gamma-ray log, bulk-density data, neutron porosities, compressional-wave acoustic velocities, and deep-reading electrical resistivities. Inferred occurrence of gas hydrates is shaded.

These oil-bearing zones have been described extensively by Werner (1987) and are estimated to contain more than approximately 6 million metric tons of in-place oil.

The occurrence of natural-gas hydrate on the North Slope of Alaska was confirmed in 1972 with data from the Northwest Eileen State-2 well in the northwestern part of Prudhoe Bay oil field. Studies of pressurized core samples, downhole logs, and the results of formation production testing have confirmed the occurrence of three gas-hydrate-bearing stratigraphic units in the Northwest Eileen State-2 well (reviewed by Collett, 1993). Gas hydrates are also inferred to occur in an additional 50 exploratory and production wells in northern Alaska, based on downhole-log responses calibrated to the known gas-hydrate occurrences in the Northwest Eileen State-2 well. Many of these wells have multiple gas-hydrate-bearing units, with individual occurrences ranging from 3 to 30 m thick. Most of these well-log-inferred gas hydrates occur in six laterally continuous sandstone and conglomerate units; all these gas hydrates are restricted geographically to the area overlying the eastern part of Kuparuk River oil field and the western part of Prudhoe Bay oil field (Figures 10 and 11). Each of the six gas-hydrate-bearing sedimentary units has been assigned a reference letter (units A through F); unit A is stratigraphically the deepest (Figure 11). Three-dimensional seismic surveys and downhole logs from wells in the western part of Prudhoe Bay oil field indicate the presence of several large free-gas accumulations trapped stratigraphically downdip below four of the log-inferred gas-hydrate units (Figures 10 and 11; units A through D). The total mapped area of all six gas-hydrate occurrences is about 1643 km^2; the areal extent of the individual units range from 3 to 404 km^2. The volume of gas within the gas hydrates of the Prudhoe Bay–Kuparuk River area is estimated to be about 1.0–1.2 trillion m^3, or about twice the volume of conventional gas in Prudhoe Bay field (Collett, 1993).

Mackenzie River Delta of Canada

Assessments of gas-hydrate occurrences in the Mackenzie Delta–Beaufort Sea area have been made mainly on the basis of data obtained during hydrocarbon exploration in the past three decades (reviewed by Judge et al., 1994). A database presented by Smith and Judge (1993) summarizes a series of unpublished consultant studies that investigated well-log data from 146 exploration wells in the Mackenzie Delta area. In total, 25 wells (17%) were identified as containing possible or probable gas hydrates (Figure 12). All of these inferred gas hydrates occur in clastic sedimentary rocks of the

Kugmallit, Mackenzie Bay, and Iperk sequences (Dixon et al., 1992). Two of the occurrences were associated with ice-bearing permafrost; the others were beneath the permafrost interval. The frequency of gas-hydrate occurrence in offshore wells was greater, with possible or probable gas hydrates identified in 35 of 55 wells (63%).

The JAPEX/JNOC/GSC Mallik 2L-38 gas-hydrate research well was designed to investigate the occurrence of in-situ natural-gas hydrates in the Mallik area of the Mackenzie River Delta of Canada (Figure 12) (Dallimore et al., 1999). The Mallik 2L-38 gas-hydrate research well was near the existing Mallik L-38 well, which had been drilled by Imperial Oil in 1972 (Bily and Dick, 1974). As described in Collett and Dallimore (1998), the Mallik L-38 well is believed to have encountered at least 10 significant gas-hydrate-bearing stratigraphic units in the depth interval of 810.1 to 1102.3 m. Bily and Dick (1974) concluded that each of the gas-hydrate-bearing units in the Mallik L-38 well contained substantial amounts of gas hydrate. However, no attempt was made to quantify the amount of gas hydrate or associated free gas that may have been trapped in the log-inferred gas-hydrate occurrences.

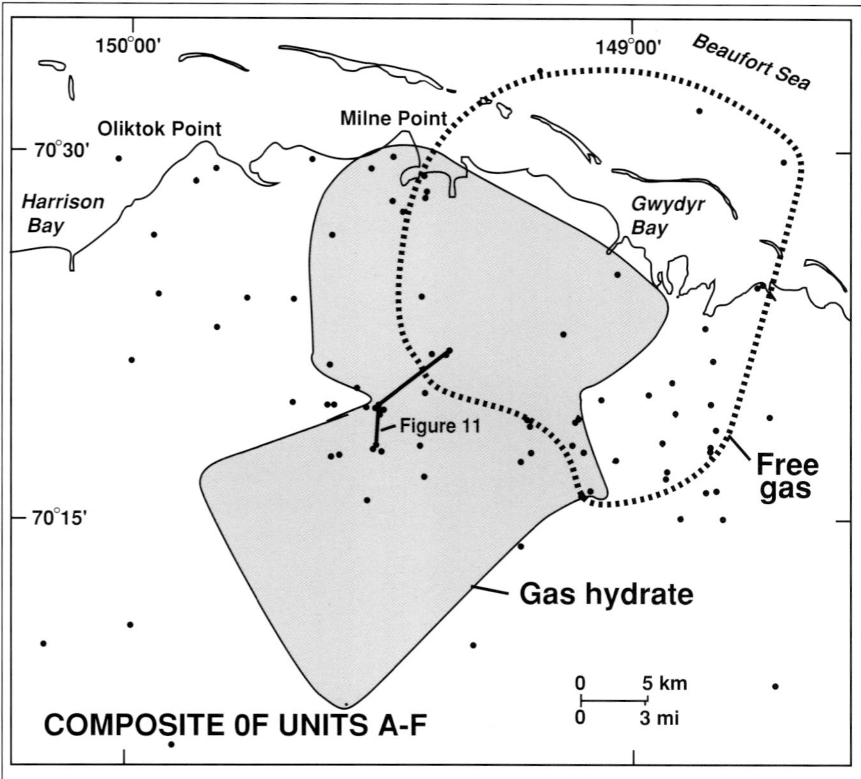

Figure 10. Composite map of all six gas-hydrate/free-gas units (units A-F) from the Prudhoe Bay–Kuparuk River area in northern Alaska. Also shown is the location of the cross section in Figure 11 (modified from Collett, 1993).

In drilling the Mallik 2L-38 well, major emphasis was placed on coring the log-inferred gas-hydrate intervals identified in the Mallik L-38 well. Thirteen coring runs were attempted with a variety of coring systems. Approximately 37 m of core was recovered from the gas-hydrate interval (878–944 m) in the Mallik 2L-38 well (Dallimore et al., 1999). Pore-space gas hydrate and several forms of visible gas hydrate were observed in a variety of sediment types.

Data from downhole logging in the Mallik L-38 and 2L-38 (Figure 13) wells and formation-production testing in the Mallik L-38 well have been used to assess local geology, permafrost, and gas-hydrate conditions. In the upper 1500 m, three stratigraphic sequences have been identified using reflection seismic records and well data (Jenner et al., 1999): These include the Iperk sequence (0–337.6 m), the Mackenzie Bay sequence (337.6–918.1 m), and the Kugmallit sequence (918.1 m–bottom of hole). The Iperk sequence appears to be composed almost entirely of coarse-grained sandy sediments. Previous coring experience (Dallimore and Matthews, 1997) indicates that the Iperk sediments are unconsolidated. The Mackenzie Bay sequence is also dominated by sand, with a distinct fining-upward section near its upper contact with the Iperk sequence. The Kugmallit sequence (>918 m) consists of interbedded sandstone and siltstone. Drill cuttings and drilling records suggest that the grain cementation in the Mackenzie Bay and Kugmallit sequences is quite variable. The base of ice-bearing permafrost in the Mallik 2L-38 well is estimated at about 640 m, on the basis of available well-log information.

The well-log-inferred gas-hydrate occurrence in the Mallik 2L-38 well occupies the depth interval between 888.84 and 1101.09 m (Figures 13 and 14); however, not all of this interval is occupied by gas hydrate. The cored and logged gas-hydrate occurrences in the Mallik 2L-38 well (Figure 14) exhibit deep electrical-resistivity measurements ranging from 10 to 100 ohm-m and compressional-wave acoustic velocities (V_p) ranging from 2.5 to 3.6 km/s. In addition, the measured shear-wave acoustic velocities (V_s) of the confirmed gas-hydrate-bearing units in the Mallik 2L-38 well range from 1.1 to 2.0 km/s.

Bily and Dick (1974) originally interpreted the presence of free gas in contact with gas hydrate on the basis of spontaneous-potential well-log responses in several intervals of the Mallik L-38 well. They also speculated that rapid pressure responses during a production test (production test-1: 1098–1101 m) in a suspected free-gas unit are evidence of highly permeable free-gas-bearing sediments. Acoustic transit-time log data from the Mallik 2L-38 well confirmed the occurrence of a relatively thin

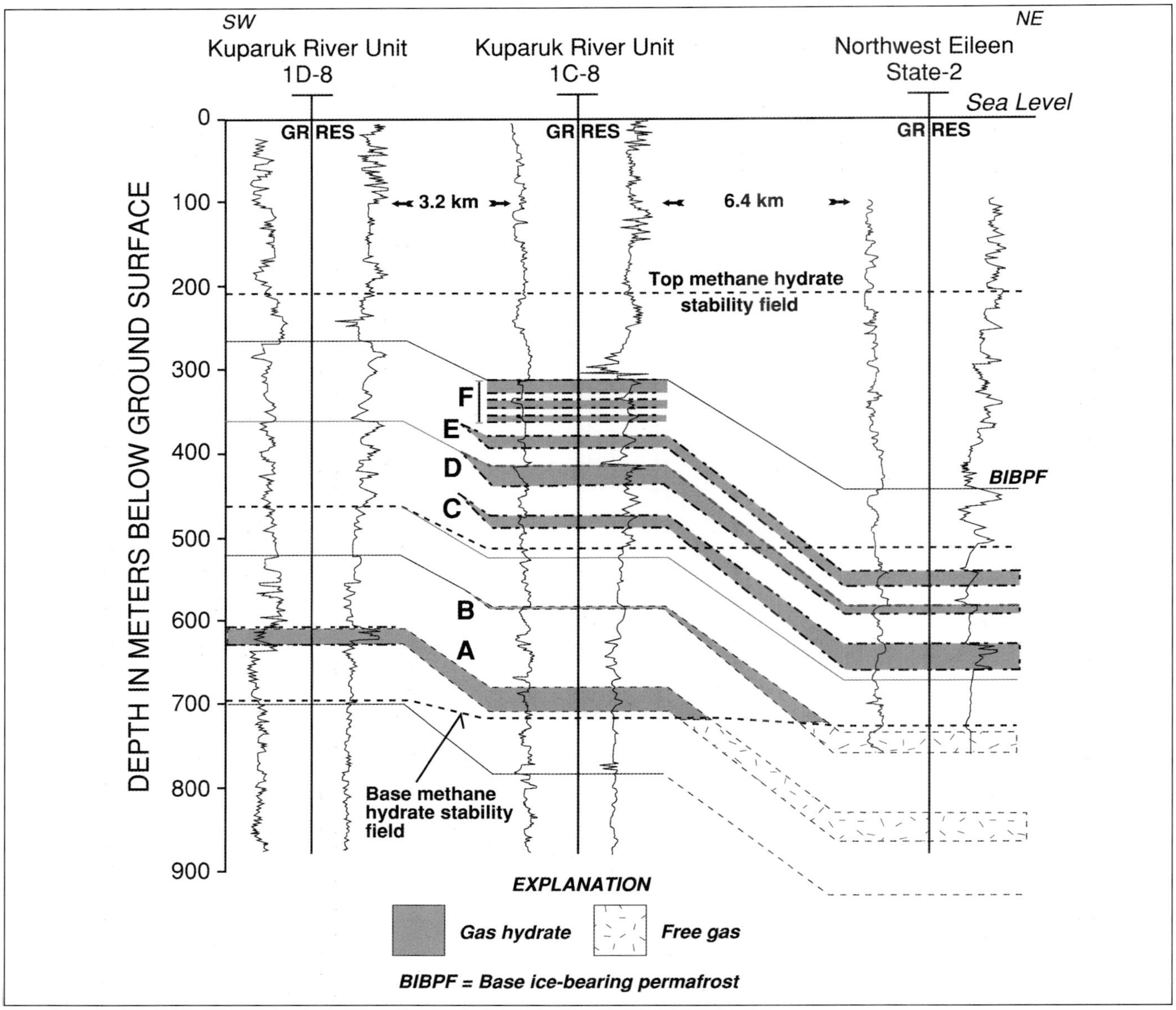

Figure 11. Cross section showing the lateral and vertical extent of gas hydrates and underlying free-gas occurrences in the Prudhoe Bay–Kuparuk River area in northern Alaska. See Figure 10 for location of cross section. The gas-hydrate-bearing units are identified with reference letters A through F (modified from Collett, 1993).

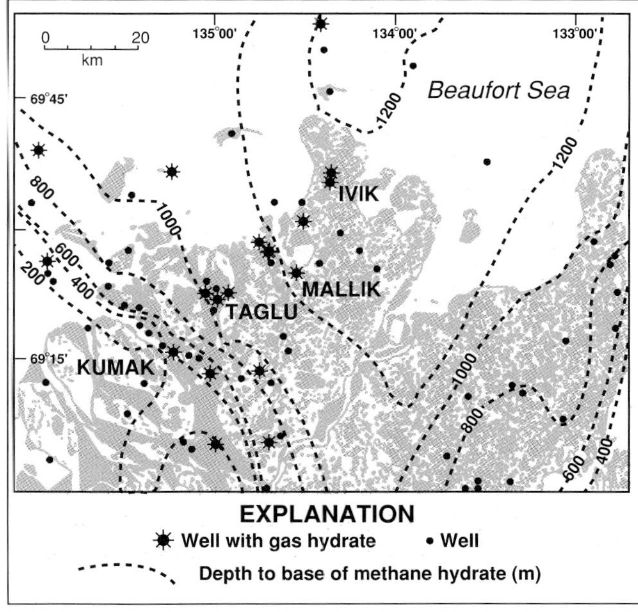

Figure 12. Map of part of the Mackenzie Delta region showing the calculated depth to the base of the methane-hydrate stability zone (modified from Judge and Majorowicz, 1992). Exploration wells with well-log-inferred gas-hydrate occurrences are shown. Contours are in meters.

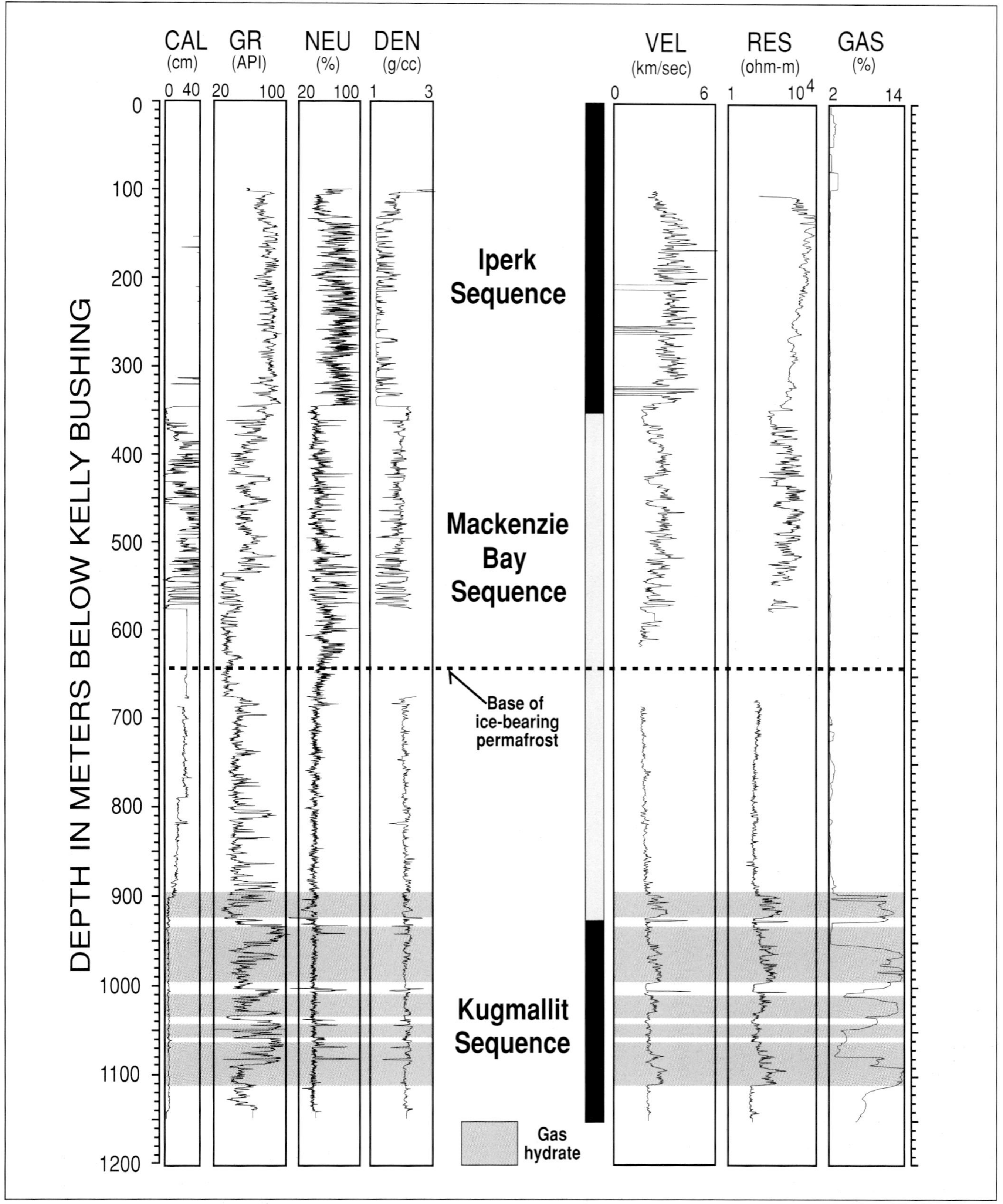

Figure 13. Well-log display for the Mallik 2L-38 well, Mackenzie Delta, showing gas-hydrate occurrences (shaded) and interpreted geology. Data shown include the caliper log (CAL), natural gamma-ray log (GR), neutron porosity (NEU), bulk density (DEN), compressional-wave acoustic velocity (VEL), deep-reading electrical resistivity (RES), and total gas from the mud log (GAS).

free-gas zone (1100.0–1101.9 m) at the base of the deepest downhole log-inferred gas hydrate. As shown in Figure 14, the log-measured compressional-shear-wave velocity ratios (V_p / V_s) below 1.8 are indicative of free-gas-bearing sediment.

ENERGY-RESOURCE POTENTIAL OF GAS HYDRATES

The remaining portion of this paper deals with the assessment of the geologic, engineering, economic, and political factors that control the ultimate resource potential of gas hydrates.

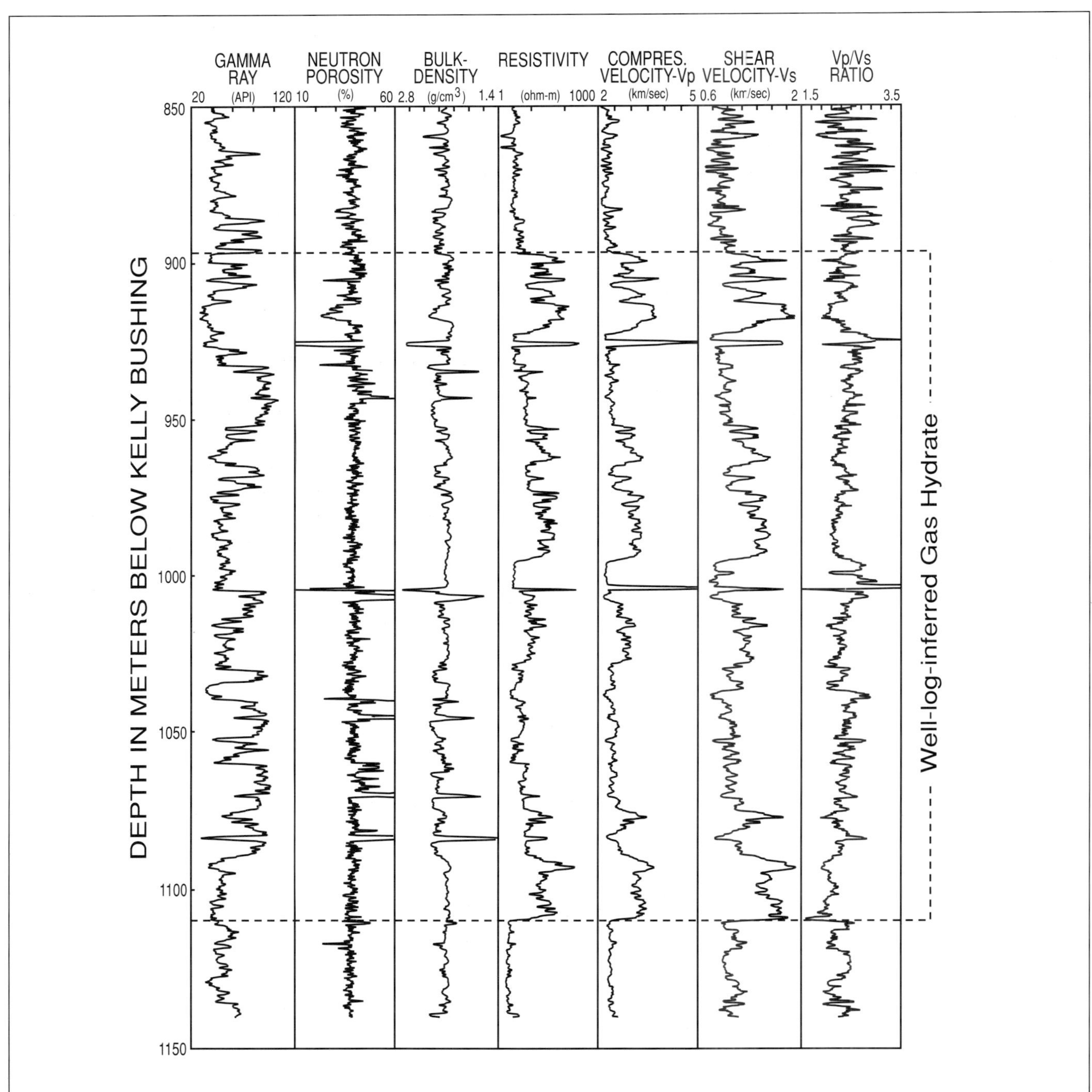

Figure 14. Enlarged portion of Figure 13 showing downhole log data from the Mallik 2L-38 well. Data shown include the natural gamma-ray log, neutron porosity, bulk density, deep-reading electrical resistivity, compressional- and shear-wave acoustic velocities, and compressional-shear-wave acoustic velocity ratios (V_p/V_s). Also shown is the depth of the downhole log-inferred gas-hydrate-bearing stratigraphic interval.

Where Do Gas Hydrates Occur?

The geologic occurrence of gas hydrates has been known since the mid-1960s, when gas-hydrate accumulations were discovered in Russia (reviewed by Makogon, 1981). As discussed in the previous section of this paper, gas hydrates are widespread in permafrost regions and beneath the sea in sediment of outer continental margins (reviewed by Kvenvolden, 1993). Cold surface temperatures at high latitudes on earth are conducive to the development of onshore permafrost and gas hydrate in the subsurface. Onshore gas hydrates (Figure 4) are known to be present in the West Siberian Basin (Makogon et al., 1972) and are believed to occur in other permafrost areas of northern Russia, including the Timan-Pechora province, the eastern Siberian Craton, and the northeastern Siberia and Kamchatka areas (Cherskiy et al., 1985).

Permafrost-associated gas hydrates are also present in the North American Arctic. Direct evidence for gas hydrates on the North Slope of Alaska comes from a core test in the Northwest Eileen State-2 well. Indirect evidence comes from drilling and industry open-hole well logs that suggest the presence of numerous gas-hydrate layers in the area of the Prudhoe Bay and Kuparuk River oil fields. Well-log responses attributed to the presence of gas hydrates have been obtained in about one-fifth of the wells drilled in the Mackenzie Delta, and more than half of the wells in the Arctic Islands are inferred to contain gas hydrates (Judge and Majorowicz, 1992). The recently completed Mallik 2L-38 gas-hydrate research well confirmed the presence of a relatively thick, highly concentrated gas-hydrate accumulation on Richards Island in the outer portion of the Mackenzie River Delta (Dallimore et al., 1999).

The presence of gas hydrates in offshore continental margins (Figure 4) has been inferred mainly from anomalous seismic reflectors (i.e., BSRs) that coincide with the predicted phase boundary at the base of the gas-hydrate stability zone. Gas hydrates have been recovered in gravity cores within 10 m of the seafloor in sediment of the Gulf of Mexico (Brooks et al., 1986), the offshore portion of the Eel River Basin of California (Brooks et al., 1991), the Black Sea (Yefremova and Zhizhchenko, 1974), the Caspian Sea (Ginsburg et al., 1992), and the Sea of Okhotsk (Ginsburg et al., 1993). Gas hydrates also have been recovered at greater subbottom depths during research coring along the southeastern coast of the United States on the Blake Ridge (Kvenvolden and Barnard, 1983; Shipboard Scientific Party, 1996), in the Gulf of Mexico (Shipboard Scientific Party, 1986), in the Cascadia Basin near Oregon (Shipboard Scientific Party, 1994), the Middle America Trench (Kvenvolden and McDonald, 1985), offshore Peru (Kvenvolden and Kastner, 1990), and on both the eastern and western margins of Japan (Shipboard Scientific Party, 1990, 1991).

How Do Gas Hydrates Occur in Nature?

Little is known about the nature of gas-hydrate reservoirs. For example, do hydrates occur as pore-filling constituents or are they found only in massive form (Figure 15)? Information about the nature and texture of reservoired gas hydrates is needed to accurately determine the amount of gas hydrate and associated gas in a given gas-hydrate accumulation. The textural nature of gas hydrate

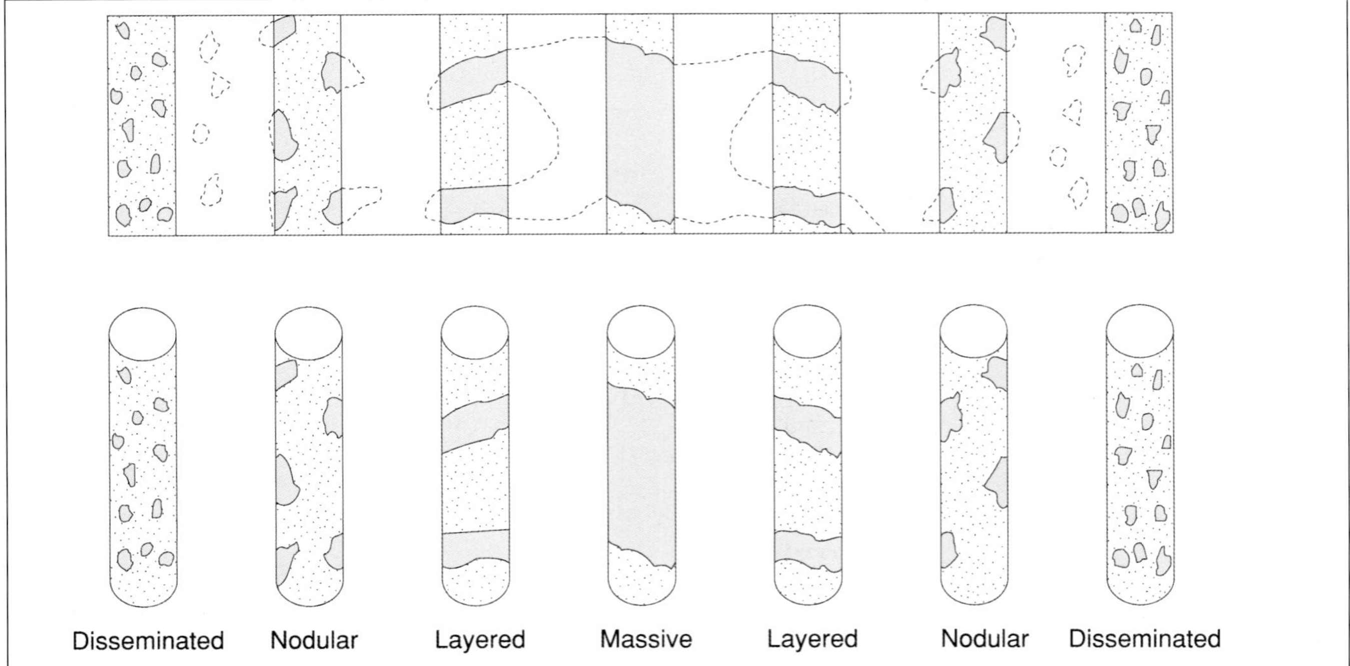

Figure 15. Various morphological forms of natural-gas-hydrate occurrence (modified from Sloan, 1998).

in the reservoir also controls the production potential and characteristics of a gas-hydrate accumulation. The physical and chemical conditions that result in different forms (disseminated, nodular, layered, massive) and distributions (uniform or heterogeneous) of gas hydrates are not understood (reviewed by Sloan, 1998). It is necessary, therefore, to systematically review descriptions of known gas-hydrate occurrences and to evaluate existing gas-hydrate reservoir models at both microscopic and macroscopic scales to assess the nature of gas-hydrate-bearing reservoirs. This section of the paper begins with a review of published gas-hydrate sample descriptions from both marine and permafrost environments, followed by an interpretive discussion of existing and proposed microscopic and macroscopic gas-hydrate reservoir models.

Recovered Gas-hydrate Samples

For this review of the nature of gas-hydrate occurrences, I have relied extensively on the offshore gas-hydrate sample database published by Booth et al. (1996). In this database, Booth et al. (1996) systematically review and describe more than 90 marine gas-hydrate samples recovered from 15 geologic regions. The individual descriptions of gas-hydrate occurrences include information on the number of recovered samples, physiographic province, tectonic setting, geographic position, water depth, subseafloor depth of the recovered sample, geothermal gradient and temperature conditions, depth to the base of the gas-hydrate stability zone, presence of a bottom-simulating seismic reflector, thickness of the gas-hydrate-bearing sedimentary interval, thickness and size of pure gas-hydrate layers and grains, habit or mode of occurrence, host-sediment lithologic description, and origin of the included gas.

In general, most of the recovered gas-hydrate samples consist of individual grains or particles, often described as inclusions or disseminated in the sedimentary section (Figure 15). Descriptions of the habitat of gas hydrates also characterize them as forming cements or nodules or as laminae and veins, which tend to be characterized by dimensions of a few centimeters or less. In several cases, thick, pure gas-hydrate layers measuring as much as 3–4 m in thickness have been sampled (DSDP Site 570; Shipboard Scientific Party, 1985). In both marine and terrestrial permafrost environments, the thickness of identified gas-hydrate-bearing sedimentary sections ranges from a few centimeters to as much as 30 m (Collett, 1993; Booth et al., 1996; Dallimore et al., 1999). Most pure gas-hydrate laminae and layers, however, are often characterized by thicknesses of millimeters to centimeters (Dallimore and Collett, 1995; Booth et al., 1996; Dallimore et al., 1999). Booth et al. (1996) concluded that gas-hydrate-bearing sedimentary sections tend to be tens of centimeters to tens of meters thick. However, thick zones of pure hydrate are relatively rare and represent only a minor constituent of potential gas-hydrate accumulations.

The Booth et al. (1996) review, along with gas-hydrate sample descriptions from the Mackenzie Delta (Dallimore and Collett, 1995; Dallimore et al., 1999) and the Blake Ridge (ODP Leg 164; Shipboard Scientific Party 1996), confirms that gas hydrates are usually uniformly distributed within sediments as mostly pore-filling constituents.

Gas-hydrate Reservoir Models

Most discussions of the nature or texture of gas-hydrate occurrences deal with macroscopic issues (reviewed by Booth et al., 1996). However, information on the occurrence of gas hydrates at the pore scale are needed, because many physical properties of gas-hydrate reservoirs are controlled by microscopic parameters (Dvorkin and Nur, 1993). Of particular concern are the acoustic nature and the fluid-flow permeability characteristics of gas-hydrate-bearing sediments (Lee et al., 1993; Dvorkin and Nur, 1993).

Dvorkin and Nur (1993), along with Ecker et al. (1996), proposed and examined two "micromechanical" models that represent the two extreme cases of gas-hydrate occurrence at the pore scale: Gas-hydrate cement grain contacts and increases the stiffness of the sediment (model-1); gas hydrate is located away from grain contacts in the "bulk" pore volume and so does not affect the stiffness of the sediment frame (model-2). Dvorkin and Nur (1993) experimentally demonstrated that even small amounts of intergranular cementation, such as proposed by model-1, can dramatically increase the stiffness of granular material. They used the intergranular gas-hydrate cementation model (model-1) to explain the occurrence of BSRs, which they attributed to a strong increase of the elastic moduli of the rock resulting from the occurrence of gas hydrates at the base of the gas-hydrate stability zone. Based on amplitude-versus-offset (AVO) analyses of the BSR on the Blake Ridge, however, Ecker et al. (1996) concluded that only reservoir model-2 could qualitatively reproduce the observed BSR, and that gas hydrates at the pore scale are located away from the intergranular contacts in large pores. Ecker et al. (1996) further concluded that the sediment above the BSR is uncemented and mechanically weak. However, they did not explain the acoustic parameters that control the occurrence of the BSR on the Blake Ridge. At this time, we must consider the conclusions of Dvorkin and Nur (1993) and Ecker et al. (1996) to be preliminary until additional laboratory field observations become available.

Before attempting to assess the volume of gas hydrate in a particular reservoir, we need to develop and define a series of reservoir models for the occurrence of gas hydrates in nature. Most reservoir models are based on simple mixing rules, in which complex multicomponent

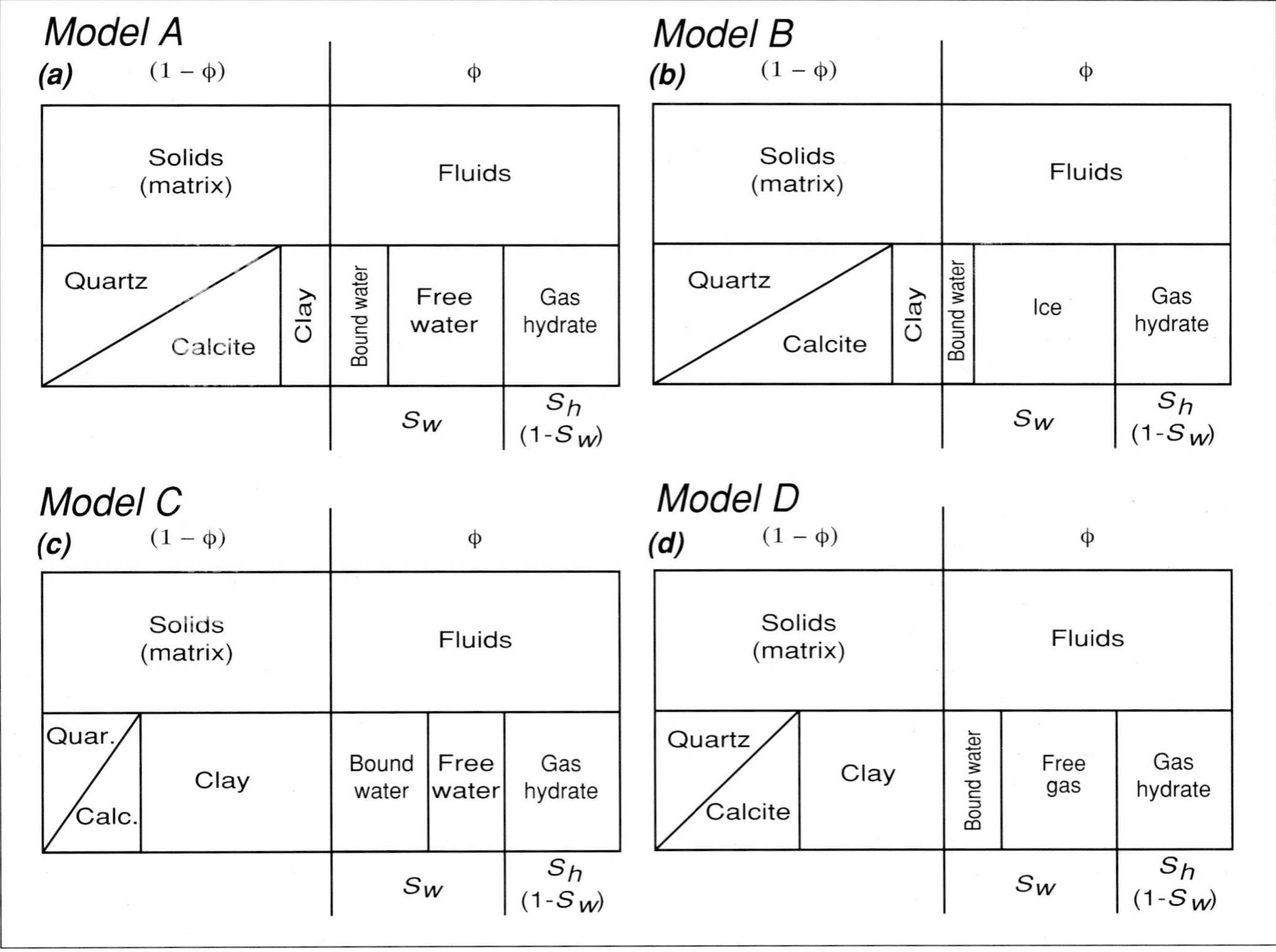

Figure 16. (a) Permafrost-associated gas-hydrate reservoir model for conditions below the base of ice-bearing permafrost (model A). (b) Permafrost-associated gas-hydrate reservoir model for conditions above the base of ice-bearing permafrost (model B). (c) Marine (clay-rich) gas-hydrate reservoir model (model C). (d) Free-gas and gas-hydrate-bearing reservoir model (model D).

systems consist of simple mixtures of rock matrix (consisting of quartz, calcite, and/or clay), water (including clay-bound and free water), and hydrocarbons (gas and/or oil). In permafrost and relatively deep marine environments, however, other reservoir constituents can include gas hydrates and permafrost ice. The first two reservoir models to be considered represent complex gas-hydrate-bearing reservoirs below (model A; Figure 16a) and above (model B; Figure 16b) the base of ice-bearing permafrost in a terrestrial setting. In both models, the sediment matrix consists of a simple mixture of quartz, calcite, and a relatively small amount of clay. Gas-hydrate reservoir models A and B assume no free-gas phase, because all the available gas is in the gas hydrate. The only difference between models A and B is that model B assumes that all the free water and some of the clay-bound water are frozen. Reservoir model C (Figure 16c) has been designed to represent a clay-rich marine gas-hydrate reservoir. Reservoir models C and A are similar, but model C assumes that the clay content of the sediment and associated volume of bound water are higher in most marine gas-hydrate reservoirs. The last gas-hydrate reservoir model to be considered may not occur in nature. Reservoir model D (Figure 4d) assumes that a free-gas phase exists and that all the available water is included in the gas hydrate. Water, being relatively abundant in nature, should not be a gas-hydrate-limiting factor in most reservoirs.

Why Do Gas Hydrates Occur in a Particular Setting?

Review of previous gas-hydrate studies indicates that formation and occurrence of gas hydrates are controlled by formation temperature, formation pore pressure, gas chemistry, pore-water salinity, availability of gas and water, gas and water migration pathways, and presence of reservoir rocks and seals (reviewed by Collett, 1995). In the following section, these geologic controls on the sta-

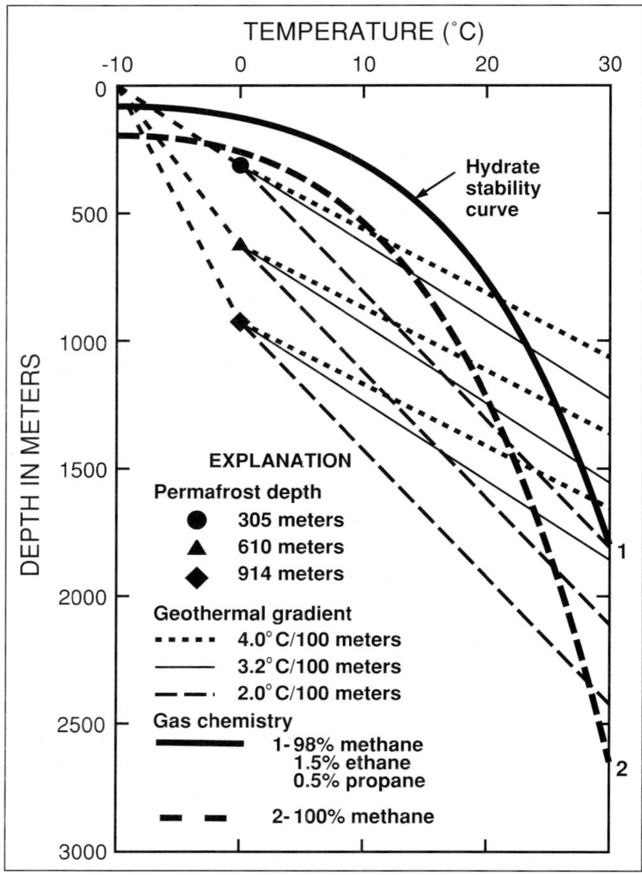

Figure 17. Graph showing the depth-temperature zone in which gas hydrates are stable in a permafrost region, assuming a 9.795 kPa/m pore-pressure gradient (modified from Holder et al., 1987).

bility and formation of gas hydrates are reviewed and assessed.

Formation Temperature, Formation Pore Pressure, and Gas Chemistry

Gas hydrates exist under a limited range of temperature and pressure conditions, such that depth and thickness of the zone of potential gas-hydrate stability can be calculated. Depicted in the temperature-depth plot of Figure 17 are a series of subsurface temperature profiles from an onshore permafrost area and two laboratory-derived gas-hydrate stability curves for different natural gases (modified from Holder et al., 1987). This gas-hydrate phase diagram (Figure 17) illustrates how variations in formation temperature and gas composition can affect thickness of the gas-hydrate stability zone. In Figure 17, the mean annual surface temperature is assumed to be –10°C. However, the depth to the base of permafrost (0°C isotherm) is varied for each temperature profile (assumed permafrost depths of 305, 610, and 914 m). Below permafrost, three geothermal gradients (4.0°C/100 m, 3.2°C/100 m, and 2.0°C/100 m) are used to project the subpermafrost temperature profiles. The two gas-hydrate stability curves represent gas hydrates with different gas chemistries. One of the curves is for a 100% methane hydrate and the other is for a hydrate that contains 98% methane, 1.5% ethane, and 0.5% propane.

The zone of potential gas-hydrate stability in Figure 17 lies in the area between the geothermal gradient and the gas-hydrate stability curve. For example, in Figure 17, which assumes a hydrostatic pore-pressure gradient, the temperature profile projected to an assumed permafrost base of 610 m intersects the 100% methane-hydrate stability curve at about 200 m, thus marking the upper boundary of the methane-hydrate stability zone. A geothermal gradient of 4.0°C/100 m projected from the base of permafrost at 610 m intersects the 100% methane-hydrate stability curve at about 1100 m. Thus, the zone of potential methane-hydrate stability is approximately 900 m thick. However, if permafrost extended to a depth of 914 m and if the geothermal gradient below permafrost is 2.0°C/100 m, the zone of potential methane-hydrate stability would be approximately 2100 m thick.

Most gas-hydrate stability studies assume that the pore-pressure gradient is hydrostatic (9.795 kPa/m or 0.433 psi/ft). Pore-pressure gradients greater than hydrostatic will correspond to higher pore pressures with depth and a thicker gas-hydrate stability zone. A pore-pressure gradient less than hydrostatic will correspond to a thinner gas-hydrate stability zone. For example, in Figure 17, which assumes a hydrostatic (9.795 kPa/m or 0.433 psi/ft) pore-pressure gradient, the thickness of the 100% methane-hydrate stability zone with a 610-m permafrost depth and a subpermafrost geothermal gradient of 2.0°C/100 m would be about 1700 m. However, if a pore-pressure gradient of 11.311 kPa/m (0.500 psi/ft) is assumed, the thickness of the methane-hydrate stability zone would be increased to about 1850 m.

The gas-hydrate stability curves in Figure 17 were obtained from laboratory data published in Holder et al. (1987). The addition of 1.5% ethane and 0.5% propane to the pure methane gas system shifts the stability curve to the right, thus deepening the zone of potential gas-hydrate stability. For example, assuming a hydrostatic pore-pressure gradient (Figure 17), a permafrost depth of 610 m, and a subpermafrost geothermal gradient of 4.0°C/100 m, the zone of methane-hydrate (100% methane) stability would be about 900 m thick. However, the addition of ethane (1.5%) and propane (0.5%) would thicken the potential gas-hydrate stability zone to 1100 m.

Pore-water Salinity

Salt such as NaCl, when added to a gas-hydrate system, lowers the temperature at which gas hydrates form. Pore-water salts in contact with gas during gas-hydrate formation can reduce crystallization temperature by about 0.06°C for each part per thousand (ppt) of salt (Holder et al., 1987). Therefore, a pore-water salinity similar to that of seawater (32 ppt) would shift the gas-hydrate stability curves in Figure 17 to the left about 2°C and reduce the thickness of the gas-hydrate stability zone.

Availability of Gas and Water

Most naturally occurring gas hydrates are characterized by two crystal structures, known as structure I and structure II (reviewed by Sloan, 1998). The ideal gas/water ratio of structure I gas hydrate is 8/46, and the ideal gas/water ratio of structure II gas hydrate is 24/136. These ratios confirm the observation that gas hydrates contain a substantial volume of gas. The ideal gas/water ratios also indicate that a substantial amount of fresh water is stored in the gas-hydrate structure. These high gas and water concentrations demonstrate that formation of gas hydrate requires a large source of both gas and water. Thus, it becomes necessary to quantify the potential sources of gas and water when assessing a potential gas-hydrate accumulation. In previous studies, this evaluation was based on assessment of a set of minimum source-rock criteria that includes organic richness (total organic carbon), sediment thickness, and thermal maturity. The availability of large quantities of hydrocarbon gas from microbial and thermogenic sources is an important factor controlling the formation and distribution of natural-gas hydrates (Kvenvolden, 1988; Collett, 1993). Carbon-isotope analyses indicate that the methane in many oceanic hydrates is derived from microbial sources. However, molecular and isotopic analyses indicate a thermal origin for the methane in several offshore Gulf of Mexico and onshore Alaskan gas-hydrate occurrences.

Gas and Water Migration Pathways

Other factors controlling the availability of gas and water are the geologic controls on fluid migration. As previously shown, gas hydrates contain a substantial volume of gas and water that must be supplied to a developing gas-hydrate accumulation. If effective migration pathways are not available, it is unlikely that a significant volume of gas hydrates would accumulate. Therefore, geologic parameters such as rock permeability and the nature of faulting must be evaluated to determine if the required gas and water can be delivered to the potential hydrate reservoir.

Presence of Reservoir Rocks and Seals

The study of gas-hydrate samples recovered during research coring operations in oceanic sediments suggests that the physical nature of in-situ gas hydrates may be highly variable (Figure 15). Gas hydrates were observed to be (1) occupying pores of coarse-grained rocks, (2) nodules disseminated within fine-grained rocks, (3) a solid filling fractures, and (4) a massive unit composed mainly of solid gas hydrate with minor amounts of sediment. This review suggests that porous rock intervals serve as reservoir rocks where gas and water can be concentrated in the amounts necessary for gas-hydrate formation. Therefore, the presence of reservoir rocks may play a role in gas-hydrate formation, particularly in well-consolidated rock intervals. However, in marine environments at shallow depths below the seafloor, gas hydrates can mechanically displace sediments to form their own reservoir. Thus, the availability of reservoir-quality rocks may not always be a limiting factor.

The presence of effective reservoir seals or traps may play a role in gas-hydrate formation. Gas generated at depth moves upward, generally along tilted permeable carrier beds, until it either seeps at the surface or meets an impermeable barrier (trap) that stops or impedes its flow. As migrating gas accumulates below an effective seal, total gas concentrations may reach the critical amounts necessary for the formation of gas hydrates. Thus, impermeable seals can provide a mechanism by which the required gas can be concentrated within reservoir rocks.

Besides conventional reservoirs and trapping mechanisms, it is possible for gas hydrate to form its own reservoir and trap (Downey, 1984). As gas migrates into the zone of gas-hydrate stability, it may interact with the available pore water to generate gas hydrate. With the appropriate volumes of gas and water, the pore space in the reservoir rock could be completely filled, thus making the rock impermeable to further hydrocarbon migration. The plugging of gas pipelines and production tubing by gas hydrates is testimony to the sealing potential of gas hydrates (Sloan, 1998).

How Much Gas Is Trapped within Gas-hydrate Accumulations?

The amount of methane sequestered in gas hydrates is probably enormous, but estimates of the amounts are speculative and range over three orders of magnitude, from about 3114 to 7,634,000 trillion m^3 (reviewed by Kvenvolden, 1993). It is likely, however, that the amount of gas in the hydrate reservoirs of the world greatly exceeds the volume of known conventional gas reserves. Before reviewing assessments of world gas-hydrate resources, it is necessary to examine the quality and variability of gas-hydrate assessments at the accumulation and reservoir scale.

Gas Hydrates at the Reservoir and Accumulation Scale

Estimates of the amount of gas hydrates and associated gas in a given gas-hydrate accumulation can vary considerably. For example, recent estimates of the volume of gas that may be contained in gas hydrates and free gas beneath gas hydrates on the Blake Ridge range from about 70 trillion m^3 of gas in an area of 26,000 km^2 (Dickens et al., 1997) to about 80 trillion m^3 of gas in an area of 100,000 km^2 (Holbrook et al., 1996). The

Table 1. Volume of gas in the downhole log-inferred gas-hydrate occurrences at ODP sites 994, 995, 997, and 889 and in the Northwest Eileen State-2 and Mallik 2L-38 wells.

Site/Well identification	Depth of log-inferred gas hydrates (m)	Thickness of hydrate-bearing zone (m)	Sediment porosity (%)	Gas-hydrate saturation (%)	Volume of gas within hydrate per square kilometer (m^3)
Ocean Drilling Program drill sites:					
Site 994	212.0–428.8	216.8	57.0	3.3	669,970,673
Site 995	193.0–450.0	257.0	58.0	5.2	1,267,941,673
Site 997	186.4–450.9	264.5	58.1	5.8	1,449,746,073
Site 889	127.6–228.4	100.8	51.8	5.4	466,635,705
Northwest Eileen State-2 well:					
Unit C	651.5–680.5	29.0	35.6	60.9	1,030,904,796
Unit D	602.7–609.4	6.7	35.8	33.9	133,382,462
Unit E	564.0–580.8	16.8	38.6	32.6	346,928,811
			Total for Northwest Eileen State-2:		1,511,216,069
Mallik 2L-38 well:					
Hydrate unit	888.8–1101.1	212.3	31.0	44.0	4,749,066,080

difference between these estimates has been attributed to the observation that the amount of free gas directly measured in pressure-core samples (Dickens et al., 1997) from beneath gas hydrates is significantly larger than that estimated from borehole vertical seismic profile data (Holbrook et al., 1996). Other published studies indicate that gas hydrates at the crest of the Blake Ridge alone (area of about 3000 km^2) may contain more than 18 trillion m^3 of gas (Dillon and Paull, 1983). The broad range of these estimates demonstrates the need for high-resolution measurements of gas-hydrate and associated free-gas volumes in any gas-hydrate accumulation of interest.

It has been suggested that the volume of gas that may be contained in a gas-hydrate accumulation depends on five "reservoir" parameters (modified from Collett, 1993): (1) areal extent of the gas-hydrate occurrence, (2) "reservoir" thickness, (3) sediment porosity, (4) degree of gas-hydrate saturation, and (5) the hydrate gas-yield volumetric parameter, which defines how much free gas (at STP) is stored within a gas hydrate (also known as the hydrate number). The following section will assess the five "reservoir" parameters (Table 1) needed to calculate the volume of gas associated with gas hydrates on the Blake Ridge (ODP Sites 994, 995, and 997; Shipboard Scientific Party, 1996), along the Cascadia continental margin (ODP Site 889; Shipboard Scientific Party, 1994), on the North Slope of Alaska (Northwest Eileen State-2 well; Collett, 1993), and in the Mackenzie River Delta of Canada (Mallik 2L-38 well; Dallimore et al., 1999.

The following "resource" assessment (modified from Collett, 1998) has been conducted on a site-by-site basis; for each site examined, the volume of gas hydrate and associated gas in a 1-km^2 area surrounding each drill site has been calculated individually (Table 1). For this "resource" assessment, I have defined the thickness of the gas-hydrate-bearing sedimentary section at the marine and permafrost drill sites to be the total thickness of the downhole log-inferred gas-hydrate accumulation (Table 1). Average core-derived sediment porosities for the gas-hydrate-bearing reservoirs along the Blake Ridge at sites 994, 995, and 997 and along the Cascadia continental margin at site 889 range from about 52% to 58% (Table 1). The "corrected" density-log-derived sedimentary porosities for the three gas-hydrate-bearing units identified in the Northwest Eileen State-2 (units C, D, and E) range from an average value of about 36% to 39% (Table 1). The density-log-derived sediment porosities for the gas-hydrate-bearing interval in the Mallik 2L-38 well averages about 31% (Table 1). Gas-hydrate saturations at the marine drill sites (sites 994, 995, 997, and 889), calculated from available downhole logs, range from an average value of about 3% to 6% (Table 1). Gas-hydrate saturations in all three gas-hydrate-bearing units (units C, D, and E) in the Northwest Eileen State-2 well, calculated from available downhole log data, range from an average value of about 33% to 61% (Table 1). The resistivity well-log-derived gas-hydrate saturations in the Mallik 2L-38 well average about 44% (Table 1).

In this assessment, I have assumed a hydrate number of 6.325 (90% gas-filled clathrate), which corresponds to a gas yield of 164 m^3 of methane (at STP) for each 1 m^3 of gas hydrate. The log-inferred gas hydrates at sites 994, 995, and 997 on the Blake Ridge contain between 670,000,000 and 1,450,000,000 m^3 of gas/km^2 (Table 1). The volume of gas in the log-inferred gas hydrates at site 889,on the Cascadia continental margin, is about 467,000,000 m^3 of gas/km^2 (Table 1). Cumulatively, all three log-inferred gas-hydrate-bearing stratigraphic units (units C, D, and E) drilled and cored in the Northwest Eileen State-2 well may contain about 1,511,000,000 m^3 of gas in the 1-km^2 area surrounding this drill site (Table 1). It was also determined that the log-inferred gas-hydrate-bearing stratigraphic interval drilled in the Mallik 2L-38 well contains about 4,750,000,000 m^3 of gas in the 1-km^2 area surrounding the Mallik drill site (Table 1).

A close examination of the gas-hydrate saturations shown in Table 1 reveals a potential problem associated with production of gas from marine gas hydrates. Even though vast portions of the world's continental shelves appear to be underlain by gas hydrates, the concentration of hydrates in most marine accumulations appears to be very low. Low gas-hydrate concentrations will significantly affect the economic-production potential of marine gas hydrates. (Gas-hydrate production is discussed in more detail later in this chapter.)

Gas Hydrates at the World and National Scale

World estimates for the amount of natural gas in gas-hydrate deposits range from 14 trillion to 34,000 trillion m^3 for permafrost areas and from 3100 trillion to 7,600,000 trillion m^3 for oceanic sediments (modified from Kvenvolden, 1993). The estimates in Table 2 show considerable variation, but oceanic sediments seem to be a much greater resource of natural gas than continental sediments. Current estimates of the amount of methane in the world's gas-hydrate accumulations are in rough accord, at about 20,000 trillion m^3 (reviewed by Kvenvolden, 1993). If these estimates are valid, the amount of methane in gas hydrates is almost two orders of magnitude larger than the total remaining recoverable conventional methane resources, estimated to be about 250 trillion m^3 (Masters et al., 1991).

The 1995 National Assessment of United States Oil and Gas Resources, conducted by the U.S. Geological Survey, focused on assessing the undiscovered conventional and unconventional resources of crude oil and natural gas in the United States (Gautier et al., 1995). This assessment included for the first time a systematic resource appraisal of the in-place natural-gas-hydrate resources of the United States onshore and offshore regions (Collett, 1995). In this assessment, 11 gas-hydrate plays were identified in one onshore and four offshore gas-hydrate provinces. The offshore gas-hydrate provinces assessed lie within the U.S. Exclusive Economic Zone adjacent to the lower 48 states and Alaska. The only onshore province assessed was the North Slope of Alaska. In-place gas resources in the gas hydrates of the United States are estimated to range from about 3200 trillion to 19,000 trillion m^3 of gas at the 0.95 and 0.05 probability levels, respectively. Although this wide range of values shows a high degree of uncertainty, it does indicate the potential for enormous quantities of gas stored as gas hydrates. The mean in-place value for the entire United States is calculated to be about 9,000 trillion m^3 of gas.

Table 2. World estimates of the amount of gas within hydrates.

In-place Natural-gas Resources Terrestrial Gas Hydrates	
Volume (m^3)	Reference
1.4×10^{13}	Meyer (1981)
3.1×10^{13}	McIver (1981)
5.7×10^{13}	Trofimuk et al. (1977)
7.4×10^{14}	MacDonald (1990)
3.4×10^{16}	Dobrynin et al. (1981)
In-place Natural-gas Resources Oceanic Gas Hydrates	
Volume (m^3)	Reference
3.1×10^{15}	Meyer (1981)
$5–25 \times 10^{15}$	Trofimuk et al. (1977)
2×10^{16}	Kvenvolden (1988)
2.1×10^{16}	MacDonald (1990)
4×10^{16}	Kvenvolden and Claypool (1988)
7.6×10^{18}	Dobrynin et al. (1981)

Gas-hydrate Production Technology

Even though gas hydrates are known to occur in numerous marine and arctic settings, little is known about the technology necessary to produce gas hydrates. Most of the existing gas-hydrate "resource" assessments do not address the problem of gas-hydrate recoverability. Proposed methods of gas recovery from hydrates (Figure 18) usually deal with dissociating or "melting" in-situ gas hydrates by (1) heating the reservoir beyond hydrate-formation temperatures; (2) decreasing the reservoir pressure below hydrate equilibrium; or (3) injecting an inhibitor, such as methanol or glycol, into the reservoir to decrease hydrate stability conditions. Gas recovery from hydrates is hindered because the gas is in a solid form and because hydrates are usually widely dispersed in hostile arctic and deep-marine environments. First-order thermal-stimulation computer models (incorporating heat and mass balance) have been developed to evaluate hydrate gas production from hot water and steam floods. They have shown that gas can be produced from hydrates at sufficient rates to make gas hydrates a technically recoverable resource (Sloan, 1998). However, the economic cost associated with these types of enhanced gas-recovery techniques would be prohibitive. Similarly, the use of gas-hydrate inhibitors in the production of gas from

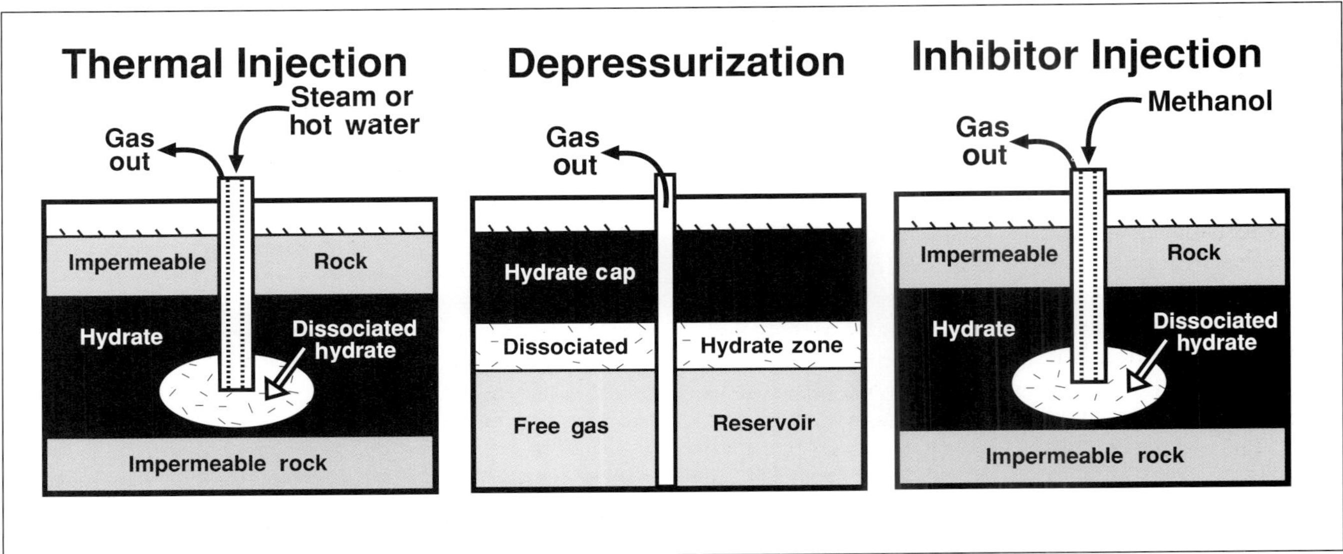

Figure 18. Schematic of proposed gas-hydrate production methods.

hydrates has been shown to be technically feasible (Sloan, 1998). However, the use of large volumes of chemicals such as methanol comes with a high economic and environmental cost. Among various techniques for production of natural gas from in-situ gas hydrates, the most economically promising is depressurization. However, extraction of gas from a gas-hydrate accumulation by depressurization will be hampered by formation of ice and/or reformation of gas hydrate resulting from the endothermic nature of gas-hydrate dissociation.

The Messoyakha gas field in the northern part of the West Siberian Basin is often used as an example of a hydrocarbon accumulation from which gas has been produced from in-situ natural-gas hydrates. Production data and other pertinent geologic information have been used to document the presence of gas hydrates in the upper part of Messoyakha field (Makogon, 1981). The production history of the field demonstrates that gas hydrates are an immediate producible source of natural gas and that production can be started and maintained by conventional methods. Long-term production from the gas-hydrate part of the Messoyakha field is presumed to have been achieved by the simple depressurization scheme. As production began from the lower free-gas portion of the field in 1969, the measured reservoir pressures followed predicted decline relations. However, by 1971, reservoir pressures began to deviate from expected values. This deviation has been attributed to liberation of free gas from dissociating gas hydrates. Throughout the production history of Messoyakha field, it is estimated that about 36% (about 5 billion m^3) of the gas withdrawn from the field has come from gas hydrates (Makogon, 1981). Recently, however, several studies have suggested that gas hydrates may not contribute significantly to gas production in Messoyakha field (reviewed by Collett and Ginsburg, 1998).

It should be noted that our current assessment of proposed methods for gas-hydrate production do not consider some of the more recently developed advanced oil and gas production schemes. For example, the usefulness of downhole heating methods such as in-situ combustion, electromagnetic heating, or downhole electrical heating has not been evaluated. In addition, advanced drilling techniques and complex downhole completions, including horizontal wells and multiple laterals, have not been considered in any comprehensive gas-hydrate production scheme. Gas-hydrate provinces with existing conventional oil and gas production may also provide an opportunity to test relatively more advanced gas-hydrate production methods. For example, in northern Alaska, existing "watered-out" production wells are being evaluated as potential sources for hot geopressured brines that will be used to thermally stimulate gas-hydrate production.

As previously noted, the low concentration of hydrates in most of the world's marine gas-hydrate occurrences raises a concern about the production technology required to produce gas from highly disseminated gas-hydrate accumulations. In addition, the host sediments also represent a significant technical challenge to potential gas-hydrate production. In most cases, marine gas hydrates have been found in clay-rich, unconsolidated sedimentary sections that exhibit little or no permeability. Most of the existing gas-hydrate production models require the establishment of reliable flow paths within the formation to allow the movement of produced gas to the wellbore and injected fluids into the gas-hydrate-bearing sediments. It is unlikely, however, that most marine sediments possess the mechanical strength to allow the generation of significant flow paths. It is possible that in basins with significant input of coarse-grained siliciclastic sediments, such as the Gulf of Mexico or along the eastern margin of India, gas hydrates may be reservoired at high concentrations in more conventional siliciclastic reservoirs. This would be more nearly analogous to the nature of gas-hydrate occurrences in onshore permafrost environments (Collett, 1993; Dallimore et al., 1999).

Motivations for Gas-hydrate Production

Significant, if not insurmountable, technical issues need to be resolved before gas hydrates can be counted as a viable option for future supplies of natural gas. In most cases, viability of an energy resource is based almost solely on economics. It is important to note, however, that in some cases, viability of a particular hydrocarbon resource can be controlled by unique local economic and nontechnical factors. For example, countries with little domestic energy production usually pay considerably more for energy needs because they rely more on imported hydrocarbons, which often come with additional tariffs and transportation expenses. Energy security is often a concern to resource-poor countries, which will often invest more money in relatively expensive unconventional domestic energy resources than energy-rich countries will. In some cases, uniqueness of a particular location, such as distance to a conventional energy resource, may lead to development of otherwise noneconomic unconventional resources. In the following section, the economic and noneconomic motivations that may eventually lead to sustained production of gas from hydrates are discussed.

Economic Motivations

Because of uncertainties about geologic settings and feasible production technology, few economic studies have been published on gas hydrates. The National Petroleum Council, in its major 1992 study of gas (National Petroleum Council, 1992), published one of the few available economic assessments of gas-hydrate production (Table 3). This information, extracted from MacDonald (1990), assessed the relative economics of gas recovery from hydrates using thermal injection and depressurization. It also benchmarks the cost of gas-hydrate production with the costs of conventional gas production on Alaska's North Slope. Since the 1992 National Petroleum Council report, the costs of conventional gas production on the North Slope and elsewhere have declined. In countries with considerable production of cheaper conventional natural gas, hydrates may soon become an economically viable energy resource in a competitive energy market.

Japan, India, and South Korea, like many other countries with little indigenous energy resources, pay a very high price for imported liquid natural gas (LNG) and oil. The high cost of imported hydrocarbon resources is one reason why, in the last few years, government agencies in Japan, India, and South Korea have begun to develop hydrate research programs to recover gas from oceanic hydrates. One of the most notable gas-hydrate projects is under way in Japan, where the Japan National Oil Corporation (JNOC), with funding from the Ministry of International Trade and Industry (MITI), has launched a five-year study to assess the domestic resource potential of natural-gas hydrates. In numerous press releases, MITI has indicated that "methane hydrates could be the next generation's source of producible domestic energy." JNOC drilled a series of gas-hydrate test wells in the Nankai Trough area, near Tokyo, in late 1999. As much as 50 trillion m^3 of gas may be stored within the gas hydrates of the Nankai Trough. In 1998, JNOC also drilled the Mallik 2L-38 gas-hydrate research well with the Geological Survey of Canada in the Mackenzie Delta of northern Canada (Dallimore et al., 1999).

Table 3. Economic study of gas-hydrate production (after National Petroleum Council, 1992).

	Thermal injection*	Depressurization	Conventional gas
Investment ($1000)	5084	3320	3150
Annual cost ($1000)	3200	2510	2000
Total production (mmcf/year)**	900	1100	1100
Production cost ($/mcf)	3.60	2.28	1.82
Break-even wellhead price ($/mcf)	4.50	2.85	2.25

* *Injection of 30,000 bbl/day of 150°C water.*
** *Assumed reservoir properties: $h = 7.3$ m, $\phi = 40\%$, $k = 600$ md*

India, like Japan, has initiated a very ambitious national gas-hydrate research program. In March 1997, the government of India announced new exploration-licensing policies which included the release of several deep-water (>400 m) lease blocks along the east coast of India between Madras and Calcutta (Kuldeep et al., 1997). Recently acquired seismic data have revealed possible evidence of widespread gas-hydrate occurrences throughout the proposed lease blocks (Hanumantha et al., 1998). Also announced was a large gas-hydrate prospect in the Andaman Sea, between India and Myanmar, which is estimated to contain as much as 6 trillion m^3 of gas. The government of India has indicated that gas hydrates are of "utmost importance to meet their growing domestic energy needs." The National Gas Hydrate Program of India calls for drilling as many as five gas-hydrate test wells.

Most recently, the U.S. Department of Energy has launched a national-level research program to assess resource potential of marine and permafrost-associated gas hydrates.

Political Motivations

The world will consume increasing volumes of natural-gas well into the twenty-first century if reliable, low-cost supplies can be discovered and exploited. In the near term, natural gas is expected to take on a greater role in power generation and transportation because of increasing pressure for cleaner fuels and reduced carbon-dioxide emissions. Gas demand is also expected to grow throughout the first half of the new century because of the expanding role of gas as a competitive transportation fuel resulting from commercial development of gas-to-liquids technology. The drive to increased reliance on natural gas will be based on economics only in part. Government regulatory and taxation policy may also dictate viability of a particular energy commodity such as gas hydrates. In the recent past, government subsidies for unconventional gas resources such as coal-bed methane contributed to their technical and economic viability. Similar forms of government support may have a significant impact on the resource viability of gas hydrates. Another noneconomic factor that may affect the resource potential of gas hydrates in a particular country is the concerns dealing with national security and dependence on foreign energy resources. The governments of many countries, including the United States, often express concerns about reliance on imported energy resources. Most certainly, the international gas-hydrate research programs of Japan, India, and South Korea have been established in part to address reliance on foreign energy resources.

Unique Motivations

The first gas-hydrate accumulations to be produced may have unique characteristics, such as location, that may make them technically and economically viable. For example, gas associated with conventional oil fields on the North Slope of Alaska is used to generate electricity in support of local field operations, such as miscible gas floods and gas-lift operations in producing oil wells, and is reinjected to maintain reservoir pressures in producing fields. In the future, gas may be used to generate steam to produce the known vast quantities of heavy oil on the North Slope. Existing and emerging operational needs for natural gas on the North Slope are outpacing discovery of new conventional resources, and at least one of the operators in Alaska is looking at gas hydrates as a potential source of gas for field operations. The North Slope of Alaska contains vast, highly concentrated gas-hydrate accumulations that may be exploited because of a unique local need for natural gas.

CONCLUSIONS

World estimates of the amount of methane sequestered in gas hydrates are enormous, but published estimates are highly speculative. It is generally believed, however, that the amount of gas in the hydrate reservoirs of the world greatly exceeds the volume of known conventional gas reserves. Until recently, relatively little work has been done to assess the availability and production potential of gas hydrates. Gas recovery from hydrates is hindered because hydrates occur as a solid in nature and are commonly widely dispersed in hostile arctic and deep-marine environments. Proposed methods of gas recovery from hydrates usually deal with dissociating in-situ gas hydrates by heating and/or depressurizing the reservoir. Among various techniques for production of natural gas from in-situ gas hydrates, the most economically promising is depressurization.

Despite the fact that relatively little is known about the ultimate resource potential of natural-gas hydrates, it is certain that gas hydrates are a vast storehouse of natural gas, and the national gas-hydrate research programs of Japan, India, and the United States will significantly contribute to our understanding of the technical challenges needed to turn this enormous resource into a economically producible reserve.

Will gas hydrates become a significant energy resource? It is unlikely that we will see significant worldwide gas production from hydrates in the next 30 to 50 years. However, in certain parts of the world characterized by unique economic and/or political motivations, gas hydrates may become a critical sustainable source of natural gas within the foreseeable future, possibly in the next five to 10 years.

REFERENCES CITED

Bily, C., and J. W. L. Dick, 1974, Natural occurring gas hydrates in the Mackenzie Delta, Northwest Territories: Bulletin of Canadian Petroleum Geology, v. 22, p. 340–352.

Booth, J. S., M. M. Rowe, and K. M. Fischer, 1996, Offshore gas hydrate sample database with an overview and preliminary analysis: U.S. Geological Survey Open-File Report 96-272, 17 p.

Brooks, J. M., B. H. Cox, W .R. Bryant, M. C. Kennicutt, R. G. Mann, and T. J. McDonald, 1986, Association of gas hydrates and oil seepage in the Gulf of Mexico: Organic Geochemistry, v. 10, p. 221–234.

Brooks, J. M., M. E. Field, and M. C. Kennicutt, 1991, Observations of gas hydrates in marine sediments, offshore northern California: Marine Geology, v. 96, p. 103–109.

Cherskiy, N. V., V. P. Tsarev, and S. P. Nikitin, 1985, Investigation and prediction of conditions of accumulation of gas resources in gas-hydrate pools: Petroleum Geology, v. 21, p. 65–89.

Collett, T. S., 1983, Detection and evaluation of natural gas hydrates from well logs, Prudhoe Bay, Alaska, *in* Proceedings of the Fourth International Conference on Permafrost: Washington, D. C., National Academy of Sciences, p. 169–174.

Collett, T. S., 1993, Natural gas hydrates of the Prudhoe Bay and Kuparuk River area, North Slope, Alaska: AAPG Bulletin, v. 77, p. 793–812.

Collett, T. S., 1995, Gas hydrate resources of the United States, *in* D. L. Gautier, G. L. Dolton, K. I. Takahashi, and K. L. Varnes, eds., 1995 National assessment of United States oil and gas resources on CD-ROM: U.S. Geological Survey Digital Data Series 30.

Collett, T. S., 1998, Well log evaluation of gas hydrate saturations: Transactions of the 39th Annual Symposium of the Society of Professional Well Log Analysts, paper MM.

Collett, T. S., K. J. Bird, K. A. Kvenvolden, and L. B. Magoon, 1988, Geologic interrelations relative to gas hydrates within the North Slope of Alaska: U.S. Geological Survey Open-File Report 88–389, 150 p.

Collett, T. S., and S. R. Dallimore, 1998, Quantitative assessment of gas hydrates in the Mallik L-38 well, Mackenzie Delta, N.W.T., Canada: Proceedings of the Eighth International Conference on Permafrost, p. 189–194.

Collett, T. S., and G. D. Ginsburg, 1998, Gas hydrates in the Messoyakha gas field of the West Siberian Basin—A re-examination of the geologic evidence: International Journal of Offshore and Polar Engineering, v. 8, p. 22–29.

Dallimore, S. R., and T. S. Collett, 1995, Intrapermafrost gas hydrates from a deep core hole in the Mackenzie Delta, Northwest Territories, Canada: Geology, v. 23, p. 527–530.

Dallimore, S. R., and J. V. Matthews, 1997, The Mackenzie Delta borehole project: Environmental Studies Research Funds Report 135, 1 CD-ROM.

Dallimore, S. R., T. Uchida, and T. S. Collett, 1999, Scientific results from JAPEX/JNOC/GSC Mallik 2L-38 gas hydrate research well, Mackenzie Delta, Northwest Territories, Canada: Geological Survey of Canada Bulletin 544, 403 p.

Dickens, G. R., C. K. Paull, P. Wallace, and the ODP Leg 164 Scientific Party, 1997, Direct measurement of in situ methane quantities in a large gas-hydrate reservoir: Nature, v. 385, p. 426–428.

Dillon, W. P., M. W. Lee, K. Fehlhaber, and D. F. Coleman, 1993, Gas hydrates on the Atlantic continental margin of the United States—Controls on concentration, *in* D. G. Howell, ed., The Future of energy gases: U.S. Geological Survey Professional Paper 1570, p. 313–330.

Dillon, W. P., and C. K. Paull, 1983, Marine gas hydrates, II. Geophysical evidence, *in* J. S. Cox, ed., Natural gas hydrates: Properties, occurrences, and recovery: London, Butterworth Publishing, p. 73–90.

Dixon, J., J. R. Dietrich, and D. H. McNeil, 1992, Upper Cretaceous to Pleistocene sequence stratigraphy of the Beaufort-Mackenzie and Banks Island areas, northwest Canada: Geological Survey of Canada Bulletin 407, p. 1–90.

Dobrynin, V. M., Yu. P. Korotajev, and D. V. Plyuschev, 1981, Gas hydrates—A possible energy resource, *in* R .F. Meyer and J. C. Olson, eds., Long-term energy resources: Boston, Pitman Publishing, p. 727–729.

Downey, M. W., 1984, Evaluating seals for hydrocarbon accumulations: AAPG Bulletin, v. 68, no. 11, p. 1752–1763.

Dvorkin, J., and A. Nur, 1993, Rock physics for characterization of gas hydrates, *in* D. G. Howell, ed., The Future of energy gases: U.S. Geological Survey Professional Paper 1570, p. 293–311.

Ecker, C., D. Lumley, J. Dvorkin, and A. Nur, 1996, Structure of hydrate sediment from seismic and rock physics: Proceedings of the Second International Conference on Natural Gas Hydrates, p. 491–498.

Field, M. E., and K. A. Kvenvolden, 1985, Gas hydrates on the northern California continental margin: Geology, v. 13, p. 517–520.

Gautier, D. L., G. L. Dolton, K. I. Takahashi, and K. L. Varnes, 1995, National assessment of United States oil and gas resources on CD-ROM: U.S. Geological Survey Digital Data Series 30.

Ginsburg, G. D., R. A. Guseinov, A. A. Dadashev, G. A. Ivanova, S. A. Kazantsev, V. A. Soloviev, Ye. V. Telepnev, R. E. Askery-Nasirov, A. D. Yesikov, V. I. Mal'tseva, Yu. G. Mashirov, and I. Yu. Shabayeva, 1992, Gas hydrates in the southern Caspian Sea: Izvestiya Akademii Nauk Serya Geologisheskaya, v. 7, p. 5–20.

Ginsburg, G. D., V. A. Soloviev, R. E. Cranston, T. D. Lorenson, and K. A. Kvenvolden, 1993, Gas hydrates from the continental slope, offshore from Sakhalin Island, Okhotsk Sea: Geo-Marine Letters, v. 13, p. 41–48.

Hanumantha, R., S. I. Reddy, R. Khanna, T. G. Rao, N. K. Thakur, and C. Subrahmanyam, 1998, Potential distribution of methane hydrates along the continental margins of India: Current Science, v. 74, p. 468–566.

Holbrook, W. S., H. Hoskins, W. T. Wood, R. A. Stephen, and the ODP Leg 164 Scientific Party, 1996, Methane hydrate, bottom-simulating reflectors, and gas bubbles: Results of vertical seismic profiles on the Blake Ridge: Science, v. 273, p. 1840–1843.

Holder, G .D., R. D. Malone, and W. F. Lawson, 1987, Effects of gas composition and geothermal properties on the thickness and depth of natural-gas-hydrate zone: Journal of Petroleum Technology, v. 39, p. 1147–1152.

Hyndman, R., G. Spence, T. Yuan, and B. Desmons, 1996, Gas hydrate on the continental slope off Vancouver Island: Proceedings of the Second International Conference on Natural Gas Hydrates, p. 485–489.

Jenner, K. A., S. R. Dallimore, I. D. Clark, D. Pare, and B. E. Medioli, 1999, Sedimentology of gas hydrate host strata from the JAPEX/JNOC/GSC Mallik 2L-38 gas hydrate research well, *in* S. R. Dallimore, T. Uchida, and T. S. Collett, eds., Scientific results from JAPEX/JNOC/GSC Mallik 2L-38 gas hydrate research well, Mackenzie Delta, Northwest Territories, Canada: Geological Survey of Canada Bulletin 544, p. 57–68.

Judge, A. S., and J. A Majorowicz,., 1992, Geothermal conditions for gas hydrate stability in the Beaufort-Mackenzie area: The global change aspect: Global and Planetary Change, v. 98, p. 251–263.

Judge, A. S., S. L. Smith, and J. Majorowicz, 1994, The current distribution and thermal stability of natural gas hydrates in the Canadian polar regions: Proceedings of the Fourth International Offshore and Polar Engineering Conference, p 307–313.

Kuldeep, C., D. K. Pandey, A. K. Pathak, and A. Sahu, 1997, Investment opportunities in deep waters of India: Proceedings of the Second International Petroleum Conference and Exhibition, Petrotech-97, p. 257–267.

Kvenvolden, K. A., 1988, Methane hydrate—A major reservoir of carbon in the shallow geosphere?: Chemical Geology, v. 71, p. 41–51.

Kvenvolden, K. A., 1993, Gas hydrates as a potential energy resource—A review of their methane content, *in* D. G. Howell, ed., The Future of energy gases: U.S. Geological Survey Professional Paper 1570, p. 555–561.

Kvenvolden, K. A., and L. A. Barnard, 1983, Hydrates of natural gas in continental margins, *in* J. S. Watkins and C. L. Drake, eds., Studies in continental margin geology: AAPG Memoir 34, p. 631–641.

Kvenvolden, K. A., and G. E. Claypool, 1988, Gas hydrates in oceanic sediment: U.S. Geological Survey Open-File Report 88-216, 50 p.

Kvenvolden, K. A., and M. Kastner, 1990, Gas hydrates of the Peruvian outer continental margin, *in* E. Suess, R. von Huene, et al., Proceedings, Ocean Drilling Program, Scientific Results, v. 112, p. 517–526.

Kvenvolden, K. A., and T .J. McDonald, 1985, Gas hydrates of the Middle America Trench, Deep Sea Drilling Project Leg 84, *in* R. von Huene, J. Aubouin, et al., Initial Reports Deep Sea Drilling Project, v. 84, p. 667– 682.

Lee, M. W., D. R. Hutchinson, W. P. Dillon, J. J. Miller, W.F. Agena, and B. A. Swift, 1993, Method of estimating the amount of in situ gas hydrates in deep marine sediments: Marine and Petroleum Geology, v. 10, p. 493–506.

MacDonald, G. T., 1990, The future of methane as an energy resource: Annual Review of Energy, v. 15, p. 53–83.

Makogon, Y .F., 1981, Hydrates of natural gas: Tulsa, Oklahoma, PennWell Publishing Company, 237 p.

Makogon, Y. F., F. A. Trebin, A. A. Trofimuk, V. P. Tsarev, and N. V. Cherskiy, 1972, Detection of a pool of natural gas in a solid (hydrate gas) state: Doklady Academy of Sciences U.S.S.R., Earth Science Section, v. 196, p. 197–200.

Masters, C. D., D. H. Root, and E. D. Attanasi, 1991, Resource constraints in petroleum production potential: Science, v. 253, p. 146–152.

McIver, R. D., 1981, Gas hydrates, *in* R. F. Meyer and J. C. Olson, eds., Long-term energy resources: Boston, Pitman Publishing, p. 713–726.

Meyer, R. F., 1981, Speculations on oil and gas resources in small fields and unconventional deposits, *in* R. F. Meyer and J. C. Olson, eds., Long-term energy resources: Boston, Pitman Publishing, p. 49–72.

National Petroleum Council, 1992, The potential for natural gas in the United States: National Petroleum Council, v. I and II, 520 p.

Ripmeester, J. A., J. S. Tse, C. I. Ratcliffe, and B. M. Powell, 1987, A new clathrate hydrate structure: Nature, v. 325, p. 135.

Shipboard Scientific Party, 1980, Site 533 and 534 (Leg 76), *in* R. E. Sheridan, et al., Proceedings, Deep Sea Drilling Project, Initial Reports, v. 76, p. 35–80.

Shipboard Scientific Party, 1985, Site 570 (Leg 84), *in* R. von Huene, et al., Proceedings, Deep Sea Drilling Project, Initial Reports, v. 67, p. 283–336.

Shipboard Scientific Party, 1986, Sites 614–624 (Leg 96), *in* A. H. Bouma, et al., Proceedings, Deep Sea Drilling Project, Initial Reports, v. 96, p. 3–424.

Shipboard Scientific Party, 1990, Site 796 (Leg 127), *in* K. Tamake, et al., Proceedings, Ocean Drilling Program, Initial Reports, v. 127, p. 247–322.

Shipboard Scientific Party, 1991, Site 808 (Leg 128), *in* A. Taira, et al., Proceedings, Ocean Drilling Program, Initial Reports, v. 131, p. 71–269.

Shipboard Scientific Party, 1994, Sites 892 and 889 (Leg 146), *in* G. K. Westbrook, et al., Proceedings, Ocean Drilling Program, Initial Reports, v. 146, p. 301–396.

Shipboard Scientific Party, 1996, Sites 994, 995, and 997 (Leg 164), *in* C. K. Paull, et al., Proceedings, Ocean Drilling Program, Initial Reports, v. 164, p. 99–623.

Sloan, E. D., 1998, Clathrate hydrates of natural gases (2nd ed.): New York, Marcel Dekker Inc., 705 p.

Smith, S. L., and A. S. Judge, 1993, Gas hydrate database for Canadian Arctic and selected east coast wells: Geological Survey of Canada Open File Report 2746, p. 1–7.

Spence, G. D., T. A. Minshull, and C. Fink, 1995, Seismic studies of methane gas hydrate, offshore Vancouver Island, *in* B. Carson, G. K. Westbrook, R. J. Musgrave, and E. Suess, eds., Proceedings of the Ocean Drilling Program, Scientific Results, v. 146, pt. 1, p. 163–174.

Trofimuk, A. A., N. V. Cherskiy, and V. P. Tsarev, 1977, The role of continental glaciation and hydrate formation on petroleum occurrences, *in* R. F. Meyer, ed., Future supply of nature-made petroleum and gas: New York, Pergamon Press, p. 919–926.

Tucholke, B. E., G. M. Bryan, and J. I. Ewing, 1977, Gas-hydrate horizons detected in seismic-profiler data from the western North Atlantic: AAPG Bulletin, v. 61, no. 5, p. 698–707.

Werner, M. R., 1987, Tertiary and Upper Cretaceous heavy oil sands, Kuparuk River area, Alaskan North Slope, *in* I. L. Tailleur and P. Weimer, eds., Alaskan North Slope geology: Bakersfield, California, Pacific Section, Society of Economic Paleontologists and Mineralogists and the Alaskan Geological Society, Book 50, v. 1, p. 109–118.

Yefremova, A. G., and B. P. Zhizhchenko, 1974, Occurrence of crystal hydrates of gas in sediments of modern marine basins: Doklady Akademii Nauk SSSR, v. 214, p. 1179–1181.

Skipper, K., 2001, Petroleum resources of Canada in the twenty-first century, *in* M. W. Downey, J. C. Threet, and W. A. Morgan, eds., Petroleum provinces of the twenty-first century: AAPG Memoir 74, p. 109–135.

Chapter 8

Petroleum Resources of Canada in the Twenty-first Century

Keith Skipper
Antrim Energy Inc., Calgary, Alberta, Canada

ABSTRACT

Remaining reserves of marketable crude oil and natural gas in Canada are more than 1.43 billion [10^9] m^3 (9 billion bbl) and 1.84 trillion [10^{12}] m^3 (65 trillion cubic feet [tcf]), respectively. These reserves enable current annual extraction rates of 127 million m^3 (800 million bbl) of oil and 170 billion m^3 (6 tcf) of natural gas, mainly from the mature Western Canada Sedimentary Basin. In the new millennium, expanded contributions to production capacity will come initially from the Mesozoic Jeanne d'Arc Basin (e.g., Hibernia and Terra Nova oil) offshore Newfoundland and from basins off Nova Scotia (e.g., Sable Island gas). In northern Alberta, additional investment in exploiting the Cretaceous oil sands will enhance the production of upgraded (synthetic) crude oil, bitumen, and heavy oil.

Notwithstanding the technical and commercial challenges, predictions of remaining exploitable resources in accessible areas exceed 5.6 trillion m^3 (200 tcf) of gas and 16 billion m^3 (100 billion bbl) of bitumen. In addition to oil sands, tight gas, and coal-bed methane in the Western Canada Sedimentary Basin, significant undeveloped resources are known in the remote Canadian Arctic islands (Sverdrup Basin), the Labrador shelf (gas), and the Beaufort Basin (gas and oil). Many of these resources will remain "orphaned," depending on environmental aspects, delivery costs, markets, and commodity prices. Current "stranded gas" in the Mackenzie Delta and the shallow offshore waters of the Beaufort Sea will be connected (via the Mackenzie Valley corridor) to the natural-gas pipeline grid serving Canadian and United States markets. Associated gas reserves (presently reinjected at Hibernia) in the Jeanne d'Arc Basin, if not connected to shore by pipeline, may be developed using either natural-gas-to-liquid conversion or compressed-gas transport technologies.

Canada's resource base is not in crisis, but the rate of conversion of the resource base to productive capacity cannot be as rapid as the resource potential might suggest.

INTRODUCTION

Canada is the world's third-largest producer of natural gas, with an 8% production share from 1.3% of the world's proven gas reserves, thirteenth-largest producer of crude oil, with 0.6% of the world's proven reserves, and has huge undeveloped future energy resources (BP, 1999; Energy Information Administration, 1999; World Energy Council, 1999; see also Canadian Association of Petroleum Producers, 1999a). To date, production in Canada has recovered more than 3.3 billion [10^9] m^3 (21 billion barrels [bbl]) of oil and 3.2 trillion [10^{12}] m^3 (114 trillion cubic feet [tcf]) of natural gas. Remaining reserves of marketable crude oil and natural gas ("booked reserves") are more than 1.4 billion m^3 (9 billion bbl) and 1.84 trillion m^3 (65 tcf), respectively. These reserves enable current annual extraction rates of 127 million m^3 (800 million bbl) of oil and 170 billion m^3 (6 tcf) of natural gas, mainly from the mature Western Canada Sedimentary Basin. Total liquids production, including conventional, synthetic (upgraded), heavy crudes, and natural-gas liquids (NGLs), reached a record 424,000 m^3/day or 2.67 million bbl per day in 1998 (Roberge, 1999). Undeveloped discovered resources, including "stranded gas" and static heavy-oil and bitumen resources (Imperial Oil, 1999a, b), are capable of sustaining greater extraction rates if economic hurdles (including availability of investment capital) and technical problems are continuously overcome. Canadian producers must compete with the industry's pursuit of high volumes of production at a low per-unit cost—an efficiency driven by volatile commodity prices.

Figure 1 shows the locations of the major sedimentary basins in Canada. Publications far too numerous to be cited here detail specific tectonic elements, stratigraphy, sedimentology, geochemistry, and resource endowment of these basins. Nevertheless, the references cited may allow the interested reader to wade into specific elements. Significant resources discovered to date in Canada are highlighted in Figure 2.

Multitudinous studies by government, industry, and individuals (see, for example, Young and Drummond, 1994) have been directed at the remaining petroleum-resource potential of Canada. These studies cover the mature producing basins (e.g., Western Canada Sedimentary Basin), Canada's new evolving areas of production (Atlantic margin, or "East Coast"), and its frontier areas (e.g., Mackenzie Delta and Beaufort Sea). Nevertheless, exploration activities and discoveries to enhance the understanding of the prospectivity in the Canadian frontiers, beyond those reported by Meneley (1986) and by Grant et al. (1986) in the inaugural Wallace E. Pratt Memorial Conference, have not occurred. Both papers remain excellent primers on which to build subsequent analysis. Indeed, the development and enhancement of early discoveries in the Canadian frontiers (including those in the Mackenzie Delta, Beaufort Sea, Arctic islands, and Grand Banks) have not progressed as much as visionaries in the early 1980s thought they might. Native people's issues, moratoriums, low prices, lack of available capital, better economic returns elsewhere in traditional basins (e.g., the Western Canada Sedimentary Basin), lack of clear Canadian federal and provincial government fiscal policies, and harsh environmental conditions, among other factors, have stalled investment.

Resource Numbers Quoted

Any discussion of ultimate resource endowments is based on many assumptions which are dynamic and continue to evolve with new petroleum discoveries, different perceptions of technology enhancements and applications, economic projections, fiscal tax and incentive

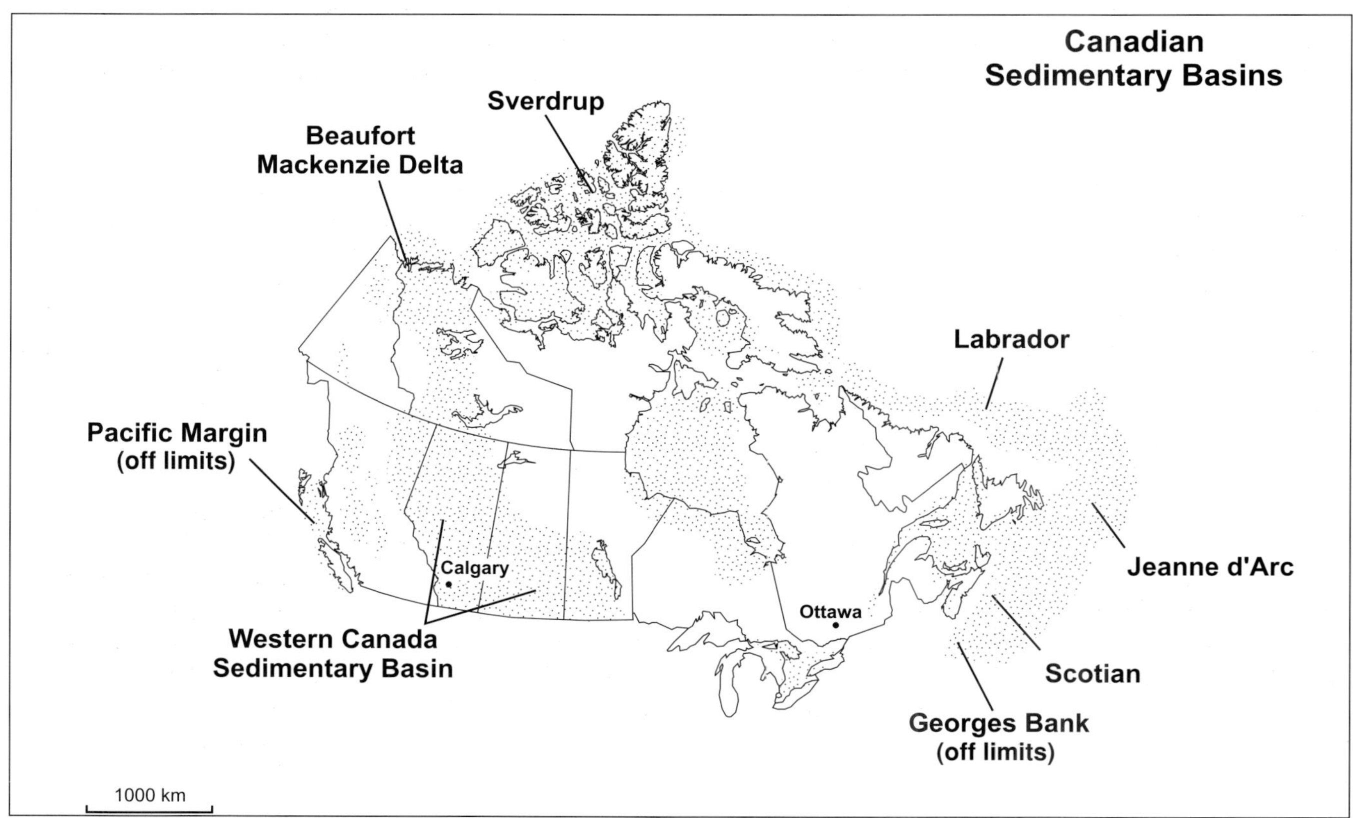

Figure 1. Sedimentary basins of Canada. Sverdrup is the major basin of interest in the Canadian Arctic islands. Labrador includes the Hopedale Basin. Jeanne d'Arc is the most important basin on the Grand Banks. The basins of the Pacific Margin and Georges Bank off eastern Canada are subject to existing drilling moratoriums.

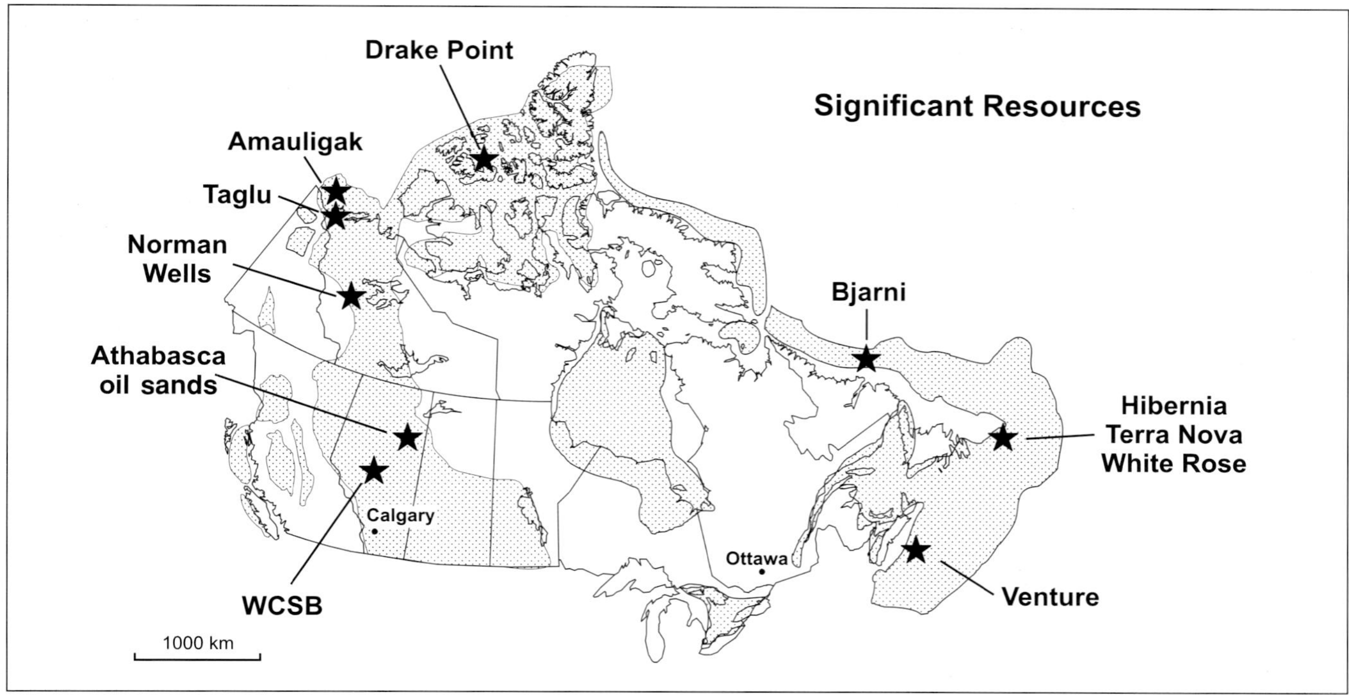

Figure 2. Location of significant resources of Canada as defined to date. Exploration activities have been conducted in all areas considered to offer potential for major discoveries. Drake Point (with 170 billion m^3, or 6 tcf, of gas) is the largest undeveloped gas discovery in Canada. Amauligak is the largest oil discovery in the Beaufort Sea. Taglu is the largest gas discovery on the Tuktoyaktuk Peninsula. The Western Canada Sedimentary Basin (WCSB) has produced most of Canada's petroleum resources to date and is the site of the Athabasca oil sands. Venture, the largest gas field on the Scotian Shelf, underpins the Sable Island gas development inaugurated in late 1999. Hibernia, in the Jeanne d'Arc Basin, is the first oil field to be developed on the Grand Banks. Bjarni is the largest gas discovery offshore Labrador.

regimes, and time of assessment. Figures quoted in this paper are from a variety of publications and Web sites, and caution is advised in the repeatability and application of these estimates. Figures assigned to Geological Survey of Canada publications frequently use probabilistic methods for estimating resource endowment; others may use Delphi approaches (Canadian Gas Potential Committee, 1997) or hybrid schemes. Estimates for some areas (e.g., Western Canada Sedimentary Basin) have a habit of increasing in magnitude over time (Drummond, 1995), but others become severely reduced as early optimism is tainted by realities of the producing environment, the commodity market, and pricing. In this paper, estimates are used wholesale as of reference date. They should be viewed as providing an order-of-magnitude estimate, as developed by informed scientists and practitioners. The figures may be useful in global analyses, comparisons, and strategic planning.

ORIGINAL IN-PLACE RESOURCES IN CANADA

The primary source document for the present status of petroleum resource assessment in Canada is a report by the National Energy Board on energy supply and demand to 2025 (National Energy Board, 1999a). For crude oil, this report also updates the compilation by Lee (1998). The latest assessment of natural-gas resources (estimated to year-end 1993) is by the Canadian Gas Potential Committee (1997), whose report is under revision. An earlier overall assessment by the Geological Survey of Canada was by Proctor et al. (1984).

Conventional Crude-oil Resources

Original oil in place in Canada is estimated at 34.4 billion m^3 (216 billion bbl), of which 9.2 billion m^3 (58 billion bbl), or 27%, is estimated to be ultimately recoverable and marketable (National Energy Board, 1999a). The figure for recoverable oil compares with an estimated 13 billion m^3 (85 billion bbl) in 1973 (McCrossan and Porter, in McCrossan, 1973, p. 704) and a speculative estimate of 9.0 billion m^3 (56 billion bbl) by Proctor et al. (1984, p. 53).

Unconventional Crude-oil Resources

Unconventional resources of crude oil consist entirely of bitumen, estimated to be 400 billion m^3 (2.5 trillion bbl), of which 49 billion m^3 (300 billion bbl), located in the Western Canada Sedimentary Basin (Figure 3), is considered to be ultimately recoverable (National Energy Board, 1999a).

Conventional Natural-gas Resources

Ultimate recoverable and marketable resources of natural gas are estimated to be in the range of 18.7 to 20.7 trillion m^3 (660–730 tcf) (National Energy Board, 1999a), depending on price. This figure compares with a 1973 estimate of 16.3 trillion m^3 (577 tcf) by McCrossan and Porter (in McCrossan, 1973, p. 704) and to an estimate of 18.2 trillion m^3 (642 tcf) by Proctor et al. (1984). However, 45–50%, or 8.5 trillion m^3 (303 tcf), is ascribed to frontier areas, which include the Arctic islands, Labrador, and the Mackenzie Delta/Beaufort Sea (see also Young and Drummond, 1994; Drummond, 1998). Because frontier resources are high cost, remote, or located in inaccessible areas which would "have no greater opportunities for exploitation than unconventional natural gas resources in the Western Canada Sedimentary Basin" (Young and Drummond, 1994, p. 10), some should be deleted from the potential marketable resource base. The Canadian Gas Potential Committee (1997) concluded on an undiscovered conventional marketable gas estimate of 3.4 trillion m^3 (122 tcf) for the Western Canada Sedimentary Basin and 1.8 trillion m^3 (63 tcf) for the frontier areas. Remaining marketable conventional gas resources are quoted as 5.2 trillion m^3 (185 tcf) in producing areas and 3 trillion m^3 (107 tcf) in nonproducing areas (Canadian Gas Potential Committee, 1997; its Tables 1-A and 1-B, p. 1). Of this total, 866 billion m^3 (30.4 tcf) was assigned to Labrador and the Sverdrup Basin; these are unlikely to be accessed in any current forecast span. The largest potential is approximately 1.6 trillion m^3 (56 tcf) in the Beaufort Basin.

Unconventional Natural-gas Resources

Unconventional gas resources in Canada include coal-bed methane (CBM), tight-gas reservoirs, shale gas, and hydrates. Comments are made here only about coal-bed methane and tight-gas reservoirs.

Great uncertainty exists as to the appropriate estimate of in-place resources of combustible natural gas (methane) associated with coal beds. Canada, particularly the Western Canada Sedimentary Basin, has vast quantities of coal measures which may offer the opportunity for methane desorption and production (Dawson, 1995). Ranges for this resource are highly variable and speculative. CBM resources of the Cretaceous upper Mannville in Alberta have been estimated at more than 4.5 trillion m^3 (160 tcf) (Langenberg et al., 1997). The Canadian Gas Potential Committee (1997) estimated recoverable resources at 135–261 tcf. Nevertheless, the National Energy Board (1999a) has assumed that a resource of 2.1 trillion m^3 (75 tcf) may be recoverable, but Young and Drummond (1994) limit the figure to 570 billion m^3 (20 tcf). If the technology works and investment is available, the recoverable CBM resource may be large (perhaps more than 2.8 trillion m^3, or 100 tcf). Further analysis and pilot projects are required to ascertain the viability of such production. CBM extraction is notorious for being site or field specific, and further work is required in Canada to justify present assumptions and projections.

However, projections on the supply of gas produced in the Western Canada Sedimentary Basin require and assume production rates from CBM of more than 283 million m^3/day (10 bcf/day) by 2025 (National Energy Board, 1999a). Tardy development of CBM production (or alternative sources of gas supply) therefore implies shortfalls against a planning scenario requiring such supply. The early development of Mackenzie Delta conventional natural gas may delay the need for early CBM or other unconventional natural-gas production, however.

Tight-gas reservoirs, particularly tight or very low permeability gas sands, are quite prevalent in the Western Canada Sedimentary Basin and have been discussed by Masters (1984) as "Deep Basin" resources. Estimates vary, but the resource may be large—175 to 3500 tcf (Canadian Gas Potential Committee, 1997, p. 3).

Production from unconventional gas resources is particularly responsive to changes in assumed levels of technological progress and commodity prices. Tax (fiscal) incentives may accelerate industry interest and investment. None of the resources discussed above is considered to be economically recoverable at this time.

PRODUCTION OF PETROLEUM IN CANADA

Cumulative Production

Estimates of cumulative production to year-end 1999 of conventional crude oil are 3 billion m^3 (18.5 billion bbl) and 397 million m^3 (2.5 billion bbl) of liquid petroleum products from oil sands. Total production of marketable natural gas is estimated at 3.2 trillion m^3 (114 tcf).

2000 Production

According to 2000 figures, Canada produces 127 million m^3 (800 million bbl) of crude oil annually, or 350,000 m^3/day (2.2 million bbl/day) (including conventional light and heavy, synthetic, bitumen, and pentanes plus), and 178 billion m^3 (6.3 tcf) of natural gas annually or 467 million m^3/day (16.5 bcf/day) per day. Exports, mainly to the United States, are approximately 85 billion m^3 (3.0 tcf) of gas annually and 222,000 m^3/day (1.4 million bbl/day) of crude oil (Canadian Association of Petroleum Producers, 1999a).

Conventional Production

Canada produces approximately 253,000 m^3/day (1.6 million bbl/day) of conventional crude oil (which includes both light and heavy conventional crude oil and condensate) (Canadian Association of Petroleum Producers, 1999b). The majority of this production comes from the Western Canada Sedimentary Basin, mainly Alberta, with minor contributions from Ontario.

Frontier Production

The Hibernia crude-oil project, located offshore Newfoundland and Canada's first large-scale offshore development, came onstream in late 1997. At year-end 1999, Hibernia was producing 23,800 m^3/day (150,000 bbl/day) of conventional light crude oil. Cohasset/Panuke field, located offshore Nova Scotia, has been in production since 1992. It was the first commercial oil development in the Atlantic Canada offshore. Cumulative production is approximately 7.3 million m^3, or 46 million bbl (Newfoundland Ocean Industries Association, 1998, section 2.3).

Production of natural gas began in November 1999 from the Sable Island project, offshore Nova Scotia. Initial rates of gas production are expected to reach 500 million cubic feet per day (mmcf/day).

In 1999, modest gas production began from Inuvialuit Petroleum Corporation's Ikhail gas development (396 million m^3, or 14 bcf of recoverable sales gas) in the Mackenzie Delta, delivering gas at a rate of 2 mmcf/day to Inuviuk (Northern Oil and Gas Directorate, 1999).

Oil Sands and Nonconventional Production

Today, 14% of the total annual oil production in Canada, or 51,587 m^3/day (325,000 bbl/day), is synthetic (or, more nearly correctly, "upgraded") crude oil produced from two large integrated oil-sands mining operations. A further 13% of annual production, or 44,000 m^3/day (280,000 bbl/day), is sold directly as bitumen (Figure 4). The production of heavy oil, bitumen, and upgraded crude accounts for 36% of current Canadian liquids supply.

CANADIAN RESERVES

Crude Oil

At year-end 1999, remaining reserves of conventional crude oil and equivalent were estimated at 714 million m^3 (4.5 billion bbl). Natural-gas liquids reserves were estimated at 103 million m^3 (650 million bbl). Remaining reserves of developed oil sands, both integrated mining and in-situ bitumen, were estimated at 635 million m^3 (4.0 billion bbl).

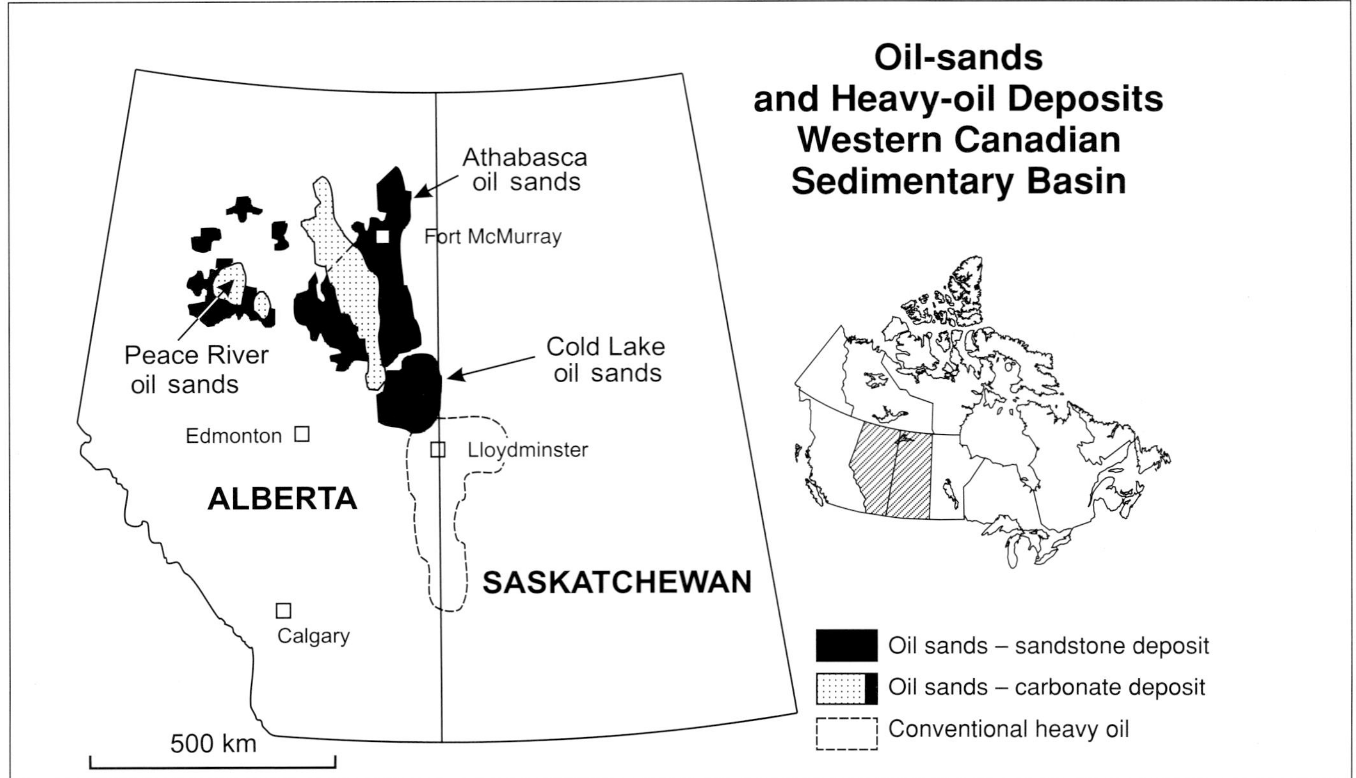

Figure 3. Location of oil-sands and heavy-oil deposits in the Western Canada Sedimentary Basin. Lloydminster is the location of major upgrading facilities. Fort McMurray is the location of all the integrated mining plants (e.g., Syncrude and Suncor).

Natural Gas

At year-end 1999, reserves of marketable natural gas were estimated at 1.84 trillion m^3 (65 tcf), consisting of 1.6 trillion m^3 (56.5 tcf), or 87% of the total, in the Western Canada Sedimentary Basin.

GAS EXPORT CAPACITY

Canada is a major exporter of natural gas to the U.S. market. Indeed, natural gas is truly a North American commodity. Present export capacity (as measured by cross-border pipeline capacity) is approximately 339 million m^3/day, or 12 bcf/day. With the completion of all currently planned facilities, including the Alliance Pipeline (see below), capacity of 452 million m^3/day, or 16 bcf/day, would be available. By 2000, export capacity from Canada was about 42% of the export capacity from the U.S. Southwest (Gulf Coast) (Energy Information Administration, 1999). Whether they will be profitable or the gas supply will be adequate to fill these pipelines is a critical issue.

RECENT AND CURRENT PROJECTS

Recent "high-impact" conventional-production developments have centered on discoveries made in the late 1960s and 1970s. The industry continues to struggle with severe operating conditions, attempts to lower its cost base, and marginal full-cycle economics. This is likely to continue to challenge investment planning and the timing of investment.

East Coast

On the East Coast (i.e., Atlantic Canada), several major projects have recently begun production, are under development, or are in advanced stages of prebuilding (see Government of Newfoundland and Labrador, 1999). These include the Hibernia oil field, in the Jeanne d'Arc Basin on the Grand Banks off Newfoundland; the Sable Offshore Energy Project, being developed to deliver natural gas from six gas fields offshore Nova Scotia (see Sable Offshore Energy Incorporated, 1999); and the Terra Nova oil-field development, also in the Jeanne d'Arc Basin. Drilling operations at White Rose field (gas and oil) in 1999 has also resulted in the filing of a plan to develop an estimated 36 million m^3 (230 million bbl) of recoverable oil.

Hibernia

The Hibernia oil field, discovered in 1979, is in 80 m (262 ft) of water, 315 km (195 mi) east-southeast of Saint John's, Newfoundland (Mackay and Tankard, 1990). It is the first development of a gravity-based platform designed to withstand direct impact with an iceberg. Cost for development has been quoted at Canadian $5.8 billion. Total oil in place is estimated at 476 million m^3 (3 billion bbl), with more than 97.6 million m^3 (615 million bbl) recoverable. Production began in late 1997 and has been building since then to a plateau of 23,800 m^3/day (150,000 bbl/d). Hibernia crude oil is 32–34° API, with low sulfur (Hibernia, 1999a). Its production is offloaded from on-site storage to shuttle tankers; it is not piped to shore because of difficulties and risk from ice scouring the seabed. The development of satellite fields, which require tieback to Hibernia later in its field life, will need buried umbilicals to drain them.

Sable Offshore Energy Project

The Sable Offshore Energy Project consists of six fields: Venture, South Venture, Thebaud, North Triumph, Glenelg, and Alma. These fields are 225 km (140 mi) offshore Nova Scotia near Sable Island. Cost of the first-phase development is Canadian $2.0 billion (Owens, 1999). The project has recoverable reserves of 100 billion m^3 (3.5 tcf). Raw gas production is estimated to begin at 500 mmcf/day, with 20,000 bbl/day of natural-gas liquids (November 1999; see Sable Offshore Energy Incorporated, 1999, and Owens, 1999).

Terra Nova

Terra Nova field, discovered in 1984, is 350 km (220 mi) east-southeast of Saint John's, Newfoundland, in 95 m (312 ft) of water, and about 35 km (22 mi) east of Hibernia. Unlike Hibernia, it is being developed via a floating production storage and offloading (FPSO) facility. Total recoverable reserves are estimated at 59 million m^3 (370 million bbl) but could reach 92 million m^3 (580 million bbl). Initial oil production is targeted for late 2001, beginning at 18,200 m^3/day (115,000 bbl/day) (Terra Nova, 1999).

White Rose

Husky Oil Ltd. drilled three successful wells in 1999 in the White Rose fields. Recoverable hydrocarbon volumes may be of 36 million m^3 (230 million bbl) of oil and 42 billion m^3 (1.5 tcf) of gas in South White Rose field. North White Rose field has 150 million bbl recoverable of natural-gas liquids and 2 tcf of gas. Initial production is expected in 2004 (Bruce et al., 1999).

Western Canada

In western Canada, major projects include additional pipeline construction to enhance the capability of delivering more natural gas to Midwest U.S. markets. Major investments are also being made in expanding existing oil-sands mining operations and the construction of new mines.

The Alliance Pipeline

The Alliance pipeline (Alliance, 1999) is a major pipeline (U.S. $3.0 billion) being constructed to move natural gas from northeastern British Columbia to the

Chicago, Illinois, area market, where it will connect into the North American gas-pipeline grid. Total distance will be 2990 km (1858 mi), the longest pipeline ever built in North America. Initial throughput will be 37.5 million m^3/day (1.325 bcf/day). The pipeline was commissioned in late 2000.

Oil-sands Upgrades, Expansion, and New Mines

Suncor and Syncrude, the two existing miners of oil sands, are expanding their present mining and upgrading operations. Suncor's Millennium project will expand its oil production to 35,000 m^3/day (220,000 bbl/day), from the current 16,600 m^3/day (105,000 bbl/day), by 2002 (Suncor, 1999). Reserve life at Suncor's new Steepbank mine is 35 years. Syncrude is extracting from its new North Mine, constructing the Aurora mine, and expects to lift its oil production from its present 35,000 m^3/day (220,000 bbl/day) to more than 76,000 m^3/day (480,000 bbl/day) by 2007, or 23.8 million m^3/year (150 million bbl/year), at a cost of Canadian $6.0 billion (Syncrude, 1999). Shell also has plans to develop the Muskeg mine, with capacity for 23,800 m^3/day (150,000 bbl/day) of oil in 2002, at a price tag of Canadian $3.4 billion. All these developments will mean that production targets of more than 80,000 m^3/day (500,000 bbl/day) may be attainable by 2005. These projects are capital intensive and are the foundation for additional supplies of Canadian petroleum to reach markets in the twenty-first century.

CANADIAN SEDIMENTARY BASINS AND PETROLEUM SYSTEMS

Canada has more than 6.5 million km^2 (2.5 million mi^2) of prospective sedimentary section in 38 unmetamorphosed basins (McCrossan, 1973, p. 4), with approximately 60% of this area onshore.

The most important and familiar basins to petroleum geologists are the Western Canada Sedimentary, Jeanne d'Arc, Scotian, Mackenzie Delta, Beaufort, and Sverdrup (Arctic islands) Basins. (Figure 1). The geographic locations of the frontier basins of Mackenzie, Beaufort, Sverdrup, Labrador, and Jeanne d'Arc have required innovative and expensive developments of drilling capabilities to evaluate them. Environments are harsh, and the costs of exploitation remain prohibitive for some. All favored basins are reasonably well known in terms of their potential petroleum endowment, and all have been the subject of exhaustive studies. Data available on all the frontier basins are arguably adequate to define the scale and nature of the resource. Substantial work has focused on locating and analyzing favorable "world-class" source rocks, and some areas worthy of potential wildcat evaluation have been suggested. However, these areas are unlikely to attract industry investment until other opportunities (both North American and foreign) appear remote and less valuable.

Western Canada Sedimentary Basin

The Western Canada Sedimentary Basin (WCSB; province code 5243 in U.S. Geological Survey, 1997), also known as the Alberta Basin, occupies an area of 1.4 million km^2 (540,000 mi^2) of southwestern Manitoba, southern Saskatchewan, Alberta, northeastern British Columbia, and the southwestern corner of the Northwest and Yukon Territories (see Mackenzie Valley, below). Petroleum resources of the basin range from natural gas and conventional light oil to bitumen (a good summary is provided by Johnson and McMillan, 1993). The basin is situated between the deformed Cordilleran belt to the west and the Precambrian (or Canadian) shield to the east (Ricketts, 1989; Barclay and Smith, 1992; Mossop and Shetsoen, 1994). The Western Canada Sedimentary Basin has been responsible for more than 90% of the 21 billion bbl of crude oil produced in Canada since 1856.

It is only in the Western Canada Sedimentary Basin that the petroleum industry comes close to a "just-in-time" inventory business, with operators able to shift rapidly from drilling for oil to gas (as occurred in 1998) and vice versa.

A total of 10 marine petroleum source rocks, ranging from Middle Devonian to Upper Cretaceous, exists in the basin. Principal hydrocarbon reservoirs occur in the Lower Cretaceous Mannville Group and subcropping Paleozoic carbonates (Creaney and Allan, 1990), including Devonian reefs (Geological Atlas of Western Canada Sedimentary Basin, 1998).

Natural Gas

Characteristics of natural-gas production from the Western Canada Sedimentary Basin include:

- Alberta's production of 5.0 tcf/year accounts for 80% of total Canadian natural-gas production.
- Recoverable reserves per well are small (Samson, 1999; Samson and Kirsch, 1999).
- The annual production rate per well declines rapidly.
- As of 1999, pipeline capacity exceeds supply.
- A high rate or intensity of drilling each year is necessary to replace produced reserves and to capture markets.
- Midstream rationalization brings greater efficiency into the use of plants, pipelines, and facilities.
- There is a gradual "infrastructure creep" into the northernmost reaches of the basin (beyond latitude 60°N).

Commensurate with the maturity of the exploration in the basin, median gas reserves per discovery well today are less than 28.3 million m^3 (1 bcf). Statistical analysis by K. J. Drummond (personal communication, 1999) indicates that this pattern is unlikely to change any time soon. Only three gas fields discovered in the last 10 years have been larger than 1.5 billion m^3 (52 bcf) when booked back to their discovery date (Samson and Kirsch, 1999). Of a quoted 4.1 trillion m^3 (148 tcf) of marketable natural gas, 82% was discovered before 1980 (Drummond, 1995), with the average field size being 161 million m^3 (5.7 bcf). Since 1980, the average field size has been slightly more than 59 million m^3, or 2.1 bcf (Young and Drummond, 1994, p. 17).

The high rates of natural-gas extraction from the basin are underpinned by 45,000 currently connected and producing wells. Gas-production rates from these wells (1997 figures) will decline from 430 million m^3/day (15.3 bcf/day) to approximately 225 million m^3/day (8.0 bcf/day) in 2001, and will contribute less than 50% of total deliverability in 2001. Wells connected since 1994 contribute about 50% of current Western Canada Sedimentary Basin production (National Energy Board, 1999b).

In 1997, the initial capability of the average producing gas well was 35.7 million m^3/day (1.26 mmcf/day) (Samson, 1999). Current rates of drilling must be increased to meet future demand (National Energy Board, 1999b, p. 21). Studies indicate rapid decline rates for a majority of wells, with a 95% probability of any new well being depleted and abandoned within four years of connection. Considerable potential for gas remains in the basin, however, with discovery trends implying that the approach of the asymptotic limit is some time off (K. J. Drummond, personal communication, 1999). To date, the basin is recognized as containing remaining identified marketable gas reserves of 1.7 trillion m^3 (60 tcf), with an ultimate potential of approximately 5.6 trillion m^3 (200 tcf), of which more than 2.8 trillion m^3 (100 tcf) has been produced. Depending on gas pricing, mean estimates of economically recoverable conventional undiscovered natural-gas resources for the basin range from 3.9 trillion m^3 (140 tcf) at a gas-plant gate price of Canadian $3.00/mcf to 849 billion m^3 (30 tcf) at a gas-plant gate price of Canadian $1.00/mcf (Conn et al., 1995).

There appears to be little in the way of new geologic concepts which will enhance these numbers, irrespective of wellhead gas prices. The basin is clearly in the drill-out phase. Gas wells connected since 1988 show decline rates of about 35% per year (National Energy Board, 1999b). However, the potential for significant contributions to Canada's natural gas remains well into the new millennium and, with midstream and other producing efficiencies, significant economic rewards await lowest-cost producers. Potential will be found in thousands of relatively small gas fields. To maintain production and reserve-to-production-cover ratio, more than 5000 gas-well completions will be required in the near term (National Energy Board, 1999b). Concerns do exist about the capacity of the Western Canada Sedimentary Basin to deliver gas in the volumes required through the next two decades, but National Energy Board consultations with industry indicate that gas from unconventional resources (such as coal-bed methane) may attain 283 million m^3/day (10 bcf/day) by 2025. Even so, some concern could be expressed about the sustainability of production at these rates. Nonconventional production would then constitute 37% of total estimated production. Conventional gas production is expected to reach its maximum in 2008.

The remaining big potential in the Western Canada Sedimentary Basin will be in the foothills and deep Devonian plays. These areas have high potential but also higher costs. Statistics suggest that 95% of all future discoveries will be fields of less than 283 million m^3 (10 bcf).

With the major decrease in field size, the intensity of exploration activity and the finding and connecting rate will impact deliverability of new gas supplies. Better prices and netbacks to industry will promote deeper drilling, and improved technologies will enhance production. However, projections made about the need to increase deep drilling in the deeper parts of the Western Canada Sedimentary Basin and the foothills imply that industry has been complacent both in the application of technology (e.g., portable 3-D) and in defining drilling targets in those areas. This may not be the case. Major players have shunned shallow gas drilling because of high overheads, low flow rates, and small additions to their reserve bases (low reserve-to-production ratios). The industry has kept up the pace to seek large prizes in the deeper portions of the Western Canada Sedimentary Basin.

Light Oil

Major characteristics of light-oil production in the Western Canada Sedimentary basin include:

- The absolute magnitude of annual production has been in decline for several years. The trend appears to be irreversible.
- The average oil well today produces less than 7.9 m^3/day (50 bbl/day) of oil. (The U. S. average in 1998 was 1.8 m^3/day or 11.5 bbl/day.)
- Light-oil production in the Western Canada Sedimentary Basin has become "noncore" to the major companies.

The conventional crude-oil resource base for the Western Canada Sedimentary Basin is about 18 billion m^3 (113 billion bbl) of oil in place, consisting of about 11.5 billion m^3 (72.5 billion bbl) of light oil and 6.5 billion m^3 (41 billion bbl) of heavy oil (National Energy Board, 1999a). Remaining established reserves of light oil are approximately 338 million m^3. Ultimate recoverable reserves are about 3.6 billion m^3 (22 billion bbl) of light oil and 1.3 billion m^3 (8 billion bbl) of heavy oil. Undiscovered recoverable resources amounted to approximately 666 million m^3 (4.2 billion bbl) of light oil and 230 million m^3 (1.45 billion bbl) of heavy oil (see also Bowers and Drummond, 1997).

Light-crude production, presently at 132,000 m^3/day (830,000 bbl/day), has been in decline for a considerable time. Irrespective of new seismic and drilling technologies, that decline will not be arrested. Many of Alberta's traditional light-oil producing fields produce with high water cuts and are tightly infilled with wells. Bowers et al. (1995) suggest that horizontal drilling could increase recovery and incremental supply in the basin on the order of 450 million m^3, or 2.8 billion bbl).

Heavy Oil

Most heavy-oil production occurs from deposits in eastern Alberta and central Saskatchewan, around Lloydminster. These deposits occur downdip of and are the southern extension of the bituminous sands deposits of northern Alberta. The oils are typically lighter than the bitumen, and some can be produced by conventional methods (e.g., with screw pumps), although recovery may be lower than those from more conventional light-oil fields (see, for example, Johnson and McMillan, 1993, p. 553–555). The reservoirs are complex multiple sand units of the Cretaceous Upper Mannville (e.g., Sparkey) at depths of 550 m (1800 ft). Porosity can be high (as much as 30%), with high permeability (as much as several darcys).

Recent heavy-oil production, although affected by the price collapse of 1998 and volatility in the "price differential" in early 1999, amounts to 88,000 m^3/day (550,000 bbl/day).

The National Energy Board (1999a) reported cumulative production to year-end 1997 of 536 million m^3 (3.3 billion bbl), with remaining established reserves and those attributed to future enhanced recovery (presumably including screw-pump technology) of 520 million m^3 (3.3 billion bbl). Ultimate recoverable resources are 1.285 billion m^3 (8.1 billion bbl), of which approximately 42% has been produced. Original oil-in-place figures are 6.575 billion m^3 (41 billion bbl), of which 19% is considered recoverable. Future demand for heavy oil (and bitumen) is assured for refinery feedstock (see Figure 4, after Imperial Oil, 1999a).

Oil Sands

In 2000, total production from oil sands was nearly 600,000 bbl/day (Figure 5), representing 27% of total Canadian crude-oil production. Because light-crude production from the Western Canada Sedimentary Basin is under rapid decline, its contribution and importance to supply are becoming of less importance. Heavy-crude extraction and the production of upgraded crude from the mining of oil sands are increasing rapidly (see the excellent summary by Singh et al., 1999). This is because the costs of producing upgraded crude oil from oil sands are now competitive with full-cycle conventional oil-production costs, particularly in North America. The resource is also essentially limitless and does not deplete, for all practical purposes, and there is a ready market.

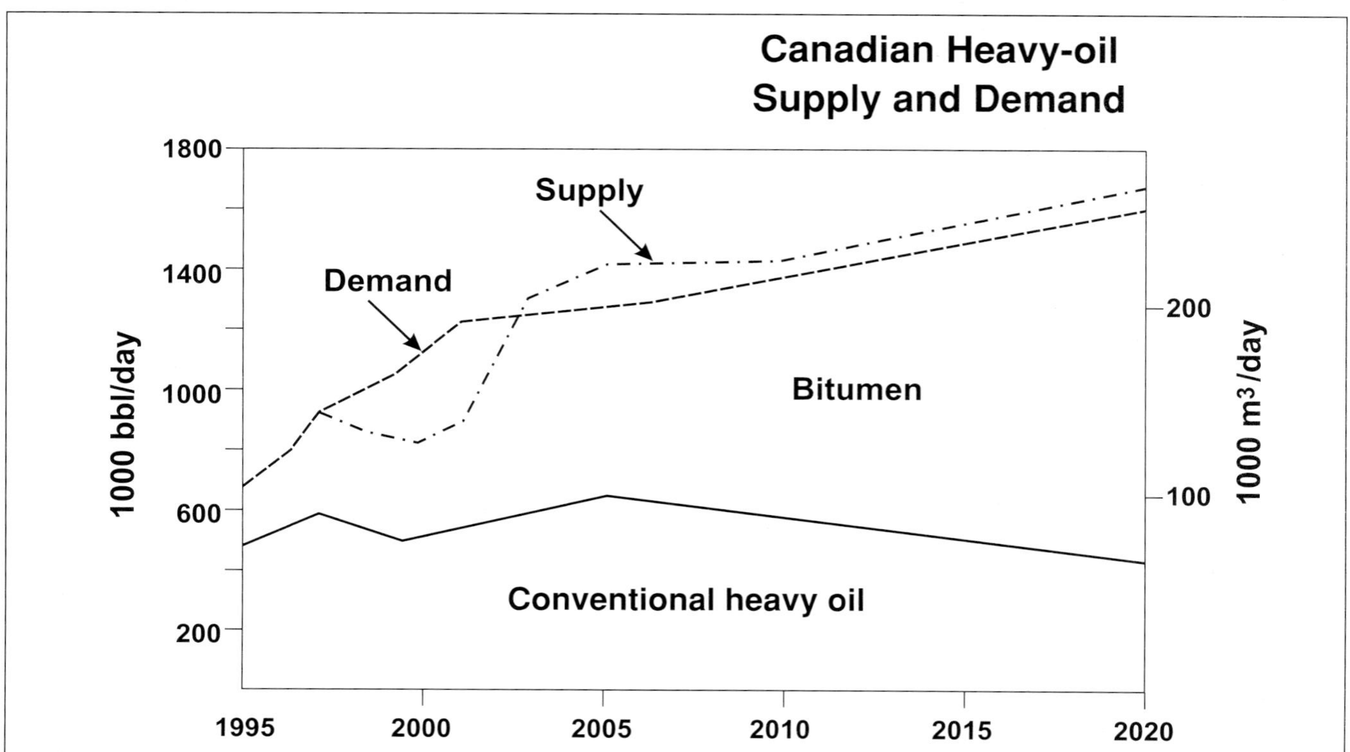

Figure 4. Canadian heavy-oil supply and demand (after Dingle, 1999a).

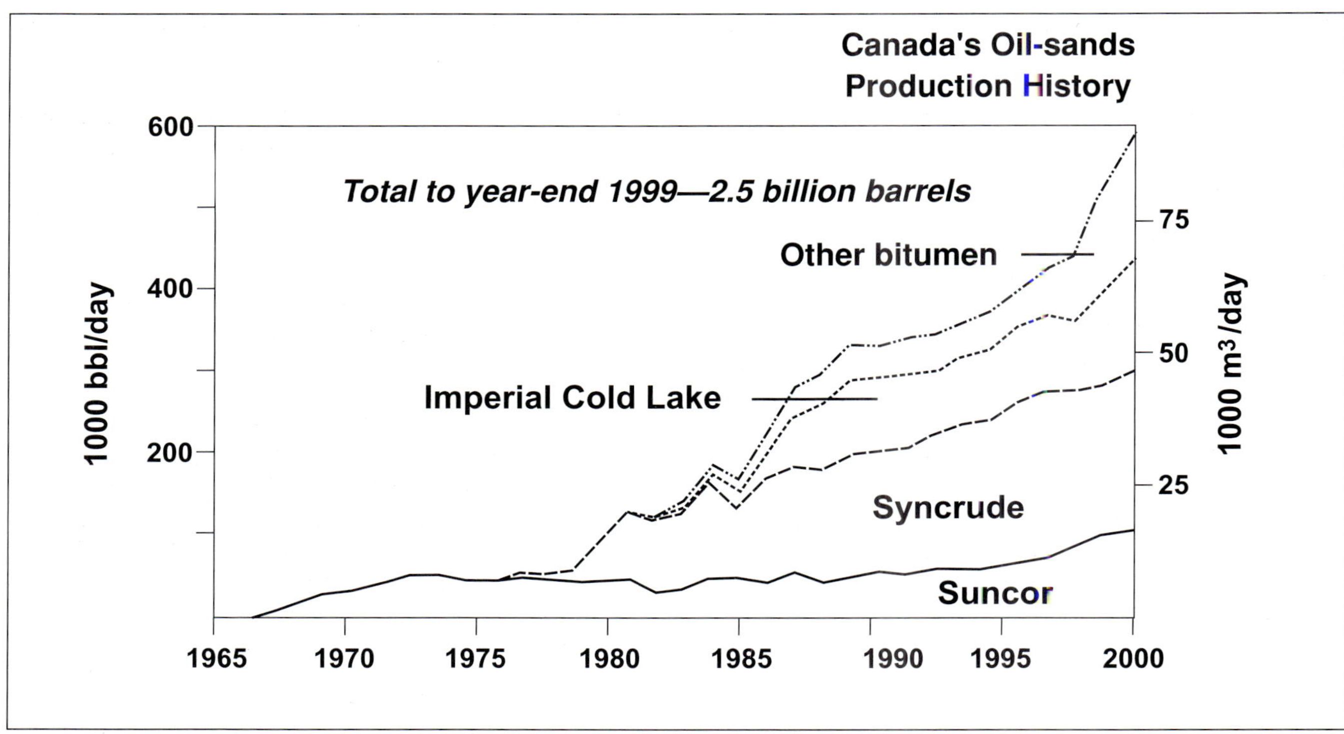

Figure 5. Canada's oil-sands production history.

Major oil-sands deposits underlie about 77,000 km^2 (29,730 mi^2) of the province of Alberta (Figure 3). They occur in Cretaceous siliciclastics and Devonian carbonates (Hills, 1974; Ranger and Pemberton, 1997; Stobl et al., 1997). Original bitumen in-place resources are reported as 1700 billion bbl, or 268 million m^3 (Imperial Oil, 1999a), of which about 9% (24 billion m^3, or 152 million bbl) is amenable to surface mining and the remainder to in-situ thermal stimulation processes at depth. The bitumen extracted from oil sands is generally less than 12° API, with reservoir bitumen saturation levels of 1% to 18%. Upgraded product, for example, Syncrude Synthetic Sweet blend, is 31°–33° API crude containing less than 0.2% sulfur and is comparable to Saudi Light. It sells for a premium to West Texas Intermediate.

The origin of the oil sands/heavy oils from most of the major Cretaceous deposits and the underlying carbonate trend has been the subject of much debate and geochemical analysis (Brooks et al., 1988). To date, conclusions are that these oils originated from a mature conventional oil generated by an unknown source. In any event, the oil migrated over a vast distance, suffering biodegradation in place and during migration.

The deposits occur in three major areas: Athabasca, Peace River, and Cold Lake (Figure 3). The Athabasca deposit accounts for 76% of the total oil in place in the oil-sands deposits, and the Peace River and Cold Lake deposits account equally for the remainder.

Athabasca.—At Athabasca, the bitumen-bearing deposits are at mining depths (overburden less than 75 m). These areas are the location of two current mining operations (Suncor and Syncrude), with others in an advance-planning stage (Shell's Muskeg). The mining involves stripping the overburden, which is saved for reclamation, and excavating the bitumen by mechanical shovel and truck or by dredging. Where overburden thickness is prohibitive for surface mining, in-situ thermal mechanisms for the recovery of bitumen from oil-sands deposits are attempted. To date, these oil sands have produced more than 396 million m^3 (2.5 billion bbl) (Figure 5).

The Upper Devonian Grosmont Formation is juxtaposed to the Athabasca tar-sand deposit in northeast Alberta. The Grosmont Formation contains approximately 50 billion m^3 (300 billion bbl) of 7° API bitumen and is a giant hydrocarbon deposit. Except for three pilot sites that were operated for a few years in the 1980s, the Grosmont reservoir has not been exploited. At present, the Grosmont is under consideration for steam and/or CO_2 injection (Luo et al., 1994). If large sinkholes or caverns filled with bitumen were ever detected in the Grosmont, these would make interesting recovery points.

The Lower Cretaceous McMurray/Wabiskaw stratigraphic interval contains approximately 143 billion m^3 (902 billion bbl) of bitumen in the Athabasca oil-sands area, north of Fort McMurray in northeastern Alberta. Of this, about 24 billion m^3 (152 billion bbl) occurs at minable depths (overburden less than 15 m or 50 ft) (Wightman and Pemberton, 1997; see also Flach, 1984).

Two major mining enterprises, Suncor (formerly Great Canadian Oil Sands, the world's first integrated oil-sands mining and upgrading plant, which began in 1967; see Suncor, 1999) and Syncrude Canada, are actively mining and upgrading the local bitumen near Fort McMurray. Syncrude Canada, onstream since 1978, is the largest operator. It currently produces 12% of Canada's light crude oil and manufactures more than 34,900 m^3/day (220,000 bbl/day) of upgraded or synthetic crude. In 1998, it produced its billionth barrel. Suncor produces approximately 16,600 m^3/day (105,000 bbl/day) of oil on an annualized basis.

Since 1986, significant progress has been made in reducing the cost of extraction, such that today, the Syncrude and Suncor mines have production and operating costs of approximately Canadian $12/bbl (U.S. $ 8.00/bbl) (Figure 6). Changes in mining machinery, with large draglines replaced by mobile truck and shovel operations, are allowing more confidence in large-scale investment in additional oil-sands plants. About 2 metric tons of oil sands is required to produce 1 bbl of light sweet synthetic (upgraded) crude oil.

Additional investment is being made in upgrader capability. It is projected that more than 50% of Canadian liquids production will come from oil sands by 2015. Issues which affect the mining of oil sands include environmental reclamation, particularly of tailings ponds; an adequate supply of diluent; the price of natural gas used as fuel; and the massive front-end investments required to take advantage of economies of scale. The product produced is not a major hurdle. Today, Syncrude makes nine products and can supply tailor-made feedstock for refineries.

Peace River.—The Peace River deposit comprises bitumen-rich sands from the Cretaceous Aptian-Albian Gething Formation, Ostracode Zone, and the Bluesky Formation, which overlie Paleozoic and older Mesozoic strata. Exploitable reserves at Peace River are contained predominantly within estuarine sands, 15–20 m (50–65 ft) thick, of the Bluesky Formation. Where combined with the underlying fluvial sands of the Ostracode Zone, net pay can reach 30 m (100 ft) in thickness. These sands occur at a depth of 550 m (1800 ft), and production of the bitumen relies on nonconventional methods. Heavy-oil (>10° API) resources in place in the Bluesky-Gething interval are estimated at almost 14 billion m^3 (90 billion bbl). Approximately 2.2 billion m^3 (13.8 billion bbl) is considered to be recoverable (Singh at al., 1999). Cyclic steam stimulation (see below) and steam-assisted gravity drainage (see below) are the usual methods to extract the bitumen, which reaches saturations as high as 88% in sandstones with porosity to 28% (Hubbard et al., 1999).

Cold Lake.—The Cold Lake oil-sands deposits occur in the Lower Cretaceous Mannville Group covering 6527 km^2 (2520 mi^2) in northeastern Alberta. Reservoir units attain thickness of more than 70 m (21 ft) at depths

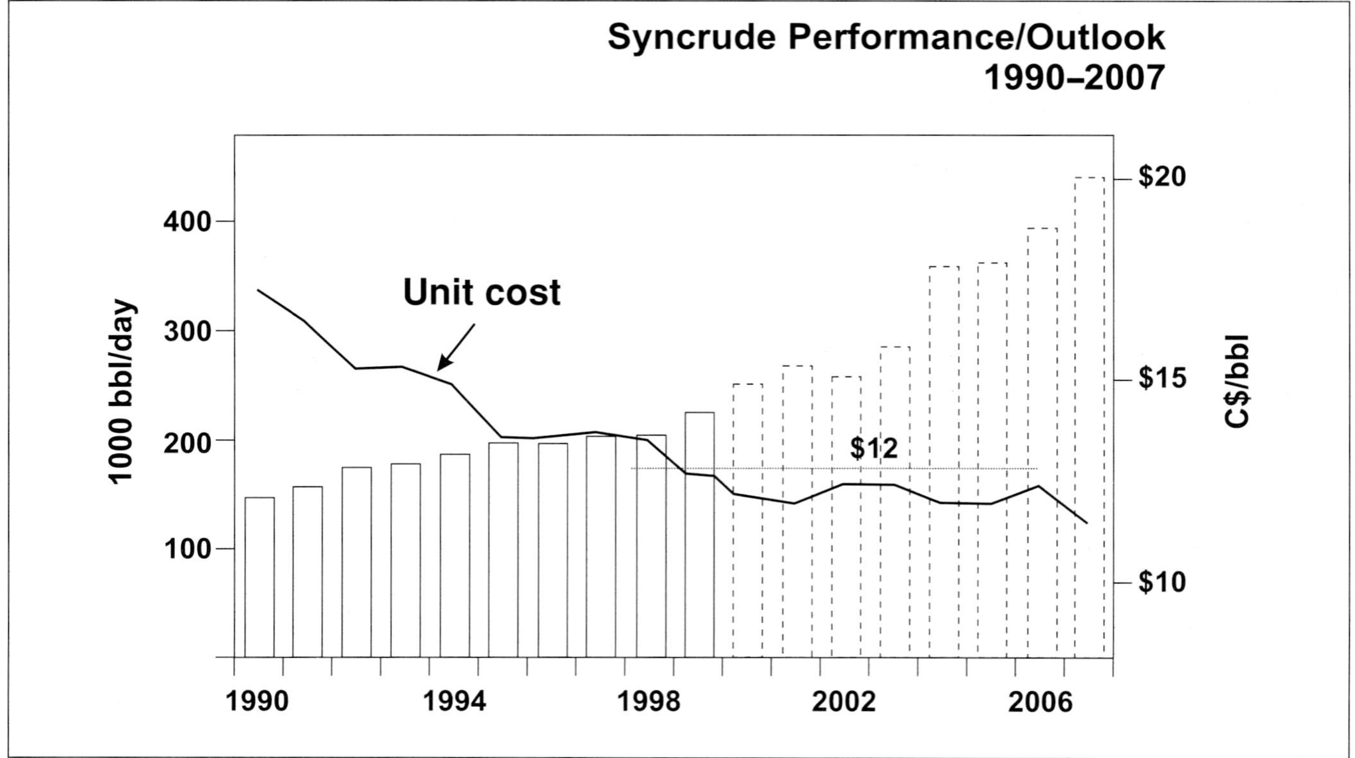

Figure 6. Syncrude performance and outlook. Note that the operating cost per unit of production has fallen to approximately Canadian $12/bbl (excluding capital recovery).

of 400–500 m (1313–1640 ft). Typical resource values are 15 million m^3 (100 million bbl) of oil in place per square mile. Cyclic steam stimulation is expected to result in recovery factors of 25–30% in better-quality reservoir areas.

In-situ recovery.—In-situ thermal recovery processes include steam-assisted gravity drainage (SAGD) technology and cyclic steam stimulation (CSS).

SAGD uses parallel horizontal injector and producer wells (Figure 7). Steam is injected into the oil-bearing formation via an upper horizontal well bore to create a "steam chamber" which progressively heats the oil. It is desirable to achieve a pressure increase in the formation that approaches the fracture gradient. The oil's viscosity is reduced such that oil flows by gravity down toward the lower horizontal well bore, which acts as the collection point. The oil is then pumped to the surface. Typical recovery rates for well pairs are 100 m^3/day (600 bbl/day) of oil. Various combinations of horizontal well configurations are being evaluated to maximize recoveries. SAGD is undergoing testing by several companies in Athabasca and Cold Lake, with promising results, but a large-scale commercial plant has yet to be announced. Typical of in-situ potentially commercial SAGD projects is the project by PanCanadian at Christina Lake, 130 km (81 mi) northeast of Lac La Biche. The project is to develop an area of 20.7 km^2 (8 mi^2) by 2005 through three phases of development, producing 8000–11,000 m^3/day (50,000–70,000 bbl/day) of crude bitumen from the McMurray Formation.

CSS is predominantly a vertical well process, with each well alternately injecting steam and producing bitumen and steam condensate. A heated zone is created whereby the bitumen can flow back to the well. Imperial Oil's Cold Lake project is the major proponent of CSS and is the flagship of in-situ projects. Cold Lake is projected to produce more than 23,800 m^3/day (150,000 bbl/day) of bitumen early in the 2000–2010 period (Imperial Oil, 1999b).

Other technological developments are on the horizon, including the Vapex process, in which solvents are recycled and reused. Significant cost reductions over SAGD may be possible and might enhance the projected ultimate recoverable resource.

Oil-sands summary.—Multiple projects will pump Canadian $30 billion in investments into development of Cold Lake, Peace River, and Athabasca. Of this, Canadian $25 billion is targeted at the Athabasca area alone. Production from oil sands is projected to rise to 50% of Canadian and 10% of North American crude production in the next 10 years. According to Dingle (1999b), "[a] solid oil sands industry base is in place today, including infrastructure, production facilities and organizations, as well as technology and know-how." For the Canadian petroleum industry, the intent is to more than offset the decline in conventional liquids production by increased

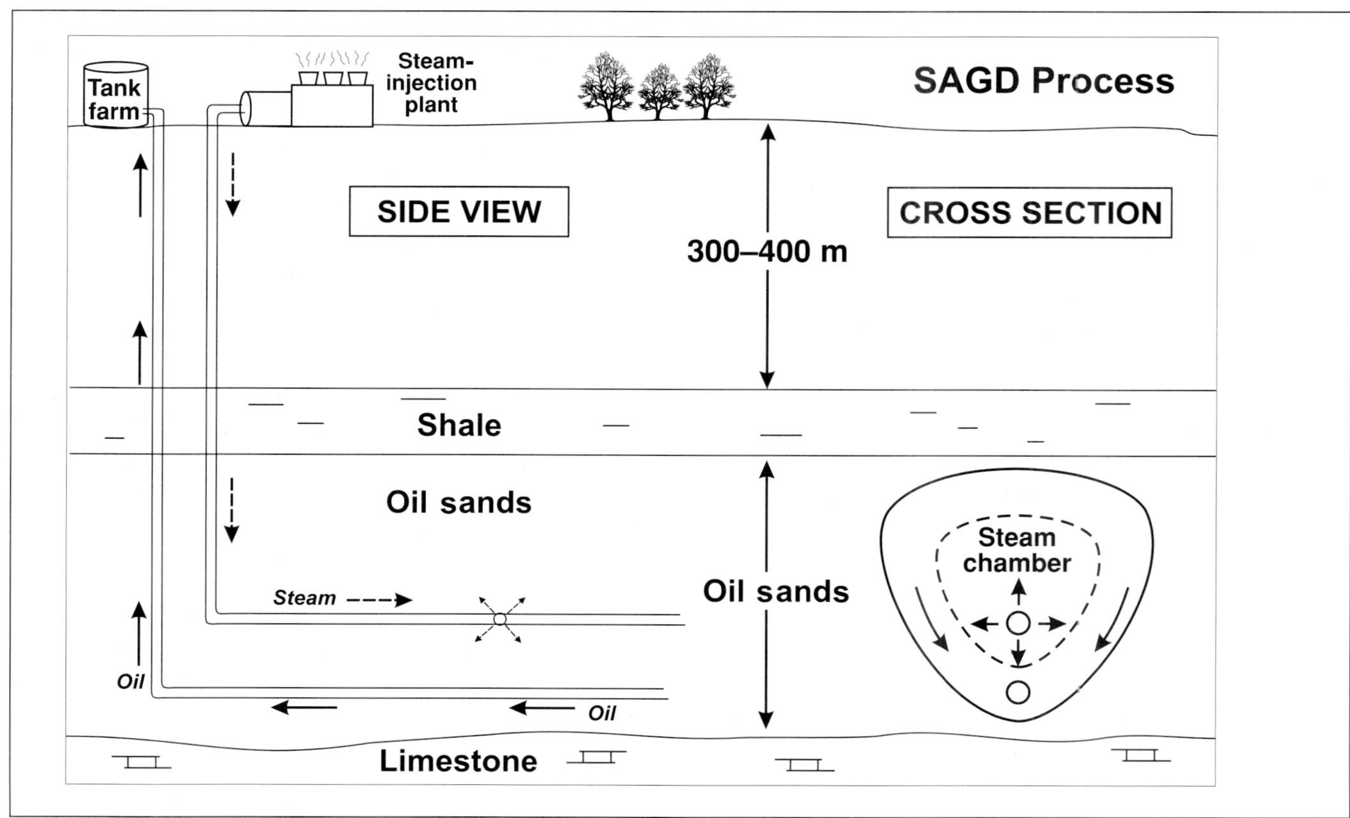

Figure 7. The SAGD process for in-situ recovery of bitumen (oil) from oil sands.

nonconventional production, primarily through growth in cost-efficient mining projects and in-situ thermal-stimulation projects.

The Western Canada Sedimentary Basin North of Latitude 60°N: Northwest Territories and Yukon Territory

Vast areas of the Northwest Territories (NWT) are only lightly explored. Excluding the Mackenzie Delta, approximately 500 exploratory wells have been drilled to seek reservoirs, mainly in Paleozoic siliciclastics (Cambrian) and carbonates (Devonian, Mississippian) and, in the southern territories, in Mesozoic (Cretaceous) sandstones.

The NWT has five regions with hydrocarbon potential: the southern territories (up to latitude 64°N), Mackenzie plain, Peel plateau, Colville Hills, and Mackenzie Delta region (Bird et al., 1995). Yukon potential is limited to the Whitehorse trough, Liard plateau, and Eagle Plain Basin (for Eagle Plain Basin, see Yukon Territorial Government, 1994b; Davidson et al., 1995). To date, most of the significant discoveries in the north are located throughout the western part of the NWT (within the northern part of the Western Canada Sedimentary Basin) and in the extreme southeastern part of the Yukon along Liard plateau (Yukon Territorial Government, 1994a; see also Price, 1995; National Energy Board, 1996). Dixon and Stasiuk (1998) summarized the hydrocarbon potential of Cambrian ("Tedji Lake") plays in the interior plains (Colville Hills) east and northeast of Norman Wells.

In the Mackenzie Valley, land issuance was under a moratorium between 1979 and 1994 because of recommendations of the Berger Commission and unsettled aboriginal land claims. Hence, even though some of the exploration plays pursued in Alberta and British Columbia extend into the NWT, there has been much less exploration than in those provinces. After the settling of aboriginal land claims, the issuance of exploration rights recommenced in the Beaufort in 1989, in the high Arctic in 1991, and in the NWT in 1994 (Morrell et al., 1995).

Gas production takes place at Pointed Mountain (mainly depleted, but to be replaced with production from recent gas discoveries at Liard). Oil production is limited to Norman Wells (for location, see Figure 2).

Norman Wells field is on the Mackenzie River, just south of the Arctic Circle, and it produces approximately 4400 m^3/day, or 28,000 bbl/day, of oil. Since 1985, the field has produced more than 23.8 million m^3 (150 million bbl) of oil, which has been piped 864 km (540 mi) to connect to infrastructure in northern Alberta. Ultimate recovery may be 37.5 million m^3 (235 million bbl). Attempts to duplicate this discovery have been futile so far.

Remaining potential in the NWT will be pursued vigorously after infrastructure is in place to the Mackenzie Delta, but the resource potential is not considered to be large.

East Coast Basins

Williams and Grant (1998) published the latest definitive tectonic assemblages map of the coastal region of Atlantic Canada. Rifting between North America and Africa that began in the Late Triassic resulted in a rifted margin off Nova Scotia and a transform margin south of the Grand Banks. Seafloor spreading thereafter produced Jurassic and Cretaceous oceanic crust south of the Grand Banks. Stretching and rifting of the crust between North America and Greenland began in the Early Cretaceous, accompanied by faulted troughs, some of which contain Cretaceous sediments, in the region surrounding the Labrador Sea and Baffin Bay. East-northeasterly seafloor spreading in the Labrador Sea began in the Late Cretaceous, but the orientation changed to north-northeasterly in the early Eocene. Seafloor spreading started in Baffin Bay in the latest Cretaceous or Paleocene but ceased in Baffin Bay and the Labrador Sea in the late Eocene. Seafloor spreading continued during the Cretaceous and Cenozoic in the North Atlantic, where the Mid-Atlantic Ridge remains an active spreading center today (Geological Map of Canada, 1999).

Bell and Campbell (1990) summarized the three main areas of interest for significant petroleum resources in Mesozoic basins along Canada's East Coast margin (Figure 8): the Scotian Shelf and Slope off Nova Scotia, the Grand Banks of Newfoundland (Jeanne d'Arc Basin), and the Labrador Shelf (Hopedale and Saglek Basins). (See also Grant et al., 1986; Meneley, 1986; Balkwill et al., 1990; and Wade and MacLean, 1990.)

Exploration enthusiasm remains high for Canada's East Coast basins, mainly on the Scotian Shelf off Nova Scotia and in the Jeanne d'Arc Basin, on the Grand Banks offshore Newfoundland (see Newfoundland Ocean Industries Association, 1998). Recent exploration commitments will cause the expenditure of more than Canadian $1.0 billion on new seismic and drilling in these areas in 2000–2005.

Canada's East Coast basins, excluding Labrador, have a gas-resource potential of more than 1.7 trillion m^3 (60 tcf) (Drummond, 1998). Of this total, approximately 283 billion m^3 (10 tcf) have been discovered. The prime area for early exploitation and further exploration success is the Scotian Shelf, primarily because development is under way and operating conditions are within technology capabilities. Investment is cautiously proceeding in other more northerly areas along this margin as operators gain confidence, comfort, and experience in harsher conditions.

Basins off Nova Scotia

Scotian Shelf.—The Scotian Basin (province code 5217, U.S. Geological Survey, 1997) is located beneath

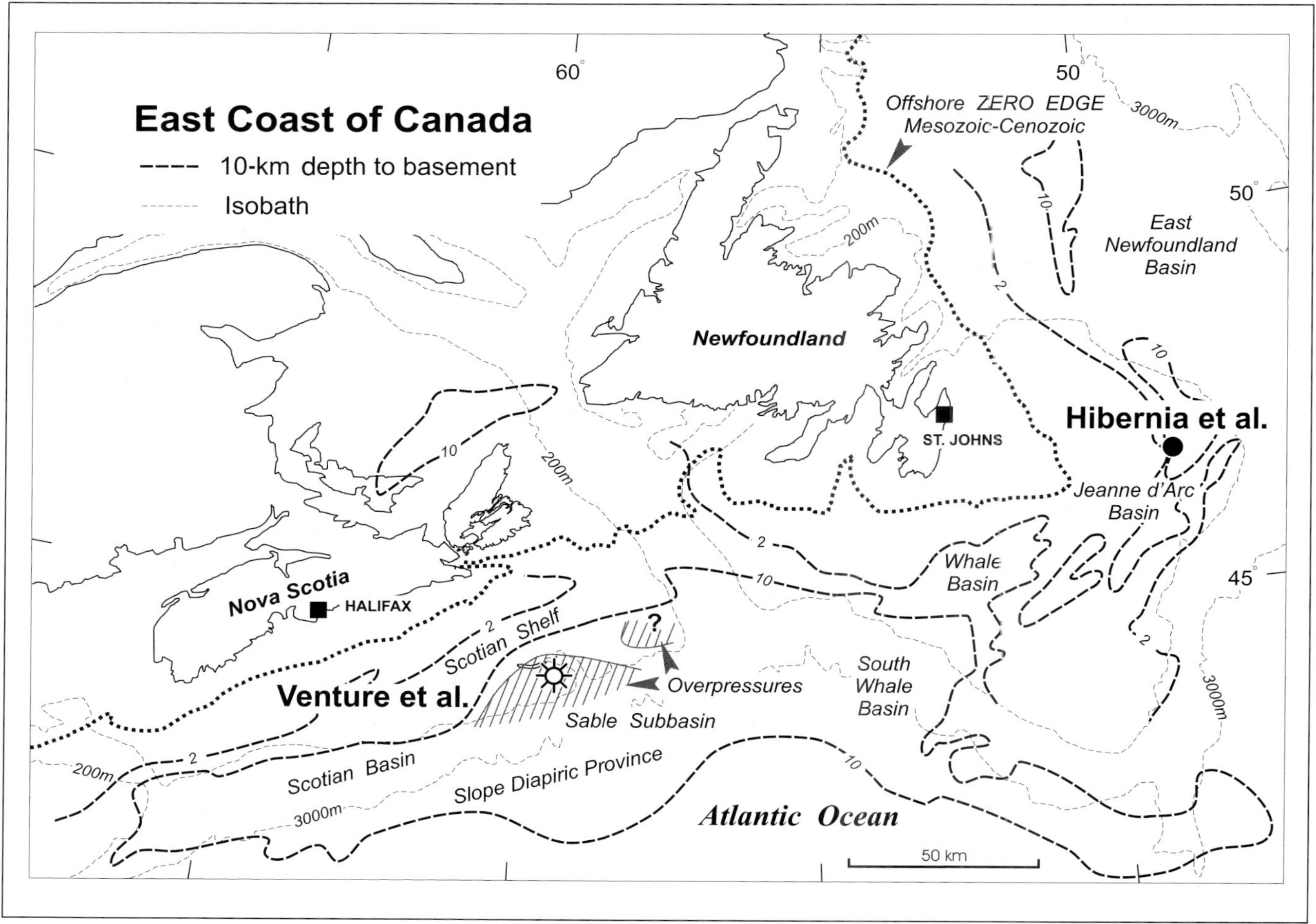

Figure 8. East Coast of Canada. Venture et al. is the location of the Sable Offshore Energy Project. Hibernia et al. includes the development at Hibernia, Terra Nova, and others (e.g., White Rose) in the Jeanne d'Arc Basin.

the outer part of the Scotian Shelf, offshore Nova Scotia (Atlantic Geoscience Centre, 1991) (Figure 8). The basin exhibits all the features of a classic passive margin: early rift deposits, a wide mantle of evaporites, carbonate platforms, giant prograding wedges of siliciclastic sediments, salt diapirism, overpressuring, hydrocarbon generation and entrapment, extensional faulting, and basinward tilting of postrift sediments. Beneath the continental shelf, the basement is faulted into a series of horsts and grabens. Locally, they are filled with Upper Triassic–Lower Jurassic synrift siliciclastics. Overpressuring is present (Drummond, 1992; Yassir and Bell, 1994) in such severity that fracture gradients may be exceeded, allowing for leakage and secondary migration of petroleum into shallow, normally pressured intervals. In Venture field, discovered in 1979, gas occurs in multiple sandstone reservoirs (Upper Jurassic to Lower Cretaceous) over a stratigraphic interval of 1600 m (5250 ft) (Drummond, 1992). The majority of accumulations discovered is restricted to siliciclastic reservoir intervals in Upper Jurassic to Upper Cretaceous rocks. Recent exploration activity along the previously disappointing Jurassic Abenaki bank-edge carbonates has shown encouraging results (PanCanadian, 1999).

The source rock for the gas discoveries on the Scotian Shelf is the Upper Jurassic–Lower Cretaceous Mic Mac (Verrill Canyon Formation) shales. Most gas discoveries appear to occur proximal to the western (updip) limit of the geochemically mature interval. Petroleum maturation and migration have occurred since the middle Tertiary. Evidence for secondary migration includes the presence of shallower reservoirs containing early-generated hydrocarbons displaced from depth by later maturation products.

Traps are dominated by growth-faulted rollover anticlines, although compaction drape features over salt pillows and other lithologies do occur.

Wade et al. (1989) concluded that the median expectation for gas resources is 512 billion m^3 (14.5 tcf) and the Canadian Gas Potential Committee (1997) estimated marketable gas potential at 455 billion m^3 (12.9 tcf), whereas the National Energy Board (1998) reported an ultimate marketable resource figure of 635 billion m^3 (18 tcf). The largest discovery to date is Venture D-23 (Canadian Gas Potential Committee, 1997), with 59.5 billion m^3 (2.1 tcf) gas in place; it is unlikely that any larger accumulation will be discovered. Discoveries of this size in deeper water may not be attractive stand-alone development candidates (Canada–Nova Scotia Offshore Petroleum Board, 1997).

Offshore Nova Scotia is an area of significant growth potential. It offers a large gas-resource base, a developing infrastructure, and proximity to large and growing gas markets in the Maritimes and the northeastern United States. Oil potential is considered insignificant, although minor oil accumulations are known (e.g., Panuke field). Verification of widespread occurrence of mature oil-prone source rocks is a prerequisite for generating new play concepts for oil.

Basins off Newfoundland

Grand Banks.— The Grand Banks basins (province code 5215, U.S. Geological Survey, 1997) cover an area of approximately 330,000 km^2 (127,000 mi^2). The water depths are relatively shallow at 60–300 m (200-1000 ft). Approximately 115 exploration wells have been drilled in this area, including 69 (60%) in the Jeanne d'Arc Basin (Bruce, 1999). Seventeen fields have been found, containing 254 billion m^3 (1.6 billion bbl) of recoverable oil and 113 billion m^3 (4 tcf) of gas (Bell and Campbell, 1990; Grant and McAlpine, 1990; Canada–Newfoundland Offshore Petroleum Board, 1999). The Flemish Pass, East Newfoundland, Whale, Horseshoe, and Carson Basins are largely unexplored. The physical environment is similar to the North Sea except for the presence of icebergs in April to July and sea ice in winter.

Reserve base is large, with significant discovered fields. Hibernia and Terra Nova fields are the largest accumulations of conventional oil remaining in Canada today. Reservoir quality is good; wells have high productivity. In 1998, the Hibernia B-16-1 well set a Canadian daily flow record for oil when it tested 8900 m^3/day (56,000 bbl/day).

The development of suitable stand-alone production facilities such as that at Hibernia and those planned for Terra Nova, White Rose, possibly Hebron/Ben Nevis, and Riverhead would allow processing of oil produced from satellite fields within 15 km (9 mi) (Taylor et al., 1991; Sinclair et al., 1992).

Jeanne d'Arc Basin.—The Jeanne d'Arc Basin is a Mesozoic failed-rift basin containing more than 20 km (65,000 ft) of sedimentary fill. The sedimentary succession ranges from Triassic to Tertiary. An overpressured zone occurs below approximately 4 km (15,200 ft) and is concentrated mainly in Jurassic rocks (Rogers and Yassir, 1993). The overpressuring possibly is related to oil generation in the Jurassic source intervals, mainly the Egret Member shales (Kimmeridgian) which contain oil-prone (type II) kerogens (Williamson et al., 1993). The Egret is recognized as a "world-class" oil source. The unit ranges in thickness from 55 m (180 ft) to more than 200 m (656 ft). Peak hydrocarbon generation occurred in the early Tertiary (50 ma) and postdates early traps. On the basis of mass-balance and hydrogen-index techniques, Fowler and McAlpine (1995) suggested that perhaps 39 billion m^3 (245 billion bbl) of oil has been generated from the Egret Member. If only 10% was captured in reservoirs and 30% was recoverable, this would give an exploration target of 1.2 billion m^3 (8 billion bbl) of oil, 75% more than that located to date. It seems appropriate to refer to the Egret-Hibernia and Egret-Avalon petroleum system when referring to the Jeanne d'Arc Basin, because Egret-sourced oil is trapped in reservoirs of the Cretaceous Berriasian to Valanginian Hibernia sandstones at a depth of 3700 m (12,136 ft) and in Cretaceous Barremian to Albian Avalon sandstones at a depth of 2400 m (7872 ft) in Hibernia field (Mackay and Tankard, 1990; Hurley et al., 1992).

Petroleum resources of the Jeanne d'Arc Basin have been lightly tapped so far, but other Mesozoic basins off Newfoundland are considered to have limited potential for additional light-crude reserves because of logistical difficulties of developing fields. Estimates of potential economic resources and technically recoverable oil are about an additional 476 million m^3 (3 billion bbl) to 634 million m^3 (4 billion bbl) more than the reserves in Hibernia (estimated by Mobil at 120 million m^3, or 750 million bbl recoverable) and Terra Nova (58–92 million m^3, or 370–580 million bbl recoverable). Extrapolations of gas potential have not been published, but as much as 283 billion m^3 (10 tcf) recoverable might be possible. Beyond those numbers, projections may be meaningless, because the comfort and experience to develop those kinds of reserves off Newfoundland must await offtake mechanisms at Hibernia and White Rose. White Rose is scheduled to be onstream in 2004.

Some observers question whether development of these resources is economically justified. In addition to the hostile environment, the decision to invest is difficult because of variations in crude gravity, the geologic complexity of some reservoirs, and the areal spread of some fields. A major challenge for the Grand Banks area is the production of associated and potentially nonassociated natural gas. Several potential investors are looking at compressed coselle technology. Each coselle (coiled pipe in a carousel-shaped container) can store about 3.1 mmcf of compressed gas; 108 coselles in one ship could store 330 mmcf. A pipeline alternative will require large volumes of natural gas to be available (say, 7–10 tcf), and threshold volumes have yet to be proved. Unless very large features (capable of containing stand-alone resources) are mapped on seismic surveys, explorers will be reluctant to drill if a gas risk exists. Costs for a 1400-km buried pipeline from the Jeanne d'Arc Basin to Nova Scotia, with a crossing of the Laurentian Channel, would be high (say, Canadian $3.5 billion). However, if large volumes of gas are discovered after further successful exploration in the coming decade, a pipeline to the large United States market will become a serious option.

Basins off Labrador

Gas discoveries were initially made off Labrador in the early 1980s (Atlantic Geoscience Center, 1989; Balkwill et al., 1990), and discovered resources are held within five Significant Discovery Licenses. The most significant accumulation discovered is at Bjarni, with more than 63.3 billion m^3 (2.23 tcf) of marketable gas reserves (this reserve estimate is larger than that in any other gas field discovered on the East Coast, e.g., the Scotian Shelf). The gas is trapped in Lower Cretaceous clastic reservoirs; the gas probably was sourced from interformational coals and lignites.

Assuming adequate per-well productivities, the estimated threshold reserves to interest investment may be on the order of 170 billion m^3 (6 tcf) of gas. The appetite for additional exploration in the next 20 years is not evident and, in any present forecast span, these resources should be considered to be orphaned. Total known gas resources in several scattered fields are approximately 141 billion m^3 (5 tcf). This is not a large accessible resource compared with some other areas of North America.

Key elements affecting the attainment of threshold reserves for marketing are:

- geologic risk. Results to date have not established major natural-gas (or oil) deposits adequate to entice stand-alone development.
- ice management. The area is a major thoroughfare for movement of icebergs south along the Labrador offshore margin. Scouring of the seabed by large icebergs poses a risk for pipeline connections and subsea wellheads.
- geotechnical elements. The seabed is littered with boulder fields produced by the melting of rafted material. These fields historically have caused difficulties in the drilling of conductor holes for offshore wells and the placing of protected subsea wellheads.
- distance from markets, market needs, and hence adequate commodity prices to provide a reasonable netback
- well costs and limited drilling and construction windows

The industry needs to achieve a high degree of comfort with operating conditions farther south, particularly on the Grand Banks, to take much interest in the known and potential gas resources off Labrador. This might occur when the more accessible resources off Nova Scotia and Grand Banks are determined to be inadequate to service northeastern United States markets. Because it is unlikely that the petroleum industry will show much interest in additional exploration to garner threshold reserves to initiate a gas-production project off Labrador, the resources will clearly remain stranded and orphaned well into the new century.

Future Potential

The discovery of new oil reserves in unproven East Coast basins relies on the presence of world-class source rocks such as the Jurassic Egret Member. There are basins on the Grand Banks (e.g., Whale Basin) where oil shows might indicate the presence of a potential Jurassic source. Future exploratory drilling in the Flemish Pass basin, situated in deeper water northeast of the Jeanne d'Arc Basin, is keenly anticipated. Significant prospects with reserve potentials of more than 80 million m^3 (500 million bbl) of oil are known both in well-defined four-way dip closures and tilted fault blocks in water depths to 1200 m (3900 ft). Reservoir objectives are below 3000 m (9800 ft) subsea (Bruce et al., 1999).

Oil potential off Nova Scotia remains speculative because there has been no encouragement that a significant oil source is preserved beneath the Scotian Shelf. Off the Nova Scotia margin, anoxic conditions or restricted circulation were not characteristic of either Late Jurassic or Cretaceous paleogeography. Tertiary siliciclastic wedges appear to be dominated, as expected, by terriginous material suitable only as a potential source of gas or wet gas and condensate.

Exploration permits have recently been awarded off Nova Scotia in water depths of 3000 m (9840 ft). There is considerable interest in the potential for the discovery of giant gas fields in the zone of assumed (Triassic Argo) salt diapirs ("Slope Diapiric Province" in Wade and MacLean, 1990, p. 200, their Figure 5.23), which occurs along and beyond the break in slope of the Scotian Shelf. For suprasalt plays, the major reservoir objective must be in the Tertiary, Upper Cretaceous, and Jurassic. Lowstand wedges in the Tertiary would be a local play, but hydrocarbon charge and the ubiquitous presence of good-quality reservoirs may be an issue. High heat transfer resulting from the presence of salt and features produced by halokinesis should bode well for maturation if suitable source intervals were present and speculative source rocks are buried to suitable depths. There is no doubt that large salt-related rollover prospects are present, but they may exist in considerable water depths (as great as 2000 m, or 6500 ft). Ponded resedimented siliciclastic facies are possible between diapiric swales (stratigraphic and "structural-stratigraphic assist" pinch-out plays). An initial significant discovery of petroleum in the "Slope Diapiric Province" will be met with intense enthusiasm because the trend is a virtual flower garden of prospects.

Exploration is now being conducted in the previously declared St. Pierre Moratorium Block south of the French islands of Saint Pierre and Miquelon (off the southern coast of Newfoundland). The Laurentian Subbasin is a deep Mesozoic extension of the Scotian Basin. MacLean and Wade (1992) assessed the area as having similar potential to other areas off Nova Scotia because evidence for good Jurassic (or other) oil source was not present. More than 300 leads were identified, and based on parameters similar to the Sable Island "model," the average expectation was for the discovery of 255 billion m^3 (9 tcf) of gas and 15.8 million m^3 (100 million bbl) of liquids. Areal size of leads (and/or stacked pays), reservoir properties, and fluid contents will be major considerations for predrill decisions and for the potential for commercial development. Mesozoic prospects, on the updip margin of the Laurentian Basin against the Burin Platform (in water depths to 200 m or 600 ft), would be of interest.

Tertiary wedge-edge plays are an exploration target, and the influx of mass-flow ("resedimented") sands may have occurred (Batemen et al., 1999), associated with, for example, the Oligocene lowstand. This event is pervasive elsewhere along the Labrador and North Atlantic margins. Charging these plays with hydrocarbons sourced from mature older stratigraphic intervals may pose a challenge, however.

The sizes of gas fields discovered to date are not large by world standards, but the area is within reach of a large continental market. If the gas finds are typical of the size of fields yet to be found and these new discoveries occur in deeper waters, then they will lie dormant for a considerable time without higher gas prices. Gas discoveries in the deeper waters off the Scotian Shelf may need to exceed 5 tcf to be attractive for development (but see Bruce et al., 1999, who quote a threshold as low as 2 tcf of recoverable sales gas). Thick pays and high-productivity wells are also required. It is unlikely that this area will drive the technology required to develop a group of smaller fields in water depths greater than about 1500 m (4950 ft).

Basins North of the Arctic Circle

Mackenzie Delta

Natural gas and oil from the Beaufort-Mackenzie Basin (province code 5239, U.S. Geological Survey, 1997) is hosted in high-productivity Tertiary and Cretaceous sandstone reservoirs (Dixon, 1995; Lane and Dietrich, 1995). More than 200 wells have been drilled in the Beaufort Basin since 1965. Currently, 62 discoveries are held under Significant Discovery Licenses (Resources, Wildlife and Economic Development, 1999) (Figure 9). The largest discovery of gas in the Mackenzie Delta is Taglu field (Figure 9), discovered by Imperial Oil Limited in 1971 (Tsang, 1990). Estimated mean marketable gas reserves are 58.6 billion m^3 (2.1 tcf) in Taglu field, 35.4 billion m^3 (1.25 tcf) in Parsons Lake field, and 13.6 billion m^3 (480 bcf) in Niglintgak field. In the Beaufort Sea, Amauligak is the largest oil and gas (associated) discovery, with mean recoverable oil of 37.3 million m^3 (235 million bbl) and mean marketable gas of 38.5 billion m^3 (1.3 tcf) (National Energy Board, 1998).

Dixon et al. (1992, 1994) estimated that the Eocene (Taglu sequence) and Oligocene (Kugmallit sequence) contain recoverable resources of 542 million m^3 (3.4 billion bbl) of oil and 1037 billion m^3 (37 tcf) of gas. These reserves and others probably have been sourced from a common Eocene source rock (Richards Formation).

Development of Mackenzie Delta gas reserves faces significant economic and technical challenges, primarily because of the long distance (1350 km, or 843 mi) from existing infrastructure. As long as lower "supply-cost" gas exists near infrastructure, the economics of Mackenzie Delta gas development remain challenging. Recent awards of exploration permits in this area (Northern Oil and Gas Directorate, 1999) and upcoming drilling activities optimistically suggest a declaration of commerciality before 2005, with an in-service pipeline constructed by 2008. This will depend on the establishment of threshold reserves, higher gas-price netbacks, reasonable transportation costs (tolls) to tie into the extended infrastructure from the northern reaches of the Western Canada Sedimentary Basin, and cooperation among producers, pipeline companies, and local communities. Extended- (and long-) reach and horizontal drilling may also play a part in their ultimate development. These technologies would provide the ability to use smaller footprints in this environmentally sensitive area to access separate fields.

The delta area and the shallow waters of the Beaufort Sea (south of the southern limit of permanent sea ice) seem to have limited light-oil potential for commercial development. The biggest discovery made to date is Amauligak, in 1984. Most researchers consider the total recoverable oil discovered in the province to be a modest 161 million m^3 (1.0 billion bbl). The potential could exceed 800 million m^3 (5 billion bbl). If scattered through several "small" fields, these resources will not be attractive to develop, based on present price trends or competitively priced upgraded crude oil produced from oil sands. During the recent National Energy Board consultations, no interest was indicated in oil production from the Mackenzie-Beaufort area before 2025 (National Energy Board, 1999a). Perhaps if natural-gas thresholds are exceeded and additional oil is discovered, oil resources may be commingled in a condensate line from the Delta area to Norman Wells. However, this does not look like an early (pre-2025) possibility. Development of Beaufort Sea discoveries in shallow water (less than 20 m, or 60 ft) may involve dry (buried) subsea completions, development pads built on artificial islands, or steel or concrete monocones designed to withstand the pressure of sea ice.

The National Energy Board (1998) estimated that 255 billion m^3 (9 tcf) of marketable natural gas has been discovered and 1.56 trillion m^3 (55 tcf) of undiscovered gas resources remain. Threshold reserves for a pipeline tie-in are likely to be proved initially in Taglu and associated fields within a 50-km radius. Initial design suggests that a high-pressure pipeline with capacity of 65 million

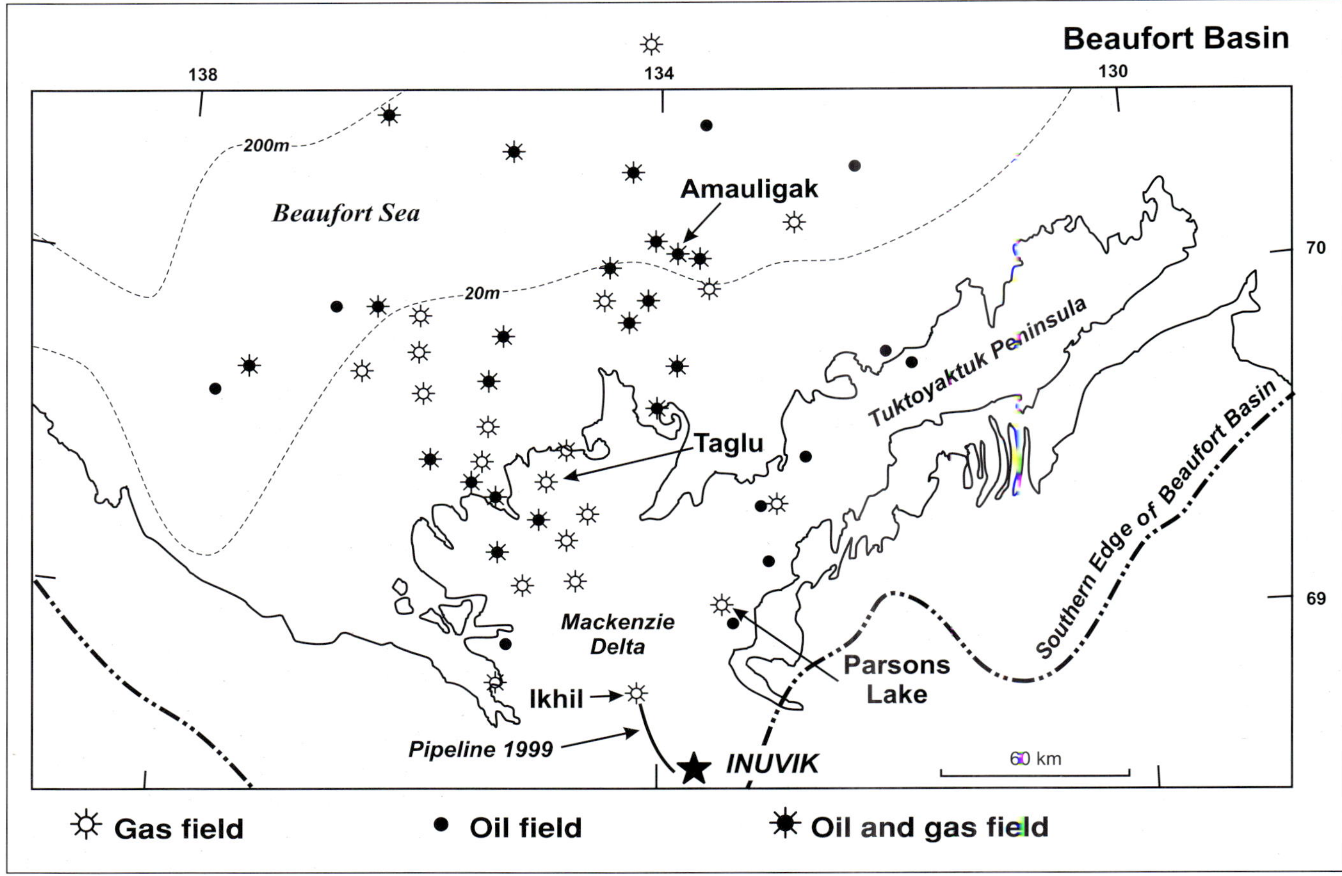

Figure 9. Discoveries in the Beaufort Basin (classification after National Energy Board, 1998). Amauligak is the largest offshore oil discovery to date. Taglu and Parsons Lake (and others) should be part of the overall future gas-development scenario. The small Ikhil gas field was commissioned in 1999 to supply gas to Inuvik.

m^3/day (2.33 bcf/day) will be required for economic thresholds to be exceeded. A separate 12-inch pipeline for natural-gas liquids and condensate connected to the existing Enbridge system via Norman Wells would be viable. Gas-throughput volumes of this magnitude suggest that threshold gas reserves approaching 566 billion m^3 (20 tcf) will be needed before there will be a commitment to construct a natural-gas export line.

The challenge of meeting sufficient reserves for pipeline commitment with throughput exceeding 56 million m^3/day (2.0 bcf/day) is not an easy one. The analysis by the Canadian Gas Potential Committee (1997, see its Figure 12.6) indicated from its projections that perhaps more than 30 new fields, each with marketable reserves of more than 8.5 billion m^3 (300 bcf), would be needed. This is a formidable investment which may take some time to assemble. For the first time in eight years, new awards were made in 1999 for additional exploration in the Delta (Northern Oil and Gas Directorate, 1999). Most National Energy Board projections, done in consultation with industry, indicate that Mackenzie Delta gas may not reach southern markets until after 2015.

Access to Alaskan gas resources on the North Slope, in conjunction with gas from the Mackenzie Delta, could possibly accelerate the construction of an export pipeline which would connect these two important future suppliers of natural gas to the Alberta and North American gas grid. Key issues are the participation of local communities and native peoples, threshold volumes for commercial development of natural gas, the associated liquids production that would "piggyback" on natural-gas development, price issues, construction costs, pipeline tolls, and environmental solutions.

Sverdrup Basin

The Sverdrup Basin (province code 5234, U.S. Geological Survey, 1997) is situated in the Arctic islands in polar latitudes between 76°N and 80°N. The Sverdrup is a lower Carboniferous to Tertiary basin. Extensional faulting affected its early history, while compressional tectonics in the Late Cretaceous to Oligocene impressed a westerly diminishing structural grain on the area.

Much of the offshore area is prone to permanent sea ice, but it was well explored during the late 1960s to mid-1980s. More than 176 exploratory wells and approximately 120,000 line-km (19,000 line-mi) of 2-D reflection seismic data have been recorded (Geological Survey of Canada, 1995). There are 19 significant discoveries (held under Significant Discovery Licenses; see Resources, Wildlife and Economic Development, 1999). The largest field and the first significant gas discovery in the Canadian Arctic islands is Drake Point (Waylett, 1990), an anticlinal trap discovered in 1969 on the Sabine Peninsula of Melville Island in the Sverdrup Basin. The field has estimated discovered marketable gas reserves of 150 billion m^3 (5.3 tcf), with potential of "reserve creep" to beyond 170 billion m^3 (6 tcf). The dry gas is reservoired in Jurassic sandstones with beach (barrier-bar) affinities. Middle Triassic to Middle Jurassic shales are potential source rocks of the gas. Characteristics of the field include a 180-m (600-ft) gas column and an areal extent of 436 km^2, (168 mi^2, or 100,000 ac). The total discovered initial volume in place is approximately 566 billion m^3 (20 tcf) of gas, including 15 discovered fields, each with more than 100 bcf in place. It may be significant that within 100 km (63 mi) of Drake Point, which is 1280 km (800 mi) from the Mackenzie Delta, are approximately 10 tcf of discovered marketable gas (Canadian Gas Potential Committee, 1997, p. 78, and its Figure 12.8).

Marketable natural-gas reserves are estimated to be 452 billion m^3 (16 tcf) (Canadian Gas Potential Committee, 1997), and in-place oil resources are estimated at 254 million m^3 (600 million bbl) (Waylett, 1990). Reservoirs found to date are mainly sandstones of the Jurassic, Triassic, and Lower Cretaceous. The Bent Horn oil field, which produced from a single well for Panarctic Oils for a number of years until 1997, has a small oil reserve in a Devonian bioherm (Meyerhoff, 1982). A major source rock for the oil and gas discoveries in the Sverdrup Basin is the bituminous ("anoxic") shales of the Schei Point Group (Triassic), which have an average 4% total organic-carbon content of marine type II algal origin. Multiple formal phases of migration have caused mixed oil and gas accumulations (Waylett and Embry, 1992).

Early optimism for the development of such stranded reserves by ice-designed vessels (gas carriers) transporting gas through the Northwest Passage has faded (Couper, 1983, p. 148). Feasibility studies have been conducted for liquefied natural-gas terminals, particularly at Bridport Inlet, Melville Inland (Lewis and Keen, 1990, p. 758). Pipeline options are probably favored. Indeed, research on construction of such a pipeline has been conducted, with submarine trenching from Drake Point to Bathurst Island and connection to the Boothia Peninsula and thence overland to markets in Ontario, Quebec, and the United States (see Strain, in Clark et al., 1997, p. 227). However, momentum and enthusiasm for Arctic islands investment have waned, and it is highly probable that these resources will remain "stranded and orphaned" well beyond the present generation or any current forecast span. Any estimated resources, therefore, should be deleted from current assessments of the Canadian marketable natural-gas resource base. The principal problems are those of logistics, environmental protection, great expense, and the limited drilling season. For the foreseeable future, these are not attractive investment areas.

The potential of gas hydrates in the Canadian Arctic, although it is the subject of recent scientific research, is not discussed here.

Areas under Moratoriums

West Coast: British Columbia Basins

The British Columbia offshore basins have been off limits since 1972. The Neogene and younger Pacific margin is structurally complex and variable. A convergent margin now exists west of Vancouver Island where the Juan de Fuca and Explorer Plates are being subducted beneath the North American Plate. A mainly transform margin extends from west of Queen Charlotte Islands to the submergent Yakutat terrane in the Gulf of Alaska. Along this segment, the Pacific Plate continues to move northward relative to North America along a dextral strike-slip fault (Geological Map of Canada, 1999).

Major sedimentary basins of the Pacific margin of Canada are the Tofino, Queen Charlotte, and Georgia Basins. These areas are considered to have the potential for 1.56 billion m^3 (9.8 billion bbl) of oil and 1.228 trillion m^3 (43.4 tcf) of gas (Hannigan et al., 1998). Dietrich (1995a, b) has detailed the petroleum-resource potential of the Queen Charlotte Basin, offshore British Columbia. Resource estimates are based on three conceptual plays involving Cretaceous and Neogene reservoirs (sandstones and conglomerates) and Jurassic and/or Tertiary source rocks. The Neogene plays of the Queen Charlotte Basin are considered to be the most prospective. Total recoverable resources are estimated at 414 million m^3 (2.6 billion bbl) of oil and 565 billion m^3 (20 tcf) of gas at median values. These numbers seem to be overly optimistic, based on earlier drilling results and the nature of the potential reservoirs, source, and charge characteristics. If the moratorium on activities is lifted, large 3-D seismic efforts, combined with high-graded locations for exploration wells, will quickly indicate the overall prospectivity of the area.

East Coast: Georges Bank

The Georges Bank Basin underlies the central and western parts of Georges Bank and the adjacent slope (see Wade, in Wade and MacLean, 1990, p. 171–190). The prospective area straddles the United States–Canada border. Wells have been drilled on the U.S. side without success, but no wells have been drilled on the Canadian side. Most of the oil and gas potential, if any, is likely to occur in the Middle Jurassic to Lower Cretaceous section. The Canadian ban on exploration activities has been extended to 2012.

PROJECTED CANADIAN PETROLEUM RESOURCES

Based on recognized resources and extrapolation of play trends, projections of total marketable supplies of conventional natural gas from Canada are on the order of 5.66 trillion m^3 (200 tcf) (Table 1). The ultimate recoverable volume for conventional natural gas, however, is very dependent on price and technology. Thus, there is much uncertainty in the numbers given for individual basins in Table 1.

Projections of light conventional crude oil indicate very limited potential outside the Beaufort Sea and the Grand Banks. Beaufort Sea projections hinge on the estimate of the Amauligak discovery of only 47.7 million m^3 (300 million bbl), insufficient for an initial development. Grand Banks (Jeanne d'Arc Basin) assessments are supported by rich Jurassic source rock (Egret) and the discoveries to date (Hibernia, etc.). Vast resources await investment and extraction in the Alberta oil sands. Only 447 million m^3 (3 billion bbl) is "booked" against current projects, but the exploitable resource is well beyond 15.9 billion m^3 (100 billion bbl).

Table 1. Estimated accessible conventional natural gas by basin.

Basin	Trillion m^3	tcf
Western Canada Sedimentary Basin	3.39	120+
Northwest Territories and Yukon (e.g., Liard plateau)	0.14	5+
Scotian Shelf	0.707	25+
Mackenzie Delta	0.850	30-
Grand Banks	0.283	10+
Total	**5.66**	**200**

CANADIAN PRODUCTIVE CAPACITY

According to most estimates (e.g., National Energy Board, 1999a), Canada is expected to provide a minimum of 4.2 trillion m^3 (150 tcf) of marketable gas supplies in the next 20 years, at present production rates. Based on current assessments, one must assume these volumes will be produced from the Western Canada Sedimentary Basin (3.4 trillion m^3, or 120 tcf, which is about the same volume as all the gas produced to date), the Scotian Shelf (509 billion m^3, or 18 tcf), and the remainder from Alberta unconventional sources (coal-bed methane, tight sands, etc.) or the Mackenzie Delta (283 billion m^3, or 10 tcf). Based on the analyses presented, one must surmise a very different situation in the year 2020 from that perceived today.

Projected crude-oil supply from Canada (Figure 10) suggests that a production rate approaching 480,000 m^3/day (2.8 million bbl/day) is attainable. Of note is the growing importance of production from oil sands with upgraded (synthetic) crude and bitumen exceeding 50% of total production by 2015. At rates of oil-sands mining and in-situ bitumen extraction of 239,000 m^3/day (1.5 million bbl/day), only 4.29 billion m^3 (27 billion bbl) would be consumed in 50 years.

Cost of Canadian Productive Capacity

No matter how we view the extensive resource base in Canada (or elsewhere, for that matter), we must be aware of the cost of supply, the cost of production capacity, and the limitations to the rate or intensity of investment in productive capacity. When viewed in a North American context, wellhead price increases for natural gas increase the resources that may be accessed at acceptable supply costs. These resources may be abundant, and once their exploitation begins, production may dampen prices, reducing the impetus for continued investment in resources with marginal economics. For oil, the large exploitable resources are the oil sands (Figure 11), where mining, handling, and processing costs can now deliver a barrel of refined product at a very competitive supply cost. The accessible supply of oil sands is enormous, but the rate and magnitude of investment will limit its production capacity. Much will remain orphaned well beyond any current forecast span. The favorable royalty and tax treatment for oil sands will depress investment enthusiasm for remote exploration and development of conventional oil discoveries in harsh, high-cost Arctic locations. For example, extraction costs for Hibernia have been calculated at Canadian $23.03 (U.S. $17.30) per barrel as spent on a 615-million-barrel recoverable case (Hibernia, 1999b). For Terra Nova (Terra Nova, 1999), capital costs as spent are Canadian $4.5 billion, and extraction costs (including operating costs) are U.S. $7.50/barrel. These figures compare with Canadian $12.00 (U.S. $8.00) per barrel for upgraded crude oil from oil-sands mining.

"Rate of Conversion" of Canadian Petroleum Resources

Industry invested more than U.S. $14.0 billion in oil and gas activities in Canada in the year 2000. Attracting, rewarding, and sustaining this scale of investment are major priorities if the rate at which Canadian resources are converted to productive capacity is to be maintained. The rate of capture or conversion of resources to productive capacity is a function of profitability, the resource habitat and its distribution, available markets, technology, financing capability, activity levels, and success, among other factors.

The capture of gas resources in the Western Canada Sedimentary Basin requires drilling at historically high

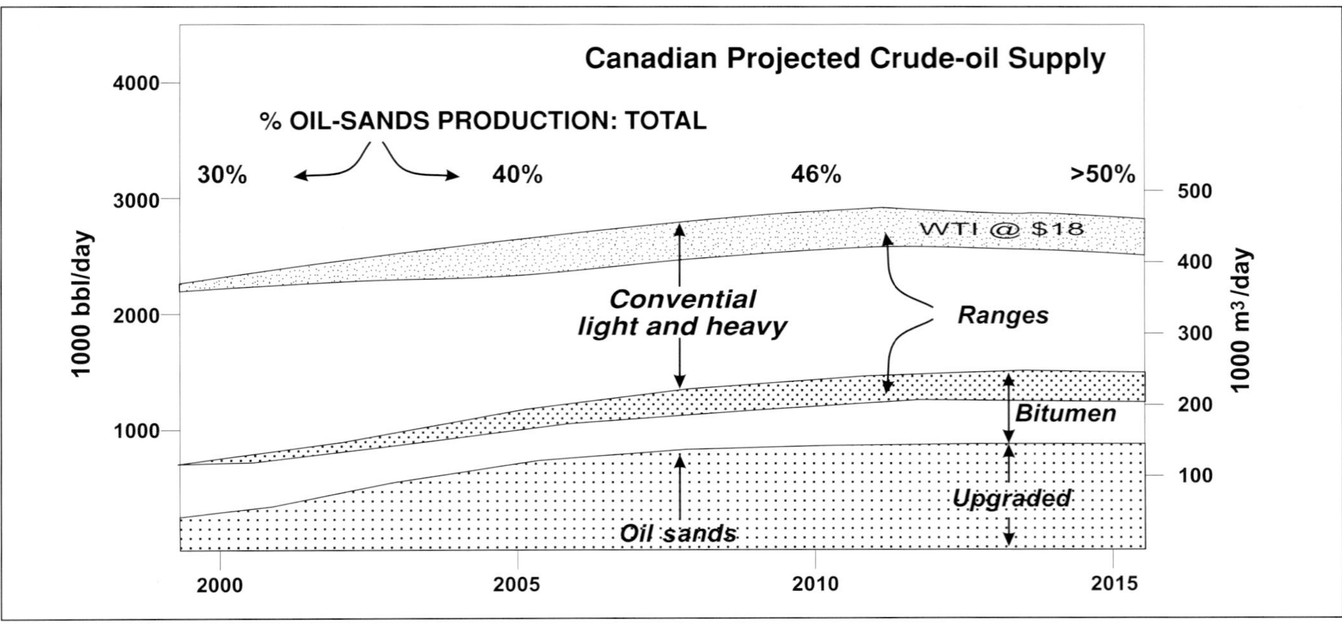

Figure 10. Projected crude-oil supply in Canada to 2015. Note the importance of oil-sands production (data modified from National Energy Board, 1999a).

levels (the connection of more than 5000 successful gas wells each year in the next decade). Based on projected market-driven requirements for Canadian gas, the industry must "book" and deplete (consume) more than 4.25 trillion m^3 (150 tcf) of natural gas in the next 25 years—and achieve a reasonable reserves-to-production ratio in 2025! The commissioning of large-scale oil-sands mining ventures must be staged because of the stress their construction places on the work force and other infrastructure. However, once in place, the productive capacity ("mine life") of a mine may extend well beyond 35 years with continual replacement and upgrading of mining and processing methods. With additional new base mines upgrading oil from oil sands at levels of 20,000 m^3/day (125,000 bbl/day), the incremental additions to Canadian oil production can be substantial.

A level of operating experience (including potential spill containment) and comfort is required for the ice-bearing waters off Newfoundland. This will result in a very cautious approach to investment and construction of production facilities "fit for purpose." The operating environment is harsh and liable to be costly. Forays into the Mackenzie Delta and Beaufort Sea will be carefully orchestrated with involvement of all interested stakeholders. Activity levels ought to be coordinated to efficiently build the threshold reserves required for pipeline connection to southern markets. All activity in Canada's emerging basins (Jeanne d'Arc, Scotian Shelf, Mackenzie Delta, and Beaufort) is relatively costly, with seasonal, logistical, and environmental limitations. Even with favorable factors, the rate of capture (or conversion) of the

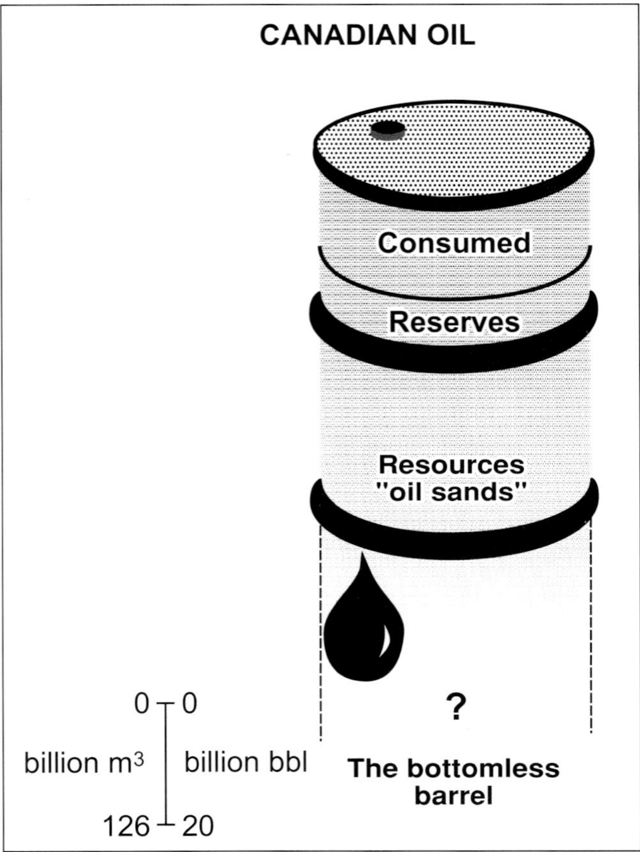

Figure 11. Canadian oil resources. The figure is drawn to highlight the large resource base captured in sands. Accessible mining resources alone can support the projected oil-production rate of 239,000 m^3/day (1.5 million bbl/day) for well beyond 100 years.

resource base in Canada cannot be as rapid as the resource potential might suggest.

DISCUSSION

Canada is blessed with substantial petroleum resources, such that *"the resource base is not in crisis"* (Kirk Osadetz, personal communication, 1998). Maintaining cost-effective and competitively priced productive capacity for natural gas, conventional light oil, upgraded (synthetic) oil, in-situ bitumen production, and heavy oil poses a serious challenge to alternative sources of hydrocarbons (e.g., tight gas, coal-bed methane, and oil shales). Supply costs for products produced from large-scale oil-sands mining at Athabasca and in-situ thermal recovery on the caliber of that at Cold Lake place a ceiling on supply costs from untapped frontier oil resources and redirect (exploration) investment for oil away from those areas. In any case, Canada has the resource base to maintain crude-oil production at more than 318,000 m^3/day (2.0 million bbl/day) into the foreseeable future because deficiencies from conventional light-crude production will be replaced by supplies of upgraded crude and bitumen. Canada may be one of the few non-OPEC countries which already has the committed investment to increase oil supplies in the next decade (Littell, 1999, his Figure 3), with output already planned to reach 452,000 m^3/day (2.8 million bbl/day) by 2005.

Natural-gas resources are sufficient to support production rates as high as 254 billion m^3/year (9 tcf/year), but that may not be sustainable or advisable. Continued high contributions will come from the Alberta portion of the Western Canada Sedimentary Basin (including the "fold belt") and its northern extensions (if the intensity of drilling is maintained), East Coast basins (principally off Nova Scotia), the Mackenzie Delta, and the shallow-water areas of the Beaufort Sea. Again, gas supplies from remote areas may proceed only if alternative supply costs and large volumes of unconventional gas in areas of existing infrastructure do not undermine projected investments.

Large amounts of Canada's petroleum resources may never be extracted. For example, natural gas is a North American continental commodity, so areas favorable for natural-gas production in Canada must compete with other areas in the North American continent such as the Mid-Continent and the Gulf of Mexico. The gas resources off Labrador, the oil and gas resources of the Arctic islands, and Arctic exploration generally (Figure 12) are areas likely to remain orphaned, perhaps throughout the first few decades of the twenty-first century. The rate of production from massive oil-sands deposits will forever

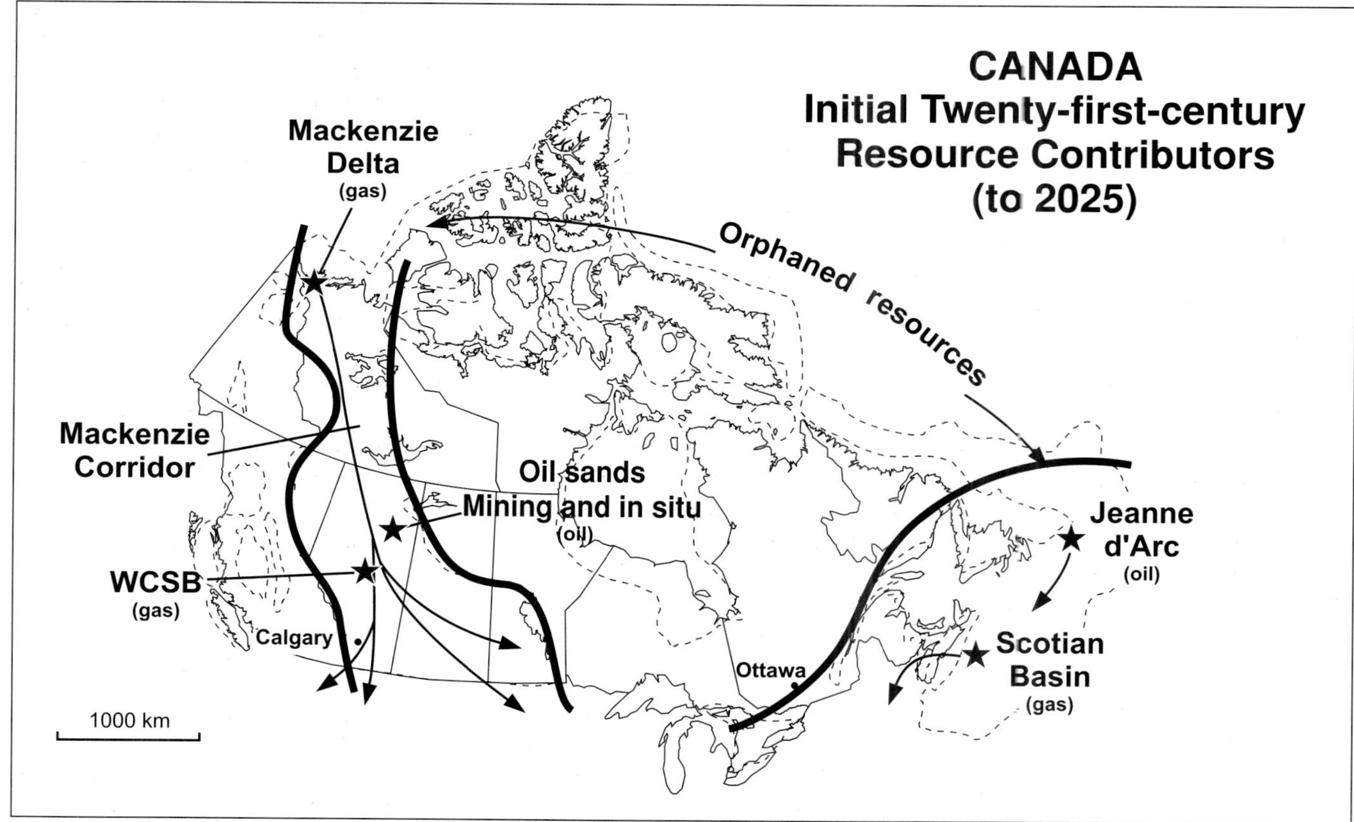

Figure 12. Initial resource contributors of Canada in the twenty-first century, highlighting the continued importance of the Western Canada Sedimentary Basin (WSCB), the oil sands, increasing production from the Scotian and Jeanne d'Arc Basins, and development of resources of the Mackenzie Delta.

fall short of potential capacity because of environmental demands, infrastructure requirements, demographic limitations in the work force, and commercial (massive-investment) considerations. A high proportion will continue to be reported as orphaned, stranded, or "static."

Unless commodity prices differ from those of the past 25 years, the investment focus for the next 25 years is already in place. New conceptual plays remain elusive in spite of new technologies to visualize the subsurface and to predict fluid contents in reservoirs "ahead of the bit." Nevertheless, significant ("trend-changing") new reserves await discovery, but these will occur initially (to 2025) only in areas of recently implanted infrastructure (e.g., Jeanne d'Arc Basin and the Scotian Shelf) and in areas of already known significant undeveloped resources requiring additional volumes to overcome development thresholds (e.g., Mackenzie Delta). Investors who are not party to existing or planned infrastructure in those areas may need to prove up initial stand-alone reserves to support alternative infrastructure or to form cooperative ventures (including unitization) to guarantee threshold recoverable resources for construction, funding, and optimal development of them. If "alliances" or other commercial arrangements are not made, substantial investment risk is present, because new add-on resources may not be brought to market for 10 to 20 years.

CONCLUSIONS

Overall, the Canadian petroleum industry is becoming a gas- and oil-sands business which will be enhanced further in the coming decades. The long-term future of the Canadian oil industry has to be in Alberta oil sands.

Initially in the twenty-first century, petroleum investment in Canada for the "big picture" and significant contributions to petroleum supply will focus on five major areas (Figure 12):

1. "mopping up" the Western Canada Sedimentary Basin. As major players vacate noncore production, smaller companies will continue the "feeding frenzy" for Cretaceous shallow gas and will probe the deeper Paleozoic potential, particularly in the fold belt and extensions north of latitude 60°N. The decline in conventional oil production will continue.
2. East Coast Scotian Shelf. The successful commissioning of the Sable Island gas project will encourage additional infrastructure expansion to serve the eastern-seaboard gas markets, while exploration and appraisal on other shallow-water shelf blocks will confirm and add to the reserve inventory. New wildcat wells will test prospects, particularly in the "diapiric zone" in deeper water.
3. East Coast Grand Banks. In these basins, particularly the Jeanne d'Arc Basin, operators will gain confidence with oil production in this harsh environment, resulting in a gradual uplift in the volumes of oil produced from this area. Effort will be directed to solving the gas-production challenge and ice management.
4. Alberta oil sands. Bitumen recovery and the manufacture of upgraded crude oil will continue to expand. Relative to frontier conventional oil, the product is low cost and competitively priced, and the resource is effectively unlimited and nondepleting.
5. Mackenzie Delta and shallow Beaufort Sea. Additional exploration will need to confirm adequate gas reserves prior to construction of a connector to the existing gas and liquid pipeline systems in the south.

With global scenarios preaching the end of low-cost oil (increasing production from OPEC), the use of abundant natural gas, alternative transport technologies, and environmental commitments (e.g., the Kyoto treaty), the first 25 years of the new century will be as challenging for the Canadian petroleum industry as the last 100 years were.

ACKNOWLEDGMENTS

I express my thanks to Grant Mossop and all members of his staff at the Geological Survey of Canada in Calgary, in particular Kirk Osadetz and Lloyd R. Snowdon, who unselfishly gave time to share thoughts and research papers with me during the formative stages of this paper. Ken Drummond freely shared with me his contributions to the assessment of Canadian resources. Andrew Nelson of Moose Oils tolerated my probing on the potential of the Canadian thrust belt and foothills. Jack Ingram kindly shared some of his thoughts on Canadian gas supply and marketing issues. The staffs of libraries of the National Energy Board and Geological Survey of Canada were very tolerant of my transgressions. My wife, Lyn, survived my Web browsing. My horse, Macs, was patient with an absent "cowboy."

KS Management Services Inc. provided funds for this project. The comments made herein are those of the author and may not reflect those of Antrim, its executive, or its directors. Brian Wyatt of Precise Drafting Services Ltd. of Calgary prepared the figures.

All errors of commission and omission remain those of the author.

REFERENCES CITED

Alliance, 1999, The pipeline: www.alliance-pipeline.com/Pipeline.html, accessed October 21, 1999.

Atlantic Geoscience Centre, 1989, East Coast basin atlas series: Labrador: Geological Survey of Canada Atlantic Geoscience Centre, 111 p.

Atlantic Geoscience Centre, 1991, East Coast basin atlas series: Scotian Shelf: Geological Survey of Canada Atlantic Geoscience Centre, 152 p.

Balkwill, H .R., N. J. McMillan, B. MacLean, G. L. Williams, and S. P. Srivastava, 1990, Geology of the Labrador Shelf, Baffin Bay, and Davis Strait, *in* M. J. Keen and G. L. Williams, eds., Geology of the continental margin of eastern Canada: Geological Survey of Canada, Geology of Canada, no. 2, p. 293–348 (also Geological Society of America, The geology of North America, v. I-1).

Barclay, J. E., and D. G. Smith, 1992, Western Canada foreland basin oil and gas plays, *in* R. W. Macqueen and D. A. Leckie, eds., Foreland basins and fold belts: AAPG Memoir 55, p. 191–228.

Bateman, J. A., E. H. Davies, J. R. Sarg, and J. W. Sneddon, 1999, Deepwater lowstand exploration plays of the Jeanne d'Arc Basin, offshore Newfoundland, Canada (abs.): AAPG Bulletin, v. 3, p. 1299.

Bell, J. S., and G. R. Campbell, 1990, Petroleum resources, *in* M. J. Keen and G. L. Williams, eds., Geology of the continental margin of eastern Canada: Geological Survey of Canada, Geology of Canada, no. 2, p. 677–720 (also Geological Society of America, The geology of North America, v. I-1).

Bird, T. J., P. R. Price, M. C. Fortier, and G. Cave, 1995, Oil and gas exploration plays of mainland Northwest Territories and Yukon, *in* J. S. Bell, T. D. Bird, T L. Hillier, and P. L. Greener, eds., Proceedings of the oil and gas forum '95—Energy from sediments: Geological Survey of Canada Open File Report 3058, p. 349–352.

Bowers, B., and K. J. Drummond, 1997, Conventional crude oil resources of the Western Canadian Sedimentary Basin: Journal of Petroleum Technology, v. 36, p.56–63.

Bowers, B., J. Bielecki., J. Hu., W. T. Wall, and K. J. Drummond, 1995, Impact of horizontal drilling on western Canadian supply of conventional crude oil: Paper 95-58, presented at the 46th Annual Meeting of the Petroleum Society of Canadian Institute of Mining, Metallurgy and Petroleum, Banff.

BP, 1999, Statistical review of world energy 1999: www.bpamoco.com/worldenergy, accessed October 26, 1999.

Brooks, P. W., M. G. Fowler and R. W. Macqueen, 1988, Biological marker and conventional organic chemistry of oil sands/heavy oils, Western Canada basin: Organic Geochemistry, v. 12, p. 519–538.

Bruce, G. C., 1999, East Coast outlook: Paper presented at CERI North American Crude Oil and NGL's Conference, Calgary, Alberta.

Canada–Nova Scotia Offshore Petroleum Board, 1997, Technical summaries of Scotian Shelf significant and commercial discoveries: 205.150.149.160/tech2.html, accessed October 31, 1999.

Canadian Association of Petroleum Producers, 1999a, Industry performance: www.capp.ca/02b.html, accessed September 27, 1999.

Canadian Association of Petroleum Producers, 1999b, Producing areas: www.capp.ca/producingareas.html, accessed September 27, 1999.

Canadian Gas Potential Committee, 1997, Natural gas potential in Canada: Canadian Gas Potential Committee, 113 p.

Clark, K., C. Hetherington, C. O'Neill, and J. Zavitz, 1997, Breaking the ice with finesse. Oil and gas exploration in the Canadian Arctic: The Arctic Institute of North America, 248 p.

Conn, R. F., R. R. Waghmare, L. Roux, and S. M. Dallaire, 1995, Preliminary estimates of recoverable gas—Future reserve additions from undiscovered natural gas resources in Western Canada Sedimentary Basin, *in* J. S. Bell, T. D. Bird, T. L. Hillier, and P. L. Greener, eds., Proceedings of the oil and gas forum '95—Energy from sediments: Geological Survey of Canada Open File Report 3058, p. 329–330.

Couper, A., 1983, The Times atlas of the oceans: Angus & Robertson Publishers, Sydney, Australia, 272 p.

Creaney, S., and J. Allan, 1990, Hydrocarbon generation and migration in the Western Canada Sedimentary Basin, *in* J. Brooks, ed., Classic petroleum provinces: Geological Society Special Publication 50, p. 189–202.

Davidson, J. A., A. P. Hamblin, and J. Dixon, 1995, Petroleum resources of the Eagle Plain Basin, Yukon Territory, Canada, *in* J. S. Bell, T. D. Bird T. L. Hillier, and P. L. Greener, eds., Proceedings of the oil and gas forum '95—Energy from sediments: Geological Survey of Canada Open File Report 3058, p. 357–359.

Dawson, F. M., 1995, Coalbed methane: A comparison between Canada and the United States: Geological Survey of Canada Bulletin 489, 60 p.

Dietrich, J. R., 1995a, Petroleum resource potential of the Queen Charlotte Basin region, West Coast Canada, *in* J. S. Bell, T. D. Bird, T. L. Hillier, and P. L. Greener, eds., Proceedings of the oil and gas forum '95—Energy from sediments: Geological Survey of Canada Open File Report 3058, p. 331– 333.

Dietrich, J. R., 1995b, Petroleum resource potential of the Queen Charlotte Basin and environs, West Coast Canada: Bulletin of Canadian Petroleum Geology, v. 43, p. 20–34.

Dingle, H. B., 1999a, Notes for remarks to Canadian Heavy Oil Association Annual General Meeting, Calgary: www.imperialoil.ca/news/sp99023.htm, accessed September 17, 1999.

Dingle, H. B., 1999b, Notes for remarks to Canadian Association of Petroleum Producers' Investment Symposium, Calgary, www.imperialoil.ca/news/sp990615.htm, accessed September 17, 1999.

Dixon, J. R., 1995, ed., Geological atlas of the Beaufort-Mackenzie area: Geological Survey of Canada Miscellaneous Report 59, 17p.

Dixon, J. R., and L. D. Stasiuk, 1998, Stratigraphy and hydrocarbon potential of Cambrian strata, Northern Interior Plains, Northwest Territories: Bulletin of Canadian Petroleum Geology, v. 46, p. 445–470.

Dixon, J., J. Dietrich, L. R. Snowdon, G. Morrell, and D. H. McNeil, 1992, Geology and petroleum potential of Upper Cretaceous and Tertiary strata, Beaufort-Mackenzie area, northwest Canada: AAPG Bulletin, v. 76, p. 927–947.

Dixon, J., G. R. Morrell, J. R. Dietrich, G. C. Taylor, R. M. Procter, R. F. Conn, S. M. Dallaire, and J. A. Christie, 1994, Petroleum resources of the Mackenzie Delta and Beaufort Sea: Geological Survey of Canada Bulletin 474, 52 p.

Drummond, K. J., 1992, Geology of Venture, a geopressured gas field, offshore Nova Scotia, *in* M. T. Halbouty, ed., Giant oil and gas fields of the decade 1978–1988: AAPG Memoir 54, p. 55–71.

Drummond, K. J., 1995, Gas reserves growth in the Western Canada Sedimentary Basin: CSPG-CWLS Symposium, Calgary, 10 p.

Drummond, K. J., 1998, East Coast gas—The big picture: CERI Eastern Canadian Natural Gas Conference, Halifax, 7 p.

Energy Information Administration, 1999, Canada (December 1998): www.eia.doe.gov/emeu/cabs/canada.html, accessed September 25, 1999.

Flach, P. D., 1984, Oil sands geology—Athabasca deposit north Alberta: Alberta Research Council Bulletin 46, 31 p.

Fowler, M. G., and McAlpine, K. D., 19995, The Egret member, a prolific Kimmeridgian source rock from offshore Eastern Canada, *in* B. J. Katz, ed., Sedimentary source rock case studies: Berlin, Springer-Verlag, 327 p.

Geological Atlas of the Western Canada Sedimentary Basin, 1998, Geological atlas of the Western Canada Sedimentary Basin: www.ags.gov.ab.ca/AGS_PUB/ATLAS_WWW/ATLAS.HTML, accessed July 26, 1999.

Geological Map of Canada, 1999, Map D1860A (CD-ROM): www.nrcan.gc.ca/ess/carto/gmc/gnotes_e.htm, accessed September 6, 1999.

Geological Survey of Canada, 1995, Arctic islands subsurface and geophysics: www.nrcan.gc.ca/~tbrent/wrasub.htm, accessed September 6, 1999.

Government of Newfoundland and Labrador, 1999, Oil and gas report, October 1999: www.gov.nf.ca/mines&en/ENERGY/Oil&Gas.htm, accessed October 31, 1999.

Grant, A. C., and McAlpine, K. D., 1990, The continental margin around Newfoundland, *in* M. J. Keen and G. L. Williams, eds., Geology of the continental margin of eastern Canada: Geological Survey of Canada, Geology of Canada, no. 2, p. 239–292 (also Geological Society of America, The geology of North America, v. I-1).

Grant, A. C., K. D. McAlpine, and J. A. Wade, 1986, The continental margin of eastern Canada: Geological framework and petroleum potential, *in* M. T. Halbouty, ed., Future petroleum provinces of the world: AAPG Memoir 40, p. 177–205.

Hannigan, P. K., J. R. Dietrich, P. J. Lee, and K. G. Osadetz, 1998, Petroleum resource potential of sedimentary basins of the Pacific margin of Canada: Geological Survey of Canada Open File Report 3629, 85 p.

Hibernia, 1999a, Oil production at Hibernia: www.hibernia.ca/product.htm, accessed October 21, 1999.

Hibernia, 1999b, People pioneering offshore excellence, extraction costs: www.hibernia.ca/p-extrac.htm, accessed September 27, 1999.

Hills, L. V., 1974, Oil sands fuel of the future: Canadian Society of Petroleum Geologists Memoir 1, 263 p.

Hubbard, S. M., S. G. Pemberton, and E. A. Howard, 1999, Regional geology and sedimentology of the basal Cretaceous Peace River oil sands deposit, north-central Alberta: Bulletin of Canadian Petroleum Geology, v. 47, p. 270–297.

Hurley, T. J., R. D. Kreisa, G. C. Taylor, W. R. L. Yates, 1992, The reservoir geology and geophysics of the Hibernia field, offshore Newfoundland, *in* M. T. Halbouty, ed., Giant oil and gas fields of the decade 1978–1988: AAPG Memoir 54, p. 35–54.

Imperial Oil, 1999a, Fact sheet: www.imperialoil.ca/investor/factbook/fb_OilSands.htm, accessed September 17, 1999.

Imperial Oil, 1999b, Overview of operations—Cold Lake: www.imperialoil.ca/investor/factbook/fb_resources-overview.htm, accessed September 17, 1999.

Johnson R. D., and N. J. McMillan, 1993, Petroleum, *in* D. F. Stott and J. D. Aitken, eds., Sedimentary cover of the craton in Canada: Geological Survey of Canada, Geology of Canada, no. 5, p. 505–562 (also Geological Society of America, The geology of North America, v. D-1).

Lane, L. S., and J. R. Dietrich, 1995, Tertiary structural evolution of the Beaufort Sea–Mackenzie Delta region, Arctic Canada: Bulletin of Canadian Petroleum Geology, v. 43, p. 293–314.

Langenberg, C. W., B. A. Rottenfusser, and R. J. H. Richardson, 1997, Coal and coalbed methane in the Mannville Group and its equivalents, Alberta, *in* S. G. Pemberton and D. P. James, eds., Petroleum geology of the Cretaceous Mannville Group, western Canada: Canadian Society of Petroleum Geologists Memoir 18, p. 475–486.

Lee, P. J. 1998, Oil resources of western Canada: Geological Survey of Canada Open File Report 3674, 142 p.

Lewis, C. F. M., and M. J. Keen, 1990, Constraints to development, *in* M. J. Keen and G. L. Williams, eds., Geology of the continental margin of eastern Canada: Geological Survey of Canada, Geology of Canada, no. 2, p. 743–823 (also Geological Society of America, The geology of North America, v. I-1).

Littell, G., 1999, The Future of oil: Paper presented to Canadian Landman's Annual Meeting, Vancouver.

Lou, P., H. G. Machel, and J. Shaw, 1994, Petrophysical properties of matrix blocks of the hetrerogeneous dolostone reservoir—the Upper Devonian Grosmont Formation, Alberta, Canada: Bulletin of Canadian Petroleum Geology, vol. 42, p. 465–481.

Mackay, A. H., and A. J. Tankard, 1990, Hibernia oil field—Canada, Jeanne d'Arc Basin, Grand Banks Offshore Newfoundland, *in* E. A. Beaumont and N. H. Foster, eds., Structural traps III: AAPG Treatise of Petroleum Geology, p. 145–176.

MacLean, B. C., and J. A. Wade, 1992, Petroleum geology of the continental margin south of the islands of St. Pierre and Miquelon, offshore Eastern Canada: Bulletin of Canadian Petroleum Geology, v. 40, p. 222–253.

Masters, J. A., 1984, Lower Cretaceous oil and gas in western Canada, *in* J. A. Masters, ed., Elmworth—Case study of a Deep Basin gas field: AAPG Memoir 38, p. 1–35.

McCrossen, R. G., 1973, The future petroleum provinces of Canada: Canadian Society of Petroleum Geologists Memoir 1, 720 p.

Meneley, R. A., 1986, Oil and gas fields in the East Coast and Arctic basins of Canada, *in* M. T. Halbouty, ed., Future petroleum provinces of the world: AAPG Memoir 40, p. 143–176.

Meyerhoff, A. A.,1982, Hydrocarbon resources in Arctic and subarctic regions, *in* A. F. Embry and H.R. Balkwill, 1982, eds., Arctic geology and geophysics: Canadian Society of Petroleum Geologists Memoir 8, p. 451–552.

Morrell, G. R., M. Fortier, P. R. Price, and R. Polt, 1995, Petroleum exploration in northern Canada: A guide to oil and gas exploration potential: Northern Oil and Gas Directorate, Indian and Northern Affairs Canada, 110 p.

Mossop, G. D., and I. Shetsoen, comps., 1994, Geological atlas of the Western Canada Sedimentary Basin: Canadian Society of Petroleum Geologists and Alberta Research Council, 510 p.

National Energy Board, 1996, A natural gas resource assessment of southeast Yukon and Northwest Territories, Canada: National Energy Board Energy Resources Branch report, 140 p.

National Energy Board, 1998, Probabilistic estimate of hydrocarbon volumes in the Mackenzie Delta and Beaufort Sea discoveries: National Energy Board report, 8 p.

National Energy Board, 1999a, Canadian energy supply and demand to 2025: National Energy Board report, 96 p.

National Energy Board, 1999b, Short-term natural gas deliverability from the Western Canada Sedimentary Basin 1998–2001: National Energy Board report, 39 p.

Newfoundland Ocean Industries Association, 1998, Harnessing the potential— Atlantic Canada's oil and gas industry: www.noia.nf.ca/ch2.html, accessed October 20, 1999.

Northern Oil and Gas Directorate, 1999, Winning bids—1999 Beaufort Sea/Mackenzie Delta call for bids: Northern Oil and Gas Bulletin, v. 6, no. 3, www.inac.gc.ca/oil/bulletin/vol6_3html, accessed September 25, 1999.

Owens, D., 1999, Gas by the turn of the century: www.soep.com/speech/owen.html, accessed October 31, 1999.

PanCanadian, 1999, Deep Panuke Carbonate gas play: Investor presentation, November 1999: www.pancanadian.ca/financial/NovPresentation/sld028.htm, accessed November 15, 1999.

Price, P. R., 1995, Petroleum resources of the Liard plateau area, southern Yukon Territory, Canada, *in* J. S. Bell, T. D. Bird, T. L. Hillier, and P. L. Greener, eds., Proceedings of the oil and gas forum '95—Energy from sediments: Geological Survey of Canada Open File Report 3058, p. 353–355.

Procter, R. M., G. C. Taylor, and J. A. Wade, 1984, Oil and natural gas resources of Canada: Geological Survey of Canada Paper 83-31, 59 p.

Ranger, M. J., and S. G. Pemberton, 1997, Elements of a stratigraphic framework for the McMurray Formation in south Athabasca area, Alberta, *in* S. G. Pemberton and D. P. James, eds., Petroleum geology of the Cretaceous Mannville Group, western Canada: Canadian Society of Petroleum Geologists Memoir 18, p. 263–291.

Ricketts, B. D., ed., 1989, Western Canada Sedimentary Basin: A case history: Canadian Society of Petroleum Geologists, 320 p.

Roberge, R .B., 1999, Canadian oil and gas survey: Calgary, PricewaterhouseCoopers, 36 p.

Rogers, A .L., and N. A. Yassir, 1993, Hydrodynamics and overpressuring in the Jeanne d'Arc Basin, offshore Newfoundland, Canada: Possible implications for hydrocarbon exploration: Bulletin of Canadian Petroleum Geology, v. 41, p. 275–289.

Resources, Wildlife and Economic Development, 1999, Minerals, Oil and Gas Division: Oil and gas: Exploration highlights: Significant Discovery Licenses (SDLs): www.gov.nt.ca/RWED/mog/html/sdl.htm, accessed September 10, 1999.

Sable Offshore Energy Incorporated, 1999, Development plan application: www.soep.com/volume1/1_1.html, accessed February 10, 1999.

Samson, L., 1999, Gas production and decline rates in the province of Alberta: Journal of Canadian Petroleum Technology, v. 38, p 60–65.

Samson, L., and M.-A. Kirsch, 1999, Alberta's drilling, gas reserve additions, and gas supply and demand: Journal of Canadian Petroleum Technology, v. 38, p 55–59.

Sinclair, I. K., K .D. McAlpine, D. F. Sherwin, N. J. McMillan, G. C. Taylor, M. E. Bast, G. R. Campbell, J. P. Hea, D. Hanao, and R. M. Procter, 1992, Petroleum resources of the Jeanne d'Arc Basin and Environs, Grand Banks, Newfoundland: Geological Survey of Canada Paper 92-8, 48 p.

Singh, S., M. P. Du Plessis, E. E. Isaacs, and R. Kerr, 1999, Cost analysis of advanced technologies for the production of heavy oil and bitumen in western Canada: World Energy Council: www.worldenergy.org/wecgeis/members_only/registered/open.plx?file=p1, accessed October 15, 1999.

Stobl, R. S., W. K. Muwais, D. M. Wightman, D. K. Cotterill, and L. Yuan, 1997, Geologic modelling of McMurray Formation reservoirs based on outcrop and subsurface analogues, *in* S. G. Pemberton and D. P. James, eds., Petroleum geology of the Cretaceous Mannville Group, western Canada: Canadian Society of Petroleum Geologists Memoir 18, p. 292–311.

Suncor, 1999, Oilsands: www.suncor.com/oilsands/, accessed October 31, 1999.

Syncrude, 1999, Syncrude 21: www.syncrude.com, accessed October 31, 1999.

Taylor, G. C., M. Best, G. R. Campbell, J. P. Hea, D. Henao, and R. M. Proctor, 1991, Petroleum resources of the Jeanne d'Arc Basin, Grand Banks of Newfoundland: Geological Survey of Canada Open File Report 2150, 13 p.

Terra Nova, 1999, What is Terra Nova?: www.terranovaproject.com/whatis/whatis.htm, accessed October 21, 1999.

Tsang, P. B., 1990, Taglu Field, *in* E. A. Beaumont and N. H. Foster, eds., Atlas of oil and gas fields: Structural traps 1: AAPG Treatise of petroleum geology, p. 191–212.

U.S. Geological Survey, 1997, Ranking of the world's oil and gas provinces by known petroleum volumes: Open-file Report 97-463, Table 1: www.certmetra.cr.usgs.gov/energy/WorldEnergy/OF97-463/97463tb11.html, accessed August 12, 1999.

Wade, J. A., and B. C. MacLean, 1990, The geology of the southeastern margin of Canada, *in* M. J. Keen and G. L. Williams, eds., Geology of the continental margin of eastern Canada: Geological Survey of Canada, Geology of Canada, no. 2, p. 167–238 (also Geological Society of America, The geology of North America, v. I-1).

Wade, J. A., G. R. Campbell, R. M. Procter, and G. C. Taylor, 1989, Petroleum resources of the Scotian Shelf: Geological Survey of Canada Paper 88-19, 26 p.

Waylett, D. C., 1990, Drake Point gas field, Canadian Arctic islands, *in* E. A. Beaumont and N. H. Foster, eds., Structural traps I: AAPG Treatise of petroleum geology, p. 77–102.

Waylett, D. C., and A. F. Embry, 1992, Hydrocarbon loss from oil and gas fields of the Sverdrup Basin, Canadian Arctic islands, *in* T. O. Vorren et al., eds., Arctic geology and petroleum potential: Amsterdam, Elsevier, p. 195–204.

Wightman, D. M., and S. G. Pemberton, 1997, The Lower Cretaceous (Aptian) McMurray Formation: An overview of the Fort McMurray area, northeastern Alberta, *in* S. G. Pemberton and D. P. James, eds., Petroleum geology of the Cretaceous Mannville Group, western Canada: Canadian Society of Petroleum Geologists Memoir 18, p. 312–344.

Williams, H., and A. C. Grant, 1998, Tectonic assemblages map, Atlantic region, Canada: Geological Survey of Canada Open File 3657, scale 1:3,000,000.

Williamson, M. A., K. DesRoches, and S. King, 1993, Overpressures and hydrocarbon migration in the Hibernia-Nautilus area of the Jeanne d'Arc Basin, offshore Newfoundland: Bulletin of Canadian Petroleum Geology, v. 41, p. 389–406.

World Energy Council, 1999, Canada: www.worldenrgy.org/wec-geis/members_only/registered/open.plx?file, accessed August 7, 1999.

Yassir, N. A., and J. S. Bell, 1994, Relationships between pore pressure, stresses, and present-day geodynamics in the Scotian Shelf, offshore eastern Canada: AAPG Bulletin, v. 78, p. 1863–1880.

Young, S. B., and K. J. Drummond, 1994, An analysis of the Canadian natural gas resource base: National Energy Board, Canada paper presented at Global Gas Resources Workshop, Vail, Colorado, 18 p.

Yukon Territorial Government, 1994a, Petroleum resource assessment of the Liard plateau area: National Energy Board report prepared for the Yukon Territorial Government, Yukon Economic Development—Oil and Gas Resources report, 55 p.

Yukon Territorial Government, 1994b, Petroleum resource assessment of the Eagle Plain Basin, Yukon Territory, Canada: National Energy Board report prepared for the Yukon Territorial Government, Yukon Economic Development—Oil and Gas Resources report, 65 p.

Bird, K. J., 2001, Alaska: A twenty-first-century petroleum province, *in* M. W. Downey, J. C. Threet, and W. A. Morgan, eds., Petroleum provinces of the twenty-first century: AAPG Memoir 74, p. 137–165.

Chapter 9

ALASKA: A TWENTY-FIRST-CENTURY PETROLEUM PROVINCE

Kenneth J. Bird
U.S. Geological Survey, Menlo Park, California, U.S.A.

ABSTRACT

Alaska, the least explored of all United States regions, is estimated to contain approximately 40% of total U.S. undiscovered, technically recoverable oil and natural-gas resources, based on the most recent U.S. Department of the Interior (U.S. Geological Survey and Minerals Management Service) estimates.

Northern Alaska, including the North Slope and adjacent Beaufort and Chukchi continental shelves, holds the lion's share of the total Alaskan endowment of more than 30 billion barrels (4.8 billion m^3) of oil and natural-gas liquids plus nearly 200 trillion cubic feet (5.7 trillion m^3) of natural gas. This geologically complex region includes prospective strata within passive-margin, rift, and foreland-basin sequences. Multiple source-rock zones have charged several regionally extensive petroleum systems. Extensional and compressional structures provide ample structural objectives. In addition, recent emphasis on stratigraphic traps has demonstrated significant resource potential in shelf and turbidite systems in Jurassic to Tertiary strata.

Despite robust potential, northern Alaska remains a risky exploration frontier—a nexus of geologic complexity, harsh economic conditions, and volatile policy issues. Its role as a major petroleum province in this century will depend on continued technological innovations, not only in exploration and drilling operations, but also in development of huge, currently unmarketable natural-gas resources. Ultimately, policy decisions will determine whether exploration of arctic Alaska will proceed.

INTRODUCTION

Alaska, America's remaining frontier, is a large and lightly populated region with abundant mineral, energy, and biologic resources. Its arctic and subarctic ecosystems are rich in scenic beauty and wilderness characteristics. As in much of the American West, land ownership is mostly public. Competing land-use interests have been and likely will continue to be a significant factor in resource development and in shaping Alaska's role as a petroleum province of the twenty-first century.

Alaska is a mountainous region, but it has many sedimentary basins. The physiographic map (Figure 1a) shows the mountainous nature of this region, a reflection of the fact that it represents the northern terminus of the American Cordillera. Only the coastal plain of northern Alaska, a continuation of the North American Interior Plains province, lies outside the cordillera. The geologic history of Alaska, as summarized by Plafker and Berg (1994), shows that this region was assembled by terrane accretion and that a great number of sedimentary basins are present. However, most early-formed basins were deformed, uplifted, and eroded in the accretion process, greatly diminishing their petroleum potential. Basins with greatest petroleum potential are those which formed during or after accretion. These basins are shown in Figure 1b. Most Alaskan basins now have been available for exploration and have undergone at least one round of leasing and drilling. Thus, considerably more is known about their petroleum potential than, for example, in the early 1980s, at the time of the first Pratt Conference (Hohler and Bischoff, 1986). Commercial petroleum production comes from the North Slope and Cook Inlet.

The basis for this analysis of Alaska's twenty-first-century petroleum potential is the assessment of undiscovered oil and gas resources conducted in 1995 by the U.S. Department of the Interior's U.S. Geological Survey and Minerals Management Service (U.S. Geological Survey National Oil and Gas Assessment Team, 1995; Sherwood, 1998). Considering only conventional resources, the mean undiscovered oil and gas resource potential of Alaska is estimated at more than 30 billion barrels (bbl) (4.8 billion m^3) of oil and natural-gas liquids and nearly 200 trillion cubic feet (tcf) (5.7 trillion m^3) of gas (Figure 2)—an amount that represents about 40% of the total undiscovered United States oil and gas endowment. Analysis by individual Alaskan region shows that the future petroleum potential lies overwhelmingly (90% oil and 84% gas) in northern Alaska (referred to as the North Slope Basin in Figure 1b). By comparison, the next most prospective area of Alaska is the Cook Inlet Basin, with a mean undiscovered resource potential (onshore and offshore) estimated at 1.4 billion bbl (0.2 billion m^3) of oil and natural-gas liquids and 2.3 tcf (65 billion m^3) of gas.

With such a clear concentration of undiscovered oil and gas resources, the focus of the remainder of this paper will be on the north Alaska region. What follows is a description of its geologic-tectonic setting, stratigraphy, exploration status, infrastructure and land ownership, petroleum discoveries and production, and petroleum systems, and a review of its petroleum potential by major reservoir interval.

GEOLOGIC AND TECTONIC SETTING OF NORTHERN ALASKA

The northern Alaska petroleum province extends about 1100 km from the Canadian border westward to the maritime Russian border and from 100 to 600 km northward from the Brooks Range to the edge of the continental shelf (Figure 3). The province is bounded on the south by the Brooks Range–Herald arch orogenic belt and on the north by a passive continental margin. The offshore part of the province is characterized in the east by the relatively narrow, 100-km-wide Beaufort continental shelf and in the west by the broad, 600-km-wide Chukchi platform.

Northern Alaska is but a part of the largest terrane comprising Alaska. By some accounts, this terrane extends several hundred kilometers eastward to include part of western Canada and more than 1000 km westward to include the Russian Chukotka peninsula (e.g., Churkin and Trexler, 1980). Its origin is controversial, with interpretations ranging from no displacement to

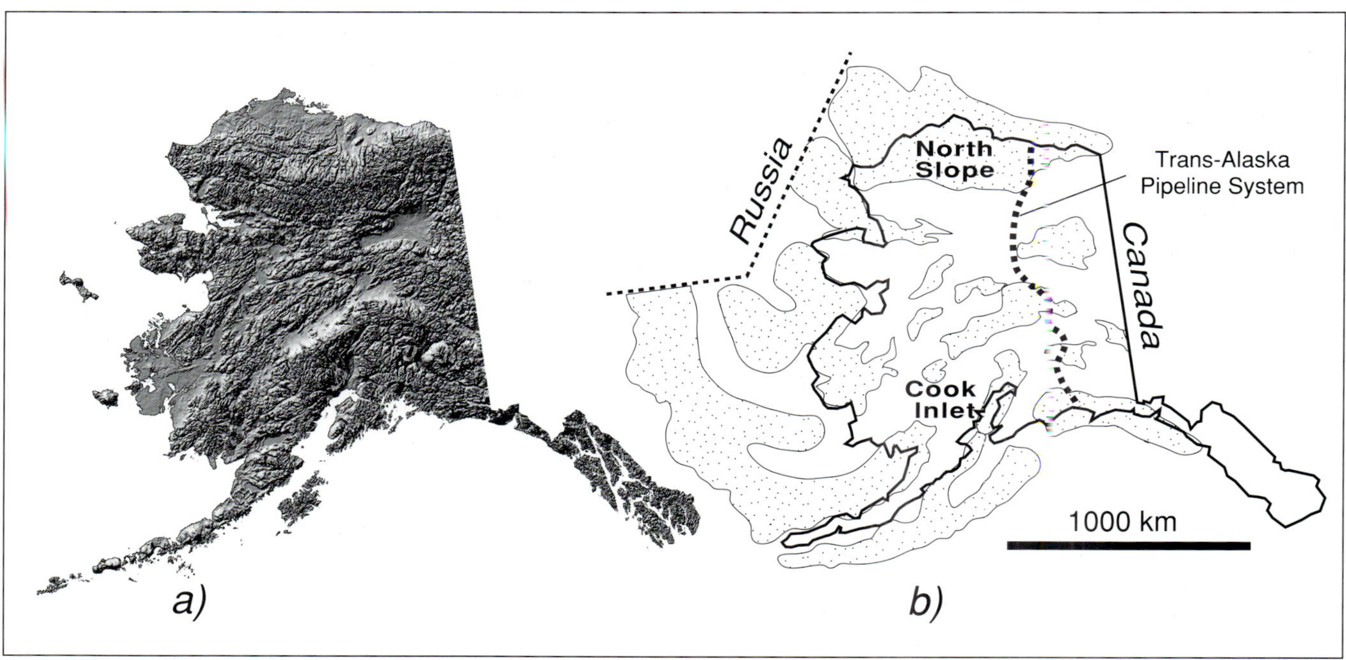

Figure 1. Maps of Alaska. (a) Shaded relief displays the mountainous nature of the region, demonstrating that all but northernmost Alaska is part of the North American Cordillera. (b) Map showing distribution of relatively intact and generally young sedimentary basins (stippled). Cook Inlet and North Slope are currently the only oil-productive basins. The Trans-Alaska Pipeline System is the transportation route linking North Slope oil fields with shipping terminal at Valdez. Basin outlines are adapted from Kirschner (1994).

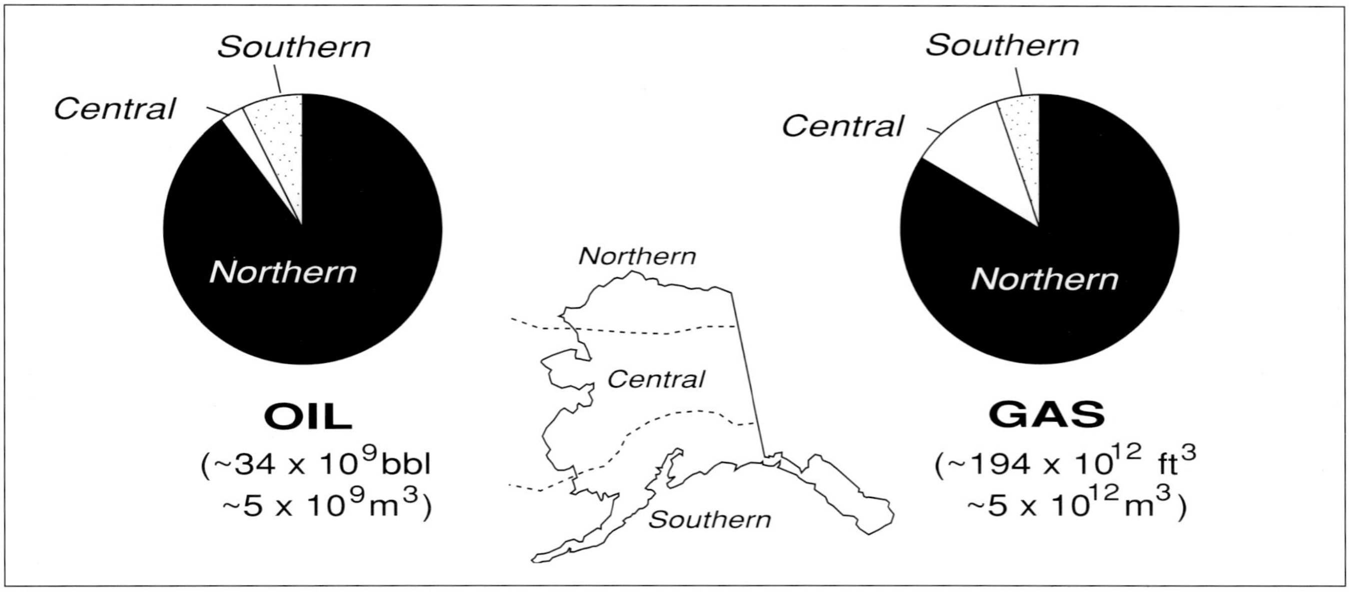

Figure 2. Diagrams showing the proportion of undiscovered, technically recoverable oil and gas resources of Alaska by regions, including onshore and offshore. Total volume of oil and natural-gas liquids and natural gas represents the sum of regional mean-value estimates from the Department of the Interior 1995 national oil and gas resource assessment (Sherwood, 1998; Gautier et al., 1995).

large-scale rotational or translational displacement along transform faults. See Lawver and Scotese (1990) and Moore et al. (1994) for summaries of these interpretations, and Lane (1997) and Grantz et al. (1998) for recent interpretations.

The major tectonic features of northern Alaska are summarized in Figure 3. Because of uncertainties in original orientation of this continental fragment, all references to directions are given in present-day coordinates. The Chukchi and Arctic platforms are remnants of a late Paleozoic to early Mesozoic south-facing continental margin. These features are separated by a north-trending structural sag, the Hanna trough, characterized by extensional normal faulting and thick sediment accumulations developed in the Devonian (?) and Mississippian. Sherwood et al. (1998) considered the Hanna trough to be a failed rift. The Barrow arch (rift shoulder) and adjacent hinge-line fault zone (rift margin) are features developed during an episode of successful rifting in the Jurassic and Early Cretaceous. The oceanic Canada basin and the flanking passive margin owe their origin to this rifting event. At the southern margin of the Arctic and Chukchi platforms and overlapping in time with rifting to the north, an arc-continent collision produced the Brooks Range, the adjacent Colville foreland basin and, presumably, the Herald arch orogenic belt. Compressional deformation in the Tertiary produced a fold-and-thrust belt that extends well north of the Brooks Range. The belt crosses the entire southern part of the foreland basin and, in the east, it extends offshore across the passive margin.

STRATIGRAPHY

The stratigraphic record of the northern Alaska province extends into the Precambrian, but rocks with petroleum potential are mostly younger than Devonian (Figure 4). The traditional grouping of the rocks into tectono-stratigraphic sequences is followed here. This scheme, proposed by Lerand (1973) and modified by later investigators, emphasizes tectonic history, provenance, and genetic relations. The sequences are described briefly here. Additional stratigraphic details are provided in later discussions of petroleum potential.

The Franklinian sequence includes Devonian and older sedimentary and some igneous rocks representing diverse origins and a complex geologic history. Although knowledge of these rocks is limited, it is known that they have everywhere been buried and metamorphosed beyond the thermal stage for oil preservation. Thus, they are generally considered economic basement.

The Mississippian to Triassic Ellesmerian sequence consists of carbonate and nonmarine to shallow-marine siliciclastic deposits. On the Arctic platform, these are continental-shelf deposits developed on a south-facing passive margin; from a northern pinch-out or erosional edge, they thicken southward to a typical thickness of about 2 km (Moore et al., 1994). Westward, these deposits thicken dramatically into the Hanna trough and then thin against the Chukchi platform. In the Hanna trough, Devonian (?) and Mississippian rocks are interpreted as synrift deposits, and younger strata as thermal sag-phase deposits (Sherwood et al., 1998). The Ellesmerian sequence includes both reservoir and source rocks. However, because the sequence is generally thin, source rocks, which lie near the top of the sequence, were not capable of petroleum generation until they were buried by Beaufortian and Brookian deposits.

The Jurassic and Lower Cretaceous (Neocomian) Beaufortian sequence (Hubbard et al., 1987) consists of synrift deposits derived locally or from the north. This is a stratigraphically complex, mud-dominated sequence with multiple unconformities and large variations in thickness. It includes source rocks and reservoir rocks. Normal faulting and development of sediment-filled grabens and half grabens, some with more than 3 km of fill (e.g., Dinkum graben; Grantz et al., 1988), occur mainly north of the present coastline. Uplift and erosion along the rift margin resulted in formation of the regional Lower Cretaceous unconformity (LCU), considered to be the breakup unconformity (Grantz and May, 1983). This unconformity progressively truncates all older units northward onto the Barrow arch. Because it truncates many reservoir and source rocks, it plays an important role in many of the largest oil fields in northern Alaska. This role includes development of enhanced porosity in subunconformity reservoirs (Jameson, 1994; Shanmugam and Higgins, 1988); provision of a migration pathway for hydrocarbons to charge multiple sub- and supraunconformity reservoirs; and juxtaposition of overlying marine mudstone source and seal rocks—pebble-shale unit and highly radioactive zone (HRZ) of the Hue Shale—with subunconformity reservoir rocks. The Barrow arch, first formed during rift-related uplift and subsequently accentuated by subsidence to the north and south, is a regional basement high and a key element in the formation of most north Alaskan oil fields. Subsidence of the rift margin and the ensuing marine transgression resulted in deposition of a blanketlike marine mudstone (pebble-shale unit) that marks the end of the Beaufortian sequence

Cretaceous and Tertiary deposits derived from the Brooks Range orogen are assigned to the Brookian sequence. These voluminous deposits filled the Colville foreland basin, overtopped the rift shoulder (Barrow arch), and built the passive margin that forms the present continental terrace north of Alaska. The Brookian sequence consists of a complex assemblage of siliciclastic strata that includes distal, condensed marine mudstone (Hue Shale); relatively deep marine basinal, slope, and outer-shelf mudstones and turbiditic sandstones (Torok, Seabee, and Canning Formations); and shallow-marine to coal-bearing nonmarine sandstone, mudstone, and conglomerate (Nanushuk Group and Sagavanirktok Formation). The lower part of the Hue Shale, an important source rock, is characterized by an interval of high gamma-ray readings known as the gamma-ray zone (GRZ), or the highly radioactive zone (HRZ) (Molenaar

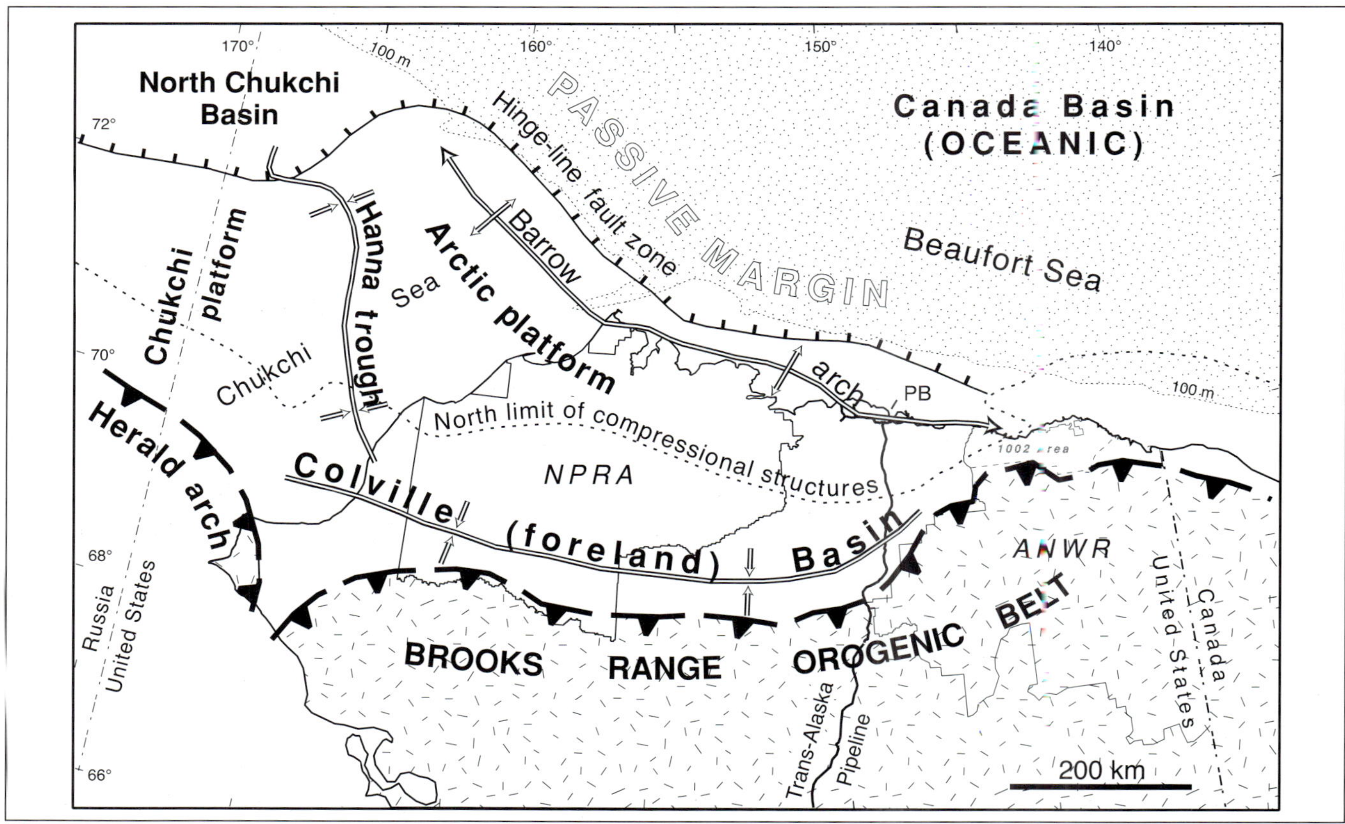

Figure 3. Map showing major tectonic features of northern Alaska. ANWR = Arctic National Wildlife Refuge; NPRA = National Petroleum Reserve–Alaska; PB = Prudhoe Bay.

et al., 1987). This unit typically occurs in a bottomset seismic facies. However, it also interfingers with downlapping clinoform strata and, in extreme cases such as the 93-Ma transgression responsible for the middle Cretaceous Seabee Formation, it may even occur in a slope position. Basin filling, including construction of the passive margin, occurred generally from west to east, with the exception of renewed Tertiary deposition in the northwestern part of the Chukchi Sea. Brookian deposits provided the overburden necessary for maturation of petroleum source rocks. Migrating depocenters related to longitudinal basin filling resulted in changing sites of petroleum generation and migration through time (e.g., Hubbard et al., 1987).

EXPLORATION, INFRASTRUCTURE, AND LAND OWNERSHIP

Oil exploration in north Alaska began in 1923 with federal field surveys at the time of creation of Naval Petroleum Reserve No. 4 (currently known as the National Petroleum Reserve, Alaska, or NPRA). The government

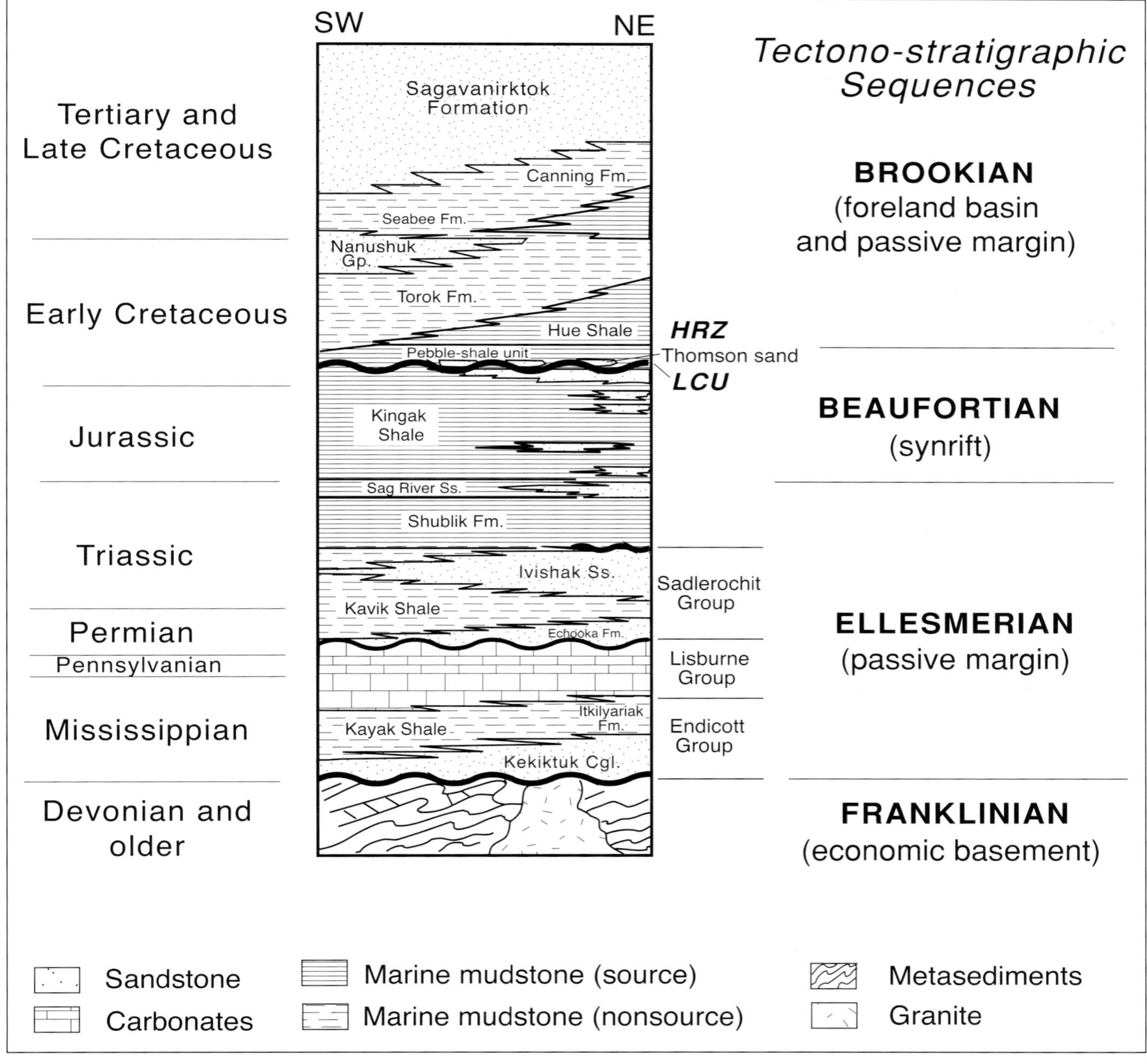

Figure 4. Generalized stratigraphic column for northern Alaska, emphasizing petroleum-prospective rocks and their tectono-stratigraphic subdivision, reflecting major stages in the tectonic development of the region. Oil and gas accumulations are found throughout the stratigraphic section. HRZ = highly radioactive zone of the Hue Shale; LCU = the regional Lower Cretaceous unconformity.

effort continued intermittently from 1923 to 1981 (Reed, 1958; Gryc, 1988). Industry exploration dates from 1958 (Jamison et al., 1980). About 400 exploratory wells have been drilled in this region (Figure 5). Three-fourths of them are located between the NPRA and the Arctic National Wildlife Refuge (ANWR), and most of those are within 100 km of the coastline. For this part of the region, exploration may be approaching a mature stage; elsewhere, exploration is at an immature stage.

Infrastructure is concentrated in the greater Prudhoe Bay area. A 48-inch- (1.2-m-) diameter oil pipeline, known as the Trans-Alaska Pipeline System, or TAPS, and its companion Dalton Highway provide the critical land-transportation link to the outside world. Remoteness, harsh arctic climate, and environmental concerns impose stringent economic conditions on north Alaska petroleum development. However, these factors also have spurred innovations such as extensive use of ice roads and ice pads for exploration, coiled tubing, extended-reach horizontal drilling, smaller and fewer development drill pads, elimination of mud pits, and subsurface disposal of drilling wastes by injection. Recently developed outlying oil fields are being operated as remote sites connected by feeder pipelines and accessible only by air or winter-ice roads. As shown in Figure 5, these feeder pipelines, which mark the outer limits of current infrastructure, now extend from the northern terminus of TAPS about 90 km west to Alpine field, 65 km east to Badami field, and about 15 km north (offshore) to Northstar field.

The federal government dominates land ownership in northern Alaska. Of the petroleum-prospective region north of the Brooks Range, federal landholdings include NPRA, the northern part of ANWR, and offshore lands beyond the state-federal three-mile boundary. Nonfederal onshore landholdings are divided between state and Native American ownership. State offshore ownership includes lands from the shoreline out to the state-federal three-mile boundary (Figure 5).

PETROLEUM DISCOVERIES AND PRODUCTION

Approximately 60 oil and gas accumulations (pools) have been discovered in northern Alaska (Figure 6a). Most are concentrated along the Barrow arch, a regional high and the focus of migrating hydrocarbons. The schematic regional cross section in Figure 6b shows the setting of the Barrow arch and the important Lower Cretaceous unconformity (LCU). This unconformity truncates many of the most important reservoir rocks along the arch; it provides a regional hydrocarbon migration pathway; and, with overlying mudstone and favorable structure, it creates a seal. In the Chukchi Sea region, the age of the main, regional-truncating unconformity is Late Jurassic (Sherwood et al., 1998). The two largest producing oil fields, Prudhoe Bay and Kuparuk River, and many smaller oil fields are combination structural-stratigraphic traps in which the LCU and overlying mudstones provide the stratigraphic-trapping component.

About one-third of north Alaskan oil and gas fields are currently producing. Many of the nonproducing fields are poorly known, single-well accumulations that may be little more than a good "show." Others, such as Ugnu field (discussed under "Brookian Topset Facies"), hold billions of barrels of heavy oil in shallow reservoirs just beneath the permafrost (Werner, 1987).

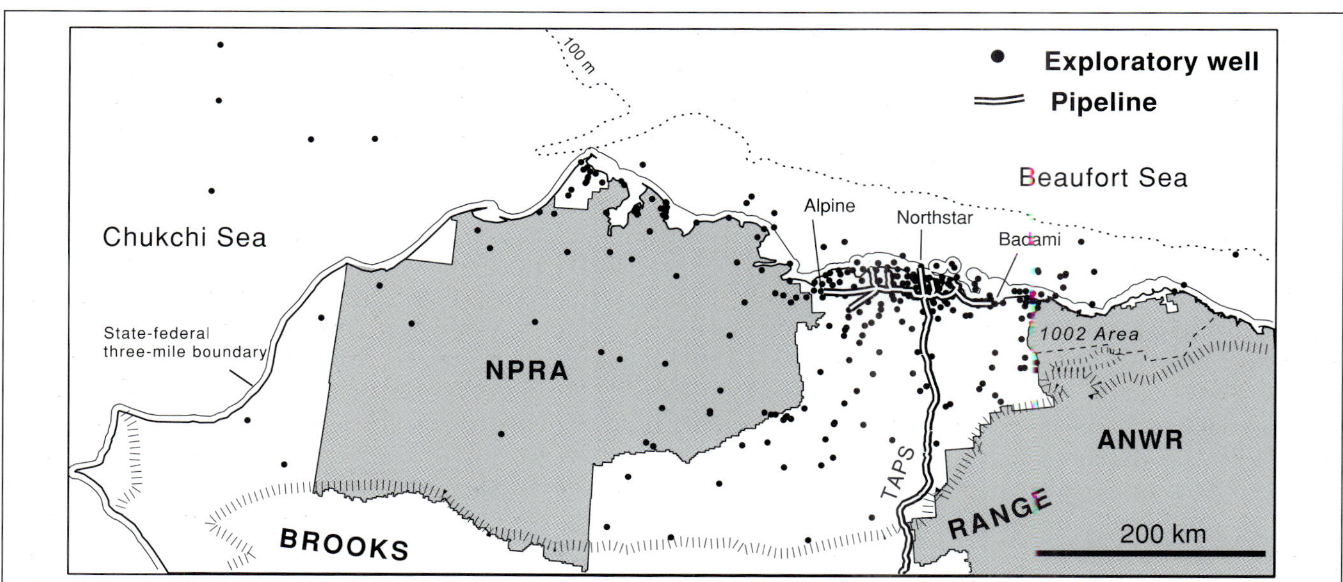

Figure 5. Map of northern Alaska showing exploratory drilling density, pipeline infrastructure, and land ownership. North of the Brooks Range, federal ownership includes NPRA, ANWR, and offshore beyond the state-federal three-mile boundary. Ownership of nonfederal lands is divided between the state and Native American organizations. TAPS = Trans-Alaska Pipeline System.

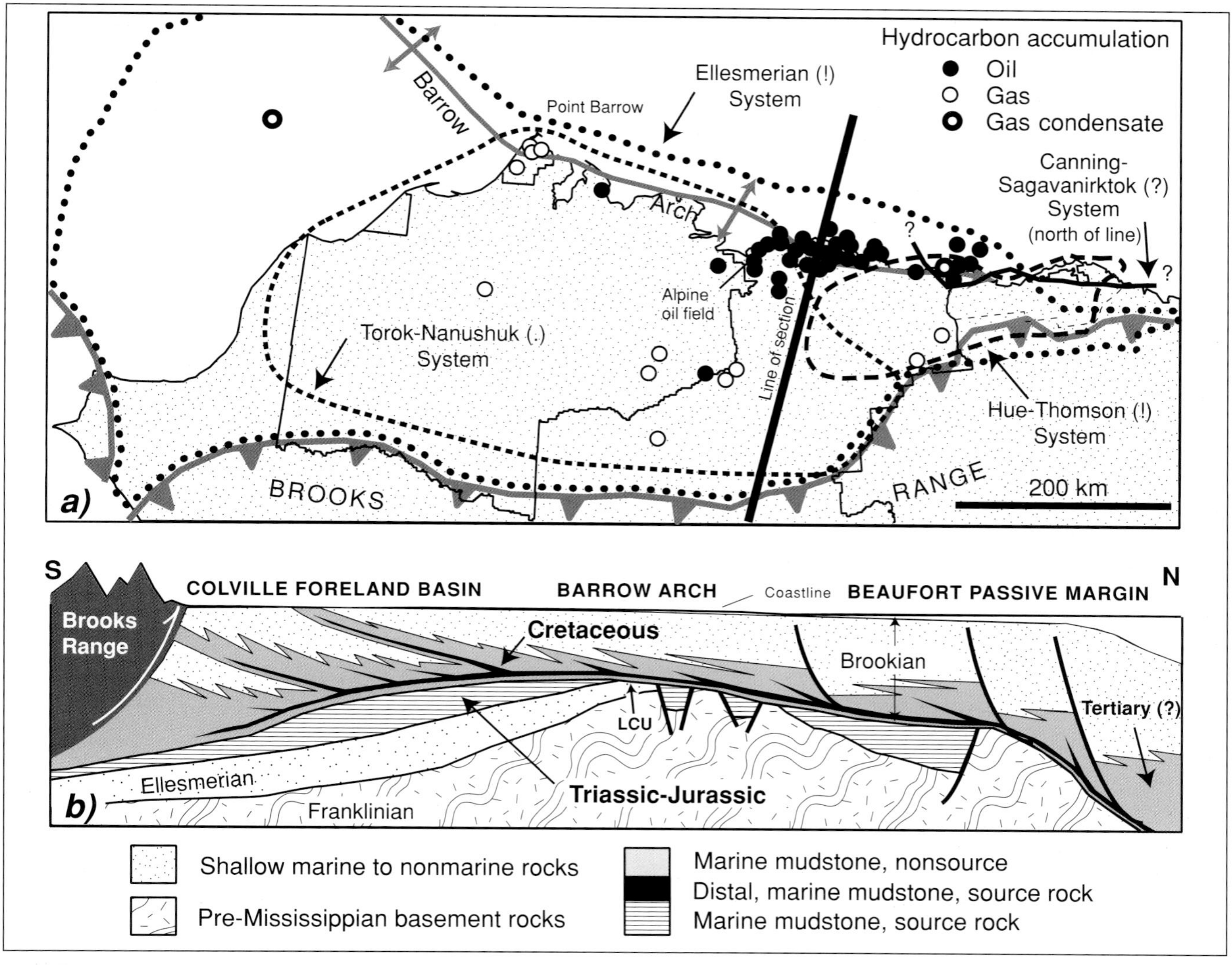

Figure 6. Hydrocarbon accumulations, petroleum systems, and source-rock intervals of northern Alaska. (a) Map showing areal distribution of four identified petroleum systems and nearly 60 oil and gas accumulations. (b) Schematic cross section showing structural and stratigraphic position of the source-rock intervals (Triassic-Jurassic, Cretaceous, and Tertiary [?]) related to these systems and stratigraphic truncation beneath the LCU, a key trapping element in many Barrow arch oil accumulations.

In spite of the uncertain size of many oil and gas accumulations, a sense of the richness of the province is gained by tabulating the estimated volume of petroleum represented by these accumulations. Table 1 shows that the total amount of in-place oil is about 60–80 billion bbl (10–13 billion m^3), the total ultimately recoverable oil is about 20 billion bbl (~3 billion m^3), and total ultimately recoverable gas is about 34 tcf (~1 trillion m^3). These volumes demonstrate the presence of a rich petroleum province and support Demaison and Huizinga's (1994) classification of this as a "supercharged" basin. This tabulation also shows that oil recovery varies greatly among sequences. The remarkably small proportion of recoverable oil in the Brookian sequence (less than 5%) is a reflection of large volumes of low-gravity oil in shallow reservoirs associated with permafrost.

Commercial oil production in north Alaska began in 1977 with completion of TAPS. This allowed the start-up of production from Prudhoe Bay field, North America's largest. By 1988, daily oil production from Prudhoe Bay field and several nearby fields peaked at slightly more than 2 million barrels (300,000 m^3). It has declined to the current rate of about 1 million barrels (160,000 m^3). Through the end of 1999, northern Alaska had produced 12.9 billion bbl (2 billion m^3) of oil and natural-gas liquids (Alaska Division of Oil and Gas, 2000). Because of lack of a transportation system, natural gas is used as a local energy source (about 4 tcf consumed thus far) or is reinjected for reservoir pressure maintenance. With current reserves of about 31 tcf (~1 trillion m^3), natural gas in this region represents a large but currently stranded resource (Alaska Division of Oil and Gas, 2000).

Table 1. Approximate volumes of discovered oil and gas in northern Alaska, listed according to tectonostratigraphic unit. In-place volumes are modified from Bird (1994) and recoverable volumes from Alaska Division of Oil and Gas (2000).

	Oil in place		Estimated ultimate recoverable			
			Oil		Gas	
	(billion bbl)	(billion m^3)	(billion bbl)	(billion m^3)	(tcf)	(trillion m^3)
Brookian	20–40	3–6	1	<1	1	30
Beaufortian	10	2	4	<1	5	140
Ellesmerian	30	5	15	2	28	790
Total	**60–80**	**10–13**	**20**	**3**	**34**	**960**

PETROLEUM SYSTEMS

Analysis of north Alaska oils and source rocks indicates that multiple source rocks have generated multiple oil types. As many as eight petroleum systems may be present, four of which have been described in the literature. Source rocks include Tertiary (Canning Formation), Cretaceous (Seabee, Torok, HRZ, pebble shale), Jurassic (Kingak), Triassic (Shublik), and Carboniferous (Lisburne) (e.g., Holba et al., 2000b). By definition, there should be at least as many petroleum systems as thermally mature source rocks. Analytic data for many north Alaskan oils have yet to be published, especially for those from more recent discoveries. Thus, more petroleum systems are likely to be present than the four that have been identified to date. It is also likely that existing systems will be modified as new information becomes available.

Petroleum systems currently identified and described in northern Alaska include the Ellesmerian (!), the Hue-Thomson (!), the Torok-Nanushuk (.), and the Canning-Sagavanirktok (?) (Figure 6) The symbol in parenthesis following each petroleum-system name is part of the classification scheme of Magoon and Dow (1994). A confident match between source rock and oil is classified as a "known" system (!), whereas increasing uncertainty in the match results in a "hypothetical" system (.) or a "speculative" system (?). Most oil in northern Alaska (more than 90%) is assigned to the Ellesmerian (!) system.

Oil

The Ellesmerian (!) petroleum system—so named because many of its reservoirs and its principal source rock occur in the Ellesmerian sequence (Magoon et al., 1987; Magoon, 1994)—contains a mixture of oil types. Ellesmerian oil is derived mostly from the Shublik Formation, with varying admixtures (in different accumulations) from the Hue Shale (HRZ) and the Kingak Shale (Seifert et al., 1980; Wicks et al., 1991; Masterson et al., 1997; Masterson et al., 2000). Oils in this hybrid system are found along the Barrow arch in reservoir rocks that range in age from Mississippian to Tertiary (Figure 6a). Mixing of oils is apparently the result of common migration pathways along rift-related normal faults and along the LCU (Figure 6b). Some mixing also may have occurred during spilling and remigration of early-accumulated oil in Prudhoe Bay field, a result of regional eastward tilting (e.g., Jones and Speers, 1976; Carman and Hardwick, 1983; Erickson and Sneider, 1997; Masterson et al., 2000). Characteristics of Ellesmerian oil are 1–2% sulfur, 20–30°API, low gas-oil ratio (less than 1500 ft^3/bbl), vanadium content greater than nickel, and a low saturate-aromatic hydrocarbon ratio (0.9–1.7). Bird (1994) provided details of this system. Sherwood et al. (1998) provide more recent information on Shublik Formation source potential in the Chukchi Sea. The regional extent of the Ellesmerian system as shown in Figure 6a is based on distribution of thermally mature Triassic and Jurassic source rocks and related petroleum accumulations as described by Bird (1994) and as modified by Magoon et al. (1999).

The geographic outlines and definitions of the other north Alaska petroleum systems (Figure 6a) are based on reports by Magoon (1994) and Magoon et al. (1999). For the Hue-Thomson system, a definitive match between oils and the Hue Shale (HRZ) source rock has been established. The Torok-Nanushuk system is postulated to have source rocks in the Brookian and/or Beaufortian sequences, and the Canning-Sagavanirktok system is postulated to have a Brookian (Eocene) source similar to that proposed for some oils in the nearby Mackenzie Delta region of Canada. Oils in these three systems have similar characteristics and are distinguished mainly on the basis of carbon isotopes and biomarkers. Common characteristics of these oils include low sulfur (less than 1%), 30–40° API, moderate gas-oil ratio (1500–7000 ft^3/bbl), moderate to high saturate-aromatic hydrocarbon ratio (1.7–3.5), and very low nickel and vanadium content or nickel content greater than vanadium (Magoon, 1994; Lillis et al., 1999; Magoon et al., 1999).

The Alpine oil field, a 400+ million barrel 1994 discovery in an Upper Jurassic reservoir at the eastern edge of NPRA, is a Kingak-sourced oil, according to Hannon et al. (2000a, 2000b), with characteristics similar to those described above. However, Kingak oil is, by definition, part of the Ellesmerian system. A large accumulation of

Kingak oil such as the Alpine and the possibility of additional similar accumulations indicate the need to reevaluate the Ellesmerian system with focus on its component oil systems.

Gas

In the analysis of north Alaskan petroleum systems, gas accumulations generally have been considered part of the same system as that of nearby oil accumulations (e.g., Figure 6a). This resulted from the fact that very little was known about natural gases in northern Alaska. Recent studies suggest that gas accumulations may have origins as complex as that of oil accumulations. Therefore, those studies call into question the practice of petroleum-system assignment based on proximity. For example, Barrow peninsula gas accumulations, composed mostly of methane, are reported by Holba et al. (2000a) to be mixtures of at least two inputs: a major dry-gas component generated at higher maturity ($R_o > 1.1–1.2$) and a minor wet-gas component generated at lower maturity (R_o ~0.6). Holba et al. (2000a) postulated that the Shublik Formation downdip to the south is the source of these gases. In apparent contrast, Burruss and Collett (2000) postulated a Cretaceous, type III kerogen source for these gases, based on isotopic similarities with gases in middle Cretaceous Brookian reservoirs in the foothills fold belt. Prudhoe-Kuparuk-area gas accumulations may be related to a late-stage gas-flushing event, according to Holba et al. (2000b). Burruss and Collett (2000) reported these to be a mixture of isotopically heavy and light gases derived from type I–II and type III organic matter. In addition, the isotopically heavy gas has an unusual composition, suggesting biodegradation or mixing with a third type of gas.

PETROLEUM POTENTIAL

With the publication of structural and stratigraphic details of the Chukchi Sea region by Sherwood et al. (1998), a more complete picture is available of north Alaska's petroleum geology. Information from that report has been combined with earlier reports to produce the following summary of the petroleum potential of northern Alaska. In the following section, each of the significant reservoir intervals from oldest to youngest is described, highlighting regionally significant tectonic, sedimentologic, and thermal-maturity characteristics; exploration status; oil and gas accumulations; and future potential.

Earlier reports include those of the Chukchi Sea (Thurston and Thiess, 1987; Grantz and May, 1988), NPRA (Bruynzeel et al., 1982; Patterson et al., 1982; Gryc, 1988), the Beaufort Shelf (Craig et al., 1985; Grantz et al., 1988), and the northern part of ANWR (ANWR Assessment Team, 1999). Hubbard et al. (1987) and Moore et al. (1994) provide regional syntheses. Thermal-maturity trends are based on data compilations by Johnsson et al. (1992, 1999) and on more recent data in the Chukchi Sea (Sherwood et al., 1998) and the northern part of ANWR (Bird et al., 1999).

Ellesmerian Sequence

Endicott Group

Stratigraphy and structure.—The Endicott Group, composed of nonmarine and marine siliciclastic rocks, represents the oldest petroleum-prospective unit in northern Alaska. These rocks unconformably overlie economic basement (the Franklinian sequence) and gradationally lie beneath carbonate platform deposits of the Lisburne Group. Nonmarine sandstone in the lower part of the Endicott, the Kekiktuk Conglomerate, is a potential reservoir rock. The Kekiktuk forms the reservoir for the 600-million-bbl Endicott and the 100-million-bbl Liberty oil fields, located just east of Prudhoe Bay (Figure 7).

Regional distribution of the Endicott Group is based largely on seismic character, because well penetrations of this unit are limited (Figure 7). Its zero edge along the Arctic and Chukchi platforms, at depths of 2000–4000 m, is a combination of depositional onlap and/or erosional truncation by the LCU. Isopach maps (Bird, 1978, 1988; Bruynzeel et al., 1982; Woidneck et al., 1987; Grantz and May, 1988) and seismic sections (Kirschner and Rycerski, 1988) show that the Endicott varies greatly in thickness in response to deposition during a period of extensional deformation. The Hanna trough—a failed rift, according to Sherwood et al. (1998)—was the site of thick accumulations of Endicott sediments. Eastward, basin sags and half grabens are present and seem to be related genetically to the Hanna trough because of general similarities in time of formation and structural trends. The onset of Endicott deposition, postulated from subsurface data, is Late Devonian (e.g., Kelley and Brosgé, 1995), but available paleontologic control documents only Mississippian ages (Witmer et al., 1981; Ravn, 1991).

The major tectonic features of the Endicott (Figure 7) consist of positive areas where the Endicott is relatively thin or absent, and several basins where it may be as much as 5000 m thick. Figure 8 shows the seismic expression of the Meade Basin, a half graben presumed to be filled with Endicott strata and to have a significant component of Kekiktuk in its lower part. Well penetrations in the Ikpikpuk-Umiat and Endicott Basins show that Kekiktuk thickness varies in more or less direct relation to that of the entire Endicott. The only direct evidence for the existence of Kekiktuk on the west flank of Hanna trough is the 600-m-thick coal-bearing sandstone and mudstone outcrop section at Kapaloak Creek (Tailleur, 1965; Moore et al., 1984), south of Cape Lisburne (Figure 7).

Source rocks.—Source potential for oil is low in the

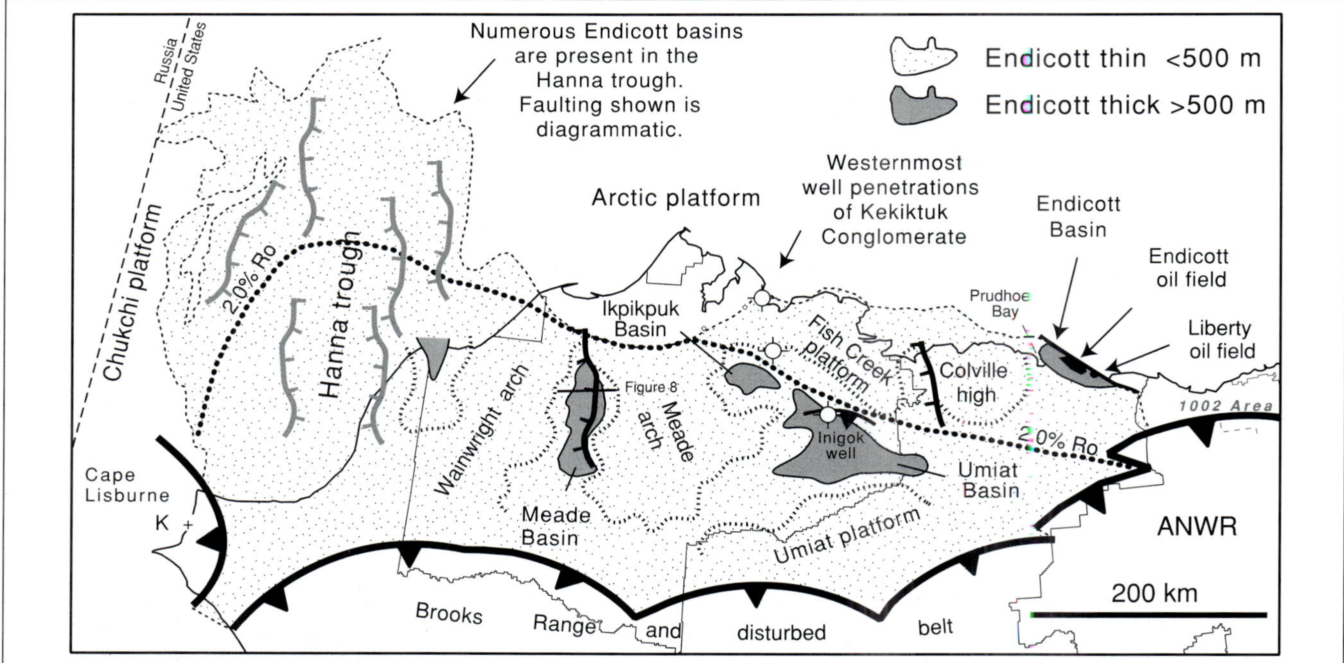

Figure 7. Map summarizing areal distribution of Endicott Group (stippled and shaded), major tectonic features, selected well control, oil accumulations, and the 2% R_o value (the lower maturation limit for liquid hydrocarbons) in the subsurface north of the Brooks Range. For clarity, the numerous wells that penetrate the Endicott in the area between NPRA and ANWR are not shown. K = Kapaloak Creek locality.

Endicott, but coal and carbonaceous mudstone in the Kekiktuk constitute a source for gas (Magoon et al., 1987; Magoon and Bird, 1988). Most of the Kekiktuk is separated stratigraphically from the main north Alaskan source rocks (Triassic–Lower Cretaceous) except in areas along the Endicott zero edge where truncation by the LCU has occurred. In those areas, hydrocarbon migration along the LCU provides a mechanism for placing hydrocarbons generated in overlying units stratigraphically downward into older units, as is the case at the Endicott oil field (Wicks et al., 1991). The truncation of the Kekiktuk by the LCU occurs along the Barrow arch and, according to Sherwood et al. (1998), along the eastern flank of the Chukchi platform, an area considered highly prospective for Kekiktuk oil accumulations.

The thermal maturity of the Kekiktuk, obtained primarily by downward projection of vitrinite-reflectance gradients in wells, shows that potential for liquid hydrocarbons (vitrinite reflectance < 2.0% R_o) is limited to a 100-km-wide zone along the Barrow arch and a somewhat broader area in the northwestern part of the Hanna trough (Figure 7). Most of the Kekiktuk lies to the south of the 2.0% R_o isograd and is therefore prospective only for natural gas.

Traps.—Structural and combination structural-stratigraphic traps occur in the Kekiktuk. These include fault traps related to extensional faulting on the Barrow arch and in the Hanna trough, compressional (transpressional) folds in east-central NPRA, and combination structural-stratigraphic traps along the Endicott zero edge related to the LCU. Traps are postulated to be related to rotated normal fault blocks, horsts, and local compressional structures associated with inversion of the Umiat Basin (Inigok well, Figure 7). The potential for traps related to erosional truncation by the LCU, similar to that at the Endicott and Liberty oil fields, lies along the outer limits of the Kekiktuk along the Barrow arch; similar potential along the western flank of the Hanna trough is related to truncation by the regional Upper Jurassic unconformity (Sherwood et al., 1998).

The Kekiktuk is composed mostly of quartz and chert, similar to the composition of most Ellesmerian sandstones. Woidneck et al. (1987) described Endicott field as having excellent reservoir characteristics in a half graben that was truncated erosionally by the LCU and sealed by overlying Cretaceous mudstones. These mudstones, the pebble-shale unit, and Hue Shale (HRZ) also are important source rocks. Their juxtaposition to the Kekiktuk provides a method for charging the reservoir.

The potential for future Kekiktuk oil discoveries is limited to those areas along the Barrow arch and eastern flank of the Chukchi platform where traps are in proximity to the LCU truncation. Away from those areas, the potential is for natural gas in deeply buried sandstones.

Lisburne Group

Stratigraphy and structure.—The Lisburne Group represents a broad carbonate ramp of Mississippian to

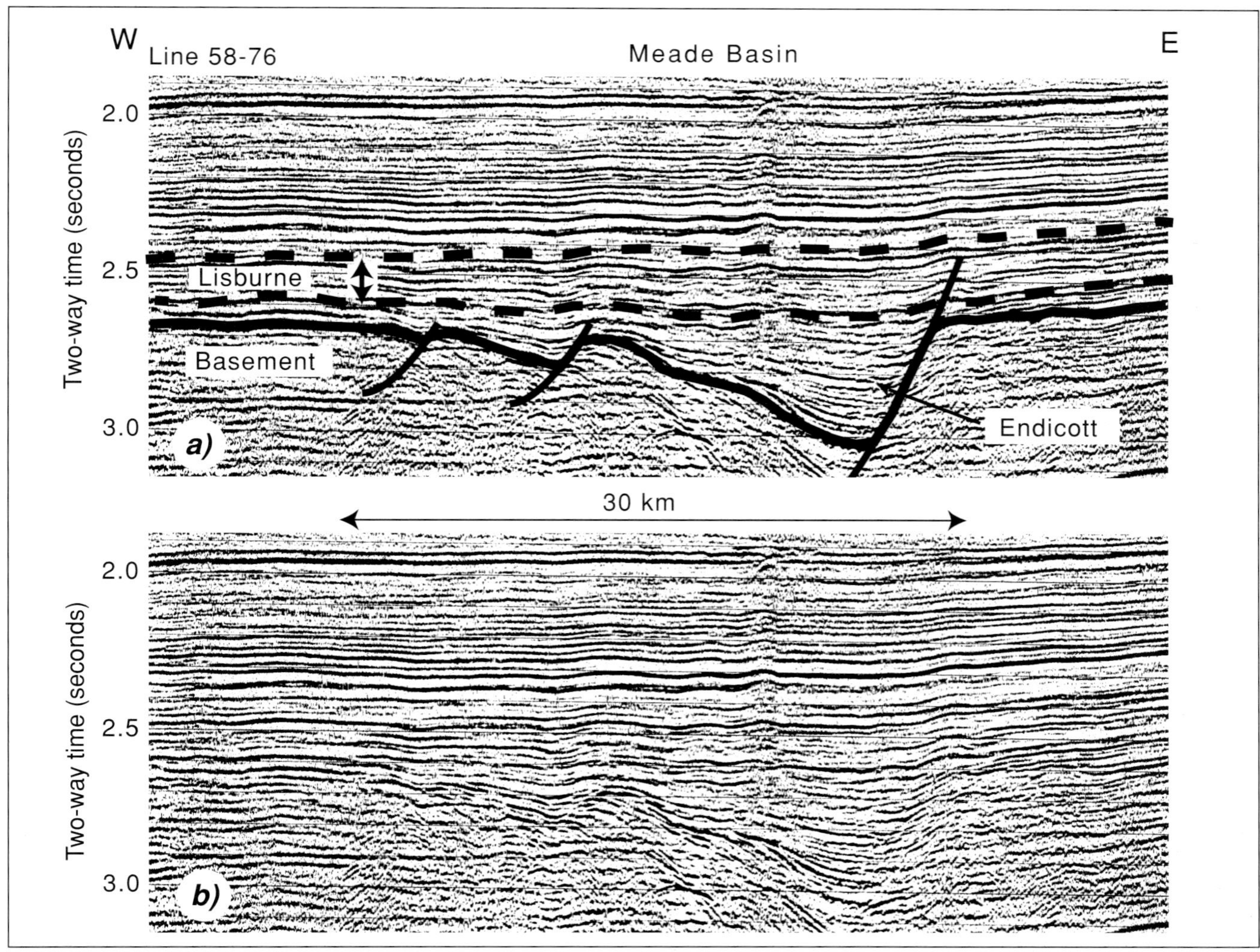

Figure 8. Meade Basin is an example of extensional basin development in northern Alaska in the Devonian (?) to Mississippian. Although no wells have penetrated this basin, its sedimentary fill is postulated by analogy with the Ikpikpuk-Umiat and Endicott Basins, where well control exists, to be coal-bearing sandstone, mudstone, and conglomerate of the Kekiktuk Conglomerate, grading upward into marine Kayak Shale, together comprising the Endicott Group. See Figure 7 for location. (a) Interpreted and (b) uninterpreted seismic section.

Early Permian age, bounded at the top by regional unconformities and at the base by a gradational contact with the Endicott Group. Its western and northern limits along the Chukchi and Arctic platforms, respectively, are characterized by depositional onlap pinch-out or erosional truncation (Figure 9). From those limits, which occur at depths of 1800–2700 m, the Lisburne gradually deepens to more than 8000 m in the Colville Basin and Hanna trough. The absence of an abrupt shelf margin and the presence of widespread shallow-marine carbonate rocks displaying gradual facies changes suggest a ramp rather than a platform configuration (Armstrong and Bird, 1976; Watts et al., 1995). The Lisburne produces oil at the 3-billion-bbl-in-place Prudhoe Bay field (Lisburne pool) and nearby at the several-million-barrel Alapah pool, a part of Endicott field (Figure 9). Both accumulations lie on the Barrow arch, and both occur in combination structural-stratigraphic traps formed during Early Cretaceous rifting.

The subsurface Lisburne Group consists mainly of limestone; lesser dolostone; and minor shale, sandstone, and evaporites (Armstrong, 1974; Armstrong and Mamet, 1974; Bird and Jordan, 1977; Jameson, 1994). Depositional patterns are complex, the result of a series of transgressions over a surface of considerable relief. This relief, consisting of basins, positive areas, and growing structures, was inherited from Endicott time. Lisburne thickness may exceed 1000 m in basins and may be 100 m or less in positive areas. Locally, extensional and transpressional deformation was active during the early part of Lisburne deposition, and basalts in the Tunalik well on the eastern flank of the Hanna trough suggest activity in the Permian.

A wealth of detail from new well and seismic data can

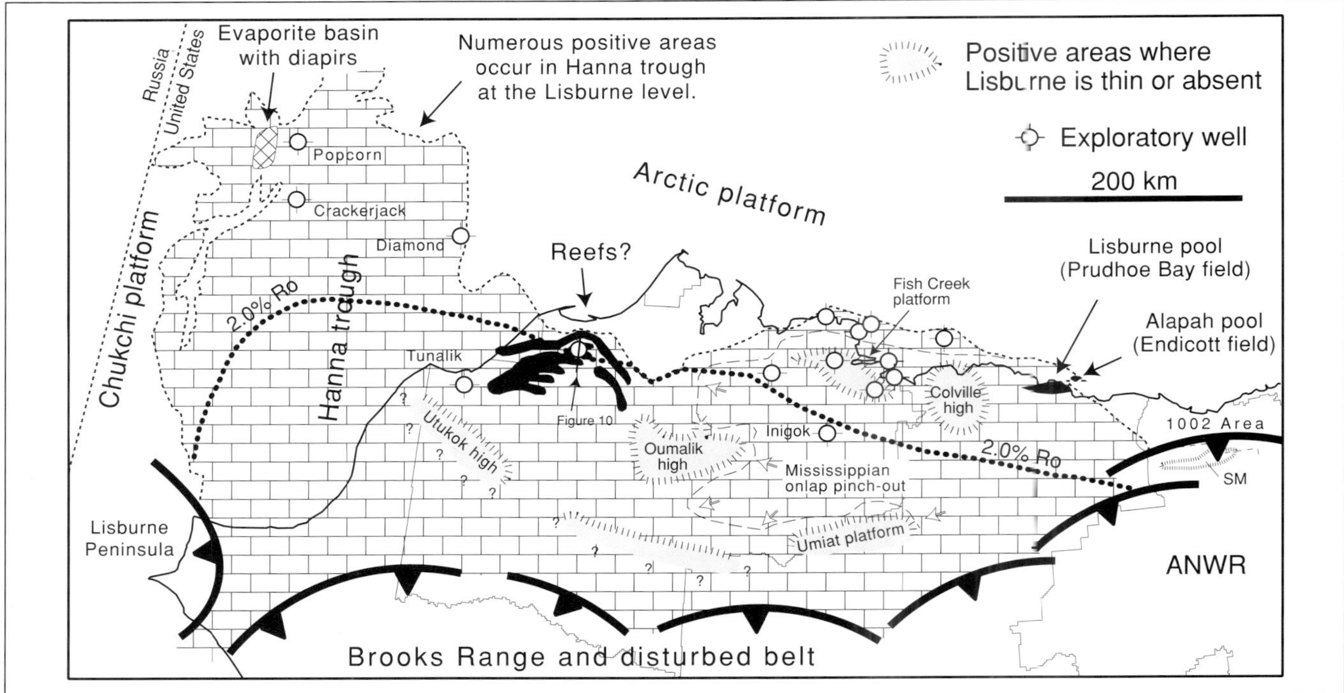

Figure 9. Map summarizing areal distribution of Mississippian to Lower Permian Lisburne Group carbonate rocks (pattern), major tectonic features, selected well control, oil accumulations, and the 2% R_o value (the lower maturation limit for liquid hydrocarbons) in the subsurface north of the Brooks Range. For clarity, the numerous wells that penetrate the Lisburne in the area between NPRA and ANWR are not shown. SM = Sadlerochit Mountains.

be added to Armstrong's (1974) regional Lisburne picture of progressive northward depositional onlap. Subsurface mapping in the NPRA (Bruynzeel et al., 1982) shows an irregular, westward onlap of the Mississippian part of the Lisburne. That pattern suggests a strong influence by the arches, platforms, and basins developed in Endicott time (Figure 7). In the northwestern NPRA, the Lisburne is predominantly Pennsylvanian and Permian, whereas in the northern Hanna trough, the Lisburne is Mississippian to Permian (Sherwood et al., 1998). On the western flank of the Hanna trough west of the Popcorn well, seismic data show a fault-bounded evaporite basin with diapiric, pillow, and withdrawal structures (Thurston and Lothamer, 1991; Sherwood et al., 1998). Although the evaporites are undated, they may be Late Mississippian (Chesterian), coeval with development of anhydrite-bearing red-bed facies and supratidal dolomites in the Prudhoe Bay region (Bird and Jordan, 1977).

Reservoir characteristics.—Reservoir characteristics in the Lisburne are known from widely scattered well penetrations (Bird and Jordan, 1977), detailed studies in the Lisburne oil field (Missman and Jameson, 1991; Jameson, 1994; Belfield, 1988), and outcrops about 100 km southeast of Lisburne field in the Sadlerochit Mountains (Figure 9; Watts et al., 1995; Dumoulin, 1999). Those studies show that calcite cementation and pressure solution have destroyed virtually all primary porosity and that most porosity is secondary, developed within dolostone formed during periods of subaerial exposure and influx of meteoric waters. The latest Mississippian (Chesterian) was a time of widespread development of dolostone in northern Alaska (Bird and Jordan, 1977), and cavernous porosity, mostly calcite filled, is observed in rocks of that age in outcrop (Dumoulin, 1999). Faults and fractures play an important role in connecting porous and permeable zones in Lisburne field, and porosity enhancement as high as 40%–50% occurs adjacent to the LCU. Therefore, the most favorable sites for Lisburne porosity appear to be positive areas which have undergone multiple periods of exposure, areas adjacent to the regional Permian or Early Cretaceous unconformities, and areas of faulting.

Source rocks.—Source rocks in the subsurface Lisburne have yet to be identified, although data are sparse (Magoon et al., 1987; Magoon and Bird, 1988). Limited source potential is consistent with observations of rare and widely scattered occurrences of solid hydrocarbons in the subsurface downdip and away from truncation by the LCU. However, oil recovered from Lisburne reservoirs in two exploratory wells on the Barrow arch is postulated to be from a Lisburne source (Hughes et al., 1985; Hughes and Holba, 1988; Lillis et al., 1999). Thermal-maturity patterns show that potential for liquid hydrocarbons (vitrinite reflectance < 2.0% R_o) in the Lisburne is limited to a 100-km-wide zone along the Barrow arch and a somewhat broader area in the northwestern part of the Hanna trough (Figure 9). Most of the Lisburne lies south of the 2.0% R_o isograd and is therefore prospective only for gas.

Traps.—Structural and combination structural-stratigraphic traps occur in the Lisburne. These include fault traps related to extensional faulting on the Barrow arch and in the Hanna trough, compressional (transpressional) folds in east-central NPRA (near the Inigok well), combination structural-stratigraphic traps along the basin margin related to the LCU, and carbonate buildups (Figure 9). The seismic expression of one of the carbonate buildups, a drilling objective of the Kugrua well, is shown in Figure 10. The origin of the buildup is not clear, based on analysis of samples. Fragments of *Paleoaplysina*, a boreal-realm Permian mound-building organism (Wilson, 1975), are present, but they do not seem abundant enough to explain the origin of the buildup.

The potential for future Lisburne oil discoveries is limited to areas along the Barrow arch and northwestern Hanna trough, especially where traps are in close proximity to the LCU truncation. Away from those areas, the potential is for natural gas in deeply buried carbonate reservoirs.

Ivishak Sandstone

Stratigraphy and structure.—The Permian and Lower Triassic Sadlerochit Group records a major influx of siliciclastic sediment deposited on the subsiding Lisburne carbonate platform. Irregular basin topography that existed during Lisburne and Endicott time had been filled in by that time, resulting in the gradual thickening southward of the Sadlerochit Group to more than 1000 m southwest of the Tunalik well (Figure 11). The northern and western limits of the Sadlerochit are characterized by a combination of erosional truncation and depositional onlap. Depth to the top of the Sadlerochit along the northern margin ranges from 1500 to 3700 m. From there, it gradually deepens to more than 7000 m in the Colville Basin and Hanna trough. Using the subsurface stratigraphic nomenclature of Jones and Speers (1976), the Sadlerochit Group consists, in upward suc-

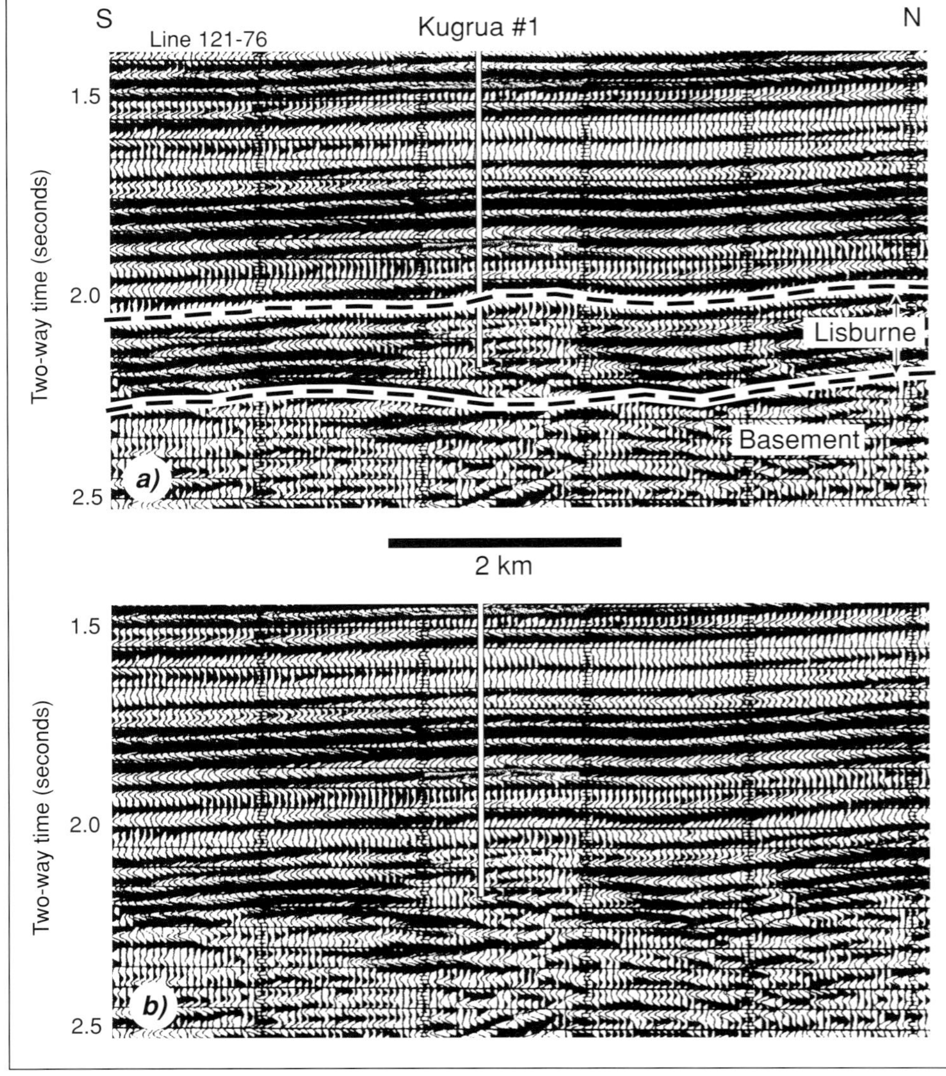

Figure 10. This seismically identified carbonate buildup in the Lisburne Group in northwestern NPRA is one of several that were originally identified as reefs (?). Feature was tested by the Kugrua well, but complete Lisburne penetration was cut short by early spring ice breakup after more than 425 m of Lower Permian and Pennsylvanian grainstone and packstone was penetrated without encountering significant reservoirs or hydrocarbon shows. (a) Interpreted and (b) uninterpreted seismic section. See Figure 9 for location.

cession, of a transgressive marine sandstone, the Echooka Formation; a marine mudstone, the Kavik Shale; and a marine and nonmarine sandstone, the Ivishak Sandstone (Figure 4).

Reservoir characteristics.—The Ivishak Sandstone is the primary reservoir of the Prudhoe Bay oil field (estimated ultimate recovery [EUR] of 13 billion bbl [2 billion m^3] of oil) and eight other oil accumulations, the largest of which is the 200-million-bbl (32-million-m^3) (EUR) offshore Northstar field that is under development (Figure 11). Along the northern basin margin, the Ivishak consists of fluvial, conglomeratic sandstone facies, interpreted as fan deltas (Detterman, 1970; Melvin and Knight, 1984; Lawton et al., 1987; McMillen and Colvin, 1987; Atkinson et al., 1988, 1990; Crowder, 1990). It grades southward to marine sandstone and ultimately to mudstone (e.g., in the Klondike well), where it is a fair to poor source rock (Sherwood et al., 1998).

A net sandstone isopach map (Figure 11) illustrates the regional Ivishak characteristics. Maximum net sandstone occurs in the north and decreases southwestward to a zero edge, located 200–300 km from the northern basin margin. It also shows a pronounced increase in net sandstone east of the longitude of Point Barrow, to a maximum of 180 m at the Prudhoe Bay oil field (Atkinson et al., 1990). Although constrained by a few widely scattered wells west of Point Barrow, the net sandstone maximum there is only slightly more than 90 m. A fan-delta origin for the Ivishak implies short-distance transport from nearby source highlands. The isopach pattern suggests that the source highlands west of Barrow were greatly diminished or were farther from the basin than the area east of Barrow.

Not only is the Ivishak very thick in Prudhoe Bay field, but its reservoir characteristics are also excellent. The reservoir is the result of both primary and secondary porosity and early filling with hydrocarbons. Its origin includes a long, initial period of shallow burial (~600 m) which maintained a loose grain-packing density without early cementation (Payne, 1987). That period was followed by uplift and partial erosion along the LCU, at which time porosity was enhanced by leaching of chert grains and siderite cement (Shanmugam and Higgins, 1988). With resumption of subsidence and burial, the trap was created, and its filling with oil occurred at about 40 Ma, when the reservoir was only at a depth of about 1500 m, compared with its present 2500-m depth of burial (Erickson and Sneider, 1997).

Source rocks.—The Ivishak is located favorably with respect to source rocks. It is directly overlain by the Shublik Formation, one of the most important source rocks in northern Alaska. Along its northern margin, normal faulting and erosional truncation related to Jurassic and Early Cretaceous rifting resulted in the favorable juxtaposition of the Ivishak with all the major source rocks—Shublik, Kingak, and HRZ. Many of the traps in that part of the region are combination structural-stratigraphic traps related to rifting.

The thermal maturity of the Ivishak shows that potential for liquid hydrocarbons (vitrinite reflectance <

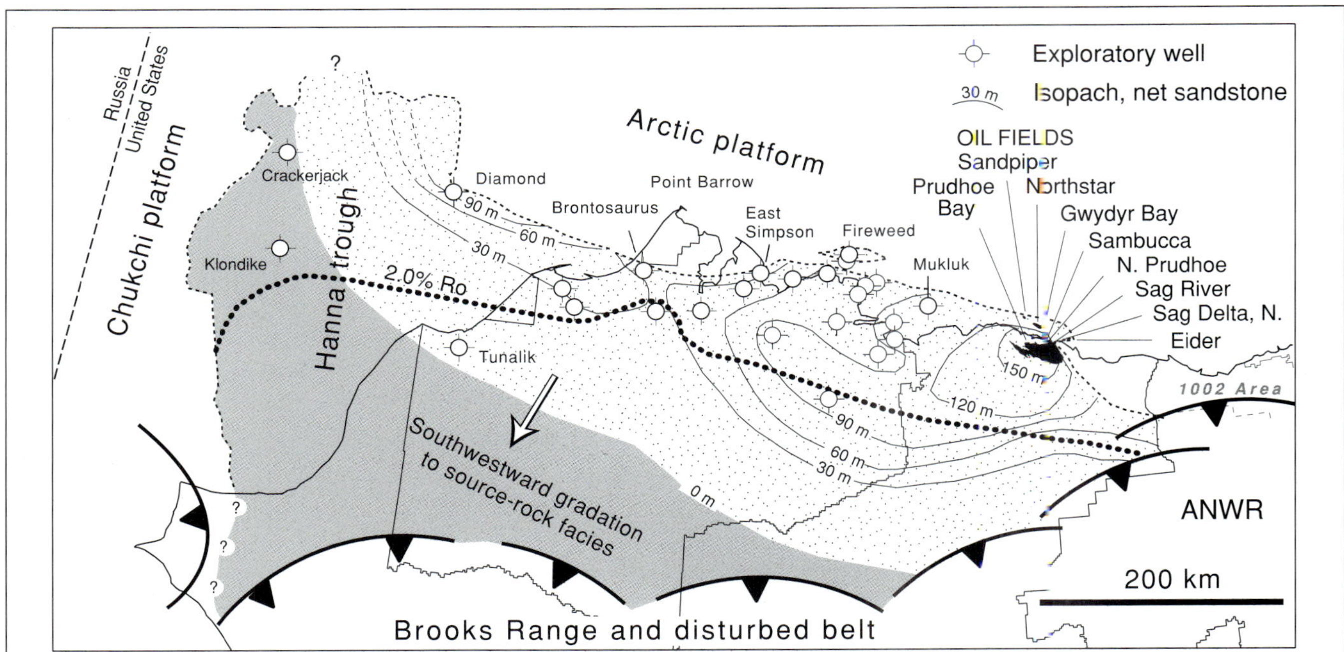

Figure 11. Map summarizing areal distribution and net thickness of sandstone of Lower Triassic Ivishak Sandstone (stippled), major tectonic features, selected well control, oil accumulations, and the 2% R_o value (the lower maturation limit for liquid hydrocarbons) in the subsurface north of the Brooks Range. For clarity, the numerous wells that penetrate the Ivishak in the area between NPRA and ANWR are not shown.

2.0% R_o) is limited to a 100-km-wide zone along the Barrow arch and a somewhat broader area in the northern part of the Hanna trough (Figure 11). Most of the Ivishak lies south of the 2.0% R_o isograd and is therefore prospective only for natural gas.

Traps.—The early and spectacular success at Prudhoe Bay made the Ivishak a primary exploration target throughout the region. However, the potential for additional large Ivishak oil discoveries now appears to be limited. The region's most promising large Prudhoe-type combination trap was tested by Mukluk (Figure 11), and just to the south, the Colville high, a large structural closure, was one of the first drilled on the Barrow arch. Ivishak was the primary objective of most wells drilled in the northeastern NPRA. Offshore, the Fireweed and nearby Antares wells tested proximal Ivishak in a graben north of the general Ivishak truncation edge. Tests near the Ivishak depositional onlap edge in the north-central NPRA include East Simpson and Brontosaurus. In the Chukchi Sea, the Diamond prospect tested Ivishak in an updip, near-truncation position. It appears that most major prospects involving the Ivishak Sandstone along the Barrow arch have been tested, with the possible exception of those along its western extent in the Chukchi Sea. The possibility of a western, Chukchi platform sand source is diminished by the absence of sand in the Crackerjack and Klondike wells (Sherwood et al., 1998). South of the Barrow arch, structural closures are nonexistent on the homoclinal south-dipping Ivishak, thus leaving only the potential for traps related to faulting or facies change. In addition, thermal maturity increases south of the Barrow arch, so that reservoir potential is diminished and gas is expected to be the only hydrocarbon phase.

Beaufortian Sequence

Stratigraphy and Structure

The Jurassic and Lower Cretaceous Beaufortian sequence is a mud-dominated succession of marine strata with numerous local sandstones deposited during a period of rifting (Figure 4). Most of the sequence is represented by the Kingak Shale, which may reach a thickness of as much as 1200 m in the onshore region, whereas age-equivalent strata in grabens offshore may be four times as thick (Hubbard et al., 1987; Grantz et al., 1988). Multiple unconformities are present. The youngest and most widespread of these is the LCU, interpreted by Grantz et al. (1988) as the breakup unconformity. Eighteen Beaufortian hydrocarbon accumulations ranging from dry gas to gas condensate to oil occur in structural, stratigraphic, and combination traps. Demonstrated source rocks are located below, within, and above the sequence, and three separate petroleum systems are documented. Potential for future discoveries is considered promising, in part because of the recent discovery of Alpine field (a nearly half-billion-barrel accumulation of high-quality oil in a stratigraphic trap), with the possibility of more such traps along trend in the NPRA.

Figure 12 summarizes the regional extent, major geologic features, and hydrocarbon accumulations of the Beaufortian sequence. South of a schematically shown rift system, the Beaufortian may be divided into two regions—an active margin and a stable margin. This useful distinction, applied in material on the Chukchi shelf by Sherwood et al. (1998), has been extended eastward across the entire region. Although the boundary between these regions is gradational, a reasonably good approximation is the onset of truncation of Jurassic strata. Figure 13 shows the salient features of these two regions in cross-section view.

Normal faulting, graben formation, and multiple unconformities characterize the active margin. It is interpreted as the rift shoulder. In this margin, south-facing normal faults are attributed to a Jurassic episode of failed rifting, whereas north-facing normal faults are attributed to an Early Cretaceous episode of successful rifting (Hubbard et al., 1987). In addition to the regional LCU, a Middle or Upper Jurassic unconformity is also widespread and accounts for the general absence of Middle Jurassic strata in the NPRA and the Chukchi shelf (Sherwood et al., 1998).

The stable margin at its progradational maximum is shown in Figure 12. It consists of a broad shelf in the north and an equally broad basin plain in the south, separated by a long linear shelf margin and slope which, in the Chukchi Sea, exhibits a dramatic hook to the south, based on mapping by Sherwood et al. (1998). Internally, diminished erosional truncation and southward clinoformal progradation characterize the stable margin. Figure 14 shows the seismic expression of a part of this margin in the western part of the NPRA. Recent evaluation of the Kingak in the NPRA (Houseknecht, 2001) shows that Kingak internal geometry is much more complex than suggested by the relatively simple map pattern and seismic section. That investigation identified four depositional sequences that show a complex, often lobate pattern of progradation with significant shifting of lobes through time.

Source Rocks and Known Traps

All Beaufortian hydrocarbon accumulations—12 oil, four gas, and two gas condensate—are located in the active margin (Figure 12). At least three petroleum systems can be demonstrated to have sourced these accumulations. Most Beaufortian oil accumulations occur in combination structural-stratigraphic traps and contain 20–30° API, high-sulfur Ellesmerian oil. The largest oil accumulation, the Kuparuk River field (2.66 billion bbl [0.42 billion m^3] EUR), was sourced predominantly from the Shublik Formation (Masterson et al., 2000). The Alpine field (429 million bbl [68 million m^3]) is a Kingak-sourced oil and the only stratigraphically trapped

Beaufortian accumulation (Hannon et al., 2000a). The Point Thomson gas-condensate field (3.5 tcf [99 billion m³] of gas and 200 million bbl [32 x10⁶ m³] of oil) is sourced from the Hue Shale (Magoon et al., 1999). The Burger gas-condensate accumulation is estimated to contain 5 tcf (142 billion m³) of gas (Sherwood and Craig, 2001). Gas accumulations on the Barrow peninsula are small (< 31 bcf [0.88 billion m³]) and apparently have multiple sources (Burruss and Collett, 2000; Holba et al., 2000a). Figure 12 shows that the downdip limit of Beaufortian liquid hydrocarbons, as inferred from the 2% R_o isograd, coincides with the ultimate Kingak shelf margin.

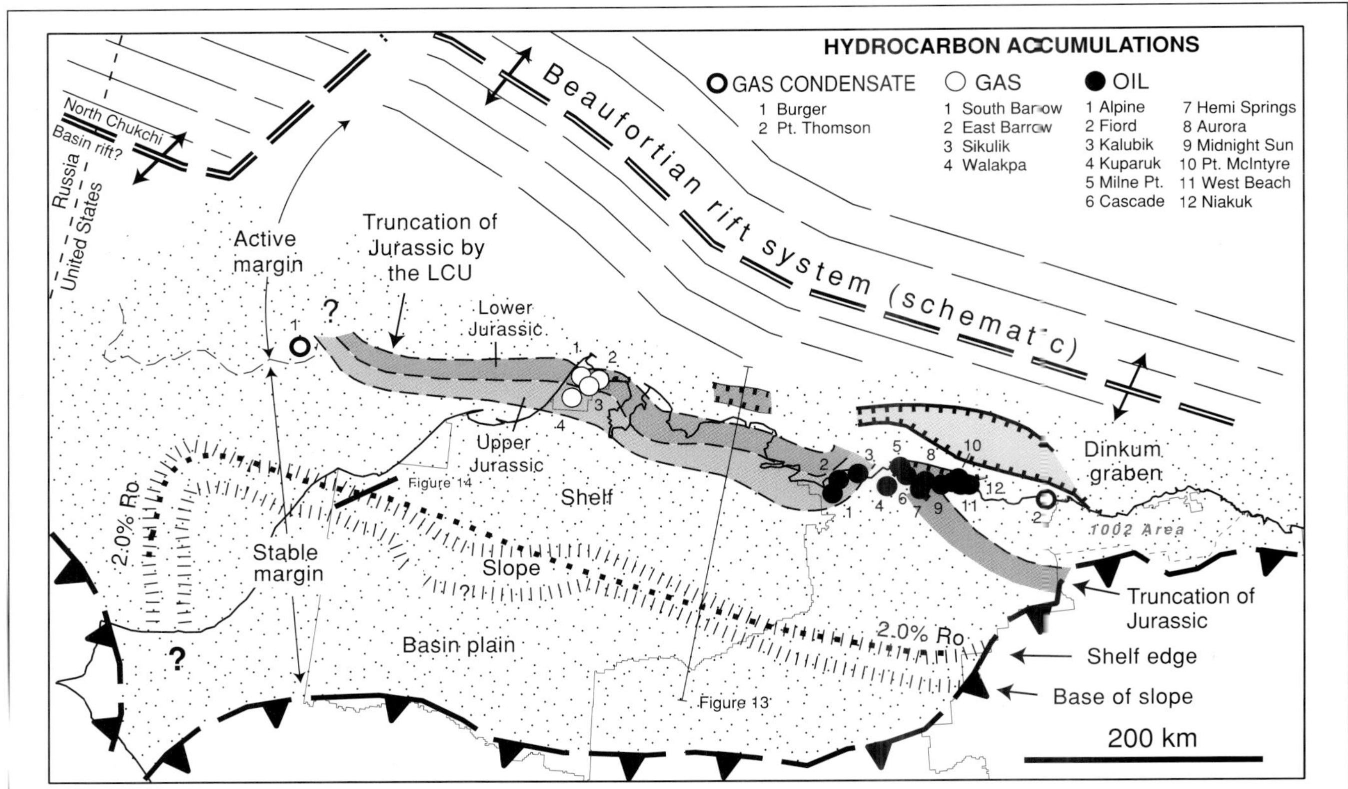

Figure 12. Map summarizing areal distribution of Jurassic and Lower Cretaceous Beaufortian sequence (stippled), major tectonic features, hydrocarbon accumulations, and the 2% R_o value (the lower maturation limit for liquid hydrocarbons) in the subsurface north of the Brooks Range. Trends of shelf edge and base of slope are those of the youngest, Lower Cretaceous (Hauterivian) part of the sequence.

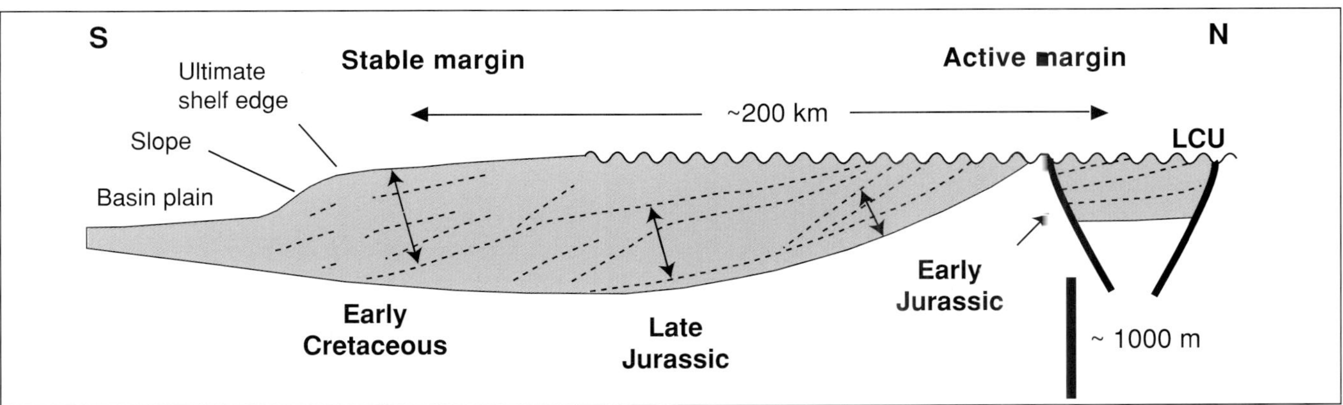

Figure 13. Schematic diagram, based on well and seismic data in the eastern part of NPRA and the adjacent offshore, showing Beaufortian (Kingak Shale) depositional style, age relations, and the distinction between "active margin" and "stable margin." Section is restored using the regional LCU as a datum. For simplicity, the 30–90 m of Beaufortian strata overlying the LCU and its correlative conformity (the pebble-shale unit, Kalubik Formation, and several locally derived sandstone units) are not included. See Figure 12 for location.

Reservoir Characteristics

Beaufortian reservoirs are sandstones that are relatively rare, only locally developed, and hard to predict. These sandstones are typically of shallow-marine origin, fine grained, and composed of quartz and chert with varying amounts of glauconite. Similar characteristics are found in the latest Triassic Sag River Sandstone (Barnes, 1987) that lies directly beneath the Kingak and is typically assigned to the Ellesmerian sequence. Clinoform seismic geometry and limited well data suggest that deeper-marine sandstones are also present in the Beaufortian. In producing fields where stratigraphic details are well known, reservoirs may display a shingled, offlapping (regressive) or onlapping (transgressive) depositional architecture, often with multiple unconformities (e.g., Masterson and Paris, 1987; Hannon et al., 2000a). Although relatively rare and only locally developed, sandstones are known to occur in all parts of the sequence—Lower and Upper Jurassic and Lower Cretaceous. Figure 15 shows examples of the log character of these sandstones. A funnel-shaped, upward-coarsening signature is common, locally culminating in a blocky, incised sandstone (the best reservoir). Sandstone prediction in the Beaufortian is a challenge—most sandstones were initially encountered by accident and their areal extent was determined by drilling. A recent example is the Alpine discovery. However, tools for identification of these types of accumulations are reported by Gingrich et al. (2000, 2001). Postdiscovery analysis at Alpine found that the presence of oil-charged reservoir was marked by anomalies in amplitude variation with offset (AVO), and elastic-impedance (EI) inversion of 3-D seismic data proved to be a good predictor of reservoir thickness.

Future Hydrocarbon Potential

Beaufortian stratigraphic and structural complexity (its synrift heritage) offers significant opportunities for future hydrocarbon discoveries. In the active margin, additional accumulations are likely in structural, stratigraphic, and combination traps, similar to those already discovered. In the stable margin, stratigraphically trapped accumulations are most likely. Most important to successful exploration will be the understanding of subtleties of reservoir development—occurrence, distribution, thickness, and quality of sandstones in shelfal and deeper-marine settings. As demonstrated in Alpine field, important exploration tools in meeting these challenges are 3-D seismic surveys and the use of AVO and EI inversion analysis. Of considerable economic importance is a sufficient understanding of hydrocarbon plumbing systems to predict the occurrence of mainly gas versus various types of oils from different petroleum systems.

Brookian Sequence

The Cretaceous and Tertiary Brookian sequence, as summarized by Bird and Molenaar (1992), is a succession of marine and nonmarine deposits shed northward from the Brooks Range orogenic belt. The sequence includes thick (>8 km) Colville foreland-basin deposits developed ahead of northward-advancing thrust sheets, relatively thin (<1–4 km) deposits over the subsiding Barrow arch rift shoulder, and thick (>8 km) Canada Basin passive-margin deposits. Colville Basin is the result of flexural subsidence produced by crustal thickening in the Brooks Range orogen in the Barremian (Early Cretaceous) (Cole et al., 1997). Brookian deposits are deformed on the south and east by compressional folding and faulting, on the west by wrench faulting, and along the passive margin on the north by growth faulting.

The hallmark seismic signature of the Brookian sequence is that of topset-clinoform-bottomset geometry (Figure 16). The lower part of the topset seismic facies, when correlated with wells and outcrops, is seen to consist primarily of marine shelf mudstone of the Torok or Canning Formation (e.g., Molenaar, 1988; Houseknecht and Schenk, 1999). This implies that sand was deposited on the delta plain and nearshore and that little if any reached the shelf edge or deeper water beyond. Yet shelf-margin and turbidite sands are well known. Relative sea-level lowstands were key to the transport of sands to the shelf edge and beyond and to the development of stratigraphic traps such as valley fill, shelf-edge sandstone, turbidite channel, and turbidite fans in the topset and clinoform facies (e.g., Morrow, 1987; Houseknecht and Schenk, 2001).

Fifteen oil and six gas accumulations have been discovered in the Brookian sequence in structural and stratigraphic traps. Demonstrated source rocks are located below and within the sequence, and four petroleum systems are documented. Potential for future discoveries is considered promising, especially in stratigraphically trapped turbidite reservoirs and, to a lesser extent, in shelf-margin sandstones. Many untested structural traps are also present. In the analysis of Brookian petroleum potential, topset facies, clinoform facies, and the fold belt are considered separately because of differences in reservoir geometry, trap types, sealing characteristics, timing, etc. The salient features of these Brookian subdivisions are summarized in Figures 17 and 18 and are described below.

Brookian Topset Facies

Depositional and tectonic features of the seismically identified topset facies are summarized in Figure 17. Rocks of this facies lie at the surface across most of the North Slope and beneath the continental shelves. Total thickness of this facies is generally less than 3000 m. The age progression of outcropping topset facies—Lower Cretaceous in the west, Upper Cretaceous and Tertiary strata in the east—reflects the longitudinal filling of the foreland basin. In the Chukchi Sea, Tertiary strata rest unconformably on Lower Cretaceous strata and thicken northward into the North Chukchi Basin, a part of the passive margin (Sherwood et al., 1998). The fold-and-thrust belt involves the southern part of the topset facies, a swath 30–200 km wide. The surface trace of the 0.6% R_o vitrinite-reflectance isograd (the top of the oil window) approximates the southern extent of this facies and indicates uplift and erosion of 3 km or more. The amount of erosion is also demonstrated by complete removal of the topset facies and exposure of the underlying clinoform facies in the foothills of the Brooks Range.

Eight oil and five gas fields are known in the topset facies (Figure 17). All oil accumulations except one are located along the Barrow arch, and all gas accumulations are located in the fold belt. The largest accumulations are the multibillion-barrel low-gravity West Sak and Ugnu oil deposits (Werner, 1987). These are Ellesmerian petroleum-system oils that were spilled from the underlying Prudhoe Bay and Kuparuk oil fields (Carman and Hardwick, 1983; Masterson et al., 2000) and were trapped in homoclinal east-dipping reservoirs sealed by small-scale normal faults, facies changes, and possibly by tar and/or permafrost. Currently, only the downdip, higher-gravity portion of the West Sak accumulation is producing (Milne Point field, Schrader Bluff pool). Smaller oil accumulations, in the size range of 10 million to perhaps 100 million bbl, are located along the Barrow arch in stratigraphic or fault traps. These include the shallow Simpson oil accumulation, trapped against a shale-filled canyon (Robinson, 1964); Tabasco, a stratigraphically trapped accumulation in a shelf-edge, incised-valley-fill reservoir

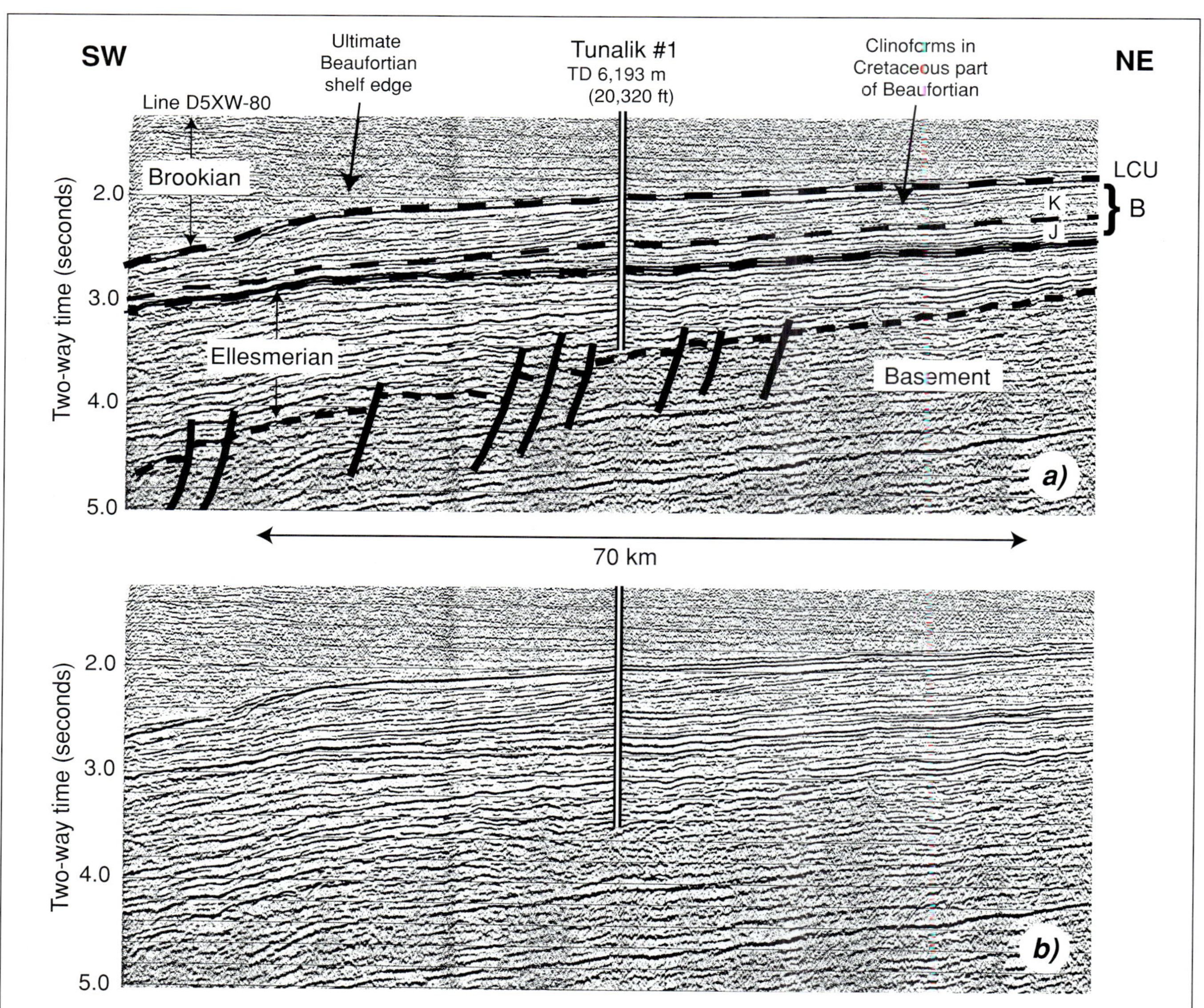

Figure 14. Seismic expression of the "stable-margin" part of the Beaufortian sequence and its well-developed ultimate shelf edge in the western part of NPRA. See Figure 12 for location. (a) Interpreted and (b) uninterpreted seismic section. In (a), B = Beaufortian sequence; K = Cretaceous; J = Jurassic; LCU = regional Lower Cretaceous unconformity.

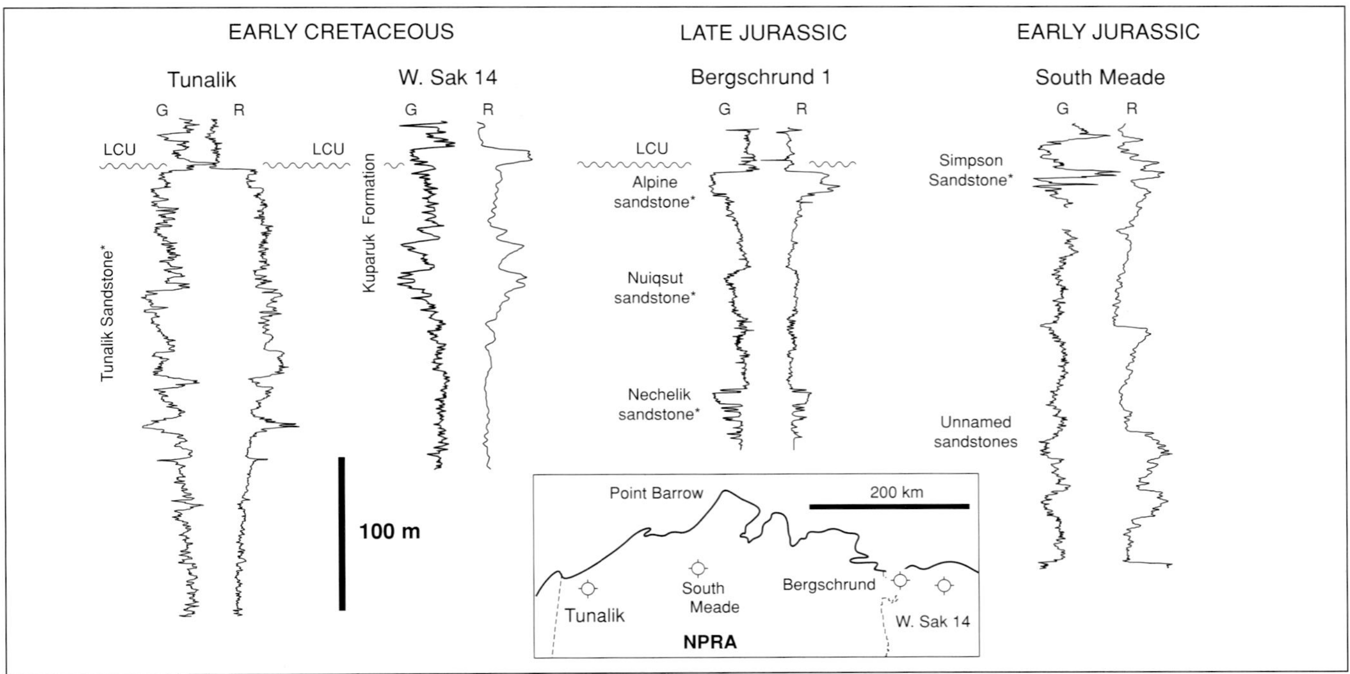

Figure 15. Examples of log character and thickness of selected Beaufortian sandstones in each of the three age-related packages of the mudstone-dominated Kingak Shale shown in Figure 13. LCU = Lower Cretaceous unconformity. Log curves: G = gamma ray, R = resistivity, * = informal stratigraphic name.

(Konkler et al., 2000); and the Fish Creek and Hammerhead accumulations, in traps formed by normal faulting (Robinson and Collins, 1959; Scherr et al., 1991). Kuvlum and Umiat (Molenaar, 1982) are anticlinal trapped oil accumulations in the fold belt. All topset-facies oil accumulations along the Barrow arch are low gravity (10–20° API), whereas those in the fold belt are high gravity (~35° API). At least three petroleum systems are represented by these accumulations (Figure 6a).

Topset-facies potential is considered significant because of the presence of a wide variety of plays and light exploration. Topset potential includes sand-rich paleovalleys, intervalley highs, and structures in the wrench-fault province; a possible sand-rich fringe or apron around a persistent high (North Chukchi high) near the western end of the Barrow arch; a broad region across the North Slope of north-trending, eastward-marching shelf edges with possible shelf-margin stratigraphic and rotational growth-fault traps and incised valleys oriented perpendicular to shelf margins; and abundant rotational growth-fault traps along and north of the Barrow arch.

Brookian Clinoform Facies

Depositional and tectonic features of the seismically identified clinoform facies are summarized in Figure 18. Rocks of this facies are generally encountered at depths of 2000–3000 m, and their thickness ranges from 300 m to more than 4000 m. The southern part of the clinoform facies is involved in the fold-and-thrust belt. The 0.6% R_o vitrinite-reflectance isograd (the top of the oil window) lies within this facies over a large part of the North Slope. It comes to the surface in the southern foothills, where uplift and erosion have removed the topset facies and exposed the clinoform facies. Slope failure and downslope transport are characteristic of the clinoform facies (Figure 16) and are observed at all scales (Houseknecht and Schenk, 1999). Several submarine canyons and a large-scale slide originate along the Barrow arch and trend toward the Canada Basin (Figure 18). Most of these features are not well defined, but two of the better known are Simpson Canyon (Robinson, 1964; Kirschner and Rycerski, 1988) and the Fish Creek Slide (Weimer, 1987; T. X. Homza, unpublished work).

Seven oil fields and one gas field are known in the clinoform facies (Figure 18). All oil accumulations are located along or adjacent to the Barrow arch; the gas accumulation is located in the fold belt. Three of these oil fields are currently producing (Badami, Tarn, and Nanuq), and a fourth (Meltwater) is nearing start-up. Sizes of these accumulations are not all publicly documented, but they appear to be in the range of 10 million to 100 million bbl. Oil gravities that range from about 25° to 40° API and limited geochemical information suggest that several petroleum systems are present. Information on Tarn field (Morris et al., 2000) and Badami field (Lemley, 1998) shows that these are complex turbidite reservoirs in confined and unconfined slope-apron systems. The complexity of these reservoirs is illustrated by Badami field, where initial estimates of about 100 million bbl of recoverable oil have been scaled back to about 10 million bbl (Alaska Division of Oil and Gas, 2000). Most wells in the fold-and-thrust belt that penetrate this facies were drilled on anticlinal structures, and gas shows commonly were encountered. The most important of these is the East Kurupa well, listed by the Alaska State Division of Oil and Gas as a gas discovery (Figure 18).

Clinoform-facies potential may be more significant than indicated by the 1995 Department of the Interior assessment (Gautier et al., 1995), when only three oil accumulations were known Since that assessment, this facies has been the focus of increased exploratory effort, resulting in discovery of four additional oil fields. Clinoform-facies potential is considered significant because of the close proximity of reservoir and source and favorable sealing. The greatest risk is the reservoir—its occurrence,

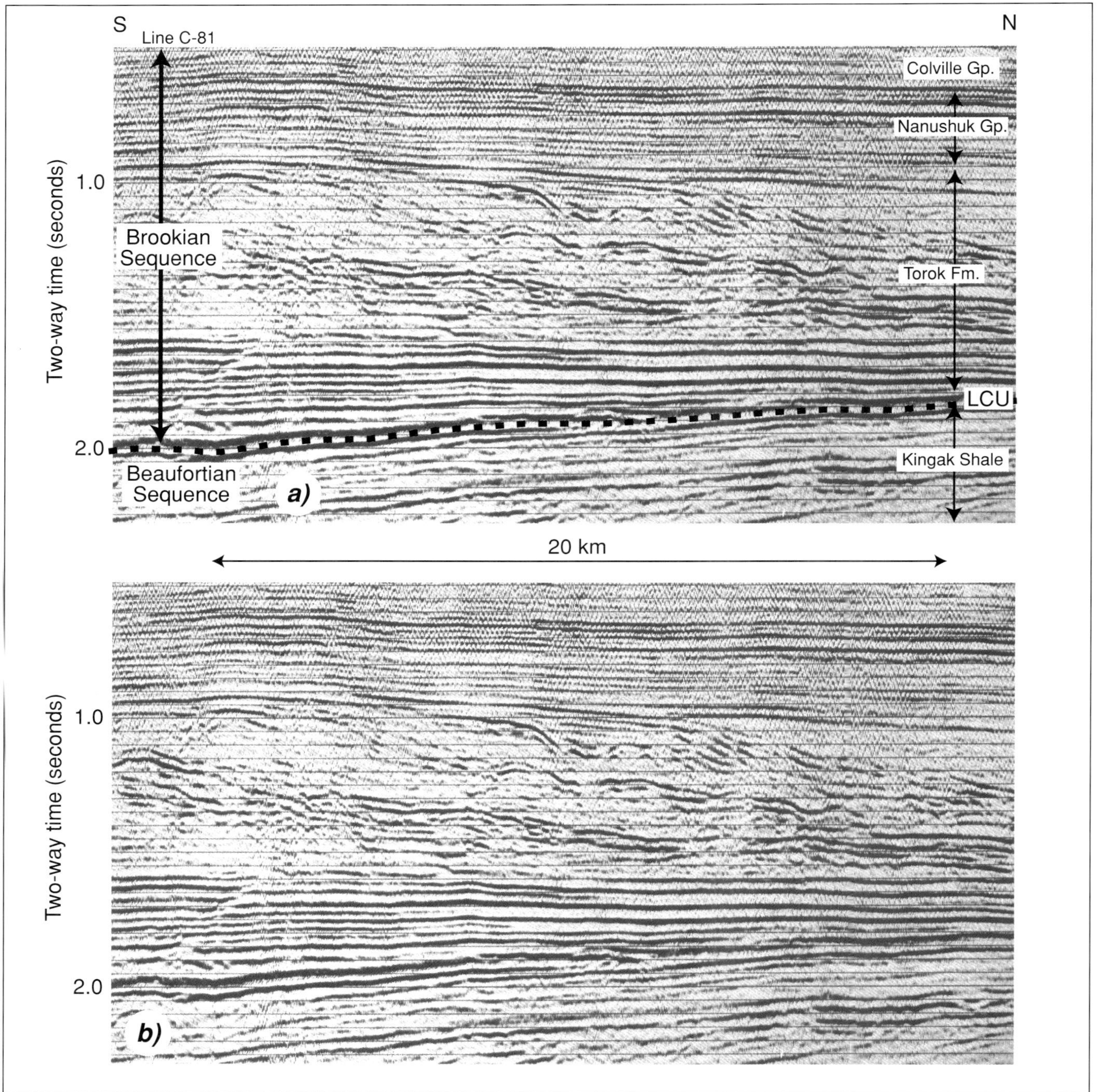

Figure 16. Seismic expression of Brookian depositional style in the eastern part of NPRA showing topset, clinoform, and bottomset stratal geometries and relation to the LCU and Beaufortian sequence. HRZ, the Brookian distal facies just above the LCU, is not resolvable on this section. (a) Interpreted and (b) uninterpreted seismic section. See Figure 17 for location.

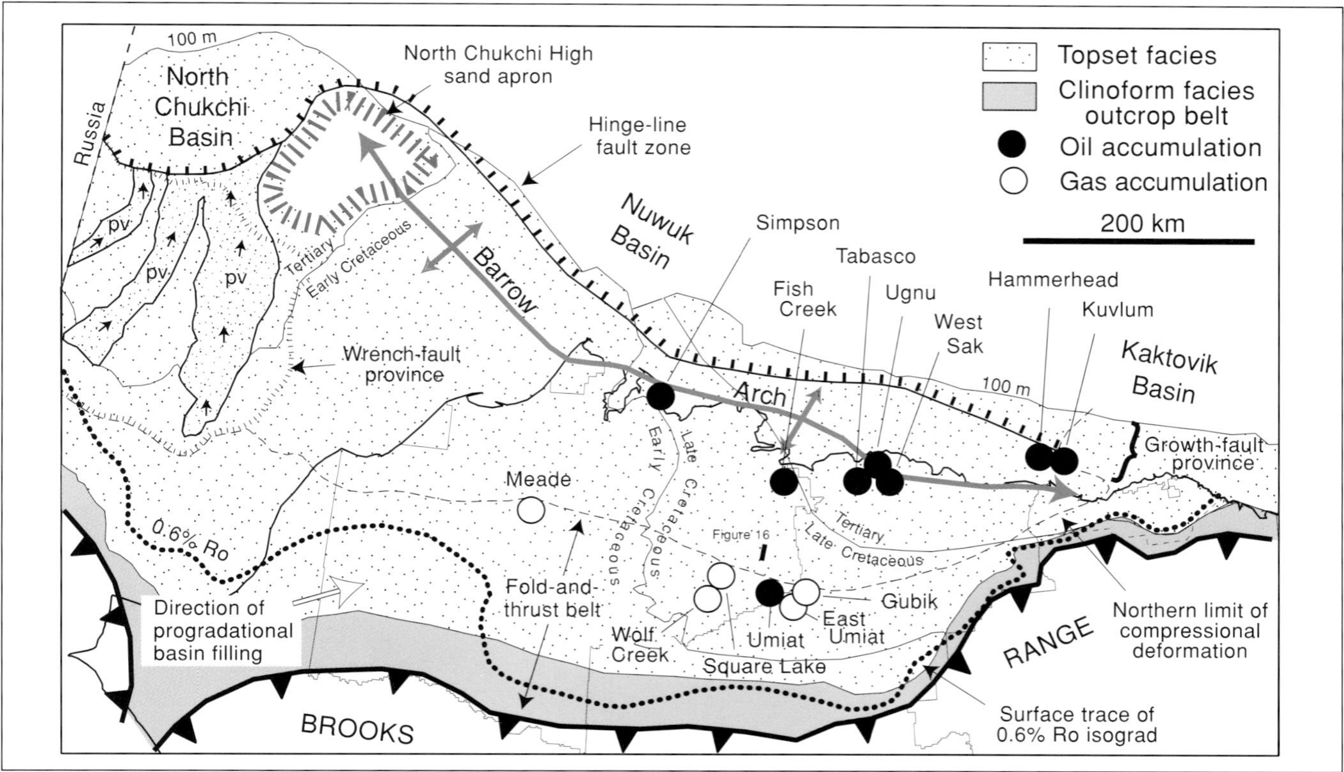

Figure 17. Summary of Brookian topset facies depositional and tectonic features, oil and gas accumulations, significant age boundaries, and the surface trace of the top of the oil window (the 0.6% R_o value). Stippling shows distribution of the topset facies as far north as the 100-m isobath. PV = paleovalley (arrows indicate direction of transport).

thickness, continuity, and loss of porosity with burial. Important questions about the clinoform facies include: Are there "sweet spots" where greater amounts of sand were deposited during lowstands? Did oil charge the lithic-rich reservoirs early enough to protect or arrest the normal rapid porosity loss by compaction? Are there intact stratigraphically trapped accumulations in the fold belt?

Brookian Fold Belt

The Brookian fold belt has exploration potential in anticlinal traps in both topset and clinoform facies. Some anticlines in the fold belt are 50–100 km long. About 35 separate structures have been tested by at least one well, and several times that many structures remain untested. Figures 17 and 18 show that five gas and two oil accumulations are known in the topset facies and one gas accumulation in the clinoform facies in the fold belt. At least three petroleum systems have been identified. In addition to the Hue-Thomson and Torok-Nanushuk systems (Figure 6a), recent studies in the fold belt south of Umiat (Houseknecht and Schenk, 2000) have discovered several occurrences of oil-saturated Brookian sandstone and solid hydrocarbon dikes that correlate in part with a Shublik source rock, indicating that yet another petroleum system is present. Important considerations in evaluation of the potential of this region are thermal maturity, trap seal integrity, timing of trap development, and reservoir character.

Timing of structural development relative to generation and migration is considered poor in the west-central part of the fold belt but fair to good in the eastern part. In the west-central fold belt, maturation/generation occurred before 100 Ma, but structures generally did not form until about 60 Ma (O'Sullivan, 1996; O'Sullivan et al., 1997). In the eastern fold belt, maturation/generation and structural development are more nearly coincident in the early to middle Tertiary (O'Sullivan et al., 1993). A possible exception to the poor-timing scenario in the west-central fold belt is the occurrence of some structures that may have formed during the Albian, at about the same time as hydrocarbon generation. Three occurrences of possible early-formed structures are shown in Figure 18 (Molenaar et al., 1988; Cole et al., 1997; Mull et al., 2000a; Mull et al., 2000b). These structures, developed mostly in the clinoform facies, may be a class of overlooked exploration targets in the fold belt.

Thrust Belt

The thrust belt is defined as the southern part of the fold-and-thrust belt where subsurface structure and stratigraphy are similar to those exposed in the adjacent

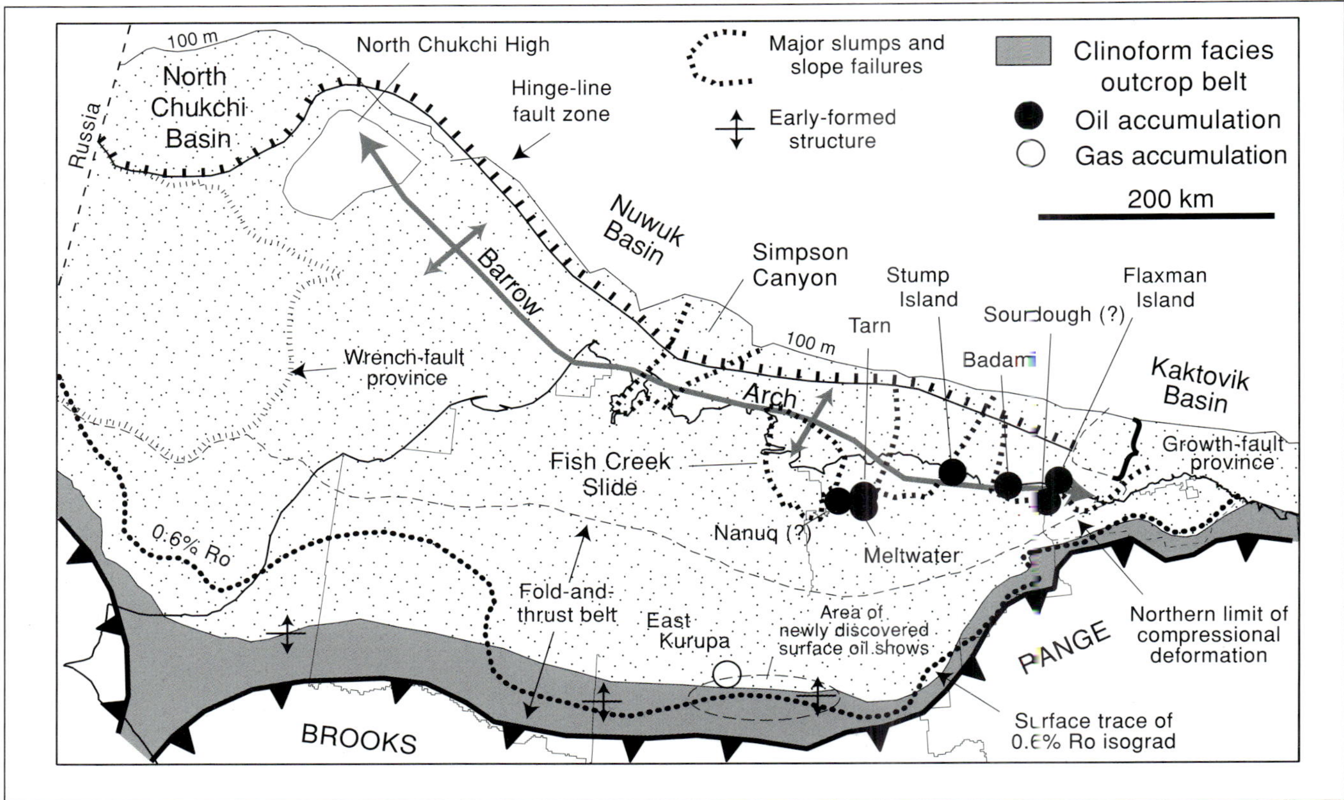

Figure 18. Summary of Brookian clinoform facies depositional and tectonic features, oil and gas accumulations, and the surface trace of the top of the oil window (the 0.6% R_o value). Stippling shows subsurface distribution of the clinoform facies as far north as the 100-m isobath.

Brooks Range. It is a region characterized by a thick succession of carbonate and associated chert and siliciclastic rocks, including some of the richest source rocks in all of Alaska, that are incorporated in large, thrust-faulted anticlines. As defined, the thrust belt is limited to a relatively narrow zone, 30–100 km wide, beneath the foothills adjacent to the northern margin of the Brooks Range (Figure 19).

At about 150°W longitude, a fundamental structural-stratigraphic distinction is made between the western and eastern parts of the thrust belt. To the west, the structural style is one of multiple-thrust imbricates of competent Lisburne Group carbonate rocks and associated strata (e.g., Mayfield et al., 1988). To the east, the structural style consists of broad, basement-involved folds characterized by high-angle reverse faulting (e.g., Wallace and Hanks, 1990). Examples of these structural styles observed in thrust-belt wells are provided in Figure 20. In recognition of these differences, this belt is typically assessed in two parts—eastern and western (Bird, 1995). Drilling in the thrust belt is limited to its central and eastern parts, where two gas fields, Kemik and Kavik, have been discovered (Figure 19). The sizes of these accumulations have not been released.

The practical definition of the thrust belt is the northern extent of subsurface structures with Ellesmerian rocks at drillable depths. Early exploratory drilling was based on misinterpreted strong reflectors in the Brookian for thrust sheets of Ellesmerian (Lisburne carbonate) strata. These early failures, shown by the smaller well symbols in Figure 19, help to constrain the northern boundary of the thrust belt—a boundary that is imaged poorly on seismic-reflection records. Thrust-belt stratigraphy is somewhat different from that of the rest of the North Slope. Ellesmerian and Beaufortian sequence rocks are mostly thin, siliceous, distal facies. A significant exception is the Lisburne Group which, throughout most of the thrust belt, remains a carbonate platform facies and is considered the main prospective reservoir in the thrust belt. In the northeast, the Ivishak and Sag River Sandstones also constitute viable reservoirs (e.g., the Kavik gas field). Brookian rocks in the thrust belt are represented by proximal, coarse clastic facies with minor reservoir potential. Fracturing may be a critical component of thrust-belt reservoir development (e.g., Hanks et al., 1997) and may enable even siliceous rocks to be potential reservoirs. Fractured lime-mudstone (Shublik Formation) apparently constitutes the reservoir of the Kemik gas accumulation.

Good to excellent source rocks occur in Mississip-

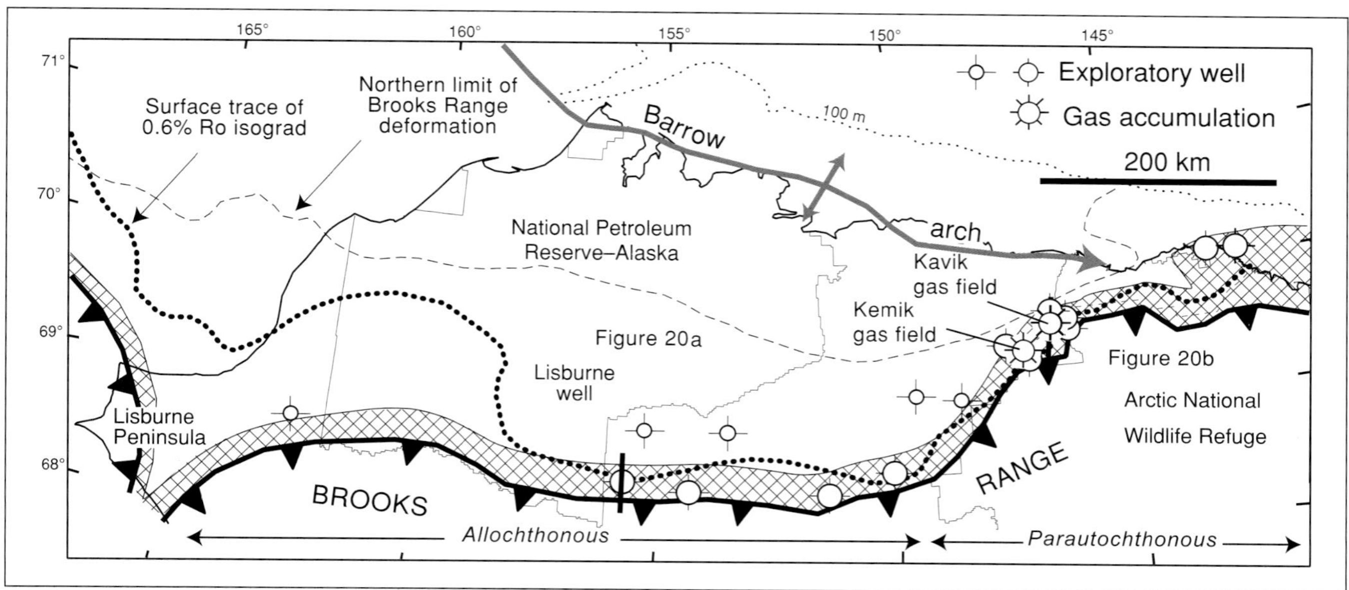

Figure 19. Location of the thrust belt (crosshatched) where subsurface structure and stratigraphy are similar to those of the adjacent Brooks Range. Allochthonous and parautochthonous designations mark regions of different structural styles and timing of deformation. Smaller well symbols show early thrust-belt exploratory failures where strong seismic reflections in the Brookian were misinterpreted as thrust sheets of Lisburne carbonate rocks. Amount of uplift and erosion along this belt is indicated by the surface trace of the top of the oil window (the 0.6% R_o value).

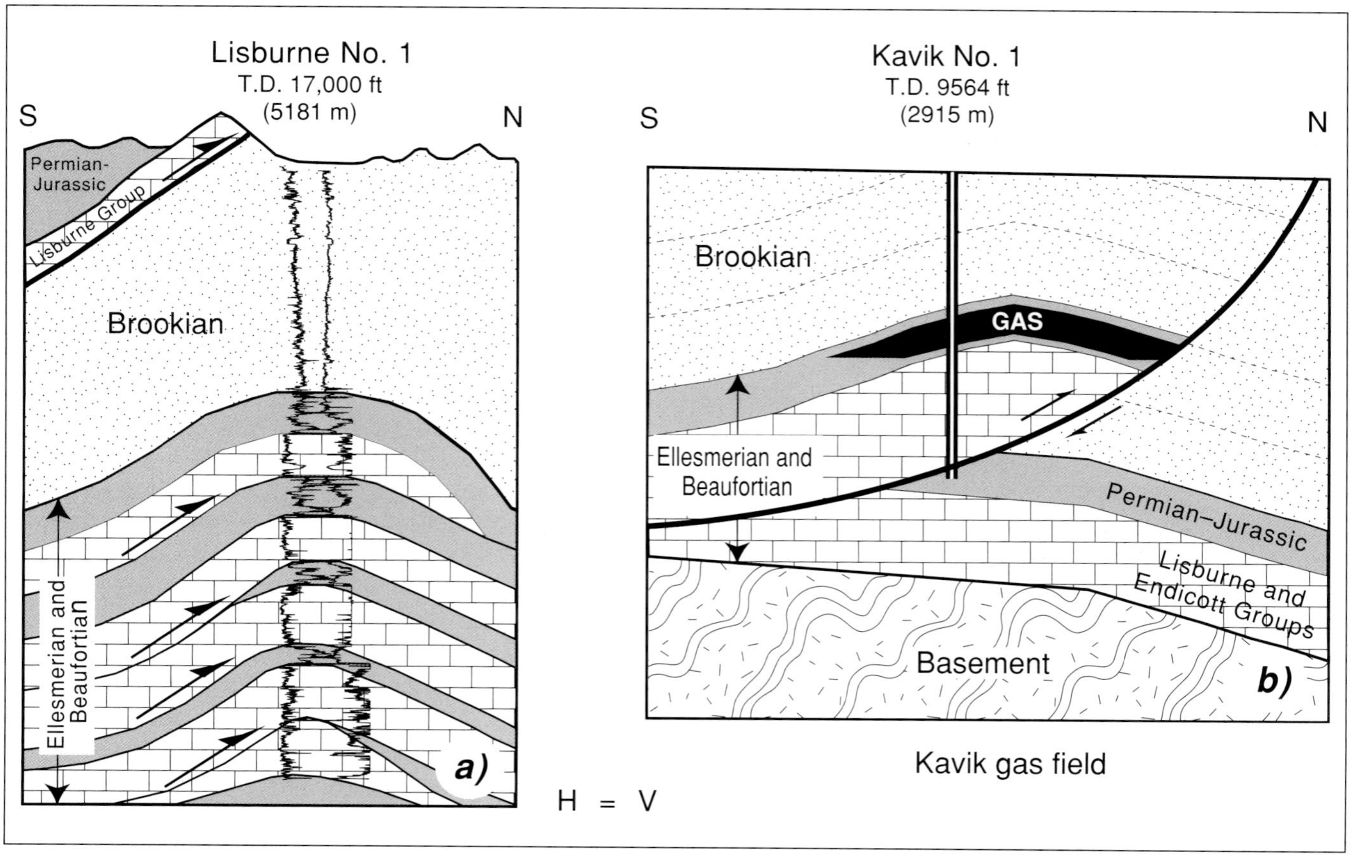

Figure 20. Comparison of Brooks Range thrust-belt structural styles. (a) Western thrust belt and (b) eastern thrust belt. See Figure 19 for locations. Sections are adapted from Cole et al. (1997) and Magoon et al. (1999), respectively.

pian, Triassic, and Jurassic strata in the central part of the thrust belt (Tailleur, 1964; Tourtelot et al., 1967; Bodnar, 1984), and solid bitumen is common in rock pores and fractures in this area. The surface trace of the top of the oil window generally lies within or north of the thrust belt (Figure 19). Thermal maturity in the west and central thrust belt is the result of some combination of tectonic and sedimentary burial (Harris et al., 1987; Cole et al., 1997), whereas in the eastern thrust belt, it is the result of mainly sedimentary burial (Bird et al., 1999).

Structural style and timing of deformation vary from west to east within the thrust belt. In the west-central area, structural style is that of multiple thrust repetitions of competent Lisburne Group carbonates. In Figure 20a, five repeats of the Lisburne are encountered in the subsurface and two more are evident in surface exposures immediately to the south (Mull et al., 1994). In this area, Cole et al. (1997) documented structural deformation in the earliest Cretaceous, early Tertiary, and perhaps middle Cretaceous. In the eastern area, structural style is that of broad, basement-involved folds with high-angle reverse faults (Wallace and Hanks, 1990; Cole et al., 1999; Moore, 1999). That style is illustrated by the structure at the Kavik gas field (Figure 20b). There, multiple episodes of Tertiary deformation (uplift and cooling) are indicated by apatite fission-track analysis (O'Sullivan et al., 1993). The earliest deformation (~45 Ma) is relatively close to the time of maximum burial and hydrocarbon generation (~55 Ma). Although subsurface information is lacking for the thrust belt on the Lisburne Peninsula (Figure 19), recent outcrop studies by T. E. Moore et al. (unpublished work) suggest that the structural style is more akin to the northeastern Brooks Range.

Thermal-maturity considerations suggest that thrust-belt potential is primarily natural gas. Some oil potential in the central part of the thrust belt is suggested by lower thermal maturities. Two possible stages of oil generation and three possible stages of deformation complicate evaluation of this potential. In the earliest Cretaceous, maturation and oil generation may have occurred as a result of burial by thrust imbricates (Harris et al., 1987). Later, in the middle Cretaceous, stratigraphic burial may have contributed to maturation and generation. Both mechanisms have played a role, but to different extents in different areas. Structural deformation occurred in the central part of the thrust belt in the earliest Cretaceous, early Tertiary, and possibly middle Cretaceous (Cole et al., 1997; Mull et al., 2000a; Mull et al., 2000b). Trap integrity and/or remigration of early-trapped oil to later-formed traps are important factors in evaluating petroleum potential of this area. In evaluating thrust-belt gas potential, one may be able to rely on late-stage gas generation and simply look for traps with adequate reservoir and seals. Those with the best seals would likely be the youngest, least-deformed structures.

SUMMARY OF NORTHERN ALASKA PETROLEUM POTENTIAL

The potential for undiscovered petroleum in northern Alaska is estimated at about 30 billion bbl (4.8 billion m^3) of oil and natural-gas liquids and about 160 tcf (4.5 trillion m^3) of natural gas (mean value, technically recoverable), based on the 1995 Department of the Interior national oil and gas assessment (Gautier et al., 1995; Sherwood, 1998). Table 2 ranks the stratigraphic and structural subdivisions in order of decreasing oil potential and expresses that potential as a percentage of the regional total. The number of plays identified and assessed in each subdivision is also shown, to give a sense of geologic variability.

By this analysis, the greatest potential in northern Alaska for oil and natural-gas liquids is in the Beaufortian and Brookian topset plays, with about one-third and one-quarter, respectively, of the undiscovered oil. The next five subdivisions (Brookian fold belt, Endicott, Ivishak, thrust belt, and Brookian clinoform) are of relatively similar potential, ranging from 11% to 6% of the total. The Lisburne is rated least prospective, with only 1% of oil potential. Potential for natural gas is estimated to be greatest in the Brookian topset and to have somewhat less potential in the thrust belt, Brookian fold belt, and Beaufortian. The least potential is in the Ivishak, Brookian clinoform, Endicott, and Lisburne.

The years since the 1995 oil and gas assessment probably mark a major turning point for exploration on the North Slope—a turn away from structural-combination-trap exploration strategy toward a stratigraphic exploration strategy guided by innovative technology—both exploration (3-D seismic, AVO, etc.) and drilling technology (ice roads, mobile rigs, directional drilling, coiled tubing, etc.). Whereas the Barrow arch has been the focus of exploration and the site of most discoveries to date, a weaker link between future discoveries and the Barrow arch is likely. A key aspect of this prediction is the occurrence of relatively "closed" petroleum systems—those (such as the Alpine system) that are isolated from the regional systems that have charged the large Barrow-arch accumulations. It is likely that a map of new discoveries compiled 10 or 25 years from now will show little relationship between field locations and the Barrow arch—except perhaps for the relationship imposed by depth to objectives. Although this change in strategy may result in a significant shift in estimated volumes of undiscovered resources, it may not significantly change the ranking of plays shown in Table 2. Clearly, the Beaufortian and Brookian topset plays would be at or near the top of a new list, and the Brookian clinoform play may be the dark horse that jumps past several plays to land in the top three or four ranking.

Table 2. Summary of future oil and gas potential of northern Alaska, expressed as a percentage of the total estimated undiscovered resources of the region, about 30 billion bbl (4.8 billion m^3) of oil and natural-gas liquids and about 160 tcf (4.5 trillion m^3) of natural gas (technically recoverable, mean value). Estimates are based on the 1995 Department of the Interior national oil and gas resource assessment.

Structural or stratigraphic subdivision	Percent of undiscovered resource		Number of plays
	Oil	Gas	
Beaufortian	31	14	5
Brookian topset	25	26	12
Brookian fold belt	11	14	3
Endicott	10	6	4
Ivishak	9	9	6
Thrust belt	7	17	3
Brookian clinoform	6	7	9
Lisburne	1	3	4

CONCLUSIONS

Northern Alaska, a demonstrated rich petroleum province, is hypothesized by government assessors to hold a significant proportion of United States undiscovered oil and gas resources. This is an extremely remote area with harsh arctic climate and limited infrastructure. The federal government owns most of the land, and energy-resource development is viewed as competing with other land uses.

This geologically complex region includes prospective strata within passive-margin, rift, and foreland-basin sequences. Outside of a relatively small area where infrastructure is concentrated and most drilling has occurred, it is lightly explored. Multiple source-rock horizons have charged several regionally extensive petroleum systems. Both extensional and compressional structures provide ample objectives. Intense historical exploration for the region's premier Triassic (Prudhoe Bay) reservoir appears to leave it with limited future potential. However, recent emphasis on stratigraphic traps has demonstrated significant resource potential in shelf and turbidite systems in Jurassic to Tertiary strata.

Despite robust potential, northern Alaska remains a risky exploration frontier—a geologically complex frontier with harsh economic conditions and volatile policy issues. Its role as a major petroleum province in the new century will depend on continued technological innovations, not only in exploration and drilling operations, but also in development of a market for huge, currently unmarketable natural-gas resources. Ultimately, government-policy decisions will determine whether exploration of the region will proceed.

ACKNOWLEDGMENTS

I am indebted to reviewers Dave Houseknecht, William A. Morgan, and Kirk Sherwood and to Margaret Keller and Les Magoon, who reviewed selected parts of the manuscript. Their thoughtful reviews and insightful comments have measurably improved this paper.

REFERENCES CITED

Alaska Division of Oil and Gas, 2000, Oil and gas annual report—2000: Anchorage, Alaska Department of Natural Resources, Division of Oil and Gas, 113 p.

ANWR Assessment Team, 1999, The oil and gas resource potential of the Arctic National Wildlife Refuge 1002 Area, Alaska: U.S. Geological Survey Open-File Report 98-34, 2 CD-ROMs.

Armstrong, A. K., 1974, Carboniferous carbonate depositional models, preliminary lithofacies and paleotectonic maps, Arctic Alaska: AAPG Bulletin, v. 58, p. 621–645.

Armstrong, A. K., and K. J. Bird, 1976, Facies and environments of deposition of Carboniferous rocks, arctic Alaska: Recent and ancient depositional environments of Alaska Symposium: Anchorage, Alaska Geological Society, p. A1–A16.

Armstrong, A. K., and B. L. Mamet, 1974, Carboniferous biostratigraphy, Prudhoe Bay State 1, Arctic Alaska: AAPG Bulletin, v. 58, p. 646–660.

Atkinson, C. D., J. H. McGowen, S. Bloch, L. L. Lundell, and P. N. Trumbly, 1990, Braidplain and deltaic reservoir, Prudhoe Bay field, Alaska, *in* J. H. Barwis, J. G. McPherson, and J. R. J. Studlick, eds., Sandstone petroleum reservoirs: New York, Springer-Verlag, p. 7–29.

Atkinson, C. D., P. N. Trumbly, and M. C. Kremer, 1988, Sedimentology and depositional environments of the Ivishak Sandstone, Prudhoe Bay field, North Slope, Alaska: SEPM Core Workshop No. 12, v. 2, p. 561–614.

Barnes, D. A., 1987, Reservoir quality in the Sag River Formation, Prudhoe Bay field, Alaska: Depositional environment and diagenesis, *in* I. Tailleur and P. Weimer, eds., Alaskan North Slope geology, v. 1: Pacific Section SEPM and Alaska Geological Society, p. 85–94.

Belfield, W. C., 1988, Characterization of a naturally fractured carbonate reservoir: Lisburne field, Prudhoe Bay, Alaska: Society of Petroleum Engineers Paper 18174, 11 p.

Bird, K. J., 1978, Generalized isopach map of the Endicott Group, eastern North Slope petroleum province, Alaska: U.S. Geological Survey Miscellaneous Field Studies Map MF-928 T, scale 1:500,000, 1 sheet.

Bird, K. J., 1988, Structure-contour and isopach maps of the National Petroleum Reserve in Alaska, *in* G. Gryc, ed., Geology and exploration of the National Petroleum Reserve in Alaska, 1974 to 1982: U.S. Geological Survey Professional Paper 1399, p. 355–377.

Bird, K. J., 1994, The Ellesmerian petroleum system, North Slope of Alaska, U.S.A., *in* L .B Magoon and W. G. Dow, eds., The petroleum system—From source to trap: AAPG Memoir 60, p. 339–358.

Bird, K. J., 1995, Northern Alaska Province (001), *in* D. L. Gautier, K. I. Takahashi, and K. L. Varnes, eds., U.S. Geological Survey 1995 national assessment of United States oil and gas resources—Results, methodology, and supporting data: U.S. Geological Survey Digital Data Series DDS-30.

Bird, K. J., and C. F. Jordan, 1977, Lisburne Group (Mississippian and Pennsylvanian), potential major hydrocarbon objective of Arctic Slope, Alaska: AAPG Bulletin, v. 61, p. 1493–1512.

Bird, K. J., and C. M. Molenaar, 1992, The North Slope foreland basin, Alaska, *in* R. Macqueen and D. Leckie, eds., Foreland basins and fold belts: AAPG Memoir 55, p. 363–393.

Bird, K. J., R. C. Burruss, and M. J. Pawlewicz, 1999, Thermal maturity, *in* ANWR Assessment Team, ed., The oil and gas resource potential of the 1002 Area, Arctic National Wildlife Refuge, Alaska: U. S. Geological Survey Open File Report 98-34, p. V1–V64.

Bodnar, D. A., 1984, Stratigraphy, age, depositional environments, and hydrocarbon source rock evaluation of the Otuk Formation, north-central Brooks Range, Alaska: M.S. thesis, University of Alaska, Fairbanks, 232 p.

Bruynzeel, J. W., E. C. Guldenzopf, and J. E. Pickard, 1982, Petroleum exploration of NPRA, 1974–1981: Tetra Tech, Inc., final report no. 8200, prepared for Husky Oil NPR Operations, Inc., 183 p.

Burruss, R. C., and T. S. Collett, 2000, Stable isotope geochemistry of natural gas, North Slope, Alaska: Evidence for multiple sources, mixing, and alteration (abs.): AAPG Annual Meeting, New Orleans, Abstracts with Program, p. A20–A21.

Carman, G. J., and P. Hardwick, 1983, Geology and regional setting of the Kuparuk oil field, Alaska: AAPG Bulletin, v. 67, p. 1014–1031.

Churkin, M. Jr., and J. H. Trexler Jr., 1980, Circum-Arctic plate accretion—Isolating part of a Pacific plate to form the nucleus of the Arctic Basin: Earth and Planetary Science Letters, v. 48, p. 356–362.

Cole, F., K. J. Bird, C. G. Mull, W. K. Wallace, W. Sassi, J. M. Murphy, and M. Lee, 1999, A balanced cross section and kinematic and thermal model across the northeastern Brooks Range mountain front, Arctic National Wildlife Refuge, Alaska, *in* ANWR Assessment Team, ed., The oil and gas resource potential of the 1002 Area, Arctic National Wildlife Refuge, Alaska: U.S. Geological Survey Open-File Report 98-34, 2 CD-ROMs, p. SM1–SM60.

Cole, F., K. J. Bird, J. Toro, F. Roure, P. B. O'Sullivan, M. J. Pawlewicz, and D. G. Howell, 1997, An integrated model for the tectonic development of the frontal Brooks Range and Colville Basin 250 km west of the Trans-Alaska Crustal Transect: Journal of Geophysical Research, v. 102, no. B9, p. 20,685–20,708.

Craig, J. D., K. W. Sherwood, and P. P. Johnson, 1985, Geologic report for the Beaufort Sea planning area, Alaska—Regional geology, petroleum geology, environmental geology: Minerals Management Service OCS Report MMS 85-0111, 192 p.

Crowder, R. K., 1990, Permian and Triassic sedimentation in the northeastern Brooks Range, Alaska: Deposition of the Sadlerochit Group: AAPG Bulletin, v. 74, p. 1351–1370.

Demaison, G., and B. J. Huizinga, 1994, Genetic classification of petroleum systems using three factors: Charge, migration, and entrapment, *in* L. B. Magoon and W. G. Dow, eds., The petroleum system—From source to trap: AAPG Memoir 60, p. 73–89.

Detterman, R. L., 1970, Sedimentary history of Sadlerochit and Shublik Formations in northeastern Alaska, *in* W. L. Adkison and M. M. Brosge, eds., Proceedings of the geologic seminar on the North Slope of Alaska: Los Angeles, AAPG Pacific Section, p. 1–13.

Dumoulin, J. A., 1999, Carboniferous and older carbonate rocks: Lithofacies, extent, and reservoir quality, *in* ANWR Assessment Team, ed., The oil and gas resource potential of the 1002 Area, Arctic National Wildlife Refuge, Alaska: U.S. Geological Survey Open-File Report 98-34, CD-ROM, p. CC1–CC33.

Erickson, J. W., and R. M. Sneider, 1997, Structural and hydrocarbon histories of the Ivishak (Sadlerochit) reservoir, Prudhoe Bay field: SPE Reservoir Engineering, v. 12, p. 18–22.

Gautier, D. L., G. L. Dolton, K. I. Takahashi, and K. L. Varnes, eds., 1995, 1995 national assessment of United States oil and gas resources—Results, methodology, and supporting data: U.S. Geological Survey Digital Data Series DDS-30, 1 CD-ROM.

Gingrich, D. A., D. Hughes, K. Martindale, M. Savic, A. Sena, D. Kerr, R. C. Hannon, and D. G. Knock, 2000, Alpine oil field: Geophysics from wildcat to development (abs.): Alaska Geological Society and Geophysical Society of Alaska, 2000 Science and Technology Conference.

Gingrich, D. A., D. G. Knock, and R. Masters, 2001, Geophysical interpretation methods applied at Alpine oil field: North Slope, Alaska: The Leading Edge, v. 20, no. 7, p. 730–738.

Grantz, A., D. L. Clark, R. L. Phillips, and S. P. Srivastava, 1998, Phanerozoic stratigraphy of Northwind Ridge, magnetic anomalies in the Canada Basin, and the geometry and timing of rifting in the Amerasia Basin, Arctic Ocean: Geological Society of America Bulletin, v. 110, p. 801–820.

Grantz, A., D. A. Dinter, and R. C. Culotta, 1987, Structure of the continental shelf north of the Arctic National Wildlife Refuge, *in* K. J. Bird and L. B. Magoon, eds., Petroleum geology of the northern part of the Arctic National Wildlife Refuge, northeastern Alaska, U.S. Geological Survey Bulletin 1778, p. 271–276.

Grantz, A., and S. D. May, 1983, Regional geology and petroleum potential of the United States Chukchi shelf north of Point Hope, *in* G. Gryc, ed., Geology and exploration of the National Petroleum Reserve in Alaska, 1974 to 1982: U.S. Geological Survey Professional Paper 1399, p. 209–230.

Grantz, A., S. D. May, and D. A. Dinter, 1988, Geologic framework, petroleum potential, and environmental geology of the United States Beaufort and northeasternmost Chukchi Seas, *in* G. Gryc, ed., Geology and exploration of the National Petroleum Reserve in Alaska, 1974 to 1982: U.S. Geological Survey Professional Paper 1399, p. 231–256.

Gryc, G., ed., 1988, Geology and exploration of the National Petroleum Reserve in Alaska, 1974 to 1982: U.S. Geological Survey Professional Paper 1399, 940 p.

Hanks, C. L., J. Lorenz, T. Lawrence, and A. P. Krumhardt, 1997, Lithologic and structural controls on natural fracture distribution and behavior within the Lisburne Group, northeastern Brooks Range and North Slope subsurface, Alaska: AAPG Bulletin, v. 81, p. 1700–1720.

Hannon, R. C., D. A. Gingrich, D. G. Knock, K. P. Helmold, W. J. Campaign, W. R. Morris, W. D. Masterson, and S. R. Redman, 2000a, The discovery and delineation of a new Alaskan reservoir: The Alpine field, North Slope, Alaska (abs.): Alaska Geological Society and Geophysical Society of Alaska, 2000 Science and Technology Conference.

Hannon, R. C., D. A. Gingrich, S. R. Redman, K. P. Helmold, W. J. Campaign, and C. B. Dotson, 2000b, The discovery and delineation of a new Alaskan reservoir: The Alpine field,

North Slope, Alaska, U.S.A. (abs.): AAPG Annual Meeting, New Orleans, Abstracts with Program, p. A63.

Harris, A. G., H. R. Lane, I. L. Tailleur, and I. Ellersieck, 1987, Conodont thermal maturation patterns in Paleozoic and Triassic rocks, northern Alaska—Geologic and exploration implications, *in* I. Tailleur and P. Weimer, eds., Alaskan North Slope geology, v. 1: Pacific Section SEPM and Alaska Geological Society, p. 181–191.

Hohler, J. J., and W. E. Bischoff, 1986, Alaska: Potential for giant fields, *in* M. T. Halbouty, ed., Future petroleum provinces of the world: AAPG Memoir 40, p. 131–142.

Holba, A. G., L. Ellis, P. Decker, G. Wilson, A. Watts, and D. Masterson, 2000a, Organic geochemistry of gas fields flanking the Avak impact structure, western NPRA, North Slope, Alaska (abs.): Alaska Geological Society and Geophysical Society of Alaska 2000 Science and Technology Conference, Anchorage.

Holba, A. G., G. Wilson, P. Decker, R. Garrard, and L. Ellis, 2000b, North Slope, Alaska petroleum systems (abs.): Alaska Geological Society and Geophysical Society of Alaska 2000 Science and Technology Conference, Anchorage.

Houseknecht, D. W., 2001, Sequence stratigraphy and sedimentology of Beaufortian strata (Jurassic–Lower Cretaceous) in the National Petroleum Reserve—Alaska (NPRA), *in* D. W. Houseknecht, ed., NPRA core workshop: Petroleum plays and systems in the National Petroleum Reserve—Alaska: SEPM Core Workshop 21, p. 57–88.

Houseknecht, D. W., and C. J. Schenk, 1999, Seismic facies analysis and hydrocarbon potential of Brookian strata, *in* ANWR Assessment Team, ed., The oil and gas resource potential of the 1002 Area, Arctic National Wildlife Refuge, Alaska: U.S. Geological Survey Open-File Report 98-34, 2 CD-ROMs., p. BS-1–BS-60.

Houseknecht, D. W., and C. J. Schenk, 2000, Sequence stratigraphic framework for oil-prospective Cretaceous turbidites, National Petroleum Reserve—Alaska (abs.): AAPG Annual Meeting, New Orleans, Abstracts with Program, p. A70.

Houseknecht, D. W., and C. J. Schenk, 2001, Depositional sequences and facies in the Torok Formation, National Petroleum Reserve—Alaska (NPRA), *in* D. W. Houseknecht, ed., NPRA core workshop: Petroleum plays and systems in the National Petroleum Reserve—Alaska: SEPM Core Workshop 21, p. 179–199.

Hubbard, R. J., S. P. Edrich, and R. P. Rattey, 1987, Geologic evolution and hydrocarbon habitat of the "Arctic Alaska Microplate," *in* I. Tailleur and P. Weimer, eds., Alaskan North Slope geology: Pacific Section SEPM and Alaska Geological Society, p. 797–830.

Hughes, W. B., and A. G. Holba, 1988, Relationship between crude oil quality and biomarker patterns: Organic Geochemistry, v. 13, p. 15–30.

Hughes, W. B., A. G. Holba, and D. E. Miller, 1985, North Slope Alaska oil-rock correlation study, *in* L. B. Magoon and G. E. Claypool, eds., Alaska North Slope oil/source rock correlation study: AAPG Studies in Geology 20, p. 379–402.

Jameson, J., 1994, Models of porosity formation and their impact on reservoir description, Lisburne field, Prudhoe Bay, Alaska: AAPG Bulletin, v. 78, p. 1651–1678.

Jamison, H. C., L. D. Brockett, and R. A. MacIntosh, 1980, Prudhoe Bay—A 10-year perspective, *in* M. T. Halbouty, ed., Giant oil and gas fields of the decade 1968–1978: AAPG Memoir 30, p. 289–314.

Johnsson, M. J., K. R. Evans, and H. A. Marshall, 1999, Thermal maturity of sedimentary rocks in Alaska: Digital resources: U.S. Geological Survey Digital Data Series DDS-54, 1 CD-ROM.

Johnsson, M. J., M. J. Pawlewicz, A. G. Harris, and Z. C. Valin, 1992, Vitrinite reflectance and conodont color alteration index data from Alaska: Data to accompany the thermal maturity map of Alaska: U.S. Geological Survey Circular Open-File Report 409, 3 computer disks.

Jones, H. P., and R. G. Speers, 1976, Permo-Triassic reservoirs of Prudhoe Bay field, North Slope, Alaska, *in* J. Braunstein, ed., North American oil and gas fields: AAPG Memoir 24, p. 23–50.

Kelley, J. S., and Brosgé, W. P., 1995, Geologic framework of a transect of the central Brooks Range: Regional relations and an alternative to the Endicott Mountains allochthon: AAPG Bulletin, v. 79, p. 1087–1116.

Kirschner, C. E., 1994, Map showing sedimentary basins of Alaska, *in* G. Plafker and H. C. Berg, eds., The geology of Alaska: Geological Society of America, The geology of North America, vol. G-1, plate 7, 1:2,500,000 scale, 1 sheet.

Kirschner, C. E., and B. A. Rycerski, 1988, Petroleum potential of representative stratigraphic and structural elements in the National Petroleum Reserve in Alaska, *in* G. Gryc, ed., Geology and exploration of the National Petroleum Reserve in Alaska, 1974 to 1982: U.S. Geological Survey Professional Paper 1399, p. 191–208.

Konkler, J. L., D. S. Hastings, D. H. McGimsey, and H. Posamentier, 2000, Tabasco sands, North Slope, Alaska: An incised valley mapped with 3-D seismic data (abs.): AAPG Annual Meeting, New Orleans, Abstracts with Program, p. A80.

Lane, L. S., 1997, Canada Basin, Arctic Ocean: Evidence against a rotational origin: Tectonics, v. 16, p. 363–387.

Lawton, T. F., G. W. Geehan, and B. J. Voorhees, 1987, Lithofacies and depositional environments of the Ivishak Formation, Prudhoe Bay field, *in* I. Tailleur and P. Weimer, eds., Alaskan North Slope geology, v. 1: Pacific Section SEPM and Alaska Geological Society, p. 61–76.

Lawver, L. A., and C. R. Scotese, 1990, A review of tectonic models for the evolution of the Canada Basin, *in* A. Grantz, G. L. Johnson, and J. F. Sweeney, eds., The Arctic Ocean region: Geological Society of America, The geology of North America, vol. L, p. 593–618.

Lemley, K., 1998, Badami oil field, North Slope Alaska (abs.): Alaska Geological Society Newsletter, v. 27, 2 p.

Lerand, M., 1973, Beaufort Sea, *in* R. G. McCrossan, ed., The future petroleum provinces of Canada—Their geology and potential: Canadian Society of Petroleum Geology Memoir 1, p. 315–386.

Lillis, P. G., M. D. Lewan, A. Warden, S. M. Monk, and J. D. King, 1999, Identification and characterization of oil types and their source rocks, *in* ANWR Assessment Team, ed., The oil and gas resource potential of the 1002 Area, Arctic National Wildlife Refuge, Alaska: U.S. Geological Survey Open-File Report 98-34, 2 CD-ROMs, p. OA1–OA56.

Magoon, L. B., 1994, The geology of known oil and gas resources by petroleum system—Onshore Alaska, *in* G. Plafker and H. C. Berg, eds., The geology of Alaska: Geological Society of America, The geology of North America, vol. G-1, p. 905–936.

Magoon, L. B., and K. J. Bird, 1988, Evaluation of petroleum source rocks in the National Petroleum Reserve in Alaska, using organic-carbon content, hydrocarbon content, visual kerogen, and vitrinite reflectance, *in* G. Gryc, ed., Geology and exploration of the National Petroleum Reserve in Alaska, 1974 to 1982: U.S. Geological Survey Professional Paper 1399, p. 381–450.

Magoon, L. B., and W. G. Dow, eds., 1994, The petroleum system—From source to trap: AAPG Memoir 60, 655 p.

Magoon, L. B., K. J. Bird, R. C. Burruss, D. Hayba, D. W. Houseknecht, M. A. Keller, P. G. Lillis, and E. L. Rowan, 1999, Evaluation of hydrocarbon charge and timing using the petroleum system, *in* ANWR Assessment Team, ed., The oil and gas resource potential of the 1002 Area, Arctic National Wildlife Refuge, Alaska: U.S. Geological Survey Open-File Report 98-34, 2 CD-ROMs, p. PS1–PS66.

Magoon, L. B., P. V. Woodward, A. C. Banet Jr., A. B. Griscom, and T. A. Daws, 1987, Thermal maturity, richness, and type of organic matter of source rock units, *in* K. J. Bird and L. B. Magoon, eds., Petroleum geology of the northern part of the Arctic National Wildlife Refuge, northeastern Alaska: U.S. Geological Survey Bulletin 1778, p. 127–179.

Masterson, W. D., and C. E. Paris, 1987, Depositional history and reservoir description of the Kuparuk River Formation, North Slope, Alaska, *in* I. Tailleur and P. Weimer, eds., Alaskan North Slope geology: Pacific Section SEPM and Alaska Geological Society, p. 95–107.

Masterson, W. D., Z. He, J. Corrigan, L. I. P. Dzou, and A. G. Holba, 2000, Petroleum migration and filling history models for the Prudhoe Bay, Kuparuk, and West Sak fields, North Slope, Alaska (abs.): AAPG Annual Meeting, New Orleans, Abstracts with Program, p. A93.

Masterson, W. D., A. Holba, and L. Dzou, 1997, Filling history of America's two largest oil fields: Prudhoe Bay and Kuparuk, North Slope, Alaska (abs.): AAPG Annual Convention Abstracts and Program, v. 6, p. A77.

Mayfield, C. F., I. L. Tailleur, and I. Ellersieck, 1988, Stratigraphy, structure, and palinspastic synthesis of the western Brooks Range, northwestern Alaska, *in* G. Gryc, ed., Geology and exploration of the National Petroleum Reserve in Alaska, 1974 to 1982: U.S. Geological Survey Professional Paper 1399, p. 143–186.

McMillen, K. J., and M. D. Colvin, 1987, Facies correlation and basin analysis of the Ivishak Formation, Arctic National Wildlife Refuge, Alaska, *in* I. Tailleur and P. Weimer, eds., Alaskan North Slope geology, v. 1: Pacific Section SEPM, p. 381–390.

Melvin, J., and A. S. Knight, 1984, Lithofacies, diagenesis and porosity of the Ivishak Formation, Prudhoe Bay area, Alaska, *in* D. A. McDonald and R. C. Surdam, eds., Clastic diagenesis: AAPG Memoir 37, p. 347–365.

Missman, R. A., and J. Jameson, 1991, An evolving description of a fractured carbonate reservoir: The Lisburne field, Prudhoe Bay, Alaska: SPE Paper 22161, p. 699–718.

Molenaar, C. M., 1982, Umiat field, an oil accumulation in a thrust-faulted anticline, North Slope of Alaska, *in* R. B. Powers, ed., Geologic studies of the Cordilleran thrust belt: Rocky Mountain Association of Geologists, p. 537–548.

Molenaar, C. M., 1988, Depositional history and seismic stratigraphy of Lower Cretaceous rocks in the National Petroleum Reserve in Alaska and adjacent areas, *in* G. Gryc, ed., Geology and exploration of the National Petroleum Reserve in Alaska, 1974 to 1982: U.S. Geological Survey Professional Paper 1399, p. 593–621.

Molenaar, C. M., K. J. Bird, and A. R. Kirk, 1987, Cretaceous and Tertiary stratigraphy of northeastern Alaska, *in* I. Tailleur and P. Weimer, eds., Alaskan North Slope Geology, v. 2: Pacific Section SEPM and Alaska Geological Society, p. 513–528.

Molenaar, C. M., R. M. Egbert, and L. F. Krystinik, 1988, Depositional facies, petrography, and reservoir potential of the Fortress Mountain Formation (Lower Cretaceous), central North Slope, Alaska, *in* G. Gryc, ed., Geology and exploration of the National Petroleum Reserve in Alaska, 1974 to 1982: U.S. Geological Survey Professional Paper 1399, p. 257–279.

Moore, T. E., 1999, Balanced cross section, Bathtub syncline to Beaufort Sea through Niguanak structural high, Arctic National Wildlife Refuge (ANWR), northeastern Alaska, *in* ANWR Assessment Team, ed., The oil and gas resource potential of the 1002 Area, Arctic National Wildlife Refuge, Alaska: U.S. Geological Survey Open-File Report 98-34, 2 CD-ROMs, p. BC1–BC61.

Moore, T. E., T. H. Nilsen, A. Grantz, and I. L. Tailleur, 1984, Mississippian marine and nonmarine strata, Lisburne Peninsula, Alaska, *in* K. M. Reed, and S. Bartsch-Winkler, eds., The United States Geological Survey in Alaska: Accomplishments during 1982: U. S. Geological Survey Circular 939, p. 17–21.

Moore, T. E., W. K. Wallace, K. J. Bird, S. M. Karl, C. G. Mull, and J. T. Dillon, 1994, Geology of northern Alaska, *in* G. Plafker and H. C. Berg, eds., The geology of Alaska: Geological Society of America, The geology of North America, vol. G-1, p. 49–140.

Morris, W. R., D. S. Hastings, S. E. Moothart, K. P. Helmold, and M. J. Faust, 2000, Sequence stratigraphic development and depositional framework of deep water slope apron systems, Tarn reservoir, North Slope Alaska (abs.): AAPG Annual Meeting, New Orleans, Abstracts with Program, p. A101.

Morrow, H. B., 1987, Seismic subsequences in the foothills fold belt, NPRA, Alaska (abs.), *in* I. Tailleur and P. Weimer, eds., Alaskan North Slope geology: Pacific Section SEPM and Alaska Geological Society, p. 479.

Mull, C. G., E. E. Harris, R. R. Reifenstuhl, and D. L. LePain, 2000a, Hydrocarbon source rock potential in the southern Colville Basin, east-central Brooks Range foothills belt (abs.): Alaska Geological Society and Geophysical Society of Alaska 2000 Science and Technology Conference.

Mull, C. G., E. E. Harris, R. R. Reifenstuhl, and T. E. Moore, 2000b, Geologic map of the Coke Basin–Kukpowruk Rivers area, DeLong Mountains D2 and D3 quadrangles, Alaska: Alaska Division of Geological and Geophysical Surveys Report of Investigation RI 2000-2, scale 1:63,360, 1 sheet.

Mull, C. G., T. E. Moore, E. E. Harris, and I. L. Tailleur, 1994, Preliminary geologic map of the Killik River quadrangle: U.S. Geological Survey Open-File Report 94-679, scale 1:125,000.

O'Sullivan, P. B., 1996, Late Mesozoic and Cenozoic thermotectonic evolution of the Colville Basin, North Slope, Alaska, *in* M. J. Johnsson and D. G. Howell, eds., Thermal evolution of sedimentary basins in Alaska: U.S. Geological Survey Bulletin 2142, p. 45–79.

O'Sullivan, P. B., P. F. Green, S. C. Bergman, J. Decker, I. R. Duddy, A. J. W. Gleadow, and D. L. Turner, 1993, Multiple phases of Tertiary uplift in the Arctic National Wildlife Refuge, Alaska, revealed by apatite fission track analysis: AAPG Bulletin, v. 77, p. 359–385.

O'Sullivan, P. B., J. M. Murphy, and A. E. Blythe, 1997, Late Mesozoic and Cenozoic thermotectonic evolution of the central Brooks Range and adjacent North Slope foreland basin, Alaska: Including fission track results from the Trans-Alaska Crustal Transect (TACT): Journal of Geophysical Research, v. 102, no. B9, p. 20,821–20,845.

Patterson, S. O., D. A. Sullivan, J. P. Trunz Jr., S. R. Aertker, B. R. Allard, M. D. Mangus, and A. Palenske, 1982, Technical services contract 1981 exploration program: TetraTech, Inc., Report 8202, prepared for the U.S. Geological Survey, Office of the National Petroleum Reserve–Alaska, contract no. 14-08-0001-20203; 2 v. and 5 boxes of oversized maps.

Payne, J. H., 1987, Diagnetic variations in the Permo-Triassic Ivishak Sandstone in the Prudhoe Bay field and central-northeastern National Petroleum Reserve in Alaska (NPRA), *in* I. Tailleur and P. Weimer, eds., Alaskan North Slope Geology: Bakersfield, California, Pacific Section, SEPM, v. 1, p. 77–83.

Plafker, G., and H. C. Berg, 1994, An overview of the geology and tectonic evolution of Alaska, *in* G. Plafker and H. C. Berg, eds., The geology of Alaska: Geological Society of America, The geology of North America, vol. G-1, p. 989–1021.

Ravn, R. L., 1991, Miospores of the Kekiktuk Formation (Lower Carboniferous), Endicott field area, Alaska North Slope, American Association of Stratigraphers and Palynologists Contribution Series 27, 179 p.

Reed, J. C., 1958, Exploration of Naval Petroleum Reserve No. 4 and adjacent areas, northern Alaska, 1944–53: Part 1, History of the exploration: U.S. Geological Survey Professional Paper 301, 192 p.

Robinson, F. R., 1964, Core tests, Simpson area, Alaska: U.S. Geological Survey Professional Paper 305-L, p. 645–730.

Robinson, F. M., and F. R. Collins, 1959, Core test, Sentinel Hill area and test well, Fish Creek area, Alaska: U.S. Geological Survey Professional Paper 305-I, p. 485–521.

Scherr, J., S. M. Banet, and B. J. Bascle, 1991, Correlation study of selected exploration wells from the North Slope and Beaufort Sea, Alaska: Minerals Management OCS Report MMS 91-0076, 29 p.

Seifert, W. K., J. M. Moldowan, and R. W. Jones, 1980, Application of biological marker chemistry to petroleum exploration: Proceedings of the 10th World Petroleum Congress, Bucharest, p. 425–440.

Shanmugam, G., and J. B. Higgins, 1988, Porosity enhancement from chert dissolution beneath Neocomian unconformity: Ivishak Formation, North Slope, Alaska: AAPG Bulletin, v. 72, no. 5, p. 523–535.

Sherwood, K. W., ed., 1998, Undiscovered oil and gas resources, Alaska federal offshore (as of January 1995): U.S. Minerals Management Service OCS Monograph MMS 98-0054, 531 p.

Sherwood, K. W., and J. D. Craig, 2001, Prospects for development of Alaska natural gas: A review: U.S. Department of the Interior, Minerals Management Service, 1 CD-ROM.

Sherwood, K. W., J. D. Craig, R. T. Lothamer, P. P. Johnson, and S. A. Zerwick, 1998, Chukchi shelf assessment province, *in* K. W. Sherwood, ed., Undiscovered oil and gas resources, Alaska federal offshore (as of January 1995), U.S. Minerals Management Service OCS Monograph MMS 98-0054, p. 115–196.

Tailleur, I. L., 1964, Rich oil shale from northern Alaska: U.S. Geological Survey Professional Paper 475-D, p. D131–D133.

Tailleur, I. L., 1965, Low-volatile bituminous coal of Mississippian age on the Lisburne Peninsula, northwestern Alaska: U.S. Geological Survey Professional Paper 525-B, p. B34–B38.

Thurston, D. K., and R. T. Lothamer, 1991, Seismic evidence of evaporite diapirs in the Chukchi Sea, Alaska: Geology, v. 19, p. 477–480.

Thurston, D. K., and L. A. Theiss, 1987, Geologic report for the Chukchi Sea planning area, Alaska: Minerals Management Service OCS Report MMS 87-0046, 193 p.

Tourtelot, H. A., J. R. Donnell, and I. L. Tailleur, 1967, Oil yield and chemical composition of shale from northern Alaska: Seventh World Petroleum Congress, p. 707–711.

U.S. Geological Survey National Oil and Gas Assessment Team, 1995, 1995 national assessment of United States oil and gas resources: U.S. Geological Survey Circular 1118, 20 p.

Wallace, W. K., and C. L. Hanks, 1990, Structural provinces of the northeastern Brooks Range, Arctic National Wildlife Refuge, Alaska: AAPG Bulletin, v. 74, p. 1100–1118.

Watts, K. F., A. G. Harris, R. C. Carlson, M. K. Eckstein, P. D. Gruzlovic, T. A. Imm, A. P. Krumhardt, D. K. Lasota, S. K. Morgan, P. Enos, R. Goldstein, J. A. Dumoulin, and B. Mamet, 1995, Results and synthesis of integrated geologic studies of the Carboniferous Lisburne Group of northeastern Alaska: U.S. Department of Energy Bartlesville Project Office, Final Report for 1989–1992, Contract DE-AC22-89BC14471, 433 p.

Weimer, P., 1987, Seismic stratigraphy of three areas of lower slope failure, Torok Formation, northern Alaska, *in* I. Tailleur and P. Weimer, eds., Alaskan North Slope geology: Pacific Section SEPM and Alaska Geological Society, p. 481–496.

Werner, M. R., 1987, West Sak and Ugnu sands: Low-gravity oil zones of the Kuparuk River area, Alaskan North Slope, *in* I. Tailleur and P. Weimer, eds., Alaskan North Slope geology: Pacific Section SEPM and Alaska Geological Society, v. 1, p. 109–118.

Wicks, J. L., Buckingham, M. L, and Dupree, J. H., 1991, Endicott field—U.S.A., North Slope Basin, Alaska, *in* N. H. Foster and E. A. Beaumont, compilers, Structural traps V: AAPG Treatise of petroleum geology, Atlas of oil and gas fields, p. 1–25.

Wilson, J. L., 1975, Carbonate facies in geologic history: New York, Springer-Verlag, 471 p.

Witmer, R. J., M. B. Mickey, and H. Haga, 1981, Biostratigraphic correlations of selected test wells of National Petroleum Reserve in Alaska: U.S. Geological Survey Open-File Report 81-1165, p. 89.

Woidneck, K., P. Behrman, C. Soule, and J. Wu, 1987, Reservoir description of the Endicott field, North Slope, Alaska, *in* I. Tailleur and P. Weimer, eds., Alaskan North Slope geology: Pacific Section SEPM and Alaska Geological Society, p. 43–59.

Brooks, J. R. V., S. J. Stoker, and T. D. J. Cameron, 2001, Hydrocarbon exploration opportunities of the twenty-first century in the United Kingdom, *in* M. W. Downey, J. C. Threet, and W. A. Morgan, eds., Petroleum provinces of the twenty-first century: AAPG Memoir 74, p. 167–199.

Chapter 10

Hydrocarbon Exploration Opportunities in the Twenty-first Century in the United Kingdom

J. R. V. Brooks[1]
Southampton Oceanographic Centre, Southampton, United Kingdom

S. J. Stoker
British Geological Survey, Edinburgh, United Kingdom

T. D. J. Cameron
British Geological Survey, Edinburgh, United Kingdom

ABSTRACT

Oil and gas production in the United Kingdom (U.K.) and its offshore designated area continued to rise through the 1990s and remains at record levels. This paper charts the developments and major new plays of the last decade of the twentieth century and highlights the opportunities and challenges for the twenty-first century.

Through a continuing program of licensing, the U.K. Department of Trade and Industry has successfully maintained industry interest in U.K. exploration and production despite lower oil prices and continuing competition from other oil and gas provinces in the world. A recent licensing round (the seventeenth) that offered acreage in the underexplored basins along the Atlantic margin northwest of Britain attracted many applications, partly because it followed discovery of oil in a deep-water Paleocene fan play there in 1992. Similarly, licensing rounds that focused on the mature North Sea basins have proved to be very encouraging. Improved methods in fault seal analysis have had a significant impact on well success rates in the mature basins. Onshore, exploration activity has increased significantly after the seventh and eighth licensing rounds, with a success rate of more than 50% for some plays. The U.K.'s coal-bed methane potential continues to be assessed.

The greatest potential for major new discoveries lies along the Atlantic margin, with significant undiscovered hydrocarbons likely in predominantly Cretaceous and Tertiary reservoirs. Subtle stratigraphic and structural trap plays will continue to have a key role in the mature North Sea provinces. Focus on older, deeper reservoirs will extend the geographic range of successful plays there.

[1] Formerly of Oil and Gas Directorate, Department of Trade and Industry, London, U.K.

INTRODUCTION

After 36 years of offshore exploration, the United Kingdom (U.K.) remains net self-sufficient in both oil and gas. At the end of 1998, the U.K. had a record number of 204 offshore and 35 onshore fields in production, which consisted of 136 oil, 87 gas, and 16 condensate fields (Department of Trade and Industry, 1999).[2] These fields had produced 0.133 billion metric tons (t) (0.974 billion barrels [bbl]) of oil and condensate and 96 billion cubic meters (m^3) (3.4 trillion cubic feet [tcf]) of gas in 1998, accounting for 1.7% of the contribution of industry to the U.K.'s national accounts (gross value added). The total cumulative production to the end of 1998 was 2.306 billion t (16.9 billion bbl) of oil and condensate and 1383 billion m^3 (46.3 tcf) of gas. Maximum remaining discovered reserves are estimated to be 1.8 billion t (13.2 billion bbl) of oil and 1795 billion m^3 (63.4 tcf) of gas (Department of Trade and Industry, 1999). The U.K.'s undiscovered recoverable reserves are estimated to lie in the range of 0.275–2.55 billion t (2.0–18.7 billion bbl) of oil and 440–1595 billion m^3 (15.5–56.3 tcf) of gas (Department of Trade and Industry, 1999). The United Kingdom Offshore Operators Association (1996) conservatively estimated in 1996 that there will be a sustained aggregate of 11 to 15 new field developments in the U.K. offshore designated area each year until at least 2020. This estimate may need to be revised upward if sufficient gas discoveries are made in the Atlantic margin to warrant construction of new gas-gathering infrastructure there.

At the previous Wallace E. Pratt Memorial Conference, in 1984, Brooks (1986) reviewed the status of hydrocarbon exploration on the U.K. continental shelf. Despite cumulative production of 18.8 billion bbl of oil equivalent (BBOE) between 1983 and 1998, remaining recoverable U.K. reserves have diminished by only 0.7 BBOE (Figure 1). This is because the U.K. has almost exactly replaced production by newly established reserves during the intervening years (Figure 2). Most of the production has been replaced by adding reserves from 168 new offshore oil and gas fields since 1983, although a component represents improved recovery from the 36 offshore fields that were already producing then. The immediate challenge facing the oil industry will be to discover sufficient new hydrocarbons to enable the U.K. to remain net self-sufficient in oil and gas through the early decades of the twenty-first century.

This paper provides an updated summary of the remaining and future petroleum potential of the U.K. and its surrounding designated waters. Some brief historical facts are included here, but Brennand et al. (1998) should be consulted for a comprehensive historical review. Although most of the larger oil and gas fields in the U.K. sector of the North Sea have already been discovered, the rate at which new reserves continue to be found remains remarkably healthy (Brennand et al., 1998). The distribution of oil, gas, and condensate fields in the U.K. and its offshore designated area is illustrated in Figure 3.

The oil and gas industry in the U.K. is operated entirely by the private sector in a competitive manner. Exploration has proceeded at a rapid rate, stimulated and controlled to a significant extent by the frequency and regularity of licensing rounds. The Department of Trade and Industry's (DTI) discretionary award system has elicited a keen response to license offers and remains very highly regarded (Brennand et al., 1998). Recent offshore licensing rounds have alternated between offering acreage in frontier areas such as the Atlantic margin and blocks in the more mature North Sea area. In the latter area, the main aim of some operators is to locate small satellite oil fields (as high as 50 million bbl) that lie close to existing infrastructure. As part of this objective, the "finder-well" concept of drilling a well for minimum possible costs without compromising safety or technical needs may be imperative. Reviews of offshore fallow blocks, in which there has been no drilling for at least six years, were initiated by DTI in 1996 and 1999. These reviews have proved to be very successful in encouraging licensees to bring forward exploration plans or to surrender acreage for relicensing.

Supplementing the private sector's prospect-evaluation activity, the Department of Energy (now incorporated into DTI) supported the British Geological Survey in systematic surveying of the U.K. continental shelf between 1967 and 1990. The resulting series of 328 geologic and geophysical maps is complemented by a series of U.K. Offshore Regional Reports.[3] Together, these constitute a unique and comprehensive geologic data set and evaluation. This work has provided an invaluable contribution to the successful petroleum exploration of the offshore U.K. (Walmsley, 1987).

Weakening oil prices in the 1990s precipitated the Cost Reduction Initiative for the New Era (CRINE) and successor initiatives. Oil and gas industry capital and operating costs have been cut by more than 30% by improving competitiveness while not compromising safety or environmental standards. An Oil and Gas Industry Task Force, chaired by the minister for energy and competitiveness in Europe, reported in 1999 on new strategies for cost reduction and raising competitiveness of the U.K. industry. An early by-product of this initiative has been the establishment of Internet-based mechanisms for license trading to stimulate investment in the U.K. industry. This is referred to as Licence Information

[2] A portion of the most recent volume is available at http://www.databydesign.co.uk/energy.

[3] For availability, see http://www.bgs.ac.uk/products/uk-books&reports/uk-offshore-reps.htm.

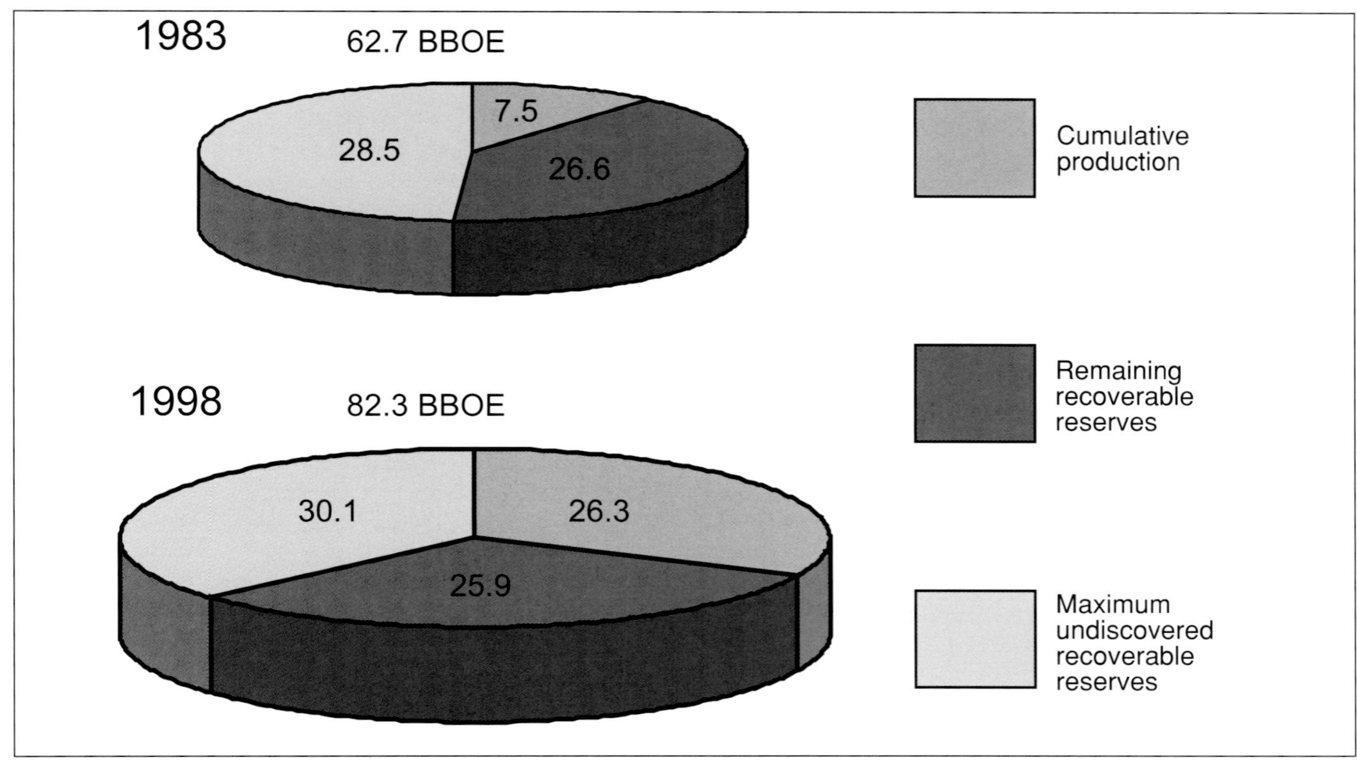

Figure 1. Changes in the U.K. hydrocarbon resource base between 1983 and 1998.

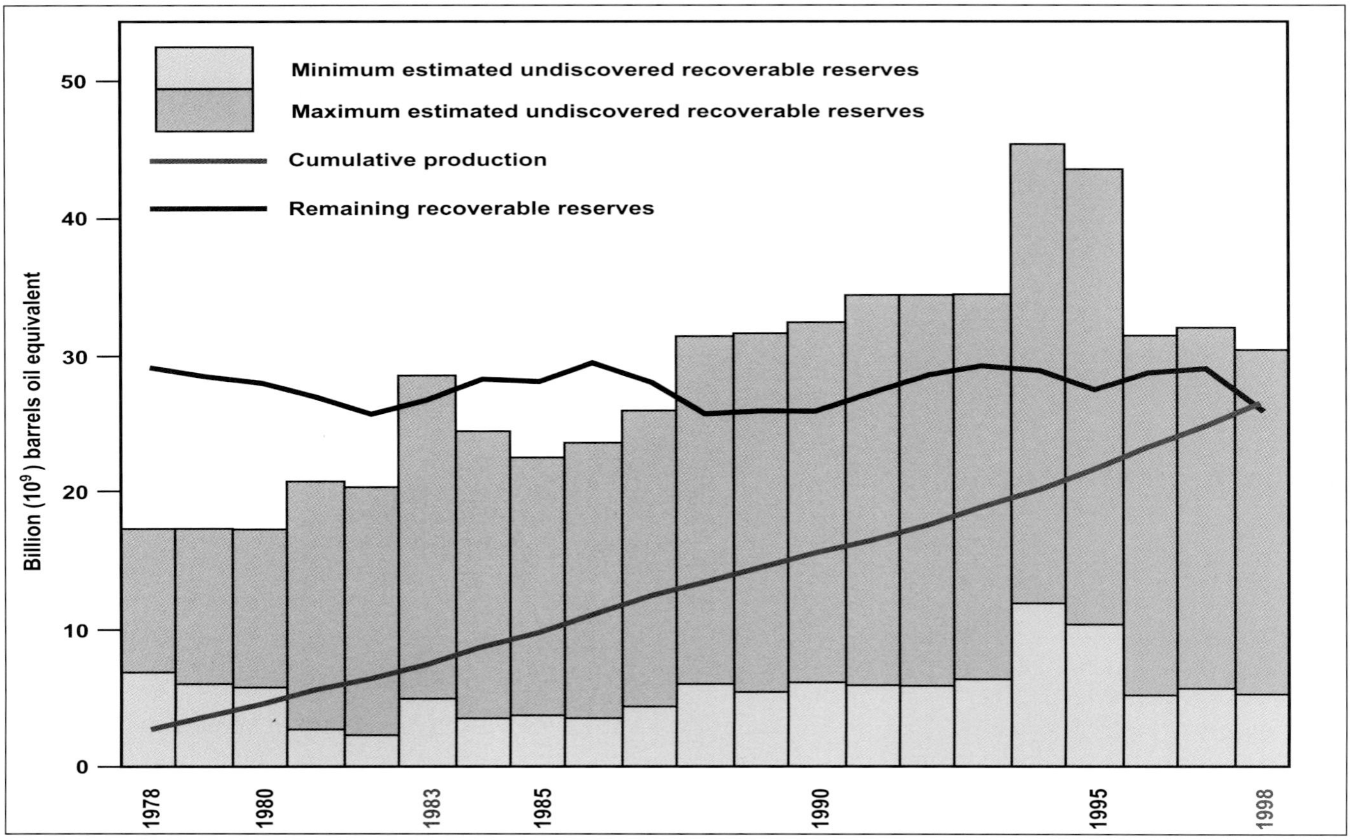

Figure 2. Historical summary of U.K. cumulative hydrocarbon production and reserve estimates between 1978 and 1998.

for Trading (LIFT).[4] A more recent development is the Digital Energy Atlas and Library (DEAL)[5] data catalogue for the U.K. offshore oil and gas industry. All these initiatives and future ones are designed to maintain the U.K.'s position as a world-class operating environment. DTI is committed to ensuring that the regulatory regime remains flexible and responsive to the changing needs of the hydrocarbon industry and of the U.K.

The U.K. oil and gas industry has a good track record in minimizing the environmental impact of its activities. Environmental awareness begins at the license-application stage, when applicants are required to demonstrate that they have considered the sensitivities of the areas they have applied for. They must show that they have environmental-management policies in effect to ensure that all activity under license will be carried out under satisfactory standards of care. Legislation introduced in 1997 and enhanced in 1998 requires environmental-impact assessments for individual well, development, and pipeline projects under new and existing licenses.

DTI has continued to participate in offshore environmental research and other environmental programs. For example, it supported a survey across the Hatton continental margin in 1999. The U.K. government's contribution toward addressing the possibility of climate change is focused mainly on its domestic aim of a 20% reduction in carbon-dioxide (CO_2) emissions by 2020 (Department of Trade and Industry, 1999).

EXPLORATION OPPORTUNITIES

North Sea Oil Province

Since the discovery of oil in Paleogene sandstones by well 22/18-1 (now part of Arbroath field) in 1969, the North Sea (north of 56°N) has become established as one of the world's major oil-producing regions. At the end of 1998, 133 fields had produced 2.15 billion t (15.8 billion bbl) of oil and condensate and 412 billion m^3 (14.5 tcf) of gas from the U.K. sector alone (Department of Trade and Industry, 1999). Production is from almost every conceivable clastic and carbonate sedimentary facies, and from strata ranging between Devonian and Eocene in age. Even in a mature stage for exploration, new discoveries are still being made, and the region continues to yield surprises.

The geologic history of the oil province was dominated by an episode of crustal extension and accelerated basin subsidence (mainly in the Late Jurassic but locally continuing into the earliest Cretaceous) that led to the formation of the Viking Graben, Central Graben, and Moray Firth rift systems (Figures 3, 4). Synrift, organic-rich marine sediments (Kimmeridge Clay) are the source interval for virtually all of the region's hydrocarbons. However, it was the postrift thermal subsidence following Early Cretaceous collapse of the rift system that enabled these source rocks to become mature for hydrocarbon generation along the rift axes from the Paleogene onward (Johnson and Fisher, 1998). Hydrocarbon migration has been essentially vertical, with significant lateral migration restricted to the Upper Jurassic and Tertiary successions. Consequently, most of the producing oil and gas fields lie within the geographic boundary of the mature source rocks (Figure 5).

Hydrocarbons occur in a range of prerift, synrift (Upper Jurassic to Lower Cretaceous, for the purposes of this paper), and postrift reservoirs. Note that it was only in the western Moray Firth that rifting continued into the Early Cretaceous. Other parts of the oil province entered a mildly compressive tectonic regime at about the time of the Jurassic/Cretaceous boundary (Oakman and Partington, 1998). Note also that a precursor episode of Permian to Early Triassic rifting may have affected "prerift" subsidence and sedimentation patterns along the axial graben of the North Sea.

Prerift

Prerift producing fields of the oil province can be subdivided into three categories: Paleozoic, Triassic to Lower Jurassic, and Middle Jurassic.

Paleozoic.—Those fields having reservoirs of Paleozoic (Devonian, Carboniferous, or Permian) age (Figure 4) are concentrated on tilted footwall blocks (e.g., Auk field), are adjacent to the major graben-bounding faults, or are on intrabasinal highs (e.g., Buchan field). Reservoir quality in these areas is notoriously difficult to predict, but it may be enhanced by fracturing during synrift footwall uplift or by leaching (in the case of Upper Permian carbonate reservoirs) beneath a regional base Cretaceous unconformity (Johnson and Fisher, 1998). All the traps are overlain by Upper Jurassic to Lower Cretaceous mudstones or Upper Cretaceous chalk, providing an effective top seal. The discovery as recently as 1998 of Flora field, with its Upper Carboniferous Stephanian (?)[6] reservoir, indicates that the traditional Paleozoic play has not reached maturity yet. Indeed, the reservoir potential of the Carboniferous strata remains underexplored, especially considering that their seismic response strongly resembles that of the equivalent gas-bearing succession of the North Sea gas province.

Prerift, Middle Devonian lacustrine sediments, a partial source for oil in Beatrice field in the Inner Moray Firth (Figure 5), are now thought to be widespread beneath the northern half of the oil province (Duncan and Buxton, 1995). This distribution provides a possibility that the geographic range of Paleozoic hydrocarbon

[4] For more information, see http://www.uk.lift.co.uk.

[5] For more information, see http://www.ukdeal.co.uk.

[6] European nomenclature is employed for subdivision of the Carboniferous throughout this paper. The Dinantian and basal part of the Namurian subsystems are equivalent to the Mississippian, and the remainder of the Namurian, with the Westphalian and Stephanian, is equivalent to the Pennsylvanian of American usage (see Figure 11).

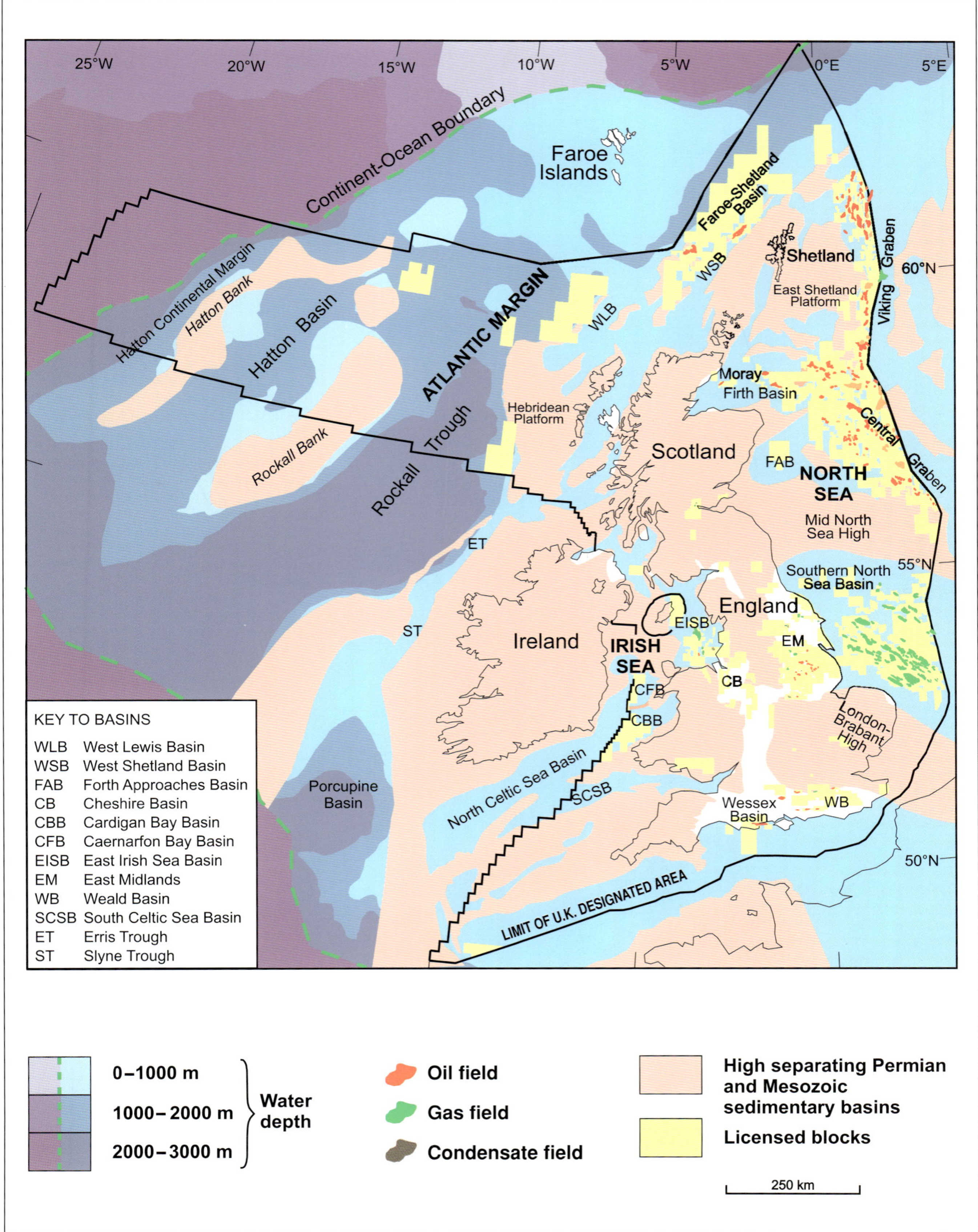

Figure 3. Location of the hydrocarbon-producing basins and frontier basins of the U.K. and its designated waters.

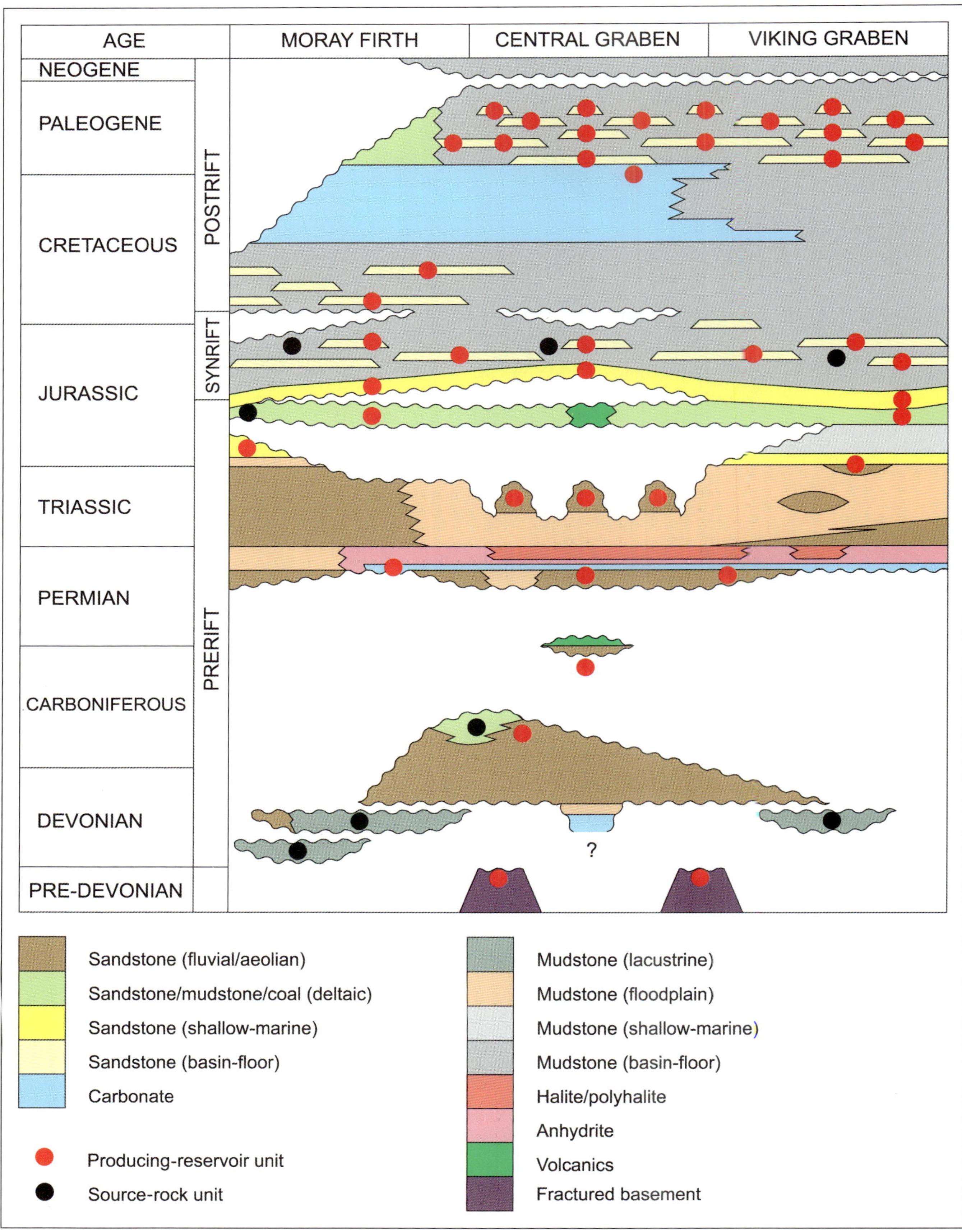

Figure 4. Simplified illustration of geographic ranges of the principal lithofacies in the North Sea oil province.

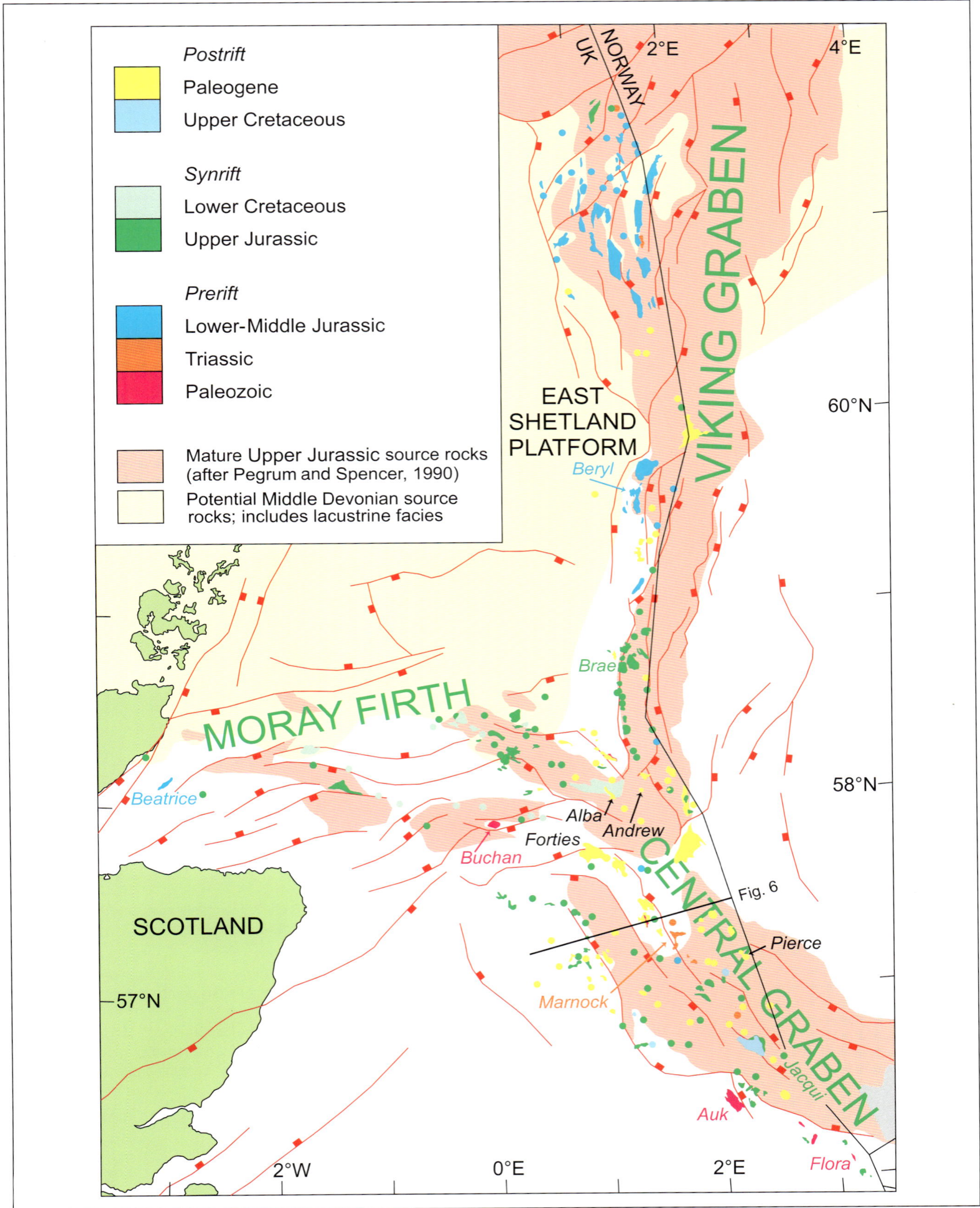

Figure 5. Age of principal reservoirs in oil and gas fields and distribution of mature Upper Jurassic and potential Middle Devonian source rock in the North Sea oil province.

reserves may extend far beyond the graben margins onto the underexplored East Shetland Platform in this northern area. Devonian subbasins that appear to be imaged by potential-field data are likely to provide the focus for initial exploration of this play.

Triassic to Lower Jurassic.—Prerift producing fields with reservoirs of Triassic to Early Jurassic age have traps either in tilted footwall blocks adjacent to the major graben (e.g., Beryl field), as subcrop closures beneath synrift or postrift strata (e.g., Marnock field), or in stacked plays beneath producing Middle or Upper Jurassic sandstones (Johnson and Fisher, 1998). These reservoirs are typically thick, highly feldspathic, fluvial channel and sheetflood sandstones. They are fine grained on the western flank of the Central Graben, where they are partly confined to topographic lows that formed in response to halokinesis of underlying Upper Permian evaporites (Figure 6). This has led to abrupt changes in reservoir thickness, elongate patterns of net sandstone, and poor connectivity between adjacent sandstone systems. Reservoir properties are highly variable. Not all the prospective structural or subcrop traps have been drilled yet, and the potential for traps defined by reservoir pinch-out is underexplored. North of a Triassic watershed at about 58°N, coarse-grained, uppermost Triassic to Lower Jurassic sandstones have excellent reservoir properties. So do the Middle Triassic eolian sandstones which are important oil reservoirs nearby in the Norwegian sector (Johnson and Fisher, 1998). However, all but the smallest of prospective structures may have already been drilled in this area.

Middle Jurassic.—The prerift, Middle Jurassic tilted fault-block play is one of the most productive in the North Sea (Figure 5), although creaming curves indicate that it may be almost fully exploited as an exploration target (Johnson and Fisher, 1998). The Middle Jurassic reservoirs were deposited within a diachronous clastic wedge, mainly in coastal or delta-plain environments in

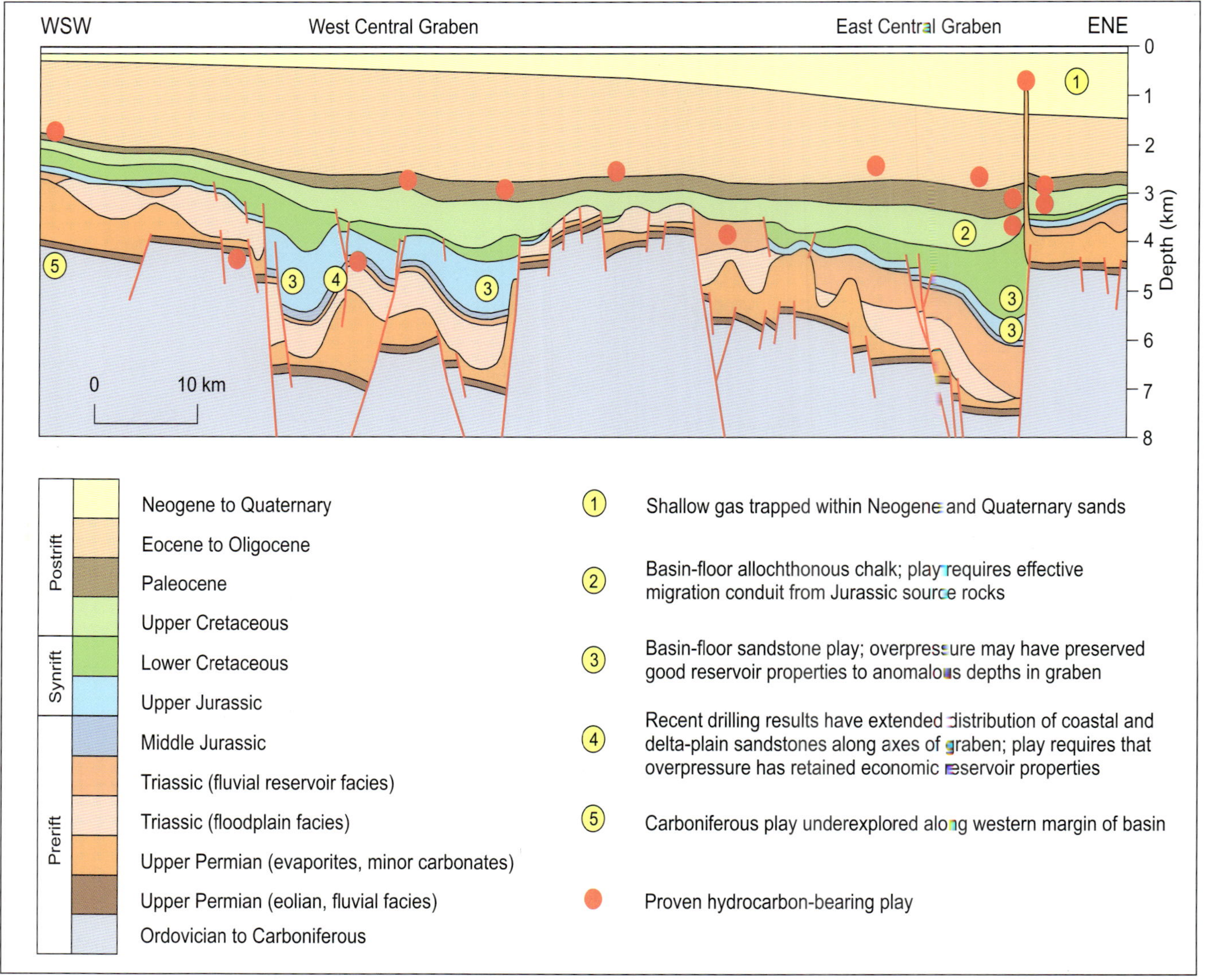

Figure 6. Schematic section illustrating principal underexplored plays in central North Sea. See Figure 5 for location.

the south, and as coastal barrier and wave-dominated deltaic facies in the north. Virtually all the footwall closures north of 58°N have already been tested, but the potential for economic reserves in hanging-wall closures is underexplored, probably because reservoir properties are perceived to deteriorate with increasing depth in this region. The opportunity to discover untapped reserves may be better in the Central Graben, although reservoir prediction is more difficult there. Nevertheless, there is a likelihood that overpressure has preserved economic porosity and permeability to anomalous depths in the graben. The most attractive prospects may be those stacked below synrift drilling targets.

Synrift

Although most of the Upper Jurassic to Lower Cretaceous synrift-producing fields (Figure 5) occur in small to medium-size traps, this exploration play is and will continue to be one of the most active in the North Sea (Figure 7). The play is largely, but not entirely, confined to the synrift graben. It owes its success to the widespread occurrence of high-quality sandstones, which directly underlie, interfinger with, or overlie mature source rocks in much of the region. However, the play has not reached maturity yet, because the distribution of reservoir facies is very complex. It reflects the interplay between local tectonic and regional eustatic controls on the rate of creation of accommodation space within each of the grabens. Contemporaneous halokinesis is another factor affecting facies distributions in the Central Graben. Later halokinesis is important as a trapping mechanism there.

The producing synrift reservoirs include both shallow-marine and deep-marine sandstones. As relative sea level rose rapidly during the early stages of rifting, retrogradational packages of shallow-marine shelf and coastal sands were deposited around the basin margins of the Outer Moray Firth and Central Graben in particular. In parts of the region, these facies continued to be deposited into the Early Cretaceous. These excellent reservoir facies pass basinward into shelf mudstones which provide a good top seal to deeper prerift hydrocarbon accumulations. In selecting drilling targets in shallow-marine sediments, the principal challenge is to develop a sequence-stratigraphic framework that assimilates all the eustatic and local tectonic influences on facies development as a means of determining the preservation potential for the reservoir sandstones (Johnson and Fisher, 1998). Nevertheless, the distribution of shoreline facies is not yet fully understood.

As rifting continued, footwall uplift along the major graben-bounding faults locally exceeded the continuing relative rise in sea level. This led to footwall erosion and mass transfer of coarse clastic debris into the adjacent, by now sediment-starved, deep-water basins. Where the clastic debris is banked against active faults, it forms small point-sourced accumulations that are often conglomeratic—the reservoir facies for several oil fields in the Viking Graben (e.g., Brae field). Where the clastic debris has been

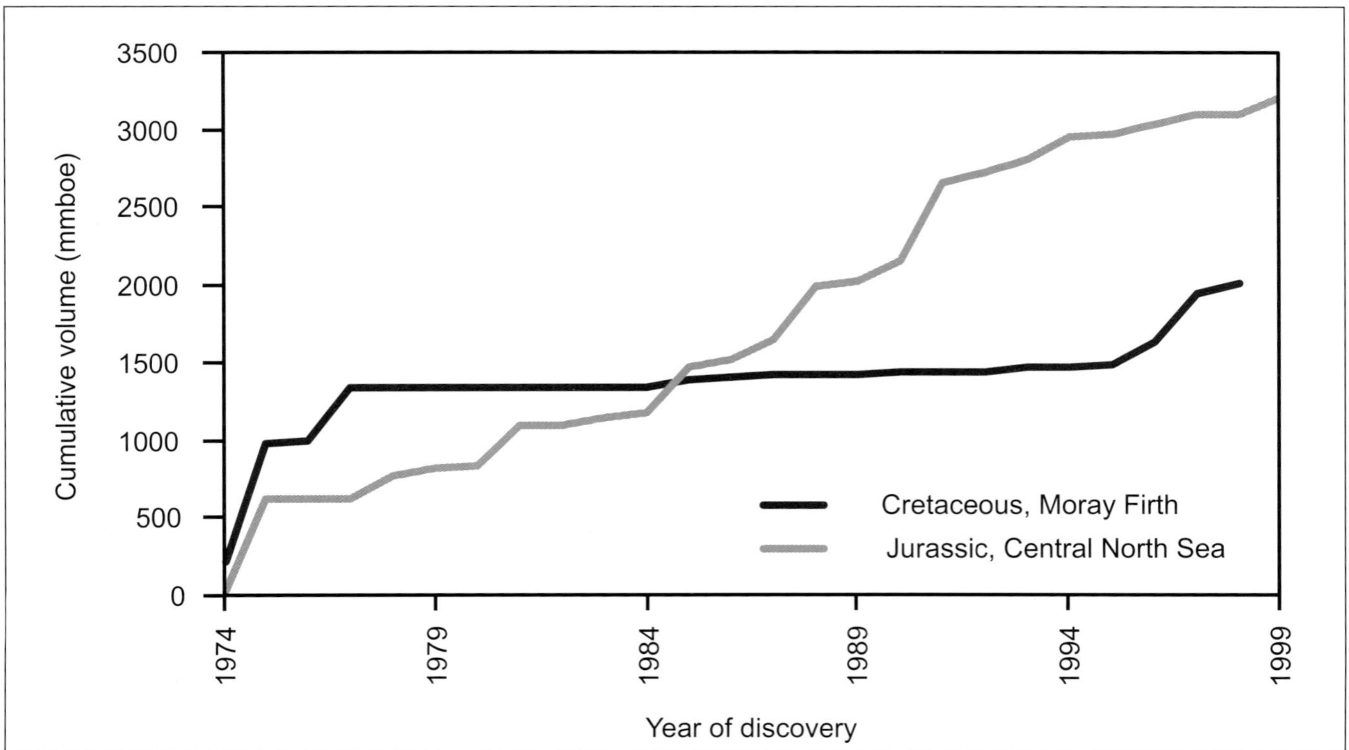

Figure 7. Creaming curves for the combined Middle and Upper Jurassic plays of the central North Sea, and Lower Cretaceous play of the Moray Firth (data supplied by Asset Geoscience Ltd.).

transported farther basinward, it forms larger, lenticular accumulations with a higher proportion of sandstone, often totally encased within basinal mudstone. The locations of the basin-floor accumulations were strongly influenced by contemporary bathymetry.

The synrift producing fields display a wide variety of trapping mechanisms, including tilted fault blocks, four-way dip closures, hanging-wall closures, stratigraphic closures, and combined structural-stratigraphic closures (Johnson and Fisher, 1998). Rifting terminated in the Viking Graben and Central Graben during the Late Jurassic. The blanket of postrift basinal mudstones deposited during the Early Cretaceous here provides a regional top seal for many drilling targets in these grabens.

The majority of synrift prospects in the Viking Graben has already been drilled. There are many synrift producing fields in the Moray Firth and Central Graben also (Figure 5), but the basinal parts of these areas remain underexplored. It is largely these areas that contain an estimated synrift component comprising 40% of the oil and 35% of the gas yet to be discovered in the oil province (Johnson and Fisher, 1998). In the Inner Moray Firth, for instance, there has been only limited exploration to date of Upper Jurassic to basal Cretaceous basin-floor sandstones. This may be because it is not widely appreciated that the contemporary source rocks (Kimmeridge Clay) are fully mature for oil generation in the deepest parts of the asymmetrical grabens that characterize this area (Oakman and Partington, 1998). Middle Devonian lacustrine sediments in the footwalls of these grabens are also likely to be mature for hydrocarbon generation. The most attractive play comprises updip, distal pinch-out of sandstone reservoirs, with encasing basinal mudstones providing lateral seals (Figure 8). It is not known whether the reservoir sandstones are widespread along the axes of the grabens or whether they are relatively localized, comprising separate accumulations derived from point sources.

Recent wells in the deepest parts of the Outer Moray Firth and Central Graben have unexpectedly extended the play fairway for thick, Upper Jurassic synrift basin-floor sandstones into this area also. Overpressure, caused mainly by high sedimentation rates of younger sediments, is expected to have preserved good reservoir quality in sandstones at anomalous depths in the basin-floor fairways. It is these sandstones that provide the gas reservoir for the recently discovered Jacqui field (Figure 5). Perhaps the greatest challenge will be to convince exploration managers that the best-quality reservoirs may be

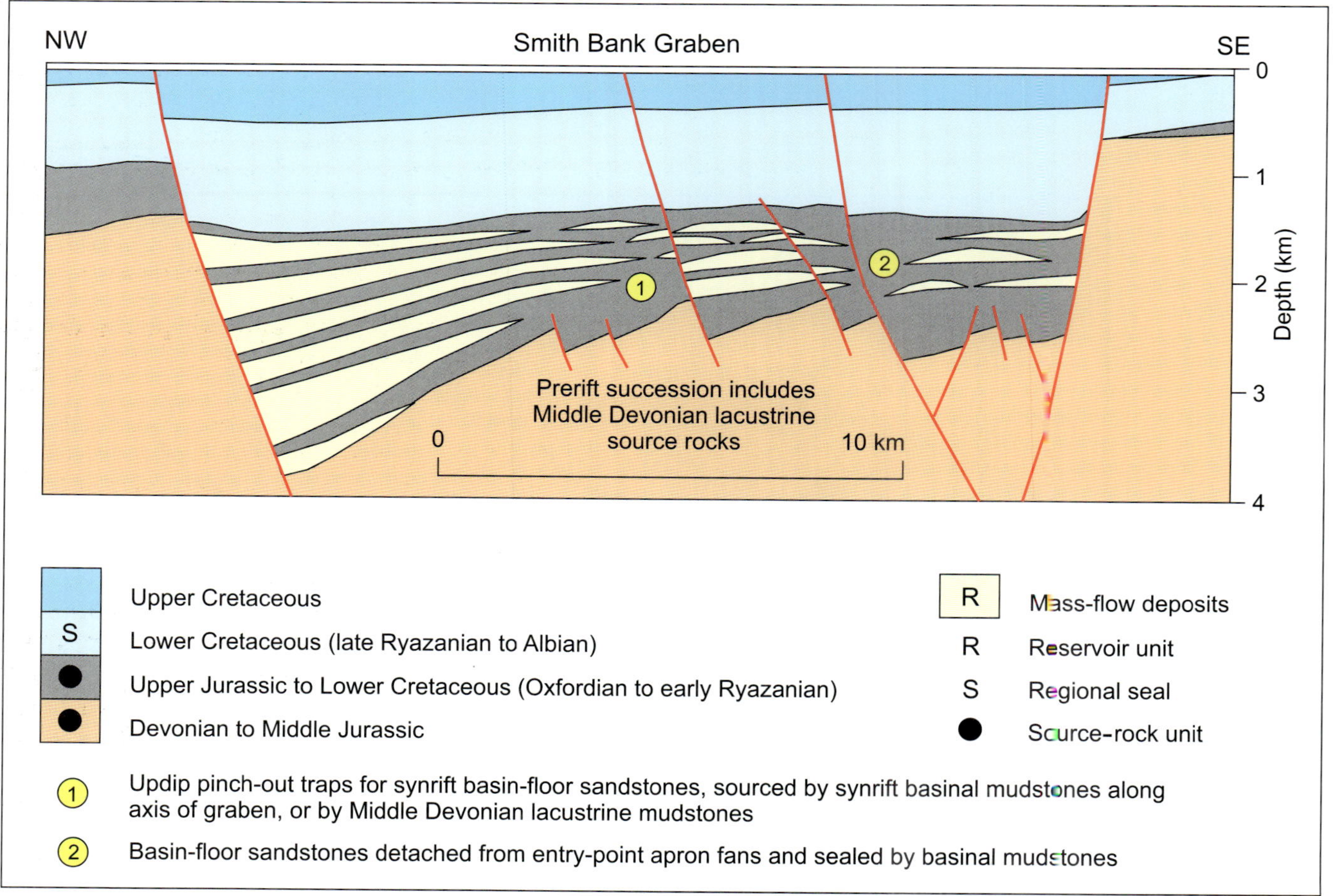

Figure 8. Schematic play diagram for Upper Jurassic to basal Cretaceous of Inner Moray Firth, North Sea oil province.

located in synclinal areas of the grabens, rather than the flank areas or intrabasinal highs that have provided the majority of drilling targets until now. Furthermore, recent drilling successes have proved that substantial hydrocarbon reserves still remain to be discovered on the hanging walls of the major faults of the region.

Tectonism continued into the Early Cretaceous in parts of the Moray Firth, where the penecontemporaneous basin-floor sandstones (e.g., Kopervik play) are proving to be attractive exploration targets. Drilling has established that seafloor bathymetry had a significant impact on depositional processes here too, with the thickest sandstones being encountered in localized trough areas within the underfilled Late Jurassic grabens (Oakman and Partington, 1998). Early successes were in structural traps, but recent discoveries have revealed the potential for significant undiscovered hydrocarbons in stratigraphic traps. Improvements in seismic resolution of the limits of the reservoir sandstones will be required to optimize the potential of this fairway.

Exploration within overpressured parts of the Central Graben has revealed that the basal synrift shallow-marine sandstones are more widespread than previously recognized (Cordey, 1993). Location of drilling targets will be more challenging there than around the basin margins, but careful sequence-stratigraphic interpretation of 3-D seismic data should yield further successes in this area.

Postrift

Postrift thermal subsidence has continued from the cessation of rifting to the present day. Subsidence patterns reflect the continuing influence of the North Sea's graben systems, although they have been modified by Late Cretaceous inversion in parts of the Central Graben. As much as 1000 m of Upper Cretaceous chalk accumulated in the latter area and in the Moray Firth, whereas the contemporaneous sediments in the Viking Graben are calcareous mudstones with no reservoir potential. The chalk also has generally poor reservoir properties. Exceptions include chalk that has been redeposited by gravity-flow processes at the base of oversteepened slopes, for example, around rising salt diapirs and near graben margins. The potential for stratigraphic entrapment within porous allocthonous chalk encased in nonporous sediment remains underexplored in basinal areas but requires an effective migration conduit from underlying source rocks and porosity preservation at depths greater than the typical chalk play. Chalk porosity may have remained anomalously high relative to depth of burial in overpressured areas of the Central Graben. Even where relatively porous, however, it has yet to be proved that significant reserves remain to be discovered from chalk reservoirs in the U.K. sector of the North Sea. All the existing fields are in the southeast part of the oil province. They are small and only marginally economic, even at $20/bbl.

Regional patterns of sedimentation changed dramatically in the early Paleocene with the influx of huge volumes of coarse clastic detritus as debris flows and/or turbidites into the basinal areas of the oil province. This detritus was being shed from the uplands of northern Scotland and from the East Shetland Platform, which were undergoing thermal uplift in response to development of the Iceland plume and opening of the North Atlantic Ocean. The resultant deep-water sandstone reservoirs of Paleocene to middle Eocene age contain about 20% of the oil province's proven hydrocarbon reserves (Pegrum and Spencer, 1990).

The Paleogene reservoirs occur in a succession of as many as 15 gently eastward-dipping, overlapping depositional sequences bounded by erosional unconformities. Each of these sequences requires a separate play evaluation (Johnson and Fisher, 1998). This is because the distribution of reservoir sandstones and potentially sealing mudstones within an individual sequence reflects the gross depositional setting of that sequence. Nevertheless, there was a progressive change from the emplacement of laterally extensive sheet sands on the basin floor during the early Paleocene, through the deposition of smaller basin-floor systems during the late Paleocene to early Eocene, to the restriction of sand bodies into narrow elongate channels intercalated within mud-dominated slope facies during the middle Eocene. Virtually all the sand systems become progressively distal toward the east or southeast. Although reservoir quality is generally very good, the overall succession displays a wide range in the thickness, geometry, orientation, and architecture of its individual sand bodies.

The Paleogene reservoirs occur in both structural and stratigraphic traps (Bain, 1993). Structural traps are located mainly over pre-Tertiary structural highs (e.g., Forties field) or are either flanking or overlying Permian salt diapirs (e.g., Andrew field, Pierce field); the latter condition is restricted to the Central Graben. Top seal is provided by regionally extensive mudstone intervals. Stratigraphic traps include mounded closures and sand pinch-outs, which can occur in a wide range of depositional settings. The pinch-out traps have a higher exploration risk of seal failure, although they have only rarely been tested in optimal drilling locations (Bain, 1993).

Since the first North Sea oil discovery, in 1969, exploration of the Paleogene play has led to the discovery of 31 fields having estimated recoverable reserves of 0.72 billion t of oil (5.3 billion bbl) and 151 billion m^3 (5.3 tcf) of gas and condensate. Nevertheless, the play will remain highly prospective, with the emphasis on locating increasingly subtle stratigraphic traps and on extending the geographic range of discoveries into basin-marginal areas such as the East Shetland Platform. Depth of burial is relatively shallow in these areas, leading to a preponderance of heavy, biodegraded oils. Nevertheless, experience gained in maximizing production from existing fields of this type in the U.K. (e.g., Alba field) and elsewhere is continuing to improve the economic potential of the many untested prospects in such areas.

The search for subtle traps is benefiting from improvements in seismic acquisition and processing, leading to better resolution, increased use of 3-D seismic interpretation, amplitude variation with offset (AVO) and seismic-attribute analysis, the application of high-resolution stratigraphic concepts, and improved models for deep-water sandstone depositional processes. Using these and other innovative techniques, the Paleogene play is likely to yield about 45% of future oil and 30% of future gas

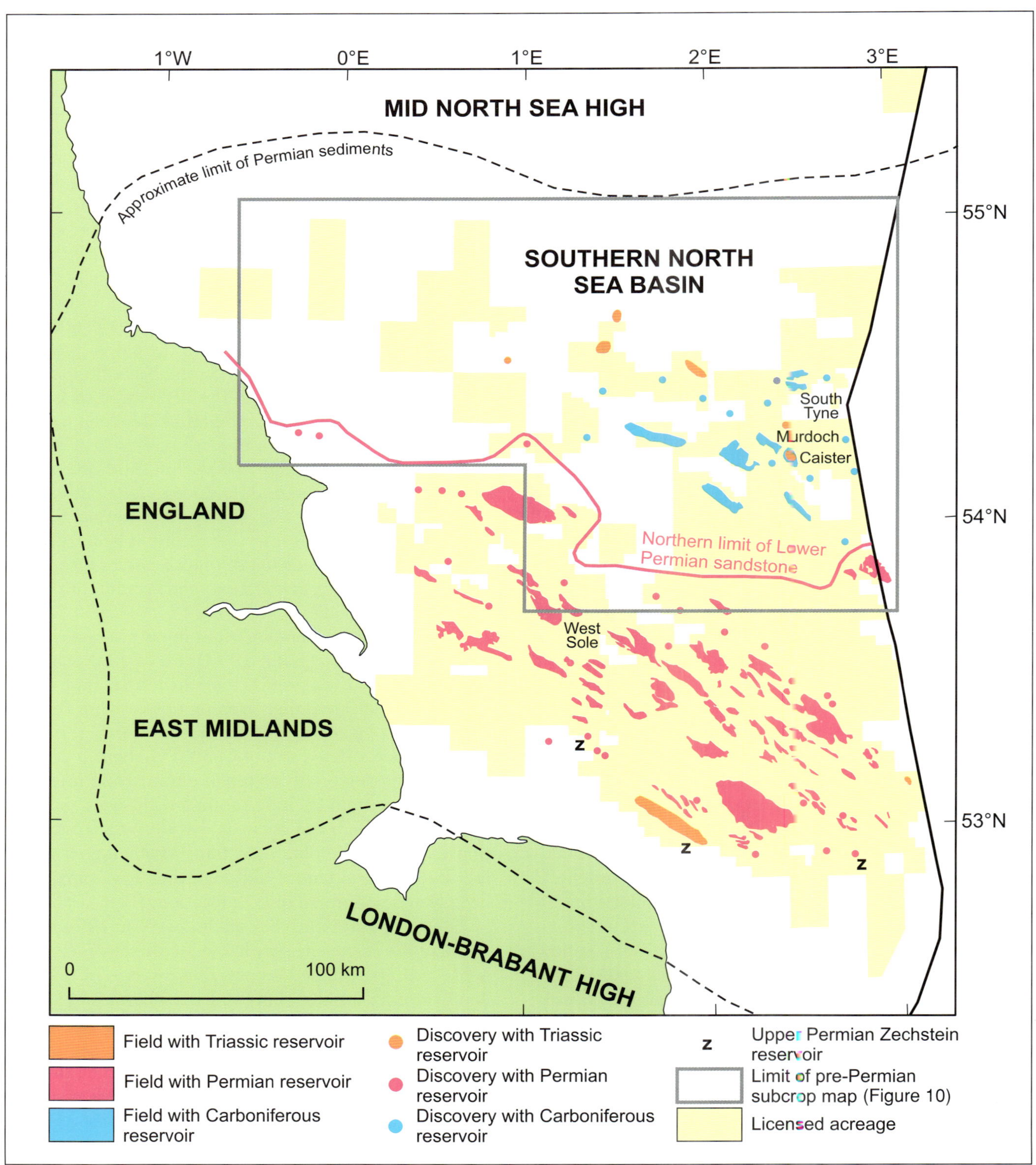

Figure 9. Fields and discoveries in the U.K. North Sea gas province.

discoveries in the North Sea's oil province (Johnson and Fisher, 1998). Overlying this play, the shallow gas that occurs within unconsolidated Neogene and Quaternary sands across large parts of the region is widely perceived as a drilling hazard but may be locally exploitable.

North Sea Gas Province

Since the first gas came ashore from West Sole field in 1967, the southern North Sea had produced about 964 billion m^3 of gas (34 tcf) from 73 gas fields by the end of 1998 (Department of Trade and Industry, 1999). About 85.5% of this production has been from Lower Permian (Rotliegend) eolian dune sandstones and 13% from Triassic fluvial sandstones (Figure 9). Upper Permian carbonates have contributed minor production from three fields. Remaining production has been from Carboniferous fluvial sandstones.

The Lower Permian (Leman Sandstone) play is restricted to the southern half of the gas province (Figure 9) because the reservoir facies passes northward into contemporaneous playa-lake mudstones and evaporites. Stretching from the U.K. eastward to the North German/Polish Plain, this fairway has ultimately recoverable reserves of about 4500 billion m^3 (159 tcf) of gas (Glennie, 1998). The Lower Permian play is continuing to evolve by integrating new techniques and technologies with the aim of recognizing ever more subtle traps. Prestack depth migration is proving to be a powerful tool in removing lateral velocity variation effects caused by the widespread presence of diapiric salt in the overburden, hence enabling increasingly accurate depth imaging of drilling targets. Other techniques that are keeping this fairway active include the application of seismic inversion to assess basin-margin plays, the evaluation of diagenetic controls on regional reservoir quality, and careful and detailed fault seal analysis. High-resolution sequence stratigraphy is now being applied to map the extent of reservoir units around the margin of the contemporaneous desert lake, and there may be potential for stratigraphic entrapment of gas in this area.

The Lower Permian reservoir sandstones and lake sediments rest on a regional base Permian unconformity. Folded and faulted Carboniferous strata beneath the unconformity include as much as 800 m of Westphalian A-C coal measures, the principal source for gas in the Southern North Sea Basin. Following strong encouragement from the U.K. Department of Energy (now DTI), exploration drilling to test the fluvial sandstones interbedded within the coal measures began during the early 1980s. This has led to the discovery of additional gas reserves in eight fields and 12 significant discoveries (Department of Trade and Industry, 1999) north of the

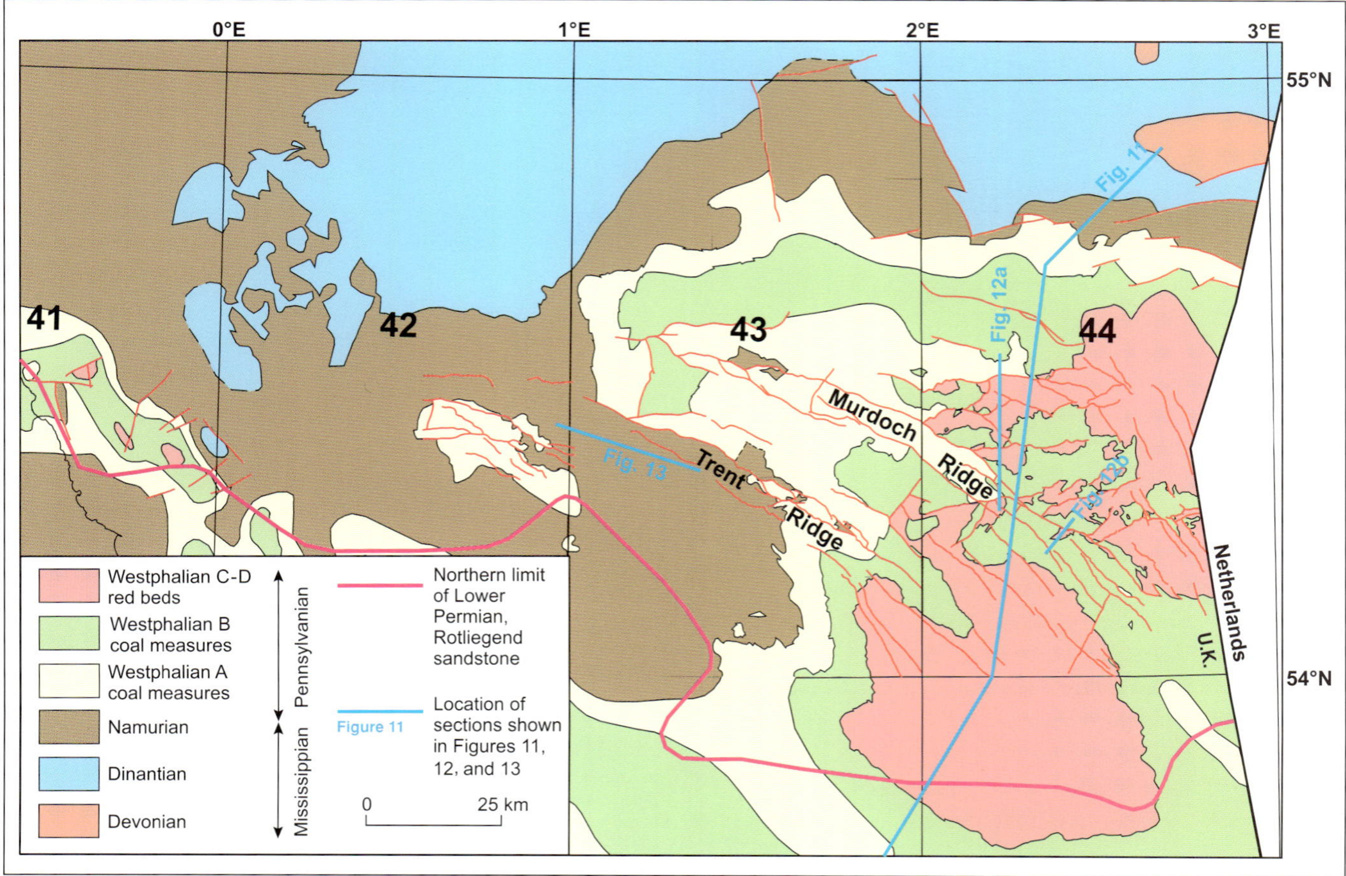

Figure 10. Subcrop of pre-Permian strata in the northern part of the North Sea gas province.

established Lower Permian play (Figure 9). Although these fields are already contributing about 5% of U.K. annual gas production, exploration of this play is still at the immature stage, with much of the Carboniferous fairway underexplored. Recently published analyses indicate that a significant component of the gas has been generated from pre-Westphalian marine source rocks (Gerling et al., 1999). This should extend the geographic range of exploration farther beyond the erosional limit of the Westphalian coal measures (Figure 10).

The structure and stratigraphy of the Carboniferous in the northern part of the Southern North Sea Basin are summarized in Figure 11. The deep structure of the basin is conjectural because seismic resolution is generally limited to within 2 km below the base Permian unconformity. Nevertheless, it is probably similar to that of the onshore U.K., where Late Devonian to Dinantian rifting established a block- and basin-deep structure and paleogeography (Leeder and Hardman, 1990) across the region. Significantly different interpretations exist for the locations of the synrift blocks and basins (e.g., Collinson et al., 1993; Cameron and Ziegler, 1997; Besly, 1998). During the early Namurian, thermal sag created basins which were filled largely with deltaic sediments by deltas prograding from the north. The remainder of the Carboniferous interval is characterized by a succession of alluvial and lacustrine beds (Westphalian A-C), passing upward into alluvial red-bed facies (Westphalian C-D). The hydrocarbon source and reservoir potential of the Carboniferous succession in the Southern North Sea Basin are summarized in Table 1.

The majority of the Carboniferous drilling targets tested so far has been defined by structural closure on the base Permian unconformity. Resting on this unconformity, the Lower Permian playa-lake sediments and overlying thick Upper Permian evaporites provide an excellent regional seal. The earliest drilling targets were major northwest-southeast-trending "pop-up" horst blocks bounded by reverse faults (Figure 11), such as the Murdoch and Trent ridges (Figure 10). This phase of exploration led to the discovery of most of the fields that are now in production or under development. It established that potentially productive fluvial sandstone reservoirs are widespread in the Westphalian C-D red beds and occur in specific successions within the late Namurian to Westphalian C coal measures and deltaic sediments (Table 1). Quirk (1997) showed that at least the upper part of the succession can be regarded as a stacked series of fluvio-deltaic sequences, with reservoir-quality sandstones having been deposited mainly during periods of

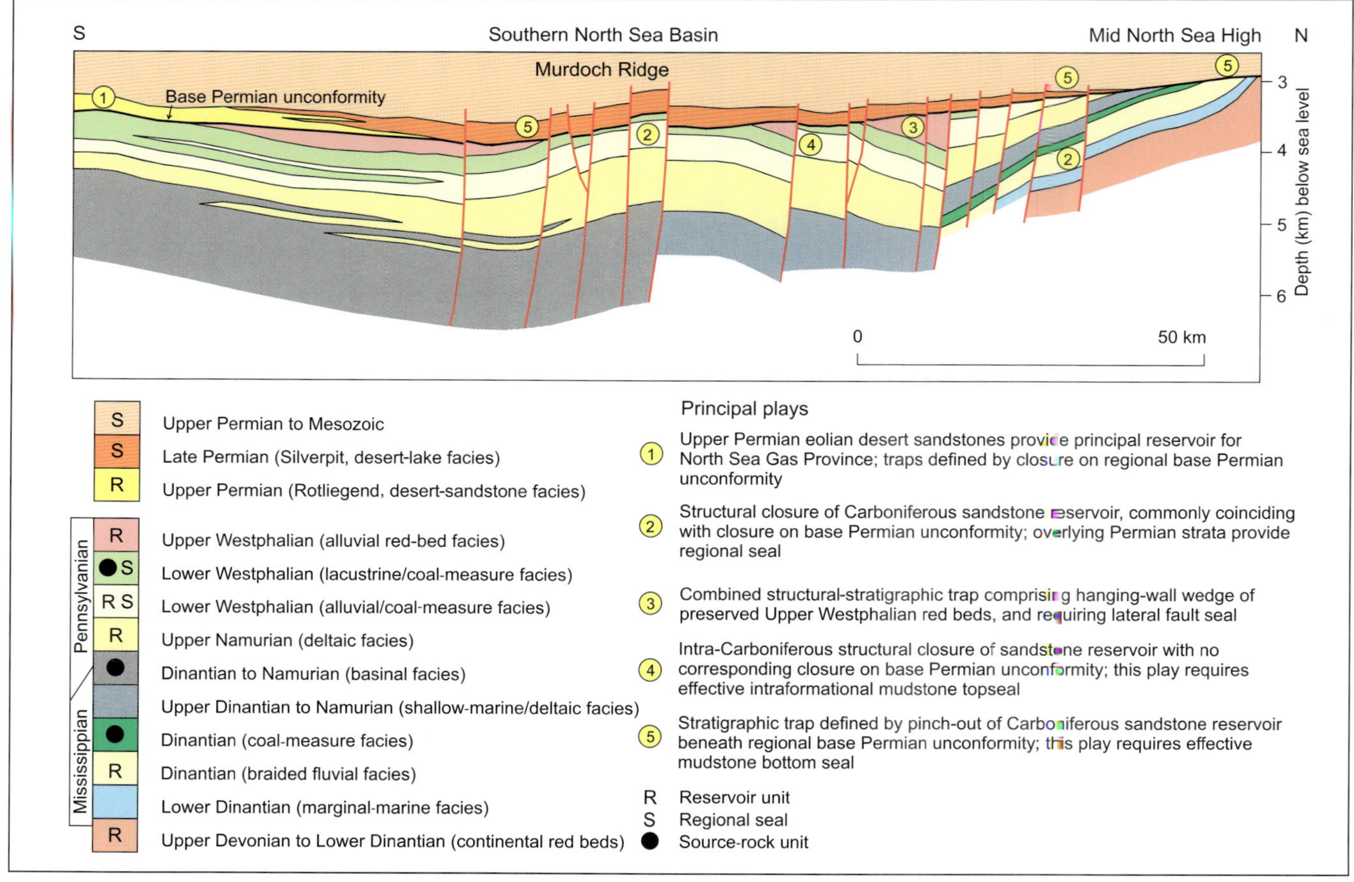

Figure 11. Schematic section summarizing principal Carboniferous plays in the North Sea gas province. See Figure 10 for location.

fluvial aggradation during the early stages of regional rise in lake level across the basin.

Challenges for the future will include extending the stratigraphic range of this sequence-stratigraphic approach to reservoir correlation, and the accurate prediction of the thickness and geometry of individual sand bodies. Quirk and Aitken (1997) have emphasized that it is all too easy to miss a gas-bearing Carboniferous reservoir within a structural closure unless a rigorous structural and stratigraphic interpretation has been carried out to delimit the subcrop limit of the target interval.

The potential for stratigraphic plays within the Carboniferous has been poorly addressed so far, mainly because of the limitations of available seismic data to resolve such plays and the need for a better understanding of the distribution and geometry of intraformational seals. South Tyne field (Figure 9) is a combined structural and stratigraphic trap found at the erosional limit of the late Westphalian C-D alluvial red beds, and it is a success story for this type of play. As 3-D seismic coverage has extended across the region, isolated wedges of the red beds have been mapped as outliers on the hanging walls of faults oriented east-west to east-northeast–west-southwest in central Quadrant 44 (Figure 10). Furthermore, the 3-D data have identified a much greater complexity of faulting within the Carboniferous interval than previously recognized from 2-D data (Oudmeyer and de Jager 1993). In almost all cases, the wedges of red beds do not occur within a structural closure at the base Permian unconformity, and hence, they require lateral fault seals (Figure 11). Numerous stratigraphic traps of a similar wedgelike geometry are expected for older Westphalian reservoirs where they form angular subcrops beneath the base of the Permian (Figure 10). Successful identification of such drilling targets requires detailed mapping of 3-D seismic data and careful fault-seal analysis.

Figure 12 illustrates one of the many undrilled prospects in the Carboniferous fairway. This prospect lies between the Murdoch and Caister gas fields, in which the productive reservoir is basal Westphalian B in age. The prospect comprises an isolated wedge of younger Westphalian C-D alluvial red beds that is preserved on the Murdoch Ridge between the gas fields. This wedge probably contains reservoir-quality sandstone beds, which are known to occur elsewhere in the region. Perhaps this wedge has not been drilled because, although it lies within the large closure on the base Permian unconformity that encloses both of the gas fields, it occurs within a local structural low at the established reservoir level.

Deeper Carboniferous plays remain speculative because the geology of the Dinantian to Namurian succession is poorly understood across much of the southern North Sea. Few wells have penetrated more than a few hundred meters of this section. A notable exception is well 43/17-2, which penetrated 2454 m of mudstone-dominated upper Dinantian to lower Namurian sediments from within a deep synrift basin (Besly, 1998). Laterally equivalent basinal mudstones are the source rocks onshore for the East Midlands oil province but have reached maturity for gas generation offshore. Basin-

Table 1. Source and reservoir potential of the Carboniferous in the North Sea gas province.

Carboniferous interval	Source rocks	Reservoirs
Westphalian C-D red beds	None	Porosity is in the range of 7–19%. The most sand-prone proximal succession occurs toward the northeast.
Westphalian A-B coal measures	Coal beds make up 5–8% of the succession and constitute the principal gas source rocks within the North Sea gas province. Local oil-prone potential is recorded.	Laterally continuous, braided channel sandstones are as much as 50 m thick and have porosity of 3–19% and permeability of 0.1–11 mD.
Namurian basinal mudstones, prodelta and deltaic sediments	Basinal mudstones provide good to excellent oil-prone source-rock potential; now mature for gas generation.	Fluvio-deltaic sandstones have an average porosity of 12–22%, with permeability greatest in tidally reworked units. Turbidite and debris-flow sandstones are poorly sampled and have low porosity and permeability where deeply buried beneath the Permian (e.g., well 43/17-2).
Dinantian synrift deposits	Coal measures preserved along the northern margin of the basin are mostly immature for gas generation.	Porosity is as high as 25% in braided river and fluvio-deltaic sandstones on the flank of the Mid North Sea High.

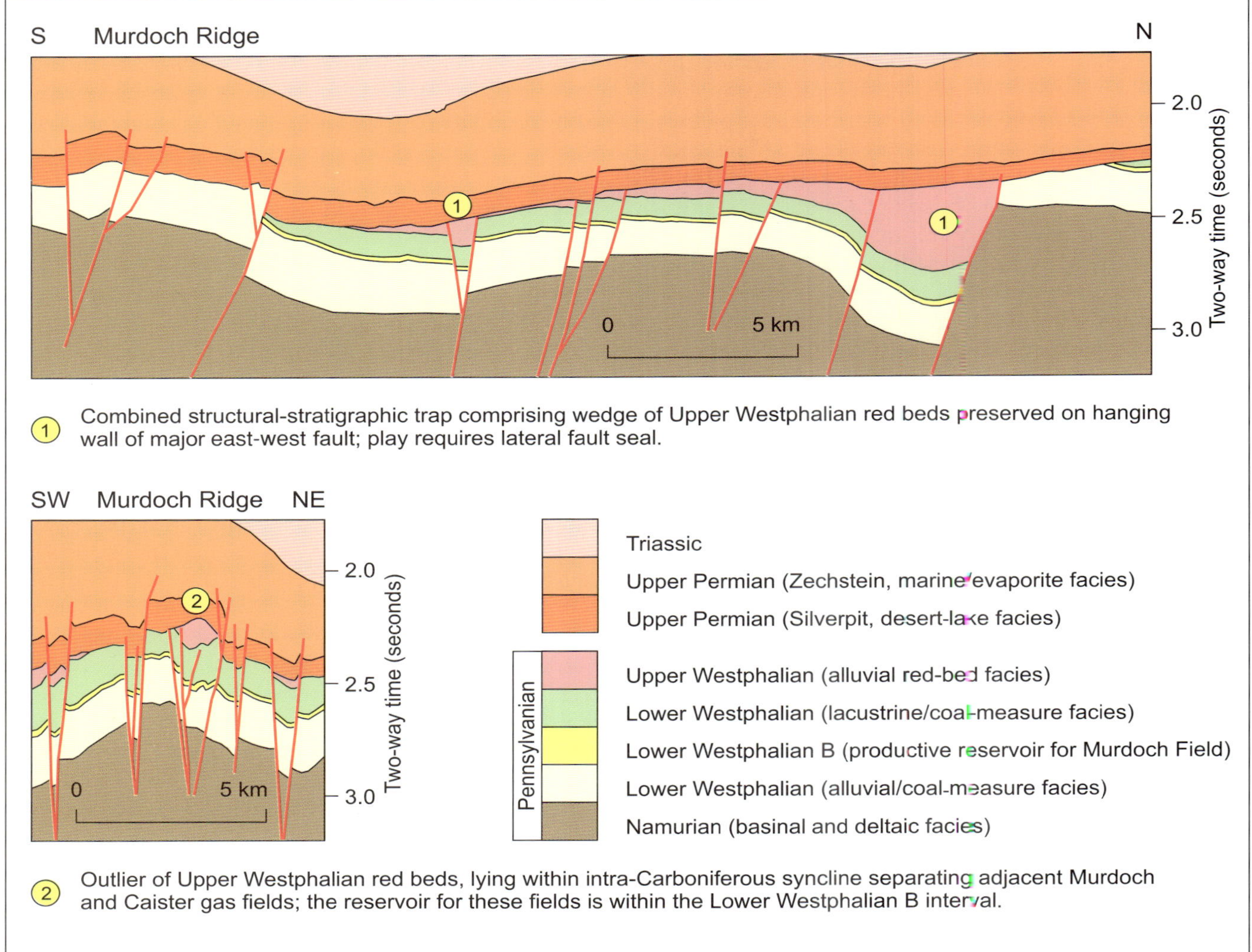

Figure 12. Geoseismic sections illustrating examples of underexplored Upper Carboniferous plays, North Sea gas province. See Figure 10 for location.

floor fan sandstones, which make up at least 10% of the succession in this well, have excellent reservoir potential, particularly if found in structural and stratigraphic traps (Figure 13).

Cameron and Ziegler (1997) suggested that significant new opportunities remain in the largely untested basal part of the Carboniferous section. In particular, they highlighted a progradational wedge play of Dinantian detrital carbonates derived from an adjacent platform succession in U.K. Quadrant 42 (Figure 13). Carbonate shelf-margin facies recorded from the U.K. onshore include carbonate shoals, buildups, talus, debris flows, and turbidites (Fraser and Gawthorpe, 1990), and secondary porosities of as much as 30% characterize carbonate debris flows where dolomitized (Gawthorpe, 1987).

Another play envisaged by Cameron and Ziegler (1997) relies on the possibility that the normally gas-prone Westphalian coal measures could have generated significant volumes of oil along the southern margin of the North Sea Basin. Their play has oil migrating from these coal measures into conventional structural traps with potentially stacked Devonian to Westphalian reservoirs.

Onshore Basins

Oil production in the U.K. began in 1919 from Hardstoft field in the East Midlands of England (Figure 14). Petroleum exploration onshore has proceeded in a cyclical manner ever since (Evans, 1990), affected by political factors and by improvements in exploratory techniques. With 35 fields currently producing about 5.2 million t of oil (38.2 million bbl) and 0.34 billion m^3 (0.012 tcf) of gas each year (Department of Trade and Industry, 1999), all the onshore area could be perceived to have been at a mature stage of exploration for many years. Nevertheless, significant discoveries are still being made, the coal-bed

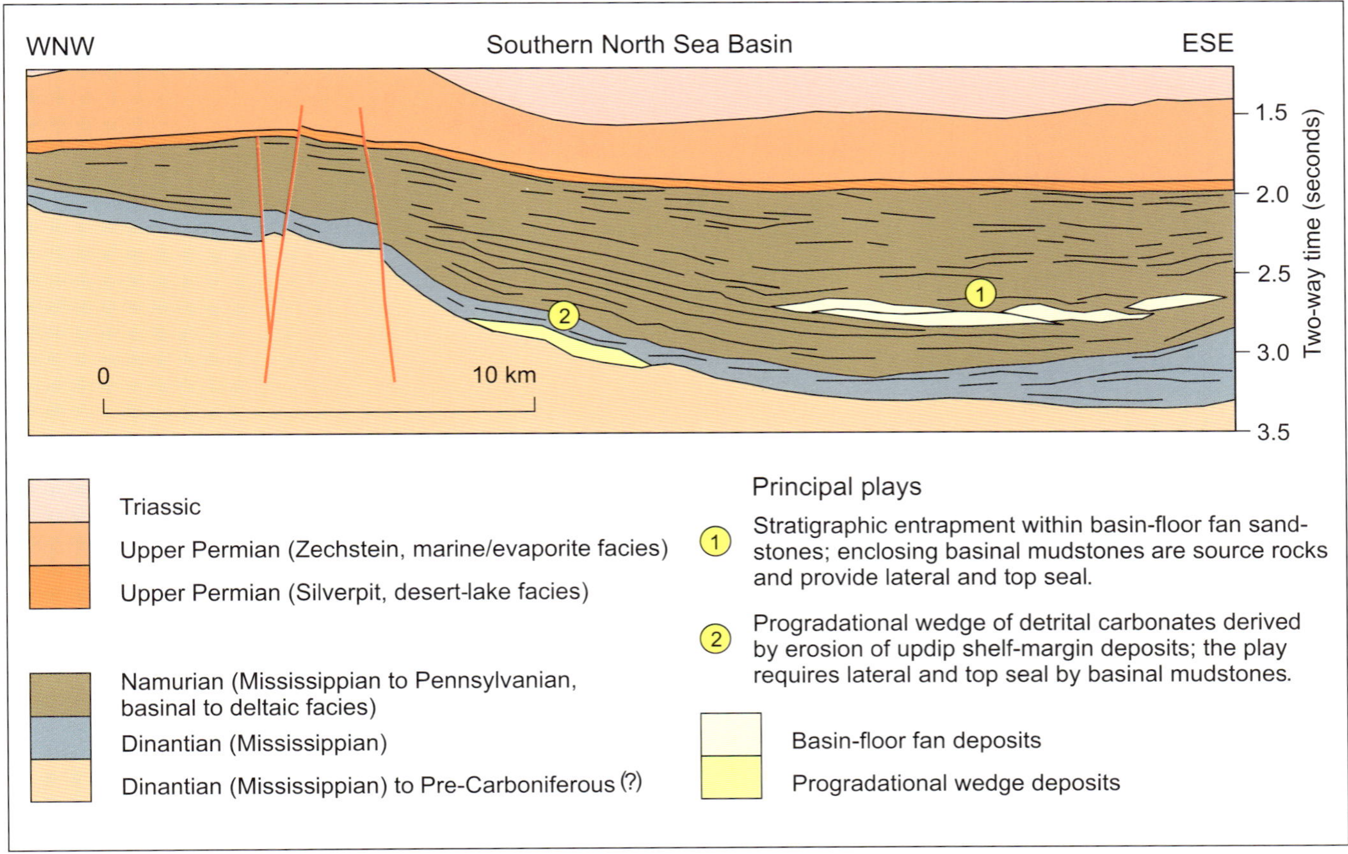

Figure 13. Geoseismic section illustrating undrilled Lower Carboniferous plays, North Sea gas province (adapted from Cameron and Ziegler, 1997). See Figure 10 for location.

methane extraction industry is in its infancy, and there is no reason to consider that innovative thinking will not yield further discoveries for many years to come.

The oil and gas fields are mainly in three geographic areas. In southern England, petroleum production is from the Weald and Wessex Basins (Figure 14) and from a range of Triassic (Sherwood Sandstone), Middle Jurassic, Upper Jurassic, and Lower Cretaceous reservoirs sourced by Lower Jurassic (Lias) oil-prone mudstones. About 90% of U.K. onshore oil production is from Wytch Farm field in the Wessex Basin (Figure 14), aided by extended-reach wells that include the 10.7-km current world record for "horizontal" stepout. The key to exploration success in southern England will continue to hinge on unraveling the relationship among the structural development of drilling targets, the timing of hydrocarbon migration into these targets, and the prediction of reservoir facies development, particularly for the Middle Jurassic oolitic limestone play.

Eastern England contains the oil province of the East Midlands (Figure 14), in which hydrocarbons sourced from lower Namurian basinal mudstones have migrated partly into Dinantian shelf-margin limestones, but mostly into younger Carboniferous sandstones. The oil of early discoveries was trapped mainly in inverted rollover anticlines. Fraser and Gawthorpe (1990) demonstrated that the hydrocarbon system includes a wider range of plays whose distribution is strongly influenced by the Dinantian rift-related tectono-stratigraphic development of the region. Not all of these plays have been fully explored yet. Recent discoveries of oil and gas have extended the East Midlands province eastward to the North Sea coastline (Figure 14), where mapping of residual Bouguer gravity anomalies has proved to be invaluable in delineating primary drilling targets. The recent successes in this area have been in upper Namurian to basal Westphalian fluvio-deltaic sandstones. Now that 3-D seismic coverage is extending across the region, there will be opportunities for defining the structural traps more clearly and for assessing the potential for stratigraphic entrapment, especially for pinch-out of basal Namurian sandstones onto the regional Dinantian structural highs.

Farther north in eastern England, there has been limited production of gas from Upper Permian carbonate reservoirs. Further exploitation of gas reserves there will depend on improvements in production technology. In northwest England, Carboniferous-sourced oil has been produced from the Lower Triassic (Sherwood Sandstone) reservoir at Formby field (Figure 14). The top seal is the Pleistocene boulder clay. Gas is being produced from an

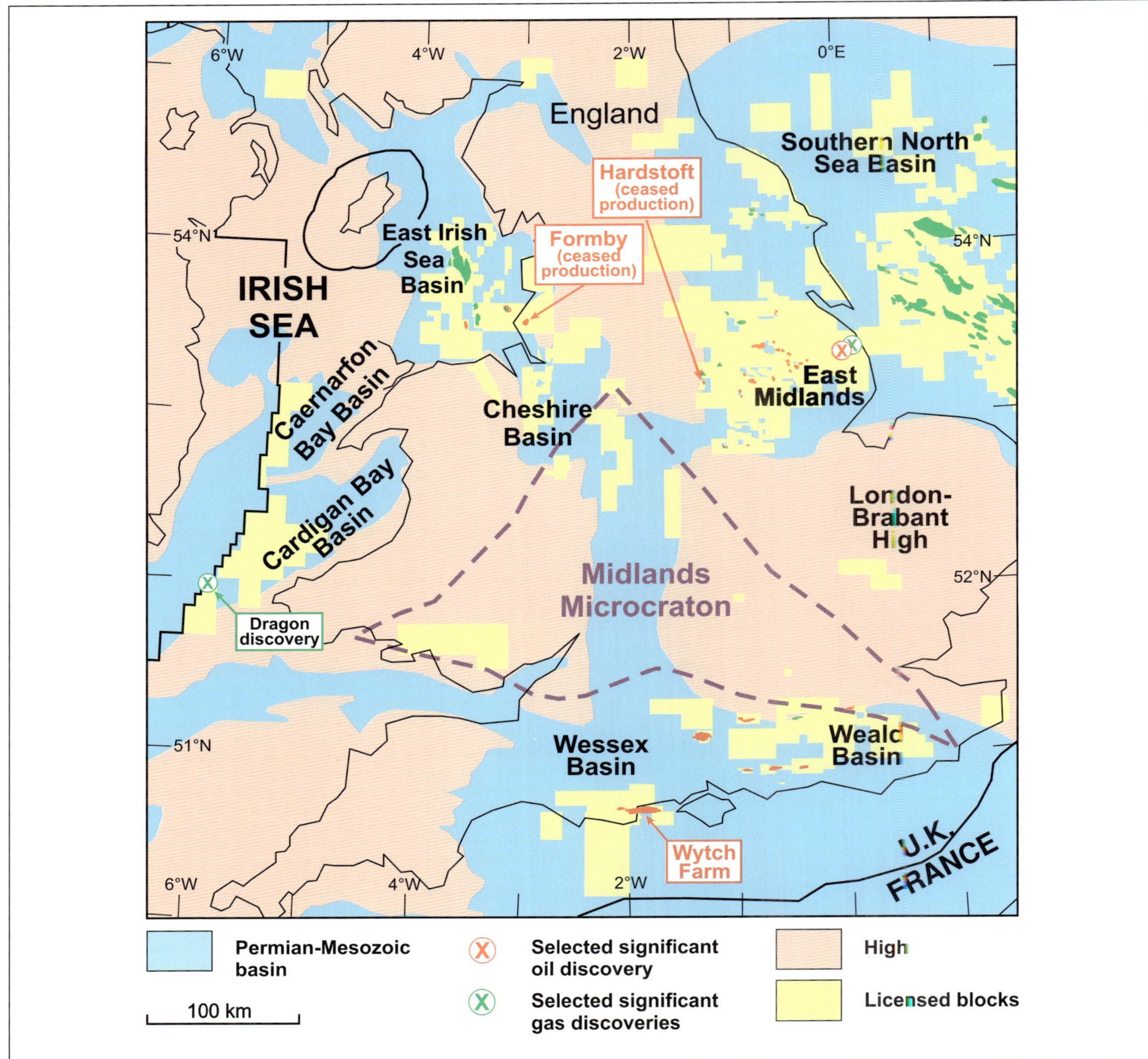

Figure 14. Hydrocarbon fields and selected hydrocarbon discoveries in the U.K. onshore and Irish Sea basins.

Upper Permian sandstone reservoir nearby, and there remain a number of undrilled leads in tilted fault-block plays, similar to those offshore in the East Irish Sea Basin.

The Cheshire Basin of northwest England and the Midlands Microcraton of central England have the best potential for expanding the geographic range of the U.K.'s onshore petroleum provinces (Figure 14). The Cheshire Basin is a Permian-Triassic half graben, partly underlain by mature, gas-prone Westphalian coal measures. Successful hydrocarbon exploration will depend on locating migration routes from these source rocks into structural traps containing reservoir-quality Lower Triassic sandstone with younger Triassic mudstone and evaporite top seals (Mikkelsen and Floodpage, 1997). The Midlands Microcraton is a triangular lower Paleozoic foreland that was bypassed by the middle Paleozoic (Caledonian) orogenic deformation which was ubiquitous to the northeast and northwest. It also lies beyond the later Paleozoic (Variscan) deformation front that traces across southern England. Relatively few exploration wells have been drilled on the microcraton, but some of these have encountered subcommercial shows of oil and gas, mainly in overlying upper Paleozoic and Mesozoic strata. Silurian marine mudstones are currently within the oil window here, whereas Cambrian to Lower Ordovician source rocks are in the gas window (Smith, 1993). Untested hydrocarbon leads in the upper Paleozoic section include pinch-out of Namurian and Westphalian sandstones onto regional highs, and four-way dip and tilt-block closures. Additional prospects in the lower Paleozoic include Silurian and Lower Cambrian marine sandstones in tilt-block traps, in particular where the hanging walls of these traps include Lower Ordovician basinal mudstone source rocks.

Coal-bed methane is already being produced in limited volumes from abandoned mines as vent gas, but there remains greater long-term potential for U.K. production from virgin coal seams. Although extraction has not been adequately tested on a commercial scale yet, Creedy (1999) has estimated that there are potentially recoverable U.K. reserves of 30 billion m^3 (1 tcf). If there is to be significant production, it will be from Westphalian seams in England and Wales and from Namurian seams in central Scotland. The principal limitation on development at present is the generally low permeability of U.K. coal seams. Until technical solutions are established for stimulating production from such rocks, the best initial prospects are in areas where mining may have already led to fracturing and enhanced permeability of the coal seams, or in structural traps. With successful resolution of this problem, unpublished data suggest that coal-bed methane could account for as much as 1% to 2% of U.K. gas production by the year 2020.

Irish Sea Basins

With 200 billion m^3 (7.1 tcf) of gas and 20 million t (147 million bbl) of oil estimated as initially recoverable reserves from six producing fields (Department of Trade and Industry, 1999), the East Irish Sea Basin (Figure 14) is in a mature exploration phase. Early Namurian basinal shales are the source rocks for these hydrocarbons. Production from all fields is from fault-bounded traps of the Lower Triassic, principally the eolian Sherwood Sandstone reservoir, which have younger Triassic continental mudstones and evaporite top seals. Future exploration will initially concentrate on extending this play, but

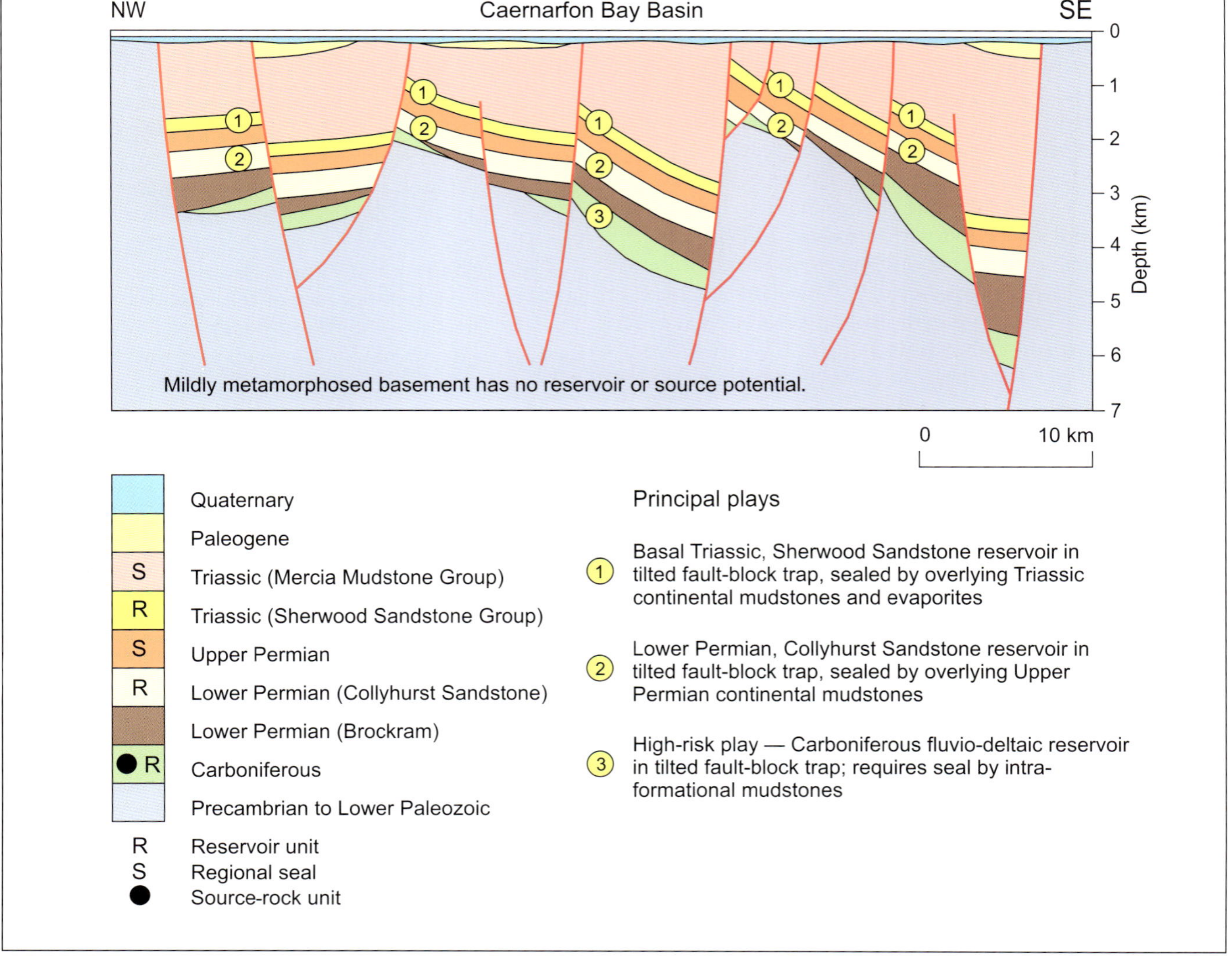

Figure 15. Schematic play diagram for Caernarfon Bay Basin, Irish Sea.

largely untested potential for gas and oil within widespread Carboniferous fluvial sandstone reservoirs also remains. This play requires intraformational mudstone seal units to be present, because there is no top seal for reservoirs subcropping the regional base Permian unconformity in the eastern part of the basin, and Carboniferous strata crop out at the seabed in the west.

The Caernarfon Bay Basin (Figure 14) contains as much as 7 km of Permian and Triassic synrift sediments in an asymmetrical graben bounded to the north and south by lower Paleozoic massifs. Only two exploration wells have been drilled, and there remain numerous undrilled targets in tilted fault-block plays (Figure 15). As in the East Irish Sea Basin, the principal target reservoir is the Lower Triassic Sherwood Sandstone, which has a top seal of younger Triassic mudstones and evaporites. Wells in the Irish sector to the west (Figure 14) have proved that prerift Westphalian coal measures are excellent hydrocarbon source rocks and are at peak maturity for gas generation there (Maddox et al., 1995). Seismic profiles clearly image these strata continuing beneath a basal Permian unconformity into at least the western part of the Caernarfon Bay Basin. The timing of gas generation presents the greatest exploration risk. Maximum burial of the source rocks and primary gas migration from them could have terminated as early as the Jurassic, whereas many of the tilted fault blocks were reactivated or created during early Tertiary inversion of the basin. However, it is also possible that a secondary gas charge occurred during regional heating associated with intrusion of Paleogene dykes, such as those that crop out nearby on the coastline of north Wales. Floodpage et al. (1999) have invoked this second phase of Paleogene hydrocarbon generation as an important factor in the charging of the East Irish Sea Basin's oil and gas fields. It is not clear yet whether aeromagnetic anomalies in the southeast Caernarfon Bay are imaging a continuation of the dyke swarm into this area too or whether they are associated with deeply buried Permian synrift volcanics. Alternatively, the fault-block traps could have been recharged by exsolution of methane from formation brines as a direct result of the Tertiary uplift (compare Doré and Jensen, 1996).

The Cardigan Bay Basin forms a continuation of Ireland's North Celtic Sea Basin (Figure 3), which has two producing gas fields, into U.K. waters. The basin comprises a southeasterly deepening half graben near the Welsh coastline; its internal structure becomes increasingly complex toward the southwest. Its Permian to Triassic synrift sediments are less than 3 km thick, but they are overlain by as much as 4 km of Jurassic strata and locally also by as much as 2 km of Paleogene fluvio-deltaic sediments. The basin has a proven petroleum system with potentially producible gas reserves at the Dragon discovery near the U.K./Eire median line (Figure 14) and oil shows in an additional three wells.

The Cardigan Bay Basin contains multiple reservoir targets (Figure 16) that include the Lower Triassic Sherwood Sandstone, Middle Jurassic shallow-marine sandstones and limestone (Great Oolite), and Upper Jurassic fluvial sandstone, the reservoir for the Dragon discovery. The most likely hydrocarbon source rocks are Lower Jurassic marine mudstones (Lias Group). These are fully mature for oil generation in the west of the U.K. sector, and are mature for gas generation nearby in the Irish sector. Gas-prone Westphalian prerift coal measures may also be present locally at depth. The Cardigan Bay Basin was subjected to two Tertiary phases of compressive uplift, whereas maximum burial that terminated primary hydrocarbon generation was probably at about the end of the Cretaceous, or earlier if Cretaceous strata, now missing, were never deposited in the basin. Despite its location on a Tertiary structure, the Dragon discovery has proved that potentially commercial volumes of hydrocarbons remained trapped at least locally in Cardigan Bay. In addition to undrilled structural traps, the basin contains untested potential for stratigraphic entrapment of hydrocarbons near synsedimentary faults, especially in the Middle Jurassic section.

Atlantic Margin

U.K.-designated waters northwest of Britain lie on the northeast Atlantic margin. They include the Faroe-Shetland Basin, with parts of the relatively unexplored Rockall Trough, Hatton Basin, and Hatton continental margin (Figure 17). Water depths in these areas locally exceed 2 km. Theoretically, this province may contain a large gas-hydrate resource, but exploitation of this resource is unlikely in the first decades of the twenty-first century. To date, the Faroe-Shetland Basin has been the main focus of conventional hydrocarbon exploration along the Atlantic margin.

The opening of the North Atlantic between Greenland and northwest Europe was accompanied by Paleogene basaltic shield volcanism across an area about 2000 km wide. The igneous activity is generally believed to have resulted from the development of the Iceland mantle plume shortly before continental breakup (White, 1989). Lavas and sills within the section are present across much of the Atlantic margin acreage, and they severely attenuate and disperse the seismic signal, degrading the seismic response at depth. This has significantly hindered our understanding of the pre-Tertiary geologic history of the region, and considerable effort is being expended toward developing a means of detailed subbasalt imaging. Seismic-refraction profiles and wide-angle seismic-reflection data have both been collected to enable models to be constructed for the crustal structure along the Atlantic margin (e.g., Joppen and White, 1990; Keser Neish, 1993; Shannon et al., 1994). Recent developments in seismic techniques are enabling more refined subbasalt imaging (e.g., Richardson et al., 1999, White et al., 1999).

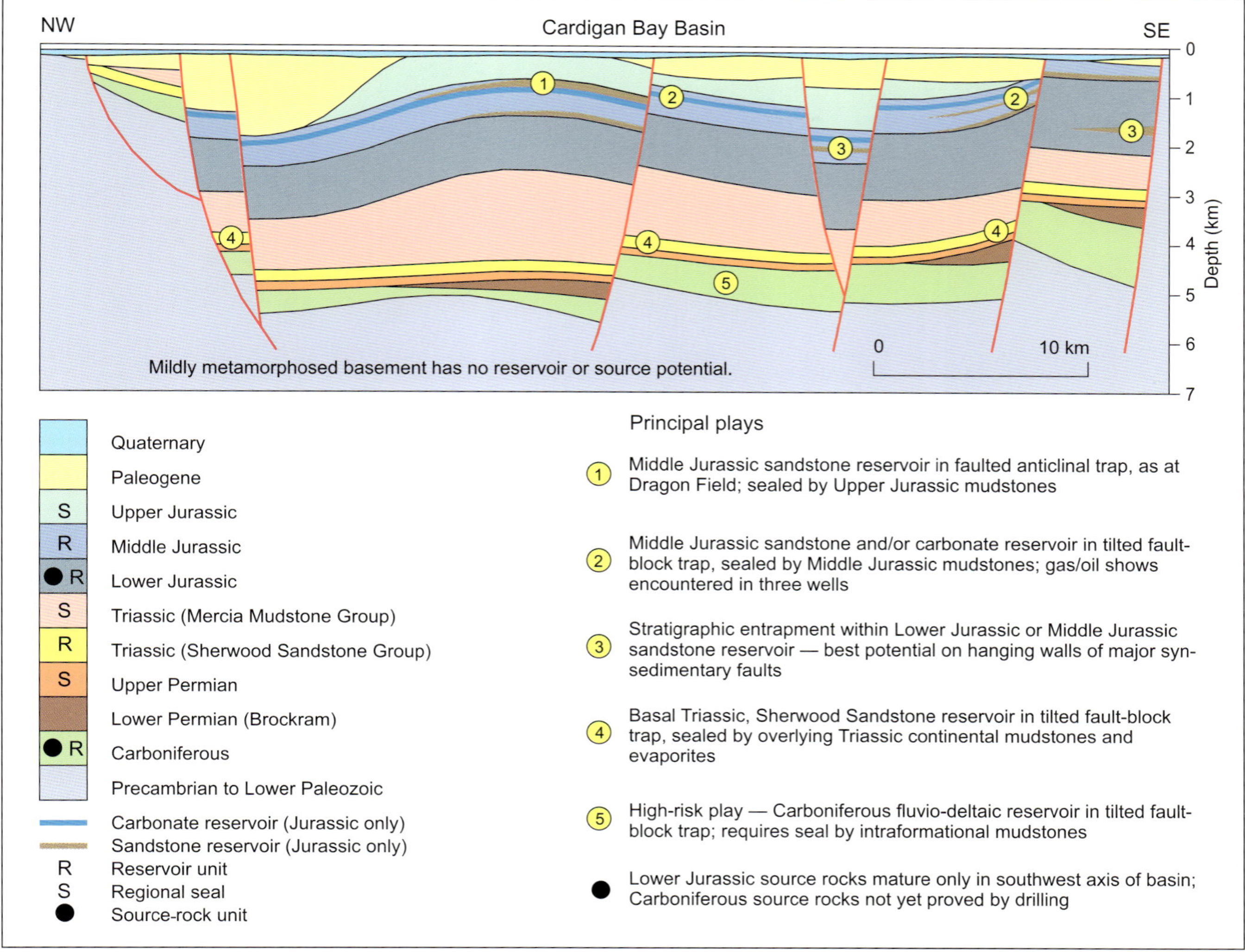

Figure 16. Schematic play diagram for Cardigan Bay Basin, Irish Sea.

As in the oil province of the North Sea, play fairways along the Atlantic margin fall into upper Paleozoic to Middle Jurassic prerift, Upper Jurassic to Lower Cretaceous synrift, and Upper Cretaceous to Tertiary postrift categories (Pegrum and Spencer, 1990; Knott et al., 1993). Tertiary inversion structures are regionally widespread (e.g., Doré et al., 1997), presenting multiple opportunities for forming structural traps. Upper Jurassic, organic-rich basinal-marine mudstones (Kimmeridge Clay) are the principal oil-prone source rocks in the Faroe-Shetland Basin, but they have not been proved yet in the Rockall Trough and adjacent Rockall Plateau. Scotchman and Broks (1999) have shown that rich oil-prone Lower Jurassic and Middle Jurassic source rocks also occur in the Atlantic margin province, and that oil in the Foinaven and Schiehallion fields (Figure 18) can be correlated to these sources. A viable petroleum system has yet to be established in the Rockall Trough and basins farther west. These areas contain the last truly frontier acreage in U.K.-designated waters.

Faroe-Shetland Basin

Although oil was found on the eastern flank of the Faroe-Shetland Basin as early as 1974, it was not until 1992 that economically viable reserves were discovered in Foinaven field (Figure 18). Recoverable oil reserves in this and the adjacent Schiehallion field are estimated at 34.4 and 79.87 million t (252 and 586 million bbl), respectively (Department of Trade and Industry, 1999). Both of these fields are using floating production systems with offshore loading because there is no existing infrastructure in the region. Much greater volumes of oil have been proved in Clair field (Figure 18), the largest oil field offshore northwest Europe. Mainly because of its very low recovery factor of 7%, the field has not proceeded to full-scale production yet (Brennand et al., 1998).

Much of the acreage along the axis of the Faroe-Shetland Basin is currently licensed; however, a large part of the acreage in the east of the basin was relinquished in 2000. In the west of the basin, the recent

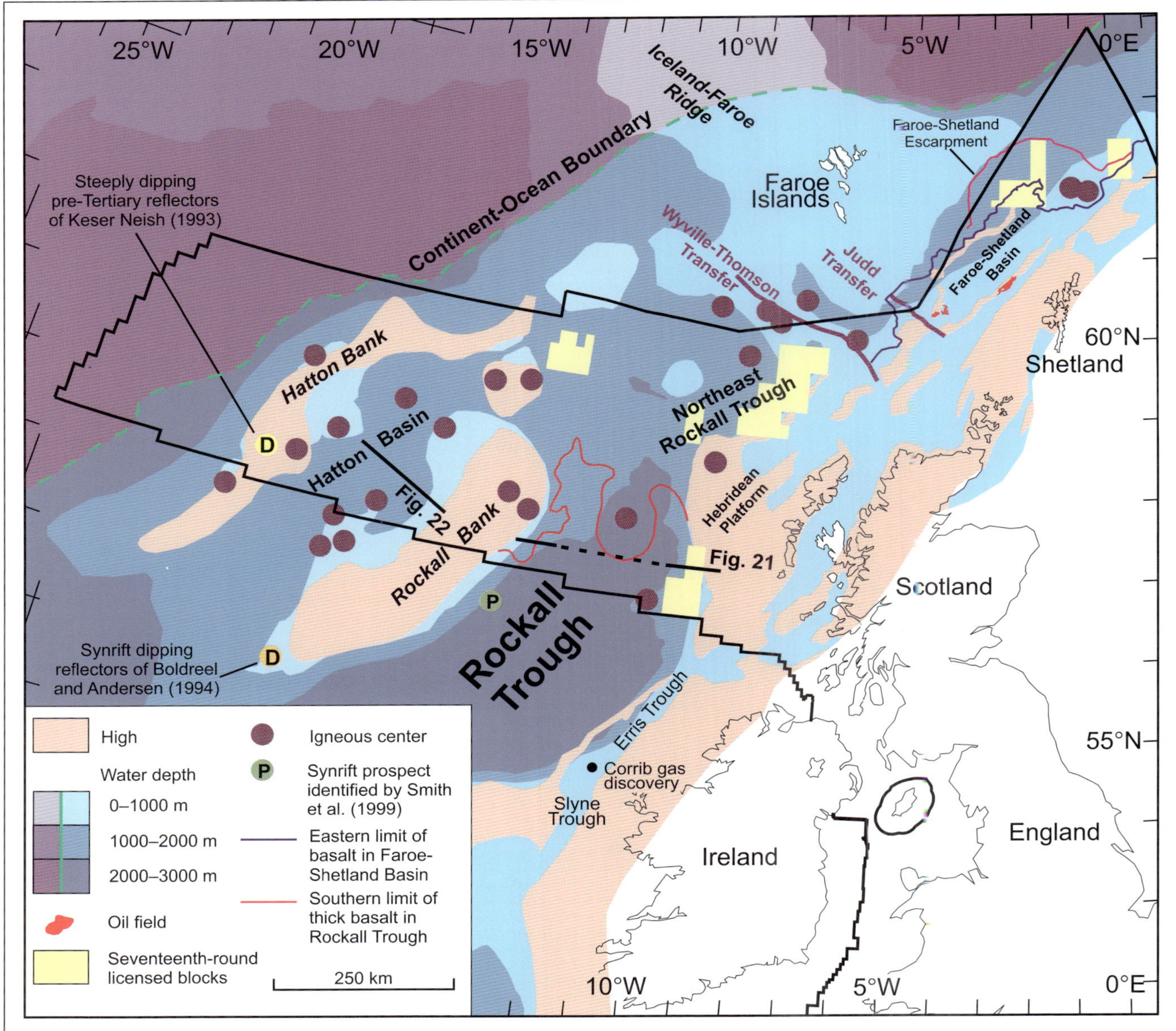

Figure 17. Location of the hydrocarbon-producing basins and frontier basins of the Atlantic margin.

(1999) agreement on a median line between the U.K. and the Faroes has provided a major stimulus to exploration, because of proximity of the newly designated acreage to proven hydrocarbon systems at Foinaven and Schiehallion fields. Paleocene and Eocene basin-floor fan reservoirs are the principal targets within the newly designated area. Prerift and synrift plays are also likely to be present there (Figure 19) and are relatively shallow, but individual prospects are currently difficult to image. This is because the Upper Cretaceous to lower Paleocene section along the northwestern flank of the Faroe-Shetland Basin is extensively intruded by Paleogene sills, which hamper seismic interpretation. Furthermore, Paleogene basalts are also present in parts of this area (Figure 18). Where they occur, wide-angle seismic data suggest that the thickness of subbasalt sediments ranges between 1.25 and 3.75 km, with the greatest thickness in the north (Richardson et al., 1999).

Prerift strata are buried too deeply to be prospective along the axis of the Faroe-Shetland Basin. Clair field sits on the Rona Ridge, which forms the eastern margin of the basin (Figure 18), and its oil reservoir is provided by fractured Precambrian to Devonian-Carboniferous rocks. Oil and gas/condensate shows have been recorded in other wells along this ridge and also on the North Rona High (Figure 18) that forms the southern margin of the basin. Middle Jurassic sandstones preserved on a fault terrace flanking the Rona Ridge have yielded oil and gas shows. Southeast of the Faroe-Shetland Basin, the Solan/Strathmore discovery (Figure 18) contains oil in overlap-

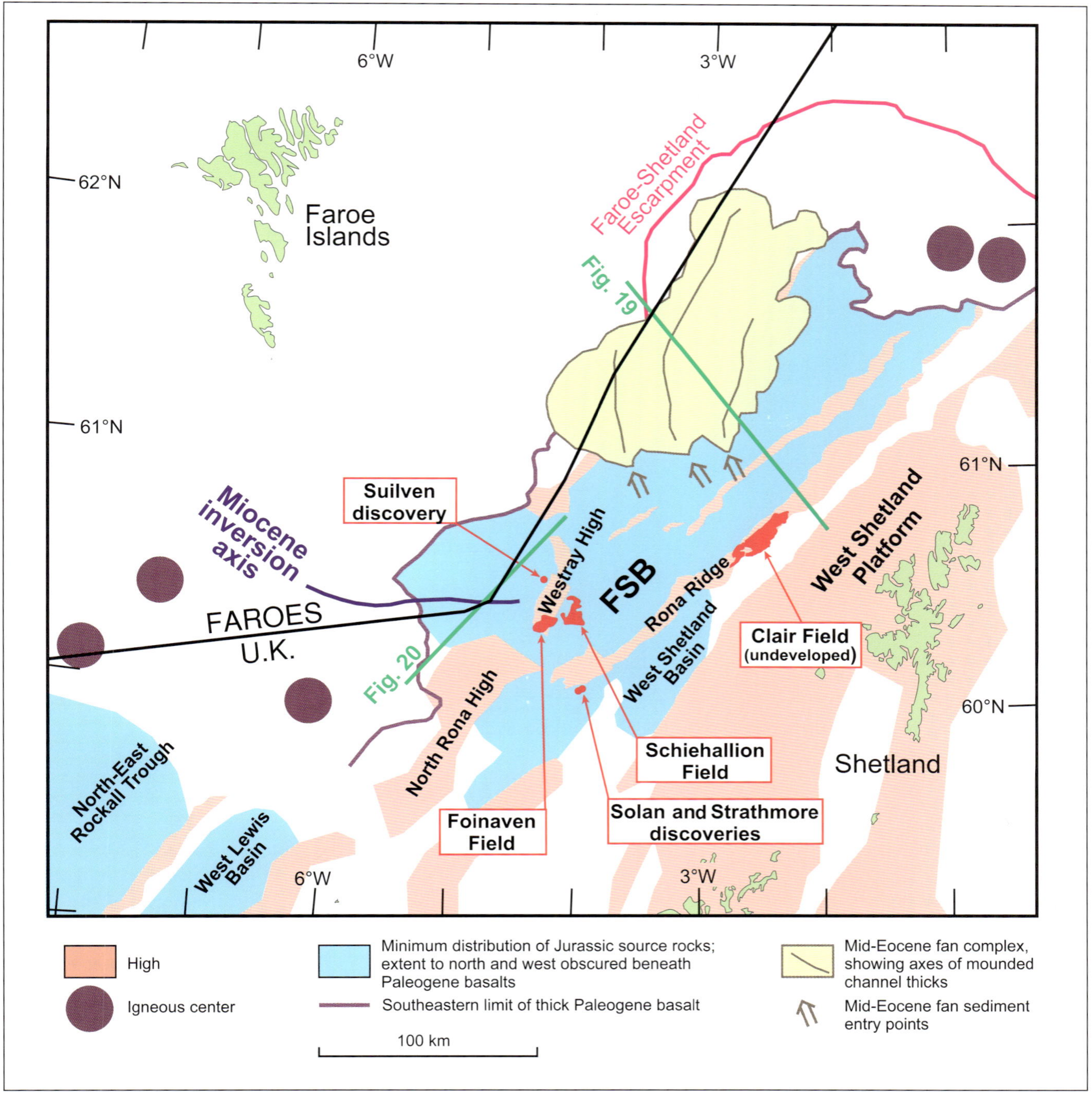

Figure 18. Location of middle Eocene fan complex, Faroe-Shetland Basin (FSB).

ping Jurassic and Triassic reservoirs within the contiguous West Shetland Basin. Prerift prospects in both basins have top seals of Upper Jurassic, Cretaceous, or Paleocene mudstones

Synrift plays are relatively lightly explored, although at least 12 wells have encountered oil or gas shows in Upper Jurassic or Lower Cretaceous sandstones within or bordering the Faroe-Shetland Basin. Two currently uneconomic gas discoveries on the eastern margin of the basin have Lower Cretaceous synrift apron-fan and basin-floor sandstone as their reservoirs. Underlying Upper Jurassic sandstones are encased in Kimmeridge Clay source rocks to provide drilling targets in combined structural-stratigraphic traps. In the southwest, Lower Cretaceous shallow-marine sandstones penetrated on the Westray High (Figure 18) constitute a synrift retrogradational shelf play, although in well 204/19-1, they are overlain by deep-water sandstone facies (Ritchie et al., 1996) and thus may locally lack a top seal. Along the axis of the Faroe-Shetland Basin, depth of burial is a prohibi-

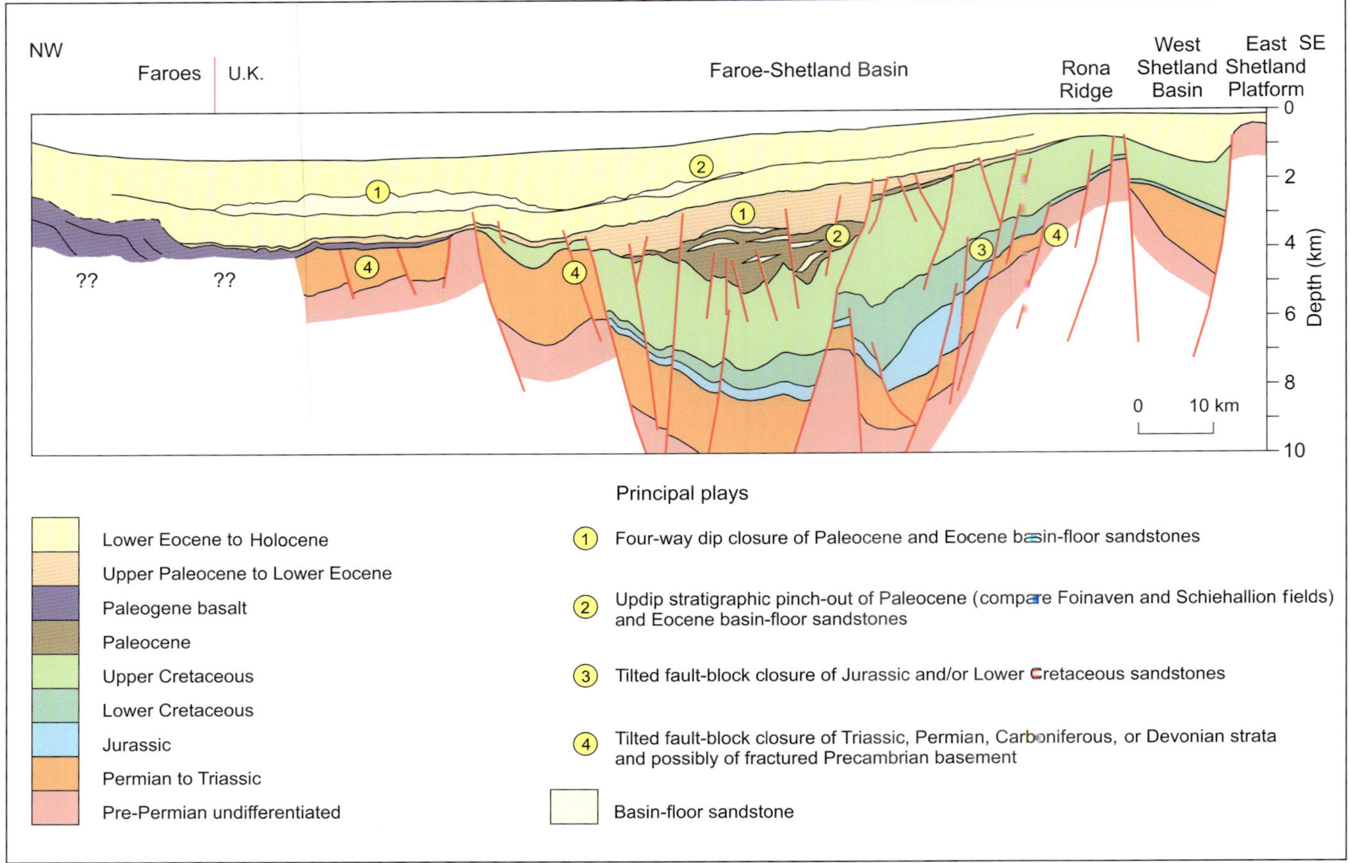

Figure 19. Schematic play diagram for the Faroe-Shetland Basin. See Figure 18 for location.

tive factor for synrift plays. Upper Cretaceous mudstones locally exceed 3800 m in thickness and provide an excellent top seal throughout the region.

Postrift Upper Cretaceous slope-apron sandstones of Cenomanian to Turonian age are locally as much as 327 m thick along the eastern flank of the Faroe-Shetland Basin, where they are stacked against the Rona Fault. Oil shows have been recorded from these sandstones, although doubts remain about the seal integrity of their boundary fault. Grant et al. (1999) speculated that if there are detached fan complexes downdip, these may offer more attractive drilling targets.

The discovery of oil in Foinaven field has completely revitalized the postrift Paleocene fan play in the region. This discovery has been followed up by the nearby Schiehallion field and Suilven discovery (Figure 18). However, the play is quite subtle, requiring accurate definition of stratigraphic pinch-out by 3-D mapping, in combination with structural closure in a zone of monoclinal basinward dip. The occurrence of an intraformational mudstone seal is also critical; mudstones associated with regional flooding surfaces have proved to be the most effective. The Paleocene interval reaches a maximum thickness of almost 3.5 km, and it contains excellent-quality sandstone reservoirs within basin-floor fan, slope fan, and shelfal facies. Porosities of more than 25% and permeabilities greater than 100 mD can be expected where the sandstones are buried less than 2.5 km below seabed (Ebdon et al., 1995). Equivalent reservoirs that occur at greater depth in the north of the Faroe-Shetland Basin are of much lower quality (Johnson and Fisher, 1998) and have yielded only gas shows. Ebdon et al. (1995) recognized eight seismically resolvable sequences within the Paleocene, whereas Mitchell et al. (1993) identified nine sequences within the same interval. The relationship between these two interpretations is complicated, because Ebdon et al. (1995) used maximum flooding surfaces to divide their sequences, whereas Mitchell et al. (1993) based their work on identifying type 1 unconformities (compare Mitchum, 1977). Both groups of authors have mapped distinct shelf, slope, and basinal systems and have compiled generalized depositional environment maps for each sequence. Their approaches are vital steps toward identify- ing potential stratigraphic targets. Mapping of seismic attributes and AVO analyses have played a critical role in current discoveries but have yet to be applied on a regional scale.

There is a particularly high potential for success in Paleocene prospects near Foinaven field within the newly designated acreage adjacent to the U.K.-Faroes median

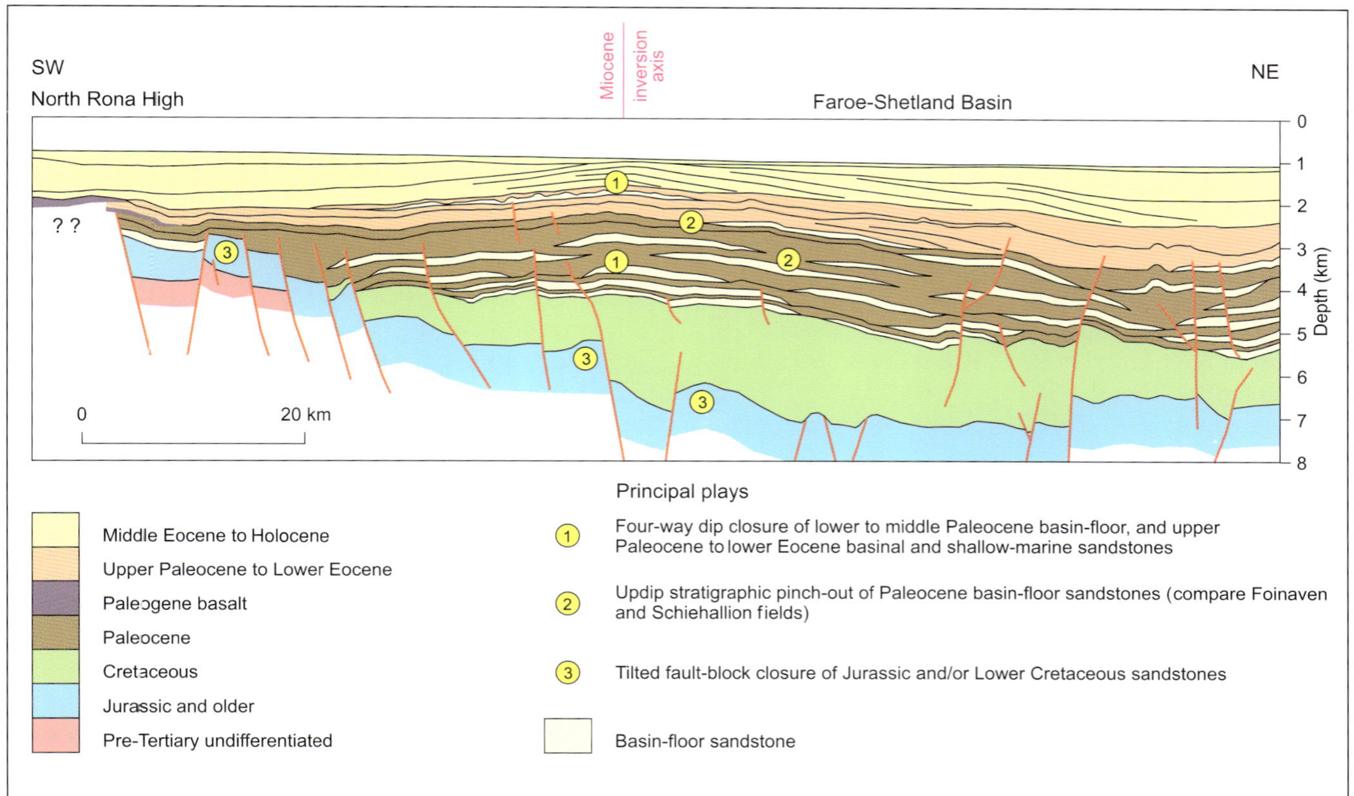

Figure 20. Schematic play diagram for newly designated acreage in the Faroe-Shetland Basin. See Figure 18 for location.

line. The crest of a major east-west-trending Oligocene-Miocene inversion axis can be traced across this area, and numerous stacked potential reservoir targets can be anticipated in four-way dip closures (Figure 20). Similar to much of the Faroe-Shetland Basin, aggradational basin-floor sandstones characterize the lower part of the Paleocene here and are likely to be sealed by both lowstand and highstand mudstones. Overlying upper Paleocene to lower Eocene sandstones include a greater proportion of shelfal and deltaic deposits, with a potentially higher risk of imperfect seals. Identification of migration pathways is of key importance in this area because of local overpressuring and a complex basin evolution (Illiffe et al., 1999).

Overlapping, linear basin-floor fans of probable middle Eocene age can be traced across the northwestern part of the Faroe-Shetland Basin and extend into the Faroes sector (Figure 18). Three principal submarine canyon entry points have been identified on the margin of the broadly northeast-southwest-trending paleoshelf, and the fan deposits appear to have flowed across the basin floor in a northerly to northeasterly direction. The fan complex is more than 100 km long and more than 50 km wide, and it attains a maximum thickness of 550 m. Internal geometry of the fans is complex, as evidenced by zones of high-amplitude layered seismic reflections bordered and locally unconformably underlain by relatively high-amplitude chaotic reflections. The principal exploration risks for this play include a risk of biodegradation because of the shallow depth of burial (about 1000 m below seabed) of the basin-floor fans and uncertainty as to the validity of migration routes through the underlying Paleocene section.

Rockall Trough

More than 80% of the Rockall Trough has never been licensed for hydrocarbon exploration, although numerous tranches of blocks were awarded in the 1997 U.K. seventeenth round of offshore licensing (Figure 17). A considerable quantity of new 2-D and 3-D seismic data has been acquired, but only one well has been drilled on the seventeenth-round acreage to date. Results from this well will not be released into the public domain until 2003. Of two older wells drilled in the area, one terminated within a thick Paleocene volcanic interval and the other drilled a Lower Cretaceous synrift section on the southeastern margin of the Rockall Trough (Musgrove and Mitchener, 1996). Three wells were drilled between 1988 and 1991 on the eastern-flanking basement horst and within an adjacent Triassic/Jurassic rift basin on the Hebridean Platform (Figure 3).

The Rockall Trough is a failed rift formed during the opening of the North Atlantic. Some authors (e.g., Tate et al., 1999) have suggested that it may be separated from the Faroe-Shetland Basin by the Wyville-Thomson Transfer Zone, although Doré et al. (1997) argued that the Judd Transfer farther north (Figure 17) is the more likely boundary. Numerous west-northwest–east-southeast to northwest-southeast-oriented lineaments traverse the Rockall Trough and offset its eastern margin; these have been linked to boundaries of pre-Caledonian basement terranes (e.g., Musgrove and Mitchener, 1996). The initiation of rifting in the trough has been variously ascribed to the Late Carboniferous, Permian, Triassic, Jurassic, and Cretaceous (e.g., Smythe, 1989, Knott et al., 1993, Musgrove and Mitchener, 1996).

Critically, if the Rockall Trough was not initiated until the Cretaceous, then there is no potential for Jurassic source rocks, and Cretaceous to Tertiary organic-rich basinal marine mudstones are required to be mature for a working petroleum system. If, instead, a proto–Rockall Trough developed as a contiguous Late Jurassic to Early Cretaceous rift with the Faroe-Shetland Basin, there may be Jurassic source rocks within one or more pre-Cretaceous basins beneath the Rockall Trough, but these basins must be of limited extent. This is because all paleomagnetic reconstructions for the Atlantic margin indicate that the pre-Cretaceous seaway was very narrow in the Rockall area (Ziegler, 1988; Knott et al., 1993; Doré et al., 1999). An alternative model presented by Brown and Beach (1999) provides some scope for significant Jurassic deposition in the region under a transtensional regime. Cole and Peachey (1999) and Nadin et al. (1999) each modeled the crustal evolution of the Rockall Trough area and concluded that an earlier pre-Cretaceous rift basin is located beneath the present-day Rockall Trough.

The western margin of the Rockall Trough comprises a series of tilted fault blocks downthrown to the east (Joppen and White, 1990), but the nature of the eastern margin is controversial because it is masked by the lower Paleogene volcanics of the Hebridean Escarpment. According to Musgrove and Mitchener (1996), the synrift succession in the trough has a westerly thickening geometry, thus implying asymmetric rifting accommodated through a series of easterly dipping faults across the entire trough. The deep seismic reflection profile WESTLINE provides evidence for such easterly facing synrift faults on the eastern margin of the trough (England and Hobbs, 1997).

Hydrocarbon exploration of the Rockall Trough is driven to a considerable extent by the recognition that upper Paleozoic and Mesozoic sections with source rocks occur in adjacent, parallel half-graben basins, such as the West Lewis Basin (Figure 18) and Erris/Slyne Trough (Figure 17). British Geological Survey boreholes on the flanks of the West Lewis Basin have proved the presence of organic-rich Middle Jurassic mudstones with excellent oil-prone source potential (Hitchen and Stoker, 1993). Rich, oil-prone Lower Jurassic source rocks have been proved in the Slyne Trough (Scotchman and Thomas, 1995). The recent Corrib gas discovery in Irish waters

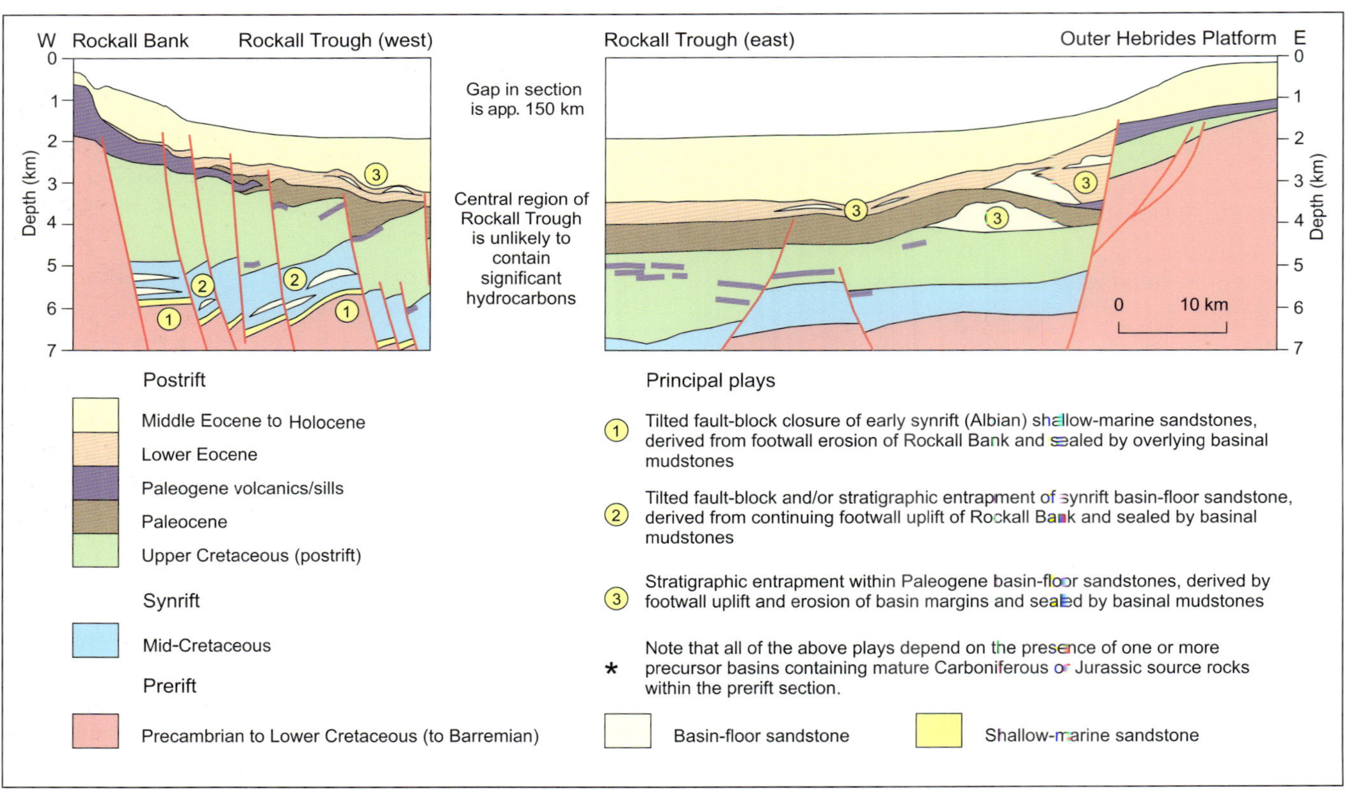

Figure 21. Schematic play diagram for the Rockall Trough. See Figure 17 for location.

(Figure 17) proves that Carboniferous coal measures also have source potential in the Slyne Trough region (Spencer and McTiernan, 1999). This discovery increases the possibility that Carboniferous source rocks occur beneath parts of the Rockall Trough also.

Figure 21 shows a range of possible exploration plays in the Rockall Trough. Lower Cretaceous synrift and Paleocene-Eocene postrift plays are generally thought to offer the best potential for significant quantities of undiscovered hydrocarbons. The potential for prerift Carboniferous to Jurassic plays is highly speculative, and such plays may be of severely limited distribution within the Rockall Trough. The southeastern part of U.K.-designated waters may contain Westphalian, Permian-Triassic, and Middle to Upper Jurassic prerift plays, as well as Paleocene-Eocene plays similar to those described by Nolan et al. (1999) from across the median line in the Irish sector. In that area, the upper part of the pre-Cretaceous section is modeled as being presently within the mid- to late oil window.

The southwestern part of U.K.-designated waters has potential for prospects in large prerift tilted fault blocks, similar to those described by Smith et al. (1999) from the Irish sector nearby. The age and nature of the reservoir, source, and seal rocks for such prospects remain entirely speculative. However, if the geologic history bears comparison with the Slyne and Erris Troughs to the southeast, then Triassic to Middle Jurassic reservoirs and Jurassic or possibly Carboniferous source rocks can be expected. There may also be potential for synrift prospects in Lower Cretaceous basin-floor sandstones there. This play requires that significant volumes of coarse siliciclastic debris were shed into this part of the Rockall Trough during contemporary footwall uplift and erosion of Rockall Bank.

The distribution of Paleocene basin-floor sandstones within the Rockall Trough is poorly known. Knott et al. (1993) speculated that such sandstones are confined to the northeastern corner of the Rockall Trough and that contemporary retrogradational shelf deposits occur locally along its southeastern margin. These authors also predicted that Eocene basin-floor sandstones are present along the eastern margin of the Rockall Trough, bordered on the east by a wide belt of shelf deposits. Both Paleocene and lower Eocene slump masses and basin-floor mounds have been observed on seismic profiles from the southeastern and western margins of the Rockall Trough, and these are presumed to be sealed by basinal mudstones.

Rockall Plateau: Hatton Basin and Hatton Continental Margin

The Hatton Basin is a northeast-southwest-trending intracontinental basin situated within the Rockall Plateau on the westernmost part of the U.K.-designated area. This basin is bounded on the west and east by Hatton Bank and Rockall Bank, respectively. Water depths across the basin are mostly between 1000 and 1300 m (Figure 17). Seismic data coverage across the Rockall Plateau remains very sparse and much of the data was acquired during the 1970s. Accordingly, the geologic history of the basin remains speculative. Recent paleogeographic reconstructions (e.g., Doré et al., 1999) suggest that the Hatton Basin was possibly initiated during the Cretaceous. However, studies of wide-angle seismic data suggest that a Carboniferous-Jurassic layer as much as 3 km thick underlies the basin nearby in the Irish sector (Jacob et al., 1995; Shannon et al., 1995). Two opposing synrift half grabens have been modeled there, separated by an intrabasinal high. Seismic profiles show a mounded slope-apron interval bordering the volcanic escarpment on the western margin of the basin (Figure 22). By analogy with the Faroe-Shetland Escarpment to the northeast, this interval can be interpreted as late Paleocene to early Eocene in age.

On those parts of the Hatton Bank where Paleogene volcanic rocks are thin or absent, a series of steeply dipping, tilted fault blocks has been imaged on seismic profiles (Keser Neish, 1993). These structures may continue westward to underlie the Hatton continental margin. Similar steeply dipping reflector packages have been identified in the Irish sector from the southeastern part of the Hatton Basin (Boldreel and Andersen, 1994). The age of these pre-Tertiary dipping strata remains open to speculation. Prograding Eocene strata penetrated by a British Geological Survey shallow borehole on the eastern flank of Rockall Bank contain Carboniferous and Jurassic/Cretaceous palynomorphs that may have been derived by reworking of upper Paleozoic and Mesozoic rocks within the Rockall Plateau area (M. S. Stoker and K. Hitchen, personal communication, 1995).

The Hatton continental margin is the westernmost sedimentary basin within U.K.-designated waters, lying adjacent to the continent-ocean boundary (Figure 3). Shannon et al. (1995) modeled the basin as comprising 0.5–1.3 km of heavily intruded Cretaceous sediments. O'Reilly and Readman (1999) suggested that a Triassic to Jurassic age is also plausible for these sediments. The basin is unconformably overlain by a thin postrift Eocene to Holocene succession.

Prerift and synrift plays remain highly speculative across the Hatton Basin to the Hatton continental margin area. If underlain by Carboniferous-Jurassic strata, the Hatton Basin has arguably the best potential for source rocks, and the presence of tilted fault blocks along its margins may provide opportunities for structural closure (Figure 22). However, Cretaceous postrift mudstones are either thin or absent, reducing the potential for effective top seal in this play. Tertiary slope-apron and basin-floor fans are likely to be present in the Hatton Basin, but their derivation from a hinterland dominated by volcanic rocks may reduce their reservoir potential.

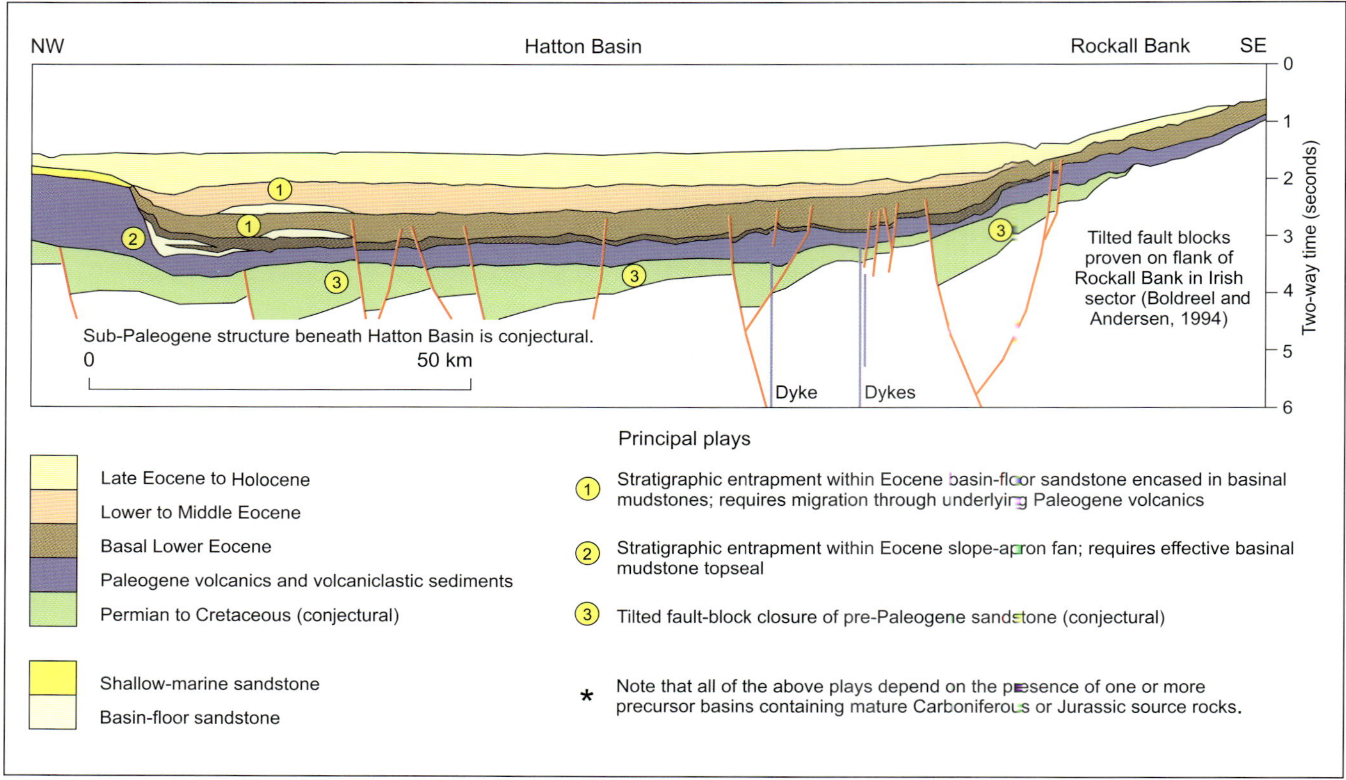

Figure 22. Schematic play diagram for the Hatton Basin. See Figure 17 for location.

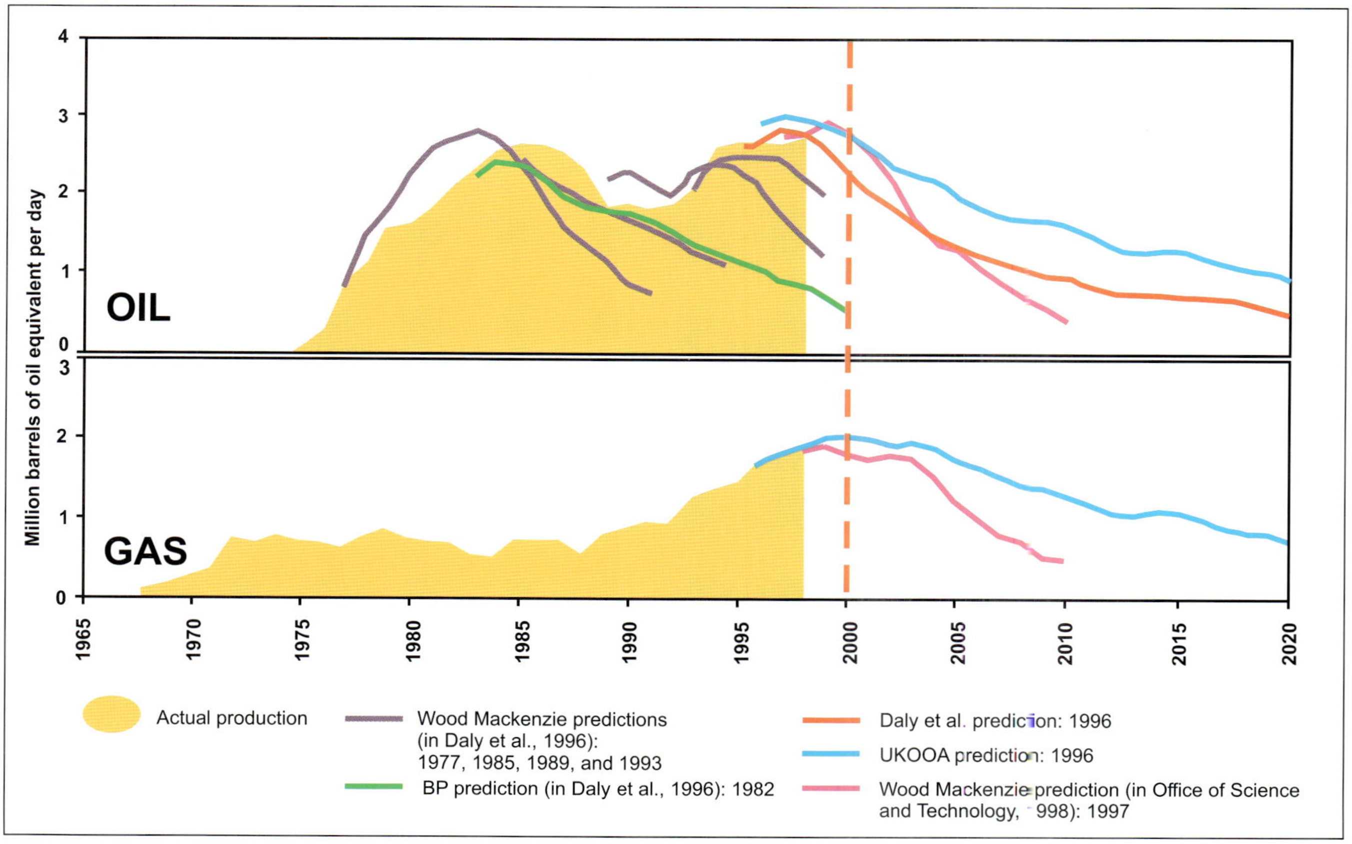

Million barrels of oil equivalent per day

OIL

GAS

1965 1970 1975 1980 1985 1990 1995 2000 2005 2010 2015 2020

Actual production

Wood Mackenzie predictions (in Daly et al., 1996): 1977, 1985, 1989, and 1993

BP prediction (in Daly et al., 1996): 1982

Daly et al. prediction: 1996

UKOOA prediction: 1996

Wood Mackenzie prediction (in Office of Science and Technology, 1998): 1997

Figure 23. Past predictions and future potential of U.K. oil and gas production (all predictions are optimum case).

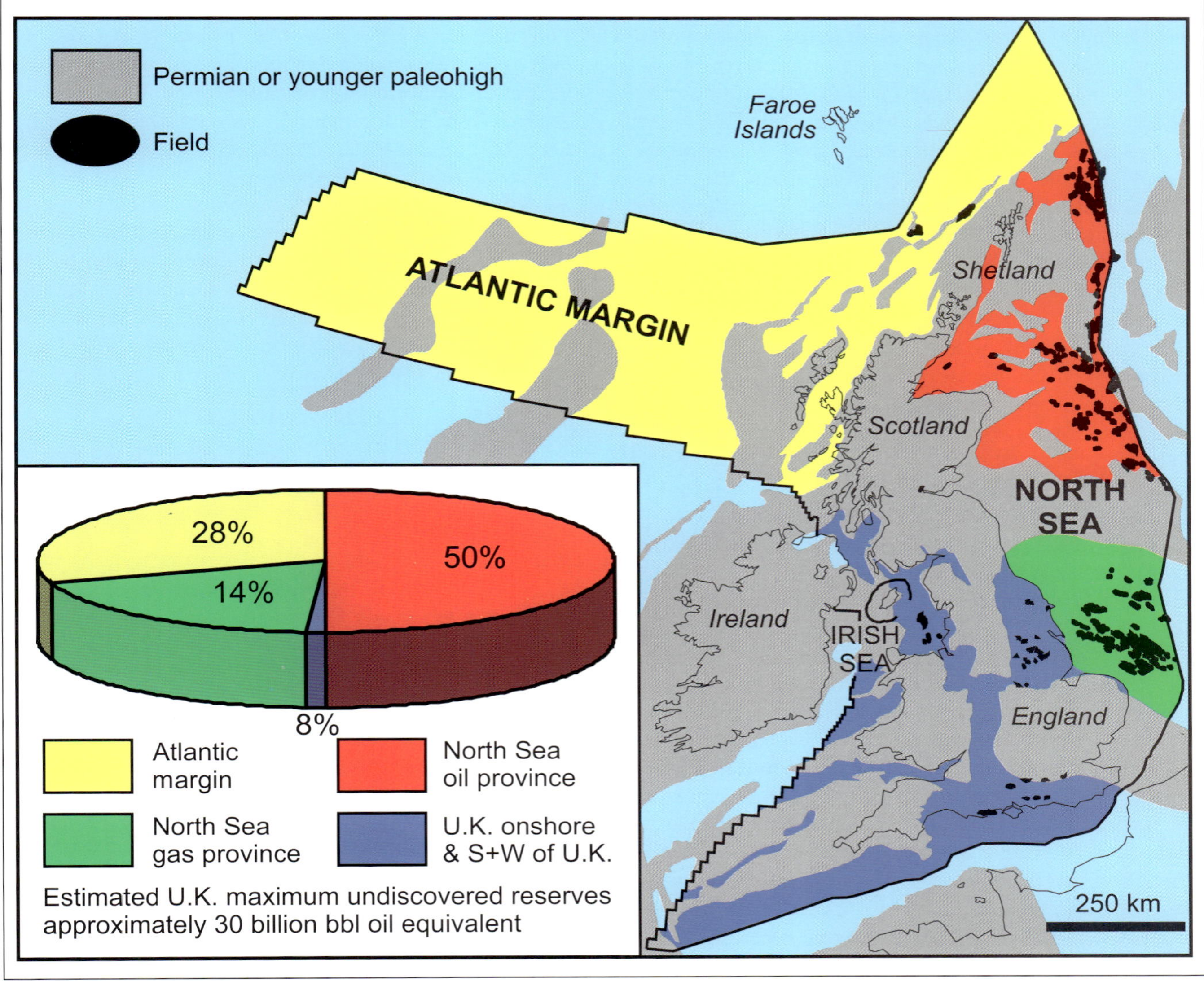

Figure 24. Proportions of maximum reserves that are estimated as remaining to be discovered in each of the U.K.'s principal hydrocarbon provinces.

SUMMARY OF OPPORTUNITIES AND CHALLENGES

The U.K. and its designated waters contain a plethora of mature, lightly explored, and frontier sedimentary basins. Historical predictions of oil and gas production have consistently underestimated the long-term resource base in the U.K. (Figure 23). This is partly because recovery factors in producing fields are continually improving, and partly it reflects an increase in the geographic range of production. However, the most significant additions to the resource base have been from the mature producing basins. This paper has demonstrated that the U.K.'s resource base will continue to increase by locating further reserves in these basins and potentially by extending production into the frontier basins of the Atlantic margin.

It is now believed that sufficient proven and undiscovered reserves remain in the mature basins of the North Sea and onshore in England to maintain the U.K.'s position as a leading hydrocarbon-producing nation into the early decades of the twenty-first century. Specifically, the Oil and Gas Industry Task Force[7] predicted that by minimizing discovery costs, U.K. production can be maintained at more than 3 million bbl of oil equivalent per day until the year 2010, partly by making an additional 5.6 billion bbl of discovered oil economic to recover. Initiatives that are being introduced to achieve this and to reduce the burdens on industry include a commitment to have one onshore and one offshore licensing round per year, with all open acreage available every two years, and a relaxation on work commitments on current licenses.

The task force predicted that more than two-thirds of the U.K.'s undiscovered oil and gas reserves lie within 50

[7] For more information, see http://www.dti.gov.uk/ogitf.

km of existing infrastructure, which provides a window of opportunity for development of marginally economic reserves. Undiscovered reserves in the North Sea oil province may be as much as 15 BBOE (Figure 24). The majority of these reserves is likely to be found in synrift Upper Jurassic to Lower Cretaceous shallow-marine and basin-floor sandstone plays and in postrift Paleogene basin-floor sandstone reservoirs. However, a significant component of discoveries will continue to be made from prerift Paleozoic, Triassic to Lower Jurassic, and Middle Jurassic plays. In the North Sea gas province, containing about 20% of undiscovered reserves, the greatest exploration challenge will be to increase the proportion of production from Carboniferous reservoirs. This will be achieved by extending the geographic range of existing discoveries, by targeting subcrops of proven reservoirs beneath the regional base Permian unconformity, and by testing hanging-wall wedges of these strata adjacent to the major intra-Carboniferous faults.

Nevertheless, it is the frontier basins of the Atlantic margin that may have the greatest potential to replace existing U.K. hydrocarbon production by the discovery of major new oil and gas fields. It is currently estimated that about 30% of the U.K.'s undiscovered reserves lie within the Atlantic margin (Figure 24), although this estimate does not include the gas potential of the Rockall Trough and Hatton Basin/continental margin. The last decade of the twentieth century saw the Faroe-Shetland Basin established as an oil-producing province, and opportunities remain for further discoveries of oil, both in the established Paleogene basin-floor sandstone play and in synrift and prerift plays along the margins of the basin. The 1990s also saw the discovery of potentially economic gas reserves in the Faroe-Shetland Basin. A realistic target for the first decade of the twenty-first century will be to discover sufficient additional reserves of gas there to encourage construction of a new gas-gathering infrastructure for the region.

The greatest remaining challenge in the Atlantic margin, however, will be to establish a viable petroleum system for the Rockall Trough and perhaps also for the Hatton Basin and Hatton continental margin. Initial exploration of these basins has led to the discovery of potentially attractive tilted fault-block plays. Viable source rocks remain to be proved, although their presence in contiguous basins to the north and south gives grounds for optimism. If economic reserves were to be discovered there, they would have the potential to maintain the U.K.'s position as a net exporter of hydrocarbons beyond the first decade of the twenty-first century.

ACKNOWLEDGMENTS

This paper was prepared as part of a contract by which the British Geological Survey supplies geologic advice to the Oil and Gas Directorate of the U.K. Department of Trade and Industry. The authors thank Tony Doré (Statoil Exploration U.K. Limited), Peter Rattey (BP Amoco Exploration), Professor Robert White (University of Cambridge), and numerous colleagues in the exploration and licensing branch of the Oil and Gas Directorate and in the British Geological Survey for reviewing drafts of the manuscript. We are grateful to Asset Geoscience Limited for supplying the data used to construct the creaming curves depicted in Figure 7. J. R. V. Brooks acknowledges permission to publish of the director, Oil and Gas Directorate, Department of Trade and Industry. S. J. Stoker and T. D. J. Cameron acknowledge permission to publish of the director, British Geological Survey (NERC).

REFERENCES CITED

Bain, J. S., 1993, Historical overview of exploration of Tertiary plays in the U.K. North Sea, *in* J. R. Parker, ed., Petroleum geology of north west Europe: Proceedings of the 4th Conference: Geological Society, London, p. 5–13.

Besly, B. M., 1998, Carboniferous, *in* K. W. Glennie, ed., Petroleum geology of the North Sea, basic concepts and recent advances, 4th edition: London, Blackwell Science Limited, p. 104–136.

Boldreel, L. O., and M. S. Andersen, 1994, Tertiary development of the Faeroe-Rockall Plateau based on reflection seismic data: Bulletin of the Geological Society of Denmark, v. 41, p. 162–180.

Brennand, T. P., B. Van Hoorn, K. H. James, and K. W. Glennie, 1998, Historical review of North Sea exploration, *in* K. W. Glennie, ed., Petroleum geology of the North Sea, basic concepts and recent advances: London, Blackwell Science Limited, p. 1–41.

Brooks, J. R .V., 1986, Future potential for hydrocarbon exploration on the United Kingdom Continental Shelf, *in* M. T. Halbouty, ed., Future petroleum provinces of the world: AAPG Memoir 40, p. 677–693.

Brown, J. L., and A. Beach, 1999, Structural modelling in the Rockall Trough: Increasing the understanding of the development of the Atlantic Margin (abs.), *in* P. F Croker and O. O'Loughlin, eds., The petroleum potential of Ireland's offshore basins: Dublin, Ireland, Petroleum Affairs Division, p. 128.

Cameron, N., and T. Ziegler, 1997, Probing the lower limits of a fairway: Further pre-Permian potential in the southern North Sea, *in* K. Ziegler, P. Turner, and S. R. Daines, eds., Petroleum geology of the southern North Sea: Future potential: Geological Society Special Publication. 123, p. 123–141.

Cole, J. E., and J. Peachey, 1999, Evidence for pre-Cretaceous rifting in the Rockall Trough: An analysis using quantitative plate tectonic modelling, *in* A. J. Fleet and S. A .R. Boldy, eds., Petroleum geology of north west Europe: Proceedings of the 5th Conference: Geological Society, London, p. 359–370.

Collinson, J. D., C. M. Jones, G. A. Blackbourn, B. M. Besly, G. M. Archard, and A. H. McMahon, 1993, Carboniferous depositional systems of the southern North Sea, *in* J. R. Parker, ed., Petroleum geology of north west Europe: Proceedings of the 4th Conference Geological Society, London, p. 677–687.

Cordey, W. G., 1993, Jurassic exploration history: A look at the past and the future, *in* J .R. Parker, ed., Petroleum geology of north west Europe: Proceedings of the 4th Conference: Geological Society, London, p. 195–198.

Creedy, D. P., 1999, Coalbed methane—the R&D needs of the U.K.: U.K. Department of Trade and Industry Report COAL R163.

Daly, M. C., M. S. Bell, and P. J. Smith, 1996, The remaining resource of the U.K. North Sea and its future development, *in* K. Glennie and A. Hurst, eds., NW Europe's hydrocarbon industry: Geological Society, London, p. 187–193.

Doré, A. G., and L. N. Jensen, 1996, The impact of late Cenozoic uplift and erosion on hydrocarbon exploration: Offshore Norway and some other uplifted basins: Global and Planetary Change, v. 12, p. 415–436.

Doré, A. G., E. R. Lundin, Ø. Birkeland, P. E. Eliassen, and L. N. Jensen, 1997, The NE Atlantic Margin: Nature, origin and potential significance for hydrocarbon exploration: Petroleum Geoscience, v. 3, p. 117–131.

Doré, A. G., E. R. Lundin, L. N. Jensen, Ø. Birkeland, P. E. Eliassen, and C. Fichler, 1999, Principal tectonic events in the evolution of the northwest European Atlantic margin, *in* A. J. Fleet and S. A. R. Boldy, eds., Petroleum Geology of north west Europe: Proceedings of the 5th Conference: Geological Society, London, p. 41–61.

Department of Trade and Industry, 1999, Development of the oil and gas resources of the United Kingdom: The Stationery Office, 206 p.

Duncan, W. I., and N. W. K. Buxton, 1995, New evidence for evaporitic Middle Devonian lacustrine sediments with hydrocarbon source potential on the East Shetland Platform, North Sea: Journal of the Geological Society, London, v. 152, p. 251–258.

Ebdon, C. C., P. J. Granger, H. D. Johnson, and A. M. Evans, 1995, Early Tertiary evolution and sequence stratigraphy of the Faeroe-Shetland Basin: Implications for hydrocarbon prospectivity, *in* R. A. Scrutton et al., eds., The tectonics, sedimentation and palaeoceanography of the North Atlantic region: Geological Society Special Publication 90, p. 51–69.

England, R. W., and R. W. Hobbs, 1997, The structure of the Rockall Trough imaged by deep seismic reflection profiling: Journal of the Geological Society, London, v. 154, p. 497–502.

Evans, W. E., 1990, Development of the petroleum exploration scene in the U.K. onshore area: A struggle over the last 75 years, *in* M. Ala et al., eds., Seventy-five years of progress in oil field science and technology: Rotterdam, A. A Balkema, p. 49–54.

Floodpage, J., J. White, and P. Newman, 1999, The hydrocarbon prospectivity of the Irish Sea: Insights from recent exploration of the Central Irish Sea, Peel and Solway Basins (abs.), *in* P. F Croker and O. O'Loughlin, eds., The petroleum potential of Ireland's offshore basins: Dublin, Ireland, Petroleum Affairs Division, p. 28–31.

Fraser, A. J., and R. L. Gawthorpe, 1990, Tectono-stratigraphic development and hydrocarbon habitat of the Carboniferous in northern England, *in* R. F. P. Hardman and J. Brooks, eds., Tectonic events responsible for Britain's oil and gas reserves: Geological Society Special Publication 55, p. 49–86.

Gawthorpe, R. L., 1987, Burial dolomitization and porosity development in a mixed carbonate-clastic sequence: An example from the Bowland Basin, northern England: Sedimentology, v. 34, p. 533–558.

Gerling, P., M. C. Geluk, F. Kockel, A. Lokhorst, G. K. Lott, and R. A. Nicholson, 1999, NW European Gas Atlas—New implications for the Carboniferous gas plays in the western part of the Southern Permian Basin, *in* A. J. Fleet and S. A. R. Boldy, eds., Petroleum Geology of north west Europe: Proceedings of the 5th Conference: Geological Society, London, p. 799–808.

Glennie, K. W., 1998, Lower Permian—Rotliegend, *in* K. W. Glennie, ed., Petroleum geology of the North Sea, basic concepts and recent advances, 4th edition: London, Blackwell Science Limited, p. 137–173.

Grant, N., A. Bouma, and A. McIntyre, 1999, The Turonian play in the Faeroe-Shetland Basin, *in* A. J. Fleet and S. A. R. Boldy, eds., Petroleum Geology of north west Europe: Proceedings of the 5th Conference: Geological Society, London, p. 661–673.

Hitchen, K., and M. S. Stoker, 1993, Mesozoic rocks from the Hebrides Shelf and implications for hydrocarbon prospectivity in the northern Rockall Trough: Marine and Petroleum Geology, v. 10, p. 246–254.

Illiffe, J. E., A. G. Robertson, G. H. F. Ward, C. Wynn, S. D. M. Pead, and N. Cameron, 1999, The importance of fluid pressures and migration to the hydrocarbon prospectivity of the Faeroe-Shetland White Zone, *in* A. J. Fleet and S. A. R. Boldy, eds., Petroleum Geology of north west Europe: Proceedings of the 5th Conference: Geological Society, London, p. 601–611.

Jacob, A. W. B., P. M. Shannon, J. Makris, F. Hauser, U. Vogt, and B. M. O'Reilly, 1995, An overview of the results of the RAPIDS seismic project, North Atlantic, *in* P. F. Croker and P. M. Shannon, eds., The petroleum geology of Ireland's offshore basins: Geological Society Special Publication 93, p. 429–431.

Johnson, H. D., and M. J. Fisher, 1998, North Sea plays: Geological controls on hydrocarbon distribution, *in* K. W. Glennie, ed., Petroleum geology of the North Sea, basic concepts and recent advances, 4th edition: London, Blackwell Science Limited, p. 463–547.

Joppen, M., and R. S. White, 1990, The structure and subsidence of Rockall Trough from two-ship seismic experiments: Journal of Geophysical Research, v. 95, p. 19821–19837.

Keser Neish, J., 1993, Seismic structure of the Hatton-Rockall area: An integrated seismic/modelling study from composite datasets, *in* J. R. Parker, ed., Petroleum geology of north west Europe: Proceedings of the 4th Conference: Geological Society, London, p. 1047–1056.

Knott, S. D., M. T. Burchell, E. J. Jolley, and A. J. Fraser, 1993, Mesozoic to Cenozoic plate reconstructions of the North Atlantic and hydrocarbon plays of the Atlantic margins, *in* J. R. Parker, ed., Petroleum Geology of north west Europe: Proceedings of the 4th Conference: Geological Society, London, p. 953–974.

Leeder, M. R., and M. Hardman, 1990, Carboniferous geology of the southern North Sea Basin and controls on hydrocarbon prospectivity, *in* R. F. P. Hardman and J. Brooks, eds., Tectonic events responsible for Britain's oil and gas reserves: Geological Society Special Publication 55, p. 87–106.

Maddox, S. J., R. Blow, and M. Hardman, 1995, Hydrocarbon prospectivity of the Central Irish Sea Basin with reference to Block 42/12, offshore Ireland, *in* P. F. Croker and P.M. Shannon, eds., The petroleum geology of Ireland's offshore basins: Geological Society Special Publication 93, p. 59–97.

Mikkelsen, P. W., and J. F. Floodpage, 1997, The hydrocarbon potential of the Cheshire Basin, *in* N. S. Meadows et al., eds., The petroleum geology of the Irish Sea and adjacent areas: Geological Society Special Publication 124, p. 161–183.

Mitchell, S. M., G. W. J. Beamish, M. V. Wood, S. J. Malacek, J. A. Armentrout, J. E. Damuth, and H. C. Olson, 1993, Paleogene sequence stratigraphic framework of the Faeroe Basin, *in* J. R. Parker, ed., Petroleum geology of north west Europe: Proceedings of the 4th Conference: Geological Society, London, p. 1011–1023.

Mitchum, R. M., 1977, Seismic stratigraphy and global changes in sea level, Part 1: Glossary of terms used in sequence stratigraphy, *in* C. E. Payton, ed., Seismic stratigraphy: Application to hydrocarbon exploration, AAPG Memoir 26, p. 205–212.

Musgrove, F. W., and B. Mitchener, 1996, Analysis of the pre-Tertiary rifting history of the Rockall Trough: Petroleum Geoscience, v. 2, p. 353–360.

Nadin, P. A., M. A. Houchen, and N. J. Kusznir, 1999, Evidence for pre-Cretaceous rifting in the Rockall Trough: Analysis using quantitative 2D structural/stratigraphic modelling, *in* A. J. Fleet and S. A. R. Boldy, eds., Petroleum geology of north west Europe: Proceedings of the 5th Conference: Geological Society, London, p. 371–378.

Nolan, C., A. McCarthy, and I. Wilson, 1999, The prospectivity of the northeast Irish Rockall Basin—A play based review (abs.), *in* P. F Croker and O. O'Loughlin, eds., The petroleum potential of Ireland's offshore basins: Dublin, Ireland, Petroleum Affairs Division, p. 112–114.

Oakman, C. D, and M. A. Partington, 1998, Cretaceous, *in* K. W. Glennie, ed., Petroleum geology of the North Sea, basic concepts and recent advances, 4th edition: London, Blackwell Science Limited, p. 294–349.

Oudmayer, B. C., and J. de Jager, 1993, Fault reactivation and oblique-slip in the southern North Sea, *in* J. R. Parker, ed., Petroleum geology of north west Europe: Proceedings of the 4th Conference: Geological Society, London, p. 1281–1290.

O'Reilly, B. M., and P. W. Readman, 1999, Thermal evolution of the lithosphere: Impact on Cretaceous/early Tertiary sedimentation patterns in the NE Atlantic and hydrocarbon prospectivity (abs.), *in* P. F Croker and O. O'Loughlin, eds., The petroleum potential of Ireland's offshore basins: Dublin, Ireland, Petroleum Affairs Division, p. 168–170.

Office of Science and Technology, 1998, Foresight: Report of the Working Group on Offshore Energies, Department of Trade and Industry.

Pegrum, R. M., and A. M. Spencer, 1990, Hydrocarbon plays in the northern North Sea, *in* J. Brooks, ed., Classic petroleum provinces: Geological Society Special Publication 50, p. 441–470.

Quirk, D. G., 1997, Sequence stratigraphy of the Westphalian in the northern part of the southern North Sea, *in* K. Ziegler, P. Turner, and S. R. Daines, eds., Petroleum geology of the southern North Sea: Future potential: Geological Society Special Publication 123, p. 153–168.

Quirk, D. G., and J. F. Aitken, 1997, The structure of the Westphalian in the northern part of the southern North Sea, *in* K. Ziegler, P. Turner, and S. R. Daines, eds., Petroleum geology of the southern North Sea: Future potential: Geological Society Special Publication 123, p. 143–152.

Richardson, K. R., R. S. White, R. W. England, and J. Fruehn, 1999, Crustal structure east of the Faroe Islands: Mapping sub-basalt sediments using wide-angle seismic data: Petroleum Geoscience, v. 5, p. 161–172.

Ritchie, J. D., R. W. Gatliff, and I. B. Riding, 1996, Stratigraphic nomenclature of the U.K. north west margin, 1. Pre-Tertiary lithostratigraphy: British Geological Survey, 193 p.

Scotchman, I. C., and T. M. Broks, 1999, Source potential of the Lower and Middle Jurassic (abs.), *in* P. F Croker and O. O'Loughlin, eds., The petroleum potential of Ireland's offshore basins: Dublin, Ireland Petroleum Affairs Division, p. 16–18.

Scotchman, I. C., and J. R. W. Thomas, 1995, Maturity and hydrocarbon generation in the Slyne Trough, north-west Ireland, *in* P. F. Croker and P. M Shannon, eds., The petroleum geology of Ireland's offshore basins: Geological Society Special Publication 93, p. 385–411.

Shannon, P. M., A. W. B. Jacob, J. Makris, B. O'Reilly, F. Hauser, and U. Vogt, 1994, Basin development and petroleum prospectivity of the Rockall and Hatton region, *in* P. F. Croker and P. M. Shannon, eds., The petroleum geology of Ireland's offshore basins: Geological Society Special Publication 93, p. 435–457.

Shannon, P. M., A. W. B. Jacob, J Makris, B. O'Reilly, F. Hauser, and U. Vogt, 1995, Basin evolution in the Rockall region, North Atlantic: First Break, v. 12, p. 515–522.

Smith, N. J. P., 1993, The case for exploration of deep plays in the Variscan fold belt and its foreland, *in* J. R. Parker, ed., Petroleum geology of north west Europe: Proceedings of the 4th Conference: London, Geological Society, p. 667–675.

Smith, S. G., A. Walsh, and J. M. Perry, 1999, Western Irish Rockall margin: Evaluating an exploration frontier (abs.), *in* P. F Croker and O. O'Loughlin, eds., The petroleum potential of Ireland's offshore basins: Dublin, Ireland, Petroleum Affairs Division, p. 108–110.

Smythe, D. K., 1989, Rockall Trough—Cretaceous or late Palaeozoic?: Scottish Journal of Geology, v. 25, p. 5–43.

Spencer, A. M., and B. McTiernan, 1999, Petroleum systems offshore western Ireland in an Atlantic margin context (abs.), *in* P. F Croker and O. O'Loughlin, eds., The petroleum potential of Ireland's offshore basins: Dublin, Ireland, Petroleum Affairs Division, p. 116–119.

Tate, M. P., C. D. Dodd, and N. T. Grant, 1999, The northeast Rockall Basin and its significance in the evolution of the Rockall-Faeroes/East Greenland rift system, *in* A. J. Fleet and S. A. R. Boldy, eds., Petroleum geology of north west Europe: Proceedings of the 5th Conference: Geological Society, London, p. 391–406.

United Kingdom Offshore Operators Association, 1996, Towards 2020: A study to assess the potential oil and gas production from the U.K. offshore: United Kingdom Offshore Operators Association Ltd., 25 p.

Walmsley, P. J., 1987, The British Geological Survey contribution to the exploration of the continental shelf: Journal of the Geological Society, London, v. 144, p. 207–212.

White, R. S., 1989, Initiation of the Iceland plume and opening of the North Atlantic, *in* A. J. Tankard and H .R. Balkwill, eds., Extensional tectonics and stratigraphy of the North Atlantic margins, p. 149–154.

White, R. S., J. Fruehn, K. R. Richardson, E. Cullen, W. Kirk, J. R. Smallwood, and C. Latkiewicz, 1999, Faeroes Large Aperture Research Experiment (FLARE): Imaging through basalt, *in* A. J. Fleet and S. A. R. Boldy, eds., Petroleum geology of north west Europe: Proceedings of the 5th Conference: Geological Society, London, p. 1243–1252.

Ziegler, P. A., 1988, The evolution of the Arctic–North Atlantic and the Western Tethys: AAPG Memoir 43, 198 p.

Meissner, F. F., and M. R. Thomasson, 2001, Exploration opportunities in the Greater Rocky Mountain Region, U.S.A., *in* M. W. Downey, J. C. Threet, and W. A. Morgan, eds., Petroleum provinces of the twenty-first century: AAPG Memoir 74, p. 201–239.

Chapter 11

Exploration Opportunities in the Greater Rocky Mountain Region, U.S.A.

Fred F. Meissner
Colorado School of Mines, Golden, Colorado, U.S.A.

M. Ray Thomasson
Thomasson Partner Associates Inc., Denver, Colorado, U.S.A.

ABSTRACT

The Greater Rocky Mountain Region covers approximately one-fifth of the contiguous 48 states. The United States Geologic Survey recognizes 22 "provinces" or "areas" in the region, of which 18 produce oil or gas. It estimates that 10.4 billion barrels of oil and condensate and 259.8 trillion cubic feet of gas remain to be discovered in the region.

Known source and reservoir rocks extend throughout a thick sedimentary section ranging from Precambrian to Tertiary in age. Virtually every conceivable type of tectonic and sedimentary environment known is present in some areas of the region. Additional oil production from established plays is not expected to be large. Greatest potential exists in unconventional plays and in sparsely drilled deeper sections of individual basins.

Gas discoveries will be more important than oil discoveries. Most of the gas potential is related to sources in coal-bearing Cretaceous and Tertiary sediments. Much of this gas will be found in coal-bed reservoirs or in low-permeability sandstones. The deeper and less-explored parts of many basins will contain gas because of advanced thermal maturity of the sediments.

Much of the potential production will fall in the middle and lower ranges of "Masters' resource triangle," which, in the case of the Rocky Mountain region, has a broader base than many other areas of the world. Exploration and development will be greatly influenced by technical, economic, and political factors. Examples of recent significant discoveries that may serve as analogs for the future include Jonah, Cave Gulch, Cedar Hills, Drunkard's Wash, and the Powder River Basin coal-bed methane fields. Many of these "discoveries" are unconventional accumulations and have resulted from the application of new technology to areas of previously abandoned, noncommercial, or subcommercial wells.

INTRODUCTION

The Greater Rocky Mountain Region (GRMR) is located in the central-western part of the contiguous 48 states (Figure 1). It extends north-south from the Canadian to the Mexican borders (1000 mi, or 1600 km) and east-west from the northern Great Plains to the central Great Basin (800 mi, or 1290 km). It contains about one-fifth of the total land area of the contiguous 48 states. It incorporates all or parts of the states of Idaho, Montana, North Dakota, South Dakota, Wyoming, Utah, Nevada, Nebraska, Colorado, New Mexico, and Arizona.

Exploration potential for the region was last published in a "province-type" assessment by the American Association of Petroleum Geologists in AAPG Memoir 15 (Cram, 1971). This memoir contained detailed papers on 15 geologic provinces and three regional summary papers contributed by 18 authors. Most of the basic geology considered in the papers is still pertinent. However, since 1971, thousands of wells have been completed, hundreds of fields have been discovered that cumulatively contain several billion barrels of oil and trillions of cubic feet of gas, and numerous papers have been written concerning various parts of the region. Better understanding has also been obtained on such subjects as the occurrence of source rocks, the features which determine a viable reservoir, the distribution and functioning of petroleum systems, and the subtleties of structure. Great progress has also been made in geophysics, wireline logging, drilling, and completion techniques. These have a significant impact on establishing a new and larger potential resource base.

This paper updates current knowledge of the exploration and development potential of this vast and geologically heterogeneous region. It must necessarily be an overview biased by the authors' knowledge, experience, and interests. Every geoscientist working in the region has ideas on what is important in characterizing an area or a prospect, where the next prospect or significant play is going to be made, and its potential. Collectively, this is a powerful force for new exploration, and it is impossible to predict what discoveries will be made in the future in technology and actual fields. The best we can do is to assess what we now know and project the use of this knowledge into a future trend.

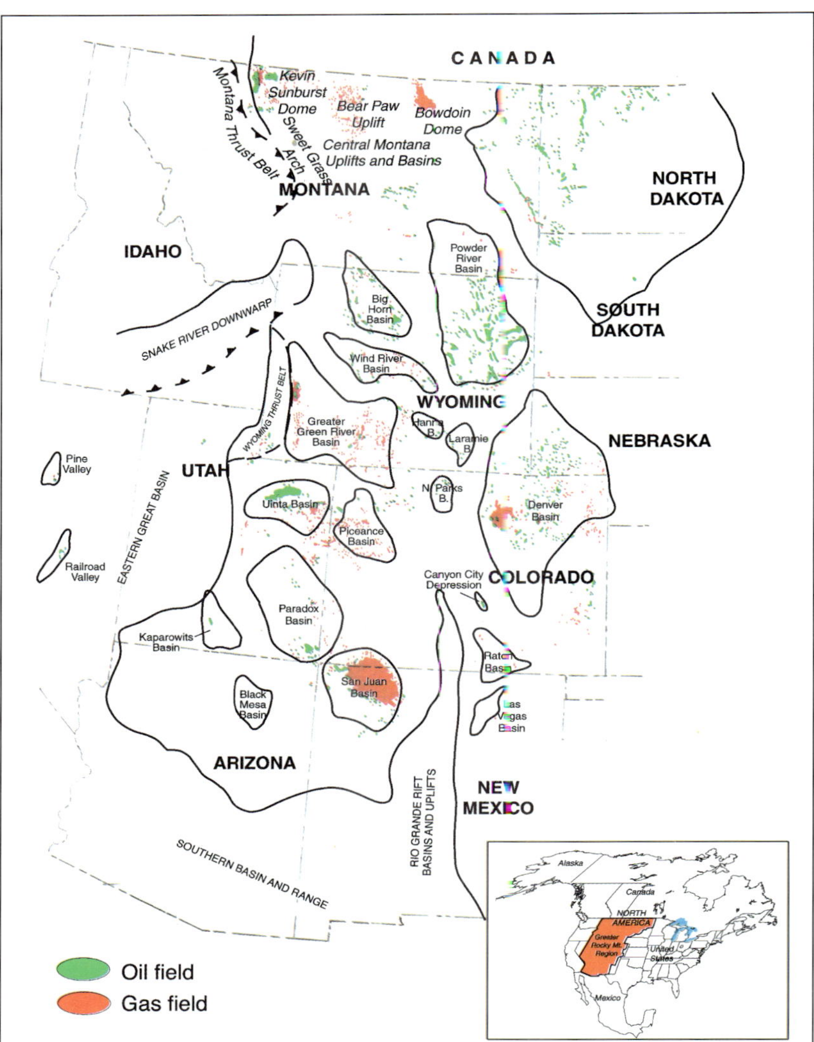

Figure 1. Maps showing location of the Greater Rocky Mountain Region (GRMR) and distribution of oil and gas fields.

HISTORY AND RESOURCE ESTIMATION

The region has a long history of oil and gas production, commencing with drilling in 1878 and establishment of production at the Florence oil field in 1881. This field lies within the Cañon City embayment, a small subbasin near the southwest corner of the Denver Basin (Kupfer, 1999). Although not recognized at the time, Florence field, which produces from a fractured shale reservoir associated with a mature source rock, is in a synclinal position and was the first of the "deep-basin" or "basin-center" oil accumulations characteristic of many subsequently discovered and prospective accumulations in other Rocky Mountain basins.

Since initial production in 1881, hundreds of significant oil and gas fields have been discovered in a wide variety of structural and stratigraphic settings (Figure 1). These include seven fields with known resources of greater than 500 million barrels (bbl) of oil or oil equivalent (Carmalt and St. John, 1986). The U.S. Geological Survey (1997) has divided the region into 22 provinces (see Figure 2 and Appendix, Figure A-1 and Table A-1). Eighteen of these provinces have produced oil and gas and are credited in the U.S. Geological Survey 1997 assessment with estimated median-value *known resources* of 17 billion bbl of petroleum liquids consisting of oil and natural-gas liquids and 80 trillion cubic feet (tcf) of gas (Appendix, Table A-2). These values represented 8.5% of U.S. oil and 8.8% of gas.

Since estimates of *future resources* in the GRMR were presented in AAPG Memoir 25 (Ambrose, 1971; Curry, 1971), several entities have made estimates of future oil and gas discoveries and developments. The most recent of these include studies made by the U.S. Geological Survey National Oil and Gas Resource Assessment Team (1995), the Potential Gas Committee (1999) of the Potential Gas Agency, the Gas Research Institute (1999), and others investigating the total in-place volume of coalbed methane (Tyler et al., 1997). A generalized summary of conclusions reached in these studies is presented in Table 1. The conclusions are supported by more detailed summaries presented for individual geologic areas in the Appendix. Table 1 also includes results of an assessment made by an AAPG "expert panel" in 1992 (Thomasson, 1992).

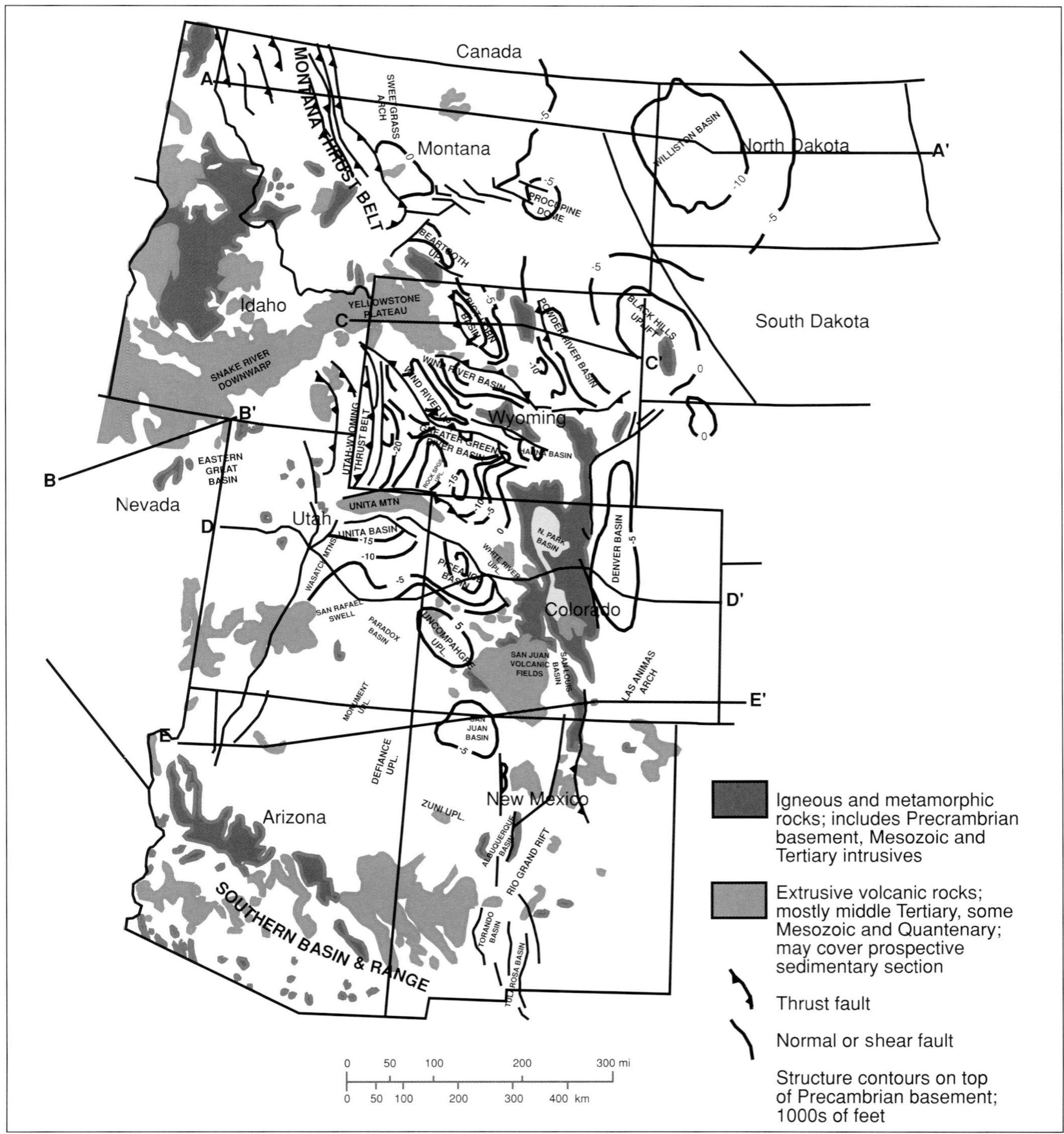

Figure 2. Geologic map of the GRMR showing areas of igneous or metamorphic basement and surface volcanic cover, structure contours at the base of the sedimentary section, and location of cross sections depicted in Figure 3 (after Cohee [1982] and Muehlberger [1992] in part).

Table 1. Summary of potential resource estimates.

	Oil (billion bbl) AAPG[1]	Oil + NGL (billion bbl) USGS[2]	Gas (tcf)			
			USGS[2]	PGA[3]	GRI	TBEG
"Conventional"	15.40	7.61	0.53			
"Traditional"				150.02		
"Continuous" (sandstone, fractured shale, chalk)		2.81	203.85			
Coal-bed methane			25.44	41.72	38.92[2] (477[3])	534[4]
TOTALS:	15.40	10.42	259.82	191.74		

[1] Median value, economically recoverable undiscovered conventional resource (excluding the Pedragosa Basin, southwest New Mexico).
[2] Median value of technically recoverable resources.
[3] Mean value by statistical aggregation of probable, possible, and speculative economically and technically recoverable resources.
[4] In-place resource. Includes deeply buried coals that may not be producible at the present state of technology because of absence of open cleats that provide permeability.

SOURCES: AAPG "Expert Panel," as reported in Thomasson, 1992; U.S. Geological Survey (USGS) National Resource Assessment Team, 1995; Potential Gas Committee (PGA) 1999; Gas Research Institute (GRI), 1999; Texas Bureau of Economic Geology (TBEG), as reported by Tyler et al., 1997.

The categories and estimates made by the various entities are not all compatible. They use different categories and different time frames. However, the bottom-line estimates should serve as a guide for an appreciation of target size. All the estimates indicate that a substantial amount of oil and gas remains to be found in the region. The importance of gas is obvious. According to the most recent Potential Gas Agency estimates, 24.4% of future additions to gas reserves in the lower 48 states will come from the GRMR. The 1992 estimate made by the AAPG "expert panel" predicts a larger volume of oil than the 1995 U.S. Geological Survey estimate. This may reflect more intimate practical knowledge of the region and of the impact of new technology, as well as a higher degree of optimism on the part of qualified scientists in the oil and gas industry.

GEOLOGIC FRAMEWORK

A generalized map (Figure 2) and five simplified east-west cross sections (Figure 3) are representative of major geologic features of the GRMR.

With a few notable exceptions, virtually every type of tectonic and sedimentary environment known to the science of geology is present in some province or area of the GRMR. Although some aspects of certain areas, basins, and uplifts are similar, many of them are unique. Because this paper is an overview of the entire region, only generalities will be presented and discussed.

Generalized Geologic History

Major episodes of structural and sedimentary activity in the GRMR are shown in the series of maps and schematic cross sections contained in Figure 4.

In the early Paleozoic, sedimentation was dominated by a large north-south-trending miogeosynclinal basin that formed on a passive continental margin located in the western part of the GRMR (Figure 4a). More than 30,000 ft (9200 m) of sediment was deposited in the geosyncline. Cambrian rocks are largely sandstones. Ordovician through Devonian rocks are largely carbonates. Throughout the early Paleozoic, continental-shelf edge and slope topography characterized the western margin of the geosyncline. In the Devonian, a well-defined carbonate shelf edge was present in this area. The western side of the GRMR considered in this paper is defined by this shelf edge and adjacent basinal slope.

In the Mississippian, the western side of the GRMR was affected by the Antler orogeny (Figure 4b). The slope and deep-water oceanic facies (the "western facies") of the lower Paleozoic sedimentary section were thrust eastward over upper Paleozoic shelf carbonates by the Roberts Mountain ("Roberts") thrust, and a large north-south-trending foredeep basin was created. Many of the shales and siliciclastic rocks of the overthrust western facies have high organic-carbon content and constitute oil-prone source rocks. Siliciclastic Mississippian rocks derived from erosion of the Antler geanticline were shed eastward into the foredeep basin, where they interfinger with deeper-water organic-rich shales of the Chainman Shale Formation. A prominent carbonate shelf edge was present on the eastern margin of the foredeep and contributed limestone turbidites moving westward into this north-south trough. Mostly shallow-water carbonates were deposited on the stable shelf farther east. The Williston Basin was an area of subsidence and received a thicker section of sedimentary fill, including an evaporite section, overlain by Upper Mississippian cyclic fluvial, estuarine, and shallow-marine sediments.

Periodic base-level changes controlled periods of nondeposition or erosion during the early and middle Paleozoic. Regional pre-Pennsylvanian erosion was responsible for the absence of rocks of these ages in much of the lower to middle Paleozoic Transcontinental arch.

The Ancestral Rockies orogeny produced a series of uplifts and basins in the central and southern GRMR during the Permian-Pennsylvanian (Figure 4c). Much of the lower Paleozoic section was removed over the uplifts, and clastic debris was deposited in adjacent areas. A cyclic section of marine to continental sediments was deposited in more distant areas. Black shales associated with these cycles represent a deepening or transgression and generally constitute rich oil-prone source rocks. A thick section of salt and cyclic interbedded black shales surrounded by "reefy" shelf carbonates was deposited in the Paradox Basin.

There was no notable regional tectonic activity during the Triassic or Jurassic. Local igneous intrusion occurred in the Great Basin portion of northeastern Nevada. Although some relatively thin marine sections are present, most of the Triassic-Jurassic section is represented by continental red beds and eolian sheet sands.

A major period of structural and sedimentary activity occurred in the Cretaceous (Figure 4d). Eastward thrusting associated with the Sevier thrust belt created a large foredeep geosyncline that extended from the Gulf of Mexico across the GRMR into Canada and Alaska. There was igneous intrusive activity in the future eastern Great Basin area west of the Sevier thrust belt. A thick section consisting predominantly of sandstones and shales was deposited in the Cretaceous Cordilleran geosyncline. This section contains source rocks and reservoirs that are associated with much of the historic oil and gas production in the GRMR. The Cretaceous Cordilleran geosyncline probably contains most of the hydrocarbons yet to be discovered in the GRMR.

The Paleocene-Eocene Laramide orogeny created most of the structural pattern characterizing the present central and northern GRMR (Figure 4e). Modern mountain ranges of the central Rocky Mountains were created and much of the sedimentary cover was eroded, exposing crystalline igneous and metamorphic basement. Clastic material derived from the uplifts was deposited in adjacent basins. Loading by lower Tertiary rocks deposited in the basins is largely responsible for triggering hydrocarbon generation in older source rocks. Oil-prone source rocks were deposited in lacustrine environments present in the Piceance, Uinta, Green River, and Wind River Basins. Similar sediments were deposited in lakes west of the Sevier uplift in the eastern Great Basin. Gas-prone coal measures were deposited in the Wind River and Powder River Basins. Scattered igneous intrusion occurred during this time period and was associated with hydrothermal activity, especially in central Colorado.

Volcanic extrusive rocks were deposited in certain areas of the GRMR during the Oligocene (Figure 2). This event may have influenced source-rock maturity in these areas; however, prospective sedimentary sections may be preserved in some rather large areas beneath volcanic cover.

During the late Tertiary, regional tectonic compression that controlled most of the structural development of the GRMR during the Cretaceous and early Tertiary was replaced by a period of extensional rifting that affected the western and southern parts of the area (Figure 4f). Listric normal faulting created a series of generally north-south-trending uplifts and basins that characterize the present Basin and Range physiographic province in Nevada, southern Arizona, and New Mexico. The Río Grande Rift valley and adjacent basins of central New Mexico are related to this event. As much as 10,000 feet of material eroded from the mountain ranges was deposited in these rift basins. Extrusive igneous rocks were deposited in the Snake River downwarp of southern Idaho (Figure 2) and in a few other scattered places during the latest Tertiary and into the Quaternary. Prospective sections of sedimentary strata are undoubtedly present beneath this volcanic cover.

Structure

The GRMR has a wide range of structural provinces. Basin types include (1) passive continental-margin miogeosynclines (lower Paleozoic of the eastern Great Basin), (2) subsiding cratonic depressions (Williston Basin), (3) foreland basins (Antler and Cretaceous Cordilleran geosynclines), (4) intermontane and perimontane basins (Ancestral Rockies and Laramide Basins), and (5) extensional rift grabens and half grabens (individual basins in the Basin and Range, Snake River downwarp, and Río Grande Rift provinces).

Structural styles represented in the various structural settings include (1) compressive "thin-skin" thrusts and folds (Powers, 1982); (2) compressive, high-angle, thick-skinned reverse faults associated with basin margins and force-fold features (Lowell, 1983; Stone, 1993); (3) extensional, listric, and high-angle normal faults creating horst-block uplifts and grabens (Wernicke, 1981; Gans and Miller, 1983; Effimoff and Pinezich, 1986; Camilleri, 1992); and (4) lateral shear zones (Stone, 1969) that have created subsidiary folds and fractured reservoirs and have influenced reservoir compartmentalization (Sonnenberg and Weimer, 1993; Warner, 1997). Other types of structure include those created by salt movement (Baars and Stevenson, 1981; Witkind, 1982) and solution (Parker, 1967; Swenson, 1967; Garner, 1997) and by meteor impacts (Brenan et al., 1975; Bridges, 1987; Koeberl and Anderson, 1996; Forsman et al., 1996; Kriens et al., 1999; Stone, 1999).

Major periods of structural deformation occurred in the Mississippian (Antler orogeny), Permian-Pennsylvanian (Ancestral Rockies), Cretaceous (Sevier orogeny), latest Cretaceous–early Tertiary (Laramide orogeny), and Miocene-Pliocene (Basin and Range/Río Grande Rift).

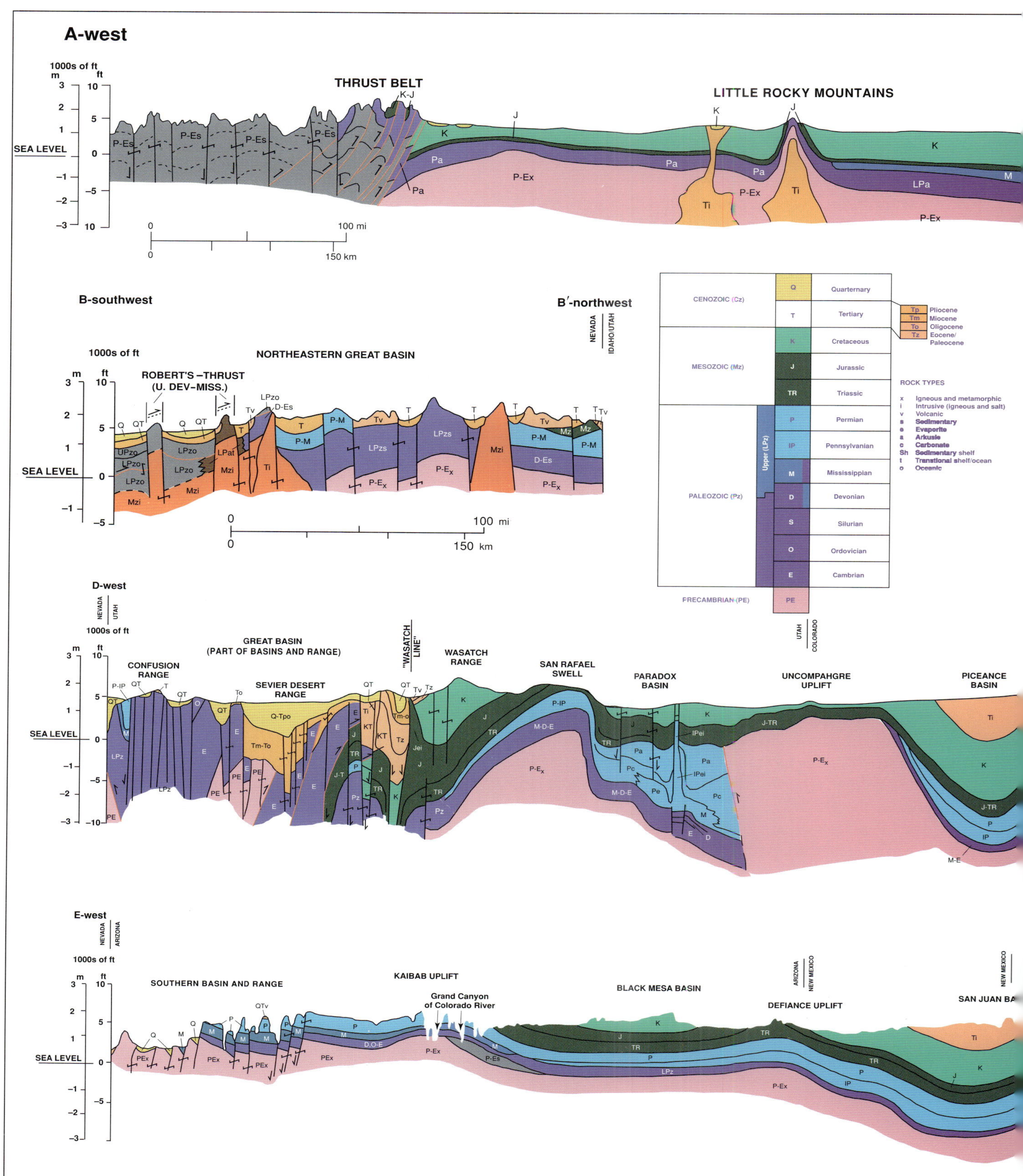

Figure 3. Cross sections through the GRMR (simplified from sections in the Northern Great Plains Region, Northern Rocky Mountain Region, Southern Rocky Mountain Region, and Southwest Region maps in the AAPG Geological Highway Map series).

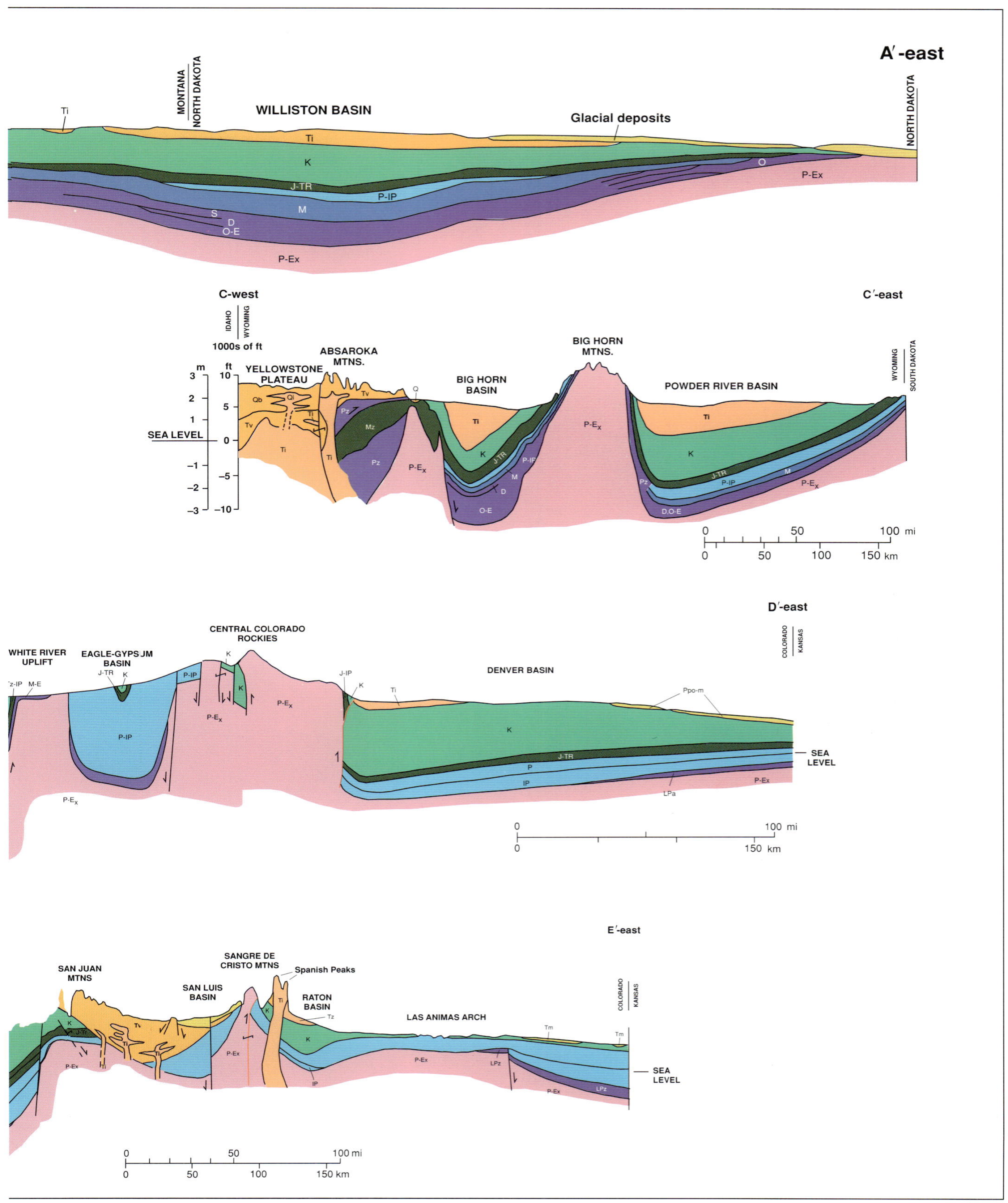
A′-east
MONTANA
NORTH DAKOTA
WILLISTON BASIN
Glacial deposits
NORTH DAKOTA
Ti
K
J-TR
P-IP
M
S
D
O-E
O
P-Ex
C-west
C′-east
IDAHO
WYOMING
1000s of ft
m
ft
SEA LEVEL
YELLOWSTONE PLATEAU
ABSAROKA MTNS.
BIG HORN BASIN
BIG HORN MTNS.
POWDER RIVER BASIN
WYOMING
SOUTH DAKOTA
Qb
Tv
Mz
Pz
D.O-E
0
50
100 mi
0
50
100
150 km
D′-east
COLORADO
KANSAS
WHITE RIVER UPLIFT
EAGLE-GYPSUM BASIN
CENTRAL COLORADO ROCKIES
DENVER BASIN
J-IP
Ppo-m
P
IP
LPa
SEA LEVEL
100 mi
150 km
E′-east
SAN JUAN MTNS
SAN LUIS BASIN
SANGRE DE CRISTO MTNS
Spanish Peaks
RATON BASIN
LAS ANIMAS ARCH
COLORADO
KANSAS
Tz
Tm
LPz
SEA LEVEL
50
100 mi
0
50
100
150 km

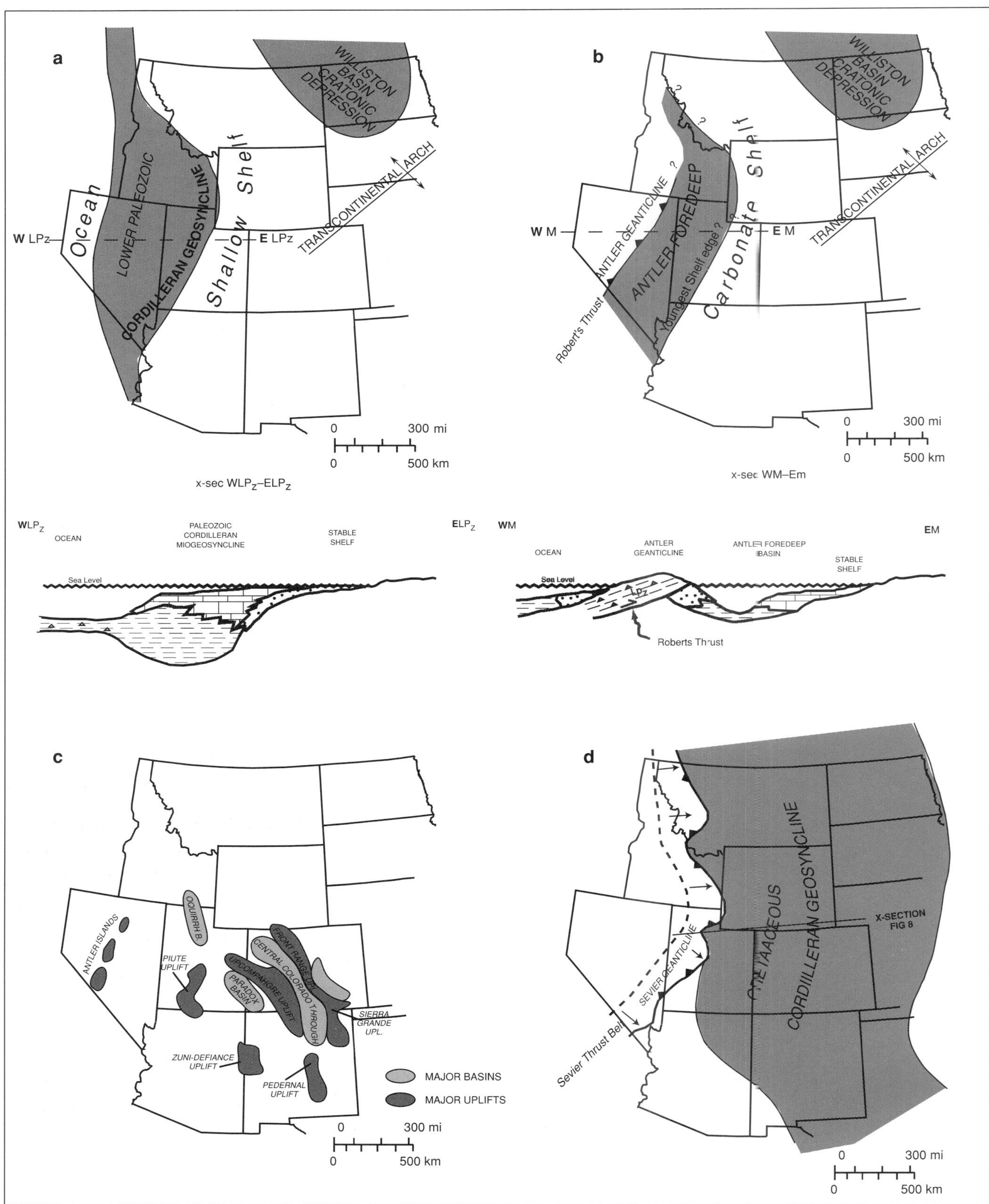

Figure 4. Generalized depositional and structural history of the GRMR at key intervals of geologic time: (a) early Paleozoic, (b) Mississippian, (c) Pennsylvanian-Permian, (d) Cretaceous, (e) Eocene- Paleocene, (f) Quaternary and Pliocene-Miocene.

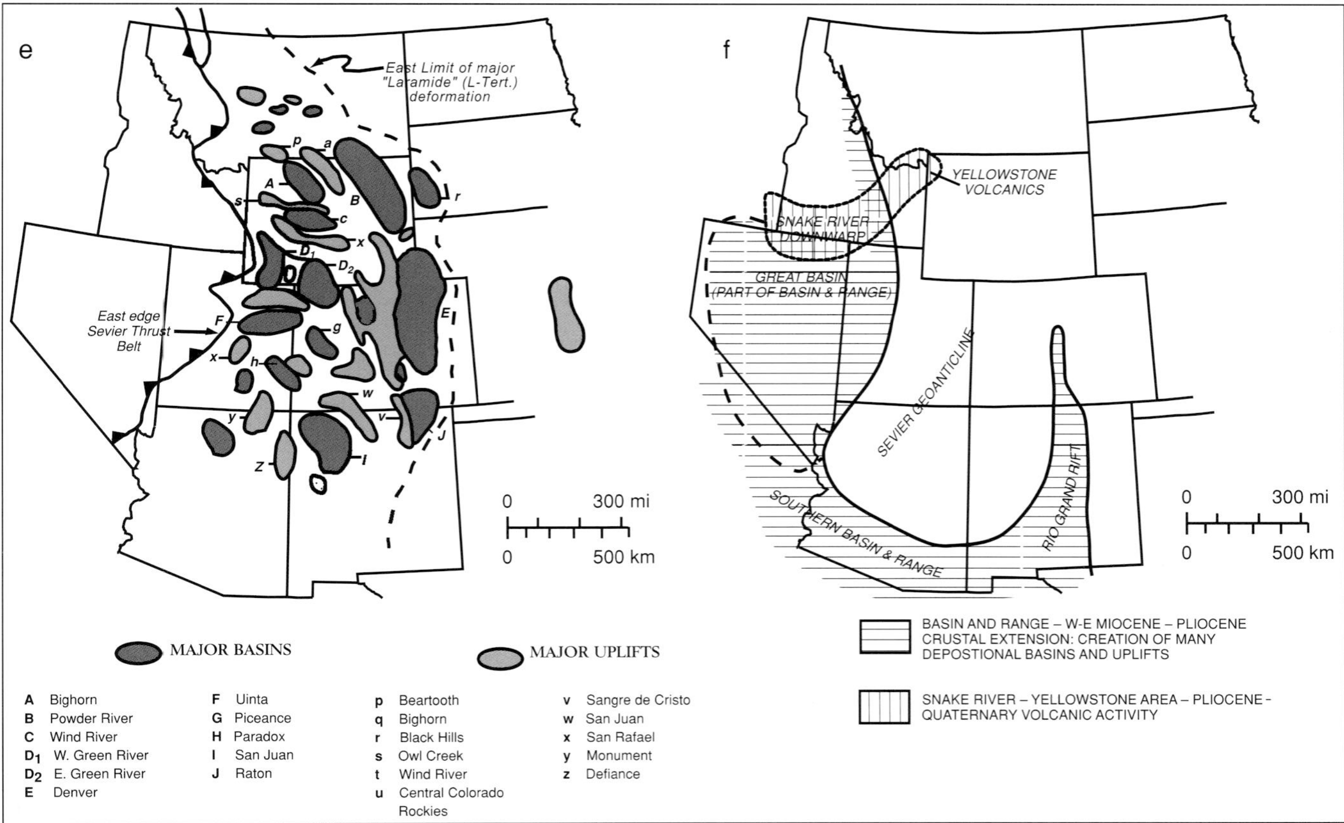

Figure 4 (cont'd.).

Significant features related to these episodes are shown in Figure 4a-e.

Stratigraphy, Depositional Environments, and Rock Types

The sedimentary section present in the GRMR ranges in age from Precambrian to Holocene (Mallory, 1972), but not all geologic periods are represented in all areas. Depositional environments represented in the stratigraphic record at various places include deep and shallow marine, lacustrine, and terrestrial. Lithologies include coarse and fine-grained siliciclastics, dolomites, limestones, salts, and anhydrites.

Intrusive and extrusive igneous rocks are present in local areas and are Jurassic, Cretaceous, Tertiary, and Quaternary in age. Most sub–upper Tertiary igneous rocks are acidic in composition, whereas those of Pliocene to Holocene age are basaltic. Layered extrusive rocks are generally intercalated with sediments and serve as cover over some large areas, and are reservoirs in certain oil fields of the Great Basin.

PETROLEUM SYSTEMS

An understanding of petroleum systems and how they operate can be of benefit in assessing discovery potential in a region (Magoon and Dow, 1994; Smith 1994). At least 30 distinct petroleum systems exist in the GRMR. Some systems are unique to a given basin or province, and some are common to several. Identified source rocks have charged reservoirs in a variety of traps through lateral and vertical migration paths.

Source Rocks and Reservoirs

The occurrence of source rocks and reservoirs in relation to geologic age and orogenic events that controlled major depositional sequences related to cratonic onlap, offlap, or erosion is summarized in Figure 5 and in Table 2. Known source rocks and productive reservoirs range in age from Precambrian through Tertiary. The spectrum of source rocks includes those that contain all of the basic kerogen types: oil-prone type I, oil-prone type II, and gas-prone type III.

Hydrocarbon production has been established from a wide variety of reservoir rocks, including sandstones, limestones, and dolomites with matrix porosity and permeability; fractured dense carbonates, shales, and igneous rocks; and coals. Considerable exploration opportunity appears to exist in ventures targeting nonconventional reservoirs in low-permeability "tight" rocks, fractured rocks, and coals. These are described in more detail.

Tight-gas Reservoirs

Tight-gas reservoirs, found primarily in Cretaceous

sandstones and chalks, generally have permeabilities that are too low to permit economic production rates using conventional completion techniques. Their characteristics and occurrence have been the subject of considerable study (e.g., Spencer, 1989; Spencer and Mast, 1986).

Chalks in the Niobrara Formation produce biogenic gas in shallow, low-relief structures on the east flank of the Denver Basin (Pollastro and Scholle, 1986). Porosities range from 25% to 50% and permeabilities range from 0.01 md to 16 md, with an average of about 1 md in the depth range of 1000–3200 ft (300–975 m). The first commercial wells drilled with air and completed open hole had initial potentials of 20–60 thousand cubic feet (mcf) (850–1700 m^3) of gas per day. Fracture-stimulation techniques and low-cost drilling and completion techniques have greatly enhanced the economics of production. Porosity and permeability decrease beyond reasonable productive reservoir limits at depths greater than about 4000 ft (1220 m). However, there appears to be a large prospective area extending along the east flank of the Denver Basin and into central South Dakota, where objective horizons are sufficiently shallow.

Matrix permeability and porosity in sandstones generally decrease with increasing burial depth and approach values that would seem to indicate uneconomic production capability (Law et al., 1986; Schmoker, 1997). However, substantial gas production has been established from so-called tight sandstones that generally have permeabilities of 0.5 md or less. Sandstone lithologies in the Cretaceous and lower Tertiary are highly variable (Coalson, 1989). Most are characteristically fine to very fine grained and consist dominantly of quartz; however, lithic fragments, feldspar grains, and primary depositional clays may be present in significant percentages. Grain size and grain composition depend on the lithology of the source area and on distance and transport mechanism to the site of deposition. Depositional facies are an important control on reservoir distribution. Although more porous and permeable sandstones may be developed in such facies as channels, point bars, and overbank deposits, they are also related to marine shorelines and bars, where substantial winnowing of fine grains and clays has taken place. These facies are often associated with sweet spots in deep-basin gas accumulations.

Although primary depositional composition is highly important, all Rocky Mountain low-permeability sandstones have been subjected to extensive diagenesis (Byrnes, 1996). Principal processes that have controlled present-day porosity and permeability include (1) grain rearrangement, (2) plastic and brittle deformation, (3) quartz pressure solution suturing, (4) quartz and calcite cementation, (5) dissolution of lithic rock and feldspar grains, and (6) precipitation of authigenic clays. These processes have resulted in destruction of much of the original intergranular porosity and have left dissolved grain porosity, clay-filled pores, and sheetlike connecting intergranular pore throats that are extremely susceptible to stress constriction. Irreducible water saturations may be unusually high. Studies of specific formations in certain basins have reached generally similar conclusions (Law et al., 1986; Pittman and Sprunt, 1986; Weimer et al., 1986). Even though permeability is low, porosities may range from 5% to 20% and provide ample storage capacity for gas. Modern methods of well stimulation by artificial hydraulic fracturing have allowed successful economic exploitation of these types of reservoirs. Tight sandstones constitute one of the most important reservoir targets for future exploration in the GRMR.

Fractured Reservoirs

Many, if not most, of the productive oil and gas reservoirs in the GRMR are fractured to some degree. Fracturing enhances the productive capability of reservoirs with effective matrix porosity and poor matrix permeability through the addition of fracture permeability. Many carbonate reservoirs in the GRMR could not be exploited economically without the presence of natural permeability-enhancing fractures. Natural fracturing is an essential characteristic in tight-sandstone reservoirs found in a deep-basin setting (Pitman and Sprunt, 1986). Fracturing also creates reservoir porosity and permeability in rocks with negligible amounts of matrix porosity and permeability, such as dense carbonates, shales, and igneous rocks. Many reservoirs of this type in the GRMR are found in mature hydrocarbon-saturated source rocks and represent an indigenous accumulation. Fracturing in source rocks and other low-matrix-permeability rocks in the deep-basin setting is believed to have been initiated by hydrocarbon generation producing overpressures (Meissner, 1974; Bredehoeft et al., 1994). We believe that fractured reservoirs in source rocks and tight-gas sands in the deep-basin settings of several Rocky Mountain basins are a significant exploration target for discovery of future reserves.

Coal-bed Methane Reservoirs

Cretaceous and Lower Tertiary coals provide one of the major future exploration and exploitation targets in the GRMR. They serve as sources of indigenous gas found within the coals as well as sources of expelled gas that charge other reservoir types (Rightmire and Choate, 1986). Coals have peculiar and unorthodox reservoir properties when compared with reservoirs with conventional matrix porosity and permeability. An understanding of these properties and characteristics should aid in predicting likely places for exploration.

Basic principles of thermal gas generation and storage, according to the model proposed by Juntgen and Karweil (1966), are shown in the graphs in Figure 6. Significant thermal generation of methane starts when a coal reaches a rank of about 37.8% volatile matter, equivalent to a vitrinite reflectance (R_o) of about 0.73%. Not all ther-

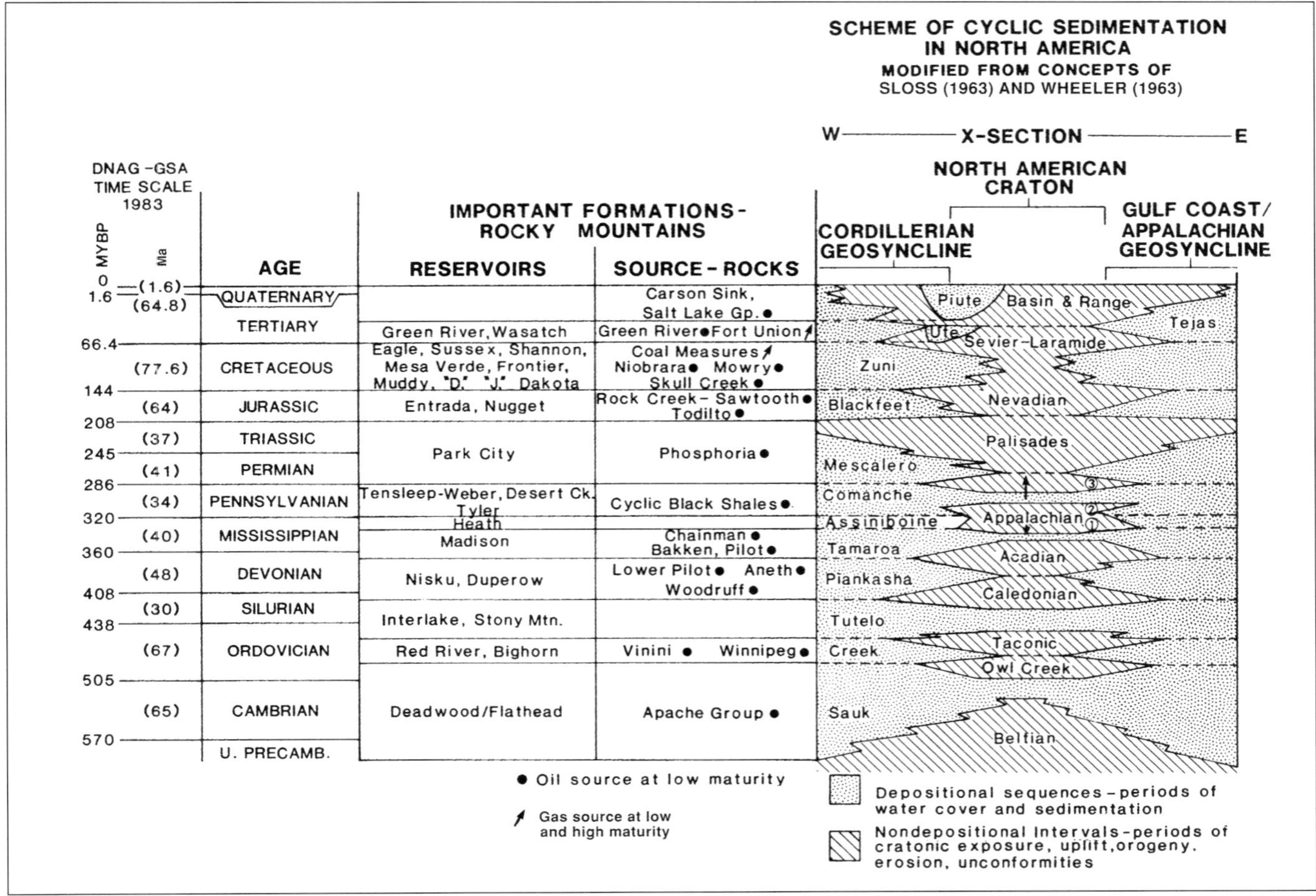

Figure 5. Stratigraphic chart showing ages and names of significant productive oil and gas reservoirs and source rocks, with their stratigraphic position in major depositional sequences (Meissner et al., 1984).

mally generated methane may be expelled. Coals have a significant capability of retaining or storing gas by processes of molecular adsorption on the coal kerogen and conventional volume storage in coal micropores (Figure 6a). Saturation storage capacities depend on coal rank, temperature, and pressure (Figure 6b). If pressure in a gas-saturated coal is lowered, storage capacity becomes less and gas is released for expulsion/migration. This is a primary mechanism for the production of coal-bed methane through wells that lower pressure in a coal-bed reservoir.

In addition to thermally generated gas, characterized by high concentrations of the heavy carbon isotope ^{13}C, many coal beds have been found to contain variable proportions of methane enriched in the "light" carbon isotope ^{12}C that is generally believed to be of biogenic origin (Rice, 1993; Scott, 1993). Biogenic gases found in coal are thought to have been generated by methane-generating bacteria that were introduced from the surface by dynamic groundwater and have subsequently used the coal as an energy source (Scott et al., 1991). Biogenic gases may charge low-rank coals that have not generated thermal methane. Although storage capacities of low-rank coals are not shown in Figure 6, the trends shown suggest that these coals may have very high storage capacities capable of storing large amounts of bacterially generated gas, particularly in microporosity.

Most shallow coal beds have developed a "cleat," or extensional fracture system (Close, 1993). Production of gas from a coal involves molecular diffusion from the coal matrix into the fracture system, followed by Darcy-type movement through the fracture system into a well bore. Production of coal-bed methane is generally characterized by modest rates and long life. An unusual feature of reservoir behavior is the fact that production rate characteristically increases with time before decline begins. Even though the coal matrix may be more or less saturated with gas, the fracture system in most shallow coals is filled with groundwater, and these must be "dewatered" before maximum production potential is achieved. Dewatering generally involves handling and disposing of large volumes of water. This disposal affects economics and may have an environmental impact. The cleat system found in shallow coals provides a critical element of permeability that controls economic rates of production. The effectiveness of the cleat system appears to diminish and disappear at depths of 4000–5000 ft (1220–1525 m). Even though more deeply buried coals may contain large volumes of gas, their exploitation is limited by absence of an effective cleat system. Developing a technology to exploit this resource will create a significant opportunity for exploration and development.

Types of Traps and Accumulations

Oil and gas production in the GRMR has been obtained from virtually every type of "trap" and "accumulation" setting known to the science of petroleum geology and includes a wide variety of "classical," "traditional," or "conventional" types as well as those considered "unorthodox" or "unconventional." Figure 7 is a simple diagrammatic sketch that depicts trap types and classifications as used by the U.S. Geological Survey National Oil and Gas Resource Assessment Team (1995) and subsequently modified by Surdam (1997a) and the authors of this paper. Figure 7 is particularly applicable to oil and gas accumulations in the Rocky Mountain region.

Table 2. Major productive petroleum systems.

Source rock	Age	Reservoir rock	Age	Basin(s)/area(s)
Green River Formation* (lake-core facies)	Eocene-Paleocene	Wasatch–North Horn (lake-margin, fluvial facies)	Eocene-Paleocene	Uinta, Western Green River
Sheep Pass Formation* (lake-core facies)	Eocene	Fractured Sheep Pass and Tertiary volcanics; Ely Limestone and Devonian-Silurian dolomite	Eocene, Oligocene, Pennsylvanian, Devonian, Silurian	Railroad Valley in eastern Great Basin
Tongue River Coal***	Paleocene	Tongue River Coal	Paleocene	Powder River
Oil prone**: Sharron Springs Shale Niobrara Formation Mowry Shale Skull Creek Shale	lower-Upper and Lower Cretaceous	Fractured source-rock sandstones in Mesaverde, Frontier, Gallup, Muddy, "D" and "J," Dakota, Lakota Formations	Upper and Lower Cretaceous	Sweetgrass Arch, Bighorn, Wind River, Powder River, Laramie Green River, Piceance, Uinta, North Park, Denver, San Juan
Gas prone***: Coals in Fruitland, Raton, Vermejo, Mesaverde, Ferron Formations	Upper and Lower Cretaceous	Sandstones and coals in Fruitland, Pictured Cliffs, Raton, Vermejo, Mesaverde, Ferron Formations	Upper and Lower Cretaceous	Wasatch Plateau Green River, Uinta, San Juan, Raton
Phosphoria Formation**	Permian	Park City carbonates Tensleep and Weber sandstones, middle and lower Paleozoic carbonates and sandstones. Crystalline basement	Permian Pennsylvanian, Mississipian, Ordovician, Cambrian, Precambrian	Bighorn, Wind River, Green River, Piceance
Cyclic black shales**	Permian, Pennsylvanian	Minnelusa and Leo sandstones Permian and Pennsylvanian limestones, dolomites, and sandstones	Permian, Pennsylvanian	Powder River, Denver
Cyclic black shales**	Pennsylvanian	Fractured shale (Cane Creek Shale) carbonate mounds	Pennsylvanian	Paradox
Cyclic black shales* in Tyler and Heath Formations	Lower Pennsylvanian–Upper Mississippian	Sandstones in Tyler and Heath Formations, Amsden limestones	Pennsylvanian–Upper Mississippian	Williston, Central Montana
Lower Mississippian and Bakken Shales**	Mississippian and uppermost Devonian	Fractured Bakken, Madison Group limestones, and dolomites. Sanish Sand, Nisku carbonates	Mississippian, Upper Devonian	Williston
Chainman Shales**	Mississippian	Tertiary volcanics, Devonian-Silurian dolomites	Oligocene, Devonian-Silurian	Railroad and Pine Valleys in eastern Great Basin
Upper Red River Formation**	Ordovician	Interlake, Stony Mountain, Red River limestones and dolomites	Silurian, Ordovician	Williston

*Type I kerogen, **Type II kerogen, ***Type III kerogen.

"Conventional Traps"

Conventional traps (called "discrete-type" accumulations by the U.S. Geological Survey) are depicted in a typical setting on the shallower flanks of the basin. Although not shown, combination-type traps consisting of structural and stratigraphic elements also belong to this realm. Structural traps include simple anticlinal closures related to such processes as thrusting, draping, salt movement, or solution. Stratigraphic traps include those related to subcrops, lateral facies changes, and erosional topography. Hydrocarbon accumulations in most of these settings are underlain by water, which regionally saturates the reservoir. The oil and gas present in these accumulations are in a static state and therefore are "trapped." In contrast, the groundwater surrounding or underlying the hydrocarbons is generally in a dynamic state. This situation often affects the hydrocarbon accumulations, for instance, by tilting oil-water contacts on "unclosed" structural noses (e.g., on the Billings Nose in the Williston Basin; see Berg et al., 1994).

"Unconventional Accumulations"

The GRMR contains a number of significant "unconventional" types of accumulations. Examples include those associated with gas production from coal beds and those associated with "deep-basin," "basin-center," or "continuous-type" accumulations. These last three names are used somewhat interchangeably; however, not all continuous-type accumulations are related to the same processes that cause deep-basin or basin-center accumulations. The U.S. Geological Survey has used the term *continuous-type* to cover all accumulations of a more nearly regional nature with no clear evidence of a controlling classical trap style. Lignite coals containing biogenic methane and shallow accumulations of biogenic gas in shallow Cretaceous sandstones and chalks that are present in central and eastern Montana, central South Dakota, and eastern Colorado are clearly not the same type of accumulation as those commonly associated with the term *deep basin* or *basin center*. All these unconventional types of accumulations are of particular interest because they have major exploration potential in the Rocky Mountain region. They can also serve as "type-locality" analog examples for the rest of the world.

A basin-center accumulation is shown in the bottom of the syncline depicted in the cross section in Figure 7. This type of deposit is characterized by hydrocarbon-saturated reservoirs that are either significantly overpressured or underpressured with respect to normal pressure gradients. The accumulations are within or adjacent to mature source rocks that either are actively generating at

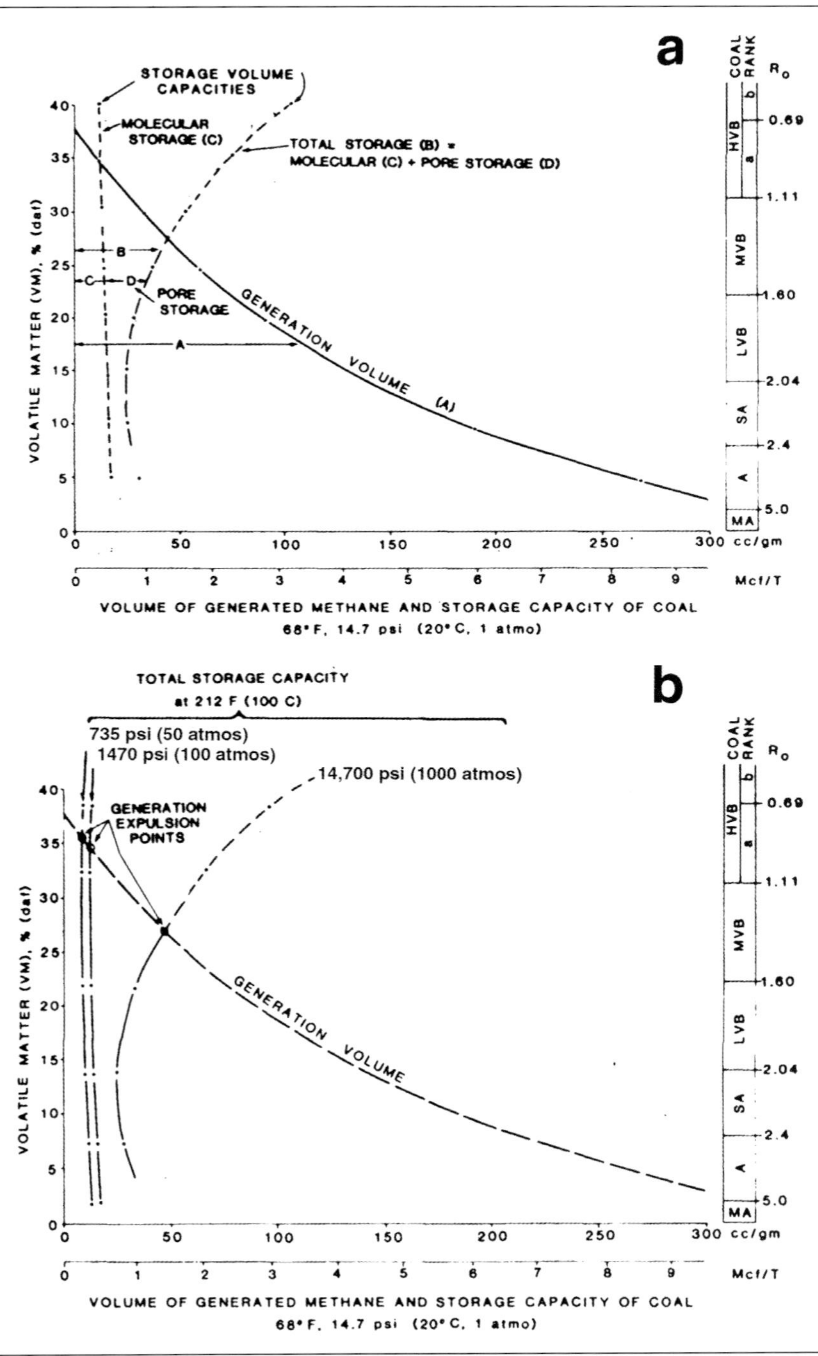

Figure 6. Generation and storage of gas in coals as a function of coal rank: (a) theoretical generation with adsorption and pore-storage volumes at 14.7 psia and 68°F (1 atmo., 20°C); (b) theoretical generation and storage volumes at variable pressures and constant 68°F (20°C) (Meissner, 1984).

high rates or have ceased such generation in the relatively recent geologic past. Fractured reservoirs may be present within the source rock. Outward migration from the source rock may also have charged reservoirs with matrix porosity; however, because of great burial depths, the matrix reservoirs are generally characterized by low porosities and permeabilities. Poor matrix permeabilities are commonly enhanced by fracturing. The accumulations have no basal hydrocarbon-water contacts but are characterized by high water saturations regionally updip within reservoirs found in the same general stratigraphic interval. The species of hydrocarbons present (oil or gas) is controlled by both the type of source rocks and their thermal maturity.

The ubiquitous nature of total hydrocarbon saturation in reservoirs associated with mature source rocks indicates an area of supercharge and high migrational impedance. Conventional "trapped" accumulations are generally found along updip migration paths, suggesting that deep-basin accumulations leak excess hydrocarbon charge. Traditional concepts of petroleum geology consider that hydrocarbon entrapment is controlled by a "seal" at which migration stops or is impeded by capillary seal capacity, and many investigators have attempted to explain the updip limit of deep-basin hydrocarbon accumulations by various types of seals. Although updip limits of deep-basin accumulations may be controlled by capillary-entry or fracture-opening pressure, they may also reflect dynamic "backup" in a region where generation rate exceeds that of migration. The volume of hydrocarbon saturation may not actually be "trapped." Instead, it is a transient accumulation that may dissipate with time.

Basin-center accumulations generally involve large rock volumes and contain extremely large volumes of hydrocarbons. However, because of generally low reservoir porosity and permeability, most of these accumulations may be characterized by noncommercial or subcommercial production at the present state of exploitation technology. Local accumulations characterized by enhanced reservoir properties are present in a regional basin-center setting; these constitute what are termed "sweet spots," where commercial production may be established. Sweet spots may be localized in areas of enhanced fracture or matrix permeability and porosity, either in the source-rock unit associated with the accumulation or in a more nearly regional low-quality reservoir that was charged by the source rock.

Cretaceous Petroleum Systems

Petroleum systems present in the Cretaceous section of the GRMR have been some of the most important to historic production and probably constitute the major contributor to future development and discovery. The distribution of both oil- and gas-prone source rocks and the reservoirs they may charge in the syntectonic depositional sequence that filled the Cretaceous Cordilleran geosyncline (Figure 4d) is shown in Figure 8. Oil-prone source rocks containing type II kerogen are present in minor cycles of transgression or basin deepening in the marine lower part of the section. Gas-prone source rocks are associated with humic coal measures containing type III kerogen that are present in regressive cycles of terrestrial sedimentation that originated from the uplifted thrust belt on the west side of the geosyncline. Sandstone

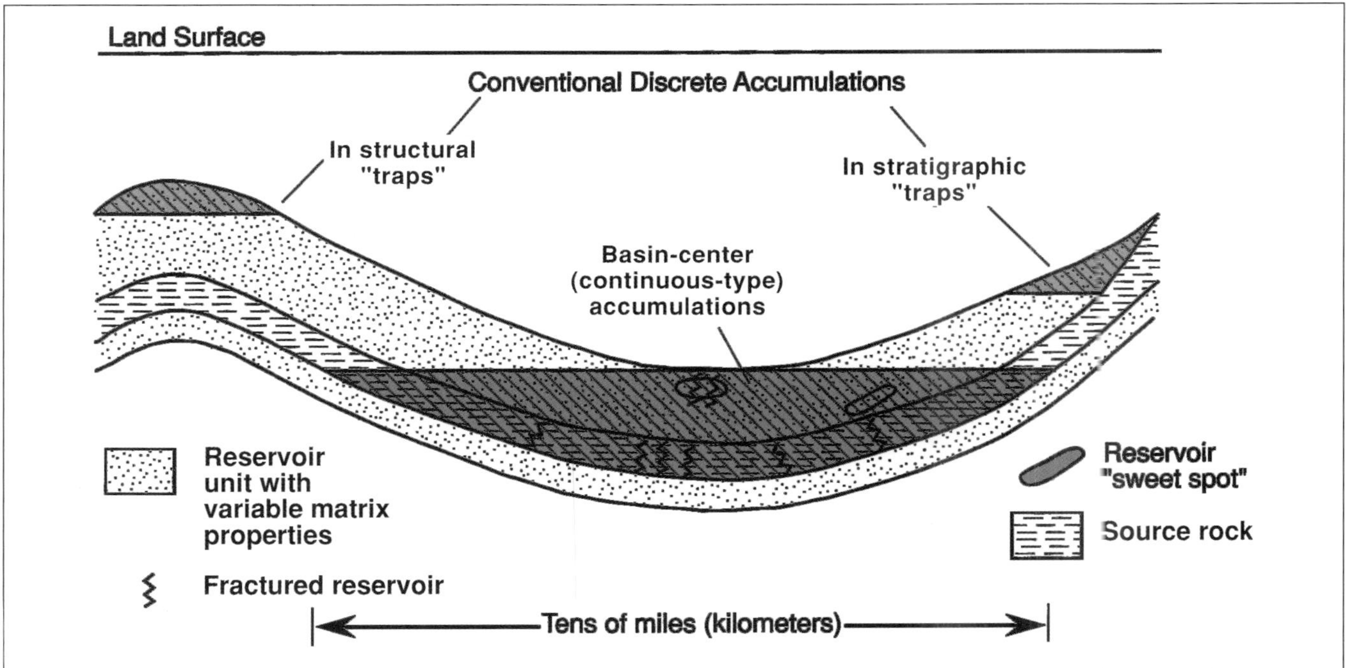

Figure 7. Simple classification of oil and gas accumulations found in the GRMR (modified from U.S. Geological Survey National Oil and Gas Resource Assessment Team, 1995, its Figure 5, and Surdam, 1997a).

reservoirs are associated mostly with marine coastal inter-deltaic, delta complexes, and fluvial systems.

In the early Tertiary, the Cretaceous section was structurally deformed by the Laramide orogeny. Sediments were eroded from mountain uplifts and buried to varying depths by lower Tertiary sediments derived from adjacent uplifts (Figure 8). Most of the maturity patterns present in Cretaceous source rocks were the result of Tertiary burial in the deepest part of the Laramide basins. Actual depths to maturity, however, are affected by present-day and paleo heat-flow conditions. Maturity in the lower part of the section along the western margin of the geosyncline was produced by burial beneath thrust plates or as a consequence of the thick Upper Cretaceous section.

Figure 9 shows the distribution of oil-prone source rocks in the lower part of the Cretaceous. These rocks are responsible for most of the oil found in Cretaceous sandstone, fractured source-rock shale, and limestone reservoirs. Although these rocks are basically oil prone, they have generated gas where they have been more deeply buried in the centers of some of the Rocky Mountain basins or exposed to unusually high temperatures.

The depositional distribution of gas-prone humic coal measures in the Cretaceous and lowermost Tertiary is shown in Figure 10. The thickness and extent of these coals constitute a major concentration of source rocks for the generation of world-class gas accumulations.

SIGNIFICANCE OF FLUID PRESSURE REGIMES AND HYDRODYNAMICS

Abnormally low "underpressure" and high "overpressure" are characteristic of many areas of the GRMR (Figure 11). Anomalous pressures in the region are caused by two basic processes: (1) topographically driven groundwater flow (pervasive groundwater phase in the reservoir/aquifer) and (2) hydrocarbon generation and migration (pervasive hydrocarbon phase in reservoir).

Groundwater-dominated Fluid Systems

Because of the variation in outcrop topography between mountain uplifts and low-lying basins and sufficient amounts of precipitation and surface-water flow, most subsurface reservoir sections in the GRMR are hydrodynamically active (Figure 11). This activity is commonly demonstrated by the presence of potentiometric gradients that are inclined toward the direction of groundwater flow. Active groundwater movement has influenced migration paths from source rocks to sites of accumulation and has caused commonly observed tilted oil-water contacts (Hubbert, 1953; Dahlberg, 1982). The attitude of these contacts has affected the distribution of hydrocarbon accumulations on "closed" and "unclosed" structures (Murray, 1959; Vincelette and Chittum, 1981; DeMis, 1987; Berg et al., 1994) as well as in stratigraphically controlled traps (Moore, 1984; Meissner, 1988). It has also greatly influenced the critical seal capacity of traps, with groundwater flow from the reservoir toward the seal reducing normal seal capacity. Conversely, flow from the seal toward the reservoir enlarges seal capacity (Stone and Hoeger, 1973; Berg, 1975; Schowalter,1976; Larber, 1981; Linn, 1981). Groundwater flow has also been shown to alter the subsurface thermal regime (Willet and Chapman, 1987) and to control coal-bed methane saturation and composition (Oldaker, 1991, Scott et al., 1996).

Hydrocarbon-dominated Fluid Systems

Anomalous areas of both overpressure and underpressure that are characterized by the presence of a predominantly hydrocarbon-saturated reservoir fluid system have been found in the deeper parts of several basins (Figure 12). These pressure anomalies form "cells" around a "core" of mature source rocks. Overpressures in this setting are believed to have been created during active hydrocarbon generation by volume changes produced during the conversion of immature solid kerogen into potentially expellable fluid hydrocarbons and kerogen residue (Meissner, 1974, 1980; Momper, 1980; Law, 1984; Law and Dickinson, 1985; Spencer, 1987; Bredehoeft et al., 1994).

Figure 13 depicts a typical example of overpressure in the Cretaceous Mesaverde Group of the eastern Green River Basin of southwestern Wyoming. Pressures are related to active gas generation in the bottom of the basin from coals contained in the Mesaverde (McPeek, 1981: Meissner, 1987). The area of source-rock maturity for the actively generating coals corresponds to the area of overpressures where pore-fluid pressure gradients exceed 0.45 psi/ft (10.2 kPa/m). Although Mesaverde sandstones in the area of overpressure are pervasively gas saturated, commercial production has been obtained only in sweet spots associated with cleaner sandstones in marine shoreface or bar facies at the top of the Mesaverde.

The amount of overpressure created by hydrocarbon generation is governed by Darcy's law and depends mostly on generation/expulsion rates, effective permeabilities, and capillary-barrier or fracture-leakage permeabilities. Pressure may be transferred from the source rock to a reservoir along the expulsion/migration path (Martinsen, 1997). Although hydrocarbon-phase overpressures may be maintained by stratigraphically or diagenetically controlled "capillary seals," the presence of generation overpressure is most likely transient. In the transient case, the pervasive presence of hydrocarbons in a generation/expulsion cell is produced by a dynamic migration bottleneck rather than by the presence of a classical static hydrocarbon "trap." Hydrocarbon generation is simply overwhelming the migration "pipeline" (Law et al., 1986).

Underpressures associated with areas of pervasive deep-basin gas saturation may theoretically be caused by

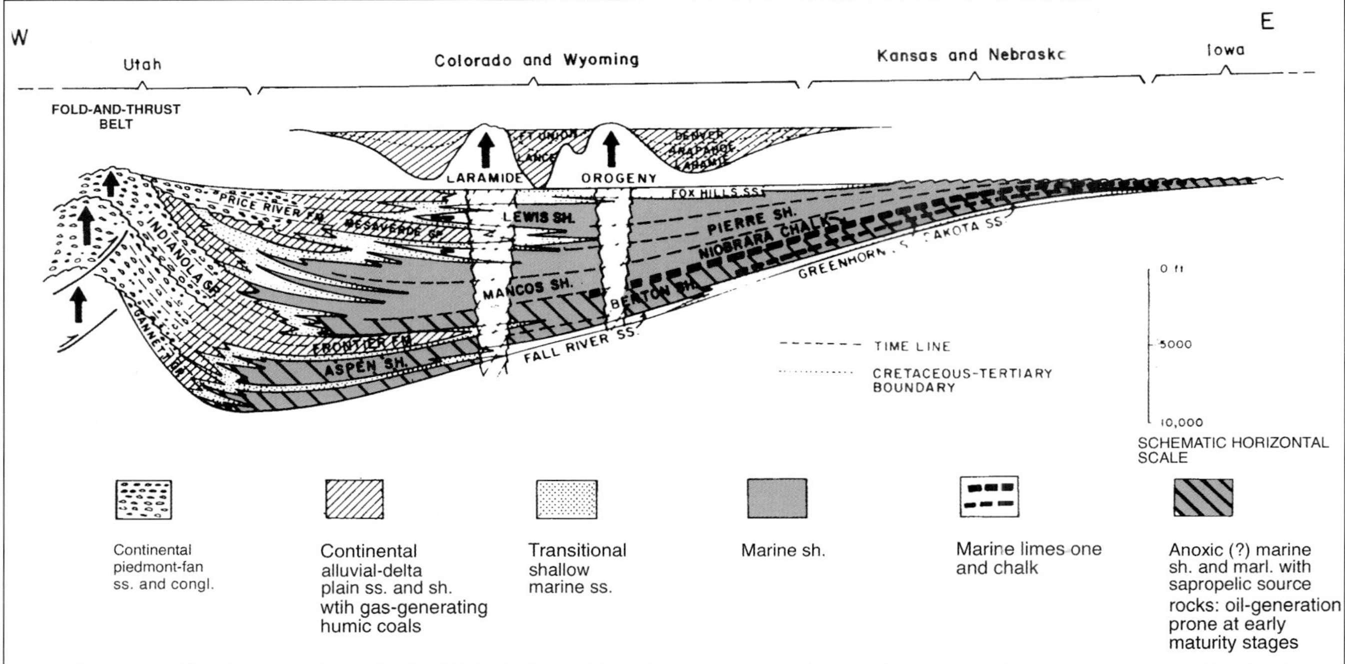

Figure 8. Schematic cross section through the Cretaceous geosyncline showing the distribution of lithofacies, depositional environments, and oil/gas source rocks. Also shown is the distribution of Tertiary rocks superimposed on the Cretaceous as a result of Laramide structural movement (after Kauffman, 1977, in Meissner et al., 1984).

a number of mechanisms, including (1) gas-volume contraction produced by temperature reduction related to changing heat flow or uplift, (2) gas readsorption from adjacent matrix reservoirs into coals when the temperature is lowered (Meissner, 1987), (3) elastic porosity dilation produced during uplift and erosion (Bachu and Undershultz, 1995), and (4) readjustment of pressure when the gradient of a gas column is reequilibrated to shallower conditions produced during uplift (Surdam, 1997b). Underpressures may also be created in both deep-basin oil and gas accumulations by the transient migration process. When active high-rate generation ceases, overpressure created by the process will diminish if the hydrocarbons are able to leak off and migrate away from their deep-basin position. As leak-off occurs, formation water will be imbibed into formation porosity, and this leads to underpressure in the region formerly characterized by overpressure. Leak-off may be accomplished by conventional phase migration or by diffusion. Diffusion may be particularly important in the case of gas (Krooss et al., 1992; Nelson and Simmons, 1992, 1995). If pressures and migration are time transient, underpressures will remain until all mechanically unstable hydrocarbons (i.e., those not in a state of static equilibrium associated with values of minimum potential energy within the constraints of a continuously permeable reservoir/carrier system) have left the synclinal deep-basin position or are stabilized in conventional traps. When this occurs, the dynamic pressure regime will return to normal (Meissner, 1985).

An example of underpressures related to pervasive deep-basin gas accumulation is found in the San Juan Basin of southwestern Colorado and northeastern New Mexico (Figure 14). Deep-basin gas accumulation there is associated with the area of thermal generation in coals of the Mesaverde Group, as is the case for overpressured gas accumulation in the Green River Basin. Maturation models (Bond, 1984; Meissner, 1987; Law, 1992) indicate that maturity was achieved in the coals as a result of an abnormal Oligocene heating event related to igneous activity in surrounding areas. Gas generation began about 35 Ma and ended about 15 Ma. As is the present situation in the actively generating Green River Basin, a large gas-saturated overpressure cell was undoubtedly present in the San Juan Basin at that time. A present-day potentiometric surface map for the Mesaverde is shown in Figure 14b. Potentiometric contours, which indicate elevations to which groundwater should rise based on the pressures measured in the reservoir, are substantially below ground-surface elevation and demonstrate the presence of underpressure. Closed contours in the deep part of the basin are of minimum value and represent the existence of a "potentiometric sink" toward which groundwater is flowing. Although several processes may contribute to this phenomenon, we believe the main mechanism creating the underpressured potentiometric sink is the loss of gas from deep-basin gas accumulation by diffusion and the replacement of pore volume by water imbibition from the updip area.

The sequence of events described above for the time-transient behavior of anomalous pressures related to hy-

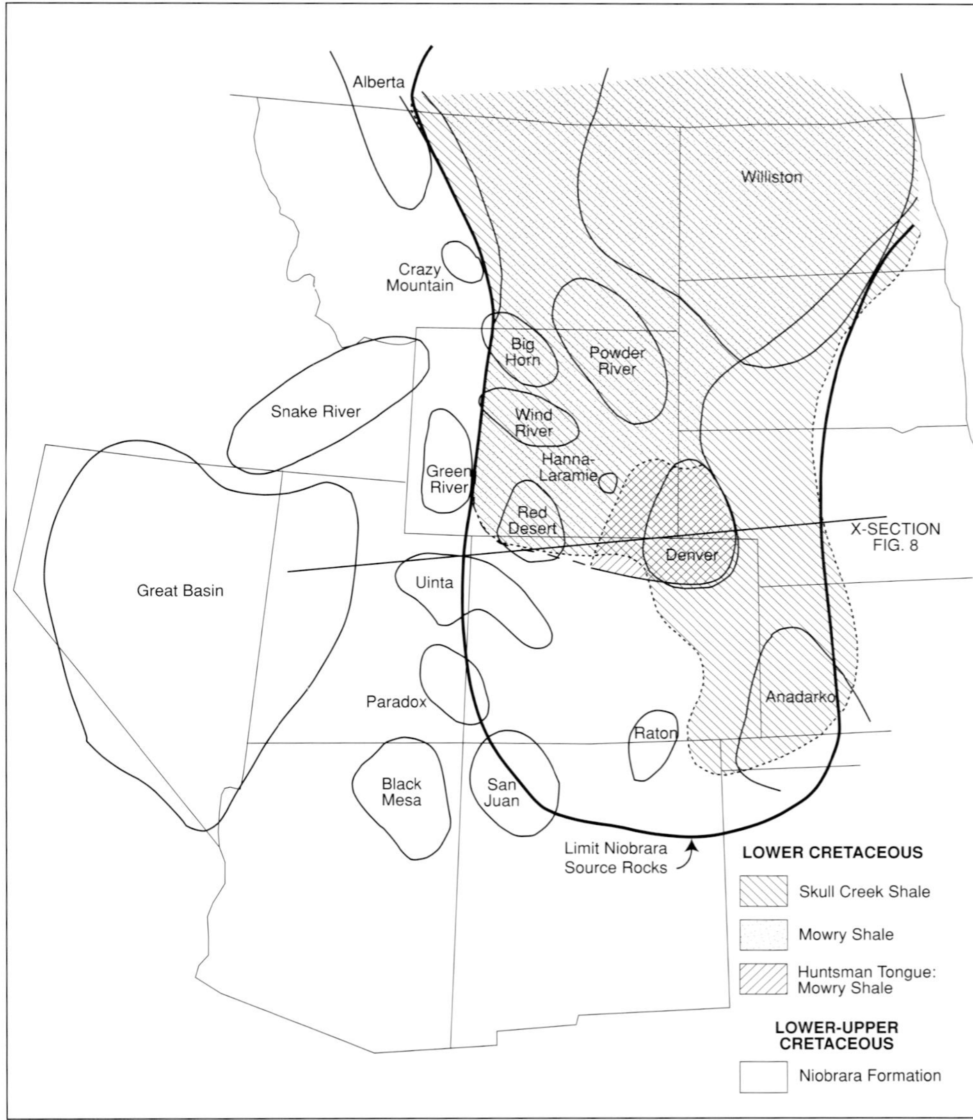

Figure 9. Distribution of Cretaceous oil-prone source rocks in the GRMR. These rocks generate oil at lower stages of maturity and gas at higher stages (Meissner et al., 1984).

drocarbon generation represents a cycle of anomalous pressure buildup and decay (Figure 15). As long as the pressure cycle related to this phenomenon is in a state of overpressure or underpressure or in the geologically short time between these states when pressure is normal, economically viable accumulations of the "deep-basin type" may exist in the anomalous pressure cell.

EXPLORATION POTENTIAL

Based on assessments made by various entities (Table 1 and Appendix) and our own exploration optimism, we believe the following situations offer significant opportunity for future discoveries of oil and gas in the GRMR.

Continuous-type Accumulations, Including Deep-basin and Shallow, Tight-gas Reservoirs

Prospective regions for discoveries in this category are shown in Figure 16. Areas shown in the northern Great Plains of Montana, North Dakota, and South Dakota include shallow Upper Cretaceous tight sandstones and chalks that are not considered to be deep-basin accumulations (Rice and Shurr, 1980; Rice and Spencer, 1995). Estimates of undiscovered technically recoverable biogenic methane from the sandstones are as high as 91 tcf (Dyman et al., 1995). Exploration and development for this resource beyond a few fields on pronounced uplifts have been hampered by the difficulty in recognizing productive intervals and areas and the lack of pipeline infrastructure (Hester, 1999).

Prospective deep-basin accumulations containing tight-sand reservoirs (also shown in Figure 16) are believed to exist in the Cretaceous section of most of the Laramide Rocky Mountain basins where burial has been sufficient to cause generation from available source rocks. This setting offers perhaps the greatest potential for discovery and development in the entire GRMR.

Although general areas of deep-basin accumulation have been identified, most development to date has been in sweet spots. Other sweet spots remain to be discovered. More importantly, new technology is being developed to economically exploit lower-grade reserves, redefining the criteria for a sweet spot. Such advancement will greatly expand the gas-resource base.

Coal-bed Methane

The development of coal-bed methane has been one of the most significant and ongoing plays developed in the last few years (Schwochow, 1991; Murray and Schwochow, 1997; Schwochow and Murray, 1999). The largest area of development has been in Fruitland coals of the

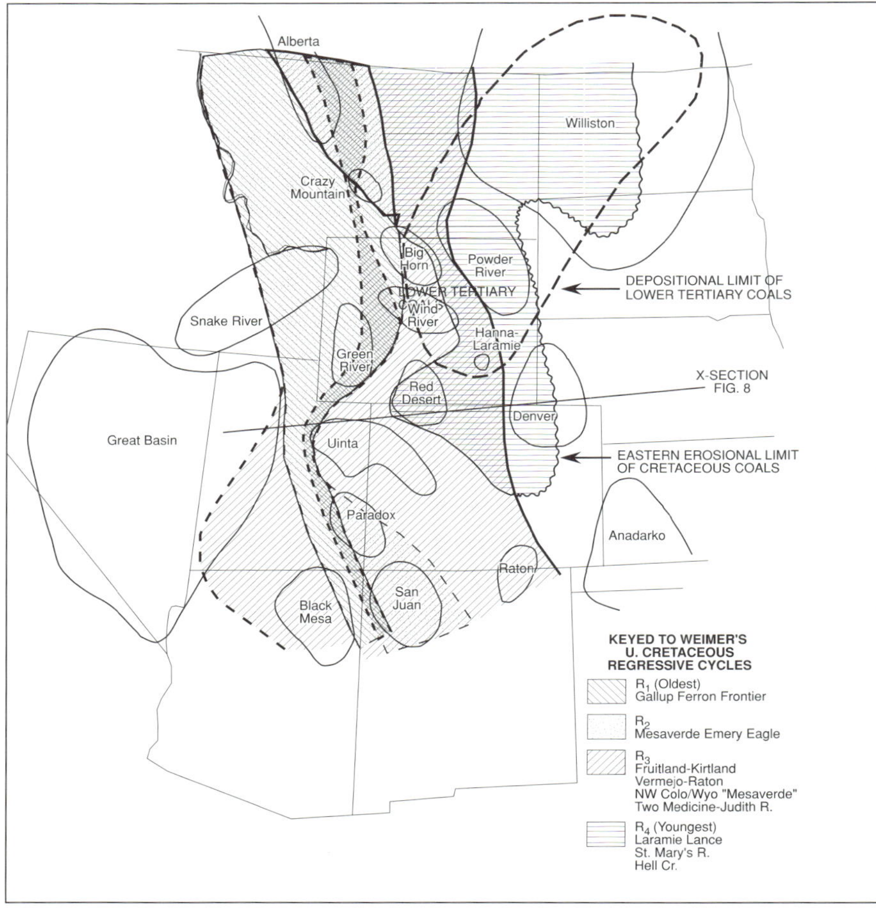

Figure 10. Distribution of Cretaceous and lower Tertiary gas-generating coal measures in the GRMR. Cretaceous units are keyed to Weimer's (1960) regressive cycles (Meissner et al., 1984).

San Juan Basin (Fassett, 1998). However, several other areas continue to experience exploration and development (e.g., LaMarre and Burns, 1997), and future plays undoubtedly will result in establishment of significant resources (Scott, 1999).

The distribution and rank of coals in the GRMR and the volumes of in-place gas they are estimated to contain are shown in Figure 17. The distribution of rank in shallow coals appears to be influenced in part by proximity to igneous activity associated with the Río Grande Rift. In general, higher-rank coals have the capability of thermally generating and storing larger amounts of gas; however, the importance of biogenic gases in lower-rank coals must also be considered. Major plays could be developed wherever sufficient gas saturation and productive capacity are found. A large part of the estimated deep in-place coal-bed methane resource is in low-permeability coals that are not economically exploitable. Technology that would enable economic production to be established in those settings may be developed.

Compartmentalized and Thin Oil Reservoirs

Many recognized pay zones have been encountered that do not permit

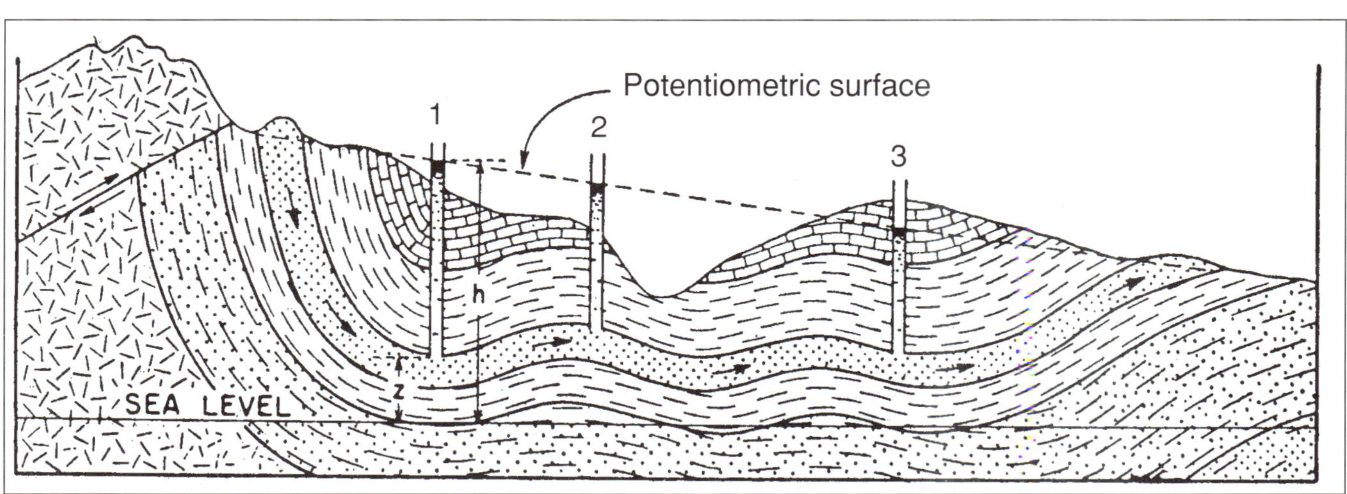

Figure 11. Cross section depicting regional flow of groundwater through an aquifer-reservoir from higher to lower outcrop elevation and a related potentiometric surface (from Hubbert, 1953). Because groundwater is capable of rising to the height (h) of the potentiometric surface in any well drilled into the reservoir, the relation of the potentiometric surface to ground elevation indicates that wells 1 and 2 will be overpressured in the reservoir, whereas well 3 will be underpressured.

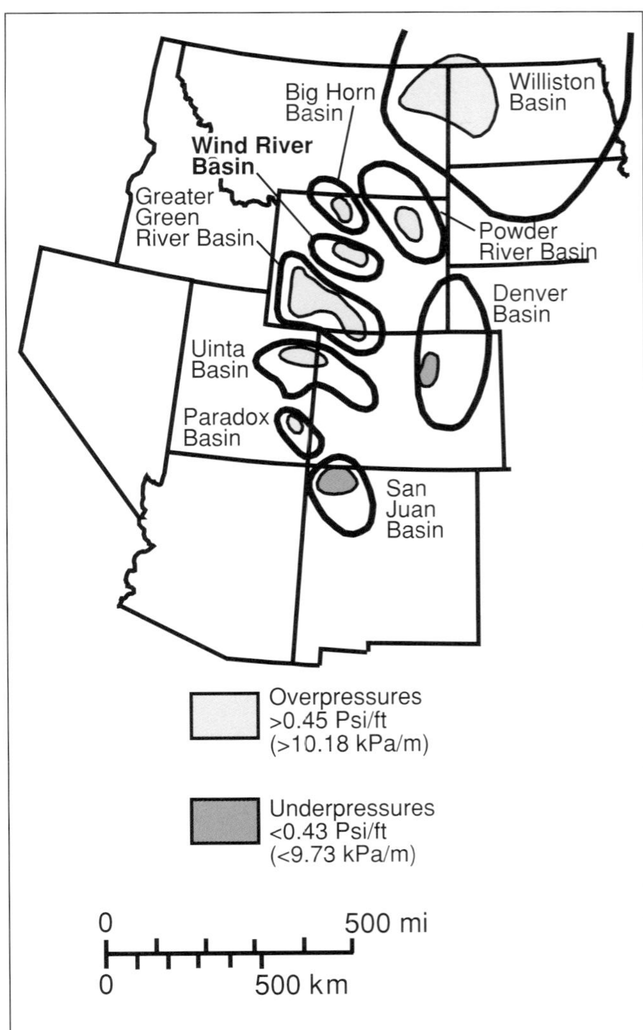

Figure 12. Basins in the GRMR with anomalous reservoir pressures associated with deep-basin-type oil or gas accumulations.

establishment of economic production rates because of limited drainage area resulting from reservoir heterogeneity or because pays which do contain potentially economic permeability are too thin. Horizontal-drilling techniques have proved to be successful in commercially developing these resources in some areas. There appear to be many areas where horizontal drilling has not been tried, and considerable opportunity for developing reserves by that method may exist.

Fractured Oil-bearing Reservoirs

Fractured reservoirs have long been of historical significance in the GRMR. Production from fractured reservoirs is widely scattered and is found in several formations, including the lower Tertiary Green River and Wasatch Formations in the Uinta Basin (Lucas and Drexler, 1975, 1976; Narr and Curry, 1982); the Cretaceous Niobrara Formation and its equivalents in the Denver, Powder River, North Park, Piceance, and San Juan Basins (Vincelette and Foster, 1992; Sonnenberg and Weimer, 1993); Pennsylvanian black shales in the Paradox Basin (Hite et al., 1984); and the uppermost Devonian and lowermost Mississippian Bakken Formation in the Williston Basin (Meissner, 1974; LeFever, 1991; Hansen and Long, 1991). All these occurrences are indigenous accumulations in or adjacent to mature source rocks. Many occur in deep-basin settings and are associated with generation overpressure. Historically, most of the early production was established in vertical wells. Many of those wells were subeconomic, and offset well success was unpredictable.

In the last few years, the industry has made considerable progress in evaluating, predicting, and exploiting the occurrence of fractured reservoirs (Hoak et al., 1997; Hoak, 1998; Nelson, 1998). Development efforts using horizontal-drilling techniques (Figure 18) have been successful in the Niobrara Formation at Silo field in the Denver Basin (Montgomery, 1991a, 1991b; Campbell et al. 1992), the Cane Creek Shale in Bartlett Flat field in the Paradox Basin (Morgan, 1992a, 1992b; Grummon, 1992), and the Bakken Formation in the so-called Fairway Trend of the southern Williston Basin (LeFever, 1991; Hansen and Long, 1991), where more than 200 wells have been drilled since 1987.

Confirmed Source Rocks with Little or No Production

Cyclic black shales of Permian and Pennsylvanian age in the northern Denver Basin, Nebraska, have been identified as thin but very rich oil-prone source rocks (Clayton and King, 1984). Drilling in the area has encountered numerous oil shows, and several small fields have been found. Carbonate reservoir permeabilities have generally been low, and productive rates, therefore, have been marginal. Formation pressures are subnormal, and drilling with conventional overbalanced mud systems has undoubtedly caused formation damage. The area is underexplored in general. Additional drilling may find better reservoirs, and advanced drilling techniques may improve production rates. Horizontal drilling has been attempted but has not yet been economically successful.

Excellent oil-prone source rocks are known to exist in lower Paleozoic (Vinini, Woodruff, Pilot Shales), middle Paleozoic (Chainman Formation, Delle Member of the Woodman Formation), and lower Tertiary (Sheep Pass, Elko Formations) sections in the eastern Great Basin of western Utah and eastern Nevada (Poole and Claypool, 1984; Sandberg and Gutschick, 1984). Several small but prolific oil fields have been discovered; however, the richness and distribution of viable source rocks suggest that a large potential for future discoveries is present in the area. As an example, the Mississippian Chainman Shale, penetrated in a deep well near the center of Railroad Valley Basin (see Figure 1), was found to be 2454 ft (748 m)

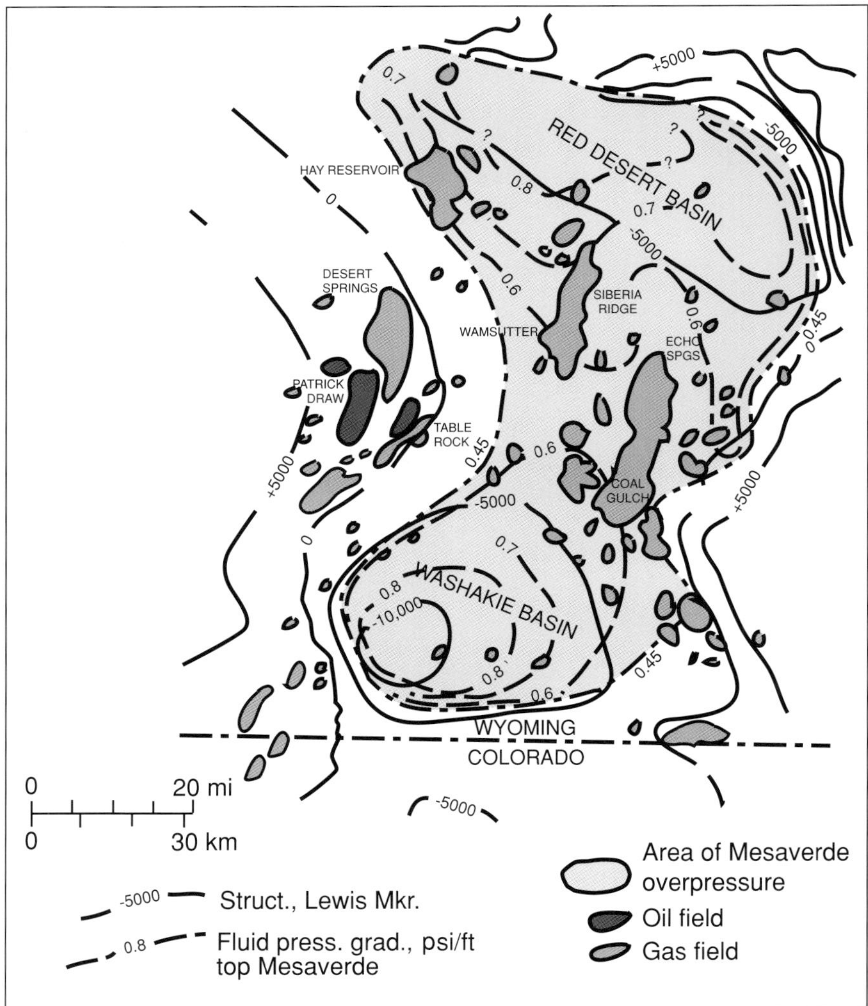

Figure 13. Structure, distribution of anomalous overpressures, and gas production in the Mesaverde (Cretaceous) section of the eastern Green River Basin (after McPeek, 1981).

thick and to have an organic-carbon content ranging from 1.0% to 5.2%, with an average of 2.7% (French, 1994). The Chainman section there is entirely within the oil window (R_o 0.8 at the top and 1.25 at the bottom), and maturity was achieved by burial during the relatively recent Basin and Range phase of structural development (Meissner, 1995). This evidence suggests a much larger potential for accumulation than has been found in related oil fields, which have established volumes of only 30–40 million bbl.

Prospects beneath Volcanic or Overthrust Cover

Several large areas in the GRMR contain middle to upper Tertiary and Quaternary extrusive volcanic rocks that cover an underlying prospective sedimentary section (Figures 2, 3b-d). These include the following volcanic "fields": San Juan (southwest Colorado; see Gries, 1985, 1989), northwest Wyoming, Marysvale (central Utah), Snake River downwarp (southern Idaho), and certain areas of the Basin and Range (western Utah, eastern Nevada, southern Arizona, and New Mexico). Although the underlying sedimentary section may contain viable petroleum systems, exploration has been hampered by the masking nature of the volcanics and the general inability of existing seismic techniques to image the section beneath them. Ability to prospect in these areas may depend on newly developing technology.

The margins of many of the Rocky Mountain Laramide basins are characterized by high-angle thrust faults that hide underlying structure (Figures 2, 3d-e; see Gries, 1983). Closed anticlines may exist, and the faults themselves may provide a sealing element to upturned beds or structural noses. There are several examples of oil and gas fields beneath these faults, including the recently discovered Cave Gulch field (discussed later).

APPLICATIONS OF NEW TECHNOLOGY

The application of new technology has been an important element in discovering and establishing new oil and gas reserves in the GRMR. Many recent developments have been made in areas where the presence of hydrocarbon saturation was known but could not be economically produced with capabilities existing then. Considerable effort is being made in developing new concepts and techniques in almost every category that affects exploration, development, and production (Crow, 1996; Coalson et al., 1997). These developments have been so extensive and pervasive that we will present only a short summary of those we consider to be most significant.

Basic Geology

Constant advances have been made in understanding basic petroleum geology, both as a fundamental science applicable on a global scale and as specific regional cases. As applied to the GRMR, these include such items as structure (Baars and Stevenson, 1981; Powers, 1982; Gans and Miller, 1983; Stone, 1993; Koeberl and Anderson, 1996), stratigraphy (Weimer, 1988; Dolson, 1994), source-rock presence and maturity (Woodward et al., 1984), and reservoir development and behavior (Weimer, 1988; Goolsby and Longman, 1988; Coalson, 1989;

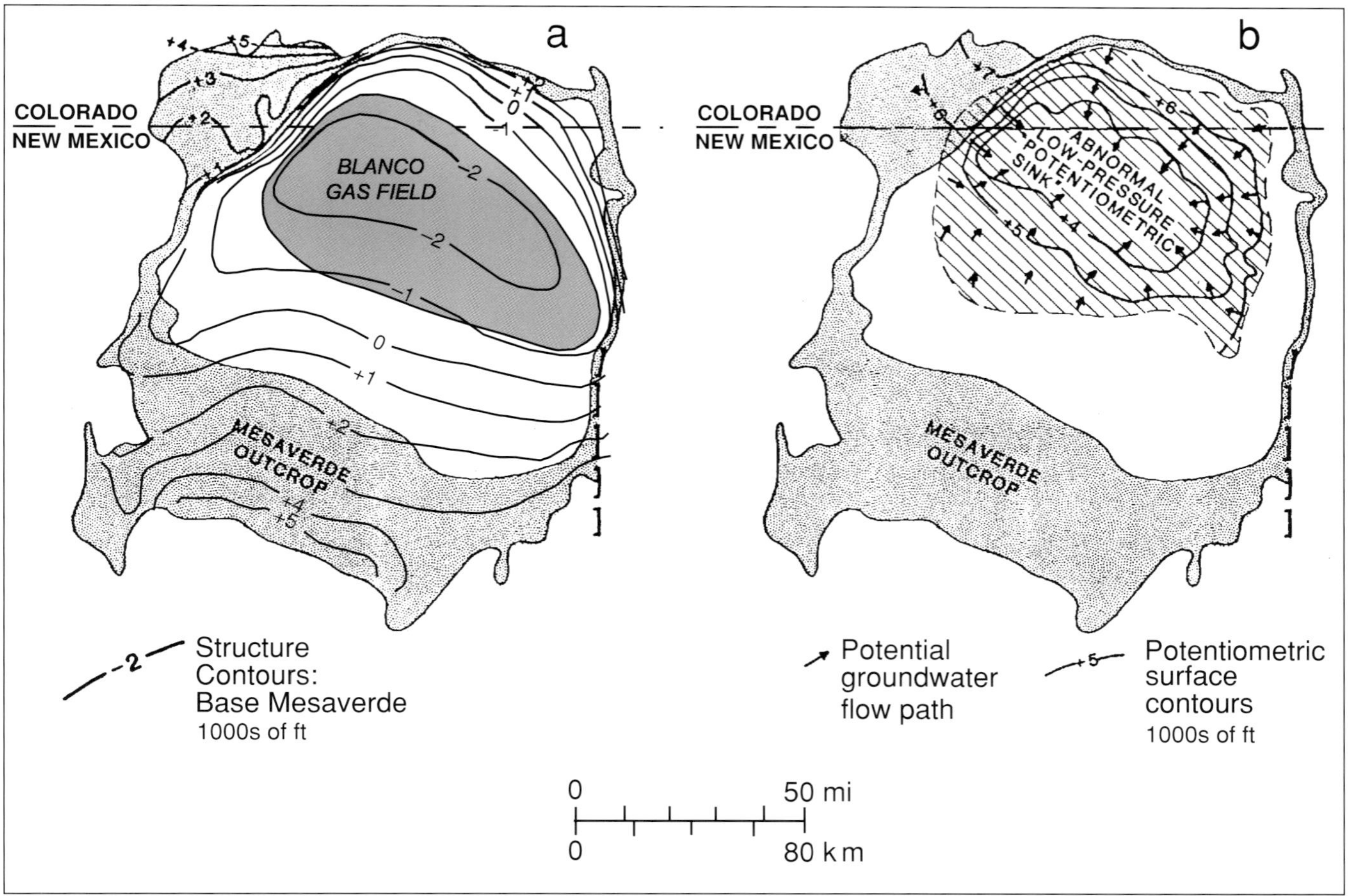

Figure 14. Maps depicting distribution of anomalous underpressures and gas production in the Mesaverde (Cretaceous) section of the San Juan Basin (after Berry, 1959). (a) Structure at the base of the Mesaverde and gas production. (b) Potentiometric surface map of the Mesaverde showing the area of an anomalous underpressured "potentiometric sink" and potential flow paths for groundwater.

Dolson, 1994; Slatt, 1998a, b; Kuuskraa, 1999). The latter is directly related to the development of Cave Gulch and Jonah gas fields (discussed later in this paper), where the understanding of limited drainage in thick, stacked pay intervals is important.

Seismic Techniques

The use of modern seismic acquisition and processing techniques has had a profound influence on exploration and development in recent years (Gries and Dyer, 1985; Ray, 1995; Rocky Mountain Association of Geologists, 1996, 1997, 1998, 1999). Because most Rocky Mountain reservoirs have low porosity, "bright-spot" technology has not been notably successful in directly identifying hydrocarbon accumulations. However, high-frequency 2-D, 2-D swath, and 3-D seismic methods, coupled with enhanced understanding of reflection amplitude, anisotropy, character, interval velocity, and apparent dip attitude, have resulted in numerous discoveries. Unusual or complex structures, such as overthrust areas, shear zones, and meteor impact features, have been interpreted through the use of analog geologic models. Similarly,

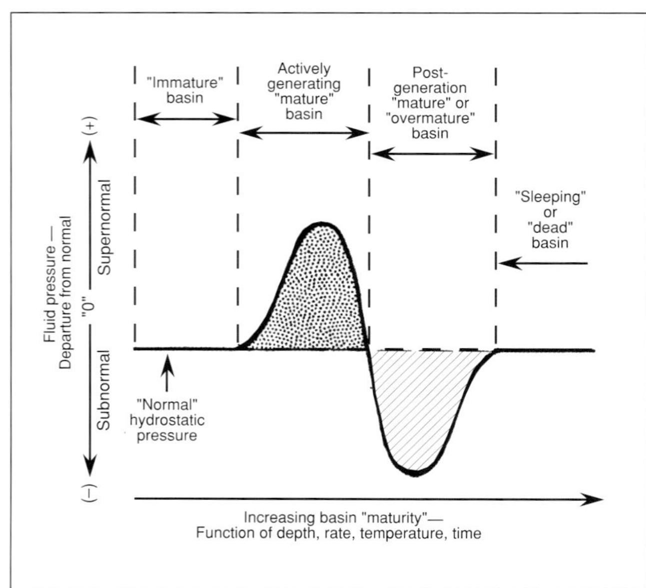

Figure 15. Evolution of fluid pressures in relation to hydrocarbon generation and migration in the setting of a deep-basin accumulation (Meissner, 1987).

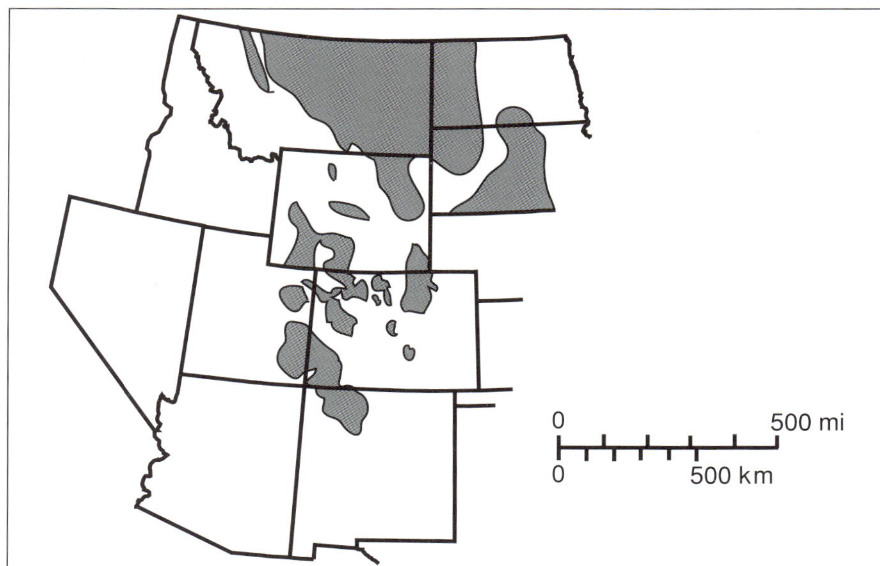

Figure 16. Shaded areas have potential for continuous-type gas production, exclusive of coal-bed methane (U.S. Geological Survey National Oil and Gas Resource Assessment Team, 1995).

reefs, mounds, channels, and other stratigraphic features have been identified. Current and ongoing advances in seismic techniques and interpretation will have a great impact on future exploration and development.

Data Management

The thousands of wells drilled in the GRMR have generated an extremely large amount of basic geologic and engineering data. Modern computer techniques have been and are being developed to sort, analyze, and plot this large volume of information. This continuing development will undoubtedly aid in identifying exploration prospects and development projects.

Drilling, Evaluation, and Completion

Advances in well drilling, evaluation, and completion technology have had significant impact on exploration and development. Horizontal wells offer great promise for exploiting reservoirs that are thin, have low permeability, are compartmentalized, are fractured, or contain viscous oil. Although the chief application of horizontal drilling in the Rocky Mountain region has been in developing fractured reservoirs, opportunity exists in many other reservoir types and conditions (Lacy et al., 1992; LeFever, 1992; Schmoker et al., 1992; Nydegger, 1992; Amateis and Hall, 1997). Many Rocky Mountain reservoirs

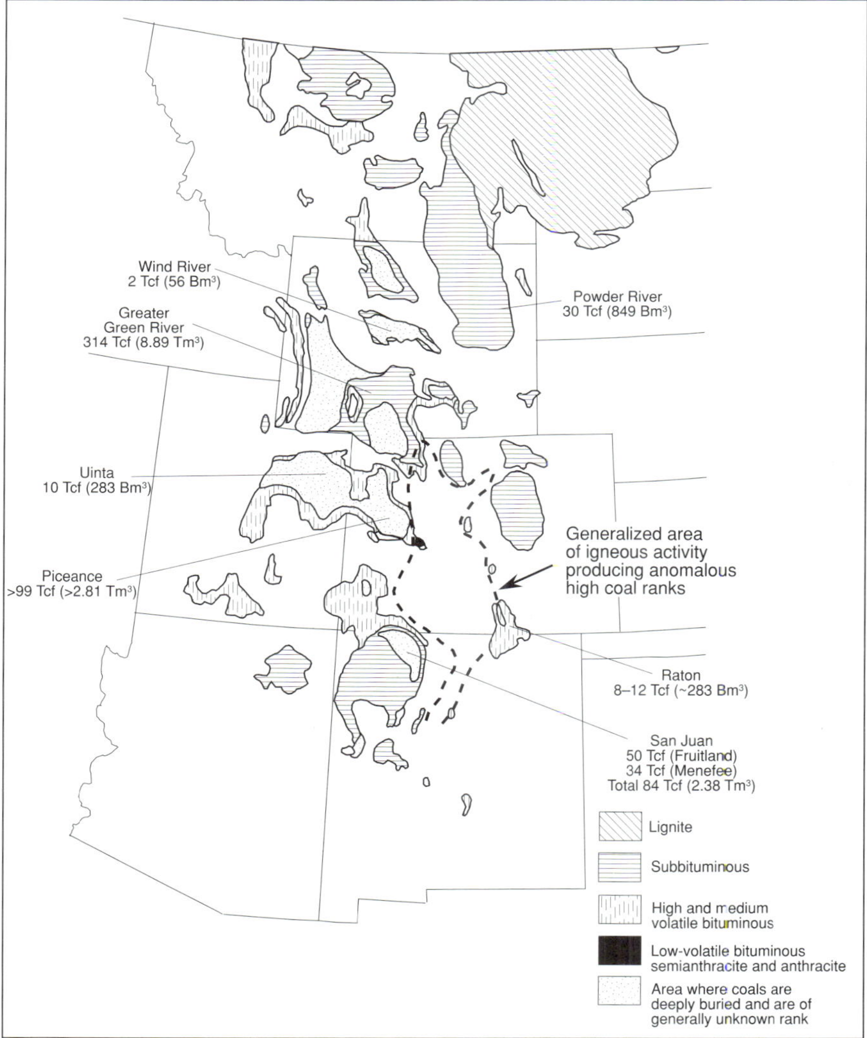

Figure 17. Major coal basins in the GRMR, with an assessment of rank at shallow depths and estimated contained volumes of gas (after Meissner, 1984; Scott, 1999).

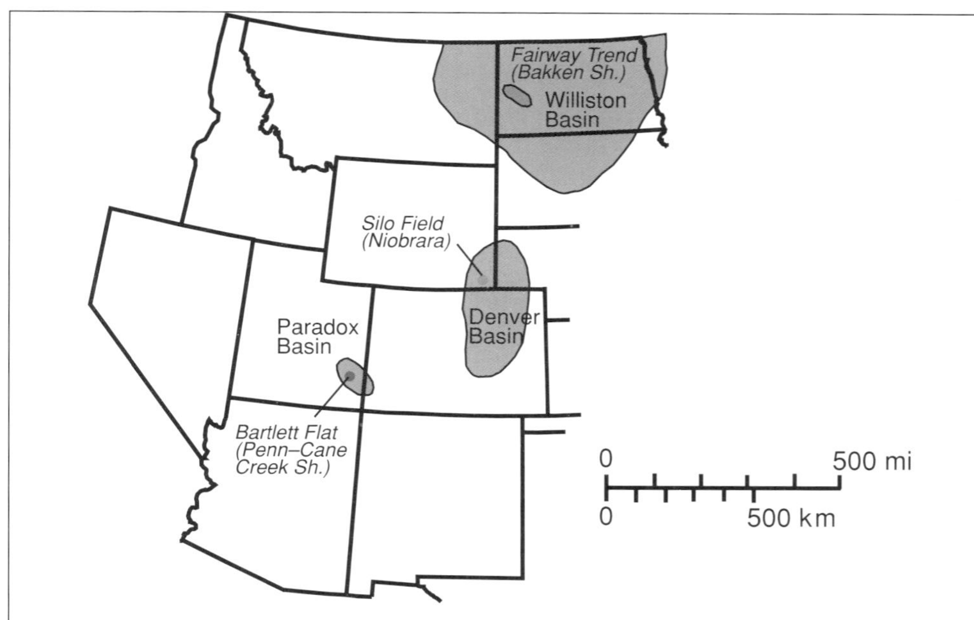

Figure 18. Areas where production from fractured source rocks has been established by horizontal drilling.

are substantially underpressured and have been penetrated with wells using overbalanced mud systems. Many of these reservoirs have undergone extensive reservoir damage that can be minimized by drilling with recently developed underbalanced mud- and flow-control drilling systems. Use of downhole motors and slim-hole drilling has lowered drilling expenses and has increased profits in the Denver Basin and has made some uneconomic reservoirs viable development targets.

Development and use of formation imaging logs have aided the identification of depositional and structural features (Bourke et al., 1989; Serra, 1989; Seiler et al., 1990). These logs have proved to be especially useful in evaluation of fractured reservoirs. Better understanding of log behavior in low-resistivity, low-contrast formations has led to better evaluation of potentially productive intervals in new or existing wells (Dolly and Mullarkey, 1996).

Hydraulic fracturing of low-permeability reservoirs in the Rocky Mountain area has produced economic production rates. Considerable progress has been made in designing less expensive and more efficient techniques, and improvements continue. Hydraulic fracture stimulation has proved to be successful in development of coal-bed methane (Ely et al., 1988). Cavity enlargement ("cavitation") has also proved to be a viable technique for enhancing production of coal-bed methane (Palmer et al., 1992).

POLITICAL AND ECONOMIC CONSIDERATIONS

Although the potential for discovering and developing substantial reserves in the GRMR obviously exists, whether it will occur in the future depends heavily on political and economic conditions. Agricultural areas are generally owned privately but may have government mineral rights. Some potentially productive areas are Native American tribal lands. Much of the GRMR is public land, administered by the Bureau of Land Management or the U.S. Forest Service. Some potentially prospective regions have been excluded from exploration as designated wilderness or national parks and monuments (e.g., northern Montana thrust belt, Escalante–Grand Staircase area of the Colorado Plateau in southern Utah). In addition, environmental restrictions and areas containing protected and endangered animal and plant species limit access and operations. Many prospective areas are in remote locations away from industry infrastructure. This is particularly important in the case of gas pipelines.

The Resource Pyramid

The concept of a resource triangle applied to an assessment of the economic viability of existing petroleum deposits was first proposed by Masters (1979). It was subsequently adapted by Thomasson (1982) and modified by Kuuskraa and Schmoker (1998) into a resource pyramid, as depicted in Figure 19. The apex of the pyramid represents a relatively small amount of oil or gas in very rich, easily found and exploited fields that have highly favorable economics. Most of the total available resource lies in the lower part of the pyramid, in accumulations that are leaner and less easily found and exploitable, associated with poor or unprofitable economics. At any given time, the ability to move downward from the apex of the pyramid depends on the product price and the finding and production costs. Increasing technical capability gives the explorationist and the exploitationist the ability to discover commercial oil and gas from leaner accumulations.

In the case of oil, several recent plays have demonstrated the trend toward exploiting resources in the lower part of the pyramid. Examples of this are production established from fractured reservoirs in limestones in the Niobrara Formation at Silo field in the southeast Wyoming part of the Denver Basin, shales and siltstones of

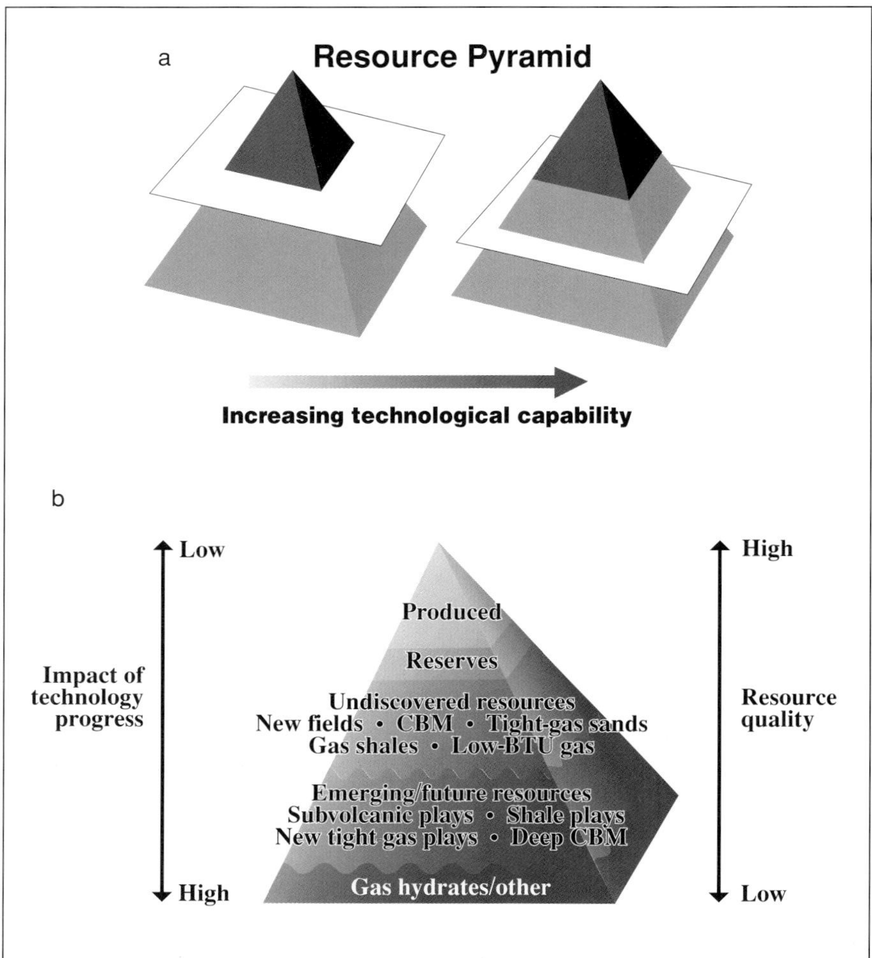

Figure 19. Resource pyramids. (a) Pyramid showing amount of recoverable resource as a function of technological capability to exploit the resource. (b) Resource pyramid for gas in the GRMR showing the position of major future exploration potential with respect to reservoir types and plays and their relation to resource quality, and the impact that future technology may have on the ability to exploit the available resource. CBM = coal-bed methane.

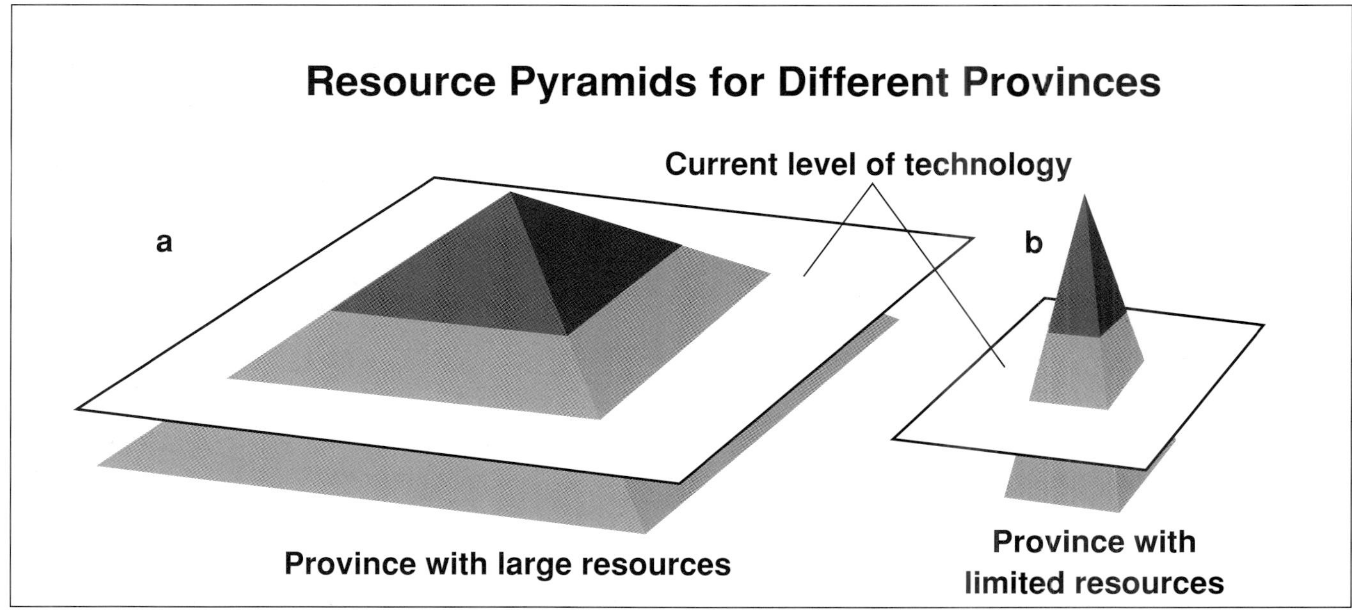

Figure 20. Resource pyramids with different base dimensions. (a) Pyramid with large amount of lower-quality resource. This type of pyramid is considered representative of the Rocky Mountain region. (b) Pyramid with small amount of low-quality resource. This pyramid is considered to be representative of many of the world's productive areas.

the Bakken Formation Fairway Trend in the North Dakota part of the Williston Basin, and shale in the Cane Creek Member of the Paradox Formation in the Paradox Basin of eastern Utah. All of these examples represent basin-center-type accumulations developed in mature source rocks that were probably fractured by overpressures developed during active generation. Fractured reservoirs in Bakken and Cane Creek are substantially overpressured; those in the Niobrara at Silo field are slightly underpressured.

We see very significant additional opportunities for such plays in the San Juan, Uinta, Powder River, Denver, Paradox, and Williston Basins and many of the Basin and Range basins. With further technological advances, an increasing volume of rock will become attractive for effective exploration and economic exploitation.

The shape of the resource pyramid shown in Figure 20a clearly describes a much larger potential resource base than that shown in Figure 20b; i.e., the volume-to-height ratio is greater in Figure 20a and increases exponentially downward from the apex. Because of the extremely large coal-bed gas resources (Figure 17) and the abundance of basin-centered or continuous-type oil and gas accumulations (Figures 12, 16) that are generally associated with large but poor-quality accumulations, the Rocky Mountain region is characterized best by the middle portion of the much broader pyramid of Figure 20a. A position in the middle of the broader pyramid is highly favorable for exploiting large reserves, if technology for doing so is available and economic considerations are favorable. Most of the oil and gas postulated by the U.S. Geological Survey to be technically available for future exploration and exploitation is contained in what it terms "unconventional" continuous-type accumulations in the middle part of the broader resource pyramid. These accumulations are generally characterized by "tight" matrix porosity and/or fractured reservoirs, anomalous pressures, large areas of complete hydrocarbon saturation, and coal-bed methane.

EXAMPLES OF RECENT SIGNIFICANT DISCOVERIES—ANALOGS FOR THE FUTURE

Although an assessment of the resource pyramid appropriate for the GRMR demonstrates a basis for predicting a large amount of potentially discoverable and exploitable hydrocarbon resources, further evidence justifying this prediction may be based on the recent history of exploration and development, wherein five "giant" fields containing more than 100 million bbl of oil or 1 tcf of gas have been discovered (in most cases, rediscovered) in the mid- to late 1990s, largely through application of new technology. These fields (Figure 21) include (1) an oil field with a hydrodynamically tilted oil-water contact localized in a thin pay on the flank of a large regional anticline (Cedar Hills field, Williston Basin), (2) an overpressured gas field where reserves were derived from an underlying and downdip regional basin-center accumulation (Jonah field, Green River Basin), (3) a field containing thick columns of gas and condensate localized in a complex anticline hidden beneath a thrust plate (Cave Gulch field, Wind River Basin), (4) a coal-bed methane field containing a mix of thermal and biogenic gas in a high-volatile coal (Greater Drunkard's Wash field, western Uinta Basin), and (5) a coal-bed methane field containing biogenic gas in a low-rank coal (Tongue River coals, Powder River Basin). Details relating to geologic concepts and applied technology that led to the discovery or redevelopment of these fields are described in the following sections.

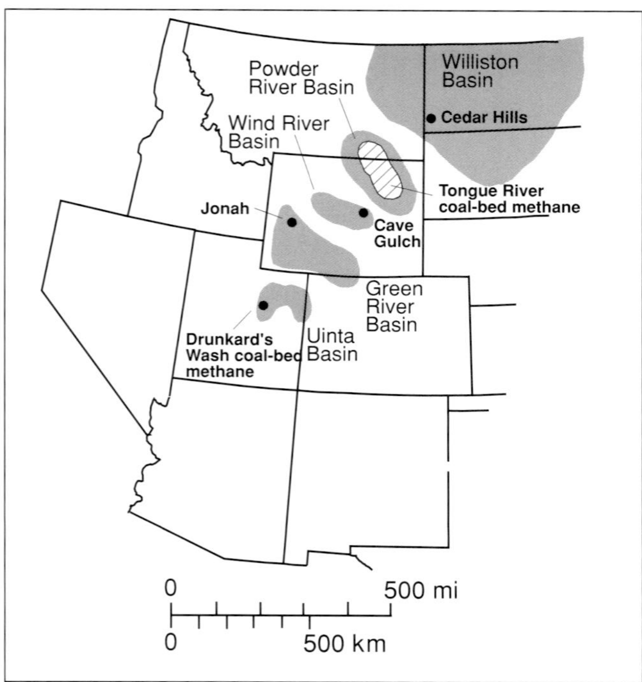

Figure 21. Map showing locations of recent major oil and gas discoveries that may serve as examples of the potential for discovery of future accumulations which may be found in the GRMR.

Cedar Hills Oil Field

A large oil field has recently been "discovered" in the Ordovician Red River "B" zone in the Williston Basin. Production at Cedar Hills field is from a conventional reservoir but is localized in a somewhat unconventional hydrodynamic trap. The oil at Cedar Hills field is being exploited using horizontal-drilling technology (Montgomery, 1997). The reservoir is a thin porous interval that had previously been penetrated by several completed and abandoned oil wells associated with noneconomic rates of production. A porous interval (the Ordovician Red River "B" zone) is somewhat variable in thickness and extends over a very large area where oil is trapped hydrodynamically.

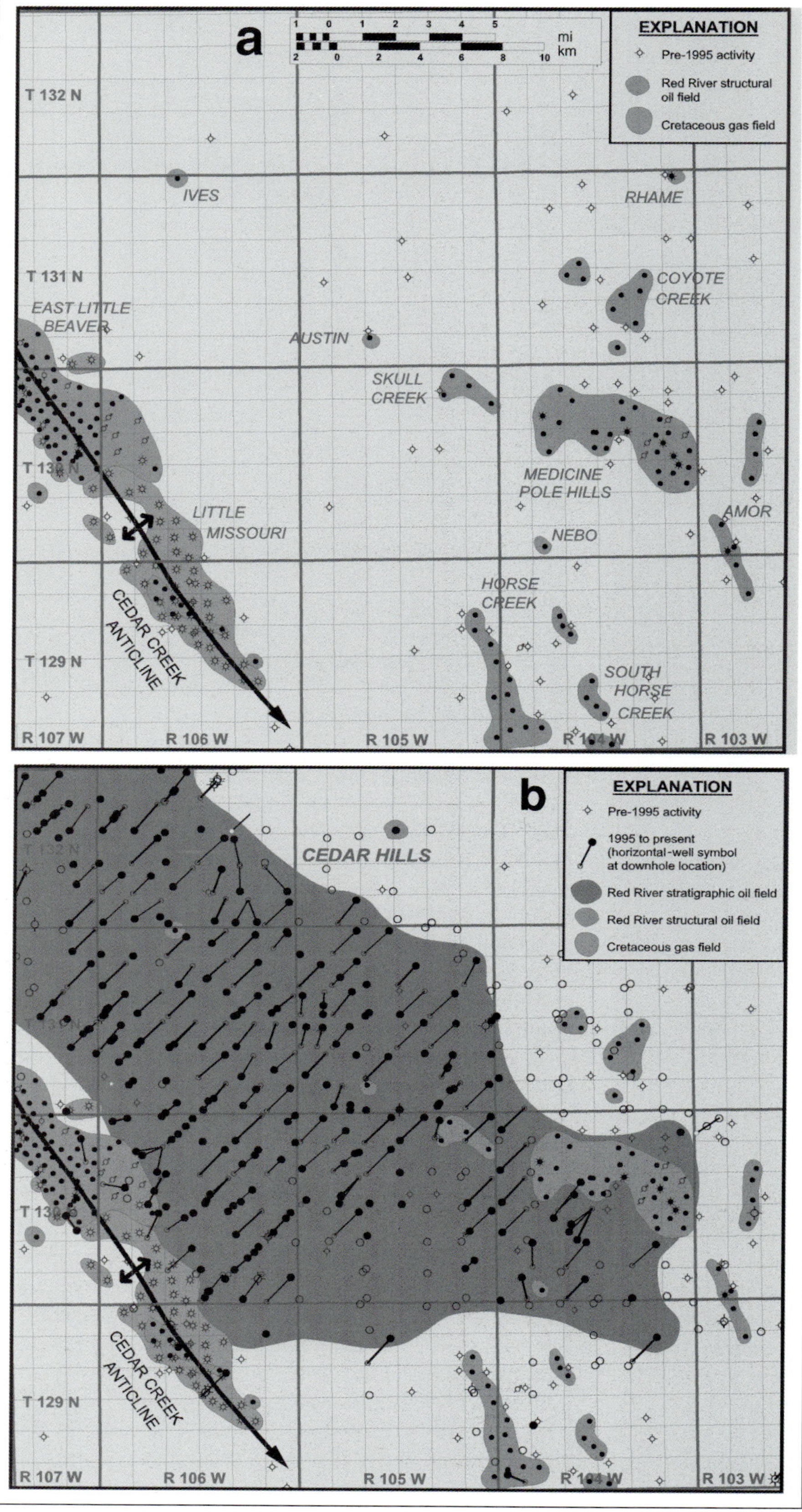

Figure 22. Cedar Hills field, Williston Basin, Bowman County, North Dakota. Productive area and well penetrations (a) before 1995, (b) 1995 to present. The EUR added since 1995 is more than 100 million bbl of oil.

Figure 22a shows the drilling status for the Cedar Hills area prior to 1995. Figure 22b shows the approximate outline in the same area of the newly horizontally exploited Red River "B" zone. Approximately 170 horizontal wells have been drilled, resulting in 150 productive completions ranging in true vertical depth of 8800 to 9500 ft. The total estimated ultimate recovery is now expected to be more than 130 million bbl of oil.

Jonah Gas Field

The Jonah gas field is a structurally controlled sweet spot in the basin-centered gas area of the Green River Basin of Wyoming (Warner, 1997, 1998). By year-end 2000, it contained more than 335 wells, with a per-well average producible reserve of 6 billion cubic feet (bcf) of gas (Ed Warner and Snyder Oil Company, personal communication, 2001). Figure 23a shows the situation at Jonah field in 1993. The field was discovered in 1975 by the Davis Oil 1 Wardell Federal, which had an initial flow rate of 303 mcf of gas/day (MCFGD) and 2 bbl of oil/day (BOPD). It was rediscovered in 1985 by the Home Petroleum 1-4 Jonah Federal, which tested at an initial rate of 470 MCFGD. It was rediscovered in 1993 by the McMurry Oil 1-5 Jonah Federal, which tested at an initial rate of 3.7 million cubic feet of gas/day (MMCFGD) and 40 BOPD. The increase in production rates is attributed to improved technology in the form of better stimulation techniques. Completion technology continues to improve so that today the initial flow rate in a pay section similar to that found in the McMurry Oil 1-5 Jonah Federal could be 10–12 MMCFGD. The average estimated ultimate recovery (EUR) per well was 2 bcf before 1994. Improved fracturing technology has raised the EUR steadily since 1992. For 1997, it was 6.7 bcf (Esphahanian et al., 1998). The current configuration of the field is shown in Figure 23b.

Current development has not yet established the eastern limit of the field, and a recent step-out by Amoco has extended production more than four miles. Jonah field has a gas column of approximately 3000 ft. Gas is trapped laterally and updip by a set of shear zones that raises a geopressured sweet spot about 3000 ft stratigraphically high to the top of regional deep-basin-type overpressuring, with essentially no vertical displacement of the main reservoir section. The use of 3-D seismic has been of particular benefit in recognizing and mapping the controlling faults.

The shear zones that seal the sweet spot laterally and updip originate in the basement and have almost no vertical throw. However, the field itself is highly broken by faults which control anomalous overpressure displacements of as much as 600 ft in separated fault blocks within the field. This extensive internal faulting and fracturing have allowed gas to migrate vertically upward through tight rocks to a shallower reservoir, which has experienced less diagenesis and thus has higher porosity than the underlying rocks associated with source-rock gas generation. Current development indicates a recoverable volume of at least 1 tcf of gas. The field appears to have as much as 12–18 tcf of gas in place (Ed Warner, personal communication, 2000). With a 25% recovery, ultimate production may amount to as much as 3–5 tcf of gas.

Cave Gulch Gas Field

The locations and status of wells drilled in the Cave Gulch–Waltman area of the Wind River Basin in Wyoming prior to 1994 are shown in Figure 24a. The area may ultimately contain more than the 1 tcf of new gas reserves added since 1994, primarily because of a different geologic concept, understood by Larry McPeek, originator of the project that led to the rediscovery and new development. McPeek recognized that even though Cave Gulch was a relatively small structural closure under the Owl Creek thrust plate, the fluvial depositional regime of the Fort Union (lower Tertiary) and Lance (uppermost Cretaceous) Formations would allow stacking of a very thick package of sandstones containing complexly compart-

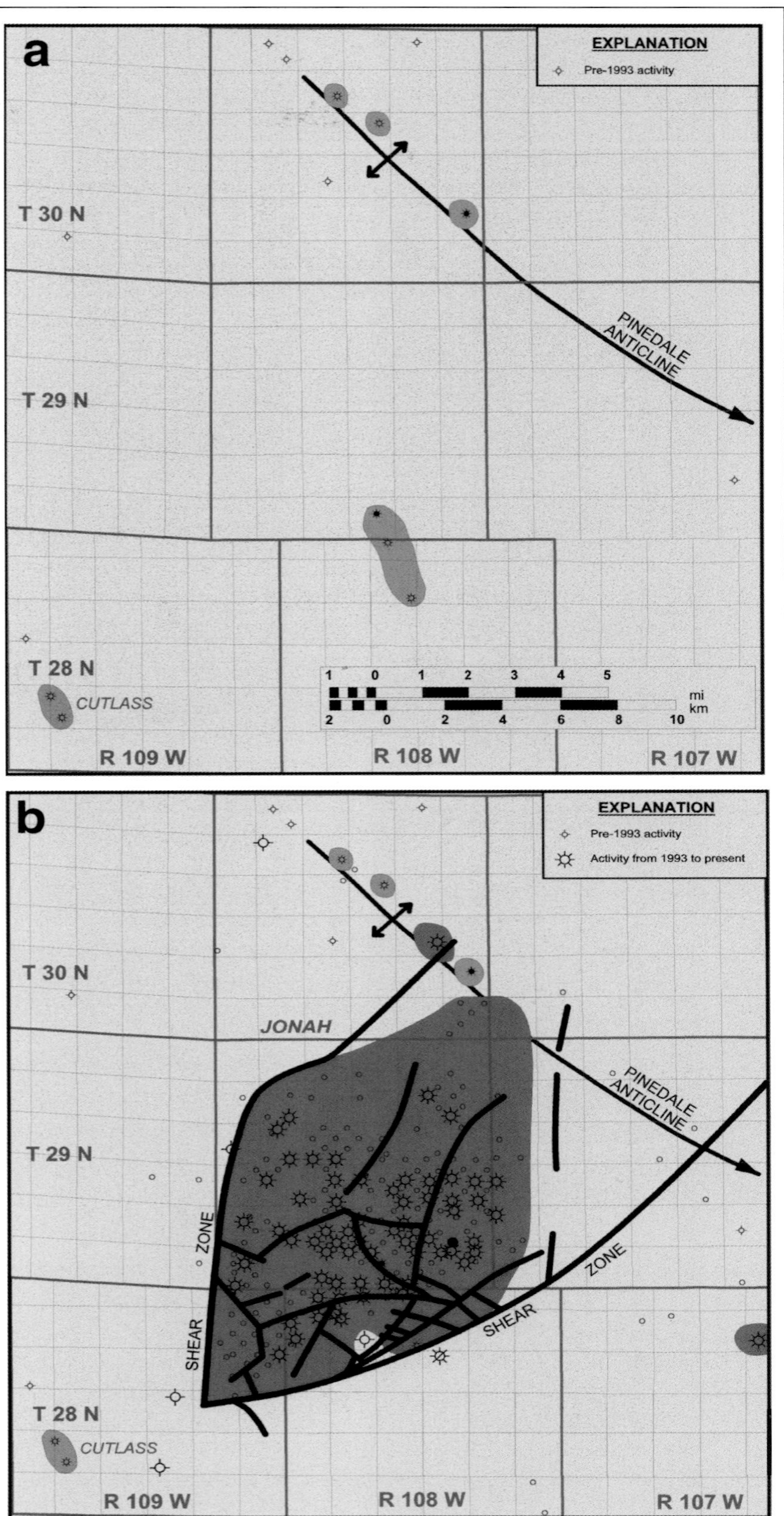

Figure 23. Jonah field, Green River Basin, Sublette County, Wyoming. Productive area and well penetrations (a) before 1993, (b) 1993 to present. The EUR added since 1993 is more than 1 tcf of gas.

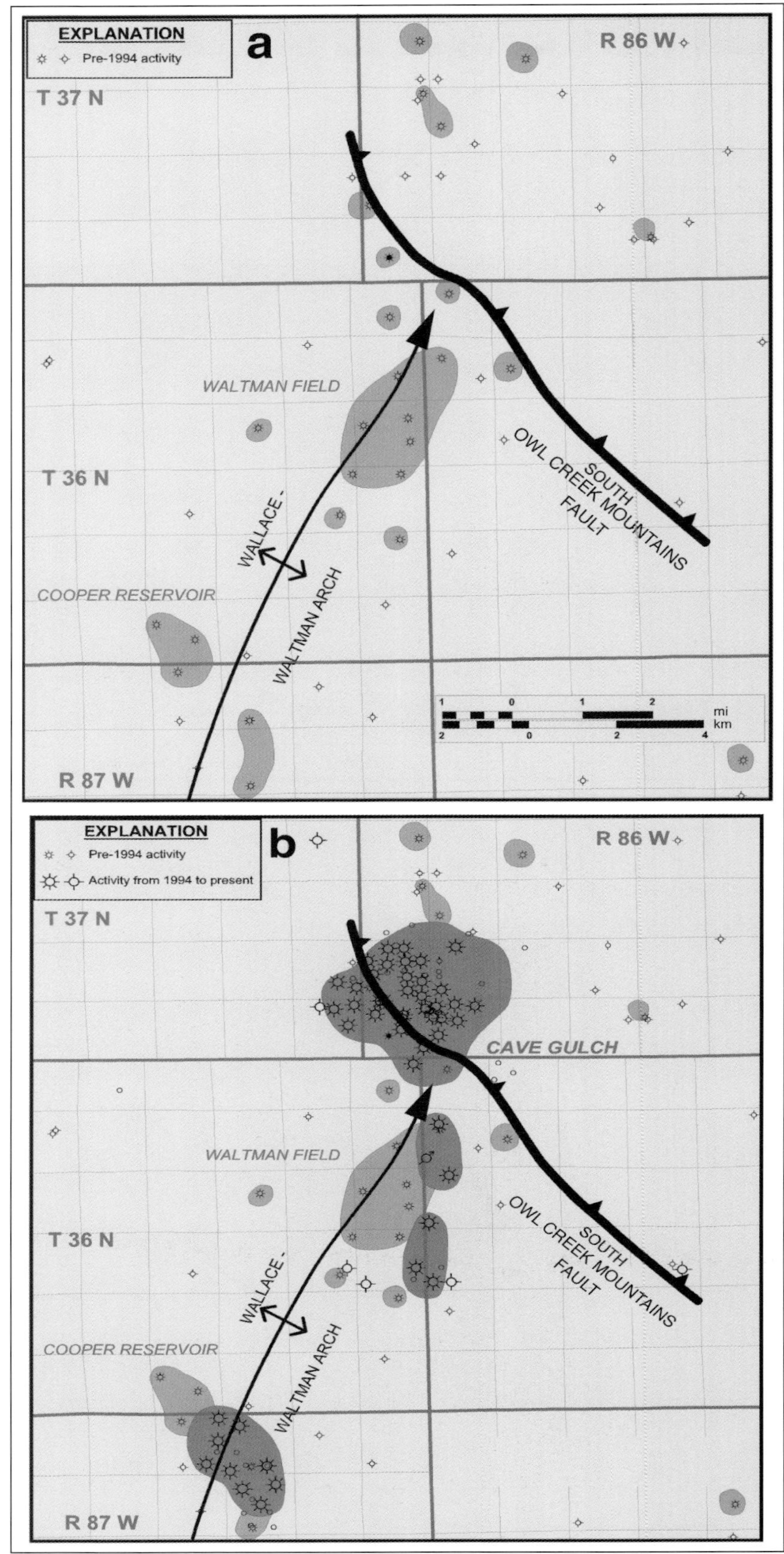

Figure 24. Cave Gulch field, Wind River Basin, Natrona County, Wyoming. Productive area and well penetrations (a) before 1994, (b) 1994 to present. The EUR added since 1993 is more than 1 tcf of gas.

mentalized reservoirs and very limited drainage areas, favoring tight well spacing and multiple twins[1]. For instance, in one 160-acre area, 14 wells have been completed from two surface locations. In addition, McPeek recognized that the deeper sands, which had proved to be noncommercial in wells drilled prior to 1994, were prospective on paleostructural highs where early gas accumulation should reduce or eliminate diagenetic destruction of reservoir quality in deeper reservoirs. New fracturing technology has also played a significant role in successfully stimulating these deeper zones.

Since 1994, 48 shallow wells (3000–10,000 ft) and six deep wells (17,000+ ft) have been drilled in Cave Gulch field (Figure 24b). Current development has established a maximum net pay section of approximately 1300 ft. One deep well blew out at calculated absolute open flow of 1 bcf of gas/day.

Drunkard's Wash Coal-bed Methane Field

Figure 25a shows the Greater Drunkard's Wash gas area in the Uinta Basin prior to 1993. Today, 240 wells are producing in the area, with a drilling program of approximately 60 wells/year projected for the foreseeable future. In 1993, the Buzzard's Bench coal-bed methane field had only one well. It now has more than 40. It appears that the entire area between Drunkard's Wash and Buzzard's Bench fields eventually will form a continuous field (see Figure 25b).

According to Lamarre and Burns (1999), the average coal thickness in Drunkard's Wash field is 24 ft (8 m). One well that has been producing for more than five years from a 28-ft-thick coal has a cumulative production of 3.5 bcf of gas and is producing 1461 MCFGD and 369 bbl of water/day (BWPD). The first 33 wells have produced for more than 65 months, and their per-well daily production averages slightly less than 1 MMCFGD and 85 BWPD. The average per-well daily gas production has increased by 380% and water production has decreased by 80%. None of these wells declined after 5.5 years of continuous production. Considering the current and projected number of development

[1] "Twin" = two wells, drilled from the same surface location and completed in different zones.

wells, it seems reasonable to assume that the Greater Drunkard's Wash area, including Buzzard's Bench, may eventually produce as much as 3 tcf of gas.

Powder River Basin Coal-bed Play

Another coal-bed methane giant gas field is being developed rapidly in the Powder River Basin. Figure 26a shows wells drilled prior to 1994; Figure 26b approximates current well development. However, with 41 wells being drilled and plans for as many as 1000 wells, this play is in a state of explosive development. We believe that this giant field developing in the Tongue River Coal Member of the Paleocene Fort Union Formation, which ranges from 300 to 1200 ft in depth, will establish reserves in excess of 1 tcf of gas. Estimates of 7–12 tcf of gas have been made (Montgomery, 1999).

SUMMARY AND CONCLUSIONS

The GRMR is a large, geologically heterogeneous area containing numerous basins and uplifts. Although it contains a wide variety of structures that were generated at several times, those that were produced in the early and late Tertiary are the most significant commercially. Sedimentary rocks representing certain time periods are absent in some areas. However, numerous oil-prone and gas-prone source rocks and prospective reservoirs ranging from Precambrian to Tertiary in age are present in one area or another. These have contributed to the presence of a large number and variety of petroleum systems. Productive and prospective reservoirs include a spectrum of carbonates and sandstones containing matrix porosity and permeability, as well as fracture-type and coal-bed methane reservoirs. Various investigating entities have estimated the potential for future producible hydrocarbon discoveries to be 10.4 to 15.4 billion bbl of petroleum liquids and 192 to 260 tcf of gas.

Known hydrocarbon accumulations include those controlled by conventional structural, stratigraphic, and combination trap types that often are affected significantly by dynamic groundwater

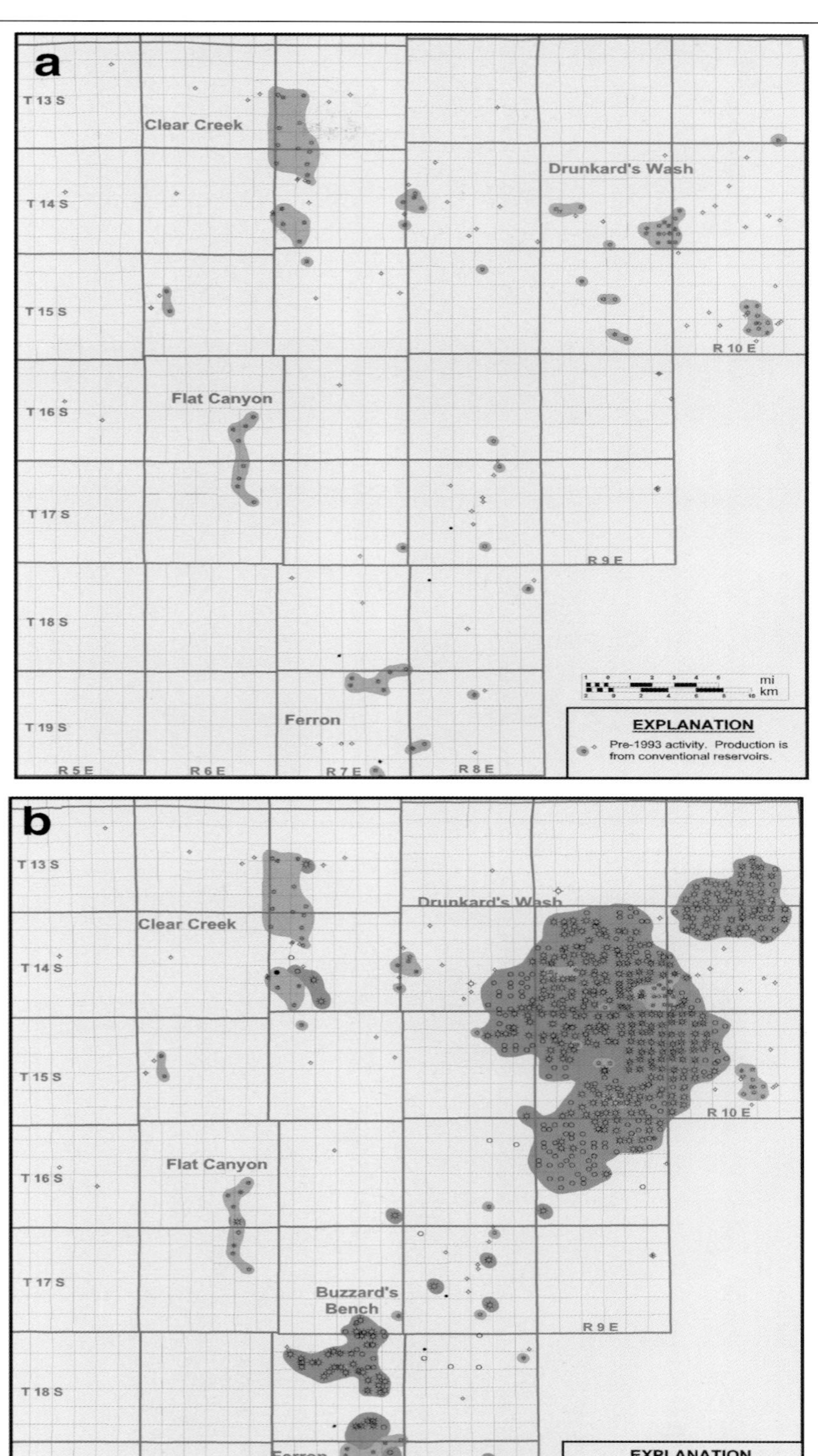

Figure 25. Drunkard's Wash coal-bed methane field, Wasatch plateau, Nevada. Productive area and well penetrations (a) before 1993, (b) 1993 to present. Pre-1993 production shown was all from conventional sandstone reservoirs. The added production is primarily from coals with an EUR of 1–3 tcf of gas.

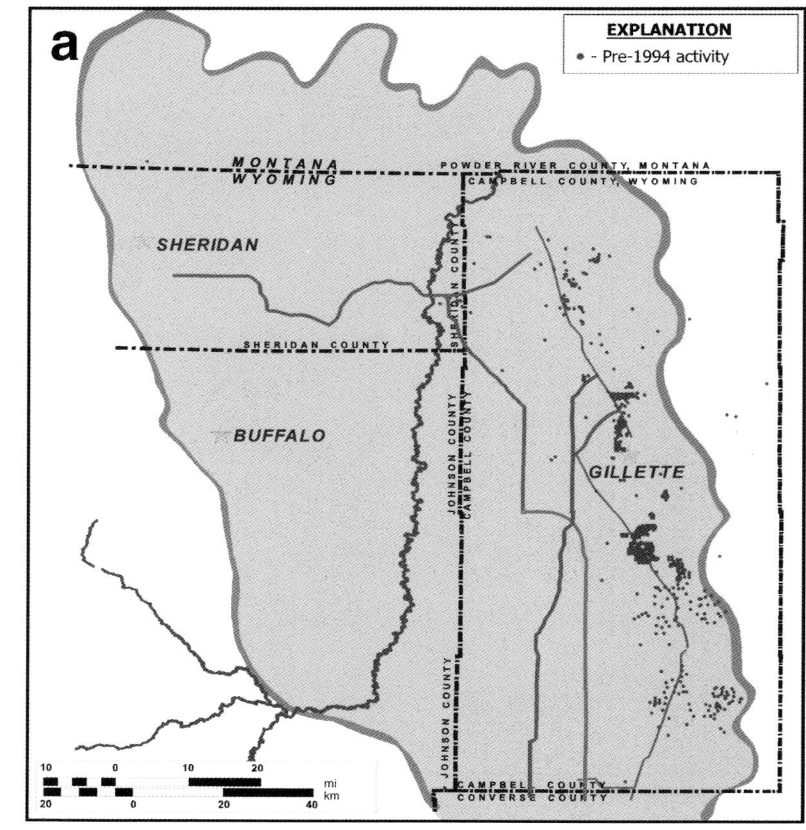

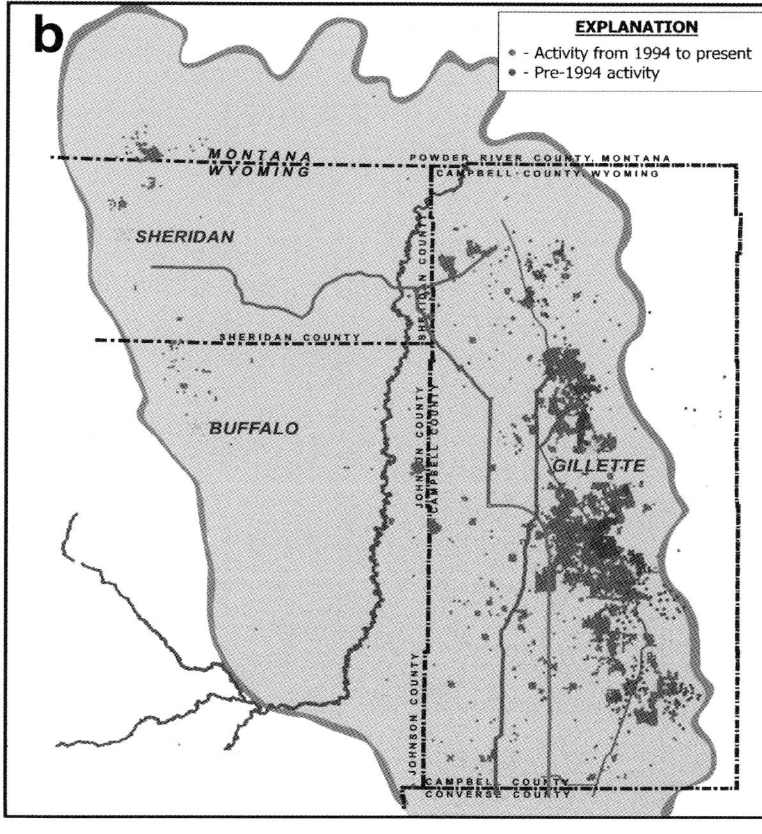

Figure 26. Powder River Basin coal-bed methane development, Campbell, Johnson, and Sheridan Counties, Wyoming, and Powder River County, Montana. Productive area and well penetrations (a) before 1994, (b)1994 to present. The EUR added since 1994 is more than 1 tcf of gas. Basin potential is 7–12 tcf of gas.

flow. Of particular significance to further exploration and development potential is a class of unconventional accumulations associated with pervasive regional hydrocarbon saturation, general absence of free water, and presence of abnormally high or low fluid pressures. These accumulations may be dynamic and transient in nature. They commonly occur in low-permeability or fractured reservoirs associated with mature source rocks in the deeper parts of typical Rocky Mountain basins. Petroleum systems in the Cretaceous and lower Tertiary section will be major contributors to future hydrocarbon production, and gas will be of particular importance because of the large number of coal measures present.

Much of the potential hydrocarbon resource remaining to be discovered and developed is characterized as representing the largest overall volume in the resource pyramid. The exploitation of this resource will depend heavily on the price of the product and the application of new and developing technology that will lower the cost of exploration and enable economically attractive development.

The presence of five large oil and gas fields that have been developed in the last seven years may be taken as an indication of the type and magnitude of remaining resource type and potential. Most of these so-called discoveries represent hydrocarbon accumulations that were previously known through prior exploration but had little economic significance until the development of geologic understanding, drilling, evaluation, and completion technology rendered them economically viable.

APPENDIX

Estimated amounts of oil and gas that may be found and developed in the GRMR.

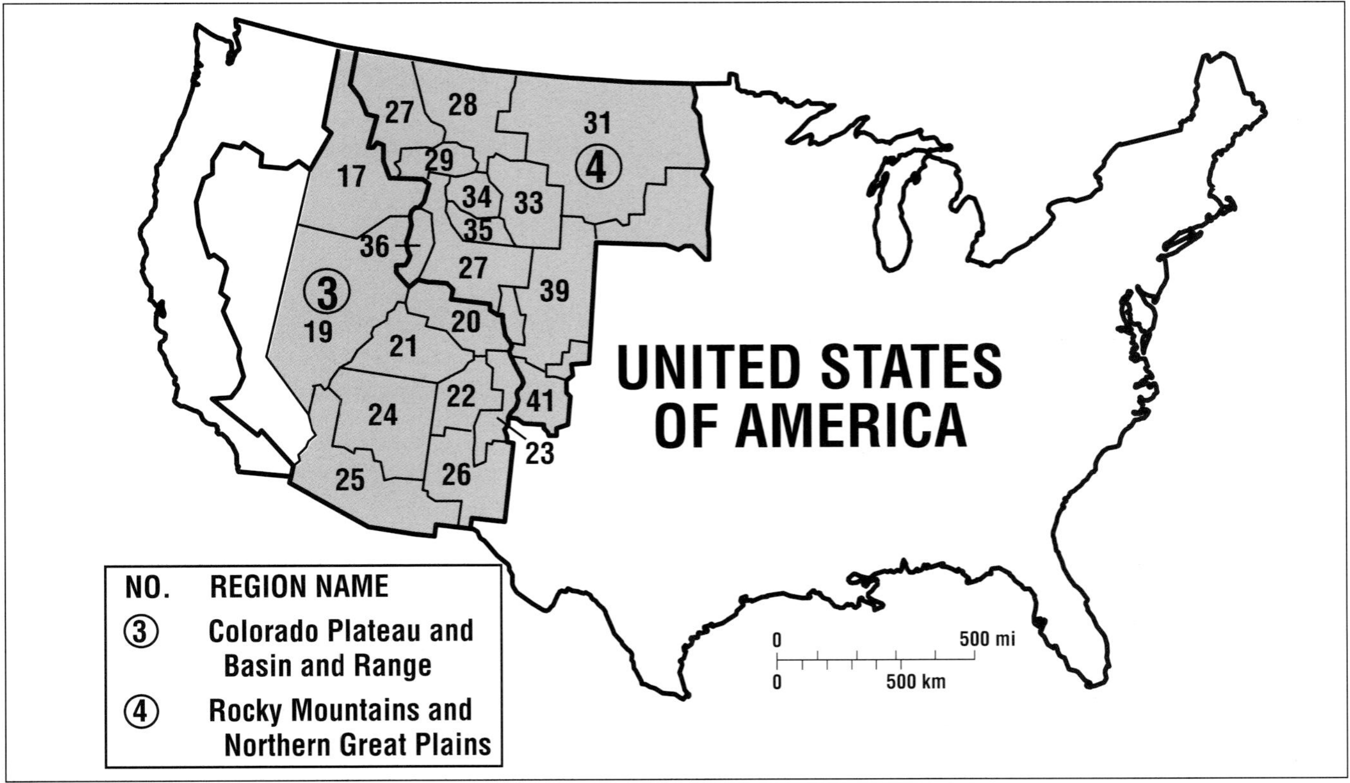

Figure A-1. Index map showing the Greater Rocky Mountain Region and its geologic province boundaries as defined by the U.S. Geologic Survey (1995). Index numbers are keyed to province names in Tables A-1–A-3.

Table A-1. Greater Rocky Mountain Region U.S. Geological Survey province names (from U.S. Geological Survey National Oil and Gas Resource Assessment Team, 1995, Table 2, p. 14).

Province number	Province name	Province number	Province name
17	Idaho–Snake River Downwarp	29	Southwest Montana*
19	Eastern Great Basin*	31	Williston Basin, U.S.*
20	Uinta-Piceance Basin*	33	Powder River Basin*
21	Paradox Basin*	34	Big Horn Basin*
22	San Juan Basin*	35	Wind River Basin*
23	Albuquerque–Santa Fe Rift	36	Wyoming Thrust Belt*
24	Northern Arizona*	37	Southwestern Wyoming*
25	Southern Arizona–Southwestern New Mexico	38	Park Basins*
26	South-central New Mexico	39	Denver Basin*
27	Montana Thrust Belt*	40	Las Animas Arch*
28	North-central Montana*	41	Raton Basin–Sierra Grande Uplift

* Provinces with oil and/or gas production

Table A-2. Known volumes of oil, natural-gas liquids (NGL), and gas in Greater Rocky Mountain Region geologic provinces (from Kett et al., 1997, Table 1).

		Median estimated values				Percent of GRMR reserves	
Province number (Figure A-1)	Name	Oil (billion bbl)	NGL (billion bbl)	Total liquids (billion bbl)	Gas (tcf)	Total liquids (billion bbl)	Gas (tcf)
19	Eastern Great Basin	0.05	0	0.1	0.05	0.3	0.1
20	Uinta-Piceance Basins	1.6	0.1	1.7	4.7	10.0	5.9
21	Paradox Basin	0.6	0.1	0.7	0.9	4.1	1.1
22	San Juan Basin	0.3	1.4	1.7	38.2	10.0	47.8
24	Northern Arizona	0.05	0	0.1	0.05	0.3	0.1
27	Montana Thrust Belt	0.05	0	0.1	0.05	0.3	0.1
28	North-central Montana	0.5	0.05	0.6	2.3	3.2	2.9
29	Southwestern Montana	0.05	0.05	0.1	0.2	0.6	0.3
31	Williston Basin, U.S.	2.1	0.2	2.3	2.4	13.5	3.0
33	Powder River Basin	2.8	0.2	3	2.6	17.6	3.3
34	Big Horn Basin	2.7	0.05	2.8	1.8	16.2	2.3
35	Wind River Basin	0.5	0.05	0.6	2.8	3.2	3.5
36	Wyoming Thrust Belt	0.2	0.7	0.9	4	5.3	5.0
37	Southwestern Wyoming	0.8	0.4	1.2	16.3	7.1	20.4
38	Park Basins	0.05	0	0.05	0.05	0.3	0.1
39	Denver Basin	0.9	0.3	1.2	3.5	7.1	4.4
40	Las Animas Arch	0.1	0.05	0.2	0.1	0.9	0.1
	Total GRMR:	**13.4**	**3.7**	**17**	**80**	**100.0**	**100.0**
	Total U.S.:	**171.7**	**29.4**	**201.1**	**908.6**		
GRMR percent of total U.S.:			7.8%	12.6%	8.5%	8.8%	

Table A-3. Greater Rocky Mountain Region estimated undiscovered technically recoverable resources.

Province number (Figure A-1)	Province name	Oil (billion bbl)			Gas (tcf)			NGL (billion bbl)		
		F_{95}*	F_5*	Mean	F_{95}*	F_5*	Mean	F_{95}*	F_5*	Mean
Estimates of undiscovered technically recoverable convention oil, NGL, and gas resources by province (U.S. Geological Survey National Oil and Gas Resource Assessment Team, 1995, Table 2, p. 12)										
17	Idaho–Snake River Downwarp	0.00	0.01	<0.01	0.00	0.09	0.01	0.00	0.00	0.00
19	Eastern Great Basin	0.06	1.34	0.49	0.01	1.12	0.33	0.00	0.02	0.01
20	Uinta-Piceance Basin	0.04	0.59	0.21	1.94	9.56	4.54	0.01	0.21	0.07
21	Paradox Basin	0.11	0.61	0.31	0.93	3.48	2.02	0.03	0.17	0.09
22	San Juan Basin	0.07	0.29	0.16	0.52	1.53	0.98	0.01	0.04	0.02
23	Albuquerque–Río Grande Rift	0.00	0.15	0.05	0.00	1.29	0.36	0.00	0.06	0.02
24	Northern Arizona	0.00	0.32	0.07	0.00	0.96	0.17	0.00	0.08	0.01
25	Southern Arizona–Southwestern New Mexico	0.00	0.06	0.02	<0.01	0.53	0.20	<0.01	0.05	0.02
26	South-central New Mexico	0.00	0.00	0.00	0.00	0.00	0.00	0.00	0.00	0.00
27	Montana Thrust Belt	0.00	0.01	<0.01	0.00	8.53	1.93	0.00	0.03	0.01
28	Central Montana	0.13	0.42	0.27	0.04	1.37	0.84	<0.01	<0.01	<0.01
29	Southwest Montana	0.00	0.13	0.03	0.12	0.76	0.40	<0.01	0.01	<0.01
31	Williston Basin	0.25	1.17	0.65	0.89	2.61	1.69	0.07	0.22	0.14
33	Powder River Basin	0.70	3.88	1.94	0.66	2.88	1.60	0.04	0.17	0.09
34	Big Horn Basin	0.08	0.87	0.39	0.24	1.19	0.62	<0.01	0.02	0.01
35	Wind River Basin	0.05	0.32	0.16	57.00	2.21	1.24	0.01	0.02	0.01
36	Wyoming Thrust Belt	0.21	1.16	0.63	5.55	16.60	10.68	0.51	1.84	1.13
37	Southwestern Wyoming	0.04	0.40	0.17	0.70	2.86	1.58	0.01	0.04	0.02
38	Park Basins	<0.01	0.11	0.03	<0.01	0.07	0.02	0.00	<0.01	0.00
39	Denver Basin	0.09	0.42	0.23	0.34	1.38	0.75	0.01	0.04	0.02
40	Las Animas Arch	0.04	0.28	0.14	0.20	1.07	0.53	0.01	0.03	0.01
41	Raton Basin–Sierra Grand Uplift	0.00	0.00	0.00	0.00	0.11	0.04	0.00	<0.01	<0.01
	TOTAL	**3.72**	**8.93**	**5.93**	**20.70**	**43.88**	**30.53**	**0.95**	**2.59**	**1.68**
Technically recoverable resources estimated for continuous-type plays in sandstones, shales, and chalks (U.S. Geological Survey National Oil and Gas Resource Assessment Team, 1995, Table 4, p. 18)										
20	Uinta-Piceance Basin	0.059	0.139	0.094	11.55	23.38	16.74	63	139	96
21	Paradox Basin	0.061	0.139	0.242	0.05	0.48	0.19	0	0	0
22	San Juan Basin	0.068	0.597	0.189	10.66	36.84	21.15		2	1
28	Central Montana	0	0.394	0	19.92	79.03	43.16	0	0	0
31	Williston Basin	0.097	0.283	0.167	0.08	0.24	0.14	0	0	0
37	Southwestern Wyoming	0	0	0	55.95	213.51	119.30	810	3104	1733
39	Denver Basin	0.139	0.502	0.285	1.49	5.69	3.16			
	TOTAL	**0.520**	**1.635**	**0.977**	**116.74**	**323.91**	**203.85**	**873**	**3244**	**1829**
Technically recoverable resources estimated for continuous-type plays in coal beds (U.S. Geological Survey National Oil and Gas Resource Assessment Team, 1995, Table 5, p. 19)										
20	Uinta Basin				1.86	4.82	3.21			
20	Piceance Basin				5.47	10.09	7.49			
22	San Juan Basin				5.76	9.67	7.53			
33	Powder River Basin				0.32	2.90	1.11			
35	Wind River Basin				0.22	0.72	0.43			
37	Southwestern Wyoming				0.83	7.66	3.89			
41	Raton Basin				1.39	2.23	1.78			
	TOTAL				**18.97**	**33.59**	**25.44**			

*F_{95}= 95% probability and F_5= 5% probability of the occurrence of at least the amount tabulated.

Table A-4. Summary of Greater Rocky Mountain Region estimated undiscovered technically recoverable resources listed in Table A-3.

Resource category	Oil (billion bbl)			NGL (billion bbl)			Total liquid HCs* (billion bbl)			Gas (tcf)		
	F_{95}*	F_5*	Mean	F_{95}*	F_5*	Mean	F_{95}*	F_5*	Mean	F_{95}*	F_5*	Mean
Conventional resources	3.72	8.93	5.93	0.95	2.59	1.68	4.67	11.52	7.61	20.70	43.88	30.53
Continuous-type plays in sandstone, shales, carbonates	0.52	1.64	0.98	0.87	3.24	1.83	1.39	4.88	2.81	116.74	323.91	203.85
Cotinuous–type plays in coal										18.97	33.59	25.44
TOTAL	**4.24**	**10.57**	**6.91**	**1.82**	**5.83**	**3.51**	**6.06**	**16.40**	**10.42**	**156.41**	**401.38**	**259.82**

*F_{95}= 95% probability and F_5= 5% probability of the occurrence of at least the amount tabulated.

Table A-5. Texas Bureau of Economic Geology estimates of in-place coal-bed methane* (compiled from Tyler et al., 1997, Figure 1, p. 8).

Coal basin	Estimated in-place resource (tcf)			
	Minimum	Maximum	Range	Average
Raton Basin	8	12	8–12	10.0
San Juan Basin	84	84	84	84.0
Piceance Basin	84	84	>84	84.0
Uinta Basin	8.3	10.6	8.3–10.6	9.5
Greater Green River Basin	314	314	314	314.0
Wind River Basin	2	2	2	2.0
Powder River Basin	30	30	30	30.0
TOTAL	**530.3**	**536.6**	**530.3.–536.6**	**533.5**

*This estimate includes total in-place resources, including those in coals deeper than 4000 ft which may not be producible with present technology because of the absence of open cleats that provide permeability.

Table A-6. Greater Rocky Mountain Region estimated coal-bed methane resources (based on Gas Research Institute, 1999). Symbols for coal ranks listed in order of increasing rank: Lig = lignite; Sb = subbituminous; Hvb = high-volatile bituminous; Mvb = medium-volatile bituminous; Lvb = low-volatile bituminous; An = anthracite.

Coal basin or deposit	State(s)	Coal-bearing formations or groups	Area (mi^2)	In-place coal (billions short tons)	Coal rank	Coal-bed gas (tcf)	
						In place	Recoverable
San Juan	Colo., N.M.	Fruitland Menefee	7500	358	Hvb-Lvb	84	11.567
Piceance and Uinta	Colo., Utah	Mesaverde Blackhawk, Ferron	6700 14,450	289 65	Hvb-An Sb-Hvb	99 10	5.528
Powder River	Wyo., Mont.	Fort Union, Wasatch	25,800	1300	Lig-Sb	39	9.329
Raton	Colo., N.M.	Raton, Vermejo	2200	52	Hvb-Mvb	10	3.493
Greater Green River	Wyo., Colo.	Fort Union, Lance, Mesaverde, Frontier	21,000	287	Sb-Hvb	214	2.500
Kaiparowits Plateau	Utah	Straight Cliffs	1650	62	Sb-Hvb	10	—
Wind River	Wyo.	Fort Union, Mesaverde	8100	87	Lig-Hvb	6	2.450
Bighorn	Wyo., Mont.	Fort Union, Mesaverde		28	Lig-Hvb	3	0.825
Denver	Colo., Wyo.	Denver, Laramie		51.8	Lig-Sb	2	0.300
Henry Mountains	Utah	Ferron		1.1	Sb-Hvb	—	2.745
Black Mesa	Ariz.	Dakota, Mesaverde	3200	21	Sb-Hvb	—	0.180
					TOTAL	**477**	**38.917**

REFERENCES CITED

Amateis, L. J., and S. Hall, 1997, Drilling multilaterals in a complex carbonate reservoir, Aneth Field, Utah, *in* E. B. Coalson, J. C. Osmond, and E. T. Williams, eds., Innovative applications of petroleum technology in the Rocky Mountain Area: Rocky Mountain Association of Geologists, p. 125–137.

Ambrose, L. L. Jr., 1971, Summary of possible future petroleum potential, Region 3, Western Rocky Mountains, *in* I. H. Cram, ed., Future petroleum provinces of the United States—Their geology and potential: AAPG Memoir 15, v. 1, p. 406–412.

Baars, D. L., and G. M. Stevenson, 1981, Tectonic evolution of the Paradox Basin, Utah and Colorado, *in* D. L. Wiegand, ed., Geology of the Paradox Basin: Rocky Mountain Association of Geologists, p. 23–31.

Bachu, S., and J. R. Underschultz, 1995, Large scale underpressuring in the Mississippian-Cretaceous succession, southwestern Alberta Basin: AAPG Bulletin, v. 79, p. 989–1004.

Berg, R. R., 1975, Capillary pressure in stratigraphic traps: AAPG Bulletin, v. 59, p. 939–956.

Berg, R. R., W. D. DeMis, and A. R. Mitsdarffer, 1994, Hydrodynamic effects on Mission Canyon (Mississippian) oil accumulations, Billings Nose area, North Dakota: AAPG Bulletin, v. 78, p. 501–518.

Berry, F. A. F., 1959, Hydrodynamics and chemistry of the Jurassic and Cretaceous Systems in the San Juan Basin, northwestern New Mexico and southwestern Colorado: Ph.D. thesis, Stanford University, Stanford, California.

Bond, W. A., 1984, Application of Lopatin's method to determine burial history, evolution of the geothermal gradient, and timing of hydrocarbon generation in Cretaceous source rocks in the San Juan Basin, northwestern New Mexico and southwestern Colorado, *in* J. Woodward, F. F. Meissner, and J. L. Clayton, eds., Hydrocarbon source rocks in the Greater Rocky Mountain Region: Rocky Mountain Association of Geologists, p. 433–447.

Bourke, L., P. Delfiner, J. C. Trouiller, T. Fett, M. Grace, S. Luthi, O. Serra, and E. Standen, 1989, Using formation microscanner images: Schlumberger—The Technical Review, v. 37, p. 16–40.

Bredehoeft, J. D., J. B. Wesley, and T. D. Fouch, 1994, Simulations of the origin of fluid pressure, fracture generation and the movement of fluids in the Uinta Basin, Utah: AAPG Bulletin, v. 78, p. 1729–1747.

Brenan, R. L., B. J. Peterson, and H. J. Smith, 1975, The origin of Red Wing Creek structure, McKenzie County, North Dakota: Wyoming Geological Association Earth Science Bulletin, v. 8, p. 1–41.

Bridges, L. W., 1987, Red Wing Creek field, North Dakota—A growth faulted or meteoritic impact structure, *in* J. A. Peterson, D. M. Kent, S. B. Anderson, R. H. Pilatzke, and M. W. Longman, eds., Williston Basin—Anatomy of a cratonic oil province: Rocky Mountain Association of Petroleum Geologists, p. 433–440.

Byrnes, A. P., 1997, Reservoir characteristics of low-permeability sandstones in the Rocky Mountains: The Mountain Geologist, v. 34, p. 39–51.

Camilleri, P., 1992, Extensional geometry of a part of the northwestern flank of the northern Grant Range, Nevada: Influences on its evolution: The Mountain Geologist, v. 29, p. 75–84.

Campbell, M., T. Heinzler, and V. C. Voss, 1992, Evolution of horizontal drilling in Silo field, Niobrara Formation, D.J. Basin, *in* G. Nydegger, general chairman, Proceedings of the horizontal drilling symposium—Domestic and international case studies: Rocky Mountain Association of Geologists.

Carmalt, S. W., and F. St. John, 1986, Giant oil and gas fields, *in* M. T. Halbouty, ed., Future petroleum provinces of the world: AAPG Memoir 40, p. 17–40.

Clayton, J. R., and J. D. King, 1984, Organic geochemistry of Pennsylvanian-Permian oils and black shales—northern Denver Basin (abs.): AAPG Bulletin, v. 68, p. 463.

Close, J. C., 1993 Natural fractures in coal, *in* B. E. Law and D.D. Rice, eds., Hydrocarbons from coal: AAPG Studies in Geology 38, p. 119–132.

Coalson, E. B., ed., 1989, Petrogeneses and petrophysics of selected sandstone reservoirs of the Rocky Mountain region: Rocky Mountain Association of Geologists, 353 p.

Coalson, E. B., J. C. Osmond, and E. T. Williams, eds., 1997, Innovative applications of petroleum technology in the Rocky Mountain Area: Rocky Mountain Association of Geologists, 255 p.

Cohee, G. V., chairman, 1962, Tectonic map of the United States: U.S. Geological Survey and AAPG, scale 1:2,500,00.

Cram, I. H., ed., 1971, Future petroleum provinces of the United States—Their geology and potential: AAPG Memoir 15, 1496 p.

Crow, P., 1996, U.S. independents' technology transfer initiative mushrooming: Oil & Gas Journal, August 19, p. 24–30.

Curtis, J. B., 1997, Natural gas resources of the Rocky Mountains and considerations for future supply: The Mountain Geologist, v. 34, p. 3–6.

Curry, W. H., 1971, Summary of possible future petroleum potential, Region 4, Northern Rocky Mountains, *in* I. H. Cram, ed., Future petroleum provinces of the United States—Their geology and potential: AAPG Memoir 15, p. 538–546.

Dalberg, E. C., 1982, Applied hydrodynamics in petroleum exploration: New York, Springer-Verlag, 161 p.

DeMis, W. D., 1987, Hydrodynamic trapping in Mission Canyon reservoirs—Elkhorn Ranch Field, North Dakota, *in* C. G. Carlson and J. E. Christopher, eds., Fifth International Williston Basin Symposium: Saskatchewan Geological Society, p. 217–225.

Dolly, E. D., and J. C. Mullarkey, eds., 1996, Hydrocarbon production from low contrast, low resistivity reservoirs, Rocky Mountain and Mid-Continent Regions: Rocky Mountain Association of Geologists, 290 p.

Dolson, J. C., chief ed., 1994, Unconformity-related hydrocarbons in sedimentary sequences—A guidebook for petroleum exploration and exploitation in clastic and carbonate sediments: Rocky Mountain Association of Geologists, 298 p.

Dyman, T. S., J. A. Peterson, J. W. Schmoker, C. W. Spencer, D. D. Rice, K. W. Porter, D. A. Lopez, T. J. Heck, and W. R. Beeman, 1995, Assessment of undiscovered resources in petroleum plays: U.S. portion of Williston Basin and north-central Montana: Seventh International Williston Basin Symposium, p. 323–339.

Effimoff, I., and A. R. Pinezich, 1986, Tertiary structural development of selected basins: Basin and Range province, northeastern Nevada: Geological Society of America Special Paper 208, p. 31–42.

Ely, S. W., S. A. Holditch, and R. H. Carter, 1988, Improved hydraulic fracturing strategy for Fruitland Formation coal-bed methane recovery, San Juan Basin, New Mexico and Colorado, *in* J. E. Fassett, ed., Geology and coal-bed methane resources of the northern San Juan Basin, Colorado and New Mexico: Rocky Mountain Association of Geologists, p. 155–158.

Esphahanian, C., J. Johnson, and J. Stabenau, 1998, Evolution of completion practices—Jonah field, Sublette County, Wyoming, *in* Developing a better understanding of basin centered gas plays: Consortium Meeting for the Emerging Resources in the Greater Green River Basin, p. 161–179.

Fassett, J. E., ed., 1988, Geology and coal-bed methane resources of the Northern San Juan Basin, Colorado and New Mexico: Rocky Mountain Association of Geologists, 351 p.

Forsman, N. F, T. R. Gerlach, and N. L. Anderson, 1996, Impact origin of the Newporte structure, Williston Basin, North Dakota: AAPG Bulletin, v. 80, p. 721–730.

French, D. E., 1994, Results of drilling and testing at the Meridian #32-29 Spencer-Federal, Railroad Valley, Nye County, Nevada, *in* R. A. Schalla and E. H. Johnson, eds., Oil fields of the Great Basin: Nevada Petroleum Society, p. 171–178.

Gans, P. B., and E. L. Miller, 1983, Style of mid-Tertiary extension in east-central Nevada: Utah Geological and Mineral Survey Special Studies 59, p. 107–139.

Garner, G. L., 1997, Seismic and subsurface evidence of local sedimentary infill of massive salt solution "micro-basins" in the Middle Jurassic Carmel Formation, central Utah: The Mountain Geologist, v. 4, p. 63–72.

Gautier, D. L., G. L. Dolton, K. I. Takahashi, and K. L. Varnes, 1995, 1995 national assessment of United States oil and gas resources—Results, methodology and supporting data: U.S. Geological Survey Digital Data Series DDS-30, CD-ROM.

Gas Research Institute, 1999, North American coalbed methane resource map: Chicago, Gas Research Institute.

Goolsby, S. M., and M. W. Longman, eds., 1988, Occurrence and petrophysical properties of carbonate reservoirs in the Rocky Mountain region: Rocky Mountain Association of Geologists, 474 p.

Gries, R., 1983, Oil and gas prospecting beneath Precambrian of foreland thrust plates in Rocky Mountains: AAPG Bulletin, v. 67, p. 1–28.

Gries, R. R., 1985, San Juan Sag—Cretaceous rocks in a volcanic-covered basin, south-central Colorado: The Mountain Geologist, v. 22, p. 167–179.

Gries, R. R., 1989, San Juan Sag—Oil and gas exploration in a newly discovered basin beneath the San Juan volcanic field, *in* J. C. Lorenz and S. G. Lucas, eds., Energy frontiers in the Rockies: Albuquerque Geological Society, p. 69–78.

Gries, R. R., and R. C. Dyer, eds., 1985, Seismic exploration of the Rocky Mountain region: Rocky Mountain Association of Geologists, 300 p.

Grummon, M. L., 1992, Cane Creek Shale Member of the Paradox Formation—Exploiting a self-sourced reservoir with horizontal well bores (abs.), *in* G. L. Nydegger, general chairman, Horizontal drilling symposium—Domestic and international case studies: Rocky Mountain Association of Geologists.

Hansen, W. B., and G. I. W. Long, 1991, Bakken production and potential in the U.S. and Canada—Can the fairway be defined?, *in* W. B. Hansen, ed., Geology and horizontal drilling of the Bakken Formation: Montana Geological Society, p. 69–88.

Hester, T .C., 1999, Prediction of gas production using well logs, Cretaceous of north-central Montana: The Mountain Geologist, v. 36, p. 85–98.

Hite, R. J., D. E. Anders, and T. G. Ging, 1984, Organic-rich source rocks of Pennsylvanian age in the Paradox Basin of Utah and Colorado, *in* J. Woodward, F. F. Meissner, and J. L. Clayton, eds., Hydrocarbon source rocks in the greater Rocky Mountain Region: Rocky Mountain Association of Geologists, p. 255–274.

Hoak, T. E., symposium coordinator, 1998, Fractured reservoirs—Practical exploration and development strategies: Rocky Mountain Association of Geologists symposium, January 19–20, Denver, Colorado, extended abstracts, 300 p.

Hoak, T. E., A. L. Klawitter, and P. K. Blomquist, eds., 1997, Fractured reservoirs—Their characterization and modeling: Rocky Mountain Association of Geologists, 230 p.

Hubbert, M .K., 1953, Entrapment of petroleum under hydrodynamic conditions: AAPG Bulletin, v. 37, p. 1954–2026.

Juntgen, J., and H. Karwiel, 1966, Gasbildung und Gasspeicherung in Steinkohlenflozen, parts I and II: Erdol and Kohle, Erdgas, Petrochemie, v. 19, p. 251–258, 339–344.

Kauffman, E. G., 1977, Geological and biological overview—Western Interior Cretaceous Basin: Mountain Geologist, v. 14, p. 75–99.

Klett, T. R., T. S. Ahlbrandt, J. W. Schomoker, and G. L. Dolton, 1997, Ranking of the world's oil and gas provinces by known petroleum volumes: U.S. Geological Survey Open File Report 97–463, CD-ROM (also http://greenwood.cr.usgs.gov/energy/worldenergy/of97–463).

Koeberl, C., and R. R. Anderson, 1996, Manson and company—Impact structures in the United States, *in* C. Koeberl and R. R. Anderson, eds., The Manson impact structure, Iowa: Anatomy of an impact crater: Geological Society of America Special Paper 302, p. 29.

Kriens, B. J., E. M. Shoemaker, and K. E. Herkenhoff, 1999, Geology of Upheaval Dome impact structure, southeast Utah: Journal of Geophysical Research (Planets), v. 104, no. E8, p. 18,867–18,887.

Krooss, B. M., D. Leythaeuser, and R. G. Schaefer, 1992, The quantification of diffuse hydrocarbon losses through cap rocks of natural gas reservoirs—A re-evaluation: AAPG Bulletin, v. 76, p. 403–406.

Kupfer, D. H., 1999, Colorado's early oil happenings—History of the Florence oil field: The Outcrop, v. 48, n. 10, p. 6, 20.

Kuuskraa, V. A., 1999, Emerging gas resources and technology, *in* Future of coal bed methane in the Rocky Mountain region symposium: Rocky Mountain Association of Geologists, Petroleum Technology Transfer Council, and Gas Research Institute.

Kuuskraa, V. A., and J. W. Schmoker, 1998, Diverse gas plays lurk in gas resource pyramid: Oil & Gas Journal, June 8, p. 123–130.

Lacy, S. L., W. D. Ding, and S. D. Joshi, 1992, Perspectives on horizontal wells in the Rocky Mountain region, *in* J. W. Schmoker, E. B. Coalson, and C. A. Brown, eds., Geological studies relevant to horizontal drilling—Examples from western North America: Rocky Mountain Association of Geologists, p. 25–32.

Lamarre, R. A., and T. D. Burns, 1997, Drunkard's Wash Unit—Coalbed methane production from Ferron coals in east-central Utah, *in* E. B. Coalson, J. C. Osmond, and E. T. Williams, eds., Innovative applications of petroleum technology in the Rocky Mountain Area: Rocky Mountain Association of Geologists, p. 47–60.

Lamarre, R. A., and T. D. Burns, 1999, Drunkard's Wash Unit—Production characteristics of an expanding coalbed methane field in east-central Utah, *in* Future of coal bed methane in the Rocky Mountain region symposium: Rocky Mountain Association of Geologists, Petroleum Technology Transfer Council, and Gas Research Institute.

Larber, G. M., 1981, Hydrodynamic effect on oil accumulation in a stratigraphic trap, Kitty field, Powder River Basin, Wyoming: M.S. thesis, Texas A&M University, College Station, Texas, 180 p.

Law, B. E., 1984, Relationships of source-rock, thermal maturity, and overpressuring to gas generation and occurrence in low-permeability Upper Cretaceous and lower Tertiary Rocks, Greater Green River Basin, Wyoming, Colorado and Utah, *in* J. Woodward, F. F. Meissner, and J. L Clayton, eds., Hydrocarbon source rocks of the Greater Rocky Mountain region: Rocky Mountain Association of Geologists, p. 469–490.

Law, B. E., 1992, Thermal maturity patterns of Cretaceous and Tertiary rocks, San Juan Basin, Colorado and New Mexico: Geological Society of America Bulletin, v. 104, p. 192–207.

Law, B. E., and Dickinson, W. W., 1985, Conceptual model for origin of abnormally pressured gas accumulations in low permeability reservoirs: AAPG Bulletin, v. 69, p. 1295–1304.

Law, B. E., R. M. Pollastro, and C. W. Keighin, 1986, Geologic characterization of low-permeability gas reservoirs in selected wells, Greater Green River Basin, Wyoming, Colorado, and Utah, *in* C. W. Spencer and R. F. Mast, eds., Geology of tight gas reservoirs: AAPG Studies in Geology 24, p. 253–269.

LeFever, J. A., 1991, History of oil production from the Bakken Formation, North Dakota, in W. B. Hansen, ed., Geology and horizontal drilling of the Bakken Formation: Montana Geological Society, p. 3–17.

LeFever, J. A., 1992, Horizontal drilling in the Williston Basin, United States and Canada, *in* J. W. Schmoker, E. B. Coalson, and C. A. Brown, eds., Geological studies relevant to horizontal drilling—Examples from western North America: Rocky Mountain Association of Geologists, p. 177–197.

Linn, J. T. C., 1981, Hydrodynamic flow in Lower Cretaceous Muddy Sandstone, Gas Draw field, Powder River Basin, Wyoming: The Mountain Geologist. v. 18, p. 78–87.

Lowell, J. D., ed., 1983, Rocky Mountain foreland basins and uplifts: Rocky Mountain Association of Geologists, 392 p.

Lucas, P. T., and J. M. Drexler, 1975, Altamont-Bluebell—A major fractured and overpressured stratigraphic trap, Uinta Basin, Utah, *in* D. W. Bolyard, ed., Deep drilling frontiers symposium: Rocky Mountain Association of Geologists, p. 265–274.

Lucas, P. T., and J. M. Drexler, 1976, Altamont-Bluebell: A major naturally fractured stratigraphic trap, Uinta Basin, Utah, *in* J. Braunstein, ed., North American oil and gas fields: AAPG Memoir 24, p. 121–135.

Magoon, L. B., and W. G. Dow, 1994, The petroleum system—From source to trap, *in* L. B. Magoon and W. G. Dow, eds., The Petroleum system—From source to trap: AAPG Memoir 60, p. 3–24.

Mallory, W. W., ed., 1972, Geologic atlas of the Rocky Mountain Region: Rocky Mountain Association of Geologists, 331 p.

Martinsen, R. S., 1997, Stratigraphic controls on the development and distribution of fluid pressure compartments, *in* R. C. Surdam, ed., 1997, Seals, traps and the petroleum system: AAPG Memoir 67, p. 223–241.

Masters, J. A., 1979, Deep basin gas trap, western Canada: AAPG Bulletin, v. 63, p. 152–181.

McPeek, L .A., 1981, Green River Basin—A developing giant gas supply from deep over-pressured Upper Cretaceous sandstone: AAPG Bulletin, v. 65, p. 1078–1098.

Meissner, F. F., 1974, Petroleum geology of the Bakken Formation, Williston Basin, North Dakota and Montana, *in* D. Rehrig, ed., Economic geology of the Williston Basin: Williston Basin symposium: Montana Geological Society, p. 207– 227.

Meissner, F. F., 1980, Examples of abnormal pressure produced by hydrocarbon generation (abs.): AAPG Bulletin, v. 64, p. 749.

Meissner, F. F., 1984, Cretaceous and lower Tertiary coals as sources for gas accumulations in the Rocky Mountain area, *in* J. Woodward, F. F. Meissner, and J. L. Clayton, eds., Hydrocarbon source rocks in the Greater Rocky Mountain Region: Rocky Mountain Association of Geologists, p. 410–431.

Meissner, F. F., 1985, Pressure cycles related to gas generation in coals and their relation to Deep Basin gas accumulations (abs.): AAPG Bulletin, v. 69, p. 856.

Meissner, F. F., 1987, Mechanisms and patterns of gas generation, storage, expulsion-migration, and accumulation associated with coal measures, Green River and San Juan Basins, Rocky Mountain region, U.S.A., *in* B. Dolegez, ed., Migration of hydrocarbons in sedimentary basins: Paris, Editions Technip, p. 79–112.

Meissner, F. F., 1988, Basin analysis in oil/gas exploration with special consideration of fluid pressure and hydrodynamic controls—Examples from the Williston, Green River and San Juan Basins, *in* notes accompanying a continuing-education course, "Principles and practices of formation pressures and applied hydrogeology in petroleum exploration and basin evolution": Rocky Mountain Association of Geologists, Denver, April 4–6, 91 p.

Meissner, F. F., 1995, Pattern of maturity in source rocks of the Chainman Formation, central Railroad Valley, Nye County, Nevada, and its relation to oil migration and accumulation, *in* M. W. Hanson, chief ed., Mississippian source rocks in the Antler Basin of Nevada, and associated structural and stratigraphic traps: Nevada Petroleum Society Field Trip Guidebook, p. 65–74.

Meissner, F. F., J. Woodward, and J. L. Clayton, 1984, Stratigraphic relationships and source rocks in the Greater Rocky Mountain Region, *in* J. Woodward, F. F. Meissner, and J .L. Clayton, eds., Hydrocarbon source rocks in the Greater Rocky Mountain region:, Rocky Mountain Association of Geologists, p. 1–34.

Momper, J. A., 1980, Generation of abnormal pressure through organic matter transformations (abs.): AAPG Bulletin, v. 64, p. 753.

Montgomery, S. L., 1991a, Horizontal drilling in the Niobrara–northern D-J Basin, part 1—Geology and drilling history, *in* Petroleum frontiers: Denver, Petroleum Information Corp., v. 8, 50 p.

Montgomery, S. L., 1991b, Horizontal drilling in the Niobrara–northern D-J Basin, part 2—Silo field, *in* Petroleum frontiers: Denver, Petroleum Information Corp., v. 8, 48 p.

Montgomery, S. L., 1997, Horizontal drilling in the Ordovician Red River "B," Williston Basin, *in* E. B. Coalson, J. C. Osmond, and E. T. Williams, eds., Innovative applications of petroleum technology in the Rocky Mountain area: Rocky Mountain Association of Geologists, p. 111–124.

Montgomery, S. L., 1999, Powder River Basin, Wyoming—An expanding coalbed methane (CBM) play: AAPG Bulletin, v. 83, p. 1207–1222.

Moore, R. W., 1984, Hydrodynamic control on oil entrapment in channel sandstones of the Dakota Sandstone, South Cole Creek field, Converse County, Wyoming: The Mountain Geologist, v. 21, p. 105–113.

Morgan, C. D., 1992a, Horizontal drilling potential of the Cane Creek Shale, Paradox Formation, Utah, *in* J. W. Schmoker, E. B. Coalson, and C. A. Brown, eds., Geological studies relevant to horizontal drilling—Examples from western North America: Rocky Mountain Association of Geologists, p. 257–266.

Morgan, C. D., 1992b, "Cane Creek" exploration play area, Emery, Grand and San Juan Counties: Utah Geological Survey Open File Report 232, 5 p.

Muehlberger, W. R., comp., 1992, Tectonic map of North America—Southwest sheet: AAPG, scale 1:5,000,000.

Murray, D. K., and S. D. Schwochow, 1997, Coalbed gas development in the Rockies—Analogues for the World, *in* E. B. Coalson, J. C. Osmond, and E. T. Williams, eds., Innovative applications of petroleum technology in the Rocky Mountain area: Rocky Mountain Association of Geologists, p. 31–46.

Murray, G. H., 1959, Examples of hydrodynamics in the Williston Basin at Poplar and North Tioga fields: AAPG Geological Record, Rocky Mountain Section, p. 55–60.

Narr, W., and J. B. Currie, 1982, Origin of fracture porosity—Example from Altamont field, Utah: AAPG Bulletin, v. 66, p. 1231–1247.

Nelson, J. S., and E. C. Simmons, 1992, The quantification of diffuse hydrocarbon losses through cap rocks of natural gas reservoirs—A re-evaluation: Discussion: AAPG Bulletin, v. 76, p. 1839–1841.

Nelson, J. A., and E. C. Simmons, 1995, Diffusion of methane and ethane through the reservoir cap rock—Implications for the timing and duration of catagenesis: AAPG Bulletin, v. 79, p. 1064–1074.

Nelson, R. A., 1998, Overview of research approaches to understanding fractured reservoirs (abs.), *in* T. E. Hoak, symposium coordinator, Fractured reservoirs—Practical exploration and development strategies: Rocky Mountain Association of Geologists, p. 13–14.

Nydegger, G. L., general chairman, 1992, Horizontal drilling symposium—Domestic and international case studies: Rocky Mountain Association of Geologists.

Oldaker, P. R.,1991, Hydrogeology of the Fruitland Formation, San Juan Basin, Colorado and New Mexico, *in* S. D. Schwochow, ed., Coalbed methane of western North America: Rocky Mountain Association of Geologists, p. 61–66.

Palmer, I. D., M. J. Mavor, J. P. Seidle, J. L. Spitler, and R. F. Voltz, 1992, Open hole cavity completions in coalbed methane wells in the San Juan Basin: Society of Petroleum Engineers Paper 24906.

Parker, J., 1967, Salt solution and subsidence structures: AAPG Bulletin, v. 67, p. 1929–1947.

Pitman, J. K., and E. S. Sprunt, 1986, Origin and distribution of fractures in lower Tertiary and Upper Cretaceous rocks, Piceance Basin, Colorado, and their relation to the occurrence of hydrocarbons, *in* C. W. Spencer and R. F. Mast, eds., Geology of tight gas reservoirs: AAPG Studies in Geology 24, p. 221–233.

Pollastro, R. M., and P. A. Scholle, 1986, Exploration and development of hydrocarbons from low-permeability chalks—An example from the Upper Cretaceous Niobrara Formation, Rocky Mountain region, *in* C. W. Spencer and R. F. Mast, eds., Geology of tight gas reservoirs: AAPG Studies in Geology 24, p. 129–141.

Poole, F. G., and G. E. Claypool, 1984, Petroleum source rock potential and crude-oil correlation in the Great Basin, *in* J. Woodward, F. F. Meissner, and J. L. Clayton, eds., Hydrocarbon source rocks in the Greater Rocky Mountain region: Rocky Mountain Association of Geologists, p. 231–253.

Potential Gas Committee, 1999, Potential supply of natural gas in the United States (December 31, 1998): Potential Gas Agency, Colorado School of Mines, Golden, Colorado, 196 p.

Powers, R. B., ed., 1982, Geologic studies of the Cordilleran thrust belt: Rocky Mountain Association of Geologists, 875 p.

Ray, R. R., chief ed., 1995, High definition seismic—2-D, 2-D swath and 3-D case histories: Rocky Mountain Association of Geologists, 214 p.

Rice, D. L, 1993, Composition and origins of coalbed gas, *in* B. E. Law and D. D. Rice, eds., Hydrocarbons from coal: AAPG Studies in Geology 38, p. 159–184.

Rice, D. D., and W. G. Shurr, 1980, Shallow, low-permeability reservoirs of the northern Great Plains—An assessment of their natural resources: AAPG Bulletin, v. 64, p. 969–987.

Rice, D. D., and C. W. Spencer, 1995, Northern Great Plains shallow biogenic gas, *in* D. L. Gautier, G. I. Dolton, K. I. Takahashi, and K. L. Varnes, eds., 1995 National assessment of United States oil and gas resources—Results, methodology, and supporting data: U.S. Geological Survey Digital Data Series DDS-30 CD-ROM.

Rightmire, C. T., and R. Choate, 1986, Coal-bed methane and tight gas sands interrelationships, *in* C. W. Spencer and R. F. Mast, eds., Geology of tight gas reservoirs: AAPG Studies in Geology 24, p. 87–111.

Rocky Mountain Association of Geologists, 1996, Expanded abstracts from the Second Annual Seismic Symposium: Rocky Mountain Association of Geologists.

Rocky Mountain Association of Geologists, 1997, Expanded abstracts from the Third Annual Seismic Symposium: Rocky Mountain Association of Geologists.

Rocky Mountain Association of Geologists, 1998, Expanded abstracts from the Fourth Annual Seismic Symposium: Rocky Mountain Association of Geologists.

Rocky Mountain Association of Geologists, 1999, Expanded abstracts from the Fifth Annual Seismic Symposium: Rocky Mountain Association of Geologists.

Sandberg, C. A., and R. C. Gutschick, 1984, Distribution, microfauna, and source-rock potential of Mississippian Delle Phosphatic Member of the Woodman Formation and equivalents, Utah and adjacent states, *in* J. Woodward, F. F. Meissner, and J. L. Clayton, eds., Hydrocarbon source rocks in the Greater Rocky Mountain region: Rocky Mountain Association of Geologists, p. 135–159.

Schmoker, J. W., 1997, Porosity predictions in deeply buried sandstones, with examples from Cretaceous formations of the Rocky Mountain region, *in* T. S. Dyman, D. D. Rice, and W. A. Westcott, eds., Geologic controls of natural gas in deep sedimentary basins in the United States: U.S. Geological Survey Bulletin 2146-H, p. 77–104.

Schmoker, J. W., E. B. Coalson, and C. A. Brown, eds., 1992, Geological studies relevant to horizontal drilling—Examples from western North America: Rocky Mountain Association of Geologists, 284 p.

Schowalter, T. T., 1976, Mechanics of secondary migration and hydrocarbon entrapment: AAPG Bulletin, v. 63, p. 723–760.

Schwochow, S. D., ed., 1991, Coalbed methane of western North America: Rocky Mountain Association of Geologists, 336 p.

Schwochow, S. D., and D. K. Murray, 1999, Overview of coal bed methane activity and recoverable resources in the Rocky Mountain Region, *in* Future of coal bed methane in the Rocky Mountain region symposium: Rocky Mountain Association of Geologists, Petroleum Technology Transfer Council, and Gas Research Institute.

Scott, A. R., 1993, Composition and origins of coalbed gas, *in* D. A. Thompson, ed., Proceedings from the 1993 International Coalbed Methane Symposium: University of Alabama, Tuscaloosa, v. 1, p. 207–222.

Scott, A. R., 1999, Review of Rocky Mountain coal basins—Where are the next coalbed methane plays?, *in* Future of coal bed methane in the Rocky Mountain region symposium: Rocky Mountain Association of Geologists, Petroleum Technology Transfer Council, and Gas Research Institute.

Scott, A. R., W. R. Kaiser, and W. B. Ayers Jr., 1991, Composition, distribution, and origin of Fruitland Formation and Pictured Cliffs Sandstone gases, San Juan Basin, Colorado and New Mexico, *in* J. E. Fassett, ed., 1988, Geology and coal-bed methane resources of the northern San Juan Basin, Colorado and New Mexico: Rocky Mountain Association of Geologists, p. 93–108.

Scott, A. R., W. R. Kaiser, R. Tyler, D. S. Hamilton, and R. J. Finley, 1996, Geologic and hydrologic factors affecting coalbed methane producibility in the San Juan, Greater Green River, and Piceance Basins: Coalbed Methane Review, no. 5, p. 6–10.

Seiler, D., C. Edmiston, D. Torres, and J. Goetz, 1990, Field performance of a new borehole televiewer tool and associated image processing techniques: Society of Professional Well Log Analysts 31st Annual Symposium Transactions, paper H.

Serra, O., 1989, Formation MicroScanner image interpretation: Schlumberger brochure SMP 7028.

Slatt, R. M., ed., 1998a, Compartmentalized reservoirs in Rocky Mountain basins: Rocky Mountain Association of Geologists, 250 p.

Slatt, R. M., 1998b, Forward: Compartmentalized reservoirs—The exception or the rule?, *in* R. M Slatt, ed., Compartmentalized reservoirs in Rocky Mountain basins: Rocky Mountain Association of Geologists, p. v–vi.

Sloss, L. L., 1963, Sequences in the cratonic interior of North America: Geological Society of America Bulletin, v. 74, p. 93–114.

Smith, J. T., 1994, Petroleum system logic as an exploration tool in a frontier setting, in L.B. Magoon and W. G. Dow, eds., The petroleum system—From source to trap: AAPG Memoir 60, p. 25–49.

Sonnenberg, S. A., and R. J. Weimer, 1993, Oil production from Niobrara Formation, Silo field, Wyoming—Fracturing associated with a possible wrench fault system (?): The Mountain Geologist, v. 30, p. 39–53.

Spencer, C. W., 1987, Hydrocarbon generation as a mechanism for overpressuring in the Rocky Mountain region: AAPG Bulletin, v. 71, p. 368–388.

Spencer, C. W., 1989, Review of characteristics of low-permeability gas reservoirs in western United States: AAPG Bulletin, v. 73, p. 613–629.

Spencer, C. W., and R. F. Mast, eds., 1986, Geology of tight gas reservoirs: AAPG Studies in Geology 24, 299 p.

Stone, D. S., 1969, Wrench faulting and Rocky Mountain tectonics: The Mountain Geologist, v. 6, p. 67–79.

Stone, D. S., 1993, Basement-involved thrust-generated folds as seismically imaged in the subsurface of the central Rocky Mountain foreland, *in* C. J. Schmidt, R. B. Chase, and E. A. Erslev, eds., Laramide basement deformation in the Rocky Mountain foreland of the western United States: Geological Society of America Special Paper 280, p. 271–318.

Stone, D. S., 1999, Cloud Creek—A possible impact structure on the Casper arch, Wyoming: The Mountain Geologist, v. 36, p. 211–234.

Stone, D. S., and R. L. Hoeger, 1973, Importance of hydrodynamic factor in formation of Lower Cretaceous combination traps, Big Muddy–South Glenrock area, Wyoming: AAPG Bulletin, v. 57, p. 1714–1733.

Surdam, R. C., 1997a, A new paradigm for gas exploration in anomalously pressured "tight gas sands" in Rocky Mountain Laramide basins, *in* R. C. Surdam, ed., Seals, traps and the petroleum system: AAPG Memoir 67, p. 283–298.

Surdam, R. C., 1997b, New exploration strategy for anomalously pressured reservoirs in Rocky Mountain basins: Gas Research Institute, Gas Tips, winter 1996–1997, p. 4–8.

Swenson, R. E., 1967, Nisku fields, Montana: AAPG Bulletin, v. 67, p. 1948–1958.

Thomasson, M. R., 1982, Synergism in exploration, *in* K. C. Jain and J. P. de Figueirido, eds., Concepts and techniques in oil and gas exploration: Society of Exploration Geophysicists, p. 3–12.

Thomasson, M. R., moderator, 1992, Proceedings of the Department of Energy (DOE) Resources Dephi Panel discussion considering evaluations prepared by the American Association of Petroleum Geologists, August 31–September 1, Austin, Texas.

Tyler, R., W. R Kaiser, A. R. Scott, and D. S. Hamilton, 1997, The potential for coalbed gas exploration and production in the Greater Green River Basin, southwest Wyoming and northwest Colorado: The Mountain Geologist, v. 34, p. 7–24.

U.S. Geological Survey National Oil and Gas Resource Assessment Team, 1995, 1995 national assessment of United States oil and gas resources: U.S. Geological Survey Circular 1118, 20 p.

Vincelette, R. R., and W. E. Chittum, 1981, Exploration for oil accumulations in Entrada Sandstone, San Juan Basin, New Mexico: AAPG Bulletin, v. 65, p. 2446–2570.

Vincelette, R. R., and N. H. Foster, 1992, Fractured Niobrara of northwestern Colorado, *in* J. W. Schmoker, E. B. Coalson, and C. A. Brown, eds., Geological studies relevant to horizontal drilling—Examples from western North America: Rocky Mountain Association of Geologists, p. 227–242.

Warner, E. M., 1997, Geology of Jonah field, a major gas accumulation in the Upper Cretaceous Lance Formation, Sublette County, Wyoming, *in* E. B. Coalson, J. C. Osmond, and E. T. Williams, eds., Innovative applications of petroleum technology in the Rocky Mountain area: Rocky Mountain Association of Geologists, p. 1–12.

Warner, E .M., 1998, Structural geology and pressure compartmentalization of Jonah field, Sublette County, Wyoming, *in* R. M. Slatt, ed., 1998, Compartmentalized reservoirs in Rocky Mountain basins: Rocky Mountain Association of Geologists, p. 29–46.

Weimer, R. J., 1960, Upper Cretaceous stratigraphy, Rocky Mountain area: AAPG Bulletin, v. 44, p. 1–20.

Weimer, R. J., 1988, Sequence stratigraphy—The Western Interior Cretaceous Basin: Rocky Mountain Association of Geologists, videotape.

Weimer, R. J., S. A. Sonnenberg, and G. B. C. Young, 1986, Wattenberg field, Denver Basin, *in* C. W. Spencer and R. F. Mast, eds., Geology of tight gas reservoirs: AAPG Studies in Geology 24, p. 143–164.

Wernicke, B., 1981, Low-angle normal faults in the Basin and Range province: Nappe tectonics in an extending orogen: Nature, v. 291, p. 645–648.

Wheeler, H. E., 1963, Post-Sauk and pre-Absaroka Paleozoic stratigraphic patterns in North America: AAPG Bulletin, v. 47, p. 1497–1526.

Willet, S. D., and D. S. Chapman, 1987, Temperatures, fluid flow and thermal history of the Uintah Basin, *in* B. Dolegez, ed., Migration of hydrocarbons in sedimentary basins: Paris, Editions Technip, p. 533–551.

Witkind, I. J., 1982, Salt diapirism in central Utah, *in* D. L. Nielson, ed., Overthrust belt of Utah: Utah Geological Association Publication 10, p. 13–30.

Woodward, J., F. F. Meissner, and J. L. Clayton, eds., 1984, Hydrocarbon source rocks in the Greater Rocky Mountain region: Rocky Mountain Association of Geologists, 557 p.

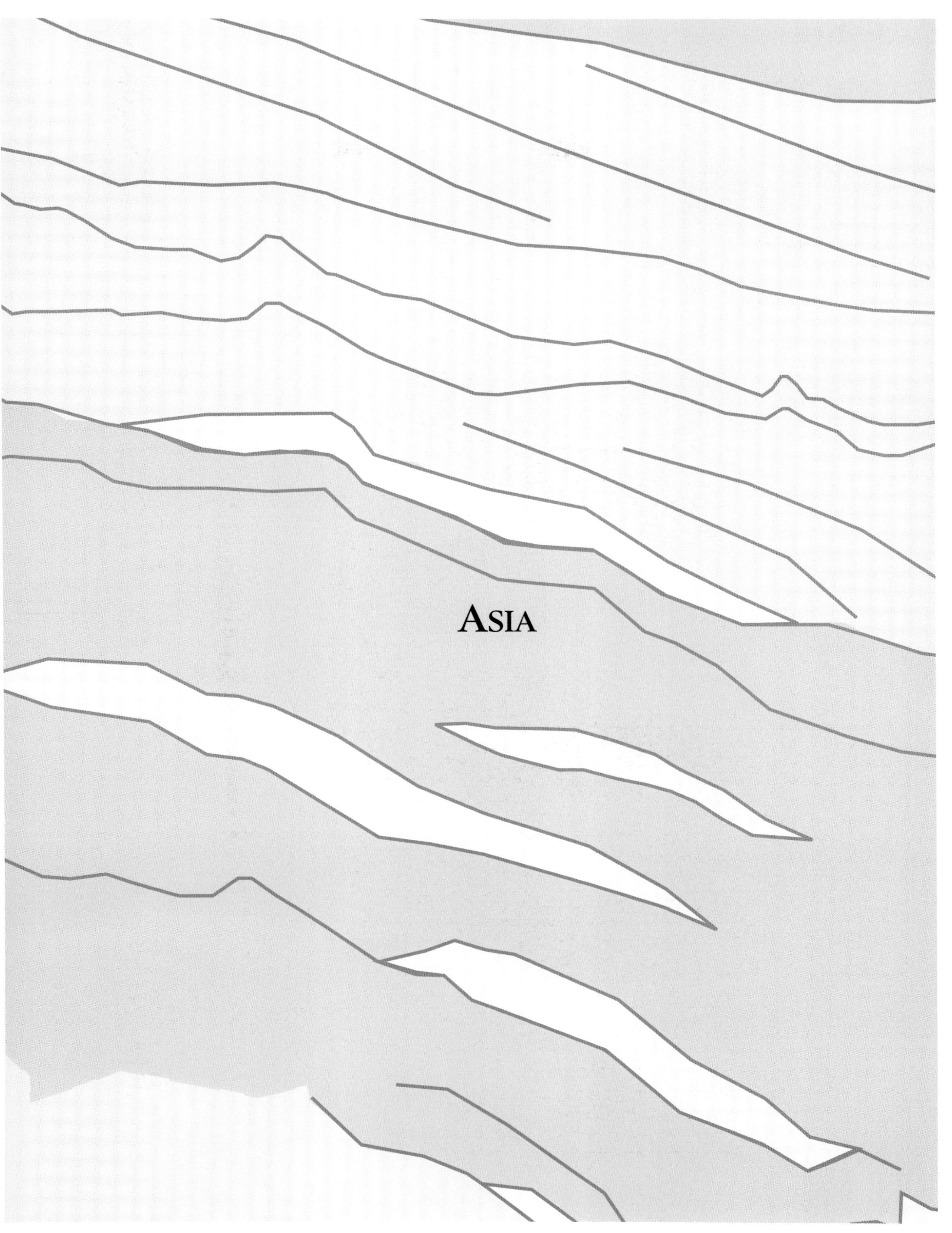

ASIA

Effimoff, I., 2001, Future hydrocarbon potential of Kazakhstan, *in* M. W. Downey, J. C. Threet, and W. A. Morgan, eds., Petroleum provinces of the twenty-first century: AAPG Memoir 74, p. 243–258.

Chapter 12

Future Hydrocarbon Potential of Kazakhstan

Igor Effimoff

Pennzoil Caspian Corporation, subsidiary of Devon Energy Corporation, Baku, Azerbaijan

ABSTRACT

Most of Kazakhstan's oil and gas reserves have not been developed. Many areas remain under- or unexplored. It is expected that a considerable portion of Kazakhstan's potential oil and gas reserves will be located offshore in the Caspian Sea. Future reserves will be discovered and developed using international technology that was unavailable to the nation during the Soviet period. Those reserves will be commercially justified with better access to world markets.

About 250 oil and gas accumulations have been identified in Kazakhstan, of which 107 have been developed. Four basins/provinces, considered to have the most potential for undiscovered reserves, are (in order of descending potential):

1. North Caspian (Peri-Caspian) Basin, also referred to as the Pri- or Pre-Caspian Basin
2. Middle Caspian Basin–Mangyshlak province
3. South Turgay Basin
4. North Ustyurt Basin

Kazakhstan's cumulative oil production is estimated to be 3.8 billion barrels (bbl); its identified reserves are approximately 13.2 billion bbl. Probabilistic estimates of undiscovered oil/condensate are in the range of 14 billion bbl (P_{95}) to 89 billion bbl (P_5), with the mode (most likely) at 26 billion bbl. Potential gas reserves are equally impressive, with identified reserves of 83 trillion cubic feet (tcf). Probabilistic estimates of undiscovered gas are in the range of 66 tcf (P_{95}) to 495 tcf (P_5), with the mode (most likely) at 124 tcf. The bulk of undiscovered oil/condensate and gas reserves is estimated to be attributable to Kazakhstan's portion of the North Caspian (Peri-Caspian) basin.

INTRODUCTION

Kazakhstan is one of several former Soviet Union nations in the Caspian Sea region (including Azerbaijan, Russia, Turkmenistan, and Uzbekistan) that are major energy producers (Figure 1). Hydrocarbon production in those nations likely will increase with additional investment, technology, and development of new export outlets.

Most of the oil and gas reserves in Kazakhstan and bordering regions have not been developed, and many areas remain under- or unexplored. A sizable portion of Kazakhstan's resources likely will be located in the offshore Caspian using international state-of-the-art technology, some of which was unavailable to the nation during the Soviet period (Effimoff, 1999).

Oil and gas basins/provinces and regions of Kazakhstan are depicted in Figure 2. Provinces and regions are administrative divisions, but they generally reflect the area's geology or type of production.

Figure 1. Kazakhstan location map.

HISTORY OF OIL AND GAS RESERVES AND PRODUCTION

Oil in Kazakhstan has been known anecdotally for centuries, as evidenced by many localities near the Caspian incorporating the word *munai* in their name. *Munai* is the Kazakh word for oil. A. Bekovitch-Cherkasky, the head of a military expedition sent by Russian Czar Peter the Great to Khiva, provided some of the earliest written documentation of oil deposits in present-day Kazakhstan. In 1717, the expedition crossed the border of Guryev province and compiled geographic information, including data on local oil seeps. Drilling for oil began in the 1890s, and the first commercial oil production was established in 1911 from a depth of 225 m in the Dossor area (Myzaprova, 1998).

To date, about 250 oil and gas accumulations have been identified in Kazakhstan, of which 107 have been developed (Myzaprova, 1998). Cumulative oil production is approximately 3 billion bbl; identified reserves are approximately 17 billion bbl. Probabilistic estimates of undiscovered oil are in the range of 14 billion (P_{95}) to 89 billion (P_5) bbl. The mode (most likely) is 26 billion bbl. The bulk of oil reserves is attributable to the Kazakh portion of the North Caspian (Peri-Caspian) Basin (the basin is also referred to in the literature as the Pri-Caspian or Pre-Caspian Basin). Potential gas reserves are equally impressive, with identified reserves of 83.7 tcf. Probabilistic estimates of undiscovered gas are in the range of 66 tcf (P_{95}) to 495 tcf (P_5). The mode (most likely) is 124 tcf. The majority of potential gas reserves is also expected to be found in the North Caspian Basin (Ulmishek and Masters, 1993; Energy Information Administration 1999b, c, d).

Kazakhstan is the second-largest oil producer among the former Soviet republics, after Russia. In 1997, Kazakhstan produced 521,000 bbl oil/day (BOPD), down from 530,000 BOPD in 1992 but up from 414,000 BOPD in 1995. Gas production in 1997 was 605 million cubic feet gas/day (MMCFGD), down from 795 MMCFGD in 1992 but up from 410 MMCFGD in 1996 (Energy Information Administration, 1999b,c).

Almost half of Kazakhstan's oil production comes from three large onshore fields: Tengiz, Uzen, and Karachaganak. In 1993, Chevron concluded a $20 billion joint venture (Tengizchevroil) to develop the Tengiz oil field, which has 6–9 billion bbl of estimated recoverable oil reserves. Tengizchevroil exports about 170,000 BOPD through the Russian pipeline system; by barge and rail to the Baltic; and by ship, pipeline, and rail to the Black Sea. Given adequate export outlets, Chevron believes it can reach peak production of 750,000 BOPD from the field by 2010. The Caspian Pipeline Consortium (CPC) will export Tengiz oil to world markets via a 900-mile, $2.3 billion oil-export pipeline connecting to the Russian Black Sea port of Novorosiisk.

Kazakhstan has undertaken several reforms to develop its hydrocarbon potential, including privatizing some existing energy concerns. However, Kazakhstan needs to resolve two major issues for it to increase oil production further. Development of its offshore potential in the Caspian Sea has been slowed by a dispute over ownership rights. This disagreement is associated with a broader debate between Caspian Sea region states over the treatment of the Caspian Sea under international law. The other major issue is the development of export routes to bring Kazakhstan's oil to world markets. Under the former Soviet Union, all of Kazakhstan's oil was exported through the Russian pipeline system. Kazakhstan still views Russia as a viable export option, and the existing export pipeline to Russia was expanded in 1999. In addition, the CPC pipeline will also pass through Russia en route to the Black Sea. Other oil-export pipeline options from the Caspian Sea region are also being explored.

More than 40% of Kazakhstan's 83.7 tcf of natural-gas reserves are located in the giant Karachaganak field, which is in the northwest of the country, on trend with Russia's Orenburg field. Liquids production from the field is expected to exceed 300,000 bbl/day by 2006, and the output will be exported using the CPC pipeline. Gas

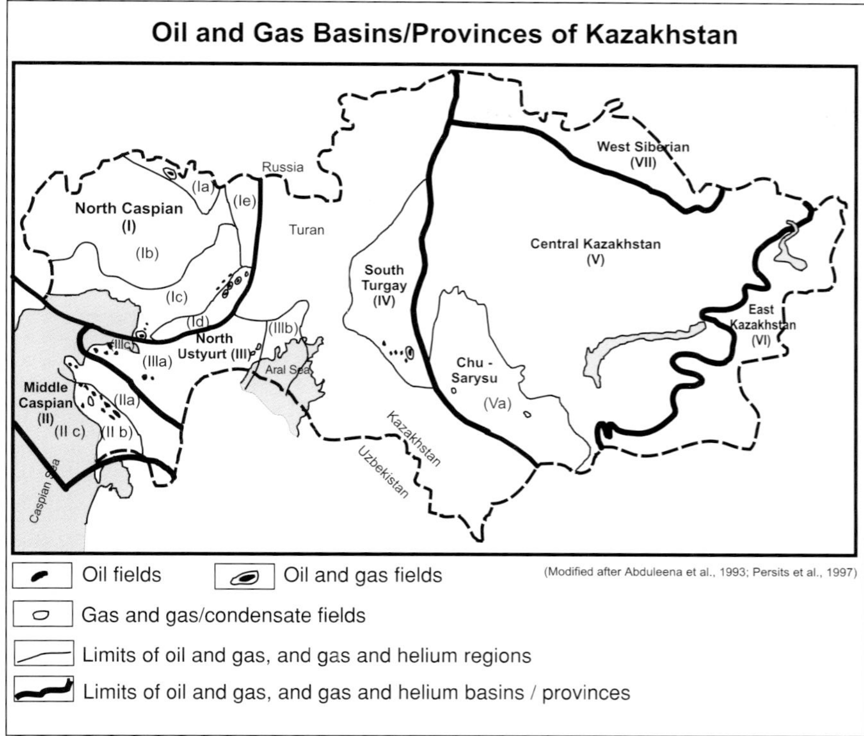

Figure 2. Oil and gas basins/provinces of Kazakhstan (modified after Abduleena et al., 1993; Persits et al., 1997).

I. North Caspian (Peri-Caspian) Basin and oil and gas province: (a) northwest oil and gas region, (b) central oil and gas region, (c) Astrakhan-Aktubinsk oil and gas region, (d) South Emba oil and gas region, (e) Ural foredeep oil and gas region.

II. Middle Caspian Basin–Mangyshlak province; (a) Mangyshlak oil and gas region, (b) South Mangyshlak oil and gas region, (c) Middle Caspian oil and gas region.

III. North Ustyurt Basin (part of Turan oil and gas province): (a) North Ustyurt oil and gas region, (b) Central Aral oil and gas region, (c) Buzachi oil and gas region.

IV. South Turgay Basin (part of Turan oil and gas province).

V. Central gas and helium province: (a) Chu-Sarysu Basin and gas-helium region.

VI. East Kazakhstan gas and helium province.

VII. West Siberian oil and gas province.

output should reach more than 883 billion cubic feet (bcf) annually.

Kazakhstan's other significant gas production comes from Tengiz, Zhanazhol, and Urikhtau fields. Associated gas production at Tengiz field rose to more than 50 bcf in 1997, and the field will become the second-largest producer of gas in Kazakhstan. The undeveloped offshore areas are also believed to hold large amounts of gas. Although some of these fields are near the Russian gas-pipeline system, they are not linked to it, and Russia's Gazprom is a potential competitor with Central Asian gas on world markets. Kazakhstan must either negotiate to connect its fields with the existing Russian gas-pipeline system or develop new ways of transporting gas to markets.

In general, the Kazakh gas sector faces a lack of infrastructure, especially pipelines. Although six gas pipelines connect Kazakhstan to other Central Asian republics and Russia, gas-producing areas in western Kazakhstan are not connected to consuming areas, such as the populous southeast and industrial north. Construction of an internal pipeline is under consideration to transport gas from Kazakhstan's western field to all oblasts in Kazakhstan (Energy Information Administration, 1999a).

ASSESSMENT OF UNDISCOVERED RESERVES

Ulmishek and Masters (1993) provided an excellent assessment of undiscovered oil and gas resources in countries of the former Soviet Union (FSU), including Kazakhstan. The assessments were made by participants of the World Energy Resources Program of the U.S. Geological Survey, using a modified Delphi (subjective) method and based on multiyear studies of the geology of FSU basins and exploration results (Masters et al., 1998). Kazakhstan's reserves and undiscovered resources allocated on a basin-by-basin basis are presented in Table 1.

Four basins/provinces (Figure 2) considered to have the most potential for undiscovered reserves (primarily on propensity for oil and secondarily for natural gas) are (in order of descending potential):

1. North Caspian (Peri-Caspian) Basin, also referred to as the Pri- or Pre-Caspian Basin
2. Middle Caspian Basin–Mangyshlak province
3. South Turgay Basin
4. North Ustyurt Basin

In view of the significance of the oil and gas potential of the North Caspian Basin (Table 1), the following discussion will focus on the North Caspian (Peri-Caspian) Basin, with shorter discussions on the other three basins/provinces. For completeness, the Chu-Sarysu Basin is described briefly. The basins mentioned are depicted on the geologic map of western and central Kazakhstan (Figure 3).

North Caspian (Peri-Caspian) Basin

The North Caspian Basin possesses the most significant potential for the discovery of oil and gas in Kazakhstan. Russia has about 20% of the basin, and Kazakhstan

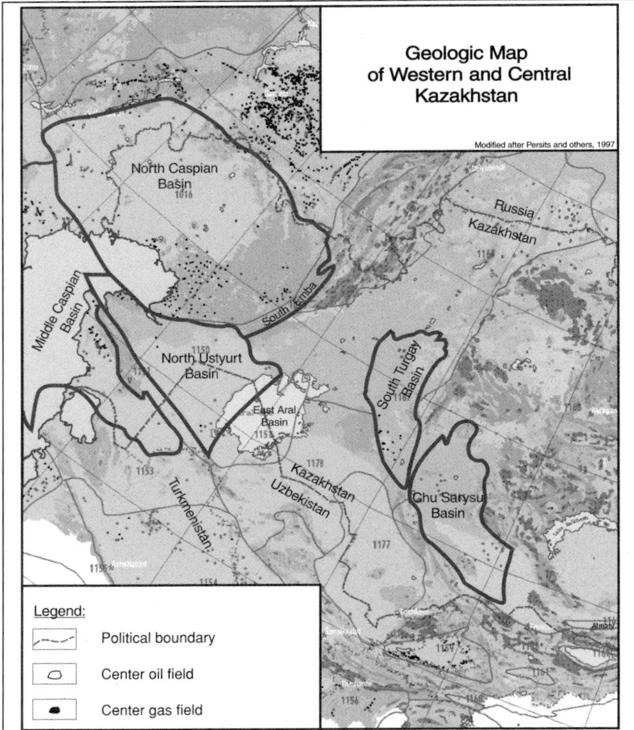

Figure 3. Geologic map of western and central Kazakhstan (modified after Persits et al., 1997). Basins discussed are indicated.

has the other approximately 80%. The basin covers approximately 550,000 km^2 and represents a deeply subsided portion of the Russian platform bound on all sides by deep-seated regional faults, which accommodated Paleozoic and Mesozoic subsidence (Figure 4). Basement faulting is documented throughout the basin but is most prevalent along the southern and eastern margins. Although basin formation can be inferred to have begun during the Riphean (Proterozoic), subsidence can be documented to have started in the Devonian.

The Permian (Kungurian) salt section divides the basin's hydrocarbon systems and stratigraphy into a presalt section and a postsalt section (Figure 5). Presalt hydrocarbon accumulations are localized primarily in Devonian to Permian carbonates in stratigraphic and/or basement-related structural traps. Hydrocarbons in the postsalt section are located in Mesozoic clastic reservoirs, generally in salt-related structural traps, that often have a strong stratigraphic component (Dongaryan, 1990; Pronichva and Savvinova, 1980).

At its center, the basin may contain more than 20 km of sediments ranging in age from Riphean to Quaternary (Figure 6). An oceanic crust may underlie the center of the basin. Siliciclastics prevail in the section from the Riphean to the Late Devonian. The organic-rich facies of the Domanik formation, which is the main source rock for hydrocarbons in Paleozoic reservoirs, was deposited at the center of the basin during the Middle Devonian. This

Table 1. Petroleum Resources of Kazakhstan.

Oil (Billion Barrels)

Basin / Province	Cumulative Production	Identified Reserves	Undiscovered Resources 95%	Undiscovered Resources Mode	Undiscovered Resources 5%	Mean
North Caspian (Peri-Caspian)	0.8	12.8	10	23	80	36
Middle Caspian–Mangyshiak	2.1	1.6	2	1	2	2
South Turgay	0	0.7	1	1	5	3
North Ustyurt	0.3	1.8	1	1	2	1
Chu-Sarysu	0.0	0.0	0	0	0	0
Total	**3.2**	**16.9**	**14**	**26**	**89**	**42**

Natural Gas (Trillions of Cubic Feet)

Basin / Province	Cumulative Production	Identified Reserves	Undiscovered Resources 95%	Undiscovered Resources Mode	Undiscovered Resources 5%	Mean
North Caspian (Peri-Caspian)		80.0	63	118	480	209
Middle Caspian–Mangyshiak	6.5	1.3	1	2	8	5
South Turgay	0	negligible	0	1	1	1
North Ustyurt	0.6	0.4	1	1	2	1
Chu-Sarysu	0.0	2.0	1	2	4	2
Total	**7.1**	**83.7**	**66**	**124**	**495**	**218**

Modified after Ulmishek and Masters, 1993

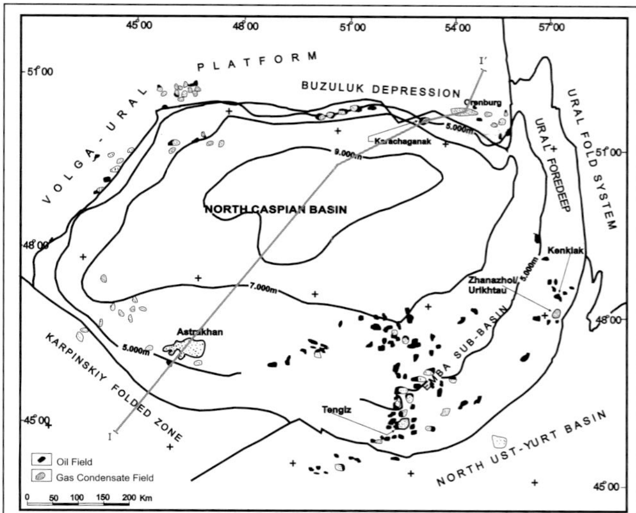

Figure 4. North Caspian Basin: structure map on top of pre-Kungurian (subsalt) section showing oil and gas fields. Major fields are noted.

early siliciclastic sequence is primarily overlain by Upper Devonian through Lower Permian carbonates and minor siliciclastics along the basin margins, sourced mainly from the east-southeast and secondarily from the northwest. Most of the large fields in the basin occur in reefs and/or carbonate debris of the Carboniferous to Permian that ring the basin. Reservoir facies include high-energy fringing reefs and banks, low-energy platform interior facies, and slope deposits. Production usually is associated with zones that have been enhanced by fracturing and/or development of karst. Examples of such fields are Tengiz (Figure 7), Karachaganak (Figure 8), and Zhanazhol (Figure 9). There is some production from Carboniferous siliciclastic facies.

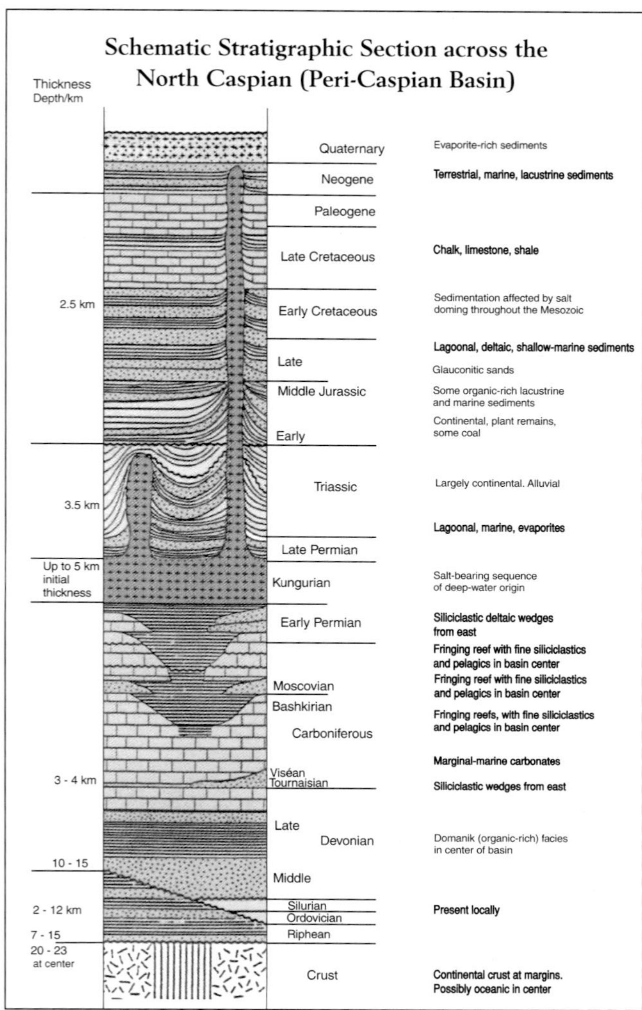

Figure 6. Schematic stratigraphic section across the North Caspian (Peri-Caspian) Basin.

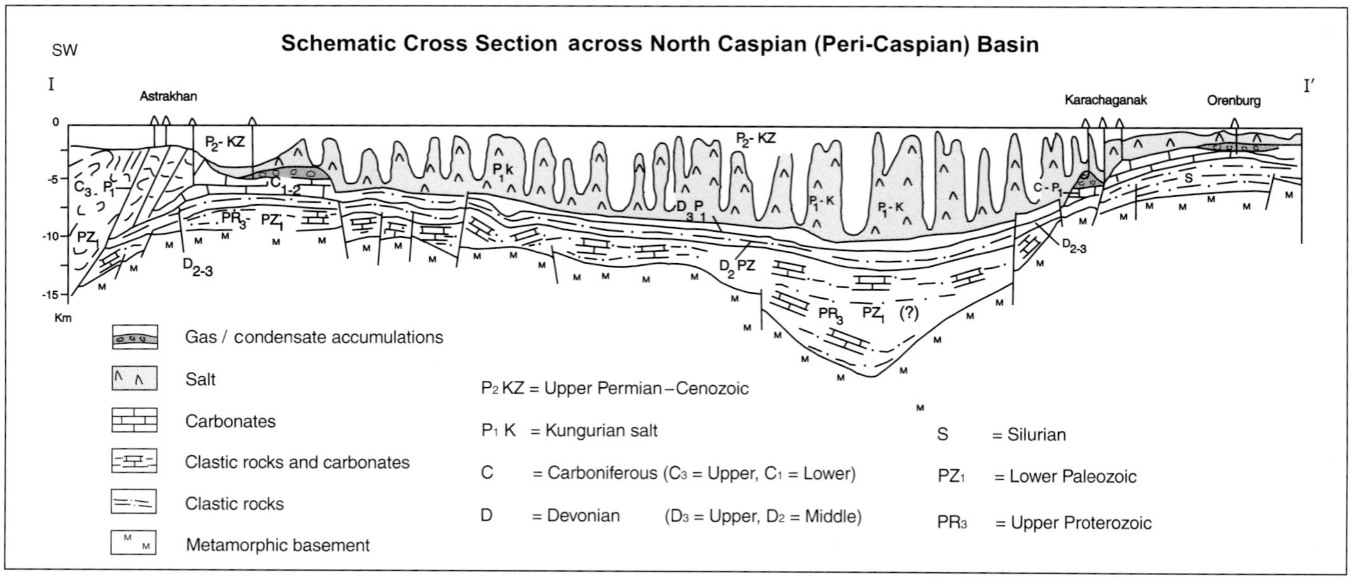

Figure 5. Schematic cross section across North Caspian (Peri-Caspian) Basin. See Figure 4 for location of cross section I–I′.

A Lower to middle Permian (Kungurian) salt-bearing sequence is believed to be of deep-water origin and as much as 2500–3000 m thick at the time of deposition. This salt is diapric and therefore varies greatly in thickness. Postsalt stratigraphy consists primarily of continental and lacustrine facies from the Late Permian through Early Jurassic, grading into shallow-marine siliciclastics in the Early Cretaceous. Middle Jurassic shales which accumulated in anoxic environments are considered to be source rocks for oil and gas accumulations in the postsalt sequence. Chalks, limes, and shales dominate from the Late Cretaceous to the Neogene, followed by a mixed stratigraphy of continental and shallow-marine sequence in the Neogene. The Quaternary consists of evaporite-rich sediments. It should be noted that postsalt stratigraphy, particularly during the Mesozoic, is influenced by movement of the Kungurian salt. Postsalt reservoirs are Middle to Upper Jurassic and Lower Cretaceous sandstones, such as at Kenkiak field (Figure 10) (Dalyan and Posadskaya, 1972; Kochariyants et al., 1979; Zamarenov et al., 1986; Bagrintseva and Belozerova, 1987; Trochimenko, 1987; Ivanov, 1988; Vladimirova and Maltseva, 1990; Demidov, 1992, 1996; Arabadzhi et al., 1993; Shlygin et al., 1997, Tarkanov and Bezborodova, 1992).

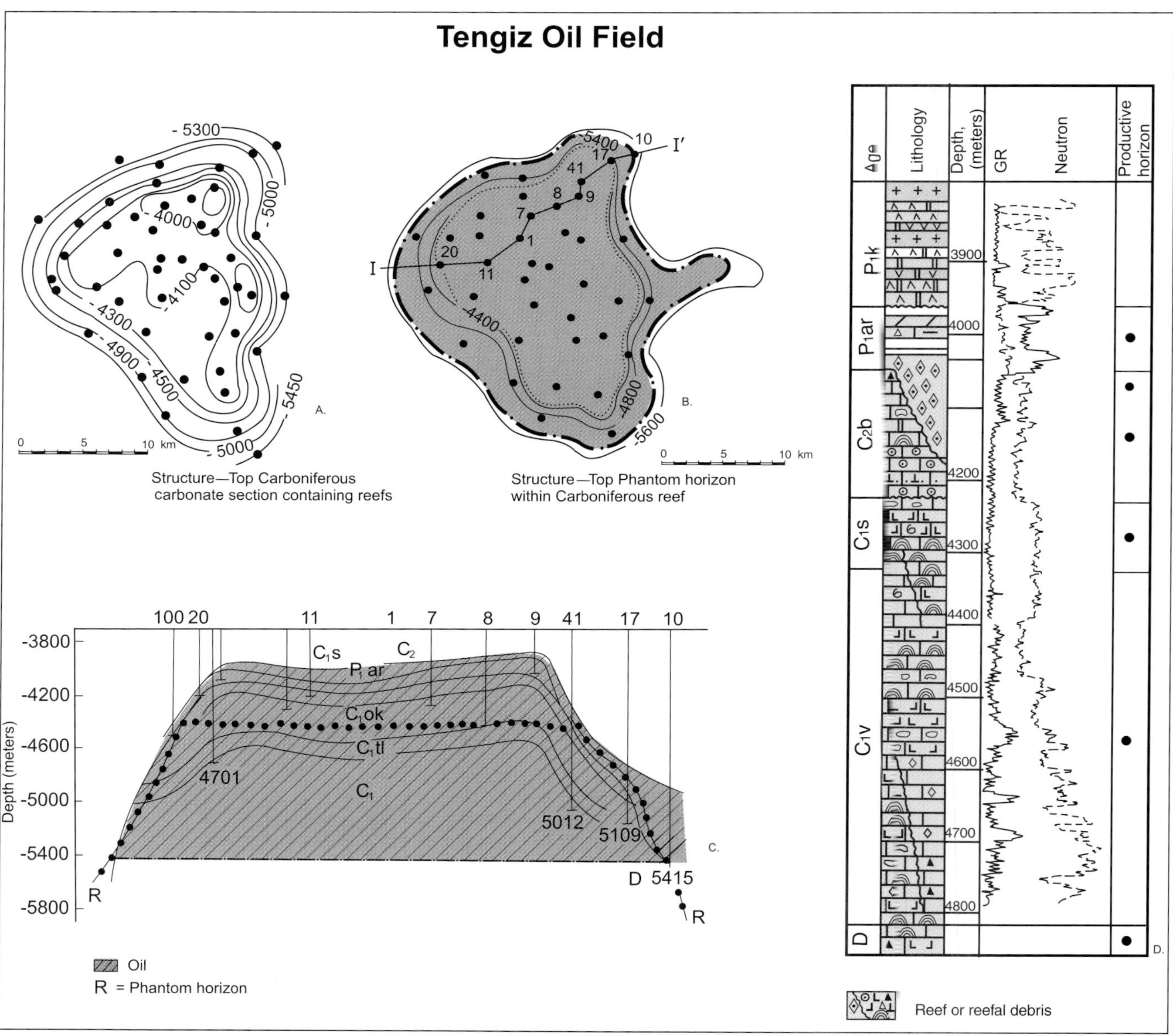

Figure 7. Tengiz oil field (modified after Abduleena et al., 1993). (a) Structure map on top of Carboniferous carbonate section containing reefs. (b) Structural map on a "correlation-marker" horizon within the Carboniferous reef section with oil-water contact indicated. (c) Composite stratigraphic section for Tengiz field. (d) Lithographic log. Permian: P_1k = Kungurian, P_2ar = Artinskian. Carboniferous: C_2b = Bashkirian, C_1s = Serpukhovian, C_1v = Viséan. Devonian undifferentiated: D.

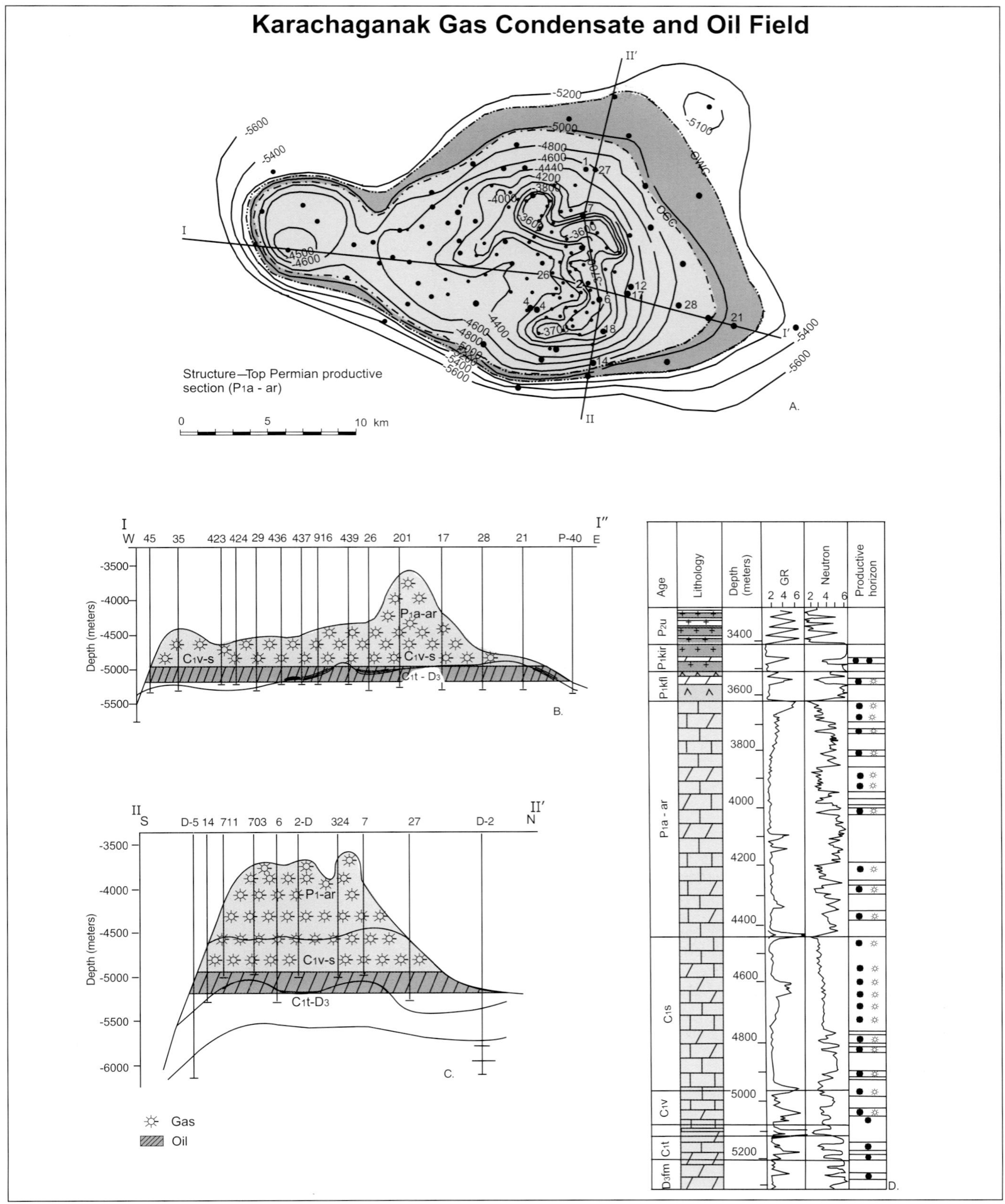

Figure 8. Karachaganak gas condensate and oil field (modified after Abduleena et al., 1993). (a) Structural map on top of Permian productive section; top and base of oil rim outlined. (b) and (c) Composite stratigraphic section for Karachaganak field. (d) Lithographic log. Permian: P_2u = Ufimian, P_1k = Kungurian, P_1a–ar = Asselian-Artinskian. Carboniferous: C_1s = Serpukhovian, C_1v = Viséan, C_1t = Tournaisian. Devonian: D_3fm = Famennian.

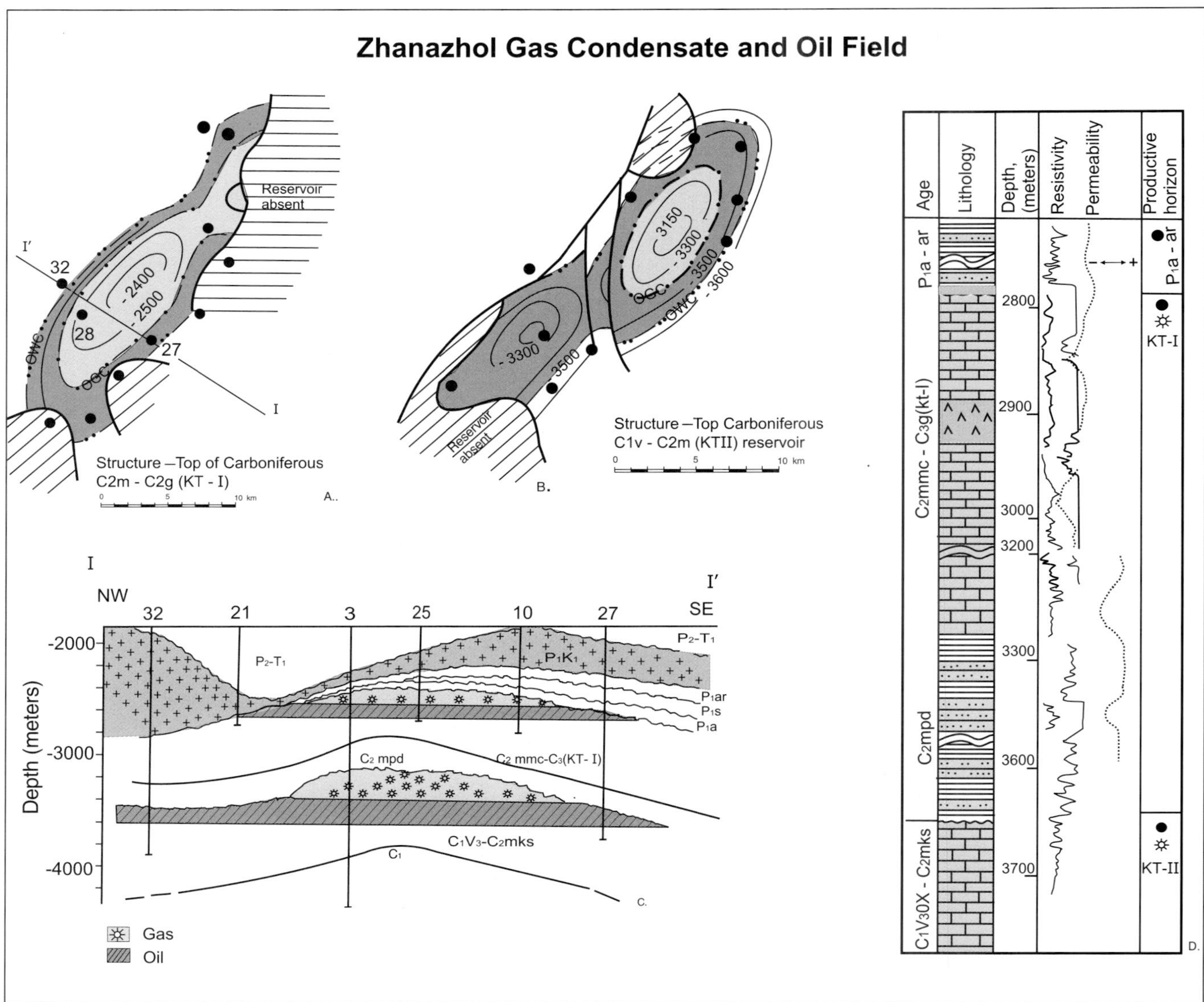

Figure 9. Zhanazhol gas condensate and oil field (modified after Abduleena et al., 1993). (a) Structural map on top of Carboniferous C_2m-C_2g (Moscovian-Gzelian) carbonate reservoir (KT-1). (b) Structural map on top of Carboniferous C_1v-C_2m (Viséan-Moscovian) sandstone reservoir (KT-II) (c) Composite stratigraphic section for Zhanazol field. (d) Lithographic log. Permian: P_1a = Artinskian. Carboniferous: C_2mmc-C_3g = Moscovian (Myachkovskian)-Gzelian, C_2mpd = Moscovian (Poldoskian), C_1v_3ox-C_2mks = Viséan (Okskian-Kashirian).

The basin has a relatively low thermal gradient in the eastern half with gradients of 1–2°C/100 m, increasing to 2–3° in the western half and to the southwest. The highest thermal gradients are in the southwesternmost portion, immediately west of Atyrau on the Caspian Sea, reaching 4.5–4.8°C/100 m. Generally, onset of hydrocarbon generation for presalt source rocks is considered to be Late Permian/Triassic and for postsalt source rocks Late Cretaceous/Paleogene (Egorova, 1979; Kiryukhin et al., 1984; Svetlakova, 1987; Kalinko et al., 1993; Medvedeva et al., 1993).

During the past two decades, several supergiant oil and gas-condensate fields (Tengiz, 1979; Karachaganak, 1979; and Zhanazhol, 1984) and some smaller but significant fields were discovered in rocks beneath the thick Permian salt of the basin. The basin is still in an immature stage of exploration because of the great depths to potential reservoirs, high reservoir pressures, and high sulfur content in the gas. The potential of the basin remains high for the presalt Paleozoic carbonate rocks and associated reefs and siliciclastic fans on the margins. Carboniferous and Lower Permian siliciclastic fans are widespread along the eastern and southern margins of the basin. A shallow Mesozoic postsalt salt-dome play has been explored for many years and still possesses a significant petroleum potential.

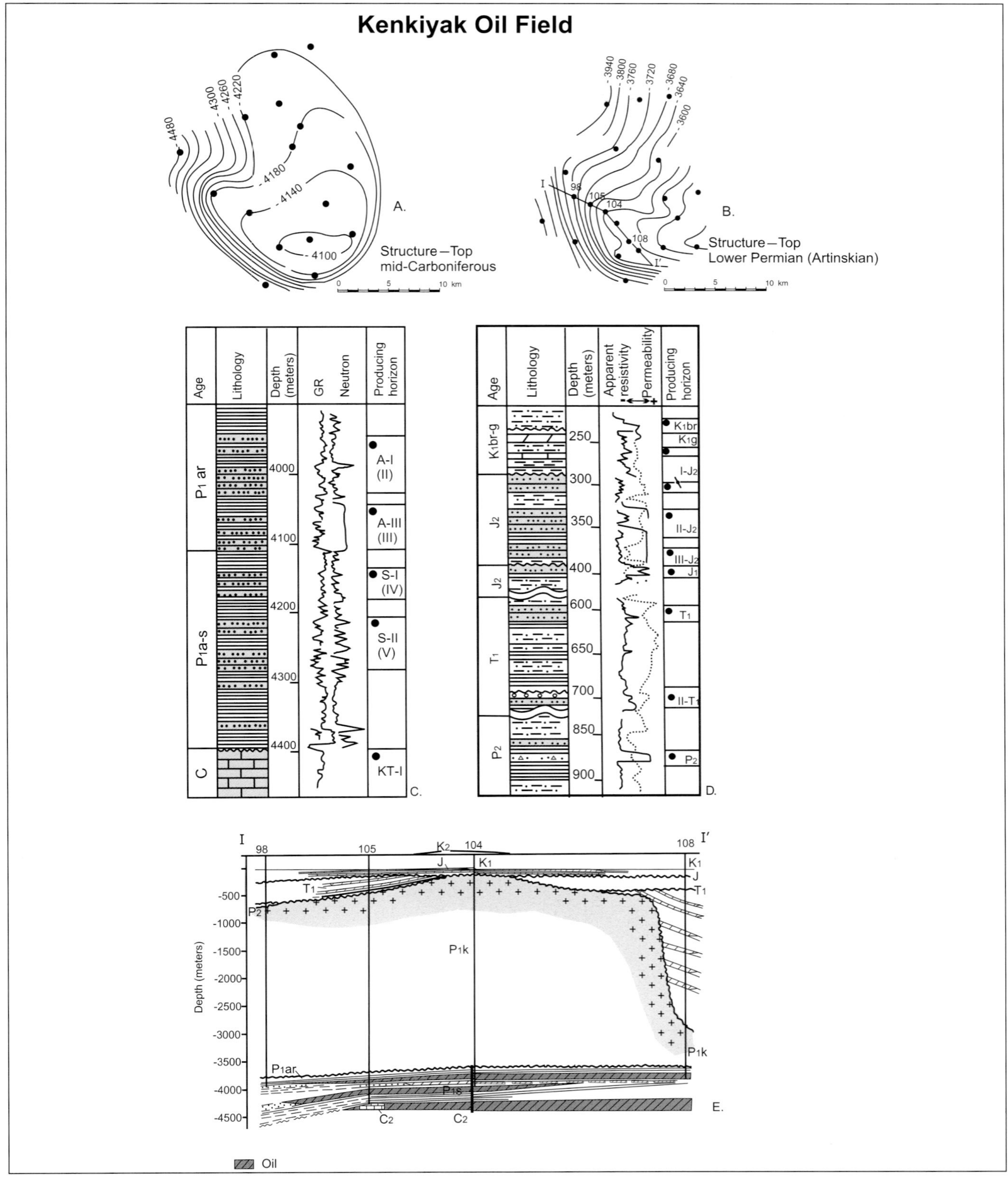

Figure 10. Kenkiyak oil field (modified after Abduleena et al., 1993). (a) Structural map on top of the middle Carboniferous. (b) Structural map on top of the Lower Permian (Artinskian) reservoir. (c) and (d) Lithographic logs. Cretaceous: K_1br-g = Barremian. Jurassic: J_1 = Dogger, J_2 = Lias. Triassic: T_1 = Scythian. Permian: P_2 = Upper Permian, P_1ar = Artinskian, P_1a-s = Asselian-Sakmarian. Carboniferous: C. (e) Composite stratigraphic section for Kenkiyak field.

Middle Caspian Basin–Mangyshlak Province

The Middle Caspian Basin–Mangyshlak province of Kazakhstan covers about 75,000 km^2 (about 35,000 km^2 is onshore and about 40,000 km^2 is offshore) (Figures 2, 3). It is part of the North Caucasus–Mangyshlak petroleum province, which extends from the Azov-Kuban Basin, in Russia, eastward across the middle of the Caspian Sea onto the onshore Mangyshlak trough in western Kazakhstan. The onshore areas of the province have been explored extensively; gas dominates in the Azov-Kuban basin, whereas most of the Middle Caspian Basin–Mangyshlak trough is oil prone (Ulmishek and Masters, 1993b).

The principal part of undiscovered resources for the Middle Caspian Basin–Mangyshlak province probably is located offshore in the unexplored central Caspian Sea of

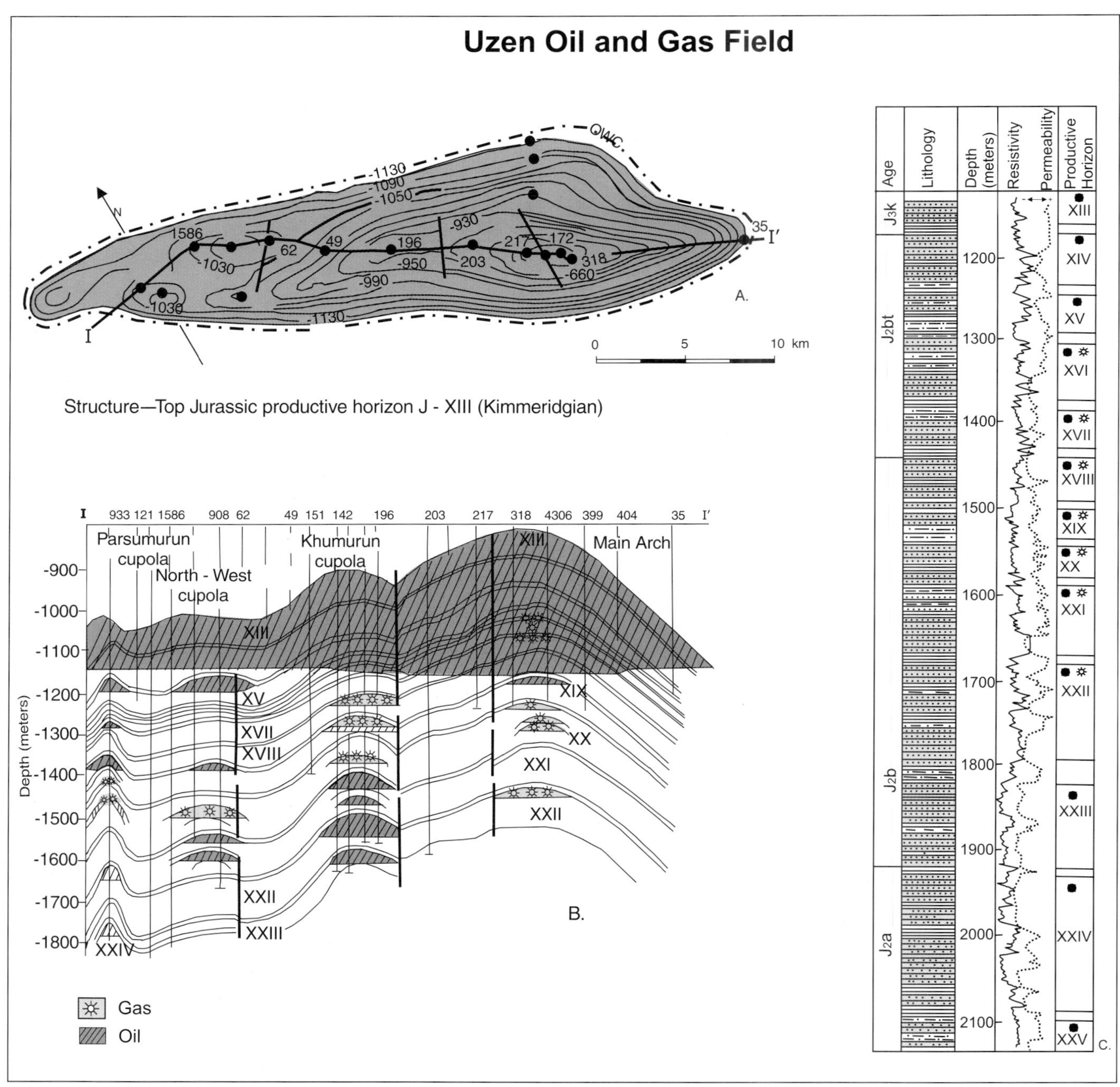

Figure 11. Uzen oil and gas field Jurassic pays (modified after Abduleena et al., 1993). (a) Structural map on top of Jurassic productive horizon J-XIII (Kimmeridgian). Oil-pool outline indicated. (b) Cross section through Uzen field showing pays and culminations (cupolas) along strike at the Jurassic level. (c) Composite lithologic log for the Jurassic producing section. J_3k_3 = Kimmeridgian, J_2bt = Bathonian, J_2b = Bojacian, J_2a = Aalenian.

the Middle Caspian Basin, where Jurassic and Lower Cretaceous siliciclastic rocks in structural traps are the prime exploration targets. Uzen (Figures 11, 12) and Zhetybay are two onshore giant fields, which produce from Middle Jurassic through lower Upper Cretaceous fluvial to deltaic siliciclastics and would serve as analogs for the offshore (Maximova, 1987a, b; Abduleena et al., 1993). The main reservoirs in these two fields are Middle and Upper Jurassic shallow-marine to nonmarine sandstones. Traps comprise drapes over Permian-Triassic asymmetric horst features. These features are modified by strike-slip faulting and attendant anticline development.

Source rocks in the area are possibly Permian shales, the Middle Triassic Olenek shale, and Lower Jurassic black shales. Hydrocarbon generation is likely to have begun in the Cretaceous. Some oils are highly paraffinic, and gas is often sour.

South Turgay Basin

The South Turgay Basin covers 160,000 km^2 and lies entirely in Kazakhstan (Figures 2, 3). Exploration began in the early 1980s and revealed flat-lying Tertiary and Cretaceous rocks underlain by a Lower-Middle Jurassic rift system. About a dozen oil and gas fields have been discovered, but most reserves are concentrated in Upper Jurassic and Cretaceous (Neocomian) siliciclastic rocks in the large Kumkol field (Figure 13), which has reserves of 50–100 million bbl oil equivalent (BOE) (Maximova, 1987a, b; Abduleena et al., 1993).

The basin is lightly explored, with the northern half essentially undrilled. However, drilling of several struc-

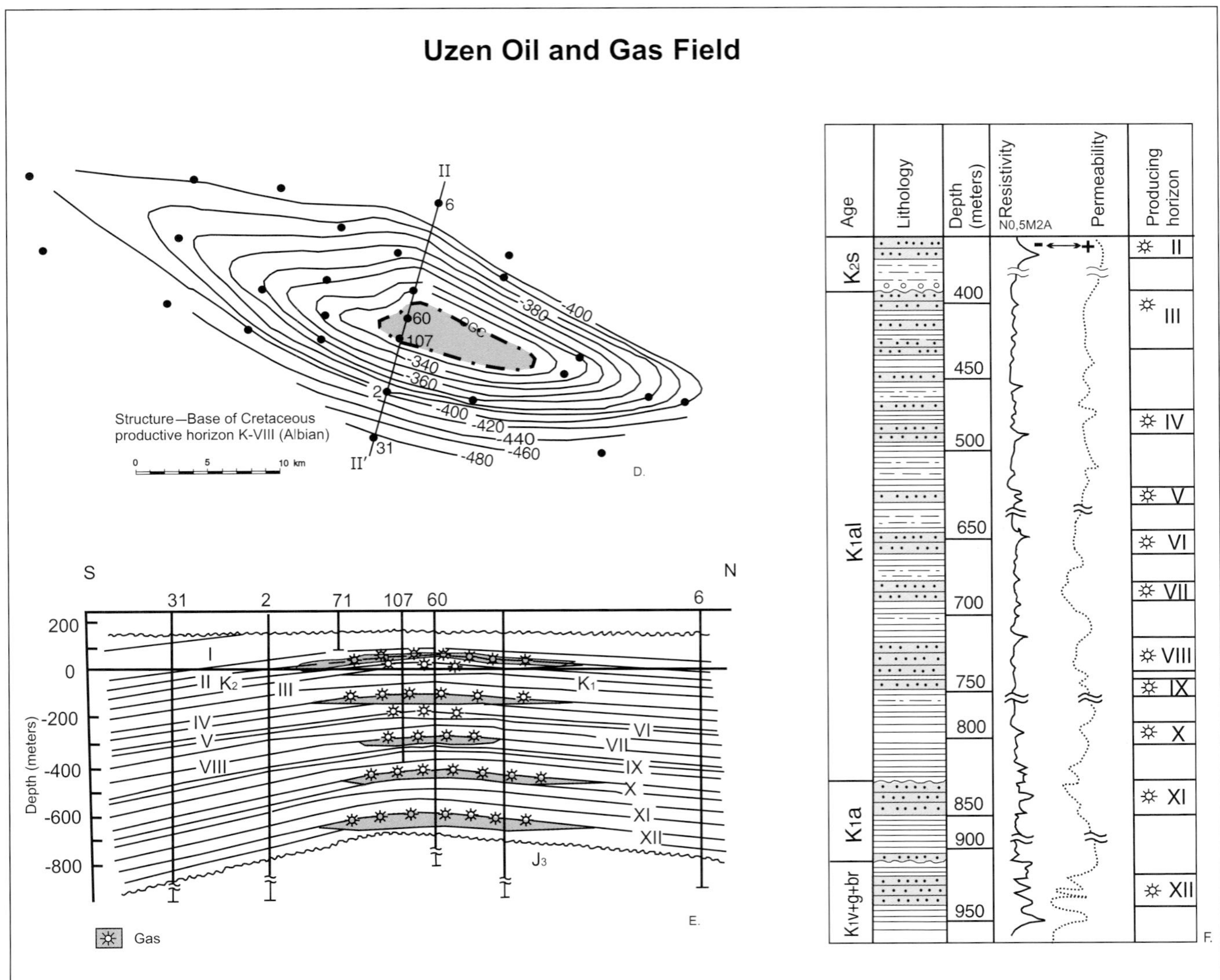

Figure 12. Uzen oil and gas field Cretaceous pays (modified after Abduleena et al., 1993). (d) Structural map at the base of Cretaceous productive horizon K-VIII (Albian). Gas-pool outline is indicated. (e) Cross section through Uzen field showing Cretaceous pays. (f) Composite lithologic log for the Cretaceous producing section. K_2S = Senonian, K_1at = Albian, K_1a = Aptian, $K_1v+g+br$ = Neocomian.

tures, similar to Kumkol field, has not resulted in significant discoveries. Much of the remaining potential is probably in stratigraphic and structural traps in the Lower-Middle Jurassic to Lower Cretaceous siliciclastic sequences on north-south-oriented horsts in the rift that developed prior to hydrocarbon migration. Migration is considered to have begun in the Cretaceous to early Tertiary. Middle to Upper Jurassic shales have high total-organic content exceeding 10% in 10-m zones and are good source rocks for oil. Triassic and Jurassic coal measuring as much as 50 m thick serves as gas source rock. A more speculative play is the Lower Carboniferous carbonates on the basin's margin and on tilted horst blocks in the basin (Ulmishek and Masters, 1993).

North Ustyurt Basin

The North Ustyurt Basin covers about 145,000 km^2, of which about the eastern 20% lies in Uzbekistan (Figures 2, 3). The basin occupies a microcontinent in the Hercynian accreted terrain. Thick Triassic and Jurassic through Tertiary siliciclastic sediments overlie a carbonate platform of the microcontinent. Reserves are heavy

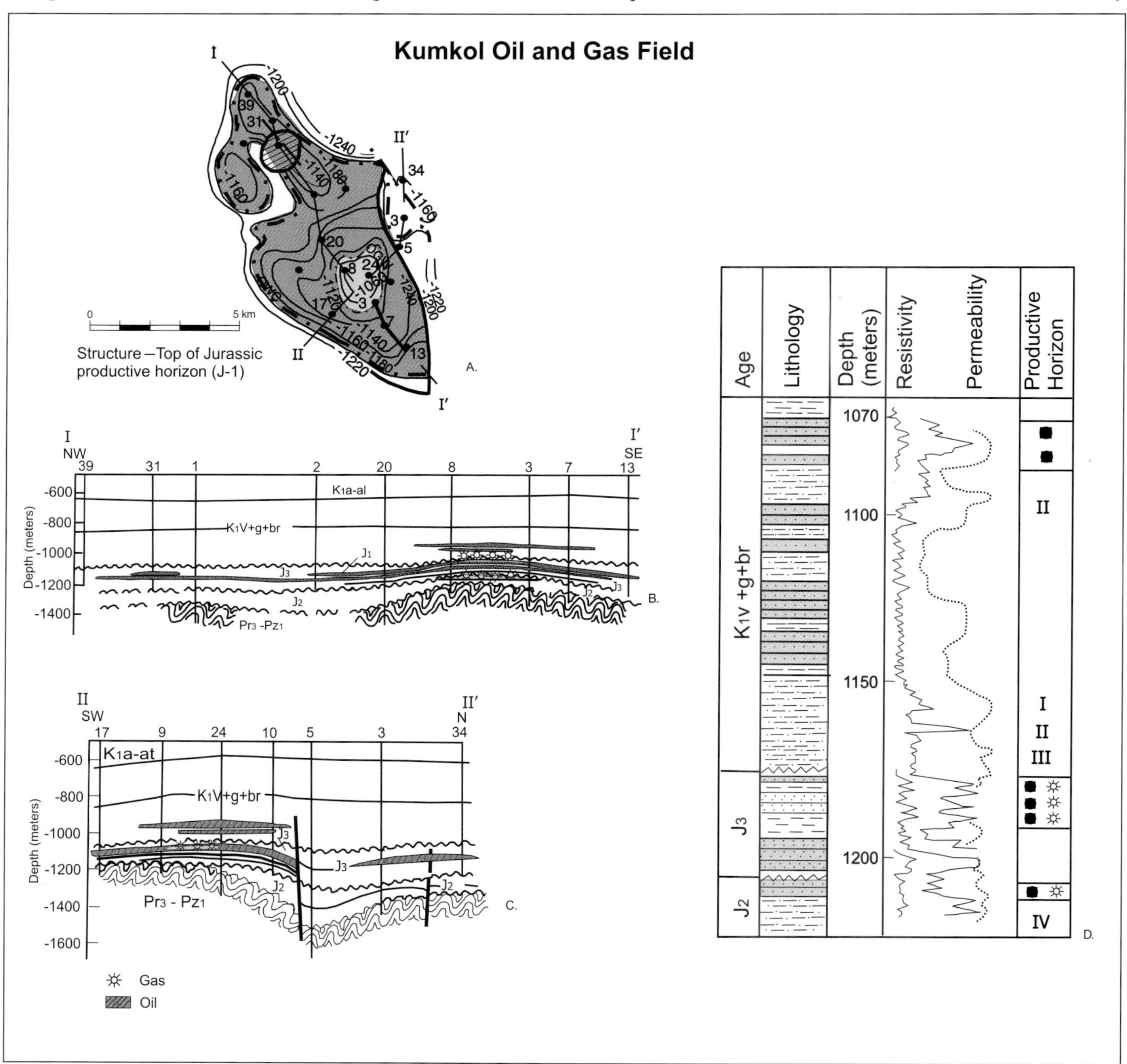

Figure 13. Kumkol oil and gas field (modified after Abduleena et al., 1993). (a) Structural map on top of Jurassic productive horizon (J-1). Outlines of oil-gas contact and oil-water contact are indicated. (b) Cross section through Kumkol field I–I′. (c) Cross section through Kumkol field II–II′. (d) Composite lithologic log for the field. Cretaceous: $K_1v+g+br$ = Valanginian, Hauterivian, Barremian. Jurassic: J_3 = Malm, J_2 = Dogger.

oil with high sulfur content (as much as 2.6%) in Jurassic and Cretaceous (Neocomian) rocks at shallow depths on the Buzachi Peninsula, on the extreme western side of the basin. Karazhanbas field (Figure 14) is the basin's largest field, with about 500 million bbl oil recoverable. Elsewhere in the basin, several small oil fields in Jurassic siliciclastics and a few gas fields of possible biogenic origin in Eocene sandstones have been discovered (Maximova, 1987a, b; Abduleena et al., 1993).

Oil source rocks are not well documented. Possibly the oil discovered has migrated from the North Caspian Basin, where it was generated from Devonian to Permian shales.

The Jurassic to Tertiary sequence is moderately explored, and its potential for undiscovered reserves is deemed relatively low. Future discoveries are likely to be analogous to accumulations discovered to date, also located in anticlines, often fault bounded. There is a possibility of stratigraphic pinch-out traps along the flanks of the Buzachi arch on the shore of the Caspian Sea. The deep Paleozoic section has been penetrated in only a few locations. Seismic data in the eastern part of the basin suggest the presence of reefs and basinal facies, perhaps containing source rocks. The assessment of undiscovered resources for the basin is highly uncertain but may be conservative (Ulmishek and Masters, 1993).

Chu-Sarysu Basin

Several hydrocarbon fields have been discovered in Carboniferous and Devonian rocks in the Chu-Sarysu Basin. Pridorozhnoye gas field is a typical example (Figure 15). Suspected source rocks in the Devonian to Car-

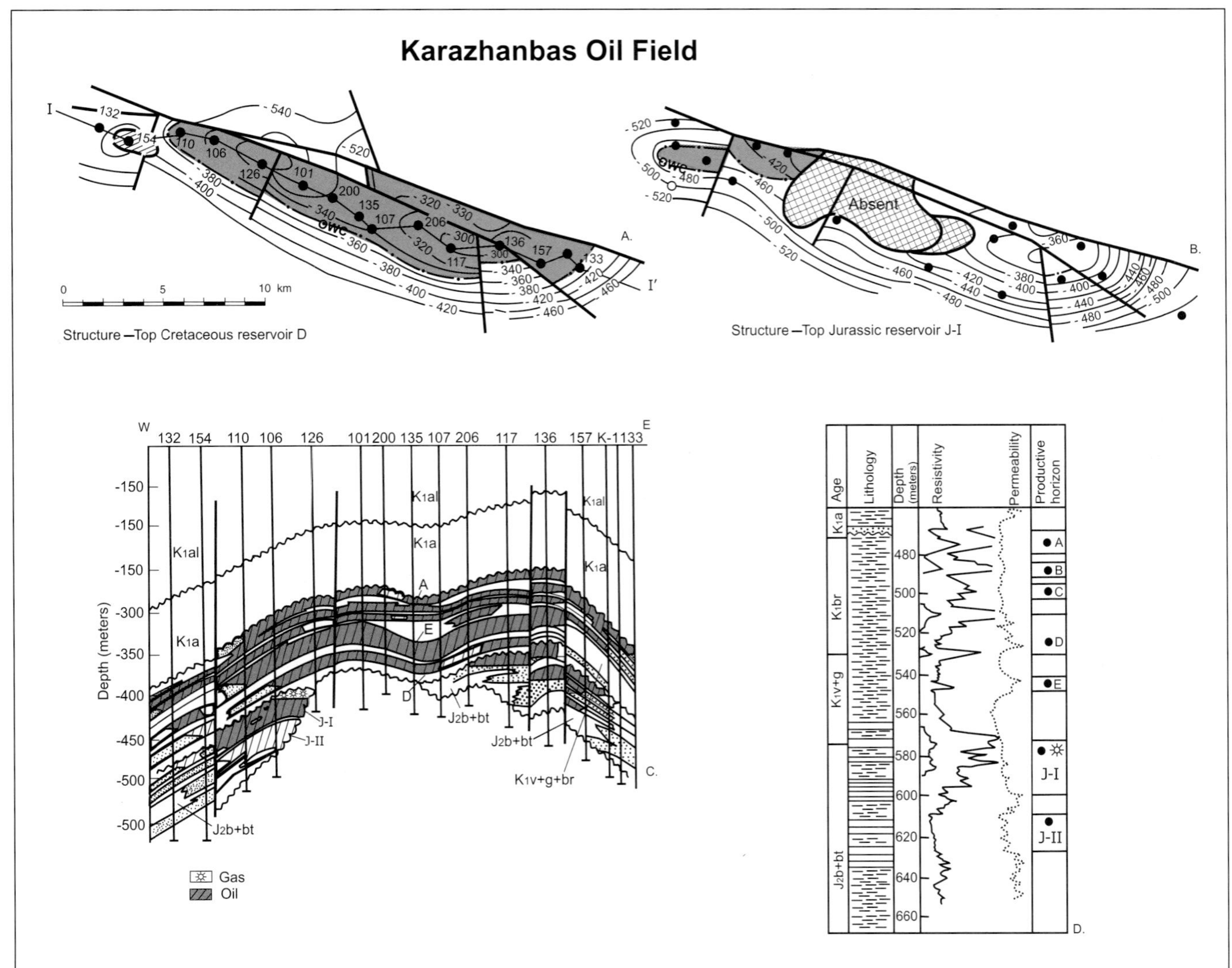

Figure 14. Karazhanbas oil field (modified after Abduleena et al., 1993). (a) Structural map on top of Cretaceous reservoir D. (b) Structural map on top of Jurassic reservoir J-1. (c) Cross section through Karazhanbas field. (d) Composite lithologic log for the field. Cretaceous: K_1a = Aptian, K_1br = Barremian, K_1v+g = Valanginian, Hauterivian. Jurassic: J_2b+bt = Bajocian, Bathonian.

boniferous (Tournaisian) section are overmature, and all discoveries have been gas. Gas in the Lower Permian reservoirs, below a salt cap, is dominantly nitrogen with high helium content. The undiscovered potential of the basin is low and related to the virtually unexplored Devonian section below the Upper Devonian salt seal (Maximova, 1987; Abduleena et al., 1993; Ulmishek and Masters, 1993).

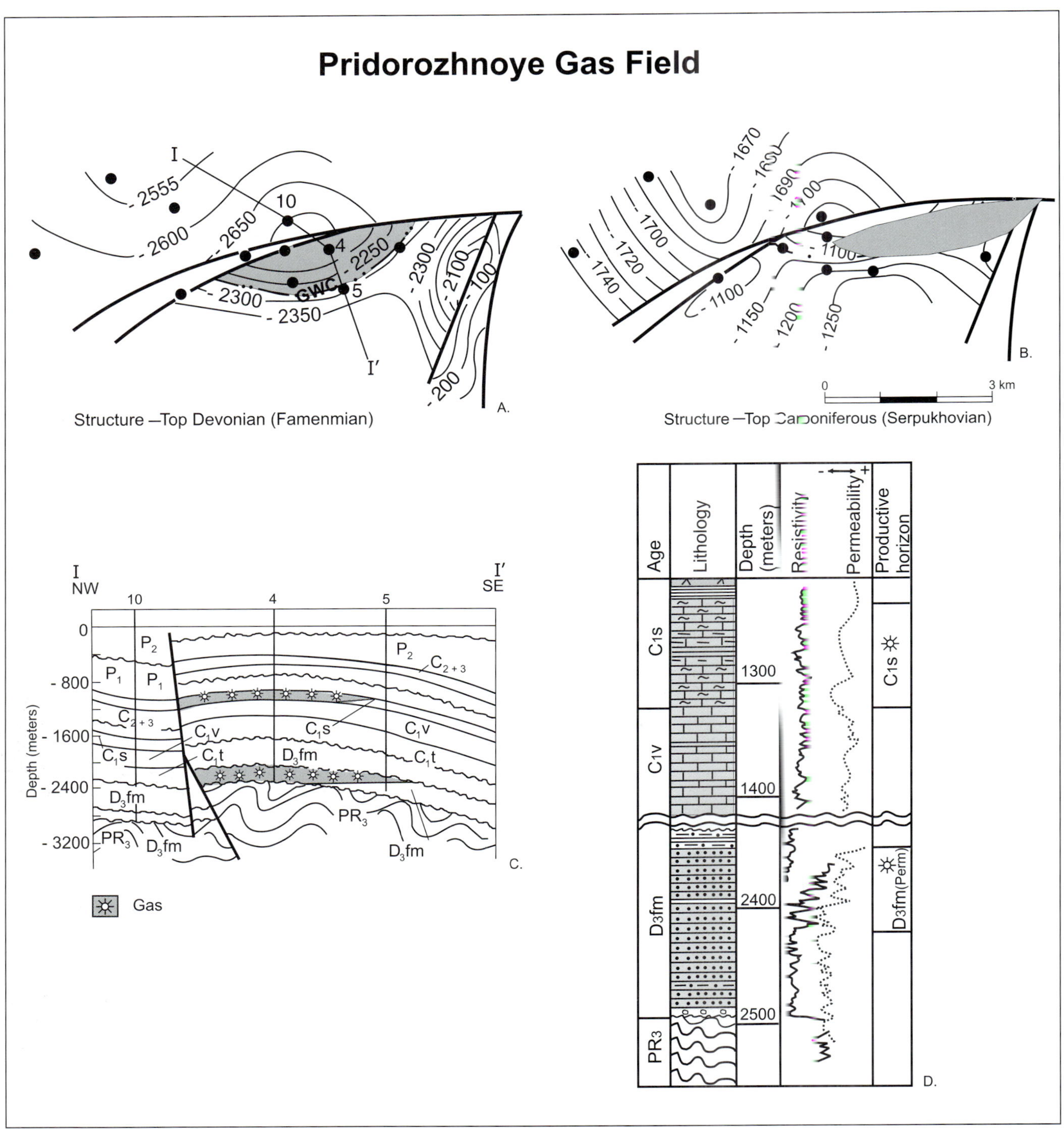

Figure 15. Pridorozhnoye gas field (modified after Abduleena et al., 1993). (a) Structural map on top of the Devonian (Famennian) reservoir. Gas-water contact is indicated. (b) Structural map on top of the Carboniferous (Serpukhovian) reservoir. (c) Cross section across Pridorozhnoye field. (d) Composite lithologic log for the field. Carboniferous: C_1s = Serpukhovian, C_1v = Viséan. Devonian: D_3fm = Famennian. Precambrian: PR_3 = Riphean.

CONCLUSIONS

Most of Kazakhstan's oil and gas reserves have not been developed. Many areas remain under- or unexplored. It is expected that a considerable portion of Kazakhstan's potential oil and gas reserves will be located offshore in the Caspian Sea. Future reserves will be discovered and developed using international technology that was unavailable to the nation during the Soviet period. Those reserves will be commercially justified with better access to world markets and improved infrastructure.

About 250 oil and gas accumulations have been identified in Kazakhstan, of which 107 have been developed. Four basins/provinces, considered to have the most potential for undiscovered reserves, are (in order of descending potential):

1. North Caspian (Peri-Caspian) Basin, also referred to as the Pri- or Pre-Caspian Basin
2. Middle Caspian Basin–Mangyshlak province
3. South Turgay Basin
4. North Ustyurt Basin

Estimates of most likely volumes of undiscovered oil/condensate and natural gas are 26 billion bbl and 124 tcf, respectively. The bulk of undiscovered oil/condensate and gas reserves is estimated to be attributable to Kazakhstan's portion of the North Caspian (Peri-Caspian) Basin.

ACKNOWLEDGMENTS

The author expresses his appreciation to Devon Energy Corporation, parent company of Pennzoil Caspian Corporation, for having granted permission to publish the manuscript. Thanks also go to Nargiz Rizazade for bibliographic research, Afa Rzayeva for technical translation, Emma Agayeva and Regina Lamy for assistance with graphics, Julie Aldrich for general assistance, and Steve Warshauer for sharing his knowledge on the geology of Kazakhstan.

REFERENCES CITED

Abduleena, A. A., E. C. Votsalevshy, and B. M. Kuandykova, 1993, Oil and gas deposits of Kazakhstan: Moscow, Nedra, 247 p.

Arabadzhi, M. S., R. S. Bezborodov, et al., 1993, Forecast of oil and gas content of the southeastern Pre-Caspian synclinorium: Moscow, Nedra.

Bagrintseva, K. I., and G .E. Belozerova, 1987, Reservoir types and properties of pre-salt deposits in the Pre-Caspian synclinorium, *in* N. A. Krylov and N. I. Nekhrikova, eds., Oil and gas content of the Pre-Caspian basin and adjacent regions: Moscow, Nauka, p. 59–64.

Dalyan, I. B., and L. S. Posadskaya, 1972, Geology and oil and gas content of the Pre-Caspian basin's east area: Almaty, Nauka, 192 p.

Demidov, V. A., 1992, Salt domes in the Pre-Caspian basin's east area and their oil and gas potential: Geology of Oil and Gas, no. 11, p. 1–5.

Demidov, V. A., 1996, On the major types of salt domes in the Pre-Caspian basin and the principles of planned locations: Geology, geophysics and development of the oil fields, no. 3.

Dongaryan, L .S., 1990, Oil and gas content of sub-salt Paleozoic in Pre-Caspian synclinorium, *in* L. S. Dongaryan, ed., Geologic basis for the formation of the Pre-Caspian oil and gas producing complex: Moscow, Nauka, p. 115–124.

Effimoff, I., 1999, Comments on the oil and gas resource base of the Caspian region, production rates and export routes: Energy Ecology Economy, no. 3–4, p. 45–59.

Egorova, R. I., 1979, Geothermal characteristics of the Pre-Caspian basin subsurface, *in* Forecast for the oil and gas content of Kazakhstan and adjacent territories: Moscow.

Energy Information Administration, 1999a, Kazakhstan: Country analysis briefs: www.eia.doe.gov/emeu/cabs/kazak.html.

Energy Information Administration, 1999b, Production (supply): Annual data: Petroleum data: www.eia.doe.gov/emeu/international/petroleu.html.

Energy Information Administration, 1999c, Production (supply): Natural gas data: www.eia.doe.gov/emeu/international/gas.html.

Energy Information Administration, 1999d, Kazakhstan—Year 1996: Country energy data report: www.eia.doe.gov/emeu/world/country/cntry_KZ.html.

Ivanov, Yu. A., 1988, Oil and gas potential of the post-salt and salt complexes of the Pre-Caspian basin: Geology of Oil and Gas, no. 7, p. 1–5.

Kalinko, M. K., G. N. Molodykh, N. I. Nemtsov, V. E. Bembeev, and V. V. Kontrovskiy, 1993, Temperature distribution on the pre-salt surface of the Pre-Caspian basin: Geology of Oil and Gas, no. 1.

Kiryukhin, L. T., D. L. Fyodorov, et al., 1984, Pre-salt oil and gas formation and distribution peculiarities in the Pre-Caspian basin: Moscow, Nedra, 144 p.

Kochariyants, S. B., T. N. Rozanova, and V. V. Paraizyan, 1979, On the possibility of Jurassic oil formation in the Pre-Caspian basin, *in* Forecast for oil and gas content of Kazakhstan and adjacent territories: Moscow.

Masters, C. D., D. H. Root, and R. M. Turner, 1998, World conventional crude oil and natural gas—Identified reserves, undiscovered resources and futures; U.S. Geological Survey Open-File Report 98-468, p. 105.

Maximova, S. P., 1987a, Oil and gas fields of the USSR: Moscow, Nedra, v. 1, p 358.

Maximova, S. P., 1987b, Oil and gas fields of the USSR: Moscow, Nedra, v. 2, p. 302.

Medvedeva, A. M., Z. E. Bulekbayav, and I. B. Daliyan, 1993, Direct proof of vertical oil migration in the eastern Pre-Caspian: Izvestiya NAS RK, no. 1.

Myzaprova, L. M., 1998, The history of oil and gas development in Kazakhstan, *in* Caspian Magazine: Almaty, KIOGA'98.

Persits, F. M., G. F. Ulmishek, and D. W. Steinshouer, 1997, Maps showing geology, oil and gas fields and geologic provinces of the former Soviet Union: U.S. Geological Survey Open-File Report 97-470E.

Pronichva, M. V., and G. N. Savvinova, 1980, Paleo-morphological analysis of oil and regions: Moscow, Nedra, 254 p.

Shlygin, D. A., N. E. Kuantaev, et al. 1997, Geochemical peculiarities of post-salt oil and gas content of the Pre-Caspian basin: Geology and exploration for the subsurface of Kazakhstan, no. 3.

Svetlakova, E. A., 1987, Formation model and principles of hydrocarbon location in the Pre-Caspian basin, *in* N. A. Krylov and N. I. Nekhrikova, eds., Oil and gas content of the Pre-Caspian basin and the adjacent regions: Moscow, Nauka, p. 151–154.

Tarkanov, M. I., and I. V. Bezborodova, 1992, Basic features of late Paleozoic paleography of the southeastern Pre-Caspian basin, *in* Southeastern Pre-Caspian synclinorium's pre-salt geology and oil and gas content: Moscow, Academy of Sciences.

Trokhimenko, M. S., 1987, Some principles of post-salt oil and bitumen fields in eastern Pre-Caspian basin, *in* N. A. Krylov and N. I. Nekhrikova, eds., Oil and gas content of the Pre-Caspian basin and the adjacent regions: Moscow, Nauka, p. 147–151.

Ulmishek, G. F., and C. D. Masters, 1993a, Oil, gas resources estimated in the former Soviet Union: Oil & Gas Journal, v. 91, no. 50, p. 59–62.

Ulmishek, G. F., and C. D. Masters, 1993b, Petroleum resources in the former Soviet Union: U.S. Geological Survey Open-File Report 93-316, 17 p.

Vladimirova, T. V., and A. K. Maltseva, 1990, The formations of the Paleozoic pre-salt deposits in the Pre-Caspian basin, *in* L. S. Dongaryan, ed., Geologic basis for the formation of the Pre-Caspian oil and gas producing complex: Moscow, Nauka, p. 32–40.

Zamarenov, A. K., et al., 1986, Sedimentation models of the pre-salt oil and gas bearing complex of the Pre-Caspian Basin: Moscow, Nedra, 137 p.

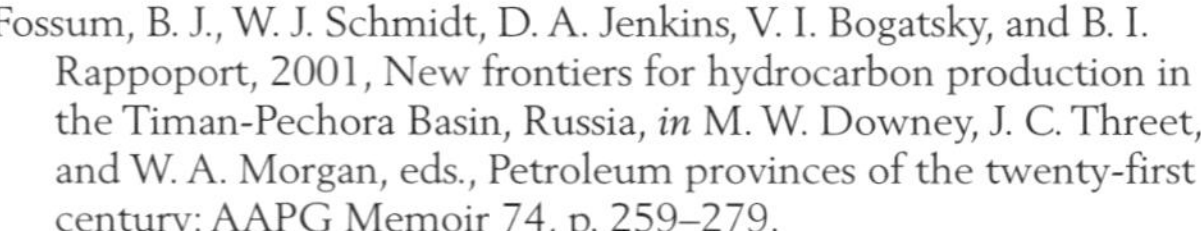
Fossum, B. J., W. J. Schmidt, D. A. Jenkins, V. I. Bogatsky, and B. I. Rappoport, 2001, New frontiers for hydrocarbon production in the Timan-Pechora Basin, Russia, *in* M. W. Downey, J. C. Threet, and W. A. Morgan, eds., Petroleum provinces of the twenty-first century: AAPG Memoir 74, p. 259–279.

Chapter 13

New Frontiers for Hydrocarbon Production in the Timan-Pechora Basin, Russia

Bret J. Fossum, William J. Schmidt, David A. Jenkins
Conoco, Inc., Houston, Texas, U.S.A.

Vladimir I. Bogatsky
Timan-Pechora Scientific Research Center, Ukhta, Komi Republic, Russia

Boris I. Rappoport
OAO Arkhangelskgeoldobycha, Arkhangelsk, Nenets Okrug, Russia

ABSTRACT

The Timan-Pechora Basin, located west of the Ural Mountains in northern Russia, is a prolific oil province covering 300,000 km^2. The basin, which has been efficiently explored, is predominantly oil prone, but significant accumulations of gas exist.

The mean discovered field size of the basin has diminished considerably since it was first explored in the 1930s. Onshore, the current exploration potential is small to moderate in terms of potential-field reserves size but large in terms of number of untested prospects and leads. Eight of the nine identified plays are in the "plateau" phase. The remaining play is in the "breakthrough" phase, but the reserves potential is small to moderate. The field-development potential of the onshore region is substantial. Overall, 80%, or 13 billion barrels, of oil reserves remain unproduced. Most of the unproduced oil reserves are found in three carbonate plays: Lower Permian, Upper Devonian Fammenian, and Lower Devonian.

Exploration potential exists in the offshore Pechora Sea, which is an immature exploration region. Four main plays, which are extensions of proven plays within the onshore portion of the basin, define the offshore exploration potential. The offshore portions of the plays are in the "active-growth" or possibly the "breakthrough" phase.

INTRODUCTION

Methodology for the Timan-Pechora Basin evaluation presented in this paper was based on petroleum-system investigation principles from Magoon and Dow (1994). Figure 1 summarizes the approach.

Four levels of petroleum investigations exist:

1. *sedimentary basin* investigations that emphasize the stratigraphic sequence and structural style of sedimentary rocks
2. *petroleum system* studies that describe the genetic relationship between a pod of active source rock and the resulting oil and gas accumulations (also known as the source rock–reservoir couplet)
3. investigations of *plays* that describe the present-day similarity of a series of present-day traps
4. studies of *prospects* that describe the individual present-day trap

Economic considerations are not important in sedimentary-basin and petroleum-system investigations, but they are essential in play and prospect investigations.

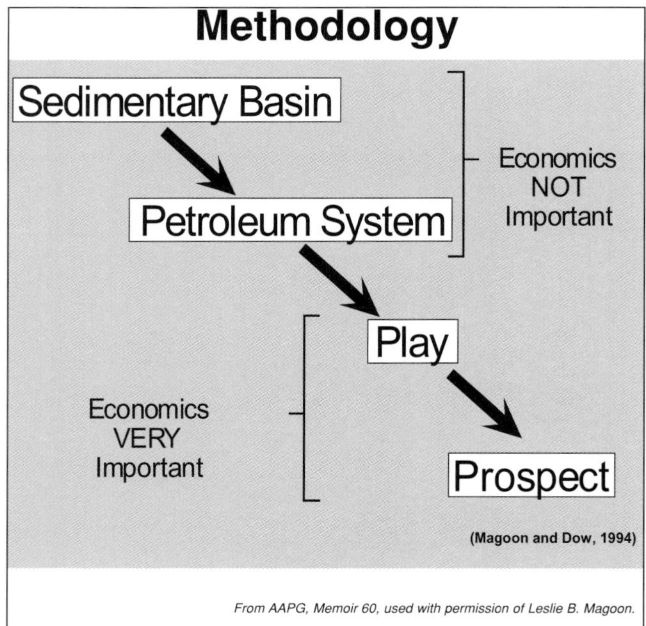

Figure 1. Outline of methodology, based on Magoon and Dow (1994).

This paper primarily describes the *sedimentary-basin*, *petroleum-system*, and *play* aspects of the oil resources in the Timan-Pechora Basin. Gas is perceived to have minimal value in much of the basin at the present time; however, the reserves are substantial (37 trillion cubic feet). Accordingly, the abundant coal and coal-bed methane resources in the basin will not be addressed. The hydrocarbon resource economic criteria are also beyond the scope of this paper.

Political boundaries (Figure 2) are important in the Timan-Pechora Basin because governing bodies and laws vary for each region. The Nenets Autonomous Okrug to the north and the Komi Republic to the south regulate the onshore portion of the basin. The Permian Oblast, in the extreme southern portion of the basin, holds minimal reserves and will not be discussed in this paper. The Murmansk District regulates the Pechora Sea.

TERMINOLOGY

In this paper, reserves are recoverable hydrocarbons. Estimated ultimate recovery (EUR) includes produced, proved, and probable reserves. In calculating barrels of oil equivalent, gas is converted to oil at a ratio of 6000 cubic feet of gas per barrel of oil. Field-size distribution (FSD) is determined from log-normal distributions of EUR field-size values. FSD values include P_{90} (90% of the values have a probability of being equal to or less than this value), P_{50} (median value of the distribution), P_{10} (10% of the values have a probability of being equal to or less than this value), and mean (probability weighted average or geometric mean of log-normal distribution).

BASIN BACKGROUND

Exploration History

The first stage of organized exploration in the Timan-Pechora Basin began in 1929 with the formation of a large geologic field team in Ukhta, Komi Republic (Aminov et al., 1993). In the following year, the team made the first onshore hydrocarbon discovery at Chibyu (EUR 4 million bbl oil), located adjacent to Ukhta (Figure 2). The first stage of exploration was completed by 1960, with the discovery of a group of oil and gas fields near Ukhta. The main hydrocarbon-bearing interval in the group of fields was the Middle Devonian and Upper Devonian Frasnian siliciclastics. The next stage of exploration, from 1960 through 1983, focused on the entire basin, including Kolguyev Island (Figure 2) in the Pechora Sea (Aminov et al., 1993). During this period, numerous oil and gas fields were discovered. Significant discoveries included Usa (EUR 1610 million bbl oil) in 1963 and Vuktyl (EUR 14.7 trillion cubic feet [tcf] gas and 515 million bbl condensate) in 1964. Usa and Vuktyl are the largest oil and gas fields, respectively, in the onshore portion of the basin. The Peschano-Ozer field (EUR 126 million bbl oil and 117 billion cubic feet [bcf] gas), on Kolguyev Island, was discovered in 1982.

The third phase of exploration occurred from 1983 to 1991 and focused in the onshore northern Nenets Okrug region and partially in the Pechora Sea. Significant discoveries were made in the Khoreyver depression, Kolva swell, and Varandey-Adzva structural zone regions (Figure 2). The first offshore discovery was made at Pomoroskoye (EUR 698 bcf gas) in 1985. Prirazlomnoye (EUR 609 million bbl oil), the largest oil field in the Pechora Sea, was discovered in 1989. Ardalin, the northernmost continuously producing field, was discovered in 1988. Since 1991, exploration activity in the Timan-Pechora Basin has been low.

Oil is the main hydrocarbon phase discovered in the Timan-Pechora Basin (approximately 68% oil versus 32% equivalent gas reserves), but significant accumulations of gas exist which are associated mostly with the foredeep region that flanks the Ural Mountains and Pay-Khoy Ridge. Other significant accumulations of gas occur in areas of gas-prone source rocks, including the northwest Shapkino-Yuraga and Laisky swells (Figure 2). The principal oil-bearing regions are in the central and northern portion of the basin in the Kolva swell, Khoreyver depression, and Varandey-Adzva structural zone (Figure 2). Approximately 230 oil and gas fields, ranging in reserve size from 1 million to 2965 million bbl oil equivalent (Vuktyl), have been discovered in the Timan-Pechora Basin.

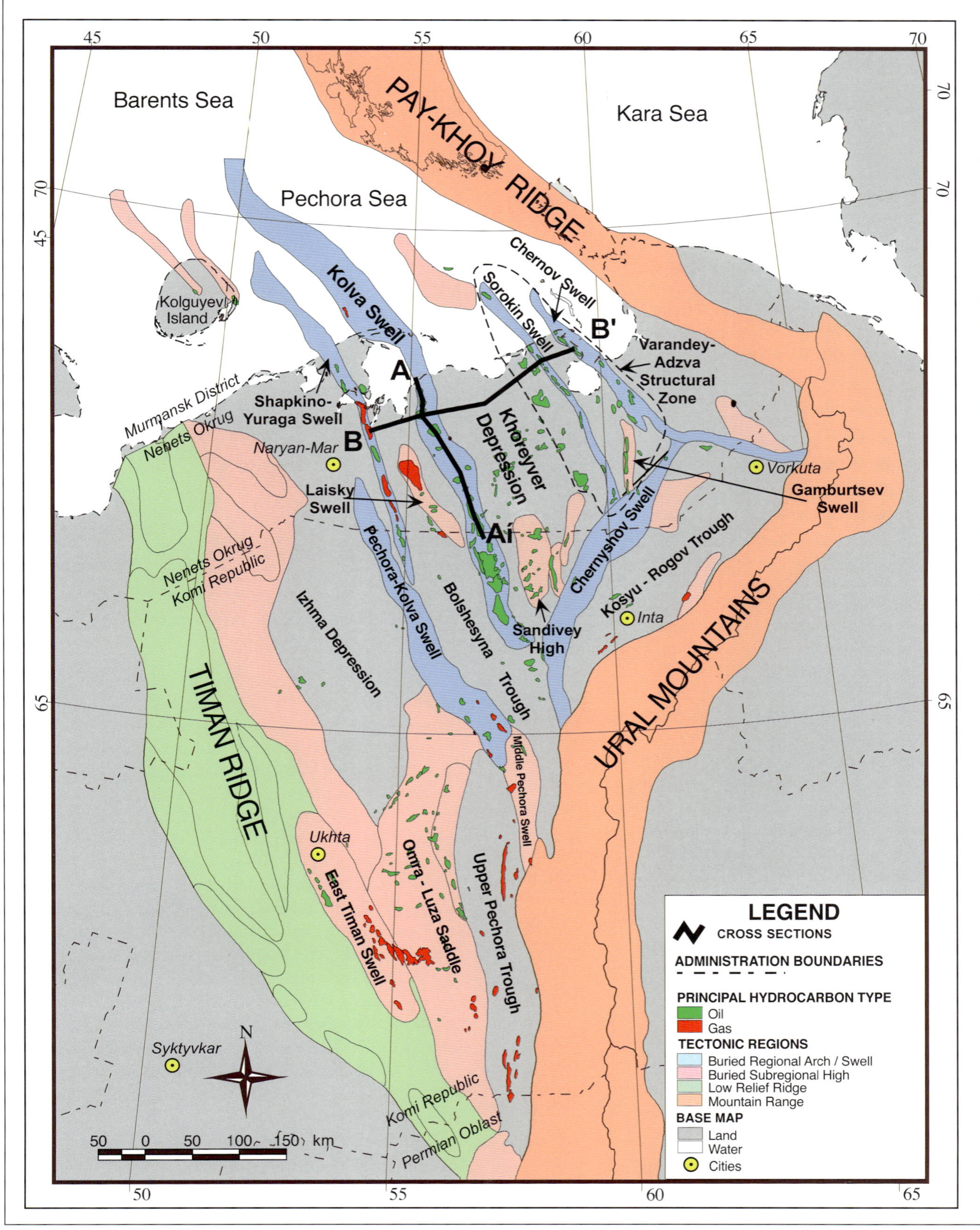

Figure 2. Timan-Pechora Basin structural features, political provinces, and cross-section locations.

Geomorphology

The geomorphology of the Timan-Pechora Basin is diverse. The northern part of the basin is situated in mostly swampy lowland with some marginal ridges known as the Malozemelsk (small land) and Bolshezemelsk (large land) Tundra. Thick permafrost tends to underlie these areas. The southern part of the basin is more elevated and extends mainly along the valley of the Pechora River, where permafrost is absent or developed locally.

Geomorphological features and biological species are important criteria considered in the positioning of pipelines, well-drilling operations, and production facilities. Barringer et al. (1998) discussed critical elements in the northern Kolva swell region, located in the northern onshore portion of the basin.

Infrastructure

An oil and gas pipeline gathering system and oil refinery in Ukhta, Komi Republic, make up the infrastructure in the Timan-Pechora Basin. The oil refinery and gas-processing plant are operational. The Usinsk-Ukhta oil pipeline (Figure 3) transports "domestic-blend" oil predominantly from the Usa, Kharyaga, and Ardalin fields in the central Nenets Okrug to refineries in the south at Yaroslov, Kirishi, and Moscow. The oil is then transported west to Baltic Sea export points in the cities of Primorsk and Porvoo. The Usinsk-Ukhta pipeline capacity is 330,000 bbl oil/day (BOPD), the Ukhta-Yaraslov capacity is 400,000 BOPD, and the Yaraslov-Kirishi capacity is 430,000 BOPD. The Kharyaga-Usinsk pipeline, completed in September 1999, provides for transportation to the south of 200,000–240,000 BOPD from fields of the Nenets Autonomous Okrug (Borovinskikh, 1998).

Market access and increased export capacity are the main concerns in developing fields in the Nenets Okrug. Because the Ukhta pipeline is at or near capacity, three transportation concepts have been suggested: (1) rebuilding the Kharyaga-Usinsk-Ukhta pipeline to increase its capacity, (2) building a new Kharyaga-Primorsk (Gulf of Finland) oil pipeline, and (3) building an offshore terminal near Varandey or Khilchuyu fields (D. E. Nevel, personal communication, 1999). Figure 4 shows two possible locations for Pechora Sea offshore single-point mooring terminals.

Figure 3. Existing pipeline infrastructure.

GEOLOGIC SETTING

Stratigraphy and Basin Fill

Figure 5 summarizes the stratigraphy and major tectonic events in the Timan-Pechora Basin, based largely on findings by Schmidt (1996) and Ressetar et al. (1997). The sedimentary fill of the Timan-Pechora Basin was dominated by tectonic controls which in turn determined the volume, fill, texture, and composition of detrital influx. Secondary controls, such as global sea level and climate, were most pronounced during episodes of reduced terrigenous supply. The fundamental late Paleozoic restructuring from passive margin to foreland basin is reflected in contrasting sedimentary composition and depositional styles below and above this boundary, which is marked by a change from platform carbonates to siliciclastics (Ressetar et al., 1997).

The lower, carbonate-dominated interval of Ordovician to middle Permian age developed on a passive margin or back-arc basin setting subjected to variations in eustatic sea level, tectonic subsidence, and sediment supply. Platform carbonates and mature, quartzose, fluvial-to-shelf sands grade eastward into outer shelf, slope and basinal mudstones, and turbidite sandstones (Ressetar et al., 1997).

In contrast, the upper foreland basin interval reflects primarily tectonically driven base-level changes and fluvially derived sediment influx. These primarily siliciclastic sediments were derived from the Ural thrust belt from the east and were rapidly deposited in alluvial-fan, fluvial, and deltaic environments. These sandstones are typically immature except where reworked by marine processes seaward of the subaerial delta margins (Ressetar et al., 1997).

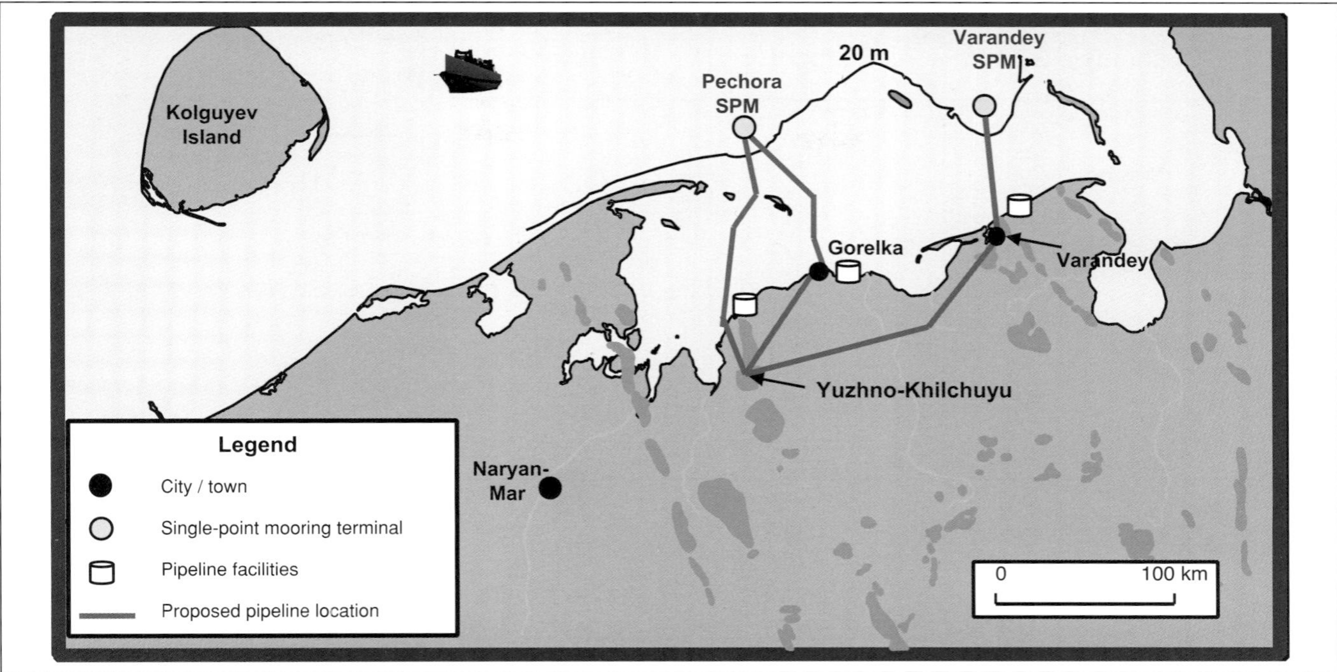

Figure 4. Proposed offshore oil terminals.

Paleozoic marine transgressions were related to expansion of the Ural seaway to the east, whereas Mesozoic transgressions originated in the Barents Sea to the north (Ressetar et al., 1997).

Structural Development

The structural development of the Timan-Pechora Basin is controlled directly by major plate boundary dynamics (Ziegler, 1988, 1989). Since the formation of the Ural Mountains in the Permian, the Timan-Pechora Basin has been in a foreland basin setting. Prior to the Permian, the basin setting varied from an accretionary margin (Cambrian) to a rift margin (Ordovician), a passive margin (Silurian), and a back-arc basin (Devonian and Early Carboniferous). The major tectonic events responsible for the present-day structural configuration of the Timan-Pechora Basin are shown in Figure 5.

As portrayed on the tectonic-features map (Figure 2), the basin is made up of a series of north-northwest-trending highs (referred to as ridges or swells) that bound intervening lows (troughs or depressions). The north-northwest structural grain is a by-product not only of the effects of the foreland basin deformation associated with the suturing of the Ural Mountains during the Permian but of older tectonic events as well. A north-northwest structural grain was developed in the basement rocks in the Cambrian as a result of the Balkalian orogeny. Periods of extension and compression in the Ordovician and Devonian resulted in structures with similar north-northwest trends. Subsequent compressive events in the Carboniferous, Permian, and Late Tertiary reactivated north-northwest-trending structures. Thus, the present-day structural grain has been active in periods of compression and extension throughout the Phanerozoic.

Two periods of tectonic activity, the Devonian to Early Carboniferous (Viséan) and the Permian, have left by far the largest impact on the area. Pulses of back-arc compression and extension resulted in a complex package of thick Devonian to Carboniferous Viséan rocks that have multiple unconformities (Figures 6 and 7) (Ziegler, 1988; Schmidt, 1996; Rappoport, 1997). The back-arc basin formed in association with the "Sakmarian" arc-trench system that lay to the east of the Timan-Pechora Basin (Ziegler, 1988, 1989). Permian inversion associated with the Uralian orogeny is responsible for the present system of ridges and troughs.

The Khoreyver depression and Kolva swell characterize the structural variations within the Timan-Pechora Basin. The Khoreyver depression has undergone very little faulting or folding throughout the Phanerozoic (Figure 7). The Kolva swell, lying to the west of the Khoreyver depression, has been tectonically active throughout this same period. The Kolva swell was one of two main rifts developed in the basin during the Devonian, namely the Kolva rift. The other was the Pechora rift, which coincides with the present Timan Ridge (Figure 2). Evidence of tectonic activity in the Kolva rift includes thick pods of Devonian sediments and multiple events of folding and faulting (Figure 7). Significant inversion of the Kolva rift took place during the Permian, resulting in the present structural high associated with the Kolva swell (Figure 7).

The sedimentation history along the Kolva swell

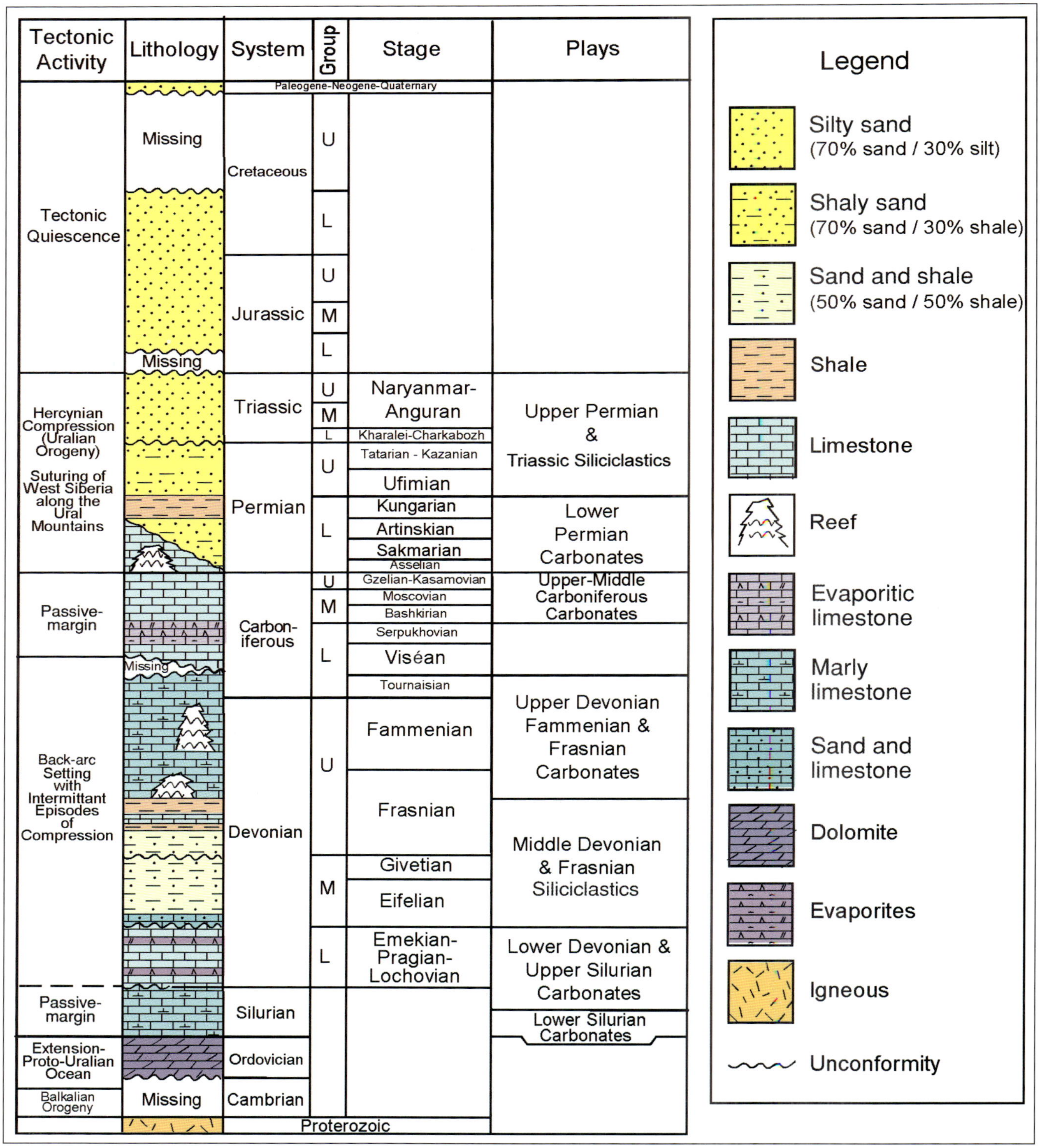

Tectonic Activity
Lithology
System
Group
Stage
Plays
Paleogene-Neogene-Quaternary
Tectonic Quiescence
Missing
Cretaceous
U
L
Jurassic
U
M
L
Missing
Hercynian Compression (Uralian Orogeny)
Suturing of West Siberia along the Ural Mountains
Triassic
U
M
L
Naryanmar-Anguran
Kharalei-Charkabozh
Upper Permian & Triassic Siliciclastics
Permian
U
Tatarian - Kazanian
Ufimian
L
Kungarian
Artinskian
Sakmarian
Asselian
Lower Permian Carbonates
Passive-margin
Carbon-iferous
U
Gzelian-Kasamovian
M
Moscovian
Bashkirian
Upper-Middle Carboniferous Carbonates
L
Serpukhovian
Viséan
Tournaisian
Missing
Back-arc Setting with Intermittant Episodes of Compression
Devonian
U
Fammenian
Frasnian
Upper Devonian Fammenian & Frasnian Carbonates
M
Givetian
Eifelian
Middle Devonian & Frasnian Siliciclastics
L
Emekian-Pragian-Lochovian
Lower Devonian & Upper Silurian Carbonates
Passive-margin
Silurian
Lower Silurian Carbonates
Extension-Proto-Uralian Ocean
Ordovician
Balkalian Orogeny
Missing
Cambrian
Proterozoic

Legend
Silty sand (70% sand / 30% silt)
Shaly sand (70% sand / 30% shale)
Sand and shale (50% sand / 50% shale)
Shale
Limestone
Reef
Evaporitic limestone
Marly limestone
Sand and limestone
Dolomite
Evaporites
Igneous
Unconformity

Figure 5. Timan-Pechora generalized stratigraphy, tectonic activity, and plays.

reflects the tectonic history. The Early Permian compressional event caused uplift of the Kolva rift, resulting in thinning of Permian sediments over the ridge (Figure 7). This is a fundamental change from the sedimentation pattern of Carboniferous and older rocks, where the Kolva swell acted as a depocenter. The change in depocenters was accompanied by a change in rock type; older carbonate rocks of Late Carboniferous age were replaced by westward-prograding siliciclastic rocks shed from the emerging Ural Mountains during the Permian.

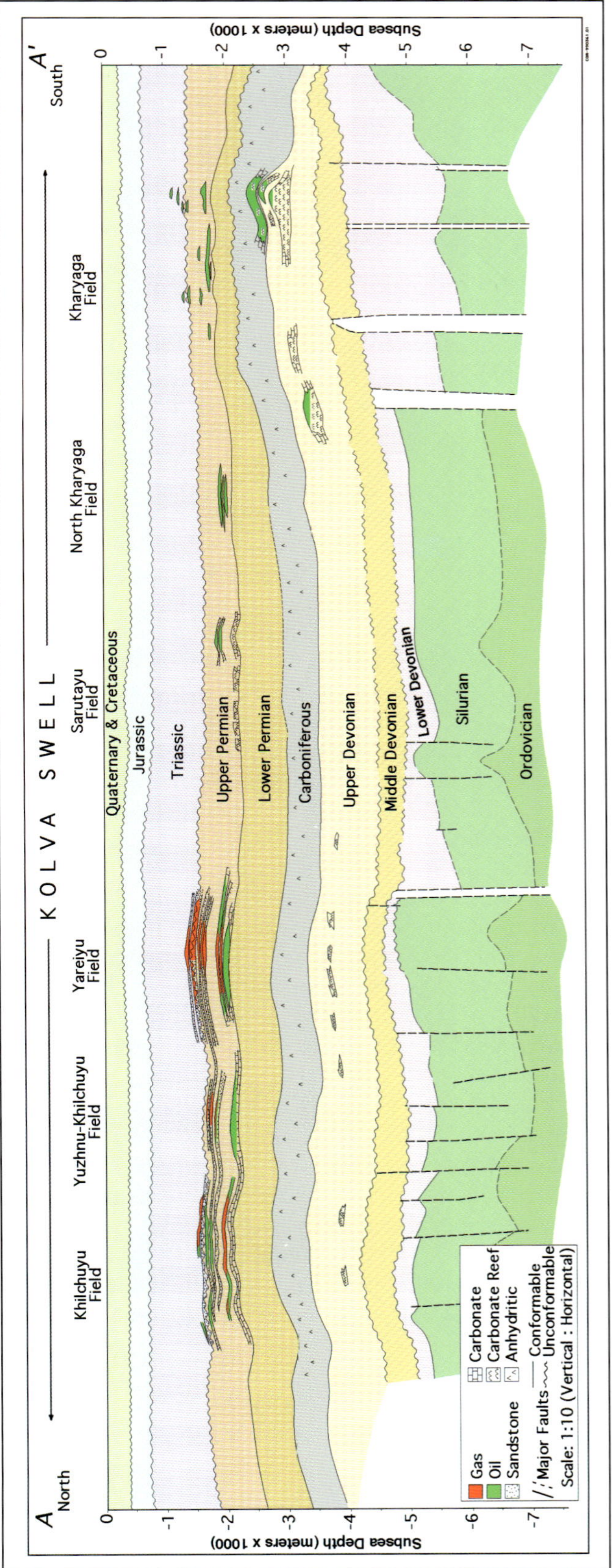

Figure 6. North-south stratigraphic cross section A-A′. See Figure 2 for location.

Based on regional seismic data in the northern portion of the Timan-Pechora Basin (Schmidt, 1996), the main phase of inversion that resulted in the north-northwest-trending swells, troughs, and depressions occurred mainly during the Early Permian (Asselian through Kungarian ages). Inversion of the Kolva swell began in the Late Carboniferous, as indicated by partial thinning of the Carboniferous section onto the Kolva swell (Figure 7). Upper Permian sediments thicken only slightly off the flanks of the Kolva swell. By the end of the Triassic, it appears that structural movement had ceased entirely, as verified by the low relief exhibited on the top Triassic surface over the Kolva swell.

Source Rocks

Five main source-rock intervals exist in the Timan-Pechora Basin (Danilevsky, 1996): (1) Middle to Upper Ordovician, Silurian, and Lower Devonian; (2) Middle Devonian (lower Frasnian); (3) Upper Devonian (lower to middle Frasnian); (4) middle Frasnian (Domanik) to Carboniferous (Tournaisian); and (5) Lower Permian.

Table 1 summarizes key characteristics of each source-rock interval where the present-day total organic carbon (TOC) is greater than 1%. Source rocks having less than 1% TOC are considered to have marginal productivity (Danilevsky, 1996).

As a means to understanding the relationship between petroleum-system elements, field size, and hydrocarbon distributions, Wavrek et al. (1997) used time-slice analysis to explain the observation by Ulmishek (1982) that a very limited number of fields account for 80% of in-place reserves (original oil in place). The analysis suggests that the most significant aspect of giant hydrocarbon accumulations is their early structural formation. Once structures were formed, the giant fields received multiple source-rock hydrocarbon charges in response to basin fill and source-rock expulsion. As an example, Usa, Vozey, and Kharyaga fields, located on the prolific Kolva swell, contain a total of approximately 3.5 billion bbl oil reserves within four plays. Assuming a 30% recovery factor, original oil in place is estimated at 11.7 billion bbl.

Migration Pathways

The characterization of hydrocarbon migration pathways is important in understanding the petroleum-system framework in the Timan-Pechora Basin. Michael and Fossum (1998) modeled regional cross sections in the northern Nenets Okrug region to understand migration pathways and trap timing. Modeling was done to determine time of hydrocarbon generation and expulsion relative to trap development and the reason for hydrocarbon emplacement in Upper and middle Carboniferous, Upper and Lower Permian, and Triassic reservoirs. Modeling focused in the Kolva swell and Khoreyver depression areas.

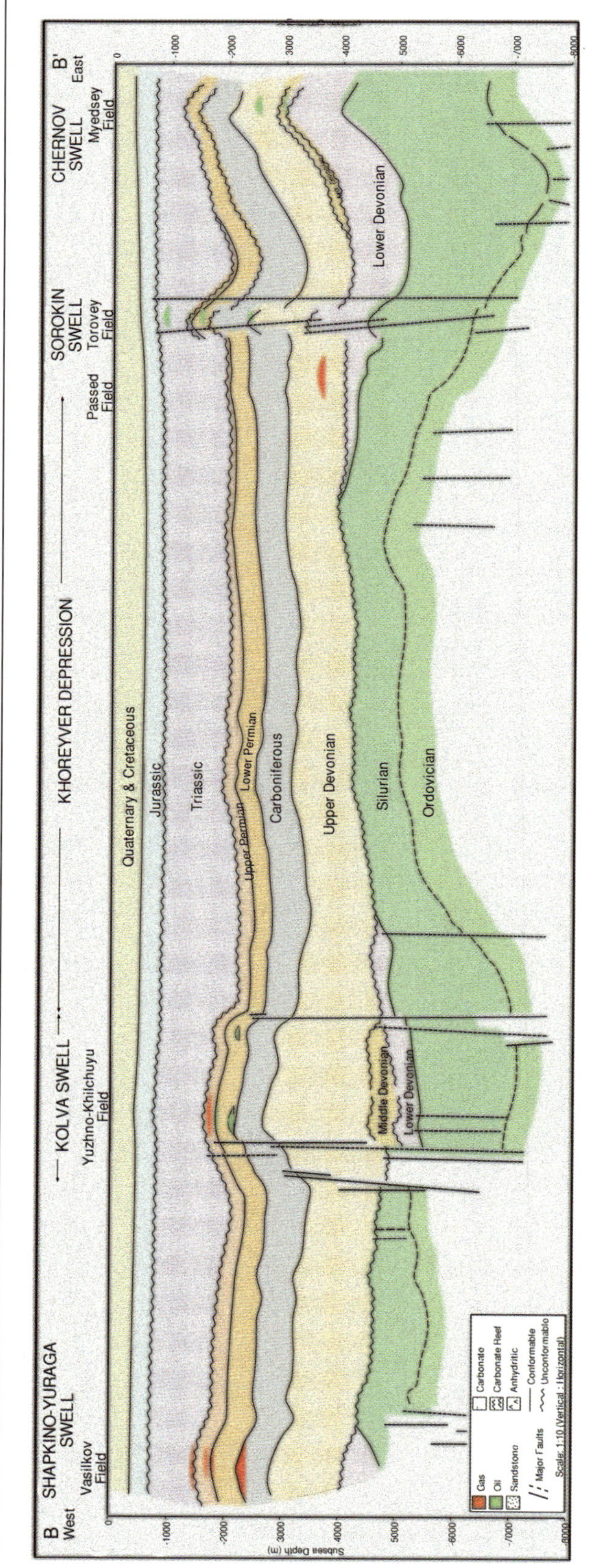

Figure 7. West-east stratigraphic cross section B-B′. See Figure 2 for location.

Results of the 2-D modeling indicated that vertical fault migration is required to explain hydrocarbon distributions in the north Timan-Pechora Basin. The vertical extent of faults controls the degree of charge in Upper Carboniferous and Permian reservoirs. The modeling explains the major distribution of oil and gas in several areas of the northern Timan-Pechora Basin and emphasizes the critical effect that migration pathways and geopressure history have in determining the distribution of hydrocarbons.

PLAY ANALYSIS AND FUTURE OIL POTENTIAL—ONSHORE NENETS OKRUG AND KOMI REPUBLIC

Summary

Onshore, the exploration potential for discovering large fields in the Timan-Pechora Basin is low to moderate, and it varies based on the geographic location and play. However, the exploration potential of discovering large reserves in numerous small accumulations is high. Borovinskikh (1998) suggested that 700 hydrocarbon fields will be discovered in the Komi Republic but that most of the fields will hold less than 75 million bbl oil equivalent in place. The discovery of many of these fields will probably be the result of the continuing application of new technologies such as 3-D seismic and integrated multidiscipline studies.

Figure 8 illustrates breakthrough, active-growth, and plateau phases of discovery history and play classification in a basin. Eight of the nine identified plays discussed below are in the plateau phase. The potential of discovering large fields in plateau-phase plays is generally low. The remaining play is in the breakthrough phase, and the potential for discovering large fields is thought to be low to moderate.

Creaming curve and play analysis indicate that approximately 16 billion bbl oil have been discovered in the basin. Play analysis indicates that future added reserves will be in small to moderate increments because the basin is in the plateau phase (Figure 9). However, possible reserves additions could be as great as 1 billion bbl in the next 10 years (Fossum, 1997). Borovinskikh (1998) calculated that potential total oil reserves in the Timan-Pechora Basin are 31.8 billion bbl, suggesting that 15.8 billion bbl oil remain, but in small accumulations. Figure 10 illustrates the decreasing mean discovered field size since hydrocarbons were discovered there in the 1930s.

Numerous wells have thoroughly tested the main play concepts and prospects. Approximately 5800 exploration wells (wildcat, step-out, and reservoir evaluation) have been drilled in the Timan-Pechora Basin. The basin was explored generally from south to north and west to east in response to the exploration expeditions men-

Table 1. Timan-Pechora Basin source-rock characteristics.

Source Rock	Principal Kerogen Type	TOC Range (%)	TOC Average (%)	Gross Thickness Range (meters)	Main Region of Deposition
Middle to Upper Ordovician, Silurian, and Lower Devonian	Type II	1–5	2	20–60	Khoreyver depression, Kolva swell, middle Pechora swell, Varandey-Adzva structural zone
Middle Devonian (lower Frasnian)	Type II (east)–Type III (west)	1– 5	2	30–60	Kolva swell, middle Pechora swell, Pechora-Kolva swell
Upper Devonian (lower to middle Frasnian)	Type II(?)	1–5	2	10–40	Pechora-Kolva swell
Middle Frasnian (Domanik) to Carboniferous (Tournaisian)	Type II (east)–Type III (west)	2–10	3	20–240	Eastern half of basin, including the East Timan swell and Omra-Luza saddle in the southwest
Lower Permian	Type III	1–5	2	50–100	Kolva, Shapkino-Yuraga, and Sorokin swells, upper Pechora trough

tioned above, which thoroughly evaluated all main exploration targets in the onshore portion of the basin. The exploration strategy implemented in the Timan-Pechora Basin resulted in high exploration efficiency, as illustrated in Figure 11. The "inefficient" curve describes a field-discovery order if the smallest field was discovered first, then the next largest, and so on. The "efficient" curve describes a field-discovery order if the largest field was discovered first, then the next smallest, and so on. The "random" curve is an average of the "inefficient" and "efficient"' field-discovery orders.

Summarizing, the exploration potential in the onshore portion of the basin has the following characteristics:

- Prospect and reserves size is limited because of mature plays.
- Although numerous prospects exist, many are deep (more than 4000 m) and are situated outside play fairways.
- Plays have a stratigraphic trap component, which may lead to greater containment uncertainty.

The oil-field development potential of the onshore Timan-Pechora Basin is substantial. Nearly 100% of the discovered reserves remain unproduced in the Nenets Autonomous Okrug, and 60% of the discovered reserves are unproduced in the Komi Republic. Overall, 80%, or 13 billion bbl, of the discovered reserves onshore remain unproduced. Figure 12 depicts annual oil and gas production since 1940, when the first oil production was recorded in Yarega field, Komi Republic. Note the severe drop in production in the early 1990s, related to a loss of operating capital during Perestroika. The subsequent production increase from 1993 through the present is caused by a slow influx of investments, both Russian and foreign, which focused on increasing production and bringing previously discovered fields on stream. One example is Ardalin field, which started producing in 1994. Cumulative production in the Timan-Pechora Basin through 1997 was 1.4 billion bbl oil and 4.0 tcf gas (Fossum, 1997).

Cumulative discovered reserves and mean field-size versus time have been depicted in Figures 13 and 14 for

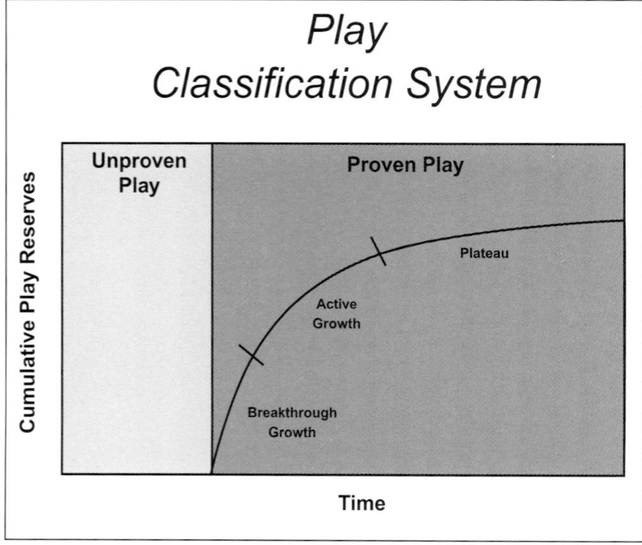

Figure 8. Play classification system.

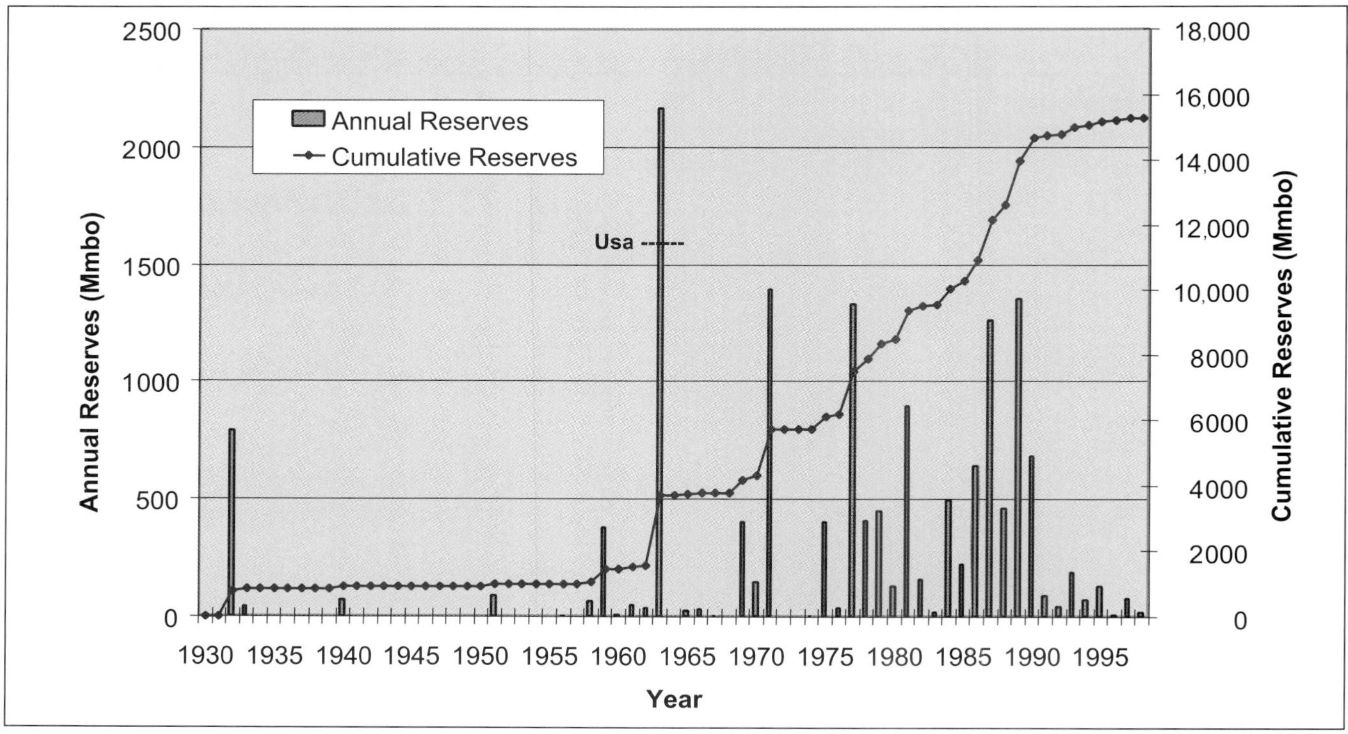

Figure 9. Timan-Pechora creaming curve for oil discoveries. Usa field, the largest oil field in the Timan-Pechora Basin, is noted. Mmbo = million bbl oil.

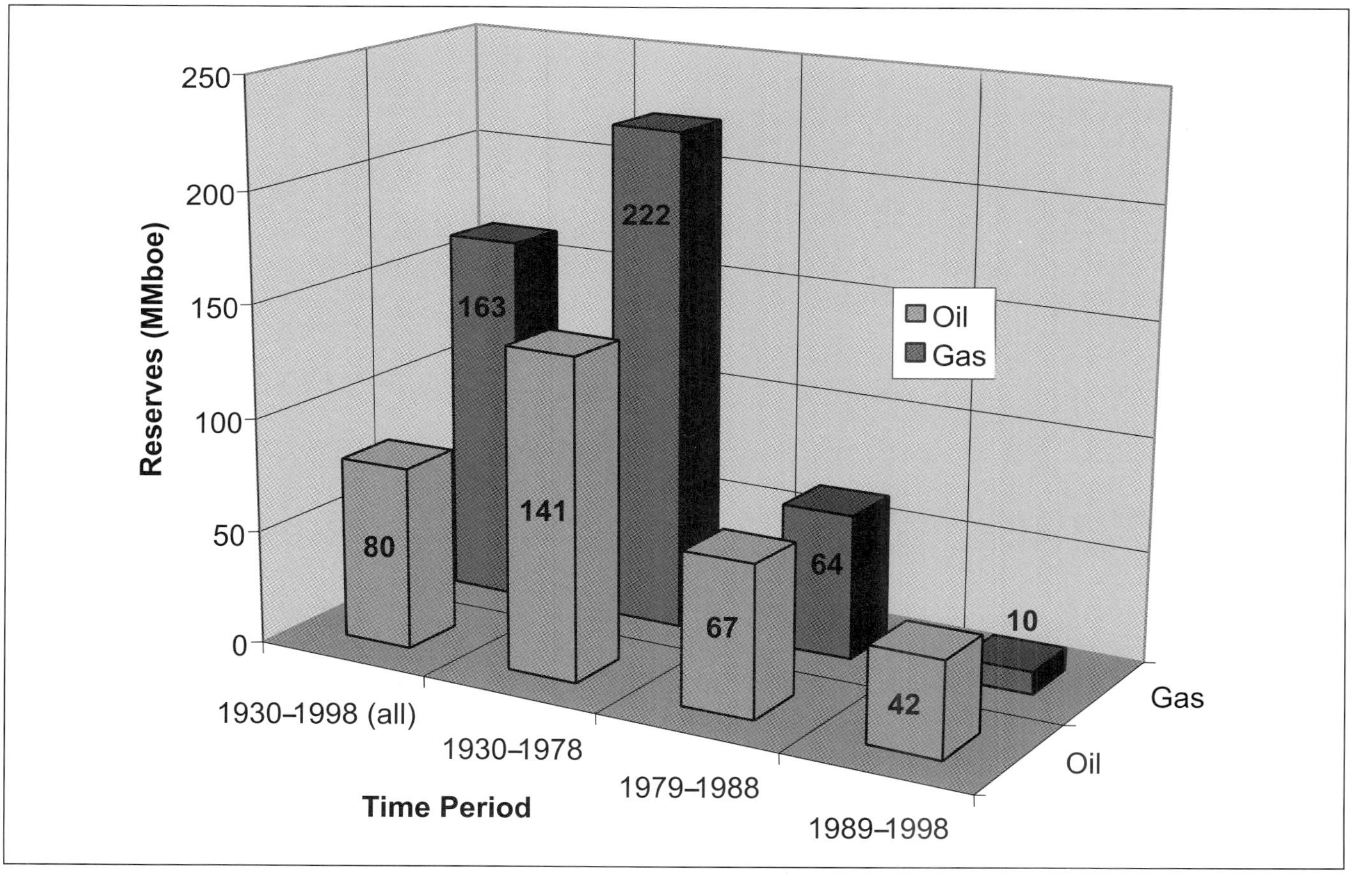

Figure 10. Total oil and gas mean field size versus time. Mmboe = million bbl oil equivalent.

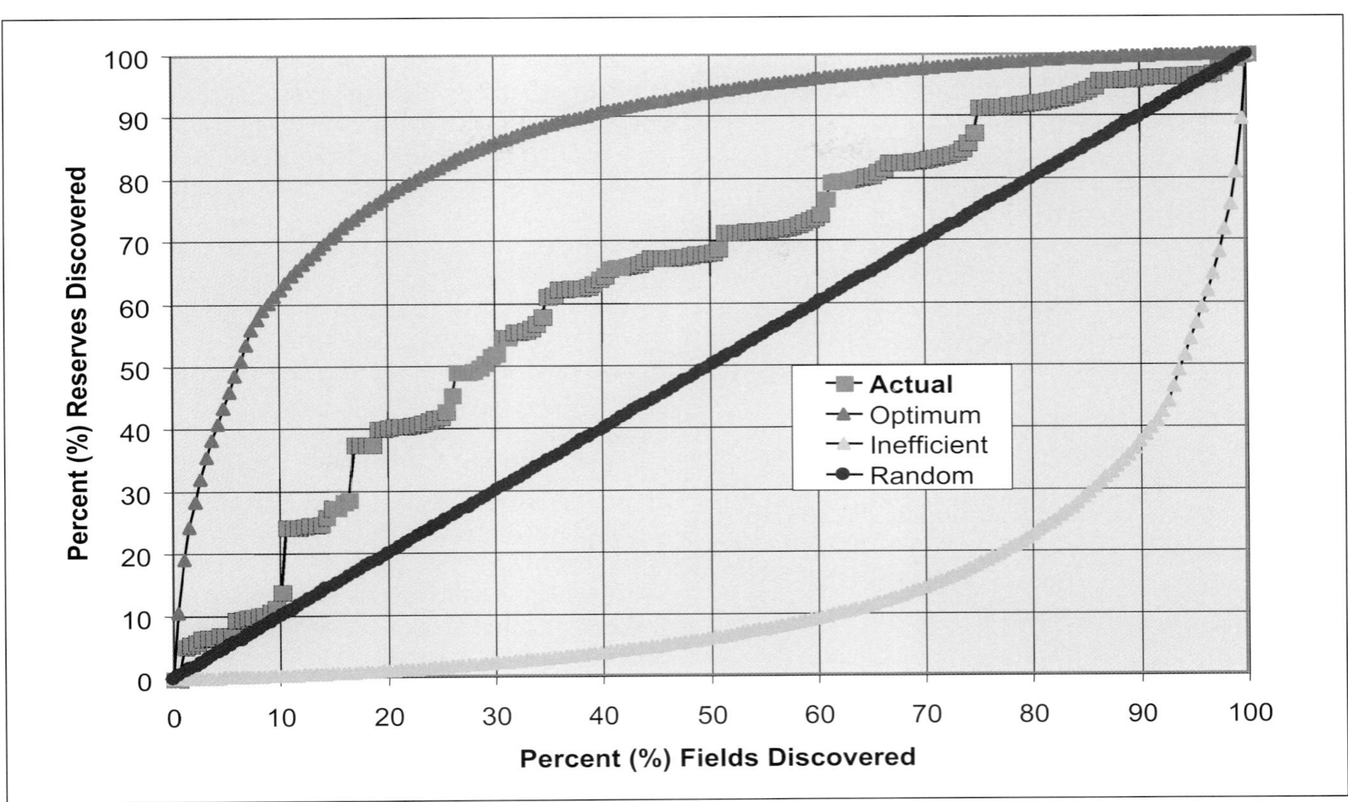

Figure 11. Timan-Pechora exploration efficiency for oil discoveries.

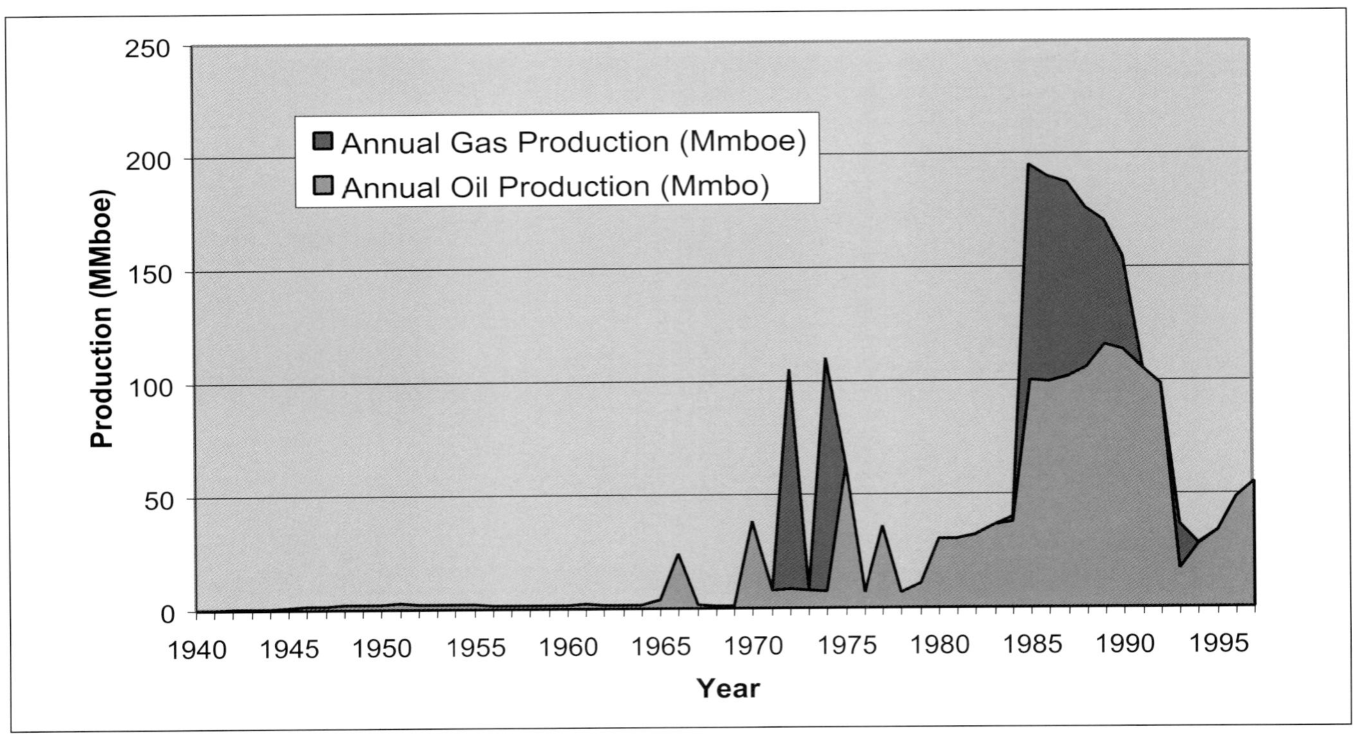

Figure 12. Timan-Pechora annual oil and gas production. Cumulative production is 1.4 billion bbl oil and 4.0 tcf gas. Mmbo = million bbl oil; Mmboe = million bbl oil equivalent.

five plays. Because of data limitations, three of the five plays (Ordovician, Silurian, and Lower Devonian carbonates; Upper Devonian carbonates; and Upper to middle Carboniferous and Lower Permian carbonates depicted in Figures 13 and 14) represent compilations of the cumulative oil-reserves discovery history and mean oil-field size versus time, respectively (Fossum, 1997).

Schematic representations of the nine identified plays are depicted in Figure 15 (cross-sections location map) and in Figures 16–21. Field-size distribution data statistics (oil only) for each play are listed at the bottom of each figure. The plays, from oldest to youngest, are:

1. Lower Silurian carbonates (Figure 16)
2. Upper Silurian and Lower Devonian carbonates (Figure 17)
3. Middle Devonian and Frasnian siliciclastics (Figure 18)
4. Upper Devonian Frasnian carbonates (Figure 19)
5. Upper Devonian Fammenian carbonates (Figure 19)
6. Upper to middle Carboniferous carbonates (Figure 20)
7. Lower Permian carbonates—structural (Figure 20)
8. Lower Permian carbonates—stratigraphic (Figure 20)
9. Upper Permian and Triassic siliciclastics (Figure 21)

A summary of each play follows (Fossum, 1997). The Ordovician carbonates play is not discussed here because of the relatively small amount of hydrocarbons discovered in this play. The only known Ordovician discovery is in Srednyaya Makarikha field, in the Khoreyver depression (Figure 2). Additional Ordovician hydrocarbon accumulations may exist but are considered insignificant. However, in the interest of completeness, Ordovician field-size statistics have been included in the reserves summary figures (Figures 9–13). Exploration risk for each play has been characterized quantitatively as low, moderate, or high, based on fundamental risk-assessment procedures documented in Rose (1992).

Lower Silurian Carbonates Play (Figure 16)

This play has moderate to high exploration risk, is in the plateau phase (Figure 13), and has small potential exploration reserves. Figure 14 illustrates the decreasing mean field size throughout time within the compiled Ordovician, Silurian, and Lower Devonian carbonates-play discovery history curve.

The Lower Silurian play is composed of vuggy, algal dolomites of the Sandivey and Veyak formations. The play is proven in the south Khoreyver depression area on the flank of the Sandivey high (Figures 2, 16). Vertical seal is provided by Upper Devonian Frasnian argillaceous shales and side seal by a pinch-out onto the structural

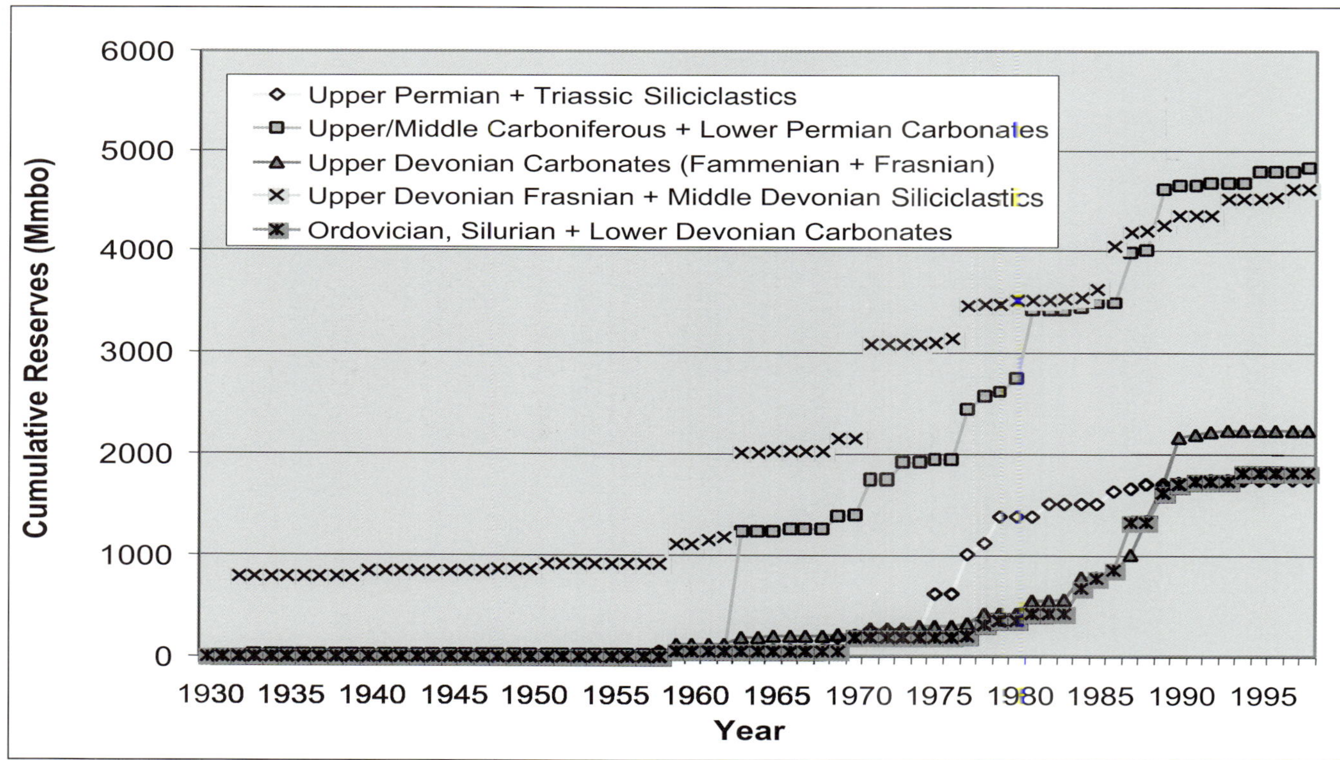

Figure 13. Cumulative discovered oil reserves by play. Because of data limitations, the Ordovician, Silurian, and Lower Devonian carbonates; Upper Devonian carbonates; and Upper to middle Carboniferous carbonates plays represent compiled cumulative field-discovery distributions. Mmbo = million bbl oil.

high and a series of reverse faults to the southwest (reverse faults not illustrated in Figure 16).

Analogous fields to the south and on trend have a Lower Silurian reserves range of 5–135 million bbl oil. The critical uncertainty for this play is reservoir quality. Porosity is poor to fair and difficult to predict. Porosity development is patchy because of dolomitization of limestone related to subaerial exposure surfaces. Hydrocarbon charge is provided by Ordovician and Silurian type II marine source rocks (Danilevsky, 1996).

Upper Silurian and Lower Devonian Carbonates Play (Figure 17)

This play has moderate to high exploration risk, is in the plateau phase (Figure 13), and has small to moderate potential exploration reserves. Figure 14 illustrates the decreasing mean field size throughout time within the compiled Ordovician, Silurian, and Lower Devonian carbonate plays.

The play is proved in two regions: the eastern Khoreyver depression and the Varandey-Adzva structural zone (Figure 2). The Roman Trebs, A. Titov, and Kolva fields, in the eastern Khoreyver depression, are composed of vuggy, dolomitic Lower Devonian limestones, which were eroded and sealed by Upper Devonian Frasnian argillaceous carbonates. Thirteen fields are situated in the Varandey-Adzva structural zone and are composed of reservoirs in Upper Silurian and Lower Devonian vuggy dolomites sealed by marls of the same age (Figure 17). Hydrocarbon charge is provided by Ordovician and Silurian type II marine source rocks (Danilevsky, 1996). Hydrocarbon quality in the Upper Silurian and Lower Devonian is a concern because the oil reserves tend to be high in paraffin, resins, and asphaltenes (Lodzhevskaya and Smolenchuk, 1998; Aminov et al., 1993).

Middle Devonian and Frasnian Siliciclastics Play (Figure 18)

This play has moderate to high exploration risk, is in the plateau phase (Figure 13), and has small to moderate potential exploration reserves.

Reservoirs of the play comprise the Middle Devonian Givetian and Eifelian siliciclastics units and the Upper Devonian Frasnian siliciclastics unit. The principal depositional environment is foreshore lower beach and shallow-marine shoals, with clay and lag deposits noted. Reservoir quality is poor to good, controlled in part by porosity reduction resulting from compaction effects. The play predominates in the Kolva swell, particularly in Usa and Vozey fields (the two largest onshore fields in the play)

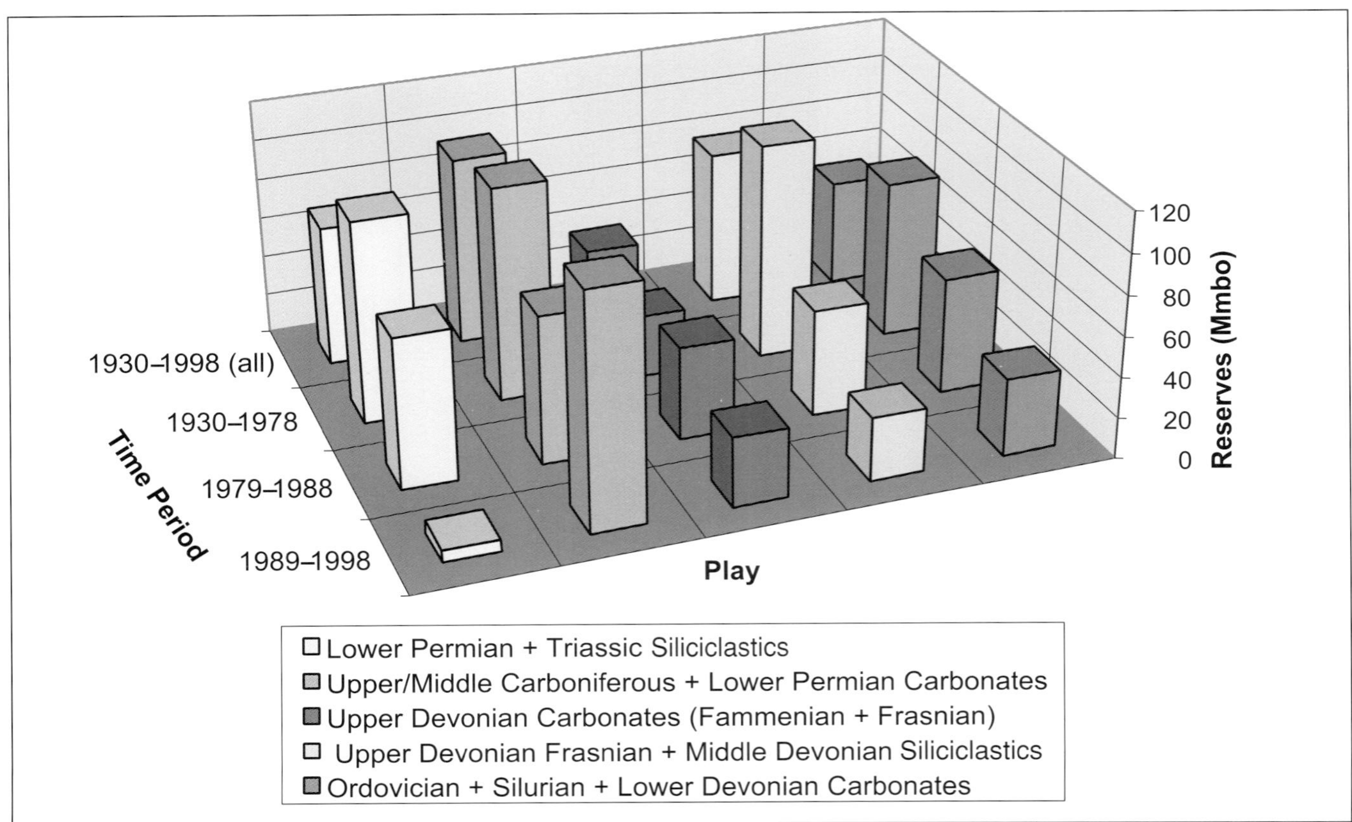

Figure 14. Mean oil-field size versus time by play. Because of data limitations, the Ordovician, Silurian, and Lower Devonian carbonates; Upper Devonian carbonates; and Upper to middle Carboniferous carbonates plays represent compiled field-size distributions. Mmbo = million bbl oil.

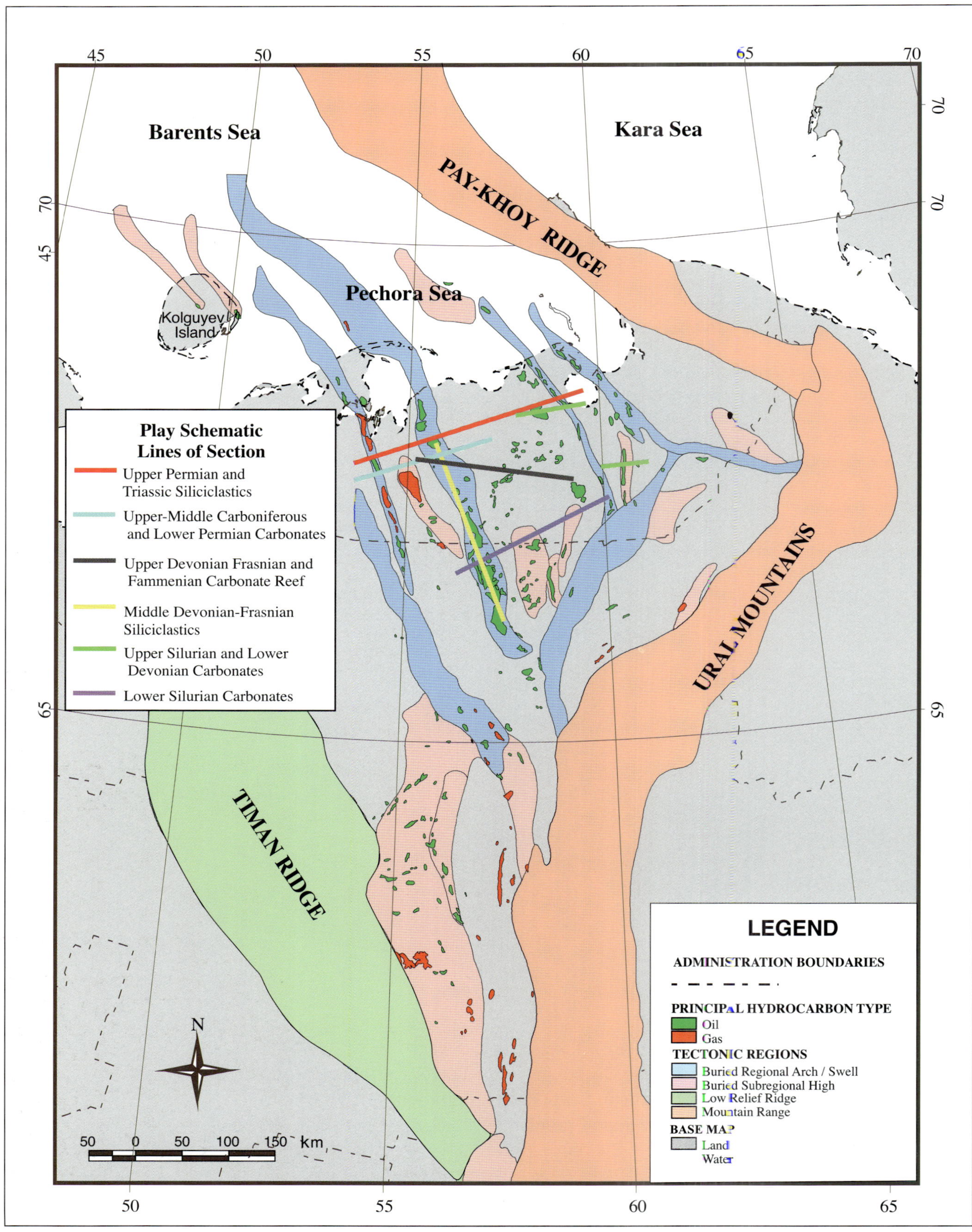

Figure 15. Index map for schematic play-type lines of cross sections.

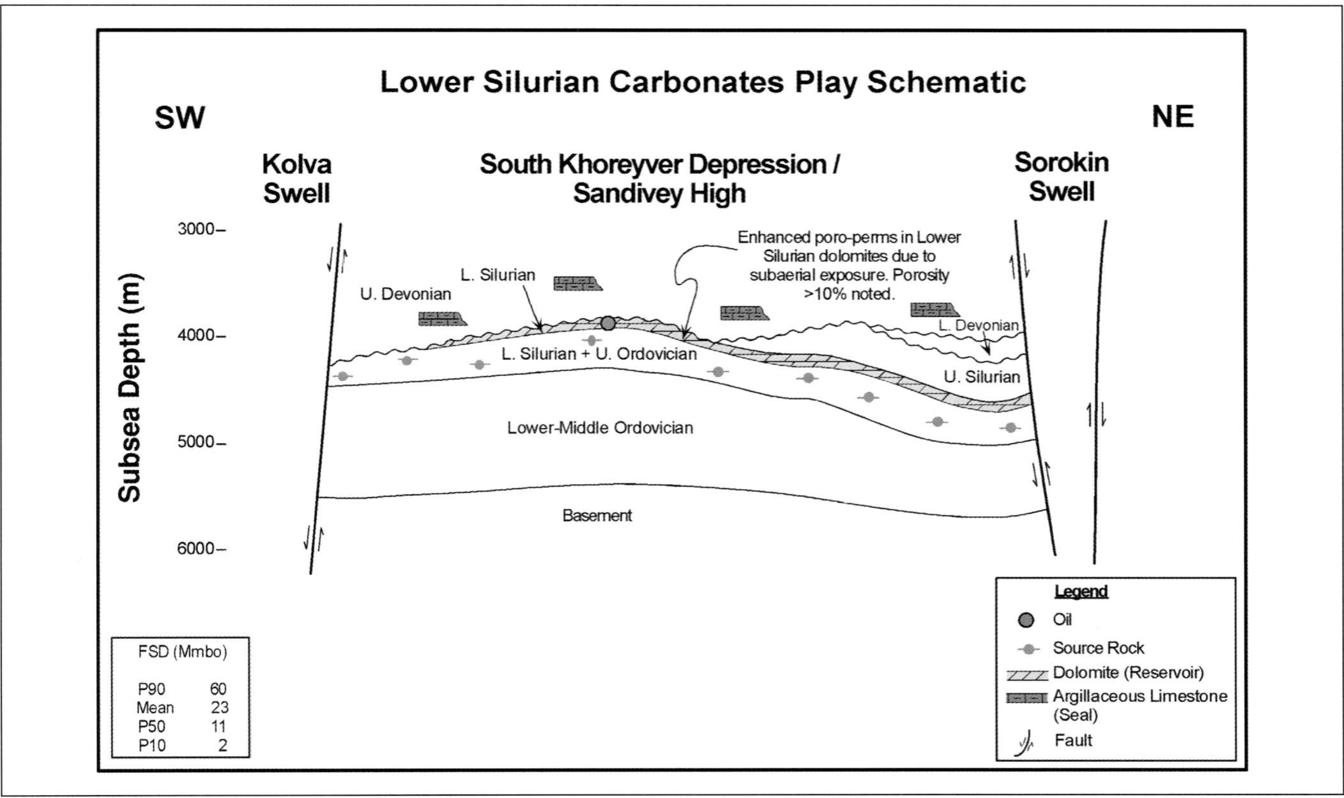

Figure 16. Lower Silurian carbonates-play schematic. See Figure 15 for line of cross section. Mmbo = million bbl oil.

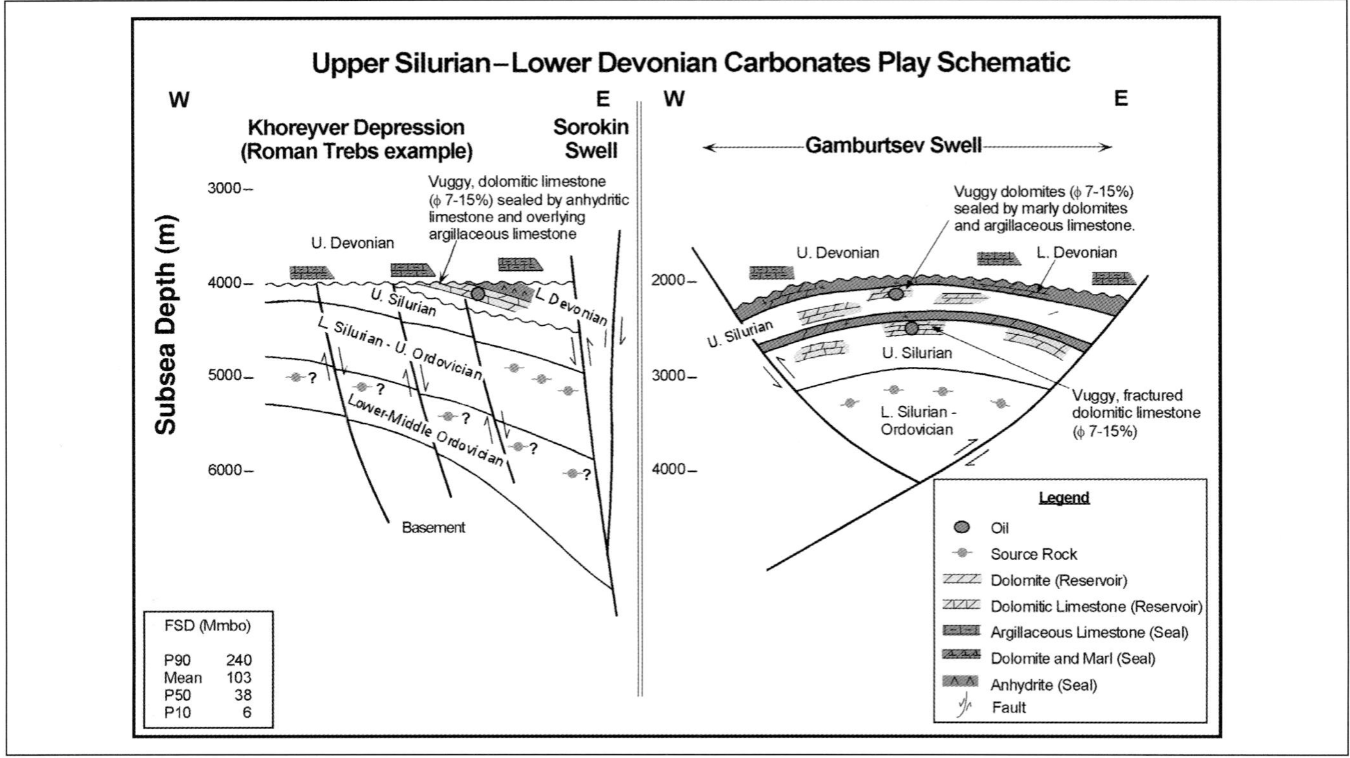

Figure 17. Upper Silurian and Lower Devonian carbonates-play schematic. See Figure 15 for line of cross section. Mmbo = million bbl oil.

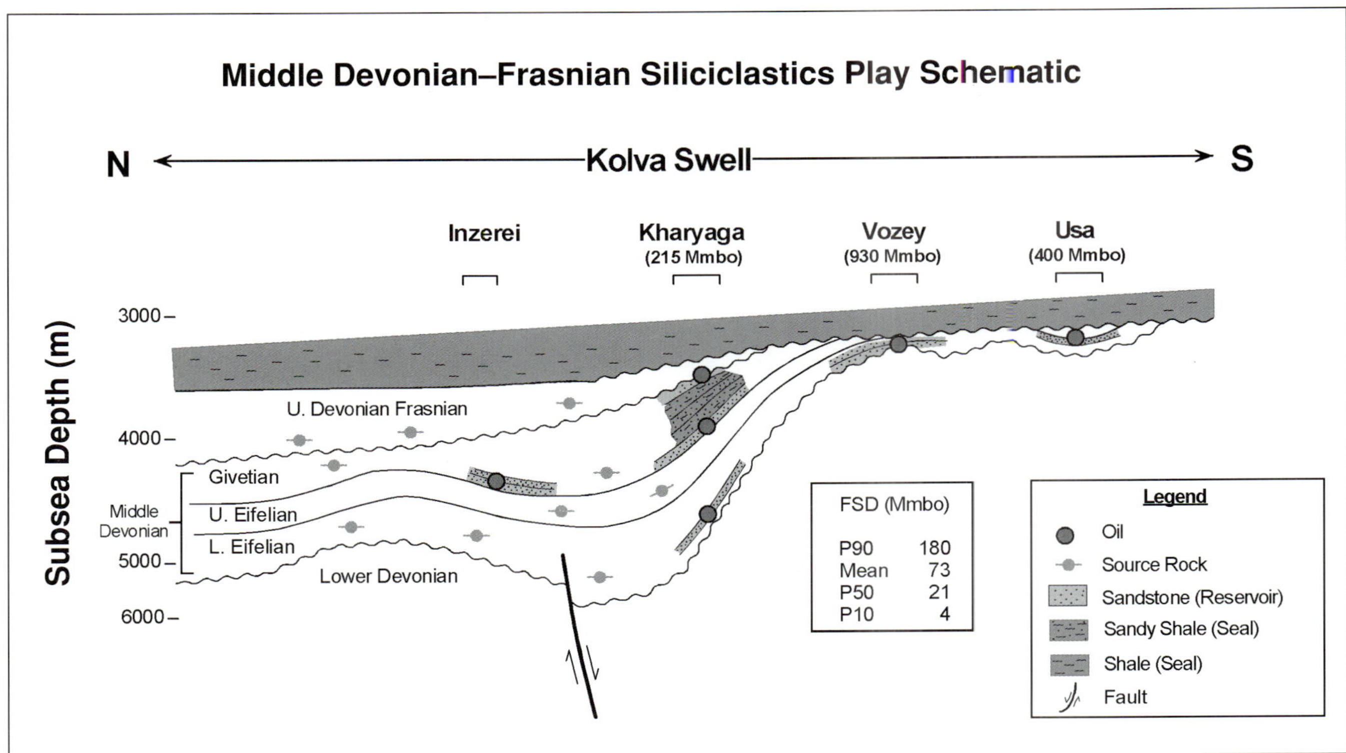

Figure 18. Middle Devonian and Frasnian siliciclastics-play schematic. See Figure 15 for line of cross section. Mmbo = million bbl oil.

and Kharyaga field. See Figure 19 for EUR reserves in the play. Porosities generally range from 8% to 11% and permeabilities from 90 to 120 md in the Kolva swell region in lower beach and shallow-marine shoal facies. Permeabilities are as high as 900 md in Kharyaga field in the shallow-marine facies (Nikonov and Bogatsky, 1996).

The principal trap type is a combination of structural and stratigraphic elements, and encasing shales provide side and top seals (Figure 18). The critical uncertainty is reservoir quality. This play is most likely sourced from Ordovician–Silurian–Lower Devonian and Middle Devonian type II source rocks (Danilevsky, 1996).

Upper Devonian Frasnian Carbonates Play (Figure 19)

The Upper Devonian Frasnian carbonate reef play has high exploration risk, is in the plateau phase (Figure 13), and has small potential exploration reserves. The Frasnian carbonate play is composed of two subplays: the Domanik and the Sirachoy-Ukhtinsky Formation reef trends (Figure 19).

The Frasnian Domanik shelfal reef trends generally north-south through the Nenets Okrug and Komi Republic region. The principal lithology is a vuggy, coralline dolomite with minor anhydrite. Porosity ranges from 7% to 15% (Nikonov and Bogatsky, 1996). The principal trap type is stratigraphic. Side seal is provided by the regionally significant, actively generating middle-shelf Domanik source-rock facies to the east and the tight, marly inner-shelf Domanik carbonates to the west (Figure 19). The Domanik type II source rock on the east, which is very prolific, is an organic-rich middle-shelf limestone with interbedded marls, argillites, and chert. The Domanik facies extends from the Pechora Sea to the southern extent of the Timan-Pechora Basin and into the Volga-Ural Basin. The Domanik is within the present-day oil window in a large portion of the basin and is generally oil prone east of the reef trend and gas prone to the west (Danilevsky, 1996). Top seal is an uncertainty because it may be composed of the potentially overlying Upper Devonian Frasnian Sirachoy-Ukhtinsky reef trend (Figure 19). However, the Upper Devonian Domanik reef play is proven to the north in Roman Trebs field. The principal critical uncertainty for this subplay is reservoir quality.

Porosity in the Frasnian Sirachoy-Ukhtinsky reef unit is poor to fair, and the principal trap type is stratigraphic. The regionally significant actively generating middle-shelf Frasnian Domanik source rock to the east and the tight, inner-shelf Frasnian carbonates to the west provide side seal. Top seal is provided by the overlying basal tight Upper Devonian Fammenian carbonates (Figure 19), and hydrocarbon charge is provided by the prolific Upper Devonian type II Domanik source rock (Danilevsky, 1996). The critical uncertainty for this subplay is reservoir quality.

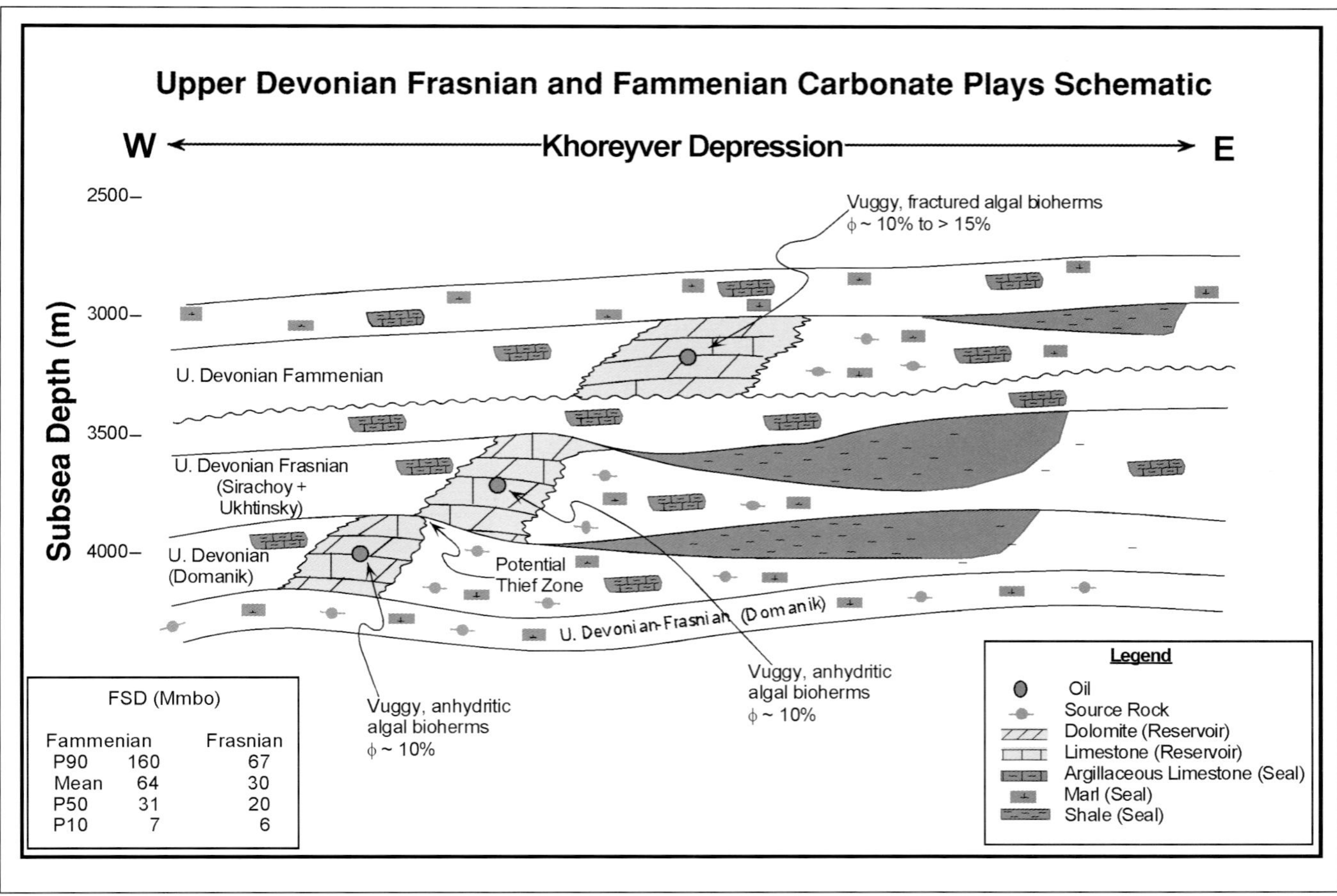

Figure 19. Upper Devonian Frasnian and Fammenian carbonates-play schematic. See Figure 15 for line of cross section. Mmbo = million bbl oil.

Upper Devonian Fammenian Carbonates Play (Figure 19)

The Upper Devonian Fammenian carbonate reef play is prolific. The play has low to moderate exploration risk, is in the plateau phase (Figure 13), and has moderate potential exploration reserves. The mean field size has remained approximately constant throughout the play development (Figure 14), presumably because of the ubiquitous similar size of the reef features.

Upper Devonian Fammenian reef carbonates have been producing in the Khoreyver depression from the Polar Lights block Ardalin field since 1996. Ardalin field is above the Arctic Circle, which makes it the northernmost continuously producing field in the Timan-Pechora Basin. Produced oil from Ardalin field is transported to the south through the Kharyaga-Usinsk-Ukhta pipeline system. Current production is 40,000 BOPD.

Fammenian reef carbonates are composed of vuggy, fractured, algal bioherms (Figure 19). Hydrocarbons in the play are sealed by Upper Devonian argillaceous carbonates and charged by Upper Devonian Domanik type II source rocks (Danilevsky, 1996) and, in some instances, by Silurian type II source rocks (Wavrek, 1997). Fammenian and Frasnian reefs typically formed on paleohighs, particularly the Bolshezemelsky basement high (Bogatsky et al., 1996; Rappoport, 1997).

Upper to Middle Carboniferous Carbonates Play (Figure 20)

The Upper to middle Carboniferous exploration play comprises fossiliferous, algal limestones sealed by interbedded, argillaceous limestones of the same age. The trap type is primarily structural; however, downdip stratigraphic traps exist (Figure 20).

The unit holds primarily gas and gas condensate to the west in the Shapkino-Yuraga swell, charged from Middle Devonian type III terrigenous gas-prone source-rock facies which are predominant in the northwest portion of the basin (Danilevsky, 1996). The Upper to middle Carboniferous carbonate unit is oil prone in the southern Kolva swell and Sandivey high, where the hydrocarbon charge is from the marine Upper Devonian Domanik type II source rock and possibly the Ordovician type II source rock (Wavrek et al., 1997).

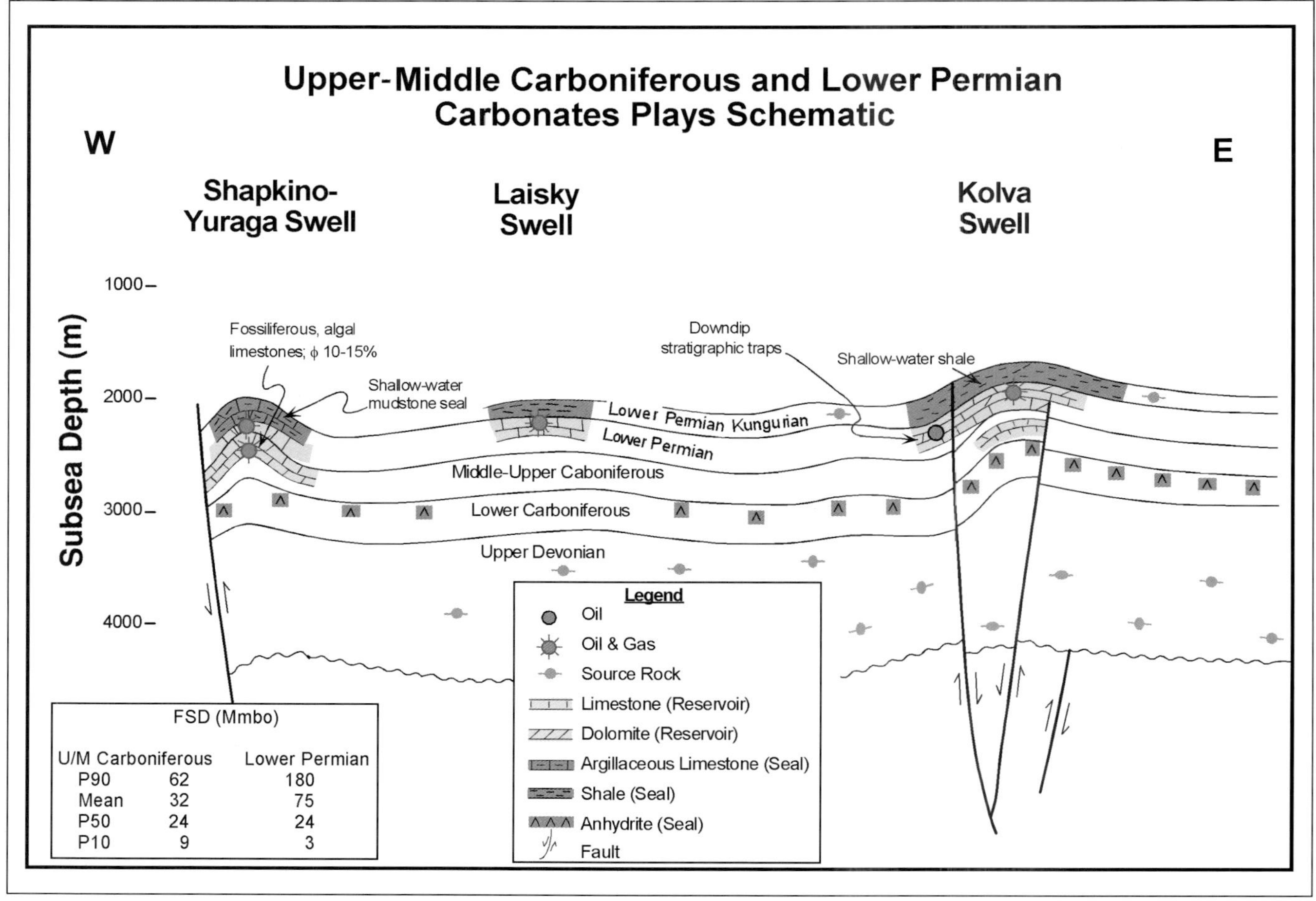

Figure 20. Upper to middle Carboniferous and Lower Permian carbonates-play schematic. See Figure 15 for line of cross section. Mmbo = million bbl oil.

The Upper to middle Carboniferous play has moderate exploration risk, is in the plateau phase (Figure 13), and has small potential exploration reserves.

Lower Permian Carbonates Play—Structural and Stratigraphic (Figure 20)

The Lower Permian exploration play is composed of two components: stratigraphic and structural (Figure 20). The structural play is situated mainly on the Kolva swell and has been drilled extensively. The play has moderate risk, is in the plateau phase (Figure 13), and has small to moderate potential exploration reserves. Figure 14 illustrates that the mean field size for this play increased in the last 10 years. This is a result of an extension of the play into the Pechora Sea with the discovery of Prirazlomnoye field (EUR 609 million bbl oil). In contrast, the mean field size for the other eight plays either decreased or remained the same during the same time period.

The primary reservoir unit in the Lower Permian carbonate play is the Asselian-Sakmarian (Figure 5). The unit is predominantly a barrier-type reef composed of bioherm structures and detrital limestone. Principal lithology is dolomitic limestone; dolomite and anhydrite are common. Porosity development is mainly secondary because of subaerial exposure and freshwater dissolution. The principal trap type is structural with a downdip stratigraphic component. The regionally extensive Lower Permian Kungarian and Artinskian marine shale provides top and side seal. Hydrocarbon charge is provided primarily by the Upper Devonian type II Domanik source rocks (Danilevsky, 1996; Wavrek et al., 1997). Critical uncertainties for the Lower Permian carbonate play include reservoir quantity and quality.

The Lower Permian carbonates stratigraphic play, which is a follow-up to the more prolific structural play, has moderate risk, is in the breakthrough phase (Figure 13), and has small to moderate potential exploration reserves. The stratigraphic play focuses on the downdip potential of Lower Permian structural features, defined primarily through the application of 3-D seismic data. In 1996, a successful well was drilled on the south flank of Yuzhno-Khilchuyu field to establish the occurrence of

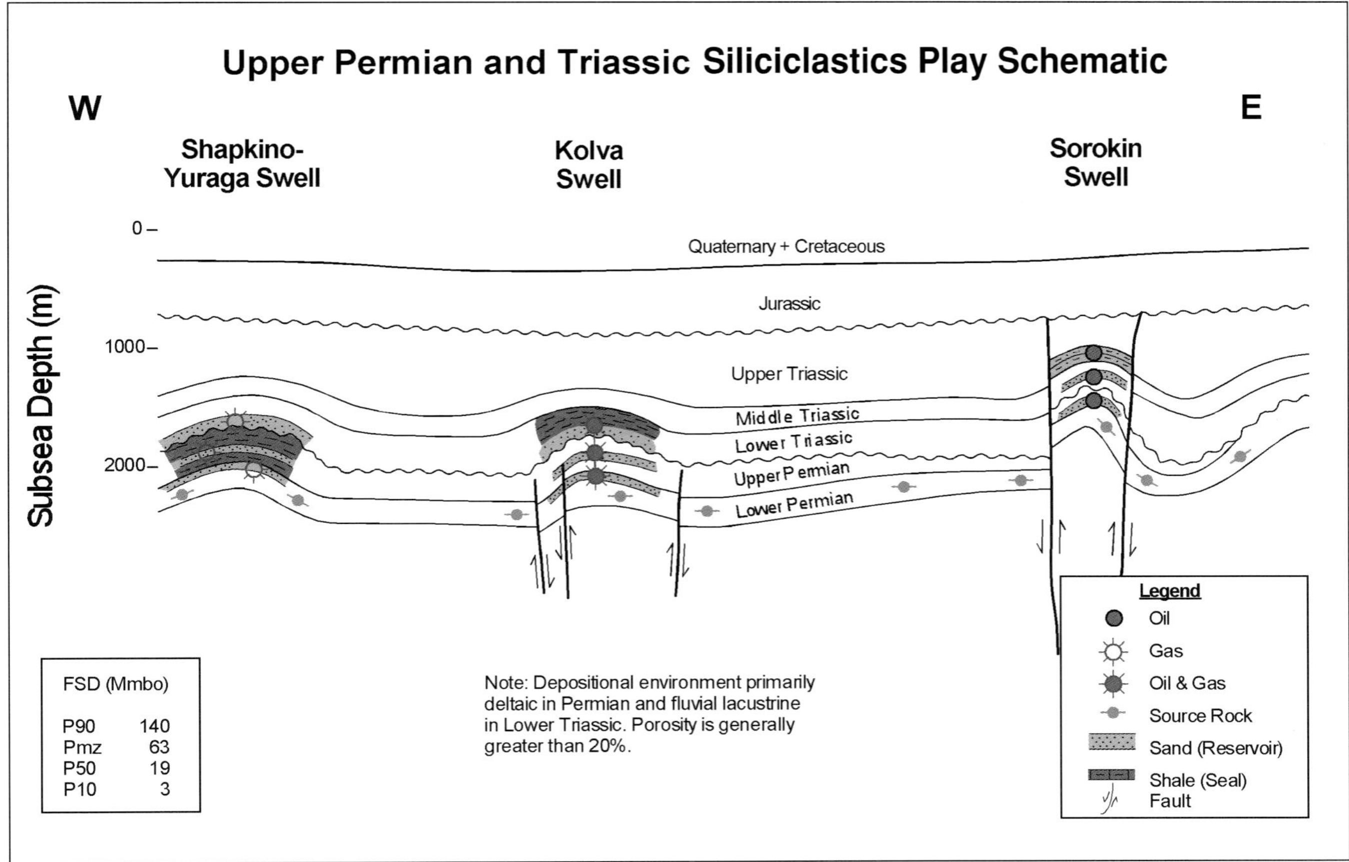

Figure 21. Upper Permian and Triassic siliciclastics-play schematic. See Figure 15 for line of cross section. Mmbo = million bbl oil.

hydrocarbons below the main oil-water contact in the field.

The downdip stratigraphic-play-type concept may be applied to other Lower Permian carbonate fields in the play fairway and to other carbonate reef or bioherm reservoirs in the Timan-Pechora Basin, if adequate data are acquired to identify the potential.

Upper Permian and Triassic Siliciclastics Play (Figure 21)

The Upper Permian and Triassic siliciclastics play reservoir facies is laterally extensive throughout the Timan-Pechora Basin, but the play fairway tends to overlay the present-day positive structural features (Figure 21). The depositional environment ranges from deltaic in the Upper Permian to fluvial and lacustrine in the Lower Triassic. Sands are generally of excellent quality (porosity >20%) but tend to be mostly thin and discontinuous, except in localized areas (Nikonov and Bogatsky, 1996). As with the Upper to middle Carboniferous carbonates play, hydrocarbon occurrence tends to be gas and condensate prone to the west in the Shapkino-Yuraga and Kolva swells and oil prone to the east in the Varandey-Adzva structural zone. This is because of variations in the source-rock organic facies (Danilevsky, 1996; Wavrek et al., 1997).

The Upper Permian and Triassic siliciclastics play has moderate risk, is in the plateau phase (Figure 13), and has small to moderate potential exploration reserves.

PLAY ANALYSIS AND FUTURE POTENTIAL—OFFSHORE PECHORA SEA, MURMANSK DISTRICT

Limited data are available for the Pechora Sea region, and therefore a detailed level of analysis is not possible. Few wells have been drilled, but exploration potential does exist. Four main plays (Lower Devonian carbonates, Upper Devonian carbonate reefs, Lower Permian carbonates, and Upper Permian and Triassic siliciclastics), which are extensions of proven plays within the onshore portion of the basin, extend into the Pechora Sea and define the offshore exploration potential (see Figures 17, 19–21 for play schematics). The offshore portions of the plays are in the active-growth phase and possibly the breakthrough phase. Creaming-curve analysis indicates that

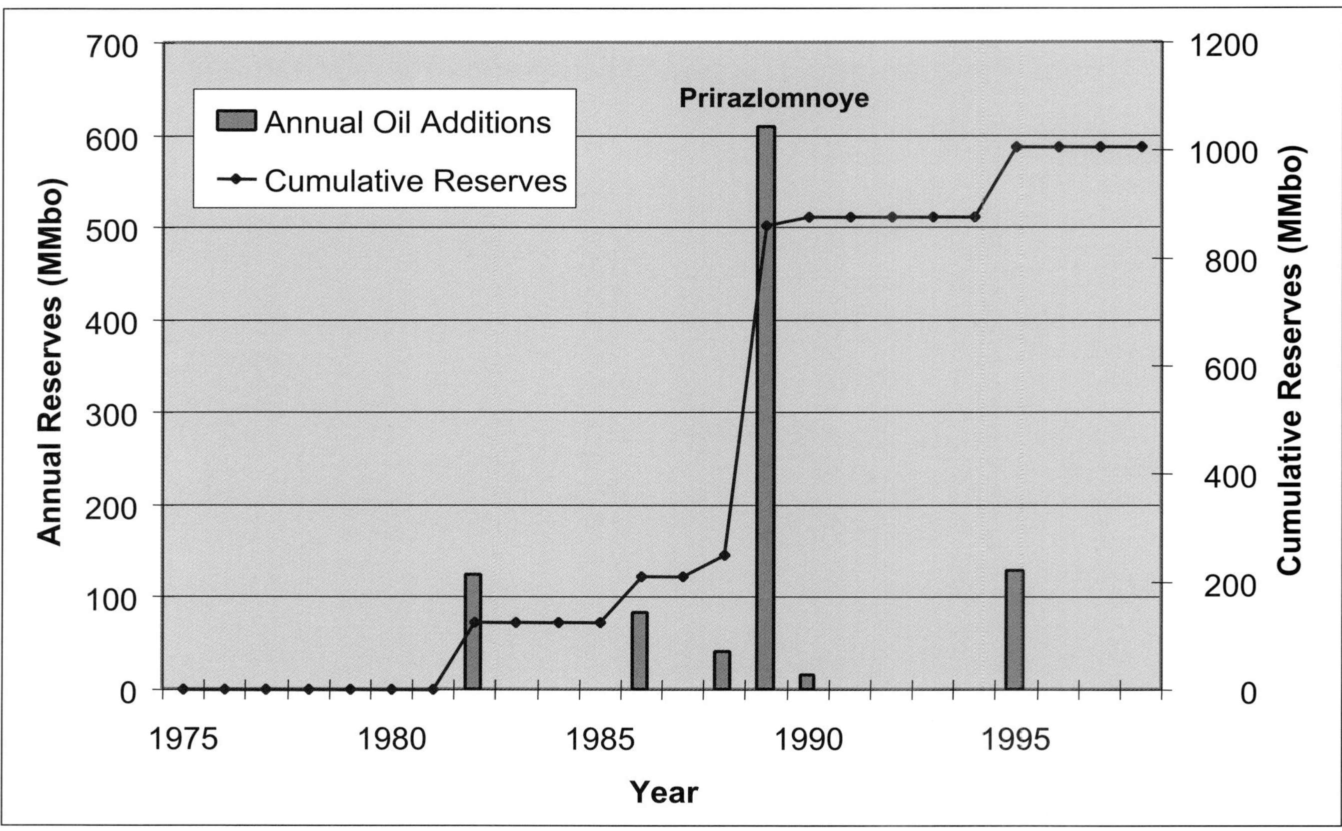

Figure 22. Pechora Sea creaming curve. Prirazlomnoye field, the largest oil field in the Pechora Sea, is noted. Mmbo = million bbl oil.

approximately 1 billion bbl oil have been discovered to date (Figure 22) and that future added reserves will be in large increments. Total undiscovered reserves might be as high as 4 billion bbl oil (Fossum, 1997). Nearly 100% of the discovered reserves remain unproduced. Critical uncertainties for Pechora Sea exploration include infrastructure, costly and hostile arctic drilling conditions (severe weather and ice), political environment, and high exploration costs.

CONCLUSIONS

Results indicate that nine proven exploration and development plays have been identified in the onshore portion of the Timan-Pechora Basin. Eight of the nine plays are in the plateau phase; one is in the breakthrough phase. The breakthrough phase was made possible by the application of new technologies such as 3-D seismic acquisition, processing, and interpretation and by the application of integration techniques in which multidisciplinary teams were used to maximize the value of acquired data.

Extracting value from the Timan-Pechora Basin lies primarily in field development and secondarily in exploration. Both field development and exploration projects require proximity to existing infrastructure to enhance value. Three plays represent the largest portion of the field development potential: the Lower Permian carbonates-structural play; the Upper Devonian Fammenian carbonates play; and the Lower Devonian carbonates play. The Upper Permian–Triassic siliciclastics play has moderate field-development potential. Overall, 80%, or 13 billion bbl, of the discovered oil reserves onshore remain unproduced. Exploration potential in the Pechora Sea exists in the same four plays.

Detailed basin analysis has indicated that 1 billion bbl oil onshore and 4 billion bbl oil in the Pechora Sea may be discovered in the next 10 years. The total future exploration potential of the basin may be as high as 15.8 billion bbl oil. Although this is a large number, the challenge for the industry (both foreign and Russian investors) is that the reserves will be in small accumulations (75 million bbl oil equivalent or less) and in a harsh operating environment where developing such small accumulations will be difficult.

ACKNOWLEDGMENTS

The authors thank the management of Conoco, Inc.; OAO Arkhangelskgeoldobycha; the Timan-Pechora Scientific Research Center; and the Environmental and Geoscience Institute (University of Utah) for permission to publish this paper. We also thank Dr. Ken C. Abdulah and Dr. Vladimir N. Vyssotski for their critical reviews of the manuscript; and Dave Warwas, Jessie Aguilar, and Debbie Hall for providing excellent graphics support.

REFERENCES CITED

Aminov, L. Z., M. Belonin, V. I. Bogatsky, A. P. Borovinskikh, V. Gaideek, Eu. Grunis, S. Danilevsky, V. Makarevich, Yu. Pankratov, O. Prischepa, Yu. Trifachov, and V. Kholodilov, 1993, Oil and gas potential and economic and geological parameters of preparation and development of hydrocarbons in the Timan-Pechora province: All-Russian Petroleum Research Geological Exploration Institute report, 92 p.

Barringer, J. J., M. Schlegel, S. Terry, and L. Zilm, 1998, Characterization of regional landforms and landscape analysis in support of an oil field development in the remote Arctic Timan-Pechora region of Russia (abs.): AAPG Bulletin, vol. 82, no. 13 (Supplement), 2 p.

Bogatsky, V. I., N. A. Bogdanov, S. L. Kostyuchenko, B. V. Senin, S. F. Sobolev, E. V. Shipilov, and V. E. Khain, 1996, Tectonic map of the Barents Sea and the northern part of European Russia: Moscow, Russian Academy of Sciences Institute of the Lithosphere.

Borovinskikh, A. P., 1998, The fuel and energy potential of Russia's European north region: The current state and development strategy: Mineral Resources of Russia, January, p. 14–31.

Danilevsky, S., 1996, Oil and gas generative potential of the oil and gas complexes within the continental portion of the Timan-Pechora sedimentary basin: Timan-Pechora Scientific Research Center report for Conoco, 13 p.

Fossum, B. J., 1997, Timan-Pechora Basin regional phase II evaluation: Conoco, unpublished internal report.

Lodzhevskaya, M. I., and T. N. Smolenchuk, 1998, Fluid dynamic aspects of oil migration and accumulation in the northeast of the Timan-Pechora Basin: Petroleum Geoscience, v. 4, p. 111–120.

Magoon, L. B., and W. G. Dow, 1994, The Petroleum system, *in* L. B. Magoon and W. G. Dow, eds., The petroleum system from source to trap: AAPG Memoir 60, p. 3–24.

Michael, G. E., and B. J. Fossum, 1998, Regional 2-D maturation models, northern Timan-Pechora Basin, Russia: Extended Abstract, AAPG Bulletin, v. 82, no. 13 (Supplement), 6 p..

Nikonov, N., and V. I. Bogatsky, 1996, Regional tectonic and lithologic characteristics of the Timan-Pechora sedimentary basin: Timan-Pechora Scientific Research Center report for Conoco, 18 p.

Rappoport, B. I., 1997, Regional geological studies to evaluate oil and gas prospects of the northern territory: OAO Arkhangelskgeoldobycha report for Conoco, 30 p.

Ressetar, R., D. Hobday, and B. Pimenov, 1997, Tectono-stratigraphic framework, *in* Petroleum systems of the Timan-Pechora basin, Russia: Energy and Geoscience Center Technical Report 97-5-20940, v. 2a, p. 1–42.

Rose, P. R., 1992, Chance of success and its use in petroleum exploration *in* R. Steinmetz, ed., The business of petroleum exploration: AAPG Treatise of Petroleum Geology, p. 71–86.

Schmidt, W. J., 1996, Tectonics and structural history, Nenets Autonomous District, Timan-Pechora Basin, Russia: Conoco, unpublished internal report, 18 p.

Ulmishek, G., 1982 Petroleum geology and resource assessment of the Timan-Pechora Basin, USSR, and the adjacent Barents-Northern Kara Shelf: U. S. Department of Energy report ANL/EES-TM-199, 195 p.

Wavrek, D. A., D. K. Curtiss, R. Ressetar, and R. Erhlich, 1997, Petroleum systems, *in* Petroleum systems of the Timan-Pechora Basin, Russia: Energy and Geoscience Institute Technical Report 97-5-20940, v. 1, p. 1–73.

Ziegler, P. A., 1988, Evolution of the Arctic–North Atlantic and the Western Tethys: AAPG Memoir 43, 198 p.

Ziegler, P. A., 1989, Evolution of Laurussia: A study in late Paleozoic plate tectonics: Boston, Kluwer Academic Publishers, 102 p.

Howes, J. V. C., 2001, Future petroleum production from Indonesia and Papua New Guinea, *in* M. W. Downey, J. C. Threet, and W. A. Morgan, eds., Petroleum provinces of the twenty-first century: AAPG Memoir 74, p. 281–286.

Chapter 14

Future Petroleum Production from Indonesia and Papua New Guinea

J. V. C. Howes
Petroknowledge.com, Christchurch, New Zealand

INTRODUCTION

Exploration for both oil and gas in deep water and a continued focus on the development of existing natural-gas resources will be important for production growth in the next two decades in Indonesia and Papua New Guinea. Overall, oil production is likely to decline, while gas production should increase. At least six deep-water areas are available for further petroleum exploration, and several large gas projects await development. Frontier exploration, innovative enhanced-recovery techniques, production from gas hydrates, and use of gas-to-liquids technology could unlock additional large petroleum volumes for exploitation in the region in the longer term. Coal-bed methane and deep, tight-gas plays are also present.

DISCOVERED RESERVES

More than 60 billion barrels oil equivalent (BOE) of ultimate reserves from more than 50 separate petroleum systems have been discovered to date in Indonesia and Papua New Guinea. Those volumes are reservoired in more than 1000 fields. Total reserve volumes discovered are weighted slightly toward gas in Indonesia and strongly toward gas in Papua New Guinea. Analysis of discovery process shows that oil exploration is mature in many areas in Indonesia (Howes and Suherman, 1995), but strong growth has occurred in gas reserves in recent years. This trend is likely to continue. Detailed analysis indicates an average of at least two significant oil and/or gas discoveries per year in the past 20 years in Indonesia in the category of greater than 50 million BOE reserves. Discovery rates in Papua New Guinea are more modest. Some basic comparative data for Indonesia and Papua New Guinea are shown in Table 1.

PETROLEUM SYSTEMS

Petroleum systems in Indonesia and Papua New Guinea have been fundamentally controlled by global tectonic events, in particular the interaction of the Eurasian, Indo-Australian, and Pacific tectonic plates (Howes, 1997). Petroleum systems on the Eurasian Plate (present in western Indonesia) are predominantly Tertiary in age, whereas those on the Indo-Australian plate (in eastern Indonesia and Papua New Guinea) are predominantly Mesozoic (Figure 1). Simpler structural styles and more accessible topography have led to a mature exploration stage in many western Indonesian systems. Petroleum systems in eastern Indonesia and Papua New Guinea, with more complex geology and more challenging topography, are less well explored in general (Figure 2) and may offer better opportunities for significant future discoveries. The maturity of the two areas is reflected by exploration statistics, which show more than 20,000 exploration and production wells in western Indonesia, compared with fewer than 2000 total wells in eastern Indonesia and Papua New Guinea.

Location, fluid type, and discovered volumes for more than 50 productive petroleum systems in the region are shown in Figure 3. The bulk of discovered reserves lies in western Indonesia. Several highly prolific oil-prone petroleum systems in that area are linked to lacustrine source rocks deposited in intracratonic to back-arc extensional settings during the extrusion of southeast Asia from Eurasia in the Eocene (Tapponier et al., 1982) (Fig-

Table 1. Basic comparative data for Indonesia and Papua New Guinea (year 2000).

	Indonesia	Papua New Guinea
Population	>200 million	<5 million
Area	>14,000 islands, 1.9 million km² land area	>600 islands, 400,000 km² land area
First exploration well	1872	1922
First oil production	1892	1992
Total wells drilled	>20,000	<400
Oil and gas fields	>1000	<30
Oil and gas discovered (EUR)	>60 billion BOE	<3 billion BOE
Oil production (barrels/day)	1.3 million	90,000
Gas production (million cubic ft/day)	8000	14
Predominant petroleum system	Tertiary (generally in extensional settings)	Jurassic-Cretaceous (fold-and-thrust belt predominant to date)
Type of contracts	Production sharing	Tax royalty

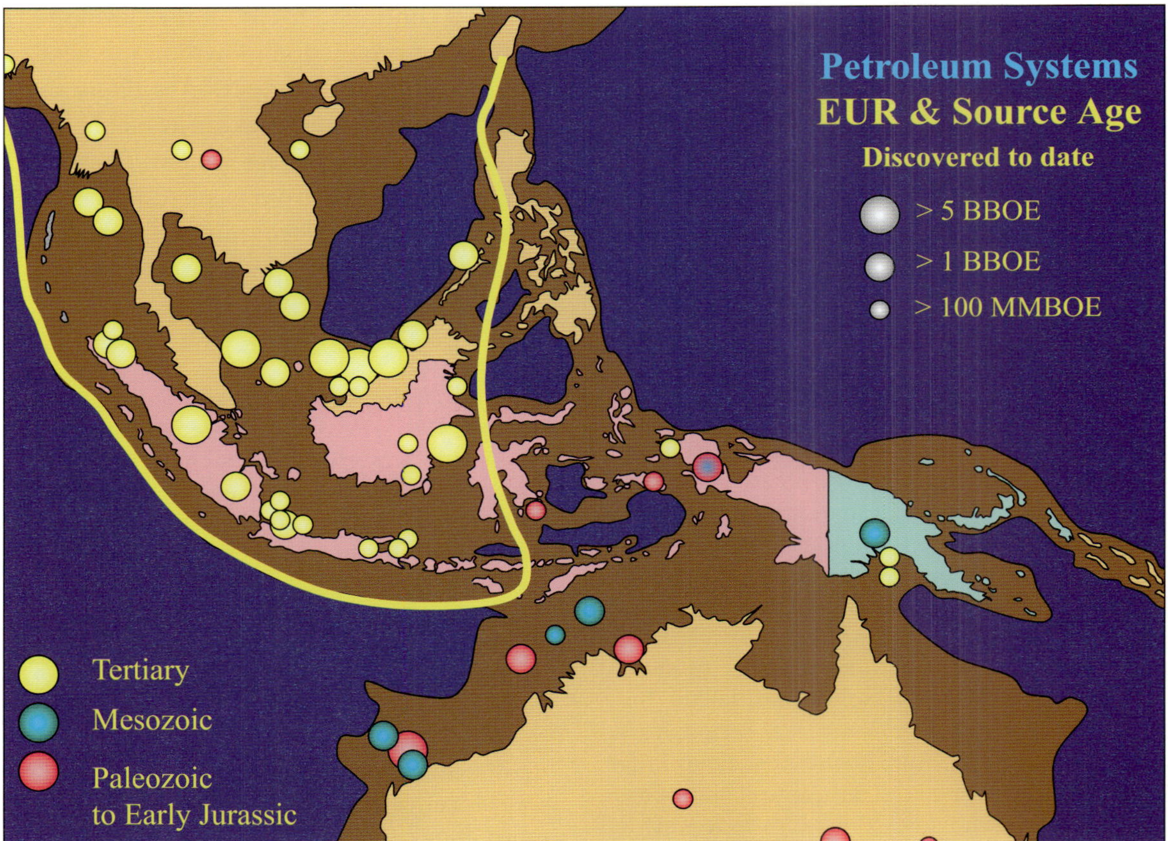

Figure 1. Age and distribution of productive petroleum systems in the southeast Asia–Australasia region. Symbols indicate estimated ultimate recoverable volumes discovered to date (EUR) in billion barrels of oil equivalent (BBOE), color-coded by principal age of petroleum system (yellow = Tertiary, blue = Middle Jurassic to middle Cretaceous, pink = Paleozoic to Early Jurassic). The land area of Indonesia is shown in pink and that of Papua New Guinea in pale blue. Other land areas are shown in light brown. Water depths less than 1000 m are shown in darker brown and greater than 1000 m in dark blue. The yellow line separates areas dominated by Tertiary petroleum systems from those dominated by Mesozoic and Paleozoic systems. The line also demarcates "western Indonesia" from "eastern Indonesia and Papua New Guinea," as referred to in the text, and coincides with the approximate eastern and southern edge of the Eurasian tectonic plate.

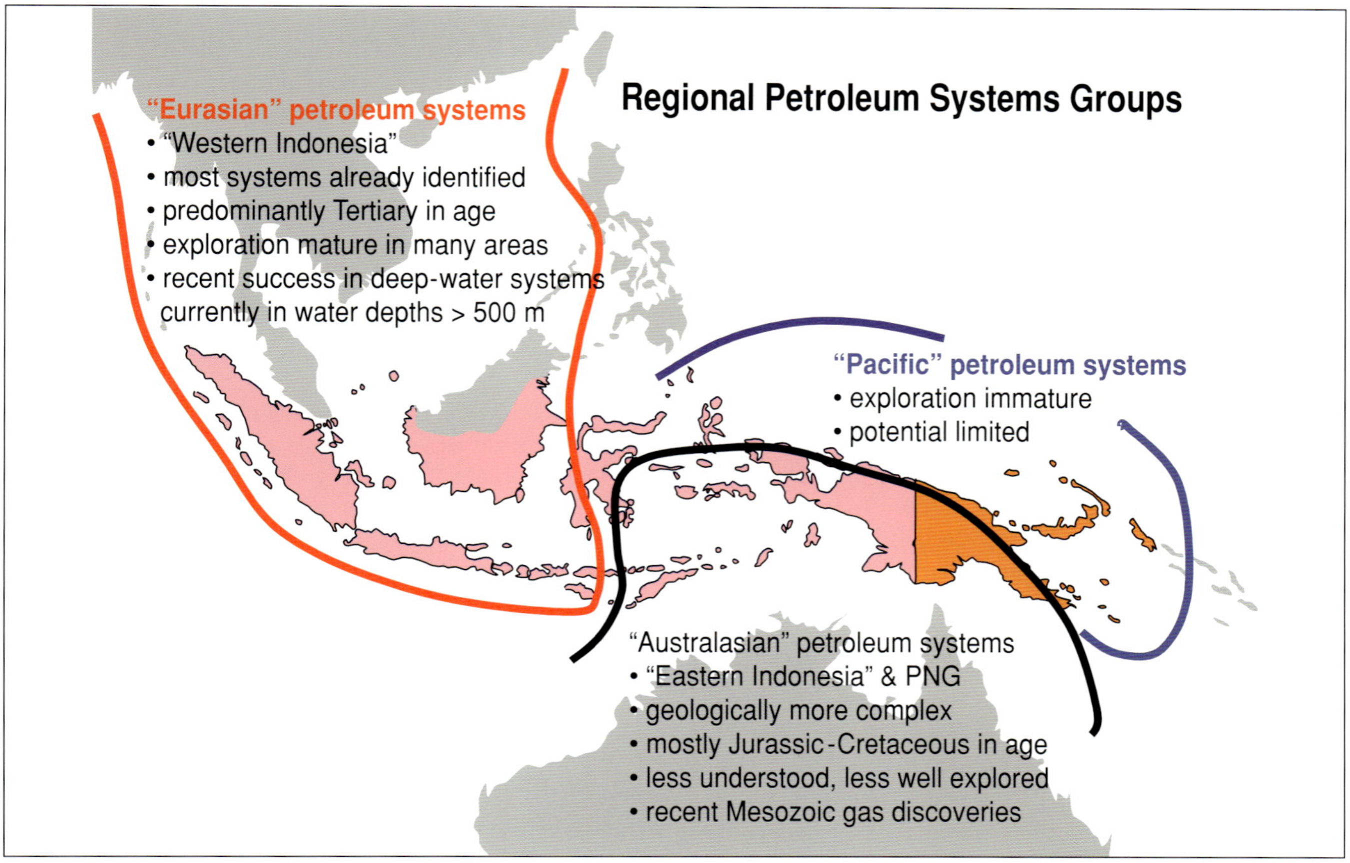

Regional Petroleum Systems Groups
"Eurasian" petroleum systems
• "Western Indonesia"
• most systems already identified
• predominantly Tertiary in age
• exploration mature in many areas
• recent success in deep-water systems currently in water depths > 500 m
"Pacific" petroleum systems
• exploration immature
• potential limited
"Australasian" petroleum systems
• "Eastern Indonesia" & PNG
• geologically more complex
• mostly Jurassic-Cretaceous in age
• less understood, less well explored
• recent Mesozoic gas discoveries

Figure 2. Location and key observations for three regional groupings of petroleum systems present in Indonesia and Papua New Guinea, referred to as "Eurasian," "Australasian," and "Pacific."

Petroleum Systems EUR and fluid type
Discovered to date
> 5 BBOE
> 1 BBOE
> 100 MMBOE
Gas prone
Oil prone
Mixed systems

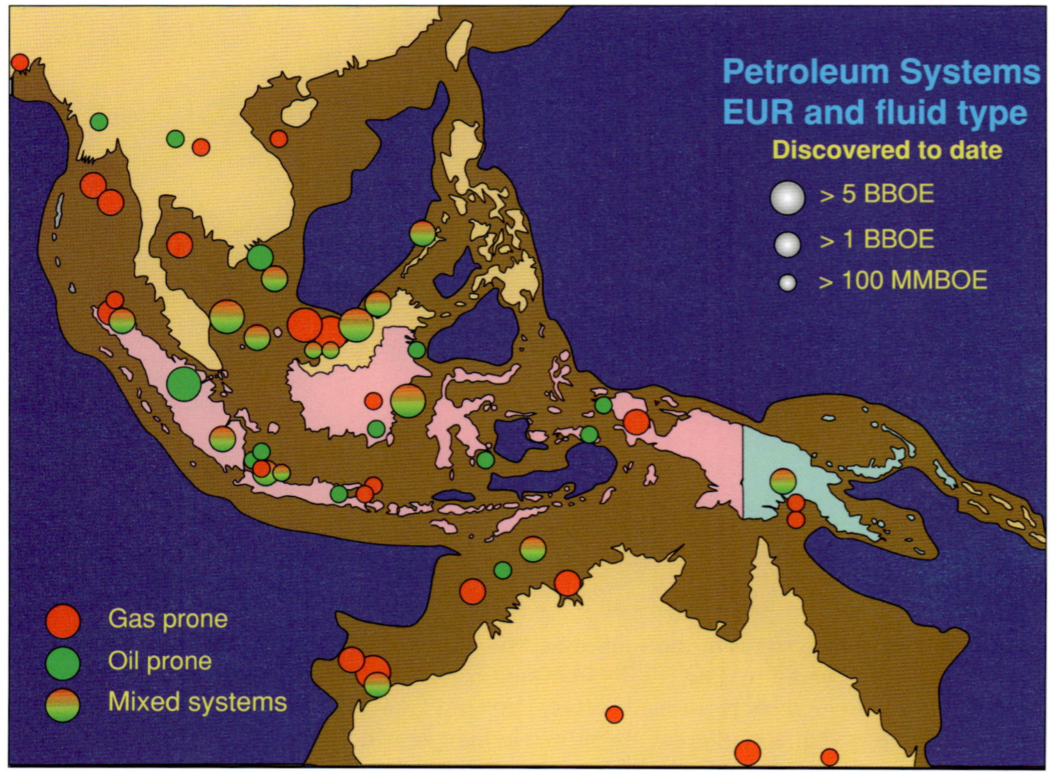

Figure 3. Fluid type of productive petroleum systems in the southeast Asia–Australasia region. Symbols indicate estimated ultimate recoverable volumes discovered to date (EUR) in billion barrels of oil equivalent (BBOE), color-coded by hydrocarbon type. The land area of Indonesia is shown in pink and that of Papua New Guinea in pale blue. Other land areas are shown in light brown. Water depths less than 1000 m are shown in darker brown and greater than 1000 m in dark blue.

ure 4). By contrast, the predominantly marine source rocks of eastern Indonesia and Papua New Guinea were deposited in rifted passive-margin settings and tend to be more gas prone. Source kitchens in both areas, usually tied to rift grabens, are often restricted in extent, and hydrocarbon-charge and migration-pathway analyses are therefore critical parts of the exploration process.

GAS PROJECTS

More than one-third of gas reserves discovered to date, or at least 12.5 billion BOE (75 trillion cubic feet), still awaits development in three areas: offshore Natuna Island ("East Natuna"), onshore and offshore Irian Jaya ("Tangguh"), and in the fold-and-thrust belt of Papua New Guinea (Figure 5). Planned infrastructure to exploit those resources, such as a possible international gas-export pipeline from Papua New Guinea to Queensland, will spur future activity in these relatively high-cost and remote areas. Significant increases in activity and production potential can also be expected in offshore deep-water areas such as the Makassar Straits in Indonesia, where recent commercial oil and gas discoveries have been made in water depths approaching 1000 m. Proximity to existing liquefied natural-gas (LNG) export facilities should ensure that deep-water Makassar Straits and offshore North Sumatra gas resources eventually will reach market.

PETROLEUM POTENTIAL

Conceptual volumes of "yet-to-be-discovered" conventional oil and gas reserves are shown in Figure 6. These volumes are subjective and unrisked, and they project additional success in proven petroleum systems and new success in presently unproductive hypothetical and speculative petroleum systems. Remaining recoverable volumes to be discovered, shown in Figure 6, total more than 20 billion BOE.

Significant reserve additions should be expected from

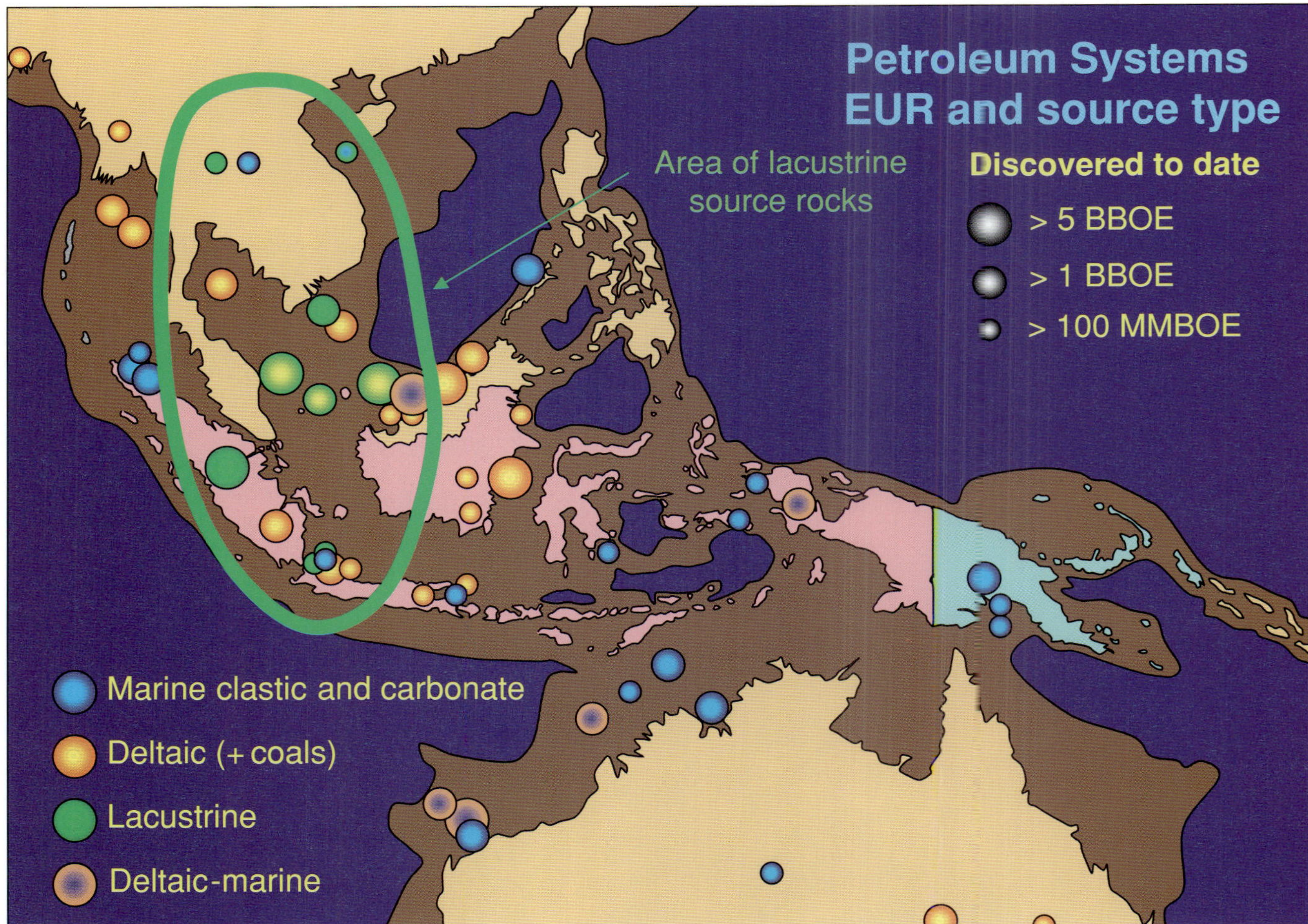

Figure 4. Source type of productive petroleum systems in the southeast Asia–Australasia region. Symbols indicate estimated ultimate recoverable volumes discovered to date (EUR) in billion barrels of oil equivalent (BBOE), color-coded by dominant source type. Mixed sources show two-toned colors. The land area of Indonesia is shown in pink and that of Papua New Guinea in pale blue. Other land areas are shown in light brown. Water depths less than 1000 m are shown in darker brown and greater than 1000 m in dark blue.

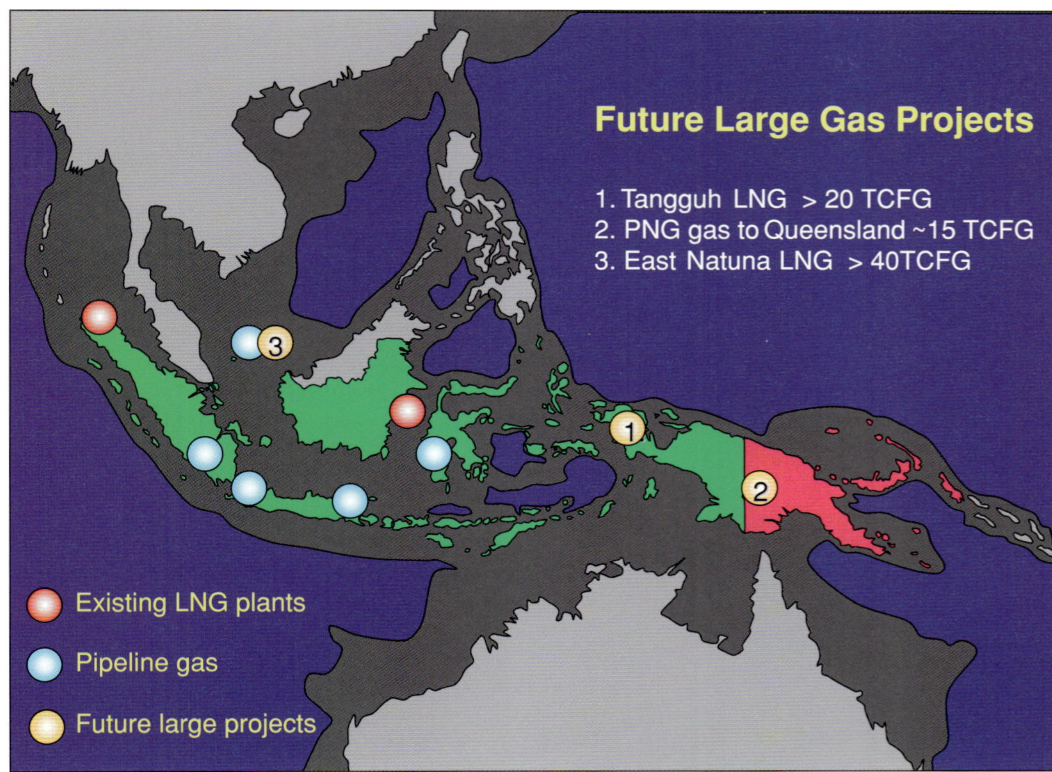

Figure 5. Location of existing and future significant natural-gas projects in Indonesia and Papua New Guinea. The land area of Indonesia is shown in green and that of Papua New Guinea in pink. Other land areas are shown in light gray. Water depths less than 1000 m are shown in darker gray and greater than 1000 m in dark blue. TCFG = trillion cubic feet of gas.

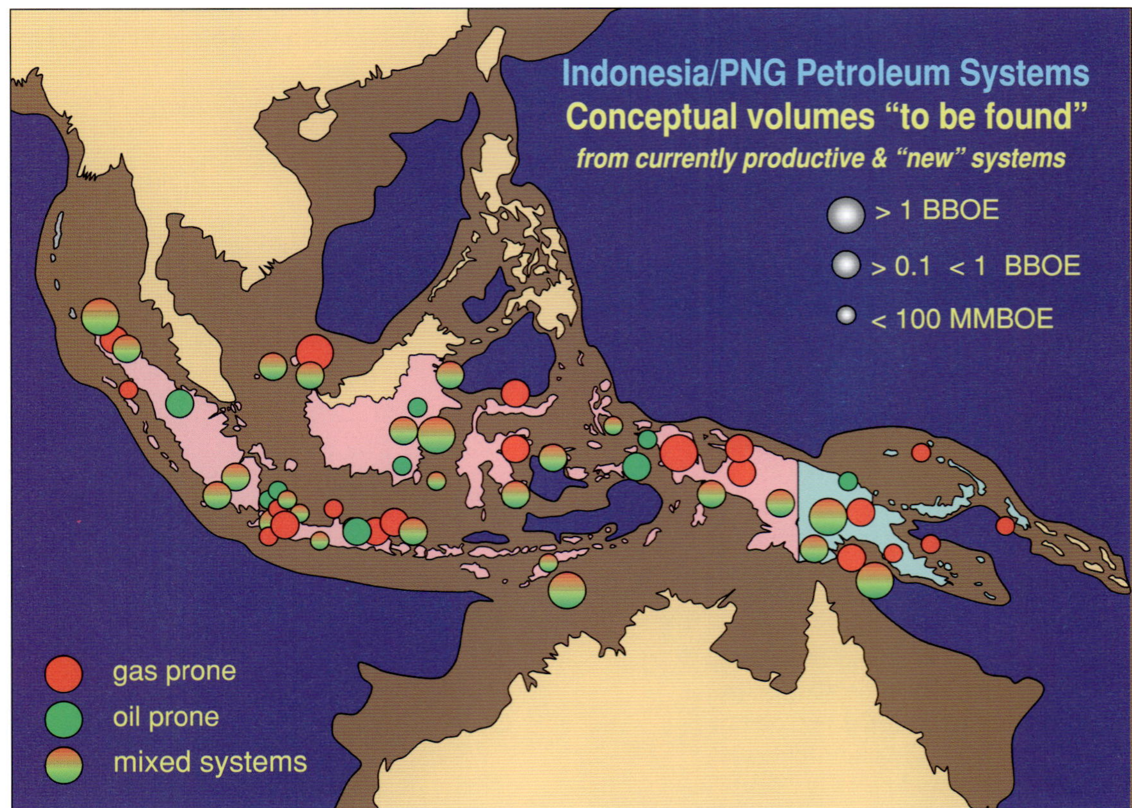

Figure 6. Distribution and possible volumes of remaining petroleum reserves to be discovered in Indonesia and Papua New Guinea, by petroleum system. Symbols indicate relative volumes to be found in each petroleum system, in billion barrels of oil equivalent (BBOE), color-coded by possible hydrocarbon type. The land area of Indonesia is shown in pink and that of Papua New Guinea in pale blue. Other land areas are shown in light brown. Water depths less than 1000 m are shown in darker brown and greater than 1000 m in dark blue.

exploration in deep water and in areas close to major natural-gas development projects. In addition, deeper pool exploration has already proved effective in some quite mature areas such as onshore East Java and South Sumatra. Other concepts to be pursued include continued exploration of fractured basement plays in Sumatra and Java, exploration beneath the current intraarc volcanic apron of Java and Nusa Tenggara, underexplored wrench-related systems in northern New Guinea, and subthrust footwall plays in Irian Jaya and Papua New Guinea fold-and-thrust belts. Because of high in-place volumes and less than optimal recovery efficiencies in some existing fields, enhanced recovery techniques could unlock significant additional reserves in Indonesia. In addition to proven petroleum systems, at least 50 lightly explored to unexplored speculative petroleum systems exist, both onshore and offshore, with the highest potential for significant new production likely in frontier areas of eastern Indonesia and Papua New Guinea. Less conventional petroleum resources that may be available for eventual exploitation include gas hydrates in deep-water areas, coal-bed methane, and deep-basin tight gas. Minor bitumen reserves are present. Development of economic gas-to-liquids technology will assist production of significant "stranded" gas reserves.

SUMMARY

Deep-water areas and those areas surrounding future major gas-sales infrastructure projects are most likely to realize the highest potential for significant increased reserves and production in the short to medium term in Indonesia and Papua New Guinea. Enhanced recovery, deeper pool drilling, and frontier exploration also will be targeted. There will be a continued shift from oil to gas production as mature oil-producing areas are depleted. Large numbers of unexplored to lightly explored petroleum systems remain, particularly in frontier areas of eastern Indonesia and Papua New Guinea. Conceptual unrisked volumes yet to be discovered total more than 20 billion BOE.

ACKNOWLEDGMENTS

Presentation of this paper at the second Wallace E. Pratt Memorial Conference was made possible with the permission and support of Atlantic Richfield Indonesia, Inc., and ARCO International Oil and Gas Company. Compilation and interpretation of regional petroleum systems data benefited greatly from the comments and encouragement of many colleagues and peers in Jakarta, Indonesia. Petroleum-systems data were compiled on a modified database supplied by IHS Energy Group (formerly Petroconsultants S.A. and IEDS). Special thanks are due to the Petroleum Division of the Department of Petroleum and Energy, Papua New Guinea, and to Paul Tiensten (director), Peter Woyengu, and Michael McWalter for their assistance during a data-gathering trip to Port Moresby in June 1999.

REFERENCES CITED

For a comprehensive set of papers on Indonesia and Papua New Guinea, readers are referred, respectively, to the *Annual Proceedings* and *Memoirs* of the Indonesian Petroleum Association (www.ipa.or.id) and to the *Proceedings* of the First, Second, Third, and Fourth Petroleum Conventions of Papua New Guinea (www.pngchamberminpet.com). In addition, the first two references listed below include many specific citations for Indonesian and regional petroleum systems.

Howes, J. V. C., 1997, Petroleum resources and petroleum systems of SE Asia, Australia, Papua New Guinea, and New Zealand, *in* J. V. C. Howes and R. A. Noble, eds., Proceedings of the Petroleum Systems of SE Asia and Australasia Conference: Jakarta, Indonesian Petroleum Association, p. 81–100.

Howes, J. V. C., and T. Suherman, 1995, Indonesian petroleum systems, reserve additions and exploration efficiency: Proceedings of the 24th Annual Convention of the Indonesian Petroleum Association, v. 1, p. 1–17.

Tapponier, P., G. Peltzer, A. LeDain, R. Armijo, and P. Cobbold, 1982, Propagating extrusion tectonics in Asia: New insights from simple experiments with plasticine: Geology, v. 10, p. 611–616.

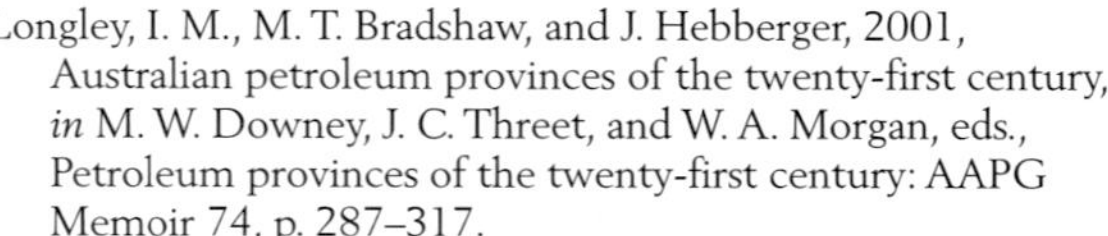
Longley, I. M., M. T. Bradshaw, and J. Hebberger, 2001,
Australian petroleum provinces of the twenty-first century,
in M. W. Downey, J. C. Threet, and W. A. Morgan, eds.,
Petroleum provinces of the twenty-first century: AAPG
Memoir 74, p. 287–317.

Chapter 15

AUSTRALIAN PETROLEUM PROVINCES OF THE TWENTY-FIRST CENTURY

Ian M. Longley
Woodside Offshore Petroleum Pty. Ltd., Perth, Western Australia, Australia

Marita T. Bradshaw
Australian Geological Survey Organisation, Canberra, Australian Capital Territory, Australia

John Hebberger
West Australian Petroleum Pty. Ltd., Perth, Western Australia, Australia

ABSTRACT

The Australian hydrocarbon exploration effort dates back to the nineteenth century, but it was not until well into the twentieth century that the first commercial field was discovered. The cumulative result of only about 3400 exploration wells drilled in 40 basins in Australia is the discovery of an estimated 6.4 billion barrels (bbl) of oil reserves, 2.1 billion bbl of condensate reserves, 136 trillion cubic feet of gas reserves, and 2.6 billion bbl of liquefied petroleum gas reserves. The majority of these hydrocarbon reserves (98%) is in the Gippsland Basin and the Cooper/Eromanga Basin and on the North West Shelf (Carnarvon, Browse, and Bonaparte Basins).

The Cooper/Eromanga and Gippsland Basins are mature exploration provinces with modest future exploration potential, but the large undeveloped reserves in the major gas fields along the North West Shelf will be developed principally for liquefied natural-gas export in the twenty-first century. The North West Shelf still has significant potential for further major discoveries, as demonstrated by its undrilled identified prospects and recent drilling success.

Other basins which offer the potential to develop into significant petroleum provinces in the twenty-first century include the basins of the Great Australian Bight and the Lord Howe Rise, which were largely overlooked during the twentieth century because of perceived excessive water depths and relatively isolated locations. These areas offer some of the best potential for undiscovered oil provinces and are the focus of current and future gazettal round opportunities.

INTRODUCTION

Australia is an island continent about 7.7 million km^2 in area, with 8.6 million km^2 of marine jurisdiction within its exclusive economic zone around the mainland and island territories (Australian Exclusive Economic Zones, or AEEZ). The exclusive economic zone off the Australian Antarctic Territory (AAT) covers another 2.5 million km^2 (Figure 1). Australia's marine jurisdiction includes another 3.8 million km^2 of legal continental shelf extending beyond the AEEZ around the mainland and island territories, and as much as an additional 1.8 million km^2 adjacent to the AAT, giving a total ocean jurisdiction of 16.7 million km^2, or more than twice the size of the continent (Commonwealth of Australia, 1999). The AAT is not considered further in this paper

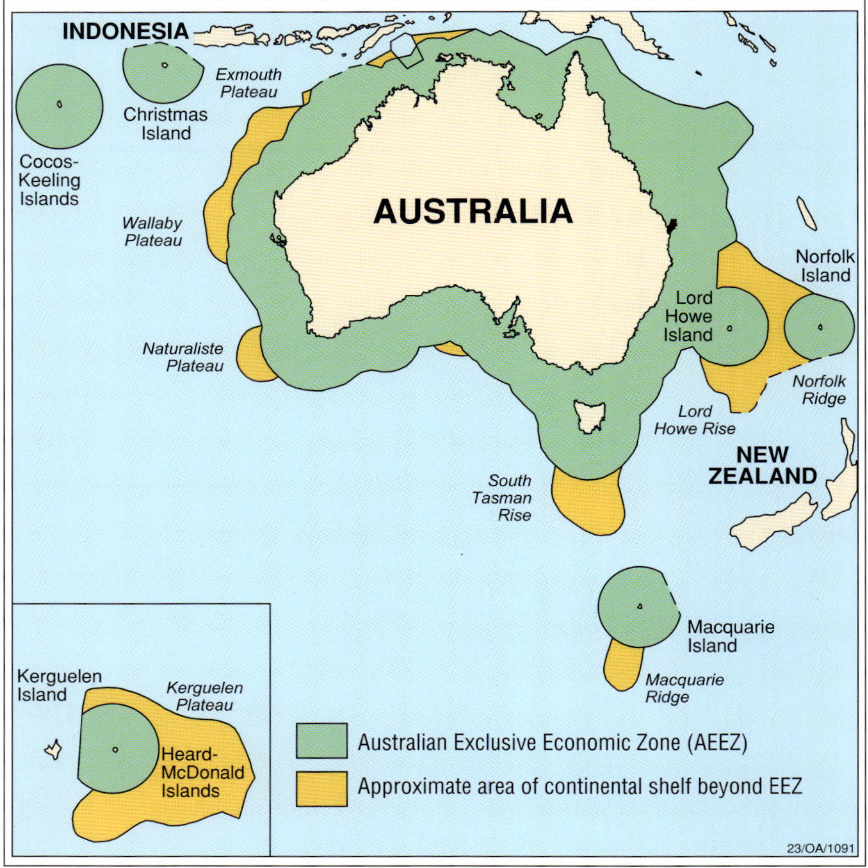

Figure 1. Australia and Australian Exclusive Economic Zones.

because the Madrid Protocol, recognized by the Australian government, prohibits the development of resources in the Antarctic region (see www.antarctic. com.au/encyclopaedia/hist/EnPrMad.html).

Across onshore Australia and in this vast ocean territory, more than 200 sedimentary basins have been described which conservatively cover about 10 million km^2. Investigations of many of the offshore basins are only in the reconnaissance stage, and advances in technology mean deep-water areas that previously were dismissed are worth serious consideration as exploration targets. The federal government has earmarked Australian $33.3 million during a four-year period to identify prospective new oil zones in frontier areas. The initial focus is the southern continental margin of the Great Australian Bight, where new regional seismic data were collected and exploration acreage was released in the 1999 bidding round (www.agso.gov.au/ marine/ar_index. html). Future releases are scheduled to include the Lord Howe Rise, Bremer Basin, and South Tasman Rise (2001–2005), followed by areas on the Kerguelen and Wallaby plateaus after 2005 (ISR, 1999).

These basins range in age from Proterozoic to Holocene and have been the focus of about 7500 exploration and development wells in the search for hydrocarbons, or about one well to every 1300 km^2 of sedimentary basin. This underlines the fact that Australia, apart from Antarctica, is the most underexplored continent on earth.

EXPLORATION HISTORY

The Australian exploration record has been described in detail by Robertson (1988), Wilkinson (1991), and Bradshaw et al. (1999). The first recorded drilling targeted at hydrocarbons began in the 1860s. Oil and gas shows in water boreholes encouraged sporadic exploration efforts throughout the first half of the twentieth century. The first commercial production of gas was in 1906 in Roma, Queensland, but it lasted for only 10 days. The first oil was recovered from a borehole in the onshore Gippsland Basin in 1924, and the first commercial condensate production was from Roma in 1928.

The first significant flow of oil in Australia was from Rough Range-1 in the onshore portion of the Northern Carnarvon Basin in 1953. Significant increased activity followed, but it was not until 1962 that the first commercial production of oil began, from the Moonie oil field. Major milestones during this modern exploration phase included the discovery of gas in the onshore Cooper Basin in 1963 (first production in 1969), discovery of major oil and gas fields in the offshore Gippsland Basin in 1965 (first production in 1969), discovery of the giant Barrow Island oil field in 1964 (first production in 1967), and discovery of the first in a series of major gas fields along the North West Shelf at North Rankin in 1971 (first production in 1984).

As summarized in Table 1 and Figures 2 and 3, an estimated 3400 exploration wells have been drilled within Australia, 23% of which are offshore. This exploration effort has discovered 14 productive hydrocarbon basins which contain an estimated 6.4 billion barrels (bbl) of oil, 2.1 billion bbl of condensate, 2.6 billion bbl of liquefied petroleum gas (LPG), and 136 trillion cubic feet (tcf) of gas. When undiscovered volume estimates of 1.9 billion bbl of oil and 32.6 tcf of gas reserves for the proven basins (after Australian Geological Survey Organisation, 1998; Bureau of Resource Sciences, 1993, 1994, 1996, 1997) are added to the discovered totals, a total hydrocarbon resource of 41.1 billion bbl of oil equivalent is calculated for Australia. The breakdown of these volume estimates by hydrocarbon phase and the status of development are shown on Figure 3.

Current methodology used by the Australian Geological Survey Organisation (AGSO) for estimating undiscovered resources in proven productive basins is described by Forman and Hinde (1985, 1986) and by Forman et al. (1992, 1993). The method is a statistical creaming-curve (declining field size as more discoveries are made) extrapolation approach based on individual play fairways. The resulting resource estimates are not ultimately recoverable reserves but refer only to conventional oil and gas accumulations that could be brought into production in the next 20–25 years.

This method gives relatively conservative estimates and is best suited for mature basins. Its focus is "on estimates of the lower limit of the petroleum potential (P95), but the real target for exploration is the small but significant chance of very much larger potential expressed at the 5% probability level" (Powell et al., 1990). Some inkling of what this prize may be is provided by an estimate of an additional 13 billion bbl of oil equivalent of "speculative" resources in remote deep-water areas (Symonds and Willcox, 1989; Willcox and Symonds, 1997). This evaluation used effective sediment volume (> 2000-m-thick kitchen areas) and volumetric yields from global analogs (Klemme, 1975).

The distribution of the discovered developed resources around Australia is summarized by basin in Figure 4; a summary of the undeveloped plus undiscovered volume estimates is shown in Figure 5. The reserves discovered and developed in the twentieth century are predominantly in the Gippsland, Carnarvon, and Cooper/Eromanga Basins, which cumulatively account for 98% of the developed reserves in Australia. The focus of activity, however, will clearly shift to the North West Shelf in the early part of the twenty-first century. This is because the Northern Carnarvon, Browse, and Bonaparte Basins contain 93% of Australia's undeveloped resources, comprising about 88 tcf of gas and 2.0 billion bbl of liquids (oil, condensate, and LPG), plus an estimated 82% of Australia's undiscovered resources in proven basins.

The discovery history of the hydrocarbon resources of Australia is shown by basin and phase in Figure 6. These data show that the rapid discovery of large volumes of oil

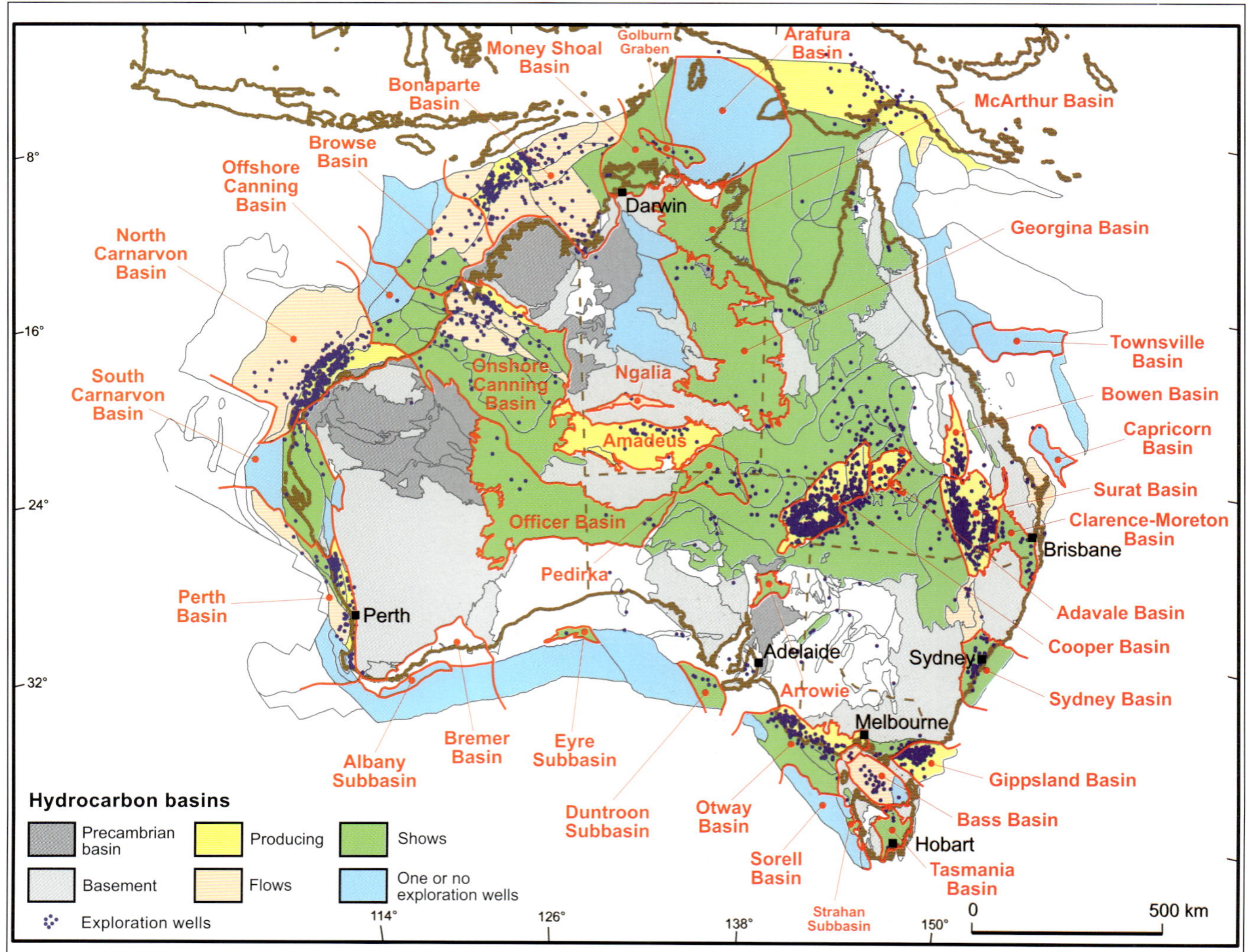

Figure 2. Australian hydrocarbon basins (producing, flows, shows).

and gas in the late 1960s in the Gippsland Basin was followed by large discoveries of predominantly gas on the North West Shelf in the 1970s and into the 1990s. These data also demonstrate that the large reserve additions in the last decade have predominantly been large gas (and some oil) discoveries on the North West Shelf.

The above data strongly indicate that the massive undeveloped gas reserves in the North West Shelf basins will dominate the exploration effort in Australia in the early part of the twenty-first century. The main undeveloped fields of this prolific gas province are described in more detail below.

PETROLEUM SUPERSYSTEMS

A broad classification of petroleum systems has been established in Australia, linking individual systems that share the same age and facies of source rock into a petroleum supersystem (Bradshaw, 1993; Bradshaw et al., 1994). This has been a useful concept in exploration and has been verified by geochemical analysis of the oils, which cluster into families mirroring the supersystem framework (GeoMark Research Inc. and Australian Geological Survey Organisation, 1996; Edwards et al., 1997, 1999). Petroleum supersystems and oil families occur regionally across basins (Figure 7) and from Australia into

Table 1: Summary of exploration drilling and discovered reserves in Australia by basin.

Basin	Age	Petroleum supersystem	Number of exploration wells	Historical hydrocarbon discovery success rate	Number of hydrocarbon discoveries	Oil status	Gas status
McArthur	Mesoproterozoic		6		0	Shows	
Amadeus	Neoproterozoic-Devonian	Certralian Larapintine	29	17%	5	Production	Production
Arrowie	Cambrian	Larapintine	1		0	Shows	
Arafura	Cambrian-Triassic	Larapintine	12		0	Recovery	
Ngaha	Cambrian-Carboniferous	Larapintine	2	50%	1		Flow
Adavale	Devonian-Carboniferous	Larapintine	11	9%	1	Production	Production
Onshore Canning	Ordovician-Cretaceous	Larapintine	164	6%	10	Production	Flow
Officer	Neoproterozoic-Permian	Larapintine	12		0	Shows	Shows
Georgina	Neoproterozoic-Ordovician	Larapintine	6	17%	1	Shows	Flow
Bonaparte	Cambrian-Holocene	Larapintine, Gondwanan, Westralian	249	16%	41	Production	Undeveloped giant fields
Perth	Silurian-Holocene	Gondwanan Austral	129	12%	16	Production	Production
Tasmania	Permian	Gondwanan	11		0	Past oil-shale production	Shows
Sydney	Permian-Triassic	Gondwanan	39	3%	1	Past oil-shale production	Coal-bed methane
Cooper/Eromanga	Carboniferous-Cretaceous	Gondwanan, Murta	933	36%	336	Production	Production
Bowen/Surat	Permian-Cretaceous	Gondwanan	815	19%	151	Production	Production
Clarence Morteon	Triassic-Cretaceous	Murta	14		0	Recovery	Shows
Pedirka	Jurassic-Cretaceous	Murto	16		0	Shows	
Gippsland	Cretacous-Holocene	Austral	314	15%	48	Production	Production
Bass	Cretacous-Holocene	Austral	28	7%	2	Flow	Flow
Otway	Jurassic-Holocene	Austral	156	16%	25	Flow	Production
Duntroon	Jurassic-Holocene	Austral	3		0	Recovery	
Eyre	Jurassic-Holocene	Austral	1		0	Show	
Carnorvon	Siluriw-Holocene	Westralian	397	23%	92	Production	Production
Offshore Canning	Devonian-Holocene	Westralian	13		0		Shows
Browse	Permian-Holocene	Westralian	58	12%	7	Flows	Undeveloped giant fields
Total			**9119**	**22%**	**797**		

other parts of the Indo-Australian Plate in Indonesia and Papua New Guinea (Bradshaw et al., 1997).

Hydrocarbon occurrence in Australia can be summarized and understood in terms of a few petroleum supersystems (Table 2). The four key Phanerozoic supersystems that provide almost all the discovered hydrocarbon resources are (1) Larapintine, characterized by lower Paleozoic marine facies deposited in tropical epicontinental seaways; (2) Gondwanan, dominated by terrestrial sediments and influenced by the Permian glaciation; (3) Westralian, the product of the Mesozoic development of Australia's northwestern margin, characterized by marine Jurassic source rocks; and (4) Austral, developed during breakup of the southern margin and having terrestrial to deltaic source rocks (Figure 7).

Of lesser importance are the Murta (Cretaceous interior sag basins with minor indigenous oil but which provide for Permian-sourced hydrocarbons) and the Capricorn (Upper Cretaceous–Cenozoic of northeast Australia, including Eocene lacustrine oil shales) supersystems. There are also Proterozoic sequences which are petroliferous in the Centralian, Urapungan, and McArthur petroleum supersystems (Table 2).

Early in the twenty-first century, the Westralian supersystem will be the dominant source of hydrocarbons—liquefied natural gas (LNG) and oil for export,

Discovered volumes November 1999				Developed volumes January 1995				Undiscovered estimates	
Oil (million bbl)	Condensate (million bbl)	LPG (million bbl)	Gas (tcf)	Oil (million bbl)	Condensate (million bbl)	LPG (million bbl)	Gas (tcf)	Million bbl oil (P_{50})	tcf gas (P_{50})
								Not assessed	Not assessed
19	6.6	4.1	0.8	19	6.4	3.3	0.5	30	0.13
								Not assessed	Not assessed
								4	0.01
								2	0.006
0	0	0	0.02	0	0	0	0.02	Not assessed	0.2
3.8	0	0	0	3.8	0	0	0	19	
								Not assessed	Not assessed
								1	
528.0	170.5	116.3	7.9	1716	0	0	0.07	520	2.4
4.2	1.4	0	0.7	4.2	1.4	0	0.7	140	0.4
								Not assessed	Not assessed
								0.1	0.004
267.5	134	173.2	8.4	266.9	105.1	153.5	6.4	68	0.8
32.7	7.8	9.9	0.7	32.7	7.7	9.8	0.6	6	0 25
								0.4	0.03
								1.5	0.03
3974.6	262.4	662.3	12.1	3797.2	238.4	654.8	9.4	180	0.7
15.4	34.7	51.2	0.3	0	0	0	0	10	0.16
0	2.5	0	0.52	0	0.5	0	0.08	100	1.8
								16	0.06
								Not assessed	Not assessed
1563.9	1075.7	828.6	70.9	1030.3	707.3	5919	24.4	585	21
								66	0.6
13.0	424.9	742	33.7	0	0	0	0	160	4
6422.	**2120.0**	**2067.6**	**196.0**	**0920.7**	**1066.6**	**1419.9**	**42.2**	**1909**	**92.6**

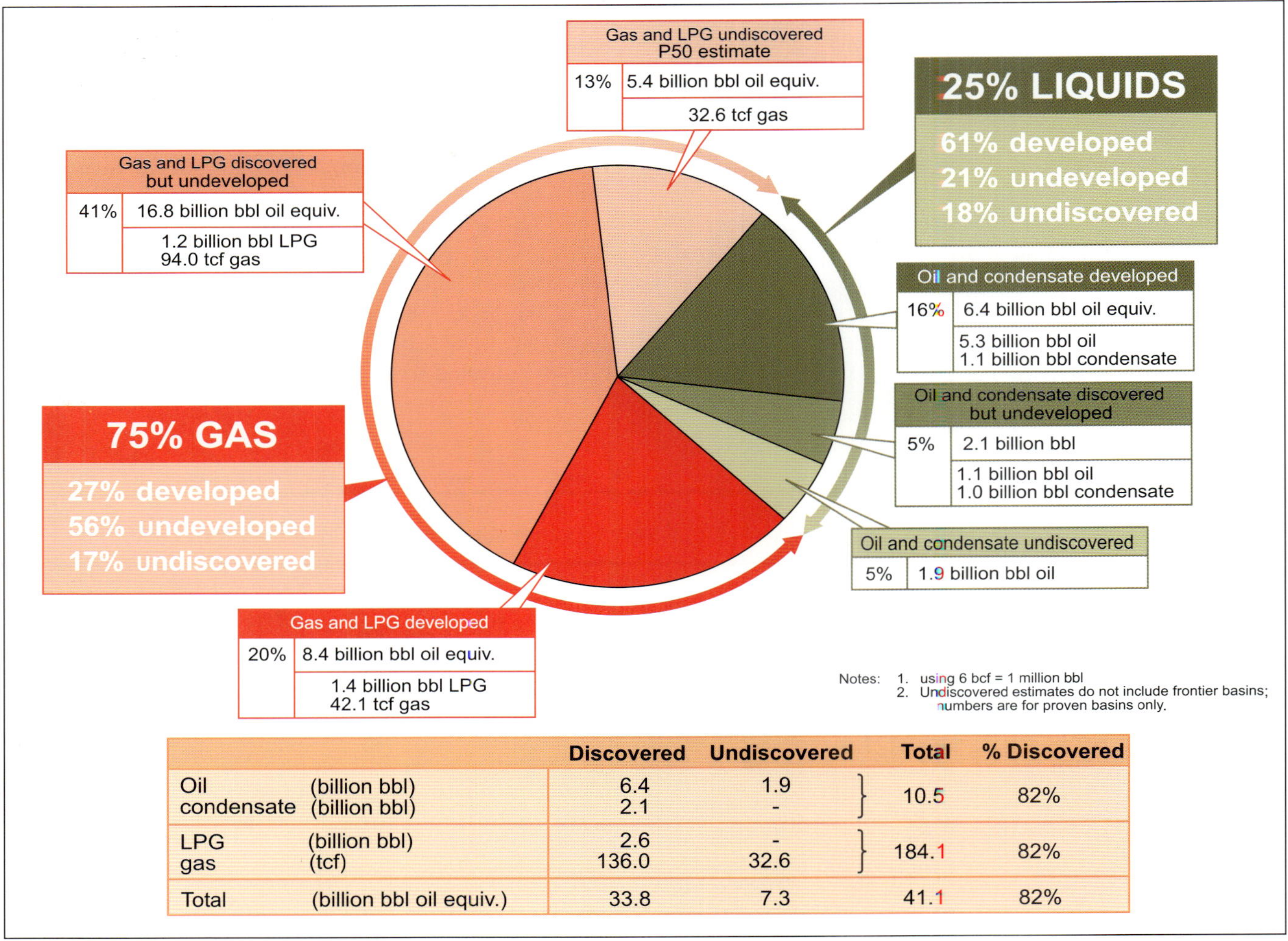

		Discovered	Undiscovered	Total	% Discovered
Oil	(billion bbl)	6.4	1.9	10.5	82%
condensate	(billion bbl)	2.1	-		
LPG	(billion bbl)	2.6	-	184.1	82%
gas	(tcf)	136.0	32.6		
Total	(billion bbl oil equiv.)	33.8	7.3	41.1	82%

Figure 3. Australian hydrocarbon resource summary by oil equivalent.

and oil and gas for domestic consumption. New finds along the southern margin and the Lord Howe Rise may cause the Austral star to rise again to replace the depleting Gippsland Basin. Gas from Gondwana and Larapintine supersystems from onshore basins will be an important component of the domestic energy mix. The contribution to the new century's energy supply from the Capricorn supersystem is unpredictable because of the compounding uncertainties of undrilled geology and environmental issues related to the Great Barrier Reef.

THE NORTH WEST SHELF OF AUSTRALIA

Introduction

The North West Shelf of Australia (Figure 8) is a Mesozoic intracratonic basin which developed into an Atlantic-style rift/drift margin after the Oxfordian (Purcell and Purcell, 1988, 1994, 1998). The margin forms one contiguous sedimentary basin (Westralian Superbasin; Yeates et al., 1987; Bradshaw et al., 1988) approximately 2400 km long by 400 km wide. It has been divided arbitrarily into different segments, namely the (Northern) Carnarvon, Offshore Canning, Browse, and Bonaparte Basins. These basins contain an estimated 2.1 billion bbl of discovered oil reserves, 1.7 billion bbl of condensate reserves, 1.7 billion bbl of LPG reserves, and 113 tcf of gas reserves (Table 1). These data indicate that the discovered resources, by oil equivalent, are 84% gas (and LPG).

Undeveloped Reserves and Twenty-first-century Activity

Much of the undeveloped resources along the North West Shelf are large gas fields (about 88 tcf gas reserves) which are the focus of exploration and development activity in the early twenty-first century. The largest nine undeveloped fields are described briefly because cumulatively they contain most likely gas reserves of 51 tcf, and they will form the nuclei for further exploration once the infrastructure has been established.

Gorgon field (Figure 9), in the Northern Carnarvon

Basin, was discovered in 1971 in water depths of 220– 270 m (Clegg et al., 1992; Sibley et al., 1999). The field comprises multipool prerift Triassic sandstone gas reservoirs in a large, 70-km-long Upper Jurassic rift-related horst block, sealed laterally and vertically by post-Triassic shales. The field has been appraised by seven wells and is covered by modern 3-D seismic data.

The trapping configuration of the deepest pools comprises faulted anticlinal traps with intra-Triassic top seal, whereas shallower pools subcrop the main (rift-onset) Oxfordian unconformity. Some sandstones also have a stratigraphic element of entrapment (Sibley et al., 1999). The field contains approximately 26–31 tcf of gas in place and a most likely reserve estimate of 9.6 tcf (an industry, not an operator, estimate) and potentially could underpin a major new LNG development (see www.gorgon.com.au). Alternatively, the field could supply gas through the existing North West Shelf Development facilities under Australian LNG (ALNG, a marketing organization formed by most of the major companies with major gas reserves on the North West Shelf; see www.australialng. com.au), because the developed Goodwyn and North Rankin fields, which supply the onshore ALNG plant, are only about 120 km to the northeast (Figure 8).

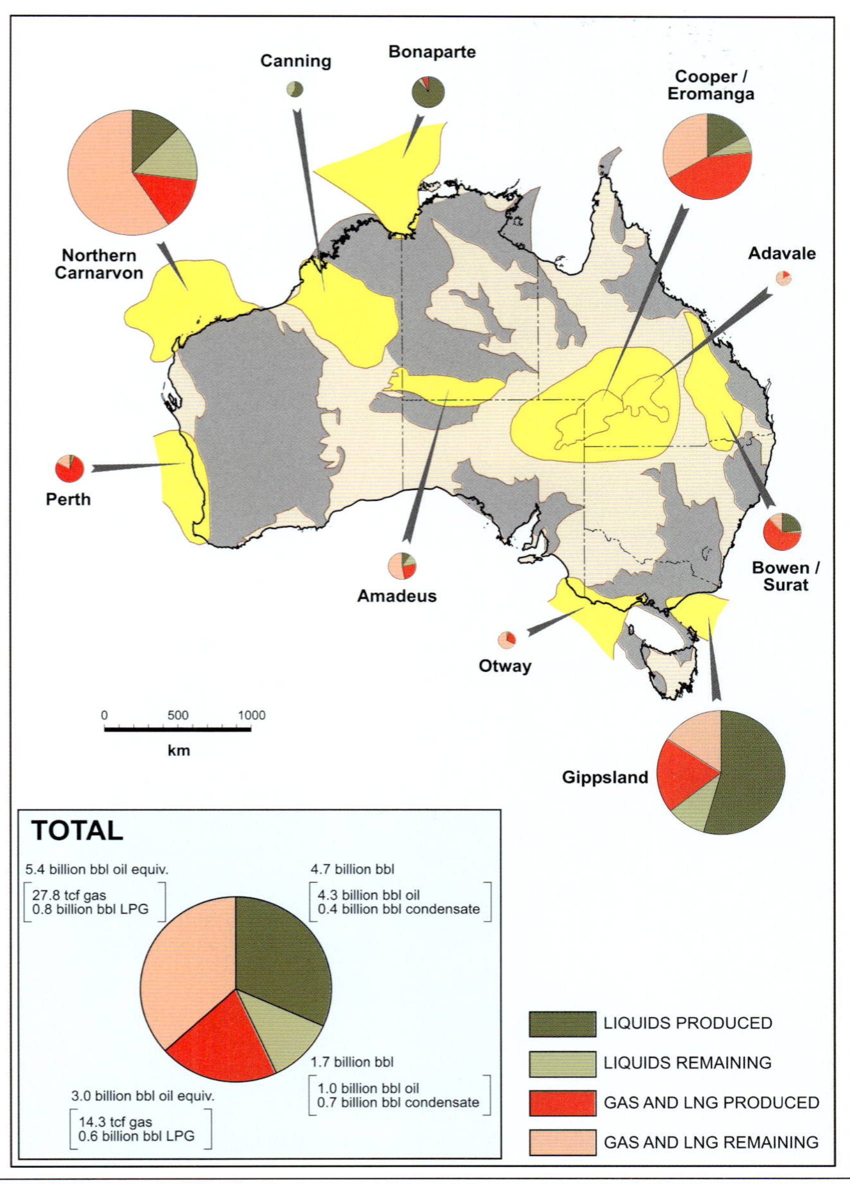

Figure 4. Summary of twentieth-century Australian developed reserves.

Scott Reef field (Figure 10), in the Browse Basin (Figure 8), was discovered in 1971 (Bint, 1998). It comprises a single-pool field, with gas reservoired within synrift Middle Jurassic sandstones sealed by postrift shales in a 55-km-long Upper Jurassic rift-related tilted horst block. The field has been appraised by two wells and is covered by a grid of 2-D seismic. The field, which contains approximately 23–40 tcf of gas in place and an estimated 13–23 tcf of gas reserves, is the largest gas field discovered on the North West Shelf. The field underlies an isolated reef atoll, which rises from relatively deep water approximately 260 km from the nearest landfall. There is an extension to the field in open water north of the atoll (water depths of 300–500 m).

Sunrise and Troubadour fields (Figure 11), approximately 400 km offshore in the Bonaparte Basin (Figure 8), comprise two separate culminations in adjacent complexly faulted anticlinal structures formed by Neogene tectonism. Sunrise field lies in 150–500 m of water; Troubadour field is in approximately 100 m of water. The reservoirs in both fields are Middle Jurassic synrift sandstones sealed by postrift shales. Although both the Sunrise-1 and Troubadour-1 discovery wells were drilled in 1974, Sunrise field is considerably larger and has been appraised by an additional four wells. Sunrise field contains approximately 9–13 tcf of gas reserves and 316–459 million bbl of condensate reserves. Plans for development are focused primarily on establishment of a new export LNG plant.

Brecknock field (Figure 12), in the Browse Basin, was discovered in 1979. It is approximately 10–20 km south of Scott Reef field in deeper water (550 m), away from the bathymetric Scott Reef atoll. The field is analogous to but smaller than Scott Reef field, comprising a single

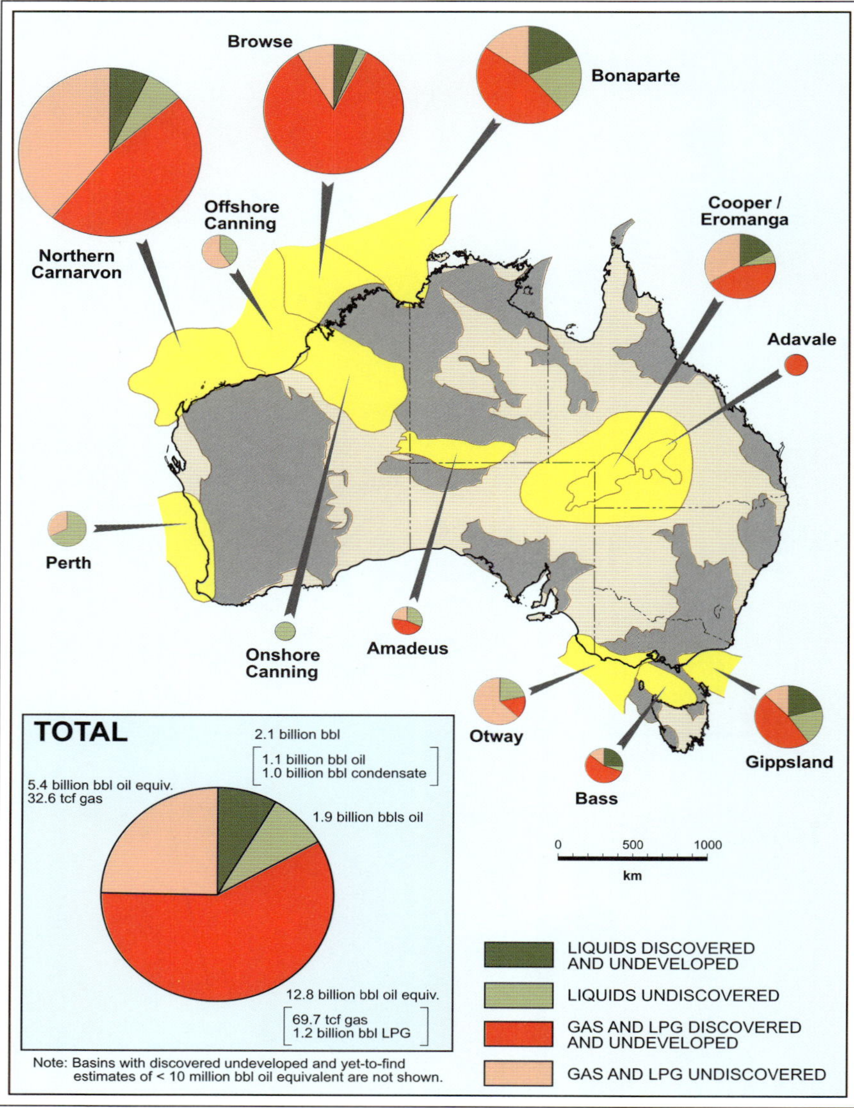

Figure 5. Summary of twenty-first-century Australian undeveloped and undiscovered reserves.

pool with gas reservoired in synrift Middle Jurassic sandstones sealed by postrift shales. The structure is a 20-km-long Upper Jurassic rift-related tilted horst block. The field is covered by 3-D seismic but has yet to be appraised by drilling. Reserves are estimated to be 7–11 tcf of gas and 123-190 million bbl of condensate. Conceptual development plans for Brecknock and the adjacent Scott Reef field comprise a new LNG project, a pipeline 850+ km long to the existing LNG plant, or a floating LNG development.

Scarborough field (Figure 13), discovered in 1979, is in 920 m of water on the Exmouth Plateau in the Northern Carnarvon Basin (Figure 8). The field, estimated to contain 4–5 tcf of gas reserves, is located approximately 180 km outboard of Gorgon gas field. Gas is reservoired in postrift Lower Cretaceous basin-floor sandstones in a complexly faulted anticlinal inversion structure formed by Late Cretaceous tectonism (Bradshaw et al., 1998).

Bayu-Undan field (Figure 14), in 70–110 m of water in the Bonaparte Basin (Figure 8), was discovered by the Bayu-1 well in 1984 (Brooks et al., 1996). The field is a faulted anticlinal structure formed by Jurassic rifting but highly modified by Neogene tectonism. Gas is reservoired in Middle Jurassic synrift sandstones. The field contains approximately 3.4–4.1 tcf of gas reserves. More important, it has a relatively large wet-gas phase, with approximately 400+ million bbl of liquid (condensate plus LPG) reserves. These liquid reserves will be developed initially as a gas-recycling development, with first production planned for late 2003 (see www.phillips66.com/bayuundan).

Chrysaor and Dionysis fields (Figure 15) are in the Northern Carnarvon Basin (Figure 8) in water depths of 810 and 1090 m, respectively (Sibley et al., 1999). Discovered in 1996 and 1997, they are located immediately north of Gorgon Field and south of the recent multi-tcf Geryon-1 gas discovery, which is in 1232 m of water. All three of these structures are covered by 3-D seismic data and contain gas in the prerift Triassic sandstones in Upper Jurassic rift-related, tilted fault blocks analogous to the larger adjacent Gorgon structure. Chrysaor and Dionysis fields combined contain an estimated 4–4.5 tcf of gas reserves that probably will be developed in conjunction with those of Gorgon field.

Undiscovered Reserves and Twenty-first-century Activity

The undiscovered reserve estimates for the North West Shelf total 1.3 billion bbl of oil and 28 tcf of gas (Table 1). Given the large undeveloped gas reserves, most of the focus in the early twenty-first century is on discovery of oil reserves and associated liquids in gas fields.

The relatively high density of exploration drilling along the inboard area of the North West Shelf implies that little probability exists for new province-scale oil or gas discoveries in those areas. Exploration effort there is concentrated on applying 3-D seismic to find smaller (predominantly oil) pools in proven play fairways and testing higher-risk trap types or deeper plays. The current limit of 3-D seismic coverage, which basically defines

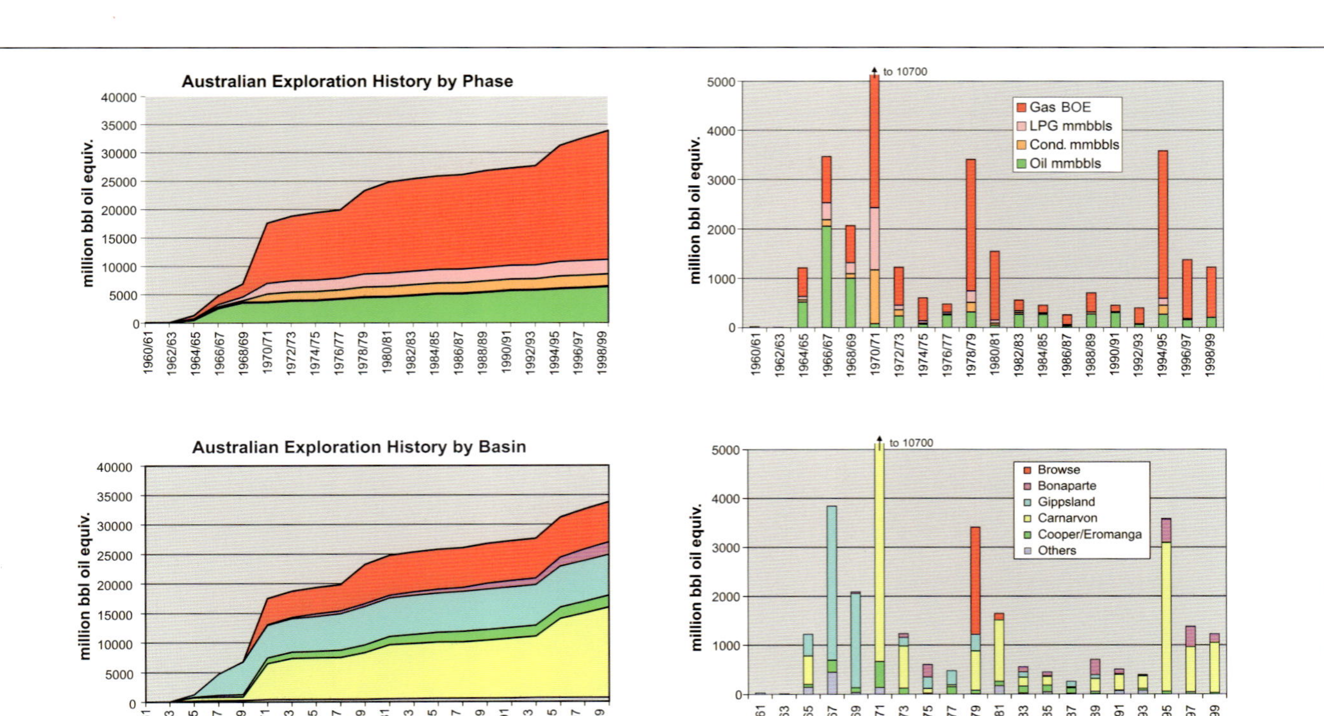

Figure 6. Cumulative and annual Australian discovery-history data by basin and phase.

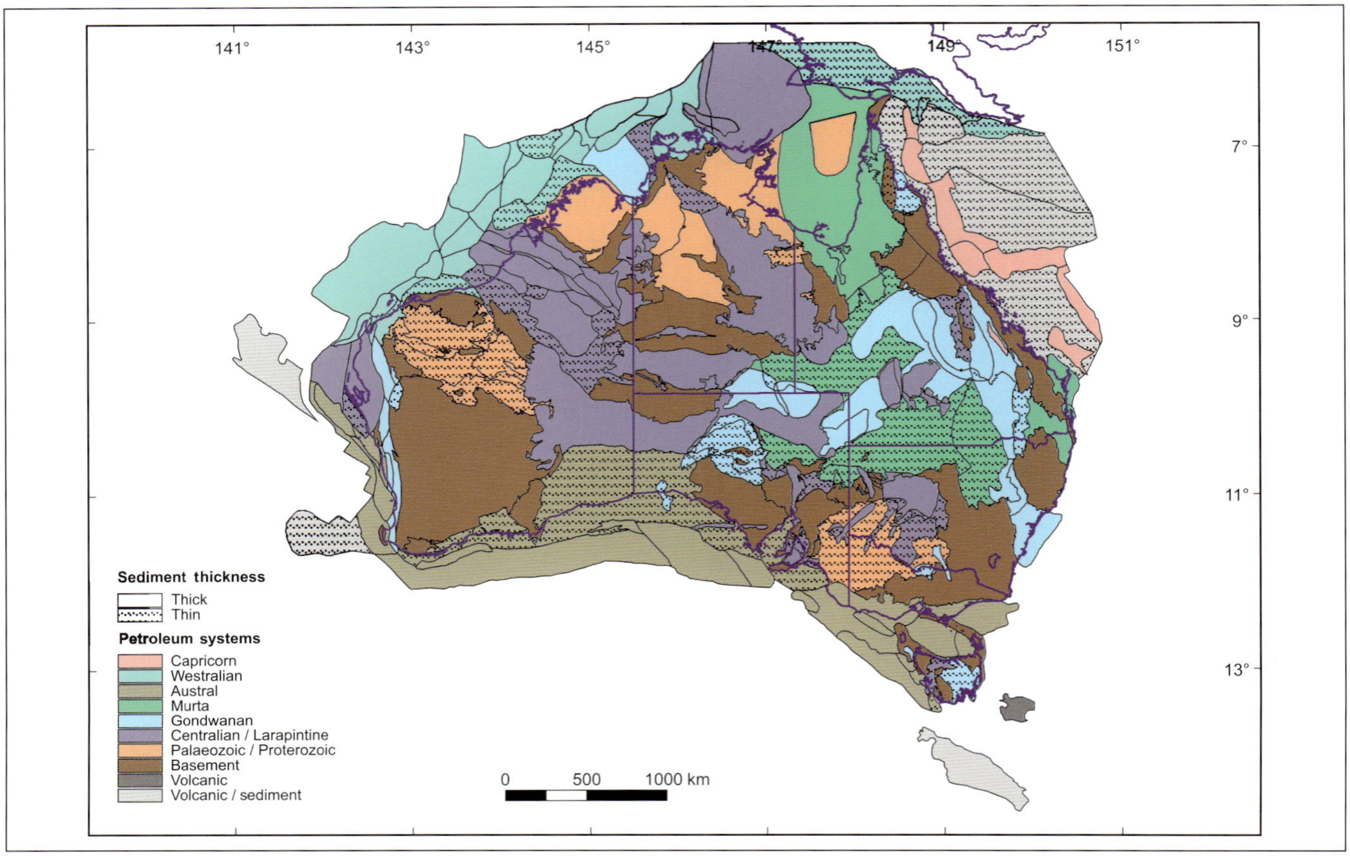

Figure 7. Australian petroleum supersystems.

Table 2: Summary of Australian petroleum supersystems.

Era	Supersystem	Subunit	Age	Source facies
Proterozoic	**McArthur** Evaporitic rifts		Mesoproterozoic 1700–1500 Ma	Lacustrine dolomite shales
	Urapungan Marine shelf and slope, foreland basin		Mesoproterozoic c. 1400 Ma	Marine shales
	Centralian Centralian Superbasin— intracratonic basin within Rodinia Marine, evaporitic, glacial	1 2 3	Neoproterozoic c. 750 Ma Neoproterozoic c. 650 Ma Neoproterozoic c. 600 Ma	Carbonates, evaporites Shales Postglacial Marine shales Postglacial Marine shales
Paleozoic	**Larapintine** Lower Palaeozoic Tropical climate Carbonates, evaporites, marine clastics	1 2 3 4	Cambrian Ordovician Middle–Late Devonian Early Carboniferous	Marine calc. shale Marine Marine carbonate Marine anoxic shale
	Gondwanan Late Carboniferous–Early Triassic Glaciation, clastics Higher plant contribution to source rocks	1 2 3	Early Permian Late Permian Earliest Triassic	Nonmarine Marine Nonmarine Marine Deltaic Marine
Mesozoic	**Westralian** Traissic-Cenozoic Breakup of northern and western margin Marine rift environments	1 2 3 4 SAHUL	Late Triassic–Early/Middle Jurassic Late Jurassic Early Cretaceous Mesozoic	Deltaic Marine, anoxic (?) Marine Marine carbonate
	Austral Late Jurassic–Cenozoic Breakup of southern and southwestern margins Terrestrial rift environments	1 2 3	Late Jurassic–Early Cretaceous Early Cretaceous Late Cretaceous	Fluviolacustrine shale Fluvial-coaly Fluviodeltaic
	Murta Cretaceous interior sag, fluviolacustrine to marine	1 2 3	Late Jurassic Neocomian Late Albian	Fluviolacustrine Lacustrine/marginal marin Anoxic marine oil shale
Cenozoic	**Capricorn** Late Cretaceous–Cenozoic rifts, northeast Australia, tropical Breakup of Coral Sea		Eocene	Lacustrine oil shales

these play areas, is shown in Figure 8. One exception for this play area is the Petrel Subbasin of the Bonaparte Basin, where a proven underexplored Paleozoic oil-charge system exists beyond the limits of current oil discoveries (Vear and DeRuig, 2000).

The underexplored deep-water areas of the North West shelf offer the best potential for significant province discoveries because only 10 wells are in water depths greater than 1000 m and all of those are in the Exmouth Plateau area of the Carnarvon Basin (Figure 8). Many large structural traps have been identified in those underexplored areas. The cumulative total unrisked volumetric estimate for these features exceeds 400 tcf. With a historical success rate of 12–23% for discovering hydrocarbons along the margin (Table 1), the yet-to-find estimates of 1.3 billion bbl of oil and 28 tcf of gas are realistic or perhaps even slightly pessimistic. The primary geologic risks in these frontier areas is a combination of reservoir presence, source presence, source effectiveness, and seal effectiveness. Recent multi-tcf discoveries of gas in the Geryon-1 and Orthrus-1 wells attest to the untapped potential in these outboard areas. These recent discoveries are play extensions to the proven gas play along the western flank of the Barrow-Dampier Subbasin. However, the underexplored deep-water area is huge, and entire untested basins are present in the Northern Beagle Subbasin of the Carnarvon Basin, Offshore Canning Basin, and Browse Basin areas (Figure 2).

THE SOUTHERN MARGIN

Australia's southern margin is the conjugate margin with Antarctica and the site of a major Mesozoic rift-valley system, which is approximately 4000 km long and contains a dozen basins that are part of the Austral petro-

Formation names	Basin	Key discoveries	Analog
Barney Creek	McArthur, Mount Isa, eastern Arafura (?)	Oil and gas shows	None
Velkerri and McMinn	McArthur, Mount Isa, Victoria River	Oil flows BMR Urapunga 4, Jamieson 1	None
Bitter Springs, Albinia	Amadeus, Ngalia, Georgina	Gas flow Magee 1	Siberia
Browne	Officer, Birrindudu, Savory	Oil and gas shows	Oman
Aralka	Amadeus, Georgina	Gas flows Ooraminna 1	
Rinkabeena	Ngalia	Davis 1	
Pertatataka, Rodda	Amadeus, Officer, Ngalia Savory, Victoria River	Dingo gas field	
Goulburn Group, Tempe	Arafura, Amadeus	Arafura 1 oil shows	Williston
Horn Valley, Goldwyer	Amadeus, Canning	Mereenie oil field	
Gogo, Ningbing	Canning, Bonaparte	Blina oil field	
Anderson, Milligans	Canning, Bonaparte	Sundown oil field	Bolivia
Irwin River, Patchawarra	Perth, Cooper, Bowen	Tirrawarra gas and oil field	
Treachery, Keyling	Bonaparte		
Blackwater Group	Bowen	Rolleston gas field	
Wagina	Perth		
Hyland Bay	Bonaparte	Petrel gas field	
Kockatea	Perth	Dongara oil and gas field	
Mungamo	Carnarvon	Rankin Trend giant gas fields	
Dingo, Flamingo	Carnarvon, Bonaparte	Barrow Island oil field	North Sea
Echuca Shoals	Bonaparte, Browse	Undan/Bayu, Cornea (?)	
	Bonaparte, Timor, Seram		
Casterton, Pretty Hill	Otway,	Katnook,	
Parmelia	Perth, Carnarvon (?)	Gage Roads	
Eumeralla	Otway	Windermere, Minerva	
Latrobe Group	Gippsland, Bass	Kingfish	
Poolawanna, Birkhead	Eromanga, Surat	Oil shows, unproven source (?)	
Murta	Eromanga	Dullingari oil field	
Toolebuc	Eromanga	Oil shows, immature source	
		Stewart oil-shale demonstration plant	Southeast Asia

leum supersystem (Figure 7). At the eastern end of the rift are the giant oil and gas fields of the Gippsland Basin and the more modest accumulations of the Otway and Bass Basins, but the western two-thirds of the rift system remains largely unexplored. Only 12 exploration wells have been drilled between Kangaroo Island and Cape Leeuwin (Duntroon Basin westward to the southwest tip of Australia, in Figure 2), an area that straddles more than 3000 km of coastline. In this region are basins covering hundreds of square kilometers, with as much as 12 km of sediments, evidence of active petroleum systems, and seismic coverage demonstrating the potential for multiple, large drilling targets. The most prospective sections are in water depths of more than 500 m and therefore have remained undrilled, but recent advances in technology have removed this barrier.

All Mesozoic basins along the southern margin share a history of continental breakup and passive-margin formation, thought to have begun in the west in the Jurassic and to have spread progressively eastward through the Cretaceous and into the Tertiary. At the eastern end of the rift system (Otway and Sorell Basins), final separation from Antarctica was delayed until the middle Tertiary and had a significant transcurrent component (Moore et al., 2000).

Unlike the breakup history on the northwestern margin, where continental slivers progressively spalled off over hundreds of million of years, the southern-margin rift system was intracontinental. This contrasting history has imparted a different character to the petroleum systems. Terrestrial source facies have been dominant in the currently explored parts of the southern margin, and this is one of the prime distinguishing characteristics of the Austral petroleum supersystem (Table 2). In the supersystem, a threefold subdivison of recovered hydrocarbons and their associated source intervals has been recognized by Edwards et al. (1999). These are (1) Austral 1—Upper Jurassic–Lower Cretaceous (Barremian) lacustrine and fluviolacustrine facies, (2) Austral 2—Lower Cretaceous (Aptian-Albian) coal-swamp to marginal-marine facies, and (3) Austral 3—Upper Cretaceous–Eocene fluvial to deltaic facies. The younger source intervals are best developed in the eastern end of the southern-margin rift system. The petroleum accumulations in the Gippsland and Bass Basins are from Austral 3 sequences, whereas most reservoired hydrocarbons in the Otway Basin are from coaly Eumeralla Austral 2 sources (Edwards et al., 1999). The older lacustrine sources (Austral 1) and entirely new marine facies can be expected to be viable source units in the frontier basins of the western and central southern margin. They are described in more detail below.

Bight Basin

The Bight Basin, in the central Great Australian Bight, contains distinct depocenters (Figures 16, 17). Earlier studies by Bien and Taylor (1981) and by Stagg et al. (1990) have been augmented by Totterdell et al. (2000), whose analyses are based on a grid of new and reprocessed seismic data combined with the limited well data. The geologic history of the region, as described by Totterdell et al. (2000), began in the Late Jurassic with deposition of fluvial and lacustrine siliciclastic sediments in half-graben systems. These synrift sediments contain potential source and reservoir units and are overlain by widespread, relatively flat-lying Berriasian to Albian sequences deposited in fluviolacustrine, deltaic, and marine environments. Late Albian marine shales provide the ductile layer in which growth faults that characterize the Cenomanian strata sole out (Figure 18). Massive shelf-margin delta complexes (as much as 5 km thick) prograded during the late Santonian to Maestrichtian in response to uplift of the eastern Australian highlands. The postrift section has been penetrated only in the most proximal positions where it is incomplete and thin, but even those positions contain examples of potential source rocks. The overlying Tertiary section is composed of transgressive siliciclastics overlain by carbonate shelf deposits.

Despite only two wells having been drilled in the Bight Basin, evidence exists of several active petroleum systems. Study of oil inclusions in Upper Jurassic sandstones intersected in Jerboa 1 indicates the presence of a paleo-oil column at least 15 m thick (Ruble et al., 1999). Edwards et al. (1999) assigned the Jerboa 1 occurrence to a fluviolacustrine source facies, whereas oil recovered from Greenley 1 in the adjacent Duntroon Basin is related to an anoxic marine source (Smith and Donaldson, 1995). Other indications of active hydrocarbon generation and migration include evidence of seepage detected from synthetic aperture radar, gas chimneys, and other possible direct hydrocarbon indicators in seismic records.

A variety of play types has been identified from seismic coverage in the Bight Basin. These include basement-involved structural traps (anticlines over basement blocks and reactivated basement faults, rotated fault blocks, and associated rollovers) and traps associated with ductile shale unit (growth faults, rollovers, and toe thrusts), as shown in Figures 18 and 19. Potential also exists for large stratigraphic traps on the basin margin where updip onlap or truncation is enhanced by compaction drape over basement topography. Other trap configurations occur in prograding wedges and incised valleys and in siliciclastic aprons abutting faults.

Bremer Basin

The Bremer Basin lies at the western end of the southern margin. It includes thin Tertiary deposits onshore and on the continental shelf, as well as Mesozoic depocenters on the slope (Hocking, 1994) such as the Albany Subbasin (Figure 2). The basin has no offshore wells, but regional seismic coverage indicates a sedimentary section at least 10 km thick in water depths of 600 to 1500 m. Stagg et al. (1990) interpreted the bulk of the sediments as being Upper Jurassic to Neocomian with a veneer of Upper Cretaceous and Tertiary sediments of variable thickness.

The Bremer Basin has the potential to contain an active Austral petroleum system (Figures 2, 7). By analogy with the Eyre Subbasin, the deeper part of the section, below the Neocomian (?) unconformity, could have good-quality lacustrine source rocks (Austral 1, Table 2). Above the unconformity, possible source units include restricted marine facies deposited in the narrow seaway between Australia and Antarctica. The known stratigraphy of adjacent basins (Perth, Eyre, Eulca) indicates potential for reservoir sandstones and intraformational shale seals in the Neocomian section, with regional seals in the Aptian and younger Cretaceous section. The major episode of trap formation occurred during wrench-fault movements at the Neocomian (?) unconformity. Trap types listed by Stagg et al. (1990) include wrench anticlines with approximately 400 m of closure (1400 m if fault dependent) in the eastern half of the basin in water depths of 700 m and more, and stratigraphic traps at the angular Neocomian (?) unconformity.

Sufficient maturation for generation has been identified as a risk in the Bremer Basin (Stagg et al. 1990). However, the proven oil column at Jerboa 1 in the Eyre Subbasin (Ruble et al., 1999), which was also dismissed as being thermally immature, now counters this argument. The oil accumulation at Jerboa 1 was breached during an episode of fault reactivation, erosion, and nondeposition in the Late Cretaceous. Some wrench anticlines in the Bremer Basin have been similarly eroded, but these are risks for specific prospects. At the basin

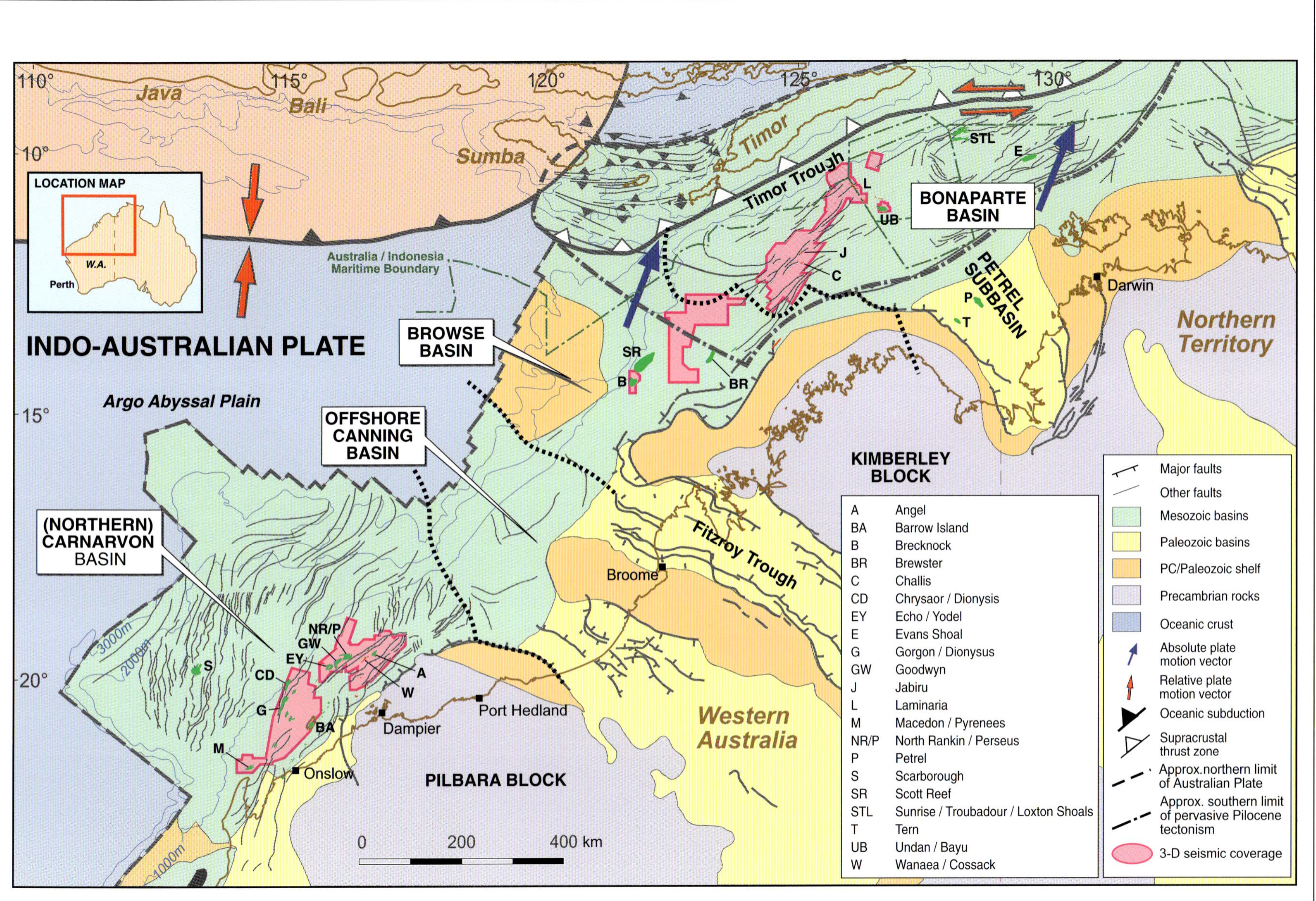

Figure 8. North West Shelf tectonic elements and field locations.

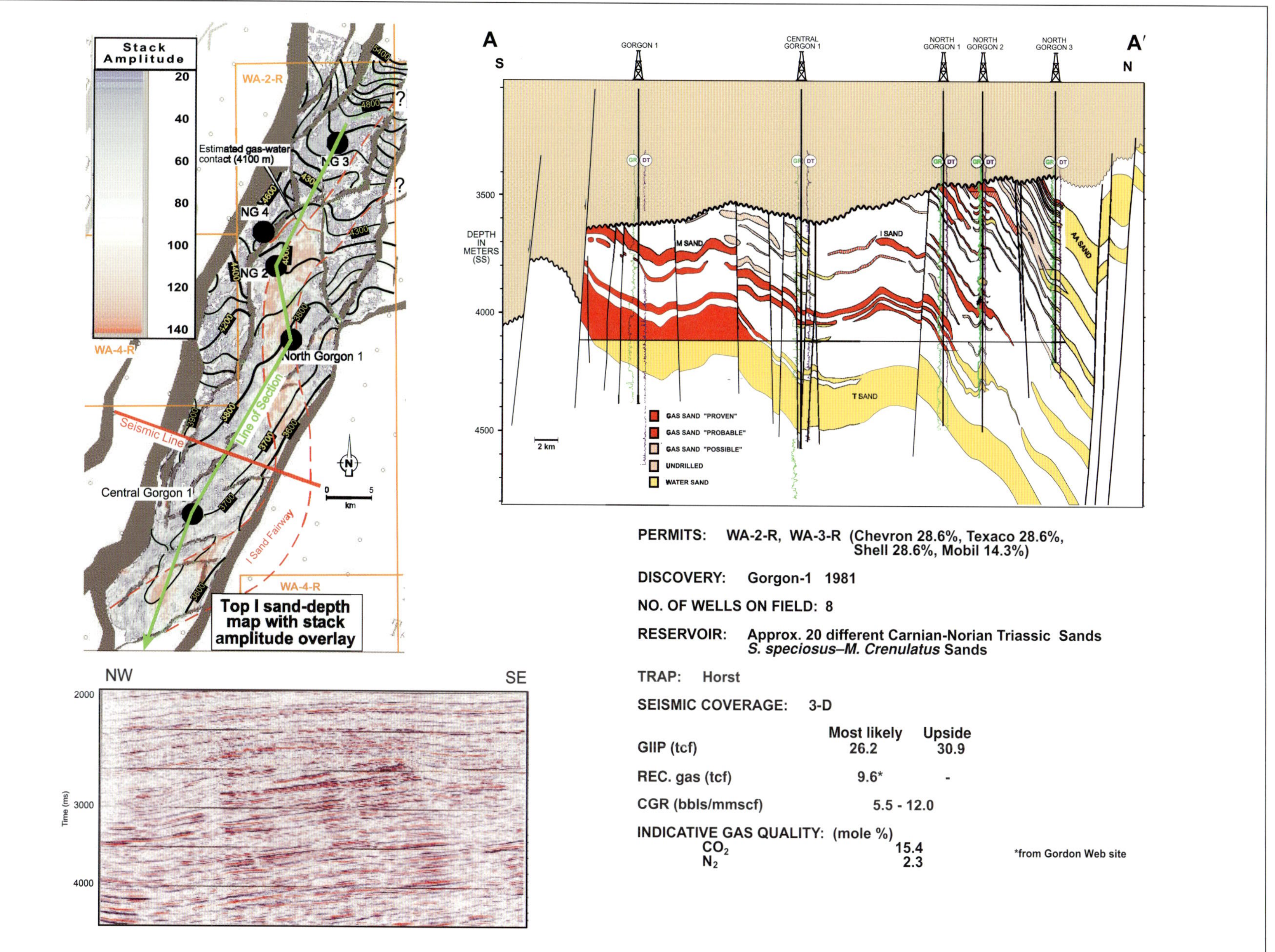

Figure 9. Gorgon field summary.

scale, there is high probability that an active petroleum system exists and that hydrocarbons have been trapped.

Sorell Basin and South Tasman Rise

West and south of Tasmania is another series of frontier basins that has petroleum potential. Along this part of the southern margin, the strike-slip character of the Australia/Antarctica separation has been a dominant influence on basin development. The Sorell Basin, offshore western Tasmania, covers about 90,000 km^2 and has sediment thicknesses of as much as 6.5 km. Seismic records show a string of subbasins trending northwest-southeast between the west Tasmanian coast and the ocean-continent boundary in water depth of more than 4000 m (Figures 2, 7). Individual depocenters are interpreted by Moore et al. (1992) as having formed at the relieving bends of a major left-lateral strike-slip fault system associated with the Australia/Antarctica breakup. Sedimentation may have commenced in the Jurassic, with southernmost depocenters being initiated somewhat later.

The lone well in the basin, Cape Sorell 1, was drilled through a sandy proximal section close to the margin of the Strahan Subbasin. It bottomed in Maestrichtian conglomerates at 3.5 km and encountered free oil near total depth. Other hydrocarbon occurrences in the Sorell Basin include wet gas of thermogenic origin in surface sediments recovered from the shelf break and upper continental slope (Hill et al., 1997). No oil-family analysis has been done, but the geology indicates potential for the Austral 1 and Austral 2 petroleum systems to be present (Edwards et al., 1999), and Austral 3 if organic-rich facies in the Upper Cretaceous to Eocene are sufficiently mature (Table 2). Trap types include early extensional faults blocks and associated drape anticlines, wrench structures developed in the Paleocene to middle Oligocene, and unconformity traps at an upper Paleocene (?) channeled surface with a relief of hundreds of meters (Moore et al., 1992). Maturation modeling indicates that the basal Cretaceous section in some depocenters entered the oil window early in the Late Cretaceous and is now capable of generating gas/condensate if organic-rich facies are present (Moore et al., 1992).

South of the Sorell Basin and about 250 km south of Tasmania is the South Tasman Rise (Figure 20). The rise is a continental fragment surrounded on three sides by Upper Cretaceous and Paleogene oceanic crust. It covers an area of 200,000 km^2 and lies in water depths of 500–4000 m. It is separated from Tasmania by an area of thinned continental crust that moved southward with Antarctic from Tasmania in the Late Cretaceous (Exon et al., 1997). The South Tasman Rise is a complex terrain made up of separate blocks underlain by different acoustic basements and with different plate-tectonic histories. The eastern block moved about 150 km south relative to Tasmania 95–65 Ma. The western block was attached to Antarctica southwest of Tasmania, then moved about 450 km southeast relative to Tasmania, mostly between 95 and 65 Ma (Royer and Rollet, 1997). The western block probably was welded to the eastern terrain in the early Eocene, thus becoming part of Australia.

Deep Sea Drilling Project (DSDP) sites 280 and 281 provide the only stratigraphic control in the region. At site 280, deep-water marine sediments were encountered above oceanic crust southwest of the rise. The well bottomed in basalt. Seismic information, however, indicates sedimentary basins on the South Tasman Rise. The largest is the Ninene Basin, which is on the western block and has a sedimentary section as much as 5 km thick. The basins are fault controlled, and the postulated sedimentary sequence includes Upper Cretaceous to lower Oligocene nonmarine and shallow-marine siliciclastics overlain by upper Oligocene and younger bathyal to pelagic chalk and ooze, 200–500 m thick in basinal areas (Hill et al., 1997).

No oil has been recovered from the South Tasman Rise, although the geology indicates that the Austral 3 petroleum system (Edwards et al., 1999) may be present, but it may not be mature. Upper Eocene mudstones intersected in DSDP site 280 are potential source rocks. Gas samples from cores indicate the presence of thermogenic hydrocarbons derived from deeply buried source rocks. There is seismic evidence of about 500 m of middle Eocene to lower Oligocene progradational marine mudstones that could provide a regional seal to Upper Cretaceous to Paleogene turbidite sandstones. Trap types include flower structures produced in the transpression regime of the strike-slip margin, where shearing continued until the earliest Miocene (23 Ma) on the western block (Hill et al., 1997).

THE EASTERN MARGIN

Australia has another petroliferous rifted margin along its eastern coast. Apart from the world-class Gippsland Basin at its southern extremity, it has been largely ignored by petroleum explorers. There are many reasons for this, including the environmental embargo on the Great Barrier Reef and adjacent areas in the northeast and the narrow continental shelf in the southeast. However, the capacity to realistically explore for and develop new resources in deeper water opens up a new frontier for exploration on the Lord Howe Rise, a large continental fragment lying between Australia, New Zealand, and New Caledonia in the Tasman Sea.

Lord Howe Rise

Lord Howe Rise covers an area of 740,000 km^2 in waters shallower than 3000 m. It is 1600 km long and 250–600 km wide with crestal depths of 750–1200 m below sea level. Widely spaced seismic profiles and 11 DSDP and Ocean Drilling Program (ODP) core sites

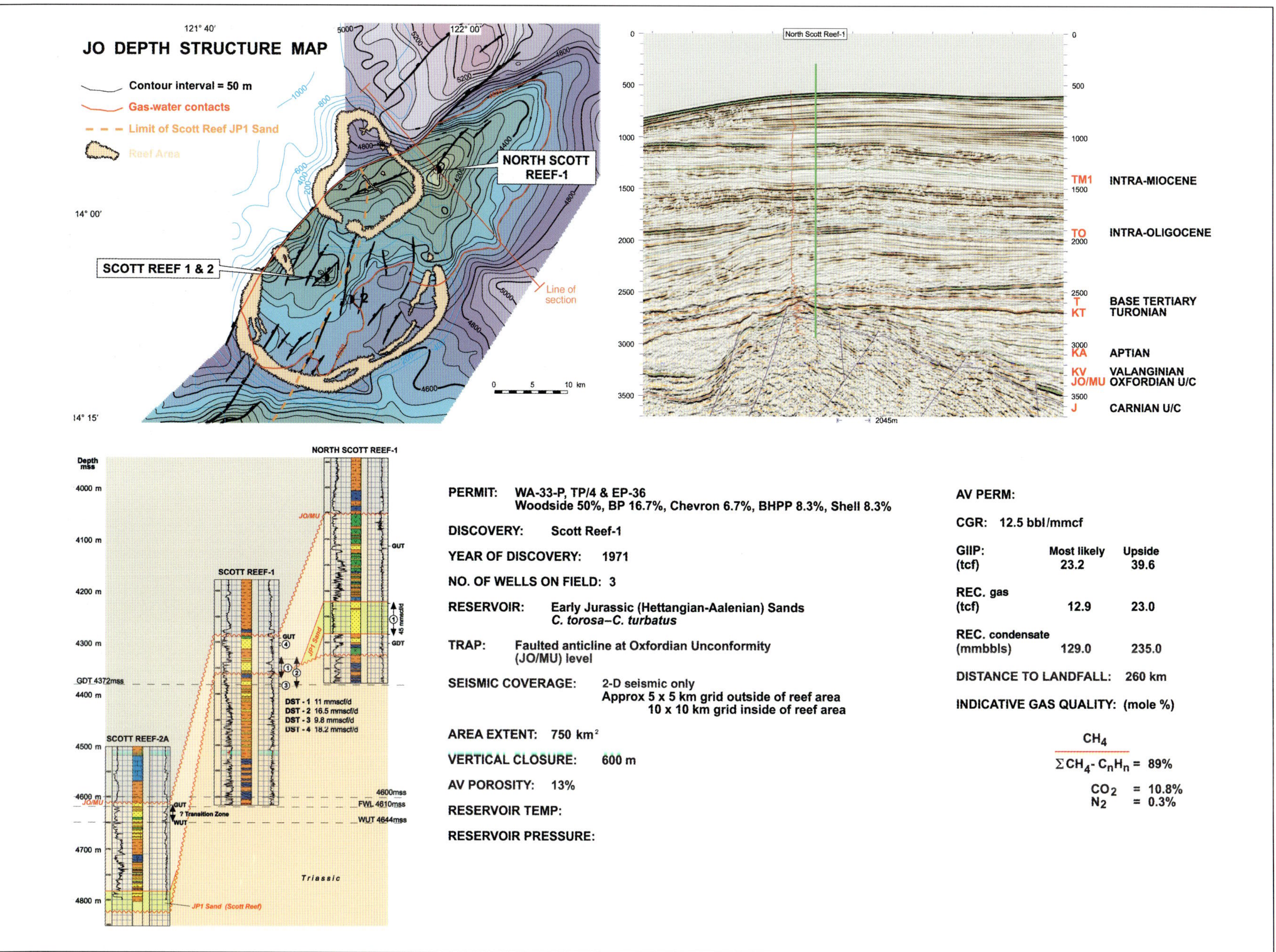

Figure 10. Scott Reef field summary.

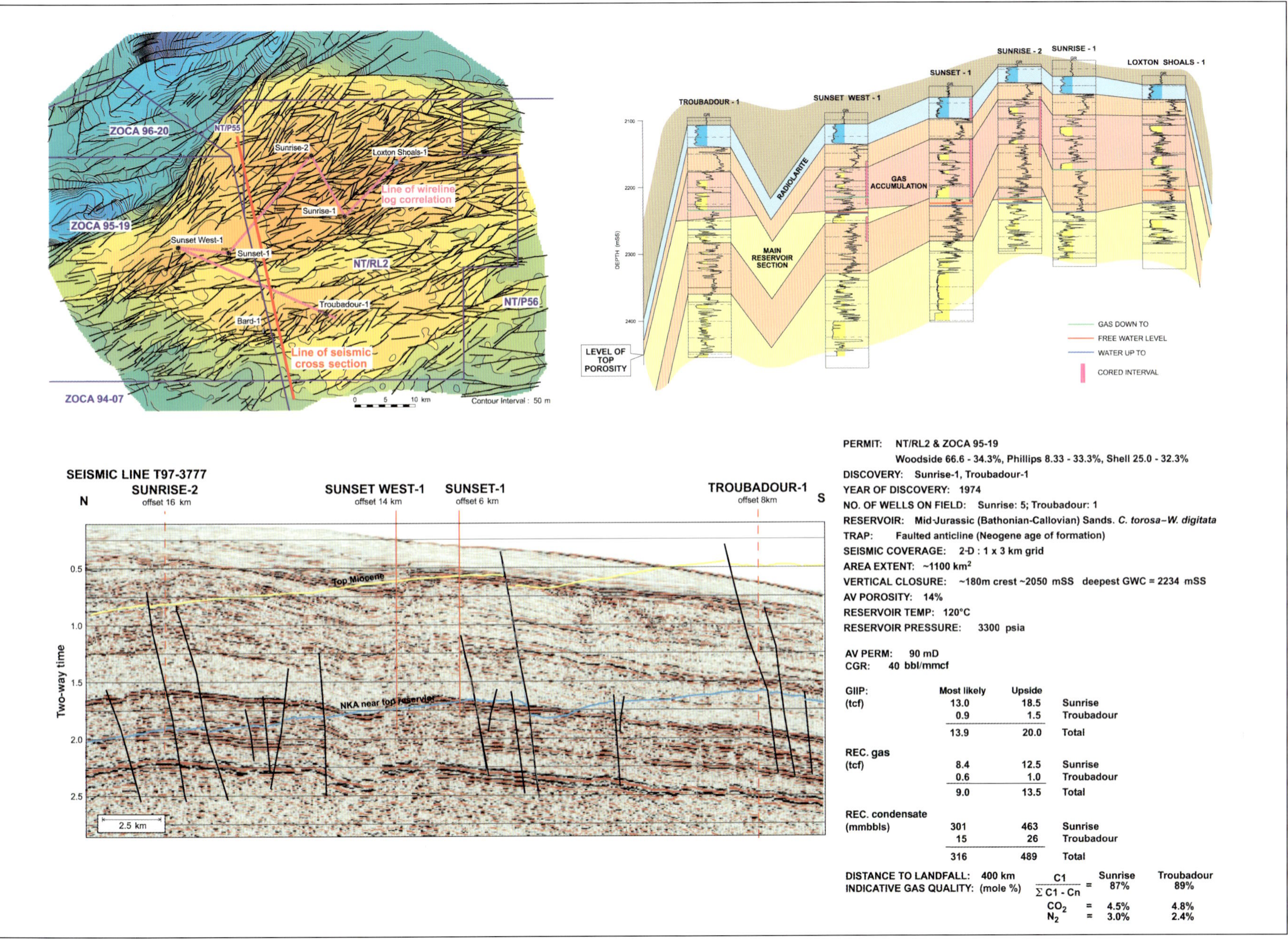

Figure 11. Sunrise and Troubadour fields summary.

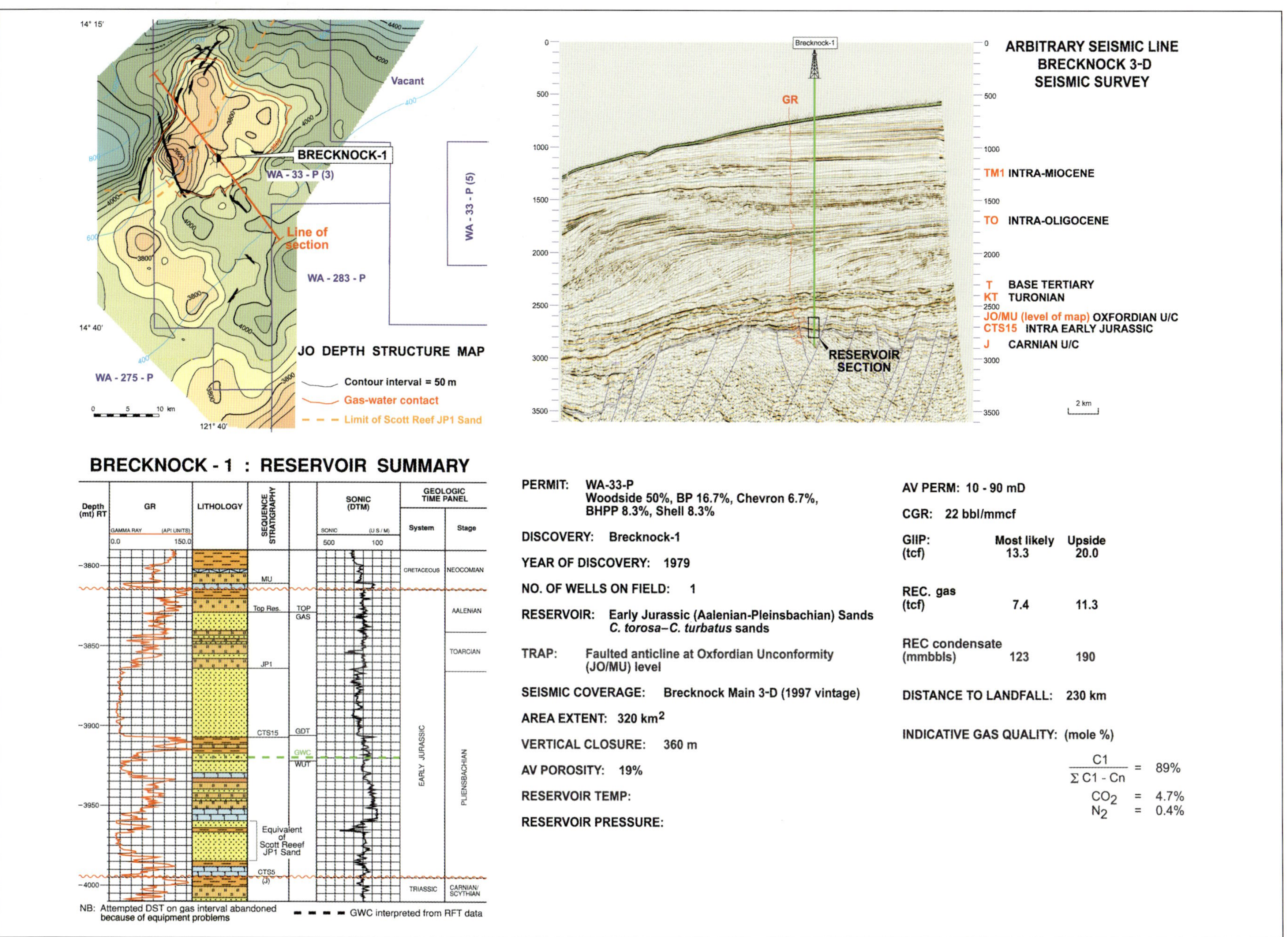

Figure 12. Brecknock field summary.

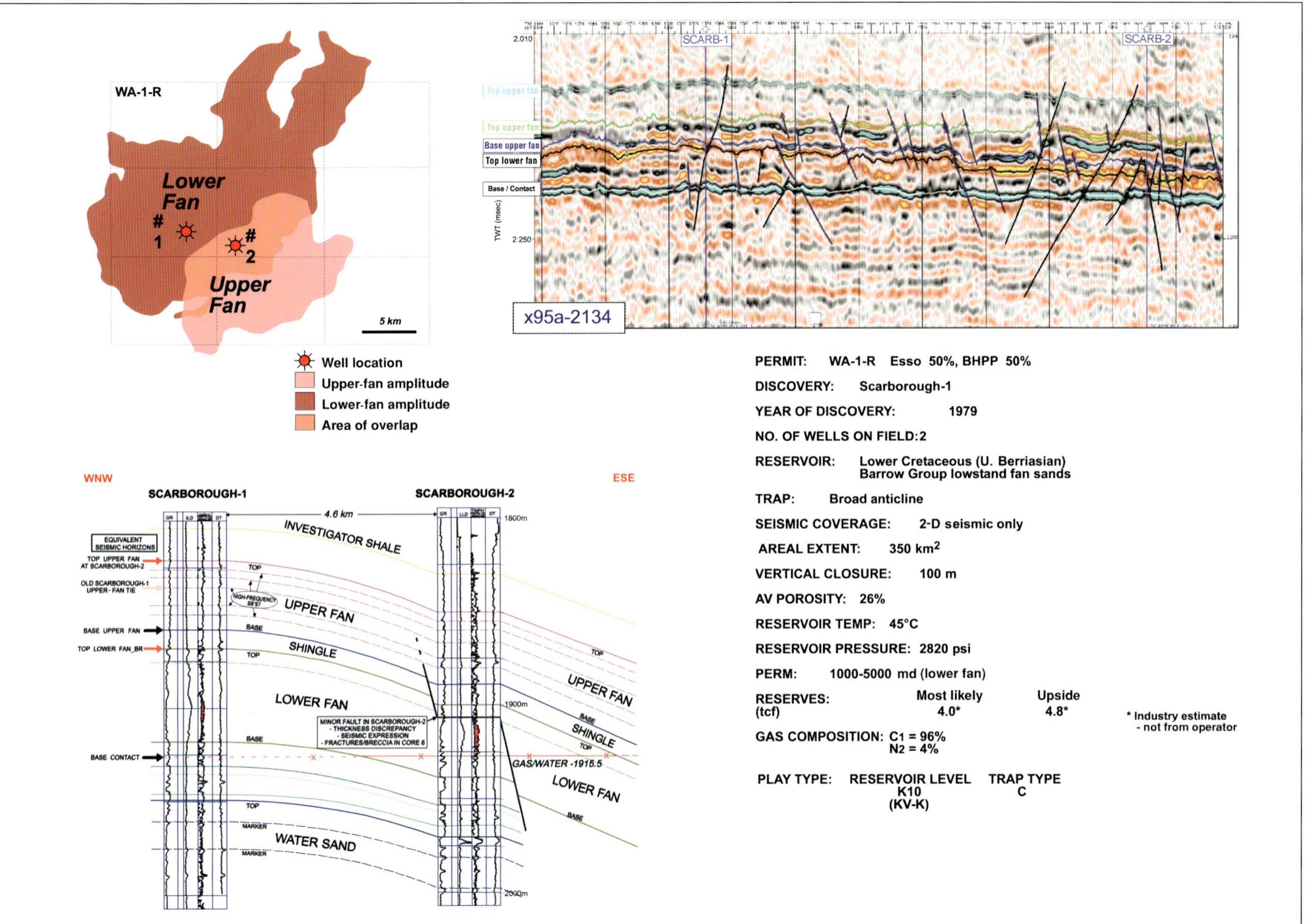

Figure 13. Scarborough field summary.

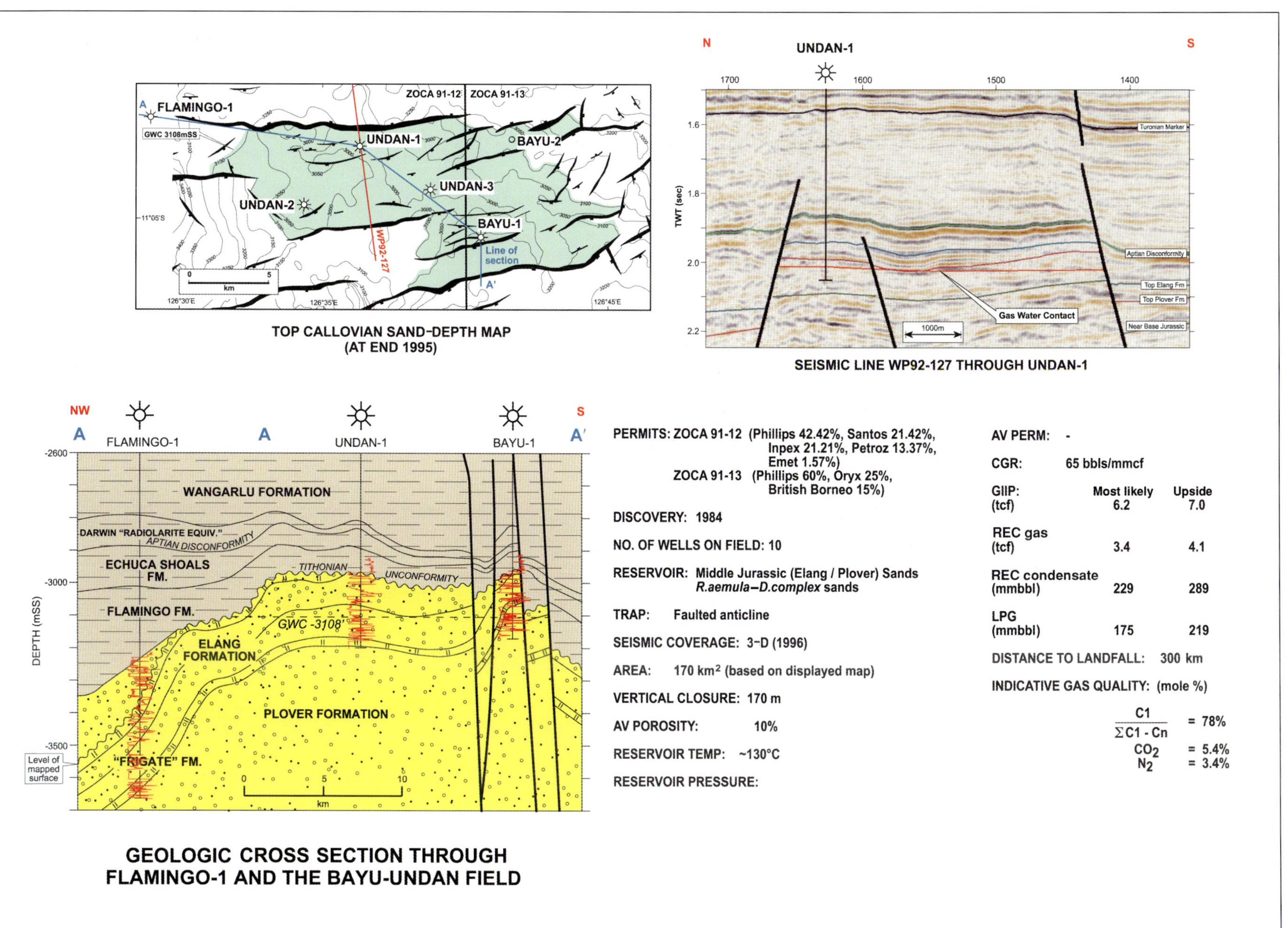

Figure 14. Bayu-Undan field summary.

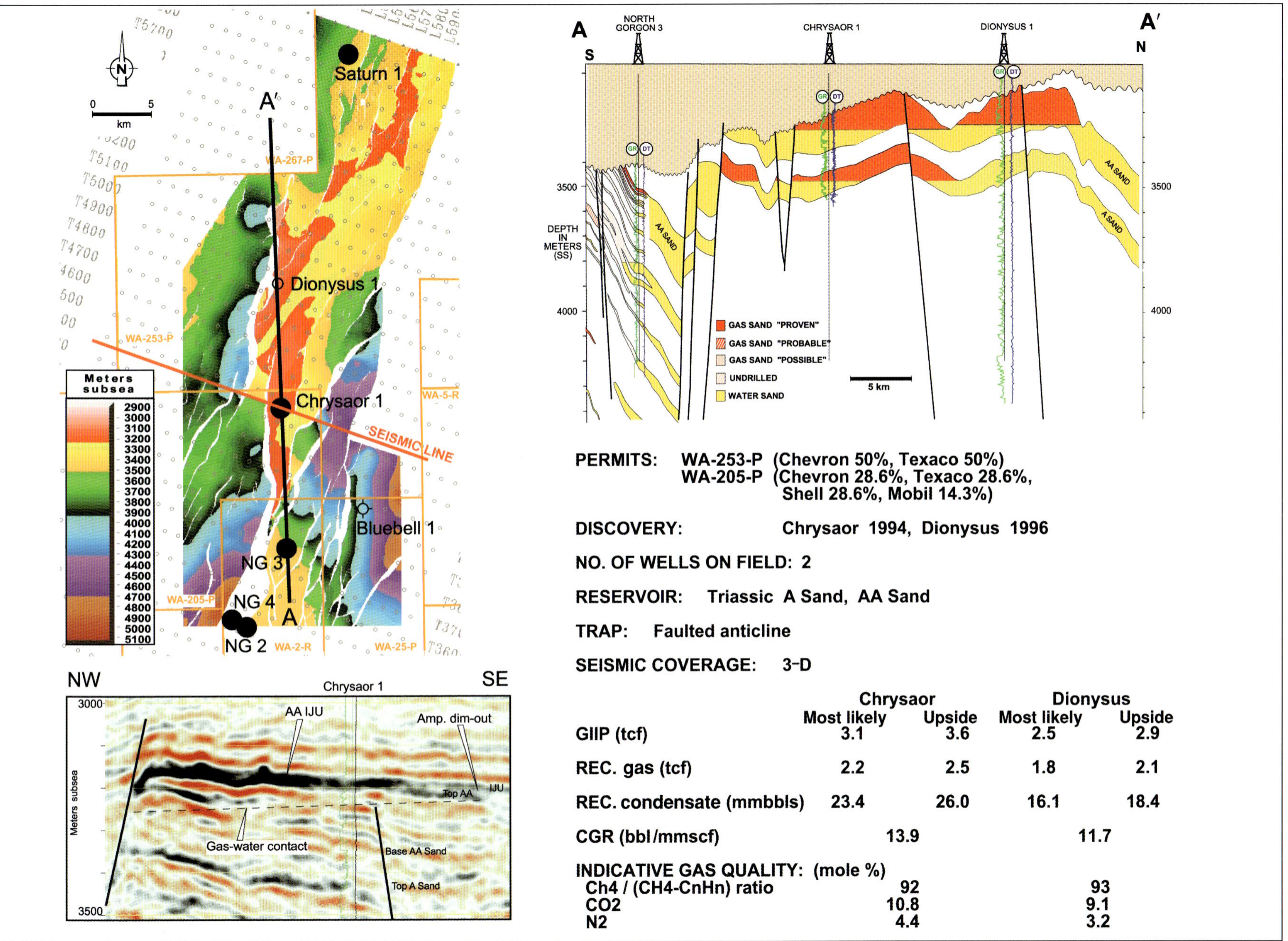

Figure 15. Chrysaor and Dionysis fields summary.

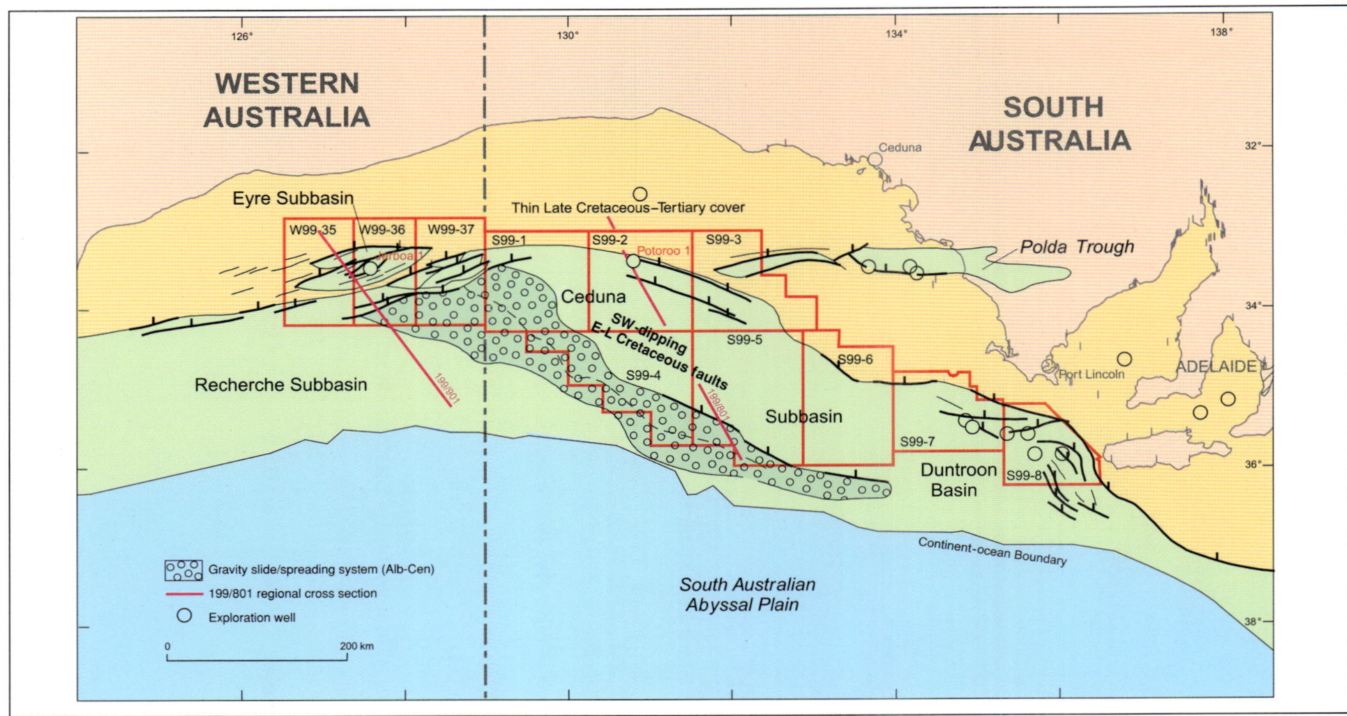

Figure 16. Southern margin regional map.

provide limited stratigraphic information. Little is known about the basement rocks because they have not been penetrated, but they are considered to comprise Paleozoic continental crust associated with the Tasman fold belt of eastern Australia. Upper Cretaceous volcanics and a condensed section of uppermost Cretaceous and Cenozoic marine sediments have been recovered, but the major part of the sedimentary section in depocenters as much as 4000 m thick remains unsampled. Lafoy et al. (1998) and Bernardel et al. (1998), however, interpreted it as Cretaceous to Paleocene (?) synrift siliciclastics with overlying Paleogene chalk and claystones. The synrift section is as much as 1500 m thick. The postrift Paleogene averages approximately 500 m thick and is overlain, above a regional middle Oligocene unconformity, by 300–700 m of Neogene section (Exon et al., 1998).

Lord Howe Rise separated from Australia in the Late Cretaceous with the opening of the Tasman Sea (Shaw, 1978). The southern part of the ridge subsided to bathyal depths in the early Eocene (DSDP site 207; McDougall and van der Lingen, 1974). However, in the north, subsidence occurred as early as the Maestrichtian (Burns, Andrew et al., 1973). A late Eocene compressive phase that produced the New Caledonian ophiolites affected the northeastern areas of Lord Howe Rise (Lafoy et al., 1994; Van de Beuque et al., 1998).

Despite lack of exploration drilling on the Lord Howe Rise, evidence exists that the sedimentary section is prospective. Bottom-simulating reflectors (BSRs) observed on seismic data indicate the presence of gas hydrates of possible thermogenic origin in an extensive area (25,000 km^2) in Australian and French territory (Exon et al. 1998; Auzende et al., 2000). Exon et al. (1998) also documented indications of free gas, including a flat spot in a horst (Figure 21). Plate-tectonic and paleogeographic reconstructions (Walley, 1992) link Lord Howe Rise with the productive Gippsland and Taranaki Basins in the Austral petroleum supersystem (Figure 22). Hydrocarbons recovered from either side of the Tasman Sea show that oil and gas fields of the Gippsland, Bass, and Taranaki Basins are derived from a similar suite of coaly source rocks deposited in fluviolacustrine to deltaic environments. Similar source facies could be expected to occur on Lord Howe Rise, along with their marine equivalents. These may include Paleocene marine shales that are the source of the small Kora discovery in the Taranaki Basin and the widespread oil seeps in the East Coast Basin of the North Island of New Zealand.

From the existing seismic coverage, Stagg et al. (1999) have produced a detailed structural-elements map of the Lord Howe Rise region (Figure 23). Highstanding basement ridges (Dampier, Monawai, Lord Howe, Fairway, Norfolk, and Three Kings) are separated by areas of Mesozoic-Cenozoic rift basins (Capel, Gower, Moore, Monawai, Fairway, and Bellona) and ocean basins (Tasman, Middleton, Lord Howe, New Caledonia, and Norfolk). A large part of the putative gas-hydrate zone, as indicated by the presence of BRS, lies in the Fairway Basin. A speculative estimate of the petroleum resources within Australian jurisdiction on Lord Howe Rise is about 4.5 billion bbl of oil equivalent (Willcox and Symonds, 1997). This estimate does not include the unconventional resource of the gas hydrates, which may be exploitable in the twenty-first century as technology advances. Exon et al. (1998) have suggested that the Lord Howe Rise gas hydrates theoretically could increase Australia's natural-gas reserves by more than an order of magnitude (> 1000 tcf) if gas concentrations are comparable to those found on the Blake Ridge off the southeast coast of the United States.

The sketchy data on which these estimates were made will soon be augmented with supporting information. AGSO, in collaboration with Institute Français de Recherche pour l'Exploitation de la Mer (IFREMER), is undertaking a major program of seabed mapping off the east coast of Australia. The French RV *l'Atalante*, in the Fairway Basin on the eastern flank of Lord Howe Rise (Figure 23), gathered multibeam sonar-swath and high-speed seismic-reflection data and magnetic and gravity profiles (Auzende et al., 2000). Chemical analyses of shallow seabed cores and pore waters from above regions where BSRs are evident have indicated the presence of ethane, propane, and higher hydrocarbons. These results suggest that gases in the southern Fairway Basin have a thermogenic component (Neville Exon, personal communication, 2001; Dickens et al., 2001).

AGSO also initiated a regional search for hydrocarbon seeps, using remote-sensing tools such as synthetic-aperture radar, and a more detailed study of any prospective areas that emerge, using airborne laser fluroesensor (ALF) and shipborne "sniffers" (continuous water-column geochemical tracers). A test of radar anomalies ("slicks") on the ZoNéCo 5 cruise by fluorometer examination of surface waters revealed nothing. However, the ZoNéCo 5 program in the Fairway Basin produced already has startling results—new seismic coverage has imaged diapiric features produced by a mobile unit that may be salt of middle to Late Cretaceous age, based on long-range seismic correlation. The paleogeographic map shown in Figure 22 depicts a narrow, restricted seaway between Lord Howe Rise and Norfolk Ridge at that time. More than 100 diapiric structures have been imaged in water depths of 2000 to 3000 m, some of which have as much as 1000 m of relief and are 30 km long (Auzende et al., 2000). The diapiric structures and the BSR are known mostly from French territory, but present indications are that they could extend well southward into the Australian part of the Fairway Basin. In this context, the anhydrite in the Eocene section of the Capricorn Basin is of some interest, as discussed below.

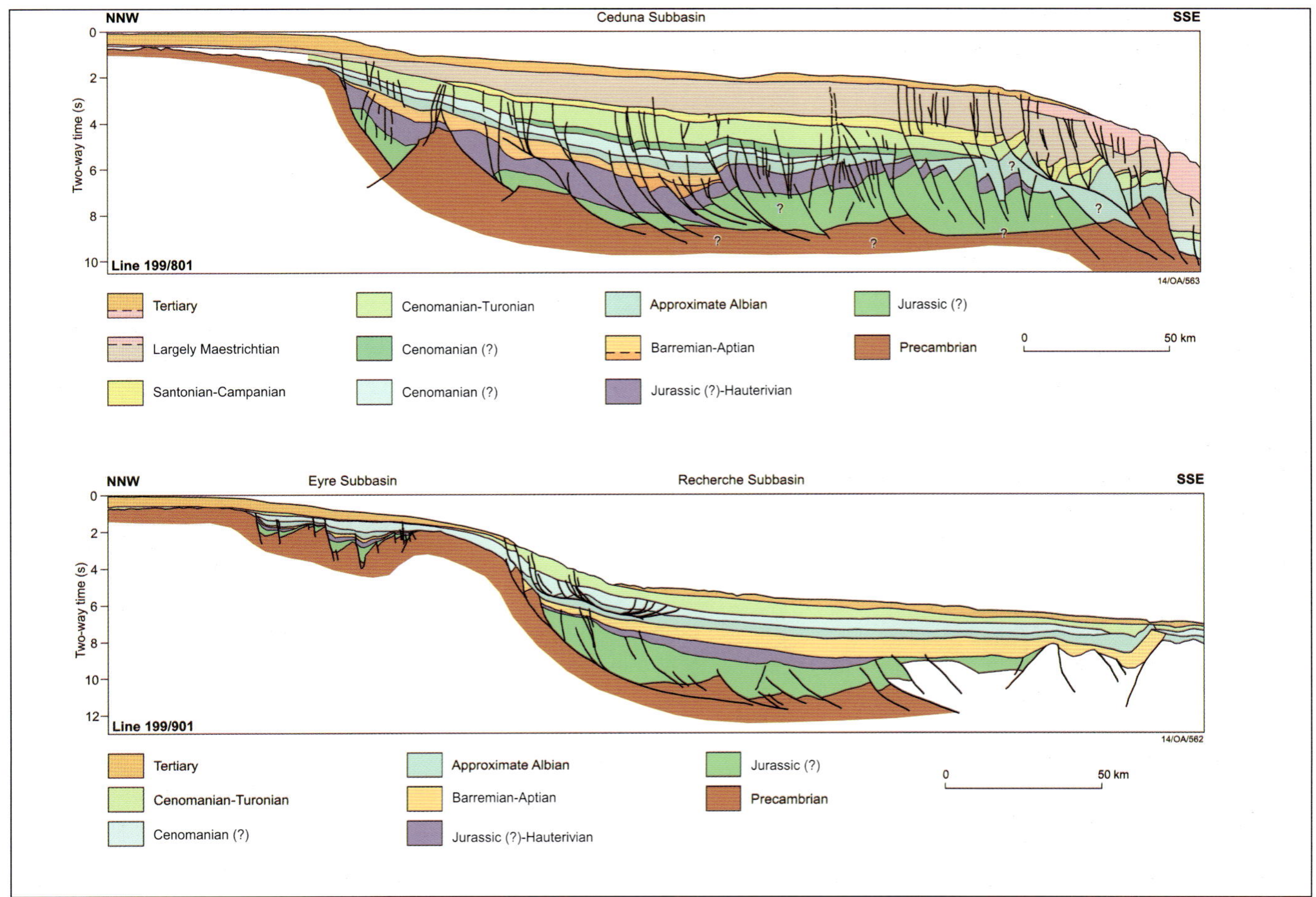

Figure 17. Great Australian Bight regional sections. See Figure 16 for locations of cross sections.

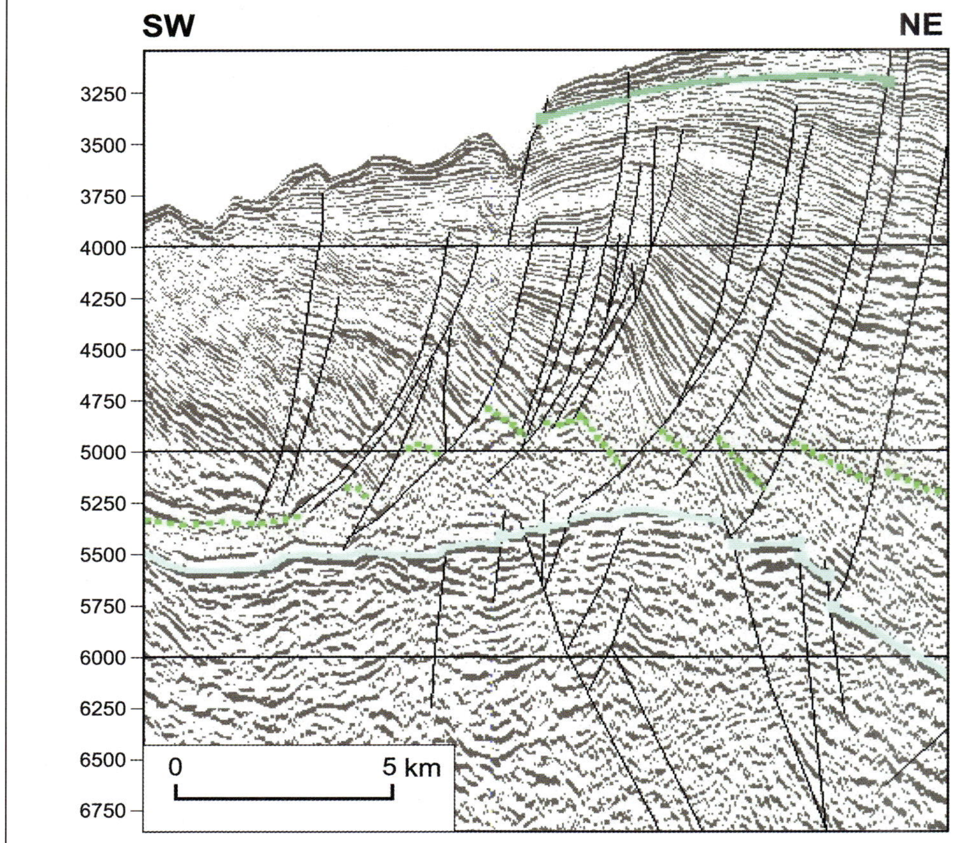

Figure 18. Seismic line for the outer Ceduna Subbasin (DWGAB survey) showing décollement within the Upper Cretaceous (Tiger supersequence = light-blue to green reflectors) overlying a deformed Cenomanian (?) succession (White Pointer supersequence). The section between the green and aqua reflectors is the Santonian to Maestrichtian deltaic Hammerhead supersequence. From Totterdell et al. (2000, their Figure 12).

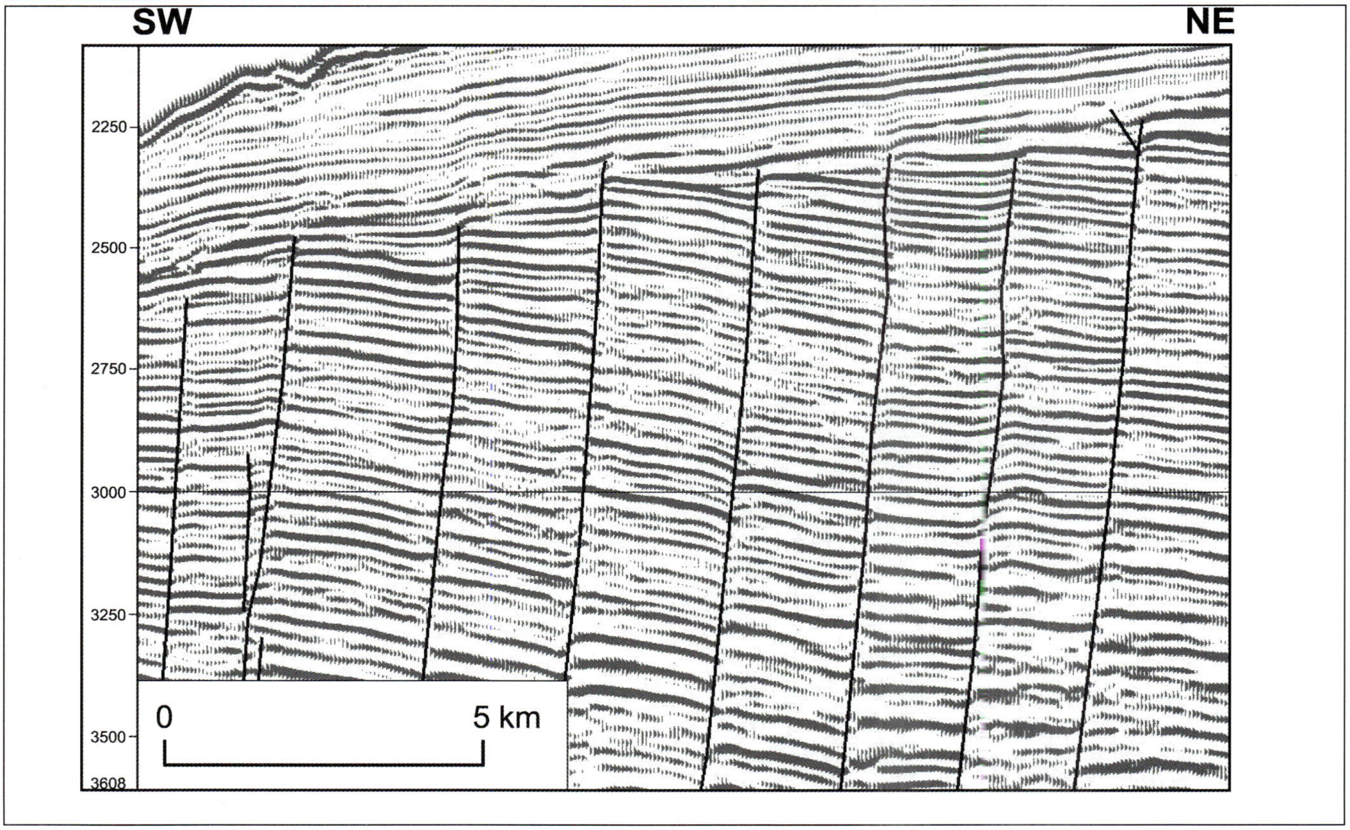

Figure 19. Closely spaced planar faults in the Upper Cretaceous (Hammerhead supersequence), southern Ceduna Subbasin (DWGAB survey). From Totterdell et al. (2000, their Figure 13).

Maryborough, Capricorn, and Townsville Basins

Arrayed along the Queensland coast are Mesozoic to Cenozoic basins that have petroleum potential (Figure 2). The most southerly is the Maryborough Basin, which covers an area of 10,000 km^2 onshore and at least 15,000 km^2 offshore, straddling Fraser Island (Hill, 1994). The entire Maryborough Basin succession is as much as 10 km thick and includes latest Triassic through to Cenozoic sediments, but it is predominantly a Lower Cretaceous depocenter, with 5 km of fine-grained marine siliciclastics and coal measures overlying more than 1 km of lowest Cretaceous intermediate volcanics. Potential source-rock facies include Upper Cretaceous marine shales, and carbonaceous shales in the Jurassic and Cretaceous coal measures. Basin inversion in the middle Cretaceous folded and faulted the thick Cretaceous section, forming major anticlines that have been the target of limited onshore drilling. Only four deep tests, all onshore, have been drilled, resulting in a gas flow (in the Gregory River 1) and gas shows. Seismic coverage interpreted by Hill (1994) indicates further potential offshore, especially southeast of Fraser Island, where the sub-Tertiary section has been folded and faulted.

The offshore basins farther north have thick Upper Cretaceous to Cenozoic sections and the potential to contain younger petroleum source rocks than the Maryborough Basin. A key clue to their prospectivity is the occurrence of Eocene lacustrine oil-shale deposits in narrow onshore basins. These constitute the main known source-rock interval for the Capricorn petroleum supersystem (Table 2). The Stuart oil-shale deposit near Gladstone is in operation, and a shipment of oil shale has been sent to Singapore for refining. However, of more importance for conventional petroleum resources in the twenty-first century is the possibility of similar Eocene oil shales occurring in offshore basins where they may be a viable mature oil source if buried by a sufficient thickness of Neogene carbonates.

Adjacent to the Maryborough Basin, across the basement high of the Bunker Ridge, is the Capricorn Basin. It is a failed continental rift associated with the formation of the Tasman Sea in the Late Cretaceous. The only two deep exploration wells in the region were drilled at the northern end of the Capricorn Basin. However, their results were obviously not encouraging—basement was reached in less than 2700 m after penetrating Upper Cretaceous–Paleocene conglomerates and red beds, Eocene sandstones, lignites and minor anhydrite, and a thick (>1000-m) marine sequence of upper Oligocene to Holocene limestone and marl. A thicker, more marine, and more prospective section is expected in deeper water (1000–3000 m) at the southeastern end of the Capricorn rift, where more than 5 km of Upper Cretaceous–Paleogene section is interpreted in association with half-graben features (Hill, 1994). The occurrence of anhydrite in the Eocene section of the Capricorn Basin lends credence to the interpretation of salt being the mobile unit in the northeast Lord Howe Rise and opens the possibility of marine as well as lacustrine source facies.

To the north of the Maryborough and Capricorn Basins lies a network of basins that partly rings the Queensland plateau (Figure 2). Their formation has been related to oblique extension along preexisting Paleozoic structural trends in the Late Jurassic (?)–Early Cretaceous (Scott, 1993; Struckmeyer et al., 1994) prior to and independent of the later opening of the Tasman and Coral Seas in the Late Cretaceous and Paleogene. The most prospective depocenter is considered to be the Townsville Basin, which underlies the east-west-trending Townsville Trough, a bathymetric low (1000–2000-m water depth) between the Marion and Queensland plateaus. The only drilling in the region has been several ODP holes which have intersected upper Miocene to Holocene sediments, but seismic coverage shows a sedimentary section as much as 6.5 km thick, made up of synrift sequences in half grabens and a thick (3.8-km) blanket of sag-phase sediments (Struckmeyer et al., 1994). North-northwest-trending faults compartmentalize the basin into a series of offset depocenters. Struckmeyer et al. (1994) interpreted three synrift sequences, the oldest of which may be Upper Jurassic to Lower Cretaceous. The remainder of the synrift section is believed to be Cretaceous to possibly Paleocene in age. It is unconformably overlain by as many as five sag-phase sequences ranging in age from Paleocene (?) to Holocene.

Potential exists for mature source rocks of various facies (fluvial, lacustrine, paralic, and marine) to have been deposited during several intervals of synrift and early sag-phase development. The complex structural history of the Townsville Basin, as described by Struckmeyer et al. (1994), has included at least two significant phases of rifting followed by wrenching and late reactivation, and has produced a variety of trap types. Some huge potential stratigraphic traps have been identified (Bureau of Resource Sciences, 1996). They could contain more than 5 billion bbl of oil, but the existence of effective charge and seal is speculative. Part of the basin underlies the Great Barrier Marine Park. It is Australian government policy not to allow petroleum exploration and development where potential exists to have an adverse impact on this World Heritage area.

THE PROMISE OF THE PALEOZOIC

In the new century, Australia's search for petroleum rightly remains focused on the Mesozoic rifted margins of the continent, but the potential of the Paleozoic should not be dismissed lightly. Some of Australia's most organic-rich rocks are Cambrian and Ordovician marine shales deposited in the warm, shallow seaways that crisscrossed the continent during the Larapintine regime (Table 2; Figure 7). Admittedly, the favorable timing of

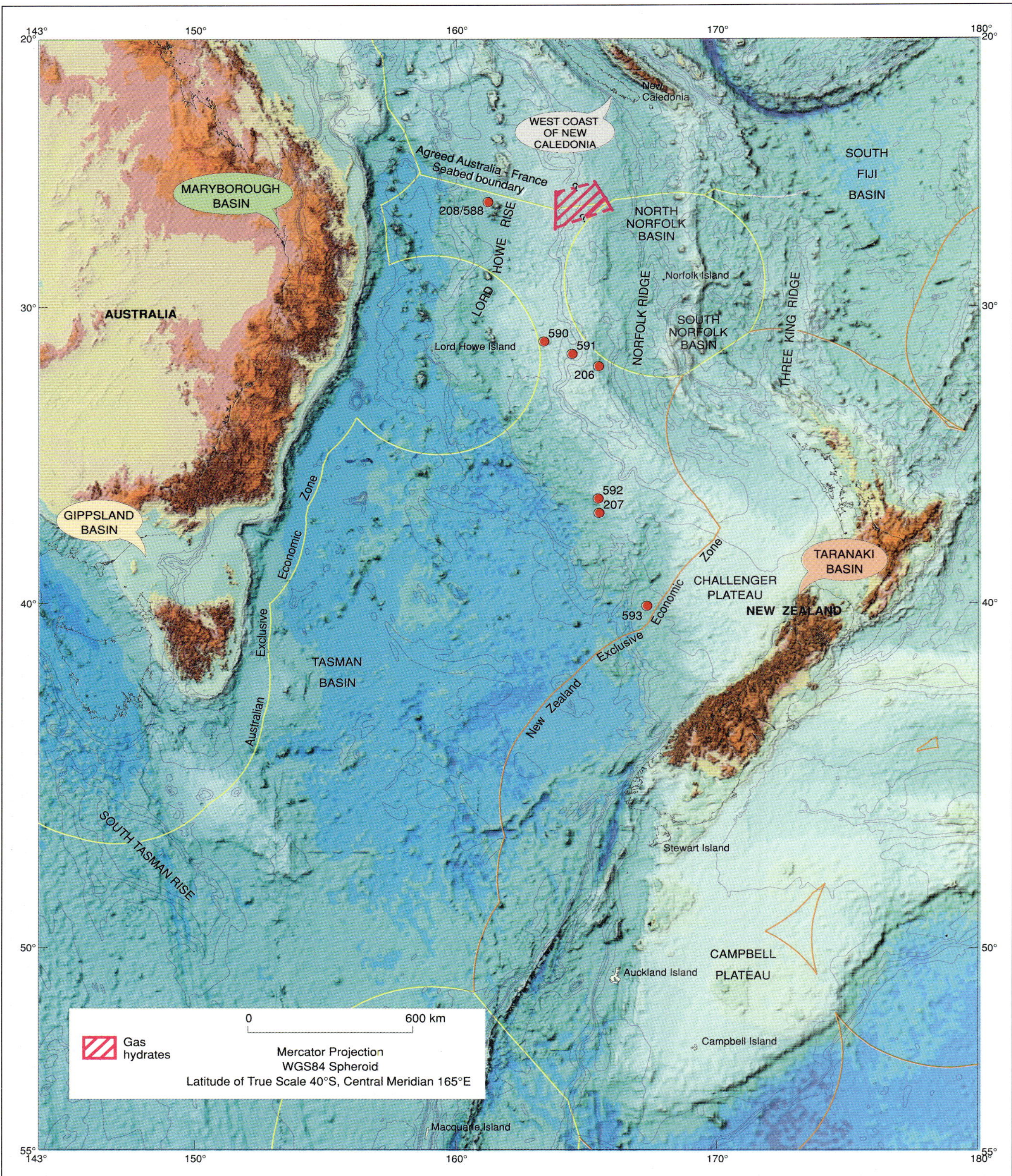

Figure 20. Eastern margin regional map, including locations of the South Tasman Rise and the gas-hydrate zone on Lord Howe Rise. The bathymetric image was prepared by I. Borissova and G. Hill by merging AGSO's 30-arc-second bathymetric data set with predicted bathymetry derived from satellite altimeter data. Onshore topography is based on Australian Surveying and Land Information Group/AGSO digital elevation model of Australia. Bathymetric contours are derived from the 1997 edition of the GEBCO Digital Atlas. ODP and DSPD holes in the Lord Howe Region are shown.

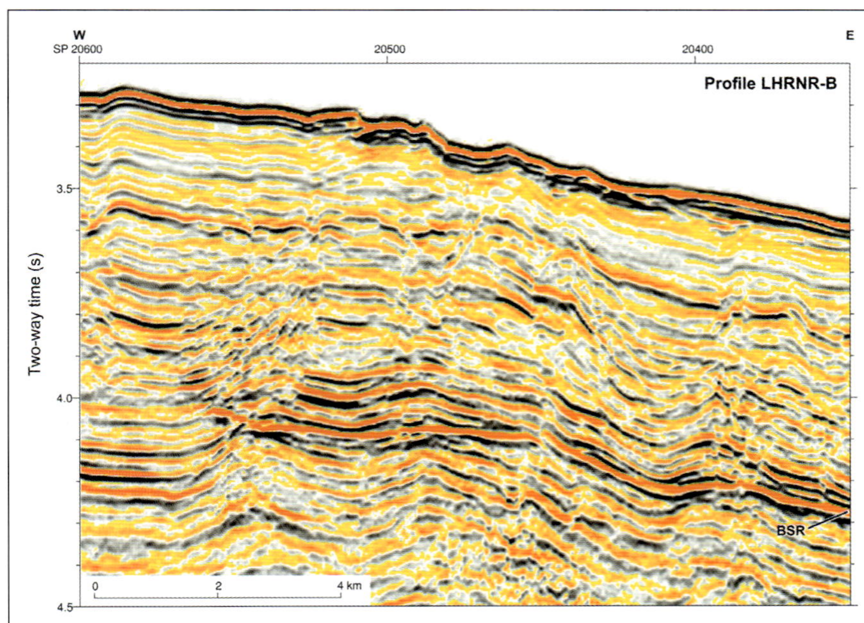

Figure 21. Bottom-simulating reflector (BSR) and flat spot in horst from Lord Howe Rise (from Exon et al., 1998, their Figure 5). The bright-red reflector on the right is a true BSR (0.73 s below seafloor), but it grades into a flat spot across the horst (0.68–0.77 s) and is not apparent farther west. The flat spot is probably a gas-water contact. Note also the bright spots extending for 0.15 s (approximately 150 m) above both the flat spot and the BSR on the right. These suggest the presence of interbedded gas and hydrate in the sedimentary column.

hydrocarbon generation and trap formation is sometimes problematic in these older basins, and the preservation of accumulations during hundreds of millions of years is a significant risk. However, 27% of the world's petroleum resources have been sourced from the Paleozoic (Klemme and Ulmishek, 1991. One of the largest onshore oil fields in Australia is Mereenie, which is sealed, sourced, and reservoired in the Lower Ordovician of the Amadeus Basin in central Australia. Other onshore basins with possible potential in lower Paleozoic sequences include the onshore Canning (King, 1999), Officer (Gravestock and Hibbert, 1991), Georgina (Lodwick and Lindsay, 1990), and the Landa Trough of the Wiso Basin (Pegum and Loeliger, 1990). The offshore Arafura Basin rates as one of the most prospective.

The Arafura Basin is located mostly offshore under the shallow waters of the Arafura Sea between northern Australia and the island of New Guinea (Figure 2). It covers about 130,000 km^2 in Australian waters and has a 10-km-thick section of Cambrian to Permian and possibly Triassic rocks (Bradshaw et al., 1990, Moore et al., 1996). Arafura 1, drilled at the eastern end of the basin, recorded more than 400 m of oil shows in Ordovician carbonates and Devonian siliciclastics. Live oil recovered from the Devonian section has been chemically typed to a Lower Cambrian marine claystone intersected 2 km deeper in the well (Moore et al., 1996; Summons et al., 1998) and thus shows that hydrocarbon generation and migration have occurred. However, Arafura 1 and all other exploration wells in the basin are in the Goulburn Graben (Figure 2), a late tectonized region where trap formation in the Triassic postdated the major phase of generation (Moore et al., 1996). North of the graben is a thick sedimentary section (as much as 7 km) of probable Paleozoic age that may contain a similar prospective Cambrian to Devonian sequence. Maturation modeling

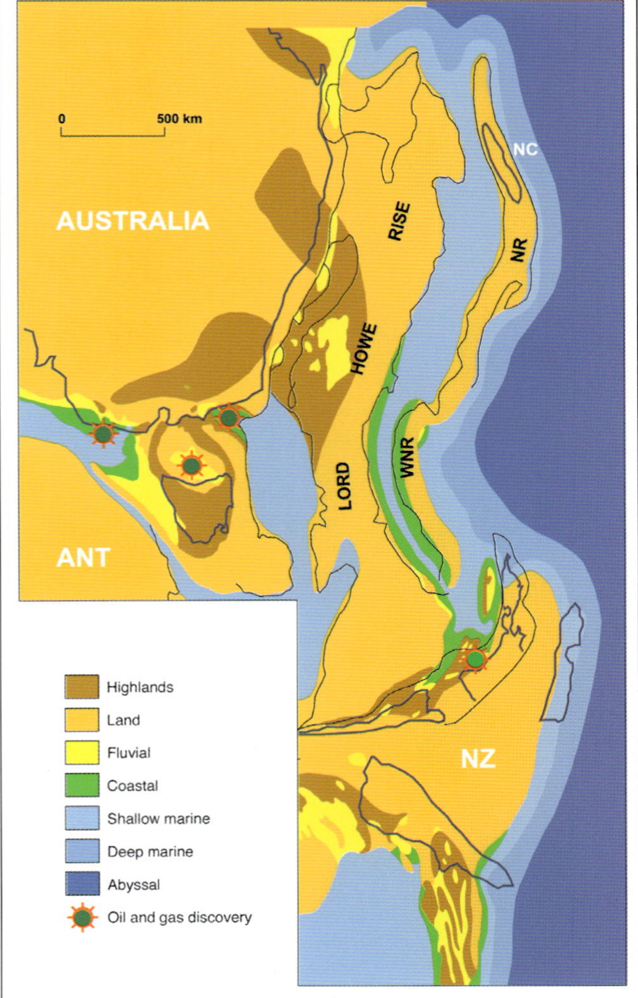

Figure 22. Eastern margin Late Cretaceous (about 95 Ma) paleogeographic map. From Walley, 1992.

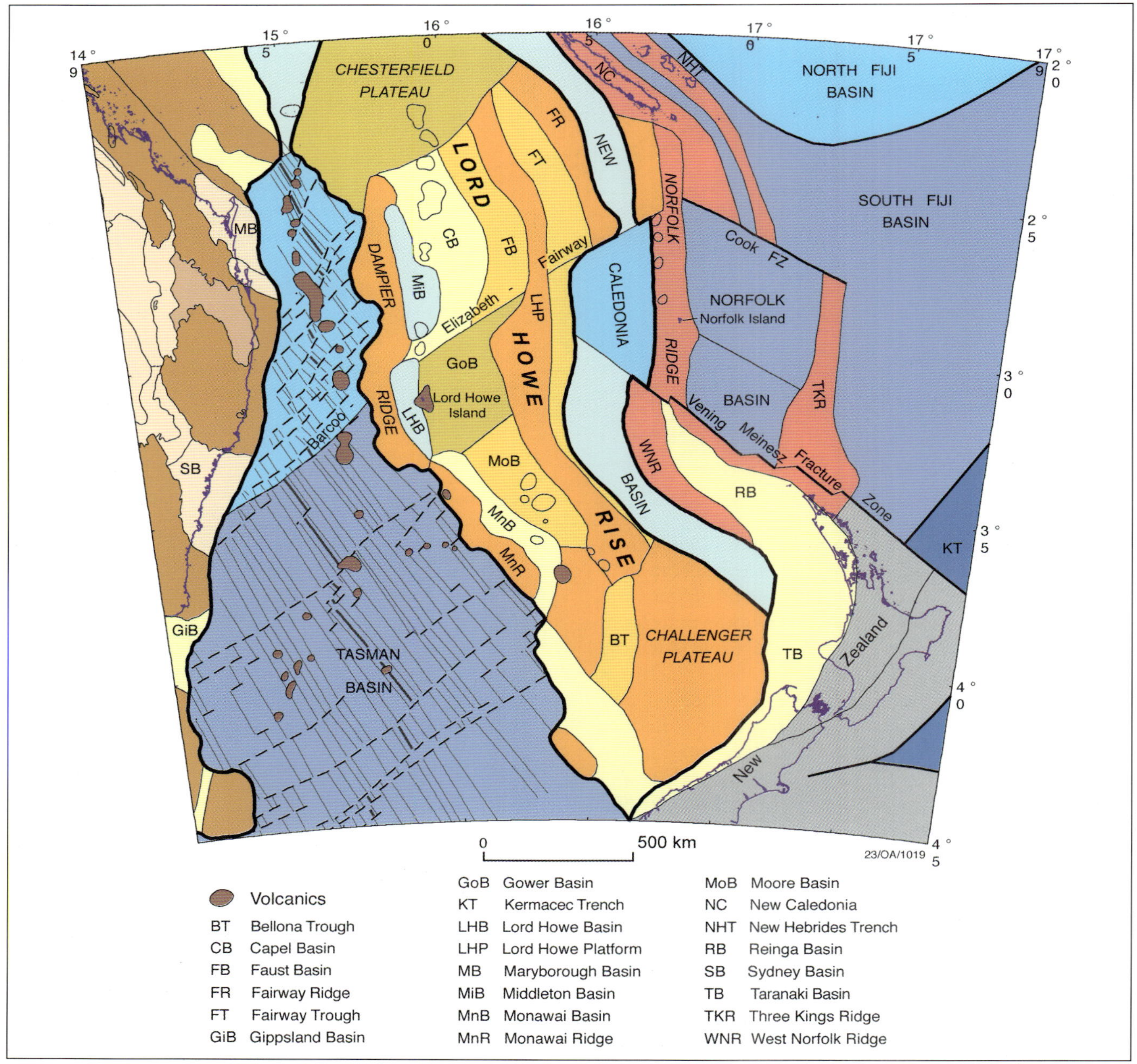

Figure 23. Structural-elements map for the Lord Howe Rise regions (from Stagg et al., 1999). Blues represent ocean basins. On the Australian continent, brown represents fold belts, beige represents Paleozoic-Mesozoic basins, and yellow represents Mesozoic-Cenozoic rift basins. On Lord Howe Rise and ridges to the east, red and orange represent highstanding basement, and tan and yellow represent Mesozoic-Cenozoic rift-basin elements. Gray is New Zealand provinces.

indicates that generation of hydrocarbons from the interpreted Cambrian-Ordovician sequence in this northern area was delayed until the Late Cretaceous (Moore et al., 1996), well after a variety of trap configurations was in place.

Apart from potential oil plays in the early Paleozoic Larapintine basins, the late Paleozoic Gondwanan supersystem (Table 2) has significant gas resources. Large undeveloped gas fields (Tern and Petrel) are in the Permian section of the Bonaparte Basin (Figure 8), and oil and gas are produced from the Permian-Triassic of the Perth Basin. In eastern Australia, much gas demand is met by production from the Cooper/Eromanga and Bowen/Surat Basins, with hydrocarbons derived from Permian coals and associated mud rocks (Boreham, 1995; Boreham and Summons, 1999). Further exploration in these basins and in other Paleozoic depocenters (Pedirka, Sydney, Gunnedah), in addition to development of coal-bed methane resources, is being encouraged in the new century by increasing demand for gas and the growth of the pipeline network.

CONCLUSIONS

For more than half of the twentieth century, Australia relied on imported petroleum, but discoveries since the 1960s supplied local demand and exports of LNG and oil. Australia's self-sufficiency in liquid hydrocarbons, which was 90% in 1986 and 66% in 1999, has been projected to increase to approximately 80% in 2001 before declining to 60% by 2009–2010 (ABARE, 1999, 2001). This reflects growth in crude-oil consumption at 1.2% per annum, offset by peak production arising from several discoveries in the 1990s (e.g., Laminaria). This decline in Australia's sufficiency in liquid-hydrocarbon production will be offset by exploitation of the vast undeveloped gas reserves of the North West Shelf. Potentially, the decline in liquid sufficiency will be slowed by exploration success from focused exploration on North West Shelf oil- and condensate-rich plays.

Nearly 100 years of effort went unrewarded in exploring onshore Australia before the first sustained commercial discoveries were made. It was not until exploration stepped offshore that giant fields were found. Do giant discoveries of the twenty-first century await us in deep-water basins surrounding Australia? Explorers may recall that offshore wells drilled in the Gippsland Basin to find the giant fields in the 1960s were the first offshore wells drilled in the world in a basin where there was no established onshore production. Perhaps the deep-water basins of Australia's southern and eastern margins will follow a similar exploration history, becoming the first deep-water discoveries made in basins which have no production on the shelf.

In the latter twenty-first century, after the vast gas reserves of the North West Shelf have been developed, we may expect to see the focus of the Australian petroleum industry shift back from the North West Shelf to the southern and eastern continental margins.

ACKNOWLEDGMENTS

We thank Steve Le Poidevin, Steve Cadman, Denis Wright, Vel Vuckovic, Eugene Petrie, Tony Stephenson, Howard Stagg, Melissa Fellows, Chris Fitzgerald, Gail Hill, Peter Hill, Rex Bates, Lindell Emerton, Roger Summons, Barry Willcox, Heike Struckmeyer, Jennifer Totterdell, Jane Blevin, Graham Logan, Dianne Edwards, Neville Montgomerie, Clinton Foster, and Neville Exon (AGSO); Peter Livingston (ISR); Agu Kantsler and Peter Moore (Woodside); and Peter Power.

MTB publishes with the permission of the chief executive officer, Australian Geological Survey Organisation.

REFERENCES CITED

ABARE, 1999, Australian energy: Market developments and projections to 2014–15: Australian Bureau of Agricultural and Resource Economics Research report 99.4, 184 p.

ABARE, 2001, Australian commodities forecasts and issues, v. 8, no. 2: Australian Bureau of Agricultural and Resource Economics.

Australian Geological Survey Organisation, 1998, Oil and gas resources of Australia 1998: Canberra, Australian Geological Survey Organisation, 168 p.

Auzende, J.-M., G. Dickens, S. Van de Beuque, N. F. Exon, C. Francois, Y. Lafoy, and O. Voutay, 2000: Thinned crust in southwest Pacific may harbor gas hydrate: EOS, v. 81 no. 17, p. 182–185.

Bint, A. N., 1988, Gas fields of the Browse Basin, *in* P. G. Purcell and R. R. Purcell, eds., 1988, The North West Shelf, Australia: Proceedings of Petroleum Exploration Society of Australia Symposium, p. 413-417.

Bein, J., and M. L. Taylor, 1981, The Eyre Sub-basin: Recent exploration results: APEA Journal, v. 21, p. 91–98.

Bernardel, G., Y. Lafoy, S. Van de Beuque, F. Missegue, and A. Nercissian, 1998, Preliminary results from AGSO Law of the Sea Cruise 206: An Australian/French collaboration deep seismic marine survey in the Lord Howe Rise/New Caledonia region: Australian Geological Survey Organisation Record, 37 p.

Boreham, C. J., 1995, Origin of petroleum in the Bowen and Surat Basins: Geochemistry revisited: APEA Journal, v. 35, p. 579–612.

Boreham, C. J., and R. E. Summons, 1999, New insights into the active petroleum systems in the Cooper and Eromanga Basins, Australia: APPEA Journal, v. 39, p. 263–296.

Bradshaw, J., R. S. Nicoll, and M. T. Bradshaw, 1990, The Cambrian to Triassic Arafura Basin, northern Australia: APEA Journal, v. 30, p. 107–127.

Bradshaw, J., J. Sayers, M. Bradshaw, R. Kneale, C. Ford, L. Spencer, and M. Lisk, 1998, Paleogeography and its impact on the petroleum systems of the North West Shelf, Australia, *in* P. G. Purcell and R. R. Purcell, eds., The sedimentary basins of Western Australia 2: Proceedings of Petroleum Exploration Society of Australia Symposium, p. 95–122.

Bradshaw, M. T., 1993, Australian petroleum systems: PESA Journal, v. 21, p. 43–53.

Bradshaw, M. T., A. N. Yeates, R. N. Benyon, A. T. Brackel, R. P. Langford, J. M. Totterdell, and M. Yeung, 1988, Paleogeographic evolution of the North West Shelf Region, *in* P. G. Purcell and R. R. Purcell, eds., 1988, The North West Shelf, Australia: Proceedings of Petroleum Exploration Society of Australia Symposium, p. 29–54.

Bradshaw, M. T., J. Bradshaw, A. Murray, D. J. Needham, L. Spencer, R. Summons, J. Wilmot, and S. Winn, 1994, Petroleum systems in west Australian basins, *in* P. G. Purcell and R. R. Purcell, eds., The sedimentary basins of Western Australia: Proceedings of Petroleum Exploration Society of Australia Symposium, 93–118.

Bradshaw, M. T., D. Edwards, J. Bradshaw, C. B. Foster, T. Loutit, B. A. McConachie, A. Moore, and R. Summons, 1997, Australian and eastern Indonesian petroleum systems: Proceedings of IPA Petroleum Systems Conference, p. 141–153.

Bradshaw, M. T., C. B. Foster, M. E. Fellows, and D. C. Rowland, 1999, The Australian search for petroleum: Patterns of discovery: APPEA Journal, v. 39, p. 12–29.

Brooks, D. M., A. K. Goody, J. B. O'Reilly, and K. L. McCarty, 1996, Bayu/Undan gas-condensate discovery: Western Timor Gap Zone of Cooperation, Area A: APPEA Journal, v. 36, p. 142–160.

Bureau of Resource Sciences, 1993, Oil and gas resources of Australia 1992: Canberra, Bureau of Resource Sciences, 128 p.

Bureau of Resource Sciences, 1994, Oil and gas resources of Australia 1993: Canberra, Bureau of Resource Sciences, 144 p.

Bureau of Resource Sciences, 1996, Oil and gas resources of Australia 1995: Canberra, Bureau of Resource Sciences, 185 p.

Bureau of Resource Sciences, 1997, Oil and gas resources of Australia 1996 : Canberra, Bureau of Resource Sciences, 157 p.

Burns, R .E., J. E. Andrews, et al., 1973, Initial Reports of the Deep Sea Drilling Project, 21: Washington, D. C., U.S. Government Printing Office, 931 p.

Clegg, L. J., M. J. Sayers, and A. M. Tait, 1992, The Gorgon gas field, *in* M. T. Halbouty, ed., Giant oil and gas fields of the decade 1978–1988: AAPG Memoir 54, p. 517–518.

Commonwealth of Australia, 1999, Australia's marine science and technology plan: Canberra, Commonwealth of Australia, 146 p.

Dickens, G., N. F. Exon, D. Holdway, Y. Lafoy, J. Auzende, G. Dunbar, and R. E. Summons, 2001, Quaternary sediment cores from the Southern Fairway Basin on the northern Lord Howe Rise (Tasman Sea): Australian Geological Survey Organisation, Record 2001/31, 29 p.

Edwards, D. S., J. Bradshaw, M. Bradshaw, C. B. Foster, J. M. Kennard, R. S. Nicoll, R. E. Summons, and J. E. Zumberge, 1997, Geochemical characteristics of Palaeozoic petroleum systems in northwestern Australia: APPEA Journal, v. 37, p. 249–277.

Edwards, D. S., H. I. M. Struckmeyer, M. T. Bradshaw, and J. E. Skinner, 1999, Geochemical characteristics of Australia's southern margin petroleum systems: APPEA Journal, v. 39, p. 297–321.

Exon, N. F., A. M. G. Moore, and P. J. Hill, 1997, Geological framework of the South Tasman Rise, south of Tasmania, and its sedimentary basins: Australian Journal of Earth Sciences, v. 44, p. 561–577.

Exon, N. F., G. R. Dickens, J.-M. Auzende, Y. Lafoy, P. A. Symonds, and S. Van de Beuque, 1998, Gas hydrates and free gas on the Lord Howe Rise, Tasman Sea: PESA Journal, v. 26, p. 148–158.

Forman, D. J., and A. L. Hinde, 1985, Improved statistical method for assessment of undiscovered petroleum resources: AAPG Bulletin, v. 69, p. 106–118.

Forman, D. J., and A. L. Hinde, 1986, Examination of the creaming methods of assessment applied to the Gippsland Basin, offshore Australia, *in* D. D. Rice, ed., Oil and gas assessment—Methods and applications: AAPG Studies in Geology 21, p. 101–110.

Forman, D. J., A. L. Hinde, and A. P. Radlinski, 1992, Assessment of undiscovered petroleum resources by the Bureau of Mineral Resources, Australia: Energy Sources, v. 14, p. 183–203.

Forman, D. J., A. L. Hinde, S. J. Cadman, A. P. Radlinski, and J. Morton, 1993, Towards assessment of plays containing migrated petroleum, *in* J. Harff and D. F. Merriam, eds., Computerized basin analysis: New York, Plenum Press, p. 275–299.

GEBCO Digital Atlas, 1997, General bathymetric chart of the oceans, 2nd release: Birkenhead, Merseyside, U. K., British Oceanographic Data Centre, under the joint auspices of the Intergovernmental Oceanographic Commission and International Hydrographic Organization, 1 computer laser optical disk.

GeoMark Research Inc. and Australian Geological Survey Organisation, 1996, Western Australian oils study: Unpublished proprietary report.

Gravestock, D. I., and J. E. Hibburt, 1991. Sequence stratigraphy of the eastern Officer and Arrowie Basins: A framework for Cambrian oil search: APEA Journal, v. 31, p. 177–190.

Hegarty, K. A., J. K. Weissel, and J. C. Mutter, 1988, Subsidence history of Australia's southern margin: Constraints on basin models: AAPG Bulletin, v. 72, p. 615–633.

Hill, P. J., 1994, Geology and geophysics of the offshore Maryborough, Capricorn and northern Tasman Basins: Results of AGSO Survey 91: Australian Geological Survey Organisation Record 1994/1, 71 p.

Hill, P. J., A. J. Meixner, A. M. G. Moore, and N. F. Exon, 1997, Structure and development of the west Tasmania offshore sedimentary basins: Results of recent marine and aeromagnetic surveys: Australian Journal of Earth Sciences, v. 44, p. 579–596.

Hocking, R. M., 1994, Subdivisions of Western Australian Neoproterozoic and Phanerozoic sedimentary basins: Geological Survey of Western Australia Record 1994/4, 84 p.

ISR, 1999, Australian offshore petroleum strategy: Canberra, Department of Industry, Science and Resources, 84 p.

King, M. R., 1998, Looma-1 reopens the Palaeozoic play in the south Canning Basin: APPEA Journal, v. 38, p. 254–277.

Klemme, H. D., 1975, Giant fields related to their geological setting: A possible guide to exploration: Bulletin of Canadian Petroleum Geology, v. 23, p. 30–66.

Klemme, H. D., and G. F. Ulmishek, 1991, Effective petroleum source rocks of the world: Stratigraphic distribution and controlling depositional factors: AAPG Bulletin, v. 75, p. 1809–1851.

Lafoy, Y., B. Pelletier, J-M. Auzende, F. Missegue, and L. Mollard, 1994, Tectonique compressive Cénozoique sur les ridges de Fairway et Lord Howe, entre Nouvelle Caledonie et Australie: Comptes Rendus Academie Sciences Paris, v. 319, series 2, p. 1063–1069.

Lafoy, Y., S. Van de Beuque, F. Missegue, A. Nercissian, and G. Bernardel, 1998, Campagne de sismique multitraces entre la marge est Australienne et le sud del'arc des Nouvelles-hébrides. Rapport de la campagne Rig Seismic 206 (21 avril–24 mai 1998): Programme FAUST (French Australian Seismic Traverse) Rapport ZoNéCo., 40 p.

Lodwick, W. R., and J. F. Lindsay, 1990, Southern Georgina Basin: A new perspective: APEA Journal, v. 30, p. 137–148.

McDougall, I., and G. J. Van Der Lingen, 1974, Age of rhyolites of the Lord Howe Rise and the evolution of the southwest Pacific Ocean: Earth and Planetary Sciences Letters, v. 21, p. 117–126.

Moore, A., J. Bradshaw, and D. Edwards, 1996, Geohistory modelling of hydrocarbon migration and trap formation in the Arafura sea: PESA Journal, v. 24, p. 35–51.

Moore, A. M .G., and H. M. J. Stagg, and M. S. Norvick, 2000, Deep-water Otway basin: A new assessment of the tectonics and hydrocarbon prospectivity: APPEA Journal, v. 40, p. 66–85.

Moore, A. M. G., J. B. Wilcox, N. F. Exon, and G. W. O'Brien, 1992, Continental shelf basins on the west Tasmania margin: APEA Journal, v. 32, p. 231–250.

Pegum, D. M., and M. Loeliger, 1990, The Lander Trough—A central Australian frontier exploration area: APEA Journal, v. 30, p. 128–136.

Powell, T. G., D. J. Wright, and E. Nicholas, 1990, Petroleum exploration and development in Australia: A BMR discussion paper: Bureau of Mineral Resources Record 1990/32, 134 p.

Purcell, P. G., and R. R. Purcell, eds., 1988, The North West Shelf, Australia: Proceedings of Petroleum Exploration Society of Australia Symposium, 651 p.

Purcell, P. G., and R. R. Purcell, eds., 1994, The sedimentary basins of Western Australia: Proceedings of Petroleum Exploration Society of Australia Symposium, 864 p.

Purcell, P. G., and R. R. Purcell, eds., 1998, The sedimentary basins of Western Australia 2: Proceedings of Petroleum Exploration Society of Australia Symposium, 743 p.

Robertson, C. S., 1988, Australia's petroleum prospects: Changing perceptions since the beginning of the century: APEA Journal, v. 28, p. 190–207.

Royer, J.-Y., and N. Rollet, 1997, Plate-tectonic setting of the Tasmanian region: Australian Journal of Earth Sciences, v. 44, p. 543–560.

Ruble, T. E., G. A. Logan, J. E. Blevin, H. I. M. Struckmeyer, M. Ahmed, and R. A. Quezada, 1999, Geochemistry of palaeo-oil in Jerboa-1, Eyre Sub-basin, Great Australian Bight: CSIRO Petroleum Confidential Report 99-060, 223 p.

Scott, D. I., 1993, Architecture of the Queensland Trough: Implications for the structure and tectonics of the northeastern Australian margin: AGSO Journal of Australian Geology and Geophysics, v. 14, p. 21–34.

Shaw, R. D., 1978, Seafloor spreading in the Tasman Sea: A Lord Howe Rise–eastern Australia reconstruction: AAPG Bulletin, v. 62, p. 75–81.

Sibley, D., F. Herkenhoff, D. Criddle, and M. McLerie, 1999, Reducing resource uncertainty using seismic amplitude analysis on the southern Rankin trend, northwest Australia: APPEA Journal, v. 39, p. 128–148.

Smith, M. A., and I. F. Donaldson, 1995, The hydrocarbon potential of the Duntroon Basin: APEA Journal, v. 35, p. 203–291.

Stagg, H. M. J., C. D. Cockshell, J. B. Willcox, A. J. Hill, D. J. L. Needham, B. Thomas, G. W. O'Brien, and L .P. Hough, 1990, Basins of the Great Australian Bight region: Geology and petroleum potential: Bureau of Mineral Continental Margins Program Folio 5, 143 p.

Stagg, H. M. J., I. Borissova, M. Alcock, and A. M. G. Moore, 1999, Tectonic provinces of the Lord Howe Rise: AGSO Research Newsletter, v. 31, November, p. 31–32.

Struckmeyer, H. I. M., P. A. Symonds, M. E. Fellows, and D. L. Scott, 1994, Structural and stratigraphic evolution of the Townsville Basin, Townsville Trough, offshore northeastern Australia: Australian Geological Survey Organisation Record 1994/50, 71 p.

Summons, R. E., M. Bradshaw, J. Crowley, D. S. Edwards, S. C. George, and J. E. Zumberge, 1998, Vagrant oils: Geochemical signposts to unrecognised petroleum systems, *in* P. G. Purcell and R. R. Purcell, eds., The sedimentary basins of Western Australia 2: Proceedings of Petroleum Exploration Society of Australia Symposium, p. 169–184

Symonds, P. A., and J. B. Willcox, 1989, Australia's petroleum potential in areas beyond an Exclusive Economic Zone: BMR Journal of Australian Geology and Geophysics, v. 11, p. 11–36.

Symonds, P. A., B. Murphy, D. C. Ramsay, K. L. Lockwood, and I. Borissova, 1998, The outer limits of Australia's resource jurisdiction off Western Australia, *in* P. G. Purcell and R. R. Purcell, eds., The sedimentary basins of Western Australia 2: Proceedings of Petroleum Exploration Society of Australia Symposium, p. 3–20.

Totterdell, J. M., J. E. Blevin, H. I. M. Struckmeyer, B. E. Bradshaw, J. B. Colwell, and J. M. Kennard, 2000, A new sequence framework for the great Australian Bight: Starting with a clean slate: APPEA Journal, v. 40, p. 95–120.

Van de Beuque, S., J.-M. Auzende, Y. Lafoy, and F. Missegue, 1998, Tectonique et volcanisme tertiarie sur la ride de Lord Howe (Sud-Oest Pacifique): Comptes Rendus Academe Sciences Paris, v. 326, series 2a, p. 663–669.

Vear, A., and M. DeRuig, 2001, Southern Bonaparte Basin revisited: The dawn of a new era of Australian exploration: AAPG International Conference, Bali, October 15–18, 2000.

Walley, A. M., 1992, Cretaceous-Cainozoic palaeogeography of the New Zealand–New Caledonia region: BMR Record 1992/011, 25 p.

Wilkinson, R., 1991, Where God never trod: Australia's oil explorers across two centuries: Sydney, David Ell Press, 482 p.

Willcox, J. B. and P. A. Symonds, 1997, Deepwater prospectivity for petroleum in the Australian region, *in* Outlook 1997—Proceedings of the National Agriculture and Resources Conference, p. 200–213.

Yeates, A. N., M. T. Bradshaw, J. M. Dickins, A. T. Brakel, N. F. Exon, R. P. Langford, S. M. Mulholland, J. M. Totterdell, and M. Yeung, 1987, The Westralian Superbasin: An Australian link with Tethys, *in* K. G. McKenzie, ed., Proceedings of Shallow Tethys 2 Symposium, p. 199–213.

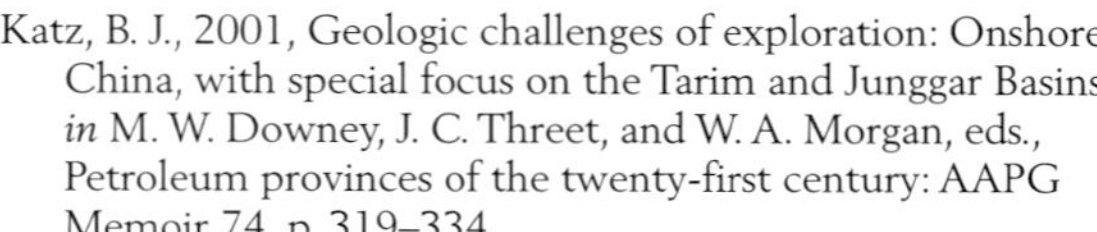
Katz, B. J., 2001, Geologic challenges of exploration: Onshore China, with special focus on the Tarim and Junggar Basins, *in* M. W. Downey, J. C. Threet, and W. A. Morgan, eds., Petroleum provinces of the twenty-first century: AAPG Memoir 74, p. 319–334.

Chapter 16

Geologic Challenges of Exploration: Onshore China, with Special Focus on the Tarim and Junggar Basins

Barry J. Katz
Texaco Group Inc., Houston, Texas, U.S.A.

ABSTRACT

Onshore China has more than 20 major sedimentary basins. Significant oil production is currently limited to only three basins—Songliao, Bohaiwan, and Junggar. Exploration results in some of the other onshore basins have been disappointing; in others, activities have been limited because of their remoteness. This paper focuses primarily on the Junggar and Tarim Basins because of their large size and potential for significant oil and gas discoveries.

Hydrocarbons discovered in the three producing basins appear to have been derived from mainly lacustrine systems and are often found in nonmarine sandstone reservoirs. The oils found are often waxy and reservoir properties are poor, with limited vertical and lateral continuity. Consequently, although field sizes may be large, flow rates from individual wells may be limited, requiring a large number of wells to capture the reserves. For example, Daqing field (Songliao Basin) had initial reserves of more than 8 billion barrels (bbl) of oil and is producing ~1.1 million bbl/day, but it contains more than 10,000 wells. Similarly, typical flow rates from vertical wells in the Shengli petroleum province (Bohaiwan Basin) are less than 700 bbl/day. Recent horizontal wells in Shengli field have shown significant improvements in production rates, with individual wells achieving rates of more than 5000 bbl/day.

An examination of the less well explored basins suggests that many of the key components of petroleum systems are present. For example, in the Tarim Basin, multiple marine and lacustrine source rocks have been identified, as have both siliciclastic and carbonate reservoirs ranging in age from Cambrian-Ordovician through Paleogene. Active seeps have also been observed in the basin and more than 200 structural targets have been identified, with more than half of them having surface expression. The primary exploration challenges in many of these basins appear to be associated with relative timing, preservation of hydrocarbon accumulations, and communication between the generative basin and the trap.

INTRODUCTION

Nearly 45% of the onshore surface area of the People's Republic of China is covered by unmetamorphosed sedimentary rock (Li Desheng, 1985). Much of this sedimentary succession is contained in more than 20 major and moderate-sized basins (Figure 1), with more than 600 sedimentary basins having been identified (Tian Zaiyi, 1990). Sediment thickness in these basins can often exceed 10 km (Li Desheng, 1985).

From a historical perspective, there is substantial evidence to suggest that these basins hold considerable exploration potential. Active oil seepage has been known for at least 4000 years (Hu Boliang, 1992). The magnitude of some of these seeps can be quite significant, with individual asphalt-covered areas in the Junggar Basin exceeding 1000 km^2 (Taner et al., 1988). In fact, the name of the village of Karamay in the Junggar Basin is derived from the "black hill" associated with an asphalt deposit that has been mined for centuries (Taner et al., 1988).

In addition to oil, the presence of natural gas has also been well documented for many centuries. For example, natural gas has been produced for commercial purposes from brine wells in the Sichuan Basin for almost 1000 years (Vogel, 1993).

Today, significant oil production onshore is limited to three basins—Songliao, Bohaiwan, and Junggar. The major field in the Songliao Basin is Daqing. The Daqing field complex currently has daily oil production of ~1.1 million barrels (bbl). The complex, which was discovered in 1959, will ultimately produce more than 8 billion bbl of oil (Meyerhoff and Willums, 1981). Within the Bohaiwan Basin, the major hydrocarbon production is from the Shengli field complex, which was discovered in 1960. Daily production from Shengli currently approaches that of Daqing. Meyerhoff and Willums (1981) estimated that the Shengli complex has reserves of ~5 billion bbl. In the Junggar Basin, the giant Karamay field complex, discovered in 1957, currently dominates the basin's production. Although initial production was limited (partially because of oil quality), the discovery of a deep light-oil pool has resulted in an increase in the field's production level. Production of ~200,000 bbl/day was targeted for the Karamay field complex for the year 2000.

Other basins have limited production or have had significant shows. For example, in the Tarim Basin, three discoveries are worth highlighting (Hu Bliang, 1992; Li Desheng, 1995). The Ke 1 well, on the southwestern margin of the basin (Figure 2), had reported initial flow rates of 10,000 bbl oil/day (BOPD) and 9.5 million cubic feet (mmcf) gas/day (MMCFGD) from lower Miocene sands. The Shacan (or Shashen) 2 well, which penetrated Ordovician dolomites, had initial flow rates of more than 7000 BOPD and 70 MMCFGD. Also from Lower Ordovician dolomites, the Tazhong 1 well, in the central part of the basin, had tested flows of more than 4000 BOPD.

Although circumstantial evidence suggests the possibility of significant hydrocarbon potential, the magnitude of the resources for individual basins (and consequently the country as a whole) is poorly known. Hsu (1994) suggested that the Tarim Basin could have generated more than 350 billion bbl of oil. Li Desheng (1996) suggested that China's in-place oil (both onshore and offshore) exceeds 500 billion bbl and that ~1160 trillion cubic feet

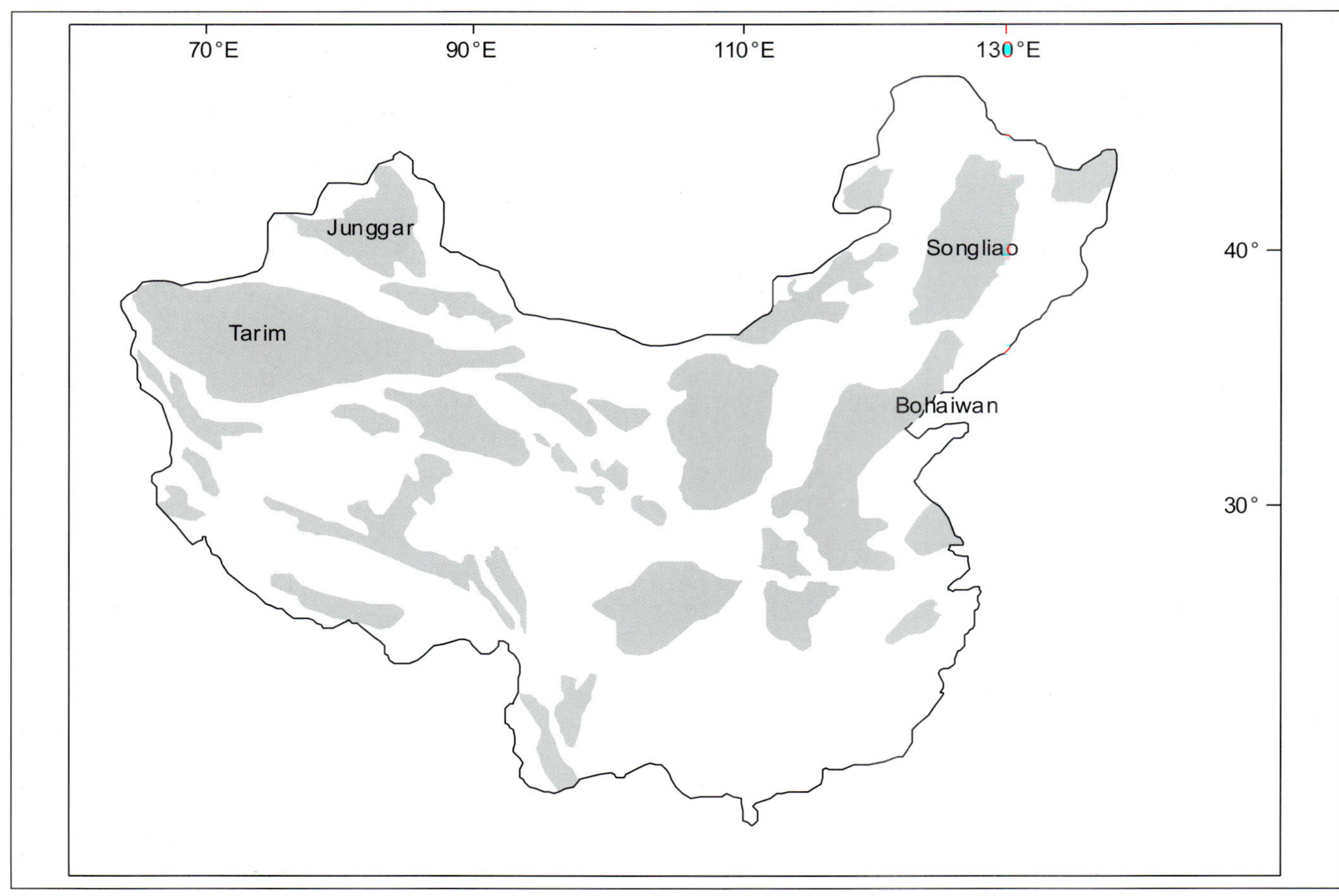

Figure 1. Distribution of major and moderate-sized sedimentary basins in China.

(tcf) of natural gas are present. But estimates of in-place oil and gas must be viewed with some degree of caution because they appear to have been based largely on hydrocarbon source-rock potential (i.e., source-rock richness and volume) rather than on a more complete integrated assessment. The U.S. Geological Survey estimated known reserves at ~52.6 billion bbl of oil and natural-gas liquids and ~57.7 tcf of natural gas (Klett et al., 1997). But even those estimates must be viewed with some degree of caution because of the wide range of published reserve estimates for individual fields. For example, published reserve estimates for Karamay field (Junggar Basin) differ by more than an order of magnitude, ranging from 219 million to more than 9 billion bbl (Carroll et al., 1992). In part, this uncertainty exists because of the limited amount of data available, the vintage of the data, the manner in which the data are interpreted, and questions of commerciality.

Regardless of the uncertainty in reserve estimates, the mere presence of oil and gas does not ensure that it will be commercial. Whether an accumulation is commercial depends on such factors as the volume of hydrocarbons present, rates of production, reservoir continuity, target depth, proximity to market, cost of infrastructure, and the value of a barrel of crude oil.

A review of exploration results in China reveals that several exploration challenges exist before commercial accumulations can be established beyond the known producing basins. These challenges include establishing the potential for initial hydrocarbon charge, the preservation of hydrocarbon accumulations, the quality of potential reservoirs, and the effectiveness of communication between source and reservoir/trap.

If one accepts the axiom that "oil and gas are where you find them," a reexamination of onshore China does appear to be warranted. It is the focus of this paper to examine aspects of hydrocarbon charge and reservoir quality and their possible implications on future exploration in the Tarim and Junggar Basins (Figures 2 and 3). Those two basins were selected because of their size and reported hydrocarbon reserve potential. Others have discussed the presence and abundance of potential structural traps in those basins as well as their different tectonic histories. For example, Xie Hong (1993) noted that more than 200 potential structural traps have been identified in the Tarim Basin, with more than half of those being identifiable through surface mapping.

HYDROCARBON CHARGE

Hydrocarbon charge is dependent on the quality and quantity of organic matter present, the extent of organic diagenesis, the relative timing of hydrocarbon generation compared with the timing of trap development, the efficiency of migration, and the preservation of the hydrocarbon accumulation.

Source-rock Potential—Tarim Basin

In the Tarim Basin, three possible source-rock intervals have often been cited: the Cambrian-Ordovician, Permian-Carboniferous, and the Upper Triassic to Middle Jurassic (Junhong Chen et al., 1996). It has also been proposed that Upper Cretaceous–Paleogene shales may be a possible source (Ulmishek, 1984). Graham et al. (1990) suggested from their limited outcrop sampling that much of the source-rock potential is limited to the Ordovician

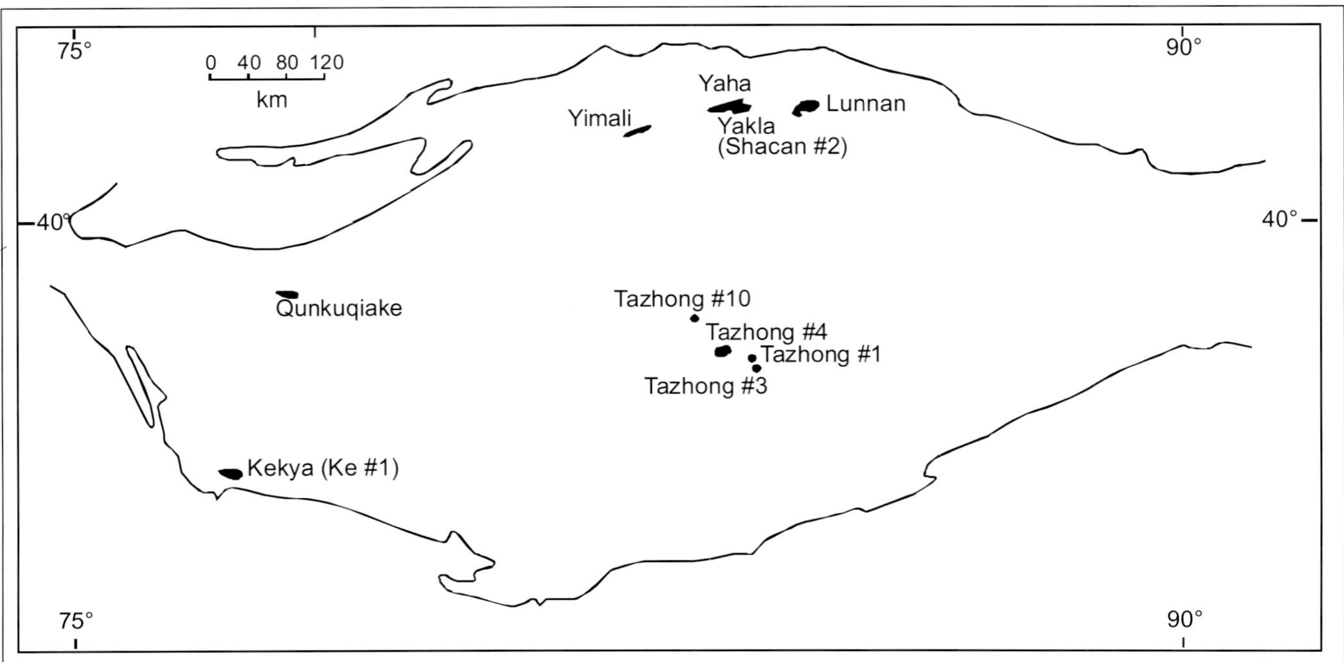

Figure 2. Index map for the Tarim Basin.

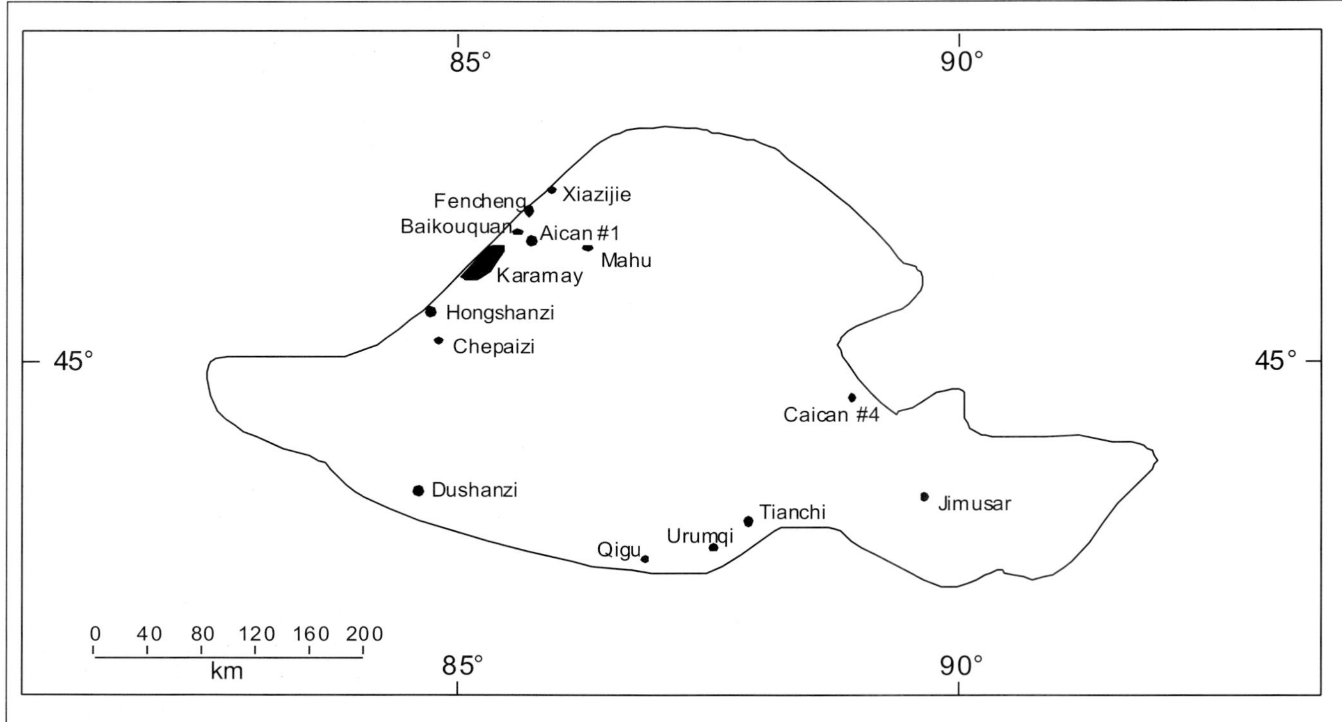

Figure 3. Index map for the Junggar Basin.

marine shales and the Upper Triassic–Middle Jurassic lacustrine and paludal sequences.

However, only limited source-rock data are available from the Tarim Basin to support those speculations. For example, Yang Bin (1991) reported that the Cambrian gray-black limestones recovered from the Kunan 1 well had organic-carbon contents that ranged from only 0.55 to 0.92 wt.%. Yang Bin further reported that outcropping Ordovician limestones and claystones also typically contained less than 1.0 wt.%, except for two samples that contained 1.17 and 1.19 wt.%. Those generally low levels of organic enrichment most probably reflect both advanced levels of thermal maturity (discussed below) and the nature of the facies examined (i.e., platformal versus basinal). At more advanced levels of thermal maturity, similar to those observed in the Lower Paleozoic sequence in the Tarim Basin, organic-carbon content could be reduced by more than 50% (Daley and Edman, 1987). Hsu (1994) suggested that the sampled facies represent shallow-water deposits rather than the more basinal facies where higher levels of organic enrichment are to be expected. In fact, Hsu suggested that a thick Lower Paleozoic euxinic basinal sequence underlies nearly half the basin at depths that cannot be reached by drilling. This basinal facies crops out along the northeastern rim of the basin, where a sequence of black shales and cherts is present.

The presence of organic-rich Lower Paleozoic sediments, however, is supported by the reported maximum organic-carbon value of 5.22 wt.% for the Cambrian-Ordovician sequence (Hu Jianyi et al., 1997). Hanson (1998), citing unpublished data, noted that organic-carbon contents of the Cambrian-Ordovician sequence display maximum values of 4 wt.% to 6 wt.%. Hanson noted, however, that much of the thick (>5 km) basinal shales, including laminated intervals, commonly contain less than 0.2 wt.% organic carbon, and he preferred an alternative depositional model for the organic-rich intervals. Hanson suggested that the organic-rich sediments were deposited in an oxygen minimum along the shelf edge or slope.

The difference between these two proposed source-rock depositional models for the lower Paleozoic of the Tarim Basin is of more than academic importance. The two models would result in dramatically different source-rock distributions and volumes. The Hsu (1994) model would provide a much wider distribution of possible source rocks than the Hanson (1998) model. It would also provide for a much greater source-rock volume. The Hanson model would suggest a distribution that largely rings the basin margin and paleohighs. Not only does the Hanson model constrain the volume of source rock, but it also restricts its geographic distribution.

Graham et al. (1990) reported a maximum organic-carbon and total hydrocarbon generation potential of 2.75 wt.% and 4.68 mg HC/g rock, respectively, for the Saergan Formation (Middle Ordovician; Figure 4). The significance of their data is that they indicated that the limited number of lower Paleozoic samples studied are gas prone, with a maximum hydrogen index of 168 mg

HC/g TOC (Figure 5). A further examination of this data set suggests that reported geochemical attributes do not appear to be representative of the original rock character and that surface weathering also may have reduced the quantity (i.e., organic carbon and generation potential) and quality of organic matter (i.e., hydrogen index).

As with the lower Paleozoic series, the amount of geochemical data available for the Permian-Carboniferous in the Tarim Basin is limited. The Permian-Carboniferous in the basin evolved from a marginal sea dominated by platform carbonates that may be 2 km thick (Li Desheng et al., 1996) to nearshore siliciclastics and finally to a nonmarine series. And although the Permian-Carboniferous has been cited as a potential source, those depositional settings are not considered very favorable for source-rock development. Limited source-rock potential does, however, appear to be present within the sequence. Li Desheng et al. (1996) cite a maximum organic-carbon content of 11.15 wt.% for that interval. Graham et al.'s (1990) data set suggests that any source-rock potential for that interval is limited to "coaly" intervals in the Kalundaer Formation (Lower Permian). Although they are organic rich (maximum reported TOC of 34.38 wt.%; Figure 4) and capable of generating significant quantities of hydrocarbons (maximum reported $S_1 + S_2$ of 26.58 mg HC/g rock; Figure 4), the studied samples are gas prone (hydrogen indices less than 150 mg HC/g TOC; Figure 5). Such rocks would not be capable of contributing significantly to the oil-resource base. Furthermore, the distribution and volumetric significance of this facies are poorly understood. It has also been speculated, however, that an Upper Permian lacustrine facies exists in the subsurface, and that those rocks could display geochemical attributes similar to organic-rich Permian lacustrine strata in the Junggar and Turpan Basins (see following discussion).

The third interval cited as a source in the Tarim Basin is in the Upper Triassic–Middle Jurassic. Organic matter in that interval is concentrated largely in a suite of coals and carbonaceous shales. Graham et al. (1990) and Hendrix et al. (1995) summarized the geochemical characteristics of those sediments. Those authors reported that the organic-carbon contents of those sediments ranges to as much as ~96 wt.% (Figure 4) and that the maximum measured hydrocarbon generation potential is 296 mg HC/g rock. Those rocks are predominantly gas prone, containing type III organic matter (Figure 5). The hydrogen indices are typically less than 250 mg HC/g TOC. Significantly, the pyrolysis gas-chromatography results suggest that those coals could generate higher molecular hydrocarbons (Hendrix et al., 1995). Pyrolytic generation of higher-molecular-weight compounds from coals, however, does not equate to an ability to expel those compounds in nature. Katz et al. (1991) noted that those higher-molecular-weight hydrocarbons would be retained in the coal until cracked and would be released as gas. A possible exception to the above discussion is a single coal sample from the Kezileinuer Formation (Middle Jurassic), which Graham et al. (1990) reported has a hydrogen index of 424 mg HC/g TOC. This sample displays a type II affinity and could, at the appropriate levels of thermal maturity, both generate and expel liquid hydrocarbons.

In addition to these coaly gas-prone facies, a Jurassic lacustrine facies has also been identified in the Tarim Basin (Hendrix et al., 1995; Ritts, 1998). Those largely oil-prone sediments average between 3 wt.% and 5 wt.% TOC, with maximum levels of organic enrichment approaching 9 wt.%.

The Upper Cretaceous–Paleogene interval, although cited as a source, appears to be organic poor, with TOC levels less than 1.0 wt.% (Ulmishek, 1984). The available data, therefore, do not support the suggestion that this stratigraphic interval is a possible source.

Source-rock Potential—Junggar Basin

The Permian sequence is thought to be the principal oil source rock for the Junggar Basin (Carroll et al., 1992). Ulmishek (1984) and Clayton et al. (1997) raised the possibility that additional effective source rocks may be present in the Junggar Basin and that some of those sources may be of only local importance or effectiveness. They suggest that those secondary source rocks may be present in the Jurassic and Tertiary sequences.

The Permian lacustrine sequence in the Junggar Basin ranks among the thickest and richest hydrocarbon source-rock sequences (Carroll, 1998). Within a gross lacustrine sequence of approximately 2000 m, Carroll et al. (1992) calculated for an 800-m net source-rock interval an average organic-carbon content and residual generation potential (S_2) for the Lucaogou Formation of 4.1 wt.% and 26.2 mg HC/g rock, respectively. Carroll et al. (1992) reported a maximum organic-carbon content exceeding 22 wt.% and a maximum generation potential approaching 200 mg HC/g rock (Figure 6). Hydrogen indices for the Lucaogou Formation typically exceed 600 mg HC/g TOC, consistent with that of type I kerogen (Figure 7).

The complete Permian lacustrine sequence records an overall evolution from a shallow evaporative lake to a deep freshwater lacustrine system associated with fluvial systems (Carroll, 1998). Carroll further suggested that the organic-rich laminated sequence of the Lucaogou Formation developed largely as a result of salinity-induced stratification and that a slow inorganic sedimentation rate may have been the primary control on organic-carbon content. He suggested that the inverse relationship between the hydrogen index and stable carbon isotope composition is supporting evidence for limited productivity within the lake basin. The underlying Jingjingzigou Formation, which was deposited under shallow lake conditions, displays, as a result of reduced preservation potential, significantly lower levels of organic enrichment (commonly less than 0.5 wt.% TOC) and lower hydrogen indices. Samples from the overlying Hongyanchi Formation often contain more than 1.0 wt.% TOC (Figure 6). The organic matter is hydrogen depleted (hydrogen indices typically less than 150 mg HC/g TOC; Figure 7) as a result of the relative abundance of vitrinite and inertinite.

The possible Jurassic source sequence in the Junggar Basin is associated with the basin's coal measures (Hendrix et al., 1995). In the Junggar Basin, Jurassic coals average 67.6 wt.% TOC and have a total hydrocarbon generation potential of 116.9 mg HC/g rock (Figure 6). Both of those averages are higher than those observed in their stratigraphic equivalents in the Tarim Basin. The coals also display a slighter higher average hydrogen index (161 vs. 127 mg HC/g TOC; Figure 7). Those differences, however, are not sufficient to suggest that the Junggar Basin coals are more oil prone (i.e., they represent an important potential gas source).

An areally and volumetrically limited Jurassic lacustrine facies most probably also exists in the Jurassic sequence of the Junggar Basin (Hendrix, 1992). Some oil-generating potential may be associated with those lacustrine strata.

No source-rock data are available to support the inferred Tertiary source in the Junggar Basin.

Thermal Maturity

Thermal maturity provides an estimate as to the extent of organic diagenesis, which relates to the extent of hydrocarbon generation and preservation.

Junhong Chen et al. (1996) reported thermal-maturity levels for the Ordovician section in the Tarim Basin equivalent to a vitrinite reflectance greater than 1.17%. Guo Jian Hua and Zhu Yangming (1995) reported that the Cambrian-Ordovician sequence in the Kunan 1 well has achieved thermal-maturity levels ranging between 1.74 and 2.04%. Hanson (1998) also reported similar maturity levels for the Cambrian-Ordovician sequence. (These vitrinite-reflectance values are estimated from bitumen reflectances.) Such elevated levels of thermal maturity suggest that the Cambrian-Ordovician sequence is largely overmature and that lower Paleozoic rocks are no longer capable of generating liquid hydrocarbons. These levels of thermal maturity, however, are consistent with the preservation of light oils and condensates. Hanson (1998), however, questioned whether the correlation between vitrinite and bitumen reflectivity is valid. He suggested that the biomarker content of some Cambrian rocks implies a lower level of thermal maturity. It is possible, however, that these slightly elevated biomarker

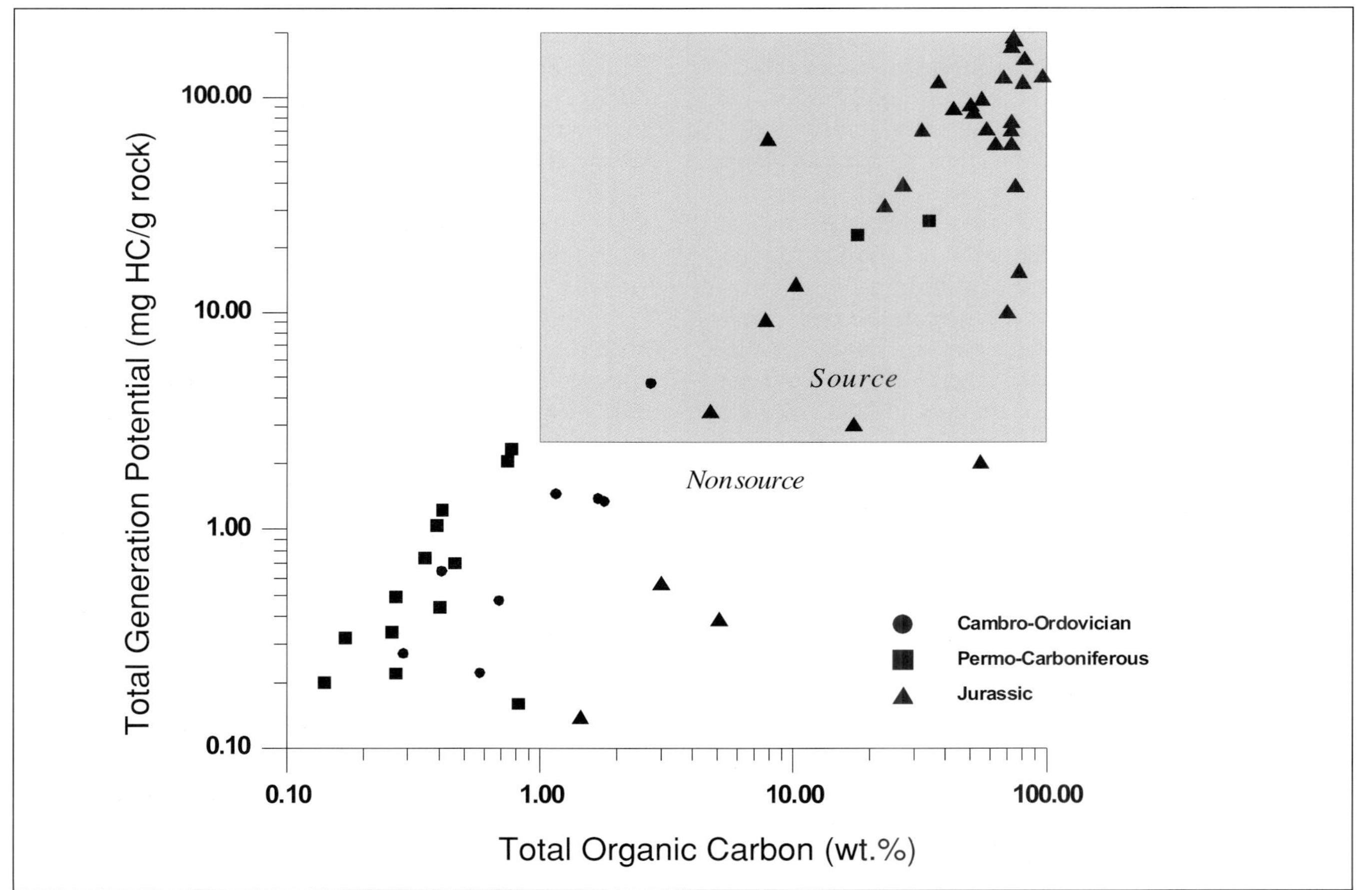

Figure 4. Generation potential versus organic-carbon content for possible source rocks in the Tarim Basin. Data sources include Graham et al. (1990) and Hendrix et al. (1995).

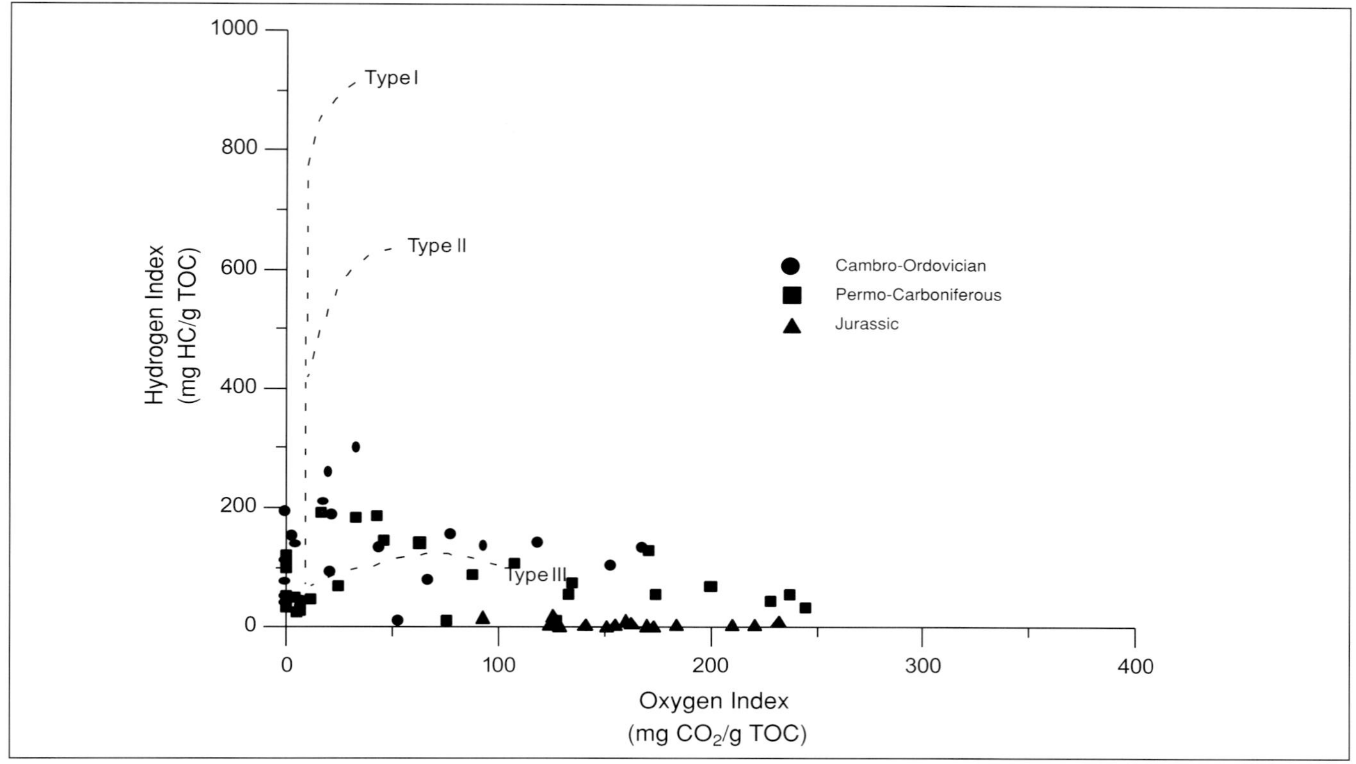

Figure 5. Modified van Krevelen–type diagram for possible source rocks in the Tarim Basin. Data sources include Graham et al. (1990) and Hendrix et al. (1995).

contents may be the result of contamination by migrated hydrocarbons.

Junhong Chen et al. (1996) suggested that the Permian-Carboniferous sequence in the Tarim Basin displays a thermal maturity equivalent to a vitrinite reflectance between 0.6 and 1.1%. The Triassic/Jurassic is immature except in the northern portion of the Tarim Basin, where Hendrix et al. (1995) reported vitrinite-reflectance values which may exceed 1.7%. They suggested that these more elevated thermal-maturity levels in the northern portion of the basin are the result of tectonic overburden caused by the stacking of thrust sheets.

King et al. (1994) examined the level of thermal maturity along the margins of the Junggar Basin. They concluded that the basin has been relatively cool since the Permian. Heat-flow levels appear to have been slightly above average during the middle Permian and have been cooling since then to their present values, which are below the global average. Available data suggest that the level of thermal maturity of the Permian sequence is highly variable, ranging from immature to overmature. For example, the Permian interval has a vitrinite reflectance between 0.79 and 1.07% in the Caican 4 well (Figure 8), but it ranges from 0.91 to 2.02% in the Aican 1 well (Figure 9). Carroll et al. (1992) reported vitrinite reflectances between 0.73 and 0.88% at their two sampled outcrop localities (Urumqi and Tianchi) in the southern Junggar Basin. The observed levels of thermal maturity at the two outcrop locations suggest that there has been a significant amount of removal of overburden, possibly on the order of 5 km.

Hendrix et al. (1995) reported that the Jurassic coals along the southern margin of the Junggar Basin are largely immature (R_o ranges between 0.45 and 0.71%).

Using the data reported by King et al. (1994), the top of the main stage of oil generation and release is located at about 3 km, with the base of the "oil window" at about 4300 m. A review of available isopach data (Ulmishek, 1984) suggests that the Permian source rock is overmature in about half of the basin (Figure 10). At these more advanced levels of thermal maturity, hydrocarbon prospectivity is largely dependent on the timing of hydrocarbon generation and the preservation of the trap. These data also imply that although the Permian lacustrine sequence was initially oil prone, large volumes of gas may be expected from this sequence.

Timing of Generation

Although thermal maturity can be measured, the timing of hydrocarbon generation can only be approximated through the use of thermal-maturation models, which integrate the effects of time and temperature on the conversion of kerogen to oil.

Hsu (1994) suggested that in the Tarim Basin, the Cambrian-Ordovician source in basinal positions generated and expelled hydrocarbons during the late Paleozoic

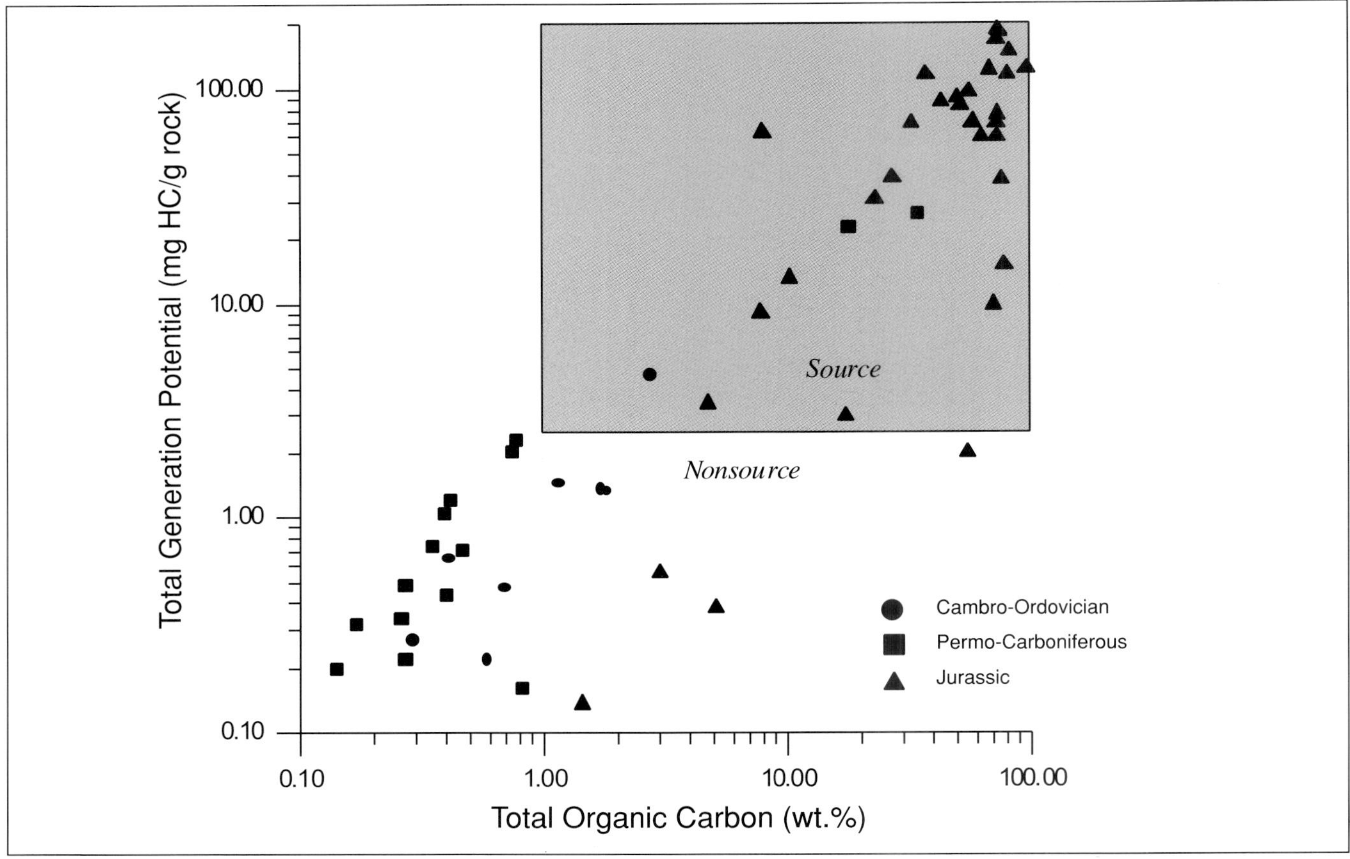

Figure 6. Generation potential versus organic-carbon content for possible source rocks in the Junggar Basin. Data sources include Graham et al. (1990), Carroll et al. (1992), and Hendrix et al. (1995).

and early Mesozoic and that generation was complete when buried by 5 km of sediment. Modeling results obtained in the vicinity of the Shacan 2 well indicate that initial generation from the Ordovician sequence would have begun during the early Carboniferous (Yang Bin, 1991). Generation was terminated and the trapped oil was uplifted during the Hercynian orogeny, which permitted biodegradation to proceed. Subsidence began again during the Mesozoic, during which a second phase of generation started. Xiao Xianming et al. (1996) also concluded that multiple generation events occurred in the central portion of the Tarim Basin in the vicinity of the Tazhong uplift. They concluded that the later phases of generation are of greater importance than the earlier phases, even though the magnitude of generation during those later episodes was significantly less. In part, this greater relative importance is a result of the breaching of earlier traps.

In the Junggar Basin, King et al. (1994) and Zhaohui Tang et al. (1997) suggested that the main stage of oil generation in the more basinal areas began early in the Triassic. Those simulations also suggest that in those basinal sequences, oil generation would have been completed prior to the close of the Jurassic (Figure 11). Lawrence (1990) further concluded that by the Cretaceous, those initially oil-prone rocks would have been generating sufficient gas to displace much of the oil that had been available to Permian-Triassic reservoirs. Those models also indicate that the Jurassic sources would have begun to contribute hydrocarbons during the Tertiary.

On the Junggar Basin margins, the generation history is much more complex. There are segments of the margin where hydrocarbon generation appears to be actively occurring. Elsewhere, the current level of thermal maturity appears to have been "frozen" because of the numerous uplift and erosional events that occurred throughout the Mesozoic and Tertiary, including a major unconformity at the base of the Tertiary. Postuplift subsidence and reburial have not commonly been sufficient for the maturation/generation process to be reinitiated. As noted above, about 5 km of overburden may have been removed. If the heat flow and/or geothermal gradients did not change significantly, at least this amount of reburial would be required for significant amounts of additional hydrocarbons to be generated. Carroll et al. (1992) further suggested that as a result of the structural history of the southern margin, hydrocarbon generation preceded the creation of potential traps formed by Neogene thrusting.

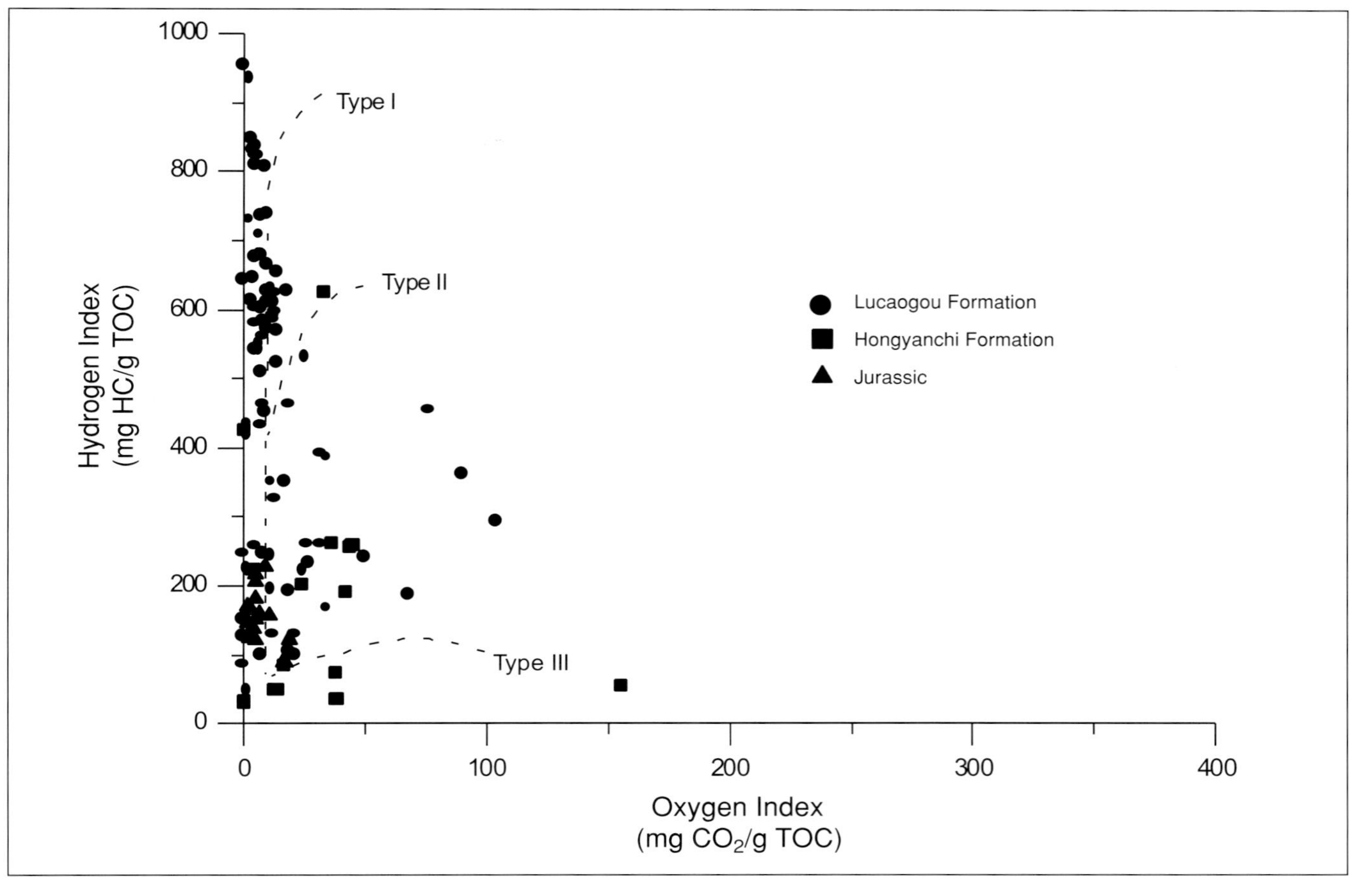

Figure 7. Modified van Krevelen–type diagram for possible source rocks in the Junggar Basin. Data sources include Graham et al. (1990), Carroll et al. (1992), and Hendrix et al. (1995).

Oil and Gas Geochemistry

Only limited source-rock data are available in the Tarim and Junggar Basins. This lack of data results from the few wells present in frontier situations, the position of wells outside the limits of the generative basin, and the often disassociated nature of the source-reservoir couplet. Nevertheless, a significant amount of information can be obtained from the geochemical character of oils and gas present in the two basins of interest.

Among the oils for which detailed geochemical data are available are those produced from the Shacan well, in the northern part of the Tarim Basin. Oils from this well, which produces from between 5363 and 5391 m from Ordovician karstified dolomites, has been genetically linked to the Cambrian-Ordovician sequence (Yang Bin, 1991). Several positive lines of evidence support this correlation. Gas chromatography reveals that the oils have a low pristane/phytane ratio (0.85/0.88) and a limited concentration of *n*-alkanes with greater than 20 carbons. Gas chromatography–mass spectrometry reveals sterane distributions dominated by C_{29} steranes (31%, 21%, and 48% for C_{27}, C_{28}, and C_{29} steranes, respectively), and the presence of C_{30} steranes. The produced oils also contain very limited quantities of trace metals (vanadium and nickel content is 2.29 ppm and 0.46 ppm, respectively) and display light carbon isotopic compositions ($\delta^{13}C$ –33.43 to –32.78‰). Negative evidence, which eliminated other possible source-rock candidates, including the nonmarine Upper Triassic to Middle-Lower Jurassic and the marine Carboniferous-Lower Permian intervals, include the absence of β-carotene, differences in sterane distributions, and the pristane/phytane ratios.

Detailed geochemical data are also available for oils and gases from Kekya field, in the southwestern part of the Tarim Basin. This field, largely a gas field, produces some light oil and condensate (API gravity ranges from 39° to 66°) from Oligocene-Miocene sandstones from between 3000 and 6450 m. The geochemical characteristics of these oils suggest that they were derived from a Permian-Carboniferous marine shale with a mixed marine and terrestrial source. As with the Shacan oil, the proposed correlation is based on both positive and negative lines of evidence. The supporting data for this assessment are the pristane/phytane ratios, which are greater than 1.5, and the saturate fraction stable carbon isotope values that are all heavier (i.e., more positive) than –30‰. These values, along with the regular sterane distribution, are more nearly consistent with the Permian-Car-

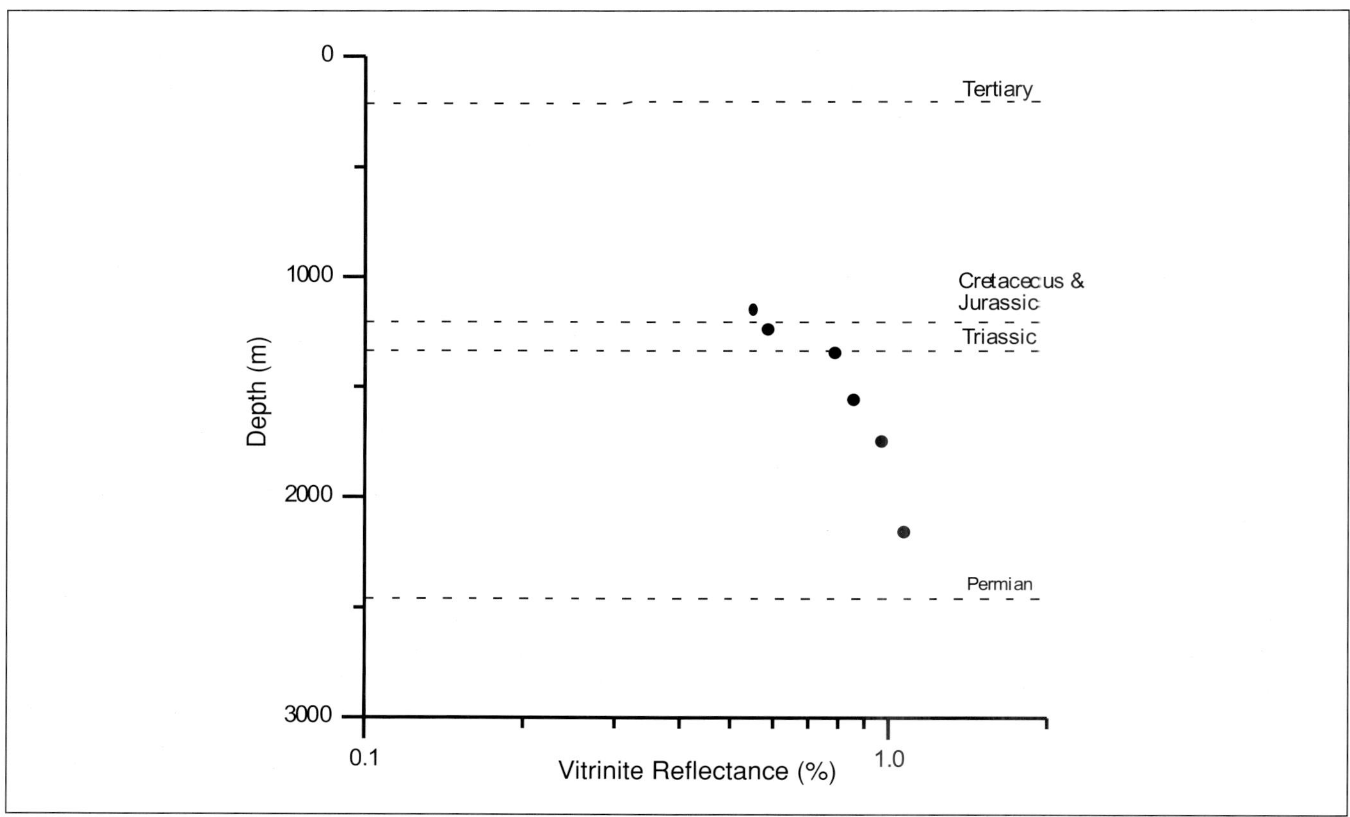

Figure 8. Vitrinite-reflectance profile for the Caican 4 well, Junggar Basin (data reported by King et al., 1994).

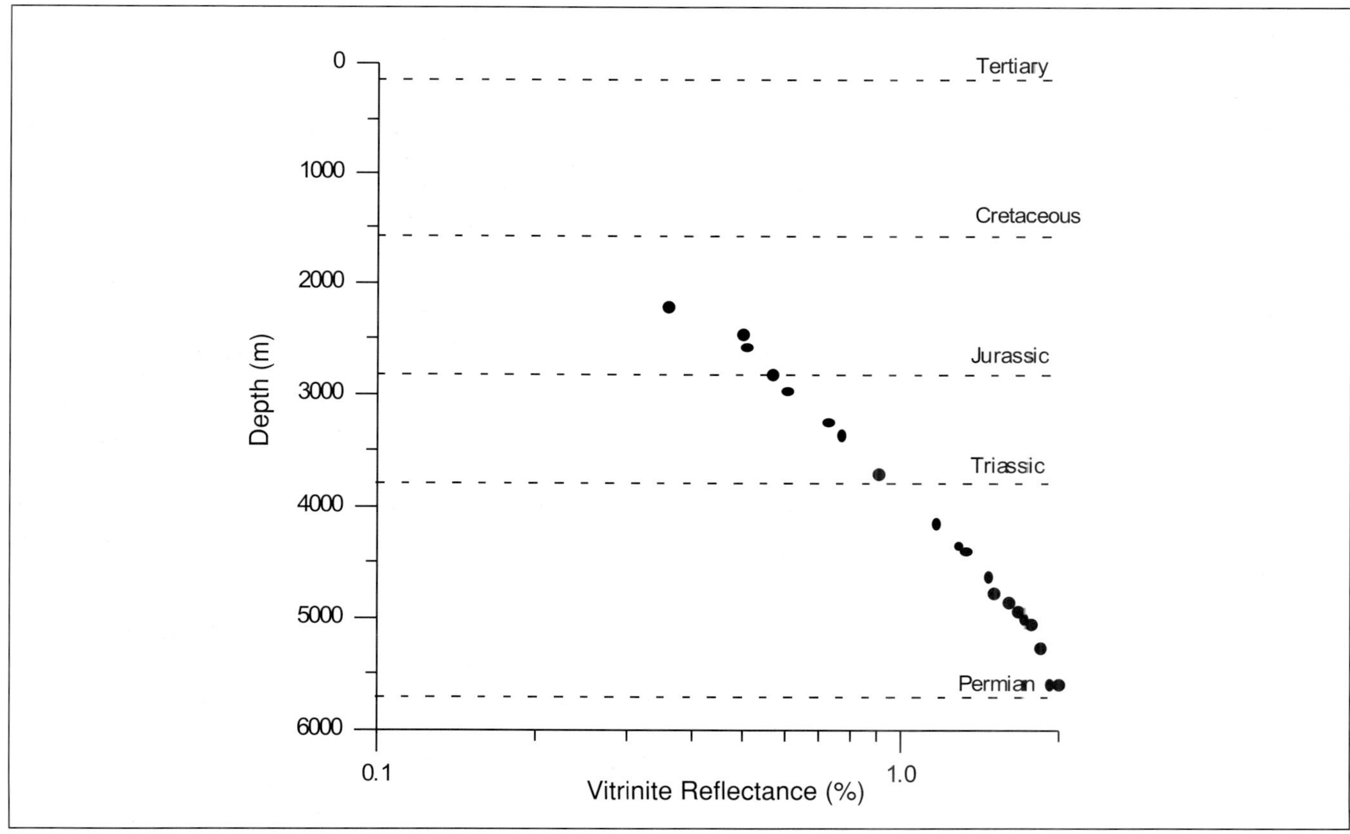

Figure 9. Vitrinite-reflectance profile for the Aican 1 well, Junggar Basin (data reported by King et al., 1994).

boniferous marine sequence than with any of the other possible source-rock sequences suggested for the basin The stratigraphic relationship between the inferred source and the reservoir suggests that vertical hydrocarbon migration has played some role.

Maowen Li et al. (1999) have further suggested that based on stable carbon isotope data, the gases and liquid hydrocarbons have been derived from different source-rock intervals. Although Maowen Li et al. suggested that the gases were derived from a Permian-Carboniferous coal-bearing sequence, the data presented are also consistent with a derivation from a more mature marine source-rock system. The light hydrocarbon stable carbon isotope data are consistent with a source-rock vitrinite reflectance greater than 1.6%. This suggests the possibility that the light hydrocarbons were derived from the Cambrian-Ordovician sequence. Additional data are required to differentiate between the two possible origins.

Maowen Li et al. (1999) presented data from Qunkuqiake field, also in the southwestern part of the Tarim Basin, and reported that hydrocarbons produced there are derived from a different source than those produced from Kekeya field (i.e., Cambrian-Ordovician rather than Permian-Carboniferous). Qunkuqiake field is, therefore, similar to the Tazhong complex, in the central part of the basin.

Hanson (1998) examined 23 oils from the Tarim Basin and concluded that at least seven oil families are present in the basin. He concluded, however, that those different oils were most probably derived from only two stratigraphic intervals: the Middle to Upper Ordovician and the Lower to Middle Jurassic. The differences in oil chemistry reflect facies variations within those stratigraphic intervals.

Data from some of the oils from the Tarim Basin suggest a high level of thermal maturity, with levels equivalent to vitrinite-reflectance values greater than 1.1% (Junhong Chen et al., 1996). The highest inferred levels of thermal maturity are associated with the northern and central parts of the Tarim Basin. Condensates produced from the Tazhong structure have a vitrinite-reflectance equivalence of ~1.9%. Those levels of thermal maturity are consistent with a derivation from the Cambrian-Ordovician sequence. Slightly lower levels of thermal maturity for the oils in Kekya field from the southwestern part of the basin are consistent with the proposed Permian-Carboniferous source for the field (Maowen Li et al., 1999). The possibility of multiple episodes of hydrocarbon-charging events in the Tarim Basin raises questions about whether a single maturity parameter can effectively represent an oil's thermal maturity (Zhao Hong et al., 1995); i.e., any estimated thermal maturity based on geochemical attributes may actually represent an "average" of the different charging episodes.

It is also interesting to note that the Shacan oil that contains a full suite of *n*-alkanes also contains a suite of 25-norhopanes (Yang Bin, 1991), which are often associated with severe biodegradation. The presence of both *n*-alkanes and the 25-norhopane series may also be used as evidence of multiple episodes of hydrocarbon generation and migration.

Clayton et al. (1997) reported geochemical data on oils from nine fields in the Junggar Basin. They suggested that at least five genetic oil types are present, representing a wide range of source-rock depositional settings. The Karamay oil type is the dominant type in the basin. It includes, in addition to oils from Karamay field, oils from Baikouquan, Fencheng, Xiazijie, Chepaizi, and Hongshanzui fields. Although Hsu (1994) proposed a lower Paleozoic marine source for at least some of the oils in Karamay field, Clayton et al. (1997) presented data that indicate a marginal, possibly saline, Permian lacustrine source. In part, this assessment is based on the relative abundance of β-carotene (β-carotene/nC_{30} greater than 1 and often exceeding 5), stable carbon isotope composition (saturate and aromatic hydrocarbons range from –30.9 to –28.4‰ and –29.0 to –27.9‰, respectively), normal sterane distribution ($C_{29} > C_{28} \geq C_{27}$), relative abundance of gammacerane (gammacerane indices typically greater than 30), and the *n*-alkane distribution. The *n*-alkane distribution includes an abundance of higher-molecular-weight (nC_{20+}) components. Lower Paleozoic marine oils lack β-carotene, are isotopically lighter, and are depleted in the longer chain *n*-alkanes. However, the distribution of *n*-alkanes within this group does display some important differences.

Clayton et al. (1997) placed Mahu field oils in a separate group and inferred a more basinal, possibly a fresher-water, lacustrine source for these oils than for the Karamay-type oils. This assessment is based on the lower relative abundance of β-carotene (β-carotene/$nC_{30} < 1$), a more depleted carbon isotopic composition (saturate and aromatic carbon isotope values range from –31.35 to –31.04‰ and –29.58 to –29.07‰, respectively), lower gammacerane indices (less than 30), and their different normal sterane distributions ($C_{29} \geq C_{28} \geq C_{27}$).

Jimusar field in the eastern Junggar Basin represents the third group of oils. Clayton et al. (1997) suggested that this group of oils was also derived from the Permian lacustrine sequence, specifically the Lacaogou Formation. It is differentiated from the group 1 oils principally by its isotopic composition. These oils are isotopically lighter. And although Clayton et al. noted that the observed isotopic variation is inconsistent with a common origin with the basin's other lacustrine oils, other lacustrine oils have been shown to display even greater isotopic variability (Katz and Mertani, 1989). Clayton et al. also suggested, on the basis of the low abundance of *n*-alkanes, that the Jimusar oils are less mature than the other oils studied from the basin. It is my opinion that the lower relative abundance of saturated hydrocarbons and the low *n*-alkane abundance are more probably a result of biodegra-

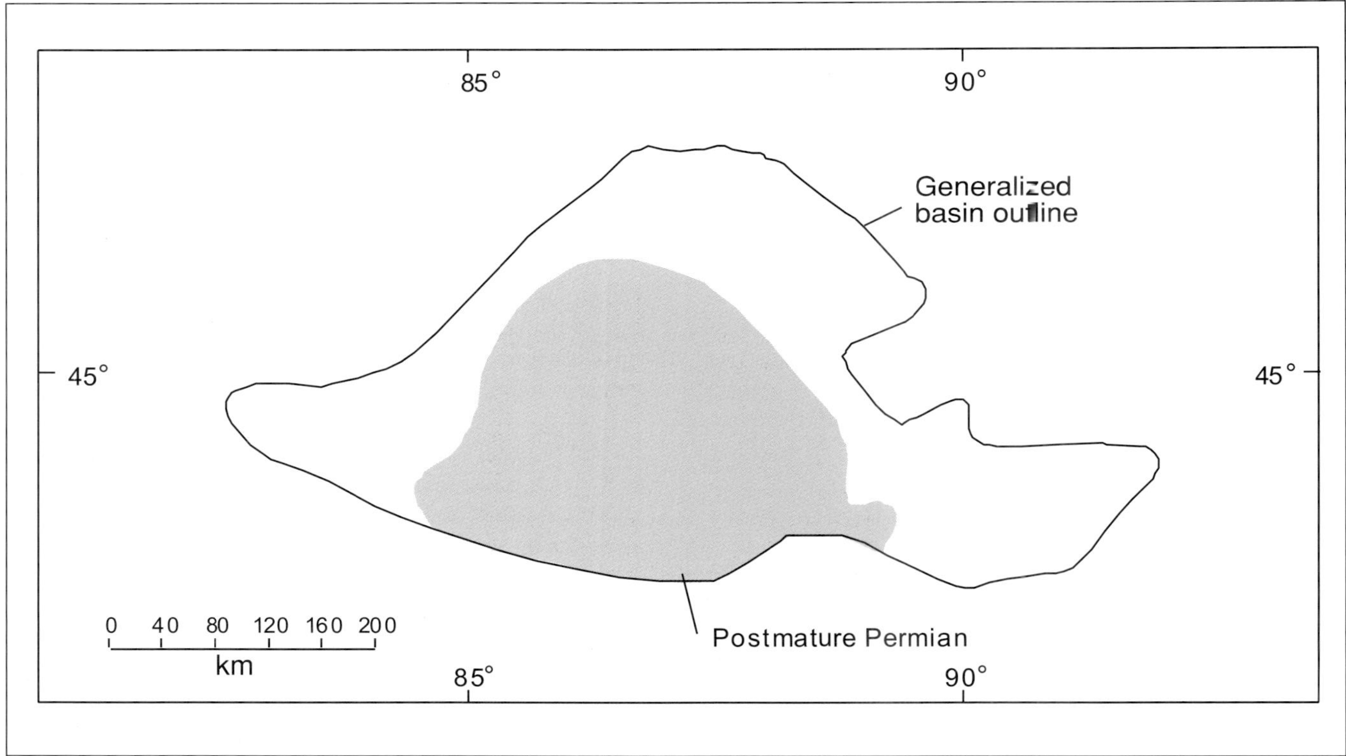

Figure 10. Generalized areal distribution of postmature Permian rocks in the Junggar Basin, based on the isopach data presented by Ulmishek (1984).

dation. Clayton et al. concluded in part that these oils were not biodegraded because of the absence of demethylated hopanes, but the absence of demethylated hopanes in an oil should not be equated to a lack of biodegradation (Peters and Moldowan, 1993).

The oils produced from Qigu and Dushanzi fields in the southern Junggar Basin are different from each other and from the other oils studied by Clayton et al. (1997). The Qigu oil displays the highest pristane/phytane ratio and the greatest relative abundance of C_{29} steranes. These characteristics are indicative of a source rich in terrestrial organic matter and would be consistent with derivation from the Jurassic coal measures. The Dushanzi oil is the isotopically heaviest oil studied by Clayton et al. ($\delta^{13}C$ of the saturated and aromatic hydrocarbon fractions are –27.63 and –25.19, respectively). This oil also has the highest relative abundance of C_{27} steranes. Possibly this oil was derived from an undefined Tertiary source rock.

Although Clayton et al. (1997) proposed that much of the Junggar Basin's oils can be correlated to the lacustrine Permian sequence, their sterane distributions are markedly different than those presented by Carroll (1998) for the extracts from the three Permian lacustrine units. Those differences could reflect either a different source than proposed or differences in analytical methods and the manner in which the sterane abundances are calculated. The available data do not permit a full examination of this observation.

RESERVOIR POTENTIAL

Potential reservoirs in China have been identified throughout the stratigraphic column as well as within altered "basement." The presence of potential reservoir rocks does not appear to be as much of a problem as producibility, reservoir continuity, and heterogeneity. For example, in Daqing field, as many as 24 separate producing zones may be present. The typical well, however, produces on the average only ~320 BOPD (Meyerhoff and Willums, 1981) because of reservoir and crude-oil properties. To compensate for these low production rates, large numbers of wells are often required. In the Daqing complex, more than 10,000 wells have been drilled and more than 3000 are actively producing.

Tarim Basin

In the Tarim Basin, potential reservoirs have been identified in the Ordovician, Carboniferous, Triassic, Paleogene, and Neogene sequences, with many of those reservoir targets at depths below 5000 m. The lower Paleozoic reservoirs are marine carbonates, whereas the remaining reservoirs are sandstones. The Carboniferous sandstones are marine, and the Mesozoic and Tertiary reservoirs are nonmarine. Little detailed information is publicly available on these different reservoir rocks.

In the Shacan 2 well (in Yakla field in the northern Tarim Basin), karsted carbonates flowed with initial production exceeding 7000 BOPD and 70 MMCFGD. The Yingmai 1 well tested at 1300 BOPD from a 23-m interval and from an Ordovician karsted dolomite (Chai Guilin et al., 1992). Fractured Ordovician limestones in Lunnan field (Lunnan 8) produced at high initial flow rates as well (5000 BOPD and 6.6 MMCFGD) (Chai Guilin et al., 1992), whereas other wells in the field had daily flow rates between 177 BOPD and 3150 BOPD. The heterogeneity of these Ordovician reservoirs appears to be typical and can be even more dramatic. On the Tazhong uplift in the central Tarim Basin, the Tazhong 1 well flow tested at ~4000 BOPD, whereas the Tazhong 3 well (approximately 25 km away) tested dry in the same interval (Li Desheng et al., 1996). No mention is made of production from preserved primary porosity.

Carboniferous, marine quartz sandstones penetrated on the Tazhong structure displayed porosity and permeability values of 15.6 % to 21.3% and 171 md to 407 md, respectively (Li Desheng et al., 1996). Carboniferous sandstones appear to be on the order of 5 to 10 m thick on the Tazhong uplift and may exceed 250 m in the northern part of the basin (Donghetang field). Those sandstones produce at rates as high as 1850 BOPD per well.

Nonmarine Triassic sandstones have flowed at initial rates exceeding 3600 BOPD in Lunnan field in the northern Tarim Basin, with similar rates being obtained in the Tazhong 4 and 10 wells (Li Desheng et al., 1996). Individual Triassic sandstones may achieve a gross thickness of as much as ~50 m.

Eocene sandstones in the northern part of the basin have also produced hydrocarbons with flow rates of more than 1200 bbl/day of condensate (Li Desheng et al., 1996) in Yaha field. In Yimali field, basal Eocene sandstones are 30 to 40 m thick.

Kekya field in the southwestern part of the basin produces largely from a series of Oligocene and Miocene sandstones. The 15 Oligocene-Miocene sandstones have effective porosities of 13.3% to 15.9% and permeabilities of 11.2 to 90.6 md (Hu Boliang, 1992). The thickest individual pay section is slightly less than 100 m. Most pay sections, however, are less than 50 m thick.

Junggar Basin

Reservoir rocks within the Junggar Basin are present largely in Permian, Triassic, and Jurassic lacustrine, fluvial, and alluvial fan deposits. Minor production has also been established from Lower Cretaceous, Oligocene, and Miocene sandstones.

Zhaohui Tang et al. (1997) suggested that the Permian, Triassic, and Jurassic sandstones are characterized by volcanic litharenites (Figure 12). Diagenetic studies by Zhaohui Tang et al. suggested that calcite cementation and authigenic clay and zeolite formation have substantially reduced porosity. Secondary porosity development varies considerably within these sandstones. Fluvial sandstones tend to display the highest average porosity and permeability values because of the retention of primary porosity through the formation of early clay coatings that prevented compaction.

The most important field in the basin is Karamay. Much of its production comes from Permian through Cretaceous alluvial fans. As would be expected in an alluvial-fan complex, these reservoirs are poorly sorted (Chang Chiyi, 1981). Individual layers have highly variable reservoir properties. Taner et al. (1988) reported permeability values for the alluvial-fan sequences ranging as high as 323 md and porosities as high as 39%. The field also produces from nearshore lacustrine deposits. Permeability in these fine- to medium-grained sandstones is significantly better, ranging between 1704 and 6783 md. Porosity values in these sandstones may reach 35%. Effective thickness of the alluvial fan reservoirs in Karamay, field may exceed 18 m (Chang Chiyi, 1981). Because of the low gravity (as low as 10° API) and biodegraded nature of the oil in the shallow pay zone (less than 500 m), production rates from individual wells in Karamay field were generally low until the discovery of the lighter, less degraded, deeper, principally Triassic pool. Wells in the shallow pay zones also experienced rapid decline, to the point that the average well was producing only 14 bbl/day. It is also interesting to note that the best-producing wells from Karamay field are located close to fault zones, especially those near intersecting faults (Taner et al., 1988).

The now abandoned Dushanzi field in the southwestern part of the Junggar Basin produced from Oligocene–Miocene reservoirs. Each of those sandstones ranges in thickness from 3 to 9 m. Effective porosity levels were between 12% and 18% (Taner et al., 1988). Individual wells had initial flow rates as high as 511 BOPD, but they rapidly declined to a few tens of barrels per day.

EXPLORATION IMPLICATIONS

An effective petroleum system requires the presence of a source, seal, reservoir, trap, and the necessary thermal maturation for hydrocarbon generation to proceed. In addition to the simple presence of these components, they must share a favorable temporal and geographic relationship. In this study, trap and seal have largely been assumed. However, as to the other critical components, the available data place some significant limitations on the remaining exploration potential of the two basins.

Tarim Basin

Within the Tarim Basin, distribution and volume of the organic-rich Ordovician facies are clearly problematic. Although the Ordovician sequence represents a significant rock volume, possibly as much as 2 million km^3

(Hsu, 1994), the available sampling suggests that the actual volume of effective source is much more limited and currently displays only limited source-rock potential. Sufficient geochemical differences are present in the oil to suggest that even those oils derived from a single stratigraphic level display facies differences suggesting that the effective source basins for each accumulation are restricted. In the most optimistic case, even if this source were universally present, it would be largely overmature. Generation appears to have begun relatively early in the basin's history, prior to the formation of many of the basin's current structural targets. Although multiple episodes of hydrocarbon generation appear to have occurred, the last episode, considered to be the most important relative to current traps, appears to have been the least significant volumetrically.

Hydrocarbons preserved in the basin would also reflect this advanced level of maturity (i.e., light oil, condensate, and large volumes of gas). The advanced level of thermal maturity would also open up the possibility of displacement of liquid hydrocarbons from early preserved traps through the late introduction of gas.

Secondary source rocks in the Tarim Basin are largely gas prone, although some lacustrine oil source-rock potential may exist. The data suggest, however, that with the exception of the northern part of the basin (i.e., the Kuche depression), much of these younger sources would be thermally immature, and these rocks would not contribute significantly to the resource base.

The best producing reservoirs in the Tarim Basin appear to be within the Ordovician karsted carbonate sequence. Prediction of those reservoirs will require a detailed understanding of their uplift and erosion history. Clearly, this history is complex, as evidenced by nearby wells from the same field having dramatically different production rates and reservoir properties. Some of the sandstones display only modest porosity and permeability values, making them better gas reservoirs than oil reservoirs. The younger Tarim Basin sandstone reservoirs can have significant production rates, but because of their nonmarine character, they display limited lateral and vertical continuity, resulting in reservoir compartmentalization.

Junggar Basin

In the Junggar Basin, a different story has evolved. The primary source rock is a lacustrine Permian shale. Although the properties of this rock have been well documented, its detailed distribution is not known. This source-rock sequence displays a wide range of thermal-maturity levels. However, more than half of the basin's area appears to have matured beyond the main stage of oil generation. In fact, a large portion of the basin has achieved levels of thermal maturity that would be consistent with gas generation.

In the central portion of the basin, it appears that the

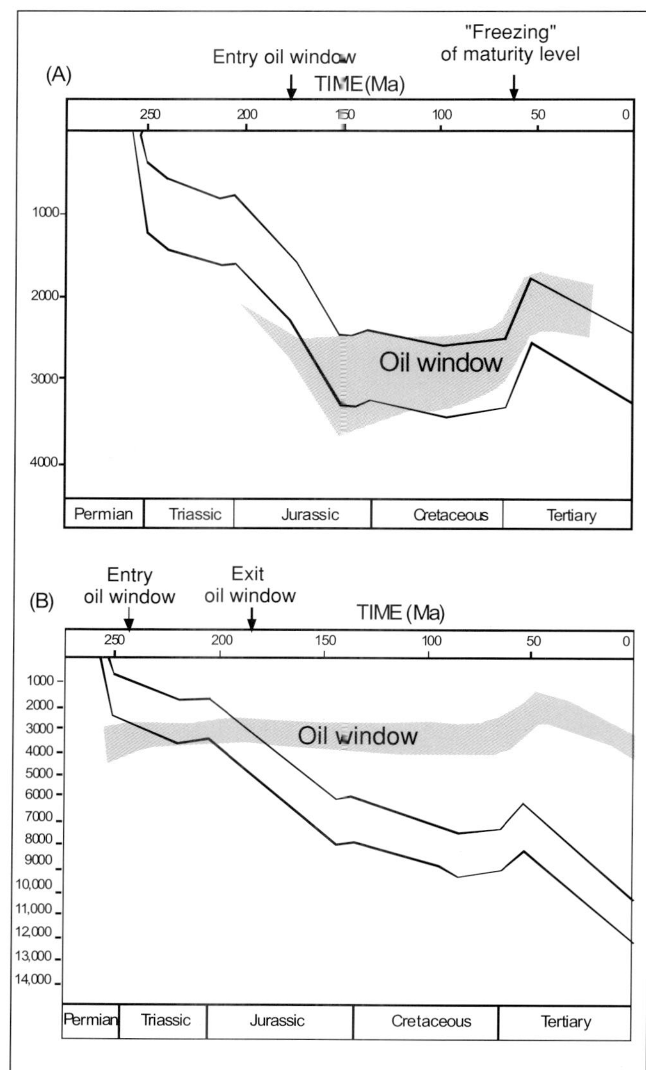

Figure 11. Thermal-maturation history of the Lucaogou Formation in the Junggar Basin along the southeastern (A) and southwestern (B) margins (after Zhaohui Tang et al., 1997).

main stage of generation would have been completed prior to the Cretaceous. On the basin margins, the generation history in the Junggar Basin is much more complex than in the center of the basin because of thrusting, uplift, and erosion. However along the basin margins, generation still appears to have occurred prior to the development of the most recent traps. The currently observed levels of thermal maturity were probably "frozen" in place because of the significant amount of erosion that has occurred. The potential for preserving some of the accumulations that may have been trapped is reduced because of the potential for breaching of the earlier traps by thrusting.

Reservoir potential in the Junggar Basin also appears to be somewhat limited. The mineralogic immaturity of the sandstones tends to result in a reduction in both po-

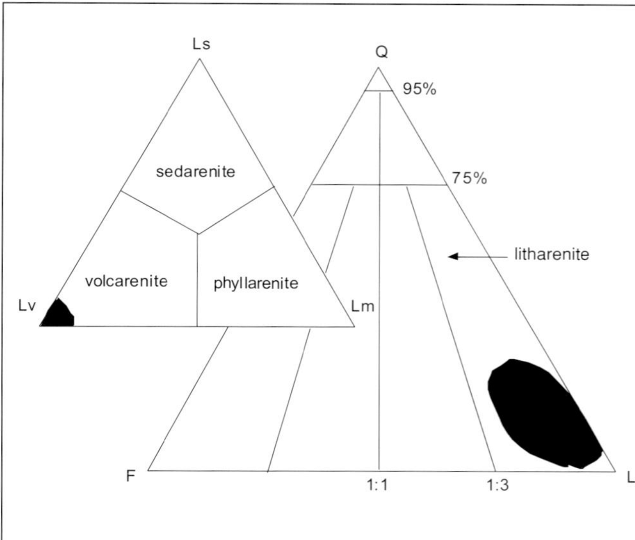

Figure 12. Classification of Permian-Triassic sandstones of the Junggar Basin (modified from Zhaohui Tang et al., 1997).

rosity and permeability. In the nearshore lacustrine sandstones, where both porosity and permeability tend to be higher, lateral and vertical reservoir continuity appear to remain a problem. Problems with potential reservoirs are complicated further by the typically very rapid drop-off in production from their initial flow rates.

Therefore, it appears that the exploration potential of these basins, which have been cited as among those with the greatest exploration potential onshore China, is significantly less than previously suggested. Although sweet spots do exist, several components of the respective petroleum systems of these basins appear to be limited and cannot be easily extrapolated beyond known limits.

SUMMARY AND CONCLUSIONS

Available geochemical data, both rock and oil data, suggest the presence of multiple source-rock zones in the Tarim and Junggar Basins. Included in the suite of source-rock candidates is the Permian lacustrine Lucaogou Formation of the Junggar Basin, which ranks among the richest and thickest. These data further suggest that there are significant facies variations in the different source intervals. These facies differences amplify the problems associated with establishing the distribution and volume of each of the source intervals. This is particularly true for the Cambrian-Ordovician source sequence in the Tarim Basin.

Even if the source rocks are widely distributed, thermal-maturity data indicate that much of the source sequences in the Tarim and Junggar Basins are currently overmature, and hydrocarbon generation has often preceded the development of the trap. The advanced level of thermal maturity suggests that there is also a potential for gas displacement of any liquid hydrocarbons that may have been retained by traps during these basins' multiple tectonic episodes. Further complicating the hydrocarbon-charge story is the potential for multiple episodes of hydrocarbon generation. Typically, the latest episodes would be most significant. Unfortunately, the generation potential remaining after the earlier episodes reduces the volume of hydrocarbons that could be generated in the Tarim and Junggar Basins.

In both the Tarim and Junggar Basins, reservoir heterogeneity and the lack of reservoir continuity place great limitations on the exploration potential of onshore China. These problems exist in both the Ordovician carbonate reservoirs of the Tarim Basin and in the Permian through Jurassic nonmarine sandstones of the Junggar Basin. Production rates from individual wells within a single field can vary from less than 200 to more than 5000 BOPD. Many of the wells also appear to experience a very rapid decline in production.

Based on these observations, it appears that the overall hydrocarbon discovery potential of these two basins is limited and that many of the previous estimates of their discovery potential are inflated. Clearly, the current fields suggest that hydrocarbon potential does exist in both of these basins and that sweet spots may exist. However, the exploration challenges and risks appear to outweigh the potential rewards. This is particularly the case when the remoteness of the basins and the lack of a local market are considered.

ACKNOWLEDGMENTS

I thank Texaco Group Inc. for permission to publish this work. This manuscript was improved through the comments of Robert K. Sawyer.

REFERENCES CITED

Carroll, A. R., 1998, Upper Permian lacustrine organic facies evolution, southern Junggar Basin, NW China: Organic Geochemistry, v. 28, p. 649–667.

Carroll, A. R., S. C. Brassell, and S. A. Graham, 1992, Upper Permian lacustrine oil shales, southern Junggar Basin, northwest China: AAPG Bulletin, v. 76, p. 1874–1902.

Chai Guilin, Wang Xiaomu, and Jin Xuezheng, 1992, Petroleum geology and oil potential of Tarim Basin, west China, *in* Proceedings of the Thirteenth World Petroleum Congress: Chichester, England, John Wiley & Sons, v. 2, p. 15–23.

Chang Chiyi, 1981, Alluvial-fan coarse clastic reservoirs in Karamay, *in* J. F. Mason, ed., Petroleum geology in China: Tulsa, Oklahoma, PennWell Books, p. 154–170.

Clayton, J. L, Jianqiang Yang, J. D. King, P. G. Lillis, and A. Warden, 1997, Geochemistry of oils from the Junggar Basin, northwest China: AAPG Bulletin, v. 81, p. 1926–1944.

Daley, A. R., and J. D. Edman, 1987, Loss of organic carbon from source rocks during thermal maturation (abs.): AAPG Bulletin, v. 71, p. 546.

Graham, S. A., A. Brassell, A. R. Carroll, X. Xiao, G. Demaison, C .L. Mcknight, Y. Liang, J. Chu, and M. S. Hendrix, 1990, Characteristics of selected petroleum source rocks, Xianjiang Uygur Autonomous Region, northwest China: AAPG Bulletin, v. 74, p. 493–512.

Guo Jian Hua and Zhu Yangming, 1995, Sedimentary features and source rocks of Cambrian-Ordovician systems in well Kunan No. 1: Acta Petrol Sinica, v. 16, no. 2, p. 8–15 (in Chinese).

Hanson, A. D., 1998, Organic geochemistry and petroleum geology, tectonics and basin analysis of southern Tarim and northern Qaidam Basins, northwest China: Ph.D. dissertation, Stanford University, Stanford, California, 565 p.

Hendrix, M. S., 1992, Sedimentary basin analysis and petroleum potential of Mesozoic strata, northwest China: Ph.D. dissertation, Stanford University, Stanford, California, 388 p.

Hendrix, M. S., S. C. Brassell, A. R. Carroll, and S. A. Graham, 1995, Sedimentology, organic geochemistry, and petroleum potential of Jurassic coal measures: Tarim, Junggar, and Turpan Basins, northwest China: AAPG Bulletin, v. 79, p. 929–959.

Hsu, K. J., 1994, Buried-euxinic-basin model sets Tarim Basin potential: Oil & Gas Journal, v. 92, no. 48, p. 51–60.

Hu Boliang, 1992, Petroleum geology and prospects of the Tarim (Talimu) Basin, China, *in* M. T. Halbouty, ed., Giant oil and gas fields of the decade 1978–1988: AAPG Memoir 54, p. 493–510.

Hu Jianyi, Zhou Xingxi, Xu Shubao, and Li Qiming, 1997, New oil and gas exploration frontiers and resource potential of the old cratons in NE Asia, in Proceedings, 30th International Geological Congress: Utrecht, Netherlands, VSP, v. 18, p. 45–56.

Junhong Chen, Jiamo Fu, Guoying Sheng, Dehan Liu, and Jianjun Zhang, 1996, Diamonoid hydrocarbon ratios: Novel maturity indices for highly mature crude oils: Organic Geochemistry, v. 25, p. 179–190.

Katz, B. J., P. A. Kelley, R. A. Royle, and T. Jorjorian, 1991, Hydrocarbon products of coal as revealed by pyrolysis-gas chromatography: Organic Geochemistry, v. 17, p. 711–722.

Katz, B. J., and B. Mertani, 1989, Central Sumatra—A geochemical paradox: Indonesian Petroleum Association, 18th Annual Convention Proceedings, v. 1, p. 403–425.

King, J. D., Jianqiang Yang, and Fan Pu, 1994, Thermal history of the periphery of the Junggar Basin, northwestern China: Organic Geochemistry, v. 21, p. 393–405.

Klett, T. R., T. S. Ahlbrandt, J. W. Schmoker, and G. L. Dolton, 1997, Ranking of the world's oil and gas provinces by known petroleum volumes: U.S. Geological Survey Open File Report 97–463, CD-ROM.

Lawrence, S. R., 1990. Aspects of the petroleum geology of the Junggar Basin, northwest China, *in* J. Brooks, ed., Classic petroleum provinces: Geological Society (London) Special Publication 50, p. 545–557.

Li Desheng, 1985, Tectonic types of oil and gas basins in China: Oil and Gas in China (selected papers from Acta Petrolei Sinica), v. 1, p. 1–19.

Li Desheng, 1995, Hydrocarbon occurrences in the petroliferous basins of western China: Marine and Petroleum Geology, v. 12, p. 26–34.

Li Desheng, 1996, Basic characteristics of oil and gas basins in China: Journal of Southeast Asian Earth Sciences, v. 13, p. 299–304.

Li Desheng, Liang Digang, Jia Chengzao, Wang Gang, Wu Qizhi, and He Dengfa, 1996, Hydrocarbon accumulations in the Tarim Basin, China: AAPG Bulletin, v. 80, p. 1587–1603.

Maowen Li, Renzi Lin, Yongsheng Liao, L. R. Snowdon, Peilong Wang, and Peilong Li., 1999, Organic geochemistry of oils and condensates in the Kekya field, southwest depression of the Tarim Basin (China): Organic Geochemistry, v. 30, p. 15–37.

Meyerhoff, A. A., and J.-O. Willums, 1981, Petroleum in the People's Republic of China, 1949-1979, *in* M. T. Halbouty, ed., Energy resources of the Pacific region: Tulsa, Oklahoma, AAPG, p. 195–214.

Peters, K. E., and J. M. Moldowan, 1993, The Biomarker guide—Interpreting molecular fossils in petroleum and ancient sediments: Englewood Cliffs, New Jersey, Prentice Hall, 363 p.

Ritts, B. D., 1998, Mesozoic tectonics and sedimentation, and petroleum systems of the Qaidam and Tarim Basins, NW China: Ph.D. dissertation, Stanford University, Stanford, California, 691 p.

Taner, I., M. Kamen-Kaye, and A. A. Meyerhoff, 1988, Petroleum in the Junggar Basin, northwestern China: Journal of Southeast Asian Earth Sciences, v. 2, p. 163–174.

Tian Zaiyi, 1990, The formation and distribution of Mesozoic-Cenozoic sedimentary basins in China: Journal of Petroleum Geology, v. 13, p. 19–34.

Ulmishek, G., 1984,. Geology and petroleum resources of basins in western China: Argonne National Laboratory Technical Report ANL/ES-146, 131 p.

Vogel, H. U., 1993, The great well of China: Scientific American, v. 268, no. 6, p. 116–121.

Yang Bin, 1991, Geochemical characteristics of oil from well Shacan 2 in the Tarim Basin: Journal of Southeast Asian Earth Sciences, v. 5, p. 401–406.

Xiao Xianming, Liu Dehan, and Fu Jiamo, 1996, Multiple phases of hydrocarbon generation and migration in the Tazhong petroleum system of the Tarim Basin, People's Republic of China: Organic Geochemistry, v. 25, p. 191–197.

Xie Hong, 1993, Petroleum geology of China—Tarim Basin: Xinjiang, China, Xinjiang Artistic Photography Publishing House, 160 p.

Zhao Hong, Wang Peirong, Chen Qi, Yao Huanxin, and Zhu Junzhang, 1995, Adamantane and its application to studying the maturity of crude oils from Tarim Basin: Journal of the Jianghan Petroleum Institute, v. 17, p. 24–30.

Zhaohui Tang, J. Parnell, and A. H. Ruffell, 1994, Deposition and diagenesis of the lacustrine-fluvial Cangfanggou Group (uppermost Permian to Lower Triassic), southern Junggar Basin, NW China: A contribution from sequence stratigraphy: Journal of Paleolimnology, v. 11, p. 67–90.

Zhaohui Tang, J. Parnell, and F. J. Longstaffe, 1997, Diagenesis and reservoir potential of Permian-Triassic fluvial/lacustrine sandstones in the southern Junggar Basin, northwestern China: AAPG Bulletin, v. 81, p. 1843–1865.

Central and South America

Guzmán, A. E., and B. Márquez-Domínguez, 2001, The Gulf of Mexico Basin south of the border, *in* M. W. Downey, J. C. Threet, and W. A. Morgan, eds., Petroleum provinces of the twenty-first century: AAPG Memoir 74, p. 337–351.

Chapter 17

THE GULF OF MEXICO BASIN SOUTH OF THE BORDER: *THE* PETROLEUM PROVINCE OF THE TWENTY-FIRST CENTURY

Alfredo E. Guzmán
PEMEX Exploracion y Producción, Coordinación de Estrategias de Exploración
Villahermosa, Tabasco, Mexico

Benjamín Márquez-Domínguez
PEMEX Exploracion y Producción, Activo de Exploración Misantla-Golfo
Poza Rica, Veracruz, Mexico

ABSTRACT

The Mexican portion of the Gulf of Mexico Basin (MGOM) extends onshore into several oil- and/or gas-producing basins: Burgos, Tampico-Misantla, Veracruz, and Sureste (the Sureste Basin includes the Salina del Istmo, Comalcalco-Chiapas-Tabasco, Macuspana, Sonda de Campeche, and Litoral de Tabasco provinces). To the east, the MGOM includes the nonproducing Plataforma de Yucatán. The deep-water Gulf of Mexico has been subdivided into eight provinces: Franja Distensiva, Delta del Río Bravo, Franja de Sal Alóctona, Cinturón Plegado de Perdido, Cordilleras Mexicanas, Cañon de Veracruz, Salina del Golfo Profundo, and Planicie Abisal. Based on the petroleum systems and exploration history of these provinces, most undiscovered reserves are likely to be found in the deep water of the MGOM.

INTRODUCTION

The Mexican portion of the Gulf of Mexico is limited on the north by the maritime United States–Mexico border; on the west and south by the states of Tamaulipas, Veracruz, and Tabasco; and on the east by the states of Campeche and Yucatán and the maritime Cuba–Mexico border (Figure 1). Geologically, the Mexican portion of the Gulf of Mexico Basin (MGOM) extends onshore (Figure 2) into several oil- and/or gas-producing basins: Burgos on the northwest, Tampico-Misantla on the west, and Veracruz and Sureste on the south. (The Sureste Basin includes three Tertiary provinces formed within the Mesozoic basin: Salina del Istmo, Comalcalco [which includes the Mesozoic province of Chiapas-Tabasco], and Macuspana, and two offshore Mesozoic provinces: Sonda de Campeche and Litoral de Tabasco.) On the east, the MGOM includes the Plataforma de Yucatán, limited by the Caribbean. The deep-water Gulf of Mexico may be subdivided into eight provinces: Franja Distensiva, Delta del Río Bravo, Franja de Sal Alóctona, Cinturón Plegado de Perdido, Cordilleras Mexicanas, Cañon de Veracruz, Salina del Golfo Profundo, and Planicie Abisal.

In this paper, we succinctly describe the petroleum geology of these basins and of the deep-water MGOM (>200-m depth). We expect that a large percentage of

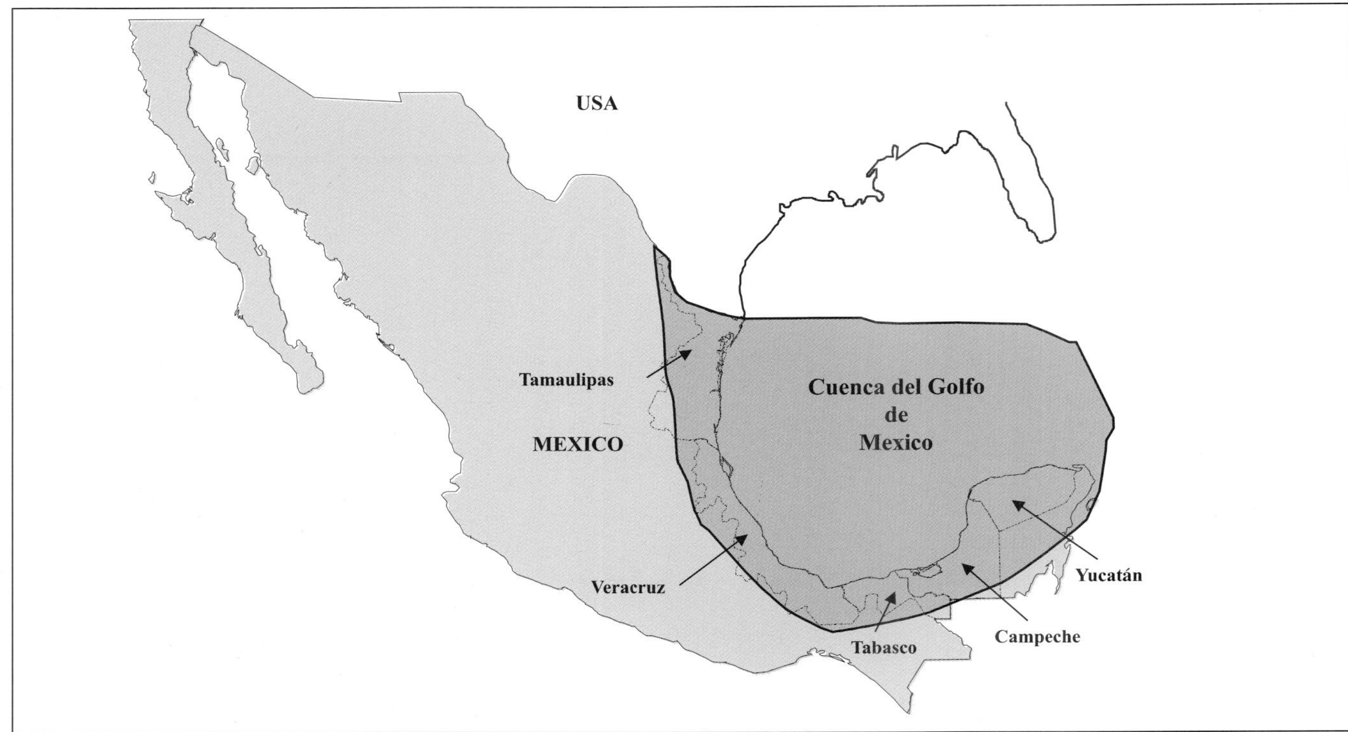

Figure 1. The Mexican portion of the Gulf of Mexico.

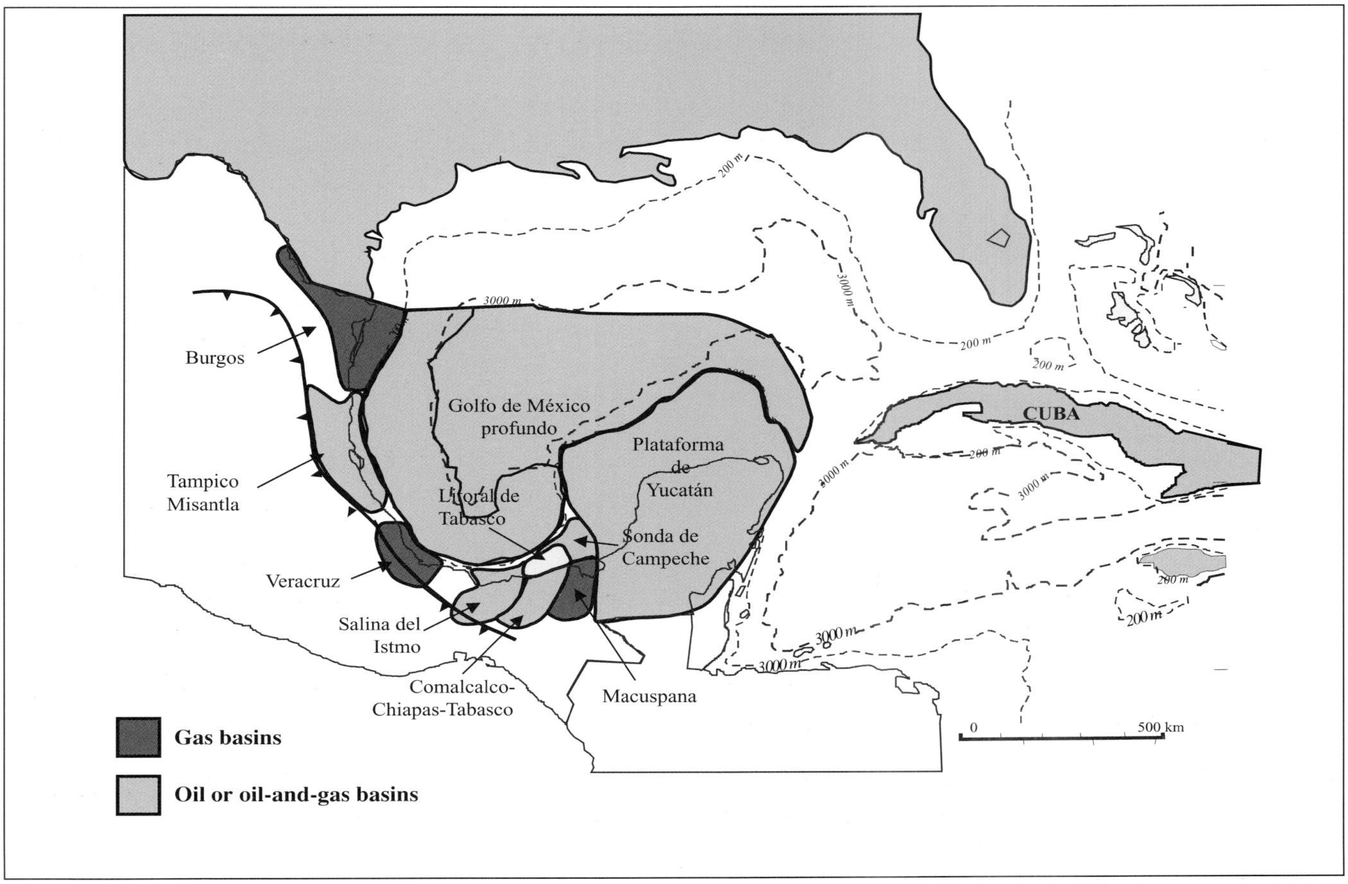

Figure 2. Provinces of the Mexican portion of the Gulf of Mexico Basin (MGOM).

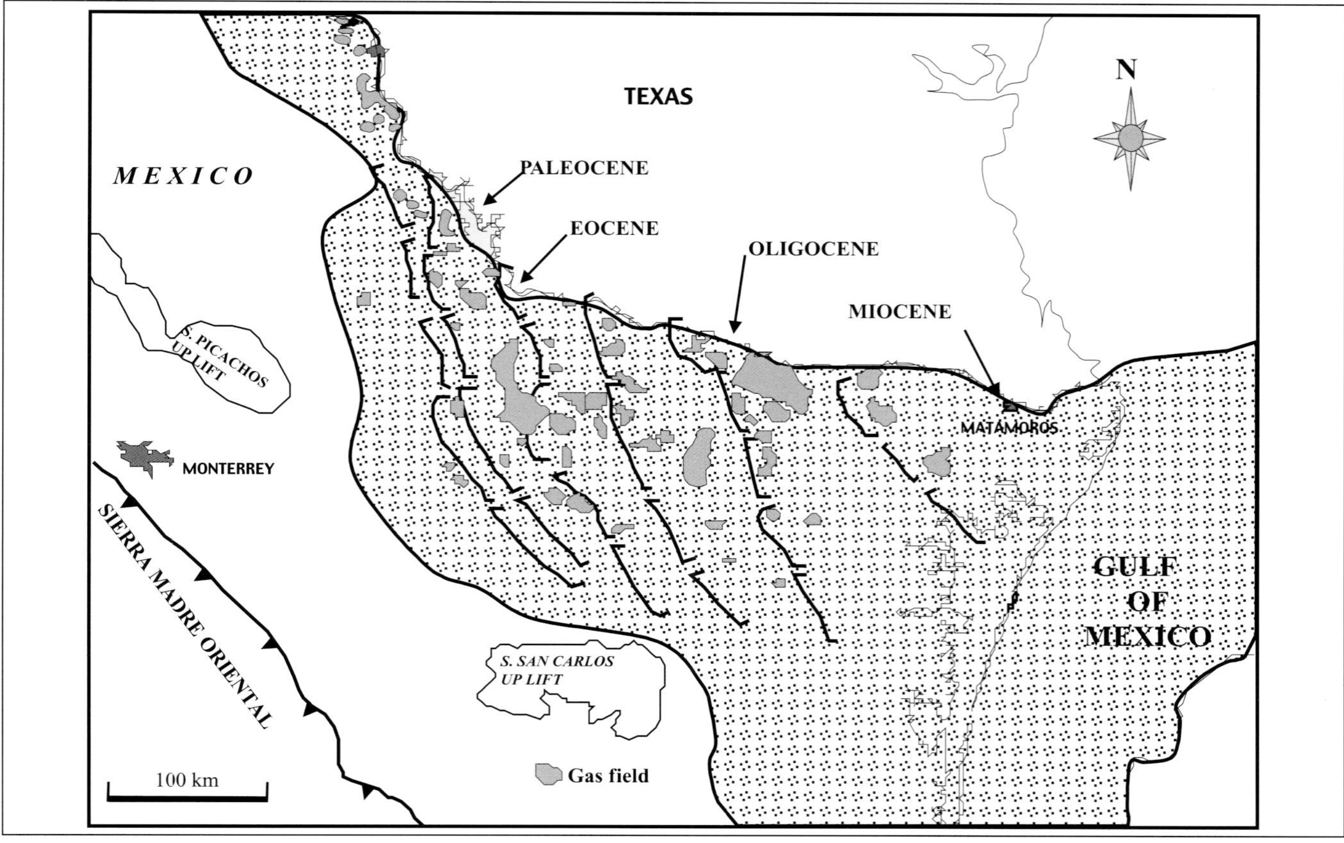

Figure 3. Burgos Basin. Ages and arrows refer to the ages of the play trends (flexure trends).

the oil and gas that will be produced in the twenty-first century will come from reserves yet to be found in the MGOM megaprovince.

PETROLEUM GEOLOGY OF THE MEXICAN PORTION OF THE GULF OF MEXICO BASIN

Burgos Basin

The Burgos Basin (Cuenca de Burgos) covers about 70,000 km^2, including part of the continental shelf to 200-m water depth (Figure 3). The basin is the northwesternmost extension into Mexico of the MGOM and is a continuation to the south of the Rio Grande embayment of south Texas. It consists of a thick (>10-km) wedge of Paleocene to Miocene siliciclastics deposited in a passive margin under extensional conditions. Most plays are continuous with those of the U.S. side of the basin. Production, mainly of nonassociated dry and wet gas, started in 1945. Since then, 187 fields have been discovered, 84 of which are still active today. Since 1994, through the extensive application of new petroleum-geoscience concepts and new technology, the main fields have been rejuvenated from a minimum gas output of 183 million cubic feet (mmcf) per day in late 1993 (after having reached 620 mmcf/day in 1970) to more than 1000 mmcf/day at the end of 1999 (Figure 4). Cumulative output is 6268 billion cubic feet (bcf), and total reserves are 7352 bcf.

The rejuvenation project for the basin, currently under way, calls for acquiring 9850 km^2 of 3-D seismic, drilling 186 exploratory wells, adding 6750 bcf of new gas reserves (P_{50}), and drilling more than 1500 development wells, thus increasing average daily gas output to 1400 mmcf by 2004.

Tampico-Misantla Basin

The Tampico-Misantla Basin (Cuenca de Tampico-Misantla), which covers about 50,000 km^2 onshore and offshore (Figure 5), has been producing mostly heavy oil since 1904, when renowned Mexican geologist Ezequiel Ordóñez discovered the first oil province in Mexico—the Ebano-Panuco district. This district has produced more than 1 billion barrels (bbl) from fractured Upper Cretaceous basinal carbonates. The basin also produces from upper Kimmeridgian oolitic carbonates and Lower Cretaceous chalks, which in Tamaulipas-Constituciones, San Andrés, and Arenque fields (the latter offshore) have produced 257, 443, and 152 million bbl oil equivalent (MMBOE), respectively.

The basin is producing approximately 85,000 bbl

oil/day (BOPD), after having reached a maximum of 600,000 BOPD in 1921. Remaining reserves are 1727 MMBOE, excluding those in the Paleocañón de Chicontepec (described below). Through the use of 3-D seismic, many new opportunities have been identified in the traditional producing plays. Those opportunities, mostly in the offshore part of the basin, have characteristics similar to the producing plays, but because the expected production from them is medium to heavy oil, prospects are undrilled, waiting for better prices for those products.

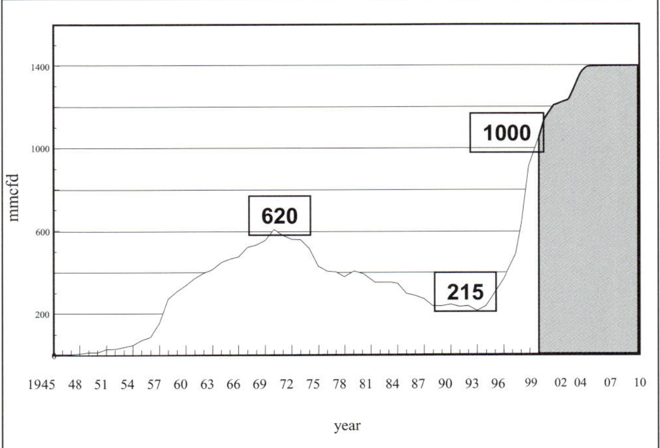

Figure 4. Burgos Basin production profile.

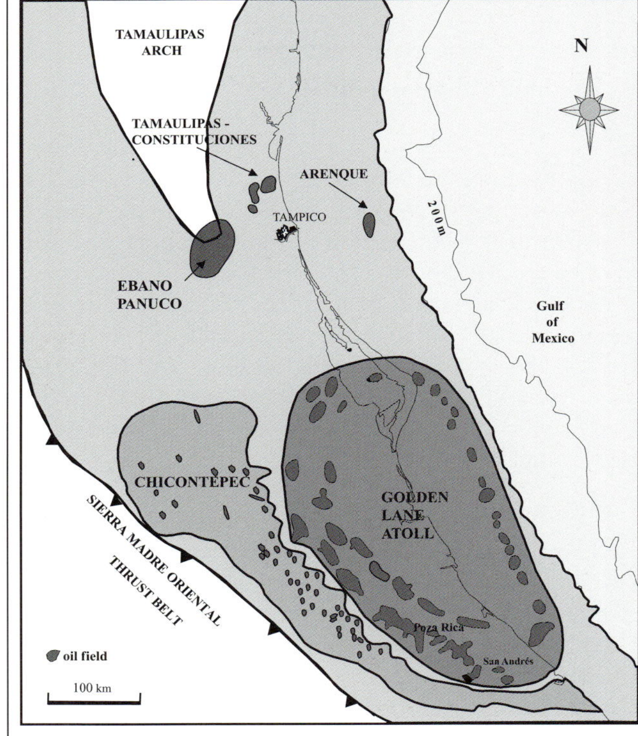

Figure 5. Tampico-Misantla Basin. Chicontepec is a Paleogene foredeep. The Golden Lane is a middle Cretaceous atoll with fields in the reef and talus facies. Ebano-Panuco produces from fractured Upper Cretaceous basinal carbonates. Tamaulipas-Constituciones, San Andrés, and Arenque fields produce from upper Kimmeridgian oolitic facies.

In the southern part of the basin, production was established in 1908 in what is known as the "Golden Lane." Since the discovery of its southern and offshore extensions, it has produced more than 1500 MMBOE from middle Cretaceous reef facies that form an atoll developed on the Tuxpan platform (Figure 6). Rimming the fields of the Golden Lane and basinward of them is a second fairway that produces from platform-derived gravity flows. The famous stratigraphic trap that became Poza Rica field, with a cumulative production of 1725 MMBOE, is the main field in this play.

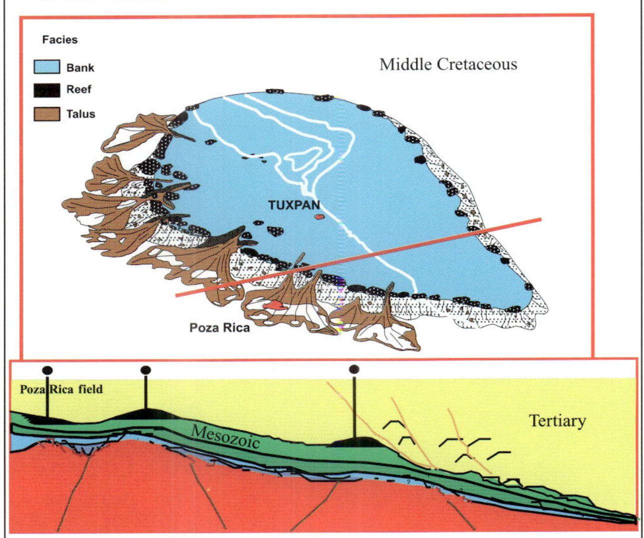

Figure 6. Main plays of the Golden Lane.

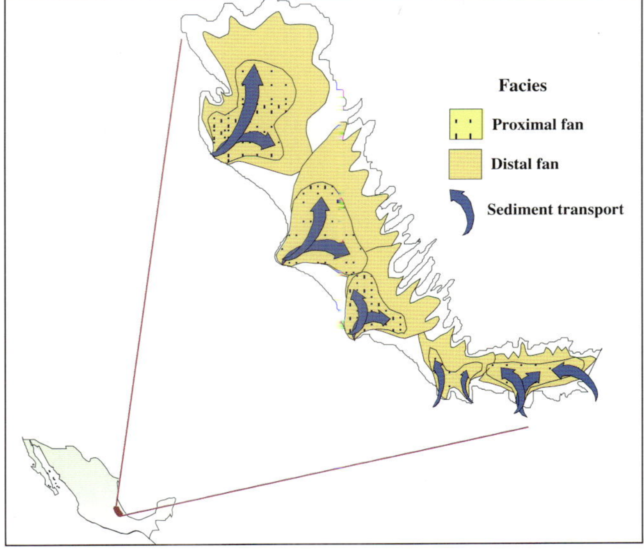

Figure 7. Paleocañon de Chicontepec depositional model.

The Paleocañon de Chicontepec, encompassing 3000 km^2 on the western side of the Golden Lane (Figure 7), developed as a foredeep filled with upper Paleocene to lower Eocene deep-water siliciclastics. The original oil and gas in place within those rocks were 139,000 million bbl and 49 trillion cubic feet (tcf), respectively, with total reserves of 13,993 million bbl oil and 28 tcf dry gas. Plans are being drawn to develop those resources after the year 2001, beginning with the northern sector, which is rich in light oil. New drilling technology, along with 3-D seismic and new geoscience concepts such as sequence stratigraphy, has allowed these formerly submarginal resources (only 141 MMBOE of cumulative production) to have rates of return attractive enough to warrant development.

In the foothills of the Sierra Madre Oriental folded thrust belt that bounds the Tampico-Misantla Basin to the west and thus the MGOM, a new province (Figure 8) is being tested that may have a significant impact in the twenty-first century. This province continues to the south into the Veracruz Basin, where it already produces oil and wet sour gas. Through the use of reprocessed old 2-D seismic, several attractive shallow structures have been identified and are expected to be gas prone, based on old well tests and geochemical modeling. Because gas demand in Mexico is expected to increase rapidly in industrial and electric-generation sectors because of concerns and economic growth, exploration for nonassociated gas is being actively favored over exploration for other types of hydrocarbons.

Veracruz Basin

The Veracruz Basin (Cuenca de Veracruz) has an area of 24,000 km^2, including part of the offshore shallow shelf. Production started in 1956, and since then, 15 fields have been discovered in two habitats (Figure 9).

The Cordoba platform, which is the buried leading

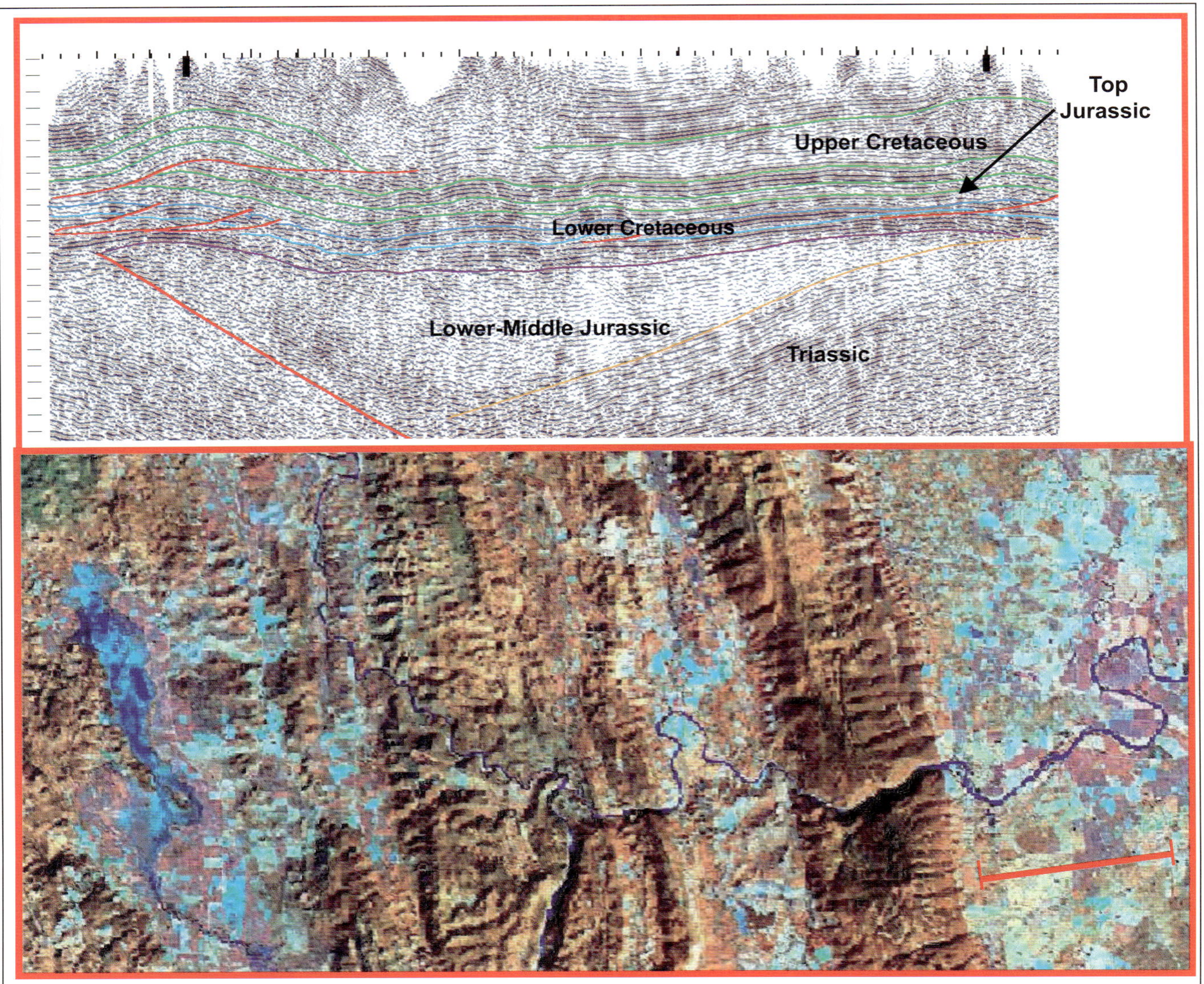

Figure 8. Eastern Sierra Madre Oriental folded thrust-belt foothills.

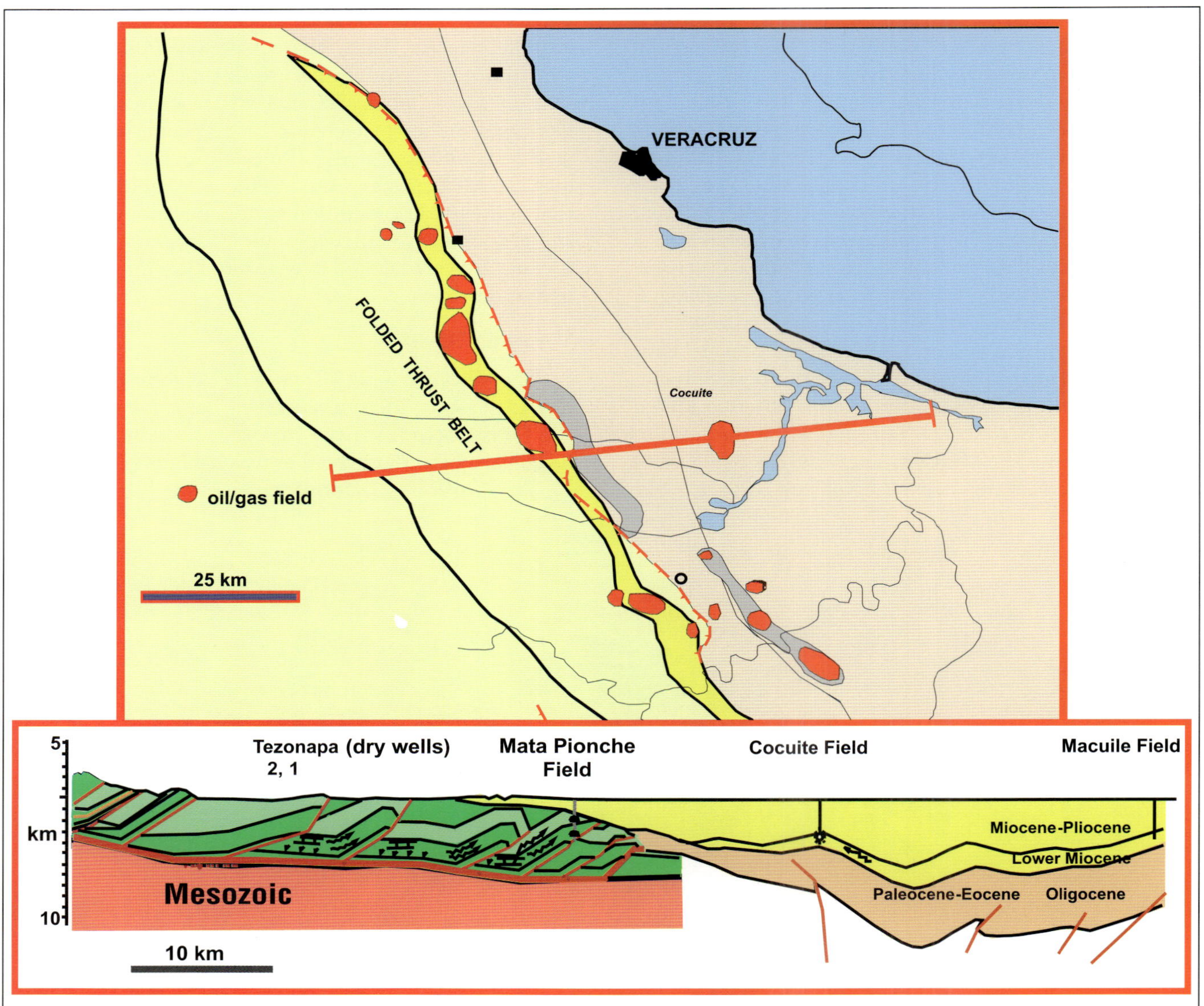

Figure 9. Veracruz Basin.

edge of the Sierra Madre Oriental folded thrust belt, is one of the principal producing areas. It consists of middle to Upper Cretaceous limestones that produce medium to heavy oil and wet sour gas from 11 fields with cumulative production of 72 million bbl of oil and 682 bcf of gas.

The other main producing area is the Tertiary trough, a depocenter filled with syntectonic conglomerates, sands, and shales deposited during erosion of the Sierra Madre uplift and deformed by neovolcanic emplacement. It has produced 136 bcf of dry sweet gas from four fields. Those sediments extend offshore to the continental shelf, where only three wildcats were drilled more than 25 years ago.

Total remaining reserves are only slightly more than 1 tcf of gas. However, through reprocessing of vintage 2-D seismic, acquisition of new 2-D and 3-D seismic, and application of new paradigms (such as testing low-contrast, low-resistivity sands), important new opportunities have been identified in both areas. This has encouraged the integration of ambitious plans to actively explore the gas-prone plays of the basin. Particularly attractive are offshore leads and prospects (Figure 10). Plans call for acquisition of more than 2000 km^2 of 3-D seismic and 4500 line-km of 2-D onshore seismic, and for drilling more than 40 exploratory wells (both onshore and offshore), with the expectation of bringing gas production up to 400 mmcf/day by the year 2005, from the 174 mmcf/day being produced today.

Sureste Basins

The Sureste Basins (Cuencas del Sureste), which encompass approximately 60,000 km^2 (Figure 11), have been the main producing area of Mexico since the mid-

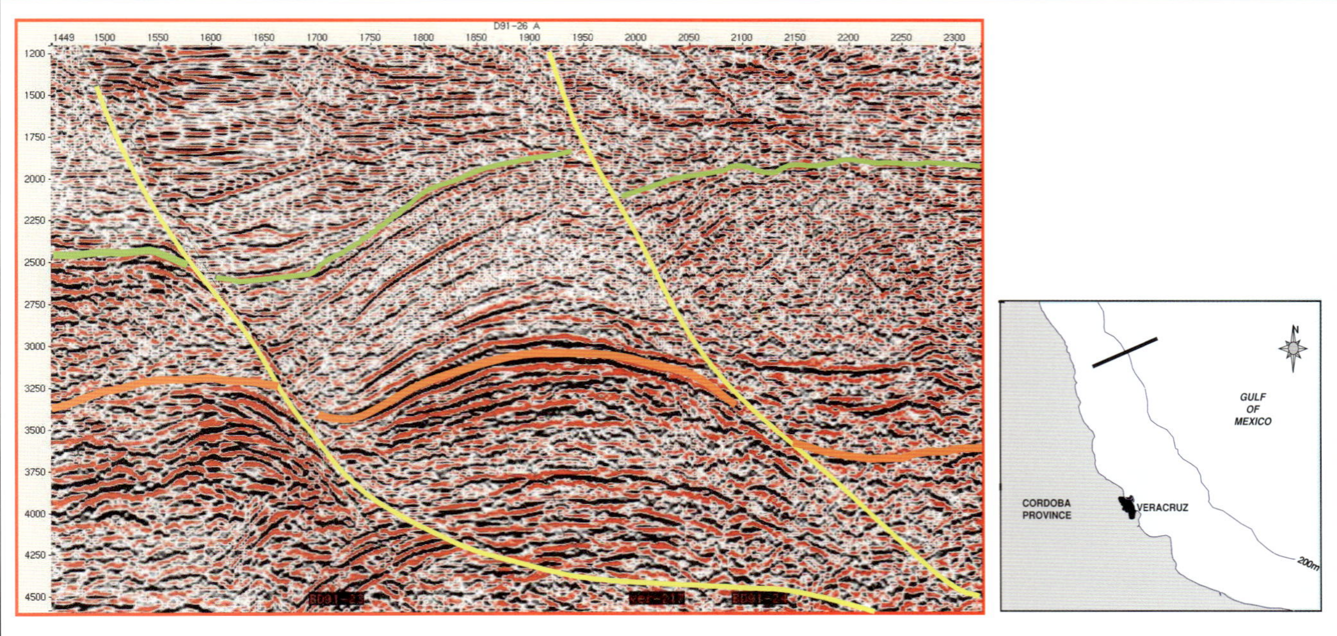

Figure 10. Seismic section, middle Miocene, offshore Veracruz.

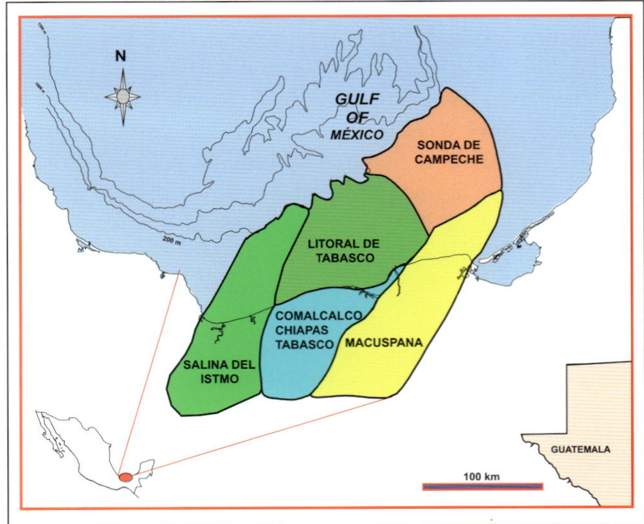

Figure 11. Sureste Basin provinces.

1970s, when the Mesozoic onshore light-oil province commonly known as Chiapas-Tabasco was discovered. The extremely prolific offshore province known as the Sonda de Campeche, producer of heavy oil in its northeastern sector, was identified in the late 1970s. Since then, 7808 million bbl of oil and 21.75 tcf of gas have been produced onshore, and 12,212 million bbl of oil and 7.55 tcf of gas have been produced offshore. Total reserves onshore are approximately 6048 million bbl of oil and 19.6 tcf of gas; offshore reserves are 21,292 million bbl of oil and 14.08 tcf of gas. The Salina del Istmo, Comacalco, and Macuspana provinces are Tertiary depocenters, with Comacalco being more or less coincident with the Mesozoic Chiapas-Tabasco light-oil province. To the north, the Sureste basins have two distinct Mesozoic offshore provinces: the northeastern, heavy-oil Sonda de Campeche, and the southwestern Litoral de Tabasco, producer of extra-light oil, condensate, and gas.

Salina del Istmo

Hydrocarbons have been extracted from the Sureste basins since the early 1900s, when production was established from shallow Tertiary siliciclastics deposited within the Salina del Istmo, Comalcalco, and Macuspana depocenters. The Salina del Istmo is in the western part of the area and encompasses approximately 15,000 km^2, about half of which is offshore (Figure 12). As its name implies, this basin was a salt depocenter. The Tertiary basin is a salt-intruded siliciclastic pile that produces light to medium oil from plays that overlay, abut against, or underlay salt of Jurassic origin. Fifty-two fields have been discovered, 28 of them still producing approximately 44,000 BOPD and 54 MMCFGD. All fields except one produce from the Tertiary. Six fields have produced more than 100 million bbl of oil. Cinco Presidentes is the most important, with production of 285 million bbl of oil and 401 bcf of gas. Cumulative production is approximately 1596 million bbl of oil and 1.9 tcf of gas; remaining reserves are 618 million bbl of oil and 658 bcf of gas.

Through application of 3-D seismic and recently developed concepts of salt emplacement, withdrawal, and associated deformation, more than 300 new exploration opportunities have been identified in this Tertiary province, particularly in the offshore extension, which has had very minor light-oil production. Application of new imaging exploration technology will allow identification of Tertiary subsalt opportunities and is expected to aid in the quest for continuation of production from the Mesozoic. In this area, the Mesozoic could have characteristics similar to the coeval section in the neighboring Chiapas-Tabasco area.

Chiapas-Tabasco and Comalcalco

The Mesozoic Chiapas-Tabasco province, discovered in 1972 and developed in the mid-1970s, and the Tertiary Comalcalco province have a combined area of approximately 13,000 km^2, which corresponds to the central part of the onshore Sureste basins (Figure 13). The offshore extension is the Litoral de Tabasco area and part of the Sonda de Campeche. Production comes from 57 fields (of 75 discovered), most of them producing from the deep, Upper Jurassic to middle Cretaceous carbonates that were deformed by salt movement and Tertiary transpression. Eleven of those fields have cumulative production of more than 100 million bbl of oil. The Bermudez complex, having produced 2225 million bbl of oil and 2894 mmcf of gas, and Jujo-Tecominoacán, having produced 810 million bbl of oil and 1049 mmcf of gas, are the most important. This region will continue to provide important volumes of light to medium oil and associated gas in the twenty-first century, particularly through application of production optimization and identification of exploration opportunities. Cumulative production for the province is 6207 million bbl of oil and 14.6 tcf of gas; remaining reserves are 5367 million bbl of oil and 17.2 tcf of gas.

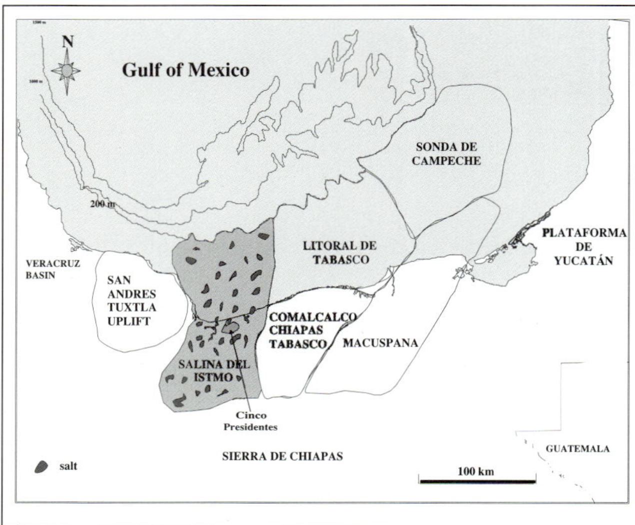

Figure 12. Salina del Istmo province.

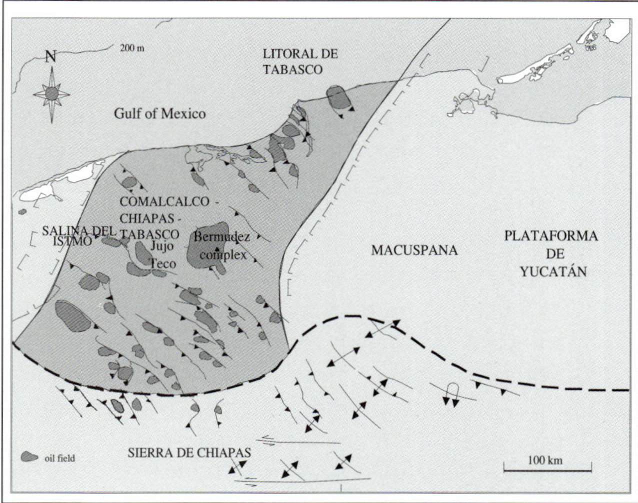

Figure 13. Comalcalco and Chiapas-Tabasco provinces. Most of the production comes from Cretaceous carbonates.

Macuspana

The Macuspana Tertiary province in the east (Figure 14) encompasses approximately 11,000 km^2; about a quarter of the province is offshore. It has been a producer of shallow (<3000 m) nonassociated sweet gas since the late 1950s. As with many of the previously described basins, application of new technologies and concepts has permitted identification of a large number of new gas-exploration opportunities. The application of an ambitious investment program that calls for acquisition of 3145 km^2 of 3-D seismic, the drilling of 100 exploration wells and more than 210 development wells, and the construction of new facilities will allow rejuvenation of the basin by bringing gas production up to 800 mmcf/day by the year 2004, from the present output of 170 mmcf/day from 13 fields (of 36 discovered). Previous maximum gas production for Macuspana was 720 mmcf/day, reached in 1975.

Unlike the reservoir rocks of Burgos and Veracruz, the fluviodeltaic and shelf sandstones in Macuspana tend to

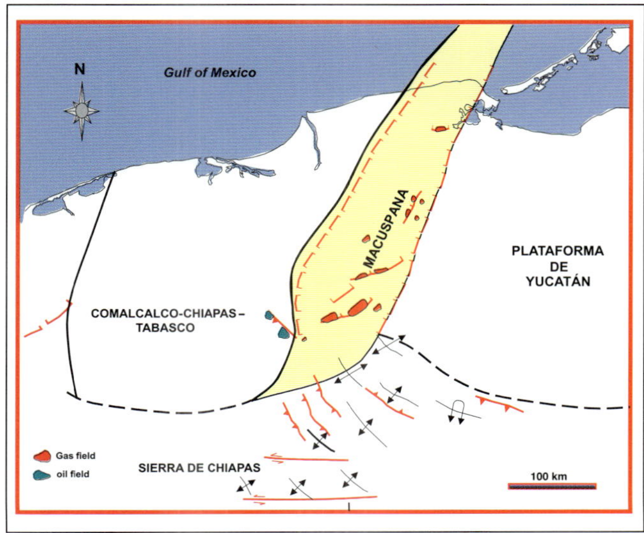

Figure 14. Macuspana Tertiary province.

be of better quality because they are "cleaner," younger (Miocene to Pliocene), and less compacted. Traps are both stratigraphic and structural, the latter being mostly rollover anticlines associated with extensional deformation. Although no production has been established in the offshore extension, very attractive structures have been identified that will begin to be tested in the next few years. Cumulative gas production for the province is 5.2 tcf; remaining gas reserves are approximately 1.8 tcf.

Sonda de Campeche

The Sonda de Campeche province (Figure 15) was discovered in 1976 in waters less than 100 m deep. Since then, 24 fields (18 producing) have been discovered, with cumulative production of 12,131 million bbl of oil and 7.37 tcf of gas. Reserves are 19,862 million bbl of oil and 11.4 tcf of gas. Most reservoirs are in Upper Cretaceous to lower Paleocene talus breccias and Upper Jurassic oolitic sediments; only two are in Oxfordian quartz sandstone dunes. The province covers 15,000 km^2 and is by far Mexico's most prolific. The Cantarell supergiant field alone (Figure 16), with remaining reserves of 11,936 million bbl of oil and 5.17 tcf of gas, produces 1.5 million BOPD and 532 MMCFGD of the 2.3 million BOPD and 1.5 BCFGD that the province generates. In the most ambitious project of the Mexican oil industry, the field is being optimized through drilling of 200 new wells, addition of new facilities and, in early 2000, injection of 1 bcf/day of nitrogen to maintain reservoir pressure. These actions will allow extraction of 2.2 million BOPD. In early 1999, the exploration of a repeated section below this field confirmed a new, totally independent reservoir with lighter oil (28° versus 24° API in the upthrown block), with total reserves of 1300 million bbl of oil.

The Ku-Maloob-Zaap (K-M-Z) complex to the northwest, with cumulative production of 1304 million bbl of oil and 680 bcf of gas and daily production of 285,000 bbl of oil, is the second most important heavy-oil field in the area. To the southwest, Abkatún field has produced 1968 million bbl of oil and 1601 bcf of gas. It is producing 153,000 BOPD and 136 MMCFGD. In addition to Abkatún, the best light-oil-producing fields are Pol (cumulative production of 788 million bbl of oil and 723 mmcf of gas; daily production of 103,000 bbl of oil and 110 mmcf of gas), Chuc (cumulative production of 453 million bbl of oil and 489 mmcf of gas; daily production of 121,000 bbl of oil and 173 mmcf of gas), and Caan (cumulative production of 391 million bbl of oil and 686 mmcf of gas; daily production of 192,000 bbl of oil and 314 mmcf of gas).

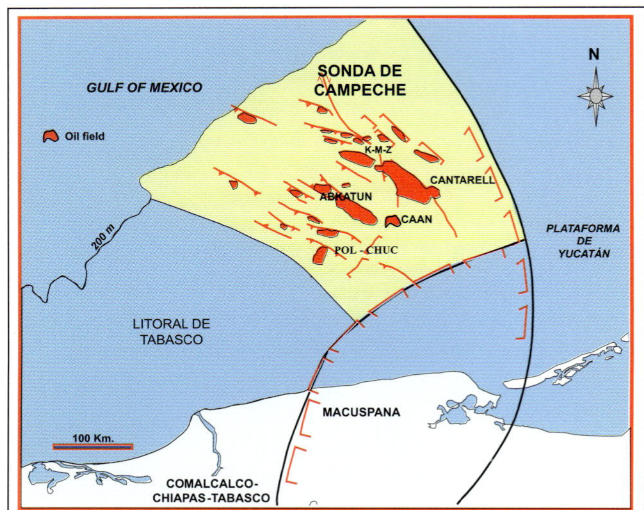

Figure 15. Sonda de Campeche province.

Although there has been little exploration focused on the Tertiary, reprocessing of the extensive 3-D seismic obtained in the area has identified several leads and prospects in the siliciclastic younger section that will be tested in the near future.

Litoral de Tabasco

The Litoral de Tabasco province (covering approximately 7000 km^2) was discovered in 1979 (Figure 17). It produces approximately 80,000 bbl/day of extra-light oil and 171 mmcf/day of associated gas from three fields (of 16 discovered). The reservoirs are Cretaceous to Paleocene breccias and Jurassic shelf facies on the east, and mostly basinal Jurassic to Cretaceous carbonates on the west. All are deformed by compressional and salt tectonics. Cumulative production is 81 million bbl of oil and 183 bcf of gas; total reserves are approximately 1430 MMBOE and 2.7 tcf of gas. The undeveloped fields are beginning to be developed in the year 2001 in a program that will increase gas production by another 1000 mmcf by the year 2008. This area has almost total 3-D seismic coverage, which has allowed identification of more than 200 exploration opportunities, in the Mesozoic and Tertiary, that will provide a large portfolio for the twenty-first century.

Plataforma de Yucatán

The Yucatán platform (Figure 1) is a very stable shelf that covers approximately 300,000 km^2. It developed in the Cretaceous and lacks the highly productive source rocks of the Late Jurassic. Although a petroleum system sourced by Lower Cretaceous supratidal facies has been documented to the south in the Sierra de Chiapas and Guatemala, it has not been confirmed in Yucatán. Thus it does not seem likely that large quantities of oil will come from this region in the twenty-first century.

Deep-water Gulf of Mexico Basin

The Mexican sector of the Gulf of Mexico basin under waters deeper than 200 m (Cuenca del Golfo de México Profundo) spans an area of approximately 530,000 km^2. Only two wells have been drilled in the area. However, based on about 28,000 line-km of recently acquired 2-D seismic, eight petroleum provinces,

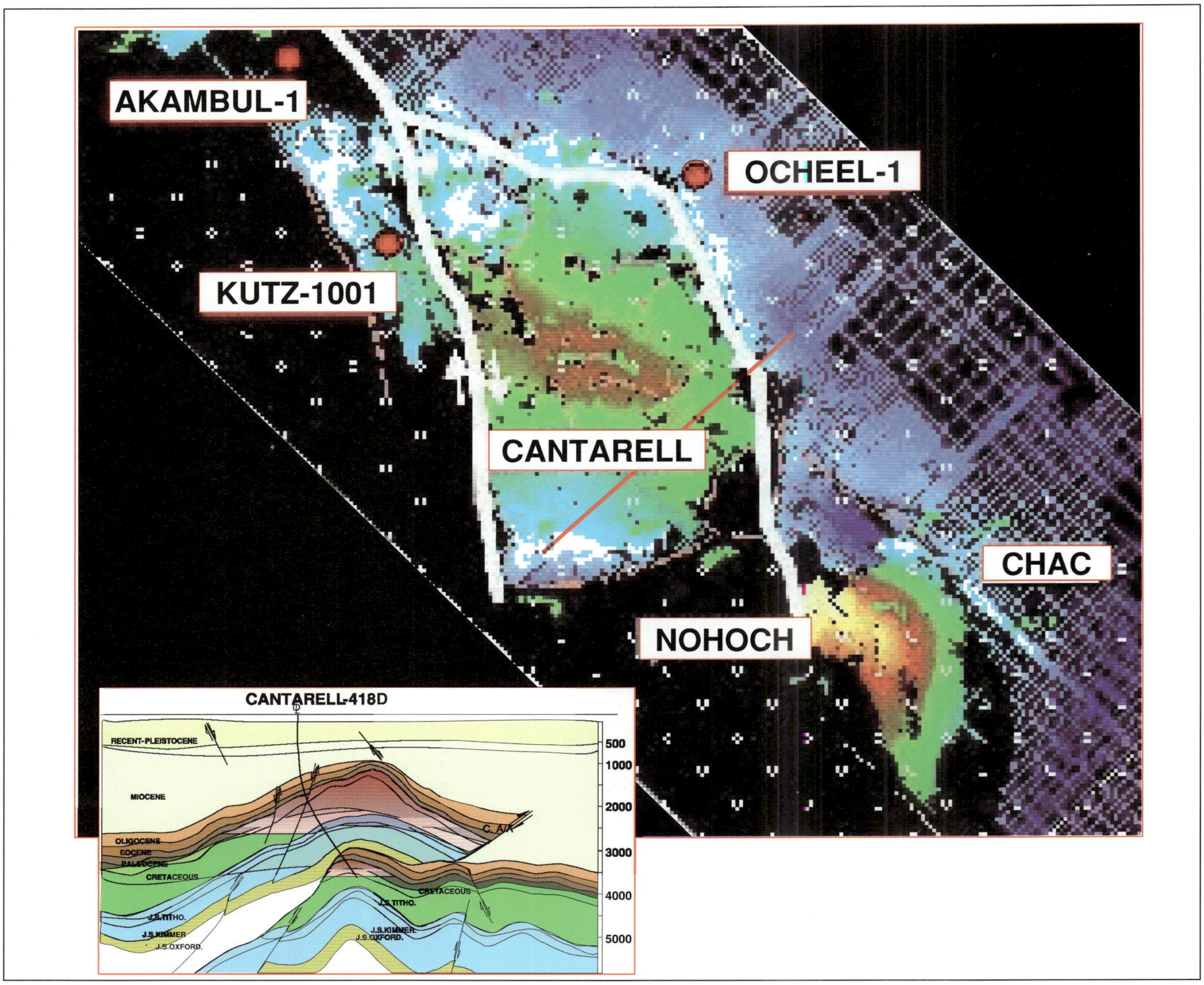

Figure 16. Cantarell field. Structure map is on top of Upper Cretaceous–Paleocene "breccia." Red line is location of the cross section.

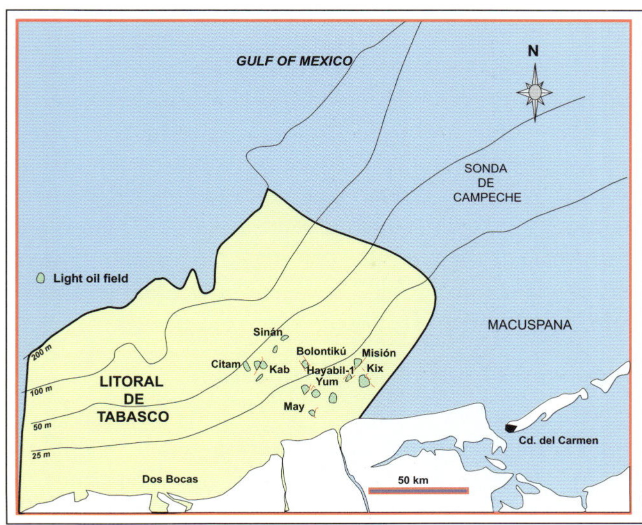

Figure 17. Litoral de Tabasco province.

six of them prospective, have been identified (Figure 18). The definition of these provinces is based mostly on their tectonic characteristics

Franja Distensiva

An extensional domain characterized by listric down to the basin faulting forms a fairway that parallels the coastline in front of the Burgos, Tampico-Misantla, and Veracruz Basins. The faults sole out within the Tertiary siliciclastic section forming large rollover structures, many of them expanded by syndepositional growth (Figure 19). The structures identified so far tend to be quite large. From seismic evidence, geochemical modeling, and sampling of a few wells and of the sea-bottom bed, the expected hydrocarbons are gas and/or very light oils. Reservoir rocks include deltaic and shelf siliciclastics and talus and basinal deep-water turbidites.

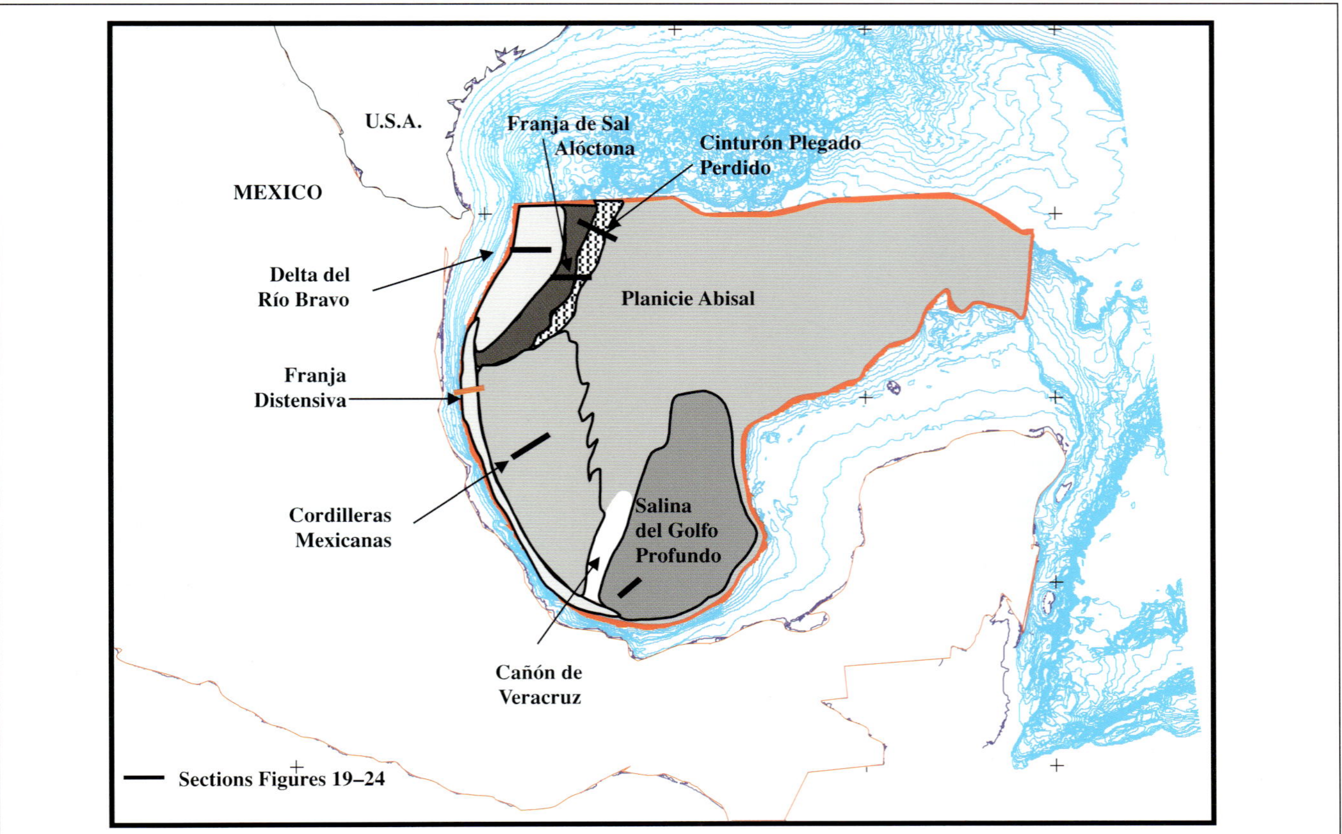

Figure 18. Deep-water Gulf of Mexico Basin provinces. Black bars show locations of sections in Figures 19–24.

Figure 19. Seismic section showing the Neogene Franja Distensiva province. For location, see Figure 18.

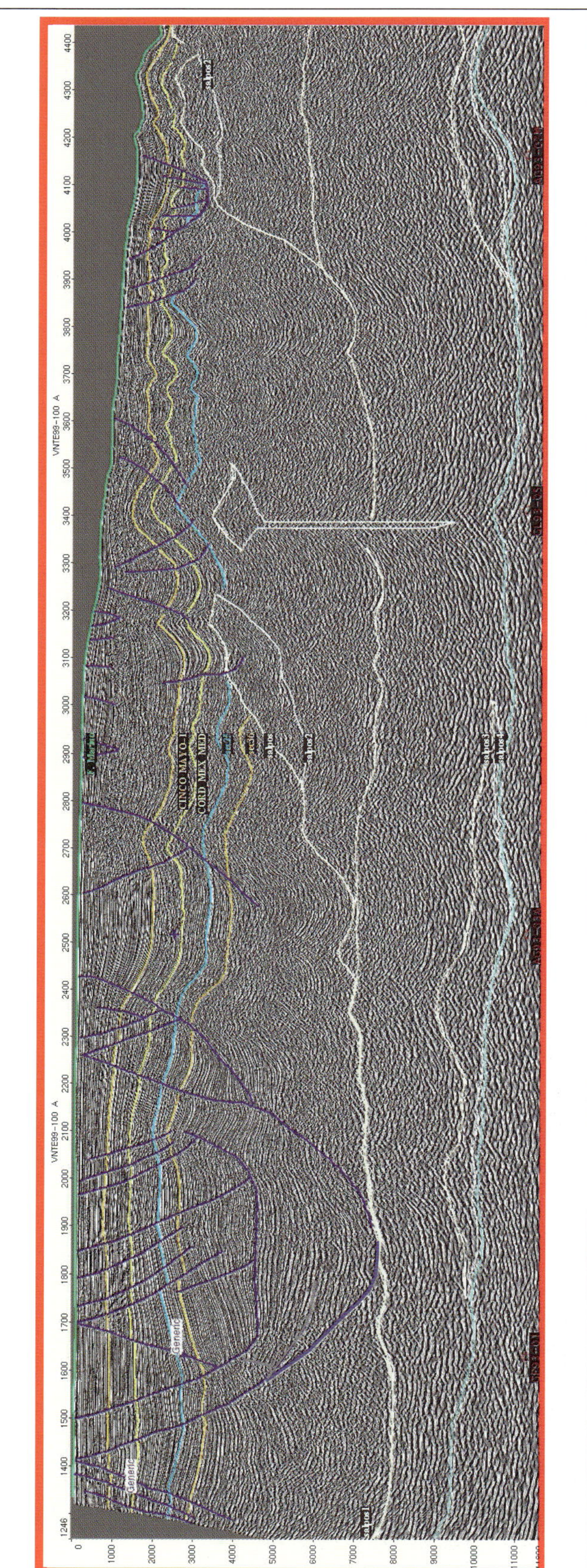

Figure 20. Section through the Delta del Río Bravo province. For location, see Figure 18.

Delta del Río Bravo

A Miocene to Pliocene depocenter has been identified in front of the delta of the Río Bravo del Norte (Figure 20). This feature originated when salt was evacuated to the east, creating structures associated with salt displacement. Based on extrapolation of Burgos and Padre Island–area data, expected hydrocarbons are mostly gas. Reservoir rocks include distal deltaic and shelf sands deposited by the ancient Río Bravo.

Franja de Sal Alóctona

This area, in the northwestern sector of the deep-water Gulf of Mexico Basin between water depths of 1000 to 3000 m, is dominated by salt sheets, canopies, and diapirs evacuated from the west (Figure 21). The area appears to be primarily gas prone. Reservoir rocks are assumed to be deep-water turbidites.

Cinturón Plegado de Perdido

Downdip from the Franja de Sal Alóctona is a folded and thrusted belt of Mesozoic rocks formed by salt emplacement and gravity detachment at the top of the Jurassic salt. Structures appear to be cored by salt and are elongated, very large (>40 km), and narrow (Figure 22). This belt underlies water between 2000 and 3500 m deep. Recently, an industry consortium tested a structure within the belt in the Alaminos Canyon area on the U.S. side that, according to some sources, was successful. Expected hydrocarbons are mostly oil, and reservoir rocks are expected to be Mesozoic, fractured, deep-water carbonates and Tertiary siliciclastic turbidites.

Cordilleras Mexicanas

Downdip from the Franja Distensiva, in front of the southern Burgos, Tampico-Misantla, and Veracruz Basins, is a wide, compressionally deformed belt known as Cordilleras Mexicanas. It extends for 500 km, covering about 70,000 km^2, in water depths of 1000–3000 m. Formed as a result of accommodation of extensional deformation updip, the displacement along a Tertiary glide plane resulted in very long (some more than 40 km), narrow anticlines (Figure 23). The structures appear to be confined to the Tertiary section. The expected hydrocarbons are light to medium oils. Reservoir rocks are postulated to be deep-water turbiditic sandstones.

Cañón de Veracruz

The Cañón de Veracruz is a physiographic feature, a submarine canyon, separating the Cordilleras Mexicanas to the west and the deep Gulf of Mexico Basin to the east. At present, it appears to have little petroleum potential.

Salina del Golfo Profundo

This province is the downdip extension of the Salina

del Istmo. In the area, both the Mesozoic and Tertiary sections are affected by salt displacement, resulting in diapirs, sheets, and canopies that have created a large number of exploration opportunities (Figure 24). Based on results of a recent well drilled in the basin, the expected hydrocarbons are heavy to medium oils. However, there could be lighter products in the older, deeper rocks.

Planicie Abisal

The central abyssal part of the basin, with water depths of more than 3000 m, covers more than 100,000 km^2. The area has little structural relief, and thus its oil potential would appear to be limited unless stratigraphic trapping is proved to exist.

CONCLUSIONS

The Mexican side of the Gulf of Mexico Basin has been producing oil and gas since the beginning of the twentieth century. All these hydrocarbons have been extracted from the onshore and shallow offshore parts of the basin (<200-m water depth), and so it would appear

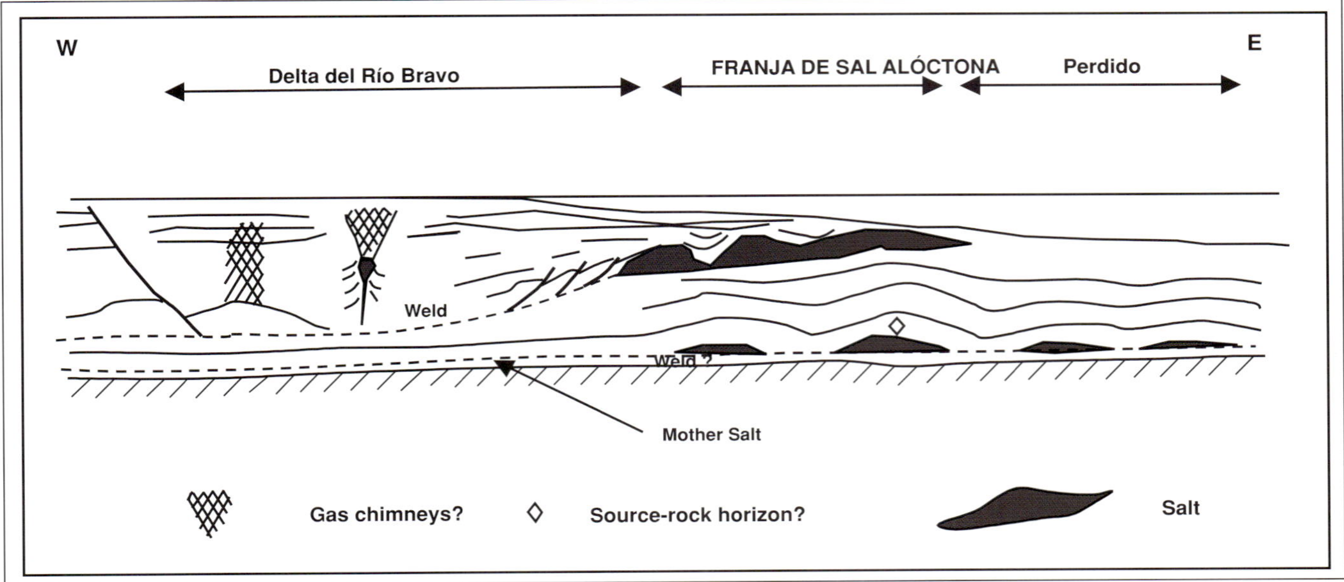

Figure 21. Section through the Franja de Sal Alóctona province. For location, see Figure 18.

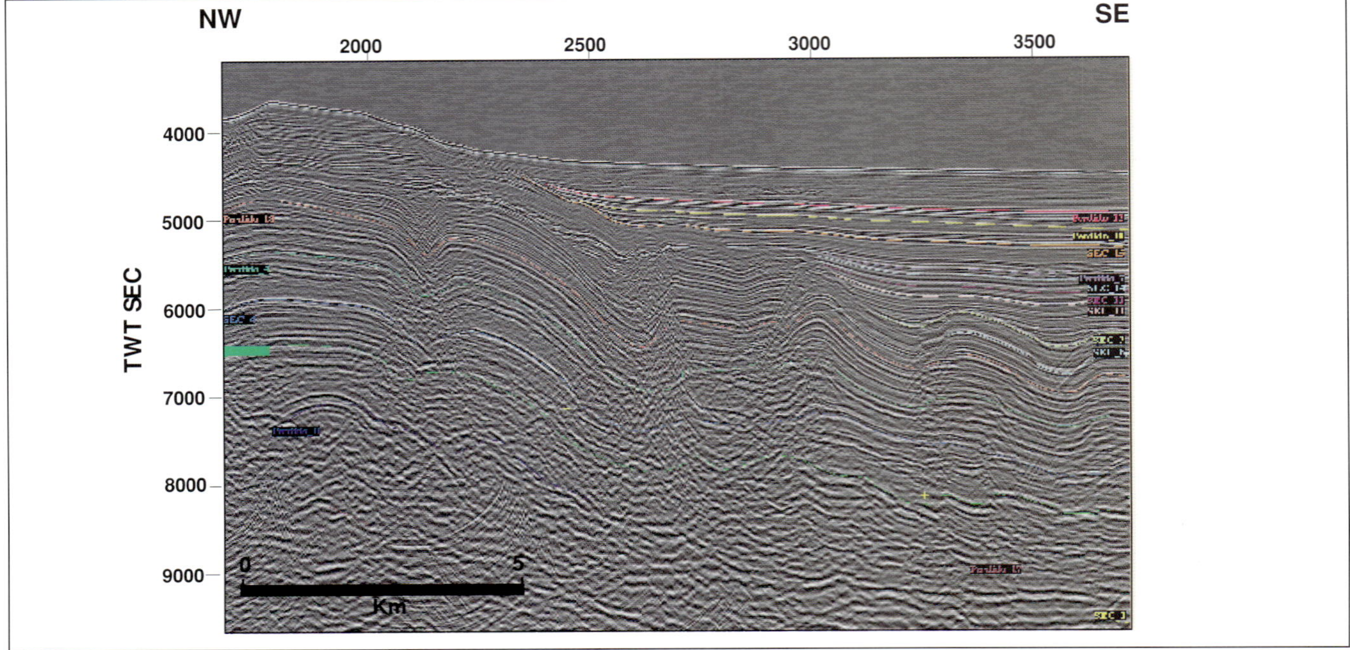

Figure 22. Seismic section through the Cinturón Plegado de Perdido province. The folded rocks are mostly Mesozoic. For location, see Figure 18.

that most future production will come from deep water. The truth is that through application of vigorous investments, technology, and new concepts and methodology, the traditional and not so traditional provinces of the basin (such as Chicontepec, the shallow offshore areas outside Campeche, and others) will carry the burden of most of the production in the early twenty-first century. It is also true that deep-water provinces of the basin hold potential that, although not tested yet, appears to be quite large, and that the technology to explore and pro-

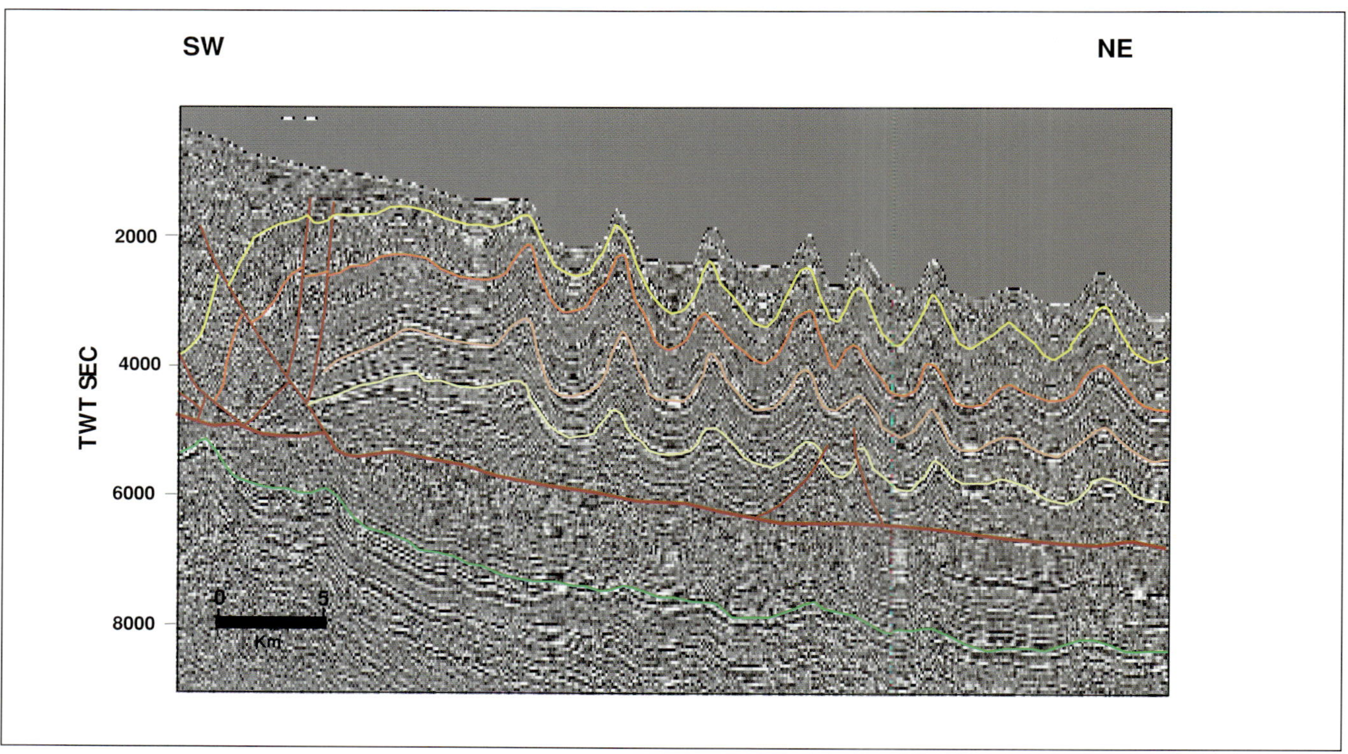

Figure 23. Seismic section through the Cordilleras Mexicanas province. The folded rocks involve only the Tertiary. For location, see Figure 18.

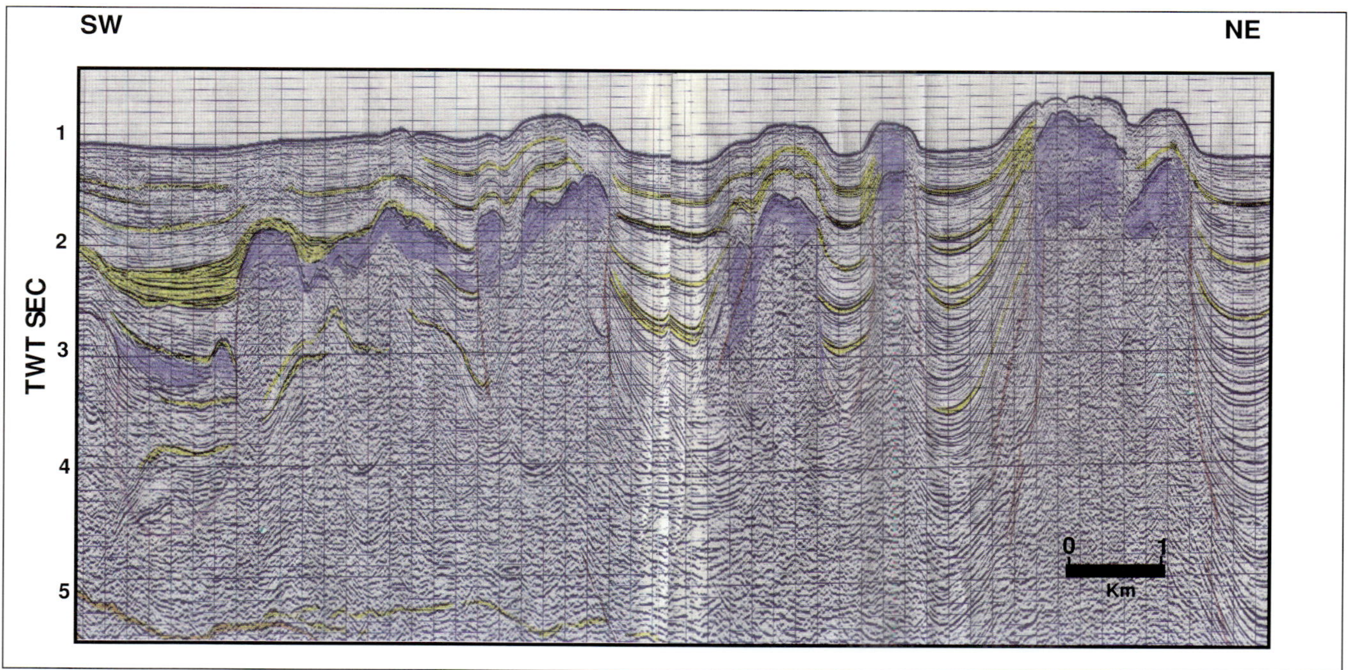

Figure 24. Seismic section through the Salina del Golfo Profundo province. The bluish-purple horizon is salt. For location, see Figure 18.

duce in deep water is already available. These factors indicate that the MGOM could very well be *the* petroleum province of the twenty-first century.

ACKNOWLEDGMENTS

We acknowledge PEMEX Exploration and Production for permission and support to present this paper, and PEMEX Exploration Asset groups for the figures included herein.

Audemard, F. E., and I. C. Serrano, 2001, Future petroliferous provinces of Venezuela, *in* M. W. Downey, J. C. Threet, and W. A. Morgan, eds., Petroleum provinces of the twenty-first century: AAPG Memoir 74, p. 353–372.

Chapter 18

FUTURE PETROLIFEROUS PROVINCES OF VENEZUELA

F. E. Audemard
PDVSA Petróleo, Caracas, Venezuela

I. C. Serrano
PDVSA Petróleo, Caracas, Venezuela

ABSTRACT

Recent regional studies have indicated that a resource potential greater than 40 billion barrels oil equivalent remains to be found in Venezuela. Evidence for new petroliferous provinces that hold this potential are discussed in this paper.

Venezuela is a showcase for the exceptional foredeep basins of northern South America, along with outstanding oil-source rocks and reservoirs associated with unconformities within these foredeeps.

In the southwest of the country, sandstones above and below the foredeep unconformity form potential stratigraphic traps, a possible western extension of the prolific Eastern Venezuelan Basin. This would be an exceptional area to evaluate weathered/fractured basement plus Jurassic fills of half grabens located below the passive-margin sequence. Also in the west, the northern and southern flanks of the Mérida Andes, with 70 oil seeps, remain virtually unexplored.

Other areas with significant petroleum potential are the 7600 m (25,000 ft) of mostly Neogene sediments offshore the Orinoco Delta, where five wells have tested 5 trillion cubic feet of gas plus condensate; a diapir wall 160 km (100 miles) in the middle of the Eastern Venezuelan Basin with three major fields; and the 110-km-long (70-mile-long) downthrown Anaco inverted structure, tested in two localities. The mountain fronts to the north are being drilled to evaluate the northern extension of the giant Furrial trend and a new thrust play to the northwest.

The 150,000-km^2 (60,000-mile^2) offshore area has only 50 wildcats, most drilled as tests for conventional traps. However, complex strike-slip structures, stratigraphic traps, and deep-water plays could exist. Oil seeps and shows in wells indicate that this is an oil-prone area.

INTRODUCTION

As Wallace Pratt used to say, "Oil is found in the minds of men," and many explorers have proposed ideas that led to the discovery of giant oil and gas fields in Venezuela. Some did not have the chance to understand the real size of the giant traps because of acreage limitations or insufficient well control. Most of the giant Venezuelan fields known today were discovered prior to 1960 (Mencher et al., 1953; Smith, 1956; Miller et al., 1958;

Salvador and Stainforth, 1968; Martínez, 1995). However, recent studies (Stephan, 1982; James, 1985; Aymard et al., 1990; Audemard, 1991; Lugo, 1991; Erlich and Barrett, 1992; Duval et al., 1994; Di Croce, 1995; Parnaud et al., 1995; Hung, 1997; Ysaccis, 1997) have provided new insights into the petroleum potential of Venezuela that allow us to propose some additional ideas, using 11 examples.

In spite of an exploration and production history that started in the nineteenth century in Venezuela, mainly around oil seeps, 75% of the sedimentary basins still remain underdrilled. This means that of a total of 500,000 km^2 (195,000 mile^2), 375,000 km^2 (145,000 mile^2) could still hold undiscovered accumulations (Figure 1). In addition, at least 20% of the areas under production have not been drilled to economic "basement."

Venezuela's cumulative oil production is close to 50 billion barrels (bbl), and proven oil reserves total 72 billion bbl. Cumulative gas production is 69 trillion cubic feet (tcf), and proven gas reserves are 147 tcf. At least 30 giant oil fields and five giant gas fields have been discovered. The largest single accumulation known is on the rim of the most prolific foredeep basin, where more than 1 trillion bbl of oil is in place. The magnitude of these numbers serves to illustrate Venezuela's hydrocarbon richness.

Despite these impressive figures, explorationists continue their search for new hydrocarbons in areas with potential for giant fields with reserves larger than 500 million bbl of oil or 3 tcf of gas. In this paper, evidence of these future petroliferous provinces is shown using reflection seismic profiles (Figure 1).

EVOLUTION OF ACTIVE AND PASSIVE MARGINS

We present two very distinctive perspectives to illustrate part of the present potential of the Venezuelan sedimentary basins. The first examines the configuration and stacking of the different basins that prevailed during the Phanerozoic; the second is devoted to the major source rocks feeding the petroleum systems active during the evolution of these basins.

Our principal focus is on the Upper Cretaceous source beds as well as the evolution of northern South America since the deposition of these rich intervals. In general, a broad passive margin containing these source rocks was progressively matured because of flexural loading and subsequent emplacement of a foredeep related to the interaction between the South American and Caribbean Plates.

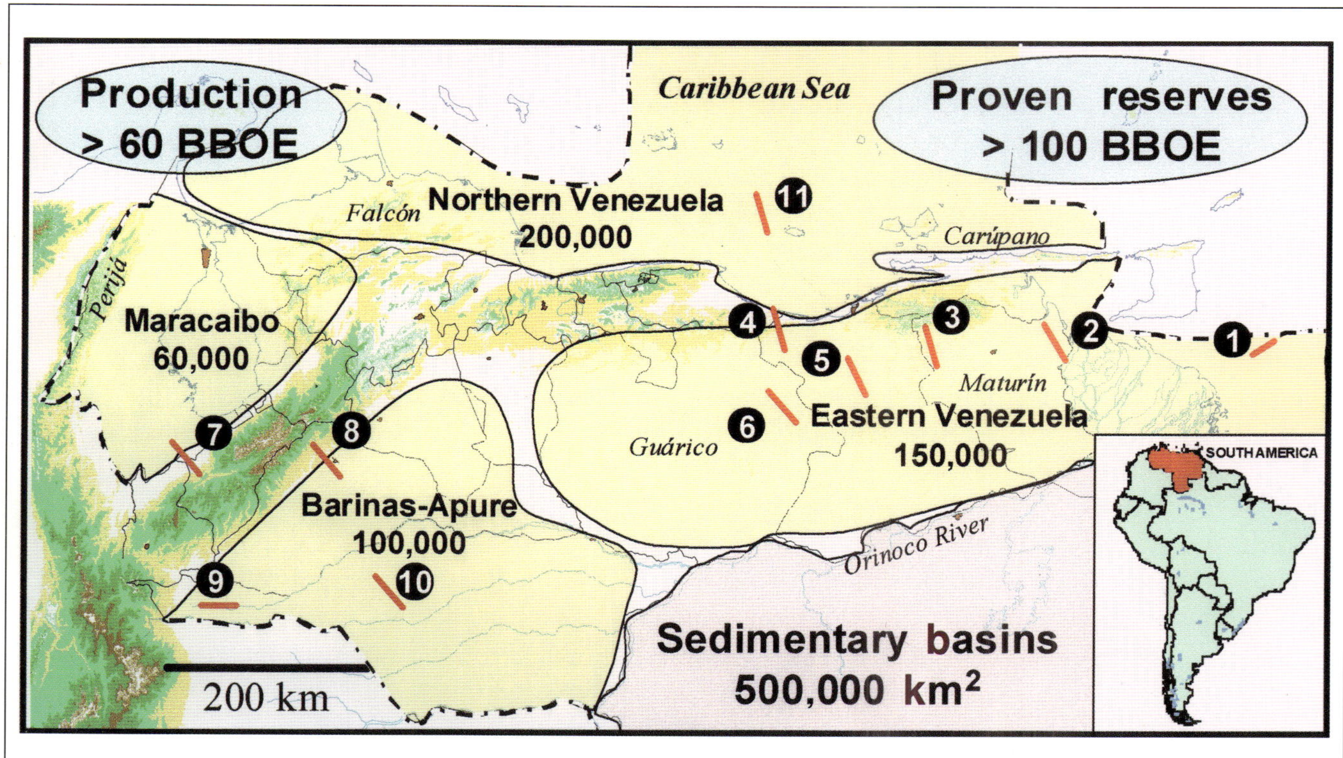

Figure 1. The areas shown are the major onshore and offshore basins (numbers correspond to areal extent of the basins in square kilometers). Cumulative production and proven reserves of Venezuela are indicated in billion bbl oil equivalent (BBOE). Locations of seismic lines used to illustrate future petroliferous provinces are in red and are numbered.

Figure 2 shows this evolutionary path across northern South America. An eastward migration of the foredeeps occurred from the Late Cretaceous (western Venezuela) to Holocene (eastern Venezuela, near the Orinoco Delta). The figure (modified from Audemard and Lugo, 1996) highlights a dual passive-margin setting (Tethys and Atlantic). This division has been considered because, in the Venezuelan literature, the oldest

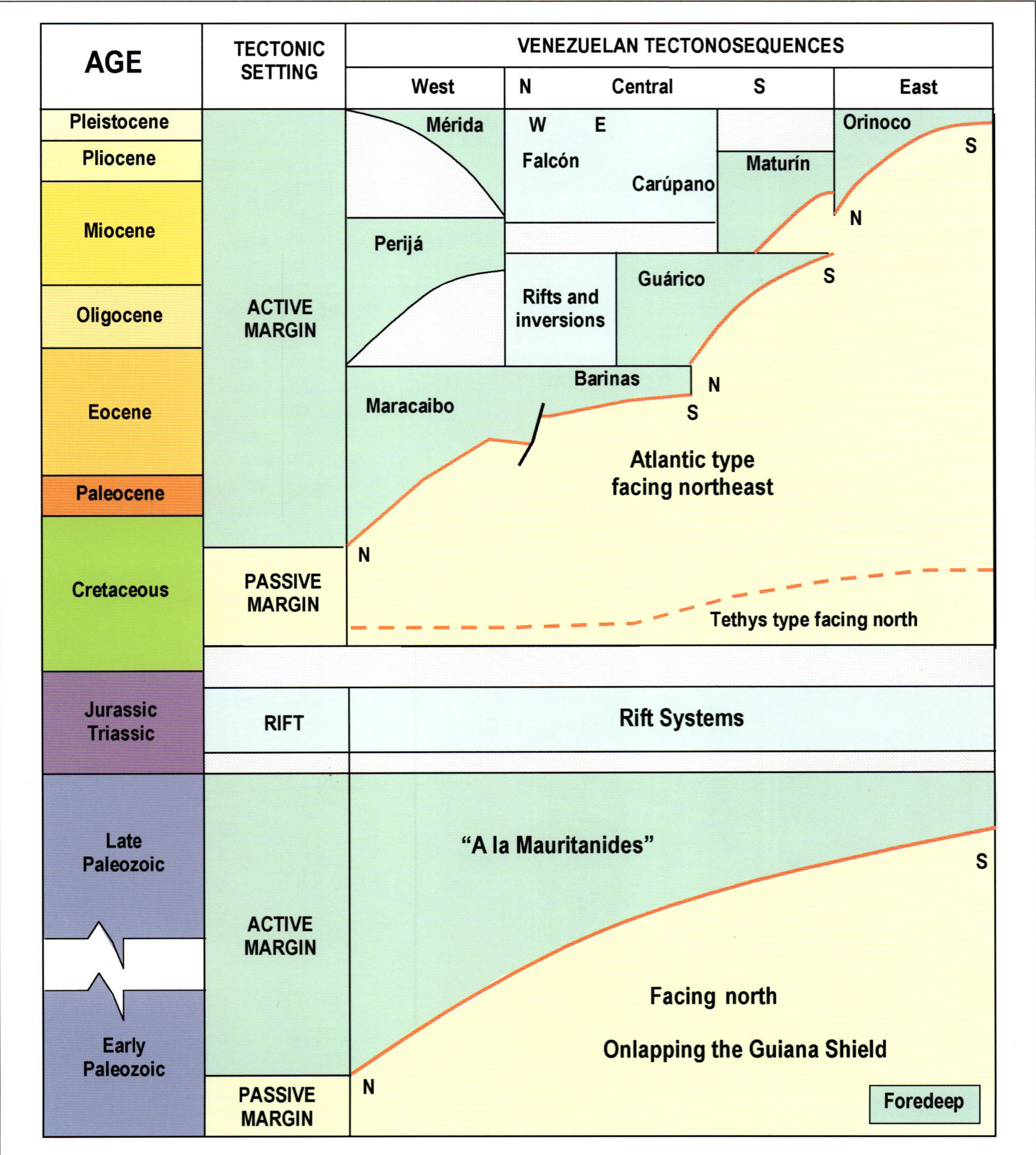

Figure 2. The diagram shows various settings: a combination of a Paleozoic passive and active margin, and two Mesozoic passive margins (Tethys and Atlantic) superimposed by a Late Cretaceous to Holocene eastward foredeep migration. Modified from Audemard and Lugo (1996) and Lugo and Audemard (1996).

rocks associated with the Mesozoic passive margin are Barremian. However, in Trinidad and Colombia, older Cretaceous and Jurassic rocks are part of the succession where no major break in sedimentation has been reported. Those age differences could be reconciled by interpreting the stack of sediments and metasediments observed along the Coastal Ranges in northern Venezuela as part of the distal foredeep and slope sediments of a passive margin and the thrusted oceanic crust (ophiolites) as portions of a folded belt developed from the frontal segments of the passive margin. This dual configuration is deduced from a major angular unconformity, probably induced by salt tectonics, well developed and visible on seismic profiles in offshore French Guyana. The nature of the unconformity probably is related to the salt, but it also could well be related to the shift from the Tethys to the Atlantic opening in this part of the planet. This implies that we are in pursuit of an unconformity, which becomes a major migration pathway across the Lower Cretaceous strata.

On the other hand, in a classic or idealized foredeep scheme, as shown in Figure 3 (Bally, 1989), we would need to consider a second breakup unconformity which will facilitate oil migration downsection. It is possible to interpret the existence of other source rocks older than the Upper Cretaceous. These additional potential source rocks might be distributed somewhat differently from the La Luna–Querecual pattern because of a slight change in the passive-margin configuration.

These potential new hydrocarbon sources can be inferred from very distinct points of view. The amount of oil in situ in the Orinoco Oil Belt does not satisfy the mass balance for the Upper Cretaceous source. A second marine source is needed. However, because the oil trapped in the Orinoco Oil Belt has certainly arrived early, then a deeper, distal source is a viable option. A second aspect pertains to the fact that the amount of oil in place forces the consideration that at least one-third of the oil leaving the source beds has been biodegraded, which means the system needs to have more oil available.

Figures 4 and 5 illustrate a third option for generation of hydrocarbons by means of overlaying the idealized foredeep model. The Tertiary foredeep is underlain by the late Paleozoic foredeep and does not directly encroach on the Precambrian shield. The leading edge of the Paleozoic folded belt (Figure 4) reached a position closer to the shield relative to that of the Tertiary. As a consequence, a narrower, elongate Paleozoic foreland basin is still preserved at a very shallow depth. The frontal folded structures that have been partially eroded remain unexplored. No effort has been made to explore all the units under the Cretaceous passive-margin unconformity, despite oil shows in some cores of Paleozoic rocks.

Within the Paleozoic folded belt, the Silurian-Devonian shales may have oil potential. These rocks are locally exposed in the Merida Andes and in northern Perija. This implies that they belong to an early Paleozoic passive margin and their thicknesses are equivalent to the Cretaceous units reported as being very prolific source beds.

VENEZUELA'S PETROLEUM SYSTEMS

One of the reasons Venezuela is rich in hydrocarbons is the widespread distribution of mature source rocks in

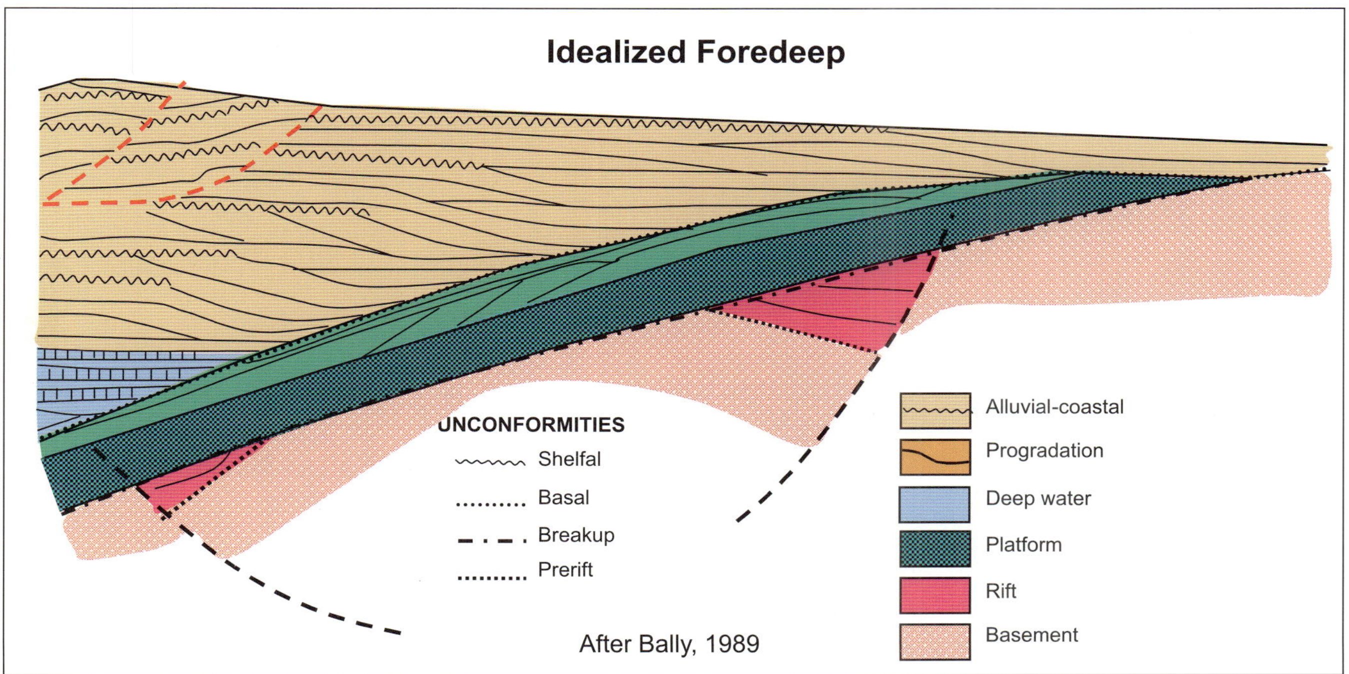

Figure 3. Idealized scheme of a foredeep.

the sedimentary column and across the country (Audemard et al., 1997). In addition, source rocks matured at different times because of the overlapping development of foredeeps and rift basins, resulting in multiple episodes of hydrocarbon migration.

Although the most prolific oil and gas source rocks identified so far in Venezuela were deposited during the Late Cretaceous (Hedberg, 1931) on a passive margin developed as a consequence of the opening of the Atlantic Ocean, many other important Cretaceous and Tertiary source rocks are present.

Organic matter in Upper Cretaceous source rocks (La Luna and Querecual Formations) is mainly marine type II kerogen, with minor amounts of type III. Algal-rich calcareous shales and shales extended across the northern part of Venezuela during that time. Their original total-

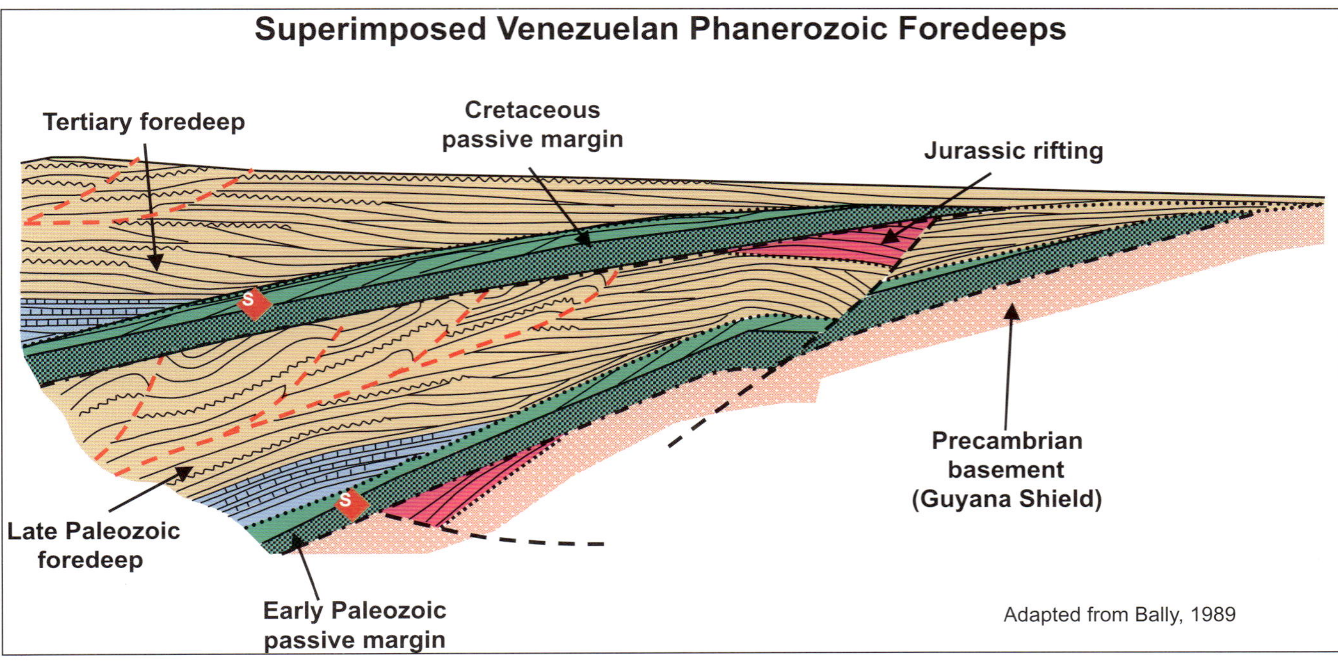

Figure 4. Idealized cross section showing Paleozoic and Mesozoic-Cenozoic passive- and active-margin relationships. S = source rocks.

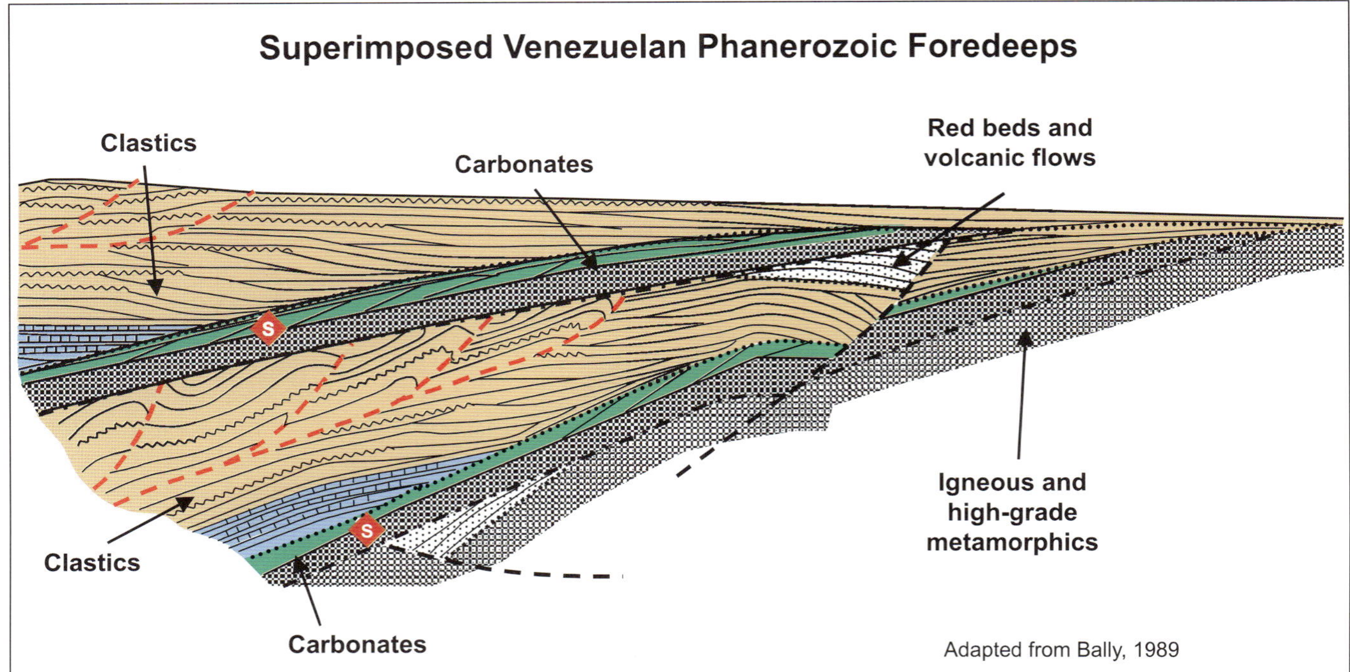

Figure 5. Same as Figure 4 but indicating lithologic relationships. S = source rocks.

organic-carbon (TOC) content was as high as 10%, especially in the La Luna Formation, and their thicknesses range from more than 195 m (650 ft) in southwestern Venezuela (Blaser and White, 1984) to 120 m (400 ft) in the east. Hydrogen indices can reach as much as 500 mg/g of total organic carbon.

In the Maracaibo Basin, the Lower Cretaceous Machiques Member is similar to the La Luna Formation. It is more than 50 m (150 ft) thick and is present toward the Perija Mountain front. Paleocene and Eocene coals and carbonaceous shales deposited in the northwest of the Andes have generated oil and gas, which are reservoired in Tertiary formations. Eastward of this basin, lower Eocene shales contain enough TOC to generate hydrocarbons, although actual hydrocarbon generation has not been demonstrated yet. In the Barinas Apure Basin, the Upper Cretaceous source rocks can account for hydrocarbons already found.

Across the Eastern Venezuelan Basin, Upper Cretaceous source rocks have been described in addition to the Oligocene and Miocene type III source rocks. They consist of marine to terrestrial shales and coals, with thicknesses that range from 30 m (150 ft) to 100 m (330 ft), and TOC as high as 11%.

Figure 6 shows the possible areas of hydrocarbon generation through time of Upper Cretaceous source rocks. Hydrocarbon generation and expulsion started during the middle Eocene in the west and continue today in the east along the Orinoco and Maturin foredeeps and the Caribbean Plate accretionary prism. Only in the south, toward the Guyana shield, are possible source rocks immature. Nevertheless, migration in coastal and deltaic sandstones and along extensive unconformities resulted in huge reserves in the Orinoco Oil Belt (Audemard et al., 1993).

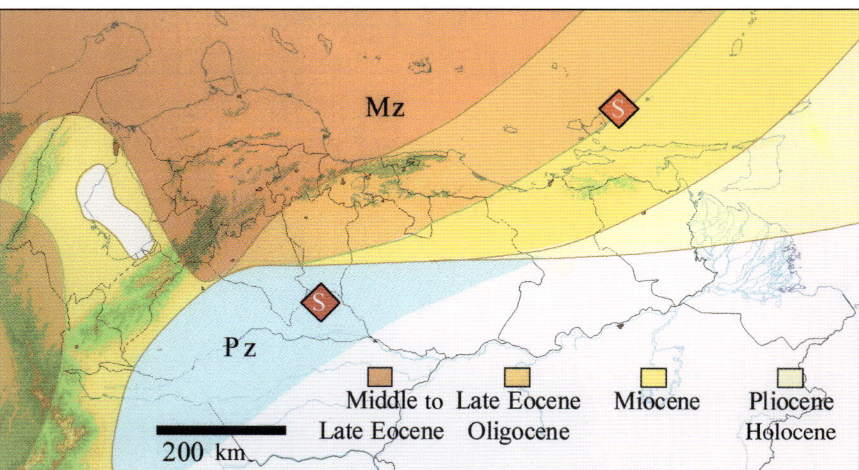

Figure 6. Upper Cretaceous source rocks and their simplified Eocene to present-day generating areas (Talukdar and Marcano, 1994; Erlich and Barrett, 1992; Audemard et al., 1993; Chigne and Hernandez, 1993; Talukdar et al., 1986; Parnaud et al., 1995). Notice the tentative extent of Paleozoic generating areas. Mz = Mesozoic, Pz = Paleozoic, S = source rocks.

Source rocks in the offshore basins and Falcón are Eocene to middle Miocene type II–III shaly source rocks, able to generate oil and gas (Boesi and Goddard, 1991). These rocks were deposited in rift basins during development of the active margin (Figure 7).

Triassic and Jurassic source rocks also may be present in the Cocinas trough of the Goajira peninsula because of deposition of marine lacustrine shales in restricted basins similar to those described in central Venezuela (Bartok, 1993).

The amount of oil that has been retained in the many traps formed along the passive and active margins during their evolution is an issue critical to the understanding of the large petroliferous provinces.

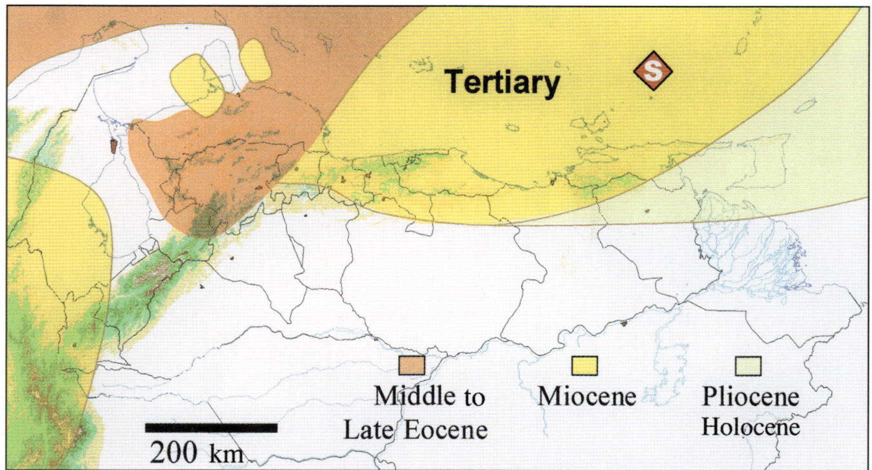

Figure 7. Tertiary source rocks and their simplified Eocene to present-day generating areas. S = source rocks.

FUTURE EXPLORATION PLAYS

Eleven plays have been selected to provide a range of underexplored play types that may exist in Venezuelan basins. They will be described from east to west. Many of the most relevant trap styles and potential source locations are shown on seismic profiles, and evidence from wells and outcrops is presented also.

Delta Platform Play (Seismic Line 1)

The Delta Platform play (Figure 8; Table 1) is constrained to the most recent part of the foredeep. It is offshore, close to the Orinoco Delta slope break, where many northwest-trending normal faults occurred (Figure 9). Each of the upthrown sides of these normal faulted blocks is a potential trap, with more than 7600 m (25,000 ft) of Miocene and Pliocene-Pleistocene sandstones and shales. Faulting dies out above the condensed Paleogene section. Some faults also cut through the Cretaceous source rocks and may provide a migration pathway. Present-day oil and gas generation and expulsion from Miocene and Cretaceous source rocks favor this area, where several fields have already been discovered in Trinidad. A discussion of the geologic setting of this province can be found in Di Croce (1995).

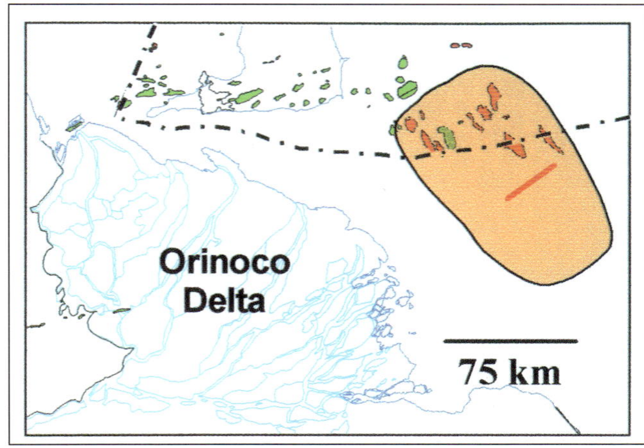

Figure 8. Location of Delta Platform play and seismic line 1.

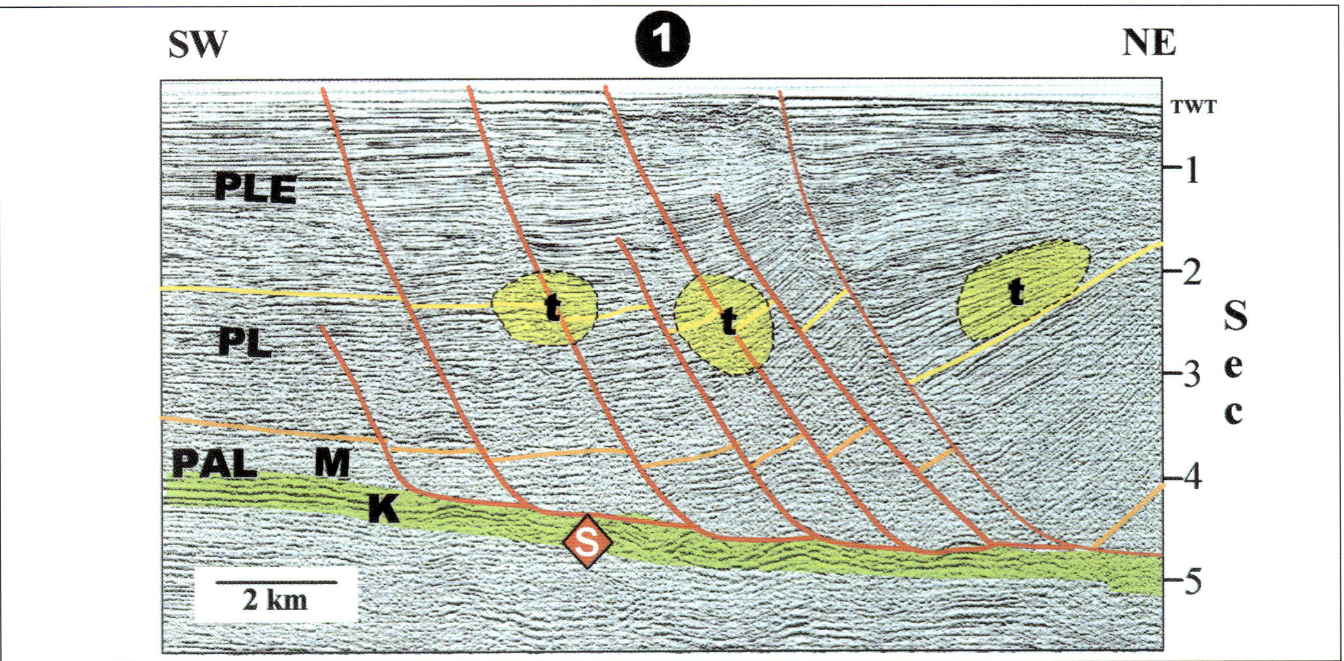

Figure 9. Seismic line 1 (Figure 8): Normal faults cutting a thick Tertiary section create a potential trap. Each footwall of a normal fault is a potential trap. Faults also may provide migration pathways from Cretaceous source rocks. K = Cretaceous, Pal = Paleogene, M = Miocene, PL = Pliocene, PLE = Pleistocene, S = source rock, t = trap. See Figure 8 for location of seismic line 1.

Table 1. Delta Platform Play.

Area	3000–5000 km^2
Wildcats drilled	Five in Venezuela, more than 20 in Trinidad
Number of discoveries	Three fields (Tajali, Loran, and Cocuina) in Venezuela, 12 fields in Trinidad
Type of traps	Hanging wall and footwall of normal faulted blocks, tilted blocks
Main reservoirs	Cretaceous continental sandstones and deep-water Pliocene-Pleistocene sandstones
Source rocks	Upper Cretaceous shales, Miocene shales
Hydrocarbon	Gas, condensates, and minor types amounts of liquids at depths less than 4000 m (13,000 ft)
Critical aspects	Reservoir distribution

Diapir Belt Play (Seismic Line 2)

The Diapir Belt play (Figures 10, 11; Table 2) is a belt of mud diapirs with two ridges extending parallel to the mountain front of the Eastern Interior Mountain Range. On both sides of these ridges, onlapping Miocene and Pliocene sandstones form traps that have been proved successfully in two fields—Pedernales in Venezuela and El Soldado in Trinidad. Hedberg (1950) described the presence of mud volcanoes and associated seeps along this belt. The shales that form the ridges were deposited during the early Miocene and remobilized during Pliocene compression. A Miocene to Holocene phase of generation and expulsion favors the presence of hydrocarbons in these traps. Although significant hydrocarbon charge is a critical issue, oil seeps associated with the mud volcanoes indicate that hydrocarbons have been generated.

Reverse faults caused by the collapse of normal faulting between the two ridges are another type of trap in this setting.

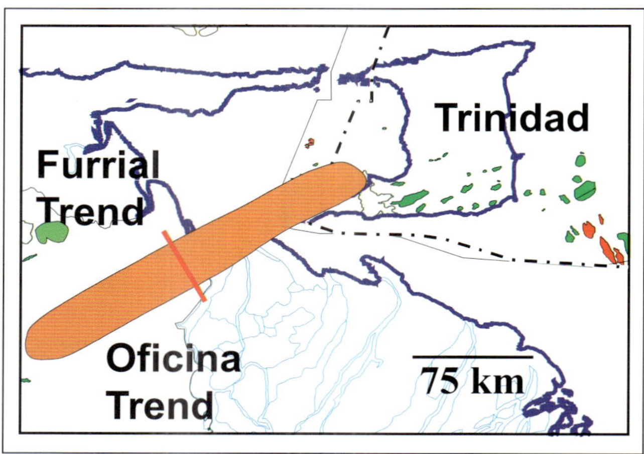

Figure 10. Location of Diapir Belt play and seismic line 2.

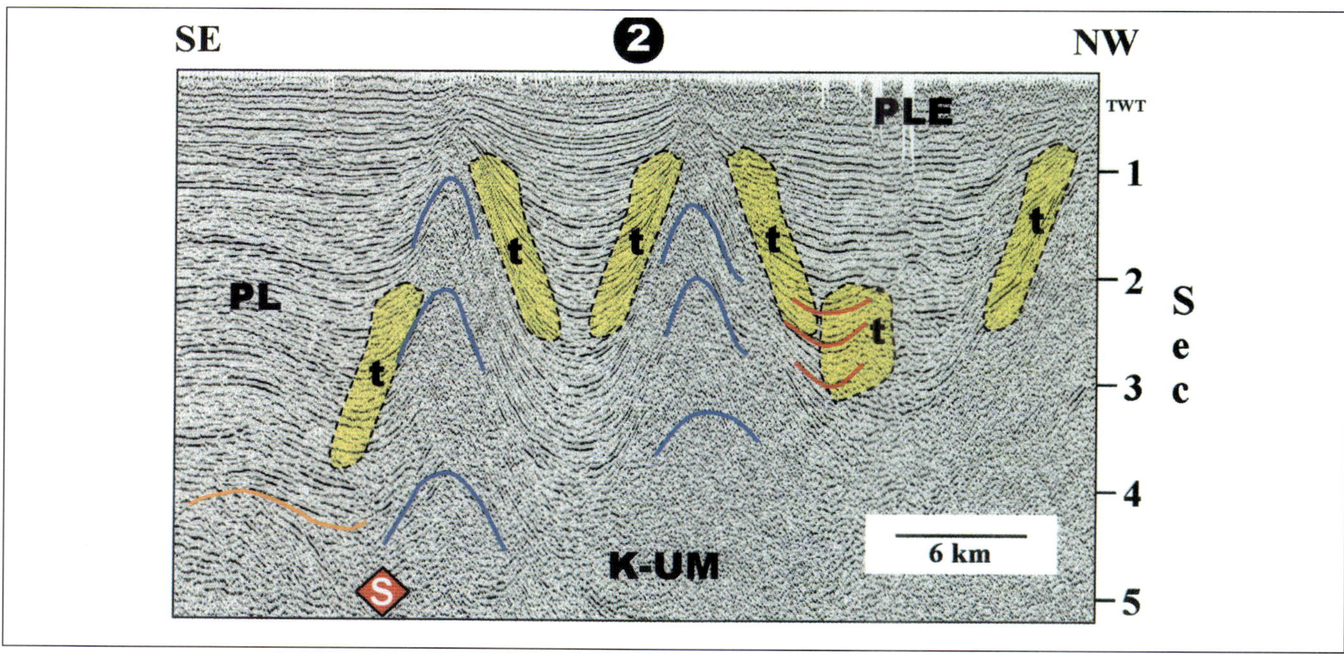

Figure 11. Seismic line 2 (Figure 10): A mud-diapir belt comprising two ridges. Onlapping sediments and collapsed reverse faults create the traps. Red lines are reverse faults. K-UM = Cretaceous to upper Miocene, PL = Pliocene, PLE = Pleistocene, S = source rock, t = trap. See Figure 10 for location of seismic line 2.

Table 2. Diapir Belt Play.

Area	1500–3000 km^2
Wildcats drilled	None for this play
Number of discoveries	1 (Pedernales field) in Venezuela, 1 (El Soldado field) in Trinidad
Type of traps	Onlaps on diapir wall, reverse-faulted blocks
Main reservoirs	Miocene to Pliocene sandstones
Source rocks	Upper Cretaceous shales, Miocene shales
Hydrocarbon types	Light oil and associated gas
Critical aspects	Significant hydrocarbon charge

Southern Eastern Mountain Range Play (Seismic Line 3)

The Southern Eastern Mountain Range play (Figures 12, 13; Table 3) represents the northern extension of the Furrial trend. The style, size, and distribution of the anticlinal structures are part of the transition to the hinterlands of the Serrania del Interior in the segment where no metamorphic rocks have been identified to date.

All of these structures have a greater probability of containing hydrocarbons with higher gas-to-oil ratios, than the Furrial trend.

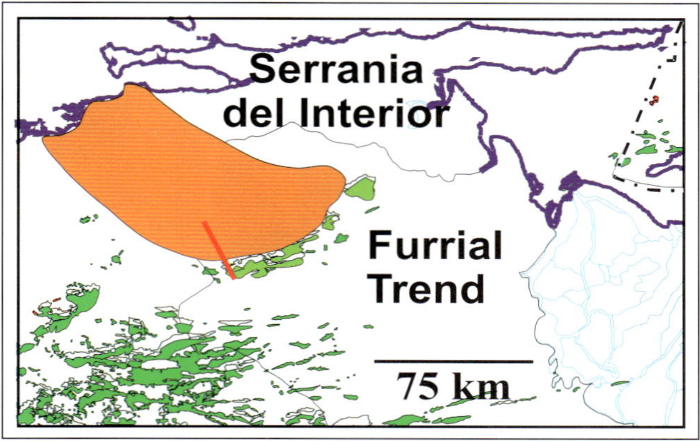

Figure 12. Location of Southern Eastern Mountain Range play and seismic line 3.

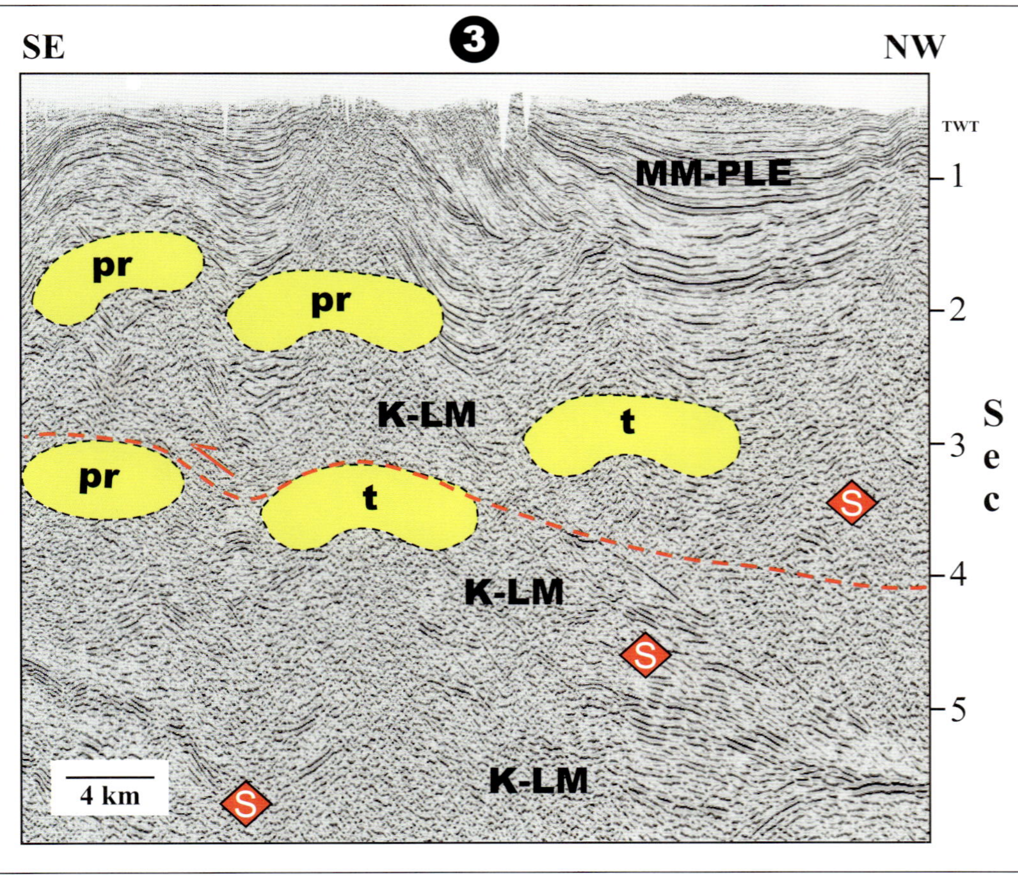

Figure 13. Seismic line 3 (Figure 12): A Cretaceous to lower Miocene section containing source rocks and reservoirs is repeated three times because of the occurrence of two décollement zones. Anticlines associated with thrusting of the two lowermost sections remain underexplored and are north of the Furrial trend. The red arrow marks the décollement zone. K-LM = Cretaceous to lower Miocene, MM-PLE = middle Miocene to Pleistocene, S = source rock, t = trap, pr = proven reservoir. See Figure 12 for location of seismic line 3.

Table 3. Southern Eastern Mountain Range Play.

Area	3000–5000 km^2
Wildcats drilled	None
Number of discoveries	None. The giant Furrial trend is downdip.
Type of traps	Thrusts
Main reservoirs	Cretaceous to Miocene sandstones
Source rocks	Upper Cretaceous shales, Miocene shales
Hydrocarbon type	Gas, condensate, and light oil
Critical aspects	Reservoir integrity

Southern Central Mountain Range Play (Seismic Line 4)

The Southern Central Mountain Range play (Figures 14, 15; Table 4) is a series of antiformal structures preserved beneath the overridden Serrania del Interior (Blin et al., 1988). The area where the allochthonous terranes are less than 3000 m (10,000 ft) thick is targetable in the search for gas, oil, and condensate accumulations. A potential risk exists of thermal cracking of oil, similar to the Yucal Placer area (Figure 14), because of high heat flow in the northern Central Guarico Basin. The reason for this flow is unknown, but it could result in a gradient of increasing liquid hydrocarbons away from that heat anomaly. The transition to the south from gas to liquid in the northern part of the Guarico Basin is abrupt and can be identified over a distance of less than 8 km (5 miles). The sizes of untested structures in this play are such that they could hold a giant accumulation. However, the quality of proven reservoirs to the south is low.

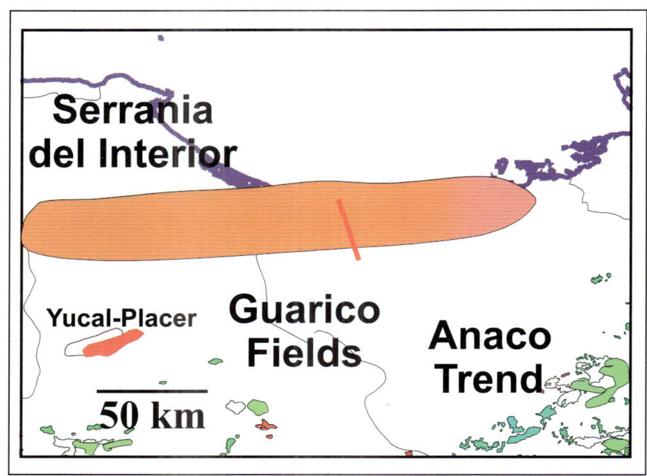

Figure 14. Location of Southern Central Mountain Range play and seismic line 4.

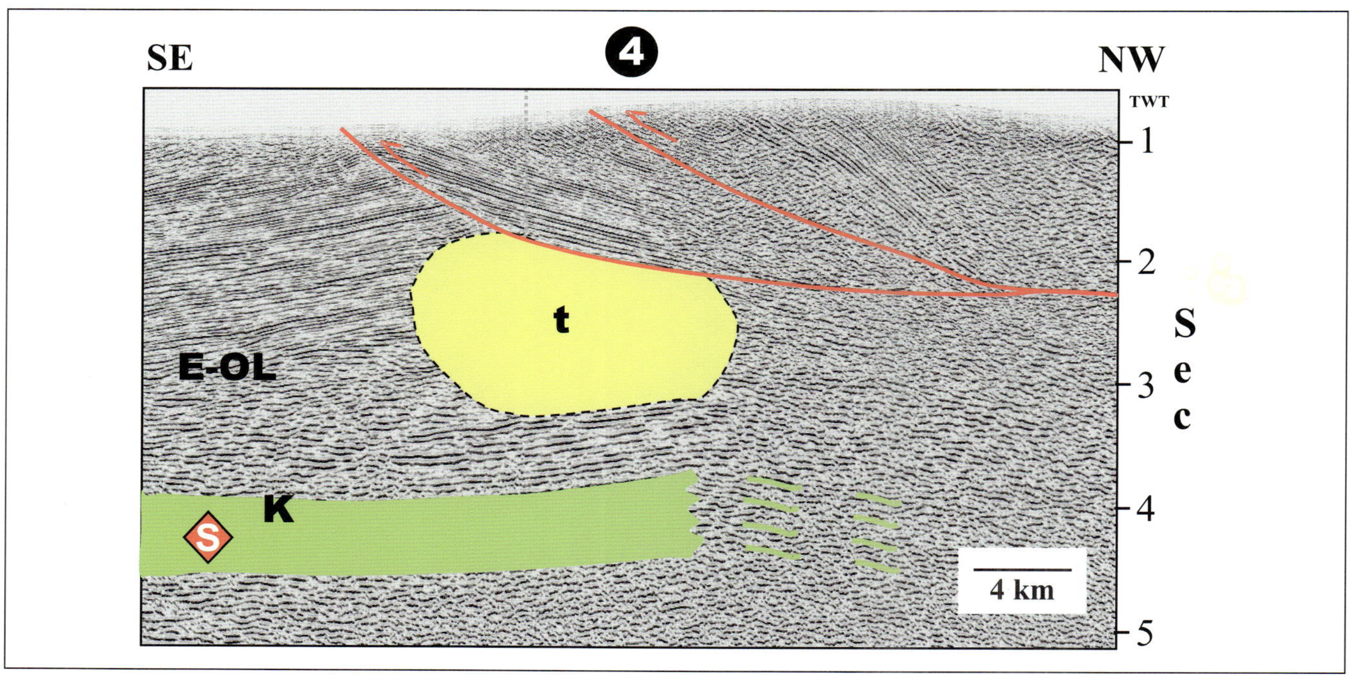

Figure 15. Seismic line 4 (Figure 14): Subthrusted structural highs with Eocene-Oligocene sandstones as the main targets. The red arrow marks a reverse fault. K = Cretaceous, E-OL = Eocene to Oligocene, S = source rock, t = trap.

Table 4. Southern Central Mountain Range Play.

Area	5000–7000 km^2
Wildcats drilled	None
Type of traps	Subthrusting anticlines
Main reservoirs	Cretaceous, Eocene, and Oligocene sandstones
Source rocks	Upper Cretaceous shales, Oligocene and Miocene shales
Hydrocarbon types	Gas, condensate, and light oil
Critical aspects	Trap-charge timing, reservoir integrity

Anaco Trend South Subthrust Play (Seismic Line 5)

The Anaco trend has been drilled since the 1930s, but only two wells have penetrated the subthrust, which has Oligocene and Miocene oil-bearing sandstones (Figures 16, 17; Table 5). The sandstones constitute the reservoirs in the hanging wall of the inverted graben in existence since the Cretaceous (Murany, 1972).

The very prolific Anaco trend has been drained consistently from the inverted footwall, but exploration along the hanging wall has been restricted to two wells which cut across the main fault plane. In addition, the Greater Oficina play, located to the south, may extend into this area. The transition between previously mentioned styles and the Oficina trend is presently underexplored, but several discoveries have been made.

If we assume an areal richness for this area similar to that found in surrounding areas, giant fields may be present.

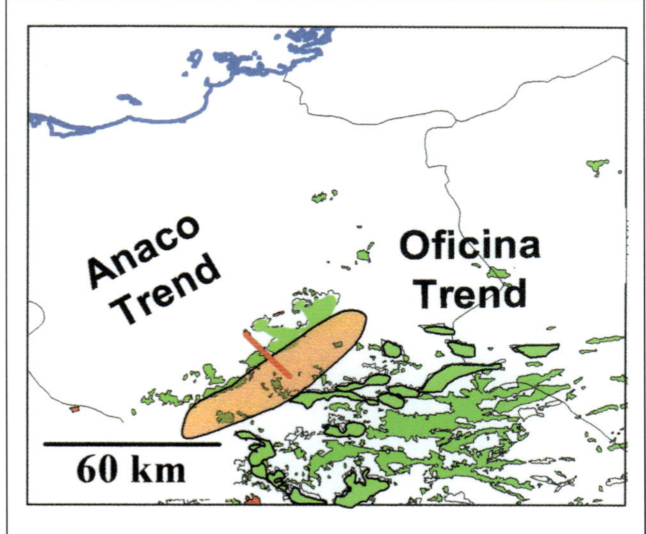

Figure 16. Location of Anaco Trend South Subthrust play and seismic line 5.

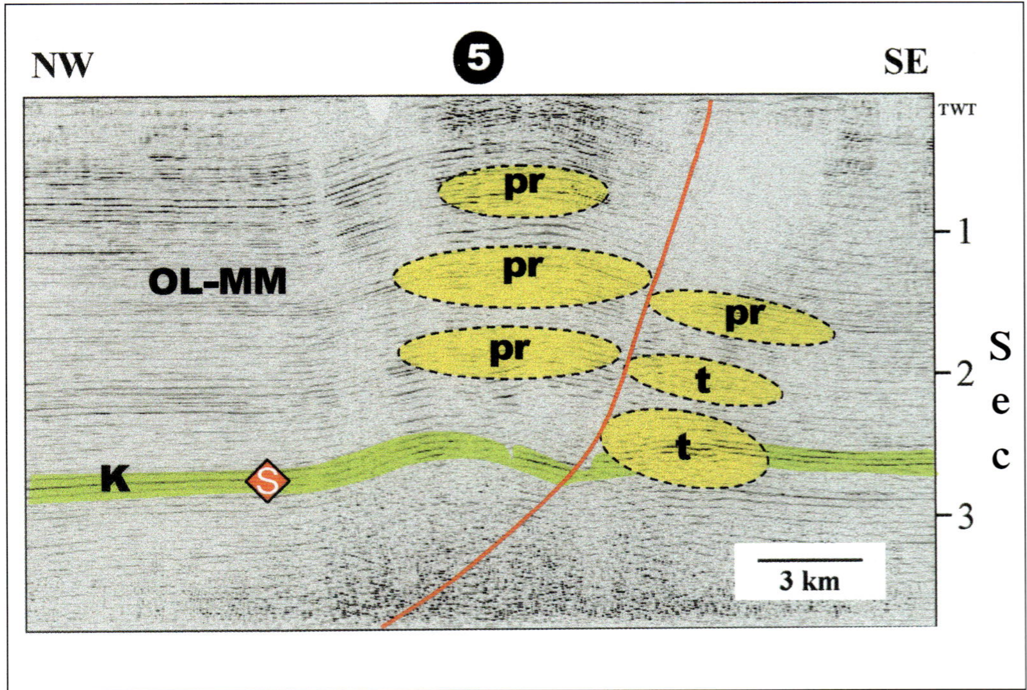

Figure 17. Seismic line 5 (Figure 16): Three-D seismic data will allow better imaging of the hanging wall of the Anaco trend, where Cretaceous (K) to Oligocene–middle Miocene (OL-MM) reservoirs can hold giant accumulations of gas, condensate, and light oil. S = source rocks, t = trap, pr = proven reservoir.

Table 5. Anaco Trend South Subthrust Play

Area	300–500 km^2
Wildcats drilled	Two
Number of discoveries	Two
Type of traps	Hanging wall of inverted graben-subthrust
Main reservoirs	Cretaceous, Oligocene to Miocene sandstones
Source rocks	Upper Cretaceous shales, Oligocene and Miocene shales
Hydrocarbon types	Gas, condensate, and light oil
Critical aspects	Reservoir quality

Espino Graben Play (Seismic Line 6)

The Espino Graben play (Figures 18, 19; Table 6) is a major Mesozoic half-graben system that cuts the Eastern and Barinas Basins transversely. This system ranges from 20 to 35 km (12–22 miles) wide and extends for more than 500 km (310 miles). The level of erosion and inversion is relatively low, but it certainly is associated with tilted blocks rotated during the extensional phase. The graben infills and rotated sedimentary units are located at depths less than 2000 m (6500 ft) (Feo Codecido et al., 1984), which suggests that favorable reservoir characteristics may be present. The largest oil accumulation of the world (the Orinoco Oil Belt) is slightly younger and south of this trend. Part of the extra-heavy oil in the Orinoco Oil Belt could be derived from potential source beds found in this structural system.

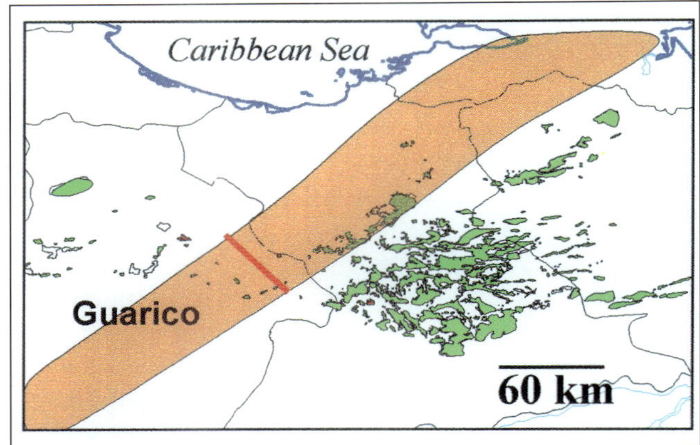

Figure 18. Location of Espiño Graben play and seismic line 6.

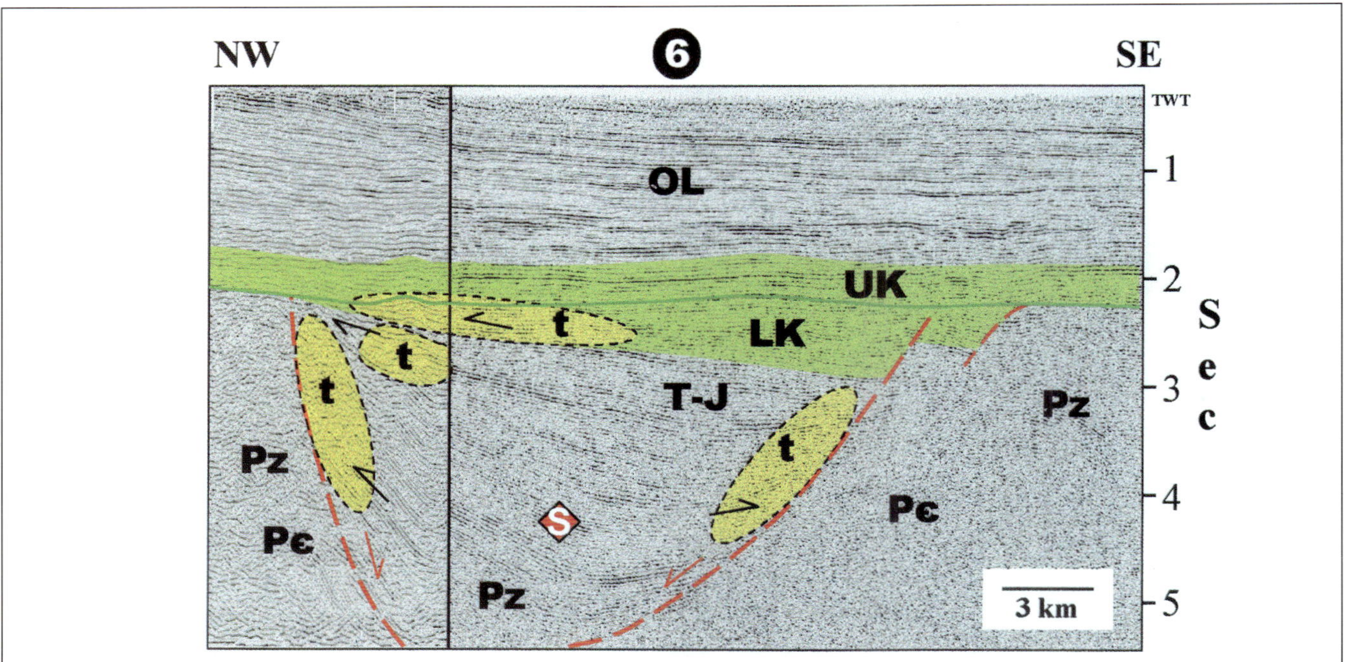

Figure 19. Seismic line 6 (Figure 18): Onlaps against Precambrian and Paleozoic basement and subunconformity toplaps constitute the main traps. Black arrows mark onlap; red arrow marks normal fault. PC = Precambrian, PZ = Paleozoic, T-J= Triassic to Jurassic, LK = Lower Cretaceous, UK = Upper Cretaceous, OL = Oligocene, S = source rock, t = trap.

Table 6. Espino Graben Play.

Area	10,000–18,000 km^2
Wildcats drilled	Two deep wells
Number of discoveries	None
Type of traps	Remigrated oil in partially inverted normal fault setting; onlaps and toplaps
Main reservoirs and seals	Paleozoic, Triassic to Jurassic, and Cretaceous sandstones
Source rocks	Triassic to Jurassic lacustrine shales, Silurian-Devonian source beds
Hydrocarbon type	Light oil
Critical aspects	Reservoir integrity for Paleozoic reservoirs, source-rock richness

North Andean Mountain Front Play (Seismic Line 7)

Large key imbricate thrusts are juxtaposed in triangle zones aligned along the northern front of the Merida Andes (Figures 20, 21; Table 7). At least four thrust sheets containing the La Luna source beds are interpreted from good-quality seismic profiles. The reservoirs are usually sandstones under the La Luna beds, but the overlying lower Tertiary sandstones are also a potential reservoirs where the triangle zones are inactive and are crosscut by deeper thrusts. All these structures are potential giant fields.

At least 50 active oil seeps are located along the roof thrust defined by the Colon shales, above the La Luna Formation. These seeps have been mapped along the northern flank of the Andes at elevations ranging from sea level to 3000 m (10,000 ft).

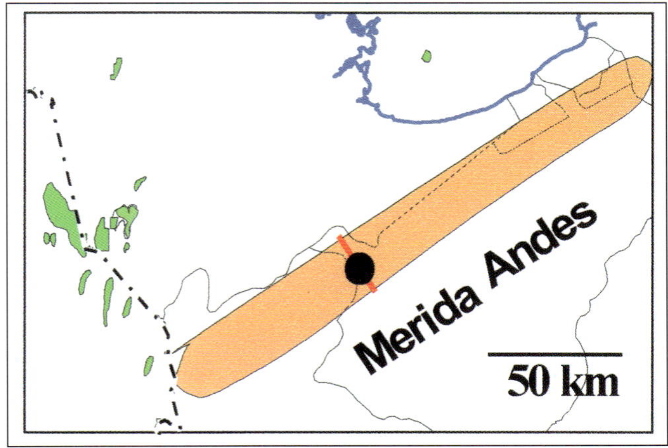

Figure 20. Location of Northern Andean Mountain Front play and seismic line 7.

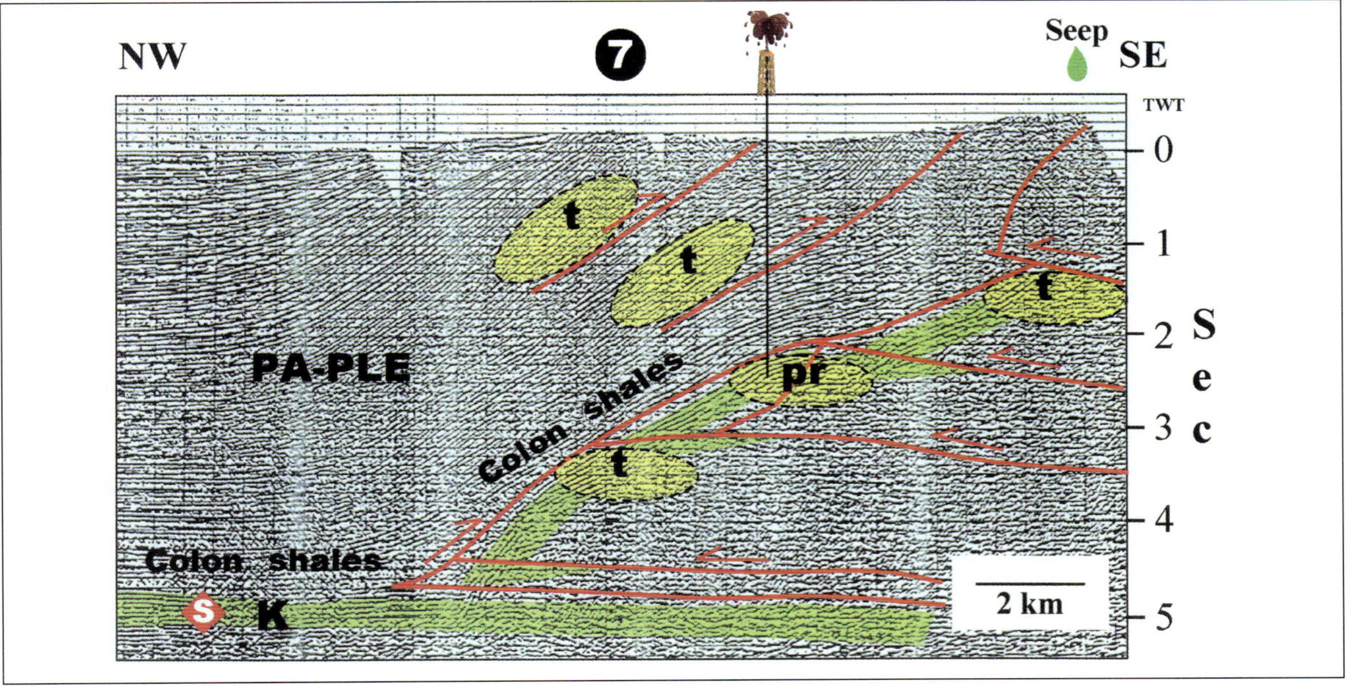

Figure 21. Seismic line 7 (Figure 20): Discovery-well location in imbricated thrusts of a North Andean triangle zone. Red arrows mark reverse faults. K= Cretaceous, PA-PLE = Paleocene to Pleistocene, S = source rock, t= trap, pr = proven reservoir.

Table 7. North Andean Mountain Front Play.

Area	3000–5000 km^2
Wildcats drilled	Four
Number of discoveries	One
Type of traps	Thrusts and back thrusts in triangle zones
Main reservoirs	Cretaceous sandstones
Source rocks	Upper Cretaceous calcareous shales, Paleogene carbonaceous shales
Hydrocarbon types	Light oil, condensate, and gas
Critical aspects	Reservoir depths

South Andean Mountain Front Play (Seismic Line 8)

The South Andean Mountain Front play (Figures 22, 23; Table 8) is the result of the evolution of the Lara Nappe accretionary prism and the formation to the south of the Eocene foredeep, where most of the oil was forced to migrate southeastward toward the Guyana shield. Most of the important traps were originally half grabens developed on the inner segments of the Cretaceous passive margin. Most of the half grabens are post-Cretaceous but predate the pre-Eocene compression. Some of them were partially reactivated during the Tertiary, mostly during the north-vergent Pliocene Andean compression. This later event was responsible for creating the diversity of traps distributed along this front. The folded structures are several kilometers in length and approximately 3 km (2 miles) in width.

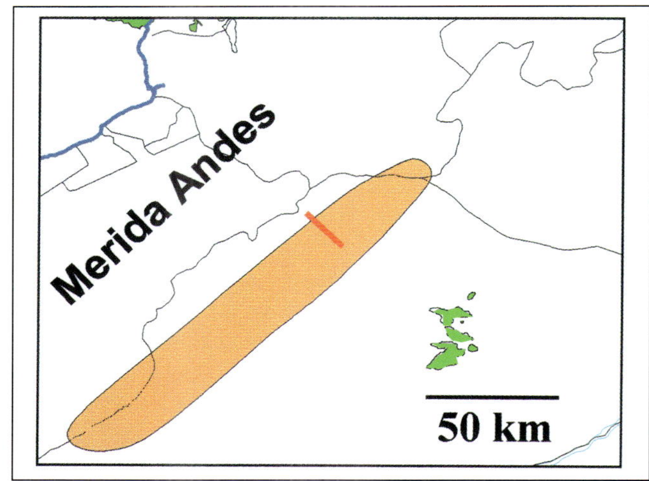

Figure 22. Location of Southern Andean Mountain Front play and seismic line 8.

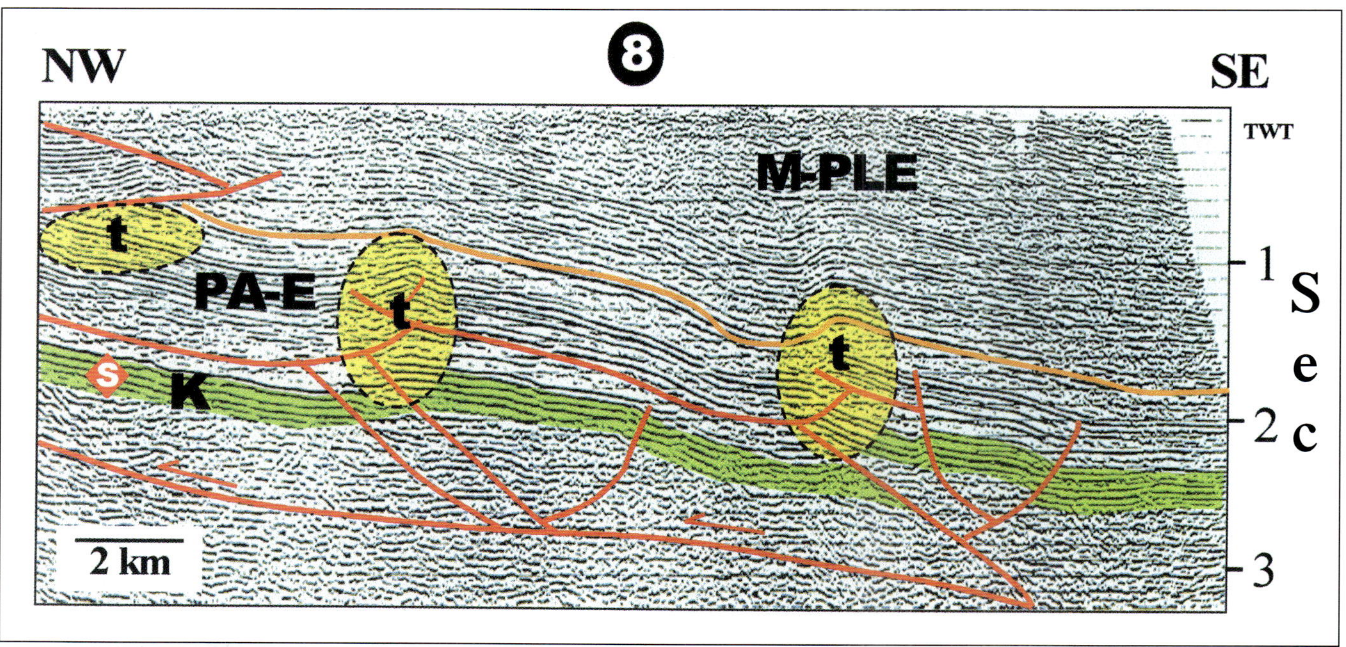

Figure 23. Seismic line 8 (Figure 22): Main targets are anticlines associated with north-verging thrusts that occurred during Andean deformation. K = Cretaceous, PA-E = Paleocene to Eocene, M-PLE = Miocene to Pleistocene, t = trap, S = source rock.

Table 8. South Andean Mountain Front Play.

Area	5000 km^2
Wildcats drilled	Two
Number of discoveries	None
Type of traps	Anticlines caused by thrusting verging to north
Main reservoirs	Cretaceous calcareous sandstones, Eocene to Oligocene sandstones
Source rocks	Upper Cretaceous calcareous shales
Hydrocarbon types	Gas and light oil
Critical aspects	Charge caused by remigration

Barinas-Apure Stratigraphic Traps Play (Seismic Line 9)

The Barinas-Apure Stratigraphic Traps play (Figures 24, 25; Table 9) includes a variety of stratigraphic traps: onlap of Paleocene to Eocene sandstones on Cretaceous shales, Cretaceous sandstones interbedded with shales subcropping below Oligocene shales, upper Eocene to Oligocene-Miocene sandstones onlapping on Cambrian to Precambrian rocks, and Upper Cretaceous deltaic sandstones which prograded to the north. Wells drilled to test structural traps may have tested some of these concepts, such as the wells in Arauca field in Colombia and La Victoria field in Venezuela. The former produced light oil and gas from Cretaceous porous, calcareous sandstones; from Paleocene and Eocene sandstones onlapping Cretaceous shales, and at pinch-outs of upper Eocene to Oligocene sandstones (C. Urbina, personal communication, 1999).

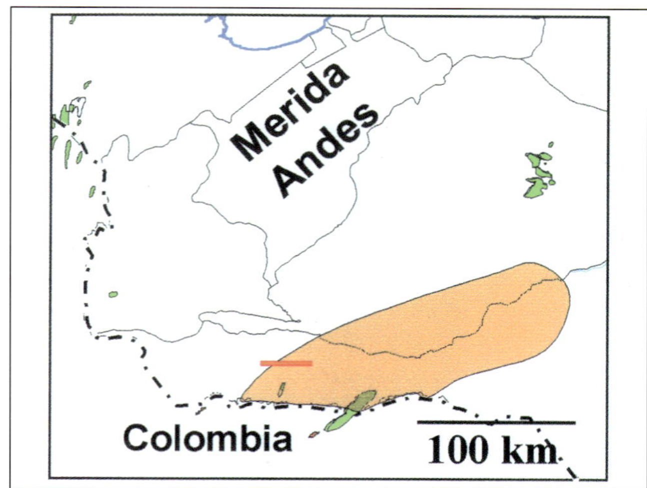

Figure 24. Location of Barinas-Apure Stratigraphic Traps play and seismic line 9.

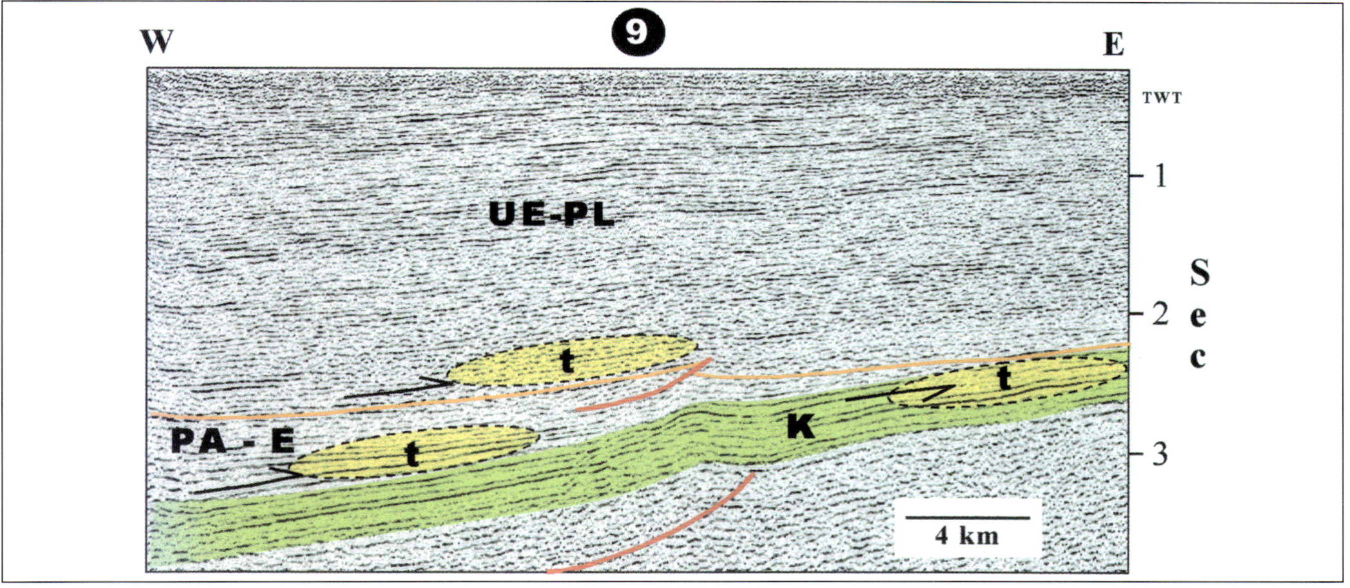

Figure 25. Seismic line 9 (Figure 24): Onlaps and subunconformity traps similar to the shallow Orinoco Oil Belt stratigraphic traps but filled with light oil because they are at a greater depth and are protected from bacterial degradation. Black arrow marks onlap and truncation. K = Cretaceous, PA-E = Paleocene to Eocene, UE–PL= upper Eocene to Pliocene, t= trap.

Table 9. Barinas-Apure Stratigraphic Traps Play.

Area	2000–2500 km^2
Wildcats drilled	None targeting these plays
Number of discoveries	At least two while drilling structural traps
Type of traps	Onlaps, subunconformity truncations, and prograding sandstones
Main reservoirs	Cretaceous calcareous sandstones, Paleocene to Eocene sandstones, upper Eocene and Oligocene-Miocene sandstones
Source rocks	Upper Cretaceous calcareous shales
Hydrocarbon types	Light oil and associated gas
Critical aspects	Mapping of trap and lateral seal

Pre-Cretaceous Weathered Zone Play (Seismic Line 10)

The peculiar traps in the Pre-Cretaceous Weathered Zone play (Figures 26, 27; Table 10), associated with an unconformable surface, were formed during the Triassic to the Jurassic. They are thought to be part of the peneplained surfaces underlying the prerift unconformity of the Tethyan passive margin. They formed through chemical weathering of siliciclastic cements or partial destruction of metasediments. They have been associated with fractured reservoirs (Smith, 1956). Good examples of this play are the basement reservoirs of the giant La Paz and Mara fields, and the very old Totumo field from northwestern Maracaibo Basin. Appreciable quantities of hydrocarbons have been produced from these fields. Similar plays could be present at depths less than 3000 m (10,000 ft) in the Barinas Basin, as well as along the belt

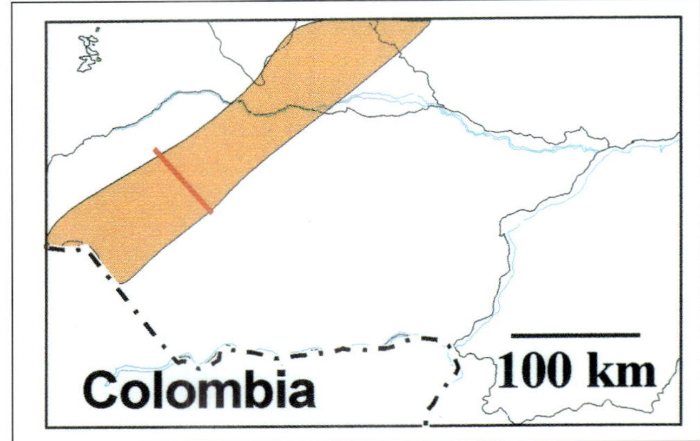

Figure 26. Location of Pre-Cretaceous Weathered Zone play and seismic line 10.

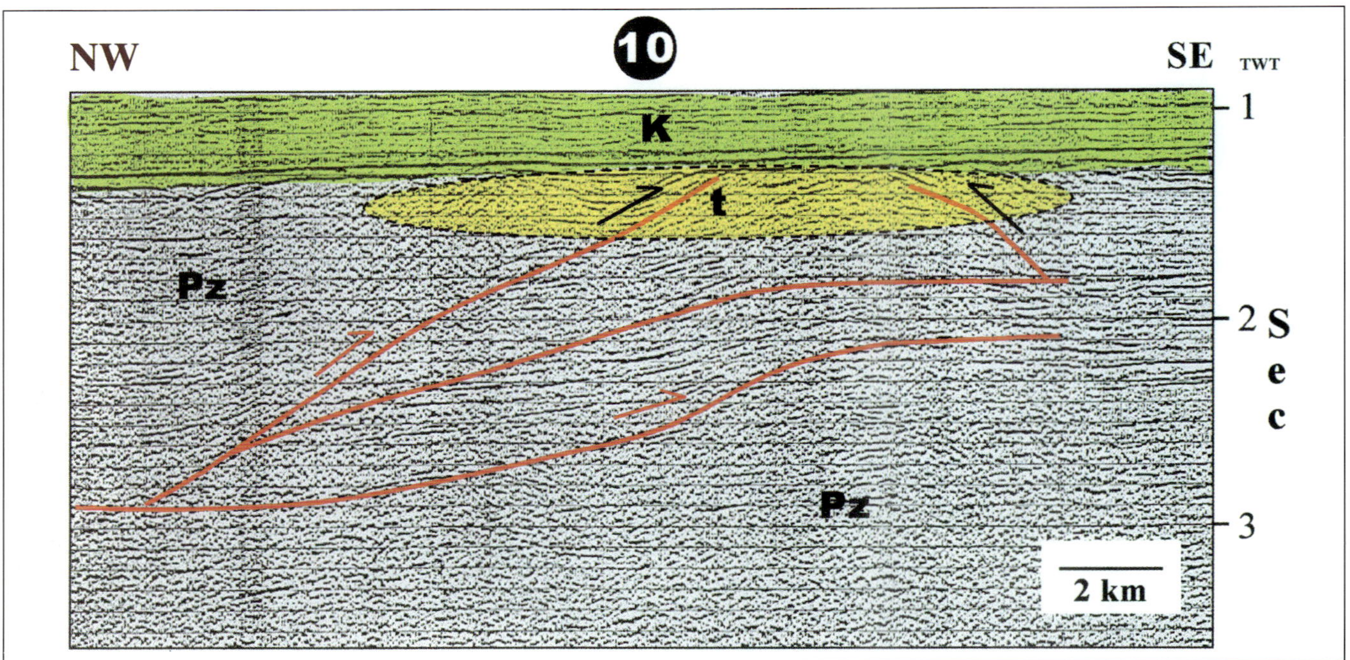

Figure 27. Seismic line 10 (Figure 26): Deformation front of the Paleozoic orogene. Long exposure of Paleozoic beds before Cretaceous sedimentation enhanced porosity and permeability of subunconformity layers. Red arrows mark reverse faults. PZ = Paleozoic, K = Cretaceous, t = trap.

Table 10. Pre-Cretaceous Weathered Zone Play.

Area	Unknown
Wildcats drilled	Unknown
Number of discoveries	Totumo, Río de Oro, La Paz, and Mara fields
Type of traps	Stratigraphic
Main reservoirs	Weathered-fractured Paleozoic siliciclastics or metasediments
Source rocks	Silurian-Devonian calcareous shales, Triassic-Jurassic shales
Hydrocarbon types	Light to medium oil
Critical aspects	Source rock, lateral seal, target depths

corresponding to the inner segment of the Paleozoic foredeep that encroached on the Guyana shield.

Northern Offshore Venezuela Basin Play (Seismic Line 11)

The most important type of trap in the Northern Offshore Venezuela Basin play (Figures 28, 29; Table 11) is the late early Miocene to middle Miocene inversion anticlines (Audemard, 1991; Macellari, 1995). This type of trap is the product of reactivation of the Paleogene to early Miocene half-graben system during Neogene transpression. Traps comprise folded Paleogene to lower Miocene discrete deep-marine sandstones interbedded with shales that form internal seals. Sealing shales result in compartmentalized hydrocarbon accumulations, each of which may have its own gas-oil-water contacts and pressure distribution. In this type of play, fracture zones and fault systems associated with the Paleogene extension could provide the main migration pathways for hydrocarbons sourced from Eocene to middle Miocene shales (Ysaccis, 1997).

Positive results have already been reported from the drilling of an inversion structure to the northwest of Tortuga Island. A well located on the flank of that anticline encountered 43° API oil in lower Miocene reservoirs.

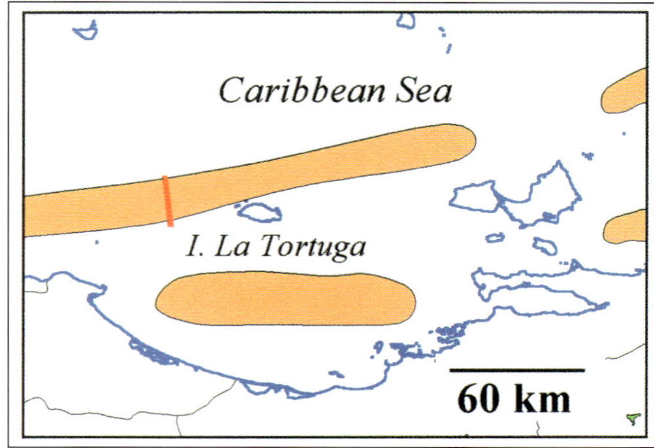

Figure 28. Location of Northern Offshore Venezuela Basin play and seismic line 11.

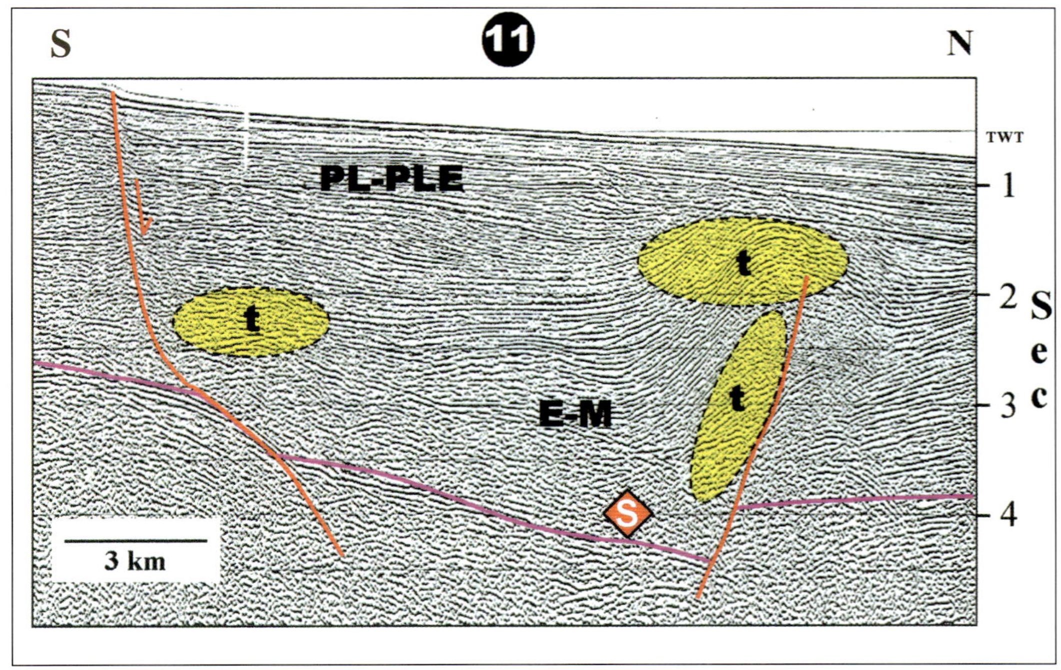

Figure 29. Seismic line 11 (Figure 28): Typical image of an offshore rift basin associated with evolution of the active margin. An Eocene (E) to Pleistocene (PLE) section fills the basin. Main source rocks (S) are in the lowermost Eocene to Miocene section. Traps (t) are rotated normal fault blocks, inverted grabens, and pinch-outs.

Table 11. Northern Offshore Venezuela Basin Play.

Area	30,000–40,000 km^2 in four areas offshore Venezuela
Wildcats drilled	Approximately 50 offshore and 50 onshore, concentrated in four areas
Number of discoveries	Six fields onshore and eight offshore
Type of traps	Rotated normal faulted blocks, inverted grabens, pinch-outs
Main reservoirs	Paleogene to Miocene turbiditic to deltaic sandstones, Mesozoic metamorphic basement
Source rocks	Eocene to middle Miocene shales
Hydrocarbon types	Light oil, condensates, and gas
Critical aspects	Reservoir presence and quality as well as probable oil-charge limitation

SUMMARY

Eleven underexplored provinces in Venezuela can hold potential giant fields (Figures 30, 31). These provinces range from extensive weathered Paleozoic basement in the west to still little-explored thick Triassic-Jurassic grabens and Pleistocene deep-water sandstones offshore and in the Eastern Venezuelan Basin. A variety of structural and stratigraphic traps, comprising continental to deep-water sandstones and local shallow-water carbonates, is present in these provinces.

Paleozoic and Mesozoic-Cenozoic active and passive margins favored sedimentation of two main source rocks, Silurian-Devonian and Upper Cretaceous calcareous shales, that matured through time. Additionally, Tertiary source rocks are present in rift basins and foredeeps.

Gas and light-oil accumulations are most common in the northern part of the country and east of Venezuela.

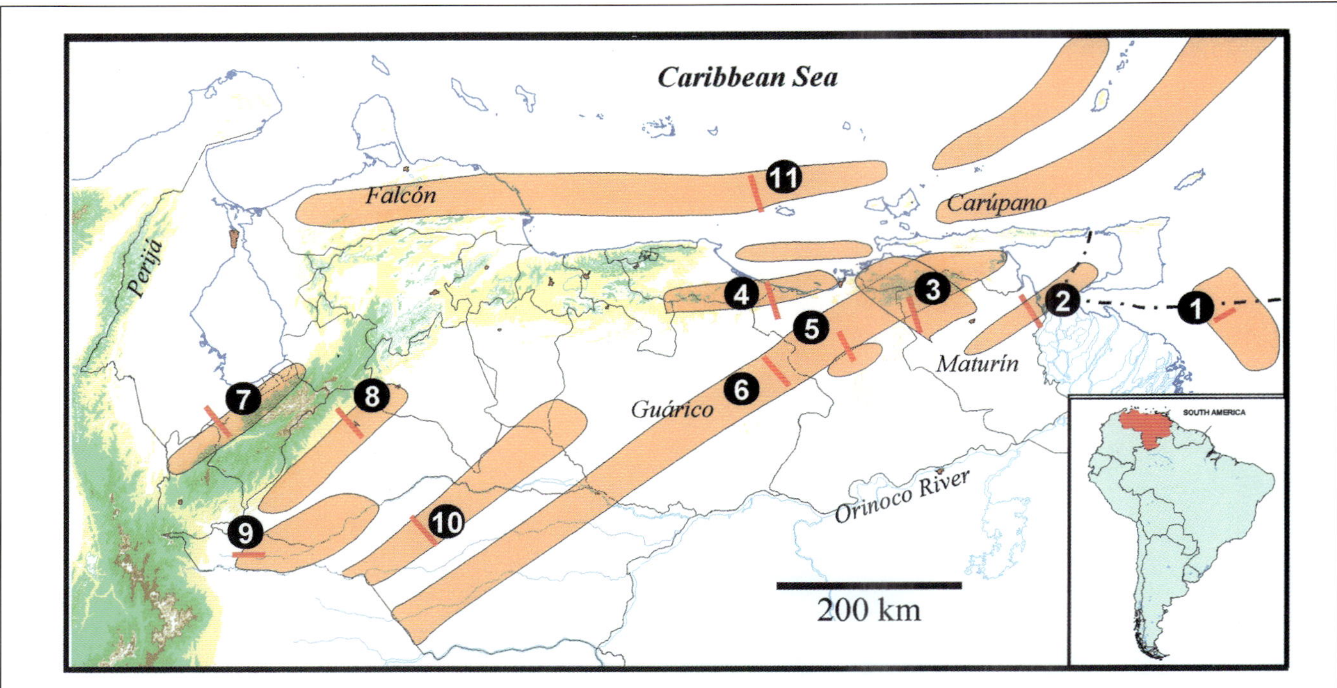

Figure 30. Future petroliferous provinces of Venezuela.

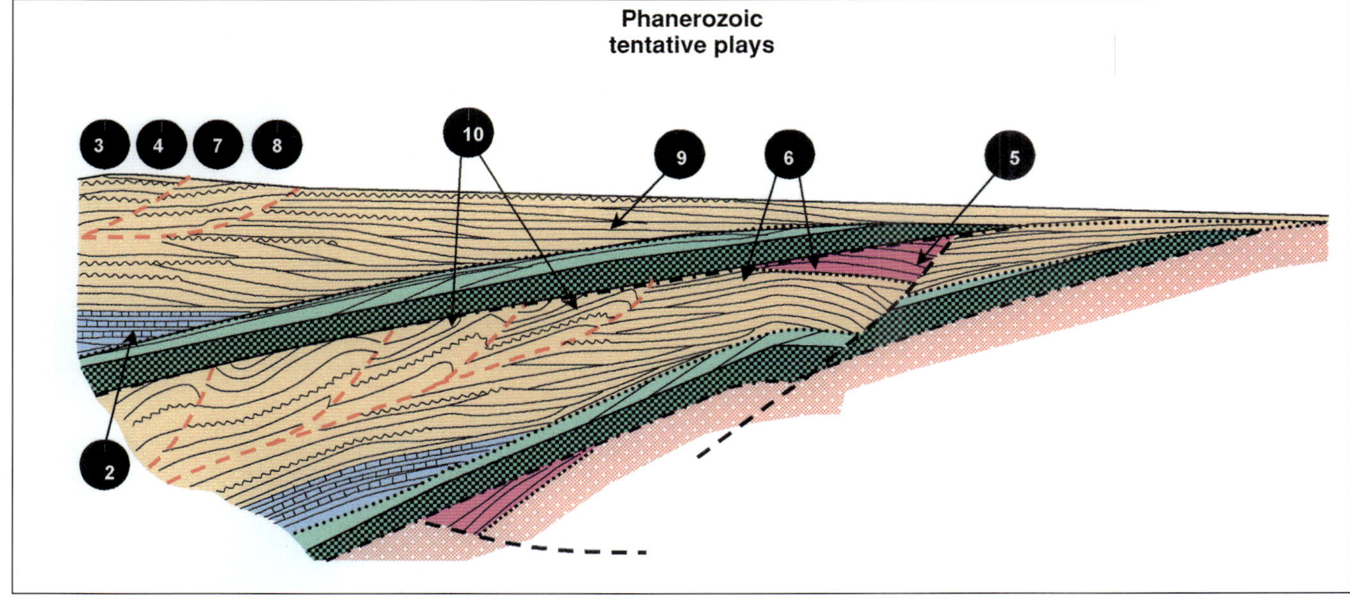

Figure 31. Schematic cross section showing possible plays. Numbers correspond to the provinces indicated in Figure 30. Play 11 is the only one not represented in the model.

Accumulations grade to medium oil and then to heavy oil toward the Guyana shield.

Venezuela has numerous unexplored and underexplored exploration opportunities ready to test in the new millennium.

ACKNOWLEDGMENTS

We greatly appreciate permission granted by PDVSA Oil and Gas to publish this paper. Technical discussions with all our colleagues of the PDVSA Regional Exploration Study Team helped to unravel the range of Venezuelan hydrocarbon potential. We also thank Christopher White for helpful suggestions and editing of this paper, and our technical staff for thoughtful comments.

REFERENCES CITED

Audemard, F. E., 1991, Tectonics of western Venezuela: Ph.D. thesis, Rice University, Houston, Texas, 245 p.

Audemard, F. E., and J. Lugo, 1996, Petroleum geology of Venezuela: AAPG/SVG International Conference course notes, 320 p.

Audemard, F. E., N. Audemard, E. Rodríguez, and N. Jordán, 1993, Generation and migration of heavy oil from Guárico and southern Maturín subbasins (abs.): AAPG Bulletin, v. 77, p. 303.

Audemard, F. E., E. Cabrera, J. Di Croce, and A. Melendez, 1997, Sinopsis de la Geología de Venezuela, *in* Léxico Estratigráfico de Venezuela, III Edition: Caracas, M. J. Editores C. A., v. 1, p. 18–28.

Aymard, R., L. Pimentel, P. Eitz, P. López, A. Chaouch, J. Navarro, J. Mijares, and J. G. Pereira, 1990, Geological integration and evaluation of northern Monagas, Eastern Venezuelan Basin, *in* J. Brooks, ed., Classic petroleum provinces: Geological Society of London Special Publication 50, p. 37–53.

Bally, A. W., 1989, Phanerozoic basins of North America, *in* A. W. Bally and A. R. Palmer, eds., The geology of North America—An overview: The geology of North America, v. A, p. 397–446.

Bartok, P., 1993, Prebreakup geology of the Gulf of Mexico–Caribbean: Its relation to Triassic and Jurassic rift systems of the region: Tectonics, v. 12, p. 441–559.

Blaser, R., and C. White, 1984, Source rock and carbonization study, Maracaibo Basin, Venezuela, *in* G. Demaison and R. J. Murris, eds., Petroleum geochemistry and basin evolution: AAPG Memoir 35, p. 229–254.

Blin, B., E. Cabrera, J. F. Stephan, and C. Beck, 1988, Estructura del Frente de Montañas de la Cadena Caribe Occidental al Este de Guarumen-Venezuela: IV Venezuelan Geophysical Congress, p. 457–466.

Boesi, T., and D. Goddard, 1991, A new geologic model related to the distribution of hydrocarbon source rocks in the Falcón Basin, northwestern Venezuela, *in* K. T. Biddle, ed., Active margins: AAPG Memoir 48, p. 303–319.

Chigne, N., and L. Hernández, 1993, Guafita field, Barinas-Apure Basin, Apure State, Venezuela, *in* N. H. Foster and E. A. Beaumont, eds., Structural traps VIII: AAPG Treatise of petroleum geology, p. 231–250.

Di Croce, J., 1995, Eastern Venezuela Basin: Sequence stratigraphy and structural evolution: Ph.D. thesis, Rice University, Houston, Texas, 225 p.

Duval, B., C. Cramez, and G. Valdez, 1994, Campos Gigantes de Finales de los 80 Asociados con Subduccion Tipo A en Suramerica: Boletín Sociedad Venezdana de Geólogos, v. 19, p. 20–40.

Erlich, R. N., and S. F. Barrett, 1992, Petroleum geology of the Eastern Venezuela Foreland Basin, *in* R. W. Macqueen and D. A. Leckie, eds., Foreland basins and fold belts: AAPG Memoir 55, p. 341–362.

Feo-Codecido, G., F. D. Smith Jr., N. Aboud, and E. Di Giacomo, 1984, Basement and Paleozoic rocks of the Venezuelan Llanos Basin, *in* W. E. Bonini, R. B. Hargraves, and R. Shagam, eds., The Caribbean–South American plate boundary and regional tectonics: Geological Society of America Memoir 162, p. 175–188.

Hedberg, H .D., 1931, Cretaceous limestones as petroleum source rocks in northwestern Venezuela: AAPG Bulletin, v. 15, p. 229–244.

Hedberg, H. D., 1950, Geology of the Eastern Venezuela Basin (Anzoátegui, Monagas-Sucre–eastern Guárico portion): Geological Society of America Bulletin, v. 61, p. 1173–1216.

Hung, E. J., 1997 Foredeep and thrust belt interpretation of the Maturin Sub-basin, Eastern Venezuela Basin: Master's thesis, Rice University, Houston, Texas, 125 p.

James, K., 1985, Marco Tectónico, Estilos Estructurales y Habitat de los Hidrocarburos Cretácicos, Venezuela: VI Venezuelan Geological Congress, p. 2452–2469.

Lugo, J., 1991, Cretaceous to Neogene tectonic control on sedimentation: Maracaibo Basin, Venezuela: Ph.D. thesis, University of Texas at Austin, 219 p.

Lugo, J., and F. E Audemard, 1996, The Venezuelan foredeeps. Part I: Stratigraphy (abs.): AAPG Annual Convention Proceedings, p. A. 86.

Macellari, C. E., 1995, Cenozoic sedimentation and tectonics of the southwestern Caribbean Pull-Apart Basin, Venezuela and Colombia, *in* A. J. Tankard, S. R. Suarez, and H. J. Welsink, Petroleum basins of South America: AAPG Memoir 62, p. 757–780.

Martínez, A., 1995, Cronología del Petróleo Venezolano 1943–1993: Caracas, Ediciones Cepet, 462 p.

Mencher, E., H. J. Fichter, H. H. Renz, W. E. Wallis, H. H. Renz, J. M. Patterson,, and R. H. Robie, 1953, Geology of Venezuela and its oil fields: AAPG Bulletin, v. 37, p. 690–777.

Miller, J., K. L. Edwards, P. P. Wolcott, H. W. Anisgard, R. Martin, and H. Anderegg, 1958, Habitat of oil in the Maracaibo Basin, Venezuela: AAPG Bulletin, v. 42, p. 601–640.

Murany, E. E., 1972, Tectonic basis for the Anaco fault: AAPG Bulletin, v. 56, p. 860–870.

Parnaud, F., Y. Gou, J.-C. Pascual, I.Truskowski, O. Gallango, H. Passalacqua, and F. Roure, 1995, Petroleum geology of the central part of the Eastern Venezuelan Basin, *in* A. J. Tankard, S. R. Suarez, and H. J. Welsink, Petroleum basins of South America: AAPG Memoir 62, p. 741–756.

Salvador, A., and R. M. Stainforth, 1968, Clues in Venezuela to the geology of Trinidad and vice-versa: IV Caribbean Geological Conference, Trinidad, 1965, p. 31–40.

Smith, J. E., 1956, Basement reservoir of La Paz–Mara oil fields, western Venezuela: AAPG Bulletin, v. 40, p. 380–387.

Stephan, J. F., 1982, Evolution Géodynamique du Domaine Caraibe, Andes et Chaine Caraibe Sur la Transversale de Barquisimeto, Venezuela: Thése d'etat, Université de Paris IV, 512 p.

Talukdar, S., O. Gallango, and M. Chin-A-Lien, 1986, Generation and migration of the hydrocarbons in the Maracaibo Basin, Venezuela: An integrated basin study: Organic Geochemistry, v. 10, p. 261–279.

Talukdar, S., and F. Marcano, 1994, Petroleum systems of the Maracaibo Basin, Venezuela, *in* L. B. Magoon and W. G. Dow, eds., The petroleum system—from source to trap: AAPG Memoir 60, p. 463–481.

Ysaccis, R., 1997, Tertiary evolution of the northeastern Venezuela offshore: Ph.D. thesis, Rice University, Houston, Texas, 285 p.

Urien, C. M., 2001, Present and future petroleum provinces of southern South America, *in* M. W. Downey, J. C. Threet, and W. A. Morgan, eds., Petroleum provinces of the twenty-first century: AAPG Memoir 74, p. 373–402.

Chapter 19

Present and Future Petroleum Provinces of Southern South America

Carlos M. Urien
Consultant, Buenos Aires, Argentina

ABSTRACT

Forty-two sedimentary basins are found in southern South America. They include intraplate and orogenic plate-margin types. Only seven of those basins produce commercial oil and gas. Six of them are in the foreland folded trend (the Subandean Belt), and one is an intracratonic basin located on the Patagonian Platform.

In most of those basins, commercial production began in the early part of the twentieth century. Since then, exploration and development work has been concentrated mostly in areas with established productive plays, postponing exploration efforts in the rest of the region.

Regional surveys have detected potential petroleum systems in more than one-third of the nonproducing basins. Moreover, in those basins with existing production, underexplored frontier areas still wait to be tested by the drill.

In this context, we can define three types of areas with exploration potential: (1) basins that are yet to be explored by the drill, (2) nonproductive basins insufficiently explored, and (3) plays in producing basins that have not yet been tested.

Giant fields continue to be discovered in provinces such as the intracratonic rifts, fold belts, and subtle stratigraphic traps. This same concept can be applied to the exploration challenge of the virtually untested South Atlantic margin.

INTRODUCTION

Southern South America extends from 14° to 54° south latitude, covering a surface of about 7.8 million km^2. Six countries (Argentina, Bolivia, Chile, Brazil, Paraguay, and Uruguay) are in the southern part of South America, with climates ranging from wet subtropical to temperate, desert, and subantarctic. Phanerozoic sedimentary basins are set in the Central Cratonic Platform, on the stable Atlantic margin, and on the Pacific active margin. Bounding the Andean oriental suture are the foreland basins, or Sub-Andean fold belt, and farther east, intracratonic-type basins lie on the Chaco Pampas plains and Patagonia Plateau (onshore and offshore).

Forty-five sedimentary basins lie in this region (Figures 1 and 2), covering a gross surface of more than 6.9 million km^2. Only seven have commercial oil and gas production from a net area of about 7000 km^2 of a gross area of 1.5 million km^2. Those seven basins are:

1) Santa Cruz Basin, Bolivia
2) Tarija Basin, south Bolivia–northwest Argentina
3) Orán-Chaco Basin, northwest Argentina
4) Cacheuta (Cuyo) Basin, Argentina
5) Neuquén Basin, Argentina
6) San Jorge Basin, Argentina
7) Austral-Magallanes Basin, Argentina-Chile

Oil production is approximately 482,000 barrels (bbl) oil/day (BOPD). Gas production is approximately 1.12 billion cubic feet (bcf) gas day (BCFGD).

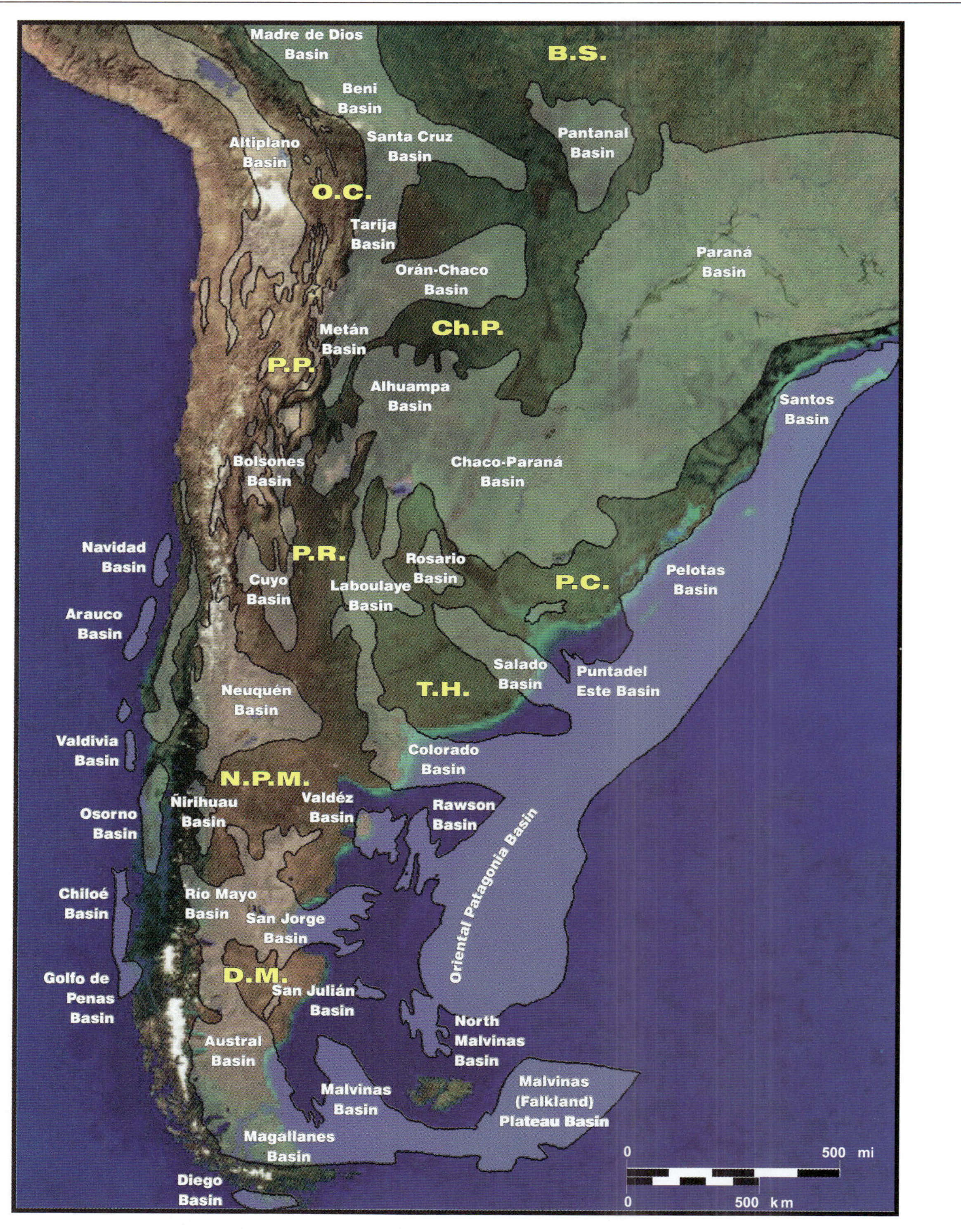

Figure 1. Relief map of southern South America. The sedimentary basins (light colors) are, from east to west, Atlantic margin, intracratonic, and foreland basins. Farther west is the Puna Plateau, bounded by the Andean belt, in central Peru, Bolivia, Chile, and Argentina. On this plateau are the intraarc and forearc basins created by the Cenozoic Andean tectonic episodes. Shields and positive areas are Brazilian shield (B.S.) and the Plata-Riveira craton (P.C.). In the map center are the uplifted Precambrian Pampean and Transpampean ranges (P.R.) that merge with the Altiplano, or Puna Plateau (P.P.); to the east is the Cordillera Oriental (O.C.). To the south are the North Patagonia Massif (N.P.M.) and Deseado Massif (D.M.), exposing the Patagonian Terrane uplifted basement. The Andes are in the west.

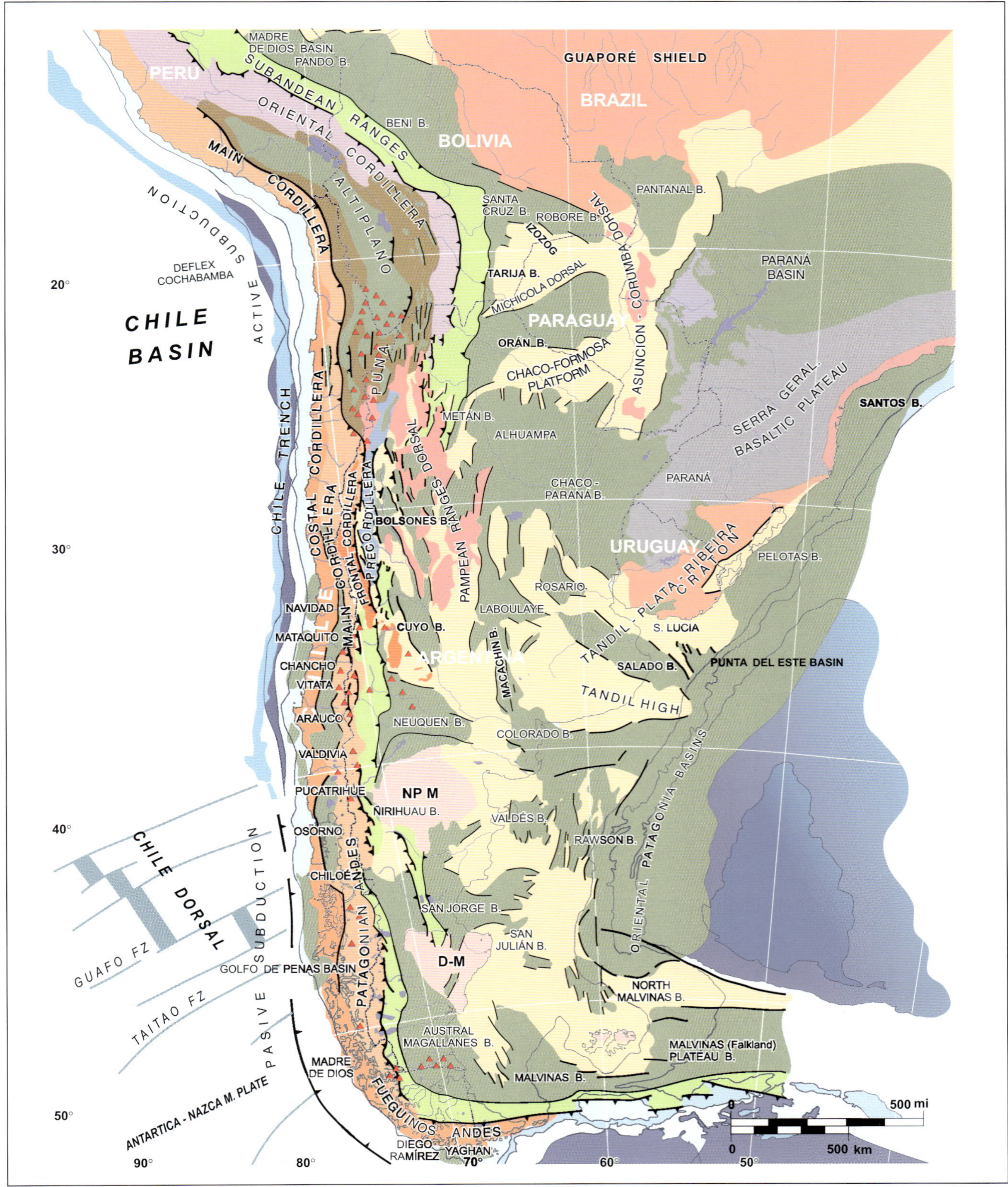

Figure 2. Southern South America tectonic setting. This map shows the main tectonic features: shield, cratons, basement and raised massifs, basins, the folded belt, and the Pre-Cordillera and Puna Plateau (Altiplano). The narrow north-south Andean orogenic belt bounds the western side of the continent, where intraarc and forearc basins are located. Overlying this belt are active and inactive Quaternary volcanoes associated with the circumpacific active margin.

Commercial use of hydrocarbons in the area began at the end of the nineteenth century, when private enterprise started to exploit oil seeps for distillation into kerosene. At the beginning of the twentieth century, local independents freely operated oil concessions (then known as "mines") with full property and disposal rights on the oil produced. Later, Standard Oil, Ultramar, and Royal Dutch Shell began to prospect for oil, particularly after the 1907 commercial oil discovery in Comodoro Rivadavia City, Argentina, in what came to be known as the San Jorge Gulf Basin. It is perhaps the oldest oil-producing basin in this part of the continent and has yielded more than 3244 million bbl oil.

Argentina

From 1928 onward, practically all exploration and production activity in Argentina was kept in the hands of the state oil company, Yacimientos Petrolíferos Fiscales (YPF). In the mid-1940s, a new state-owned company, Gas del Estado, was created to monopolize all gas surface-distribution activity (pipelines, treatment plants, distribution networks, etc.). Neighboring countries imitated this policy of state monopoly of the oil industry with the creation of YPFB in Bolivia, Petrobras in Brazil, Enap in Chile, and Ancap in Uruguay. Many years later, these national oil monopolies began to grant some exploitation and later exploration concessions to private oil companies through different types of service contracts.

In 1958, Argentina launched an ambitious program, La Batalla del Petróleo ("the Battle for Oil"), directed toward securing self-sufficiency in oil by awarding exploration and production contracts to foreign private companies. During that period, known as the "the Golden Years of Oil," many wildcatters came to work in Argentina. Agip, Union Oil, Tenneco, Cities Service (later Oxy), and Pan-American (Amoco), among other companies, started operations at that time. In fact, many of the large oil fields and the peak in oil production resulted from that effort. Unfortunately, the host government canceled some of those contracts.

New legislation was introduced to regulate exploration and production of hydrocarbons. From 1967 onward, as a result of open bidding, the governing administration signed exploration contracts with major oil companies, covering a total of 29.8 million acres. During that period, YPF signed production-services contracts with private firms.

In 1983, when a new democratic government was elected, Plan Houston for the exploration of selected blocks was introduced. In 1989, Plan Argentina was launched, opening all onshore and offshore basins for exploration and exploitation. This ushered in a new era in which private companies own and manage oil and gas fields, and a new local oil industry began to grow. This plan is still in force.

When YPF was privatized in 1995, all upstream and downstream oil and gas activities were deregulated. Pipelines and refineries were transferred to private consortiums. This deregulation allowed companies to trade, swap, and export hydrocarbons. A new oil and gas pipeline network was designed particularly to export oil and gas to Chile, Brazil, and Uruguay (Figure 3). In this scenario, Argentina, as well as neighboring countries, started out on a new exploration phase to increase oil and gas reserves.

The Falkland Islands (Malvinas) Administration opened exploration blocks in the exclusion area, including joint ventures with Repsol-YPF S.A.

Since 1907, more than 32,000 wells have been drilled in Argentina, distributed in six producing basins. In total, 470 commercial fields have been discovered, of which 390 are still in production with about 13,102 producing wells. Those discoveries have led to an estimate ultimate recovery of more than 11,757 million bbl of oil and 54.825 trillion cubic feet (tcf) of gas. Those figures are based on the evaluation of the commercial fields but not the evaluation of productive trends or untested horizons. Considering areas still awaiting exploration, oil and gas reserves in the present producing basins could be increased remarkably, especially for gas.

Bolivia

The first oil discovery in Bolivia dates from 1924, with the Bermejo field discovery in the Subandino region. Since then, 53 fields have been discovered, producing mainly gas, most of it exported to Argentina. Because of logistic and market considerations, commercial oil and gas production is centralized in the Central Subandino (Santa Cruz) and South Subandino, i.e., the Santa Cruz and Tarija Basins.

A few years ago, Bolivia began a plan of global deregulation of its oil industry. YPFB was converted into a controlling entity that transfers projects to private enterprise, either partially (in association) or totally. Under this plan, Bolivia sells minor exploitation areas and opens up exploration areas to private enterprise.

Chile

In Chile, hydrocarbons production is concentrated in the Magallanes Province, around the Magellan Straits and Tierra del Fuego. The oil and gas production comes from 37 fields with a cumulative production of about 450 million bbl of oil.

Gas

One incentive for exploration in the region is the Santa Cruz de la Sierra (Bolivia)–Sao Paulo (Brazil) gas pipeline, which captures practically all the gas produced in the Central Subandino. Another incentive is Brazil's broad plan for associations with Petrobras for exploration and concessions to private enterprise. Apart from Venezuela, Brazil today has the greatest attraction in South

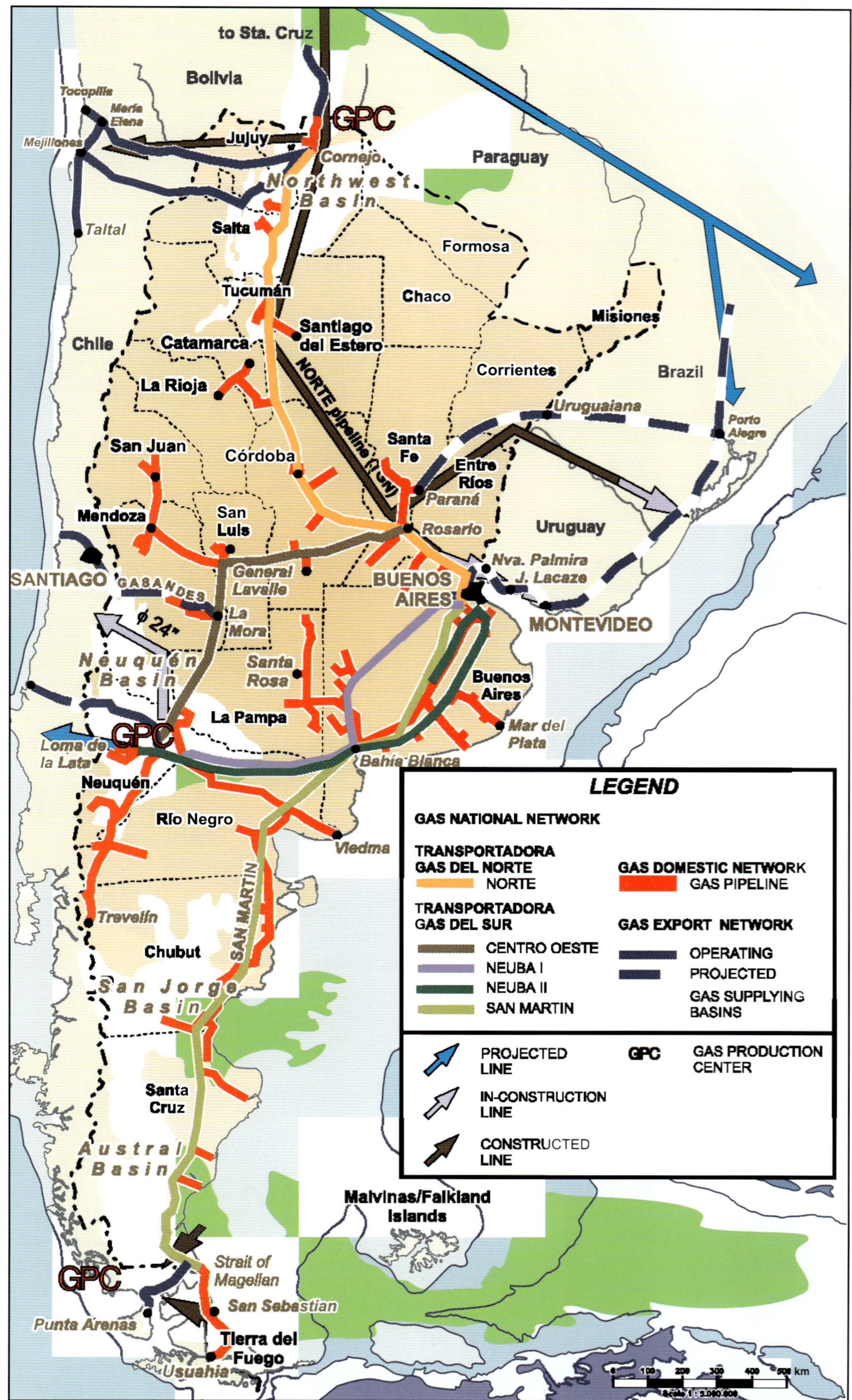

Figure 3. Argentina's main domestic distribution and export gas pipelines.

America for foreign investment in the petroleum industry. Chile follows because of its exploitation contracts with ENAP in Tierra del Fuego. In addition, new areas for hydrocarbon exploration have emerged in this part of the world, opening up new vistas for the oil industry in the twenty-first century.

The remaining Southern Cone government oil monopolies (Chile, Bolivia, and Brazil) are firmly committed to turning over production, transport, commercial sale, and distribution of gas to private enterprise. The potential for expanding domestic consumption and the demand in neighboring countries with insufficient or no fossil fuels (such as Chile, Brazil, Paraguay, and Uruguay) have turned gas into a most attractive business.

EXPLORATION OVERVIEW

In many of the partly explored, nonproducing basins, suitable hydrocarbon source rocks have been found in sequences of Ordovician, Silurian-Devonian, late Paleozoic, Triassic, Jurassic, Cretaceous, and early Tertiary age. Although potential petroleum systems have been found in many of those basins, no commercial discoveries have been reported. Commercial production from Paleozoic sequences is found only in the Central Subandean Belt in Bolivia and northern Argentina. Other basins with Paleozoic sequences and with potential hydrocarbons source are still almost unexplored, e.g., Chaco-Paraná, Bolsones, Central Patagonia, and Atlantic Margin.

Giant oil fields with recoverable reserves of about 3.5 billion bbl of oil have been developed in three of the producing basins. The basins with the highest cumulative oil production in Argentina are San Jorge (3244 million bbl), Neuquén (2463 million bbl), and Cuyo (1267 million bbl), representing only 35% of the inland basin area. In addition, 80% of the exploratory wells in Argentina were drilled there.

A cursory examination of the discovery statistics shows that in the majority of the fields discovered, the recoverable oil ranges between 10 million and 30 million bbl, but some fields have 200 million to 300 million bbl of recoverable oil.

During the last few years, deep drilling in mature areas (e.g., Bolivian Subandean and Argentinean Tarija and San Jorge Basins) has significantly enhanced known reserves. In the last 10 years, pools with more than 250 million bbl of oil and 600 bcf of gas have been discovered in the Tarija, Neuquén, San Jorge, and Magallanes Basins (e.g., Aguarague San Pedrito and Margarita fields in northwest Tarija Basin, and Filo Morado, El Trapial, El Portón, and Sierra Chata fields in Neuquén Basin).

In the areas in production, some zones are practically untested by the drill. A new opportunity is open to explorers, even in apparently mature areas.

Several databases have been assembled from studies and geologic surveys performed by the state-owned oil enterprises, national geologic surveys, and private companies. However, modern geochemical concepts and new seismic data are scarce and do not, on their own, enable an accurate definition and delineation of the real hydrocarbon-generating potential in this extensive and environmentally favorable territory.

This situation offers a challenge in this new century to be met with advanced technology to resolve the many unsolved problems that have hampered and delayed (or prevented) the opening of exploratory frontiers.

Gas, which during the last four decades was considered nonprofitable and thus was avoided or disregarded, now has a deregulated market and has become a coveted asset. New gas opportunities have emerged, forcing a reconsideration of past exploration concepts. New technology and a new and expanded market have led to more discoveries in new and old geologic provinces.

These concepts should help to encourage exploration by companies on the lookout for challenging opportunities in areas where risks are reasonable and operating conditions and the market and political climates are favorable.

One of the key issues in the assessment of basins is, of course, identification of petroleum systems in the stratigraphic record. The region discussed in this paper has quite varied basin settings and stratigraphic evolution and, therefore, diverse oil habitats. It is my hope that this paper will help readers to assess the most attractive opportunities.

BASIN TYPES

The basins in southern Bolivia, Argentina, Chile, and Uruguay have a total area of more than 7 million km^2 and are intraplate basins or plate-margin basins.

Intraplate basins (nonorogenic setting) are divided into three types: (1) intracratonic basins; (2) rift, or "pull-apart," basins; and (3) passive continental margins (divergent Atlantic-margin type).

Plate-margin basins are related to compressional orogeny (active convergent plate-margin settings) and are of two types: (1) basins related to Andean-type tectonic (subduction "A"), i.e., foreland (intermontane basins), and (2) basins related to subduction (subduction "B"), i.e., either intraarc basins or forearc basins.

INTRACRATONIC BASINS (FIGURE 4)

The nonorogenic or interplate settings include two principal classes: (1) the cratonic or intracratonic basins, entirely on continental crust, and the "pull-apart" basins developed at any stage between incipient rifting in the craton on wholly continental crust; and (2) passive Atlantic continental margins at the transition between oceanic and continental crust.

These basins are located in the stable Brazilian Shield. They are underlain by thick continental crust and have

INTRACRATONIC BASINS		
BASIN	Area 1000 km^2	Thickness Avrg. (m)
Paraná	1400	3700
Chaco-Paraná	510	4500
Alhuampa	50	3800
South Córdoba	60	3200
Rosario	20	2000
Macachin	40	3000
Valdés-Rawson (*)	60	4200
San Jorge (**)	170	5800
San Julián (*)	15	2500
North Malvinas (*)	90	4200
Malvinas Plateau (*)	230	3800
(*) Offshore (**) Onshore and offshore		

Figure 4. Intracratonic basins. The formation of these basins is related to areas of thermal subsidence in the western Gondwana cratonic and pseudocratonic attached terranes. The Chaco-Paraná and Central Pampas Basins have had a polyhistoric tectonic-sedimentary evolution. The Patagonian Platform rifts were products of Middle Jurassic to Cretaceous transtensional events preceding the Gondwana breakup.

very gentle subsidence, low sedimentation rates, and low heat flow. During times when circulation in epeiric seas on the craton was restricted by fold belts that surrounded the craton, a warm climate favored the accumulation and preservation of organic materials, especially at times of lowest sedimentation rates. Prolonged periods of transgression during the Devonian, Carboniferous, Mesozoic, and Cenozoic favored deposition of widespread source rocks, which unconformably overlie extensive cratonic areas.

Since the Late Cambrian–Silurian, areas of thermal subsidence developed in the western Gondwana continent cratons, commonly known as the Guaporé Shield, Pampean Ranges, Plata-Ribeira, Brazilian Shield, and North Patagonia Somuncura and Deseado Massifs. Sedimentary fill is made up of lower Paleozoic, Silurian-Devonian, upper Paleozoic, Triassic, Jurassic, and Cretaceous sequences. Cenozoic sediments are thin and occur as remnants in a few scattered localities. The most important intracratonic basins are (1) Paraná and Chaco-Paraná Basin (and Alhuampa), (2) Pampas and Patagonian Platform Rift Basins, and (3) San Jorge Basin.

PARANÁ AND CHACO-PARANÁ BASIN (FIGURE 5)

This basin extends over southeast Brazil, Paraguay, northern Uruguay, and northern Argentina, covering an area of about 2.4 million km^2.

The "Chaco-Paraná Complex" is the largest onshore basin, and its maximum thickness is in central-north Argentina, southern Brazil, and northern Uruguay. Deep structures, such as internal highs and antithetic folds related to wrench zones, form suitable closures. Both Chaco (Alhuampa) and North Paraná have good organic-matter content and fair to good kerogen maturation in Devonian and Permian-Carboniferous shaly beds. However, only noncommercial oil shows have been reported from study wells in these areas.

The north-south-trending grabenlike Rosario, Laboulaye (General Levalle), and Macachín Basins are in the Central Pampas Plains in Argentina. In those basins, very little drilling has been done, with no success. With the exception of old seismic surveys, little is known about the basins. The difference between those rift basins and the Chaco-Paraná is that the rifts may have been formed

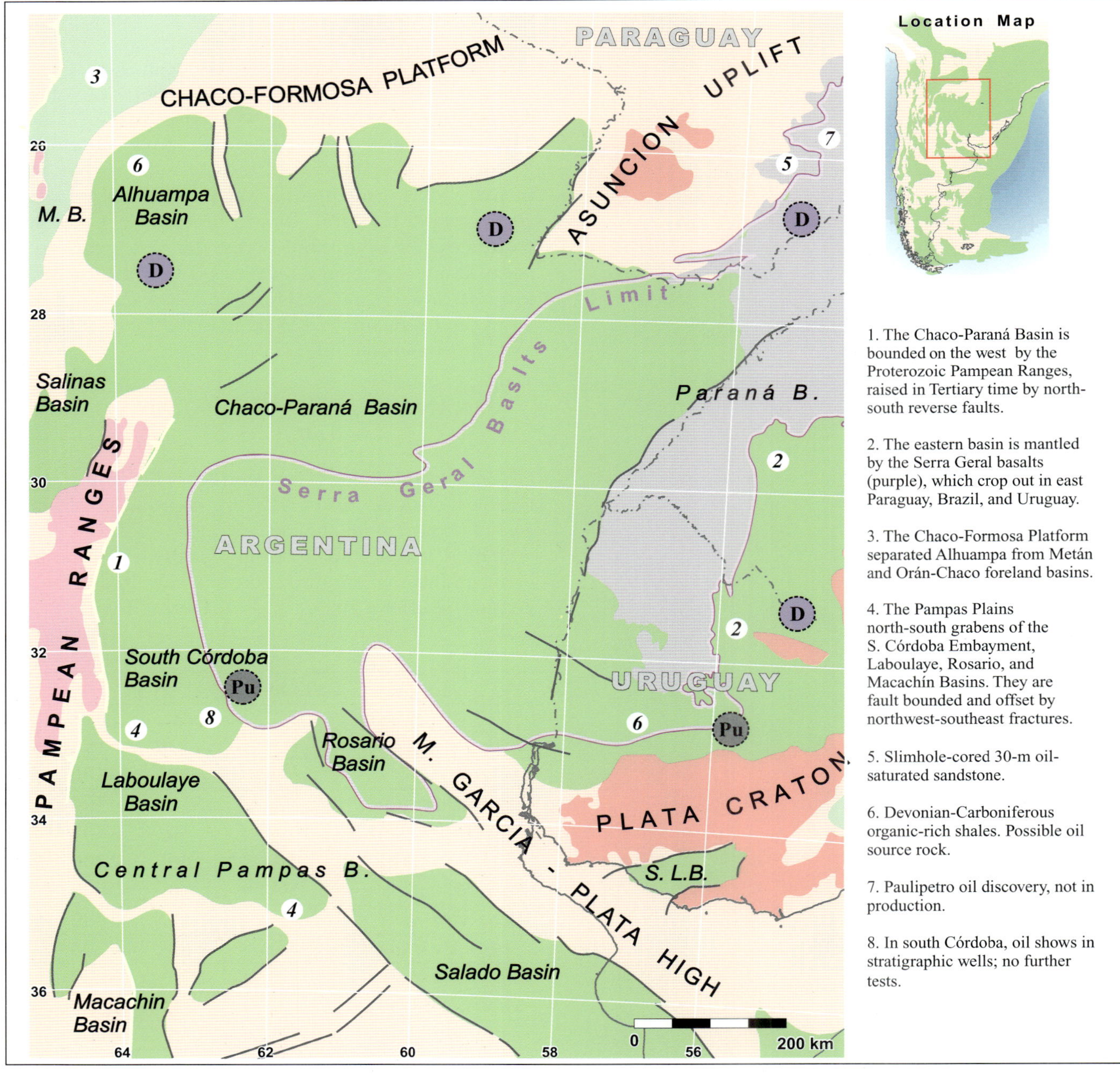

Figure 5. Chaco-Paraná and Central Pampas Basins. Set on Precambrian basement, these basins are bounded by the Proterozoic Pampean ranges. The eastern basin is mantled by the Serra Geral basalts, which crop out only in east Paraguay, Brazil, and Uruguay. In the Central Pampas Plains, north-south grabens, such as the South Córdoba Embayment and Rosario, Laboulaye, and Macachín Basins, are fault bounded and offset by northwest-southeast structural lineaments. The Chaco-Formosa Platform separates the Alhuampa Basin from the Metán and Orán-Chaco Basins in the foreland. D = Devonian petroleum system; Pu = Late Permian petroleum system. Closed circles are proven petroleum systems; dotted circles are possible petroleum systems.

as a result of Late Jurassic–Cretaceous reactivated faulting of the Precambrian basement containing remnants of late Paleozoic and Triassic depocenters. These "multigenerational basins" were formed as a result of post-Hercynian tectonic episodes at the time of the Gondwana breakup. They are filled with Cretaceous and Cenozoic red-bed sequences.

Sedimentation in the Chaco-Paraná Basin complex through the Paleozoic and Mesozoic reached a maximum total thickness surpassing 7000 m. This intracratonic basin, during its early inception stages, was probably connected with the Panthalassian sea and perhaps with the Congo Basin in Africa (Cuvette Central), and was formed on a Proterozoic basement framed by fold-and-thrust belts oriented dominantly southwest-northeast. Thermal subsidence and sediment accumulation started at about the Late Ordovician, when subsidence was the result of intraplate transtensional releasing of the compressional stress caused by the collision between Gondwana and the precordillera terrane.

During the Phanerozoic, the southwestern Gondwana margin was an area of a persistent convergent motion between the continent and the Panthalassian oceanic floor, which favored the progressive closure of the Paraná Basin up to its complete isolation within the craton's mass.

Megasequences

Three megasequences define the basin's stratigraphic framework. The overall record ranges from 450 to 65 Ma, with regional hiatuses separating the Ordovician-Silurian, Devonian (Paraná), and Upper Carboniferous–Lower Triassic (Gondwana I) successions. Gondwana II and Gondwana III are represented entirely by Mesozoic continental sedimentary successions and associated basic lava flows (Serra Geral basalts).

Gondwana I

Ordovician-Silurian

The Ordovician-Silurian units are predominantly siliciclastic, as much as 300 m thick, and include basal conglomerates and predominantly fluvial, sandy beds which indicate southwest-trending paleocurrents. Overlying these are diamictites with clasts and boulders of various origins. In general, this diamictite unit is as much as 20 m thick and is widespread across the basin, defining an important stratigraphic marker related to Late Ordovician glaciation that affected large areas of the Gondwana supercontinent. The cycle is completed by a shaly-fossiliferous package. These shales record the maximum transgression of this succession and have a total thickness that exceeds 1000 m.

Devonian

Devonian megasequences consist of an 800-m-thick basal conglomerate and blanketlike medium- to coarse-grained quartzose sandstones. The maximum marine flooding event of the entire Devonian cycle occurred in the Emsian and was followed by a lowstand system of sandy delta complexes which prograded from the northeastern basin margin. The final succession of this megasequence comprises shales rich in macrofossils (Lange, 1967).

The Devonian shales are medium- to fair-quality source rocks, with a maximum total-organic-carbon (TOC) content of about 4% (minimum 1%). Hydrogen indexes are generally low, possibly because of advanced maturation of its dominantly woody organic content. The shales are overmature in most of the eastern basin, probably because of high heat flows from the Mesozoic magmatic episodes.

Widespread organic-rich sequences are recognized mainly in the Devonian and are less common in the Upper Permian successions.

Upper Carboniferous–Lower Triassic

In the Mississippian, a global cooling resulted in the presence of a large, regional ice cap and a strong decrease of sediment accumulation. Sedimentation in the Paraná Basin resumed during an interglacial period in the Westphalian. The Westphalian Itararé Group (Gondwana I) is a thick and complex package of sedimentary rocks starting with periglacial marine and terrestrial beds in an association of diamictites, conglomerates, sandstones, and shales. Many of the deposits are the result of gravity-flow processes. Three major glacial cycles reflect probable climatic changes during the glacial regime, defined by southward-trending offlapping, which accompanied a progressive basin expansion. In the Late Carboniferous and Permian, a climatic warming, marked by the presence of Glossopteris flora, resulted in a eustatic sea-level rise, which was recorded as a transgressive sedimentation cycle that includes deltaic successions that are widespread throughout the eastern and western flanks of the basin. The alternation of sandy-shaly packages resulted from the interaction between tidal currents and waves that deposited prograding sands and reworked delta lobes. The coastal systems, lagoons, and marshes contain coal beds intercalated with transitional siliciclastic sediments. Southern Brazil is the only producer of this coal.

The Upper Permian highstand sequences (Irati Formation) (see Table 1) are represented by shales, mudstones, and bituminous shales in the southern half of the Paraná Basin and by limestone and shale and subordinate evaporites in the northern part. A restricted marine environment has been interpreted for this unit. Du Toit (1927) correlated the Irati Formation with the Whitehill Formation of southern Africa.

It seems that this shallow embayment sequence represents the last marine flooding in the Paraná Basin. This supersequence ends with an aggradational complex of red beds that marks the definitive continentalization of the Paraná Basin.

The Permian Río Bonito Irati black shales (source rocks) and Palermo sandstones (reservoir rocks) form the exploratory play. These bituminous shales are present in the southern Paraná Basin and have a total-organic-carbon content averaging 1.8% with a maximum of 24%. Kerogen is oil prone and predominantly algal in origin, with lipid-rich composition. This source is probably immature, even in the deepest depocenters. Oils related to the Irati shales probably were generated locally by abnormal heat flow influenced by the Cretaceous intrusives. Gas and condensate shows from wells drilled in the central part of the basin were collected from the Itararé sandstones and have gravity ranges of 22° to 33° API. Heavy fractions showed geochemical correlation with organic extracts from the Devonian Ponta Grossa shales.

Late Permian to Early Triassic sedimentation patterns show a marked regional regressive tendency (Gondwana II). Fluvial sandstones and lacustrine shales in the central Paraná Basin are fringed by fluvial-eolian deposits which advanced basinward at the dawn of the Mesozoic.

Gondwana II and III

The Middle to Late Triassic fauna of reptiles and mammals shows also an important correlation with analogous sequences in southern Africa.

The Upper Jurassic–Neocomian lower portion of this supersequence is characterized by a spectacular package of Botucatu eolian sandstones made up of fine- to medium-grained quartzose sands. They cover an exceptionally large area (more than 1,300,000 km^2), yet they always retain the same lithologic characteristics. They were deposited during the maximum expansion of the depositional basin in the Mesozoic. The Botucatu Formation is representative of the widespread desertification of Gondwana in the Mesozoic. Toward the top of the formation, the sandstones are intercalated with basic lava flows which perhaps marked the initial stages of continental rupture. Together, the eolian sandstones and lava mark the end of the Gondwana epoch.

The Neocomian extrusive event culminated with the extrusion of widespread Serra Geral lava flows. In Brazil, these volcanics are more than 1500 m thick and overlie most of the sediments of the Paraná Basin. They have an intricate pattern of dikes and sills and are dated 138–127 Ma. The basaltic sequences barely extend only to the Argentine Pampas.

The Aptian-Maestrichtian units comprise alternations of sands and conglomerates, which are massive or have cross-stratification and cut-and-fill features, and are as much as 250 m thick. In Brazil, Paraguay, and Uruguay, these sediments are the by-product of the basin inversion that exposed the Serra Geral basalts, Paleozoic units, and Precambrian basement.

During the Maestrichtian-Aptian in Argentina, a shallow, short-lived epeiric sea flooded most of the Chaco-Pampas plains up to the Orán-Metán Basins in the northwest.

Petroleum Systems

Throughout the Chaco-Paraná basinal complex, petroleum systems of middle to late Paleozoic age are present. Some wells tested light oil, although noncommercial, from rocks of that time period near the easternmost border of Paraguay and in southern Brazil.

In 1996, Petrobras discovered the first commercial gas accumulation in the basin, associated with the Ponta Grossa–Itararé play. Oils recovered from wells drilled in the southern basin had gravity values of 22° to 33° API, revealing a positive geochemical correlation with the organic shales of the Irati Formation.

In Argentina, as well as in Uruguay, suitable source rocks are present, but no commercial shows have been reported from exploratory boreholes. However, the exploratory effort seems minimal in relation to the vastness of the territory.

Despite the presence of source rocks and reservoirs and the possibility of large-sized traps, this basin poses technical exploratory problems. The tremendous thickness of the basaltic plateau and the basin inversion represent a real challenge in seismic imaging of the subsurface. As long as the acquisition of subsurface information continues to be a difficult task, this will mean a serious limitation to exploration. On the other hand, production from this basin should easily find a market, and therefore, there should be continued interest in exploration.

Patagonian Platform Basins (Figure 6)

During the Late Jurassic to Early Cretaceous in the Central Patagonia terrane, "basin-and-range"-style deformation was associated with east-west-trending transform faults or zones (Uliana et al., 1989; Rapela, 1990).

During the Late Jurassic, an attenuation of the Tobífera extrusive events (Uliana et al., 1985), associated with persistent extensional faulting, reactivated half grabens in which subsidence exceeded sediment supply and a high base l0evel helped to promote the lacustrine regime. This created conditions that favored the production of kerogens.

The stratigraphic successions in the active central rift depocenters suggest a local volcano-sedimentary supply and a poorly integrated drainage system. Intermontane depressions were closed basins with limited or no centrifugal drainage, and they contained lakes that favored concentration of shales rich in organic matter. Most of those depocenters remained deeply buried.

Sedimentary continental depocenters accumulated sequences consisting of thick Middle to Late Jurassic micritic limestones, intercalated with olive-gray siltstones and mudstones. Proximal facies were made up of oolitic grainstones and stromalithic boundstones, intercalated with sandstones (Cañadón Asfalto, A. Bandera–C. Guadal, and D-129 Formations) (Table 1). Because of the lack of accurate subsurface data, the stratigraphic correlation among those depocenters is ambiguous. Some of them in sectors of the Patagonian Platform were exposed as a result of Cenozoic inversion associated with the North Patagonia and Deseado Massifs.

The Upper Jurassic lacustrine deposits overlie the acidic rocks of the Bahía Laura–Chon Aike and Lonco Trapial eruptive-pyroclastic complex (more than 1250 m thick). Carbonized tree trunks and gymnosperm flora, evidence of conifer forests, suggest a mild and relatively warm climate that favored the accumulation of organic matter. These units alternate with reworked pyroclastics in a long-lived lacustrine environment.

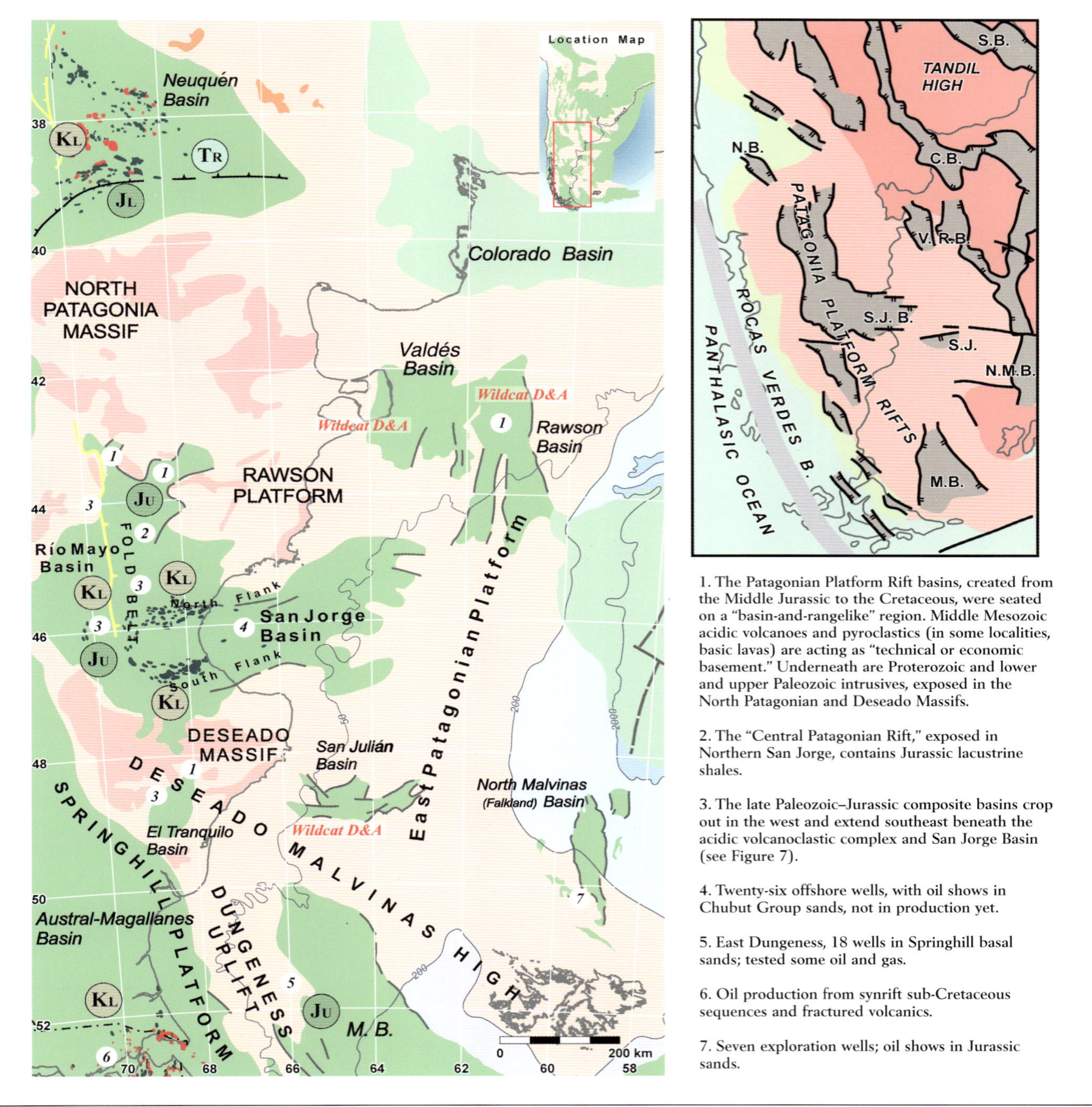

Figure 6. Central Patagonian intracratonic basins. San Jorge, Valdés-Rawson, San Julián, North Malvinas (Falkland), and El Tranquilo rift basins are seated on the Patagonian Platform basin-and-rangelike region. Middle Mesozoic acidic volcanics and pyroclastics and Proterozoic and early and late Paleozoic intrusives are exposed in the North Patagonian and Deseado Massifs. The San Jorge Basin is part of the north-south-trending "Central Patagonia rifts" where Triassic-Jurassic lakes were concentrated. Tr = Triassic petroleum system; Ju = Jurassic petroleum system; Kl = Cretaceous petroleum system.

Table 1. Stratigraphy of the intracratonic basins.

Period	Epoch	Chaco	Chaco-Paraná	Paraná	Rawson-Valdés	San Jorge	Tranquilo	San Julián
Quaternary	Pleistocene	Alluvium	Pampa	Alluvium		Alluvium	Marine bed	Marine bed
Tertiary	Pliocene	U. Chaco G.				Plioceno		
				Cocheira				
	Miocene		Parana G.			S. Cruz		
		L. Chaco G.						
	Oligocene		F. Bentos			O. Basaltos		
						Patagoniano		
	Eocene					Basaltos		
						Sarmiento		
	Paleocene		M. Boedo			R. Chico		
Cretaceous	Upper	Pirgua G.				Salamanca	L. Palacios	L. Palacios
			Yerua			Y. Terbol		
			Arata			C. Rivadavia	Baquero	
				Bauru		M. Carmen		
				Caiua				
	Neocomian		S. Geral	S. Geral			B. Grande	B. Grande
			Tacuarembo	Botucatu		D-129		
Jurassic						C. Guadal		
						A. A. Bandera		
	Main					C. Asfalto	B. Laura	B. Laura
						L. Trapial	L. Trapial	L. Trapial
	Dogger						B. Pobre	
	Lias					Liassic	R. Blamca	R. Blamca
Triassic	Upper		B. Vista	R. Do. Sul.				
Permian	Upper			R. Rastro		E. Tranquilo	E. Tranquilo	E. Tranquilo
				Teresinha				
				Salta			L. Golondrina	L. Golondrina
	Lower			Irati				
Carboniferous				Palermo				
	Upper			R. Bonito				
	Lower			Itarare		L. Juanita	L. Juanita	L. Juanita
Devonian	Upper	Jollin		P. Grosa	S. Grande G.		L. Modesta	L. Modesta
		Michicola				Granites		
		Rincon		Furnas				
	Lower	Cabure				R. Lacteo		
Precambrian		Precambrian	Precambrian	Precambrian	Precambrian			

San Jorge Basin (Figures 6 and 7)

The San Jorge Basin is the most important intracratonic Patagonian basin. It is a transtensional, fault-bounded basin surrounded by uplifted "neocratons" known as the North Patagonian (or Somuncurá) and Deseado Massifs, which are part of the Chilenia and Patagonia terrane, attached to the Gondwana continent.

In the Central Patagonia Platform (i.e., south Chubut and north Santa Cruz provinces), more than 1000 m of Middle to Late Jurassic organic-rich shales were concentrated in an intricate regional paleogeographic mosaic composed of different, but perhaps interconnected, lacustrine complexes, merging with fluvial and alluvial fan systems, localized basaltic and ignimbrite flows, and subordinate pyroclastic realms that might represent the declining stages of the "Tobífera" episodes.

In the main San Jorge Basin, the most widespread Upper Jurassic and basal Neocomian lacustrine deposits are concentrated in the western sector, where they are known as "Pelitas Laminares" (Las Heras Group or Aguada Bandera and Cerro Guadal Formations [Table 1]; Lesta and Ferello, 1972). These successions contain a basal section of grayish-green sandstones, interbedded with black shales and siltstones, that grade into a middle section dominated by gray to black pyritic shales that are excellent source rocks.

During this period in the northern part of the San Jorge Basin, expansion and contraction of lacustrine depositional environments resulted in the Cañadón Asfalto and equivalent sequences. These sequences contain spores and nonmarine charophytes and ostracodes, indicating a Late Jurassic–Neocomian age (Van Niewenhuise and Ormiston, 1989).

Overlying the Upper Jurassic successions are the fluvio-lacustrine and deltaic sequences of the D-129 Formation, a Lower Cretaceous equivalent of the nonmarine Río Mayer, upper Katterfeld and, most likely, Coyaike Formations present in the Magallanes Basin in Santa Cruz and Chile.

During the Kimmeridgian to Berriasian, the depocenters were flooded and a stagnant lacustrine environment, propitious for the accumulation of more than 3250 m of varvelike bituminous and carbonaceous shales, was common. These shales alternate with distal stacked sandstones. The last depositional episode occurred during the Neocomian and resulted in a series of mature alluvial plains and deltas that documents the fluvial offlapping and final filling of the lacustrine basins. Fluvial units are divided into four gross sandstone packages comprising laterally and vertically discontinuous sandstones.

Traps in the Patagonian Platform basins are formed by normal, nearly vertical-faulted anticlines cut by regional wrench faults. Continental successions of the Patagonian Platform basins have yielded more than 6000 million bbl of oil from giant fields oriented east-west with the structural grain.

The main commercial production is concentrated in the Lower Cretaceous fluvial Chubut Group along the north and south basin flanks. This entirely fluvial reservoir produces from four stacked sandstone complexes, which have yielded 3244 million bbl of oil from 180,000 acres. More than 4000 exploration wells have been drilled, with average depths of 1500 m.

Hydrocarbon traps are controlled by normal and antithetic faults. The basin has a circular outline within which are circular depocenters. The western pools seem to be small, but this may be because of scattered exploration in this remote area. The eastern offshore basin has a few scattered discoveries and has been explored only sparsely, especially considering the new technologies now available.

In the central basin, deep drilling has yielded condensate and gas from thermal cracking of the same oil sources. These deep-wet-gas discoveries represent a new exploratory-frontier challenge.

Additional reserves can be discovered in this basin because:

- Deep zones have not been explored thoroughly. Recent new oil discoveries detected low-permeability fractured sandstones containing light oil and gas.
- In the West Río Mayo Basin, the marine Upper Jurassic and Neocomian (and probably also upper Paleozoic formations) have good exploratory potential that has not been evaluated properly.
- The characteristics of the offshore extension of the proven onshore reservoirs are identical and suggest a potential of about 630–945 million bbl of oil or equivalent (perhaps more than 1.6 tcf of gas).

RIFT BASINS

The rift basins are north-trending faulted grabens in the Pampas Plains and Patagonian Platform extending beneath the Atlantic margin. They are the end members of a continuum of geotectonic processes, and they represent the evolution of midplate or intraplate basins, which originated in a cratonic terrane and, in late end stages, could become underlain by transitional continental-oceanic crust. Early rifting preserved a good record of the sedimentary sequences.

The evolution from earliest or incipient rifting to a fully developed passive margin may be divided into four stages of basin development:

1) incipient and aborted rift systems
2) rifts which are ocean-facing failed arms of triple junctions
3) young passive-margin basins
4) Atlantic-margin basins

Incipient and Aborted Rift Systems

Incipient and aborted rift systems are those that never developed an oceanic crust. They include the Central Pampas basins; the Cordoba Embayment; the Rosario, Laboulaye, and Macachin Basins; and, in the Patagonian Platform, the La Juanita, La Matilde, and Cañadon Asfalto Basins. The Valdés-Rawson, San Julián, and North Malvinas (Falklands) lie offshore beneath the Atlantic continental shelf.

Only five exploratory wells have been drilled in this region; no well had positive results, but petroleum systems are still probable. Borehole data have shown that sedimentary sequences offshore are quite similar to those exposed or drilled onshore. In this tectonic setting, source, reservoir, and seal rocks were deposited; overburden, in most cases, should have resulted in mature source rocks, but exploration is still in its early stages.

Rifts Which Are Ocean-facing Failed Arms of Triple Junctions (Figure 8)

In this stage, the fault system has opened so that an early ocean begins to develop as two arms of a triple junction begin to spread apart. The third, or failed, arm does not spread, but strikes (perpendicularly) into the continent, facing the newly developing ocean. Rivers flow down the rifted depression of the failed arm and develop deltas that prograde into the future ocean, and the failed arm continues to be a locus of subsidence and sedimentation. Source beds formed during the early stage become buried and continue to evolve toward maturity. The two basins having this origin, the Salado and Colorado, are bounded by faulted crystalline and metamorphic basement and run perpendicular to the continental margin. They owe their origin to tensional tectonism that originated during early stages of the separation of Africa from

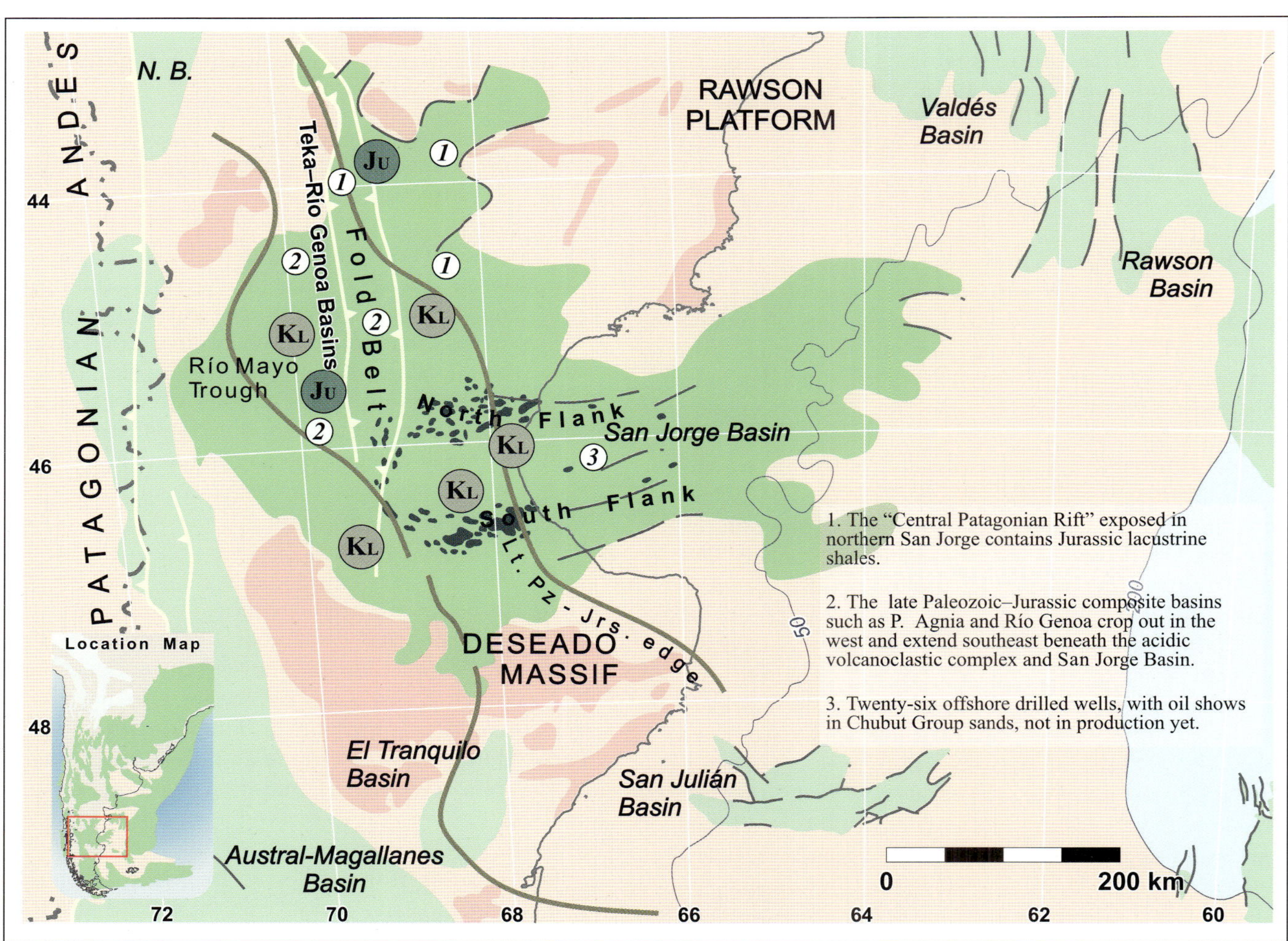

Figure 7. San Jorge Basin. The basin is bounded on the west by the exposed fold belt (San Bernardo). This inversion is exposed in the west in the Upper Paleozoic–Jurassic sequences in the Pampa de Agnia and Río Genoa Basins, which also extend beneath the San Jorge. In the north and south, the basin is shaped by wrench faulting. These transtensional lineaments are associated with the Gondwana breakup. In the west are the foreland Río Mayo and Austral-Magallanes Basins. They extend beyond the Chile border, alternating with magmatic-arc volcanics. Farther west along the Patagonian Andes, the metamorphosed Devonian La Lancha Formation and Lower Cretaceous granites are exposed. Ju = Jurassic petroleum system; Kl = Late Cretaceous petroleum system.

South America. This had occurred in successive tectonic phases since the Early Jurassic.

Young Passive-margin Basins (Figure 8)

Young passive-margin basins are closely associated with failed arms. They include young passive margins, rifts, or "pull-apart" basins, which had already begun to spread and developed into a transitional oceanic crust formation during rapid thermal subsidence. Continental source beds (alluvial and mainly lacustrine synrift facies) deposited during the early stage of rifting were buried and probably matured by the sedimentary load during the continental-margin formation, which began with a marginal sag stage. Circulation of marine waters was still quite restricted, anoxia and organic productivity prevailed, and new source beds may have developed within the sedimentary sequence deposited during this young marine stage. The widespread Upper Jurassic and Lower Cretaceous black shales of the southern South Atlantic and Weddel Sea region between Madagascar, Australia, and Antarctica were probably deposited in this type of setting. These sequences, which later extended beyond the Río Grande and Walvis Ridge between Brazil and Equatorial Africa, were deposited in the same tectonic setting and yield commercial oil in the deep Campos, Angola, Niger, and Congo Basins.

JOIDES boreholes in the M. Ewing High detected Oxfordian sapropelic shales ranked as good oil-source rocks. Along the southwestern margin, no shows have been reported from Lower Cretaceous sequences. On the continental slope in Argentina, piston coring from outcrops of the same age recovered samples with organic-rich shales.

In the Eastern Malvinas, South Malvinas, and Magallanes Basins, similar transgressive sequences (Springhill equivalents) and Upper Jurassic and basal Neocomian siliciclastics have yielded hydrocarbons. Wells drilled on the Dungeness High and in the Eastern Austral–Magallanes Basins have had positive tests and have yielded commercial production.

The geothermal gradients of fully developed passive margins tend to be low because heat flow decreases away from the spreading center. Thus, basins of this type may have two paleothermal gradients: an older, higher gradient when the basin was close to the spreading center, and a younger, lower gradient during deposition of the sedimentary prism. Hence, maturation will result from deep burial by the accumulating sediments on the subsiding outer margin (slope and rise).

Seismic data show that possible marine highstand events onlap the basal half-rift units and grade into the outer sectors of the Salado and Colorado Basins beneath the continental slope and rise. These little-known young passive-margin depocenters are unexplored. Paleoclimatic reconstructions support the inference of potential that would justify further systematic exploration.

Atlantic-margin Basins (Figures 9 and 10)

After the Gondwana breakup and the formation of the Proto-Atlantic ocean, two plates were defined: the African and the South American (Figure 8). As the spread of the continental plates continued, the oceanic crust widened, developing true "Atlantic-type" passive continental margins. A thick prism of siliciclastic sediments prograded over the "postrift unconformity." Supply of sediments was high at that time. The principal source beds for hydrocarbons may be in the underlying rift-bounded grabens or in the postbreakup rift stage of the new ocean.

These western Atlantic basins are, from the Río Grande Rise south, the Pelotas, Salado–Punta del Este, and Colorado Basins; the eastern Valdés-Rawson Basins; the Eastern Patagonia continental slope and rise; and the northeastern arm of the North Malvinas (Falklands) Basin.

Some of these basins lie on the southeastern Brazilian-Uruguay-Plata Shield, Gondwanides Belt. The rest lie on the Patagonian terrane, whose basement is made up of Proterozoic granitoids, upper Paleozoic sequences, and Jurassic acidic extrusive.

Beneath the Colorado Cretaceous Rift sequences lies the Permian-Carboniferous Ventania Belt, connected with the Cape Folds. This belt has proven source-rock potential and has yielded light oil in tests from a recently drilled offshore borehole. This discovery is considered noncommercial. This is the only Paleozoic oil detected so far in the western-southern Atlantic-margin basins.

Source beds which developed during the passive-margin sag stage that generated a sedimentary prism more than 5000 m thick are likely to be localized in depocenters that favored a restricted environment (Figure 10). Although organic productivity along the margins may have been high during marine stages, oxygen minimum layers may have existed on the continental slopes and rise, especially during times of low sedimentation (high sea-level stands).

On the easternmost continental terrace, as well as in the south Malvinas Basin (Figure 9), Tertiary deltaic and turbiditic sequences more than 3000 m thick have been identified on seismic. In the Malvinas Basin, Upper Cretaceous–lower Tertiary flyschlike beds were deposited under reducing conditions and could be potential generators of hydrocarbons as a result of adequate heat flow. The Ciclón x-1 well, near the Fagnano-Burdwood trend, has good hydrocarbon shows in Tertiary beds, but reservoirs could be a problem.

Summary

In the intracratonic basins, organic-rich rocks developed during transgressive episodes, but low heat flow and relatively thin sedimentary successions may have prevented rapid maturation. Rift-bounded grabens, aulacogens, and failed arms ("pseudoaulacogens") are the less-explored basins in the regions discussed above.

The western Atlantic-margin basins (Figure 10)—early rift, failed arms, and regional half grabens—are very immature as far as exploration is concerned. They are in the best and most predictable tectonic setting for hydrocarbon source beds because of their early-restricted circulation and generally above-average heat flow. Continuous or renewed depocenters may have provided sufficient overburden, reservoirs, and seals for maturation and preservation. Tensional tectonics reactivated fault systems that allow migratory fairways and trapping, as in the case of the San Jorge Basin, which accounts for important hydrocarbon accumulations in pull-apart basins within passive-margin sedimentary prisms. In case of successful exploration, those areas are destined to be highly prized petroleum provinces.

The hydrocarbon systems of the offshore intracratonic basins (Salado, Colorado, and Valdés-Rawson) are not well rated, considering their type of sedimentary fill (synrift fluviatile and lacustrine potentially organic-rich beds). Near the outer continental slope, shallow-marine organic shales in highstand sequences onlap the basal continental beds. They could be an attractive exploratory

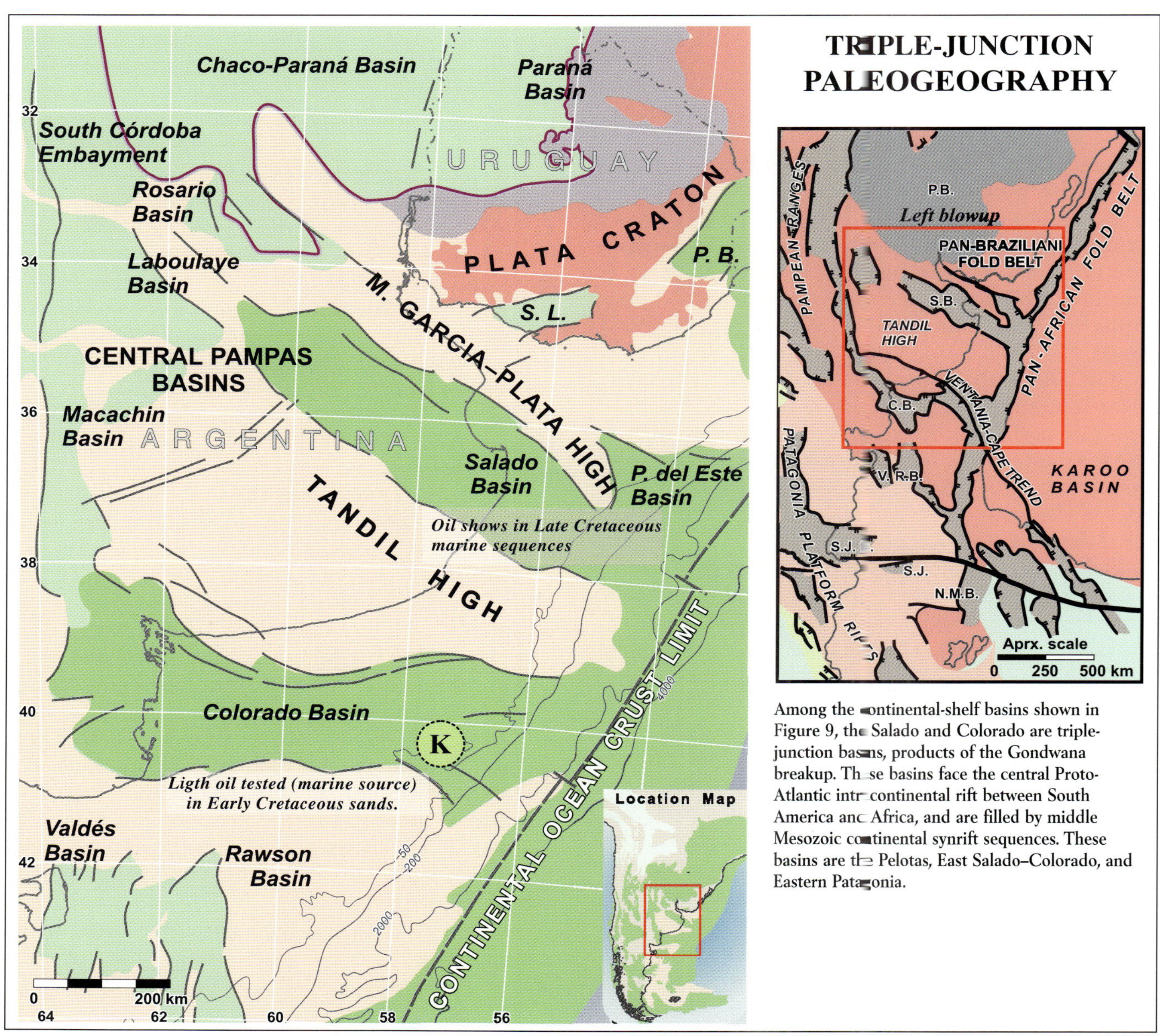

Among the continental-shelf basins shown in Figure 9, the Salado and Colorado are triple-junction basins, products of the Gondwana breakup. These basins face the central Proto-Atlantic intracontinental rift between South America and Africa, and are filled by middle Mesozoic continental synrift sequences. These basins are the Pelotas, East Salado–Colorado, and Eastern Patagonia.

Figure 8. Salado and Colorado ocean-facing failed-arms basins. These are continental-shelf basins (see Figure 9 for locations). The Salado and Colorado Basins are triple-junction basins, conjugate with the West African Orange-Namibia Basins and separated from them by the Gondwana breakup. These basins face the central Proto-Atlantic intracontinental rift inserted between the American and African Plates. Initial sedimentary fill was contributed by middle Mesozoic continental synrift terrigenous sequences. K = Cretaceous petroleum system.

lead. By analog with other South Atlantic basins with similar stratigraphic and structural sequences, reserves of approximately 100 billion bbl of oil could be expected. The Atlantic Province deserves a fair to good rating for future exploration.

PLATE-MARGIN BASINS

Plate-margin basins are related to compressional orogeny bounded by the megasutures that resulted from the collision of the Pacific and South America converging plates during the last 300 million years. The collisions on the western continental-plate margin resulted in the successive accretion of terranes in the Proterozoic and, later, from the early Paleozoic, the Panthalassian convergent oceanic (active) plate. Those basins are related to Andean megasuture-type, or "A," subduction and Benioff megasuture-type, or "B," subduction.

Basins Related to "A" Subduction

These were formed in association with Andean-type subduction related to the collision of the American continental plates and the subducting Pacific oceanic crust. They are foreland basins, aulacogens, and intermontane basins, with the intermontane basins being the "end member" of this geodynamic evolution.

Foreland Basins (Figure 11)

The basins that lie along the "South American Andean Backbone," locally called Subandean basins, are bounded by a megasuture-type "A" that is the eastern limit of the Andes fold belt.

Platform Incised Rifts

Situated on the west pericratonic foreland platform are north-south-oriented fault-bounded troughs formed by rotation of basement blocks. These depocenters, partially connected with the sea, were filled with marine and continental Carboniferous, Permian, and Triassic sequences. In the Bolsones Basin, Triassic and Jurassic sequences were laid down in association with acidic volcanics and pyroclastics. Remarkable amounts of organic-rich sediments were deposited in the region.

ATLANTIC-MARGIN BASINS		
BASIN	Area 1000 km^2	Thickness Avrg.(m)
Punta del Este	22	4100
Salado	90	4200
Colorado	126	4500
Eastern Patagonia	2800	5000

Figure 9. Atlantic-margin basins. These are the Pelotas, Punta del Este, Salado, Colorado, Valdés-Rawson, San Julián, Oriental Patagonia, and North Malvinas (Falkland) Basins. N.P.M. = North Patagonia Massif; D.M. = Deseado Massif.

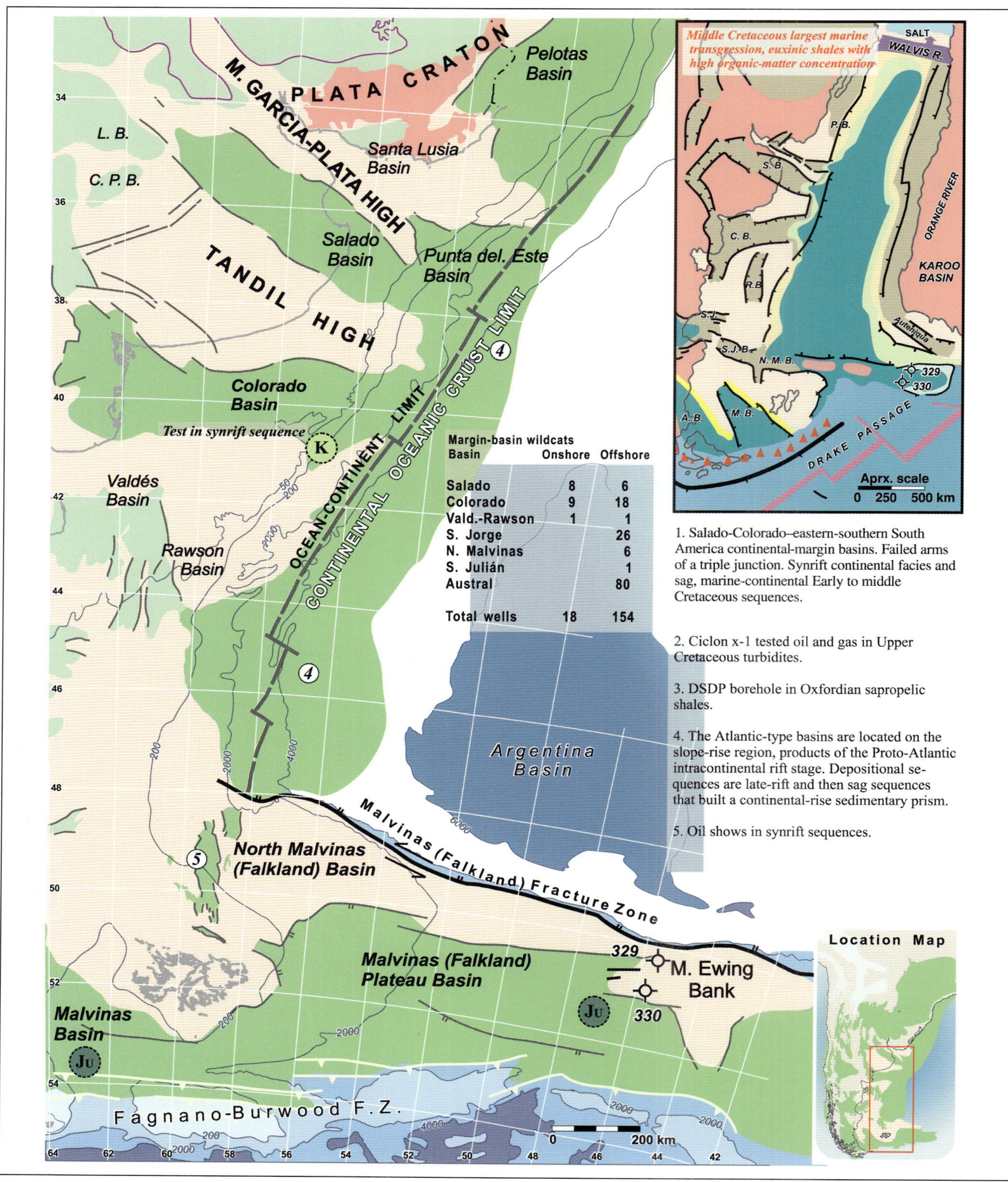

Figure 10. Southern-southwestern Atlantic-margin basins. The Atlantic-type basins lie on the continental-slope rise, a sedimentary prism where more than 4000 m of sediments filled the shifting depocenters. In their core are remnants of intracratonic basins. The Atlantic-margin basins lie on a transitional crust where basic lavas alternate with continental siliciclastic and nonsiliciclastic (probably lacustrine) sequences. Late Neocomian shallow-marine units were deposited on flooded lowlands and later were mantled by sag sequences during spreading episodes. Ju = Late Jurassic petroleum system; K = Cretaceous petroleum system.

Prospective sequences have been tested by drilling only in the Cuyo Cacheuta Subbasin and Eastern Neuquén Basin. Ultimate reserves from these rifts sum to more than 550 million bbl of oil. With the exception of the Cacheuta Subbasin, the "Intermontane Bolsones" are virtually unexplored.

In southern Patagonia of Argentina and Chile (Tierra del Fuego), Late Triassic–Jurassic north-trending grabens were filled with continental sequences alternating with acidic volcanics. Most of them have tested oil.

The flexured, thermally subsiding basinal complex was faulted in east-west-trending wrench systems where early sedimentation was predominantly continental. In the foredeep, thick packages of sediments accumulated on continental crust between the mobile magmatic arc and the central craton. The early foredeep stage accompanied the beginning of compressional folding and subduction of Pacific oceanic crust.

These foreland basins are among the most prolific hydrocarbon producers on the continent. Predominantly transgressive, marine organic-rich shales were deposited in depocenters. Heat flow was low to average. Later stages involved rapid sediment influx from the adjacent mobile belt during the Late Cretaceous and Tertiary. Initially, sediments were deposited in deep-water settings and, later, in marine and nonmarine environments.

From the late Paleozoic, the Gondwana western border, or Panthallasian margin, consisted of volcanically active lowlands, where a narrow sea, about 5000 km long, circumscribed the convergent margin. Discontinuous forearc basins with sills were episodically open to marine circulation from the Pacific, creating anoxic bottom conditions. This long foreland trend extended from the Guayaquil Gulf down to Antarctica.

During the Triassic and Tithonian-Neocomian, the richest organic-matter sequences were concentrated in this segment of the Subandean basins. This area has perhaps the most favorable thermal history in western South America. Most of the kerogen-rich deposits and active petroleum systems are concentrated in the Cuyo and Neuquén to Austral (Magellan) Basins.

Northwest Bolivia and Argentina (Figure 12)

During the Late Silurian–Devonian in the central Subandean belt (i.e., from southern Peru down through Bolivia to northern Argentina), a remarkable concentration of oil source rocks was deposited.

The early stages of the foreland belt extended south to the Chaco-Alhuampa Basins, where Silurian-Devonian sequences consist of a similar sedimentary succession. Eastward, the inner Paraná Basin may have connected the Chaco-Paraná Basin and the Panthalassian ocean. The Silurian-Devonian interval is characterized by a relatively simple quartzite-siltstone-shale association, with gradual variations in thickness and regular changes of facies and lithologic types. Along the eastern side of these basins, there is a subtle onlap to the peneplaned Guaporé craton. Regional reconstructions suggest the presence of depocenters morphologically different but tectonically connected, separated by submerged sills (Sempere, 1995; Starck, 1995).

Little is known about Silurian source rocks in Bolivia and north Argentina (Kirusillas Formation and equivalents) because the unit has not been reached frequently by the drill. The available data indicate that the Silurian-Devonian is dominated by mature shales, with low organic-matter content (average about 0.5%, exceptionally 1%). In contrast, the Devonian displays large concentrations of organic-rich marine shales in the Madre de Dios Basin, south Peru and Bolivia (Pando Plains). In the Subandean Boomerang Hills and in the Bolivian and Argentina Tarija and Alhuampa Basins, the organic content of the unit improves markedly (Moretti et al., 1994). Average TOC values range from 1% to 3%, with maximum values of more than 5%. This good-quality organic matter is situated in the thermal-maturity window, generating oil, condensate, and gas.

The basal Santa Rosa Formation (see Table 2) is a transgressive system of sandstones and shales which grades upward into the low-energy, fine-grained succession of the Los Monos and Tonono Shales, consisting of thick, black, micaceous, and laminated shales. These beds are often described as bituminous or carbonaceous, and they are intercalated locally with heterogeneous coarse-grained facies (Huamampampa Formation). These units are widely distributed, reach more than 1000 m in thickness, and are considered the most effective Devonian source-rock sequences. The horizons with brachiopods and trilobites suggest a temperate to cold-water community, deposited about 60°S paleolatitude. These faunal associations suggest connections between the Malvinas–South Africa and Antarctica provinces (Starck, 1995). Invertebrate distribution indicates that the Middle to Late Devonian was a period of maximum marine incursion and interconnection of depocenters with low relief.

In the northwest, the Los Monos Formation is held to be the Paleozoic source rock par excellence (Moretti et al., 1994; Dunn et al., 1995; Moretti, 1997), responsible for the sourcing of Devonian reservoirs. Recent information indicates that light oils in some Carboniferous sequences seem to be linked genetically to this source rock (Disalvo and Villar, 1998, 1999).

In the south basin, nonmarine deposits grade to the northeast into laminated or massive sandstones, reflecting high-energy wave action. These sedimentary prisms were probably built with sediments derived from the western orogenic belt. Sandstones and conglomerates, which are predominant near the Eastern Cordillera Basin margin, are replaced by shales in the Chaco Plains and Alhuampa (Copo and Rincón Formations).

Aramayo Flores (1987) indicated a west-northeast-trending grain-size gradient in this sector of the fold belt, where the best reservoir rocks appear to be concentrated

Good reservoir properties related to fracture trends associated with secondary permeability are found in the Devonian sandstones (Santa Rosa and Huamampampa Formations). In southern Bolivia and northwestern Argentina, drilling yielded commercial gas from Devonian reservoirs in the Macueta, San Antonio, and Aguaragüe Ranges structural trend.

In Bolivia, the main producing district is located from Santa Cruz de la Sierra south to Tarija, and in northwest Argentina. The first commercial discovery in Bolivia was Monteagudo field in 1924. In Argentina, the first commercial discovery was in 1928 at Vespucio field, which had an initial flow rate of 285 BOPD. In Bolivia, accumulated reserves have reached more than 500 million bbl oil and 27 tcf gas. In the Argentine portion, 18 fields were discovered, with recoverable reserves of more than 330 million bbl oil and 8.6 tcf gas.

In Argentina, the Upper Cretaceous (Campanian to Danian) muddy limestones and shales of the Salta

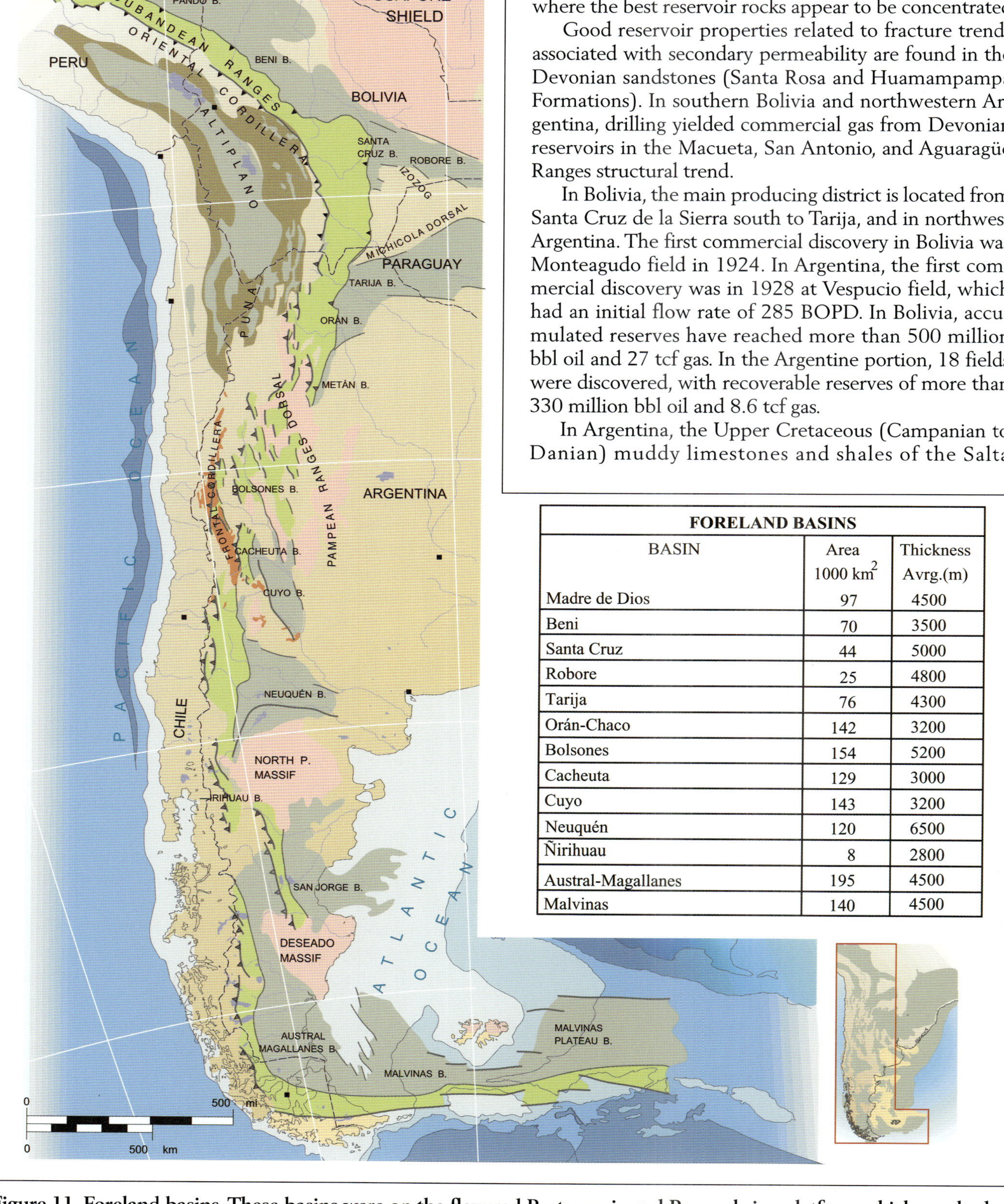

FORELAND BASINS		
BASIN	Area 1000 km^2	Thickness Avrg.(m)
Madre de Dios	97	4500
Beni	70	3500
Santa Cruz	44	5000
Robore	25	4800
Tarija	76	4300
Orán-Chaco	142	3200
Bolsones	154	5200
Cacheuta	129	3000
Cuyo	143	3200
Neuquén	120	6500
Ñirihuau	8	2800
Austral-Magallanes	195	4500
Malvinas	140	4500

Figure 11. Foreland basins. These basins were on the flexured Proterozoic and Precambrian platform which resulted in the western Foreland Belt. The western limit was an active magmatic arc that restricted marine circulation. The Cenozoic tectonic episodes raised the "Cretaceous batholith," building the Andean fold-and-fault belt.

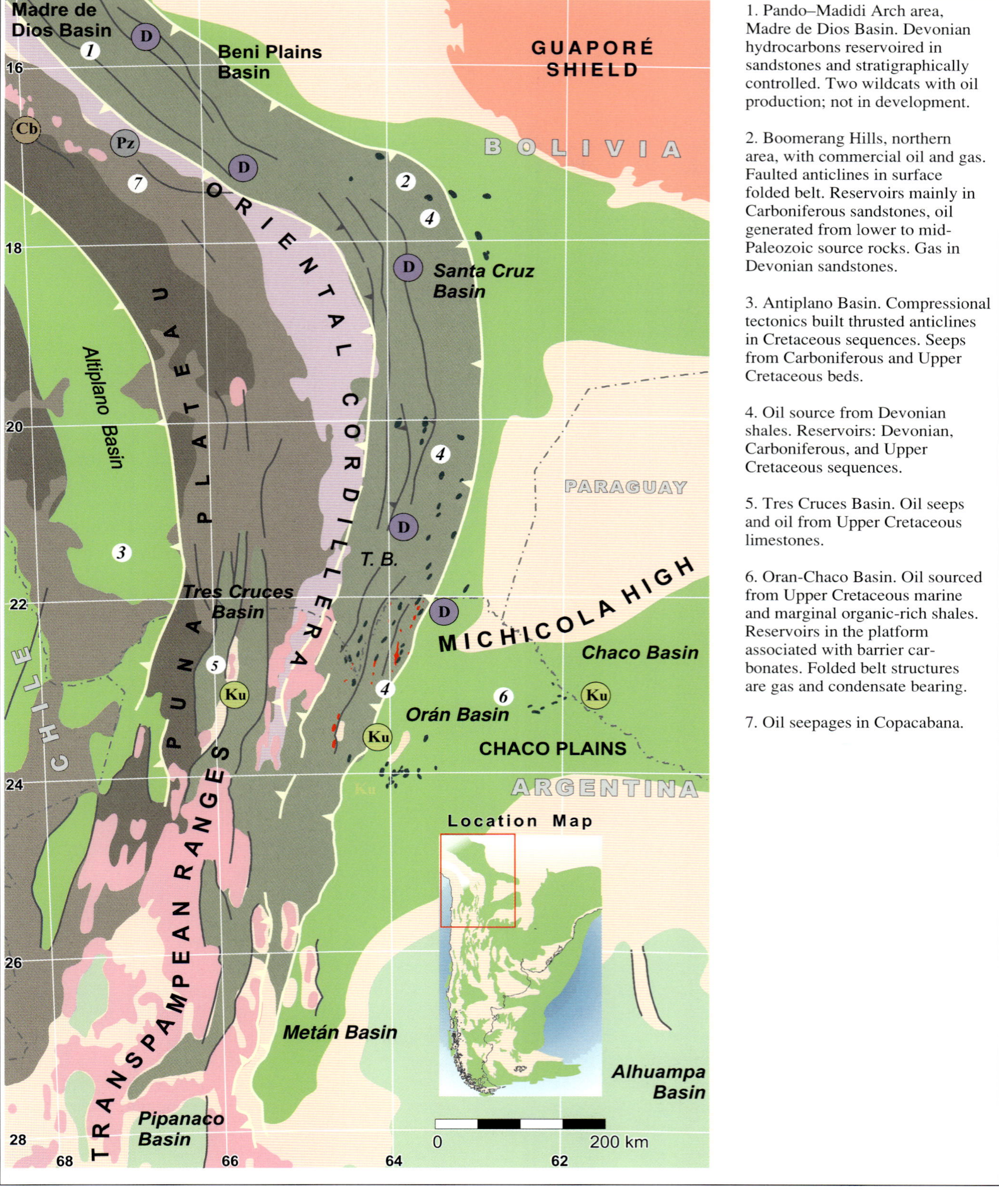

Figure 12. Northwest basins. These basins surround the Guaporé shield. A low-oxygen marine depocenter was formed, mainly in the middle Paleozoic. Late Cretaceous and Tertiary overburden promoted oil-source maturation and migration. The Oriental Cordillera is the western limit of the Subandean folded belt. Pz = Paleozoic petroleum system; D = Devonian petroleum system; Cb = Carbonaceous petroleum system; Ku = Late Cretaceous petroleum system.

Group are considered to be the oil and gas source. The Danian Olmedo black shales and Yacoraite mudstones and limestones are the main source rocks for Orán-Metán as well as potential source rocks in southern Altiplano basins.

In the Orán and Chaco Basins, Martínez del Tineo, Puesto Guardián, Pozo Escondido, and other fields are located on a regional strike-slip trend that probably favored oil migration and trapping in antithetic, fault-controlled anticlines, as well as in stratigraphic plays along the basin rim. The main reservoirs are predominantly in coastal barrier and deltaic sandstones and in limestones.

In the folded belt are Caimancito and Río Colorado–Valle Morado fields, which also produce oil and gas from Upper Cretaceous boundstones. As in the northern Subandean folded trend (from Venezuela to Peru, as well as south Bolivia), the possibilities of finding substantial new reserves of oil and gas are extremely high. The only limitations to exploration are logistic.

In the Argentine portion of the Tarija/Orán-Metán Basins, cumulative production has reached more than 209 million bbl oil and 5.6 tcf gas from 42 fields that cover about 2706 km^2.

Between the south Altiplano (Puna rifts) and Orán-Metán basins lie the north intermontane grabens—Arenal, Pipanaco, and Talampaya—filled with upper Paleozoic, Triassic, and Tertiary continental sequences. Some of the sequences have remarkable source rocks, but no hydrocarbon shows have been reported. This tectonic trend extends into the Puna Plateau, but there are no relevant data from this area.

The Late Cretaceous sequences have not yet yielded commercial production in the Puna Plateau, Tres Cruces Basin (Argentina), and Altiplano Basins (Bolivia), where the Yacoraite and Puca Formations seem to have good source-rock properties. However, oil seeps and shows have been reported from this region.

Cuyo and Bolsones (Figure 13)

In west-central Argentina, foreland platform rifting resulted in north-south "intermontane basins." Sedimentary sequences in those basins were exposed by the late Tertiary tectonic inversion controlled by the north-south faulted and rotated basement blocks of the central and western Pampean ranges. Proven and probable petroleum systems exist in this region in predominantly fluvial and lacustrine sequences, although present commercial production comes only from the Mendoza Cacheuta Subbasin. Apart from this extremely centralized, proven system, there are large areas in the "Paganzo Basin" of Carboniferous and Triassic oil source rocks. Remnants of these sequences are also exposed in the west, in the Precordillera (Calingasta-Uspallata piggyback basins), San Juan (Rincón Blanco, Marayes, and Bermejo Subbasins), and La Rioja (Ischigualasto), and west of the San Luis Province.

A global cooling took place during the Carboniferous and Permian, and an ice cap was located in southern Brazil, Uruguay, and central Argentina south of paleolatitude 60°S. Periglacial sediments, diamictites, varves, sandstones, and shales were deposited throughout the region.

Despite these periglacial conditions, in the Paganzo Basin, an extensive array of coastal-marine (lagoons), deltaic, and lacustrine sequences was deposited. In most of these, organic-rich shales have floral associations that indicate environments of tundra to taiga type, with predominantly rainy summer seasons that favored proliferation of peat, swamp, and marsh deposits. Outcrops of Carboniferous rocks display these sequences as unconformably overlying the Precambrian basement. Geochemistry data from these sequences indicate excellent source richness (TOC from 1% to 24%) and potential for generation of gas and waxy oils (Villar and López Gamundi, 1991).

In the Late Permian, warmer conditions prevailed in the region, creating drastic seasonal changes with extreme arid conditions that favored the deposition of continental red beds. These conditions continued until the Late Triassic. In the west, an active acidic magmatism rained volcanic ashes and pyroclastics throughout the region. Farther south, volcanic activity spread lava and pyroclastic material over southern Cuyo, central Pampas, and Patagonian regions.

During the Late Triassic, renewed uplift brought about the formation of closed centripetal basins that favored the deposition of alluvial fans and the formation of alluvial plains with lakes where Casbras, Potrerillos, Cacheuta, and Río Blanco Triassic sequences (Table 2) were laid down. The concentration of organic matter is greater in the upper Potrerillos–Cacheuta Formations and their time equivalents in the Paganzo Basin (the Upper Triassic Los Rastros and Ischihualasto Formations). The organic facies include mainly plant detritus contained in thin coal or carbonaceous shale layers alternating with coarse siliciclastic fluvial beds.

Studies point to a temperate to hot paleoclimate influenced by intensity of rainfall, implying that development of layers of coal and bituminous shales in the Triassic "Cacheuta" Series is closely linked to climate variations (Volkheimer, 1969; Stipanicic, 1979; Stipanicic and Bonaparte, 1979). Along with sedimentary and paleozoologic evidence, this type of association suggests the presence of a flooded landscape, dominated by freshwater marshes and bogs. The well-preserved lamination, absence of evidence of subaerial exposure, association of plants with abundant algae, and the continuous vertical succession observed in the "Cacheuta" shales suggest accumulation in a permanent subaqueous environment. All the evidence points to a very great and deep lacustrine system, presumably stratified, so that anaerobic conditions prevailed on the bottom.

Cacheuta shales generally show an average organic content of 4% and peaks of more than 12% of algae-amorphous type I or I–II material. A notable characteristic of the Cacheuta shales is their low thermal maturity, which explains why only a minor proportion of the excellent petroleum potential of its kerogen has effectively matured into hydrocarbons in localized deep portions of the main depocenters. This scenario is likewise responsible for the abundance of low-maturity, waxy, and high-viscosity oils, with little subsequent migration, that indicate early generation from source rocks.

The Triassic stages are represented by the Río Blanco siltstones and sandstones that show an upward-shallowing succession and a substantial decrease in preserved organic matter.

The presence of widespread organic-rich black shales implies that a petroleum system is present in the Bolsones region. Maturation and structural conditions, along with reservoir distribution, suggest that commercial concentrations of hydrocarbons, probably gas, should exist in this region.

Geographically, this region occupies an advantageous situation with regard to consumer markets in Argentina and in neighboring countries. This region is a frontier with great potential.

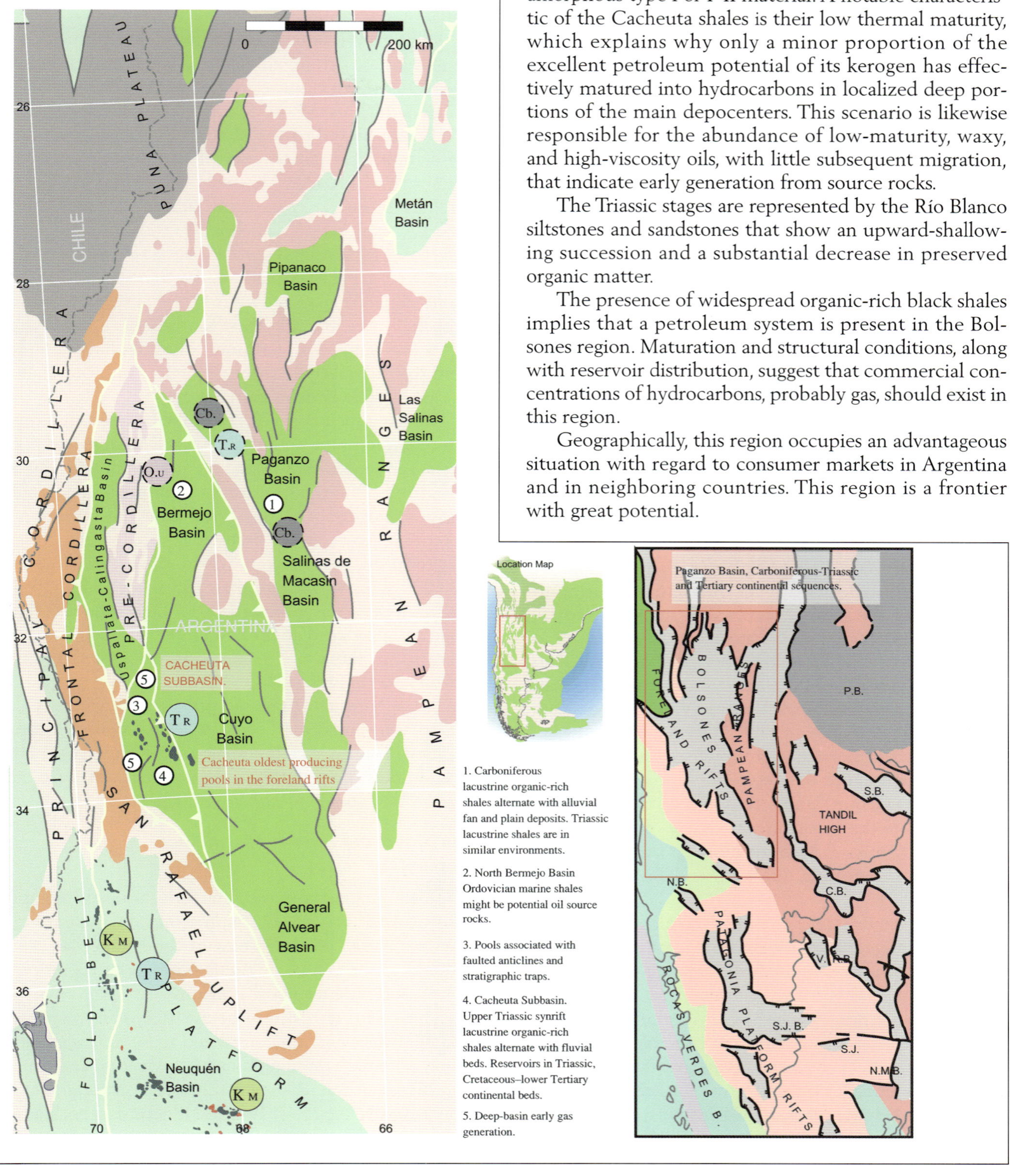

Figure 13. Cuyo-Bolsones basins. In the central-western Argentina Andes and pre-Andes foothills, intermontane valleys ("bolsones") are separated by Precambrian ridges. All the basins follow a north-south trend imposed by Tertiary tectonics. These are the Paganzo; Pipanaco; Arenal; Bermejo; Talampaya; Chilesito; Las Salinas; Salinas de Macasin; Bebedero; San Luis; Uspallata-Calingasta; Río Blanco; Cacheuta, or Cuyo; and General Alvear Basins. Ou = Late Ordovician petroleum system; Cb = Carbonaceous petroleum system; Tp = Triassic petroleum system; Km = middle Cretaceous petroleum system.

Neuquén Basin (Figure 14)

The Neuquén Basin is a typical foreland basin. This funnel-shaped basin is shaped by east-west transtensional fractures, which resulted in the formation of depocenters throughout the Mesozoic-Tertiary. Four petroleum systems are defined in this basin, and they have produced commercial hydrocarbons since 1920. Through the last 42 years, this basin has been the center of exploration and development activity because of its strategic geographic location.

Late Triassic–Early Jurassic continental sequences were also laid down in half grabens under similar structural and depositional conditions as the Cuyo-Bolsones basins. The Puesto Kaufman Formation, or "Pre–Cuyo Group" lacustrine shales (Table 2), rich in humic organic matter, have been detected by the drill and have yielded significant amounts of commercial oil. The Cacheuta Subbasin and eastern Neuquén are the only areas with proven Triassic petroleum systems. Similar sequences have been identified in the southern embayment and in the north basin (Llantenes, Remoredo, and P. Morada Formations), which suggests unexplored leads in this region.

During the Early Jurassic, a high sea level flooded the region, developing a ring of platform siliciclastics and basinal black shales. Sea-level fluctuations during the remainder of the Jurassic resulted in stacked reservoirs along the basin margins and paleohighs. At the end of the Jurassic, evaporites (Auquilco Formation) and fluvial and eolian red beds (Tordillo Formation) accumulated and are associated laterally with a large volume of volcanic facies in the magmatic-arc zone. The evaporitic and siliciclastic members make up a group of sequences that indicates an expansion of the depositional area in relation to older (Cuyo Group) units. This is particularly marked with regard to the basinwide spreading of the upper sequences of the basal Mendoza Group (or Cycle) basal sandstones (Quebrada del Sapo and Tordillo Formations).

The reestablishment of an effective communication with the Pacific through the magmatic arc and a sea-level rise favored a reflooding of the basin, again under anoxic conditions suitable for the accumulation and preservation of organic matter. The Vaca Muerta "Kimmeridge Clay" is the principal source rock of the Neuquén Basin. During the early Neocomian, platform facies (shallow, shelfal calcareous Quintuco Formation) were restricted to a narrow belt, whereas the platform carbonates, dolomites, and associated evaporites of the younger Loma Montosa Formation are widespread in the east.

The Vaca Muerta Formation contains a basal section of grayish-green sandstones, stratified with black shales and siltstones, that grades into a middle section dominated by gray to black lutites with pyrites which have an excellent source-rock potential. Progressive progradation led to the expansion of the shelf and slope, resulting in the formation of a coastal sabhka setting and the accumulation of evaporites.

Deposition of euxinic shales decreased with increased deposition of littoral siliciclastic facies and fluvial red beds. During maximum progradation in the late Valanginian, a thick section of nonmarine and littoral siliciclastics accumulated in the shallower areas of the basin (Mulichinco Formation equivalent to the carbonates in the south of Mendoza known as the Chachao Formation). In the last 15 years, these sandstones have been the object of exploration, which led to the discovery of new reservoirs with reserves exceeding 81.7 million bbl oil and 3.2 tcf gas (Filo Morado–El Portón, Sierra Chata, and Loma de las Yeguas fields). Additional reserves are still expected to be found in these sandstones.

In the Mendoza piedmont, oil must have migrated from the older Agrio marine sequences into the shallow-marine Roca-Loncoche Formation.

The Neuquén Basin can be ranked highest in terms of future exploration potential. Stacked marine and continental sequences, rich in organic matter, provide mature gas-prone objectives. In terms of strategic geographic position, this basin is considered the most important in Argentina. According to productive areas and amounts of oil produced, it is also rated as number one. More than 200,000 acres with more than 185 fields and 2619 producing wells are in production. By December 2000, the basin had produced about 2463 million bbl oil and 13 tcf gas (proven remaining reserves are about 1353 million bbl oil and 1.34 tcf gas).

In the Ñirihuau Basin in northwest Patagonia, Oligocene marine and coastal beds, as well as similar sequences in the Chilean Navidad embayment, have organic intervals, although no commercial oil has been tested from the marine "Patagonia Formation" sequences.

Río Mayo and Austral-Magallanes Basins (Figure 15)

Through central-west Patagonia, widespread upper Paleozoic to Jurassic coastal, marine, and lacustrine sequences were laid down east of the Patagonian Andes foothills in the Pampa de Agnia–Río Genoa Basins of the Tepuel and Nueva Lubecka regions.

These sequences have been studied from outcrops and boreholes and are considered to be potential petroleum systems. The main axis of the folded and faulted belt dips south, underneath the west San Jorge and Río Mayo Basins. Above it lie the Neocomian coastal and marine sequences of the Tres Lagunas, Cerro Katterfeld and, farther south, the equivalent Springhill and Río Mayer Formations. In the Austral Basin, thick, organic-rich black shales accumulated throughout the latest Jurassic-Neocomian. In the Austral Basin, these sequences are oil and gas bearing. In the west San Jorge Basin, tested wells in these sequences were noncommercial.

This region should have fair exploration potential in view of the type of potential petroleum systems, but it has been disregarded because of little detailed exploratory work and its remote location.

Throughout the Cretaceous and early Tertiary in this region, successive transgressive-regressive episodes in the foreland caused the formation of several petroleum systems. In the Austral Basin, the Springhill coastal and marine organic-rich shales and Neocomian Río Mayer marine shales are the most important source rocks. Gas and condensate are found predominantly in Springhill sandstones. In Argentina and Chile, producing pools are associated with basement highs, but the real trapping control is stratigraphic. In this producing trend, additional pools still await discovery, but only through detailed interpretation of high-resolution seismic will it be possible to detect such subtle plays. The shallow-marine Tertiary beds (Dorotea and Río Turbio Formations) are rich in coal and could be a potential source. In fact, boreholes in the west-basin piedmont (pre-Cordilleran area) tested gas during drilling of those units.

When the potential of other reservoirs is considered,

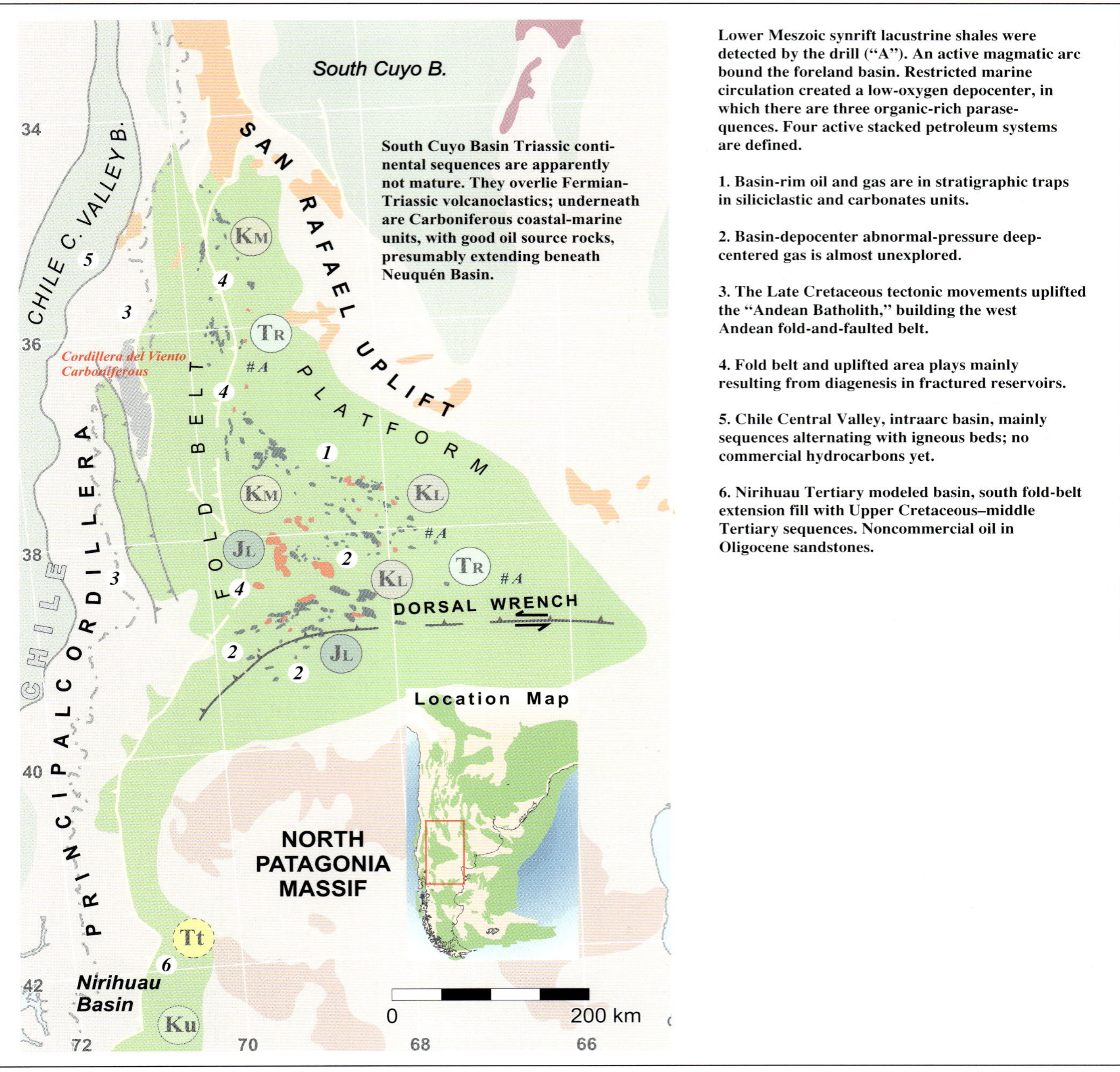

Figure 14. Neuquén Basin. This basin is in western Argentina in the Cordillera foothills, along with the western Agrio–South Mendoza folded belt and Loncopue graben. The basin is triangular and is formed by strike-slip faulting which offsets eastern-basin basement blocks. Tr = Triassic petroleum system; Jl = Late Jurassic petroleum system; Km = middle Cretaceous petroleum system; Kl = Late Cretaceous petroleum system; Tt = Tertiary petroleum system.

the "Springhill Platform or Producing Trend" in the eastern basin and the hinge line still has an important upside potential that deserves exploration, particularly with the application of new exploration techniques. This already has been proved through recent discoveries of Puesto Bravo and P. Peter fields, in which oil has been found in Upper Cretaceous and lower Tertiary siliciclastic reservoirs outside the traditional Springhill sandstones.

Pre-Springhill "basement grabens" are almost unexplored. Drilling results show that the Upper Jurassic volcanic Lemaire Formation (Tobífera) is interbedded with organic-rich shales dominated by a mixture of humic and sapropelic components from Santa Cruz in the north to Tierra del Fuego. Reserves may be small per structure, but so far, little or nothing is known about them, except in southern Chile (in the Gaviota, Angostura, and Bandurria rifted trends).

In west Tierra del Fuego, identical sedimentary units have been drilled in the north-south rifts of Lago Carmen, Gaviota, and Bandurria, testing light oil from fluvial sandstones and fractured volcanics. In addition, other structures have been detected by seismic in offshore and onshore areas. Offshore Tierra del Fuego and the Dungeness High may have additional prospectivity.

In the foothills, attractive folded and faulted structures are exposed. Sparsely drilled, most of them tested gas in Tertiary-Cretaceous siliciclastic intervals. Late Cretaceous siliciclastic sediments sourced from the west have excellent exploration potential "on structure."

Exploration and development have been concentrated mainly on the east "Platform Area." The 30,000 acres in production had yielded, by December 2000, about 488 million bbl oil and 5.582 tcf gas and have reserves of more than 188 million bbl oil and 7.3 tcf gas. In the southwest, the Austral Basin has an exciting exploratory potential.

In the southeast, the offshore Austral and Malvinas Basins merge in the Fagnano-Burdwood Bank through an onshore extension of the Andes Australes folded and faulted belt. Geologic characteristics of the offshore Upper Jurassic, Cretaceous, and Tertiary sequences are similar to those in onshore areas. The presence of folded and faulted structures, multiple reservoirs, and source rocks makes offshore petroleum systems probable.

From a production standpoint, this area seems highly risky because of severe weather and oceanic conditions and high exploration and development costs. Perhaps future exploration results in the onshore trend will help to define offshore prospectivity more precisely.

Aulacogens

Closely associated with the evolution of the southern Andean fold belts are the Malvinas and Malvinas Plateau aulacogens striking continentward from the fold belt. These are failed arms remaining from the earlier opening phase of the tectonic cycle that ended formation of the Weddel sea. The source beds were originally deposited in the failed arm and matured in the Malvinas Basin. They provide prolific hydrocarbon sources for reservoir rocks deposited during the foreland cycle.

Basins Related to "B" Subduction

These basins are related to the compressional megasuture associated with the "B" subduction zone, in which oceanic crust is subducted under continental crust. Their tectonic settings are intraarc and forearc basins.

Intraarc Basins (Figure 16)

Intraarc basins formed behind magmatic arcs. It can be difficult to define source-rock settings in this type of basin. Organic material varies, depending on geographic conditions, and dilution by rapid sedimentation may be high. Because heat flow is generally high, maturation may take place under relatively shallow burial. Basins with source rocks deposited in this paleotectonic setting are in central and southern Chile.

Forearc Basins (Figure 16)

Forearc basins lie between a magmatic arc and an associated trench (the "Chile Trench," in the case of the Antofagasta Basin). Heat flow in these basins is lower than in intraarc basins. Distribution of organic matter depends on local conditions. Sedimentation rates are high and may be excessive. This group of compressional-type basins is also associated with collision of two plates in which strike-slip faulting could be present. Good source rocks can develop in such basins because they are in tectonically isolated blocks.

Summary

Forearc and intraarc basins associated with magmatic arcs can contain good source rocks. Back-arc basins found in Chilean territory, which have high heat flow, can have mature source rocks at relatively shallow depths, but accumulation of organic material in such basins is not predictable. Collision-type basins have a good chance of developing source rocks in a restricted environment, but hydrocarbon maturation, preservation, and migration are difficult to predict.

SUMMARY

Most of the oil and gas ultimate recoverable reserves (18.5 billion bbl oil and 110 tcf gas) in Bolivia, Argentina, and Chile is in foreland basins. Concentrated in those basins are many of the best source rocks, reservoirs, seals, and traps. Most foreland basins have multiple petroleum systems and are far from being mature from an exploration and production standpoint.

Only 35% of the estimated potential reserves has been discovered in these basins. Hydrocarbon production always starts from the "easy zones," i.e., shallow intervals or obvious classic structures. The most significant new discoveries in nontraditional oil habitats have resulted from exploratory efforts in the last 20 years, when the industry began to search for new pools in folded belts, stratigraphic traps, deep basins, and other previously disregarded areas.

One of the main limitations in the exploration of the Neuquén Basin was an excess of gassy reservoirs. Despite the Neuquén Basin's having the largest number of gas pipelines in that area of the continent, a great part of the region has remained isolated from markets. Seasonal winter-summer consumer peaks have hindered rational management of oil/gas production. Deregulation of the gas market and freedom to export gas have caused gas-field development to expand. This has led to the expansion of gas exploration, even in areas apparently already developed.

Regions described in this paper include a variety of basins that have received exploratory attention governed by demand, accessibility, and evidence of the existence of active petroleum systems.

The majority of the discoveries was made in the early stages of the industry after indications that we now consider almost romantic: the presence of seepage, anticlines observed in outcrops, and even wells drilled for water that produced oil. Exploration expansion was achieved on the basis of "hot spots" and with methods that now appear rudimentary and technology that can be classed only as elementary. After World War II, methods and techniques registered great advances, and this new technology helped to expand the producing trends but did not find many new productive areas.

In the 1960s and 1970s, frontiers were expanded cautiously. Not only were virgin basins studied, but more sophisticated problems also were tackled, and offshore seismic studies were initiated. It was only in the 1980s, however, that important discoveries were made, favored by a change in the political climate as state monopolies began to be deregulated.

Today, the free market for petroleum in the Southern Cone area makes for a more aggressive type of entrepreneur, attracted by energy markets and transportation systems in full development.

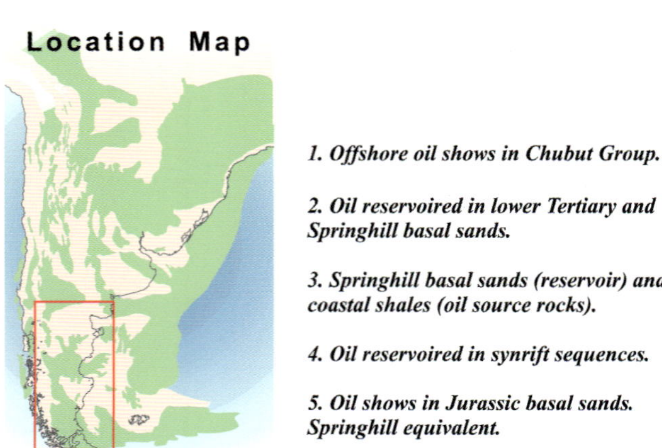

Figure 15. Southern Patagonian basins. These are the San Jorge and El Tranquilo intracratonic basins, the Tepuel–Nueva Lubeka and Río Genoa late Paleozoic–Jurassic foreland basins, and the Río Mayo and Austral-Magallanes foreland Cretaceous-Tertiary basins. Ju = Late Jurassic petroleum system; Km = middle Cretaceous petroleum system; Kl = Late Cretaceous petroleum system.

Table 2. Stratigraphy of the foreland basins.

Period	Epoch	Magallanes	San Jorge	Neuquen	Cuyo	Bolsones	Northwest	Tarija	Santa Cruz	Pando
Quaternary	Pleistocene	Basalts alluvium	Basalts alluvium	Quaternary basalts	L. Pilona	Alluvial				
Tertiary	Pliocene	Plioceno	P. Basaltos	Pliocene Tertiary	Marino		U. Chaco Gr. R. Guanaco	Quequende	Quequende	Quequende
	Miocene	Patagonia	S. Cruz	A. Piedra	C. Violaceo Papagayos	Calchaqui G.	L. Chaco Gr.	Yecua	Yecua	
	Oligocene	R. Leona	Basalts Patagoniano				Q. J. Maria Anta	Petaca	Petaca	
	Eocene	C. Bola	Basalts Sarmiento		D. Largo		R. Seco Tranquitas Lumbrera			
	Paleocene	Dorotea U.C. Cazador	Lefipan R. Chico	Coihueco		Lagarcito	M. Gordo Mealla Olmedo		Bala	Bala
Cretaceous	Upper	L.C. Cazador A. Vista Anita C. Toro	Salamanca B. Barreal Y. Terbol C. Rivadavia	Roca Neuquen G.		Basalts	Yacoraite Lecho Pirgua G. L. Curtiembres L. Yesera	Tacuru	Flora Eslabon Beu	Flora Eslabon Beu
	Middle	R. Mayer	M. Carmen	Rayoso Huitrin U. Agrio		Gigante G.	E. Cadillal			
	Neocomian	Princon Springhill	Katterfeld D-129	Mulichinco L. Montosa	P. Bardas					
Jurassic	Miam	Tobifera	C. Guadal A. A. Bandera	V. Muerta Tordillo Auquilco	Barrancas	Basalts				
	Dogger		L. Trapial C. Ferrarorn	Lotena P. Rodasa Lajas						
	Lias		Ostrarena	L. Molles P. Morada El Freno						
Triassic	Upper			Pre-Cuyo	Paganzo III R. Blanco Cacheuta	Paganzo III L. Colorados Ischigualasto	Ipaguasu	Ipaguasu		
	Lower			P. Flores	Potrerillos R. Mendoza	Rastros Ischichuca	Vitiacua Cangapi	Vitiacua Cangapi		
Permian	Upper		E. Tranquilo	Choiyoi G. Granite	Choiyoi G. Paganzo II	A. Pena G. Paganzo II				
	Lower		R. Genoa			Patquia Tupe			Copacabana	Copacabana
Carbo-niferous	Upper		L. Juanita		Paganzo I	Paganzo I	S. Telmo	S. Telmo	S. Telmo	
	Lower			E. Imperial	E. Imperial	S. Eduardo Guandacol Maliman	L. Peñas Tarija Tupambi	Escarpment Tarija Tupambi	L. Peñas Tarija Tupambi	Retama
Devonian	Upper						Iquiri L. Monos Huamapampa	Iquiri L. Monos Huamapampa	Iquiri L. Monos Huamapampa	Tequeje
	Middle		Granites				Icla	Icla	Icla	
	Lower		R. Lacteo				S. Rosa Guallabillas	S. Rosa	S. Rosa	
Silu-rian							Kirusillas Zapla	Lipeon Zapla	Kirusillas S. Benito Cuchupunata	Tenere Enadere
Ordovi-cian					Lsts-S. Juan Villavicencio		Ordovicico		Avispas Putintiri	
Precambrian					Precambrian		Precambrian	Precambrian	Precambrian	Precambrian

The start of the twenty-first century provides us with these favorable factors:

- technological advances
- basins with proven hydrocarbons, with new areas to expand frontiers
- geographically favorably situated virgin basins
- favorable legal conditions
- availability of capital
- medium-size and independent enterprises willing to expand their areas of activity

This combination of factors offers the potential of a brilliant future for petroleum exploration and development in southern South America in the twenty-first century.

REFERENCES CITED

Aramayo Flores, F., 1987, Cuencas petroleras del noroeste Argentino, *in* Evaluación de formaciones en Argentina: Buenos Aires, Schlumberger, p. 1–2, 9–15.

Disalvo, A., and H. J. Villar, 1998, The petroleum system of post-Devonian reservoirs in the Eastern Tarija Basin, Chaco Plain, Argentina, *in* Sixth Latin American Congress on Organic Chemistry, CD-ROM, 5 p.

Disalvo, A., and H. J. Villar, 1999, Los sistemas petrolíferos del área oriental de la Cuenca Paleozoica noroeste, *in* IV Congreso de Exploración y Desarrollo de Hidrocarburos: Instituto Agentino del Petroleo y Gas, p. 89–100.

FOREARC BASINS		
BASIN	Area 1000 km^2	Thickness Avrg.(m)
Tarapaca	8	400
Valparaiso	-	2000
Navidad	-	-
Valdivia	-	-
Chiloé	-	-
Golfo de Penas	-	-
Madre de Dios	-	-
Diego Ramírez	-	-

INTRAARC BASINS		
BASIN	Area 1000 km^2	Thickness Avrg.(m)
Titicaca	70	6000
Altiplano	-	-
Antofagasta	18	3000
Valle Central	-	-
Tenuco	21	2500
Ancud	21	3500
Arauco	23	2000

Figure 16. Andean intraarc and forearc basins. Intraarc and forearc basins are within raised Andes Mesozoic-Cenozoic sequences, alternating with igneous rocks that remain as relicts. During the Tertiary Andean tectonic pulses, shallow-marine and continental sequences filled these depressions.

Dunn, J. F., K. G. Hartshorn, and P. W. Hartshorn, 1995, Structural styles and hydrocarbon potential of the Sub-Andean thrust belt of southern Bolivia, *in* A. J. Tankard, R. Suarez S., and H. J. Welsink, eds., Petroleum basins of South America: AAPG Memoir 62, p. 523–543.

Du Toit, A. L., 1927, A geological comparison of South America with South Africa: Washington, D. C., Carnegie Institute, 157 p.

Lange, F. W., 1967, Biostratigraphic subdivision and correlation of the Devonian in Paraná Basin: Boletin Paranenase de Geociencias, v. 21/22, p. 63–98.

Legarreta, L., and C. A. Gulisano, 1989, Análisis estratigráfico secuencial de la cuenca Neuquina (Triásico Superior–Terciario inferior), *in* G. A. Chebli and L. A. Spalletti, eds., Cuencas Sedimentarias Argentinas: Instituto Superior de Correlación Geológic, Universidad Nacional de Tucumán, Serie Correlacion no. 6, p. 221–243.

Lesta, P. J., and R. Ferello, 1972, Región extraandina del Chubut y norte de Santa Cruz, *in* A. F. Leanza, ed., Geología Regional Argentina: Córdoba, Argentina, Academia Nacional de Ciencias, p. 601–653.

Moretti, I., 1997, Maturation and migration of hydrocarbons: South and central Subandean Zone of Bolivia, *in* Oil and gas exploration and production in fold and thrust belts: Veracruz, AAPG/AMPG, p. 1–5.

Moretti, I., E. Diaz Martínez, G. Montemuro, E. Aguilera, and M. Perez, 1994, Las rocas madre de Bolivia y su potencial petrolífero: Subandino–Madre de Dios–Chaco: Revista técnica de YPFB, v. 15, no. 3–4, p. 293–317.

Rapela, C. W., 1990, El magmatismo gondwánico y la megafractura de Gastre, *in* Onceavo Congreso Geológico Argentino, Actas I, p. 113-116.

Riccardi, A. C., S. E. Damborenea, M. O. Manceñido, R. Scasso, S. Lanes, and M. P. Iglesias, 1997, Primer registro de Triásico marino fosilífero de la Argentina: Asociación Geológica Argentina Revista, v. 52, p. 228–234.

Sempere, T., 1995, Phanerozoic evolution of Bolivia and adjacent regions, *in* A. J. Tankard, R. Suarez S., and H. J. Welsink, eds., Petroleum basins of South America: AAPG Memoir 62, p. 207–230.

Starck, D., 1995, Silurian-Jurassic stratigraphy and basin evolution of northwestern Argentina, *in* A. J. Tankard, R. Suarez S., and H. J. Welsink, eds., Petroleum basins of South America: AAPG Memoir 62, p. 207–230.

Stipanicic, P. N., 1979, El Triásico del Valle del Río de Los Patos (San Juan), *in* J. M. C. Turner, ed., II° Simposio de Geología Regional Argentina: Córdoba, Argentina, Academia Nacional de Ciencias, v. 1, p. 695–744.

Stipanicic, P. N., and J. Bonaparte, 1979, Cuenca triásica de Ischigualasto–Villa Unión, Provincia de La Roja y San Juan, *in* J. M. C. Turner, ed., II° Simposio de Geología Regional Argentina: Córdoba, Argentina, Academia Nacional de Ciencias, v. 1, p. 523–575.

Tankard, A. J., et al., 1995, Structural and tectonic controls of basin evolution in southwestern Gondwana during the Phanerozoic, *in* A. J. Tankard, R. Suarez S., and H. J. Welsink, eds., Petroleum basins of South America: AAPG Memoir 62, p. 5–52.

Turic, M., 1999, Un desafio argentino: La exploracion de frontera en el siglo que se inicia: Boletín de Informaciones Petroleras, v. 59, p. 17–35.

Ugarte, F. R., 1966, La cuenca compuesta carbonífera-jurásica de la Patagonia meridional: Universidad de la Patagonia San Juan Bosco, Ciencias Geológicas, v. 2, p. 37–68.

Uliana, M. A., K. T. Biddle, and J. J. Cerdan, 1989, Mesozoic extension and the formation of Argentina sedimentary basins, *in* A. J. Tankard and H. R. Balkwill, eds., Extensional tectonics and stratigraphy of the North Atlantic margin: AAPG Memoir 46, p. 599–613.

Uliana, M. A., K. T. Biddle, D. Phelps, and D. Gust, 1985, Significado del vulcanismo y extensión mesojurásicas en el extremo meridional de Sudamérica: Revista Asociación Geológica Argentina, v. 40, p. 221–253.

Urien, C. M., and J. J. Zambrano, 1994, Petroleum systems in the Neuquén Basin, Argentina, *in* L. B. Magoon and W. G. Dow, eds., The petroleum system—From source to trap: AAPG Memoir 60, p. 513–534.

Van Niewenhuise, D. S., and A. R. Ormiston, 1989, A model for the origin of source-rich lacustrine facies, San Jorge Basin, Argentina, *in* I Congreso de Exploración y Desarrollo de Hidrocarburos: Instituto Agentino del Petroleo y Gas, p. 853–884.

Villar H. J., O. López Gamundi, and W. Püttmann, 1991, Characterization of kerogen from Triassic oil shales of the Calingasta Valley, Cuyo Basin, Argentina: 15th International Meeting on Organic Chemistry, p. 171–172.

Volkheimer, W., 1969, Paleoclimatic evolution in Argentina and relations with other regions of Gondwana, *in* Earth Sciences 2: Gondwana stratigraphy: Paris, UNESCO, p. 551–587.

Yrigoyen, M. R., 1983, Reseña sobre los conocimientos y la exploración de hidrocarburos en Argentina antes de 1907: Petrotecnia, Revista Instituto Argentino del Petróleo, March, p. 32–38; April, p. 36–41.

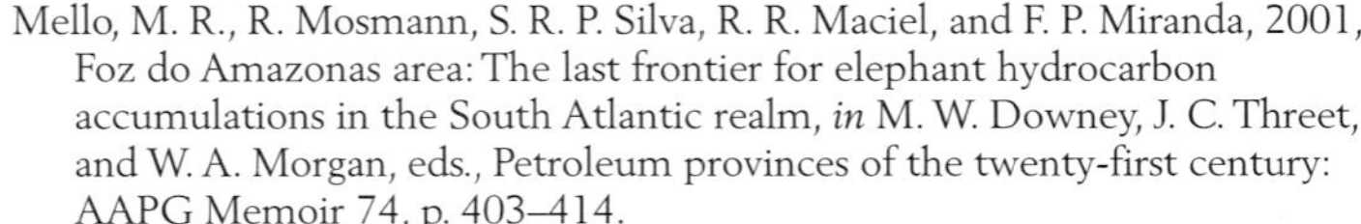
Mello, M. R., R. Mosmann, S. R. P. Silva, R. R. Maciel, and F. P. Miranda, 2001, Foz do Amazonas area: The last frontier for elephant hydrocarbon accumulations in the South Atlantic realm, *in* M. W. Downey, J. C. Threet, and W. A. Morgan, eds., Petroleum provinces of the twenty-first century: AAPG Memoir 74, p. 403–414.

Chapter 20

FOZ DO AMAZONAS AREA: THE LAST FRONTIER FOR ELEPHANT HYDROCARBON ACCUMULATIONS IN THE SOUTH ATLANTIC REALM

M. R. Mello
Petrobrás
Rio de Janeiro, Brazil

S. R. P. Silva
Petrobrás
Rio de Janeiro, Brazil

F. P. Miranda
Petrobrás
Rio de Janeiro, Brazil

R. Mosmann
Consultant
Rio de Janeiro, Brazil

R. R. Maciel
Petrobrás
Rio de Janeiro, Brazil

ABSTRACT

In most areas of the South Atlantic, exploration is far from being mature because exploration in ultradeep water has just begun. Geochemical data suggest that similar organic-rich facies and oil types are found across the South Atlantic petroleum provinces and that they have undergone similar tectonic-stratigraphic evolution, allowing the application of a unified model for hydrocarbon provenance in basins on both sides of the Atlantic. Such similarities, when interpreted in a paleogeographic context, can help reveal details of unexplored petroleum systems.

This study, based on an integrated multidisciplinary approach and using technologies ranging from remote sensing to molecular geochemistry, suggests that the Foz do Amazonas (Amazon River) area is one of the most promising oil/gas-prone provinces in the Brazilian continental margin. The tectonic-stratigraphic framework and regional facies variations of Upper Cretaceous and Tertiary source rocks are consistent with a marine carbonate and marine deltaic origin for source-rock deposition. Source rocks are related to Upper Cretaceous anoxic global events (Cenomanian to Turonian) and to an extensive fluvial-deltaic complex that has been in existence since the Miocene. The Niger Delta oil province is a comparable petroleum system analog.

INTRODUCTION

As a result of the giant oil discoveries in deep-water reservoirs in the Niger Delta during the past few years, most major oil companies are actively investigating the presence of similar petroleum systems in deep-water areas of the Amazon Delta. Although the Foz do Amazonas and Niger Delta Basins, which lie on opposite sides of the South Atlantic, are traditionally considered as independent entities, evolving geologic knowledge suggests that they share structural, stratigraphic, and geochemical elements. The major uncertainty regarding the Foz do Amazonas' huge hydrocarbon potential is the presence or absence of prolific source-rock systems in deep-water areas.

The Amazon Delta is considered important because it has the same oil types and similar petroleum systems compared with the most prolific deltaic basins in the world, yet there has been minimal deep-water exploration in this basin. In addition, the occurrence of several shallow-water, subcommercial accumulations of oil and gas and the appearance of natural oil seeps closely associated with listric and strike-slip faults that reach the surface indicate the presence of at least two active petroleum systems in the area.

The Foz do Amazonas Basin is in the northern area of the Brazilian equatorial margin and covers an area greater than 360,000 km^2 landward of the 3000-m isobath (Figure 1). The basin is directly related to the breakup of the African and American Plates during the Aptian-Cenomanian and can be classified as a typical divergent, mature, Atlantic-type continental margin (Asmus and Ponte, 1973).

This paper describes how a petroleum-system approach, supported by multidisciplinary technology involving seismic, geochemical, and remote-sensing data, is useful in the prediction and selection of future deep-water exploratory areas that may directly impact the development of new hydrocarbon frontiers in the South Atlantic realm.

GENERAL GEOLOGY

The Foz do Amazonas province can be subdivided into several geologic subprovinces. The most important are the Pará and Amapá platforms, Caciporé half-graben systems and Amazon River mouth system (Figure 1). The tectono-sedimentary history of the basin started during the Triassic with deposition in the Caciporé area of pre-rift siliciclastic sediments of the Calçoene Formation (Brandão and Feijó, 1994) (Figure 2).

The rift siliciclastic stage was a direct result of the South Atlantic opening during the Aptian-Cenomanian (e.g., Castro et al., 1978). The process was associated with basement-involved block-rotated faulting (half grabens tilting toward the southwest) and a regional unconformity overlain by a drift sequence. The rift sediments are composed of continental to marine shales and sandstones of the Caciporé Formation which were deposited in half grabens. The maximum thickness of the Caciporé Formation is approximately 6000 m (Silva and Rodarte, 1989; Silva et al., 1999).

The rift phase ceased when seafloor spreading started. The succeeding drifting stage is characterized by flexural

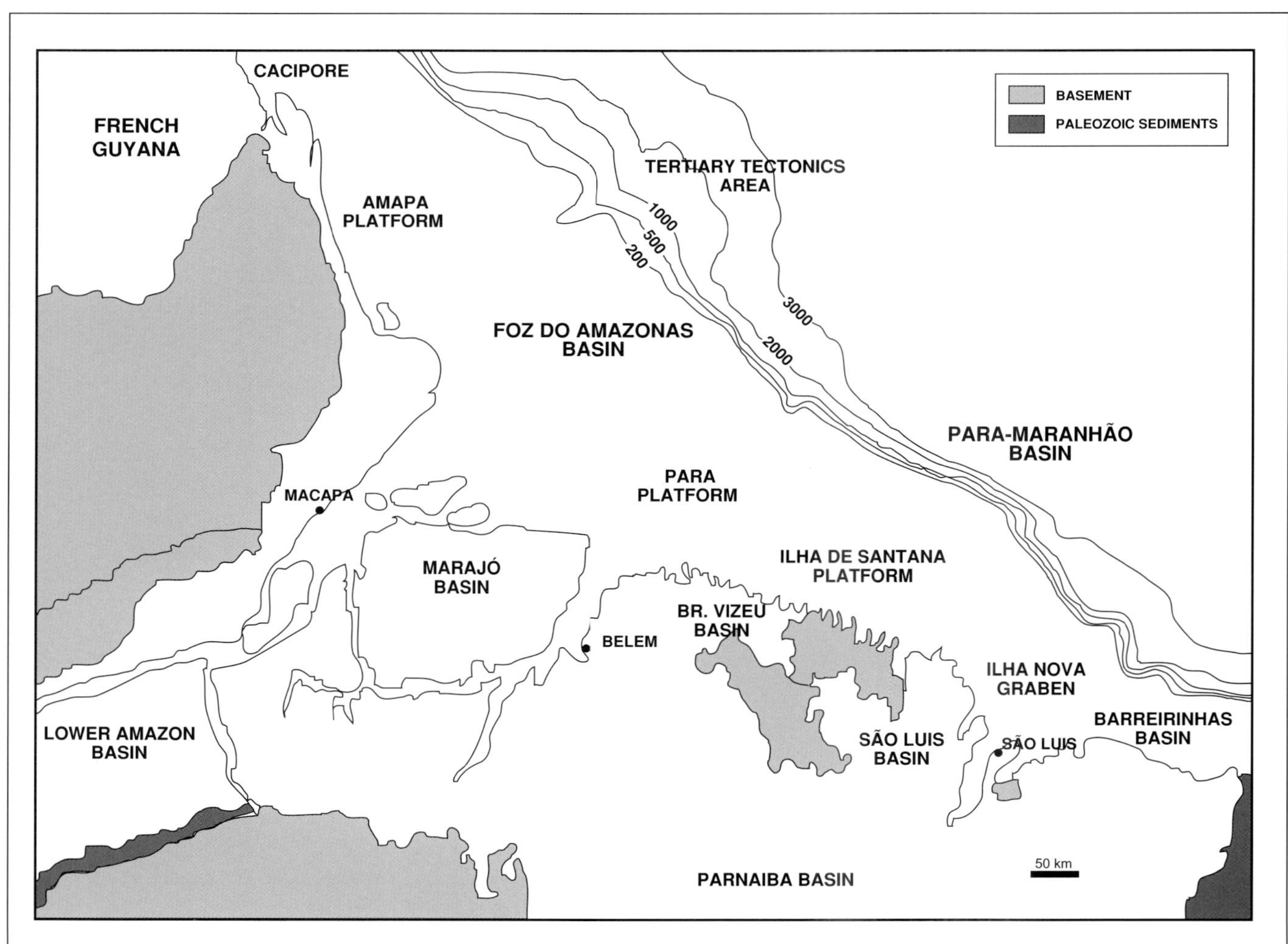

Figure 1. Location map of the Foz do Amazonas area.

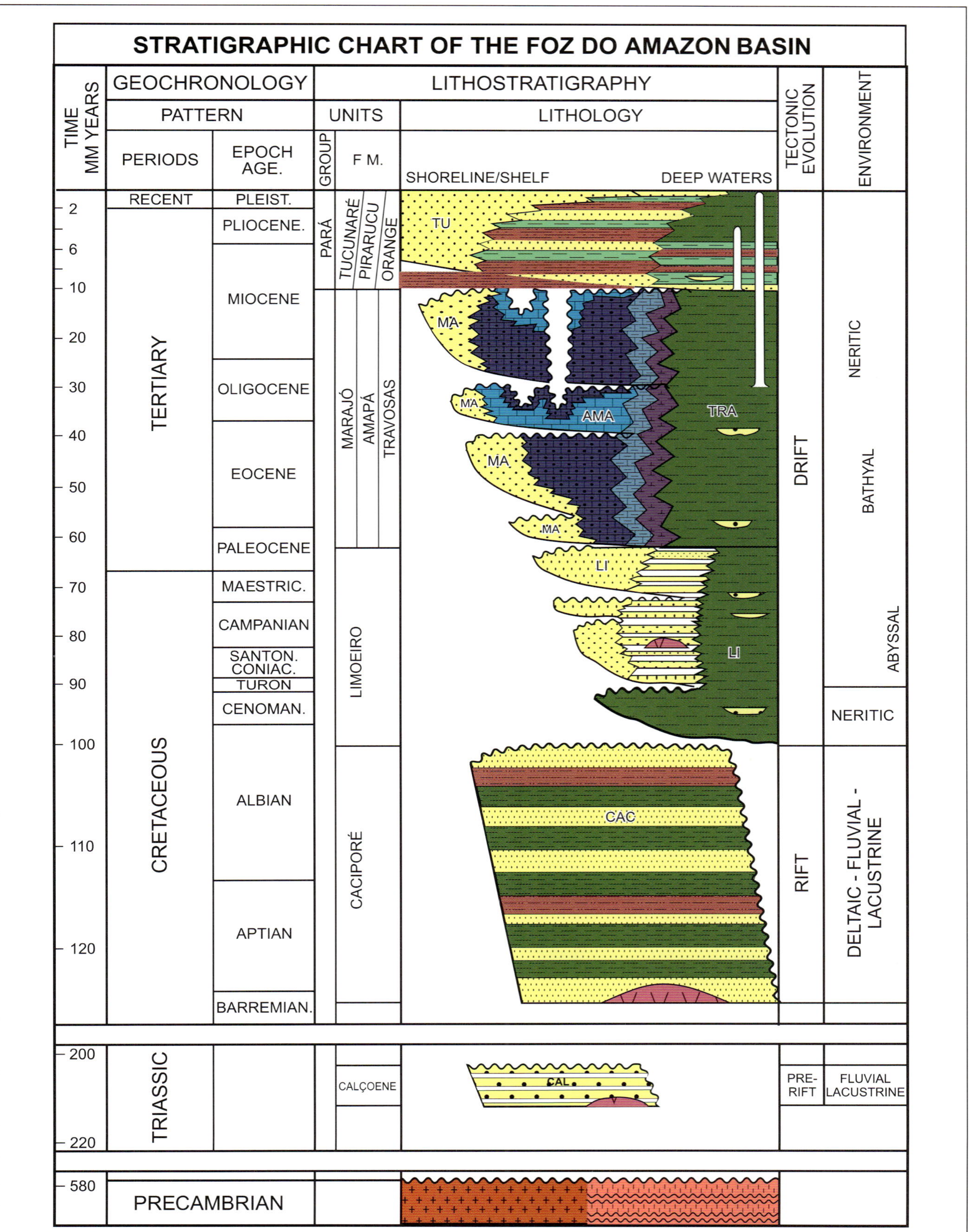

Figure 2. Stratigraphic chart of the Foz do Amazonas Basin (modified from Brandão and Feijó, 1994).

subsidence of the margin without conspicuous faulting. Three main drift sequences overlie the rift sediments: an Upper Cretaceous dominantly prograding siliciclastic continental to marine sequence; a Paleocene–lower Miocene marine carbonate platform; and an upper Miocene–Holocene very thick, prograding siliciclastic sequence related to Amazon River sedimentation (e.g., (Silva and Rodarte, 1989; Silva et al., 1999) (Figures 2–4).

The Upper Cretaceous Limoeiro Formation comprises the dominantly progradational siliciclastic continental to marine succession. It ranges from 2500 to 3000 m in thickness and is composed of shales (basal transgressive unit) and an overlying sandstone-shale succession.

The Paleocene–lower Miocene marine carbonate platform comprises a dominantly transgressive sequence (65–10 Ma) and is composed mainly of carbonates and marls of the Amapá Formation (Figure 2). This sequence is characterized by platform and slope carbonate sediments deposited in a neritic to upper bathyal environment and appears to be associated with conditions of tectonic quiescence. Near the Tertiary shelf break, salt tectonism is represented by listric detached faults soling out at the Upper Cretaceous regional unconformity (Silva et al., 1999).

Overlying the Paleocene–lower Miocene marine carbonate platform is the younger, very thick (as much as 9000 m), progradational siliciclastic marine succession of the Tucunaré, Pirarucú, and Orange Formations (Pará Group), related to Amazon River sedimentation (Silva et al., 1999) (Figure 4). A structural and seismic-stratigraphic analysis shows that most of the basin during that time was affected by very intense gravity tectonics caused by rapid siliciclastic sedimentation (e.g., Silva et al., 1999). Castro et al. (1978) identified growth faults as the major structural style of the basin and interpreted the compressional features as diapiric structures. Recently, based on new seismic data, Silva et al. (1999) identified the diapirs as compressional structures (folds and thrust faults) and defined a new tectonic model for the area

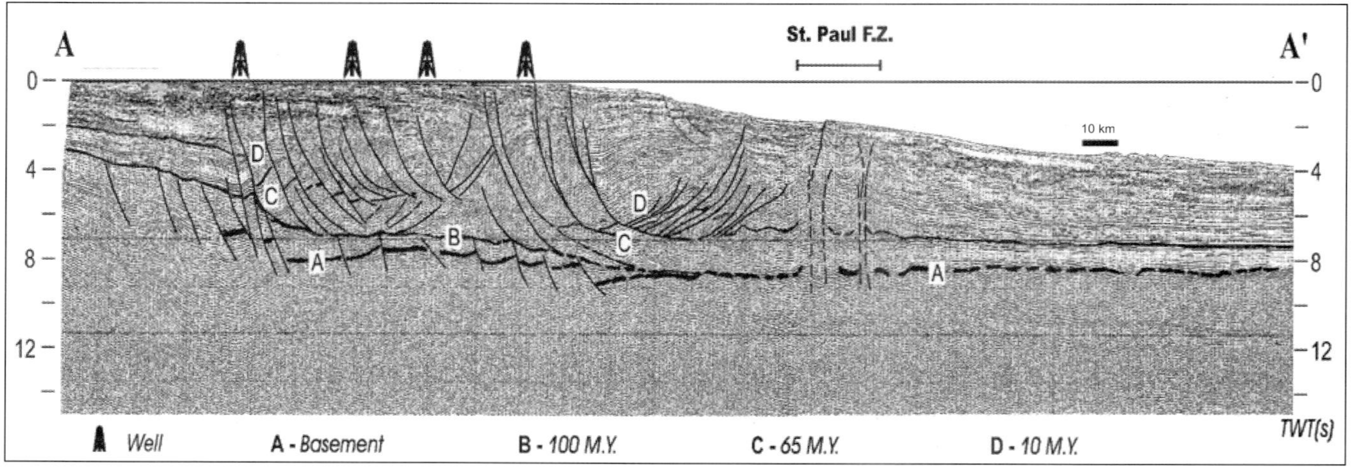

Figure 3. Multichannel seismic section typical of the Foz do Amazonas Basin, illustrating the carbonate platform, growth faults along the shelf margin, and thrust faults (Silva et al., 1999). For location, see Figure 6.

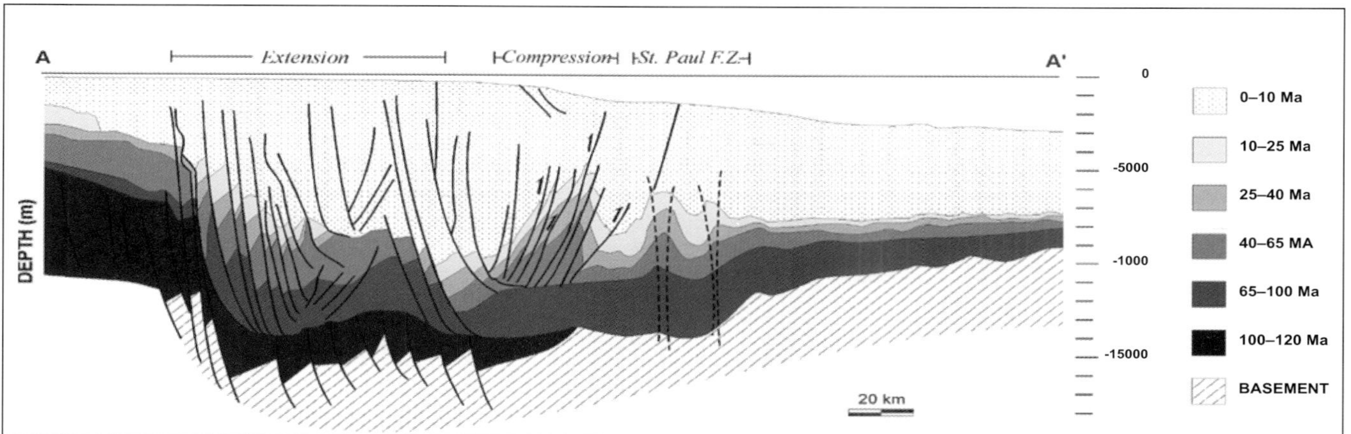

Figure 4. Depth-converted representative geologic section (based on Figure 3 seismic section) of the central part of the Foz do Amazonas Basin, showing two areas of extension and compression (Silva et al., 1999). For location, see Figure 6.

based on gravity gliding and spreading (compare Galloway, 1986).

PETROLEUM SYSTEM

In the past decade, a multidisciplinary approach involving geochemical, geologic, geophysical, and microbiostratigraphic research has greatly enhanced the level of understanding of some of the most representative petroleum systems in the Brazilian sedimentary basins (Mello et al., 1991, 1994, 1995).

The petroleum-system concept shifts emphasis from the rock to the fluid system of discovered petroleum occurrences (Magoon et al., 1990). The petroleum-system approach is the best method to evaluate exploration risk, even in unexplored provinces such as the deep-water Foz do Amazonas Basin. When exploration risk is evaluated in a sedimentary basin, it is mandatory to investigate three independent variables: hydrocarbon charge, trap, and timing.

Because petroleum is proof of the presence of a system, this is the place to start an investigation. First, representative oil samples from the Foz do Amazonas and the Niger Delta were analyzed. After a series of geochemical analyses, the data were used for oil-oil correlations to determine the number of oil types in the provinces. Generally, there are at least as many petroleum systems as there are oil types, unless there has been mixing of oils or a change in organic facies in the source rock.

The recognition of similar petroleum types in the Amazon and Niger Deltas in the Brazilian and West Africa margins (Figure 5) fosters the understanding of the hydrocarbon source potential in the Foz do Amazonas because the Niger Delta oil province has approximately 125 billion barrels (bbl) of oil in place (Doust and Omatsola, 1990; Mello et al., 1991), whereas only subcommercial accumulations have been discovered in the Foz do Amazonas.

TECTONO-STRATIGRAPHIC SETTING

The main reservoirs and traps present in the Foz do Amazonas province are lower to mid-Albian syntectonic siliciclastic rocks of the Caciporé Formation, Paleocene-Miocene biocalcarenites of the carbonate platform of the Amapá Formation, and turbiditic sandstones of the upper Miocene–Holocene Pirarucú Orange Formations (Silva and Rodarte, 1989; Miranda et al., 1998) (Figures 2–4, 6–7). They are found throughout most of the province (Silva et al., 1999) (Figure 4).

Structural and stratigraphic traps were created as a result of the Cenozoic gravitational tectonics, which gave

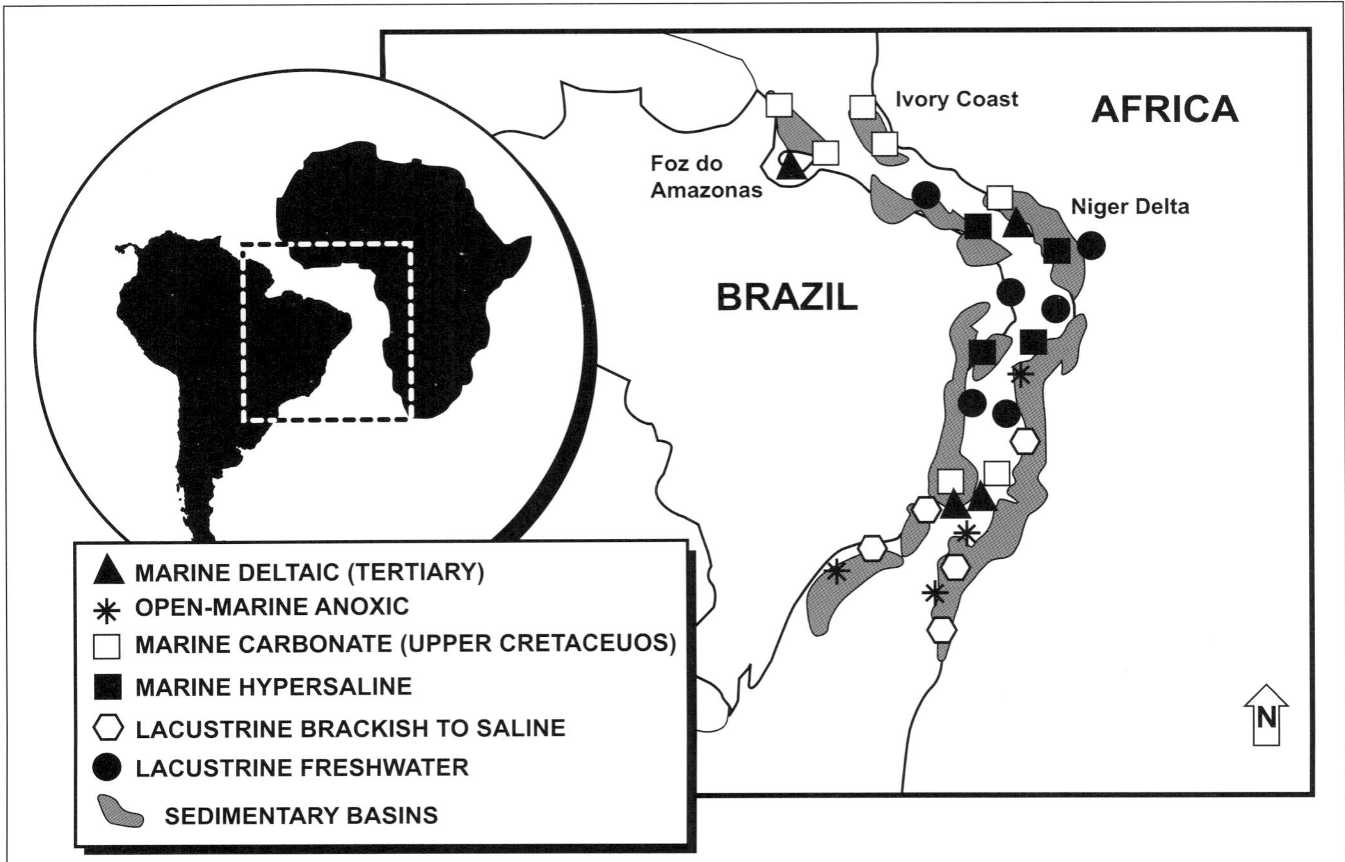

Figure 5. Map showing oil-type distribution across the South Atlantic realm.

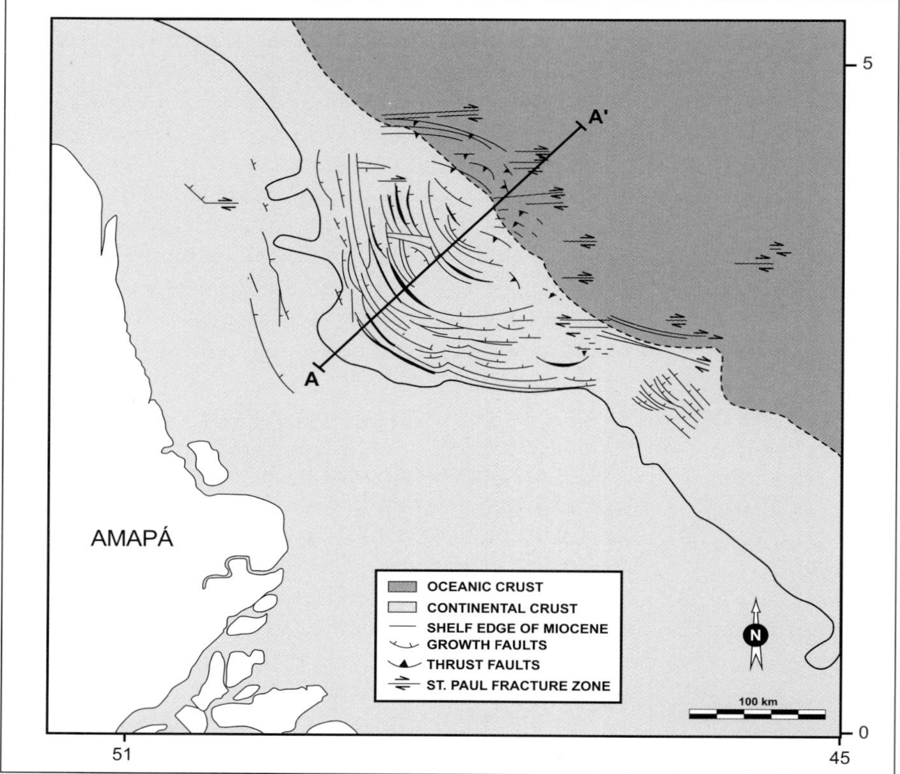

Figure 6. Miocene growth faults near the shelf edge and northeast-verging thrust faults (from Miranda et al., 1999; Silva et al., 1999). See Figures 3 and 4 for section A-A′.

rise to listric normal, thrust, and strike-slip faults that locally affected the Cenomanian-Holocene deposits. These faults are controlled by salt tectonism caused by the sediment loading of upper Miocene–Holocene sediments sliding over an unconformity (Silva and Maciel, 1998) (Figure 4). As can be observed from a Miocene fault map, growth faults located near the Miocene shelf edge and thrust faults verge toward the northeast (Silva et al., 1999) (Figure 6).

A similar assemblage of gravity-tectonic structures with their related traps is seen in profiles across the Niger Delta continental slope (Galloway, 1986; Doust and Omatsola, 1990). Thus, the Foz do Amazonas province has structural-stratigraphic features similar to those in the Niger Delta or any classic delta in the world. The young, rapid sediment loading contributed to suppression of thermal evolution of source rocks in the Amazon deltaic area, resulting in the basin's being oil prone no matter the depth of the Upper Cretaceous and Tertiary source rocks.

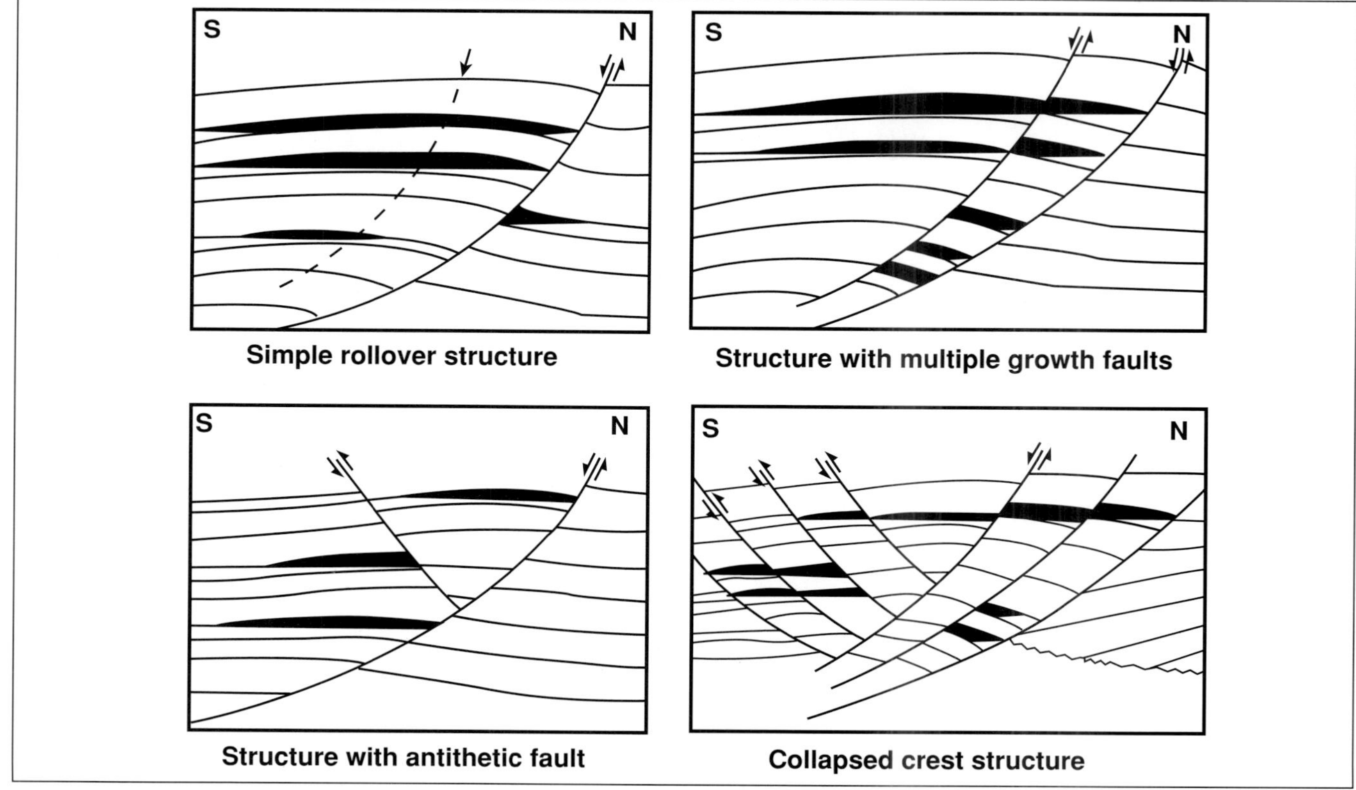

Figure 7. Reservoir structural styles present in the Niger and Foz do Amazonas Deltas (Requejo et al., 1995).

The hydrocarbon accumulations expected in this type of environment are controlled by pinch-out of sandstone turbidite reservoirs or by regional eastward dip resulting in closure against listric faults, as observed in the Niger Delta (Requejo et al., 1995) (Figures 3 and 4).

HYDROCARBON CHARACTERIZATION AND SOURCE ROCKS

Detailed molecular geochemical studies were used to identify and characterize oil types in the Foz do Amazonas and to correlate them with their Niger Delta counterparts. The results of these studies were used to assess depositional paleoenvironments of coeval source rocks, and to perform correlation of oil and source rock to predict the existence of active, prolific petroleum systems in deep waters of the Foz do Amazonas Basin. This approach was based mainly on distributions and concentrations of biological markers (Moldowan et al., 1990; 1994; Peters and Moldowan, 1993; Mello, 1988, 1993; Mello et al., 1988a, 1988b, 1989, 1995). Application of this methodology was undertaken involving oil samples pooled in reservoirs ranging in age from Early Cretaceous to Tertiary.

Age- and depositional-related biomarkers are, as the name implies, compounds whose appearance in oils and source-rock extracts is restricted to certain depositional environments and geologic time periods. Thus, their presence in oils and source-rock extracts can be of use to constrain, within limits, the depositional environments and age of source rocks. Determination of the distributions of the age-related biomarkers could be achieved only by using sophisticated gas chromatography– mass spectrometry–mass spectrometry (GC-MS-MS) techniques (Peters and Moldowan, 1993; Mello et al., 2001).

The use of specific biomarkers such as C_{26} steranes, oleananes, 2- and 3-methyl steranes, dinosterane-type steroids, and C_{30} steranes has allowed geochemists not only to distinguish different origins of oils related to specific organic facies in one particular basin but also to determine the approximate age of each organic facies (Moldowan et al., 1990, 1994; Mello, 1988b; Mello et al., 2001). This is of extreme importance when determining the petroleum systems that are active in each basin.

Results of the biological-marker investigation revealed similarities and differences among the oil samples studied. The data point to an origin from two major types of hypoxic-anoxic environments (Mello et al., 1988b, 1989, 1991, 1995) (Figures 8–13), ranging in age from Late Cretaceous to Miocene. The two types are (1) Upper

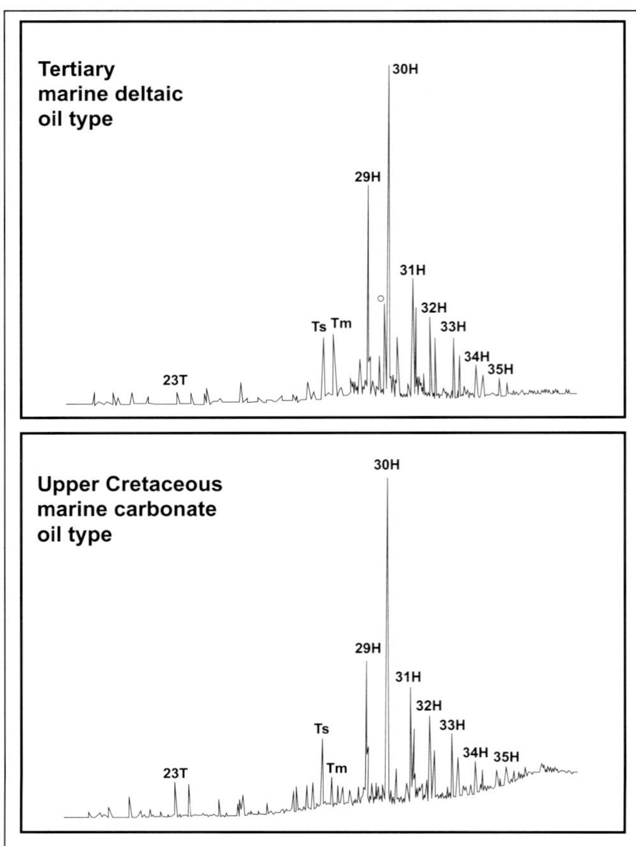

Figure 8. Mass chromatograms of mass/charge 191 showing representative hopane distribution in Niger Delta oil types (Requejo et al., 1995).

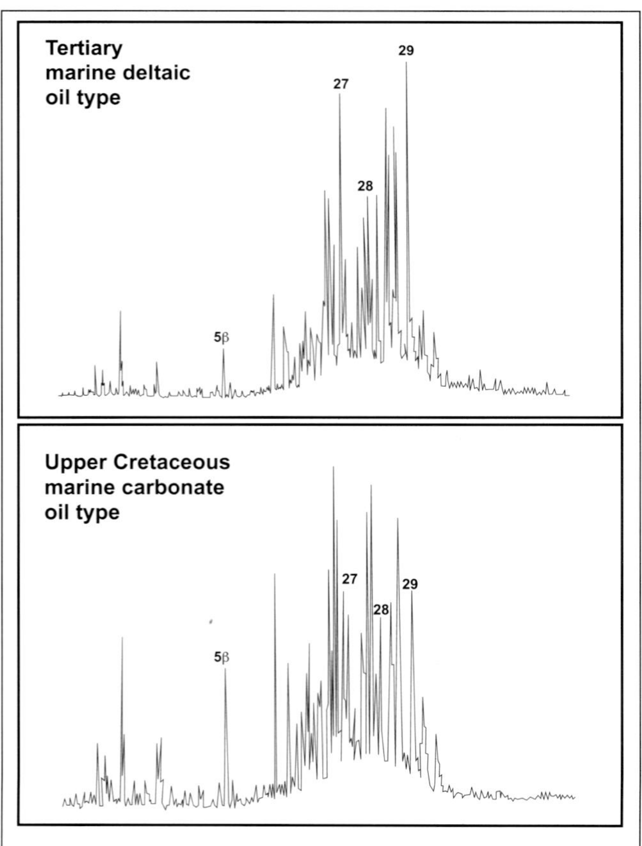

Figure 9. Mass chromatograms of mass/charge 217 showing representative sterane distribution in Niger Delta oil types (Requejo et al., 1995).

Cretaceous marine carbonate and (2) Tertiary marine deltaic with predominance of siliciclastic or carbonate lithology (Mello, 1988). Each oil type is discussed separately in the following sections.

Upper Cretaceous Marine Carbonate Oil Type

Oils from this group have been recovered from small subcommercial accumulations in the Niger Delta (Benue Trough; Requejo et al., 1995) and the Foz do Amazonas area (Caciporé to Pará-Maranhão; Mello, 1988; Mello et al., 1995) (Figure 5). They are derived from upper Albian–Cenomanian marls and calcareous black-shale source rocks deposited in a marine environment (for example, Limoeiro Formation in the Foz do Amazonas Basin and Upper Cretaceous in the Niger Delta area; Mello, 1988; Requejo et al., 1995) (Figure 5). These oils in the Foz do Amazonas are pooled mainly in lower to middle Albian syntectonic siliciclastic rocks of the Caciporé Formation and in Paleocene to Miocene biocalcarenites of the Amapá Formation. Initial charge of the reservoirs began during the Eocene from deep, distant "oil kitchens" along lateral migration pathways associated with regional unconformities and along major transform fault systems. These oils are characterized by naphtenic-aromatic, high-sulfur contents with low gas-oil ratios (Mello, 1988). As can be observed in Figures 13 and 14,

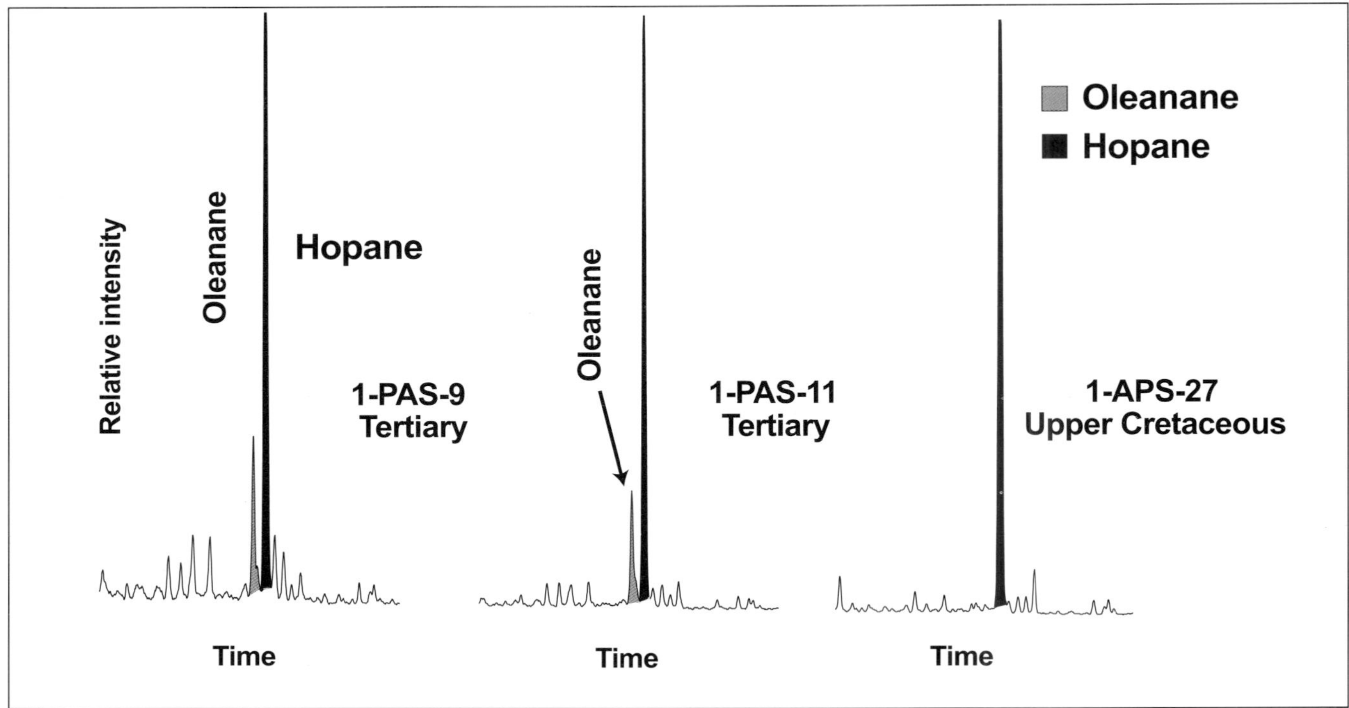

Figure 10. GC-MS-MS (m/z 412–191) traces showing representative age-related oleanane distribution in Foz do Amazonas oil types (Mello, 1988).

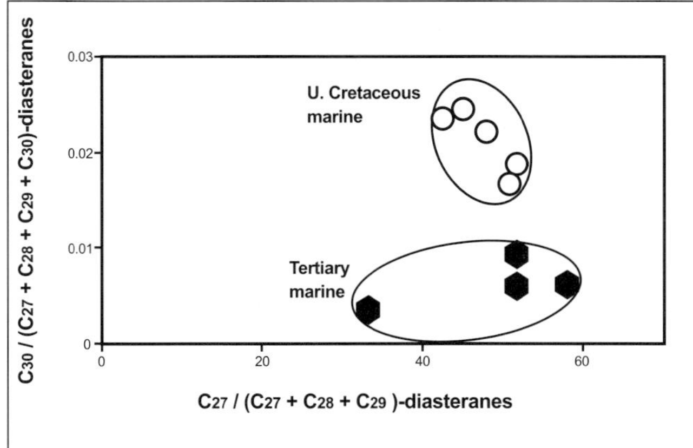

Figure 11. Crossplot of diasteranes biomarkers from Foz do Amazonas oil types.

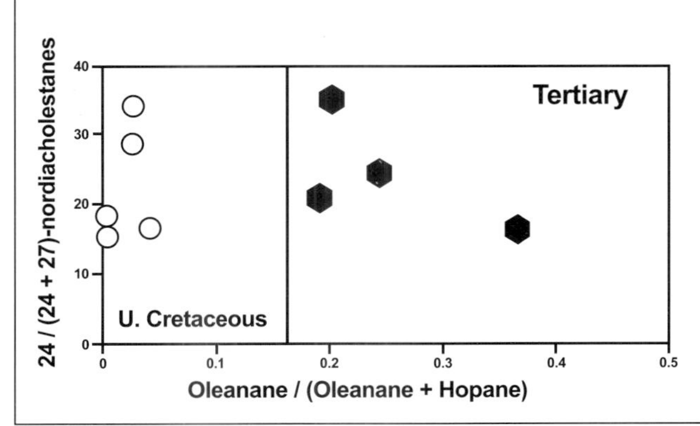

Figure 12. Crossplot of nordiacholestanes ratios versus oleanane/(oleanane + hopane) for Foz do Amazonas oil types.

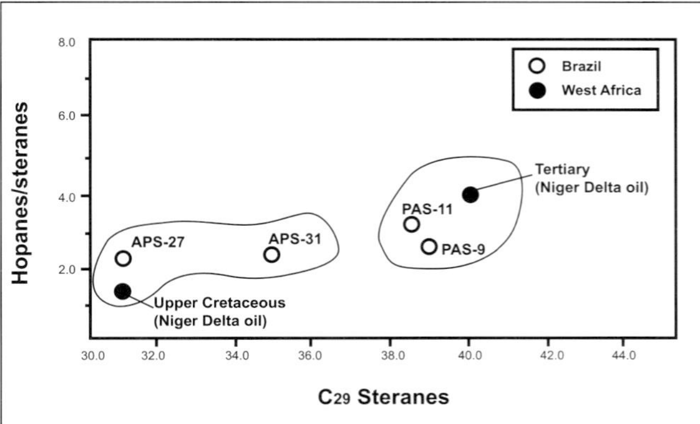

Figure 13. Crossplot of hopane/sterane ratios versus C_{29} sterane content for Niger Delta and Foz do Amazonas oil types.

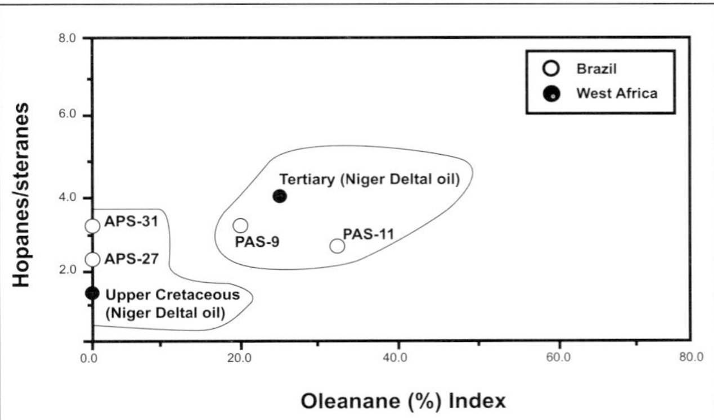

Figure 14. Crossplot of oleanane index versus C_{29} sterane content for Niger Delta and Foz do Amazonas oil types.

there is a good correlation between some of the oils from Foz do Amazonas and oils from the Niger Delta.

The source rock of the Upper Cretaceous oil derived from marine carbonates was identified in the Limoeiro Formation. It comprises approximately 50–150 m of thinly laminated, Albian-Turonian calcareous dark-gray shales ($CaCO_3$ as high as 18%; Mello, 1988; Mello et al., 1988a). Total organic carbon (TOC) averages 2–5% and locally is as high as 4% (Mello, 1988; Mello et al., 1988a). Hydrogen index values, consistent with a type II kerogen, range as high as 600-mg hydrocarbon/g organic carbon. Based on organic petrology, lipid-rich material mainly of algal and bacterial origin predominates (average 70% of amorphous organic matter; Mello, 1988; Mello et al., 1988a). Although the published data show an immature stage of thermal evolution for these rocks in the Foz do Amazonas, their good hydrocarbon source potential (as high as 30 kg hydrocarbon/ton of rock; Mello, 1988), combined with greater thermal maturity resulting from deeper burial far offshore, was sufficient to yield the oils present in the Foz do Amazonas area.

Tertiary Marine Deltaic Oil Type with Predominance of Siliciclastic or Carbonate Lithology

Oils from this group have been recovered in the Pará-Maranhão area in the Foz do Amazonas province and in the Niger Delta Basin in equatorial West Africa (Mello et al., 1991) (Figure 5). They are correlated with lower Tertiary dark-gray shale deposited in marine environments (Travosas Formation in the Amazonas Basin; Agbada and Akata Formations in the Niger Delta Basin; Mello et al., 1988a, 1991; Doust and Omatsola, 1990; Requejo et al., 1995). The oils are in reservoirs ranging in age from Eocene to Miocene which were charged during the late Miocene from deep, distant "oil kitchens" along mainly lateral migration pathways associated with regional unconformities (Mello, 1988; Doust and Omatsola, 1990). These oils are characterized by high parafinic content, low sulfur, very high API gravity (44°–46°), and high gas-oil ratios (Mello, 1988).

The Tertiary marine deltaic oil-source rocks have not been drilled. Mello (1988) identified a very thin Oligocene organic-rich interval composed of thinly laminated marls in the Amapá Formation ($CaCO_3$ as high as 54%; Mello, 1988; Mello et al., 1995). TOC averages 3% (Mello, 1988). Hydrogen index values, consistent with a type II kerogen, range as high as 400 mg hydrocarbon/g of organic carbon (Mello, 1988; Mello et al., 1995). Although the interval shows an immature stage of thermal maturation, its proximal position, along with its good correlation with Tertiary oil (Mello, 1988; Mello et al., 1995), suggests that this facies, if buried offshore to depths as great as 6000 m, can have the appropriate thermal history to yield Tertiary oils.

In summary, hydrocarbons sourced by the Upper Cretaceous marine carbonate and Tertiary marine deltaic petroleum systems in the Foz do Amazonas Basin indicate the presence of at least two active petroleum systems, with their respective kitchens located deep offshore. Because of rapid upper Miocene–Holocene sedimentary loading in the basin, the oils were probably expelled only during the Miocene, when they charged lower to middle Albian syntectonic siliciclastic rocks of the Caciporé Formation, Paleocene-Miocene biocalcarenites of the Amapá Formation, and extensive Tertiary turbiditic sandstones of the upper Miocene–Holocene Orange Formation. The structural and stratigraphic traps were created as a result of the Cenozoic gravitational tectonics, which gave rise to listric normal, thrust, and strike-slip faults. The listric fault system, in combination with several regional unconformities, acted as a migration pathway for hydrocarbons generated in the Tertiary and Upper Cretaceous sequences. The presence of biodegraded marine-carbonate-sourced oils in the Foz do Amazonas Basin with different thermal evolution profiles indicates the occurrence of more than one migration and biodegradation event during successive stages of reservoir filling (Mello, 1988). Such episodes appear to have been related to major changes in sea level, which would have controlled the influx-discharge cycles of meteoric waters in reservoirs (Soldan et al., 1995). In contrast, the marine-deltaic-sourced oils do not show any signs of biodegradation (Mello, 1988).

REMOTE SENSING APPLIED TO DETECTION OF NATURAL OIL SEEPS

Natural oil seeps historically have provided valuable information about petroleum systems. Foremost, they indicate the presence of mature hydrocarbon source

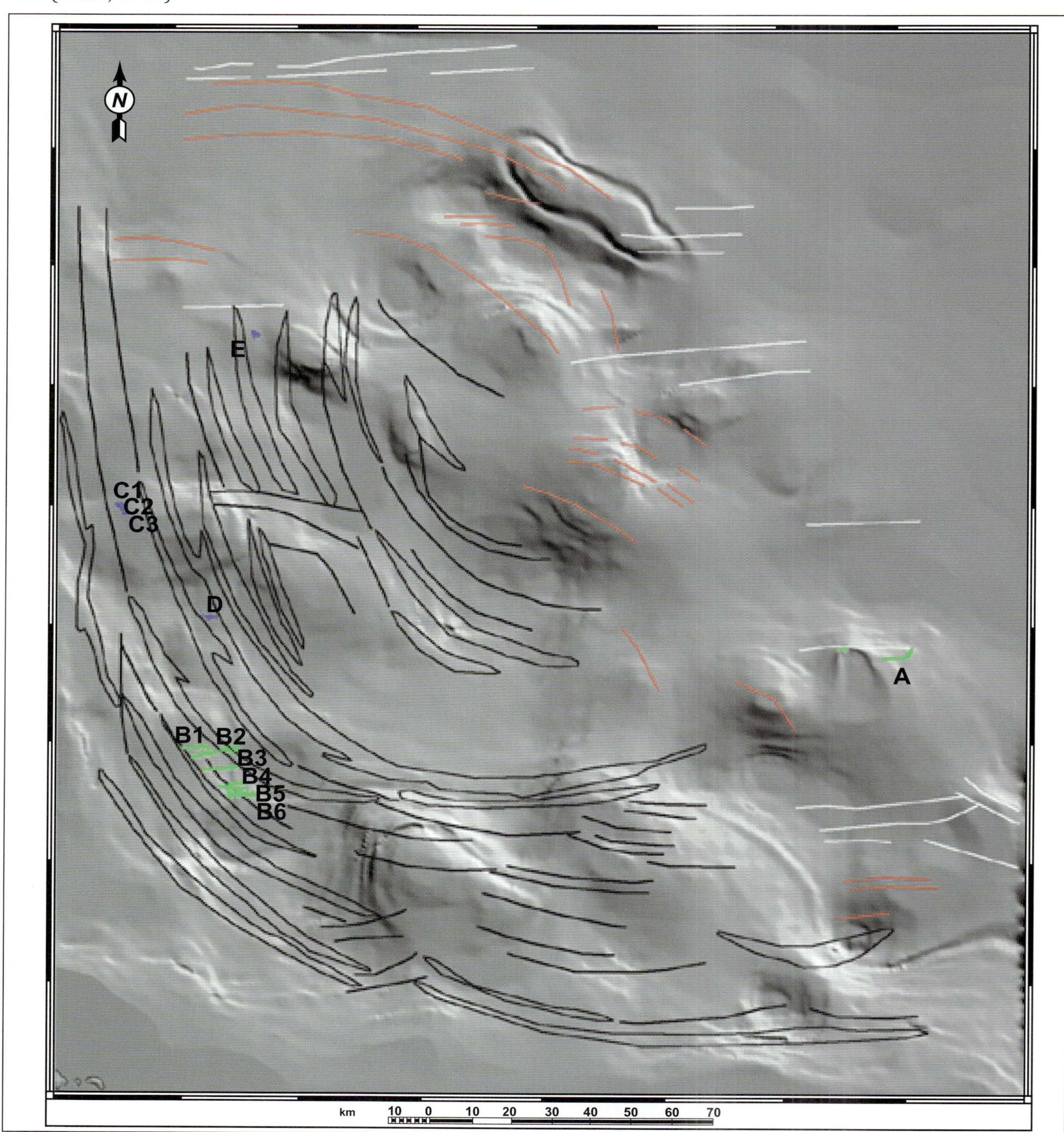

Figure 15. Miocene growth faults near the shelf edge, northeast-verging thrust faults, and interpreted natural oil slicks (from Miranda et al., 1999). Dark lines signify growth faults, red lines signify trust faults, and white lines signify strike-slip faults. Short lines labeled A, B, C, D, and E are natural oil slicks that could indicate migration pathways.

rocks and an active migration pathway in the area, without which there can be no petroleum accumulations. The high cost of deep offshore exploration has made identification of oil seeps a well-accepted risk-assessment methodology (Miranda et al., 1998).

Along the vast continental margin of the Foz do Amazonas area, identification of sea-surface hydrocarbon slicks using RADARSAT-1 imagery was undertaken to predict presence of active petroleum generation in the area. Natural oil seeps linked to structural features, such as normal (listric) and strike-slip faults along the deep-water mouth of the Amazon River, offshore Pará and Amapá states, emphasize the importance of the Miocene salt tectonism for the formation of migration pathways in the area (Miranda et al., 1998) (Figure 15).

In part of the Foz do Amazonas area, RADARSAT-1 images were acquired during a one-year period. The methodology (Miranda et al., 1998) proved to be effective because it showed that thin, elongated patches of smooth surface, interpreted as natural oil slicks, were closely associated with normal (listric) and strike-slip faults (Figure 15).

The physical mechanism that allows detection of oil seeps using RADARSAT-1 images is the dampening of capillary waves present on the ocean surface. These waves, only a few centimeters in wavelength, produce backscattering of the incident radar pulse resulting from Bragg scattering (Johannessen et al., 1994). As a result, ocean regions containing oil are dark on the images, in contrast to the background radar signal. However, with regard to radar imaging, the suppression of the wave backscatter is, unfortunately, not unique to the presence of oil. False targets include wind shadow, biological surfactants, local upwelling, shallow marine vegetation, and areas of heavy rainfall. Our approach circumvents these pitfalls by using the Unsupervised Semivariogram Textural Classifier (USTC) textural classification algorithm (Miranda et al., 1998) that enhances areas of smooth texture on RADARSAT-1 images. Oceanic databases and modeling results for winds, waves, and tidal elevation were used to provide information about environmental conditions at the time of RADARSAT-1 data acquisition. This information is crucial for discriminating between smooth texture areas caused by natural oil slicks and those having other causes. The locations of identified seeps are then combined with seismic and bathymetry data to identify geologic controls on natural hydrocarbon seepage (Miranda et al., 1997, 1998).

Raster polygons defining smooth-texture regions on USTC-classified RADARSAT-1 images were merged with geophysical and bathymetric data. Seven surface slicks were identified on the RADARSAT-1 (Figure 15). The slicks range in length from 4.6 to 11.3 km, with a mean value of 7.1 km and standard deviation of 2.04 km (Figure 15).

Spatial coincidence of surface slicks and geologic structure at the top of the Travosas Formation allowed inferences about the control of Cenozoic gravitational tectonics on natural seepage (Figure 15). Most surface slicks originated in the extensional domain from sources near listric normal faults (Figure 15). Only one surface oil slick was found in the compressional domain at greater water depths.

These results may be used to identify promising acreage for oil exploration. They also indicate potential locations for collection of sea-bottom cores for geochemical analysis.

CONCLUSIONS

An integration of molecular oil geochemistry, structural-stratigraphic, and remote-sensing data allowed us to construct models for two petroleum systems in the Foz do Amazonas area. The hydrocarbons sourced by the Upper Cretaceous marine carbonate and Tertiary marine deltaic petroleum systems in the Foz do Amazonas Basin indicate the presence of at least two active petroleum systems, with their respective kitchens buried deeply offshore. These data have also shown the similarity of the Foz do Amazonas and Niger Delta petroleum systems.

In addition, the presence of similar gravity tectonics in the Foz do Amazonas area and the Niger Delta suggests that the Foz do Amazonas area has the potential to be one of the most prolific petroleum provinces in the South Atlantic realm.

ACKNOWLEDGMENTS

We thank Drs. W. E. Galloway and Henrique L. B. Penteado for their helpful comments and suggestions during the review of this manuscript. We extend our thanks to Petrobrás and Drs. José Coutinho Barbosa and Irani Varella for their support and for permission to present and publish this paper in the 2000 Pratt II Conference. We thank C. M. Bentz (Petrobrás-Cenpes), C. H. Beisl (Petrobrás-Cenpes), J. A. Lorenzzetti (INPE), C. E. S. Araújo (INPE), and C. L. da Silva Jr. (Oceansat) for image processing and interpretation of RADARSAT and NOAA-AVHRR data. We also thank Ricardo Marins for all the artwork.

REFERENCES CITED

Asmus, H. E., and F. C. Ponte, 1973,The Brazilian marginal basins, *in* A. E. Nair and F. G. Stehli, eds., The ocean basins and margins: v. 1, The South Atlantic: New York, Plenum Press, p. 87–132.

Brandão, J. A. S., and F. J. Feijó, 1994, Bacia da Foz do Amazonas: Boletim de Geociências da Petrobrás, v. 8, p. 91–99.

Castro, J. C., K. Miura, and J. A. E. Braga, 1978, Stratigraphic and structural framework of the Foz do Amazonas Basin: 10th Annual Offshore Technology Conference, p. 1843–1847.

Doust, H., and E. Omatsola, 1990, Niger Delta, *in* J. D. Edwards and P. A. Santogrossi, eds., Divergent/passive margin basins: AAPG Memoir 48, p. 201–252.

Galloway, W. E. 1986, Growth faults and fault-related structures of prograding terrigenous clastic continental margins: Gulf Coast Association of Geological Societies Transactions, v. 36, p. 121–128.

Johannessen, J. A., G. Digranes, H. Espedal, O. M. Johannessen, P. Samuel, D. Browne, and P. Vachon, 1994, SAR ocean feature catalogue: ESA SP-1174, 106 p.

Magoon, L. B., 1990. The petroleum systems—A classification scheme for research, exploration, resource assessment, *in* L. B. Magoon, ed., Petroleum systems of the United States: U.S. Geological Survey Bulletin 1912, p. 2–10.

Mello, M. R., 1988, Geochemical and molecular studies of the depositional environments of source rocks and their derived oils from the Brazilian marginal basins: Ph.D. thesis, Bristol University, 240 p.

Mello, M. R., 1993, The hydrocarbon source potential of Brazil and West Africa marginal basins: Third International Congress of the Brazilian Geophysical Society, Expanded Abstracts, v. 2, p. 1306–1307.

Mello, M. R., P. C. Gaglianone, S. C. Brassel, and J. R. Maxwell, 1988a, Geochemical and biological marker assessment of depositional environment using Brazilian offshore oils: Marine and Petroleum Geology, v. 5, p. 205–223.

Mello, M. R., N. Telnaes, P. C. Gaglianone, M. I. Chicarelli, S. C. Brassel, and J. R. Maxwell, 1988b, Organic geochemical characterisation of depositional palaeoenvironments of source rocks and oils in Brazilian marginal basins, *in* L. Mattavelli and L. Novelli, eds., Advances in organic geochemistry 1987: Oxford, England, Pergamon Press, p. 31–45.

Mello, M. R., E. A. M. Koutsoukos, M. B. Hart, S. C. Brassel, and J. R. Maxwell, 1989, Late Cretaceous anoxic events in Brazilian continental margin: Organic Geochemistry, v. 14, p. 529–542.

Mello, M. R., W. U. Mohriak, E. A M. Koutsoukos, and J. C. A. Figueira, 1991, Brazilian and West African oils: Generation, migration, accumulation and correlation: Proceedings of the Thirteenth World Petroleum Congress, p. 153–164.

Mello, M. R., W. U. Mohriak, E. A. M. Koutsoukos, and G. Bacoccoli, 1994, Selected petroleum systems in Brazil, *in* L. B. Magoon and W. G. Dow, eds., The petroleum system—From source to trap: AAPG Memoir 60, p. 499–512.

Mello, M. R., N. Telnaes, and J. R. Maxwell, 1995, The hydrocarbon source potential in the Brazilian marginal basins: A geochemical and paleoenvironmental assessment, *in* A. Huc, ed., Paleogeography, paleoclimate, and source rocks: AAPG Studies in Geology 40, p. 233–272.

Mello, M. R., J. M. Moldowan, J. Dahl, and R. Requejo, 2001, Biological markers applied to the petroleum system approach, *in* M. R. Mello and B. Katz, eds., Petroleum systems of the South Atlantic: AAPG Memoir 73, p. 41–51.

Miranda, F. P., L. E. N. Fonseca, C. H. Beisl, A. Rosenqvist, and M. D. M. A. M. Figueiredo, 1997, Seasonal mapping of flooding extent in the vicinity of the Balbina Dam (Central Amazonia) using RADARSAT-1 and JERS-1 SAR data: Proceedings of the International Symposium Geomatics in the Era of RADARSAT (GER'97).

Miranda, F. P., C. M. Bentz, C. H. Beisl, J. A. Lorenzzetti, C. E. S. Araújo, and C. L. Silva Jr., 1998, Application of unsupervised semivariogram textural classification of RADARSAT-1 data for the detection of natural oil seeps offshore the Amazon Mouth: Proceedings of the RADARSAT ADRO Symposium, CD-ROM.

Moldowan, J. M., F. J. Fago, C. Y. Lee, S. R. Jacobson, D. S. Watt, N. E. Slougui, A. Jeganathan, and D. C. Young, 1990, Sedimentary 24-*n*-propylcholestanes, molecular fossils diagnostic of marine algae: Science, v. 247, p. 309–312.

Moldowan, J. M., J. Dahl, B. J. Huizinga, F. J. Fago, L. J. Hickey, T. M. Peakman, and D. W. Taylor, 1994, The molecular fossil record of oleanane and its relation to angiosperms: Science, v. 265, p. 768–771.

Peters, K., and J. M. Moldowan, 1993, The biomarker guide: Englewood Cliffs, New Jersey, Prentice Hall, 363 p.

Requejo, G., R. Sassen, E. Ukpabio, M. C. Kennicutt II, T. McDonald, M. C. Denous, and J. M. Brooks, 1995, Hydrocarbon source facies in the Niger Delta petroleum province as indicated by the molecular characteristics of oil: International Chemical Congress of Pacific Basins Societies Book of Abstracts, v. 7, p. 180.

Silva, S. R. P., and R. R. Maciel, 1998, Foz do Amazonas hydrocarbon systems, in M. R. Mello and P. O. Yilmaz, eds., AAPG International Conference and Exhibition, Río de Janeiro, Extended Abstracts, p. 480–481.

Silva, S. R. P, and J .B. M. Rodarte, 1989, Bacias da Foz do Amazonas e Pará-Maranhão (águas profundas): Uma analise sismoestratigrafica, tectono-sedimentar e termica: Annais I Congresso Sociedade Brasileira de Geofísica, v. 2 p. 843–852.

Silva, S. R. P., R. R. Maciel, and M. C. G. Severino, 1999, Cenozoic tectonics of Amazon Mouth Basin: Geo-Marine Letters, v. 18. p. 256–262.

Soldan, A. L., J. R. Cerqueira, J. C. Ferreira, L. A. F. Trindade, J. C. Scarton, and C. A. G. Corá, 1995, Giant deep water oil fields in Campos Basin Brazil: A geochemical approach: Revista Latino-Americana de Geoquímica Orgânica, v. 1, p. 14–27.

The Middle East and Africa

Versfelt Jr., P. L., 2001, Major hydrocarbon potential in Iran, *in* M. W. Downey, J. C. Threet, and W. A. Morgan, eds., Petroleum provinces of the twenty-first century: AAPG Memoir 74, p. 417–427.

Chapter 21

MAJOR HYDROCARBON POTENTIAL IN IRAN

Porter L. Versfelt Jr.
Versfelt & Associates, Ltd., Houston, Texas, U.S.A.

ABSTRACT

Huge hydrocarbon potential still exists in Iran, even after prolific production for many decades. Currently, Iran is operating about 28 drilling rigs (23 of them onshore) and is producing about 3.6 million barrels of oil per day and about 3 trillion cubic feet of gas per day from approximately 1200 wells in 39 fields. Since oil was discovered at Masjid-i-Suleiman in 1908, exploration has extended onshore and offshore to several provinces and regions of southwestern Iran. Oil production from the Zagros fold belt dominates Iranian hydrocarbon production. Zagros reservoirs are in Jurassic, Cretaceous, and Tertiary carbonates and siliciclastics. Despite decades of exploration and production, future exploration and field-development opportunities appear to abound, now aided by new technologies and tools.

INTRODUCTION

An Englishman named William Knox D'Arcy with an interest in Persian archaeology first learned of oil seeps at Chiah Surkh in Khuzestan in southwestern Persia (now Iran). In 1901, D'Arcy sent to Persia an English petroleum consultant who returned with a 60-year exclusive petroleum-rights agreement for most of Persia. In 1903, D'Arcy created a company and, by mid-1904, was drilling near the oil seeps at Chiah Surkh. Only noncommercial oil shows were found there. In spite of the high costs of this failed exploration, D'Arcy effectively farmed out exploration rights to Burmah Oil Company to form Concessions Syndicate Ltd.

In 1908, a successful test well was drilled at Masjid-i-Suleiman in southwestern Iran in the Miocene Asmari Limestone. At a depth of 1108 ft, this test flowed at more than 1000 barrels (bbl) of oil per day (BOPD). By the end of 1908, three wells were producing oil and some gas. This was the first commercial petroleum in the Middle East. Then in 1909, the D'Arcy-Burmah Syndicate was taken over and became the Anglo-Persian Oil Company, the predecessor of British Petroleum. In 1914, Winston Churchill arranged for the British government's purchase of the Anglo-Persian Oil Company to better safeguard a supply of oil for its navy.

In 1951, the Iranian government nationalized Anglo-Persian. In 1953, "the Consortium" (then known as the Iranian Oil Exploration and Production Company, or IOEPC) was formed by British Petroleum, Shell, Exxon, Gulf, Mobil, Chevron, Texaco, Total, ARCO, Aminoil, Getty, Signal, Continental, and Sohio.

In the 1970s, the Consortium operated more than 60 drilling rigs and produced more than 6 million BOPD in Iran, making the country one of the world's major producers in overall production, well-flow rates, reserves, and size of producing fields. Advancements in development and enhanced oil-recovery plans and designs for facilities were soon to increase production to 8.5 million BOPD. In the 24 years prior to the Islamic Revolution of the late 1970s, the Consortium produced more than 24 billion bbl of oil. Cumulative oil production is estimated at 50.9 billion bbl. Proven onshore and offshore reserves from the Zagros, Caspian, and other petroliferous provinces are estimated at 99.7 billion bbl of oil and 812.3 trillion cubic feet (tcf) of gas. Recent discoveries and large foreign investments in Iran have resulted in the possibility that production rates will reach earlier high levels. Credible production estimates by the U.S. Department of Energy's Energy Information Agency are 4.8 million BOPD by 2005 and as much as 7.3 million BOPD by 2020.

The typical oil field in the Zagros fold belt contains 5 billion bbl of oil in place, and many have recoverable reserves in excess of 1 billion bbl of oil. The flow rates of most Zagros fold-belt wells are more than 10,000 BOPD. Many wells still test from 50,000 to 100,000 BOPD. Today, about 28 drilling rigs (23 of them onshore) are operating in Iran. Production estimates range from 3.94 million BOPD (U.S. Department of Energy, Energy Information Administration) to 4.05 million BOPD (*Oil & Gas Journal*) from about 1200 wells in 36 fields, both onshore and offshore. *World Oil* now reports oil production at 3 million BOPD.

STUDY AREA

Iran has several geologic basins, but important commercial hydrocarbon production has been developed in only two: the South Caspian Sea depression in the north and the Zagros fold belt in the southwest (Figure 1; Table 1). Nonproductive basins, including the Gavkhuni, Kavir, Khur, Kopt Dagh, Lut, Makran, and Yazd Basins of central and southeastern Iran, differ in hydrocarbon prospectivity. The most likely near-future opportunities for new development and exploration rights are expected to be in the Zagros fold belt. Thus, this chapter focuses on that region.

The Alborz Mountains dominate the northern part of the country, north of which is the coast of the Caspian Sea. Central Iran consists of a central plateau bounded on the east by the Lut Block, which adjoins the Baluch Ranges to the southeast. The Zagros Mountains dominate southwestern Iran, extending from the northwest to the Makran Ranges in the south. The Persian Gulf coast is a low-relief coastal plain interrupted by salt domes. This coastal terrain and the Zagros foothills to the northeast are the focus of this discussion.

TECTONIC SETTING

The northwest-southeast-trending Zagros orogenic fold belt extends about 2000 km from southeastern Turkey to southern Iran. This fold belt was created by the collision of the Arabian Plate with part of the Eurasian margin beginning in the late Miocene and continuing today (Figure 2). This collision formed the Zagros Foreland Basin. Giant and supergiant oil fields are found there in long, asymmetrically thrusted anticlinal folds created by the Miocene compression. A collision, or "crush," zone

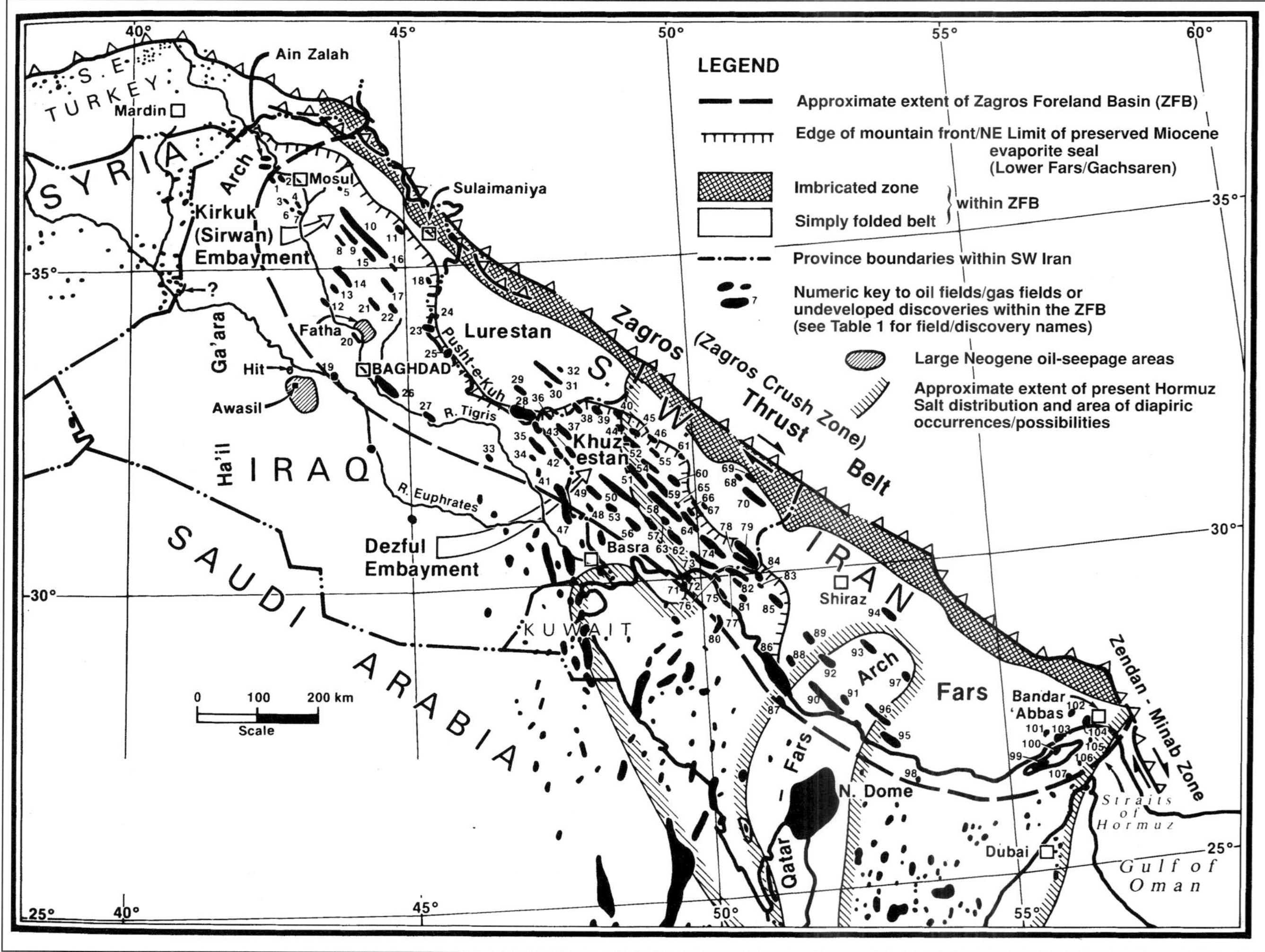

Figure 1. Regional oil and gas map showing locations, provinces, and key structural elements in the Zagros Foreland Basin. From Beydoun et al., 1992. See Table 1 for numeric key to fields and discoveries.

bounds the fold belt on the northeast. Folding intensity decreases toward the Persian Gulf, and outcrops become progressively younger in that direction.

Evolution of the Zagros orogenic fold belt began with the development of northeast-southwest-oriented sutures in the Arabian shield during the Precambrian. Collisions of the Arabian continental basement with offshore island arc terranes created the foundation for the northwest-southeast Najd fault system, along which the Zagros trend formed.

Infra-Cambrian subsidence as a result of pre-Tethys rifting led to development of the proto-Tethys realm and continental-rift basins in which thick deposits of Hormuz Salt accumulated (Figure 3). In the Ordovician, movement of the Hormuz Salt occurred along the eastern flanks of the developing basin. Widespread deposition of transgressive shale source rocks occurred during the Silurian.

The Devonian was a period of uplift and the erosion of continental-shelf strata. The collision of Laurasia and Gondwana to form Pangea began with Hercynian orogenic compression of middle to late Paleozoic strata, resulting in a regional unconformity at the Carboniferous-Permian boundary.

Breakup of Gondwana and Pangea began during the Triassic and resulted in erosional unconformities at the tops of the Triassic and Jurassic sequences. The paleo-Tethys sea was consumed, and the neo-Tethys sea was created as the Cimmerian microplate moved away from northern Arabia during rifting. A carbonate ramp formed on the north-facing passive continental margin. During Middle Jurassic extension, the Lut microcraton moved northeastward, then northward, as carbonates were deposited. Uplift, erosion, and evaporitic conditions occurred over much of this region in the Late Jurassic.

By the Late Cretaceous, the Cimmerian microplate had started to collide with the Eurasian Plate, resulting in subduction of the neo-Tethys and volcanic activity in the Tethyan sea. Widespread post-Cenomanian, post-Sarvak, and post-Turonian unconformities formed. A post-Cretaceous unconformity between the Gurpi and Pabdeh units was also widespread.

The early to middle Tertiary was a time of tectonic quiescence characterized by evaporite deposition. This was followed by north-northeast extension, magmatism, volcanism, and uplift during the Miocene. Local unconformities developed within Oligocene Asmari intervals, and widespread unconformities occurred below the Pliocene Bakhtyari Conglomerate that today crops out over much of the region.

Creation of the Zagros foreland fold-and-thrust belt left only the Persian Gulf as a remnant of the Tethys seaway. Southward thrusting over the northern Arabian passive margin created the overlying foreland basin. Loading of the foreland basin caused the Miocene Fars evaporites to facilitate Zagros foreland folding in Laurestan and Khuzestan Provinces. Farther southeast in Fars Province, the Cambrian Hormuz Salt flowed diapirically and even

Table 1. Numeric key to oil fields/gas fields or undeveloped discoveries in the Zagros Foreland Basin. See Figure 1 for locations. From Beydoun et al. (1992).

No.	Field	No.	Field	No.	Field	No.	Field
1	Sasan	28	Delhuran	55	Masjid-i-Suleiman	83	Sulabedar
2	Alan	29	Samand	56	Mansuri	84	Chillingar
3	Jawan	30	Halush	57	Shadegan	85	Nargesi
4	Qasab	31	Veyzenhar/Malah Kuh	58	Marun	86	Kuh-e-Mand
5	Demir Dagh			59	Kupal	87	Pars
6	Najmah	32	Sarkan	60	Haft Kel	88	Kuh-e-Kaki
7	Qaiyarah	33	Dujailah	61	Par-e-Siah	89	Bushgan
8	Khabbaz	34	Nur	62	Ramshir	90	Kangan
9	Bai Hassan	35	Buzurgan	63	Ramshir (gas)	91	Nar
10	Kirkuk	36	Danan	64	Agha Jari	92	Dalan
11	Chemchemal	37	Cheshmeh Khush	65	Khavizi	93	Aghar
12	Tikrit	38	Kabud	66	Karanj	94	Sarvestan
13	Saddam	39	Qaleh Nar	67	Paris	95	Lamard
14	Hamrin	40	Lab-e-Safid	68	Doudrou	96	Varavi
15	Jambur	41	Al Halfayah	69	Kuh-i-Rig	97	Bandubast
16	Kor Mor	42	Jabal Fauqui	70	Shurom	98	"T"
17	Pulkhana	43	Peydar	71	Bahrgansar	99	Salakh
18	Chiah Surkh	44	Palangan	72	Hendijan	100	Gavarzin
19	Fallujah	45	Lali	73	Rag-e-Safid	101	S. Gashu
20	Balad	46	Karun	74	Pazanan	102	West Namak
21	Injana	47	Majnoun	75	Binak	103	Suru
22	Gilabat	48	Jufeyr	76	Siah Makan	104	Sarkhun
23	Naft Khaneh/Naft-i-Shahr	49	Susangerd	77	Bibi Hakimeh	105	HD
		50	Ahwaz	78	Garangan	106	Hanquan
24	Emmam Hassan	51	Ramin	79	Gachsaran	107	Henjam
25	Tang-e-Bijar	52	Zeloi	80	Douroud/Kharg		
26	East Baghdad	53	Ab Teymour	81	Gulkhari		
27	Ahdab	54	Naft Safid	82	Kilurkarim		

breached the surface (Figure 4). More than 200 piercement salt diapirs now occur in the region, some forming mountains as much as 1200 m high and others flowing downslope as salt glaciers.

STRATIGRAPHY

In the Khuzestan foreland fold belt, the stratigraphic section ranges from middle Mesozoic to Cenozoic in age. In Fars Province, sedimentary deposits are from Precambrian to Cenozoic in age (Figure 3), and the total thickness of the stratigraphic section ranges from 10 to 15 km. The oldest rocks in Fars Province are Precambrian metamorphosed sedimentary rocks and granitic intrusives.

The Hormuz, Barut, and Zaigun Formations are Cambrian sedimentary units consisting of dolomites, shales, salt, and sandstones with a combined thickness of 3000 m. The Cambrian Lalun and Ilebeyk Formations are red sandstones and quartzites overlain by carbonates and shales. Ordovician rocks are gray to green shales and sandstones of the Zard Kuh Formation. Above the supra-Ordovician unconformity is the Carboniferous Faraghan Formation comprising dark carbonates and sandstones. Lying conformably above it are the Permian Deh Ram Group, including the dark carbonates and purple to gray shales of the Dalan Formation.

Triassic rocks consist of the Dashtak and Khaneh Formations, which are thin dolomites and sandstones with

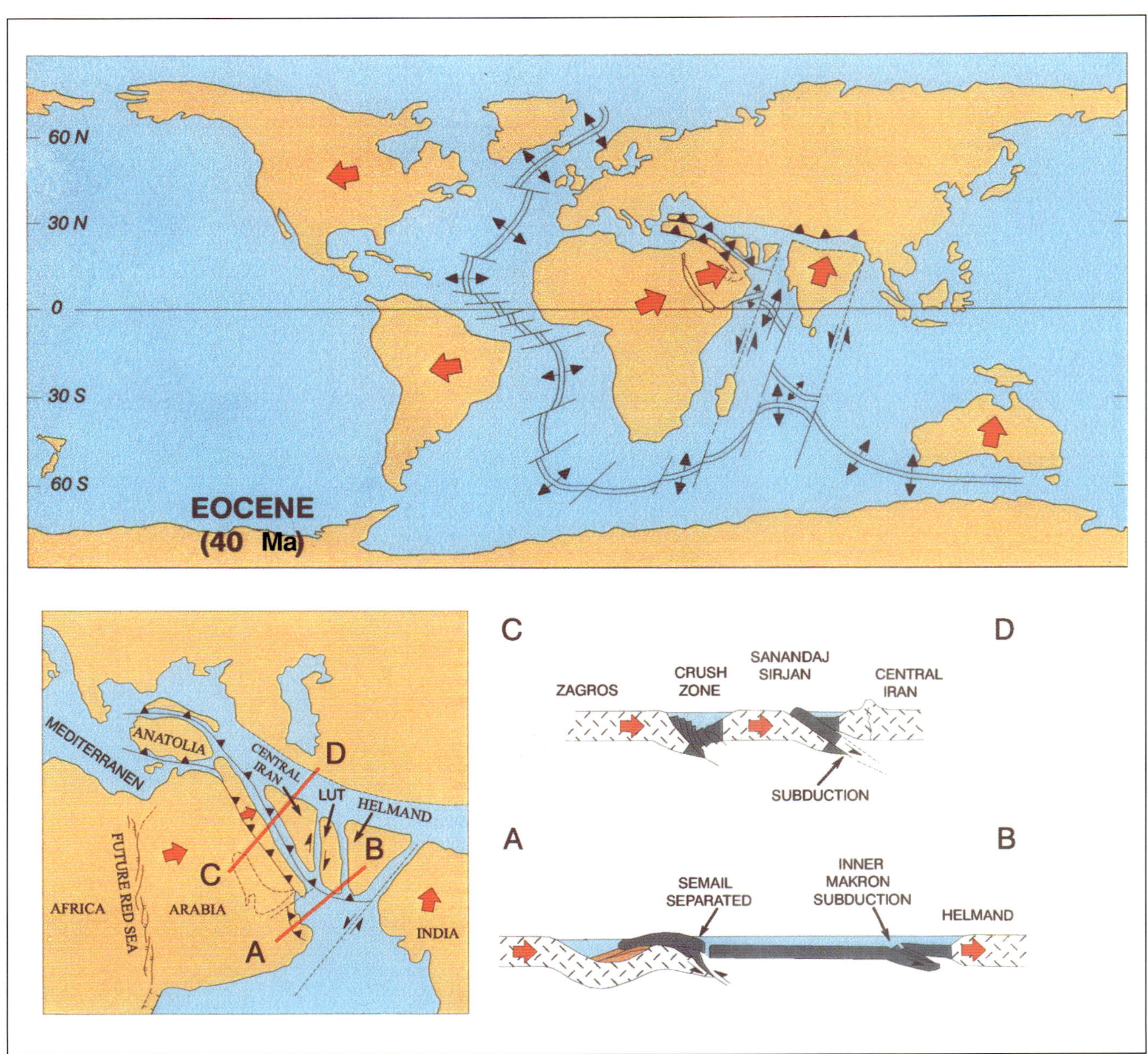

Figure 2a. Eocene plate-tectonic reconstruction showing development of the Zagros Mountains.

local anhydrite and coal. An Upper Triassic unconformity, marking the Cimmerian microplate collision with Europe, forms the lower boundary of the overlying Jurassic Surmeh Formation of the Khami Group. This formation is mainly dolomite, with local anhydrite, gypsum, and coal. Another unconformity bounds the top of the Jurassic sequence.

The Lower Cretaceous (Neocomian) sequence is represented by the Fahlyian, Gadvan, and Dariyan limestones, sandstones, and shales, which are of marine origin. An unconformity separates the Dariyan Formation from the organic-rich shales of the overlying Kazhduni Formation. Upper Cretaceous (Cenomanian-Maestrichtian) strata consist of the Bangestan Group, including the limestones of the Sarvak Formation. Separated by an unconformity are the overlying Upper Cretaceous Illam Formation carbonates and the Gurpi-Tabur Formation, an organic-rich shale with limestones and dolomites. Bangestan carbonates range from 700 to about 1000 m in thickness.

Paleocene Radhuma limestones, Pabdeh marine shales, and local Sachu evaporites rest unconformably on Maestrichtian Gurpi shales in central Khuzestan. Radhuma and Pabdeh sediments extend upward into the Eocene, which is terminated by a regional unconformity caused by the uplift associated with thrusting and folding of the Zagros Mountains. Zagros orogenesis was caused by the opening of the Red Sea and consequent northward

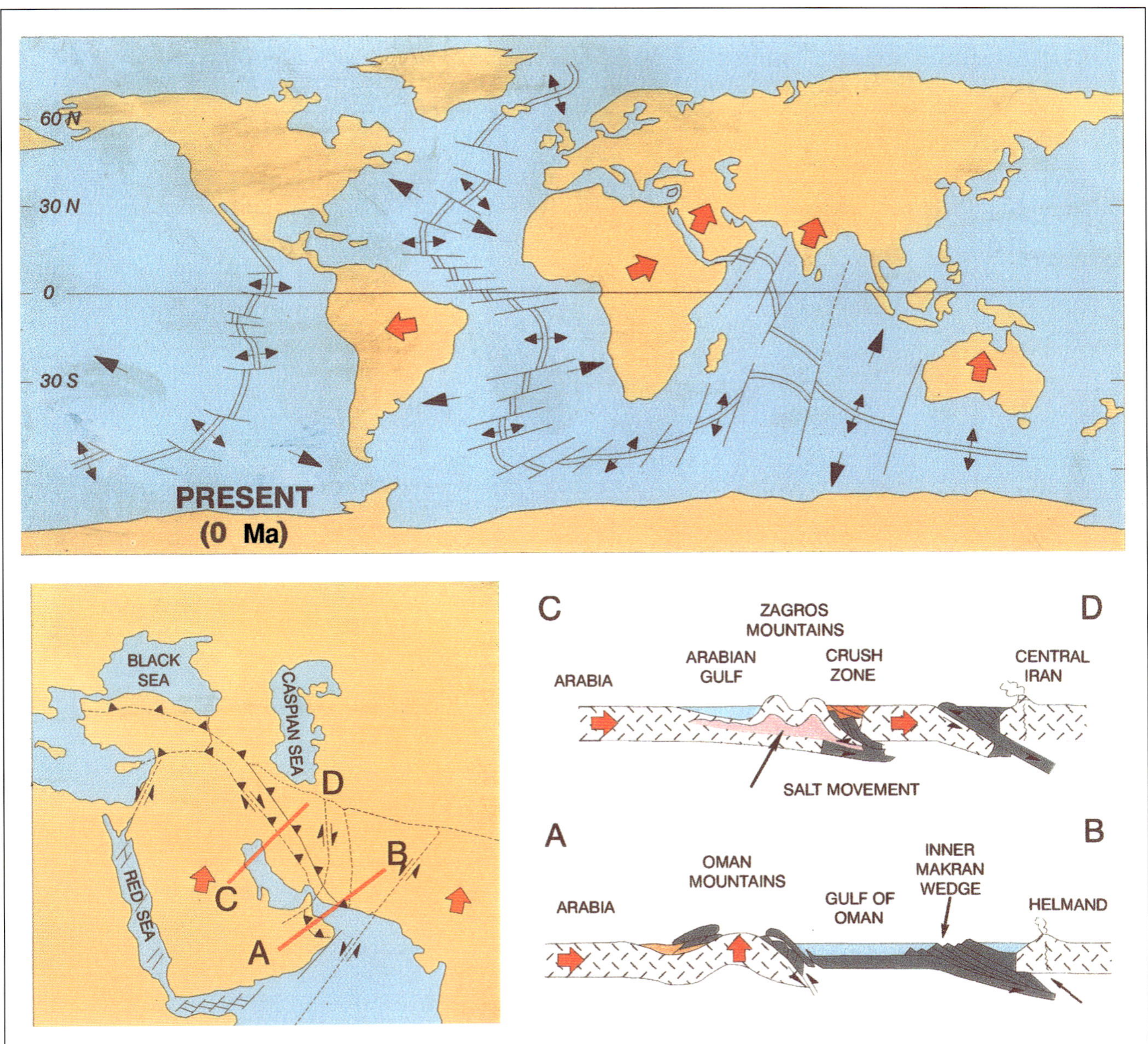

Figure 2b. Present-day plate-tectonic setting showing development of the Zagros Mountains.

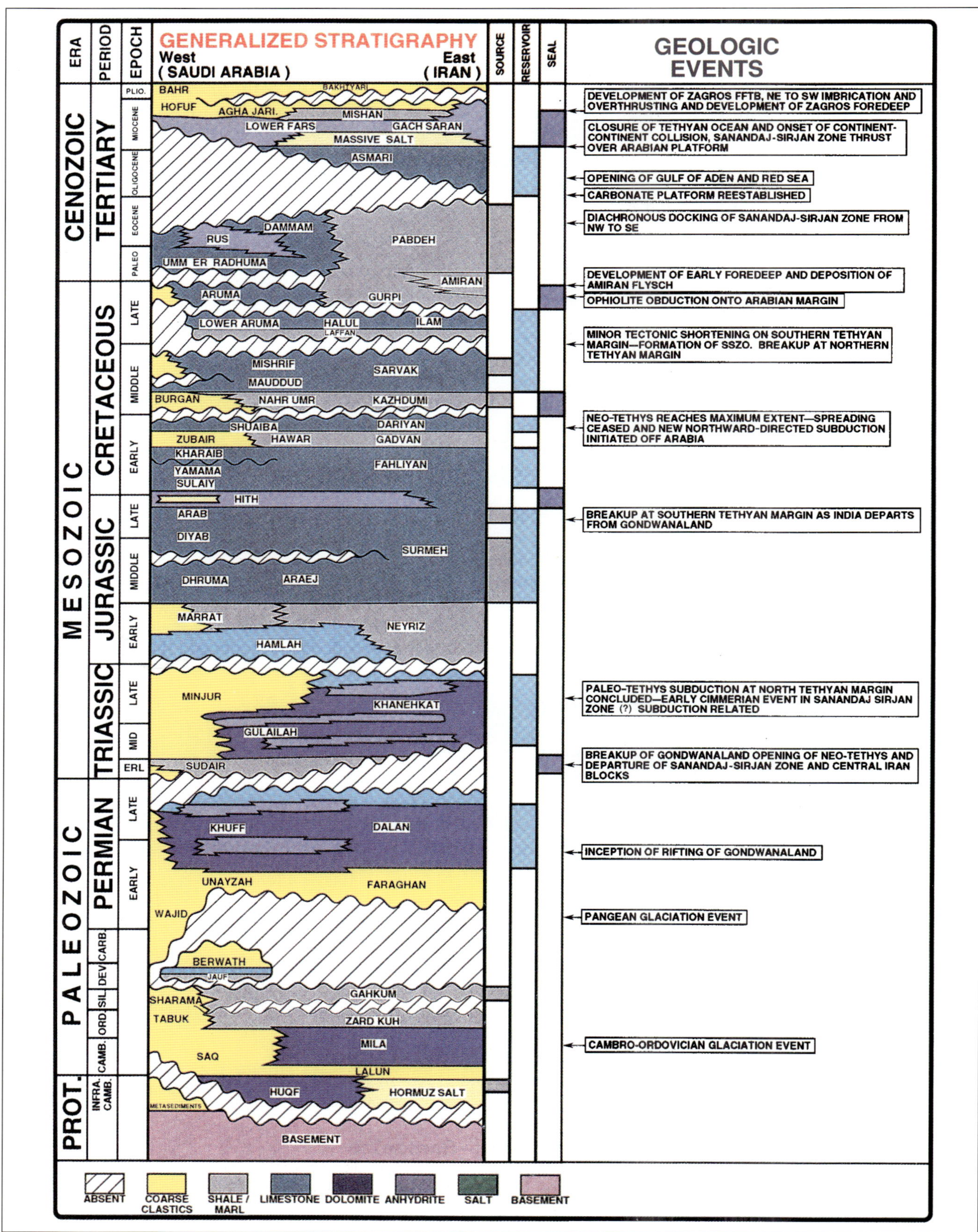

Figure 3. Stratigraphic column for the study area. From Hooper et al., 1994. Reproduced by permission of Gulf PetroLink.

collision of the Arabian Plate with the Iranian portion of Eurasia.

The Oligocene Asmari marine limestones, which include minor sandstone and shale members, are the most important hydrocarbon reservoirs in the Zagros fold belt. The Asmari carbonate and siliciclastic section ranges from about 300 to 500 m in thickness. Conformably sealing the Asmari carbonate reservoirs is the Miocene Fars Group, which consists of evaporites of the Gachsaran Formation and shales of the Razak Formation. Shaly carbonates of the Mishan Formation and sandstones and shales of the Agha Jari Formation were deposited in the shallow-marine environment of the later Miocene to early Pleistocene. The Mishan Formation is about 600 m thick, whereas the Agha Jari is about 1500–3000 m thick. Widespread Pleistocene Bakhtyari conglomerates overlie the Miocene strata. Quaternary to Holocene alluvium blanket much of the Bakhtyari and older deposits.

SOURCE ROCKS

Laurestan and Khuzestan Provinces

Cambrian supra-Hormuz Formation carbonates and siliciclastics are considered to be the earliest source rocks in the Laurestan and Khuzestan Provinces of the Zagros fold belt (Figure 5). Silurian Gahkum siliciclastics are identified as source rocks in the central regions of Khuzestan. Jurassic Sargelu Formation carbonates are sources in both Laurestan and Khuzestan Provinces.

In central Khuzestan Province, known source rocks include the Lower Cretaceous Garau and Kazhduni Formations, which are mature oil-prone shales and carbonates. The Kazhduni is considered to be the major source rock for the Dezful embayment's prolific reservoirs in the Upper Cretaceous Sarvak carbonates and the Oligocene Asmari carbonates and siliciclastics. Upper Cretaceous Gurpi organic-rich shales are also thought to be a hydrocarbon source in the same area. Total-organic-carbon (TOC) values of 6.9% have been reported in Upper Cretaceous shales. Paleocene-Eocene Pabdeh shales are organic-rich sources, primarily farther northwest in Laurestan Province.

Fars Province

Supra-Hormuz carbonates and siliciclastics are also suspected as source rocks in Fars Province and southern regions. The Silurian Ghkumm shales are considered to be overmature and gas-prone source rocks. Middle Jurassic Sargelu and Izhara carbonates are known sources. Lower Cretaceous Shu'alba-Dariyan carbonates are an oil-prone source but are less rich there than in Khuzestan Province. Most TOC values are less than 0.5%. Middle Cretaceous Shilaif Formation carbonates are source rocks in Fars Province.

Upper Cretaceous Laffan Formation shales, although organic rich, are considered to be immature. Paleocene-Oligocene Pabdeh shales contain marine algal organic matter, but they also appear to be immature.

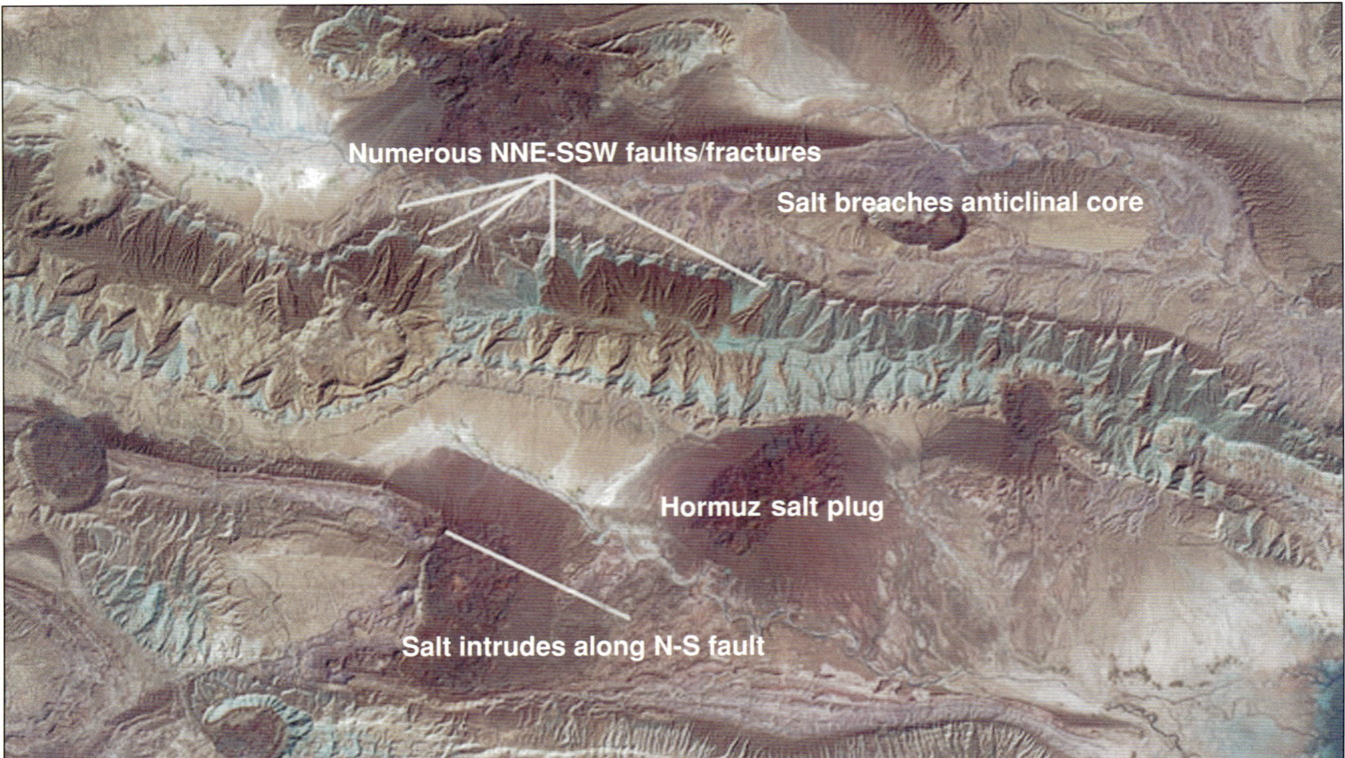

Figure 4. Satellite image showing the Hormuz salt plug.

HYDROCARBON TYPES

Oils in the Zagros fold belt range from 19.0° API at Soroosh field to 44.5° API at Naft Safied field. Cretaceous Bangestan Group (Sarvak Formation) oils range from 25° to 38° API. Oils from the Oligocene Asmari carbonates range from 28° to 36° API, whereas the Asmari sandstone oils are from 30° to 32° API. Oils at Bibi Hakimeh produced from both the Sarvak and Asmari are 29.9° API, perhaps attesting to the vertical fracture network that facilitates intercommunication of these reservoirs. The sulfur content of the Asmari reservoir oils ranges from about 1.5% to 3.5%.

RESERVOIRS AND SEALS

Khuzestan Province

Hydrocarbon accumulations are found in Permian and Jurassic carbonates in Khuzestan and Fars Provinces (Figures 1, 3, and 5). In several fields in Khuzestan Province, gas production has been established from Lower Cretaceous Khami marine carbonate reservoirs. These reservoirs are sealed by Lower Cretaceous Garau, Gadvan, and Kazhduni Formation shales.

Middle Cretaceous Sarvak marine carbonate oil-producing reservoirs (of the Bangestan Group) are sealed by Upper Cretaceous Illam carbonates and the overlying Gurpi shales. The Ahwaz, Bibi Hakimeh, Binak, Gachsaran, Marun, Ramshir, and Rag-e-Safid fields all produce from Sarvak Formation reservoirs. Hydrocarbon columns of more than 900 m have been encountered at Bibi Hakimh, and the column at Gachsaran is more than 1800 m, primarily because of extensive vertical communication through fractures and faults. Vertical reservoir communication can be demonstrated through shale beds as much as 1000 m in thickness. Fractured Upper Cretaceous Illam carbonates are also considered to be potential reservoirs.

Oligocene–lower Miocene Asmari fractured limestones, dolomites, and sandstones are the most important oil reservoirs in the Zagros fold belt. The Asmari Formation ranges from 100 to 800 m in thickness and contains several siliciclastic layers with relatively high porosity and permeability. Typical Asmari porosities range from about 5% to 15%, and permeabilities range as high as approximately 30 md. It is the fracture-and-fault network that provides the avenues of effective permeability, often exceeding several darcys. Approximately 90% of oil production in the Zagros fold belt is from Asmari reservoirs. These reservoirs are sealed by the Miocene Gachsaran evaporites, which range in thickness from 600 to 1200 m. Production has been developed in the Asmari at the Ahwaz, Agha Jari, Gachsaran, Marun, and Rag-e-Safid fields and in many other fields in the region.

Fars Province

In Fars Province, Permian Dalan Formation limestones and dolomites are established gas reservoirs. Triassic Kangan Formation limestones and dolomites are also gas reservoirs. Both the Dalan and Kangan reservoirs are sealed by the Triassic Dashtak evaporites. Lower Cretaceous Fahliyan limestones are gas reservoirs at Sallakh, Suru, and Kharg Island, and are sealed by the Kazhduni shales. Neritic limestones of the middle Cretaceous Sarvak Formation are reservoirs in Bibi Hakimeh and Servestan fields and are sealed by the middle Cretaceous Illam and Gurpi shales.

Asmari limestone and sandstone reservoirs of Oligocene age are found at Ahwaz, Bibi Hakimeh, Gulkari, Kilurkkarim, and Nangesi fields. Gachsaran Formation evaporites of the Miocene Fars Group seal these fractured reservoirs.

TRAPS

Classic anticlinal traps abound along the Zagros fold belt from Laurestan and Khuzestan to Fars Province. Secondary stratigraphic and fault-bound traps also occur on and below these folded and thrusted structures. Porosity and permeability barriers form traps on these anticlines, which are often highly fractured. In Fars Province, salt halokinesis has created a multitude of traps against the salt pillars and folds, as well as stratigraphic and facies traps along their flanks.

DEVELOPMENT POTENTIAL

Numerous subtle development possibilities exist in fields of the Zagros fold belt. In established reservoirs, many identified potential productive anticlinal targets have yet to be fully exploited. Both above and below established reservoirs are zones that are identified or suspected to have development potential. Most thrusted structures in the Zagros fold belt are asymmetrical, with the southwestern flanks being steeper. These flanks are often unexplored or left undeveloped. Below many well-established reservoirs, especially below the steeply dipping southwestern flanks, seismic data suggest the presence of undeveloped fault-bound and stratigraphic traps.

Extensive fracture networks aid primary production in most reservoirs. However, faults and low porosity and permeability barriers resulting from diagenetic and facies changes often block intrareservoir communication. Overall recovery rates for primary oil production range from 18% to 26%, with some fields having recovery rates as low as 7.5% and some as high as 42%.

Enhanced oil-recovery (EOR) techniques often require creation of multiple development cells. Secondary EOR methods were in operation as early as 1976 in sev-

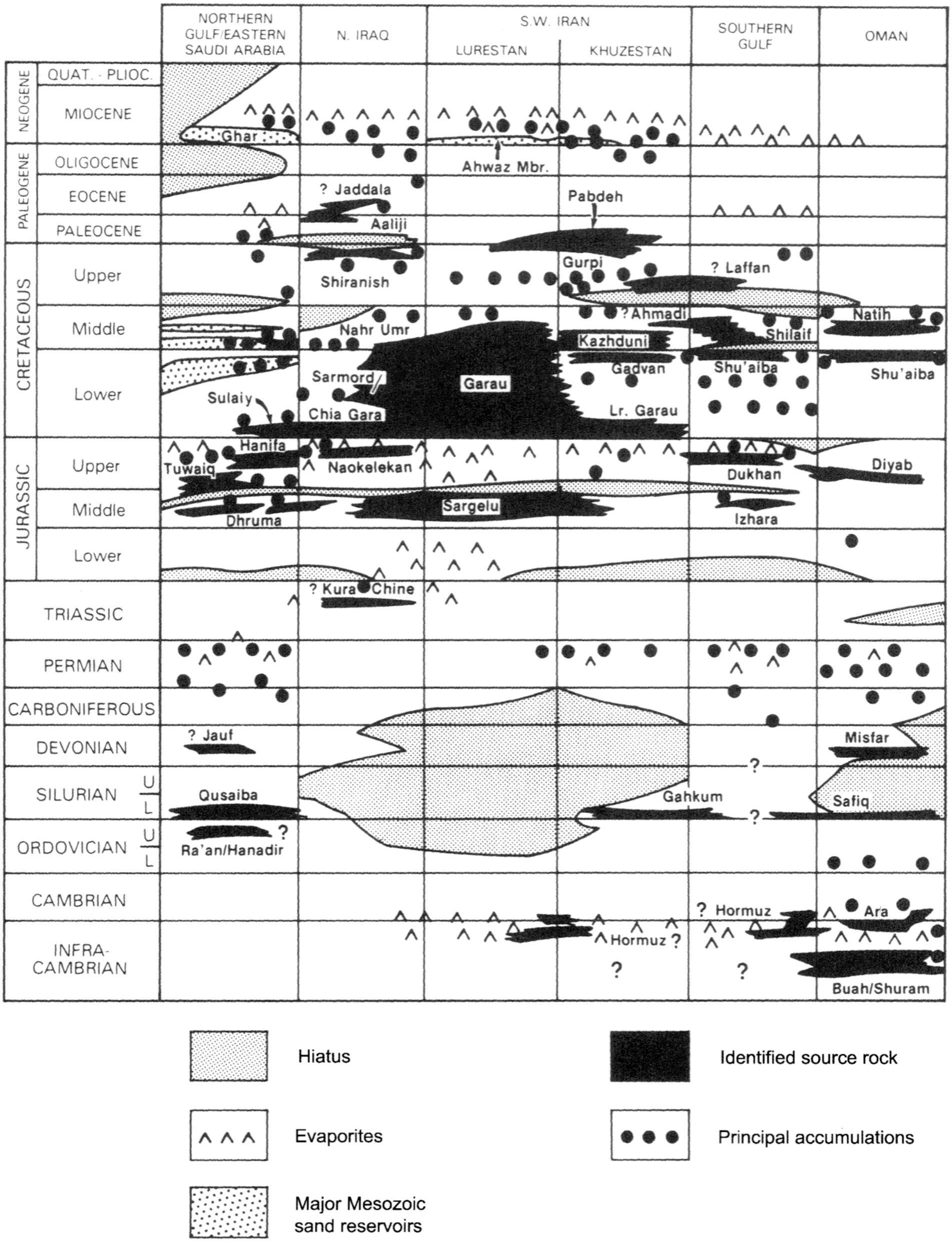

NORTHERN GULF/EASTERN SAUDI ARABIA
N. IRAQ
S.W. IRAN
LURESTAN
KHUZESTAN
SOUTHERN GULF
OMAN
NEOGENE
QUAT. - PLIOC.
MIOCENE
PALEOGENE
OLIGOCENE
EOCENE
PALEOCENE
CRETACEOUS
Upper
Middle
Lower
JURASSIC
Upper
Middle
Lower
TRIASSIC
PERMIAN
CARBONIFEROUS
DEVONIAN
SILURIAN
U
L
ORDOVICIAN
U
L
CAMBRIAN
INFRA-CAMBRIAN
Ghar
Ahwaz Mbr.
? Jaddala
Pabdeh
Aaliji
Gurpi
? Laffan
Shiranish
? Ahmadi
Natih
Nahr Umr
Shilaif
Kazhduni
Gadvan
Shu'aiba
Shu'aiba
Sarmord
Garau
Sulaiy
Chia Gara
Lr. Garau
Hanifa
Tuwaiq
Naokelekan
Dukhan
Diyab
Sargelu
Dhruma
Izhara
? Kura Chine
? Jauf
Misfar
?
Qusaiba
Gahkum
Safiq
?
?
Ra'an/Hanadir
? Hormuz
Ara
Hormuz ?
?
?
Buah/Shuram
Hiatus
Identified source rock
Evaporites
Principal accumulations
Major Mesozoic sand reservoirs

Figure 5. Stratigraphic position of the identified source rocks in the northeastern Arabian shelf region and Oman. From Beydoun et al., 1992; expanded and modified from Stoneley, 1990. Reproduced by permission of the Geological Society (London).

eral major and a few minor fields in the Zagros fold belt. Several major fields are undergoing gas injection, and more are being planned for that procedure. Additional EOR projects, ranging from simple dump waterfloods to gas injection, are possible for many fields. Exotic tertiary EOR techniques and methods, such as carbon dioxide, steam, fire floods, and microbial techniques, have the potential to recover significant quantities of remaining hydrocarbons.

EXPLORATION POTENTIAL

Despite nearly a century of exploration, the Zagros fold belt, both onshore and offshore, is still underexplored. Consider that only about 1200 wells now produce more than 3.6 million BOPD and that only a relatively few exploration wells have been drilled since the late 1970s. Early exploration began at the site of surface seeps and expanded to mapped prominent surface anticlinal folds. Surface exploration continued until the 1930s, when seismic data aided in subsurface interpretation. Even with seismic data, however, the true configuration of some structures had to be revealed by drilling.

Because most of the long anticlinal folds are asymmetrical, early exploration often missed the crest of the structures. An example is the Ahwaz field, where it took six exploration wells to discover the oil-rich crest. Modern seismic-data acquisition and processing and fault-fold interpretation models will undoubtedly improve future exploration efforts in the Zagros fold belt.

Despite decades of exploration, the potential for new discoveries is still considered to be great. Many surface structures have yet to be explored or are underexplored, including some impressive structures tens of kilometers long. In some cases, those structures were drilled in the early days without seismic data or adequate mud-logging equipment to detect hydrocarbon shows. Known surface oil and gas seeps are associated with many of those surface structures and need to be considered in the exploration prospects.

To further evaluate these onshore and offshore exploration prospects, the industry has some useful newer technologies available, including 3-D seismic techniques, satellite imagery, software applications, horizontal and multilateral completions, workover rigs, and logging tools. For example, satellite imagery today provides excellent interpretation of surface structures and tectonics, with resolution down to 1 m. This imagery can also detect previously unknown surface seeps. Offshore seeps of oil and gas can be detected with satellite radar, ultraviolet imagery, and airborne laser fluorosensor surveys. Special processing of satellite imagery onshore can show hydrocarbon indicators caused by shifts in redox conditions, associated secondary mineralization, spectral brightness of surface geology, and chlorophyll changes in vegetation. Landsat multispectral scanner (MSS), thematic mapper (TM), infrared bands, and SPOT imagery can show surface anomalies caused by subsurface hydrocarbon deposits. All of these changes are known to be caused by ascending microseeps of oil and gas.

New offshore techniques can also greatly aid exploration along the Zagros fold-belt trend in the Persian Gulf. Surface-imagery analysis of the Persian Gulf water surface can be processed to reveal oil and, in some cases, gas seeps that may lead to identification of prospective plays. Satellite radar altimetry and gravity data can be processed to reveal water-surface topographic highs and lows associated with seafloor topographic features that may aid in exploration. Offshore in the Fars Province region, several such anomalies appear to be structural features, including intrusive salt bodies that may be prospective for gas. In addition, some offshore prospective trends can be projected from onshore productive structural trends with seismic data and may lead to development of viable prospects.

As an example of the remaining exploration potential in the Zagros fold belt, in September 1999, the National Iranian Oil Company (NIOC) announced discovery of a giant field—Azadegan, with 26 billion bbl oil in place (5–6 billion bbl recoverable), west of Ahwaz. For perspective on the size of this reported discovery, 26 billion bbl oil in place is more than the total reserves of China! Another recent success has been achieved offshore in the Persian Gulf, where NIOC has 20 large gas discoveries in the South Pars region near Qatar. The South Pars gas field alone has proven reserves of about 280 tcf. In addition, Sirri C and D oil fields offshore near Abu Dhabi have total reserves of 1.2 billion bbl oil in place.

REFERENCES CONSULTED

Ala, M. A., 1974, Salt diapirism in southern Iran: AAPG Bulletin, v. 58, p. 1758–1770.

Alavi, M., 1994, Tectonics of the Zagros orogenic belt of Iran: New data and interpretations, Tectonophysics, v. 229, p. 211–238.

Beck, R., 1998, Worldwide petroleum industry outlook: Tulsa, Oklahoma, PennWell Books, p. 31–270.

Beydoun, Z. R., 1991, Arabian Plate hydrocarbon geology and potential—A plate tectonic approach: AAPG Studies in Geology no. 33, 77 p.

Beydoun, Z. R., M. W. Hughes Clarke, and R. Stonely, 1992, Petroleum in the Zagros Basin: A Late Tertiary foreland basin overprinted onto the outer edge of a vast hydrocarbon-rich Paleozoic-Mesozoic passive-margin shelf, *in* R. W. Macqueen and D. A. Leckie, eds., Foreland basins and fold belts: AAPG Memoir 55, p. 309–339.

Christian, L., 1997, Cretaceous subsurface geology of the Middle East region: GeoArabia, v. 2, p. 239–256.

Hooper, R. J., I. R. Baron, S. Agah, and R. D. Hatcher Jr., 1994, The Cenomanian to recent development of the southern Tethyan margin in Iran, *in* M. J. Al-Husseini, ed., GEO '94:

The Middle East petroleum geosciences: Selected Middle East papers from the Middle East Geoscience Conference, April 25–27: Manama, Bahrain, Gulf PetroLink, v. 2, p. 505–516, Figure 3.

James, G. A., and J. G. Wynd, 1965, Stratigraphic nomenclature of Iranian oil consortium agreement area: AAPG Bulletin, v. 49, p. 2182–2245.

Owen, E. W., 1975, Trek of the oil finders: A history of exploration for petroleum, AAPG Memoir 6, 1647 p.

Sengör, A. M. C., 1984, The Cimmeridge orogenic system and the tectonics of Eurasia: Geological Society of America Special Paper 195.

Sengör, A. M. C., 1987, Tectonic subdivisions and evolution of Asia: Bulletin of the Technical University of Istanbul, v. 40, p. 355–435.

Sengör, A. M. C., D. Altiner, A. Cin, T. Ustaomer, and K. J. Hsu, 1988, Origin and assembly of the Tethyside orogenic collage at the expense of Gondwanaland: Geological Society of London Special Publication 37, p. 119–181.

Stoneley, R., 1990, The Middle East Basin: A summary overview, *in* J. Brooks, ed., Classic petroleum provinces: The Geological Society (London) Special Publication 50, p. 293–298, Figure 2.

U. S. Department of Energy, 1996, Oil production capacity expansion costs for the Persian Gulf: Washington, D.C., U. S. Department of Energy.

Rusk, D. C., 2001, Libya: Petroleum potential of the underexplored basin centers—A twenty-first-century challenge, *in* M. W. Downey, J. C. Threet, and W. A. Morgan, eds., Petroleum provinces of the twenty-first century: AAPG Memoir 74, p. 429–452.

Chapter 22

Libya: Petroleum Potential of the Underexplored Basin Centers—A Twenty-first-century Challenge

Donald C. Rusk
Consultant, Houston, Texas, U.S.A.

ABSTRACT

Recoverable reserves in approximately 320 fields in Libya's Sirt, Ghadamis, Murzuq, and Tripolitania Basins exceed 50 billion barrels of oil and 40 trillion cubic feet of gas. Approximately 80% of these reserves were discovered prior to 1970. Since then, there has been a less active and more conservative exploration effort. Complex, subtle and, in particular, deep plays were rarely pursued during the 1970s and 1980s because of definitive imaging technologies, limited knowledge of the petroleum systems, high costs, and risk adversity.

Consequently, extensive undiscovered resources remain in Libya. These resources could be accessed if geologic and geophysical knowledge, innovation, and advanced technologies were used effectively. Three-dimensional seismic acquisition will be required to some degree for reliable trap definition and stratigraphic control.

Predictably, most of the undiscovered resources will be found in the vast, underexplored deep areas of the producing basins. Six areas are exceptional in this regard: the south Ajdabiya trough, the central Maradah graben, and the south Zallah trough–Tumayam trough in the Sirt Basin, and the central Ghadamis Basin, the central Murzuq Basin, and the offshore eastern Tripolitania Basin in the west. These highly prospective basin sectors encompass a total area of nearly 150,000 km^2, with an average well density for wells exceeding 12,000 ft of 1 well/5000 km^2.

INTRODUCTION

The exploration effort in Libya, which began in 1957, has been a phenomenal success. In the Sirt Basin (Figure 1), the drilling of 1600 new-field wildcats resulted in 250 discoveries with recoverable reserves of 45 billion barrels (bbl) of oil and 33 trillion cubic feet (tcf) of gas. These figures include 18 of the 21 giant fields in Libya, which hold reserves of 37 billion bbl of oil. In the Ghadamis Basin (including the Gheriat and Atchan Subbasins), approximately 260 exploration wells yielded 35 oil-field discoveries with an estimated 3 billion bbl of recoverable oil. The 62 wildcats drilled in the Murzuq Basin found 11 oil fields, including two giants, with reserves of approximately 2 billion bbl. The exploration effort in the offshore Tripolitania Basin has been rewarding as well. Fourteen new oil and gas-condensate fields have been discovered as a result of the drilling of about 50 wildcats. Reserves there are an estimated 2 billion bbl of oil and 8 tcf of gas. These estimates refer to activities through 1998 and include some fields categorized as marginal.

Despite this great exploration effort, the four producing basins are in the emerging stage of exploration maturity. Two aspects in particular are indicative of vast undiscovered resources in Libya and the exploration opportunities to access those resources: (1) numerous potential areas, proximal to oil-field trends where well density is extremely low; and (2) extensive areas, mostly basin centers, where valid deep objectives were reached by only a few wells.

It is noteworthy that 17 of the 21 giant oil fields and 80% of the total recoverable oil and gas were discovered prior to 1970. Since then, a less active and more conservative exploration effort has taken place. Apparently, rewards were adequate from the results of field extensions and the drilling of proven, relatively shallow plays. Complex and subtle plays (for example, low-relief structural or structural-stratigraphic traps and deep plays) were rarely pursued prior to the 1990s.

Probably the main reasons for the absence of an aggressive approach to exploration in the 1970–1990 period were lack of definitive imaging technologies (seismic acquisition and processing and other computer-related geoscience technology), limited understanding of petroleum systems, and ineffective use of sequence-stratigraphic concepts.

Today, in view of state-of-the-art technologies available for a wide range of petroleum-exploration needs and the relatively low cost to apply them, pursuit of deep plays in Libya should be a top priority. To address this objective, I have selected for evaluation six large underexplored areas with exceptional potential and, for the most part, with deep primary targets (Figure 1). However, many other promising areas are within and near the producing basins of Libya.

Three of the subject areas are in the Sirt Basin: the south part of the Ajdabiya trough, the Maradah graben, and the south part of the Zallah trough, including the adjoining Tumayam trough. The other study areas are in western Libya: the central part of the Ghadamis Basin, the central part of the Murzuq Basin, and the extreme eastern part of the Tripolitania Basin.

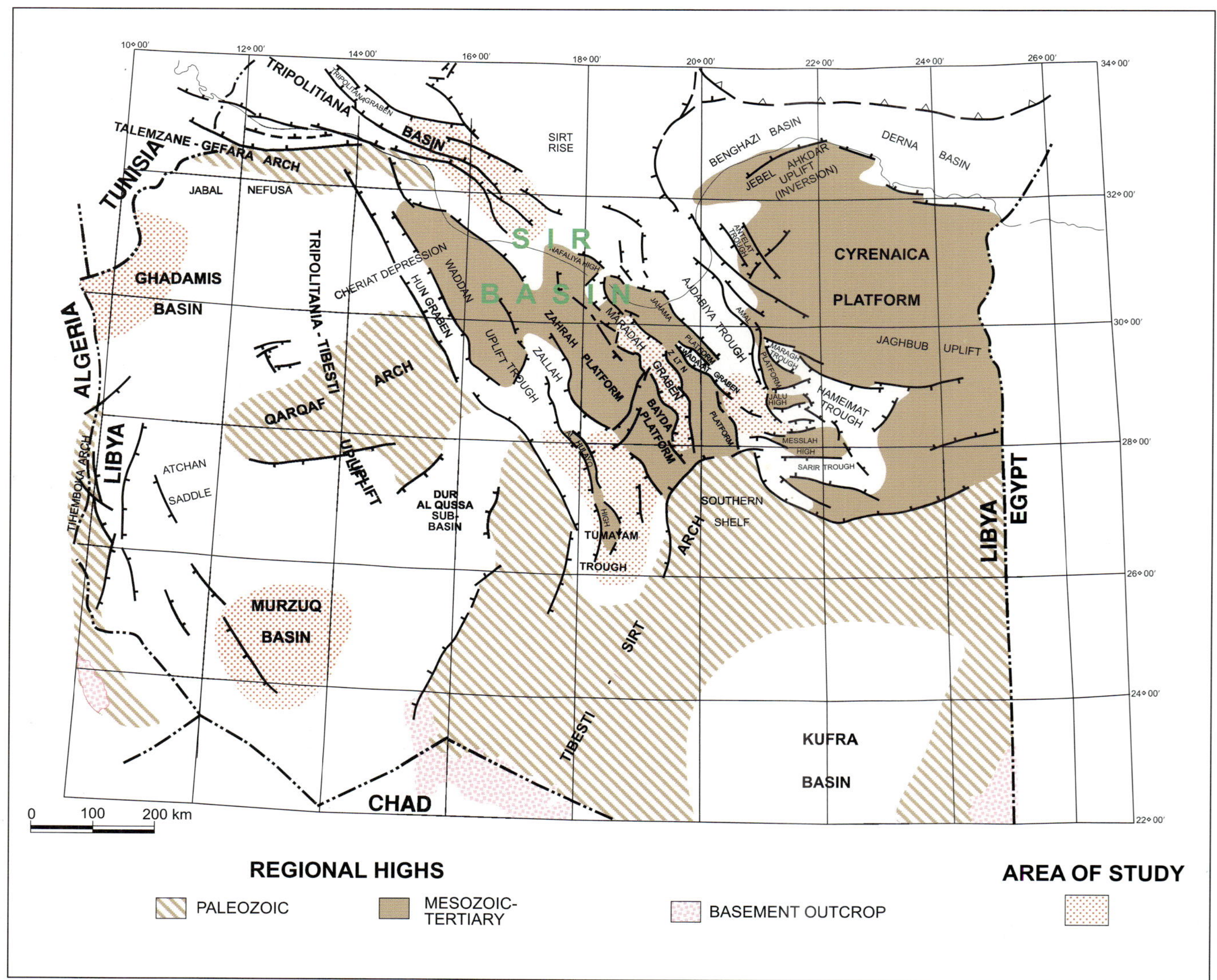

Figure 1. Generalized tectonic map of Libya showing major structural features. Also shown are six underexplored central basin or trough areas, which are the subject areas of this study.

TECTONIC SETTING

Paleozoic

Deposition of mostly continental siliciclastics during the Cambrian and marginally marine to marine siliciclastics during the Ordovician and Silurian continued essentially without interruption from Morocco to the Middle East. Uplift and erosion during the Late Silurian Caledonian orogeny initially defined the limits of the Paleozoic basins of Libya. The east-west-trending Qarqaf arch separated the Ghadamis and Murzuq Basins; the north-south-trending Sirt-Tibesti arch separated the Murzuq and Kufrah Basins and, generally, the Ghadamis Basin from the eastern Cyrenaica–Western Desert Basin (Klitzsch, 1971; Bellini and Massa, 1980).

After the dominantly marine siliciclastic deposition during the Devonian and the shallow-marine to continental deposition in the Carboniferous, widespread uplift and severe erosion during the Hercynian orogeny, particularly along the Sirt-Tibesti arch, Qarqaf arch, and Jefara uplift, further accentuated the Paleozoic basin margins.

Mesozoic

A very thick sequence of continental sediments of Triassic to Early Cretaceous age occupies the central part of Murzuq Basin. Along the Murzuq Basin margins and the nearby Qarqaf and Tibesti arches, Paleozoic and basement rocks are exposed. Gradual northward sag of the Ghadamis Basin throughout the Mesozoic resulted in continental and marine deposition, with a thickness of less than 1000 ft in the south and more than 6000 ft in the north. From the Late Permian to the Cretaceous, the extreme northern margin of the Ghadamis Basin underwent severe northward tilt, an effect of Tethyan subsidence. This resulted in a more pronounced northward increase in sedimentary thickness, with increased marine influence.

This Mesozoic depositional episode continued offshore in the Tripolitania Basin, where the thickness of post-Permian to Upper Cretaceous marine siliciclastics and carbonates may exceed 12,000 ft. Tectonic activity in the Tripolitania Basin and surrounding offshore areas during the Mesozoic was dominated by east-west-oriented dextral transtension related to movement of the African Plate relative to the Eurasian Plate (Van Houten, 1980; Anketell, 1996).

In the general area of the future Sirt Basin, the broad Sirt-Tibesti arch, with basement and Cambrian-Ordovician rocks exposed at the Hercynian surface, remained positive until the Late Jurassic. There were rare exceptions in discrete peripheral areas, where Triassic deposition occurred (the Maragh trough, for example). A variable thickness of continental siliciclastics (in the south) and marginally marine siliciclastics (in the north) of Late Jurassic to Early Cretaceous age, referred to as the Nubian sandstone, was deposited on the Hercynian surface. Nubian deposition was controlled by surface relief and, to some degree, by faulting.

In the Albian or early Cenomanian, extensional and probably transtensional faulting, followed by uplift and erosion, deformed the Sirt-Tibesti arch. This activity (the Sirt event) was a prelude to subsequent collapse of the arch (El-Alami, 1996b; Gras, 1996; Hallett and El-Ghoul, 1996; Koscec and Gherryo, 1996). The structural alignment created, which is most evident in the south and southeast, was for the most part east-west, east-southeast–west-northwest, and east-northeast–west-southwest. Consequently, the subcrop at the Sirt unconformity is a mosaic of Jurassic to Lower Cretaceous siliciclastics in grabens and half grabens, which are in depositional or fault contact with basement or Cambrian-Ordovician rocks on structural highs. Evidence of this fabric is exhibited in the Faregh, Masrab, Magid, Messlah, Jalu, and other areas in the southeast sector and is suggested by fault trends in the southern parts of the Zaltan and Bayda platforms.

The main Sirt Basin rift phase, which established the distinctive configuration of the basin, began in the Cenomanian with the collapse of the Sirt-Tibesti arch. Basically, five major grabens formed (Hun, Zallah, Maradah, Ajdabiya, and Hameimat), separated by four major platforms (Waddan, Zahrah-Bayda, Zaltan, and Amal-Jalu) (Figure 2). The orientation of these structural features was generally north-northwest–south-southeast, a fabric which persisted throughout the recurrent episodes of faulting during the Late Cretaceous and Paleocene. During this period, a great thickness of shale and subordinate carbonates and evaporites accumulated in the troughs, while a considerably reduced thickness of dominantly shallow-marine carbonates was deposited on the platforms (Barr and Weegar, 1972; Gumati and Kanes, 1985; Baird et al, 1996).

Tertiary

In the northern sector of the Ghadamis Basin, only a thin section of Tertiary shallow-marine sediments is present, and it thickens considerably northward toward the Tripolitania Basin and eastward toward the Sirt Basin. In the east on the Cyrenaican platform, deposition of thick, dominantly carbonate strata occurred.

In the Sirt Basin, from the middle Paleocene to the early Eocene, rift tectonics had less control on sedimentation, and thickness variation from trough to platform was less pronounced. From the early Eocene to the Pliocene, interior sag dynamics persisted, with a gradual eastward shift of the sag axis.

PETROLEUM SYSTEMS AND PLAYS

Summary

The petroleum systems, which have been active in

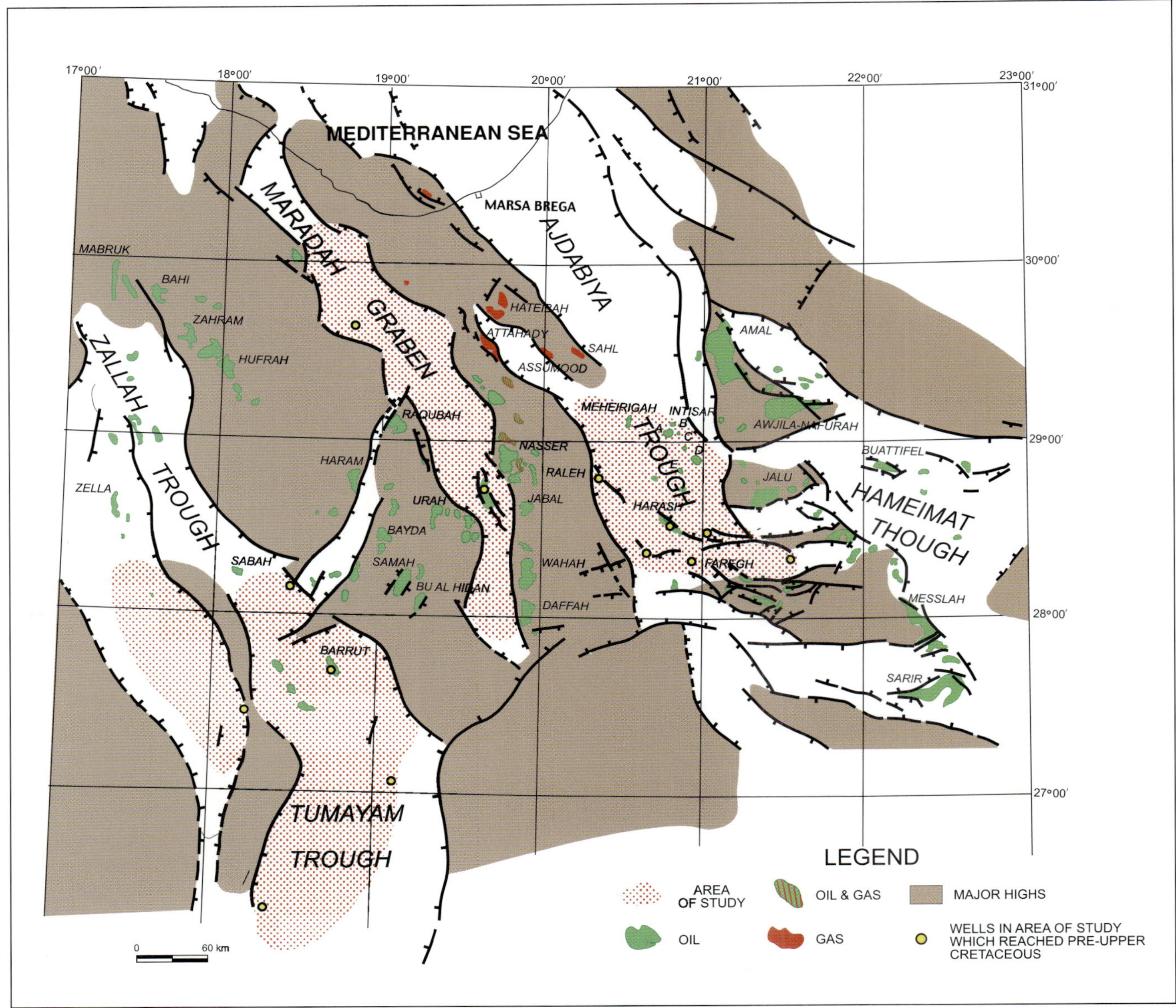

Figure 2. Structural elements of the Sirt Basin, showing oil and gas fields, areas of study, and location of wells with total depths exceeding 12,000 ft in the areas of study. The approximate size of the Sirt Basin areas of study are Ajdabiya trough, 8,500 km^2; Maradah graben, 10,000 km^2; and South Zallah trough–Tumayam trough, 25,000 km^2.

the six basin-center sectors under study, are extensive. The multiple systems in the Sirt Basin include a wide range of Cretaceous and Paleogene reservoir sequences, which were charged by three or four Cretaceous source rocks. The Ghadamis Basin petroleum systems involve Ordovician, Silurian, Devonian, and Triassic reservoirs charged by Lower Silurian and/or Middle to Upper Devonian source beds. A single petroleum system was active in the Murzuq Basin, comprising Ordovician, Silurian, and Devonian reservoirs, which were charged by Silurian source rocks (Boote et al., 1998). The Tripolitania Basin probably has a framework of several petroleum systems, which includes a wide range of Mesozoic and Tertiary formations. In the following paragraphs, the key hydrocarbon factors (reservoir, seal, source, trap, migration, and timing) will be described for each of the subject areas.

Sirt Basin

General

The underexplored sectors of the Ajdabiya trough, Maradah graben, and Zallah-Tumayam trough have important features in common: nearby oil production, only four or five exploration wells which reached sub–Upper Cretaceous horizons, a world-class source rock (the Upper Cretaceous Sirt-Rachmat shale), and large areal extent. The Ajdabiya, Maradah, and Zallah-Tumayam areas cover 8,500 km^2, 10,000 km^2, and 25,000 km^2, respectively.

Source-rock Summary (Figure 3)

The Campanian-Coniacian Sirt-Rachmat shale sequence, which includes minor amounts of carbonates (Tagrifet limestone) with variable source potential, varies in thickness from 1000 ft to more than 3000 ft in each of the three troughs (Figure 4). The total organic carbon (TOC) of this sequence ranges from 0.5% to 8%, averaging 1.5–4% (Parsons et al., 1980; Hamyouni et al., 1984; Baric et al., 1996).

The Cenomanian-Turonian Etel Formation (evaporites, shale, and minor carbonates deposited in shallow lagoonal to supratidal conditions) exhibits good source-rock characteristics, with TOC ranging from 0.6% to 6.5% in the Hameimat trough (El-Alami, 1996b). These same Etel facies, with net shale thicknesses of 200 ft to more than 1000 ft, are present in the southern Ajdabiya trough and Maradah graben (Figure 5). Therefore, they should be considered an effective source in those sectors. The source quality of the Etel shale is questionable in the southern Zallah and Tumayam troughs, where it exceeds 500 ft in a limited area only.

A third source is the Lower Cretaceous middle shale member of the Nubian Formation. Nubian lacustrine to lagoonal shale has been identified in the Hameimat trough and the adjoining Faregh and Messlah areas, where thicknesses vary from 0 to 1000 ft (Figure 6) and average TOC is approximately 3%. It is most likely a minor source in the southern part of the Ajdabiya trough. In the Maradah graben, based on only two wells (El-Hawat, 1996), the Nubian middle variegated shale member attains thicknesses ranging from 200 to 400 ft. This shale sequence was deposited in a partially anoxic, marginal-marine environment. It may have contributed some hydrocarbon to surrounding areas.

The contribution of variable quantities of oil from as many as four source units (shale or shale and carbonate) at different times of expulsion (during periods from early Oligocene to early Pliocene) has yielded several distinct crude oils in different areas. One similar characteristic of these oils is the gravity, which ranges from 36° to 40° API. More rock-oil correlation analyses and related studies are needed for more accurate determinations of regional rock-oil-timing associations.

South Ajdabiya Trough

Reservoirs.—The lower and upper sandstone members of the Upper Jurassic to Lower Cretaceous Nubian Formation are clearly the primary reservoir targets for the area (Clifford et al., 1980; Ibrahim, 1991; Abdulgader,

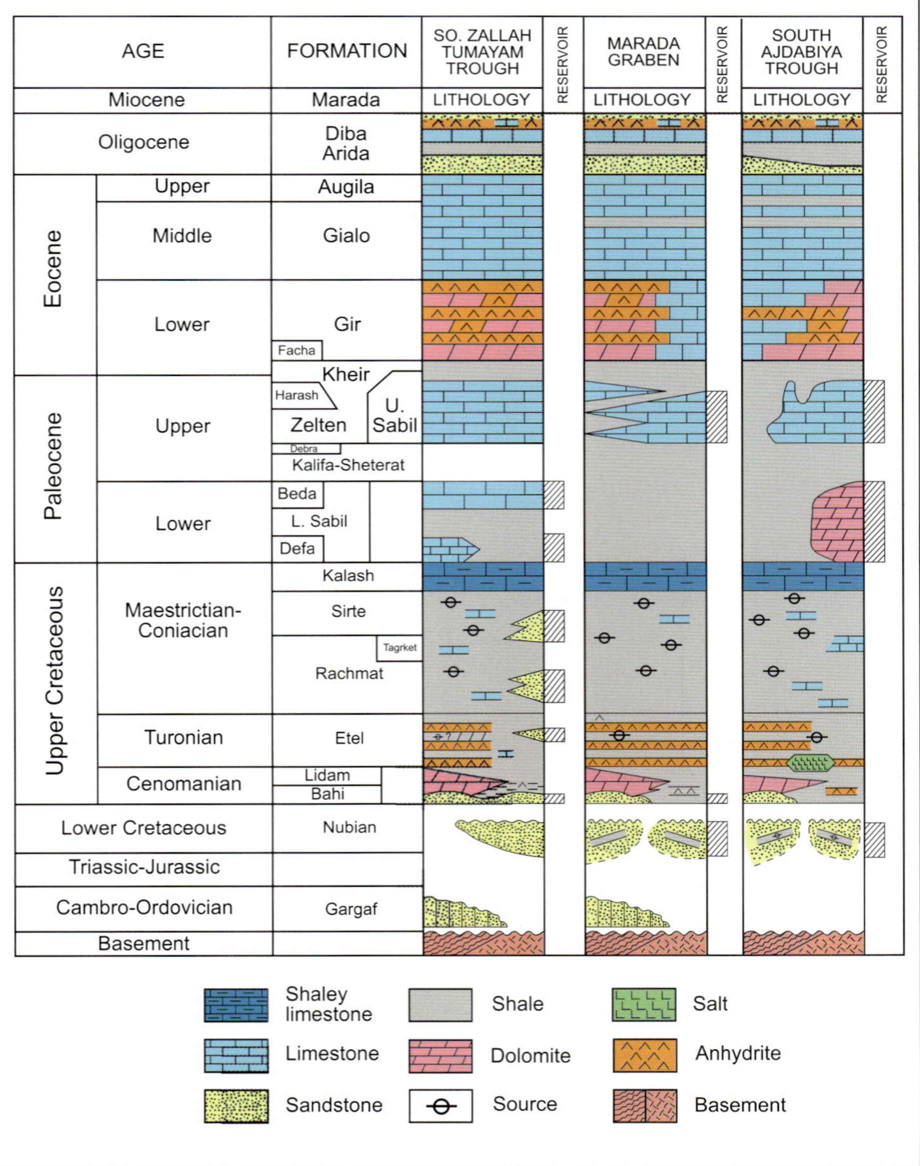

Figure 3. Generalized stratigraphic correlation chart of the Sirt Basin study areas: south Ajdabiya trough, Maradah graben, and south Zallah trough–Tumayam trough. The main reservoir and source intervals are indicated on the chart. Hachured boxes represent main reservoirs.

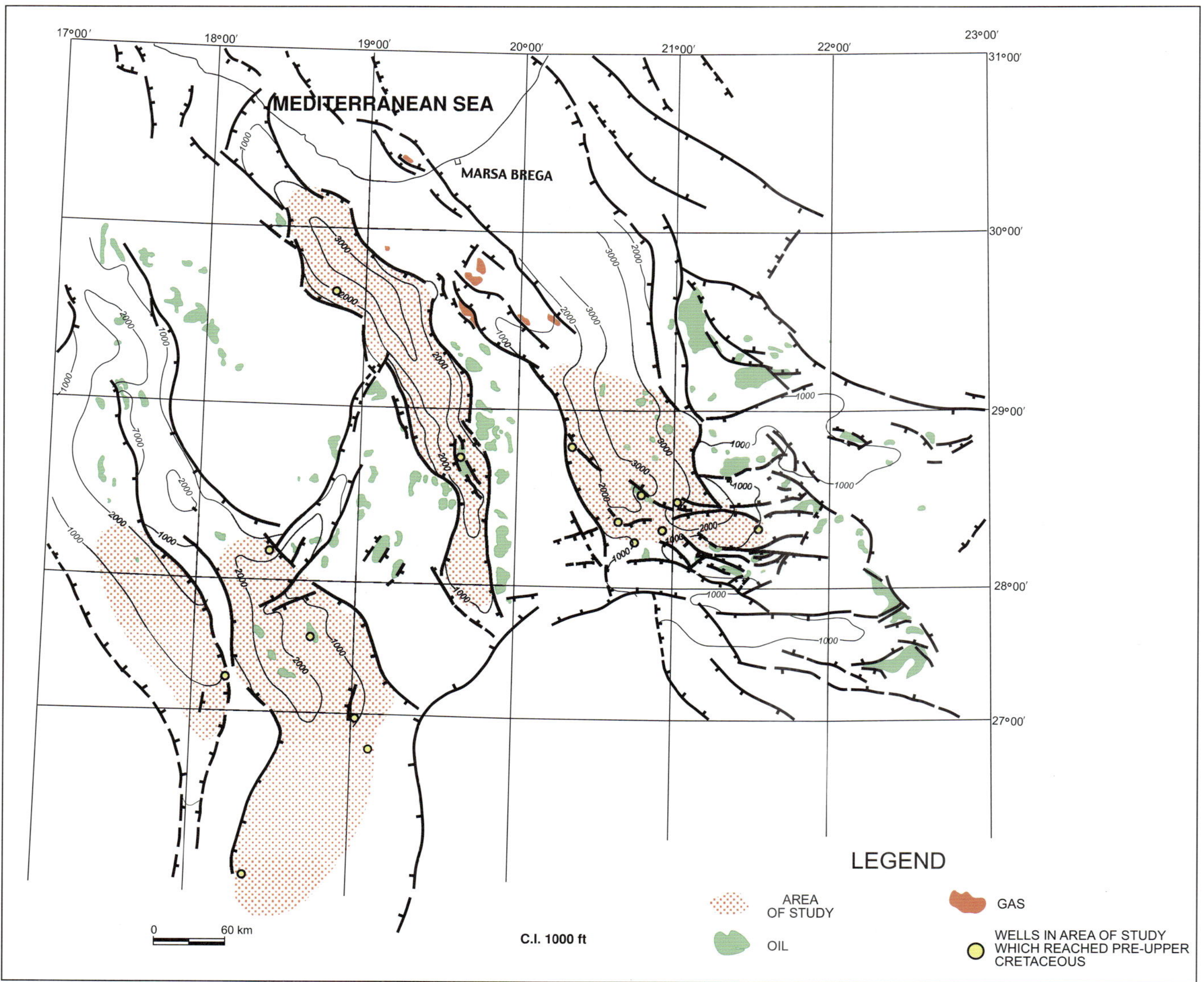

Figure 4. Net shale isopach map of Sirt and Rachmat Formations (Upper Cretaceous), Sirt Basin. Modified from Masera Corporation (1992).

1996; Mansour and Magairhy, 1996). Net sand thicknesses are estimated to range from 0 (at discrete onlap and truncation limits) to 1200 ft (Figure 6). Depth to the top Nubian ranges from 12,000 to 18,000 ft (Figure 7). Despite these depths, it is expected that average porosities will be 12–13%, with maximum porosity exceeding 20%. Average porosity at depths below 15,000 ft ranges from 12% to 13.5% in some wells in nearby Hameimat trough.

Secondary reservoir objectives are high risk in the area because of limited distribution and reservoir properties. The Bahi (Maragh) sandstone equivalent is absent or very thin in surrounding areas, with dominant siltstone and shale lithology suggestive of the Etel Formation. The Lidam dolomite, a facies of the Etel Formation in this sector of the Sirt Basin, is also very thin or absent in nearby wells. The Tagrifet limestone and equivalent Rachmat limestone beds are thin and generally argillaceous mudstones west of the Amal and Jalu highs.

Possible attractive secondary targets are Paleocene lower and upper Sabil shoal and reef limestones (Spring and Hansen, 1998). Upper Sabil shelf-edge deposition was not controlled by rift phase faulting, and the shelf extended across the southern part of the Ajdabiya trough (Figure 8). This potential reservoir is at relatively shallow depths and consequently has been the subject of exploration programs for some time. However, subtle buildups, overlooked in the past, can be imaged accurately today using state-of-the-art methods.

Seals.—Etel shale and anhydrite at the Sirt unconformity provide an effective seal for the Nubian sandstone throughout most of the area. Locally, a thin Bahi (Maragh) sandstone or Lidam dolomite sequence may directly overlie the Nubian, in which case the Nubian lacks a seal.

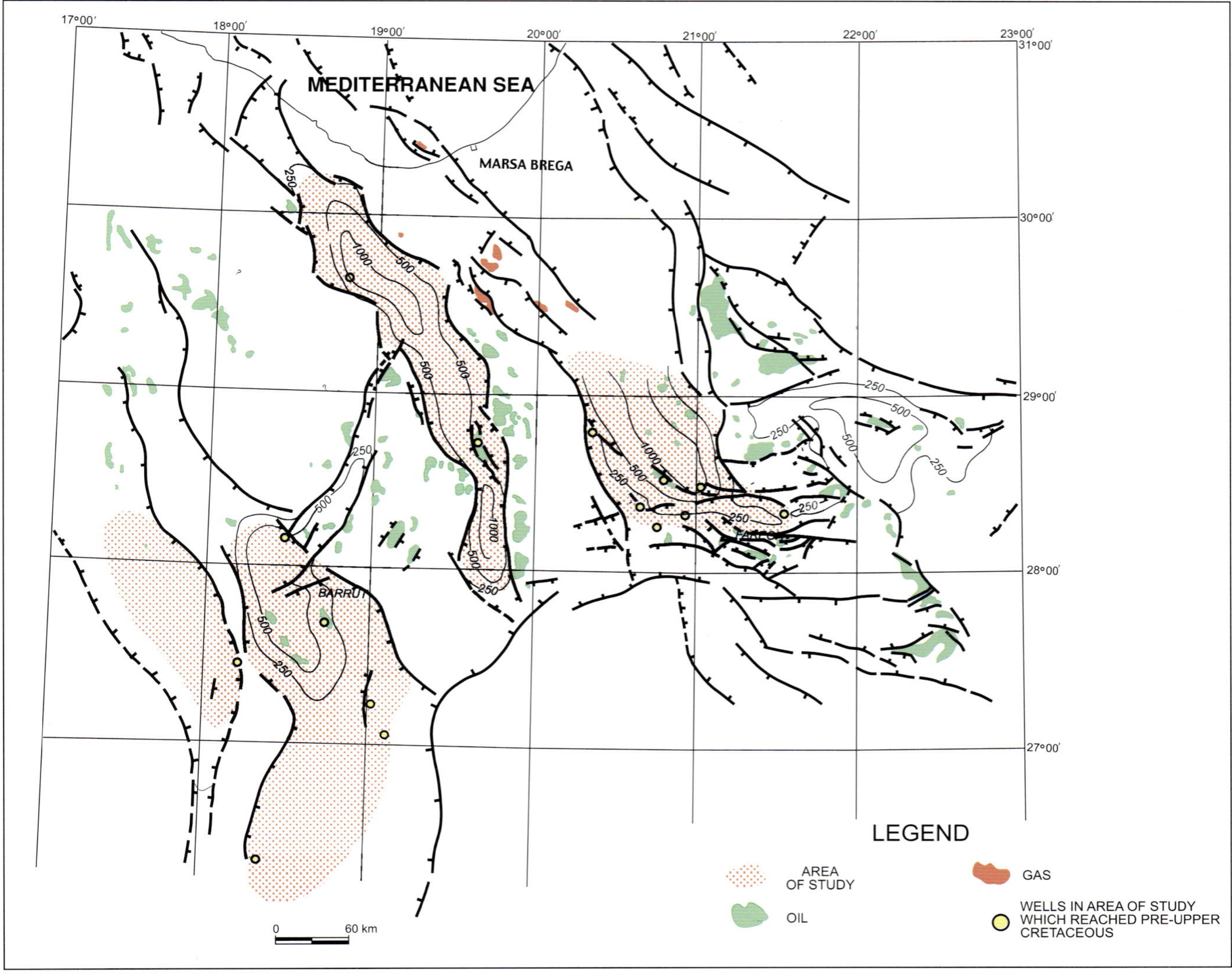

Figure 5. Net shale isopach map of the Etel Formation (Upper Cretaceous), Sirt Basin.

Sheterat and Kheir shales provide excellent seals for lower and upper Sabil carbonates, respectively.

Timing and migration.—In the southern part of the Ajdabiya trough, the peak oil-expulsion stage occurred approximately from the late Eocene to the late Pliocene from source beds of the Rachmat and Sirt Formations (Ghori and Mohamed, 1996; Roohi, 1996b; Gumati and Schamel, 1988). This stage occurred generally at depths below 11,000 ft. Because the latest significant structural and stratigraphic trap development was late Paleocene, drainage timing was ideal. The main source rocks (Sirt shale and Rachmat shale) are stratigraphically separated from the Nubian. Therefore, secondary migration would have been via faults or faults in combination with the Sirt unconformity. Migration from Etel source beds would have been accomplished by lateral drainage via the Sirt unconformity to underlying Nubian sands.

Oil from Sirt source beds reached Sabil reservoirs via vertical migration along faults and fractures.

Traps.—Trap types for Nubian reservoirs are horsts, tilted fault blocks, updip unconformity truncations, and updip terminations against basement or Cambrian-Ordovician quartzite (Figure 9). Sabil traps are usually drape anticlines over buildups with lateral permeability barriers.

Maradah Graben

Reservoirs.—The lower and upper sandstone members of the Nubian Formation are the primary reservoir targets for the area. The maximum Nubian net sand thickness in the graben is approximately 1000 ft (Figure 6). The Nubian may be absent on Cambrian-Ordovician highs, similar to the setting in the southeast Sirt Basin, but no current data support this hypothesis. Depth to the top Nubian ranges from 11,500 to 15,000 ft in the Maradah graben (Figure 7). It is expected that average porosity will be 12–13%.

Nubian thickness and porosity estimates in the Mar-

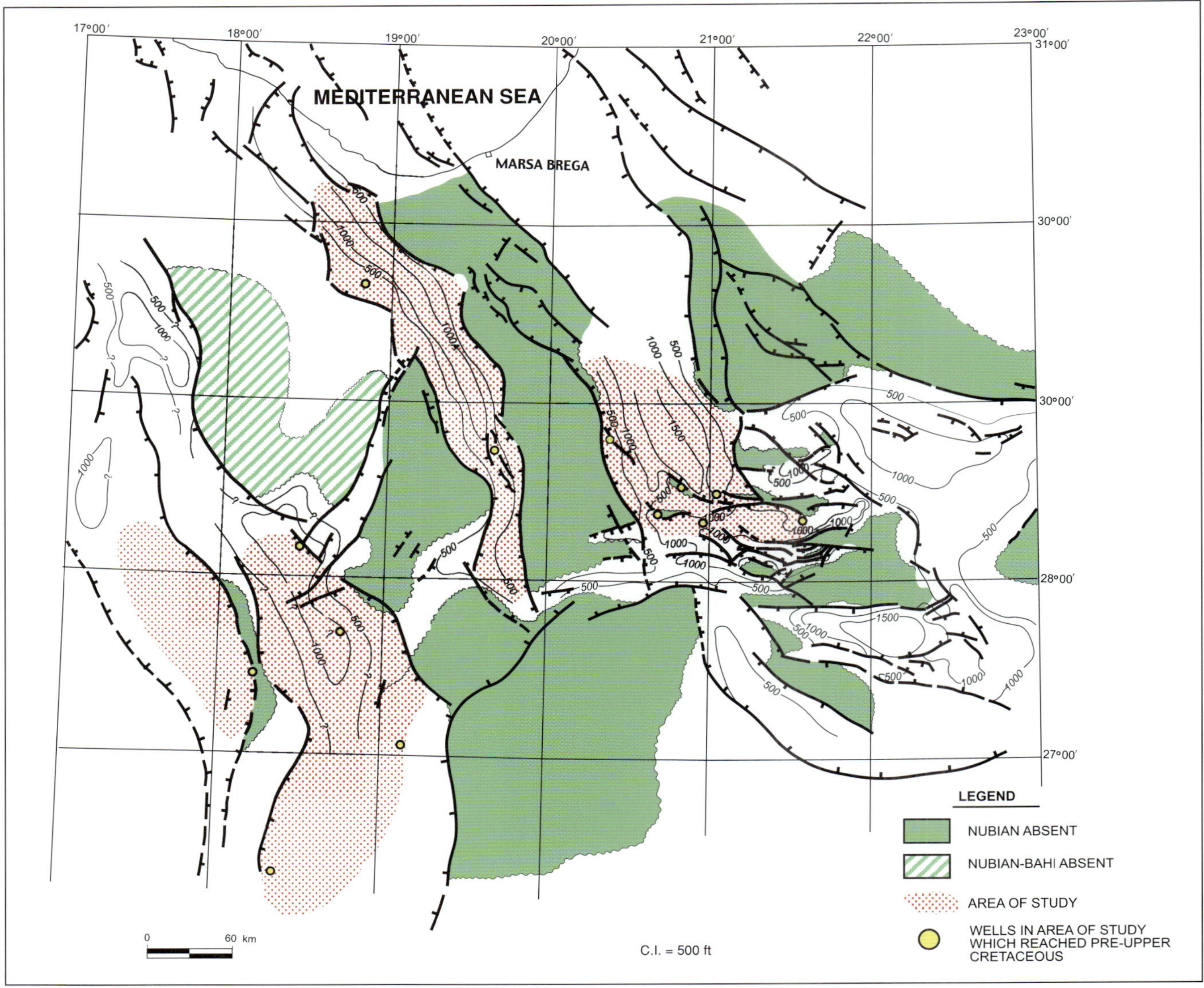

Figure 6. Net sand isopach map of the Nubian Formation (Lower Cretaceous), Sirt Basin. Areas where the Nubian is absent because of erosion or nondeposition are indicated.

adah graben are based only on regional projection and partial data from three widely separated wells (El-Hawat et al., 1996; Bonnefous, 1972): D6-NC149, in the Wadi oil field; P1-16, in the Bazuzi oil field at the northeast edge of the Zahrah platform; and V1-59, in the Bilhizan oil field in the south part of the Bayda platform (Figure 2).

Secondary reservoir objectives are few and high risk in the area because of limited distribution and poor development. The exceptions are reef and shoal carbonates of the Zaltan Formation and the Bahi sandstone. Zaltan Formation facies, consistent with the equivalent upper Sabil carbonate to the east, were not controlled by earlier faulting. A shelf margin extended across the southern part of the Maradah graben. Net thickness of the Zaltan in this area ranges from 0 in the north to more than 400 ft in the south. Depth to the top Zaltan is 7500 to 9000 ft. Because of this shallow depth, the Zaltan has been subjected to considerably more exploration than the Nubian. The basal Upper Cretaceous Bahi sandstone may attain thicknesses exceeding 600 ft in the graben. However, in places, part or all of the so-called Bahi sandstone may be Lower Cretaceous Nubian sandstone.

Seals.—Etel shale and anhydrite at the Sirt unconformity provide an effective seal for the Nubian sandstone throughout most of the area. In a few places, the Nubian may lack an effective seal because the Bahi sandstone or Lidam dolomite directly overlies it. The Etel shale-evaporite sequence is also an excellent seal for Bahi and Lidam reservoirs. The Paleocene Harash or Kheir shales provide the seals for the Zaltan carbonates.

Timing and migration.—In the Maradah trough, the peak oil-expulsion stage occurred approximately from the early Oligocene to the late Miocene for the Etel,

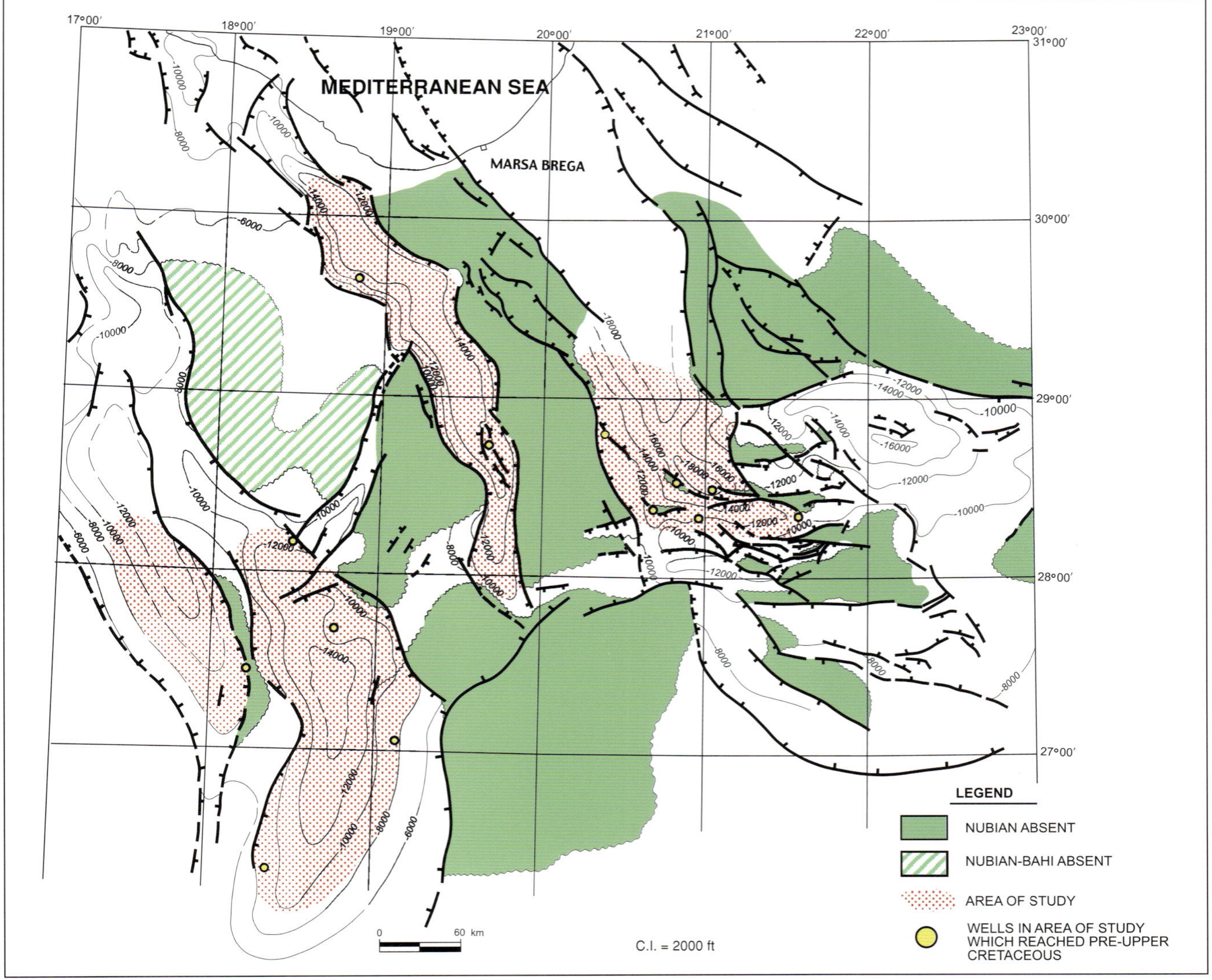

Figure 7. Structure map on the top Nubian Formation, Sirt Basin. Modified in part from Masera Corporation (1992), El-Hawat et al. (1996), and Mansour and Magairhy (1996).

Rachmat, and Sirt Formation source rocks (Roohi, 1996b). As in the case of the Ajdabiya trough, the latest significant structural and stratigraphic trap development was late Paleocene, creating ideal entrapment and retention conditions. Secondary migration from the Sirt shale and the Rachmat shale to the underlying Nubian, Bahi, and Lidam reservoirs would have required an indirect carrier system via faults or faults in combination with the Sirt unconformity. Migration from Etel source beds to overlying reservoirs would have occurred laterally via carrier beds associated with the Sirt unconformity.

Vertical migration from Sirt source beds via faults and fractures provided the charge for the overlying Zaltan reservoir.

Traps.—Trap types for Nubian reservoirs are most likely horsts, tilted fault blocks, and faulted anticlines. Combination traps also may be present, involving Nubian sandstone truncated at the Sirt unconformity or updip onlap of Nubian sandstone on the Cambrian-Ordovician surface.

Trap types for the Bahi and Lidam Formations include horsts, tilted fault blocks, drape and faulted anticlines, and pinch-outs. Expected traps for Zaltan reservoirs are reef and shoal buildups, usually in combination with drape and faulted anticlines.

Southern Zallah Trough–Tumayam Trough

Reservoirs.—The lower and upper sandstone members of the Nubian Formation and the Bahi sandstone are among the primary objectives (Schroter, 1996). In this area, it is difficult to differentiate between these two formations. Therefore, the thicknesses reported here are estimates. Nubian net sand thicknesses are estimated to range from 0 (at onlap and truncation limits) to approxi-

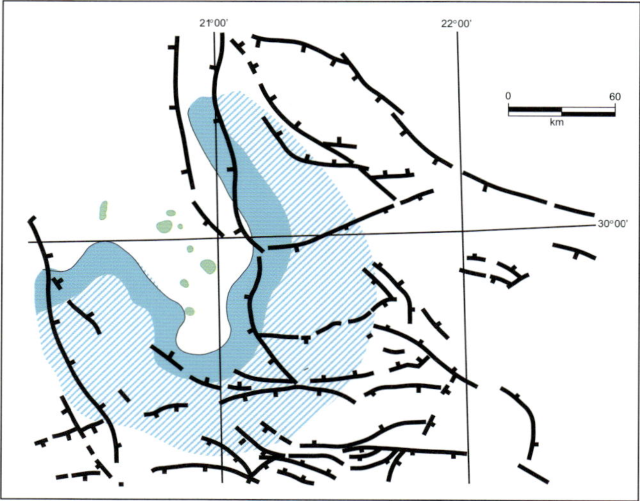

Figure 8. Approximate location of the Paleocene Upper Sabil carbonate shelf edge, Ajdabiya trough—a zone of potential reef and shoal development. Shelf slope pinnacle reef oil fields are shown.

mately 1000 ft. The Bahi sandstone is expected to be from 0 to 300 ft thick. Depth to the top Nubian and Bahi ranges from 9500 to 14,000 ft. It is expected that average porosity will be 12–14% in both formations.

Probably equally important reservoir targets are the Paleocene Defa and Beda Formations and the lower Eocene Facha high-energy carbonate facies. Barrier shoal carbonates are well developed in the Thalith, lower Beda, and upper Beda members of the Beda Formation in the northeast sector of the subject area (Bezan et al, 1996; Johnson and Nicoud, 1996; Sinha and Mriheel, 1996). Porosity in the lower and upper Beda members (Farrud sequence) ranges as high as 35%. Thickness of the Beda Formation exceeds 1000 ft, with as much as 600 ft of net porous carbonate (Figure 10). The Defa carbonate and Facha dolomite attain a net thickness of as much as 400 ft in the area. Approximately 25 wildcat wells have reached these formations in the area at depths of less than 9000 ft. However, the well density of 1 well/1000 km^2 indicates that the area is still underexplored, even at shallow levels.

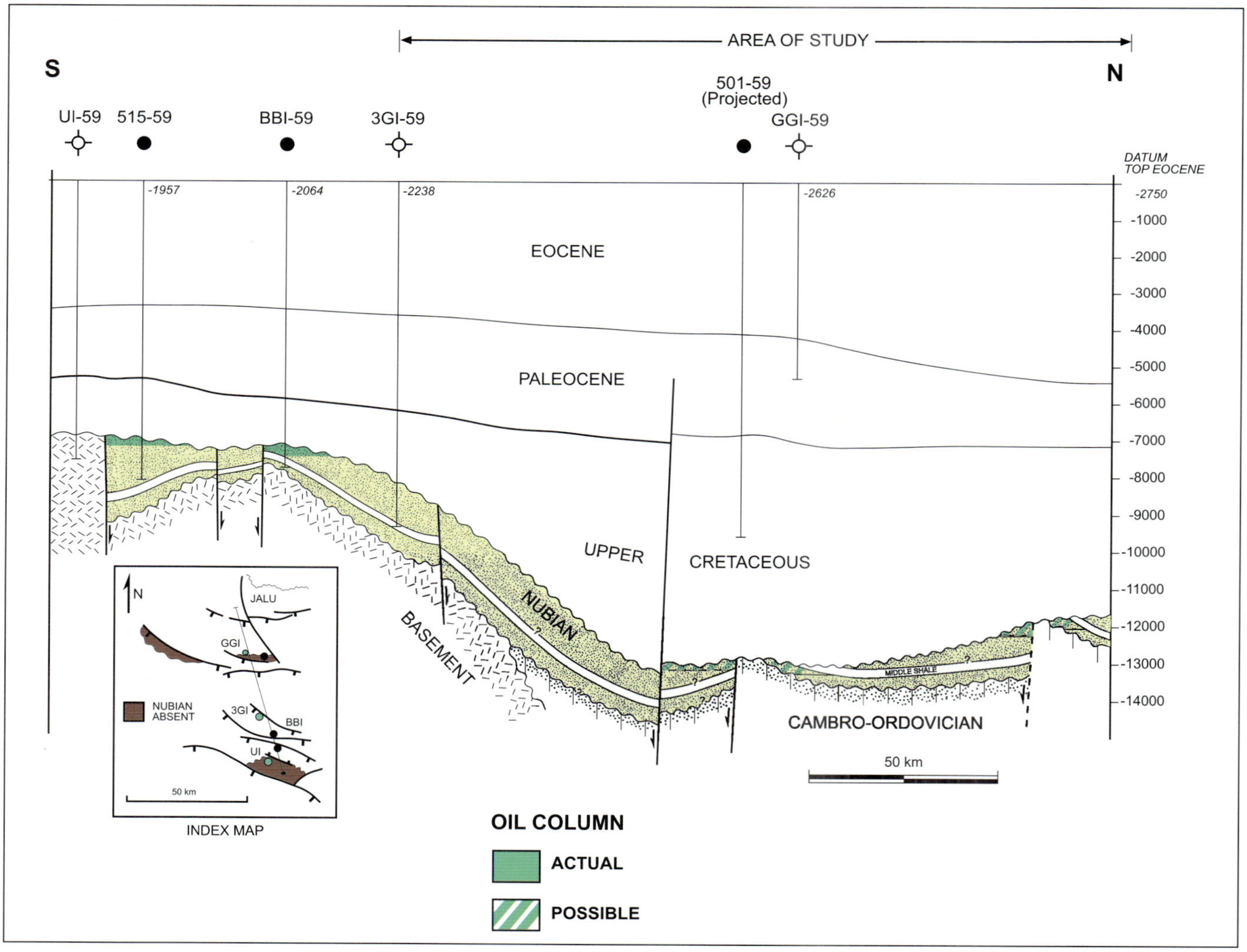

Figure 9. North-south structural cross section from the central part of the Ajdabiya trough to the Faregh oil-field area, depicting actual and inferred Nubian sandstone trap configurations.

A secondary objective that has not been pursued is a Turonian-Senonian sandstone sequence, equivalent to the Rachmat and Sirt Formations, which is developed in the southern sector of the Tumayam trough. These porous sandstone beds thicken rapidly southward from their pinch-out limits to more than 1000 net ft (Figure 11).

Seals.—Etel shale and anhydrite provide an effective seal for the Nubian and Bahi sandstones throughout most of the area. Locally, there is a slight risk that a thin Lidam dolomite sequence overlying the Nubian or Bahi would have prevented sealing. Hagfa and Khalifa shales are effective seals for Defa and Beda carbonates, and the Gir evaporites are reliable seals for the Facha dolomite. Interbedded shales should provide adequate seals for the individual Rachmat-Sirt sandstones.

Timing and migration.—In the central part of the South Zallah–Tumayan trough area, where the top of the Sirt shale is between 9000 and 11,000 ft, the main stage of oil expulsion apparently occurred throughout the Miocene. There is little doubt that the Sirt shale is the only important effective source rock in the area, based on organic richness and maturity.

Secondary migration from Sirt shale to underlying Nubian and Bahi reservoirs, as is the case throughout the Sirt Basin, requires a system of faults or faults in combination with the Sirt unconformity.

Vertical migration of oil from Sirt source beds to overlying Defa, Beda, and Facha reservoirs was accomplished via faults, fractures, and local carrier beds.

Traps.—Trap types for Nubian reservoirs are expected to be the same here as in the Maradah and Ajdabiya troughs. Trap types for Bahi sandstone should include tilted fault blocks, drape and faulted anticlines, and pinch-outs. Northerly oriented pinch-outs of the Turon-

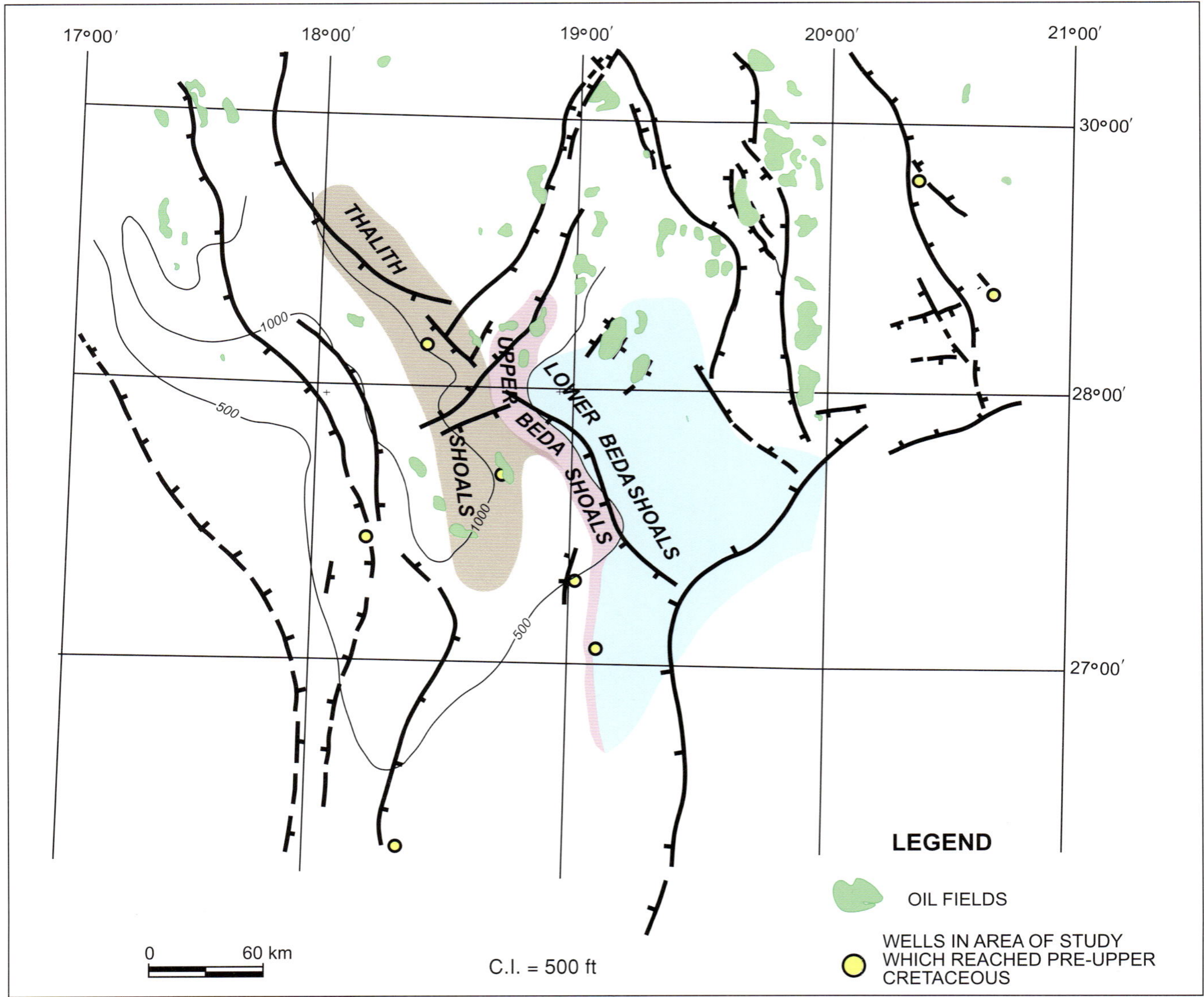

Figure 10. Isopach map of the Beda Formation, showing distribution of barrier shoal carbonate facies of the Thalith, lower Beda, and upper Beda members, south Zallah trough–Tumayam trough. After Bezan et al. (1996) and Sinha and Mriheel (1996).

ian-Senonian sandstones, in combination with dip or fault closures, are expected in the southern sector of the area. Reef and shoal buildups, in combination with anticlinal drape or faults, are the most likely traps for Defa and Beda reservoirs.

Western Libya

Central Ghadamis Basin

General.—The Ghadamis Basin area of study, which covers more than 20,000 km^2, is located in the center of the basin bordering Tunisia and Algeria (Figures 12–14a). The basin is continuous across southern Tunisia and central Algeria, covering an area of approximately 200,000 km^2. It is particularly noteworthy that in the last 10 years, an estimated 5 billion to 6 billion bbl of recoverable oil equivalent has been discovered, mainly from Devonian and Triassic sandstone reservoirs in the Algerian sector of the Ghadamis Basin. The key to these discoveries was an understanding of the plays and 3-D seismic. During that same period, there was minimal success in the Libyan sector, although geologic setting and reservoirs are essentially the same.

In the study area, 27 wildcats yielded one oil and three gas-condensate discoveries with Upper Silurian Acacus sandstone pay in the north, and two oil discoveries with Triassic and Upper Devonian Tahara sandstone pay in the central sector.

Reservoirs.—The main reservoir targets for the area are the Upper Silurian Acacus Formation and the Lower Devonian Tadrart and Kasa Formations (Figure 15) (Said, 1974; Masera Corporation, 1992; Echikh, 1998). The Acacus net sandstone thickness ranges from approximately 500 to 1300 ft (Figure 16). The Acacus average porosity is at least 16%. The Tadrart and Kasa Formations should have a net sandstone thickness of 300–700 ft and an average porosity of 14–15% in the study area. These formations, which are a more or less continuous stratigraphic succession, are at depths between 8000 and 12,500 ft (Figure 17). Only eight exploration wells, most of which were in the north, reached these objectives in the study area.

Three other sandstone reservoirs are valid objectives, but because of their shallower depths, they have been the subject of more exploratory drilling than the above formations. They are the Middle Devonian Uennin sandstone (equivalent of the F3 in Algeria), with a thickness range of 0 to 300 ft; the Upper Devonian Tahara Formation, with a net sand range of 50 to 200 ft; and the Triassic Ras Hamia Formation, with a net sandstone thickness of 200 to 700 ft. All of these sandstones have very good porosity, averaging 14–18%.

Seals.—Generally, there is an effective Acacus shale seal above the sandstone. Where it may be absent, however, the overlying Tadrart will form a combined objective with the Acacus sandstone. Shale horizons consistently provide adequate seals for Tadrart, Kasa, F3 equivalent, and Tahara sandstones. Throughout most of the area, there are effective shale, carbonate, or evaporite seals for the Ras Hamia sandstone. Because of a dominant continental siliciclastic facies above the Ras Hamia in the southern part of the area, however, a seal may be lacking.

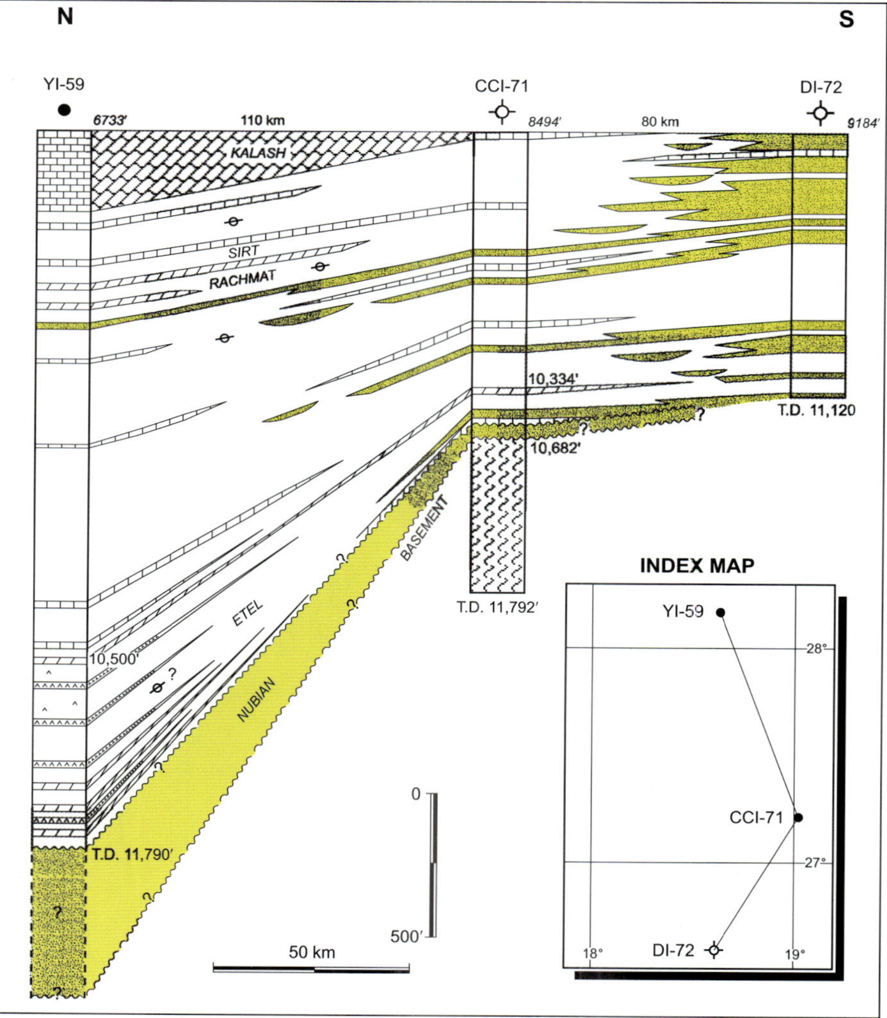

Figure 11. North-south diagrammatic correlation of the Cretaceous section of wells Y1-59, CC1-71, and D1-72. Well correlation illustrates probable relationship of the northward sandstone pinch-outs interfingering with Sirt-Rachmat shale source beds. Also shown is the interpreted Nubian sandstone correlation. Datum is the top Cretaceous.

Source rock, timing, and migration.—There are two world-class, type II source rocks distributed throughout the entire basin: the Lower Silurian Tanezzuft and the Middle to Upper Devonian Uennin Formations. The two shale formations have an average TOC of 3–5% and are approximately 1000–2000 ft thick in this prime study area.

The peak oil-generation-expulsion window (equivalent to vitrinite reflectance [R_o]of 0.8–1.3%) for both formations is approximately 8500–12,000 ft. Depths to the base of the Tanezzuft and Uennin in the area are 12,000– 14,500 ft and 8000–12,000 ft, respectively.

The main stage of oil expulsion from the Tanezzuft source probably occurred from the Late Triassic to Early Cretaceous. Oil expulsion from the Uennin source probably occurred from Early to Late Cretaceous. At present, the Tanezzuft shale is in the wet-gas to dry-gas generation stage, and the Uennin source beds are in the peak-oil to late-peak-oil stage.

In this central basin sector, structural traps were essentially established during Hercynian events, although some early development most likely occurred during the Caledonian orogeny. It is unlikely that the Albian Austrian event or the Eocene Pyrennian events, which affected major highs and coastal areas in the region, caused any significant structural modification to this sector. Consequently, traps were in place prior to migration.

Conditions for migration were optimum, in view of the short distance and vertical and lateral carrier systems from the two sources to the multiple reservoirs.

Traps.—The expected trap types are low-relief, simple, and faulted anticlines; drape anticlines over paleotopographic relief or faulted structures; unconformity truncation of the Tahara sand in the northern part of the study area; and pinch-outs of the Uennin F3 equivalent sand.

Central Murzuq Basin

General.—This underexplored basin-center area covers more than 30,000 km^2. Only four wells have been drilled there (Figure 14b), and one well, A1-NC58, is a marginal oil discovery. Within about 50 km to the north are seven small, undeveloped oil-field discoveries, with total reserves of about 150 million bbl, and one major discovery, Elephant (N1-NC174), with estimated reserves of 500 million bbl of oil. The Murzuq oil-field complex (A, B, C, H, and J-NC115 fields), with reserves of about 1 billion bbl of oil, is approximately 100 km north of the subject area. In all these discoveries, sandstones of the Ordovician Memouniat Formation are the reservoirs (Figure 13).

Reservoirs.—Main potential reservoirs for the area include the Acacus and the Lower Devonian Tadrart-Kasa sandstones, as well as the main pay in the basin, the Memouniat Formation. The net sandstone thickness of the Memouniat Formation ranges from approximately 500 to 2500 ft and has an average porosity of 10–14%. The Acacus net sandstone thickness is from 0 (at the north edge of the study area, where it is truncated) to 300 ft. The average porosity of the Acacus sandstone is approximately 15%. The Tadrart-Kasa sandstones, undifferentiated, have an estimated net thickness of as much as 200 ft and an average porosity similar to that of the Acacus. This sequence pinches out at the Caledonian surface in the northern part of the area.

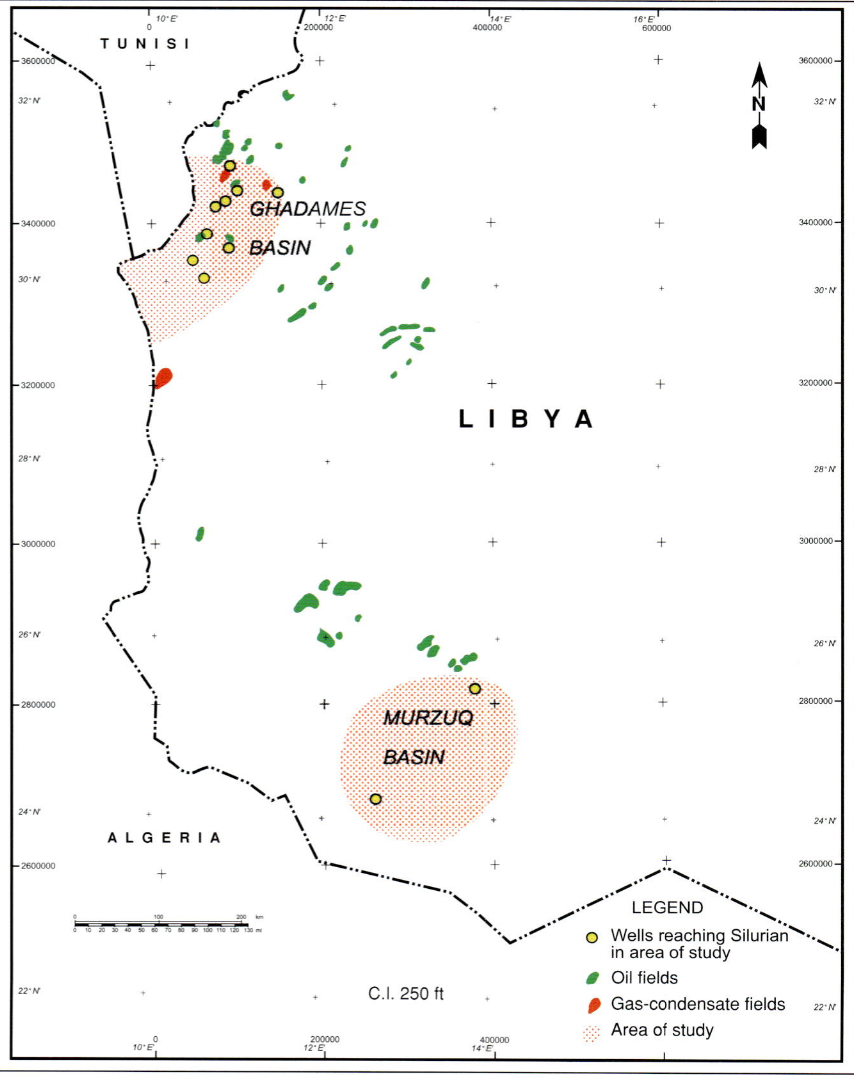

Figure 12. Location map of the Ghadamis and Murzuq Basins, showing the basin-center areas of study. The approximate size of the Ghadamis is 20,000 km^2; the approximate size of the Murzuq Basin is 30,000 km^2.

The depth to the Memouniat ranges from 8000 to 11,500 ft. The Acacus and Tadrart-Kasa are at depths of 6500 to 10,500 ft in the Murzuq Basin center (Masera Corporation, 1992).

Seals.—The Tanezzuft shale provides a reliable seal throughout the area for the Memouniat Formation. Generally, effective shale seals are interbedded with Acacus sandstone beds. In a few places, upper Acacus sandstones are overlain by Tadrart-Kasa sandstones, which could create a combined reservoir, as in the Ghadamis Basin. Uennin shale beds provide adequate seals for the Tadrart-Kasa sequence.

Source rock, timing, and migration.—The Tanezzuft shale is the only effective oil source of importance in the Murzuq Basin (Hamyouni, 1991). It is possible, however, that very minor amounts of early oil were expelled from Devonian Uennin organic-rich shale in the basin center (Meister et al., 1984). The Tanezzuft shale is 400–1600 ft thick in the study area. The average TOC is 1.8%. The peak oil-expulsion window is approximately 6500–9000 ft. Therefore, because the depth to the base Tanezzuft is from 7000 to 11,500 ft in the study area, the Tanezzuft is in peak-oil to wet-gas generation stages.

Vertical, updip, and fault pathways provided easy, short-distance pathways for migration of oil to the adjacent reservoirs. Migration apparently took place from the Early Jurassic to the Early Cretaceous, after the establishment of most, if not all, of the traps in the study area.

Traps.—Structural trap types are basically the same as those in the Ghadamis Basin center. Unconformity truncation of the Acacus and onlap pinch-out of the Tadrart-Kasa, in association with dip or fault closure, are also potential traps in the area.

Eastern Tripolitania Basin

General.—The offshore Tripolitania Basin (Gabes-Sabratha Basin) is a deep, highly faulted, elongate trough which extends from the Gulf of Gabes to the northwestern margin of the Sirt Basin. The eastern sector, which covers approximately 20,000 km^2, is essentially unexplored. To date, one dry hole has been drilled there. The oil and gas-condensate discoveries in the basin are concentrated about 100–150 km west of that area. In general, play concepts established in the productive western sector of the basin and, to some degree, in the western part of the Sirt Basin are also valid in this undrilled area (Bishop, 1988).

Reservoirs.—Based on regional projections, numerous potential reservoir suites are in this basin sector (Figure 18). The lower Eocene El Garia Formation of the Metlaoui group (Jdeir Formation), the main pay in all of the Tripolitania Basin discoveries, is obviously the most important objective in the subject area. El Garia nummulitic bank grainstone-packstone facies and equivalent or underlying dolomite and skeletal limestones (Jirani and Bilal Formations) probably have net thicknesses of as much as 600 ft in the subject area. The effective porosity range is about 5–30%, with an average of 17% in the western part of the basin. These facies pinch out toward the

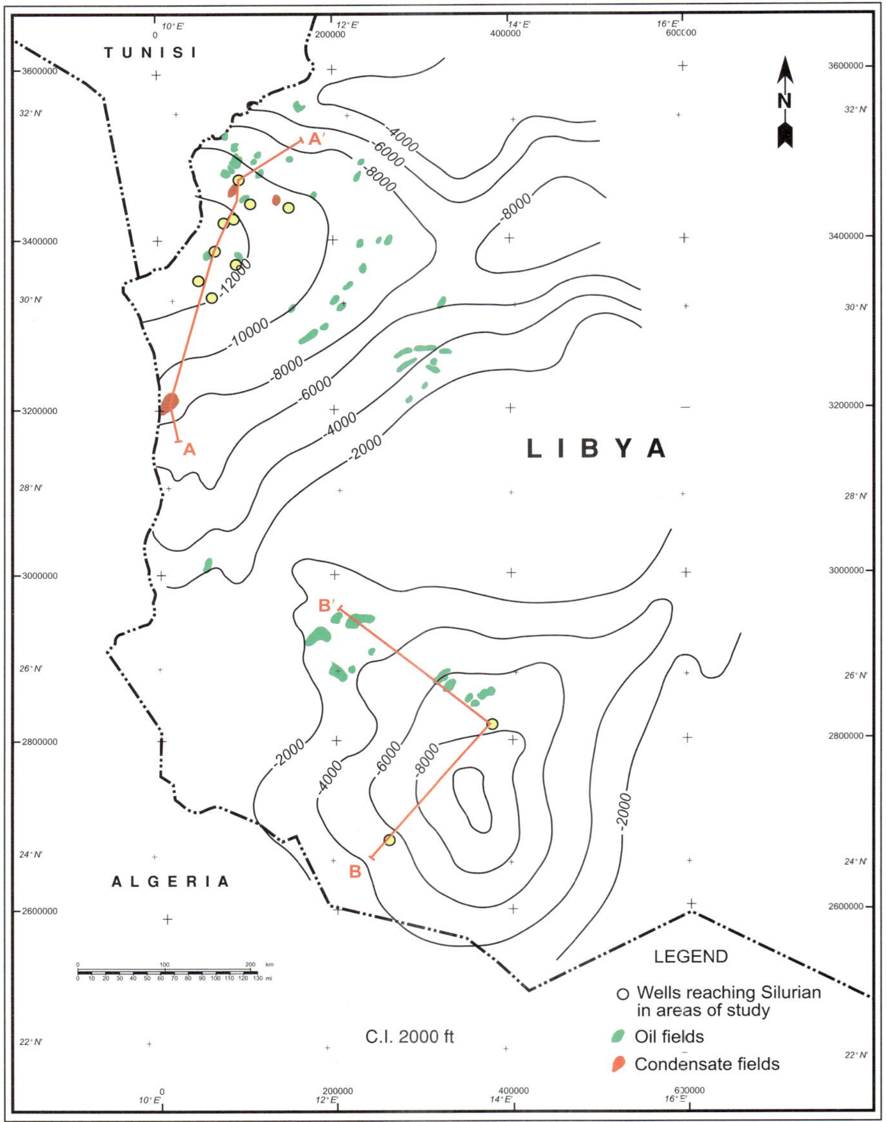

Figure 13. Structure map on the top Ordovician, Ghadamis, and Murzuq Basins, showing oil and gas fields and discoveries. Also shown are locations of cross sections A-A' and B-B' shown in Figure 14. Adapted from Masera Corporation (1992).

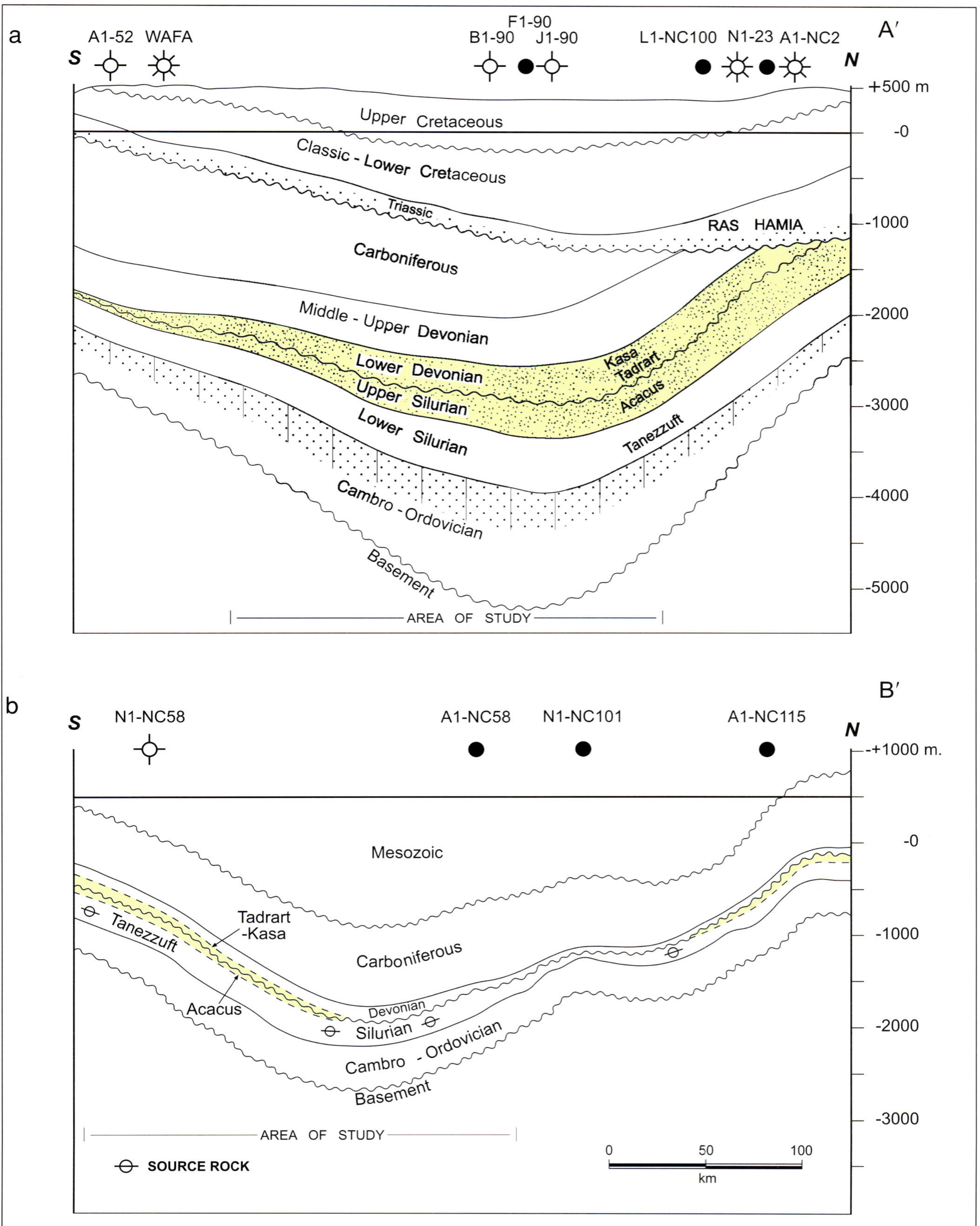

Figure 14. North-south structural cross section A-A′, Ghadamis Basin area of study, and (b) north-south structural cross section B-B′, Murzuq Basin area of study. Refer to Figure 13 for locations of cross sections.

inner shelf along the southwest margin of the study area and seaward of the shelf edge at the northern limits of the area. The top of the El Garia is at depths from 5000 ft in the southwest to 11,000 ft in the basin center (Bailey et al., 1989; Sbeta, 1990; El-Ghoul, 1991; Bernasconi et al., 1991; Loucks et al., 1998) (Figures 19, 20) .

Cretaceous reservoir considerations are speculative. However, based on stratigraphic projection from a few wells in the western part of the Tripolitania Basin and the northwestern part of the Sirt Basin, there appear to be several attractive secondary reservoir targets within the Cretaceous section. Probably the most important are the shallow-shelf skeletal limestone and dolomite facies of the Cenomanian-Turonian lower and upper Zebbag Formations. In the Libyan nomenclature, this sequence equates to the Alagah and Makhbaz Formations and the Lidam-Argub sequence. The net Upper Cretaceous porous carbonate section is estimated to thin basinward from a maximum thickness of 600 ft in the south to about 100 ft along the northern edge of the study area. These objective formations are at depths of 7500 to 15,000 ft (Figure 21).

Lower Cretaceous formations also have potential reservoir-quality facies. The shallow-marine carbonates and marginally marine sandstones of the Meloussi and Boudinar Formations and the rudist carbonates of the Serdj Formation (probable equivalents of the Turghat-Kiklah sequence) are potential targets. However, distribution and thickness are matters of speculation. Depths to Lower Cretaceous strata are 8000 to 16,000 ft.

Seals.—Shale and argillaceous limestone (mudstone-wackestone) beds provide effective seals for the underlying Cretaceous and Eocene reservoirs throughout most of the eastern sector of the basin (Figure 18).

Source, timing, and migration.—Mature organic-rich type II source beds have been identified in four formations in the basin. The best known and probably the most important is the Turonian Bahloul argillaceous limestone, with a TOC of 1–10% (Caron, 1999) (Figure 21). The Bahloul Formation is expected to have an average thickness of 100 ft in the study area. The organic-rich shale beds of the Sidi Kralif–Fahdene sequence, which have a TOC of 0.5–10% in offshore Tunisia, may be as effective as the Bahloul. The distribution and thickness of this sequence in the area of study are relatively unknown. However, on the basis of projection from a few wells to

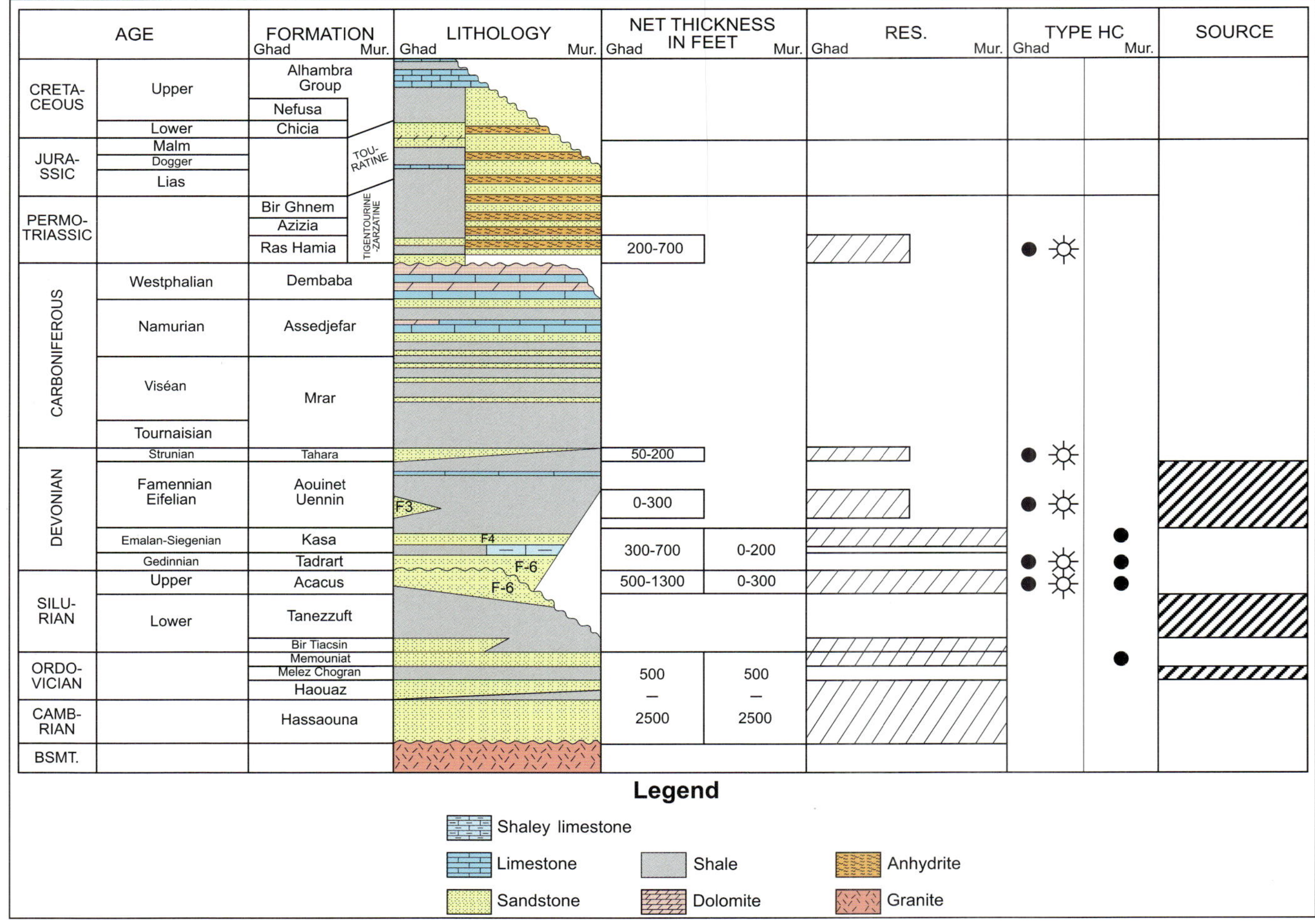

Figure 15. Generalized stratigraphic chart of Ghadamis and Murzuq Basins, showing source and potential reservoir intervals.

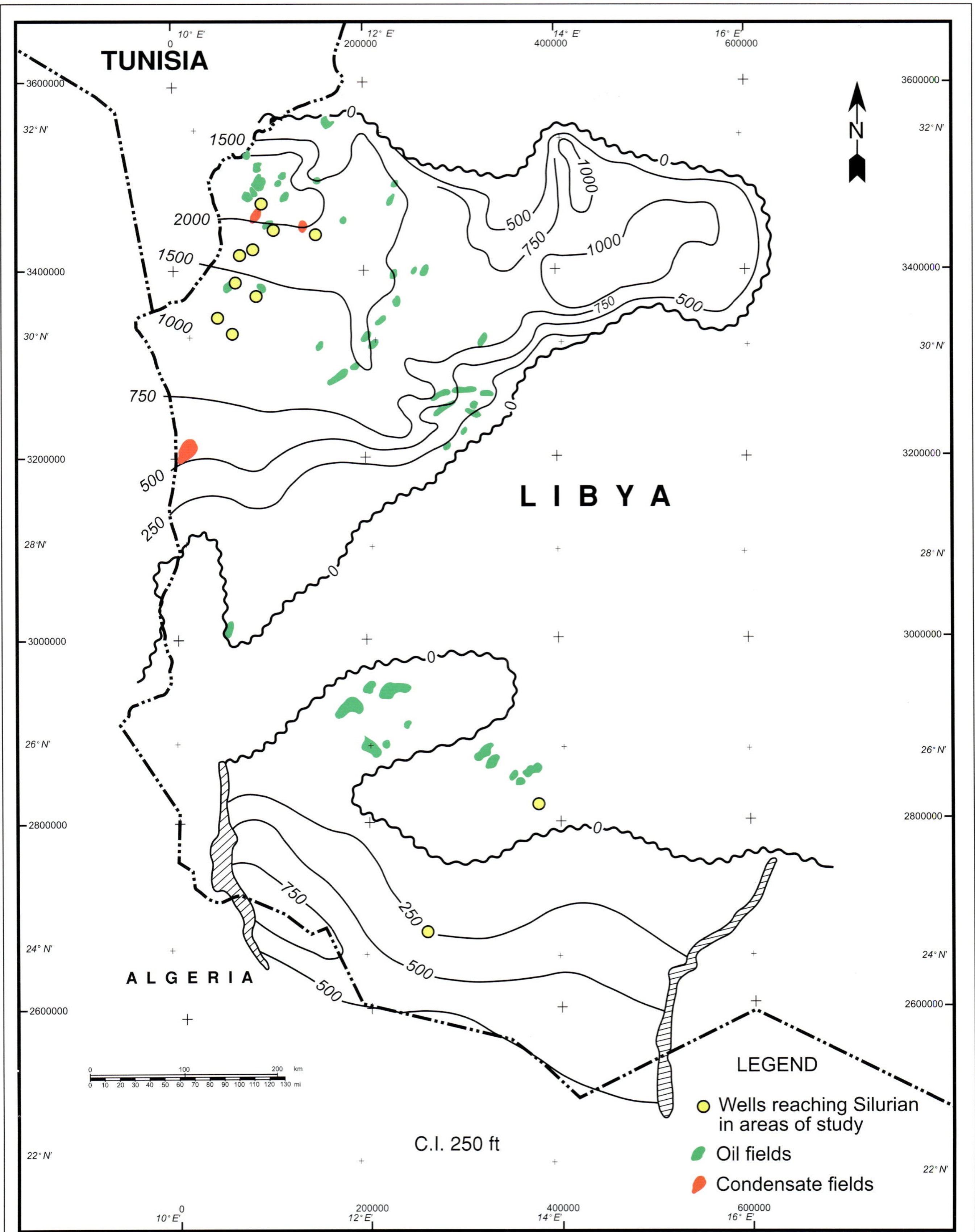

Figure 16. Isopach map of the Acacus Formation, Ghadamis and Murzuq Basins.

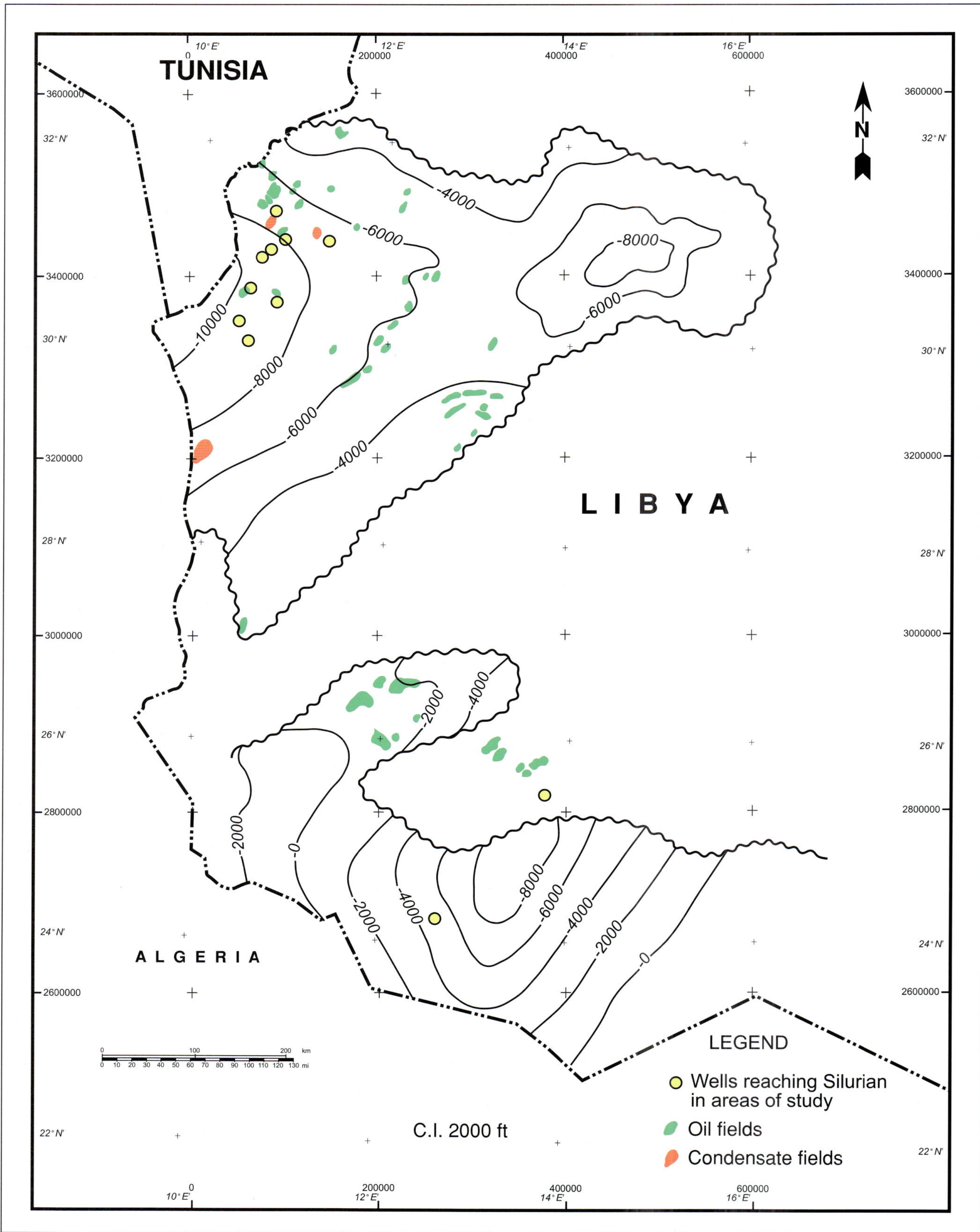

Figure 17. Structure map on the top Acacus Formation, Ghadamis and Murzuq Basins. Modified from Masera Corporation (1992).

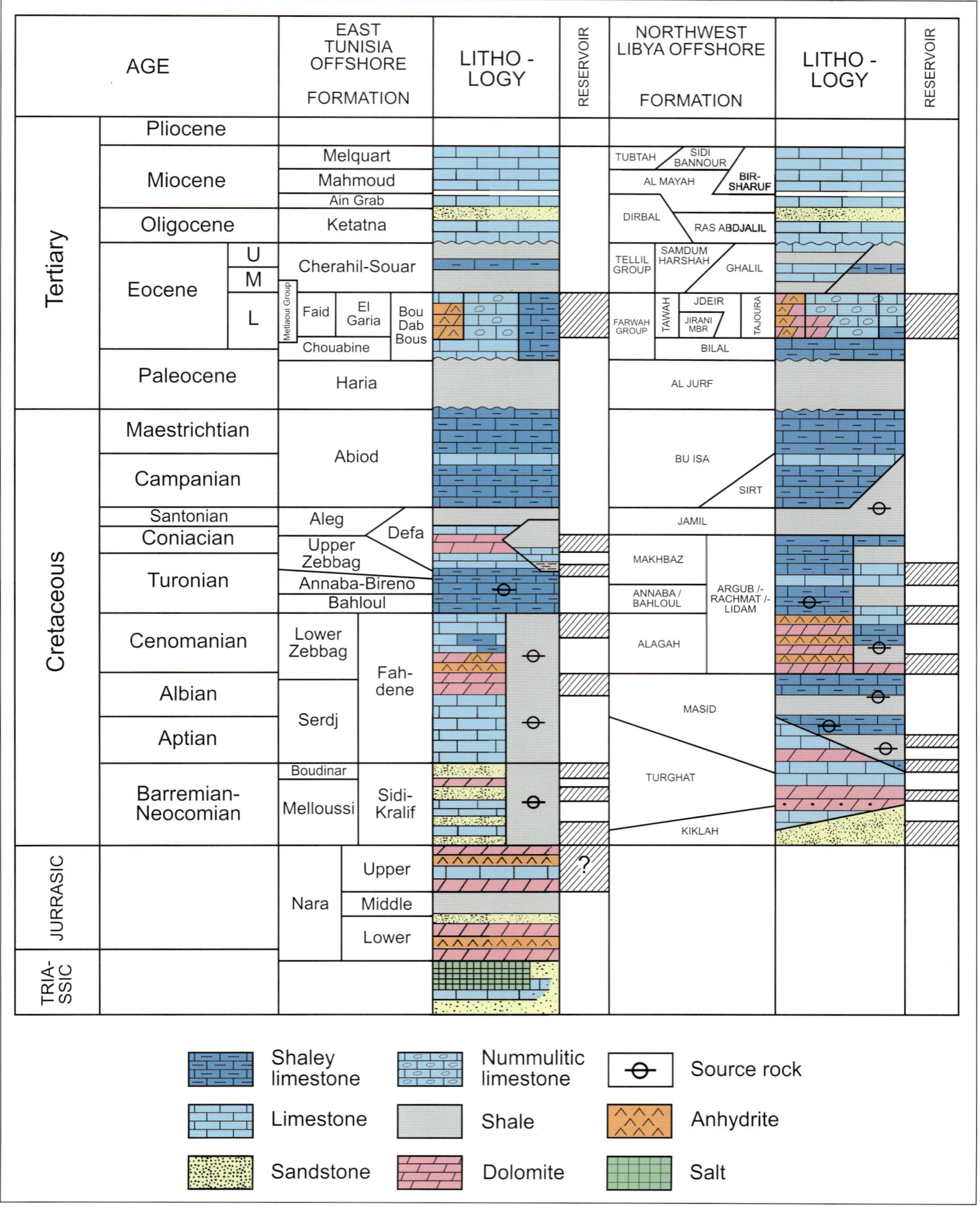

Figure 18. Stratigraphic correlation chart of formations and generalized lithologies of northwest offshore Libya and south offshore Tunisia. Also shown are the main reservoir and source units. Modified from Bishop (1988), Bernasconi et al. (1991), Sbeta (1990), and El-Ghoul (1991).

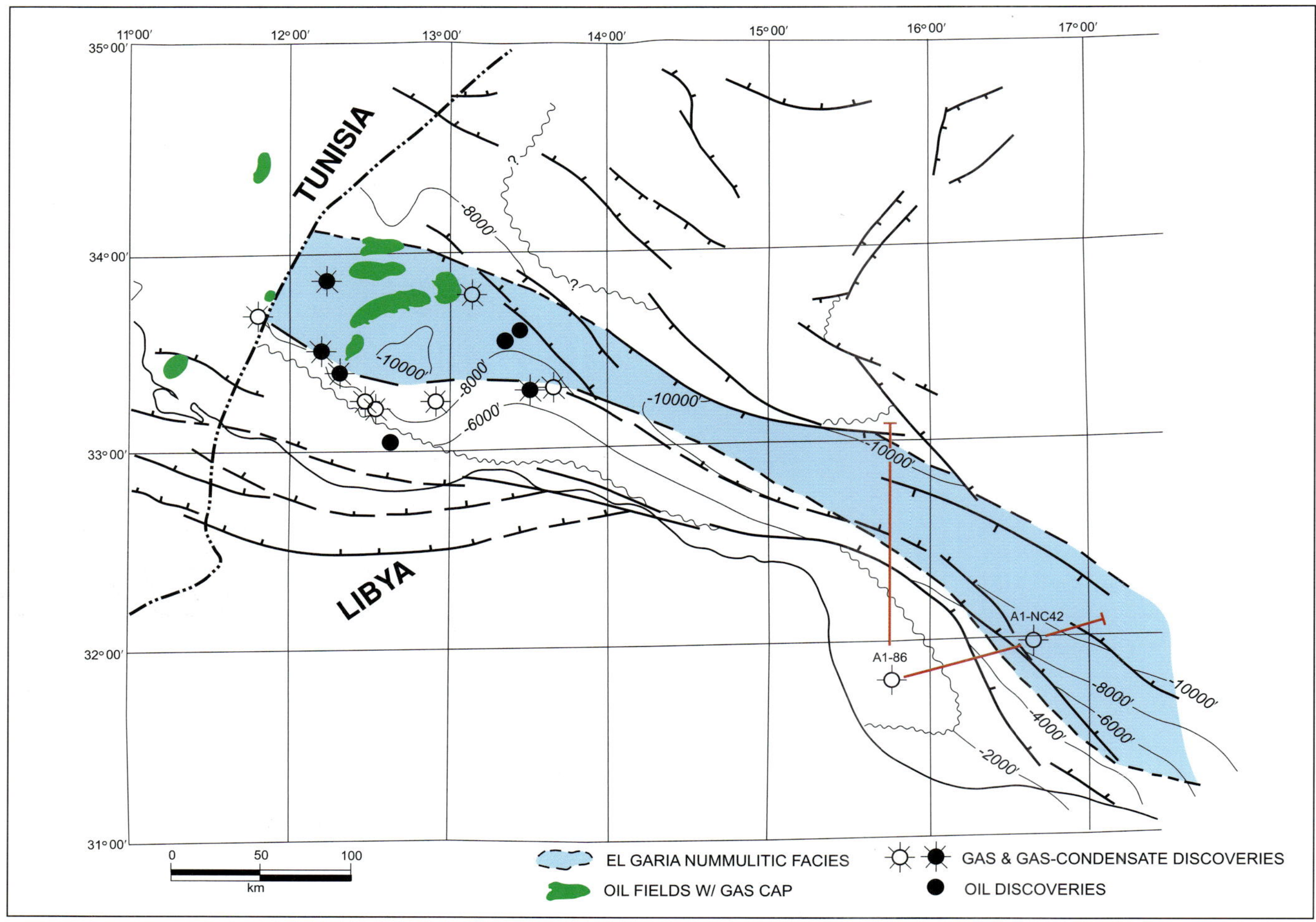

Figure 19. Structure map of the top Metlaoui group, Tripolitania Basin, showing distribution of the El Garia Formation (Jdeir) nummulitic facies.

the west, as much as approximately 400 ft can be expected in parts of the study area.

Along the extreme southwest part of the study area, the Silurian Tanezzuft shale thickens from an erosional edge on the north to more than 1000 ft at the southwest limits of the study area (Belhaj, 1996). It is estimated that Tanezzuft TOC is between 1% and 8%, based on Ghadamis Basin data.

The lower Eocene Chouabine limestone is considered to be an effective source rock in the western part of the Tripolitania Basin, although its area of peak generation is limited and it may not be present in the area of study.

The peak oil-generation-expulsion stage for the Tanezzuft shale probably occurred during the Paleogene. Peak oil generation for the Sidi Kralif-Fahdene and Bahloul Formations probably occurred from the Oligocene to the Miocene in the central part of the eastern Tripolitania Basin.

In the study area, it is likely that secondary migration was vertical or updip directly to reservoirs in some cases, and via carrier beds and faults in other cases.

Even though phases of recurrent faulting occurred throughout the Tertiary, the thick Miocene to Holocene section, with adequate shale intervals, should have preserved trap integrity in all but the southwestern quadrant. In this sector, which has a very thin Neogene section, there is a risk that late faulting could have caused seals to be breached.

Traps.—The trap types expected in the study area include faulted anticlines, horsts and tilted fault blocks, drape anticlines over carbonate buildups or faulted relief, and updip lithology or permeability pinch-outs.

CONCLUSIONS

The six underexplored basin or trough centers which are the subject of this paper have exceptional potential for major undiscovered petroleum resources.

In each of the six areas, which are peripheral to major oil and gas production, at least one well-defined petroleum system is established. These systems comprise mature, highly organic-rich source rock which provided a voluminous charge to multiple reservoirs by means of a variety of short-distance migration pathways.

In the Sirt Basin study areas, the Upper Jurassic–

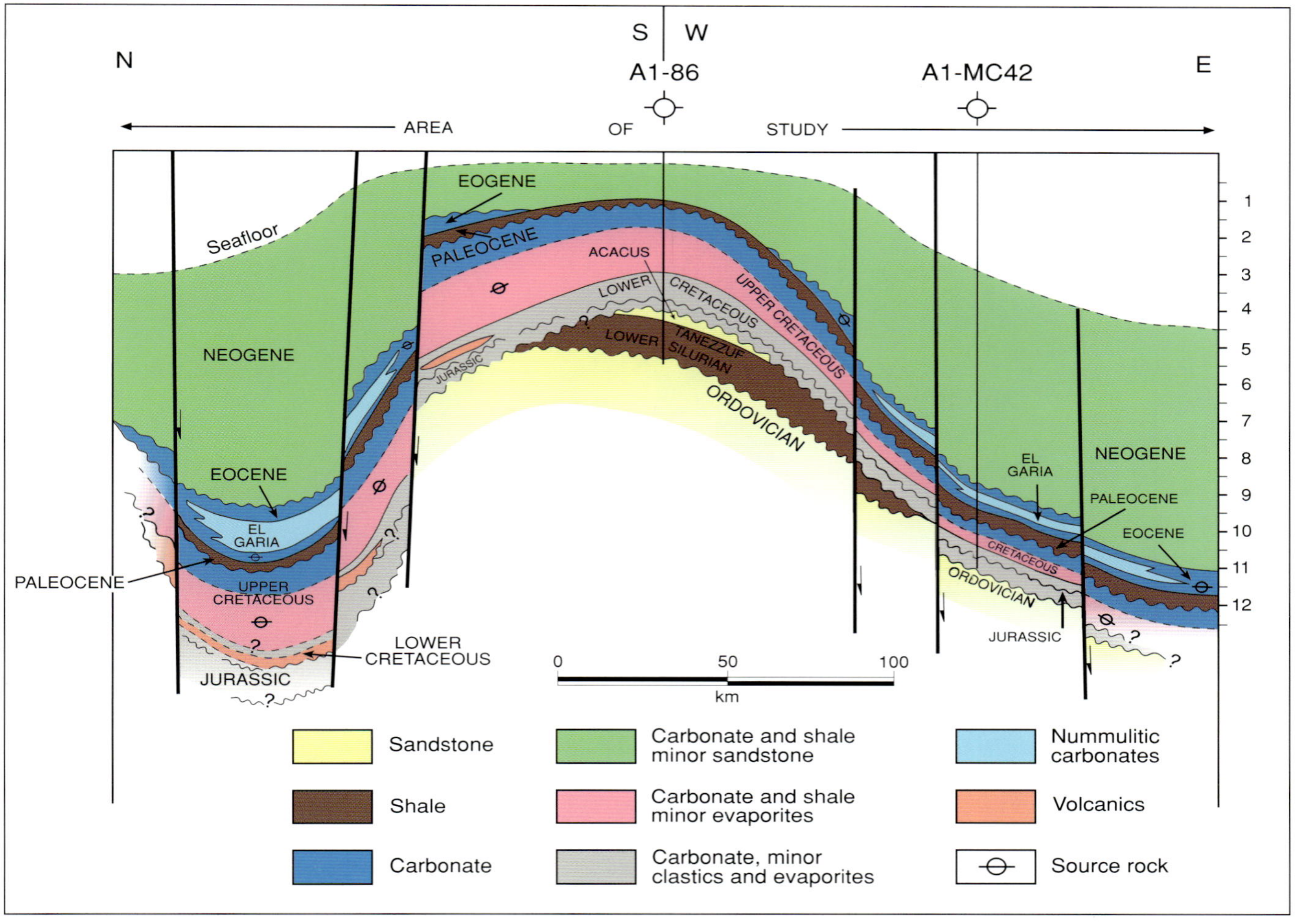

Figure 20. A diagrammatic north-south and west-east structural cross section of the Tripolitania Basin area of study. Adapted in part from Belhaj (1996).

Lower Cretaceous Nubian sandstone members should be considered primary objectives. This thick sandstone series, which is mostly at depths exceeding 12,000 ft, surprisingly has been the subject of minimal exploration to date.

In the Ghadamis and Murzuq Basins, sandstone sequences of the Upper Silurian Acacus and Lower Devonian Tadrart-Kasa Formations are definitely quality objectives, but they have not been priority targets. In the Ghadamis study area, which covers 20,000 km^2, only eight exploration wells reached the Acacus.

In the eastern Tripolitania Basin, in addition to the lower Eocene El Garia (Jdeir) nummulitic limestone, which is the major producing formation in the western part of the basin, reservoir potential includes numerous dominantly carbonate Lower and Upper Cretaceous formations.

The critical factor in determining future exploration success in the underexplored depocenters will probably be accurate trap definition. In general, at this stage in the exploration history of Libya, it is expected that the majority of the focus will be on subtle and complex trap types: low-relief faulted structures and drape anticlines, structural-stratigraphic combination traps involving facies pinch-outs, onlap terminations, and unconformity truncation. Identifying specific traps is complicated further by the fact that they are at considerable depths. Therefore, it will be necessary to adopt an integrated, interdisciplinary approach for in-depth, accurate interpretation of the specific trap or prospect. To accomplish this optimum level of trap definition, a detailed geologic database and state-of-the-art tools and methods will be required, including, for example, 3-D seismic, sequence stratigraphy, and basin-modeling concepts.

ACKNOWLEDGMENT

I extend special thanks to Paul McDaniel and John W. Shelton of Masera Corporation for their welcome assistance in the preparation of this paper and for the use of the Masera data set on Libya.

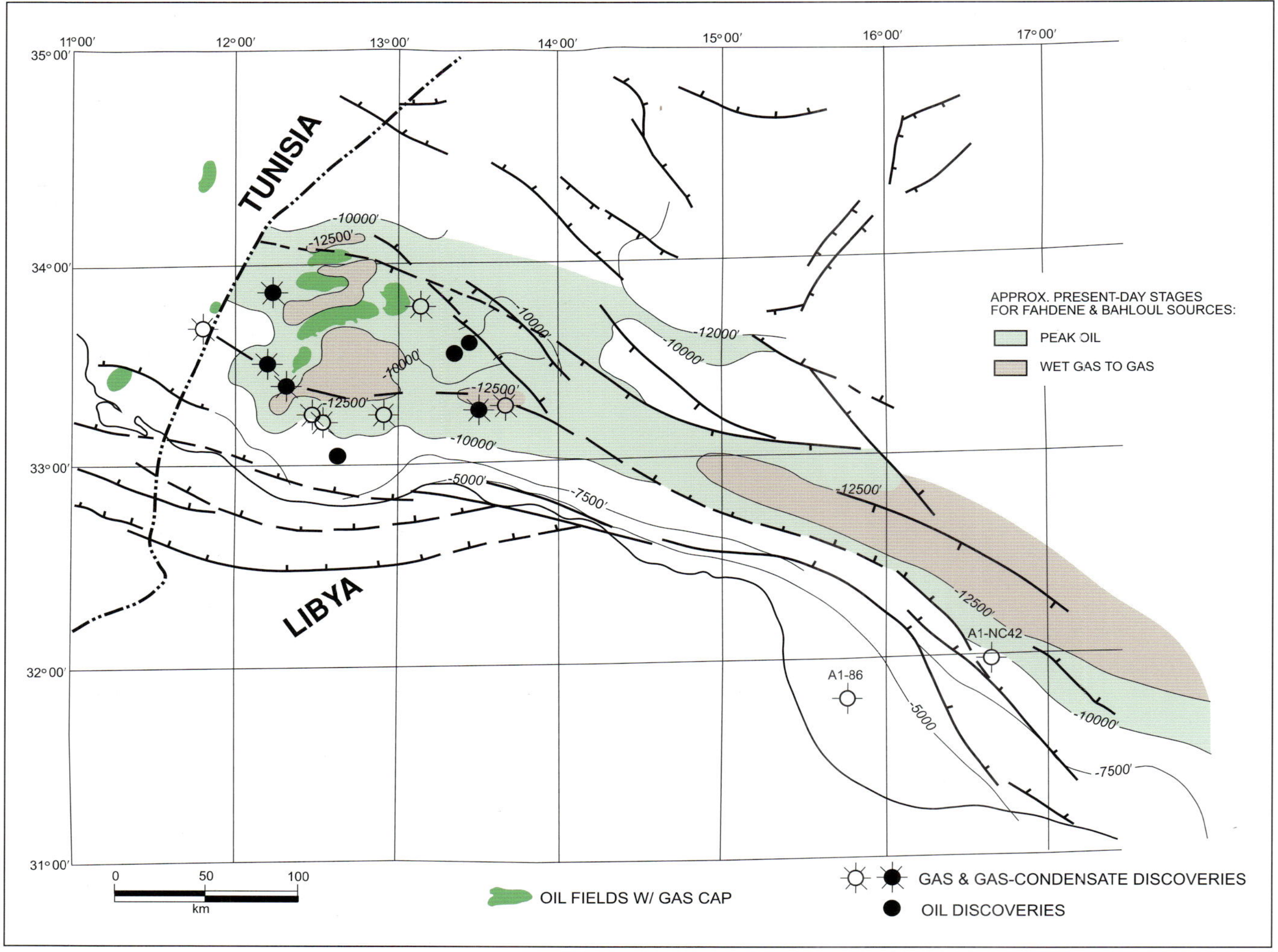

Figure 21. Structure map of the top Cretaceous, Tripolitania Basin, showing approximate zones of present-day peak oil generation of the Fahdene–Sidi Kralif and Bahloul source rocks.

REFERENCES CITED

Abdulghader, G. S., 1996, Depositional environment and diagenetic history of the Maragh formation, NE Sirt Basin, Libya, *in* M. J. Salem, A. S. El-Hawat, and A. M. Sbeta, eds., Geology of the Sirt Basin: Amsterdam, Elsevier, v. 2, p. 263–274.

Anketell, J. M., 1996, Structural history of the Sirt Basin and its relationships to the Sabratah Basin and Cyrenaican platform, northern Libya, *in* M. J. Salem, A. S. El-Hawat, and A. M. Sbeta, eds., Geology of the Sirt Basin: Amsterdam, Elsevier, v. 3, p. 57–88.

Baird, D. W., R. M. Aburawi, and N. J. L. Bailey, 1996, Geohistory and petroleum in the central Sirt Basin, *in* M. J. Salem, A. S. El-Hawat, and A. M. Sbeta, eds., Geology of the Sirt Basin: Amsterdam, Elsevier, v. 3, p. 3–56.

Bailey, H. W., G. Dungworth, M. Hardy, D. Scull, and R. D. Vaughan, 1989, A fresh approach to the Metlaoui: Actes de IIeme Journees de Géologie Tunisienne Appliquée à la Recherche des Hydrocarbures: Memoire de Enterprise Tunisienne d'Activités Petrólières 3, p. 281–308.

Baric, G., D. Spanic, and M. Maricic, 1996, Geochemical characterization of source rocks in NC 157 block (Zaltan platform), Sirt Basin, *in* M. J. Salem, A. S. El-Hawat, and A. M. Sbeta, eds., Geology of the Sirt Basin: Amsterdam, Elsevier, v. 2, p. 541–553.

Barr, F. T., and A. A. Weegar, 1972, Stratigraphic nomenclature of the Sirte Basin, Libya: Petroleum Exploration Society of Libya, 179 p.

Belhaj, F., 1996, Paleozoic and Mesozoic stratigraphy of eastern Ghadamis and western Sirt Basins, *in* M. J. Salem, A. S. El-Hawat, and A. M. Sbeta, eds., Geology of the Sirt Basin: Amsterdam, Elsevier, v. 1, p. 57–96.

Bellini, E., and D. Massa, 1980, A stratigraphic contribution to the Palaeozoic of the southern basins of Libya, *in* M. J. Salem and M. T. Busrewil, eds., Geology of Libya: London, Academic Press, p. 3–56.

Bernasconi, A., G. Poliani, and A. Dakshe, 1991, Sedimentology, petrography and diagenesis of Metlaoui Group in the offshore northwest of Tripoli, *in* M. J. Salem and M. N. Belaid, eds., The Geology of Libya: Third Symposium on the Geology of Libya, held at Tripoli, September 27–30, 1987: Amsterdam, Elsevier, v. 5, p. 1907–1928.

Bezan, A. M., F. Belhaj, and K. Hammuda, 1996, The Beda formation of the Sirt Basin, *in* M. J. Salem, A. S. El-Hawat, and A. M. Sbeta, eds., Geology of the Sirt Basin: Amsterdam, Elsevier, v. 2, p. 135–152.

Bishop, W. F., 1988, Petroleum geology of east-central Tunisia: AAPG Bulletin, v. 72, p. 1033–1058.

Bonnefous, J., 1972, Geology of the quartzitic "Gargaf Formation" in the Sirte Basin, Libya: Bulletin du Centre de Recherches de Pau, Société Nationale de Petrole Aquitaine, v. 6, p. 256–261

Boote, D. R. D., D. D. Clark-Lowes, and M. W. Traut, 1998, Palaeozoic petroleum systems of North Africa, *in* D. S. MacGregor, R. J. T. Moody, and D. D. Clark-Lowes, eds., Petroleum geology of North Africa: Geological Society of London, p. 7–68.

Caron, M., F. Robaszynski, F. Amedro, F. Baudin, J.-F. Deconinck, P. Hochuli, K. von Salis-Perch Nielsen, and N. Tribovillard, 1999, Estimation de la durée de l'événement anoxique global au passage Cenomanien/Turonien: Approche cyclostratigraphique dans la formation Bahloul en Tunisie centrale: Bulletin de la Société Géologique de France, v. 170, p. 145–160.

Clifford, H. J., R. Grund, and H. Musrati, 1980, Geology of a stratigraphic giant: Messla oil field, Libya, *in* M. T. Halbouty, ed., Giant oil and gas fields of the decade 1968–1978: AAPG Memoir 30, p. 507–524.

Echikh, K., 1998, Geology and hydrocarbon occurrences in the Ghadames Basin, Algeria, Tunisia, Libya, *in* D. S. MacGregor, R. J. T. Moody, and D. D. Clark-Lowes, eds., Petroleum geology of North Africa: Geological Society of London, p. 109–130.

El-Alami, M. A., 1996a, Petrography and reservoir quality of the Lower Cretaceous sandstone in the deep Mar trough, Sirt Basin, *in* M. J. Salem, A. S. El-Hawat, and A. M. Sbeta, eds., Geology of the Sirt Basin: Amsterdam, Elsevier, v. 2, p. 309–322.

El-Alami, M. A., 1996b, Habitat of oil in Abu Attiffel area, Sirt Basin, *in* M. J. Salem, A. S. El-Hawat, and A. M. Sbeta, eds., Geology of the Sirt Basin: Amsterdam, Elsevier, v. 2, p. 337–348.

El-Ghoul, A., 1991, A modified Farwah Group type section and its application to understanding stratigraphy and sedimentation along an E-W section through NC35A, Sabratah Basin, *in* M. J. Salem and M. N. Belaid, eds., Geology of Libya, p. 1637–1657.

El-Hawat, A. S., A. A. Missallati, A. M. Bezan, and T. M. Taleb, 1996, The Nubian sandstone in Sirt Basin and its correlatives, *in* M. J. Salem, A. S. El-Hawat, and A. M. Sbeta, eds., Geology of the Sirt Basin: Amsterdam, Elsevier, v. 2, p. 3–30.

Ghori, K. A. R., and R. A. Mohammed, 1996, The application of petroleum generation modelling to the eastern Sirt Basin, Libya, *in* M. J. Salem, A. S. El-Hawat, and A. M. Sbeta, eds., Geology of the Sirt Basin: Amsterdam, Elsevier, v. 2, p. 529–540.

Gras, R., 1996, Structural style of the southern margin of the Messlah High, *in* M. J. Salem, A. S. El-Hawat, and A. M. Sbeta, eds., Geology of the Sirt Basin: Amsterdam, Elsevier, v. 3, p. 201–210.

Gumati, Y. D., and W. H. Kanes, 1985, Early Tertiary subsidence and sedimentary facies—northern Sirte Basin, Libya: AAPG Bulletin, v. 69, p. 39–52.

Gumati, Y. D., and A. E. M. Nairn, 1991, Tectonic subsidence of the Sirte Basin, Libya: Journal of Petroleum Geology, v. 14, p. 93–102.

Gumati, Y. D., and S. Schamel, 1988, Thermal maturation history of the Sirte Basin, Libya: Journal of Petroleum Geology, v. 11, p. 205–218.

Hallett, D., and A. El-Ghoul, 1996, Oil and gas potential of the deep trough areas in the Sirt Basin, Libya, *in* M. J. Salem, A. S. El-Hawat, and A. M. Sbeta, eds., Geology of the Sirt Basin: Amsterdam, Elsevier, v. 2, p. 455–484.

Hamyouni, E. A., 1991, Petroleum source rock evaluation and timing of hydrocarbon generation, Murzuk Basin, Libya: A case study, *in* M. J. Salem and M. N. Belaid, eds., Geology of Libya, p. 183–211.

Hamyouni, E. A., I. A. Amr, M. A. Riani, A. B. El-Ghull, and S. A. Rahoma, 1984, Source and habitat of oil in Libyan basins: Presented at seminar on source and habitat of petroleum in the Arab countries, Kuwait, p. 125–178.

Ibrahim, M. W., 1991, Petroleum geology of the Sirt Group sandstones, eastern Sirt Basin, *in* M. J. Salem, M. T. Busrewil, and A. M. Ben Ashour, eds., The Geology of Libya: Third Symposium on the Geology of Libya, held at Tripoli, September 27–30, 1987: Amsterdam, Elsevier, v. 7, p. 2757–2779.

Johnson, B. A., and D. A. Nicoud, 1996, Integrated exploration for Beda Formation reservoirs in the southern Zallah trough (West Sirt Basin, Libya), *in* M. J. Salem, A. S. El-Hawat, and A. M. Sbeta, eds., Geology of the Sirt Basin: Amsterdam, Elsevier, v. 2, p. 211–222.

Klitzsch, E., 1971, The structural development of parts of North Africa since Cambrian time, *in* C. Gray, ed., Symposium on the geology of Libya: Tripoli, Faculty of Science of the University of Libya, p. 253–262.

Koscec, B. G., and Y. S. Gherryo, 1996, Geology and reservoir performance of Messlah oil field, Libya, *in* M. J. Salem, A. S. El-Hawat, and A. M. Sbeta, eds., Geology of the Sirt Basin: Amsterdam, Elsevier, v. 2, p. 365–390.

Loucks, R. G., R. T. J. Moody, J. K. Bellis, and A. A. Brown, 1998, Regional depositional setting and pore network systems of the El Garia Formation (Metlaoui group) lower Eocene, offshore Tunisia, *in* D. S. MacGregor, R. J. T. Moody, and D. D. Clark-Lowes, eds., Petroleum geology of North Africa: Geological Society of London, p. 355–374.

Mansour, A. T., and I. A. Magairhy, 1996, Petroleum geology and stratigraphy of the southeastern part of the Sirt Basin, Libya, *in* Geology of the Sirt Basin: Amsterdam, Elsevier, v. 2, p. 485–528.

Masera Corporation, 1992, Exploration geology and geophysics of Libya: Tulsa, Oklahoma, Masera Corporation, 205 p.

Meister, E. M., E. F. Ortiz, E. S. T. Pierobon, A. A. Arruda, and M. A. M. Oliveira, 1991, The origin and migration fairways of petroleum in the Murzuq Basin, Libya: An alternative exploration model, *in* M. J. Salem, M. T. Busrewil, and A. M. Ben Ashour, eds., The Geology of Libya: Third Symposium on the Geology of Libya, held at Tripoli, September 27–30, 1987: Amsterdam, Elsevier, v. 7, p. 2725–2742.

Parsons, M. G., A. M. Zagaar, and J. J. Curry, 1980, Hydrocarbon occurrence in the Sirte Basin, Libya, *in* A. D. Maill, ed., Facts and principles of world petroleum occurrence: Canadian Society of Petroleum Geology Memoir 6, p. 723–732.

Roohi, M., 1996a, A geological view of source-reservoir relationships in the western Sirt Basin, *in* M. J. Salem, A. S. El-Hawat, and A. M. Sbeta, eds., Geology of the Sirt Basin: Amsterdam, Elsevier, v. 2, p. 323–336.

Roohi, M., 1996b, Geological history and hydrocarbon migration pattern of the central Az Zahrah–Al Hufrah platform, *in* M. J. Salem, A. S. El-Hawat, and A. M. Sbeta, eds., Geology of the Sirt Basin: Amsterdam, Elsevier, v. 2, p. 435–454.

Said, F. M., 1974, Sedimentary history of the Paleozoic rocks of the Ghadames Basin in Libyan Arab Republic: Master's thesis, University of South Carolina, Columbia, 39 p.

Sbeta, A. M., 1990, Stratigraphy and lithofacies of Farwah Group and its equivalent: offshore—NW Libya: Petroleum Research Journal, v. 2, p. 42–56.

Schroter, T., 1996, Tectonic and sedimentary development of the central Zallah trough (West Sirt Basin, Libya), *in* M. J. Salem, A. S. El-Hawat, and A. M. Sbeta, eds., Geology of the Sirt Basin: Amsterdam, Elsevier, v. 3, p. 123–136.

Sinha, R. N., and I. Y. Mriheel, 1996, Evolution of subsurface Palaeocene sequence and shoal carbonates, south-central Sirt Basin, *in* M. J. Salem, A. S. El-Hawat, and A. M. Sbeta, eds., Geology of the Sirt Basin: Amsterdam, Elsevier, v. 2, p. 153–196.

Spring, D., and O. P. Hansen, 1998, The influence of platform morphology and sea level on the development of a carbonate sequence: The Harash Formation, eastern Sirt Basin, Libya, *in* D. S. MacGregor, R. J. T. Moody, and D. D. Clark-Lowes, eds., Petroleum geology of North Africa: Geological Society of London, p. 335–354.

Van Houten, B. F., 1980, Latest Jurassic–earliest Cretaceous regressive facies, northeast African craton: AAPG Bulletin, v. 64, p. 857–867.

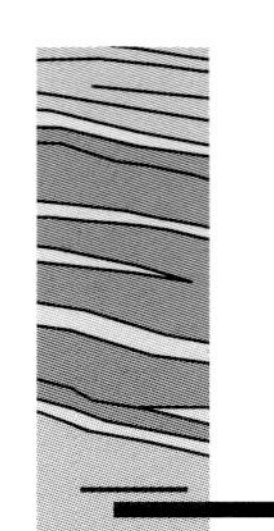

Dolson, J. C., M. V. Shann, S. Matbouly, C. Harwood, R. Rashed, and H. Hammouda, 2001, The petroleum potential of Egypt, *in* M. W. Downey, J. C. Threet, and W. A. Morgan, eds., Petroleum provinces of the twenty-first century: AAPG Memoir 74, p. 453–482.

Chapter 23

THE PETROLEUM POTENTIAL OF EGYPT

John C. Dolson
BP Amoco Egypt, Cairo, Egypt

Mark V. Shann
BP Amoco, London, United Kingdom

Sayed Matbouly
Egyptian General Petroleum Corporation Cairo, Egypt

Colin Harwood
British Gas, Cairo, Egypt

Rashed Rashed
Gulf of Suez Petroleum Company, Cairo, Egypt

Hussein Hammouda
Gulf of Suez Petroleum Company, Cairo, Egypt

ABSTRACT

Since the onshore discovery of oil in the Eastern Desert in 1886, the petroleum industry in Egypt has discovered more than 15.7 billion barrels of oil equivalent (BOE) of reserves. This paper uses an understanding of the tectono-stratigraphic history of each major basin, combined with drilling history and field-size distributions, to justify predicting that Egypt's future potential resource base will be twice what it is today.

Major reserve replacement will come from expansion of existing petroleum plays into the Mediterranean Tertiary gas trends. Additional reserve growth will result from successes using 3-D seismic methods in deeper pool exploration in and around proven fields, and for new stratigraphic plays off-structure. Examples from the Western Desert, the Gulf of Suez, and the Mediterranean demonstrate this growth potential.

More remote new exploration areas include the Komombo and other basins in Upper Egypt and the northern end of the Red Sea rift, both of which are under reevaluation by several international oil companies.

Despite a relatively complex geologic history, the geologic framework of Egypt is highly suited for oil and gas exploration. It comprises eight major tectono-stratigraphic events: (1) Paleozoic craton, (2) Jurassic rifting, (3) Cretaceous passive margin, (4) Cretaceous Syrian arc deformation and foreland transgressions, (5) Oligocene-Miocene Gulf of Suez rifting, (6) Miocene Red Sea breakup, (7) the Messinian salinity crisis, and (8) Pliocene-Pleistocene delta progradation. Each of those events created multiple reservoir and seal combinations. Source rocks occur from the Paleozoic through to the Pliocene, and petroleum is produced from Precambrian through Pleistocene reservoirs.

INTRODUCTION

Despite its petroleum-exploration history of more than 100 years, many large geographic areas in Egypt remain underexplored (Figures 1, 2). Although Egypt has the physical size of the U.S. state of Texas, it contains only 1754 exploratory tests (Figure 2). Only 245 of those wells penetrated Precambrian strata (Table 1), and many of those did not drill a complete stratigraphic section before crossing a fault into basement. Those exploratory tests have resulted in the discovery of 30 giant (>100 million barrels oil equivalent [MMBOE]) oil and gas fields, seven

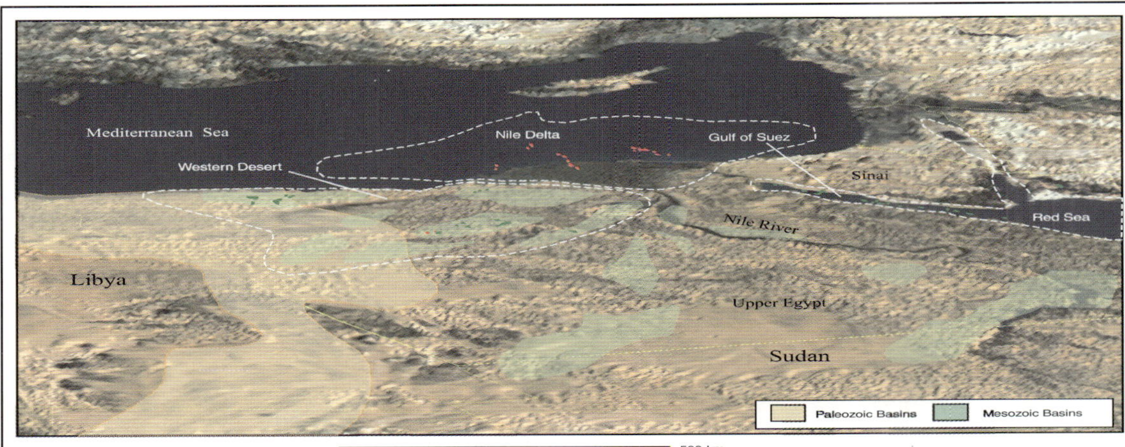

Figure 1. Landsat image of Egypt with Mesozoic, Paleozoic, and Tertiary basins and oil (green) and gas (red) fields identified.

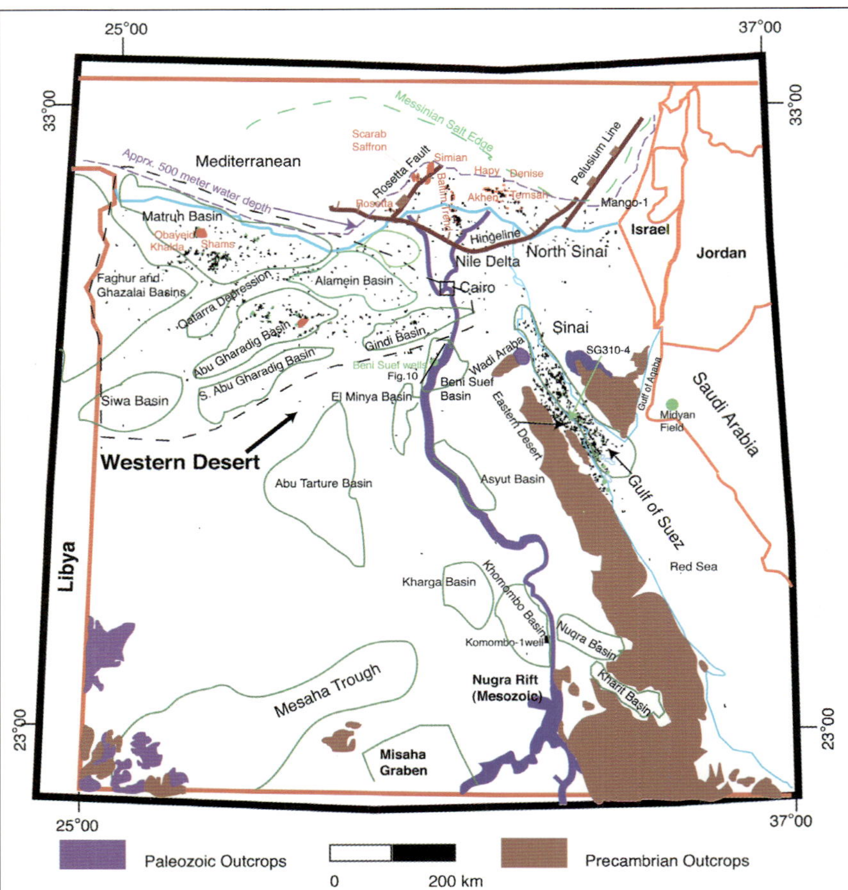

Figure 2. Location map of the study area showing all exploratory tests in Egypt. Significant wells discussed in the text are highlighted in red or indicated as text, as are the locations of giant-field discoveries in the 1990s.

of which were found in the late 1990s (Table 2). In addition, a substantial number of significant discoveries less than 100 MMBOE in size have also been made (Table 3).

Egypt's geologic history is complex and, although a full discussion is beyond the scope of this paper, a sequence-stratigraphic-based broad tectono-stratigraphic framework is presented to describe the context of the proven petroleum systems and to outline future exploration potential. Key references dealing with the petroleum systems and geology of Egypt and its surrounding areas include Dixon and Robertson (1984), Said (1990b), Sadek (1992), MacGregor et al. (1998), and Purser and Bosence (1998).

The Gulf of Suez, Nile Delta, offshore Mediterranean, and greater West-

Table 1. Exploratory penetrations in Egypt by age at total depth.

Petroleum System	Total Wells	Tertiary	Cretaceous	Jurassic	Triassic	Paleozoic	Precambrian
Western Desert	578	51	319	137	0	40	31
Nile Delta, North Sinai, Mediterranean	247	199	27	20	0	1	0
Gulf of Suez, Eastern Desert, Sinai	902	260	412	13	0	19	198
Upper Egypt	13	0	5	0	0	0	8
Red Sea	14	6	0	0	0	0	8
Totals	**1754**	**516**	**763**	**170**	0	**60**	**245**

Table 2. Giant oil and gas fields of Egypt (defined as >100 MMBOE). Data are complete to January 1, 2000 (three more giant gas fields [not shown] were discovered in the Mediterranean province in early 2001).

FIELD	COMPANY	YEAR	DRILLING PROVINCE	MMBOE	BCF GAS	DISCOVERY WELL	AGE	MAJOR TRAP	PRIMARY REFERENCE
BELAYIM MARINE	PETROBEL	1961	GULF OF SUEZ	1593	0.00	BELAYIM M-1	LANGIAN	Structural	Matbouly and Sabbagh, 1996
MORGAN OLD-SOUTH	GUPCO	1965	GULF OF SUEZ	1201	0.00	MORGAN-1	LANGIAN	Structural	Matbouly and Sabbagh, 1996
OCTOBER MAIN	GUPCO	1977	GULF OF SUEZ	848	0.00	GS195-1(OCT-A1)	CRETACEOUS	Structural	Matbouly and Sabbagh, 1996
RAMADAN	GUPCO	1974	GULF OF SUEZ	668	0.00	GS 303-1	CRETACEOUS	Structural	Matbouly and Sabbagh, 1996
SIMIAN	BRITISH GAS	1999	MEDITERRANEAN	416-666	2500-4000	SIMIAN-1	PLIOCENE	Stratigraphic	IHS Energy Group, 1999
BELAYIM LAND	PETROBEL	1955	SINAI	645	0.00	BELAYIM 112-1	SERRAVALIAN	Structural	Matbouly and Sabbagh, 1996
JULY	GUPCO	1973	GULF OF SUEZ	625	0.00	GS 311-1 (J-4)	BURDIGALIAN	Structural	Matbouly and Sabbagh, 1996
SCARAB	BRITISH GAS	1998	MEDITERRANEAN	375-466	2800.00	SCARAB-1	PLIOCENE	Structural	IHS Energy Group, 1999
TEMSAH	MOBIL	1981	MEDITERRANEAN	333-450	2000-2700	EL TEMSAH-2	SERRAVALIAN	Structural	IHS Energy Group, 1999
OBAYEID	OBAIYED	1993	WESTERN DESERT	283-366	1700-2200	OBAYID-3	JURASSIC	Structural	IHS Energy Group, 1999
ROSETTA	BRITISH GAS	1997	MEDITERRANEAN	333-416	2000-2500	ROSETTA-3	PLIOCENE	Structural	IHS Energy Group, 1999
RAS GHARIB	GPC	1938	EASTERN DESERT	357	0.00	RAS GHARIB-3	SERRAVALIAN	Combination	Matbouly and Sabbagh, 1996
SAFFRON	BRITISH GAS	1998	MEDITERRANEAN	333-416	2000-2500	SAFFRON-1	PLIOCENE	Structural	IHS Energy Group, 1999
HAPY	BP AMOCO	1997	MEDITERRANEAN	250-416	1500-2500	HAPY-1	PLIOCENE	Structural	EGPC Records
RAS BUDRAN	SUCO	1978	GULF OF SUEZ	270	0.00	EE 85-1A	CRETACEOUS	Structural	Matbouly and Sabbagh, 1996
BADRI	GUPCO	1987	GULF OF SUEZ	267	0.00	BDR-E-1	SERRAVALIAN	Structural	Matbouly and Sabbagh, 1996
DENISE	IEOC	1995	MEDITERRANEAN	125-150	750-900	DENISE-1	PLIOCENE	Structural	IHS Energy Group, 1999
ABU GHARADIG	GUPCO	1969	WESTERN DESERT	220	586.00	ABU GHARADIG-1	CRETACEOUS	Structural	Hegazy, 1992
BALTIM	IEOC	1993	MEDITERRANEAN	83-133	500-800	BALTIME-1	MESSINIAN	Combination	IHS Energy Group, 1999
ZEIT BAY	SUCO	1980	GULF OF SUEZ	215	0.00	OO 89-1	CRETACEOUS	Structural	Matbouly and Sabbagh, 1996
KHALDA	WEPCO	1971	WESTERN DESERT	213	771.00	KHALDA-1	CRETACEOUS	Structural	Hegazy, 1992
ABU MADI	IEOC	1967	NILE DELTA	209	1254.00	ABU MADI-1	MESSINIAN	Stratigraphic	Moussa and Matbouly, 1994
MORGAN OLD-NORTH	GUPCO	1965	GULF OF SUEZ	200	0.00	MORGAN-1	LANGIAN	Structural	Matbouly and Sabbagh, 1996
SHAMS	REPSOL	1997	WESTERN DESERT	176-250	1000-1500	SHAMS-2X	JURASSIC	Structural	IHS Energy Group, 1999
BED-3	SHELL	1983	WESTERN DESERT	153	847.00	BED 3-1	CRETACEOUS	Structural	Hegazy, 1992
RAS FANAR	SUCO	1978	GULF OF SUEZ	143	0.00	KK 84-1	SERRAVALIAN	Stratigraphic	Matbouly and Sabbagh, 1996
HILAL	GUPCO	1976	GULF OF SUEZ	125-140	0.00	GS 391-1	CRETACEOUS	Structural	Matbouly and Sabbagh, 1996
BAKR	GPC	1958	EASTERN DESERT	135	0.00	BAKR-6	SERRAVALIAN	Structural	Matbouly and Sabbagh, 1996
BED-2	SHELL	1982	WESTERN DESERT	132	792.10	BED 2-1	CRETACEOUS	Structural	Hegazy, 1992
SHOAB ALI	GUPCO	1977	GULF OF SUEZ	110	0.00	ALMA-2	SERRAVALIAN	Structural	Matbouly and Sabbagh, 1996
KANAYES	IEOC	1992	WESTERN DESERT	100-116	600-700	KANAYIS - 5	JURASSIC	Structural	IHS Energy Group, 1999

Table 3. Significant exploratory tests in Egypt in the 1990s.

WELL	REGION	LAT.	LONG.	YEAR	CLASS
KHARIT-1	UPPER EGYPT	23.5155	34.3507	1998	DRY HOLE
KOMOMBO-2 ST	UPPER EGYPT	24.6543	32.8083	1998	DRY HOLE
KOMOMBO-3	UPPER EGYPT	24.5995	32.7390	1998	DRY HOLE
NUQRA-1	UPPER EGYPT	24.4208	33.4982	1997	DRY HOLE
KOMOMBO-1	UPPER EGYPT	24.6665	32.8064	1997	DRY HOLE
BENI SUEF 1X	UPPER EGYPT	29.1533	30.8738	1997	DISCOVERY
BENI SUEF-4X	UPPER EGYPT	29.1533	30.8858	1998	DISCOVERY
BENI SUEF-5X	UPPER EGYPT	29.1532	30.8801	1998	DISCOVERY
ASHRAFI SW 3	GULF OF SUEZ	27.7762	33.7089	1998	DISCOVERY
E.TANKA-3 (ET-A1)	GULF OF SUEZ	28.9845	32.9443	1996	DISCOVERY
GS 184-2	GULF OF SUEZ	28.8794	33.0264	1994	DISCOVERY
RABEH-1	GULF OF SUEZ	27.2229	33.7466	1997	DISCOVERY
SG 310-4	GULF OF SUEZ	28.2532	33.2251	1998	DISCOVERY
SG 310-6A	GULF OF SUEZ	28.2498	33.2100	1999	DISCOVERY
WARDA	GULF OF SUEZ	29.1900	32.7063	1991	DISCOVERY
AKHEN-1	MEDITERRANEAN	31.9067	31.9213	1996	DISCOVERY
BALTIM E-1	MEDITERRANEAN	31.7748	31.2449	1993	DISCOVERY
DENISE-1	MEDITERRANEAN	31.8705	32.0959	1995	DISCOVERY
EL TEMSAH NW-1	MEDITERRANEAN	31.8619	32.1233	1996	DISCOVERY
HAPY-1	MEDITERRANEAN	31.9197	31.8544	1996	DISCOVERY
PFM SW-1 "ST"	MEDITERRANEAN	31.5506	32.4446	1998	DISCOVERY
ROSETTA-3	MEDITERRANEAN	31.8408	30.6262	1997	DISCOVERY
SAFFRON-1	MEDITERRANEAN	32.1047	30.5344	1998	DISCOVERY
SCARAB-1	MEDITERRANEAN	32.0474	30.6302	1998	DISCOVERY
SIMIAN-1	MEDITERRANEAN	32.1920	30.7990	1999	DISCOVERY
TAO-1	MEDITERRANEAN	31.6142	32.7733	1997	DISCOVERY
TUNA-1	MEDITERRANEAN	31.8961	32.2160	1996	DISCOVERY
MARAKIA-1	MEDITERRANEAN	31.1871	29.6316	1992	DRY HOLE
EL SAGHA-3X	WESTERN DESERT	29.7880	30.5814	1995	DISCOVERY
KANAYIS-4	WESTERN DESERT	31.0609	27.6718	1992	DISCOVERY
OBA A-3	WESTERN DESERT	31.1273	26.5693	1996	DISCOVERY
OBAYID-1	WESTERN DESERT	31.0723	27.0502	1992	DISCOVERY
QARUN A-4X	WESTERN DESERT	29.7665	30.5977	1995	DISCOVERY
S.W.QARUN-1X	WESTERN DESERT	29.7766	30.5226	1996	DISCOVERY
SHAMS S-1X	WESTERN DESERT	30.8262	26.9078	1996	DISCOVERY
SHAMS-2X	WESTERN DESERT	30.8508	26.9284	1997	DISCOVERY

ern Desert basins are the only basins proved to contain economically viable petroleum resources. At least six sedimentary basins in Upper Egypt have had little or no hydrocarbon exploration. In 1997, one of those basins, the Komombo Basin, tested live oil from Jurassic reservoirs.

The primary focus of this paper is on the basins with proven play systems where significant well and seismic data are available and where industry activity is continuing to establish significant new discoveries or has the potential to open up new trends. We provide a short overview of the potential of the Upper Egypt and Red Sea Basins, based on new data acquired in the late 1990s.

COMMENTS	COMPANY	AGE AT TD	FTD (METERS)
P & A; ESTABLISHES PRESENCE OF NEW BASIN IN EGYPT	REPSOL	PRECAMBRIAN	2307.53
P & A; ESTABLISHES PRESENCE OF NEW BASIN IN EGYPT	REPSOL	PRECAMBRIAN	2651.63
P & A; ESTABLISHES PRESENCE OF NEW BASIN IN EGYPT	REPSOL	PRECAMBRIAN	1281.32
P & A; ESTABLISHES PRESENCE OF NEW BASIN IN EGYPT	REPSOL	PRECAMBRIAN	2549.53
Tested oil in Jurassic; proves viability of new basin in Upper Egypt	REPSOL	PRECAMBRIAN	2579.00
Oil Disc. (Kharita & Bahariya); southernmost Egypt oil extension	SEAGULL	PRECAMBRIAN	3388.60
Oil well in Cretaceous; southernmost Egypt oil extension	SEAGULL	LOWER CRETACEOUS	2197.50
Oil well Bahariya & Kharita T & A.; southernmost Egypt oil extension	SEAGULL	LOWER CRETACEOUS	2256.63
Nubia oil: Significant southern extension of pay in Gulf of Suez	AGIBA	PRECAMBRIAN	1904.91
New field discovery of "downthrown" Asl sand trap: IP 10,000 BOPD	AMOCO	BURDIGALIAN	2834.50
New field discovery of "downthrown" Asl sand trap: offset flows 15,000 BOPD	GUPCO	BURDIGALIAN	3554.71
6800 BOPD Nukhul and Matulla (Nezzazat); significant southern extension of production	COPLEX	PRECAMBRIAN	
Oil well from Asl Sd &Hawara Sd (Burdigalian) flows 20,000 BOPD	GUPCO	BURDIGALIAN	3230.72
Oil well from Asl Sd &Hawara Sd: significant small field discovery on new fault block	GUPCO	BURDIGALIAN	4462.05
40+ MMBO oil-field discovery beyond limits of Belayim Salt top seal from Kareem and Rudeis Formations	BRITISH GAS	LANGHIAN	2752.21
New Serravalian field discovery (350–700 bcf)	AMOCO	BURDIGALIAN	4421.82
New Messinian valley-fill trend extension (500–800 bcf)	IEOC	SERRAVALIAN	3911.92
Pliocene gas discovery (750–900 bcf)	IEOC	PLIOCENE	2402.93
Significant Serravalian pay extension	IEOC	BURDIGALIAN	4019.81
Giant Pliocene gas discovery (1500–2500 bcf)	AMOCO	PLIOCENE	1861.93
Pliocene gas discovery Port Fouad SE (350–500 bcf)	PETROBEL	PLIOCENE	3985.98
Giant Pliocene gas discovery (1500–2000 bcf)	BRITISH GAS	PLIOCENE	1943.92
Giant Pliocene gas discovery (2000–2500 bcf)	BRITISH GAS	PLIOCENE	2374.89
Giant Pliocene gas discovery (2250–2700 bcf)	BRITISH GAS	PLIOCENE	2099.97
Giant Pliocene gas discovery (2500–4000 bcf)	BRITISH GAS	PLIOCENE	2264.86
Significant Pliocene gas discovery (250–550 bcf)	AMOCO	PLIOCENE	2719.90
Significant Pliocene gas discovery (350–500 bcf)	IEOC	PLIOCENE	1259.98
P & A; Tested 9 liters of oil in the Cretaceous; northern extension of Western Desert Cretaceous into offshore Mediterranean	SHELL	LOWER CRETACEOUS	4366.96
Qarun field discovery, Bahariya (80–100 MMBOE)	PHOENIX	LOWER CRETACEOUS	2936.60
Discovery as gas and condensate from Khatatba (Jurassic) 19.2 MMSCFGD & 1300 BCPD; 77.3 MMBOE	NORSK HYDRO	JURASSIC	4765.01
Recovered gas/condensate from Paleozoic strata	SHELL	CARBONIFEROUS	4132.89
Giant Jurassic and Lower Cretaceous field discovery: (1700–2200 bcf)	SHELL	JURASSIC	5038.10
Significant southern extension of Qarun field pay	PHOENIX	LOWER CRETACEOUS	2939.96
Extension of Qarun field pay	APACHE	LOWER CRETACEOUS	3349.89
Discovery gas/condensate in Kharita (380–500 bcf)	KHALDA	JURASSIC	4114.60
Significant Jurassic discovery (1000–1500 bcf)	REPSOL	JURASSIC	

METHODOLOGY AND PRIOR WORK

The undiscovered reserves ("yet-to-find" numbers) of potential giant fields that were used to postulate a minimum resource doubling for Egypt in the coming decades are based on the integration of two analyses: field-size distribution and drilling-success statistics, interpreted in a petroleum-systems context.

We have used a historical exploratory-drilling database made available by Egyptian General Petroleum Corporation, in conjunction with published field data from more than 251 fields (Hegazy, 1992; Moussa and Matbouly, 1994; Matbouly and Sabbagh, 1996), to understand the exploratory-drilling history by basin and play type. Post-1990 field-size data from the IHS Energy Group (London, U.K.) have been added to supplement missing data. Additional reserve data are available from El-Banbi (1999). Reserve numbers presented in this paper (Table 4) are 15% to 20% larger than those reported in El-Banbi (1999), which contains production data current through the end of 1997. Our larger reserve estimate results from inclusion of several significant new discoveries in 1998 and 1999.

This combination of historical drilling data and field data has provided field-size distributions for recoverable reserves estimates from which "yet-to-find" numbers of giant fields in each basin or trend have been derived, using the assumption that field-size distributions follow log-normal trends. This technique of estimating "yet to find" by statistical analysis of log-normal plots for known discoveries follows published work by Capen (1992), Smith and Jones (1992), Root and Attanasi (1993), and Drew and Schuenemeyer (1993). Historical drilling-success rates by basin and trend have also been used to assess which plays are "emerging" in Egypt and which are matured and apparently "played out."

The trend assessments are discussed in terms of discoveries from structural, combination, or stratigraphic traps, along with the age of the major producing zones. Structural discoveries are those consisting of four-way closures or fault-bounded traps. Combination traps involve at least one overriding component of stratigraphic seal, and stratigraphic traps are independent of any significant structural closure.

Drilling fairways, per nomenclature already established by the Egyptian General Petroleum Corporation, include the Gulf of Suez, Eastern Desert, Sinai, North Sinai, Nile Delta, and Western Desert (Figure 2). An exploratory well is defined in Egypt as any well targeting a new trap, as demonstrated by variant pressures, fluid levels, fault blocks, or stratigraphic horizons, or a well that is more than 2 km away from known production. Hence, many of the exploratory tests used for statistical analysis in this paper could be viewed in other countries as appraisal or field-delineation wells or as deeper pool tests. Further differentiation of those well types has not been undertaken in this paper, although where possible, the bulk of the reserves found in a new-field discovery has been assigned to the original exploratory test in the field. Where data were available, reserves have been designated by stratigraphic zone in multistoried pay traps, and an appropriate discovery well has been assigned to each newly discovered zone. Gas caps and associated gas in oil fields are not included as a basis for the "yet-to-find" reserve estimates.

Age control (Figure 3) is derived from extensive chronostratigraphic synthesis of the literature, much of which, however, is very general in nature. The most complete synthesis for Tertiary age dating is in Krebs et al. (1996, 1997), from outcrops fringing the western Sinai peninsula, summarized in chart format by Wescott et al. (1998). Ages presented in this outcrop synthesis have been constrained by paleomagnetic data and biostratigraphic correlation for the major unconformities and flooding surfaces (Miller, 1977). They provide the basic framework for the event stratigraphy shown on Figure 3 for Tertiary strata. We have also used pre-Tertiary correlations and paleogeographic maps based on the work of Kerdany and Cherif (1990), Klitzsch (1990), Said (1990a, c), and Boote et al. (1998) to illustrate a broad pattern of deposition in the Paleozoic, Mesozoic, and Tertiary.

Data provided by Repsol-YPF from the Komombo-1 exploratory test in Upper Egypt added significantly to our understanding of Upper Egypt. Additional key references include Wycisk (1990, 1994), Hegazy (1992), Taha (1992), Moussa and Matbouly (1994), Matbouly and Sabbagh (1996), and Morris and Tarling (1996b). The chronostratigraphic age dates for major series and stage boundaries shown in Figure 3 are derived from Haq et al. (1988).

Structural and stratigraphic synthesis of the Gulf of Suez is taken from Patton et al. (1994) and Schutz (1994), and recent concepts in sequence stratigraphy for the Gulf of Suez Tertiary synrift section are covered by Gawthorpe et al. (1990), Dolson et al. (1996), Ramzy et al. (1996), and Sharp et al. (1998). The Nile Delta and Mediterranean age dating is taken from Moussa and Matbouly (1994) and Harwood et al. (1998), supplemented by unpublished work by BP Amoco on the combined structural and stratigraphic basin evolution. Higher-resolution age analysis of the Tertiary section in the Western Desert is largely unavailable. Philobbos and Purser (1993) and Purser and Philobbos (1993) provide chronostratigraphic data from the southern Gulf of Suez and Red Sea.

Prior regional summaries of Egypt's geologic history include Robertson and Dixon (1984), Sestini (1984), Said (1990b), Halbouty and El-Baz (1992), Morris and Tarling (1996b), and MacGregor et al. (1998). Those references provide valuable insight into the major tectonic events that can explain the pattern of unconformity development shown in Figure 3.

Last, we present some recent case histories of successful exploration in each major basin which highlight the critical technical or economic issues driving significant new-field discoveries.

TECTONO-STRATIGRAPHIC HISTORY

The tectono-stratigraphic history of Egypt is recorded in eight major tectono-stratigraphic successions, which are summarized in a chronostratigraphic chart (Figure 3). Each episode created reservoir, source, and seal facies combinations controlling the hydrocarbon prospectivity of each basin.

Paleozoic Craton

There have been 245 wells drilled to Precambrian rocks in Egypt (Table 1). This limited well data, plus some outcrop exposures (Figure 2), provide control for a very generalized history of the Paleozoic of Egypt. Prior to the Triassic and Jurassic rifting which resulted in the breakup of the Pangea megacontinent, Egypt consisted of a low-relief alluvial plain dipping north and westward toward cratonic sags developed in Libya and along the proto-Mediterranean (Figure 4). Shallow-marine carbonates and siliciclastics generally increase in thickness northward in Egypt, with dominantly fluvial-alluvial lithofacies present in the south.

Most of the facies encountered are light-colored sandstones, glauconitic sandstones, and gray or red shales formed as paleosols, sabkhas, or well-oxygenated marine environments. Shelfal marine Silurian black shales, however, have been encountered in the Western Desert. Western Desert Paleozoic stratal thicknesses exceed 2500 m in the Siwa Basin. The Silurian and Devonian shales interbedded in this thick depocenter are proven source facies in the age-equivalent Tenzuft Shales in western Libya (Hegazy, 1992).

Paleozoic strata are absent in large areas in Egypt because of erosion during Triassic and Jurassic rift episodes and onlap around preexisting basement highs. Production from Paleozoic strata has been minimal, but the Paleozoic section still remains to be explored systematically.

Jurassic Rifting

In the Late Triassic through Jurassic, a series of rift basins formed during the breakup of Pangea which eventually resulted in the opening of the proto-Mediterranean Tethyan basin (Figures 4, 5). The sedimentary record of Triassic and Jurassic strata reflects a typical three-phase rift development of (1) rift initiation, (2) rift climax, and (3) postrift sag (Prosser, 1993). Jurassic strata are thickest in the northeastern corner of the Western Desert, where synrift graben fill exceeds 2500 m.

Rift grabens in northern Egypt are oriented perpendicular to the published divergent plate vectors between the African and European plates (Figure 5). The direction of other Jurassic rift systems across northern Africa may reflect inherited trends from a pre-Hercynian Paleozoic structural grain. The northwest-southeast orientation of the newly discovered Mesozoic rifts in southern Egypt may have another genesis entirely.

Upper Triassic and Lower Jurassic strata record an early rift phase of nonmarine and shallow-marine sediments that thin progressively southward away from the Tethyan margin. In the south, these strata are dominated by sandstone, red shale, and thin anhydrite deposited in rift-bounded fluvial, lacustrine, and sabkha environments. Northward, toward the proto-Mediterranean, time-equivalent rift-fill lithofacies contain progressively more carbonate and marine shale. Early movement of rift-related structures created paleostructural highs onto which many of these strata onlap.

By the Middle Jurassic, during the rift-climax phase, fault blocks had fully developed and progressively deeper marine strata were deposited. The carbonate-prone Masajid Formation contains black marine shales that form locally thick source rocks in the Western Desert. Toward the south, there is a change to progressively more nonmarine facies which contain carbonaceous shales and coal of the Khataba Formation, also considered proven source-rock facies.

In addition, Schull (1988) and Taha (1992) documented the development of additional Mesozoic rift basins in Upper Egypt and Sudan which appear to contain dominantly nonmarine lacustrine sediments. These basins were probably not physically connected to the Tethyan open-marine rift systems in the Western Desert and North Sinai areas. Repsol successfully proved the presence of three of those basins (Nuqra, Kharit, and Komombo; Figure 2) with exploratory wells drilled in the 1990s (Table 3). The northwest-southeast orientation of those basins may indicate rifting associated with breakup of the Afro-Arabian Plate, possibly as far south as Yemen (M. Winfield, personal communication, 1999) in trends which remain poorly understood.

Cretaceous Passive Margin

By the Early Cretaceous, an extended period of thermal sag associated with wide passive-margin development occurred across the northern margin of the African plate, resulting in a mixed siliciclastic and carbonate system. Local unconformities between the Lower Cretaceous and Upper Jurassic, plus the vertical transition from marine shale to lignite and carbonaceous shale near the base of the Alam el-Bueib (AEB) Formation, seem to indicate some local continuation of rift episodes into the Early Cretaceous. The AEB Formation is progressively overlain by a pattern of micrite and oolitic limestone cycles that suggests episodic series of transgressions and regressions related to regional sea-level oscillations. The AEB Formation contains proven marine carbonate source rocks. Age-equivalent strata in the Nubia Formation to the south generally consist of red shale and coarse-grained reservoir sandstones deposited in fluvial environments. Between the two areas, Darwish (1992) documented Nubia Formation shallow-marine and sabkha environments along western Sinai outcrops. The Aptian Alamein Dolomite marks a period of widespread sea-level rise which provides a useful flooding-surface marker horizon throughout the Western Desert and North Sinai. Age-equivalent marine strata in the Komombo Basin (Figure 3) indicate that this Aptian transgression reached into southern Egypt (Figure 4).

Continued thermal subsidence throughout the Western Desert was accompanied by south-directed transgressions across the stable carbonate shelf that resulted in the deposition of additional widespread source rocks

Table 4. Egypt reserves by age and petroleum system.

GULF OF SUEZ, EASTERN DESERT, SINAI	MMBOE (GAS, COND, OIL)	MMBOE (ALL)	CONDENSATE OIL (MMBO)	(MMBO)	ASSOCIATED GAS (BCF)	GAS (BCF)	GAS CAP (BCF)
TERTIARY	5951.58	6651.19	5913.85	0.00	226.40	2720.64	1477.00
CRETACEOUS	2682.19	2972.98	2680.53	0.00	10.00	1119.70	625.00
JURASSIC	0.00	0.00	0.00	0.00	0.00	0.00	0.00
TRIASSIC	0.00	0.00	0.00	0.00	0.00	0.00	0.00
PALEOZOIC	0.00	0.00	0.00	0.00	0.00	0.00	0.00
PRECAMBRIAN	0.60	0.60	0.60	0.00	0.00	0.00	0.00
SUBTOTAL	**8634.37**	**9624.76**	**8594.97**	**0.00**	**236.40**	**3840.34**	**2102.00**
WESTERN DESERT	**MMBOE (GAS, COND, OIL)**	**MMBOE (ALL)**	**CONDENSATE OIL (MMBO)**	**(MMBO)**	**ASSOCIATED GAS (BCF)**	**GAS (BCF)**	**GAS CAP (BCF)**
TERTIARY	5.00	5.33	5.00	0.00	0.00	0.00	2.00
CRETACEOUS	1623.65	1727.20	935.25	43.10	3871.83	621.30	0.00
JURASSIC	966.52	966.52	92.35	109.50	4588.00	0.00	0.00
TRIASSIC	0.00	0.00	0.00	0.00	0.00	0.00	0.00
PALEOZOIC	5.50	5.50	0.50	0.00	30.00	0.00	0.00
PRECAMBRIAN	0.00	0.00	0.00	0.00	0.00	0.00	0.00
SUBTOTAL	**2600.67**	**2704.55**	**1033.10**	**152.60**	**8489.83**	**621.30**	**2.00**
NILE DELTA, MEDITERRANEAN, NORTH SINAI	**MMBOE (GAS, COND, OIL)**	**MMBOE (ALL)**	**CONDENSATE OIL (MMBO)**	**(MMBO)**	**ASSOCIATED GAS (BCF)**	**GAS (BCF)**	**GAS CAP (BCF)**
TERTIARY	4499.65	4499.65	35.48	183.94	25681.42	0.00	0.00
CRETACEOUS	7.78	7.78	3.00	0.00	28.68	0.00	0.00
JURASSIC	0.00	0.00	0.00	0.00	0.00	0.00	0.00
TRIASSIC	0.00	0.00	0.00	0.00	0.00	0.00	0.00
PALEOZOIC	0.00	0.00	0.00	0.00	0.00	0.00	0.00
PRECAMBRIAN	0.00	0.00	0.00	0.00	0.00	0.00	0.00
SUBTOTAL	**4507.43**	**4507.43**	**38.48**	**183.94**	**25710.10**	**0.00**	**0.00**
UPPER EGYPT (BENI SUEF)	**MMBOE (GAS, COND, OIL)**	**MMBOE (ALL)**	**CONDENSATE OIL (MMBO)**	**(MMBO)**	**ASSOCIATED GAS (BCF)**	**GAS (BCF)**	**GAS CAP (BCF)**
TERTIARY	0.00	0.00	0.00	0.00	0.00	0.00	0.00
CRETACEOUS	10.00	10.00	10.00	0.00	0.00	0.00	0.00
JURASSIC	0.00	0.00	0.00	0.00	0.00	0.00	0.00
TRIASSIC	0.00	0.00	0.00	0.00	0.00	0.00	0.00
PALEOZOIC	0.00	0.00	0.00	0.00	0.00	0.00	0.00
PRECAMBRIAN	0.00	0.00	0.00	0.00	0.00	0.00	0.00
SUBTOTAL	**10.00**	**10.00**	**10.00**	**0.00**	**0.00**	**0.00**	**0.00**
TOTAL ALL BASINS	15752.47	16846.75	9676.54	336.54	34436.33	4395.64	3329.00
TOTAL LIQUIDS	10013.08						
TOTAL GAS	41001.97						

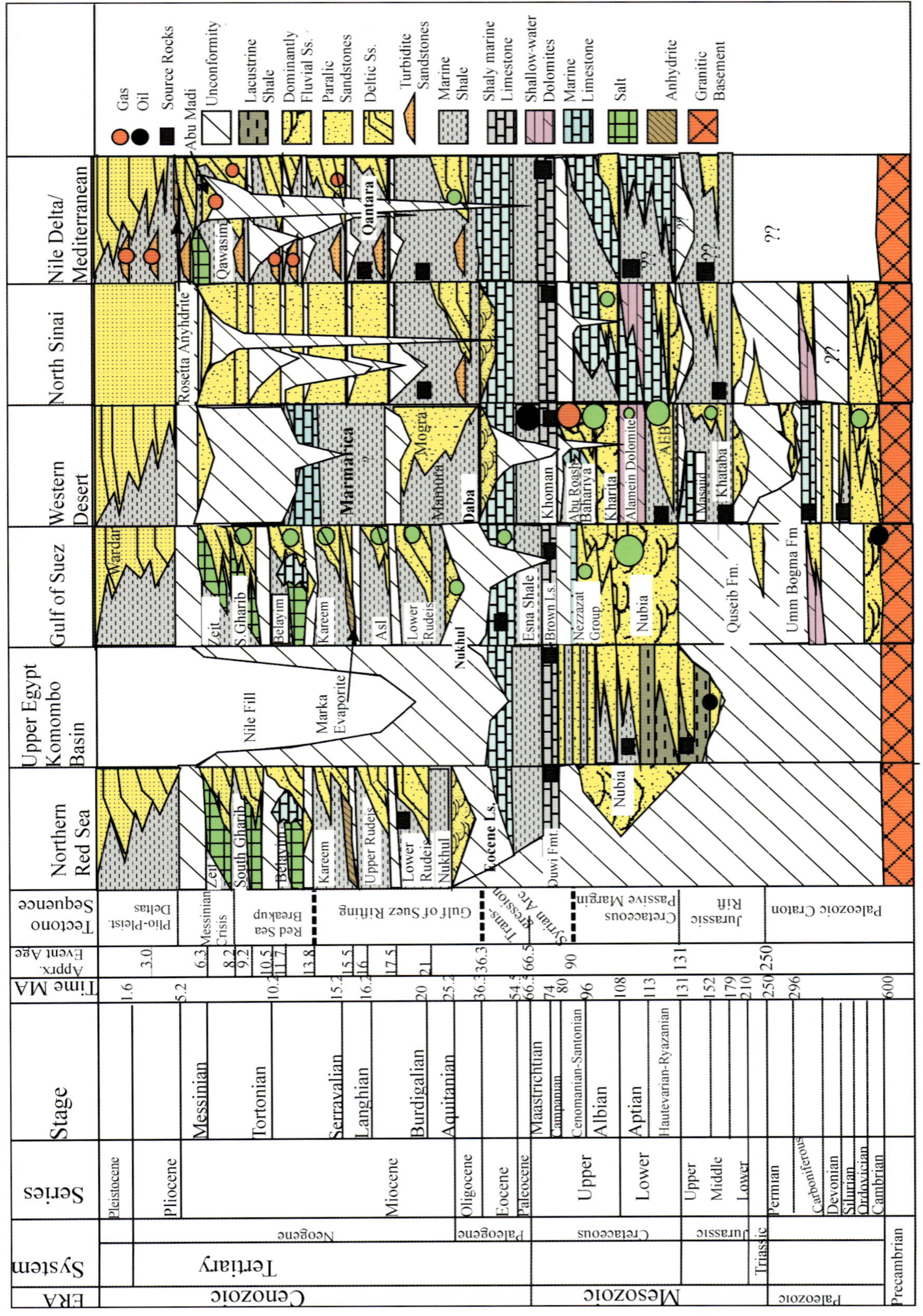

Figure 3. Stratigraphic correlation panel of Egypt. Major tectono-stratigraphic breaks are highlighted on the chart. See text for discussion.

in the Kharita and Bahariya Formations. Time-equivalent strata to the south are dominantly fluvial (Nubia Formation) and shallow marine (lower Nezzazat Group Raha Formation).

The northern edge of the Cretaceous carbonate platform system is in the Nile Delta area, where it has been mapped as the Cretaceous hinge line. The carbonate margin crops out in southern Israel as locally thick carbonate slope breccia of the Talme Yafe Formation (Bein and Weiler, 1976). To the north of this carbonate margin, deep-marine carbonates and shale were deposited across the present offshore Nile Delta and Mediterranean Sea areas.

Syrian Arc Deformation and Foreland Transgression

The onset of Tethys closure between the European and African plates during the Cenomanian through Turonian resulted in regional uplift, characterized by rift-basin inversion throughout the Western Desert, resulting in a series of "Syrian Arc" northeast-southwest-trending folds (Moustafa and Khalil, 1990; Moustafa et al., 1998). This structural deformation resulted in development of unconformities across inverted structural crests. Deposition locally continued off-structure with onlap along the flanks of these evolving highs. This important tectonic event continued episodically through the Santonian as the African plate moved northward, colliding with the European plate and completely cutting off the Tethyan seaway.

Turonian through Santonian carbonates of the Abu Roash Formation show widespread high-frequency cyclicity. Some of these correlative units (Nezzazat Group, Gulf of Suez) contain oolitic limestone and grainstone that can be traced using well control as far south as the southern end of the Gulf of Suez. At the same time, deep-marine shale and limestone were deposited northward of the remnant carbonate margin across the Mediterranean and Nile Delta areas.

Syrian Arc–related structural trends form the bulk of the productive traps discovered in the Western Desert. Extension of these productive trends eastward into the onshore part of the Nile Delta has proved disappointing.

By the early Campanian, most of the structural deformation associated with the development of the Syrian Arc had ceased. A major sea-level rise (Haq et al., 1988) resulted in widespread flooding and deposition of source-rich anoxic shelfal shale and limestone of the Khoman Formation and its southern equivalent, the Brown Limestone. Where they were not removed by Tertiary uplift and erosion, these important oil-source rocks are encountered in all Egyptian basins as far south as the northern end of the Red Sea.

From the Cenomanian onward, the African plate was gradually subducted northward under the European margin, developing a wide foreland basin across northern Egypt (Figure 5). Continued transgression resulted in the deposition of Paleocene shale (Esna Shale and equivalents) and Eocene cherty carbonate and thinly laminated shale of the Thebes Formation and equivalent strata. The Campanian through Eocene interval throughout Egypt forms significant top seals to underlying reservoirs, as well as mechanical structural boundaries impacting faulting and folding geometries.

Gulf of Suez Rifting

The opening of the Gulf of Suez began in the early Oligocene and culminated with the Red Sea breakup in the Serravalian stage of the Miocene. Biostratigraphic data (Patton et al., 1994; Krebs et al., 1996; Wescott et al., 1996) indicate that extension began in the northern part of the Gulf of Suez and spread southward during the Miocene. Marine deposits of the Nukhul and lower Rudeis Formations show a Mediterranean fauna consistent with a limited seaway connection northward to the Mediterranean. Mixed northern and southern faunal assemblages occur in the Gulf of Suez during the Burdigalian (rift-climax event), showing that a full connection was developed between the Red Sea and the Mediterranean at that time. By the middle Serravalian, however, the faunal assemblages show only a southern connection to the Indian Ocean. Structural uplift across reactivated Syrian Arc structures had closed off the northern end of the Gulf of Suez (Figure 6).

Several significant basinwide unconformities occur in the Gulf of Suez. Dolson et al. (1996) and Ramzy et al. (1996) showed that those surfaces formed primarily in response to regional tectonic adjustments associated with different phases of rift evolution.

Rift-initiation deposits in the Gulf of Suez consist of conglomeratic braided and meandering stream sandstone interbedded with brick-red paleosols and playa lake lithofacies of the Abu Zenima and Nukhul Formations. Those formations grade upward into bioturbated sandstone interbedded with progressively more "coquina" and limestone intervals deposited in estuarine and shoreface environments. A series of closely spaced marine ravinement surfaces caps the initial rift fill, which is then progressively overlain by deeper-water marl and shale of the overlying lower Rudeis Formation. Isopach maps and outcrop paleocurrent directions indicate a north-to-south transport direction for most sediments of the Nukhul Formation, which is preserved in half grabens that run subparallel to the regional structural fabric of the opening rift.

Oligocene strata are only partially preserved in the Gulf of Suez, but they consist of conglomeratic sandstone and marine shale in the Western Desert and Nile Delta areas. Many of the reservoir sandstone units formed as deltas fed by the erosion of the emergent parts of the Gulf of Suez and other topographically high areas to the south. These deltaic reservoirs and associated deep-water fans represent a deep, largely untested play across much of the Nile Delta and Mediterranean Sea. In the Nile Delta, Oligocene to lower Miocene shale of the

Qantara Formation also contain some fair to good source potential.

By the early Burdigalian, the Gulf of Suez was fully developed, and parts of the basin had reached bathyal water depths. Large fan deltas continued to enter the basin through major structural transfer zones, depositing important reservoir sandstone units. A major unconformity developed at 17.5 Ma (T20 or Mid-Rudeis unconformity). This hiatus records a rapid basin shallowing in the late Burdigalian, which may have resulted from regional isostatic rebound of the basin floor (Wescott et al., 1998).

An important basinwide lowstand event in the Gulf of Suez (Asl Formation) resulted from this relative lowering of sea level, which culminated in widespread deposition of the supratidal Markha Anhydrite. The paleobathymetric shift was profound. Areas that were once at bathyal conditions shallowed in less than 2 million years to supratidal conditions, and only deep structural synclines still continued to fill with deeper-water shale (Dolson et al., 1998). During this shallowing, coarse-grained detrital sediments prograded far beyond the limits of earlier lower Rudeis deltas, reaching, in many cases, the axis of the basin.

Miocene reservoirs continue to be primary targets for basin-flank and basin-floor fan exploration. Although the Asl Sandstone lowstand event may relate to a global sea-level lowstand, it also corresponds to regional uplift across Egypt. Harms and Wray (1990) showed a marked bathymetric change in the Nile Delta from bathyal to inner neritic water depths during the same time interval (17.5–15.2 Ma) in the Qantara-1 and Qallin-1 wells.

Red Sea Breakup

The isolation of the Gulf of Suez from the Mediterranean Sea because of reactivation of the Syrian Arc structural trend may have resulted from an overall shift to north-south (versus northeast-southwest) rift extension (Meshref, 1990). The isolation of the Gulf of Suez from the Mediterranean is marked by cyclical deposition of basin-centered salts and anhydrites (Figure 6). These strata are intermixed with coarse-grained sandstone derived from point-sourced deltas along the basin flanks. The evaporites of the Belayim and South Gharib Formations form critical regional seals to many structurally trapped accumulations, and the interbedded sandstones form significant reservoirs.

By contrast, in the Nile Delta, the Serravalian to lower Tortonian strata record a strong pulse of basinward progradation of deep-water siliciclastics from south to north into the Mediterranean. The precise entry points of feeder systems remain unknown but were presumably from the proto-Nile, which would have been developed west of the Gulf of Suez uplifts and from coastal uplands along the North Sinai.

Messinian Crisis

In the Mediterranean and Nile Delta region, the frequency of erosional unconformities in the late Miocene

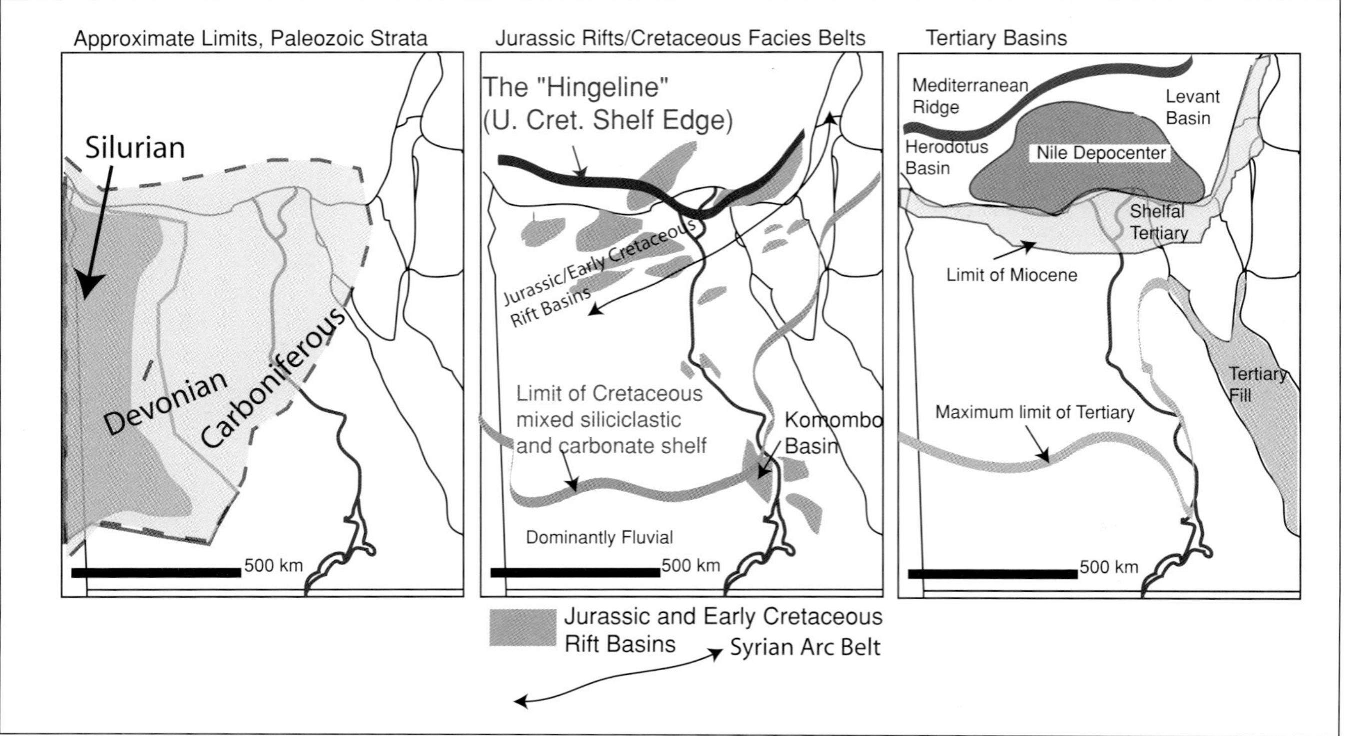

Figure 4. Paleogeographic reconstructions for major tectono-stratigraphic intervals (modified from Boote et al., 1998). See text for discussion.

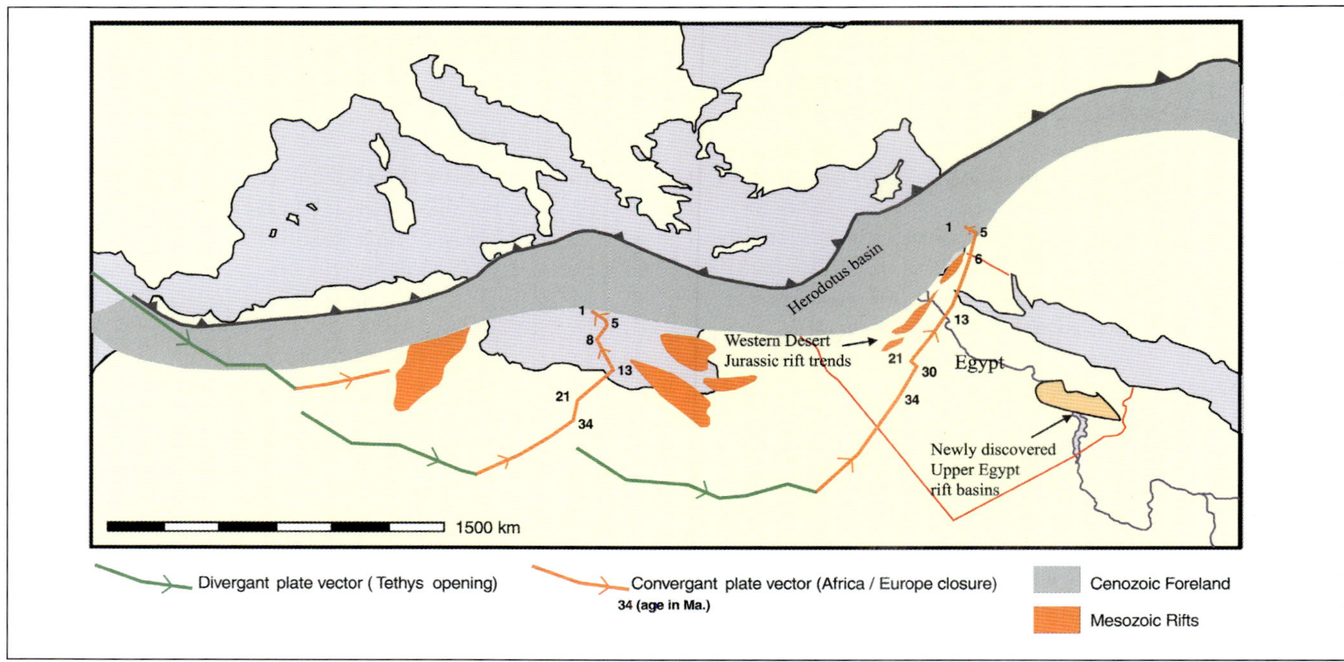

Figure 5. Summary diagram of plate reconstructions in North Africa and the Mediterranean region (modified from Morris and Tarling, 1996a).

appears to increase upward stratigraphically, culminating with the Messinian crisis at 6.7 Ma. This was essentially a tectonic-driven event related to late-stage closure of the African plate against Europe, causing the closure of the Straits of Gibraltar and evaporation of the Mediterranean Sea (Halbouty and El Baz, 1992). Basinwide lowstand deposits of salt and anhydrite occur throughout the deeper portions of the Mediterranean (see Figure 2 for limits of Messinian salt).

Compressional deformation in the Nile Delta area created wide uplifted arches and local strike-slip grabens, which were deeply incised during the evaporitic drawdown. Grand Canyon–scale incisions of these arches occurred along many coastal areas and in the proto-Nile valley (Harms and Wray, 1990). Large volumes of deltaic sediment were transported offshore through these canyons into deep-water trends that remain undrilled. As the Mediterranean finally refilled, subsequent flooding backfilled these valleys with fluvial and estuarine strata, which are productive in many combination traps. Moussa and Matbouly (1994) documented further examples of canyon incisions and petroleum traps related to the Messinian crisis. These transgressions were episodic, and numerous intra-Messinian erosional surfaces are recognizable on seismic sections. The drainage networks established around the rim of the Mediterranean provided focal points for sediment input in the Pliocene and offer numerous exploration targets.

No similar pattern of deep shelfal incision is noted in the Gulf of Suez in the equivalent strata of the Messinian Wardan Formation, although a significant hiatus developed and a long period of basin-centered salt deposition occurred at that time. In addition, drainage patterns around the Gulf of Suez were apparently diverted to the west by drainage capture into the Nile canyons.

Pliocene-Pleistocene Delta Progradation

The Pliocene section in Upper Egypt and the Gulf of Suez lies well above potential mature source rocks. The presence of intervening shale and evaporite seals has meant that the Pliocene has not been a successful exploration target. In the Nile Delta and Mediterranean areas, however, Pliocene deltaic sandstones are very significant reservoirs and have become the dominant "big play" of the 1990s. The Messinian canyons were completely overstepped and infilled by 5.6 Ma, but large volumes of sediment from the Nile Valley continued to prograde into the Mediterranean. As many as 17 sequences have been documented for the Pliocene (Harwood et al., 1998), and the associated facies shifts formed important reservoir fairways.

A switch in the plate-convergence direction toward the northwest, from approximately 5 Ma onward (Figure 5), probably caused the Nile Delta depocenter to tilt downward to the northwest into the present-day basinal low of the Herodotus Basin.

DRILLING HISTORY

Exploration drilling continues to result in significant new-field discoveries (Figure 7). Egypt's most prolific period of growth prior to the late 1990s in the Western Desert and Mediterranean was the 1955–1979 period of exploration in the Gulf of Suez. That time interval was marked by successive discoveries of four-way closures and large three-way upthrown fault-block traps easily discernible with 2-D seismic methods and, in some cases, with water bathymetry maps. The Gulf of Suez and Eastern Desert discovery rates have flattened significantly in the last 15 to 20 years, indicating a maturing of the exploratory history and/or a need to discover new trends with different technology or play concepts. The onshore Sinai has had only limited success since the discovery of the giant Belayim Land field in 1955 and has limited "running room" because of the proximity of the rift-basin margin immediately east of this drilling province.

In contrast, substantial growth has occurred in the Western Desert and Mediterranean provinces. This growth has been fueled largely by the awarding of gas rights in both basins in the late 1980s. The cumulative discovery versus time data for the Mediterranean offshore closely parallels the Gulf of Suez early discovery rate.

Although drilling statistics do not show significant hydrocarbon volumes in Upper Egypt, this area has seen some limited but significant exploration activity (five wells) in the 1990s (Table 3). This activity has established the presence of oil in the Komombo Basin with the drilling of the Komombo-1 well by Repsol in 1997. A new phase of drilling in the Red Sea may soon be under way as several companies complete interpretations of newly acquired 2-D and 3-D seismic, aeromagnetic, and gravity data in an attempt to open up new discoveries in this frontier province.

Discussion of the potential and technical challenges in each trend follows.

EXPLORATION POTENTIAL BY PETROLEUM SYSTEM

Western Desert

The Mesozoic basins of the Western Desert provide rewarding but difficult exploration opportunities. The Western Desert has proved to be challenging to understand because of its complex history of a Paleozoic margin, Jurassic rift, Early Cretaceous passive margin, and subsequent inversion during the Syrian Arc deformation. Fields are commonly segmented and complex, with multiple fluid contacts. In addition, seismic data quality degrades rapidly beneath the Alamein Dolomite; therefore, many of the productive AEB and older Jurassic reservoirs are difficult to image. Drilling has been confined almost totally to structural culminations, and the possibility of successful exploitation of numerous stratigraphic pinch-outs has not been fully tested. As a result, the distribution of discovered resources is almost exclusively in structural traps (Figure 7). Surface conditions make acquisition of seismic data difficult. Widespread 3-D seismic surveys, which have the potential to decrease risk on future drilling opportunities, have been conducted only recently, and most of those were local in extent.

Nevertheless, the Western Desert field-size distribution shows a large number of small fields, but some room does exist in the statistics for oil discoveries of 100 million barrels (bbl) and larger (Figure 8a). Indeed, the Kanayes, Shams, and Obaiyed fields all have been discovered since 1993 (Table 2; Figure 9), producing mainly from Jurassic reservoirs. The interpreted number of "yet-to-find" discoveries shown in Figure 8a is derived by assuming that one field is left to find larger than Obaiyed

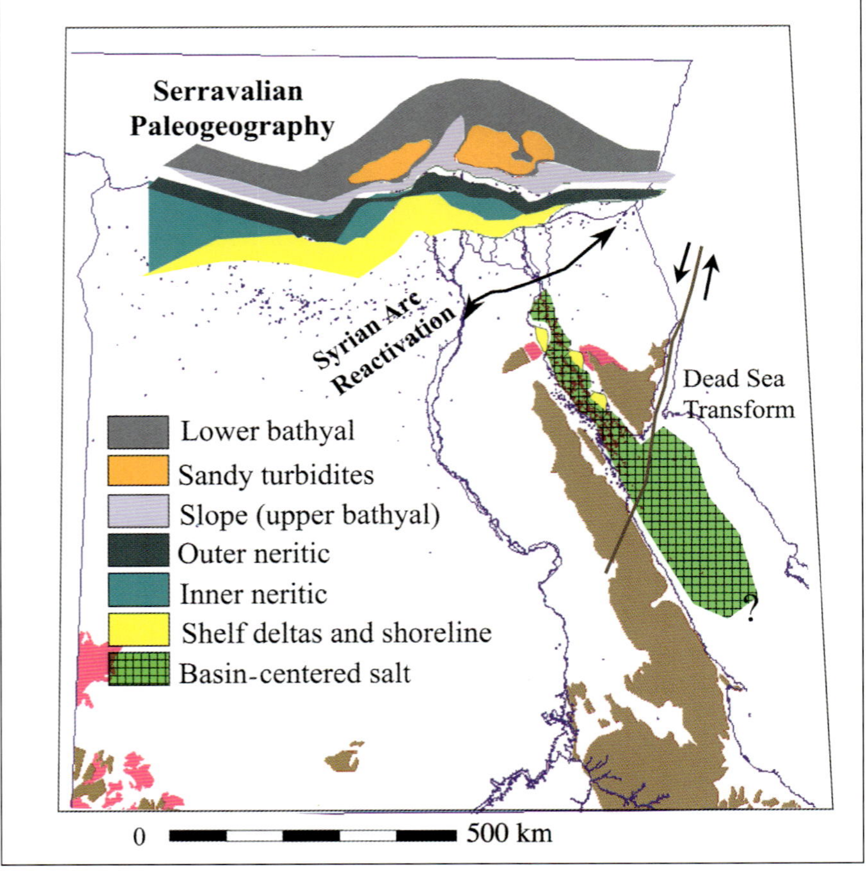

Figure 6. Simplified Serravalian (late Miocene) paleogeography. Syrian Arc structural reactivation caused development of a northeast-southwest-oriented arch which separated the Gulf of Suez and Mediterranean Basins. This event caused basinward translation of reservoir facies in the Nile Delta that has continued into the present. Concurrently, the opening of the Gulf of Aqaba stranded the Gulf of Suez, resulting in widespread evaporite deposition and development of regional seals to the underlying petroleum system.

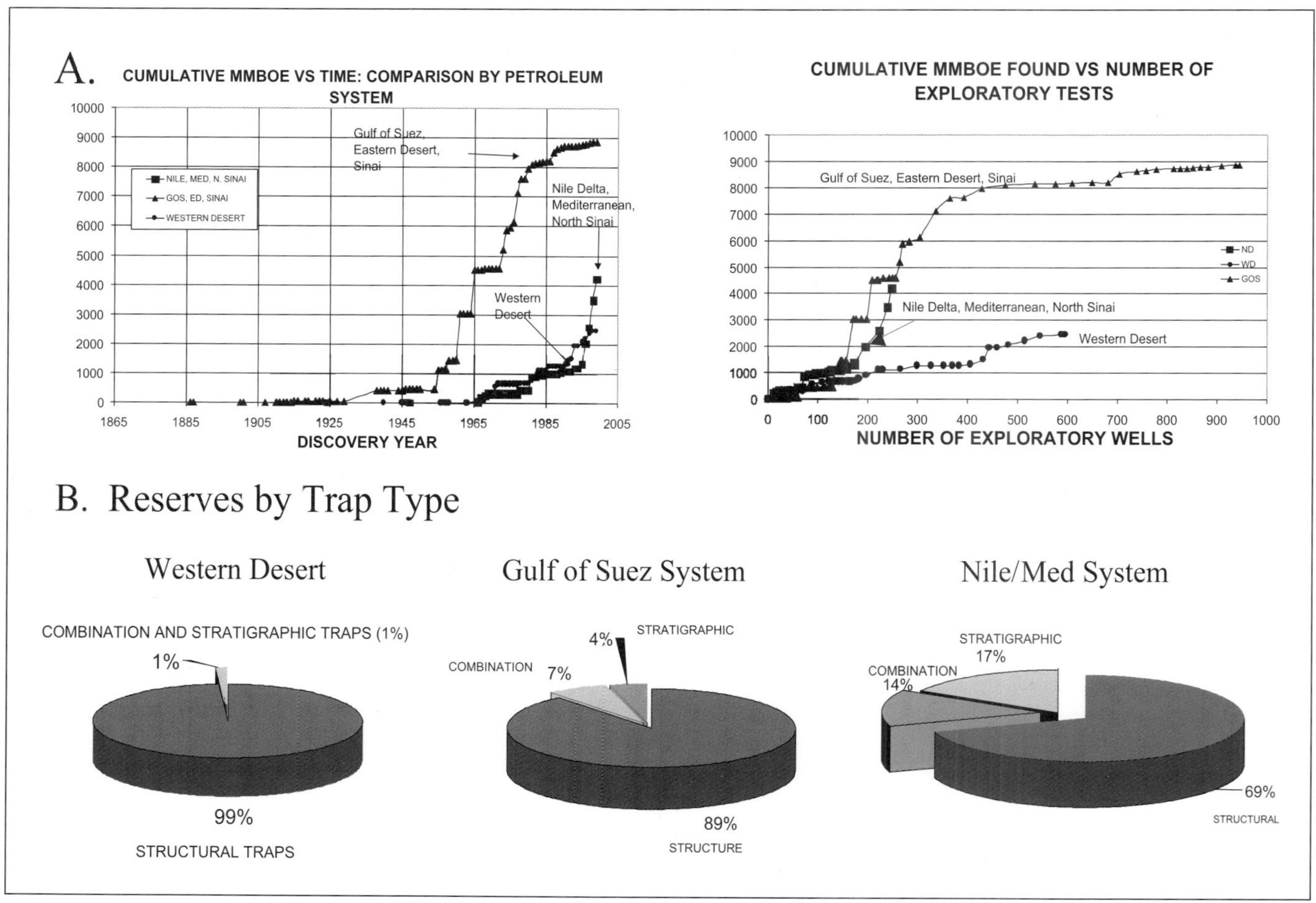

Figure 7. (a) Cumulative discovery volume by petroleum system versus time and versus number of exploratory tests. (b) Distribution of resources by trap type and petroleum system.

field, the largest yet found in the trend. In theory, because the field-size clusters shown are lumped in exponential classes on the x-axis, missing field sizes can be estimated from the difference between the diagonal line and the number of fields in each class. The diagonal line is chosen by estimating the largest field yet to be found in the trend and then reconnecting to the number of the smallest fields. The gaps are theoretical missing field numbers.

We view these interpretations as useful tools to gauge relative potential growth in each province. We acknowledge that additional assessment by play, trap, or reservoir age provides more refinement, especially if coupled with source, maturation, and migration studies which can quantify volumes of hydrocarbons expelled in each trend. Nevertheless, this method suggests that 15–33 trillion cubic feet (tcf) of gas may be left to find in the Western Desert. The location of most of those new fields would almost certainly have to come from deeper Jurassic and/or Paleozoic objectives or new stratigraphic trap concepts drilled on flank and in basinal areas.

Table 5 provides some summary data (courtesy of the Egyptian General Petroleum Corporation) which shows the potential volumes of oil and gas generated in each subbasin in the greater Western Desert. The presence of the source facies shown has been confirmed by a large number of wells (Hegazy, 1992), but the eastward extent of Paleozoic source rocks into the Natrun Basin remains conjectural. In Table 5, an assumption has been made that the thickness of source-bearing interval will be uniform across each basin area. The "expected recoverable" oil and gas numbers shown come from using a subjective and somewhat conservative 7% of the generated hydrocarbons actually migrating out of the basin toward potential traps. In contrast, Barker and Dickey (1984) used a similar number of 10% to estimate the volume of migrated hydrocarbons in Saudi Arabia. Whereas this table contains much speculative data, maturation profiles have been generated for each basin which constrain the expulsion predictions, and the petroleum volumes generated are significant.

These data suggest that 26.4 billion bbl of oil and 21 BBOE gas (144 tcf) may have migrated toward traps in the Western Desert basins (for a total of 47.4 BBOE migrated hydrocarbons). The 578 exploratory tests drilled in this large region have discovered 1.533 billion bbl of oil and condensate and 9.113 tcf of gas (2.7 BBOE;

Table 4). Using the ranges of potential missing field sizes shown in Figure 8, the 15–33 tcf (2.5–5.5 BBOE) "yet-to-find" number is well below the migrated-hydrocarbon estimate of 47.4 BBOE shown in Table 5. Even if the volumes shown in Table 5 were off by a factor of two, those "yet-to-find" numbers seem reasonable. We believe the missing 15–33 tcf represent long-term "upside," requiring a significant investment in technology and a willingness to pursue deeper and more difficult targets.

As discussed earlier, Paleozoic source strata extend under some of these rift grabens and thicken westward from the Western Desert into Libya. They may provide petroleum charge for lightly drilled Paleozoic reservoir zones. In the Western Desert, 208 exploratory wells penetrate the Jurassic interval, but only 71 reach the Paleozoic section (Table 1). In 1996, the OBA-A3 well established Paleozoic production in the Matruh Basin. The Paleozoic potential remains an interesting target because Devonian and Silurian source strata are important sources for oils in Libya (Boote et al., 1998; Keeley and Massoud, 1998; Traut et al., 1998).

The emerging play of the 1990s was Jurassic traps. Figure 9 illustrates the complex distribution of basins and intervening highs in the northern portion of the Western Desert near Obaiyed field. The giant Abu Gharadig field (see Figures 2 and 9 for location) is a typical inverted-rift structure developed during the Syrian Arc tectono-stratigraphic event. The poor seismic data quality beneath the Alamein Dolomite has made deep exploration difficult because the deeper inverted-rift structural trends do not mirror the overlying Syrian Arc–related structural closures.

The spatial alias shown for the 208 Jurassic penetrations in the Western Desert leaves significant room for further deep exploration and discovery, particularly given the lack of areawide 3-D surveys and high-quality deep resolution of structural and stratigraphic features. Mahmoud and Barkooky (1998) demonstrated more than 250 m of paleotopographic relief across the Obaiyed feature regionally, with rapid lateral facies changes and reservoir pinch-outs, which could set up additional stratigraphic traps.

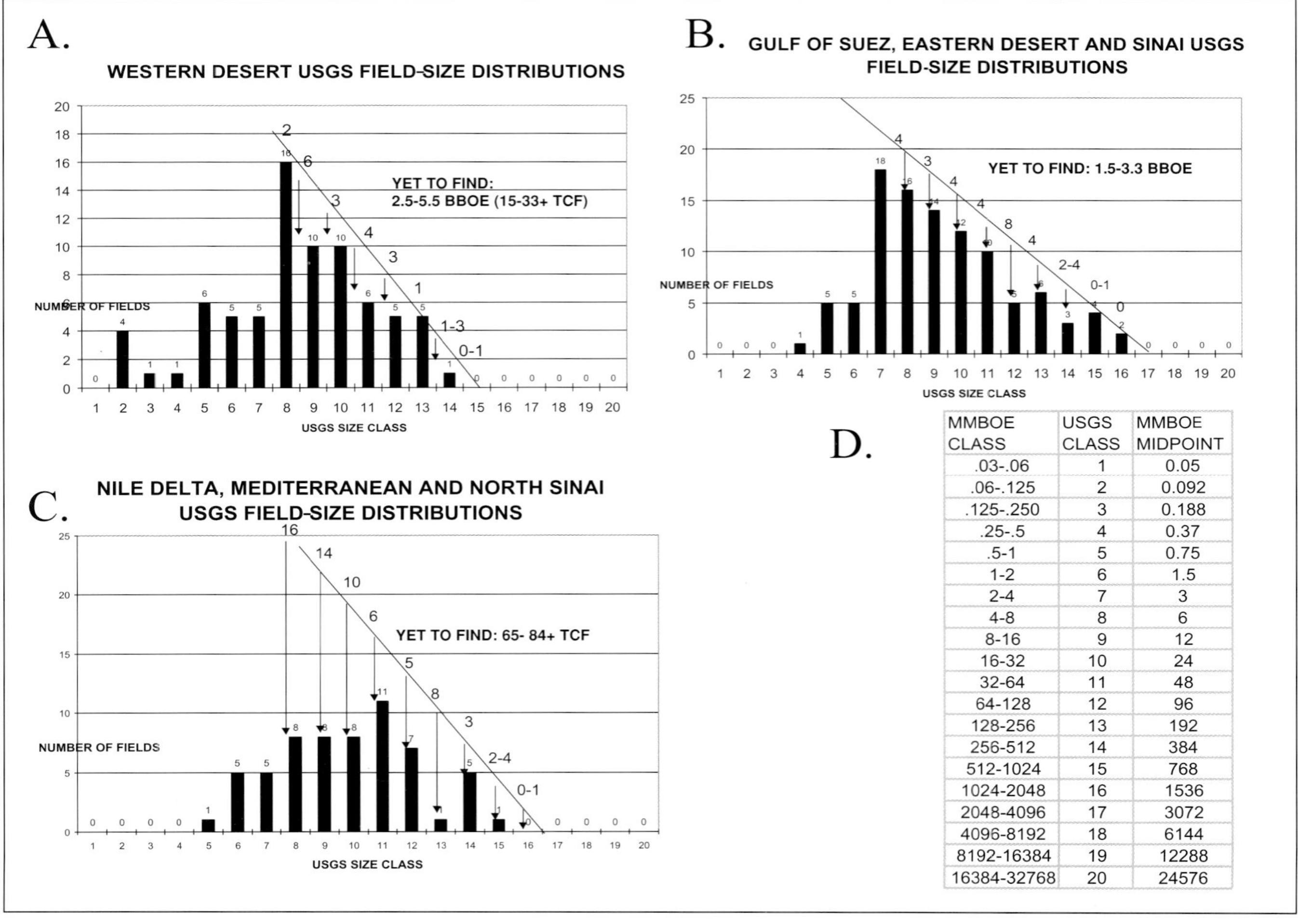

MMBOE CLASS	USGS CLASS	MMBOE MIDPOINT
.03-.06	1	0.05
.06-.125	2	0.092
.125-.250	3	0.188
.25-.5	4	0.37
.5-1	5	0.75
1-2	6	1.5
2-4	7	3
4-8	8	6
8-16	9	12
16-32	10	24
32-64	11	48
64-128	12	96
128-256	13	192
256-512	14	384
512-1024	15	768
1024-2048	16	1536
2048-4096	17	3072
4096-8192	18	6144
8192-16384	19	12288
16384-32768	20	24576

Figure 8. (a-c) U.S. Geological Survey (USGS) field-size distributions, Western Desert, Gulf of Suez, and Nile/Mediterranean region. Field classes are arranged logarithmically, so "yet-to-find" numbers of new fields are shown with arrows under the straight line. (d) USGS field-class sizes (from Drew and Schuenemeyer, 1993) showing midpoint volumes used for "yet-to-find" estimates. See text for discussion.

Southward, in the Beni Suef and Gindi Basins, recent new-field discoveries have been made in Cretaceous reservoirs (Table 3) at Qarun (Geizery et al., 1998) and Beni Suef fields (Figure 10). These discoveries have proved that hydrocarbon systems exist in these smaller basins where little or no prior successful exploration activity had occurred. The extent of the Jurassic source rocks in the deeper parts of these basins is unknown, and drilling has concentrated on the structural high along the basin boundaries.

The field-size distributions and significant reserve growth shown by the strong cumulative finding rate in the 1990s suggest that the Western Desert will continue to be an attractive exploration target in the future. Extensions into the offshore Mediterranean should also prove attractive.

Upper Egypt

Mesozoic basin exploration is not confined to the Western Desert and Northern Egypt. One of the most interesting plays, a frontier exploration play, is being tested in Upper Egypt. In the late 1990s (Table 3, Figure 2), Repsol began to explore the Komombo Basin, a Mesozoic continental rift. The rift geometry was detected from outcrop patterns and gravity data (Kamel, 1990), and five exploratory tests had been drilled by 2000. The Komombo-1 well tested 37°–39° API oil from the Jurassic and proved the existence of a working petroleum system (Taha and A/Aziz, 1998).

The Komombo well (Figure 11) penetrated nonmarine Jurassic lacustrine type II–III source rocks formed in the Middle to Late Jurassic. Additional source rocks, as well as marginal marine and marine strata, were encountered in Aptian and younger Cretaceous strata. Those Cretaceous source rocks were also present to the south in the Nuqra and Kharit Basins, but they are thermally immature. Although additional drilling has been unsuccessful so far, shows were encountered in all wells in the Komombo Basin. The nearby Nuqra and Kharit Basins wells had no oil and gas shows. The representative seismic line shown in Figure 11 over prospective leads illustrates the source, reservoir, and seal potential of this sparsely drilled province. The Komombo Basin contains a potentially ideal petroleum system, with deeply buried lacustrine source rock (3000–5000 m paleoburial depth), thick siliciclastic reservoirs, and an overlying shallow-marine shale-dominated section which can potentially form basinwide seals. Therefore, the potential exists for both structural and stratigraphic trapping in an area covering more than 2000 km^2.

Schull (1988) and Taha (1992) documented production from similar nonmarine rifts in Sudan, which have an analogous tectono-stratigraphic history, possibly related to Mesozoic rifting of the African/Arabian Plate. With further integration of data from these important new wildcats and additional seismic surveys, this new basin may one day become Egypt's newest oil province.

Gulf of Suez, Eastern Desert, and Sinai

The Gulf of Suez rifting and Red Sea breakup set up ideal petroleum systems in the Gulf of Suez and, potentially, in the Red Sea. Collectively, the Gulf of Suez, Eastern Desert, and Sinai drilling provinces are part of the Gulf of Suez petroleum system. Known source rocks are the Campanian Brown Limestone and Eocene Thebes Formations, with a minor contribution from lower Miocene shales (Robison, 1995; Barakat et al., 1997). Precambrian through Serravalian rocks produce in the basin, with most of the reserves found within Tertiary reservoirs. Most wells and producing fields are located along the crests of tilted fault blocks and four-way closures, which are charged from numerous flanking basins (Figure 12). The more than 900 exploratory wells that have been drilled in this basin continue to target structural blocks, with little basin-centered exploration. All fields appear to be filled to their sealing capacity, so the petroleum system is unlikely to be charge limited. Top seals are dominantly middle Miocene shales and evaporites of the Belayim and South Gharib Formations.

An idealized rift model and schematic petroleum-system model illustrate some of the major Miocene synrift traps of the Gulf of Suez (Figure 13). A multitude of proven play types exists in this basin. The largest fields, such as Belayim Marine, Morgan, and July fields, are rotated fault blocks with three-way and four-way closures against sealing shales or evaporites. In the late 1980s, "downthrown" traps were discovered on the flank of October field (Dolson et al., 1996). These structural and combination traps should exist elsewhere and provide an attractive future exploration target (see Table 3 for examples). Synclinal lows with deep-marine turbidite stratigraphic trap objectives remain essentially untested. Exploration for hydrocarbon accumulations in Miocene strata of the Gulf of Suez is complicated by syndepositional movement around fault blocks, which has created complicated reservoir distributions. Synchronous onlap around the high dip slopes of structures and downlap into basinal areas has created difficult well-log and seismic correlations which must be understood to predict the location of stratigraphic and combination traps which have the most "running room" to explore.

The Gulf of Suez petroleum system has a very mature drilling curve (Figure 7). Field-size distributions (Figure 8b), however, still show room for several new giant fields and large numbers of intermediate fields. These data suggest that 1.5–3.3 BBOE may yet be found, based partly on the assumption that the largest field sizes (classes 15 and 16) have already been found. Given the mature history and dense drilling in the Gulf of Suez, most new fields found there will be less than 100 MMBOE in size. Indeed, the last significant giant field (Table 2) was discovered in 1987. However, this basin is exceptionally difficult to explore seismically, with severe multiples masking real structural and stratigraphic signatures. As a result, significant structural traps (for example, Warda field, Table 3) are still being found. As seismic imaging improves, significant new reserves may be found in stratigraphic and combination traps.

Dolson et al. (1997) discussed an exploration "turnaround" in the mid-1990s through the application of integrated technology, 3-D seismic methods, and seismic multiple-suppression techniques in the Gulf of Suez. Examples of overcoming difficult seismic-imaging problems are the SG310-6 and 310-4 discoveries (Table 3, Figure 14), which flowed 10,000 and 20,000 bbl oil/day (BOPD), respectively, from the Burdigalian Hawara Sandstone (Ramzy et al., 1998). Those wells offset four dry holes, which had missed the updip termination of fault blocks. Those culminations were completely undetectable with conventional seismic processing. Multiple-removal techniques, plus dipmeter-derived dips displayed correctly on seismic depth sections using geotechnical software and computer workstations, allowed the updip traps to be imaged accurately.

We believe extremely difficult seismic imaging makes the Gulf of Suez a technologically limited basin for further growth potential. Although it is likely that the largest fields have already been found, significant gaps occur in the field-size distribution graphs (Figure 8b), particularly in the number of fields in the field-size class of sub-4–128 million bbl. Continued seismic-image enhancement will undoubtedly be the key to finding the 1.5–3.3 BBOE "yet-to-find" resources suggested by the field-size gaps shown in Figure 8b.

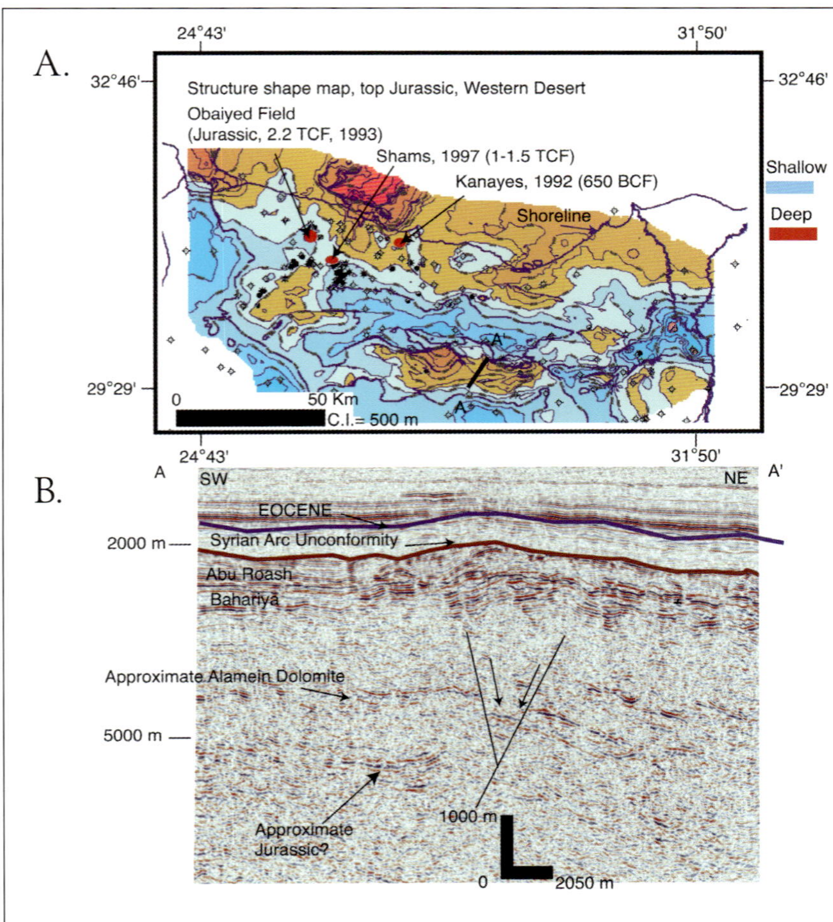

Figure 9. (a) Structural shape map, top of Jurassic, Western Desert, showing major Jurassic fields. Only 208 wells penetrate the Jurassic and older section (Table 1), and these are spatially aliased in small geographic areas. (b) Northeast-southwest arbitrary line across Abu Gharadig field [see (a) for location] illustrating a typical inverted Syrian Arc structure. Deep Jurassic structures do not mimic shallow Cretaceous objectives, and seismic image degrades rapidly beneath the Alamein Dolomite, making accurate mapping difficult.

Red Sea

The obvious southeastward extension of the Gulf of Suez productive trends has proved to be difficult, with discovery of relatively small fields in highly complicated structural blocks. However, recently acquired 3-D surveys and regional gravity and aeromagnetic data (Figure 15) illustrate the potential of this large geographic area. The Red Sea breakup phase was initiated by the development of the Gulf of Aqaba, and the southwestward extension of the Dead Sea transform zone can clearly be seen in Figure 15. This lineament continues southwestward into the northern end of the Gebel Duwi area, where it sets up a major structural transfer zone along the Egyptian side of the Red Sea coast.

Only 14 wells have been drilled in the Red Sea province, the northern half of which is shown in Figure 16. The Red Sea has more than twice the areal extent of the Gulf of Suez, which has proven reserves in excess of 8.6 BBOE (Table 4). Most significantly, Cretaceous source rocks are present along outcrop sections at Gebel Duwi immediately east of the Red Sea hills (Heath et al., 1998; Khalil et al., 1998) and are likely to be present throughout the graben areas of the Red Sea. The structural style and proven petroleum system of the Gulf of Suez should continue southward into the Red Sea, although the dominant petroleum product is likely to be gas. In addition, geochemical studies (Alsharhan and Salah, 1997) show that common Miocene source rocks have charged reservoirs in the southern end of the Gulf of Suez and eastward in Midyan field of Saudi Arabia (Figure 2).

As in the Gulf of Suez, subsalt seismic imaging poses the greatest challenge to success (Figure 15). Because multiple reservoir and seal combinations should be present in the Red Sea in Tertiary strata, significant future dis-

Table 5. Geochemical data with interpreted expelled hydrocarbons, Western Desert basins, Egypt (courtesy of Egyptian General Petroleum Corporation).

Basin	Horizon	Age	Acres	Expelled oil (bbl/acre-ft)	Expelled gas (bbl/acre-ft)	Effective source-rock thickness (ft)	Total expelled oil (BOEB)	Expected recovered oil 7% expelled	Total expelled gas (BOEB)	Expected recovered gas 7% of expelled
Shoushan	AEB	Early Cretaceous	4155775	70	110	30	8.73	0.61	13.71	0.96
Shoushan	Khataba	Jurassic	4155775	190	280	40	31.58	2.21	46.54	3.26
Shoushan	Diffah	Carboniferous	4155775	64	5	30	7.98	0.56	0.62	0.04
Shoushan	Zeitoun	Devonian	4155775	64	5	30	7.98	0.56	0.62	0.04
Shoushan	Kohla	Silurian	4155775	64	5	30	7.98	0.56	0.62	0.04
Subtotal							**64.25**	**4.5**	**62.13**	**4.35**
Alamein	AEB	Early Cretaceous	2642900	110	180	30	8.72	0.61	14.27	1
Alamein	Khataba	Jurassic	2642900	250	410	40	26.43	1.85	43.34	3.03
Alamein	Dhiffah	Carboniferous	2642900	60	10	30	4.76	0.33	0.79	0.06
Alamein	Zeitoun	Devonian	2642900	60	10	30	4.76	0.33	0.79	0.06
Alamein	Kohla	Silurian	2642900	60	10	30	4.76	0.33	0.79	0.06
Subtotal							**49.2**	**3.46**	**59.99**	**4.2**
Natrun	AEB	Early Cretaceous	3698825			30				
Natrun	Khataba	Jurassic	3698825	70	160	40	10.36	0.72	23.67	1.66
Natrun	Dhiffah	Carboniferous	3698825	70	13	30	7.77	0.54	1.44	0.1
Natrun	Zeitoun	Devonian	3698825	70	13	30	7.77	0.54	1.44	0.1
Natrun	Kohla	Silurian	3698825	70	13	30	7.77	0.54	1.44	0.1
Subtotal							**33.66**	**2.36**	**28**	**1.96**
Abu Gharadig	Abu Roash A-F	Late Cretaceous	3902600	240	50	50	46.83	3.28	9.76	0.68
Abu Gharadig	AEB	Early Cretaceous	3902600	65	180	30	7.61	0.53	21.07	1.48
Abu Gharadig	Khataba	Jurassic	3902600	160	420	40	24.98	1.75	65.56	4.59
Abu Gharadig	Dhiffah	Carboniferous	3902600	110	43	30	12.88	0.9	5.03	0.35
Abu Gharadig	Zeitoun	Devonian	3902600	110	43	30	12.88	0.9	5.03	0.35
Abu Gharadig	Kohla	Silurian	3902600	110	44	30	12.88	0.9	5.15	0.36
Subtotal							**118.05**	**8.26**	**111.61**	**7.81**
Siwa/Faghur	AEB	Early Cretaceous	14647100			30				
Siwa/Faghur	Khataba	Jurassic	14647100			40				
Siwa/Faghur	Dhiffah	Carboniferous	14647100	23	5	30	10.11	0.71	2.2	0.15
Siwa/Faghur	Zeitoun	Devonian	14647100	41	8	30	18.02	1.26	3.52	0.25
Siwa/Faghur	Kohla	Silurian	14647100	44	8	30	19.33	1.35	3.52	0.25
Subtotal							**47.46**	**3.32**	**9.23**	**0.65**
Guindi	AEB	Early Cretaceous	1661075	28	35	30	1.4	0.1	1.74	0.12
Guindi	Khataba	Jurassic	1661075	185	270	40	12.29	0.86	17.94	1.26
Guindi	Dhiffah	Carboniferous	1661075	332	61	30	16.54	1.16	3.04	0.21
Guindi	Zeitoun	Devonian	1661075	336	62	30	16.74	1.17	3.09	0.22
Guindi	Kohla	Silurian	1661075	346	65	30	17.24	1.21	3.24	0.23
Subtotal							**64.22**	**4.5**	**29.05**	**2.03**
		TOTAL					376.84	26.4	300.01	21

coveries may be made. In addition, targets drilled off the flanks of the highest structural culminations may also encounter reservoirs in pre-Tertiary strata.

Besides difficulties in subsalt imaging and reservoir prediction, an additional challenge to exploration remains the water depths of 1000+ m. This remote deep-water province will need a significant discovery to encourage widespread industry activity there.

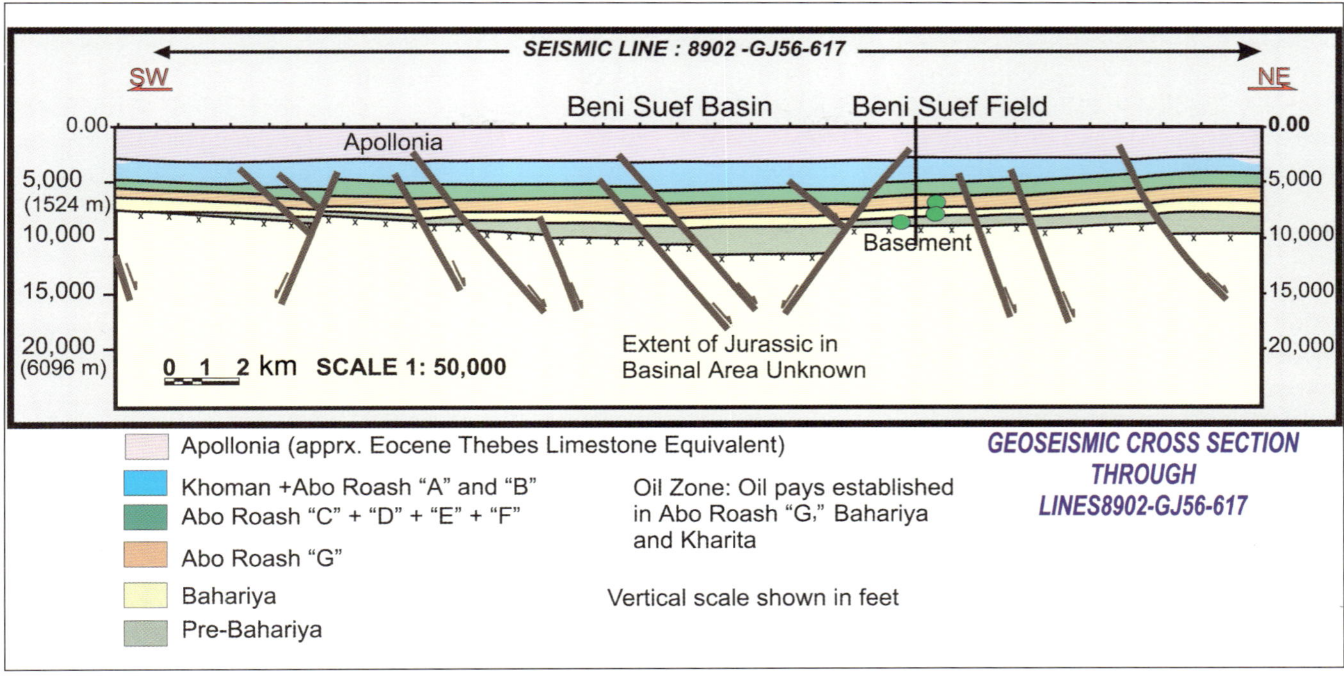

Figure 10. Northeast-southwest structural section taken from a 2-D seismic line across the Beni Suef Basin and newly discovered Beni Suef field, Upper Egypt (data and interpretation courtesy of the Apache Egypt Companies). This field is the southernmost extension of commercial pay into Upper Egypt. The Abo Roash C, D, E, and F members are Upper Cretaceous (Cenomanian) in age.

Nile Delta, Mediterranean, and North Sinai

The Nile Delta, Mediterranean, and North Sinai drilling provinces occur within a petroleum system dominated by play trends involving Pliocene turbidite fans and channels, deformation of Pliocene deltaic sandstone, Messinian valley fills, and older Miocene turbidite deposits. Most current activity is located in the offshore Mediterranean.

The North Sinai province occurs south of the Pliocene structuring limit (Cretaceous "hinge line" shown in Figure 2), and drilling has been targeted on Oligocene and older structural plays which represent northeastward extensions of Syrian Arc structures common to the Western Desert.

Charge in all three drilling provinces is from underlying Cretaceous and Jurassic source beds (Moussa and Matbouly, 1994). The North Sinai has had very limited exploratory success, with follow-up offsets to initial discoveries being unsuccessful. For instance, the Mango-1 well (Figure 2), drilled in 1986, tested 10,000 BOPD from Cretaceous sandstones on an extension of those trends into the offshore Mediterranean. It was offset with two dry holes, and the structure does not appear to be filled to spillpoint.

By contrast, the Nile Delta and Mediterranean region Tertiary targets are undergoing significant growth (Figure 7). Field-size distributions (Figure 8c) indicate very substantial room for continued large-field discoveries. Given the relatively immature exploration history of this petroleum system, the number of "yet-to-find" resources is difficult to estimate. The graph shown in Figure 8c assumes that at least one field will be found in the 512–1024-MMBOE grouping (3.07–6.14 tcf gas). In addition, we assume that a minimum of 16 new fields will be found in the 8–16-MBOE class (48–96 bcf gas). This conservative small-field estimate is based not only on the sheer size and volume of undrilled rock in this rich hydrocarbon province, but also on a large number of seismically derived leads visible from direct hydrocarbon indicators throughout the Pliocene section. The Nile Delta and Mediterranean region is geographically about half the area of the Gulf of Mexico shelf province and has similar multistoried pay potential. A value of 65–84 tcf of "yet-to-find" gas may be possible from a comparison with reserves in the Gulf of Mexico and from this simplified statistical approach.

The plays are numerous (Figure 16), and they provide a multitude of opportunities in both structural and stratigraphic traps. Tertiary reservoirs are likely to be underlain by the deep-marine equivalent facies of the same Mesozoic source zones that occur in the Western Desert, with additional sources identified in the prodelta shales of the lower Miocene to Oligocene Qantara Formation (Moussa and Matbouly, 1994).

The 17.5-Ma "mid-Rudeis" structural event caused seaward progradation of large deltas and associated turbidite deposits (TS30 and TS40 in Figure 16). By the Serravalian and Tortonian, these deltas and turbidites had prograded far offshore and now produce gas and condensate in several giant fields from structural and combination traps. The best examples are the Serravalian turbidite reservoirs of the Akhen and Temsah (2+ tcf) fields (Bertello et al., 1996) (Figure 17). Exploration for deep targets such as Temsah field has been limited in recent years because seismic data quality degrades significantly beneath Messinian evaporites and the Pliocene-Pleistocene growth-fault province. In addition, direct detection of hydrocarbons from seismic data is much more difficult in the deeper pays; hence, industry activity has focused recently on the higher-amplitude-bearing Messinian and Pliocene strata.

Favorable geologic conditions created by the Messinian crisis and Pliocene deltaic progradation tectono-stratigraphic events set up the "big plays" of the 1990s in Egypt. Underlying traps related to the Messinian crisis and drawdown occur mainly in the Baltim trend in several significant gas-condensate discoveries. Bright-spot seismic methods have provend to be successful repeatedly in the Nile Delta and Mediterranean region (Figure 18). The Baltim trend provides a typical example. The incised canyon fill of this trend is layered complexly, with multiple stacked point-bar and estuarine deposits, which form structural and stratigraphic traps within the canyon fill. The 1993 Baltim East discovery is a recent giant-field addition to Egypt's resources.

Messinian canyon cuts are not confined to the Baltim trend alone. These incisions occur throughout the Mediterranean, and additional exploration opportunities exist. Exploration for the associated downdip deltas and turbidites has not been undertaken because that play trend exists into the offshore deep water not yet tested by drilling.

Pliocene deltaic and turbidite sandstone plays currently dominate industry activity. Scarab field (Figure 19) is a trap in shelf-canyon turbidite channel sandstones. Its discovery was made possible through direct seismic hydrocarbon indicators, such as flat spots at the gas-water contacts. This 2-tcf accumulation of gas is typical of the bright-spot plays in this Mediterranean area, which were the subject of speculation as early as 1994 (Moussa and Matbouly, 1994) and have now proved to be successful.

This region offers by far the biggest short-term large-reserve growth potential. Drilling for gas targets in deep-water areas in front of the Nile Delta (water depths greater than 500 m) has only recently been attempted. In addition, oil has been encountered on the fringe of the delta in Abu Qir and Tineh fields, and the potential exists

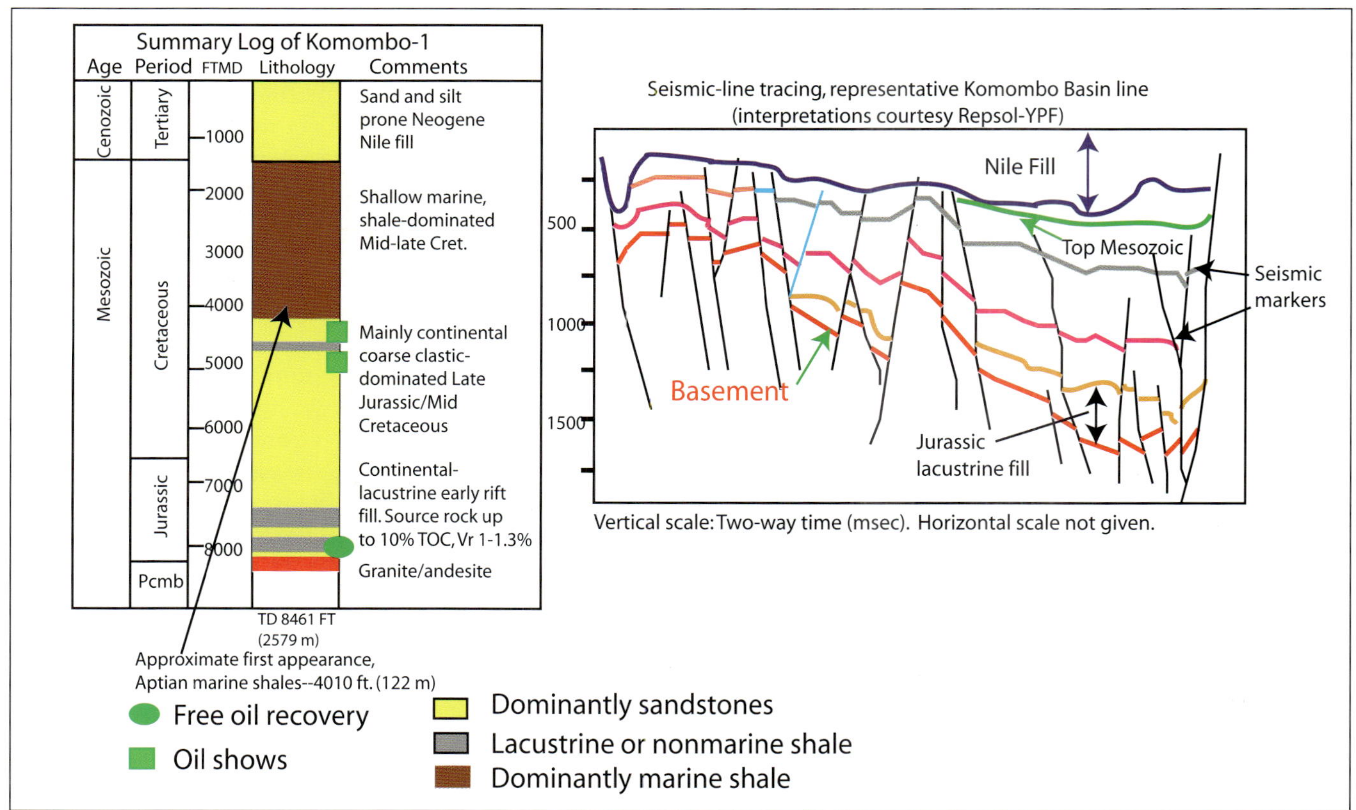

Figure 11. Representative seismic line and stratigraphic column, Komombo Basin, Upper Egypt. Patterns of onlap of Jurassic source and reservoir strata around Mesozoic rift structures are clear on this seismic profile. The Komombo-1 well proved the existence of oil in a previously unknown Mesozoic rift basin in Upper Egypt and is one of the most significant wells in Egypt. Data courtesy of Repsol-YPF. The Abo Roash C, D, E, and F members are Upper Cretaceous (Cenomanian) in age.

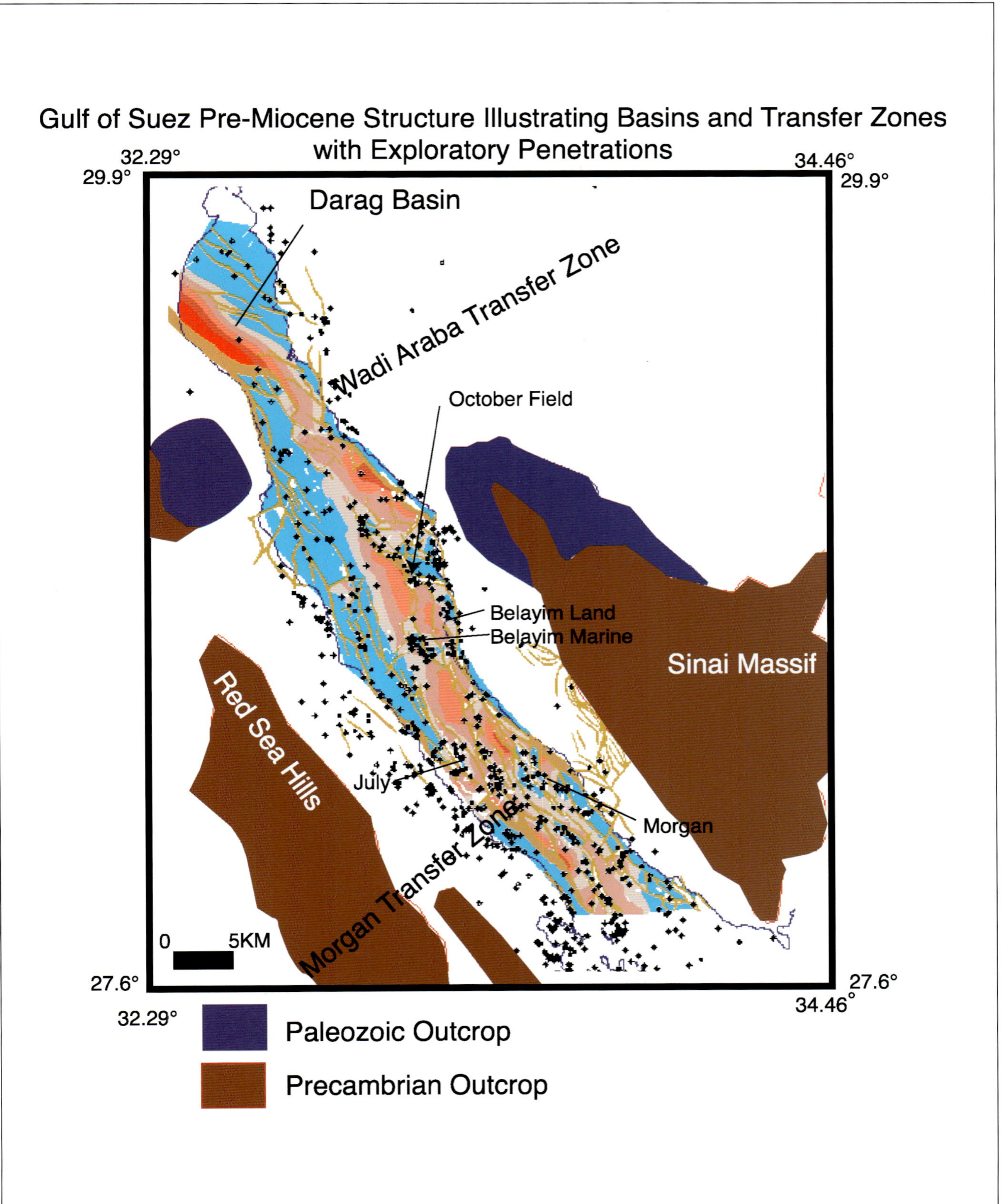

Figure 12. Gulf of Suez structural map with major faults and exploratory penetrations. Basinal areas (orange-red colors) are "kitchens" for Cretaceous and Eocene source rocks, resulting in hydrocarbon migration into flanking structural traps. Regional highs are shown in blue. Exploration in deep synclinal troughs and flank positions remains a future challenge and growth opportunity. See text for discussion.

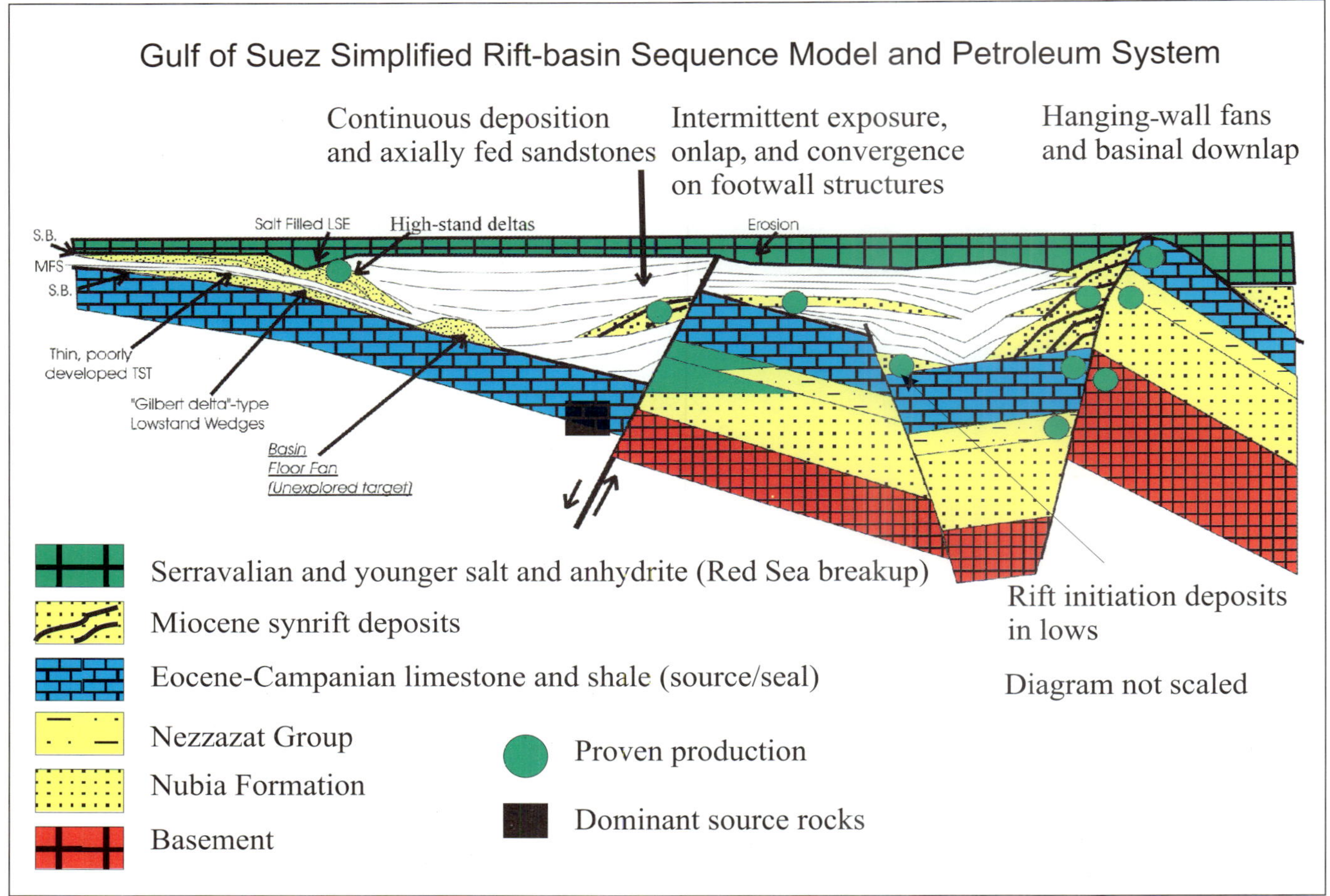

Figure 13. Simplified synrift petroleum system and Miocene sequence-stratigraphic diagram, Gulf of Suez (not to scale). Eocene and Cretaceous limestones and shales source rocks can charge reservoirs from Precambrian through Miocene strata through vertical migration along faults and through structural juxtaposition. Complex patterns of syndepositional reservoir facies preservation occur in the Miocene section. See text for discussion.

to discover oil in other parts of the Nile Delta and Mediterranean region (Moussa and Matbouly, 1994).

One of the real challenges to exploration remains the successful economic exploitation of gas resources in a deep-water setting. The successive giant-field discoveries of the 1990s in the Mediterranean have been in progressively deeper water. Additional growth will occur in water more than 500 m deep (Figure 2).

SUMMARY

When associated gas and gas-cap reserves are added to the volume of known resources in Egypt, more than 41 tcf of gas and 10 billion bbl of oil have been proved to date. Data presented in this paper suggest that additional resources of as much as 100 tcf of gas (16.6 BBOE) may remain to be discovered.

Egypt's position along the proto-Mediterranean Tethyan margin during the Mesozoic created abundant source rocks in rift grabens across the Western Desert, Nile Delta, and North Sinai regions. A probable Afro-Arabian Plate separation, still poorly understood, created nonmarine lacustrine rifts in Upper Egypt, depositing additional source rocks in large geographic areas. These remain lightly explored but have now been proved to contain light hydrocarbons. The widespread transgressions during the subsequent Cretaceous passive-margin episode added a further blanket of rich source rocks and seals over large areas. Paleocene and Eocene transgressions provide additional source and seal strata.

The Syrian Arc event inverted many of the older rift structures and basins across northern Egypt. Although this play has generally proved to be disappointing, discoveries will continue to be made in and around these structural trends.

Rifting in the Gulf of Suez and Red Sea created a world-class petroleum system which is in its mature phase of exploration in the Gulf of Suez but still provides a frontier exploration play in the Red Sea.

The Nile Delta and Mediterranean region is undergoing an early phase of exploration because of the recent awarding of gas rights in those areas. The significant num-

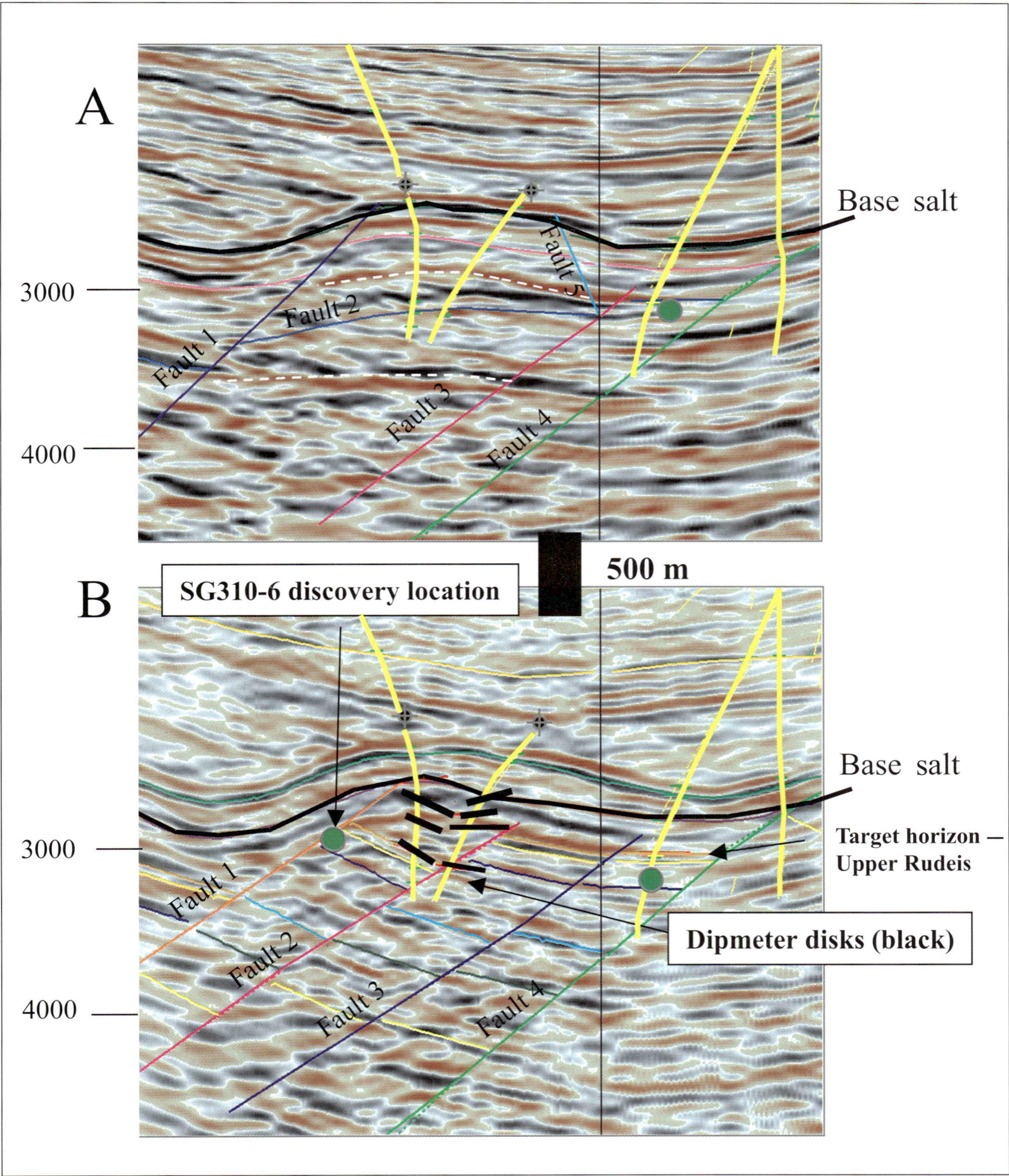

Figure 14. Example of successful step-out exploration in the Gulf of Suez. The SG310-6 well flowed 11,800 BOPD from Hawara (Burdigalian) sandstones in an attic position of a tilted fault block. An adjacent fault block (SG310-4 well, Table 3) flowed 20,000 BOPD in a similar structural position. (a) Prior seismic interpretations did not show the presence of a prospective compartment because of severe multiples originating from overlying Serravalian salts. (b) Posting of dipmeters directly on 3-D volume depth sections and successful reprocessing to remove multiples showed the updip untested fault block.

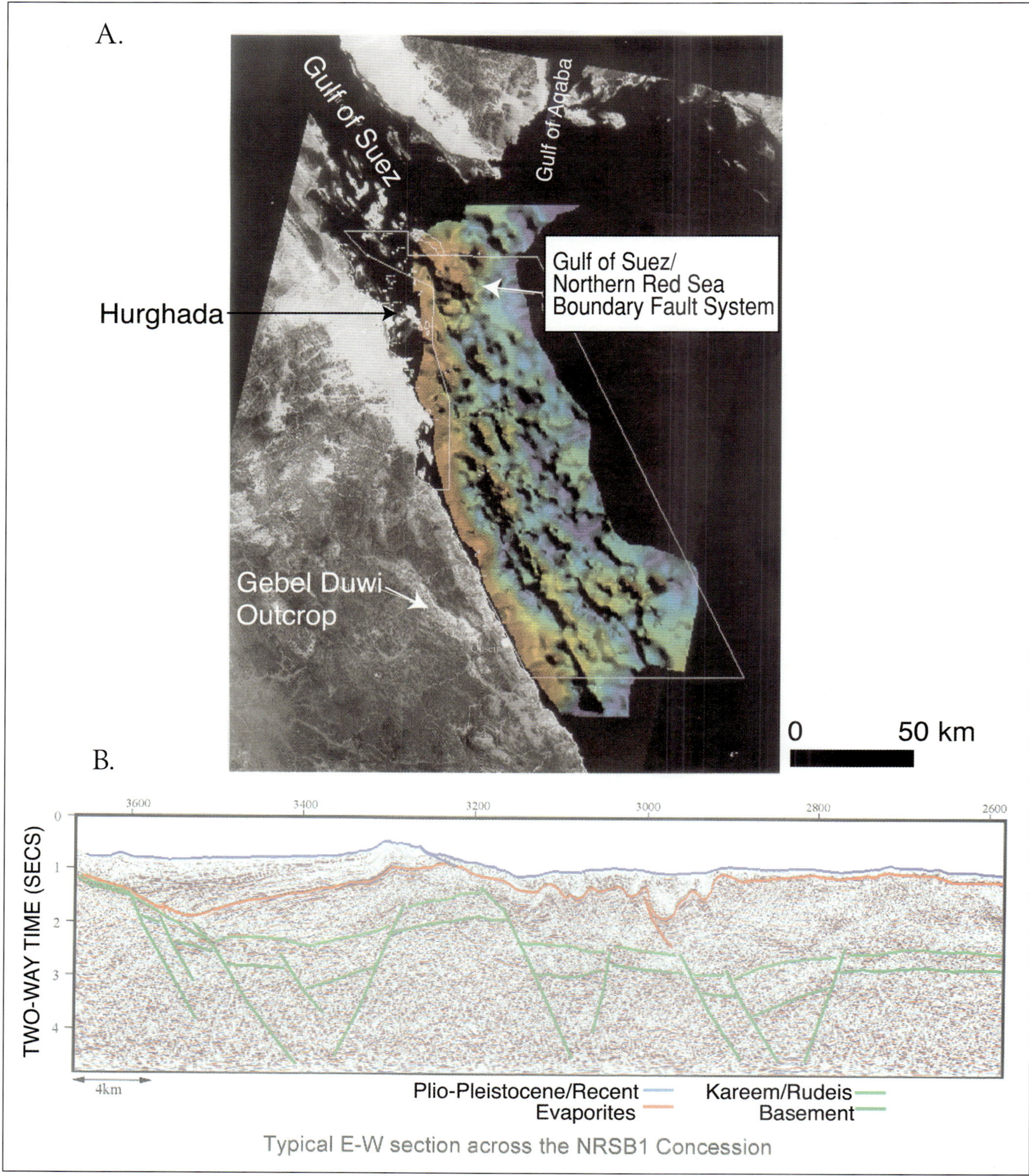

Figure 15. (a) Regional gravity interpretation of the northern portion of the Red Sea merged with the onshore Landsat image. The Gebel Duwi outcrop (highlighted) contains preserved Cretaceous source rocks and Nubia reservoirs (see Heath et al., 1998). Offshore, the shallower basement of the southern Gulf of Suez and the nearshore terrace are shown in red shading. The broad structural fabric is highlighted by the shaded relief created by illuminating from the northeast. The large number of undrilled subbasins resembling those of the Gulf of Suez is apparent. (b) Representative arbitrary seismic line across a portion of the gravity survey shown above. Data and interpretations courtesy of British Gas International and Edison Gas.

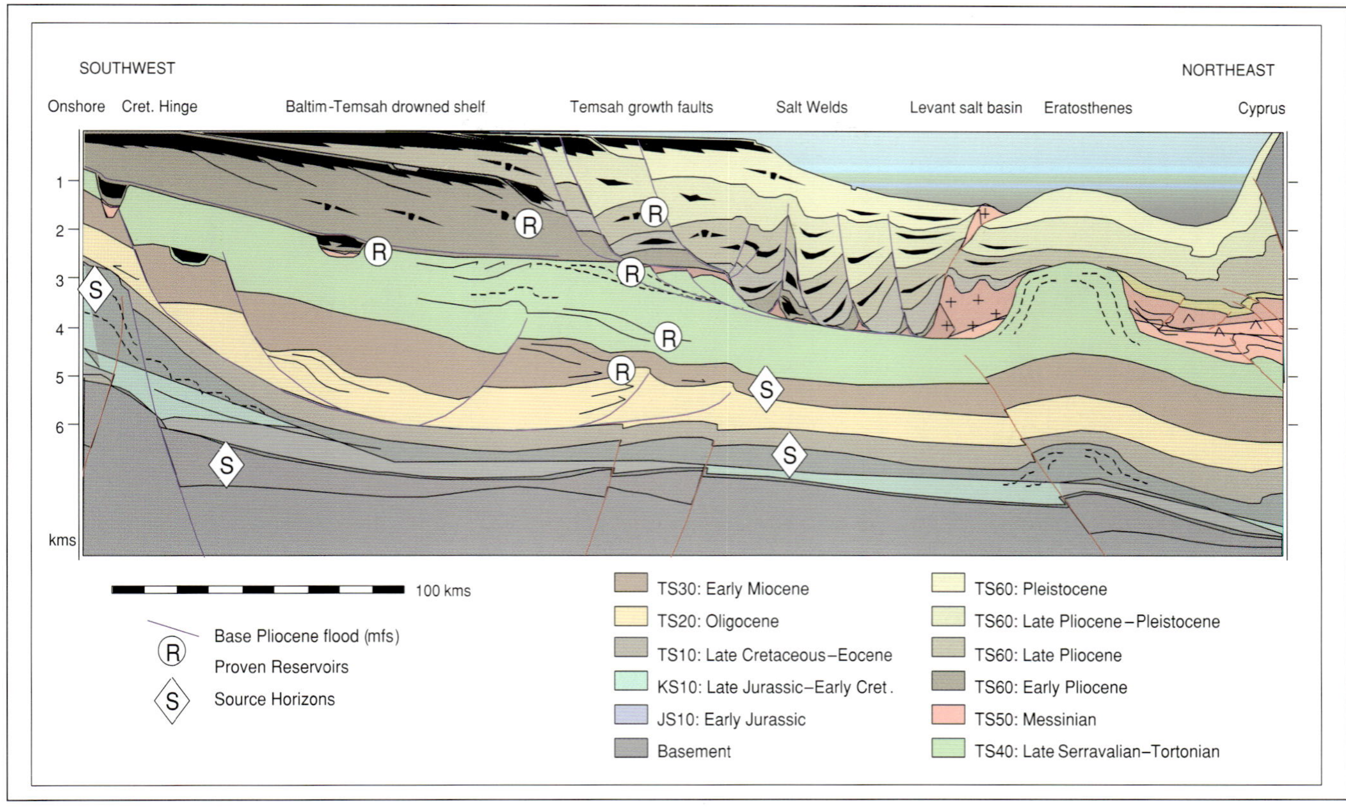

Figure 16. Schematic cross section based on regional seismic profiles across the Nile Delta and Mediterranean showing major petroleum plays. Most current activity targets the Pliocene and Messinian section, which is better imaged seismically, but deeper potential exists throughout the basin. See text for discussion.

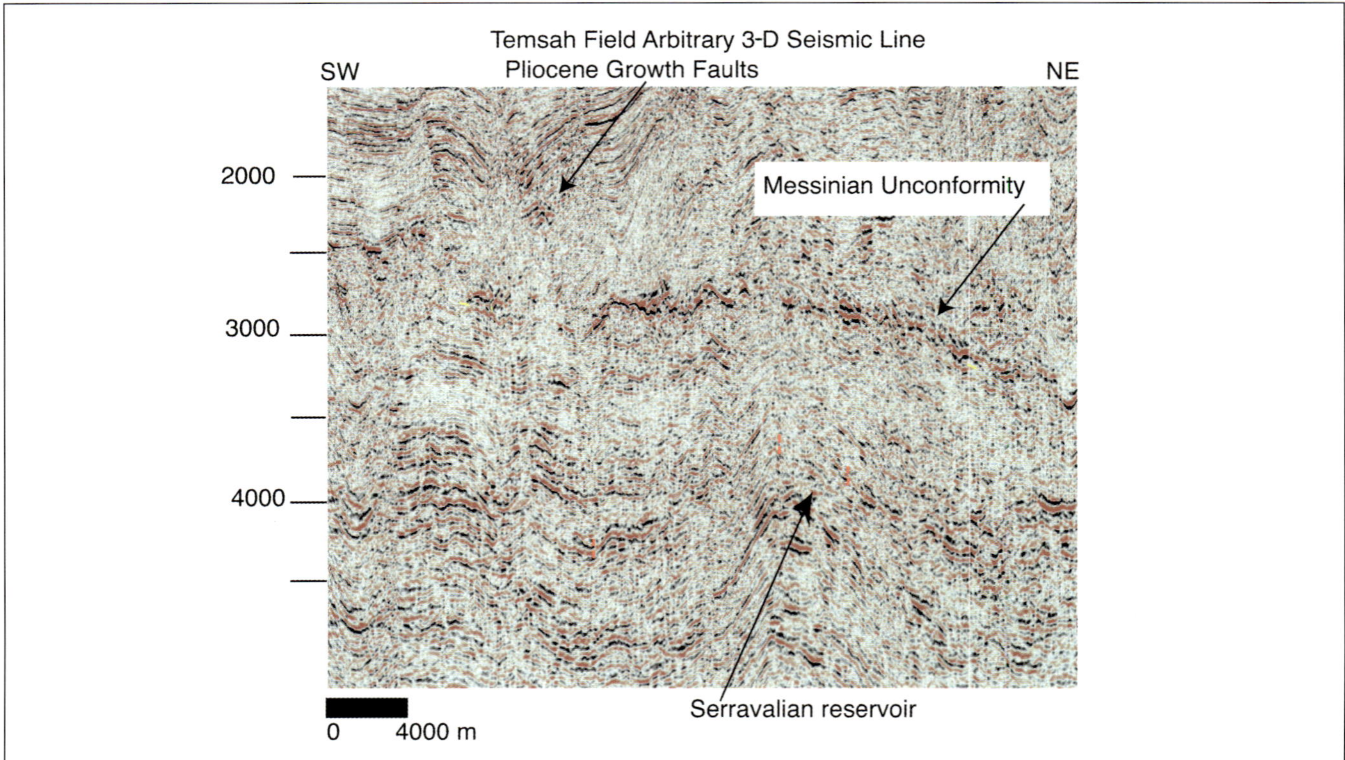

Figure 17. Arbitrary 3-D seismic profile across the giant Temsah field, Nile Delta, showing the difficult deep seismic imaging and a large thrusted structure lying beneath a Pliocene growth fault province.

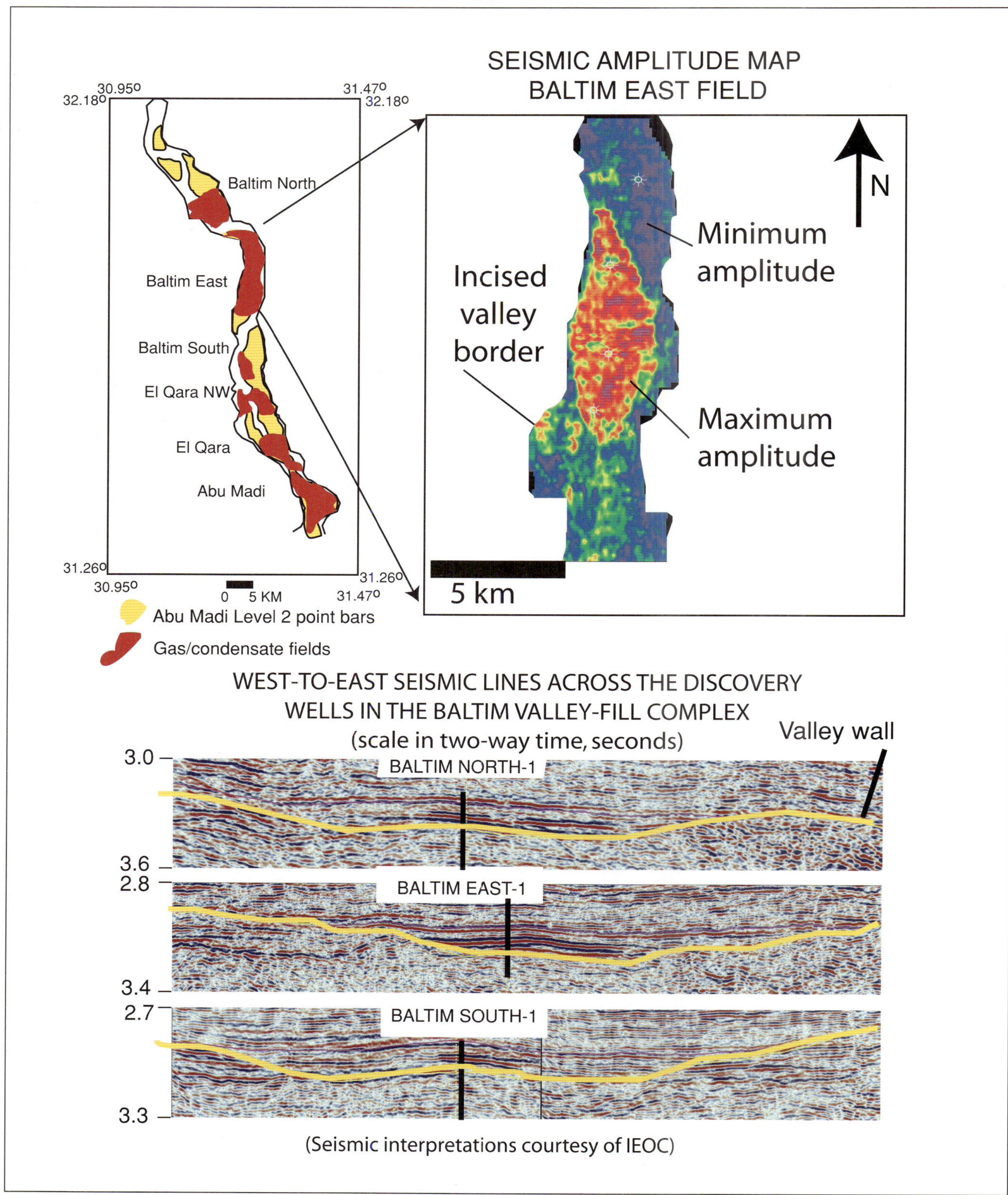

Figure 18. Seismic expression of gas pays in the Messinian incised valley-fill Baltim trend in the Nile Delta (modified from Palmieri et al., 1996, and Dalla et al., 1997). Canyon incision exceeds 1000 m in places. Fields in the trend are in small closures and fault traps encased within valley walls and are sealed dominantly by abandoned channel shales. Most production occurs in fluvial point bars and thin estuarine deposits. Seismic amplitudes have proved to be effective for identifying the gas-prone traps. Data courtesy of IEOC.

ber of giant fields discovered since 1990 attests to the exploration potential of those areas. The large volume of undrilled rock and large areal extent of this province suggests that a long period for giant-field discoveries lies ahead. Most of those discoveries will be in gas trends, but additional oil potential does exist.

Although the data presented in this paper are regional in nature, they do strongly suggest that the petroleum industry in Egypt will continue to enjoy a long period of success as the offshore Mediterranean and Jurassic and deeper Paleozoic plays in the Western Desert continue to be developed. The large volume of undrilled deep Jurassic and older rocks in the Western Desert and its offshore extensions, and the vast area of the offshore Mediterranean may provide room for that new reserve growth. The Gulf of Suez has been on a declining discovery curve since the mid-1980s and will continue to decline unless real breakthroughs are made in subsalt imaging of increasingly smaller targets and/or new concepts are proved to be viable. The Red Sea and Upper Egypt trends offer further growth potential but with multiple technical challenges, including remote locations and deep water.

"Yet-to-find" resource numbers are difficult to quantify with any degree of accuracy. But the missing field-size distributions shown in this paper, the large areal extent of rich source rocks, multiple play and basin types, and relatively sparse drilling in many trends suggest a very positive future. We believe these data show that in the coming decades, Egypt should be able to fully replace and perhaps more than replace its current proven reserve base of 15.7 BBOE.

ACKNOWLEDGMENTS

This paper would not have been possible without cooperation from numerous oil companies in Egypt, which agreed to release proprietary data, review the text,

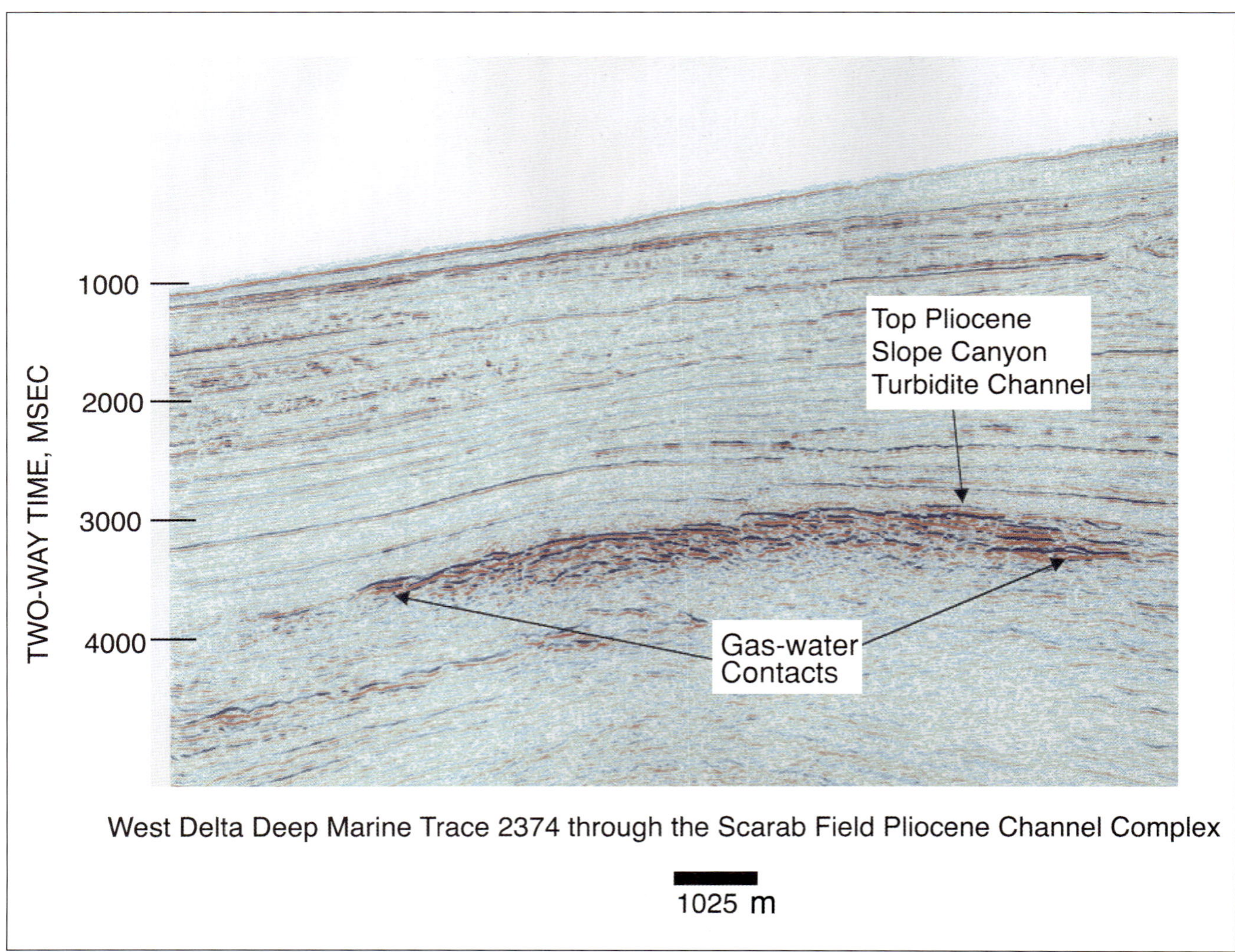

Figure 19. Scarab field discovery, Nile Delta (courtesy of British Gas International and Edison Gas). This Pliocene slope-canyon turbidite channel was found through direct hydrocarbon detection of gas-bearing amplitudes. This 3-D volume arbitrary line, shown in two-way time, is a strike view trending north-south down the channel axis.

and contribute key figures. Special thanks go to Chuck Fetzner and David Allard (Apache Corporation), Marco Boy and Marco Serazzi (International Egyptian Oil Company), and Mike Winfield and Jamie Suarez (Repsol) for contributing valuable data and ideas. Marc Gerrit, Bill Horsefield, and Marcus Schwander (Shell Egypt New Ventures) provided data from Obaiyed field. David Pivnik, Bill Chandler, and Tim Marchant provided valuable comments on the content and text. The IHS Energy Company provided critical missing field-size data and worldwide field-distribution comparisons. Lastly, we thank BP Amoco and the Egyptian General Petroleum Corporation for permission to publish this paper.

REFERENCES CITED

Alsharhan, A. S., and M. G. Salah, 1997, A common source rock for Egyptian and Saudi hydrocarbons in the Red Sea: AAPG Bulletin, v. 81, p. 1640–1659.

Barakat, A. O., A. Moustafa, M. S. El-Gayar, and J. Rullkotter, 1997, Source-dependent biomarker properties of five crude oils from the Gulf of Suez, Egypt: Organic Geochemistry, v. 27, p. 441–450.

Barker, C., and P. A. Dickey, 1984, Hydrocarbon habitat in main producing areas, Saudi Arabia: Discussion: AAPG Bulletin, v. 68, p. 108–109.

Bein, A., and Y. Weiler, 1976, The Cretaceous Talme Yafe Formation: A contour current shaped sedimentary prism of calcareous detritus at the continental margin of the Arabian craton: Sedimentology, v. 23, p. 511–532.

Bertello, F., K. Barsoum, S. Dalla, and S. Guessarian, 1996, Temsah discovery: A giant gas field in a deep sea turbidite environment, *in* M. Youssef, ed., Proceedings of the 13th Petroleum Conference: Cairo, Egypt, Egyptian General Petroleum Corporation, p. 165–180.

Boote, D. R. D., D. D. Clark-Lowes, and M. W. Traut, 1998, Paleozoic petroleum systems of North Africa, *in* D. S. MacGregor, R. T. J. Moody, and D. D. Clark-Lowes, eds., Petroleum geology of North Africa: Geological Society Special Publication 132, p. 7–68.

Capen, E., 1992, Dealing with exploration uncertainties, *in* R. Steinmetz, ed., The business of petroleum exploration: AAPG, p. 29–61.

Dalla, S., H. Harby, and M. Serazzi, 1997, Hydrocarbon exploration in a complex incised valley fill: An example from the late Messinian Abu Madi Formation (Nile Delta Basin, Egypt): The Leading Edge, v. 16, p. 1819–1824.

Darwish, M., 1992, Facies developments of the Upper Paleozoic–Lower Cretaceous sequences in the northern Galala Plateau and evidences for their hydrocarbon reservoir potentiality, northern Gulf of Suez, Egypt, *in* A. Sadek, ed., Geology of the Arab world: Cairo, Egypt, Cairo University, p. 175–214.

Dixon, J. E., and A. H. F. Robertson, 1984, The geological evolution of the eastern Mediterranean: Geological Society Special Publication 17, 836 p.

Dolson, J., O. El-Gendi, H. Sharmy, M. Fathalla, and I. Gaafar, 1996, Gulf of Suez rift basin sequence models—Part A: Miocene sequence stratigraphy and exploration significance in the greater October field area, northern Gulf of Suez, *in* M. Youssef, ed., Proceedings of the 13th Petroleum Conference: Cairo, Egypt, Egyptian General Petroleum Corporation, p. 227–241.

Dolson, J. C., B. Steer, J. Garing, G. Osborne, A. Gad, and H. Amr, 1997, 3D seismic and workstation technology brings technical revolution to the Gulf of Suez Petroleum Company: The Leading Edge, v. 16, p. 1809–1817.

Dolson, J. C., J. W. Stewart, B. Golob, O. El-Gendi, I. Gaafar, and G. Azazi, 1998, Synchronous onlap and downlap, Asl Formation lowstand wedge, October field, Gulf of Suez (abs.), *in* Integrated structural and sequence stratigraphic analysis in rift settings: AAPG Hedberg Research Conference, p. 37.

Drew, L. J., and J. H. Schuenemeyer, 1993, The evolution and use of discovery process models at the U.S. Geological Survey: AAPG Bulletin, v. 77, p. 467–478.

El-Banbi, D. H. A., 1999, Exploration, production and reserves in Egypt—Statistical data (1886–1997): Cairo, Egypt, Egyptian General Petroleum Corporation, 191 p.

Gawthorpe, R. L., J. M. Hurst, and C. P. Sladen, 1990, Evolution of Miocene footwall-derived coarse-grained deltas, Gulf of Suez, Egypt: Implications for exploration: AAPG Bulletin, v. 74, p. 1077–1086.

Geizery, M. E., I. A. Moula, S. A. Aziz, and M. Helmy, 1998, Qarun field geological model and reservoir characterization: A successful case from the Western Desert, *in* M. Eloui, ed., Proceedings of the 14th Petroleum Conference: Cairo, Egypt, gyptian General Petroleum Corporation, p. 279–297.

Halbouty, M. T., and F. El-Baz, 1992, The Mediterranean region: Its geologic history and oil and gas potentials, *in* A. Sadek, ed., Geology of the Arab world: Cairo, Egypt, Cairo University, p. 15-60.

Haq, B. U., J. Hardenbol, and P. R. Vail, 1988, Mesozoic and Cenozoic chronostratigraphy and eustatic cycles, *in* C. K. Wilgus, B. S. Hastings, C. G. S. C. Kendall, H. W. Posamentier, C. A. Ross, and J. C. V. Wagoner, eds., Sea-level changes: An Integrated approach: Society of Economic Paleontologists and Mineralogists Special Publication 42, p. 71–109.

Harms, J. C., and J. L. Wray, 1990, Nile Delta, *in* R. Said, ed., The geology of Egypt: Rotterdam, Netherlands, A. A. Balkema Publishers, p. 329–344.

Harwood, C., M. Ayyad, and N. Hodgson, 1998, The application of sequence stratigraphy in the exploration for Plio-Pleistocene hydrocarbons in the Nile Delta, *in* M. Eloui, ed., Proceedings of the 14th Petroleum Conference: Cairo, Egypt, Egyptian General Petroleum Corporation, p. 1–13.

Heath, R., S. Vanstone, J. Swallow, M. Ayyad, M. Amin, P. Huggins, R. Swift, I. Warburton, K. McClay, and A. Younnis, 1998, Renewed exploration in the offshore north Red Sea region—Egypt, *in* M. Eloui, ed., Proceedings of the 14th Petroleum Conference: Cairo, Egypt, Egyptian General Petroleum Corporation, p. 16–34.

Hegazy, A., 1992, Western Desert oil and gas fields (a comprehensive overview): Cairo, Egypt, Egyptian General Petroleum Corporation, 431 p.

Kamel, H., 1990, Gravity map, *in* R. Said, ed., The geology of Egypt:, Rotterdam, Netherlands, A. A. Balkema Publishers, p. 45–50.

Keeley, M. L., and M. S. Massoud, 1998, Tectonic controls on the petroleum geology of NE Africa, *in* D. S. MacGregor, R. T. J. Moody, and D. D. Clark-Lowes, eds., Petroleum geology of North Africa: Geological Society Special Publication 132, p. 69–78.

Kerdany, M. T., and O. H. Cherif, 1990, Mesozoic, *in* R. Said, ed., The geology of Egypt: Rotterdam, Netherlands, A. A. Balkema Publishers, p. 407–438.

Khalil, M., E. E. Nagdy, H. A. Rahman, and M. Comisso, 1998, Frontier exploration in the southern Egyptian Red Sea: Superimposed Cretaceous and Tertiary rift tectonics in the newly discovered, untested Ras Banas sub-basin, *in* M. Eloui, ed., Proceedings of the 14th Petroleum Conference: Cairo, Egypt, Egyptian General Petroleum Corporation, p. 1–15.

Klitzsch, E., 1990, Paleozoic, *in* R. Said, ed., The geology of Egypt: Rotterdam, Netherlands, A. A. Balkema Publishers, p. 393–406.

Krebs, W. N., W. A. Wescott, D. Nummedahl, I. Gaafar, G. Azazi, and S. A. Karamat, 1996, Graphic correlation and sequence stratigraphy of Neogene rocks in the Gulf of Suez, *in* M. Youssef, ed., 13th Petroleum Conference: Cairo, Egypt, Egyptian General Petroleum Corporation, p. 214–226.

Krebs, W. N., W. A. Wescott, D. Nummedahl, I. Gaafar, G. Azzazi, and S. Karamat, 1997, Graphic correlation and sequence stratigraphy of Neogene rocks in the Gulf of Suez: Bulletin of the Society of Geology, France, v. 1, p. 63–71.

MacGregor, D. S., R. T. J. Moody, and D. D. Clark-Lowes, 1998, Petroleum geology of North Africa: Geological Society Special Publication 132, 444 p.

Mahmoud, A., and A. E. Barkooky, 1998, Mesozoic valley fills incised in Paleozoic rocks: Potential exploration targets in the north Western Desert of Egypt, Obaiyed area, *in* M. Eloui, ed., Proceedings of the 14th Petroleum Conference: Cairo, Egypt, Egyptian General Petroleum Corporation, p. 84–100.

Matbouly, S., and M. E. Sabbagh, 1996, Gulf of Suez oil fields (a comprehensive overview): Cairo, Egypt, Egyptian General Petroleum Corporation, 735 p.

Meshref, W. M., 1990, Tectonic framework, *in* R. Said, ed., The geology of Egypt: Rotterdam, Netherlands, A. A. Balkema Publishers, p. 113–156.

Miller, F. X., 1977, The graphic correlation method in biostratigraphy, *in* E. G. Kauffman and J. E. Hazel, eds., Concepts and methods of biostratigraphy: Stoudsburg, Pennsylvania, Hutchinson and Ross, Inc., p. 165–186.

Morris, A., and D. H. Tarling, 1996a, Palaeomagnetism and tectonics of the Mediterranean region: An introduction, *in* A. Morris and D. H. Tarling, eds., Palaeomagnetism and tectonics of the Mediterranean region: Geological Society Special Publication 105, p. 1–17.

Morris, A., and D. H. Tarling, eds., 1996b, Palaeomagnetism and tectonics of the Mediterranean region: Geological Society Special Publication 105, 422 p.

Moussa, S. S., and S. I. Matbouly, 1994, Nile Delta and North Sinai: Fields, discoveries and hydrocarbon potentials (a comprehensive overview): Cairo, Egypt, Egyptian General Petroleum Corporation, 387 p.

Moustafa, A. R., R. E. Badrawy, and H. Gibali, 1998, Pervasive E-ENE oriented faults in northern Egypt and their effect on the development and inversion of prolific sedimentary basins, *in* M. Eloui, ed., 14th Petroleum Conference: Cairo, Egypt, Egyptian General Petroleum Corporation, p. 51–67.

Moustafa, A. R., and M. H. Khalil, 1990, Structural characteristics and tectonic evolution of north Sinai fold belts, *in* R. Said, ed., The geology of Egypt: Rotterdam, Netherlands, A. A. Balkema Publishers, p. 381–392.

Palmieri, G., H. Harby, J. A. Marini, F. Hashem, S. Dalla, and M. Shash, 1996, Baltim fields complex: An outstanding example of hydrocarbon accumulations in a fluvial Messinian incised valley, *in* M. Youssef, ed., Proceedings of the 13th Petroleum Conference: Cairo, Egypt, Egyptian General Petroleum Corporation, p. 256–269.

Patton, T. L., A. R. Moustafa, R. A. Nelson, and S. A. Abdine, 1994, Tectonic evolution and structural setting of the Gulf of Suez rift, *in* S. M. Landon, ed., Interior rift basins: AAPG Memoir 59, p. 9–56.

Philobbos, E. R., and B. H. Purser, eds., 1993, Geodynamics and sedimentation of the Red Sea–Gulf of Aden rift system: Geological Society of Egypt Special Publication 1, 456 p.

Prosser, S., 1993, Rift-related linked depositional systems and their seismic expression, *in* G. D. Williams and A. Dobb, eds., Tectonics and seismic sequence stratigraphy: Geological Society Special Publication 71, p. 35–66.

Purser, B. H., and D. W. J. Bosence, eds., 1998, Sedimentation and tectonics in rift basins, Red Sea–Gulf of Aden: London, Chapman and Hall, 663 p.

Purser, B. H., and E. R. Philobbos, 1993, The sedimentary expressions of rifting in the NW Red Sea, Egypt, *in* E. R. Philobbos and B. H. Purser, eds., Geodynamics and sedimentation of the Red Sea–Gulf of Aden rift system: Geological Society of Egypt Special Publication 1, p. 1–45.

Ramzy, M., B. Steer, F. Abu-Shadi, M. Schlorholtz, J. Mika, J. Dolson, and M. Zinger, 1996, Gulf of Suez rift basin sequence models—Part B: Miocene sequence stratigraphy and exploration significance in the central and southern Gulf of Suez, *in* M. Youssef, ed., Proceedings of the 13th Petroleum Conference: Cairo, Egypt, Egyptian General Petroleum Corporation, p. 242–256.

Ramzy, M., B. Steer, J. Thorseth, and J. Garing, 1998, Integrated exploration: The SG310-4 case history, *in* M. Eloui, ed., Proceedings of the 14th Petroleum Conference: Cairo, Egypt, Egyptian General Petroleum Corporation, p. 359–373.

Robertson, A. H., and J. E. Dixon, 1984, Introduction: Aspects of the geological evolution of the eastern Mediterranean, *in* J. E. Dixon and A. H. F. Robertson, eds., The geological evolution of the eastern Mediterranean: The Geological Society Special Publication 17, p. 1–74.

Robison, V. D., 1995, Source rock characterization of the Late Cretaceous Brown Limestone of Egypt, *in* B. Katz, ed., Petroleum source rocks: Heidelberg, Germany, Springer-Verlag, p. 265–281.

Root, D. H., and E. D. Attanasi, 1993, Small fields in the national oil and gas assessment: AAPG Bulletin, v. 77, p. 485–490.

Sadek, A., 1992, Geology of the Arab world: Cairo, Egypt, Cairo University, 541 p.

Said, R., 1990a, Cenozoic, *in* R. Said, ed., The geology of Egypt: Rotterdam, Netherlands, A. A. Balkema Publishers, p. 451–486.

Said, R., ed., 1990b, The geology of Egypt: Rotterdam, Netherlands, A. A. Balkema Publishers, 734 p.

Said, R., 1990c, Mesozoic, *in* R. Said, ed., The geology of Egypt: Rotterdam, Netherlands, A. A. Balkema Publishers, p. 439–450.

Schull, T. J., 1988, Rift basins of interior Sudan: Petroleum exploration and discovery: AAPG Bulletin, v. 72, p. 1128–1142.

Schutz, K. I., 1994, Structure and stratigraphy of the Gulf of Suez, Egypt, *in* S. M. Landon, ed., Interior rift basins: AAPG Memoir 59, p. 57–96.

Sestini, G., 1984, Tectonic and sedimentary history of the NE African margin (Egypt-Libya), *in* J. E. Dixon and A. H. F. Robertson, eds., The geological evolution of the eastern Mediterranean: Geological Society Special Publication 17, p. 161–176.

Sharp, I. R., S. Gupta, J. R. Underhill, and R. L. Gawthorpe, 1998, Synrift sequence stratigraphy and structural development of the Sinai margin, Miocene Gulf of Suez, Egypt, a field guidebook: AAPG Hedberg Conference, Integrated Structural and Sequence Stratigraphic Analysis in Rift Settings, 180 p.

Smith, M. D., and D. R. Jones, 1992, Trend analysis, *in* R. Steinmetz, ed., The business of petroleum exploration: AAPG, p. 215–236.

Taha, M. A., 1992, Mesozoic rift basins in Egypt: Their southern extension and impact on future exploration, *in* S. Abdine, ed., Proceedings of the 11th Petroleum Exploration and Production Conference: Cairo, Egypt, Egyptian General Petroleum Corporation, p. 1–19.

Taha, M. A., and H. A. Aziz, 1998, Mesozoic rifting in Upper Egypt concession, *in* S. Shaheen, conference chairman, 14th International Petroleum Conference abstract book.

Traut, M. W., D. R. D. Boote, and D. D. Clark-Lowes, 1998, Exploration history of the Paleozoic petroleum systems of North Africa, *in* D. S. MacGregor, R. T. J. Moody, and D. D. Clark-Lowes, eds., Petroleum geology of North Africa: Geological Society Special Publication 132, p. 69–78.

Wescott, W. A., W. N. Krebs, D. Nummedal, and J. C. Dolson, 1998: Field trip guidebook: Miocene sequence stratigraphy, chronostratigraphic correlation and subsurface exploration and production analogues, Gulf of Suez: Outcrop to subsurface: AAPG Hedberg Conference, Integrated Structural and Sequence Stratigraphic Analysis in Rift Settings, 150 p.

Wycisk, P., 1990, Aspects of cratonal sedimentation: Facies distribution of fluvial and shallow marine sequences in NW Sudan/SW Egypt since Silurian time: Journal of African Earth Sciences, v. 10, p. 215228.

Wycisk, P., 1994, Correlation of the major Late Jurassic–early Tertiary low and highstand cycles of southwest Egypt and northwest Sudan: Geologische Rundschau, v. 83, p. 759–772.

Konert, G., A. M. Afifi, S. A. Al-Hajri, K. de Groot, A. A. Al Naim, and H. J. Droste, 2001, Paleozoic stratigraphy and hydrocarbon habitat of the Arabian Plate, *in* M. W. Downey, J. C. Threet, and W. A. Morgan, eds., Petroleum provinces of the twenty-first century: AAPG Memoir 74, p. 483–515.

Chapter 24

Paleozoic Stratigraphy and Hydrocarbon Habitat of the Arabian Plate[1]

G. Konert
Shell International
Rijswijk, Netherlands

S. A. Al-Hajri
Saudi Aramco
Dhahran, Saudi Arabia

A. A. Al Naim
Saudi Aramco
Dhahran, Saudi Arabia

A. M. Afifi
Saudi Aramco
Dhahran, Saudi Arabia

K. de Groot
Shell International
Rijswijk, Netherlands

H. J. Droste
Petroleum Development Oman
Muscat, Sultanate of Oman

ABSTRACT

The Paleozoic section became prospective during the early 1970s when the enormous gas reserves in the Permian Khuff reservoirs were delineated in the Gulf and Zagros regions and oil was discovered in Oman. Since then, frontier exploration has targeted the Paleozoic System throughout the Middle East, driven by the need to replace oil production from maturing fields, the need to add gas reserves to meet local energy requirements, and other economic considerations.

The Paleozoic sequences were essentially deposited in continental to deep-marine siliciclastic environments at the Gondwana continental margin. Carbonates became dominant only in the Late Permian. The sediments were deposited in arid to glacial settings, reflecting the drift of the region from equatorial to high southern latitudes and back.

Following late Precambrian rifting that formed salt basins in Oman and the Arabian Gulf region, the Cambrian-Ordovician sequences were deposited on a peneplained continental platform. However, by the Late Ordovician, this margin probably differentiated into two terranes along the Zagros fault zone, as indicated by the Silurian paleogeography.

The entire region was affected by the Hercynian orogeny during the Carboniferous, which caused long-wavelength plate buckling in the north, block uplifts in the central region and regional uplift in the south, and tectonism along the Zagros fault zone. This deformation caused widespread erosion of the Devonian-Carboniferous section, and probably was caused by collision along the northern margin of Gondwana. The Paleozoic tectonic supercycle ended with the onset of breakup tectonics in the Permian and deposition of Khuff carbonates over the eastern passive margin.

A major Paleozoic petroleum system embraces reservoir seal pairs spanning the Silurian to Permian sequences. Hydrocarbons occur in a variety of traps and are sourced primarily by the Silurian "hot" shale. Hydrocarbon expulsion estimates (taking into account secondary migration losses) suggest that about 1 trillion barrels of oil equivalent may have been trapped from the Silurian "hot" shale alone. A second petroleum system occurs in areas charged from Upper Precambrian source rocks in salt basins.

Problems with deep seismic imaging and relatively tight and heterogeneous reservoirs, combined with hostile subsurface environments, pose significant challenges to exploration and development. The critical success factor is the continuous innovative effort of earth scientists and subsurface engineers to find integrated technology solutions which render the Paleozoic plays economically viable even in a low-oil-price environment.

[1] *This paper summarizes the efforts of many geoscientists from Shell, Saudi Aramco, and Petroleum Development Oman.*

INTRODUCTION

The Middle East holds estimated proven reserves of about 625 billion barrels (bbl) of crude oil and 1720 trillion (standard) cubic feet (tcf) of natural gas, approximately 64% and 34% of world reserves, respectively. This fact led Murris (1980) to describe the area as the world's richest hydrocarbon habitat. These reserves are found mainly in Mesozoic and Tertiary reservoirs in a northwest-trending zone from Oman to Turkey (Figure 1).

Crude and condensate production from these reserves was reported in 1997 to be approximately 7.4 billion bbl/year (*World Oil*, 1998). Known reserves have been developed only partly, because of remoteness from infrastructure and for strategic reasons, technical considerations, or capacity restrictions. Reserve estimates vary with project lifetime, but after initial development, reserves generally increase with technical advances, allowing higher ultimate recovery (UR). Maintaining an exploration effort of any magnitude in this environment may seem difficult to justify.

The oil industry continues to sustain an active exploration effort for several reasons. First, exploration for nonassociated gas is needed to meet local energy requirements to supply power utilities and petrochemical developments. Furthermore, additional income is derived from exploitation of gas condensates, which are not regulated by oil-production quotas. Second, exploration is required to replace produced reserves to guarantee future income and maintain production quotas. Third, exploration is needed to replace lower-value crude reserves with better-quality crudes. Fourth, some of the producing fields have reached a high level of maturity, and revitalizing producing fields is often more cost-beneficial than developing remote resources. Enhanced recovery programs may require a search for cheap local gas to optimize UR. Finally, the growing market demand continues to offer an incentive to direct frontier exploration toward increasingly

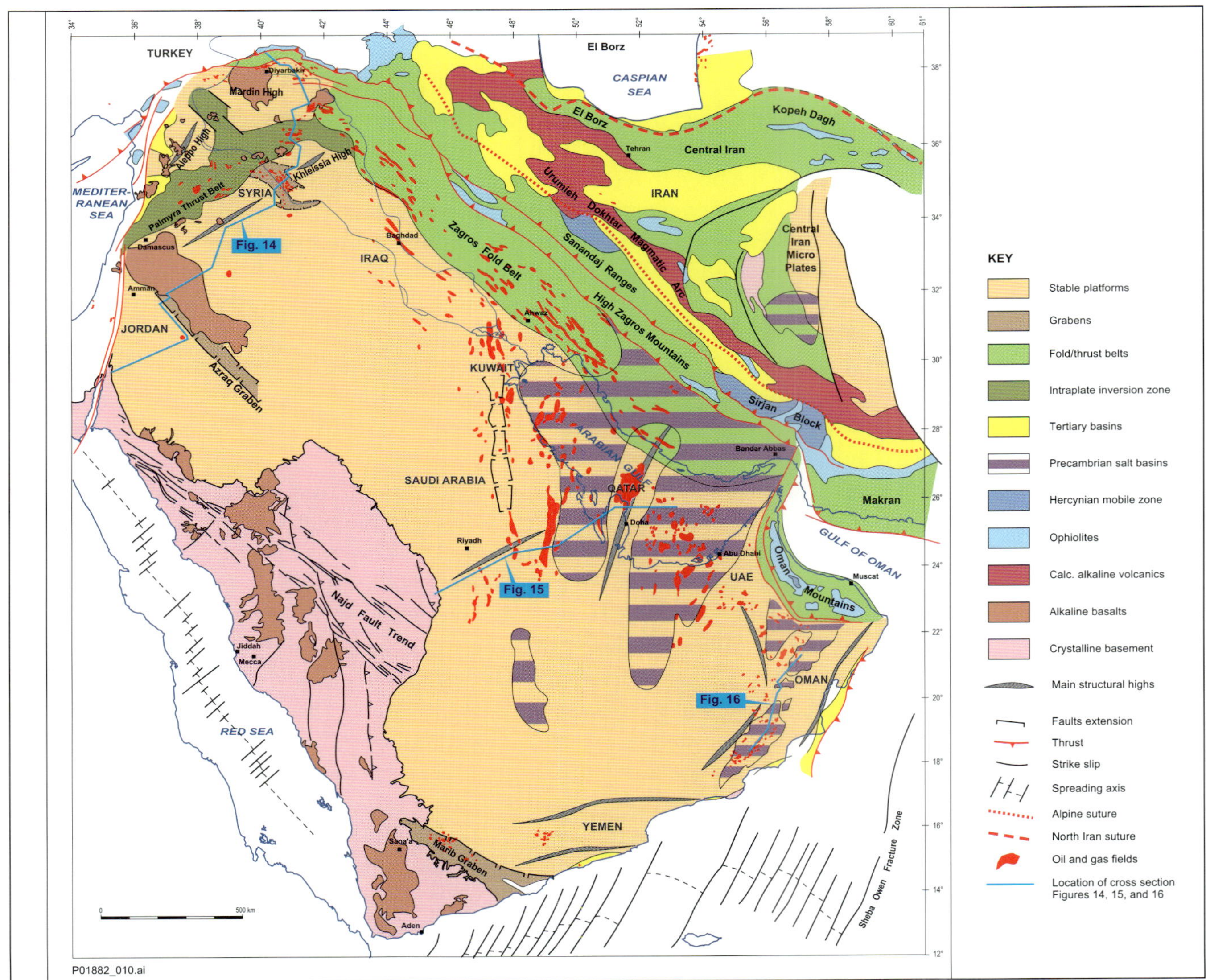

Figure 1. Major tectonic elements of the Arabian Plate and Iran.

more complex geologic settings and, during the last decade in the Middle East, toward deeper Paleozoic targets.

The Paleozoic section has been known to host significant reserves since the oil discoveries in Oman and the delineation of the giant Permian Khuff gas fields in the central Gulf and the Zagros fold belt during the early1970s. Activities, especially since the late1980s, have established the economic attractiveness of the Paleozoic petroleum systems throughout the Middle East. In particular, the highly successful campaign in Saudi Arabia for the Unayzah play established the presence of a hitherto unknown hydrocarbon province. Discoveries in Ordovician reservoirs in Jordan, Carboniferous reservoirs in Syria, Silurian and Ordovician reservoirs in Iraq, and Devonian reservoirs in Turkey, all charged essentially by Silurian source rocks, indicate widespread occurrence of Paleozoic petroleum systems.

In this paper, we describe our present understanding of the Paleozoic frontier of the Middle East because it is thought to offer the industry a major opportunity to delineate new reserves. The sequences are only lightly explored, except in Oman and central Saudi Arabia. Therefore, a discussion of the basin evolution and the hydrocarbon habitat of the Paleozoic sequences at the scale of the Arabian Plate and interior Iran remains speculative.

REGIONAL SETTING

Main Tectonic Elements

The Arabian Plate boundaries embrace all types of plate-boundary processes. They include rifting and seafloor spreading in the Red Sea and Gulf of Aden; collision and subduction along the Zagros-Bitlis suture and Makran, respectively; and transform fault activity along the Dead Sea and Owen-Sheba fracture zones (Figure 1). The Makran and Zagros convergence zones separate the Arabian Plate from the microplates in interior Iran.

The Middle East basins are underlain by Precambrian basement exposed in the Arabian shield in the west and, locally, along the Arabian Sea and in interior Iran. The Arabian platform stretches east of the shield toward the Oman and Zagros Mountains. Northward, the Arabian platform is interrupted by the intraplate Palmyra and Sinjar troughs, which were inverted during the Eocene-Miocene. The Aleppo and Mardin highs form stable blocks between this intraplate deformation zone and the alpine collision zone in Turkey.

The Zagros fold belt, High Zagros Mountains, and Sanandaj ranges form the site of Alpine orogenesis associated with the closure of Neo-Tethys, essentially starting during the Late Cretaceous (Alavi, 1994). Remnants of the Neo-Tethys ocean are found in ophiolite complexes exposed all along the trend, including the Oman Mountains, which is a zone of Late Cretaceous ophiolite obduction. The Urumieh Dokhtar volcanic arc shows evidence of subduction-related magmatic processes. The suture between the Arabian and interior Iran microplates is thought to be located just southwest of the volcanic arc and may be hidden under a linear belt of Tertiary intramontane basins.

The interior Iran microplates and the Arabian Plate are thought to have been part of Gondwana during most of the Paleozoic (Beydoun, 1993). The Paleozoic continental margin of Gondwana probably extended along the Elborz–Kopeh Dagh sutures into Turkey (Figure 1). The interior Iran microplates broke away from Gondwana during the Permian and probably docked with Eurasia during the latest Permian (Ruttner, 1993). The tectonic history of interior Iran is very complex, but the presence of large shear zones in which tectonic lenses of oceanic crust are present indicate that the area is a mosaic of smaller terranes. Paleomagnetic modeling suggests that the tectonic history of individual microplates diverged through time, involving major rotations of the Central Iran microplate (Davoudzadeh et al., 1981) before they were assembled during the Alpine orogeny.

The Precambrian basement, which underlies the Middle East Basin, consists of accreted island-arc and microcontinent terranes (Brown et al., 1989), overlain by postcratonic sediments and volcanics. Rift basins were formed during the latest Precambrian (Husseini, 1988). These rift basins were the site of salt deposition in the Arabian Gulf and Oman (Figure 1).

The main structural elements of the platform indicate the existence of several inherited mechanical weak trends. They are defined by northerly trending highs, exemplified by the Qatar Arch; northwesterly trending systems such as the Azraq and Ma'rib grabens of Mesozoic age; and northeasterly trending systems such as the South Syrian platform, Khleissia, and the Mosul trend. These trends are well expressed in the structural map of top basement (Figure 2).

The parallelism between major structures in the Precambrian basement of the Arabian shield and Phanerozoic structures is striking and suggests that rejuvenation of mechanical discontinuities in the basement played an important role in the evolution of the plate.

Figure 2 highlights the overall asymmetric nature of the basin with basement at surface in the west. The basin deepens gently in an easterly direction with maximum depth reached in a foredeep setting in front of the Zagros convergence zone. No obvious foredeep is developed along the northern plate boundary reflecting the escape tectonics of the Anatolian Plate (Turkey). Shallow basement along the Arabian Sea reflects repeated episodes of uplift associated with the breakaway and drift of the Indian subcontinent. Moreover, the northeasterly trending salt basins in Oman are well expressed. Basement trends in the Zagros follow essentially the surface structural grain.

Paleoplate Positions

The Arabian Plate is thought to have originated in the late Neoproterozoic because of accretion tectonics (Unrug, 1996). By the end of the Precambrian, the plate was located close to the equator (Figure 3) and had an east-west orientation, with Iran in the north. The area translated gradually to southern latitudes during the early Paleozoic and was accompanied by a minor counterclockwise rotation. The Levant formed the southernmost part of the area by the Late Ordovician; Oman was located in subtropical latitudes compared with present-day climate zones. During the Silurian to Late Carboniferous, the plate underwent a major clockwise rotation of about 100° and did not make significant north-south translations. By the Late Carboniferous, Oman was positioned in the south and Turkey in the north. During these rotations, the Hercynian orogeny wound down to the northwest. These movements were followed by a rapid translation of Arabia to the north during the Permian. Turkey reached the equator shortly after the end of the Paleozoic.

The journey of the Arabian Plate during the Paleozoic across the southern hemisphere may be summarized in

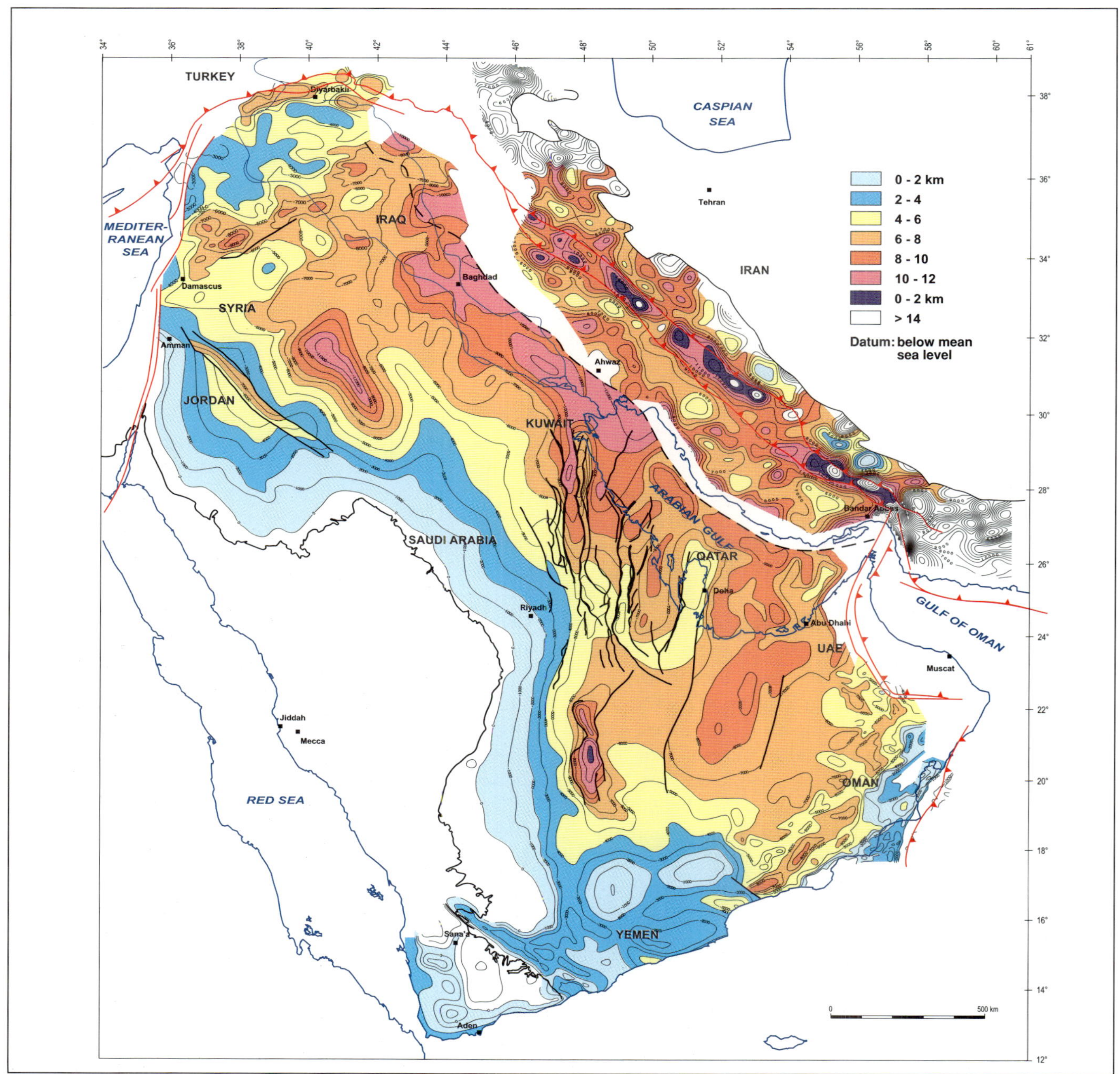

Figure 2. Tentative basement depth map. Contours in kilometers below mean sea level (partly based on modified Best et al., 1993; Buday and Jassim, 1987; Loosveld et al., 1996).

three distinct episodes: (1) Precambrian to Late Ordovician southward translation, (2) Silurian to Carboniferous clockwise rotation without north-south translation, and (3) rapid Permian northward translation.

Stratigraphic Framework

Most knowledge pertinent to the stratigraphy of the Paleozoic of Arabia stems from wells drilled over structural highs and outcrops along the present basin margins. Large areas, especially in the deeper parts of the basin, contain sparse or no well data. Most of the basin is covered by older vintages of seismic data that do not adequately image the Paleozoic section because of low-impedance contrast in the predominantly siliciclastic section, high-amplitude interbed multiples from Mesozoic and Cenozoic carbonates, and near-surface statics.

The data available for interior Iran are essentially derived from surface outcrops that were described in the 1970s and earlier. A complex geologic picture emerges. The stratigraphic resolution of the work and lithostratigraphic approach do not allow an unambiguous interpretation of the data in the context of a Middle East– scale sequence-stratigraphic model. Therefore, these data will be used only to stress key issues related to basin evolution.

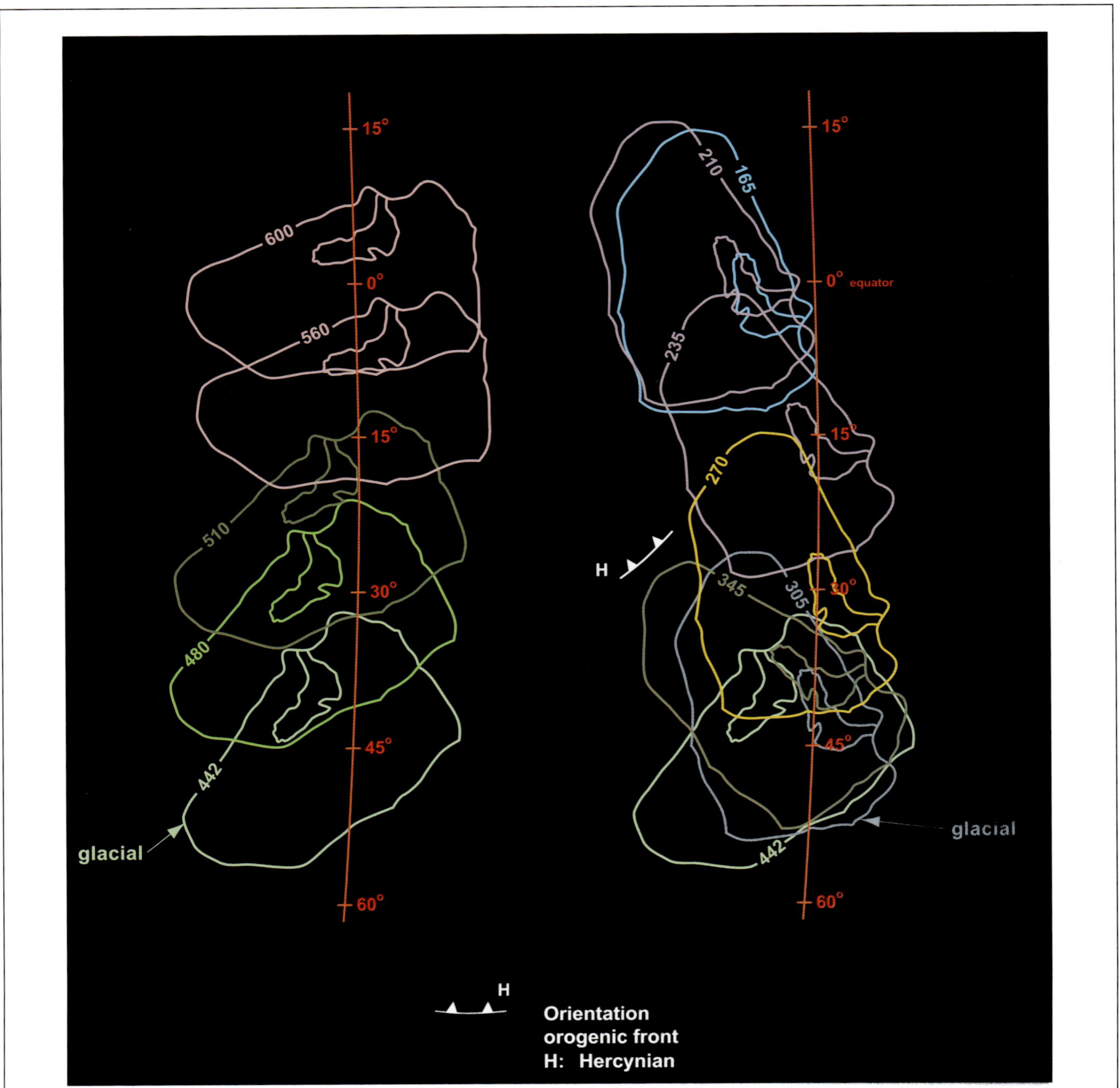

Figure 3. Paleopositions of the Arabian Plate during the Paleozoic. See time scale on Figure 5 for key to colors.

The Paleozoic sequences are dominated by siliciclastics, essentially sourced from the exposed interior of Gondwana to the west and south. Figures 4 and 5 illustrate the generalized stratigraphic framework for the Paleozoic of the Middle East, and Figure 5c includes a summary of the Precambrian section. Note that the Huqf sequences in Oman and their lateral equivalents are presently dated as Precambrian, after recent developments in chronostratigraphy (Gradstein and Ogg, 1996). Gradstein and Ogg define the base of the Cambrian at 543 Ma. This also implies that rifting which preceded basin formation is of latest Proterozoic age.

The base of the Paleozoic section comprises a massive continental sandstone unit of Early Cambrian age. The base of these sandstones is a diachronous horizon. The continental sandstones were succeeded in the north and east by the development of a shallow-marine carbonate platform of Middle Cambrian age. A return to siliciclastic braid-delta environments occurred during the Late Cambrian. These are followed by a stack of prograding braid-delta sequences in an overall transgressive setting, culminating in a major maximum flooding surface in the Middle Ordovician. The maximum flooding surface is followed by two prograding siliciclastic cycles during the Middle to Late Ordovician, interrupted by a second maximum flooding event. The close of the Ordovician is represented by a major regional unconformity caused by a large drop in sea level associated with the Ordovician glacial event. Glacial and periglacial continental to subaquatic deposits and their lateral equivalents indicate two major phases of ice advance and retreat, predominantly from the west (McClure, 1978; Vaslet, 1990).

The deglaciation phase resulted in a primary, plate-wide, maximum flooding surface of Llandoverian age, during which the prolific source rocks of the "hot" shale were deposited. A second, younger Silurian source rock was deposited in the deepest parts of the basin. The remainder of Silurian sedimentation was dominated by a prograding deltaic complex.

The latest Silurian to the latest Carboniferous period is represented poorly in the rock record. This is primarily caused by Hercynian tectonism and possibly by the increased maturity of the basin, resulting in deceleration in subsidence rates with loss of preservation potential.

The preserved Devonian is fluviatile, deltaic, and shallow marine in origin. The marine rocks include shallow carbonate platform deposits of Early Devonian age, probably deposited during maximum flooding, which are overlain uncomformably by continental siliciclastics. The Hercynian orogeny affected the basin from latest Devonian time but appears to have climaxed in the Early Carboniferous. The Carboniferous synorogenic sequences were deposited in continental to shallow-marine environments and include shallow-marine carbonates of Viséan age. The Carboniferous siliciclastics were derived mainly from erosion of older siliciclastics in uplifted areas.

During the latest Carboniferous, icehouse conditions in southern Arabia occurred, again resulting in a major drop in sea level and coeval erosion. Glacial and glaciolacustrine deposits are preserved in Oman and southern parts of Saudi Arabia.

Permian deposits are present throughout the basin, reflecting increased accommodation space related to

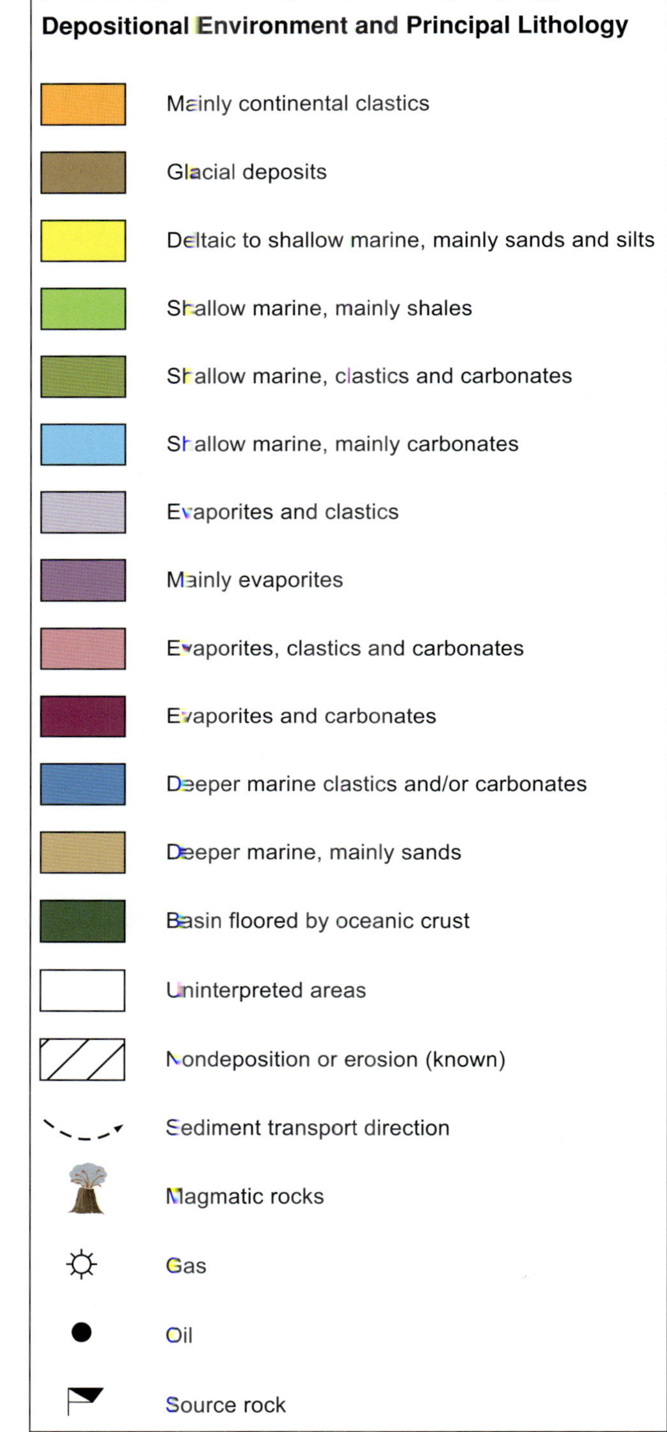

Figure 4. Key to stratigraphic diagrams and environmental maps (Figures 5–13).

stretching of the crust, which gave rise to formation of the Neo-Tethys ocean. The Lower Permian comprises siliciclastics of fluvial and eolian origin. These siliciclastics were partly deposited coeval with rift tectonics along the eastern and northern margins of the Arabian Plate. They are overlain unconformably by the platewide syndrift carbonate/evaporite platform sequences of Late Permian age.

PALEOZOIC BASIN EVOLUTION

Lower to Middle Cambrian

The Paleozoic depositional cycle started with deposition of continental siliciclastics of Early Cambrian age. These sequences unconformably overlie a peneplained, stable platform, and they were essentially derived from interior sources in Gondwana. They can be traced over the northern part of the Arabian platform, and time equivalents in similar facies are observed throughout the Iranian microplates. They are missing in the southwestern part of the platform, probably because of emergence.

The rocks consist of reddish to white variably sorted arkosic sandstones, conglomerates, and subordinate red shale. They were deposited in a system of alluvial fans grading into braid plains and braid deltas. The braid plains may include playa lake–type deposits in Oman. The distribution of these sediments in the salt basins indicates that accommodation space was generated by halokinesis.

During the latest Early Cambrian to early Middle Cambrian, the platform became inundated from the north (Figure 6). Siliciclastic tidal flats were established in marginal settings and grade basinward into low-energy carbonate and siliciclastic carbonate mixed tidal flats followed by subtidal carbonates (Amireh et al., 1994). The latter developed into a vast shallow-marine, stable carbonate platform covering most of northern Arabia and interior Iran. Locally, oolites and stromatolites have been described. Salt pseudomorphs and anhydritic dolomites indicate the temporal establishment of evaporitic conditions. These carbonates form a key seismic marker in northern Arabia. Carbonate platform environments persisted throughout the Middle Cambrian on the deeper parts of the platform, but along the basin margin, deposition returned to alluvial-fluvial environments, interrupted by subordinate marginal-marine deposits.

Continental deposits replace marginal-marine environments southward. In the salt-basin province, an angular unconformity separates the upper Lower Cambrian from the underlying sequences. This unconformity is thought to mark the onset of subsidence driven by thermal relaxation. The age of this unconformity is still controversial; it may also mark the base of the Paleozoic.

The continental deposits are interpreted as proximal alluvial-fan deposits sourced from local uplifted basin-margin highs. Basinward, these deposits interfinger with alluvial and eolian sandstones. All relict topography appears to have been leveled by the Middle Cambrian, and relatively uniform depositional conditions persisted over large areas (including central Saudi Arabia) for the first time.

Regional facies trends indicate a southwest-to-northeast transport direction, suggesting that the sediments were derived from the southwestern and southern margins. Fluvial fan conglomerates and sandstones pass northeastward into fluvial-eolian sandstones and inland sabkha deposits.

Upper Cambrian to Lower Ordovician

Increased siliciclastic influx in the Late Cambrian terminated carbonate deposition in northern Arabia and the Zagros, and a prograding siliciclastic apron was deposited conformably over the Middle Cambrian sequences (Figure 7). In interior Iran, however, carbonate deposition persisted into the Late Cambrian. In the northern and central area of the Arabian Plate, fluvial to fluviodeltaic to shallow-marine siliciclastic environments were established, sourced from exposed areas in the west. They grade eastward into distal shale-dominated marine environments in the Zagros.

In the south, the base of the Upper Cambrian is a regional unconformity. In south Oman, continental conditions persisted, and in the Ghaba and Fahud Salt Basins, a marine-influenced environment of deposition became established. The base of these sections is made up of a stack of shallow-marine to intertidal shallowing-upward cycles, consisting of alternating carbonates, sandstones, and shales. These are followed by shallow-marine mudstones deposited during maximum flooding. The section concludes with stacked braid-delta lobes separated by marine mudstones.

The Cambrian-Ordovician boundary is poorly defined in the rock record, and Upper Cambrian deposition continued uninterrupted into the earliest Ordovician.

During the later Tremadocian to Arenigian, the platform became inundated again, and deeper-marine environments became established over the basinward parts of the platform in the north, including interior Iran. Mixed siliciclastic/carbonate settings are found on the Central Iran microplates (Rickards et al., 1994). The sea invaded the basin margin, and braid-plain to braid-delta environments were overlain by coastal plain to inner neritic siliciclastic environments. The transgression involved multiple eustatic cycles, as indicated in marginal settings farther south. Here, shallow, open-marine mudstones were deposited. They are followed by a shoaling sequence before returning to mudstones. These mudstones are topped by an unconformity of late Tremadocian age, overlain by coastal sandstones, followed by a transgressive succession, culminating in maximum flooding in the earliest Arenigian. This is followed by a regression in the remainder of the Arenigian, during which a prograding

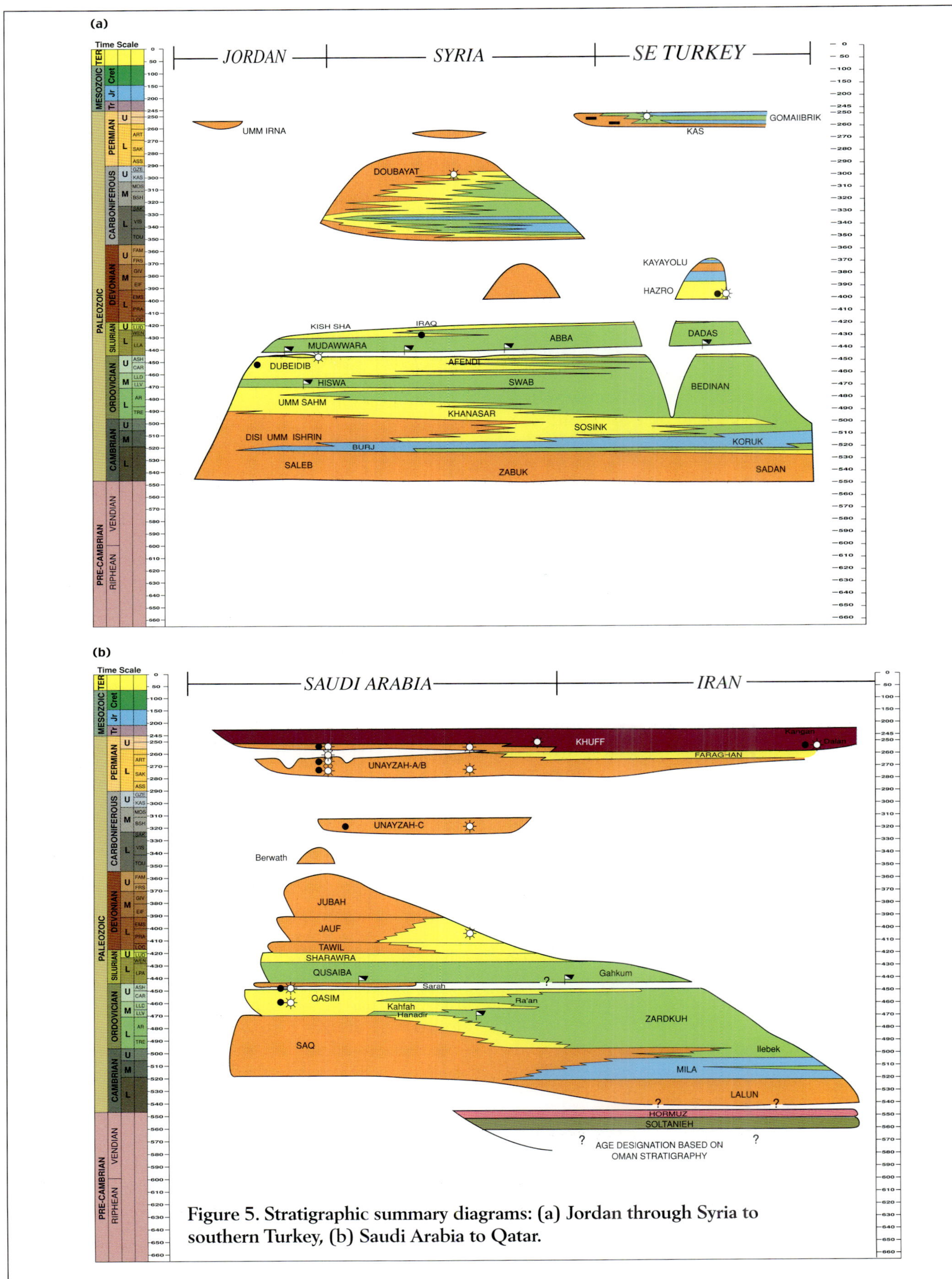

Figure 5. Stratigraphic summary diagrams: (a) Jordan through Syria to southern Turkey, (b) Saudi Arabia to Qatar.

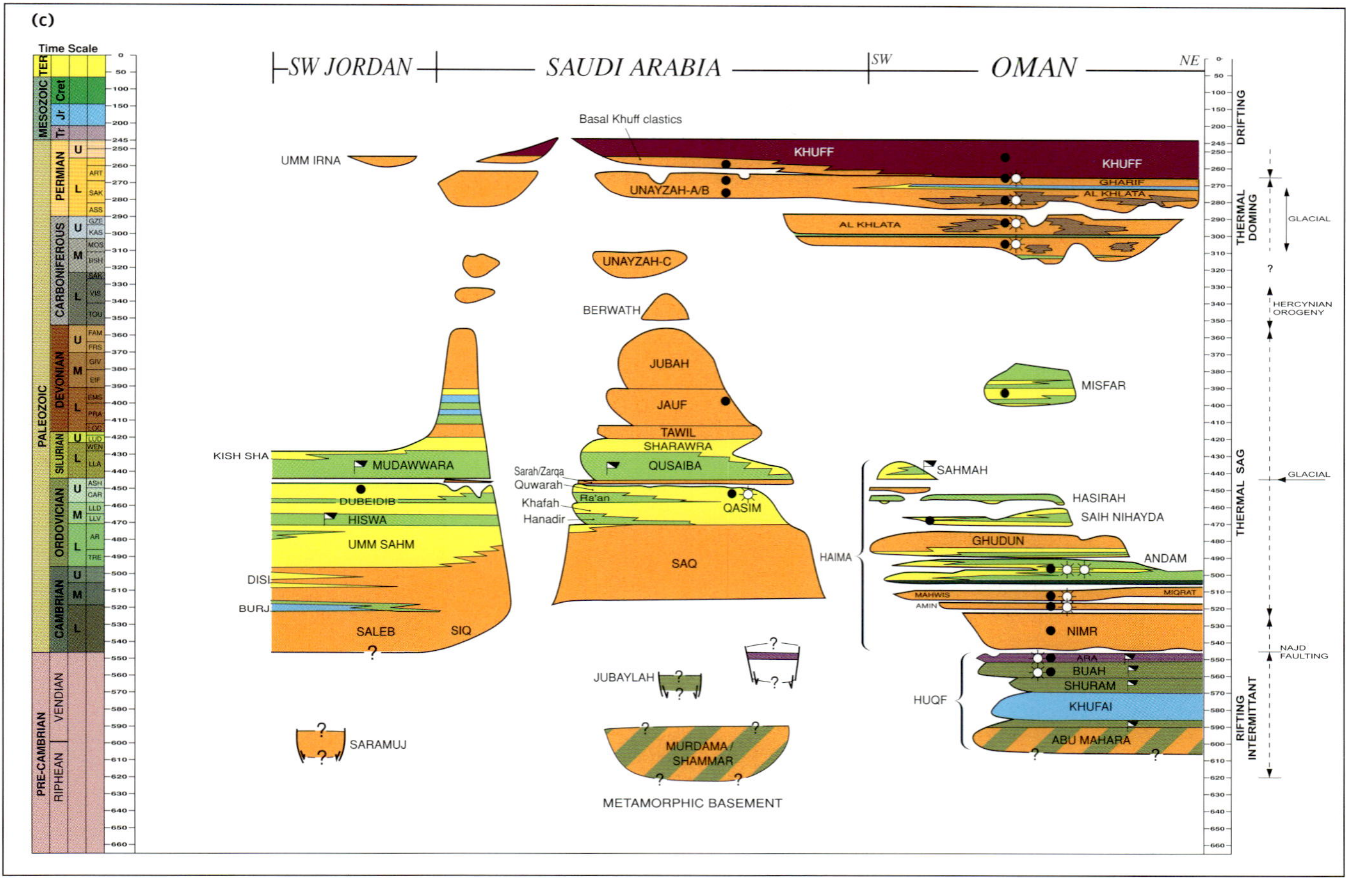

Figure 5. Stratigraphic summary diagrams: (c) Jordan through central Saudi Arabia to Oman. See Figure 4 for key to environments of deposition.

braid-delta system was deposited, consisting of massive quartz sandstones/siltstones and subordinate shales.

Middle to Upper Ordovician

A major unconformity separates the Middle from the Lower Ordovician in the south and extends into central Arabia. A thin sandy unit locally overlies this unconformity, but generally, a rapid transgression resulted in deposition of middle to outer neritic shales above the unconformity. This major maximum flooding surface is of Llanvirnian age and can be traced basinwide (Figure 8). Locally, the shale may be rich in organic material, indicating restricted water circulation in the basin for the first time.

Prograding siliciclastic aprons overlie the maximum flooding deposits near the basin margins. These Middle Ordovician sediments were deposited in inner neritic to estuarine or deltaic environments. Point sources can be recognized in Oman and northern Saudi Arabia (Figure 8).

The Middle Ordovician cycle is followed by a transgressive regressive cycle of Caradocian age. Sediments were deposited in similar environments, and the basin geometry remained apparently unchanged. Basinward, cycles are difficult to recognize because the section consists of an undifferentiated package of essentially middle to outer neritic graptolitic shales.

Time-equivalent deposits are absent in most of interior Iran; remnants have been preserved only locally (Reitz and Davoudzadeh, 1995), possibly because of Late Ordovician and younger erosion events. They are also absent in the Mardin area of southeastern Turkey, where the entire Ordovician section was progressively removed by pre-Silurian erosion (Figure 5c). Whether this is the result of tectonic processes or shelf-edge erosion associated with the fall in sea level during the close of the Ordovician remains to be resolved.

Late Ordovician Glaciation

The base of the latest Ashgillian deposits is an important unconformity which formed during the Late Ordovician glaciation of Gondwana. The polar ice cap covered sub-Saharan Africa and advanced into western Arabia in two major pulses, depositing two sequences of tillite and proglacial siliciclastics, mostly sandstones in incised valleys adjacent to the Arabian shield and southern Jordan (McClure, 1978; Vaslet, 1987, 1990) (Figure 9).

The deep-valley systems were incised to depths exceeding 500 m by glacial and fluvial processes. They have been traced into the subsurface of northern Saudi Arabia with seismic data (McGillivray and Husseini, 1992; Aoudeh and Al-Hajri, 1995). The associated major fall in relative sea level was recorded away from the glaciated areas by a sudden influx of significant amounts of fluvial to deltaic sands on top of deeper-marine sediments in parts of the basin.

Silurian

The Llandoverian saw a major phase of global warming which resulted in retreat of the glaciers. Sea level rapidly started to rise, flooding the Arabian platform (Figure 9). Shallow- to open-marine environments were established in marginal areas, and deeper-marine environments covered the inundated platform and extended southward along the narrow subsiding intrashelf trough in central Saudi Arabia (Jones and Stump, 1999). Anoxic bottom waters in the sediment-starved basin resulted in preservation of organic-rich shales, which comprise the prolific Silurian "hot" shale, one of the principal source rocks for Paleozoic hydrocarbons (Abu-Ali et al., 1991; Mahmoud et al., 1992). A second, younger source rock of possibly Wenlockian age occurs in the northern parts of the basin. The initial transgression is followed by a thick (>1000 m) coarsening-upward, progradational megasequence of shales and sandstone of Llandoverian to Peridolian age.

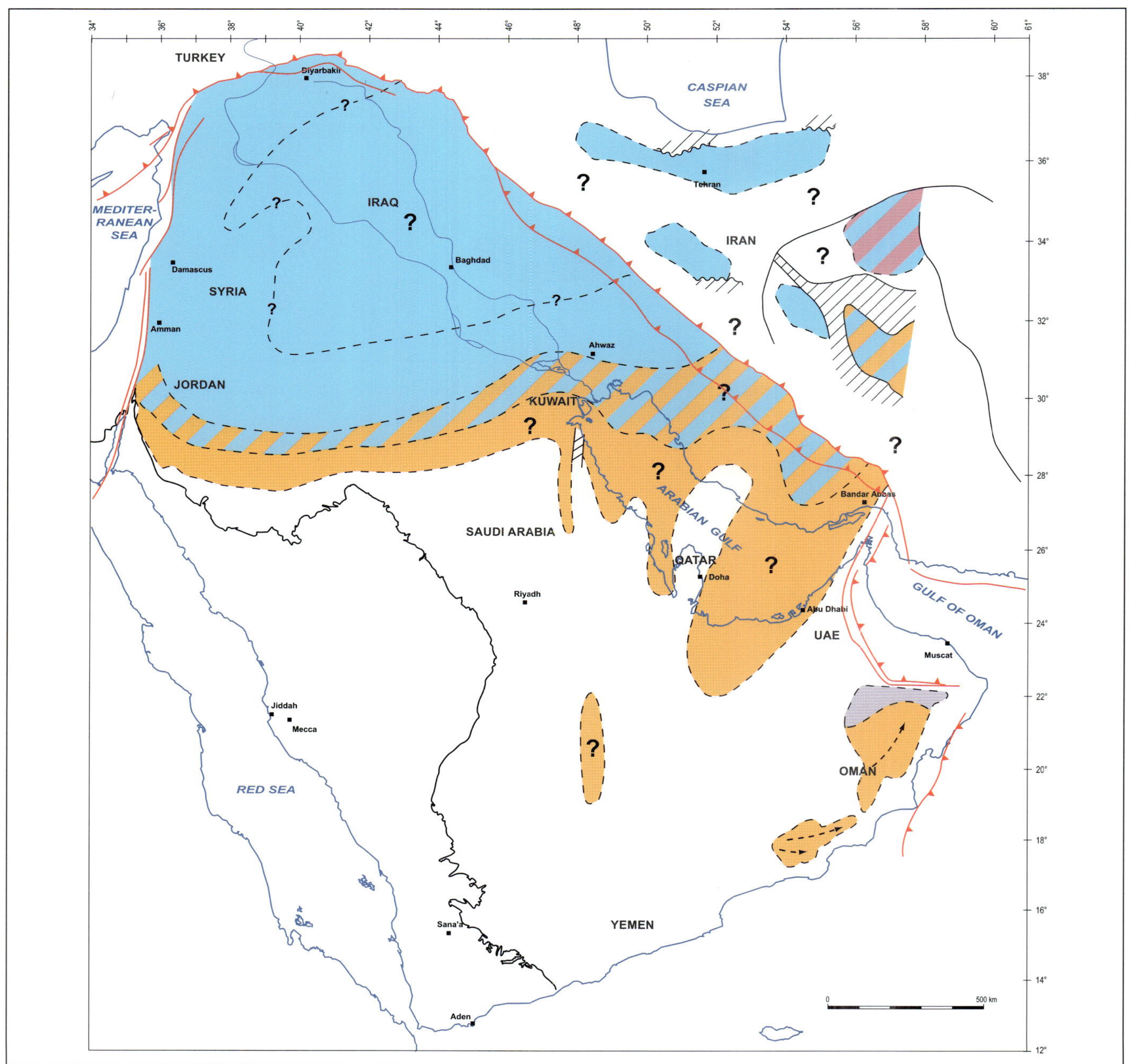

Figure 6. Middle Cambrian environments of deposition. See Figure 4 for key to environments of deposition.

Middle to outer neritic environments persisted in the north and east during the remainder of the Silurian.

In interior Iran, Lower Silurian sediments rest directly on Lower Ordovician sediments, and the entire Middle to Upper Ordovician interval is absent, except for local remnants.

Initially, coarse continental siliciclastics were deposited which laterally appear to grade into and are followed by shallow-marine carbonates. Volcanic rocks have been described from various parts of the basin. These sequences may represent the margin of Paleo-Tethys in the Kopet Dagh area and eastward, taking into account Alpine rotations associated with the Central Iran microplate (Davoudzadeh et al., 1981). In the Elborz Mountains, the Silurian was either not deposited or not preserved, and uppermost Devonian rocks rest unconformably on Cambrian–Lower Ordovician deposits, indicating uplift of the continental margin during the Devonian at the latest.

The dramatically different paleogeography of the Central Iran Basin indicates that interior Iran had been uplifted by the Early Silurian and possibly had started to follow a separate tectono-magmatic evolution from Gondwana. In contrast, similarities in the Cambrian to Lower Ordovician stratigraphy suggest that central Iran may have been an integral part of Gondwana. It follows that the Zagros fault zone may have been rejuvenated during the latest Ordovician.

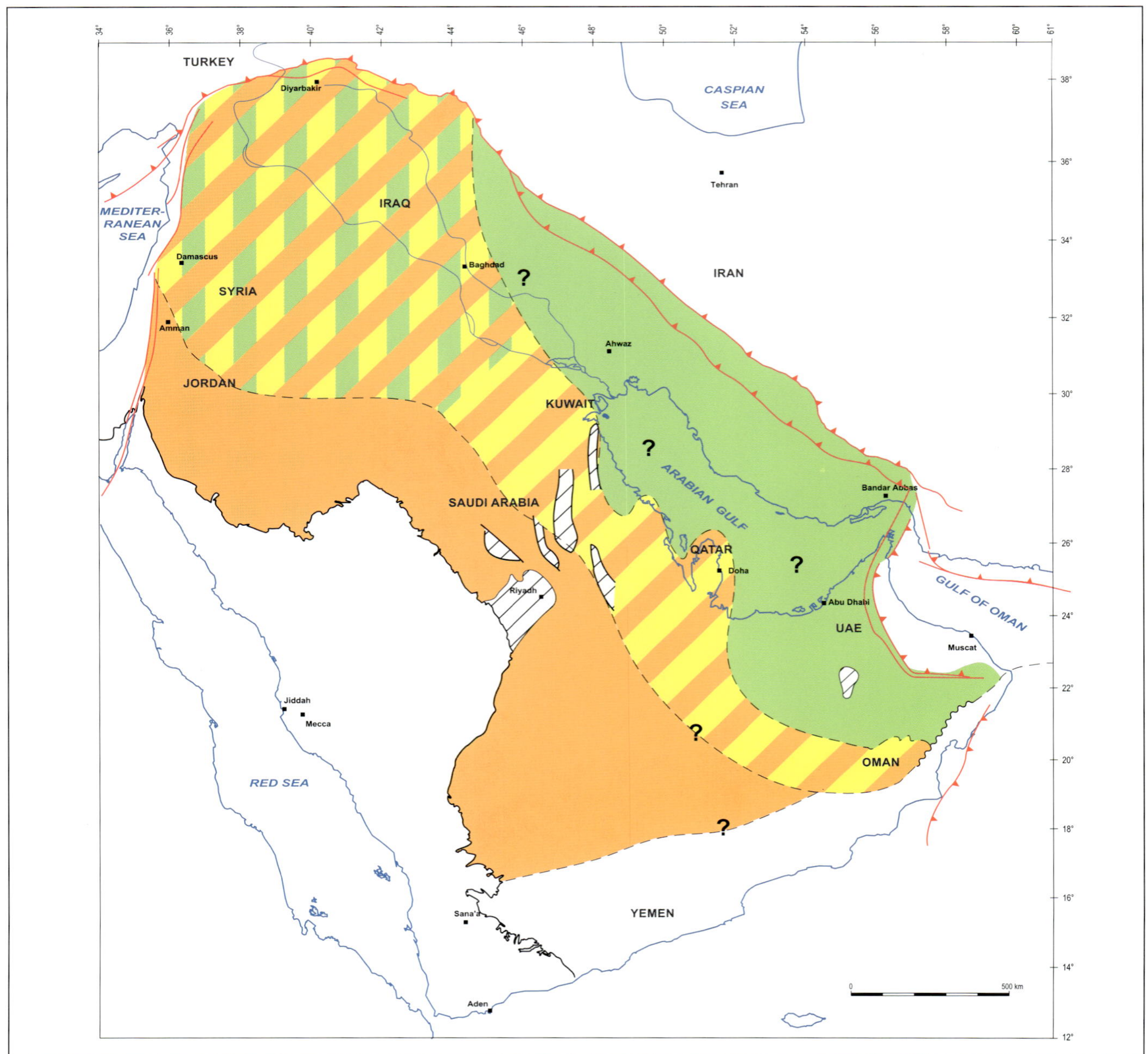

Figure 7. Late Cambrian environments of deposition. See Figure 4 for key to environments of deposition.

Silurian-Devonian

The base of the Upper Silurian–Devonian megasequence is a regional disconformity. The section is not present over large areas, probably because of Hercynian erosion (Figure 10). The most complete section has been preserved in Saudi Arabia (Al-Hajri et al., 1999). There, the cycle starts with continental siliciclastics of latest Silurian age (Figure 5a), followed by marine deposition during the Pragian (reaching into the Emsian). A large delta front became established in Saudi Arabia and Qatar. In the north, it is replaced by mixed-marine siliciclastics and carbonates.

The marine incursions of Emsian to Eifelian age also reached Oman, Iraq, Syria, and Turkey. Continental environments became established in central Arabia, Syria, and Iraq, and marginal-marine environments developed in Turkey and Oman. The marginal-marine environments include anoxic mudstones deposited in lower coastal-plain environments.

Alternating marine and continental deposits characterize the close of the Devonian in Iraq and Turkey, which are replaced southward by continental sediments. The marine sequences contain carbonate deposits.

The absence of Lower Devonian deposits in Turkey and Iraq suggests a structural-high position with respect to the depocenter in Saudi Arabia. The return of marine

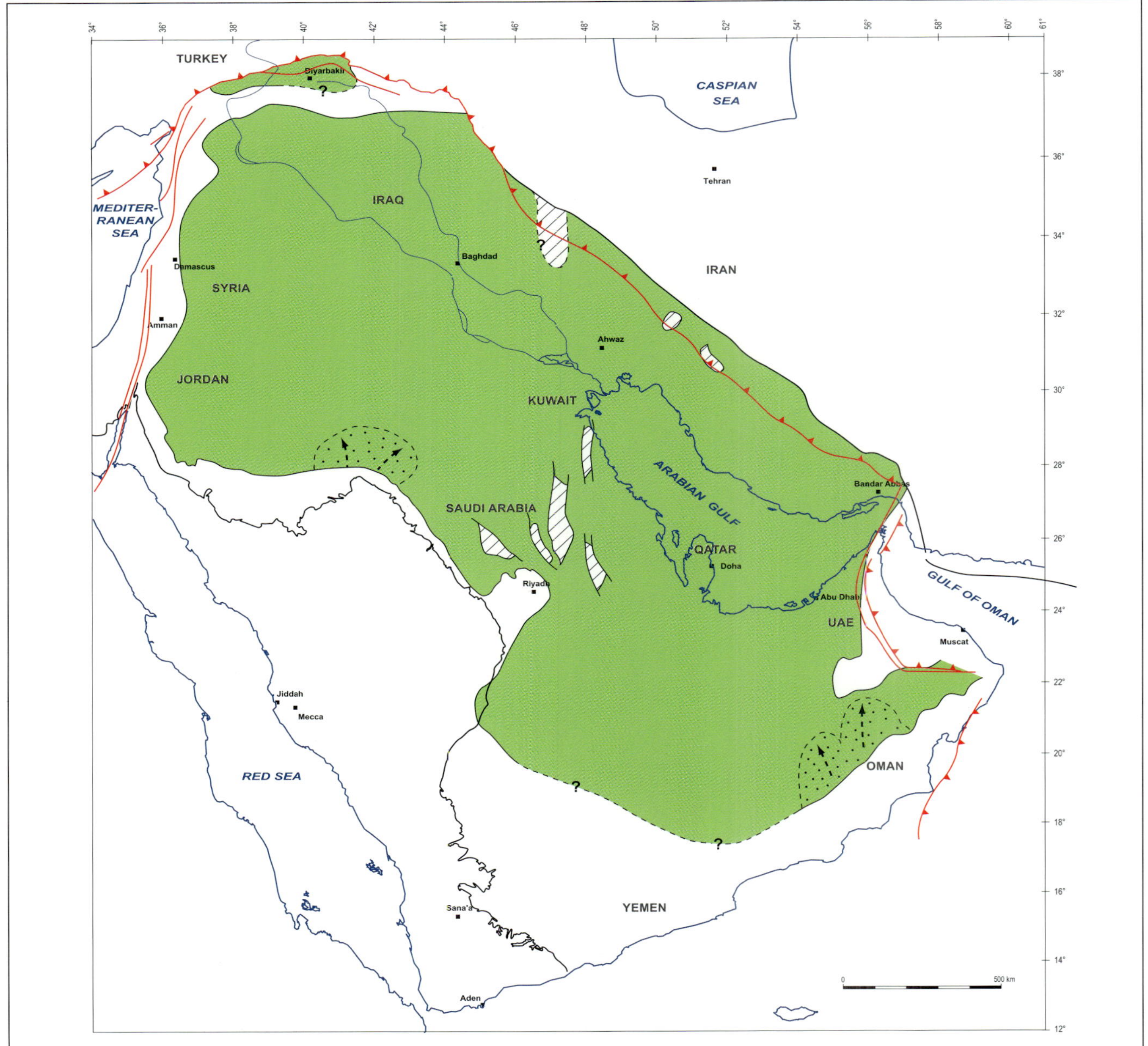

Figure 8. Middle Ordovician environments of deposition. Stippled areas indicate locations of outbuilding deltas during subsequent regressive stage. See Figure 4 for key to environments of deposition.

environments, especially during the latest Devonian in the northern region (or the preservation thereof), suggests differential downwarp of the northern margin of Gondwana. Similar relationships can be observed in interior Iran. Uppermost Devonian strata rests directly on Cambrian–Lower Ordovician sequences in the Elborz Mountains, whereas a more continuous Paleozoic section (including older Devonian) is preserved in the basin south of the Elborz Mountains (Wensink, 1991) (Figure 9). These relationships suggest that the northern margin of Gondwana became tectonically unstable at the onset of the Hercynian orogeny.

Carboniferous

The Carboniferous is largely missing because of widespread uplift and erosion during the Hercynian orogeny. In Syria, Lower Carboniferous sequences were deposited and preserved in a northeast-trending proto-Palmyra trough (Figure 11). The base of the Carboniferous is a regional unconformity that becomes angular adjacent to Hercynian uplifts.

The basal part of the Carboniferous section in Syria comprises Tournasian to earliest Viséan shallow-marine shale with subordinate sandstones/siltstones and bioclastic carbonates. Incomplete biozones are indicative of

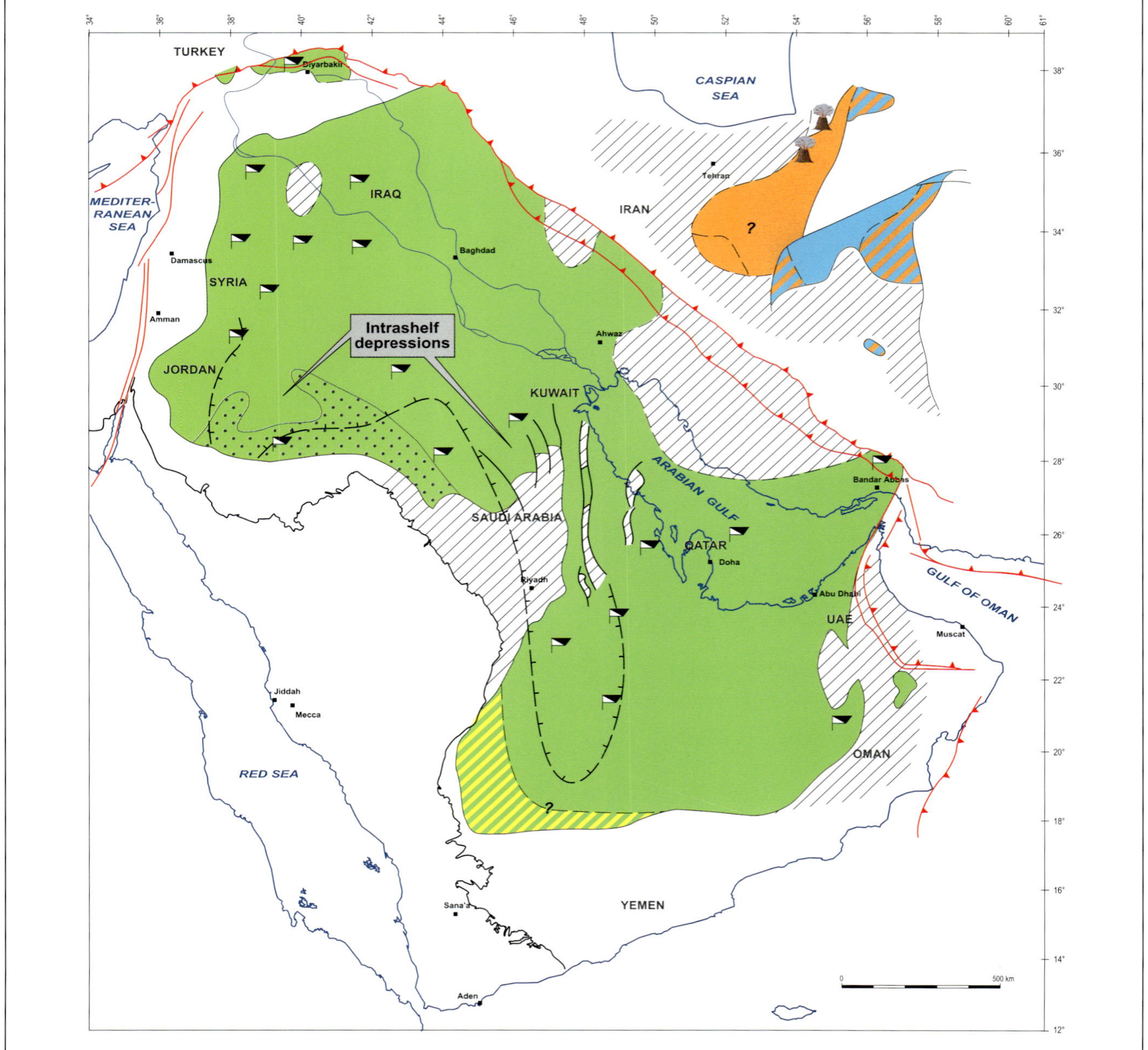

Figure 9. Early Silurian environments of deposition. Stippled area outlines area underlain by latest Ordovician glacial valleys. See Figure 4 for key to environments of deposition.

intraformational depositional hiatuses. These are followed by fully marine carbonates of Viséan age, reflecting the maximum extent of the transgression.

The overlying sequences are part of a regressive complex made up of nearshore to deltaic siliciclastics. These rocks range in age up to the Stephanian. Thinning and pinching out of the carbonates and variations in sand-shale ratios, especially of the Middle to Upper Carboniferous sequences, suggest that deposition occurred in a shallow, landlocked southwest-northeast-trending depression. This implies a major change in basin geometry, which may be attributed to the Hercynian orogeny.

Isolated Carboniferous siliciclastics have been penetrated in Saudi Arabia. They consist of poorly dated syn-Hercynian continental sandstones of the Berwath and Unayzah-C member, deposited in lows between Hercynian uplifts.

Upper Carboniferous deposits outside the proto-Palmyra depression are known from the south. Here, glacial and periglacial deposits of the Al Khlata (Helal, 1966; Braakman et al., 1982) and Juwayl Formations of Moscovian to Late Carboniferous age have been preserved (Figure 5c). The glacial deposits are related to uplifted areas southeast of Oman (Al-Belushi et al.,

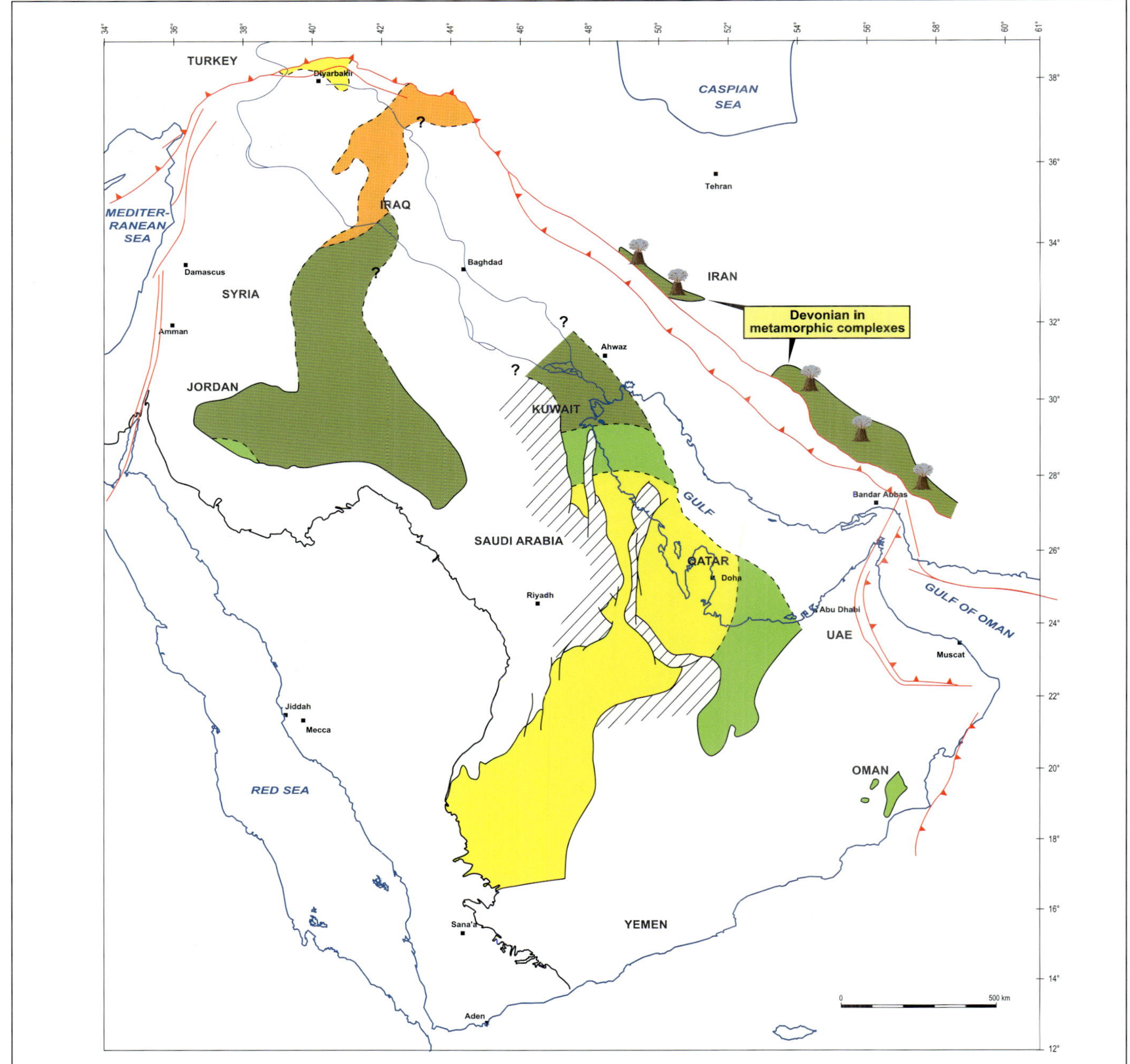

Figure 10. Devonian environments of deposition during the Emsian. See Figure 4 for key to environments of deposition.

1996). Deposition in glacial environments in Oman continued during the Early Permian.

Lower Permian

The first extensive deposits following the Hercynian orogeny are the Lower Permian siliciclastics that overlie an angular unconformity on older Paleozoic rocks and basement. These mainly continental siliciclastics are widespread, but they appear to be missing in the north and in the Oman Mountains (Figure 12). In those mountains, their absence can be explained by erosion or nondeposition associated with Permian rifting. In addition, they are missing by onlap and/or truncation over the east-northeast-trending Central Arabian arch.

Generally, the Lower Permian section is made up of braided-plain, channel-fill, and eolian sandstones and siltstones deposited under semiarid conditions (Senalp and Al-Duaiji, 1995). They are replaced basinward by braid-plain deposits, which are overlain by shallow-marine nearshore deposits to essentially shallow-marine sandstones in the Zagros (Szabo and Kheradpir, 1978). The thickness of these siliciclastics varies because of onlap on the Hercynian structures.

The Lower Permian section in Oman comprises shal-

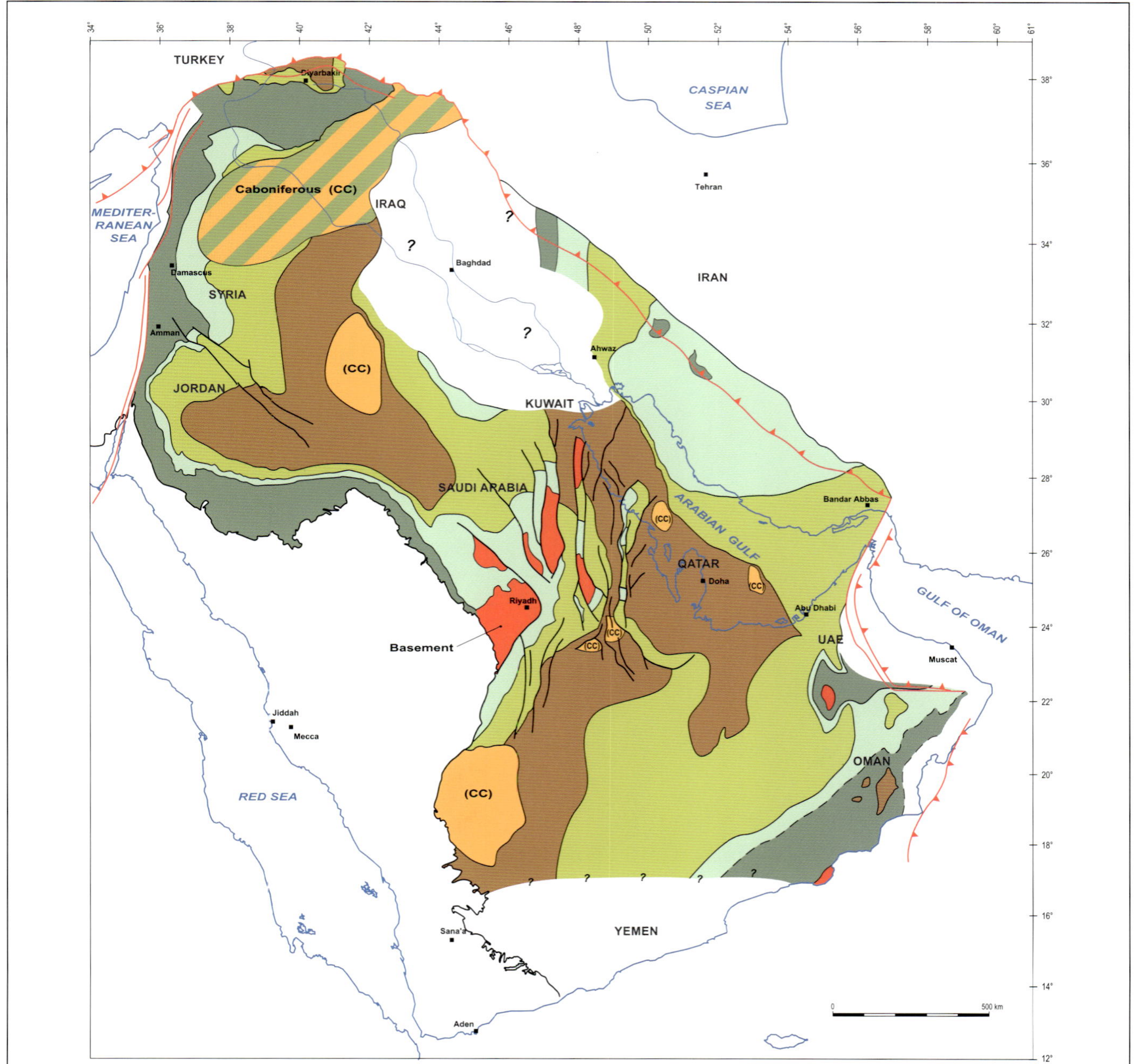

Figure 11. Carboniferous environments of deposition and Hercynian subcrop map. See Figure 4 for key to environments of deposition. See time scale on Figure 5 for colors for subcrop map.

low-marine carbonates of Sakmarian age (Figure 12). The initial transgression was recorded in the deeper part of the basin by a transgressive lag and marine mudstones, which grade laterally into alluvial and fluvial deposits. They are followed by regressive marine carbonates and their siliciclastic lateral equivalents, which are overlain by shoreface deposits followed by lower coastal-plain sediments. The latter may include lacustrine and playa deposits, suggesting diminishing basin topography. The early part of the Artinskian documents a sudden increase in sand content deposited by rivers, probably in response to uplift in the source areas associated with incipient rifting, preceding the formation of the Neo-Tethys margin.

Upper Permian

The base of the overlying Upper Permian Khuff siliciclastics and carbonates is an unconformity which marks the opening of the Neo-Thetys ocean.

The lower part of this megasequence comprises continental to marine sandstones and shales supplied from the west and deposited during the Artinksian in the basin and into the Kazanian along the basin margin. North-

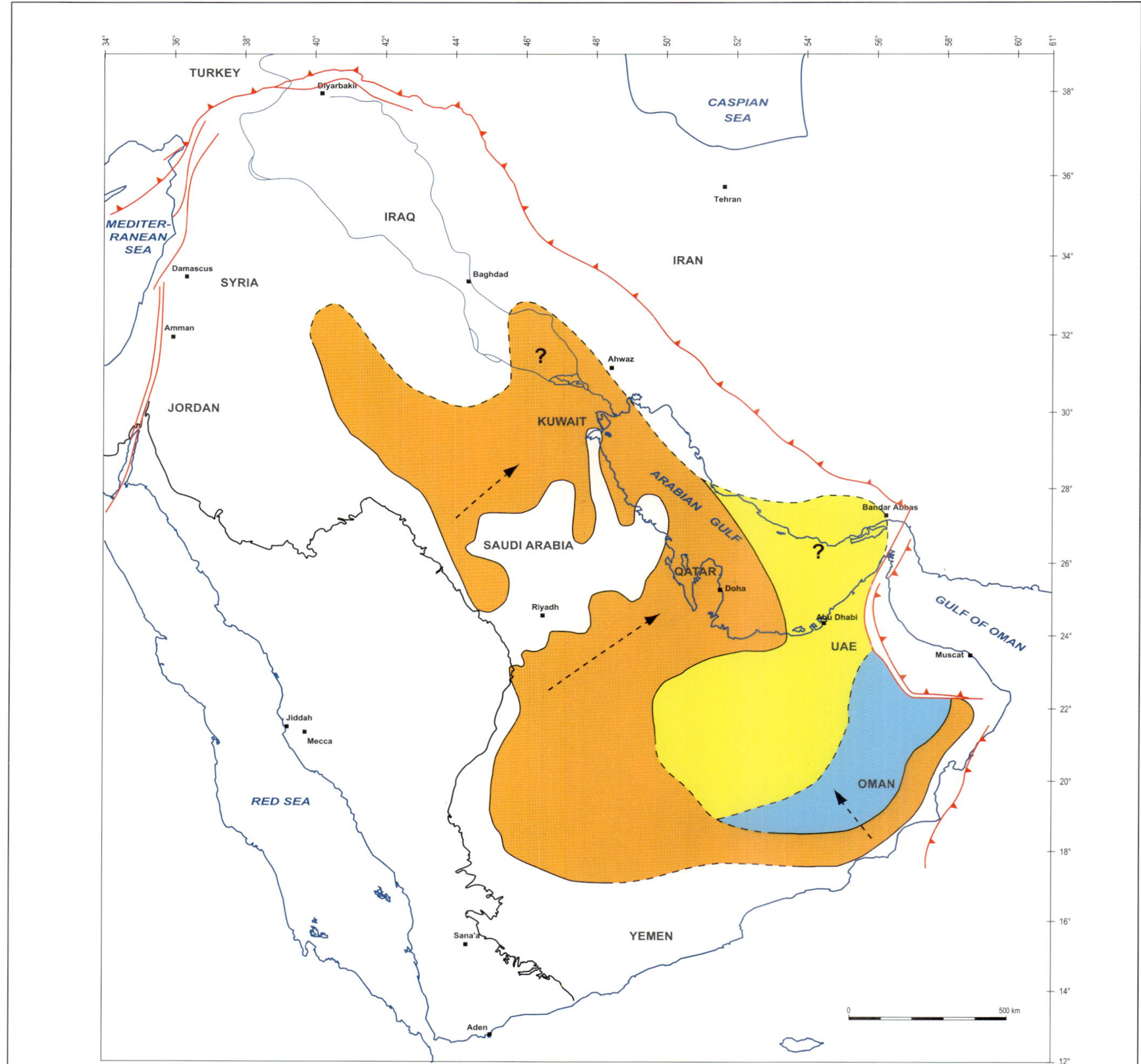

Figure 12. Early Permian environments of deposition at the close of the Sakmarian. Arrows in Saudi Arabia indicate the location of the main channel complexes in the overlying basal Khuff siliciclastics. See Figure 4 for key to environments of deposition.

ward, the continental deposits include economic coal deposits. The coals indicate wet tropical environments. A regression during the Kungurian gave rise to evaporitic, hypersaline lagoons grading into shallow-marine carbonates (Figure 13).

These lagoons were followed by deposition of extensive carbonates and anhydrites in shallow-marine to tidal-flat environments (Al-Jallal, 1995). Most of the Arabian shelf was covered by a shallow sea at that time. Restricted evaporitic environments became established on the western part of the platform, which was protected by shoals from open-marine waters in the east. Deep-marine sediments were deposited in the northeast and are preserved in the High Zagros and Oman Mountains. The Khuff Formation includes at least four depositional cycles. During maximum transgression, carbonates overstepped the siliciclastic realm and rest directly on basement over the Central Arabian arch. The close of the Paleozoic and onset of the Scythian saw the deposition of anhydritic limestones and dolomites grading into oolitic shoals and shallow-marine carbonates.

HERCYNIAN OROGENY

Stratigraphic relationships indicate tectonic instability during the latest Ordovician, the latest Devonian–

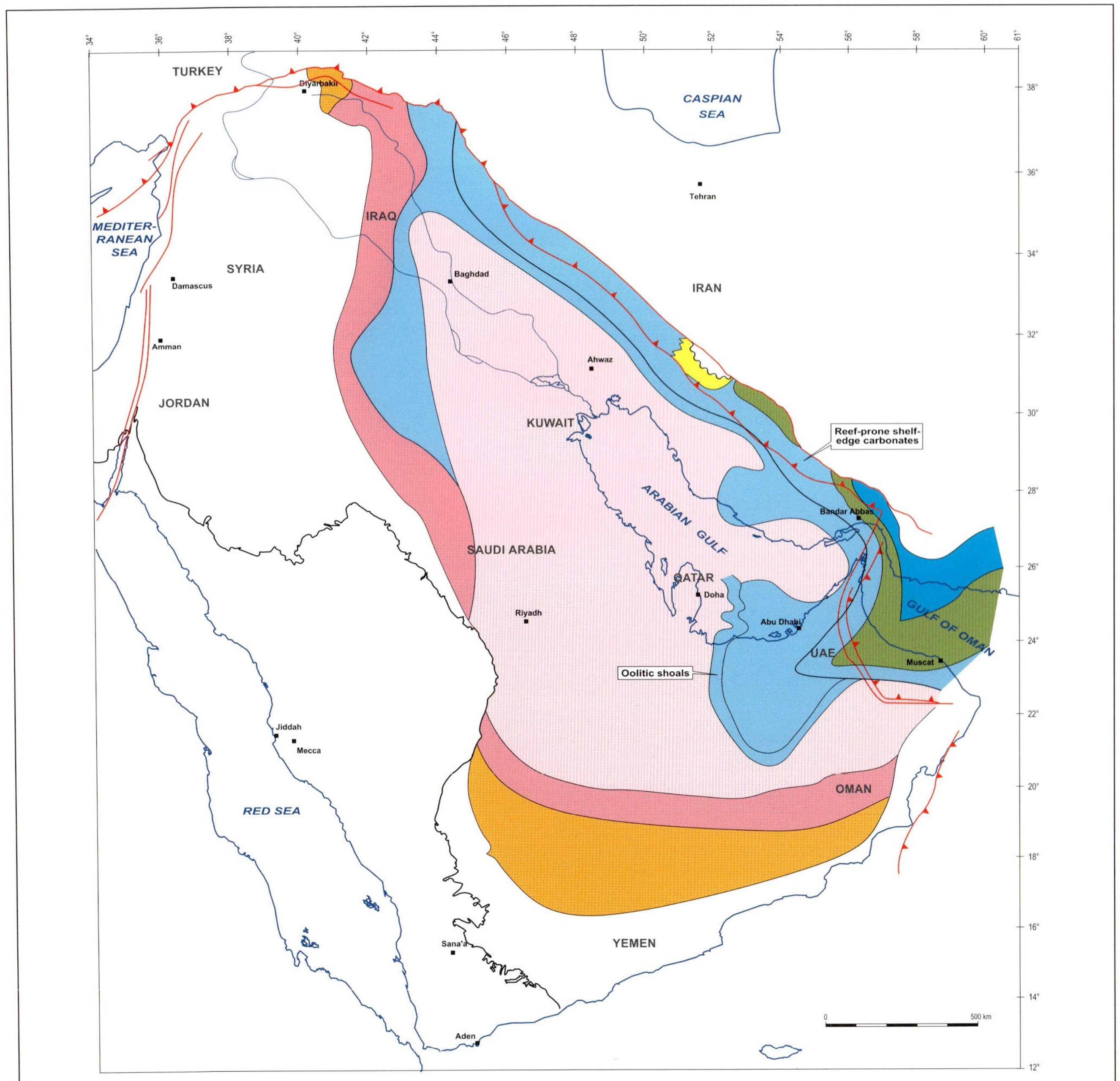

Figure 13. Late Permian environments of deposition during the Kungurian to Tartarian (modified from Al-Jallal, 1995). See Figure 4 for key to environments of deposition.

Early Carboniferous, and the Permian. The instability in the Permian is clearly related to the opening of the Neo-Tethys ocean. The Ordovician event suggests a first phase of disintegration of the northeastern margin of Gondwana.

The Carboniferous event resulted in a major change in basin geometry. The Hercynian subcrop map (Figure 11) reveals the complex regional structure which originated during the Carboniferous. This subcrop map shows a northeasterly trending basement high protruding into the basin in central Arabia—the Central Arabian arch. Facies patterns and thickness variations in Devonian-Silurian and older sequences suggest that the arch originated during the Hercynian and persisted into the Mesozoic. This high was overprinted by northerly trending basement-cored uplifts. The northwesterly trending faults in the Azraq graben were also active and are associated with large uplifts accompanied by deep erosion (Figure 14).

The proto-Palmyra and its northeasterly extension occur just south of a zone where uplift and erosion exposed Ordovician strata in the area of the Aleppo and Mardin highs. Northward, younger rock units have been preserved in the Diyarbakir Basin, suggesting that the uplift zone is a regional, east-northeast-trending foreland bulge. It is noteworthy that this bulge trends parallel to the Central Arabian arch.

Additional evidence for Hercynian tectonism stems from structural observations. Figure 14 is a geologic cross section of the structural-stratigraphic relationships in the northern Arabian Plate. The section shows that the Silurian to Cambrian sequences form one structural entity. The Devonian hiatus may result essentially from vertical movements. A major angular unconformity occurs at the base of the Carboniferous and truncated Silurian-Ordovician sequences. Moreover, sequences are folded at a regional scale prior to deposition of the Carboniferous. The axial zone appears to coincide with the South Syrian platform (compare with Figure 2). Distribution of the Carboniferous sequences suggests that the area was affected by a phase of differential uplift prior to deposition of the Triassic. Folded Silurian-Ordovician rocks occur below flat-lying basal Triassic and younger rocks farther south.

The cross section shown in Figure 15 follows the trend of the Central Arabian arch and extends from the Arabian shield across several large structures in central and eastern Saudi Arabia to the Qatar arch. Sub-Permian strata are clearly truncated by erosion below the Hercynian unconformity. This extensive erosion, particularly of the Devonian section, demonstrates that the structures were uplifted by thousands of meters during the Carboniferous (Figure 11).

The north-trending Hercynian uplifts such as Ghawar are bounded by reverse faults, indicating that the uplifts were caused by a regional, compressive-stress field.

In general, post-Hercynian pre-Permian erosion leveled the topography, but not completely, as indicated by thickness and facies variations in the Unayzah Formation. Many of the Hercynian faults bounding the major north-south uplifts were reactivated during Triassic extension and especially during Late Cretaceous compression, as indicated by the dramatic thickening of the Aruma Formation (Upper Cretaceous) on the flanks of these uplifts. It is noteworthy that not all structures shown in Figure 15 are Hercynian; the Ghawar and Qatar structures may have started to grow earlier. The Harmaliyah anticline, immediately east of Ghawar, preserves the most complete Devonian section in Arabia and is clearly post-Hercynian in origin.

Figure 16 highlights the geologic relationships in the

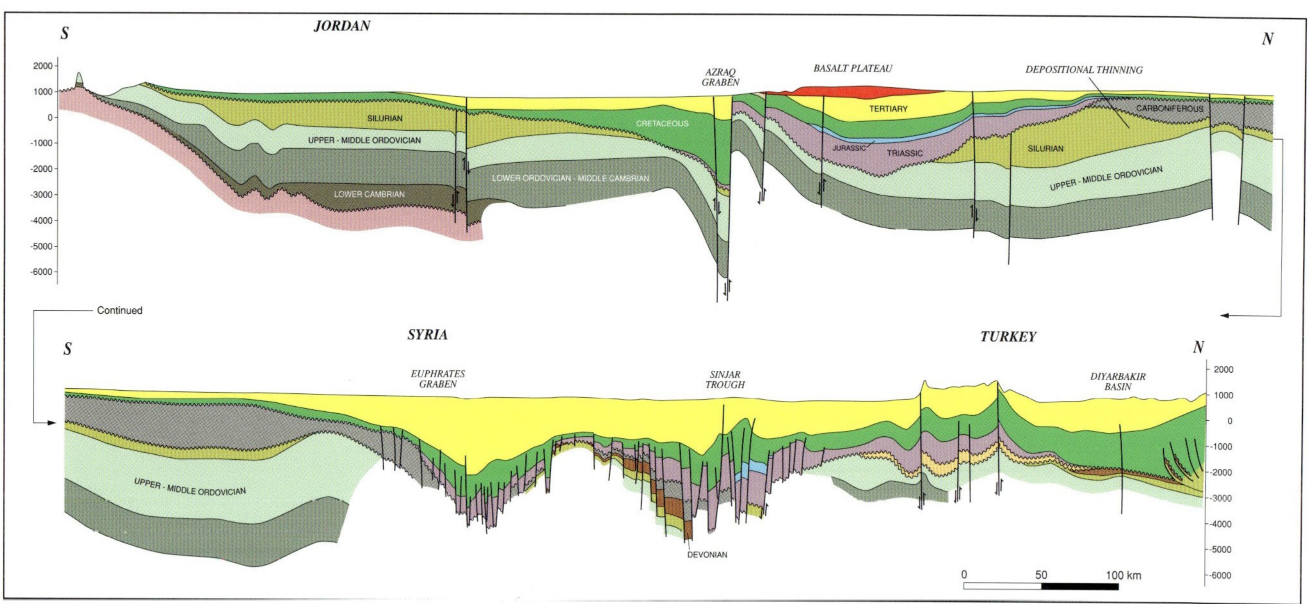

Figure 14. Geologic traverse through Jordan to Turkey. For location, see Figure 1. See time scale on Figure 5a for colors.

south. The post-Hercynian Carboniferous sequences rest generally on Ordovician or older deposits, leaving a large hiatus. Locally, some Devonian rocks are preserved. No folding or reverse faulting has been documented in Oman, suggesting that Hercynian movements were essentially vertical in nature. Note that the main phase of halokinesis affected the Cambrian-Ordovician sequences.

Fission track studies, combined with organochemical studies, carried out on rocks from Turkey to Oman indicate overburden removal on the order of kilometers over the uplifted areas. The changes in basin geometry, regional uplift, basement-cored uplifts, evidence of folding, and inversion tectonics suggest that the Arabian Plate underwent multiple phases of compression during the Hercynian orogeny. The geometry of the structures cannot have been caused by a single stress regime. The structural observations are consistent with a northwest-to-southeast-directed principal compressive-stress vector.

Further evidence for Hercynian movements, although still highly speculative, is derived from the Sanandaj-Sirjan ranges and the Oman Mountains. In the Sanandaj-Sirjan ranges, intensely folded metamorphic Devonian complexes have been found (Davoudzadeh and Weber-Diefenbach, 1987; Thiele et al., 1968) (Figure 10). These are overlain by unmetamorphosed Permian rocks. Sparse radiometric age dating indicates an Early Carboniferous age for the metamorphism (Crawford, 1977). In the Oman Mountains, the Permian rests unconformably on highly deformed and metamorphosed Lower Paleozoic rocks attributed to the Hercynian orogeny (Mann and Hanna, 1990).

The deformation and metamorphism indicate that the subsequent Zagros margin was possibly a zone of transpressional movements.

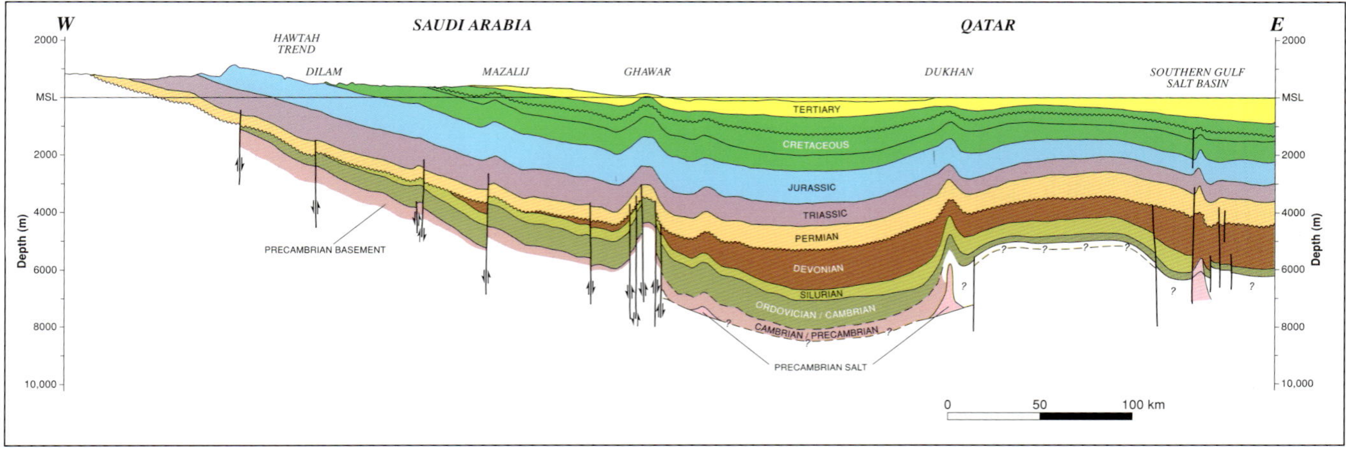

Figure 15. Geologic traverse through Saudi Arabia to Qatar (see also Alsharhan and Nairn, 1994). For location, see Figure 1. See time scale on Figure 5b for key to colors.

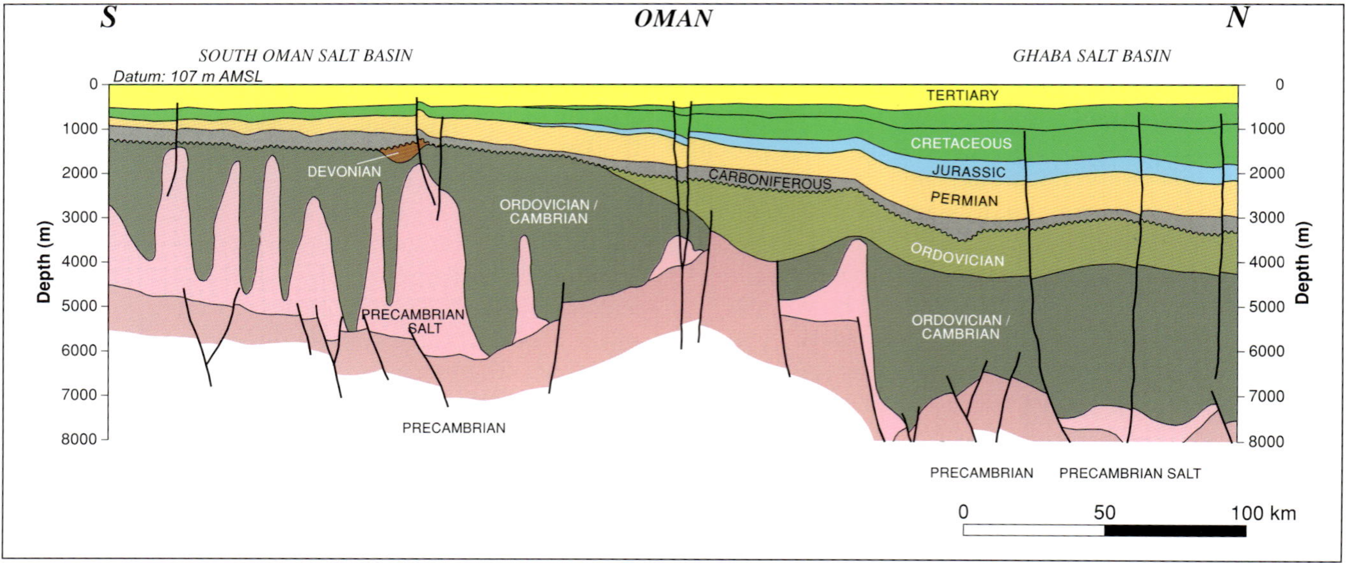

Figure 16. Geologic traverse through Oman. For location, see Figure 1. See time scale on Figure 5c for key to colors.

HYDROCARBON HABITAT

Hydrocarbon Availability

Understanding the Silurian petroleum system is one of the keys to unlocking most Paleozoic resources. Our working definition of a petroleum system differs from published definitions (Magoon and Dow, 1994). We define a petroleum system as the total space occupied by all hydrocarbons derived from one chemically distinguishable source-rock interval. Emphasis in this definition is on establishing hydrocarbon availability.

Organic-rich shales exist throughout the basin at the base of the Silurian (Figure 9). These source rocks, made up of dark gray to black shale, contain marine algae, acritarchs, and abundant chitinozoans and graptolites (Jones and Stump, 1999). Source-rock quality and thickness vary with depositional environment, as demonstrated by lack of organic matter and a shallow-marine bioturbated, sandy, micaceous claystone facies in basin-margin settings. Source-rock thickness varies from hundreds of meters in the Rub [c] al-Khali to about 50 m in the northern basin to a few meters in marginal settings.

In Jordan, Syria, and Iraq, a younger source rock of probable Wenlockian age has been documented.

Oil-to-oil and oil-to-source correlations indicate the presence of Silurian-derived fluids over a wide geographic area from Turkey to Oman and from Saudi Arabia to Qatar (Figure 17). They occur as mixtures or end-member crude. The crudes have a distinct chemical fingerprint (Grantham et al., 1987; Abu-Ali et al., 1991; Mahmoud et al., 1992; Cole et al., 1994).

Estimating availability and quality of Silurian-sourced hydrocarbons is often hindered by complex burial histories. Burial histories are complicated by multiple major phases of uplift, especially during the Hercynian. This may complicate the interpretation of vitrinite-reflectance measurements. Areas where Paleozoic rocks were deeply buried may have generated their hydrocarbons prior to the Carboniferous, and some reservoired hydrocarbons may have escaped to the surface during Hercynian deformation.

In other areas, source rocks reached the oil window only prior to Hercynian uplift, leaving only potential for gas generation during the Alpine burial phase. Therefore, critical to success is the understanding of generation histories, which can be successfully addressed only through application of inorganic paleothermometer tools.

The predicted cumulative volumes of oil and gas expelled from the Silurian "hot" shale contained within the present oil window across the depocenter range between 430 and 760 billion bbl of oil and 1540 and 2575 tcf of gas. Cumulative volumes of oil and gas expelled from the Silurian within the present gas window range between 3000 and 3600 billion bbl of oil and 21,595 and 39,200 tcf of gas. If we assume that approximately 90% of these predicted expelled volumes are not reservoired because of migration losses or various model inaccuracies, then (1) 48–83 billion bbl of oil and oil equivalents are predicted to be recoverable from the geographic area where the source rock is within the oil window, and (2) 380–439 billion bbl of oil and oil equivalent are predicted to be recoverable from the geographic area where the source rock is presently in the gas window. The Paleozoic exploration frontier, therefore, may offer a hydrocarbon initially in place of 1 trillion bbl of oil equivalent reservoired from the Silurian "hot" shale alone.

Although the Silurian "hot" shale is the principal hydrocarbon source, recent geochemical evidence from the Shamah gas field in eastern Saudi Arabia indicates that Unayzah condensates are derived from a different source rock, yet unidentified.

A second group of established petroleum systems involves upper Precambrian source rocks in Oman. They may also be present in the other Precambrian salt basins (Figure 1). Hydrocarbons derived from these source rocks have been found in reservoirs spanning the entire Phanerozoic. The hydrocarbons have been linked to several source-rock intervals deposited in prerift to synrift sequences. They include carbonate source rocks, which contain mainly type I(/II) organic matter with total-organic-carbon (TOC) contents as high as 7%. Siliciclastic source rocks are found in presalt and intrasalt settings. They have variable TOCs ranging as high as 10% and may occur in massive sections as much as 1750 m thick. They are considered a world-class source rock, with organic matter characterized by anomalously low activation energies.

Finally, a group of hydrocarbons has been defined in Oman, the so-called Q oils. These have distinct geochemical characteristics (Grantham et al., 1987). The exact source of these hydrocarbons remains to be identified, although they appear to be Precambrian in character.

Reservoirs and Seals

Stratigraphic diagrams (Figure 5) show the relationship among the main source rocks, seals, and reservoirs in the Paleozoic section. They are generalized diagrams; local exceptions are to be expected.

The Permian sandstones and carbonates contain the main reservoirs for the Silurian petroleum system, sealed by intraformational claystone/shale or by tight carbonates and evaporites. The regional seal in Saudi Arabia is shales of the Triassic, which completely separate the Silurian hydrocarbon system from the Mesozoic systems above.

The lack of seals renders little prospectivity to these Permian sequences in northern Arabia. The Carboniferous to Devonian sequences may include excellent potential reservoirs, especially the Devonian of eastern Saudi Arabia. The presence of only local seals, combined with rapid lateral facies variations, renders these sequences of limited regional prospectivity. Exceptions include structures where these reservoirs subcrop below Permian seals and where they are juxtaposed with a sealing facies across faults.

The Silurian section also includes possible reservoirs in the form of sandstones in the shale-dominated outer neritic environments. Generally, these reservoirs are thin and their quality is difficult to predict. This Silurian play was recently confirmed by discoveries in Saudi Arabia and Iraq.

The Silurian "hot" shale forms the ultimate seal to the sub-Silurian section. The sub-Silurian embraces excellent reservoirs, which also may be charged from "hot" shale. An example is Abu Jifan field in eastern Saudi Arabia, in which sandstones of Ordovician Sarah/Qasim Formation are the main reservoir.

The sub-Silurian section becomes an important target, in addition to the Permian in those regions underlain by Precambrian source rocks. The trapping potential in the Cambrian-Ordovician massive coarse siliciclastics of the basin margin sections depends on truncation, which juxtaposes reservoirs against Permian-Carboniferous or younger seals. Seal potential increases basinward in paral-

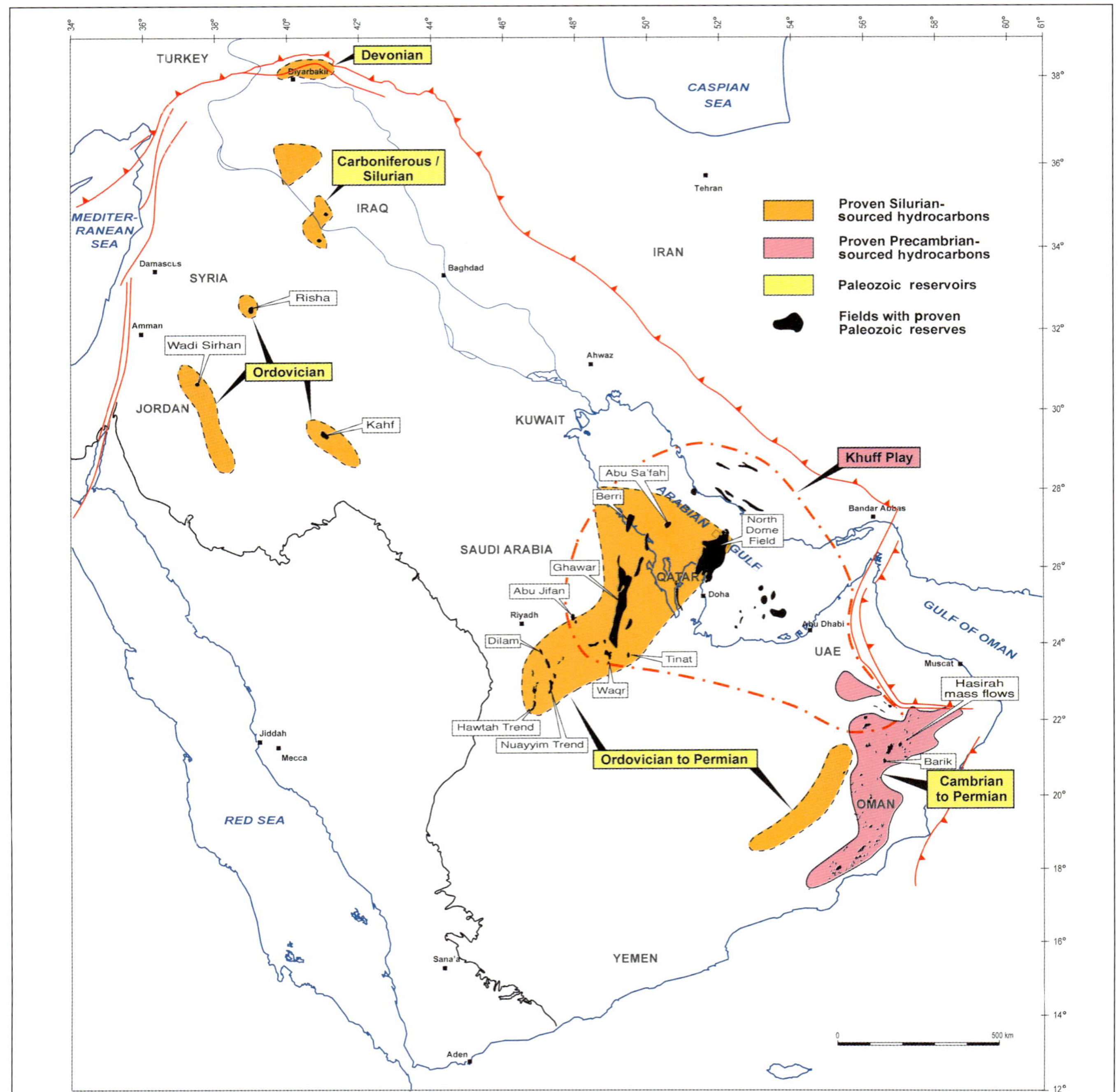

Figure 17. Map showing the proven extent of the Silurian and Precambrian petroleum systems (see text for details).

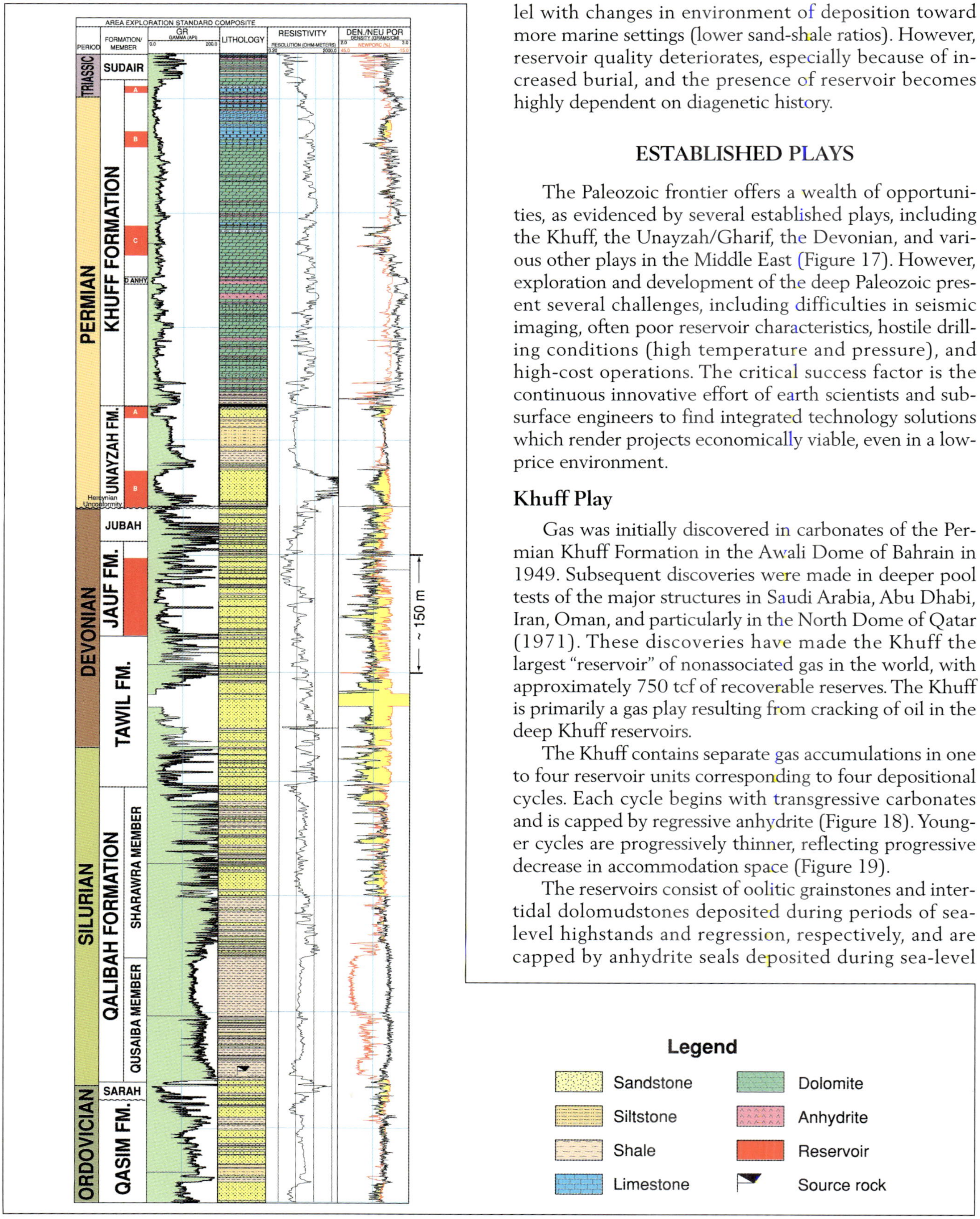

lel with changes in environment of deposition toward more marine settings (lower sand-shale ratios). However, reservoir quality deteriorates, especially because of increased burial, and the presence of reservoir becomes highly dependent on diagenetic history.

ESTABLISHED PLAYS

The Paleozoic frontier offers a wealth of opportunities, as evidenced by several established plays, including the Khuff, the Unayzah/Gharif, the Devonian, and various other plays in the Middle East (Figure 17). However, exploration and development of the deep Paleozoic present several challenges, including difficulties in seismic imaging, often poor reservoir characteristics, hostile drilling conditions (high temperature and pressure), and high-cost operations. The critical success factor is the continuous innovative effort of earth scientists and subsurface engineers to find integrated technology solutions which render projects economically viable, even in a low-price environment.

Khuff Play

Gas was initially discovered in carbonates of the Permian Khuff Formation in the Awali Dome of Bahrain in 1949. Subsequent discoveries were made in deeper pool tests of the major structures in Saudi Arabia, Abu Dhabi, Iran, Oman, and particularly in the North Dome of Qatar (1971). These discoveries have made the Khuff the largest "reservoir" of nonassociated gas in the world, with approximately 750 tcf of recoverable reserves. The Khuff is primarily a gas play resulting from cracking of oil in the deep Khuff reservoirs.

The Khuff contains separate gas accumulations in one to four reservoir units corresponding to four depositional cycles. Each cycle begins with transgressive carbonates and is capped by regressive anhydrite (Figure 18). Younger cycles are progressively thinner, reflecting progressive decrease in accommodation space (Figure 19).

The reservoirs consist of oolitic grainstones and intertidal dolomudstones deposited during periods of sea-level highstands and regression, respectively, and are capped by anhydrite seals deposited during sea-level

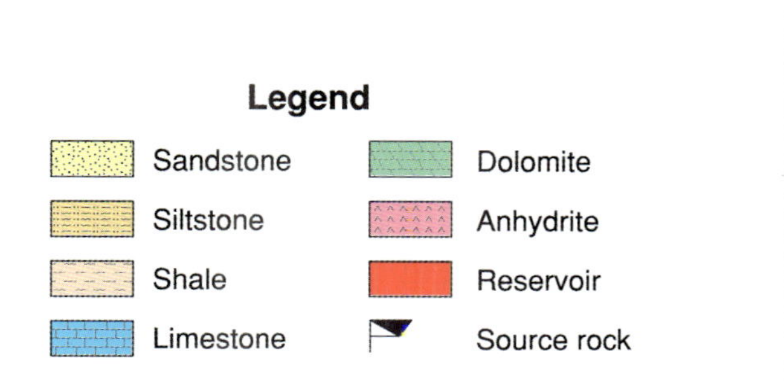

Figure 18. Composite log of the Paleozoic section in the Ghawar area in Eastern Saudi Arabia. Three of the four Khuff reservoirs are present. See Figure 4 for key to environments of deposition.

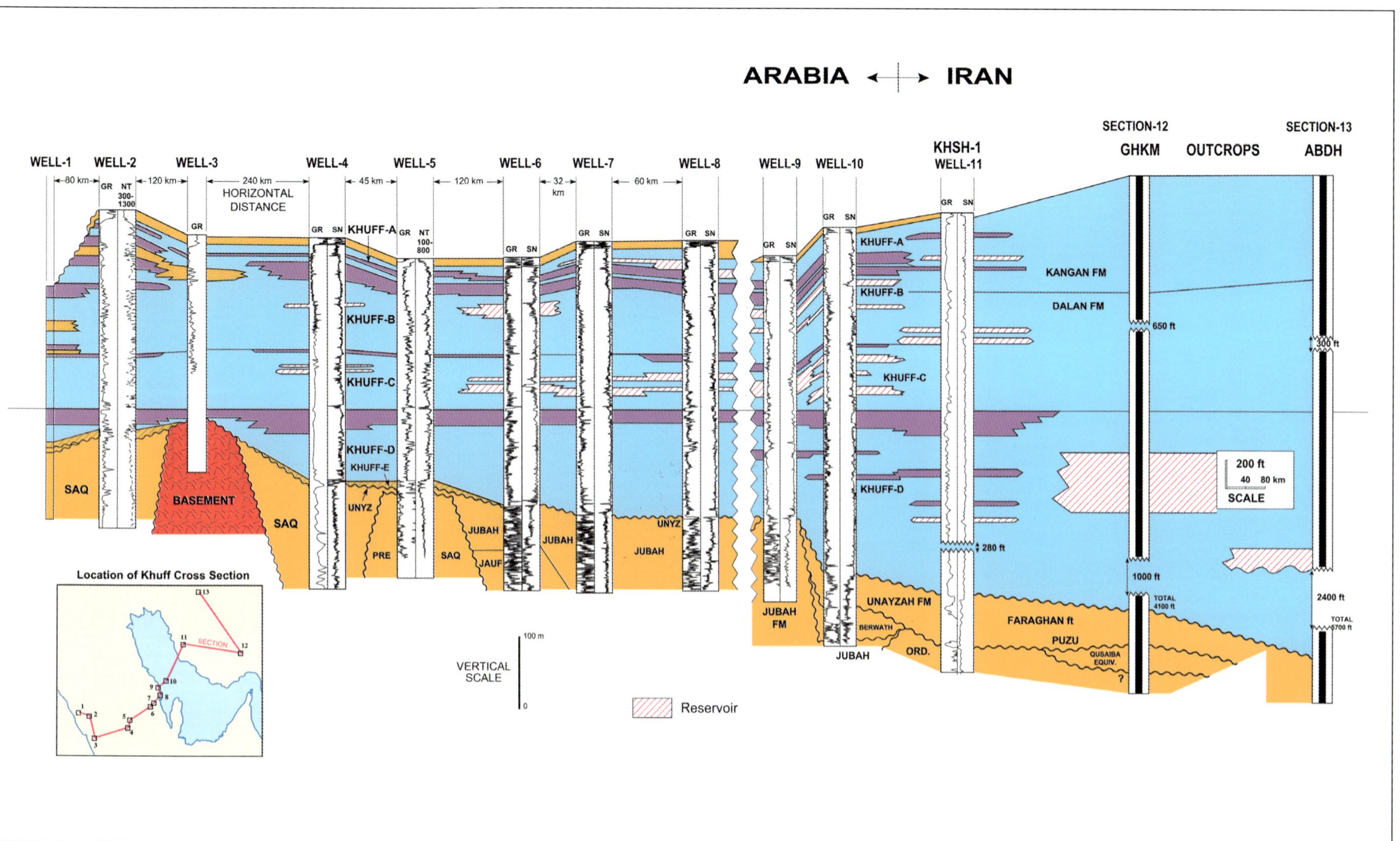

Figure 19. Stratigraphic cross section from Saudi Arabia to Iran showing reservoir development in the Khuff A-D reservoirs. Datum is the Khuff D anhydrite.

lowstands. The development of the Khuff reservoirs on this scale is related to several factors, such as the relative position on the carbonate shelf and the development of higher-energy facies on shoals that may straddle structural highs and shelf-margin reefs (Al-Jallal, 1995). The quality of the Khuff reservoirs varies from excellent to poor, depending primarily on extent of dolomitization, leaching, fracturing, and cementation (particularly by anhydrite). Leached zones often form the better portion of the reservoir. Production may be both from the matrix and from fractures, but productivity generally improves with the presence of fractures. The Khuff has considerable potential for stratigraphic traps, yet unexplored.

The Khuff play presents two major challenges: reservoir performance and gas quality.

The quality of Khuff gas varies, depending on the amounts of nonhydrocarbon gases, mainly H_2S, CO_2, and N_2. The amount of H_2S in the Khuff correlates with temperature and, consequently, with depth of the reservoirs, reflecting in-situ conversion of hydrocarbons to H_2S by thermochemical reduction of sulfate. The amounts of other gases, such as N_2 and CO_2, appear to increase with greater depth and maturity of the source rocks.

The quality of the Khuff reservoirs presents the highest risk to development because of abrupt lateral and vertical variations in porosity and permeability. The Khuff reservoirs display a wide variety of porosity types, ranging from primary intergranular to secondary oomoldic. Reservoir permeability is equally variable, depending on leaching of either matrix and cement components or extent of fracture development. For these reasons, petrophysical evaluation and geologic modeling of the Khuff reservoirs are hampered by uncertainties. Three-dimensional seismic data have proven to be the best approach for delineating zones of Khuff porosity ahead of development drilling.

For example, Figure 20 shows inverted 3-D seismic volumes from the Khuff C reservoir in the middle of the Ghawar structure. The Khuff C was divided into three 15-m-thick layers for purpose of illustration. The seismic impedance in this reservoir was calibrated with well data from nearby areas, which indicated an inverse correlation of impedance with porosity. The inverted seismic data were then used to position six development wells in the most porous zones, all of which proved successful, therefore minimizing the risk of drilling unproductive wells.

There is some uncertainty about the history and

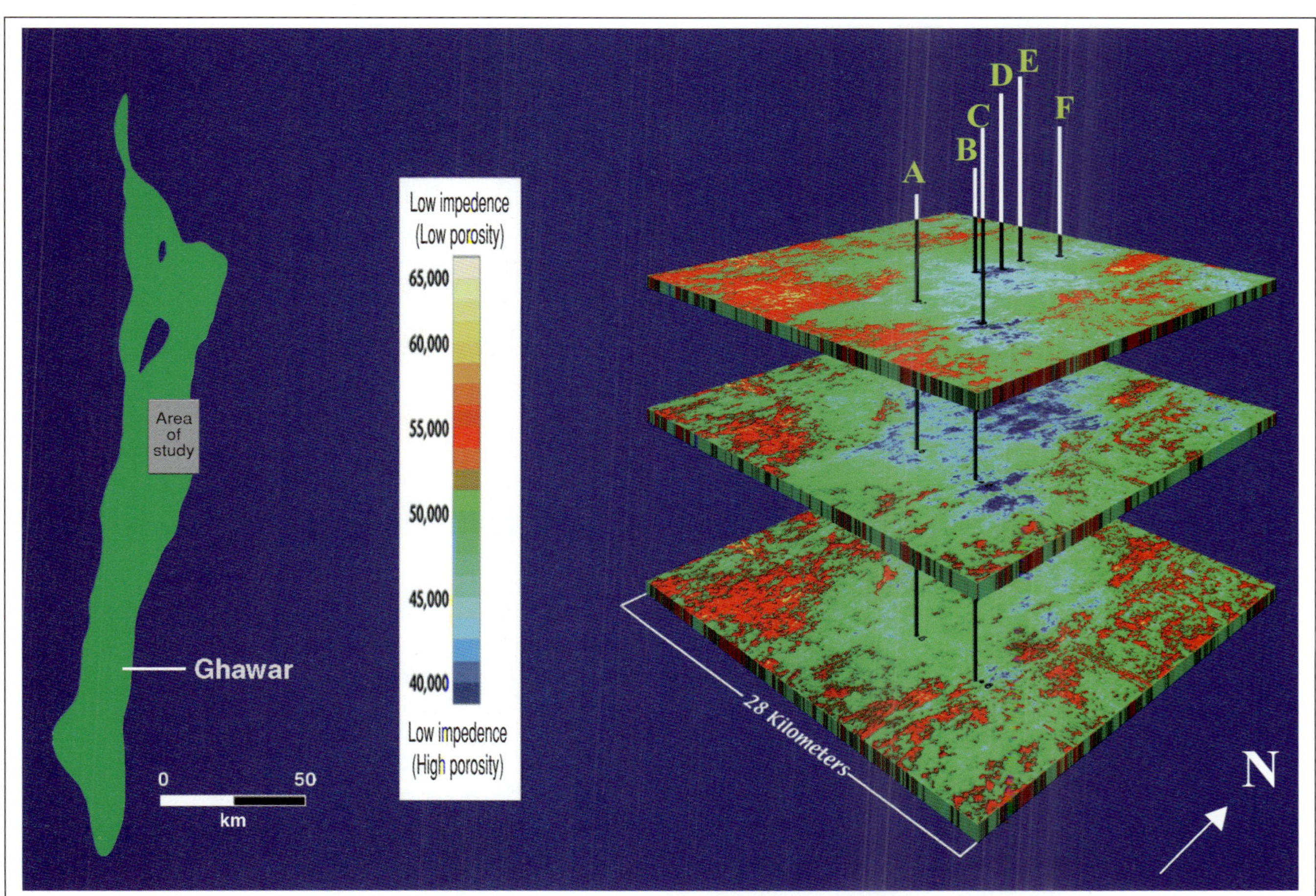

Figure 20. Perspective diagram showing 3-D seismic impedance contrast in the Khuff C reservoir used to locate six development wells.

paths of hydrocarbon migration into the Khuff, particularly in areas where shales and tight carbonates at the base of the Khuff seal hydrocarbon accumulations in the underlying Paleozoic siliciclastics. It is likely that reactivated Hercynian faults, such as those on the west flank of Ghawar (Figure 21), provided direct pathways for vertical migration into the Khuff from hydrocarbon kitchens in flanking basins.

Unayzah/Gharif Play

Oil was initially discovered in 1972 in sandstones of the Permian Gharif Formation in the Ghaba North structure in Oman, and the subsequent drilling campaign demonstrated the economic viability of the play throughout Oman. In Saudi Arabia, the potential of the Permian Unayzah Formation was confirmed in 1982 by a gas discovery in the southern part of the Ghawar structure. The Unayzah play became more significant in 1989, when superlight oil was discovered in central Saudi Arabia in the Hawtah structure. This was followed by an aggressive exploration campaign that resulted in the discovery of 18 oil and gas fields along the Hawtah and Nuayyim trends in the next 10 years. The fields are structural closures along Hercynian basement-cored uplifts that may be transpressional in origin (Simms, 1995). Moreover, the stratigraphic variability of the Unayzah Formation, influenced by paleotopography and the continental environments of deposition, lends a stratigraphic component to entrapment.

The Unayzah and overlying basal Khuff siliciclastics are composed of alluvial, fluvial, and eolian deposits that display substantial variation in facies. The Unayzah includes three sandstone reservoirs (designated informally as A, B, and C) separated by siltstone and mudstone (McGillivray and Husseini, 1992; Senalp and Al-Duaiji, 1995).

The sandstone reservoirs are laterally discontinuous, and their quality varies, depending on sorting and the amount of diagenetic quartz, kaolinite, and illite/smectite cement. Intergranular porosity ranges as high as 30% and permeability to more than 1 darcy, particularly in eolian sandstone facies. The top seal comprises transgressive shales at the base of the overlying Khuff Formation.

The Unayzah oils range from 48° to 53° API, and their gas-oil ratio (GOR) is less than 90 m^3/m^3. The low GOR is attributed to solution of methane by waters in an active hydrodynamic system driven by influx of meteoric water from outcrops along the western edge of the basin (Figure 15). The Silurian source rocks in central Saudi Arabia are immature, and the Unayzah oils were evidently generated in the deeper parts of the basin and migrated about 200 km westward toward the basin margin (Abu-Ali et al., 1991).

The acquisition of 3-D seismic surveys during field development has helped in mapping the distribution of

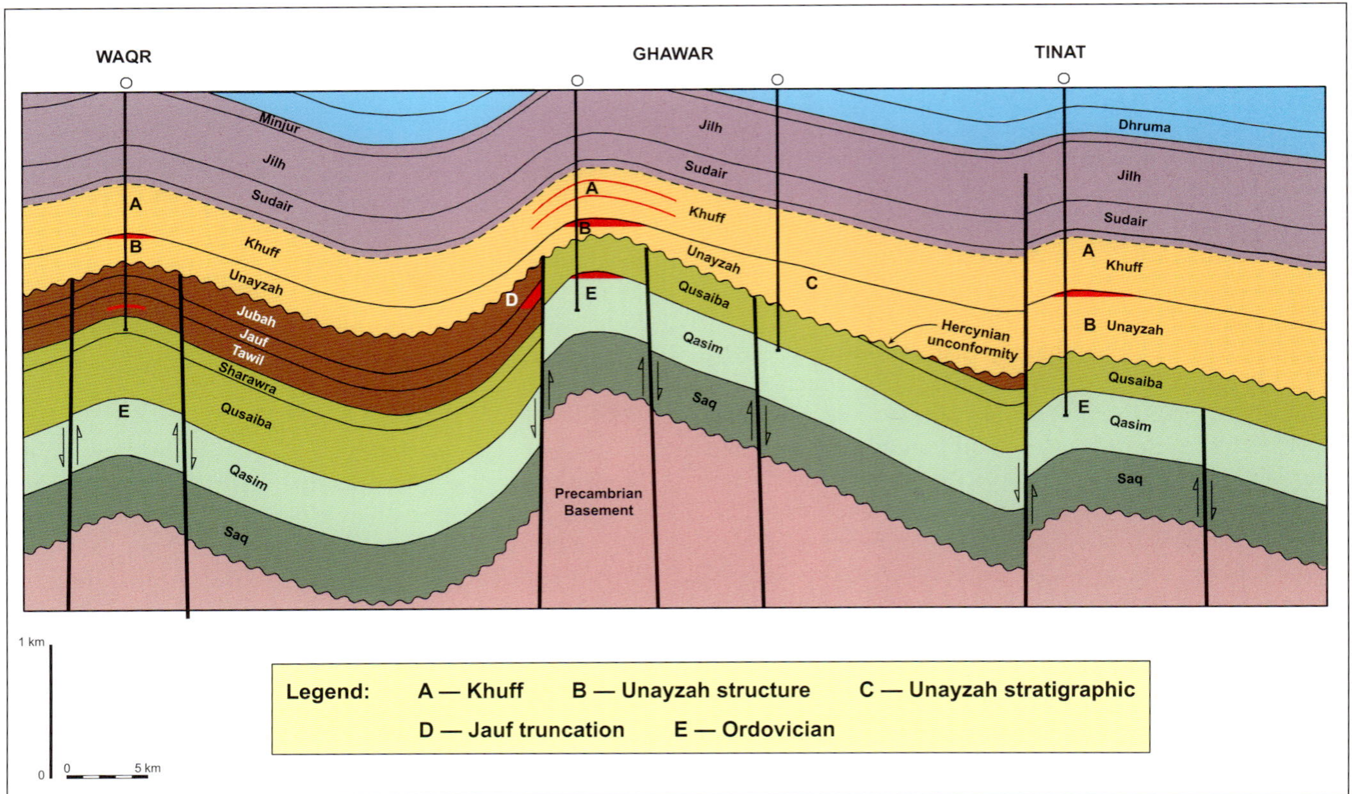

Figure 21. Schematic east-west structural cross section across the Waqr-Ghawar-Tinat fields showing the major Paleozoic plays.

the Unayzah reservoirs. Figure 22 shows the structural configuration of Hawtah field. The present structure was formed by reactivation of Hercynian faults during the Triassic. Also shown is the variation of seismic amplitudes in the Unayzah A reservoir in the field. The high seismic amplitudes correspond to eolian sandstone facies, and the low amplitudes correspond to nonreservoir siltstones and mudstones. The seismic amplitudes and other attributes were used for locating wells, which improved the success ratio of development wells from 54% to 84%. The amplitudes also show that the north Hawtah area is a stratigraphic trap resulting from updip pinch-out of the A reservoir. The 3-D surveys were used to discover other stratigraphic and combination traps, for example, in Usaylah field (Evans et al., 1997).

The Unayzah play was extended in the last decade to target gas in the deeper basin (3700+ m) near facilities in eastern Saudi Arabia (Figure 21). The gas-exploration campaign has resulted in discovery of six additional Unayzah gas/condensate fields near the Ghawar structure, such as Waqr and Tinat (Figure 21). The Unayzah deep-gas play presents additional challenges, including poor seismic imaging of the Paleozoic section and abrupt variation in reservoir quality resulting from stratigraphic and diagenetic reasons. The problems of deep seismic imaging and reservoir heterogeneity are being addressed by acquisition of high-effort (28,800-channel) 3-D seismic surveys to reduce reservoir and trap risks.

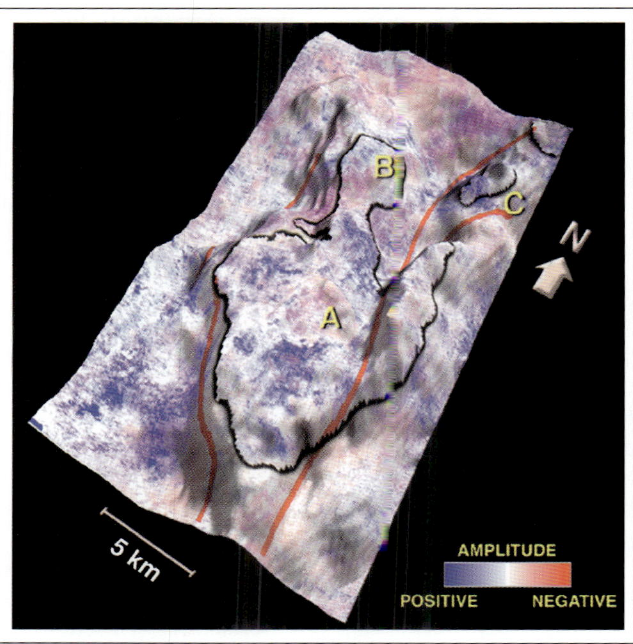

Figure 22. Perspective diagram of the Hawtah oil field, central Saudi Arabia, from 3-D seismic data, showing amplitude of the Unayzah reservoir draped over structure. The flanks of the structure are drape folds over deep Hercynian faults. The blue areas correspond to good reservoir development in eolian sandstone facies; the red areas correspond to nonreservoir facies. A = poor reservoir at the crest of the Hawtah structure; B = stratigraphic trap in the north Hawtah area defined by the updip pinch-out of the reservoir; C = Nisalah field.

Devonian Play

Gas in the Devonian Jauf sandstone was initially discovered in 1980 by a deeper pool test on the north end of the Ghawar structure. Subsequent deep tests showed that the Devonian section was mostly eroded from the crest of the structure. The discovery in 1994 of Jauf gas in a combination structural-stratigraphic trap along the flank of central Ghawar was a major exploration success, especially in light of the poor seismic imaging of the sub-Khuff section (Wender et al., 1998).

The Jauf consists of shallow-marine sandstones with relatively high porosities (as much as 25%), which is unusual, given their burial to more than 4300 m. Although sub-Khuff siliciclastics have undergone extensive silica cementation, the Jauf reservoir is weakly cemented with authigenic illite that coats grain surfaces. This early illite has apparently inhibited quartz cementation and preserved porosity. The abundant illite also lowers resistivity values because of the excess bound water and the high cation-exchange capacity of illite. This can cause pessimistic water-saturation estimates and can lead to potentially bypassed low-resistivity pay zones (Wender et al., 1998).

The cross section in Figure 21 shows the structural relationships of the Devonian Jauf play. On structures such as Ghawar that were subjected to a large amount of Hercynian growth, the Jauf Formation is eroded from the crest and preserved along the flanks. The Jauf flank play is defined by the lateral truncation of the Jauf reservoir against sealing faults or by top truncation of the reservoir by the Hercynian unconformity with top seal provided by the basal shales of the Khuff Formation. The Jauf may be preserved over the crest of low-relief structures such as Waqr, where it is a structural play.

Cambrian-Ordovician Plays

Outside Oman, the thick Silurian shales form an effective regional seal to potential hydrocarbon accumulations in the Upper Ordovician sandstones. Several discoveries have been made in structural traps, including Dilam and Abu-Jifan in central Saudi Arabia, Kahf in northern Saudi Arabia, and Wadi Sirhan 4 and Risha in Jordan (Figure 17). The hydrocarbons, mainly gas, are sourced from the overlying Silurian shales, which also act as seals.

The main challenge to the Ordovician play is the poor reservoir quality, particularly in the deeper parts of the basin.

The deep Ordovician sandstones generally have low porosities and permeabilities because of compaction and extensive cementation by quartz overgrowths during burial. Petrographic studies have shown that any significant porosity is always secondary, primarily resulting from dissolution of early intergranular carbonate cement. The early carbonate cementation was localized in areas where Hercynian uplift and erosion resulted in Permian Khuff carbonates unconformably overlying the Ordovician sandstones. The carbonate cement was probably derived from the Khuff carbonates and precipitated at shallow depths before the sandstones underwent significant compaction. The subsequent dissolution of carbonate cement preceded hydrocarbon migration into the reservoirs.

In Oman, major hydrocarbon discoveries have been made in the Cambrian-Ordovician section, despite absence of thick Silurian shales. These discoveries occur in a variety of plays. The following two examples illustrate the potential and challenges of exploring the deeper Paleozoic siliciclastics.

Hasirah Gravity-flow Play

The Caradocian Hasirah play has been defined in onshore central Oman. The sedimentary succession in this area comprises a tide-dominated sandy deltaic (or estuarine) system fed by braided rivers from an overall southerly source. These sandy sediments pass basinward into undifferentiated marine mud/claystones and interbedded, laterally discontinuous, sandy gravity-flow deposits, deposited in outer-shelf environments (Figure 23). The sandy gravity-flow deposits have excellent reservoir qualities because of reworking and rapid deposition. Porosities reach as high as 32%. Sediments were deposited in an active salt-withdrawal basin in ponded geometries and occur at an average depth of about 3000 m.

One of the critical success factors of these potential stratigraphic traps is predicting reservoir distribution and quality. The ponded nature of these sandstones, combined with their potentially erratic vertical distribution, introduces a high risk in predicting their presence using conventional technology.

Rock-property modeling indicates that the interface between the marine fines, which have high velocities, and the underlying sandstones provides a strong acoustic-impedance contrast, probably because of the highly porous nature of the sands. Peak amplitudes probably reflect the presence of hydrocarbons, which is confirmed by amplitude-versus-offset (AVO) analysis. Mapping this event confirms its discontinuous nature. Acoustic-impedance and coherency maps indicate the lateral extent of the gravity-flow system and the lobate geometry of the sands (Figure 24). Amplitude maps show the potential presence of hydrocarbons.

"Body checking" (i.e., extraction of samples with similar attributes) was applied to visualize the reservoir in its proper structural configuration (Figure 25). The result clearly shows the gravity-flow system to be limited to the axial zone of the basin southwest of the Qarat Kibrit salt

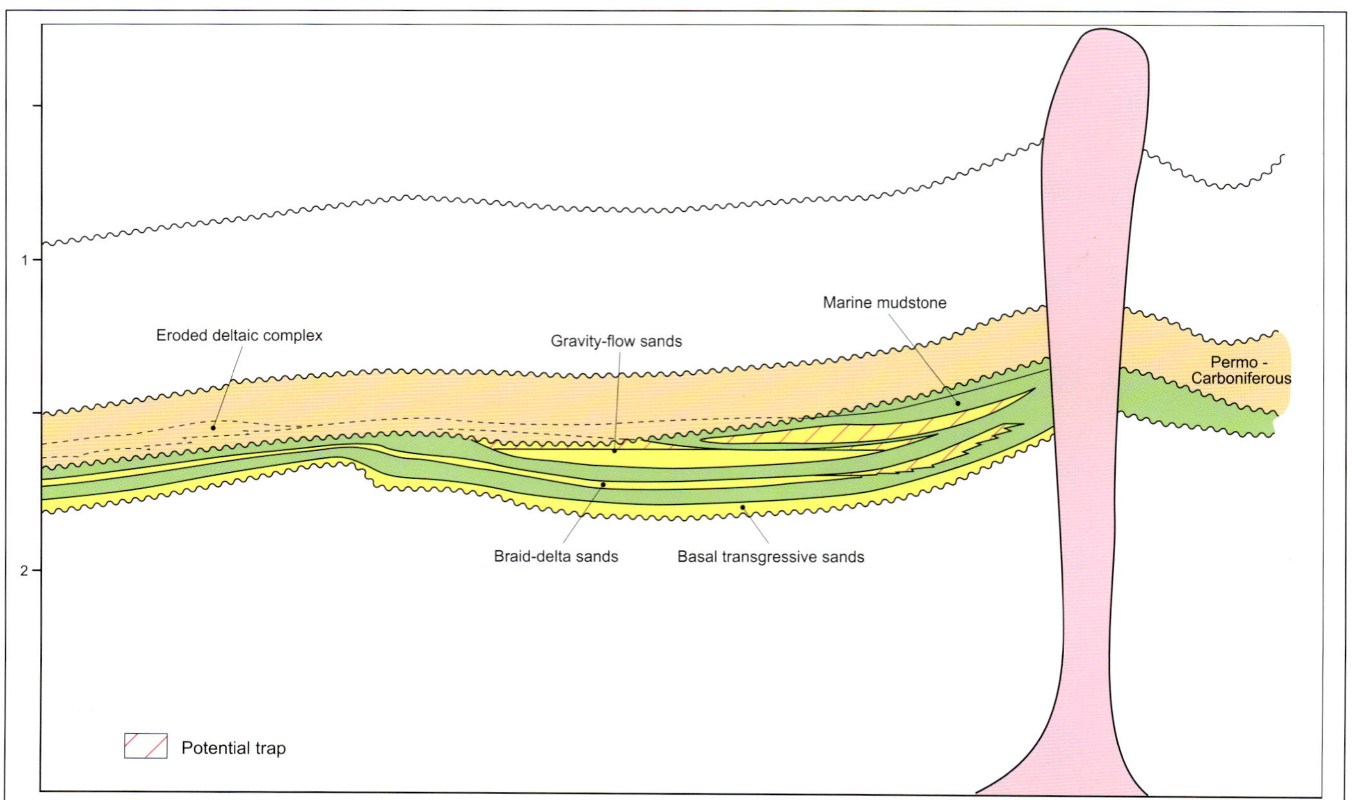

Figure 23. Conceptual diagram showing the Ordovician gravity-flow sand play.

diapir. The play clearly indicates the unconventional potential of the Paleozoic sequences, and the play may have a wide geographic extent.

Haima Deep-gas Play

The recent successful campaign for the Haima deep-gas play in onshore central Oman illustrates the challenges of exploring for hydrocarbons in the Paleozoic sequences. The initial discovery of the Haima gas/condensate resources was made in 1989. To date, about 17.6 tcf (expectation) of nonassociated gas reserves have been booked. The development project involves the supply of natural gas to an export facility for liquefied natural gas (LNG).

The gas resources occur in the Lower Ordovician Barik Sandstone at depths of more than 4 km (Figure 26). The reservoirs are found in salt-cored domes, which may be compartmentalized by faults (Figure 27). Initial reservoir pressures are about 500 bar and temperatures are 125–140°C, providing a considerable challenge in deep-well engineering. The condensate-gas ratio ranges from 0 to 950 $m^3/10^6 m^3$ (std). Hydrocarbon columns are 100 m to more than 200 m thick.

Key issues include seismic imaging and reservoir quality.

Low acoustic-impedance contrasts in the objective section, combined with a high level of multiple contamination resulting from younger reflectors with strong impedance contrasts, result in a seismic response with low signal-to-noise ratios. Thus, reservoirs and faults are difficult to image. In addition, high source-energy absorption associated with surface conditions may negatively influence seismic quality. Modeling of the reservoir is therefore dependent on neighboring reflective packages.

The reservoirs comprise sandy braid-delta sandstones and interbedded heterolithic shallow-marine facies deposited during periodic flooding events. Eustatic fluctuations gave rise to eight stacked flow units. Reservoir characteristics vary with overall position in the depositional system; average porosity and permeability are on the order of 8–10% and 1–2 md, respectively. Local variations in reservoir parameters also depend on diagenetic history and especially on the presence of an early oil charge, which inhibited quartz overgrowth and dolomite cementation. The presence of higher-quality "thief" zones complicates reservoir management because of a risk of

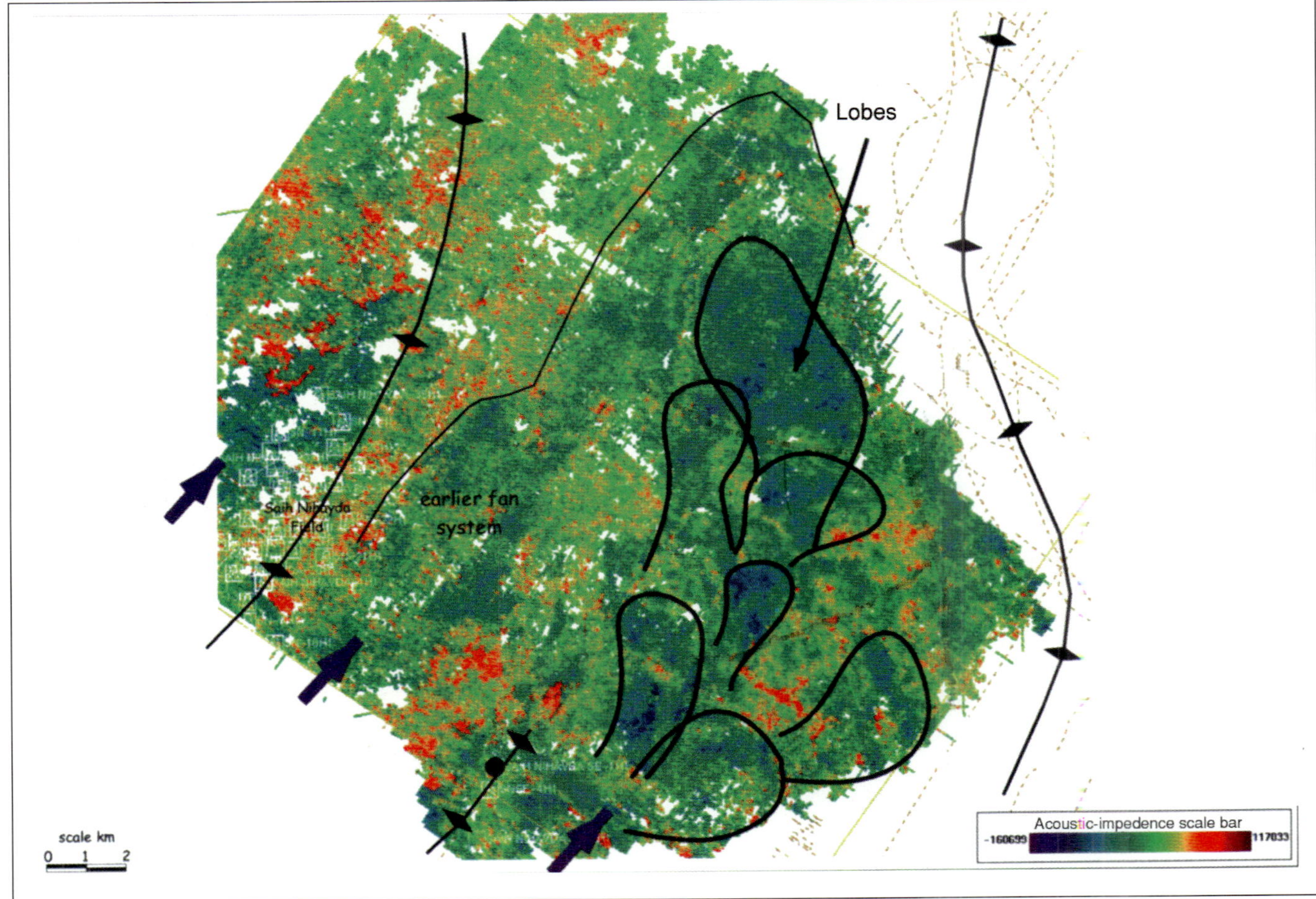

Figure 24. Acoustic-impedance time slice of the Ordovician gravity-flow deposits showing sand-prone lobes. Sands were fed from the south-southwest via paleosaddle points.

early water breakthrough. In addition, reservoir performance is highly dependent on fractures.

A dedicated team of earth scientists and subsurface engineers was given the task of finding integrated technology solutions to address the challenges of the Haima deep-gas play.

One of the solutions was the acquisition of a tailor-made 3-D survey, shot in 1998 over Barik field, to improve the structural definition. The survey used a new cost-effective acquisition approach (orthogonal wide-line acquisition) specifically designed to improve multiplicity ("fold") for deep targets (Hoetz and Duyndam, 1999). A recently developed processing sequence, specifically designed to preserve true amplitude and phase stability to allow deterministic interpretation (McGinn and Duyndam, 1998), was also applied to the data. The data quality is superior to that of the 1992 survey because of increased signal-to-noise ratio and reduced multiple contamination. The increased resolution allows more reliable fault mapping and a far more confident mapping of top reservoir, significantly reducing uncertainty (Figure 27).

Another solution was an extensive effort to increase well productivity through improved fracture stimulation, resulting in reduction in the number of wells initially required for the contractual production capacity. This involved dynamic gas-condensate reservoir modeling, combined with fracture geometry modeling to determine the preferred well configuration to drain the entire hydrocarbon column (Jones et al., 1998). The resulting design is vertical wells with single- or multiple-propped hydraulic fractures (dependent on local reservoir architecture) that cover the entire hydrocarbon column (all flow units). Development with horizontal wells was discounted because of the overall mismatch with fracture orientation. Productivity increases are among other parameters dependent on fracture width and the length of the perforated interval connected to the fracture. The studies and tests show that a productivity increase of 20–35% may be expected. Development with fracture-stimulated wells will result in significant savings in overall subsurface project cost.

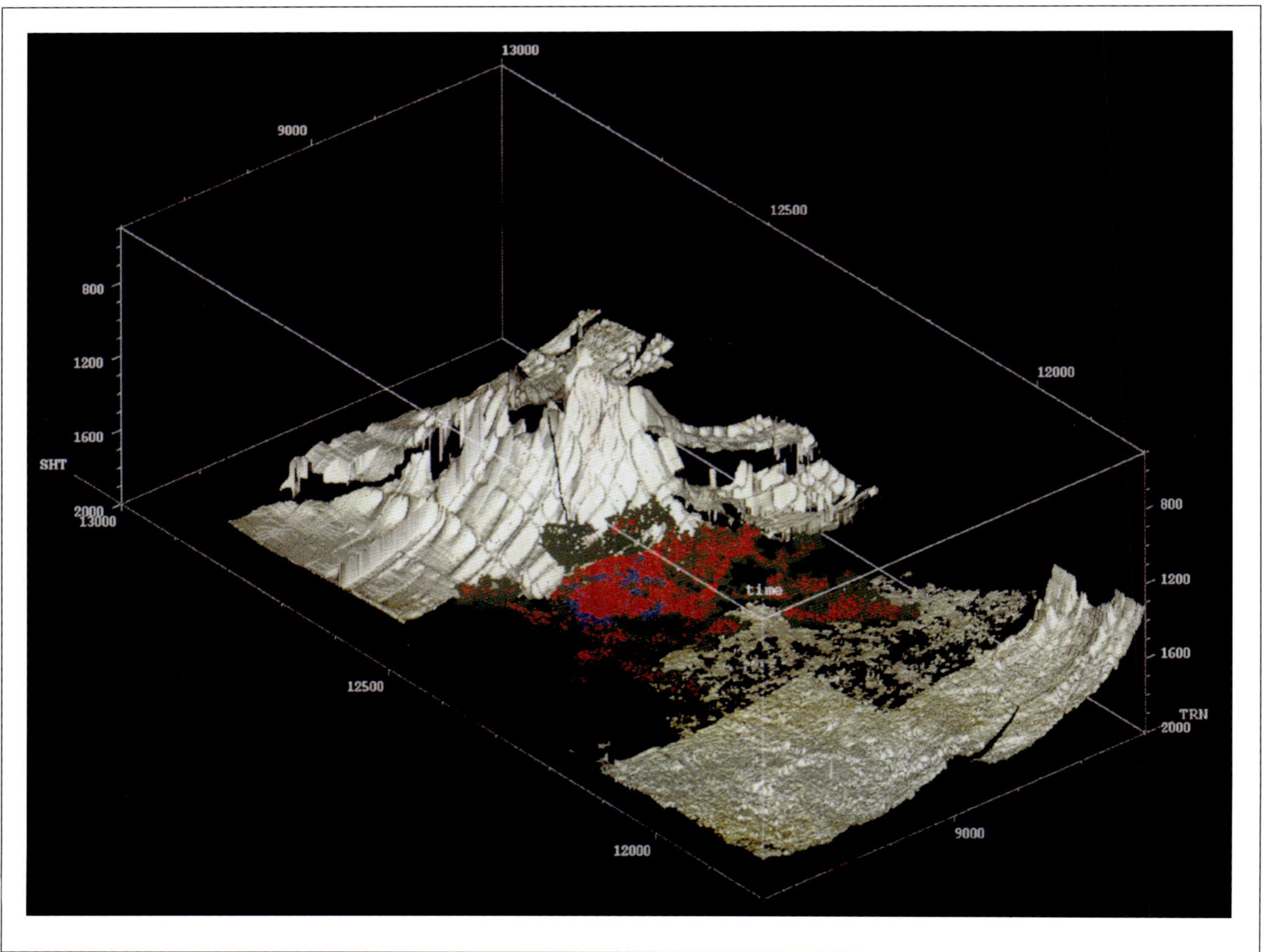

Figure 25. Body-checked gravity-flow deposits. Note that the dome structure to the north is cored by salt.

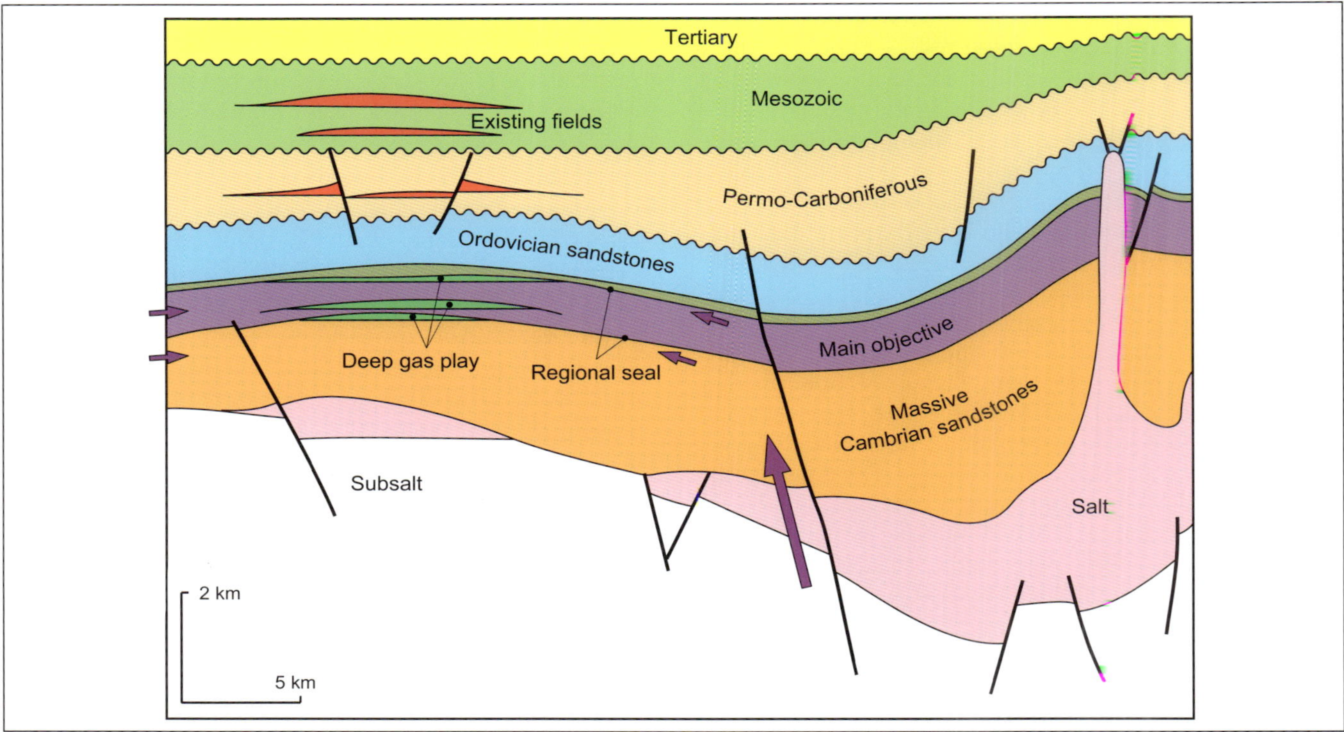

Figure 26. Conceptual diagram showing the Haima deep-gas play.

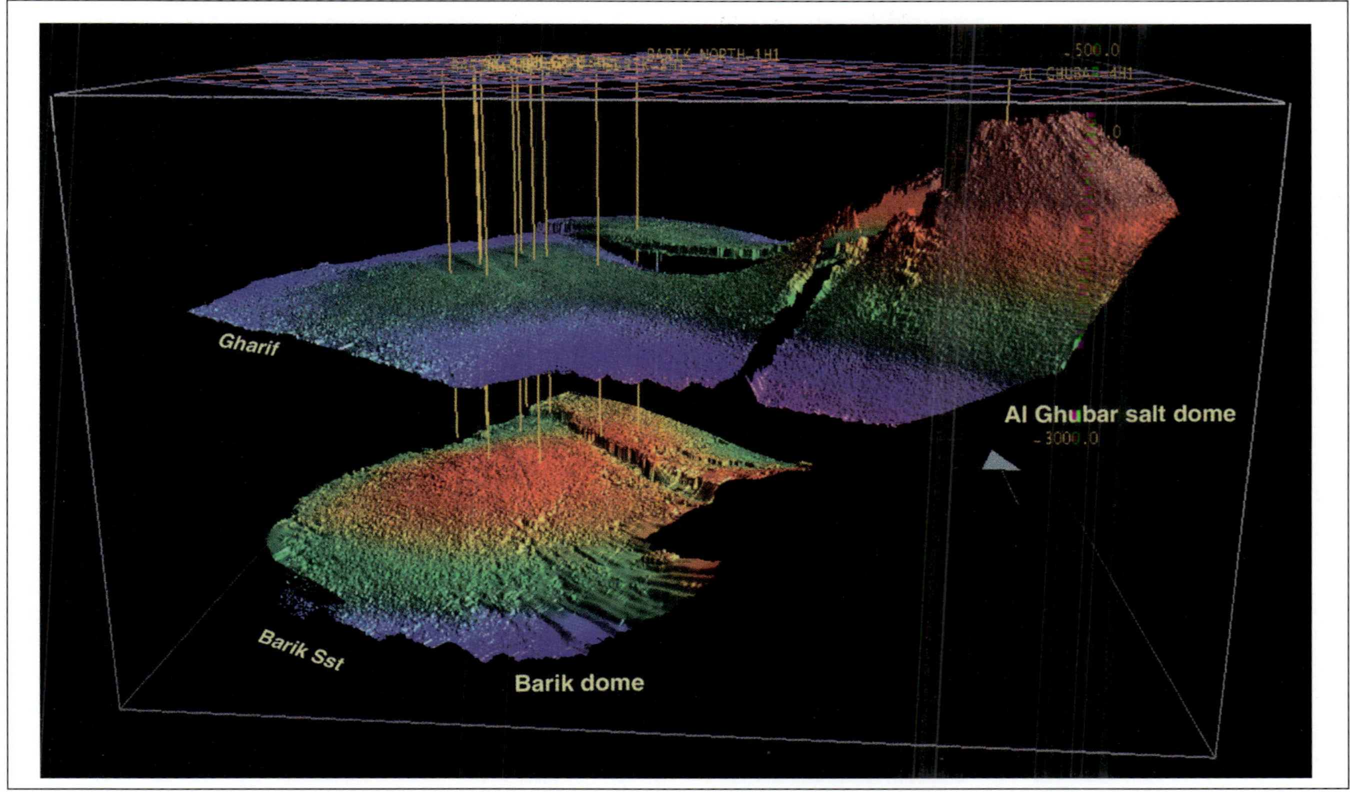

Figure 27. Structural presentation of the Barik structure and the Al Ghubar salt dome at Permian (Gharif) and Ordovician (Barik sandstone) level. The Gharif horizon represents the main reservoir in Barik field. The Barik sandstone horizon shows the structure at the deep Haima gas objective. Note that northwest-trending faults compartmentalize the structure. Colors represent traveltime.

CONCLUSIONS

The Paleozoic frontier in the Arabian Plate offers major opportunities to discover and delineate new energy reserves. The system includes multiple reservoir objectives in Cambrian–Lower Permian continental and marine siliciclastics and in Upper Permian carbonates. Hydrocarbons were derived mainly from the prolific Silurian "hot" shales, which extend over most of the basin. In addition, the Precambrian rift basins include additional source rocks in Oman and possibly elsewhere.

The Paleozoic sequences were deposited on a vast platform along the northeastern margin of Gondwana. Tectonostratigraphic relationships indicate that stable platform environments prevailed until the latest Ordovician, when the margin started to disintegrate, probably along the proto-Zagros zone.

The Hercynian orogeny was heralded by tectonic unrest at the plate margin starting during the latest Devonian, resulting in extensive intraplate deformation. This is manifested by east-northeast-trending regional upwarps in Syria, central Arabia, and Oman and by sags in the Palmyra and Rub [c] al-Khali Basins. The second manifestation of Hercynian deformation is the narrow north-trending basement-cored uplifts in central Arabia and elsewhere. The Hercynian deformation climaxed during the Carboniferous and was followed by rifting along the eastern margin during the Early Permian, which led to opening of the Neo-Tethys ocean during the Late Permian. Specifics about the pre- and syn-Hercynian tectonic history of the Arabian Plate remain to be determined.

The prospectivity of the Paleozoic section is largely determined, in addition to the sedimentary facies patterns, by the pre- and post-Hercynian burial and thermal histories, which dramatically impact reservoir quality and availability of hydrocarbons. A nontraditional approach is required to constrain thermal histories because of the complex burial/uplift history. Although porosity was largely destroyed during the deep burial of the section, it was preserved locally, either because of the presence of an early diagenetic phase or by early emplacement of hydrocarbons. Moreover, secondary porosity was selectively created in thin carrier beds by leaching during fluid flow.

Exploration and development success will depend on significant innovations to meet the challenges posed by low acoustic contrasts between the target rock units, difficult surface conditions, tight reservoirs, and deep subsurface environments. The history of hydrocarbon exploration in the Arabian Plate has yielded a wide variety of new and often unexpected hydrocarbon plays spanning the Tertiary to Precambrian section. Exploration success in these plays has been driven by creative geologists who challenged established views of play potentials of the Arabian Plate.

ACKNOWLEDGMENTS

This paper is based on the work of numerous individuals who cannot all be justly mentioned. Special thanks are due to D. Evans, A. Al-Hauwaj, M. Husseini, M. Mahmoud, A. Neville, H. McClure, J. McGillivray, A. Norton, M. Rademakers, M. Senalp, and L. Wender from Saudi Aramco, and W. O. Bement, H. G. Hoetz, P. J. F. Jeans, A. T. Jones, B. K. Levell, M. P. Ormerod, M. A. Partington, J. G. M. Raven, A. N. Richardson, P. Spaak, and W. G. Townson from Petroleum Development Oman and Shell. The authors assume full responsibility for their own conclusions. The authors are grateful to Petroleum Development Oman LLC, Saudi Aramco, Shell International Exploration and Production B.V., the Oman Ministry of Oil and Gas, and the Saudi Arabian Ministry of Petroleum and Mineral Resources for permission to publish this paper.

REFERENCES CITED

Abu-Ali, M. A., U. A. Franz, J. Shen, F. Monnier, M. D. Mahmoud, and T. M. Chambers, 1991, Hydrocarbon generation and migration in the Paleozoic sequence of Saudi Arabia: SPE Middle East Oil Show, paper 21376, p. 345–356.

Alavi, M., 1994, Tectonics of the Zagros orogenic belt of Iran: New data and interpretations: Tectonophysics, v. 229, p. 211–238.

Al-Belushi, J., K. W. Glennie, and B. P. J. Williams, 1996, Permo-Carboniferous glaciogenic Al Khlata Formation, Oman: A new hypothesis for origin of its glaciation: GeoArabia, v. 1, p. 389–404.

Al-Hajri, S. A., J. Filatof, L. E. Wender, and A. K. Norton, 1999, Stratigraphy and operational palynology of the Devonian System in Saudi Arabia: GeoArabia, v. 4, p. 53–68.

Al-Jallal, I. A., 1995, The Khuff Formation: Its reservoir potential in Saudi Arabia and other Gulf countries: Depositional and stratigraphic approach, *in* M. I. Husseini, ed., Geo '94, The Middle East petroleum geosciences: Bahrain, Gulf Petrolink (publ.), p.103–119.

Alsharhan, A. S., and A. E. M. Nairn, 1994, Geology and hydrocarbon habitat in the Arabian Basin: The Mesozoic of the state of Qatar: Geologie en Mijnbouw, v. 72, p. 265–294.

Amireh, B. S., W. Schneider, and A. M. Abed, 1994, Evolving fluvial-transitional-marine deposition through the Cambrian sequence of Jordan: Sedimentary Geology, v. 89, p. 65–90.

Aoudeh, S. M., and S. A. Al-Hajri, 1995, Regional distribution and chronostratigraphy of the Qusaiba Member of the Qalibah Formation in the Nafud Basin, northwestern Saudi Arabia, *in* M. I. Husseini, ed., Geo '94, The Middle East petroleum geosciences: Bahrain, Gulf Petrolink (publ.), p. 143–154.

Best, J. A., M. Barazangi, D. Al-Saad, T. Sawaf, and A. Gebran, 1993, Continental margin evolution of the northern Arabian platform in Syria: AAPG Bulletin, v. 77, p. 173–193.

Beydoun, Z. R., 1993, Evolution of the northeastern Arabian plate margin and shelf: Hydrocarbon habitat and conceptual future potential: Revue de l'Institut Français du Pétrole, v. 48, p. 311–345.

Buday, T., and S. Z. Jassim, 1987, The regional geology of Iraq, v. 2: Tectonism, magmatism and metamorphism: Baghdad, State Establishment of Geological Survey and Mineral Investigation, 352 p.

Braakman, J. H., B. K. Levell, J. H. Martin, T. L. Potter, and A. van Vliet, 1982, Late Palaeozoic Gondwana glaciation in Oman: Nature, v. 299, p. 48–50.

Brown, G. F., D. L. Schmidt, and A. C. Huffman Jr., 1989, Geology of the Arabian peninsula: Shield area of western Saudi Arabia: U.S. Geological Survey Professional Paper 560-A, 188 p.

Cole, G. A., M. A. Abu-Ali, S. M. Aoudeh, W. J. Carrigan, H. H. Chen, E. L. Colling, W. J. Gwathney, A. A. Al-Hajji, H. I. Halpern, P. J. Jones, S. H. Al-Sharidi, and M. H. Tobey, 1994, Organic geochemistry of the Paleozoic petroleum system of Saudi Arabia: Energy and Fuels, v. 8, p. 1425–1442.

Crawford, A. R., 1977, A summary of isotopic age data for Iran, Pakistan and India: Mémoire Hors Série Société Géologique de France, ser. h, v. 8, p. 251–260.

Davoudzadeh, M., H. Soffel, and K. Schmidt, 1981, On the rotation of the Central-East–Iran microplate: Neues Jahrbuch für Geologie und Paläontologie, Monatshefte, p. 180–192.

Davoudzadeh, M., and K. Weber-Diefenbach, 1987, Contribution to the paleogeography, stratigraphy and tectonics of the Upper Paleozoic of Iran: Neues Jahrbuch für Geologie und Paläontologie, Abhandlungen, 175, v. 2, p. 121–146.

Evans, D. S., B. H. Bahabri, and A. M. Al-Otaibi, 1997, Stratigraphic trap in the Permian Unayzah Formation, central Saudi Arabia: GeoArabia, v. 2, p. 259–278.

Gradstein, F. M., and J. Ogg, 1996, A Phanerozoic time scale: Episodes, v. 19, p. 3–5.

Grantham, P. J., G. W. M. Lijmbach, J. Posthuma, M. W. Hughes Clarke, and R. J. Willink, 1987, Origin of crude oils in Oman: Journal of Petroleum Geology, v. 11, p. 61–80.

Helal, A. H., 1966, On the occurrence and stratigraphic position of Permo-Carboniferous tillites in Saudi Arabia: Neus Jahrbuch für Geologie und Paleontologie, Monatshefte, v. 7, p. 391–415.

Hoetz, H. G., and B. P. M. Duyndam, 1999, Barik 3D reshoot: Tailormade acquisition for multiple objectives: Oral presentation, Middle East Oil Show, Society of Petroleum Engineers, Bahrain, February 20–23.

Husseini, M. I., 1988, The Arabian Infracambrian extensional system: Tectonophysics, v. 148, p. 93–103.

Jones, A. T., M. S. Al Salhi, S. M. Shidi, M. England, and R. Pongratz, 1998, Multiple hydraulic fracturing of deep gas-condensate wells in Oman: SPE Annual Technical Conference and Exhibit, paper 49100.

Jones, P. J., and T. E. Stump, 1999, Depositional and tectonic setting of the Lower Silurian hydrocarbon source rock facies, central Saudi Arabia: AAPG Bulletin, v. 83, p. 314–332.

Loosveld, R. J. H., A. Bell, and J. J. M. Terken, 1996, The Tectonic evolution of interior Oman: GeoArabia, v. 1, p. 28–51.

Magoon, L. B., and W. G. Dow, 1994, The petroleum system, *in* L. B. Magoon and W. G. Dow, eds., The petroleum system from source to trap: AAPG Memoir 60, p. 3–24.

Mann, A., and S. S. Hanna, 1990, The tectonic evolution of pre-Permian rocks, central and southeastern Oman Mountains, *in* A. H. F. Robertson, M. P. Searle, and A. C. Ries, eds., The geology and tectonics of the Oman region: Geological Society Special Publication 49, p. 307–325.

Mahmoud, M. D., D. Vaslet, and M. I. Husseini, 1992, The Lower Silurian Qalibah Formation of Saudi Arabia: An important hydrocarbon source rock: AAPG Bulletin, v. 76, p. 1491–1506.

McGillivray, J. G., and M. I. Husseini, 1992, The Paleozoic petroleum geology of central Arabia: AAPG Bulletin, v. 76, p. 1473–1490.

McClure, H. A., 1978, Early Paleozoic glaciation in Arabia: Paleogeography, Paleoclimatology, Paleoecology, v. 25, p. 315–326.

McGinn, A., and B. Duyndam, 1998, Land seismic data quality improvements: The Leading Edge, v. 17, p. 1570–1577.

Murris, R. J., 1980, Middle East: Stratigraphic evolution and oil habitat: AAPG Bulletin, v. 64, p. 597–618.

Reitz, E., and M. Davoudzadeh, 1995, Ordovician acritarchs from the Banestan, Kerman area, central Iran: Paleobiogeographical evidence for a warm water environment: Neues Jahrbuch für Geologie und Paläontologie, Monatshefte, v. 8, p. 488–500.

Rickards, R. B., M. A. Hamedi, and A. J. Wright, 1994, A new Arenig (Ordovician) graptolite fauna from the Kerman district, east-central Iran: Geological Magazine, v. 131, p. 35–42.

Ruttner, A. W., 1993, Southern borderland of Triassic Laurasia in north-east Iran: Geologische Rundschau, v. 82, p. 110–120.

Senalp, M., and A. Al-Duaiji, 1995, Stratigraphy and sedimentation of the Unayzah Reservoir, central Saudi Arabia, *in* M. I. Husseini, ed., Geo '94, The Middle East petroleum geosciences: Bahrain, Gulf Petrolink (publ.), p. 837–847.

Simms, S. C., 1995, Structural style of recently discovered oil fields, central Saudi Arabia, *in* M. I. Husseini, ed., Geo '94, The Middle East petroleum geosciences: Bahrain, Gulf Petrolink (publ.), p. 861–866.

Szabó, F., and A. Kheradpir, 1978, Permian and Triassic stratigraphy, Zagros Basin, south-west Iran: Journal of Petroleum Geology, v.1, p. 57–82.

Thiele, O. M., R. A. Assefi, A. Hushmand-Zadeh, K. Seyed-Emami, and M. Zahedi, 1968, Explanatory text of the Golpaygan quadrangle map: Geological Survey of Iran Geological Quadrangle E7, 24 p.

Unrug, R., 1996, The assembly of Gondwanaland: Episodes, v. 19, p. 11–20.

Vaslet, D., 1987, Early Paleozoic glacial deposits in Saudi Arabia, a lithostratigraphic revision: Saudi Arabian Directorate General for Mineral Resources Technical Record BRGM-TR-07-1, 24 p.

Vaslet, D., 1990, Upper Ordovician glacial deposits in Saudi Arabia: Episodes, v. 13, p. 147–161.

Wender, L. E., J. W. Bryant, M. F. Dickens, A. S. Neville, and A. M. Al-Moqbel, 1998, Paleozoic (pre-Khuff) hydrocarbon geology of the Ghawar area, eastern Saudi Arabia: GeoArabia, v. 3, p. 273–302.

Wensink, H., 1991, Late Precambrian and Paleozoic rocks of Iran and Afghanistan, in M. Moullade and A. E. Nairn, eds., The Phanerozoic geology of the world, I. The Paleozoic: New York, Elsevier Science Publishing Co., p. 147–217.

World Oil, 1998, World trends: Exploration, drilling, production: World Oil, August 98, p. 25–29.

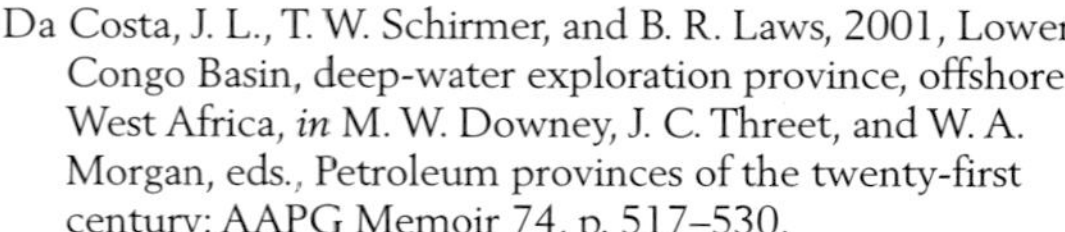
Da Costa, J. L., T. W. Schirmer, and B. R. Laws, 2001, Lower Congo Basin, deep-water exploration province, offshore West Africa, *in* M. W. Downey, J. C. Threet, and W. A. Morgan, eds., Petroleum provinces of the twenty-first century: AAPG Memoir 74, p. 517–530.

Chapter 25

Lower Congo Basin, Deep-water Exploration Province, Offshore West Africa

J. Leite Da Costa
Sonangol, Luanda, Angola

T. W. Schirmer and B. R. Laws
Chevron Overseas Petroleum Inc., Luanda, Angola

ABSTRACT

The Lower Congo Basin lies offshore the west coast of Africa between the Republic of Congo and central Angola. It covers 115,000 km^2 in water depths extending to more than 3500 m. A large number of oil and gas fields occur in the basin (14 billion barrels oil and gas equivalent produced and proved).

Two main producing trends have been discovered. The first, discovered in Block 0 in Cabinda, Angola, more than 30 years ago, produces from Cretaceous reservoirs in water depths less than 200 m.

In the past eight years, 62 exploratory wells have been drilled in the Lower Congo Basin Tertiary deep-water turbidite trend, which is associated with ancient deep-water channel deposition of the Congo River fan. At least 42 new oil fields have been discovered in the deep-water trend, in water depths between 200 and 2000 m. Three-dimensional seismic data are the key to mapping these complex turbidite channel prospects. Large areas, with numerous channels and other trap types, are undrilled in this new province. Structural traps (fault truncations, channel drape over structural highs, and salt domes) dominate. Reservoirs are complex high-quality turbidite sand systems. Source rocks occur in three separate intervals (Cretaceous Bucomazi and Iabe and Tertiary Malembo Formations).

The first field to produce from the turbidite trend is Kuito field in Block 14, Cabinda, Angola, discovered in April 1997 by Chevron (operator) and partners Sonangol, Agip, TotalFinaElf, and Petrogal. It went onstream in December 1999.

INTRODUCTION

A series of conditions makes the deep-water trend in the Lower Congo Basin one of the most competitive and successful exploration plays in the recent history of the oil business:

- recognition of geologic elements conducive to hydrocarbon occurrence in a new area, including an active, world-class petroleum system with multiple source rocks; widely distributed, excellent-quality, deep-water turbidite sandstone reservoir rocks; and a complex tectonic history with salt dynamics, producing multiple structural trends and a myriad of traps across the basin
- advances in technology which enable exploration and development of oil fields in deep-water settings
- support of the government for the petroleum industry in Angola

The Lower Congo Basin (Figure 1) is one of the hottest and most successful exploration plays in the world. In the last several years, announcements of discoveries have taken place in Blocks 14–18 in Angola and Haute Mer in the Republic of Congo. Of the 62 exploration wells drilled, 42 potential discoveries encountered testable hydrocarbons (Figure 2) and as many as 31 discovered commercially developable accumulations, suggesting that risk in the deep-water trend is low. These are remarkable numbers that have driven the excitement and competition in the basin. However, despite aggressive drilling, significant areas of the basin remain undrilled.

Several fields announced to date have been characterized as giant accumulations with recoverable reserves of 500 million barrels (bbl) of oil or greater (Girassol and Dahlia in Block 17, Kuito in Block 14). The mean size of fields discovered to date is at least 200 million bbl of recoverable reserves. Several fields are vertically stacked and grouped geographically around structural traps. These characteristics will enable production facilities to tap multiple accumulations, adding to the value of the facilities through time. Fields in this basin will benefit from recent technical advances in deep-water development technology that enable production in greater than 1000 m of water. As exploration progresses into prospective areas with water depths as great as 3500 m, this basin will provide the impetus for further advances in engineering technology to produce the reserves. The benign ocean environment characteristic of this part of the western coast of Africa will enable technology to push to greater water depths than might be practical in more adverse environments. Once infrastructure is in place in the deep water across the basin, satellite fields and smaller accumulations will become commercially viable.

Most of the basin has been covered with modern 3-D seismic data during the last five years. When combined with the powerful computer hardware and software now available to visualize and interpret the data, this has enabled a high level of understanding of the geology and prospectivity of the deep-water trend.

Because of the competitive nature of the leasing activity in the Lower Congo Basin, sparse information is published about the exploration play or the different fields discovered. The most insight into the Lower Congo Basin trend was provided at the AAPG meeting in Rio de Janeiro, Brazil, in 1998. Several papers were presented by representatives of companies exploring the basin, as well as Sonangol, the Angolan National oil company (see Barrett et al., 1998 [Chevron]; Cole et al., 1998 [BHP];

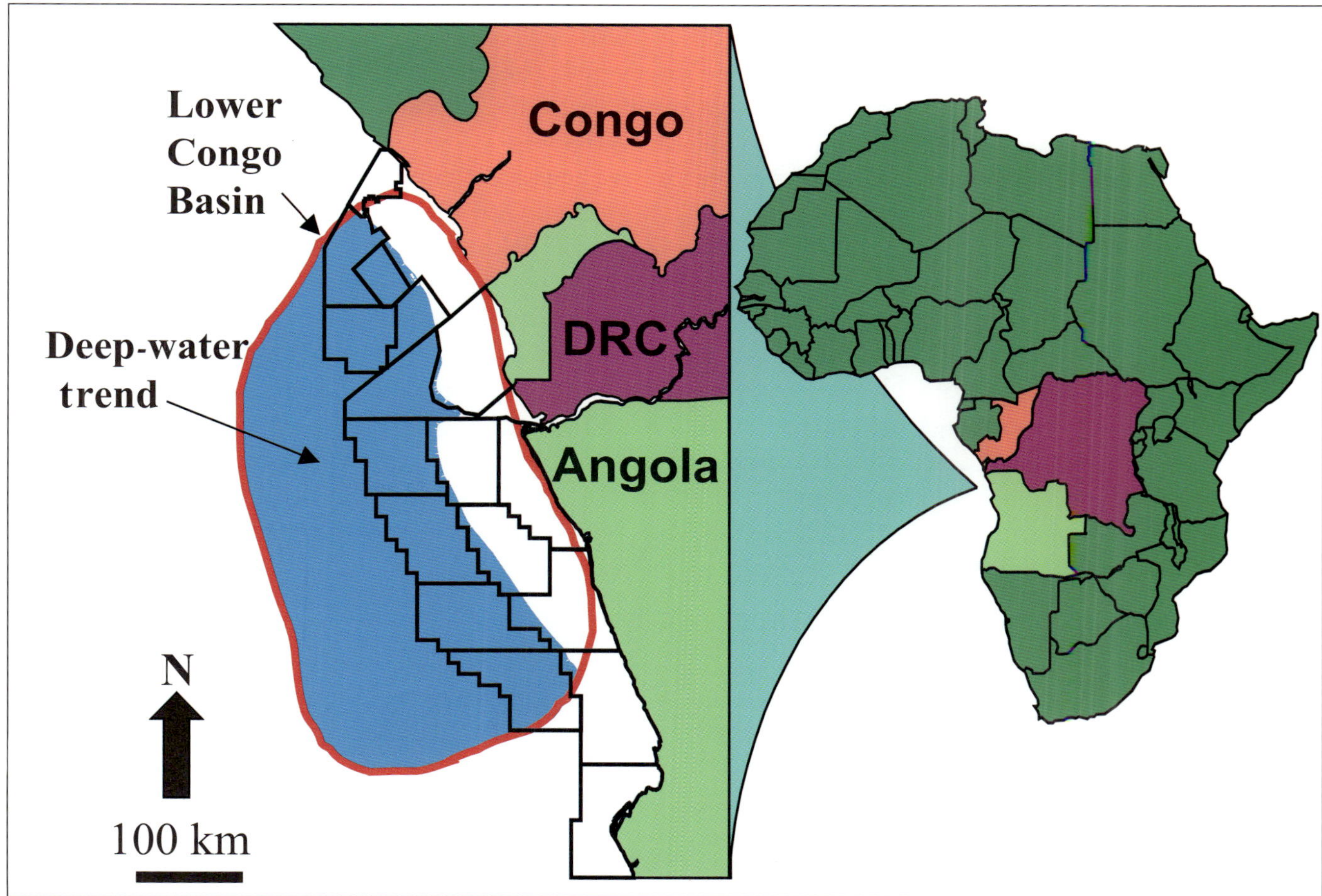

Figure 1. Index map of Africa with location of Lower Congo Basin. Deep-water exploration trend shown in blue. Outlines of offshore exploration blocks are shown in black.

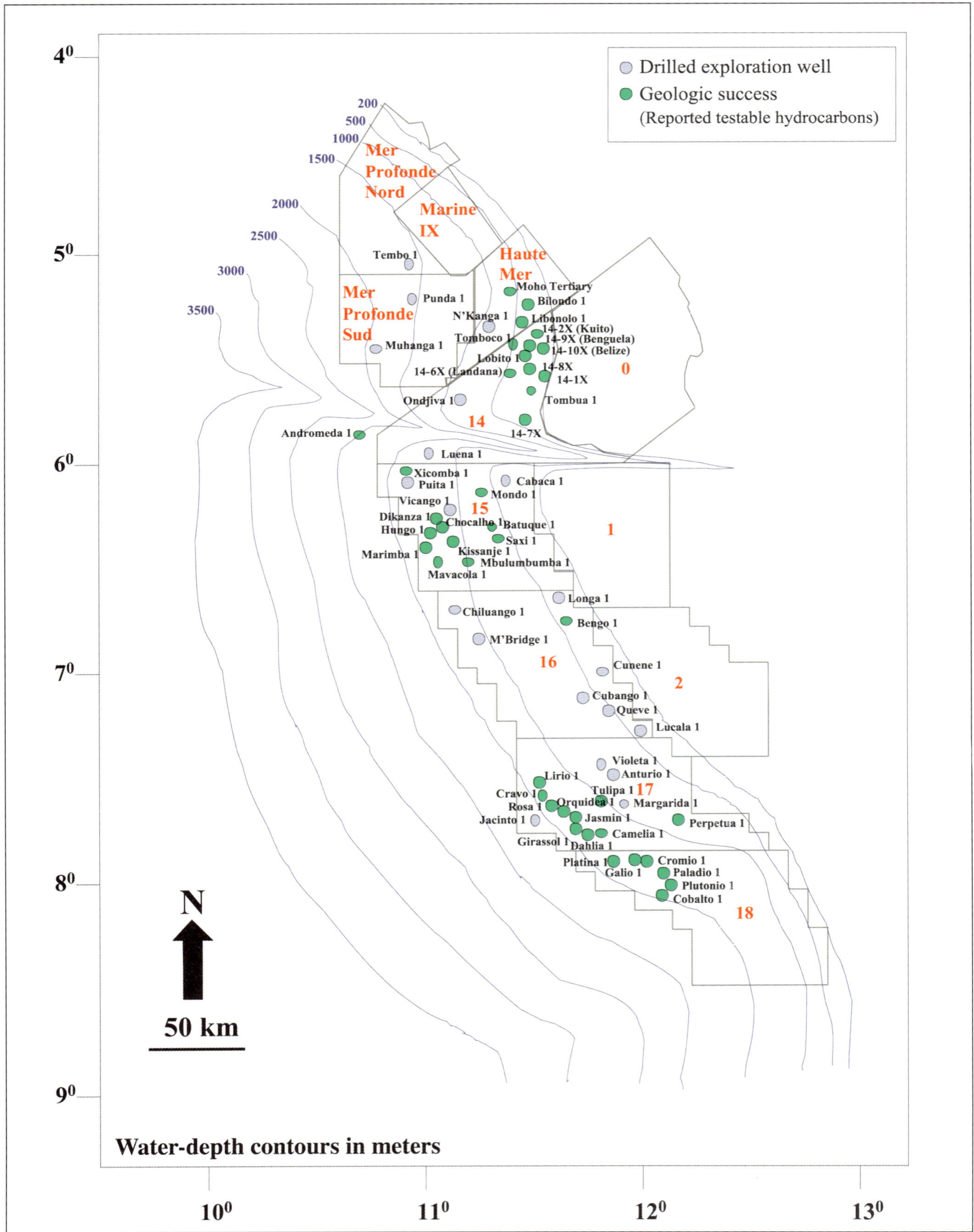

Figure 2. Water-depth contours, exploration blocks, and deep-water Tertiary exploratory wells in Lower Congo Basin.

David, 1998 [Sonangol]; Dominey and White, 1998 [Shell]; Hartman et al., 1998 [Exxon]; Marton and Tari, 1998 [Amoco]; Raillard et al., 1998 [Elf]; and Raposo and Inkollu, 1998 [Sonangol]).

This paper provides an overview of the Lower Congo Basin and a regional perspective of the deep-water turbidite exploration play, using information gleaned from industry activity and Chevron and Sonangol's experience in the basin.

GEOLOGIC SETTING

The Lower Congo Basin lies offshore the west coast of Africa between the Republic of Congo and central Angola. It covers 115,000 km^2 in water depths extending to more than 3500 m. The Lower Congo Basin is in the Congo Basin proper, a subbasin of the Aptian Salt Basin system that occurs along the western coast of Africa (Clifford, 1986).

The history of the Congo Basin can be divided into three main stages (Lehner and De Ruiter, 1977):

1. rift stage, with lacustrine and alluvial deposition in graben and half-graben structural basins (Neocomian to middle Aptian)
2. evaporite deposition stage, developed during the transition from active rifting to thermally induced crustal subsidence (Aptian)
3. subsidence stage, with regional marine deposition and active extension and salt tectonics (Albian to Holocene)

The stratigraphic succession for the Lower Congo Basin is shown in Figure 3.

Basin development along the coast of West Africa began in the late Mesozoic, resulting from rifting and separation of the South American and African continental masses during the opening of the South Atlantic Ocean (Lehner and De Ruiter, 1977). Transverse fracture zones of the Mid-Atlantic Rift segmented the rifted continental crust into a series of subbasins. The Congo Basin, one of those subbasins, extends from the Republic of Congo to central Angola. It lies between the Gabon Basin to the north and the Kwanza Basin to the south. The transition from the Congo Basin to the Kwanza Basin lies along the Ambriz spur, a northeast-southwest trend north of Luanda, the capital of Angola.

Similar basin development occurred on the opposing rifted margin—the eastern coast of South America. The Campos Basin is one of the most significant petroleum provinces where deep-water turbidites contain world-class oil fields (Pettingill, 1998a). The similarities of the Brazil margin basins (particularly the Campos and Santos Basins) to the Lower Congo Basin of Angola and the Republic of Congo are compelling.

In the rifted Congo Basin, the predominant lithologies are lacustrine silt and shale of the Bucomazi Formation (Neocomian to middle Aptian). Active tectonism during deposition of the Bucomazi resulted in changing basin geometry and stratigraphy through time. Anoxic conditions during the end of the Neocomian resulted in the deposition of a widespread organic-rich sequence in the middle of the Bucomazi Formation that is a primary source rock in the basin. Total organic content in this interval is as high as 20% (Dale et al., 1992).

By the end of the Barremian, rift activity on the Mid-Atlantic Ridge had progressed to the west, reducing the level of direct tectonic activity along the African passive margin. Final uplift and erosion produced a regional unconformity that is recognized on seismic and has been correlated in wells. The Chela Formation sandstone and shales were deposited on this unconformity during the early Aptian.

The onset of marine deposition in the Congo Basin is denoted by deposition of the Aptian Loeme Salt Formation. This extensive evaporite sequence consists of halite, potash, and local siliciclastics. An anhydrite layer marks the top of the sequence. The original stratigraphic thickness of the Loeme Salt is difficult to determine. The interval acts as the primary detachment surface for pervasive extensional faulting all along the eastern half of the basin. The interval can be seen on seismic data to be thinned and deformed along the regional detachment surface. The Loeme Salt is also involved in diapir features and complex compressional structures in the deep-water part in the western half of the basin.

Open-marine conditions continued with deposition of the Pinda Formation during the Albian. The Pinda Formation consists of a sequence of continental-shelf siliciclastics and carbonates. After deposition of the lower Pinda limestone and dolomite section, the shelf collapsed westward into a series of fault blocks, bounded by listric normal faults on the updip side. These fault blocks rode on the underlying Loeme Salt sequence, responding to sediment loading and development of regional west dip on the shelf. Pinda deposition continued and was influenced by these moving fault systems, with dramatic growth sections developed against the listric faults.

In the Cenomanian, depositional patterns changed from mostly carbonate-siliciclastic rocks of the Pinda Formation to mainly siliciclastics of the Iabe Formation. Sea level remained relatively constant, with subsidence and deposition in balance. Depositional patterns varied spatially, with nonmarine deposits to the east, nearshore and shoreface environments of the Vermelha reservoir sandstones in a band along present shallow-water areas, and shale and silts in the western part of the basin in present-day deeper water (Dale et al., 1992). The Iabe Formation has shale intervals that contain organic facies, providing an additional source rock to the stratigraphic section.

Subsidence of the passive margin of West Africa in the Congo Basin area continued through the Late Cretaceous–Eocene, with marine deposition of the Landana

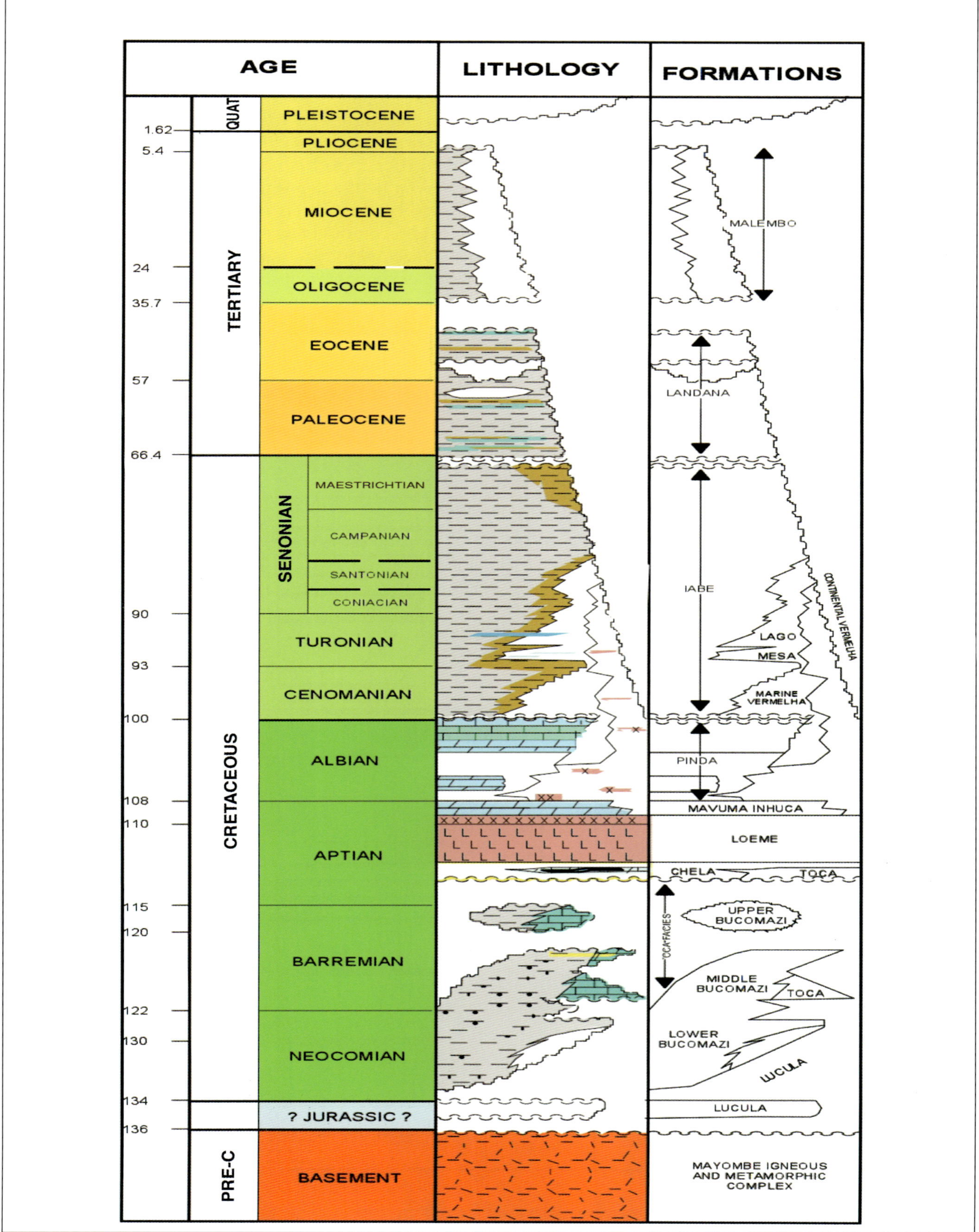

Figure 3. Stratigraphy of Lower Congo Basin based on Cabinda, Angola, well control.

Formation occurring across the basin. The Landana interval is sparsely drilled in the basin and may have turbidite sediments in the deep-water area. A major unconformity at the base of the Oligocene section, caused by a significant lowering of sea level, marks the beginning of a period of marine deposition driven by sea-level changes, which continues to the present. Continued subsidence during this phase provided significant accommodation space for a large volume of Tertiary sediment to be transported into the Congo Basin. The highest sedimentation rate occurred during the Miocene, when as much as 6000 m of section was deposited as the Malembo Formation. The large volume of sediment was deposited from the Congo River, which drains a vast area of south-central Africa.

Miocene to Holocene Turbidite and Petroleum Systems

Throughout the Miocene, the Congo River spread submarine turbidite deposits across the basin, with the vector of sedimentation varying with time within an arc from the southwest to northwest. Deposition of the deep-water turbidite facies occurs in a channel-dominated submarine fan system. Regional assessments by Chevron tie the depositional sequence to the Miocene sea-level curves (Ewins and Minck, 1998) (Figure 4) and provide a methodology to model and predict depositional geometry and sand content. Sand systems were generally deposited at sequence boundaries in cut-and-fill channels which commonly exhibit internal meander-

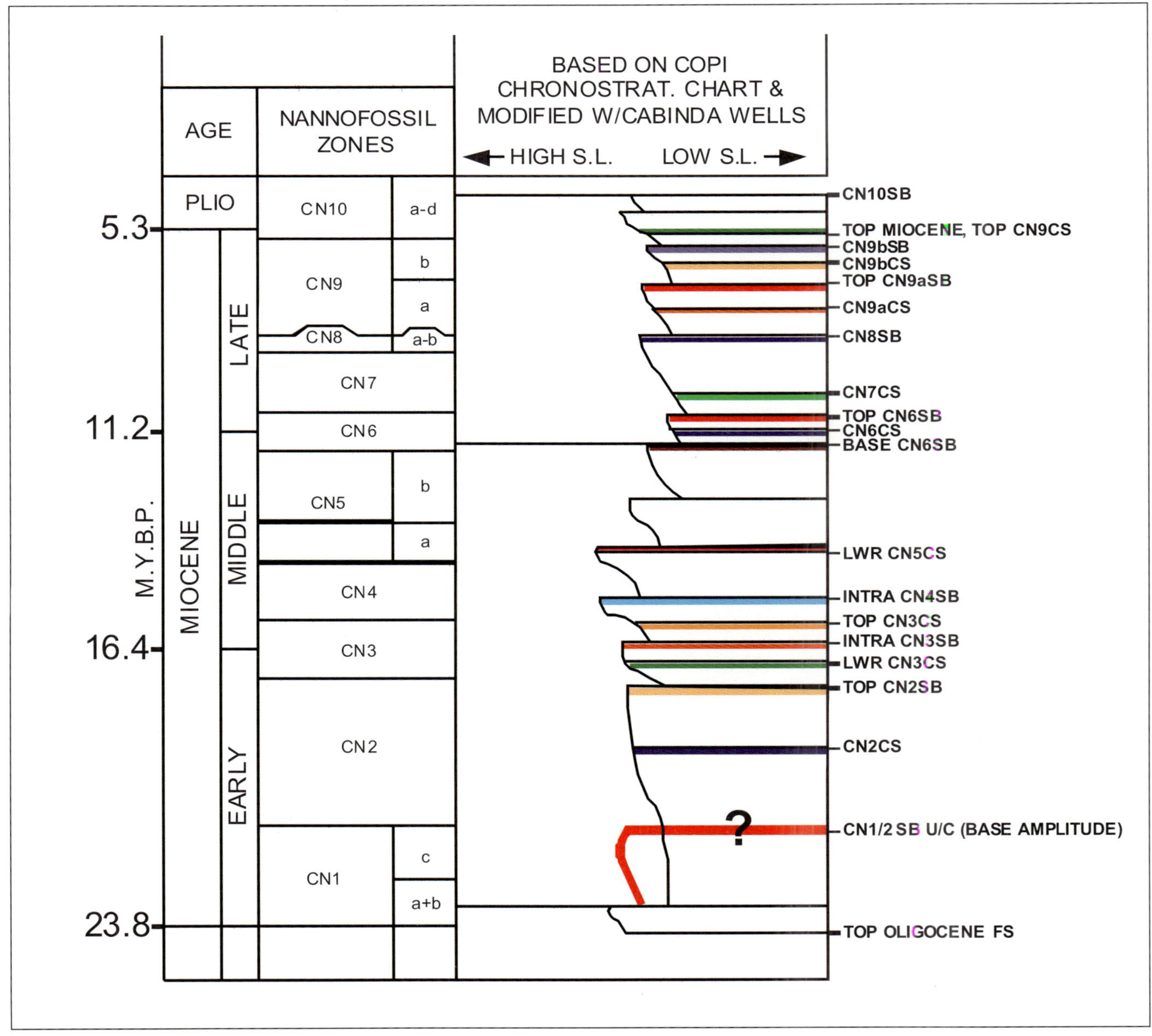

Figure 4. Chronostratigraphy of Miocene sea-level curves built from Cabinda well control. CS denotes condensed sections, SB denotes sequence boundaries, and FS denotes flooding surfaces (Ewins and Minck, 1998).

ing geometries (Figure 5) and have variable net-to-gross sand both vertically and laterally. Some channel systems exhibit more linear channel-levee geometry (Figure 5). Differential compaction is evident on seismic data in higher net-to-gross intervals and where channel meanders are stacked vertically (Figure 6). Outside of channel systems, the predominant sediment is shale, forming vertical and lateral seals to sand-filled channels. The shales of the Malembo Formation also provide a third source rock for the Congo Basin.

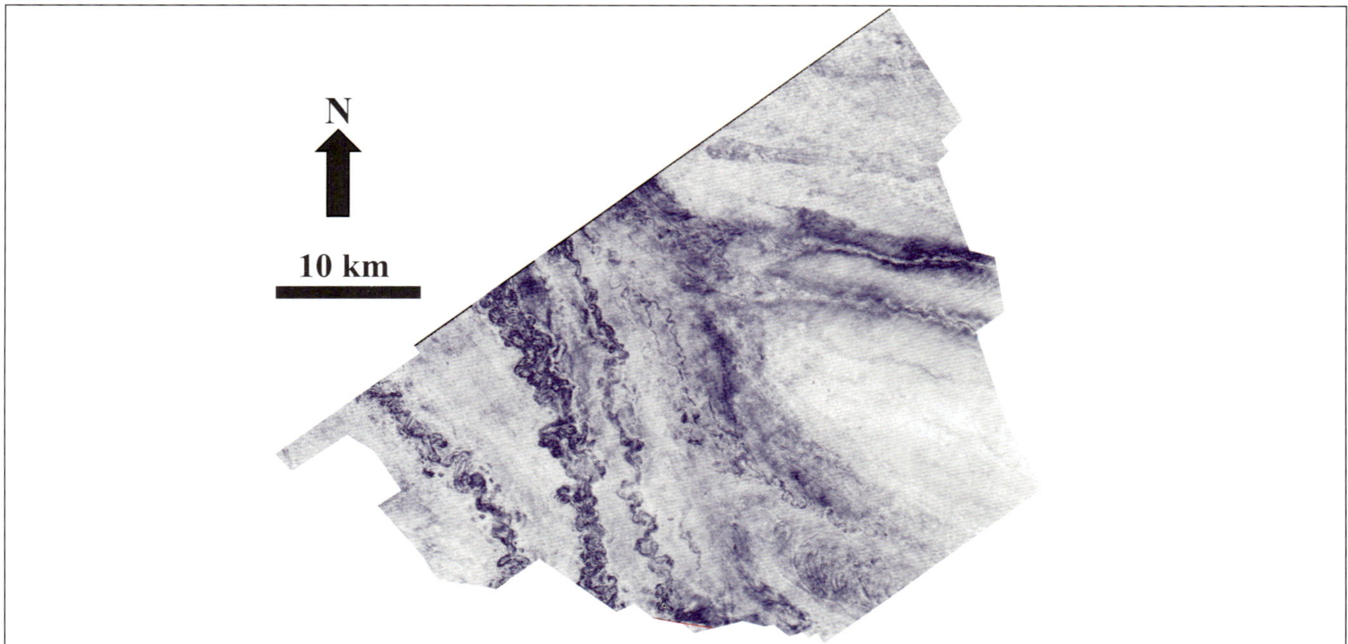

Figure 5. Amplitude extraction of Miocene sequence boundary in Block 14 illustrating depositional geometries. Note meandering channel systems in western part of data and linear channel/levee systems to east.

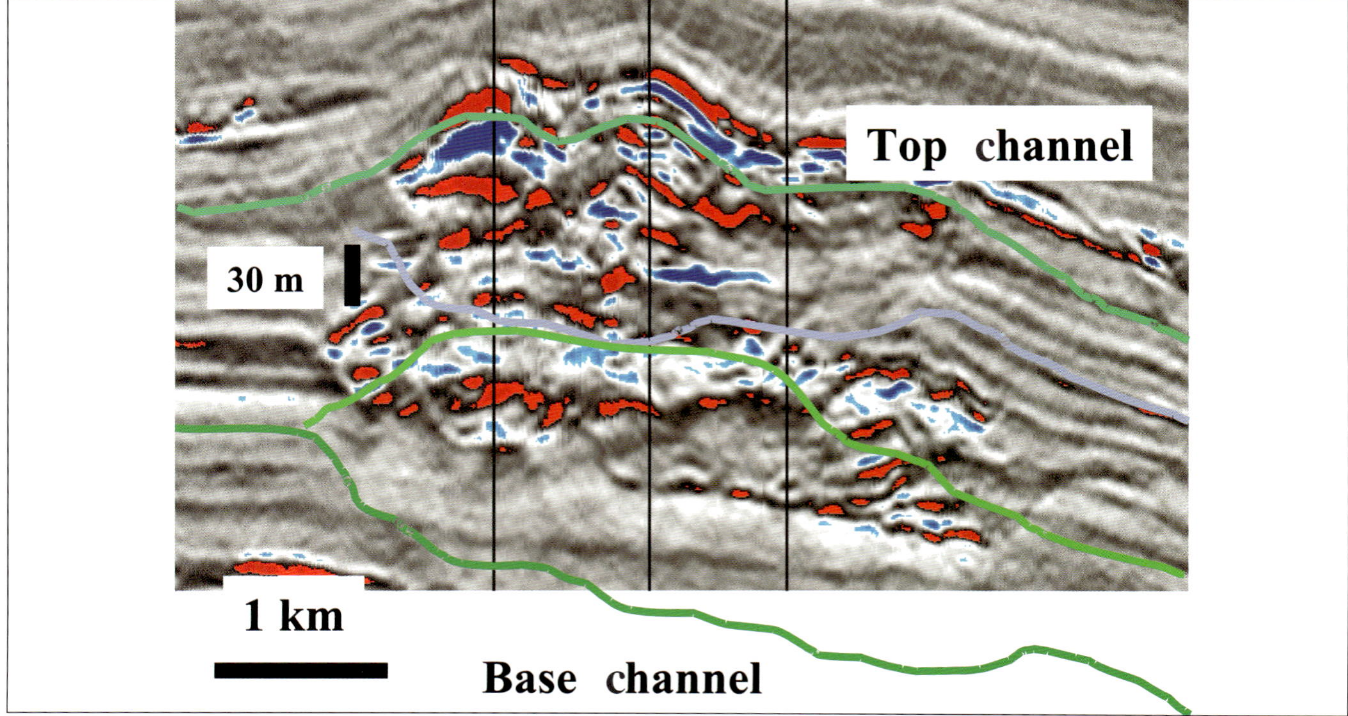

Figure 6. Seismic line through Kuito field. Line is oriented across depositional axis of channel. Note differential compaction and stacking of amplitude packages that indicate sandy turbidite channel systems.

From the late Miocene to the present, the Congo River cut large erosional canyons into the submarine fan during sea-level drops. The modern Congo Canyon provides a good analog for these features. The Congo submarine canyon has been recognized since 1886, when a seafloor cable route was surveyed (Heezen et al., 1964). The canyon runs for more than 250 km offshore (Figure 7), and is actively depositing sediments on the Angolan Abyssal Plain in 4000 m of water (Heezen et al., 1964). Cable breaks as far as 200 km from shore have occurred because of turbidity currents during times of maximum river discharge. The Miocene and Pliocene canyons were filled with pelagic sediment and sandy turbidite flows during subsequent rises in sea level.

The high sedimentation rate and continued subsidence during the Miocene caused renewed basinwide extension on the Aptian Loeme Salt detachment, with major extensional basins developing along the eastern margin of the deep-water province. This extension, which was greatest east of Block 16 and decreased to the north and south, was compensated to the west by compressional structures and salt tectonics in the western

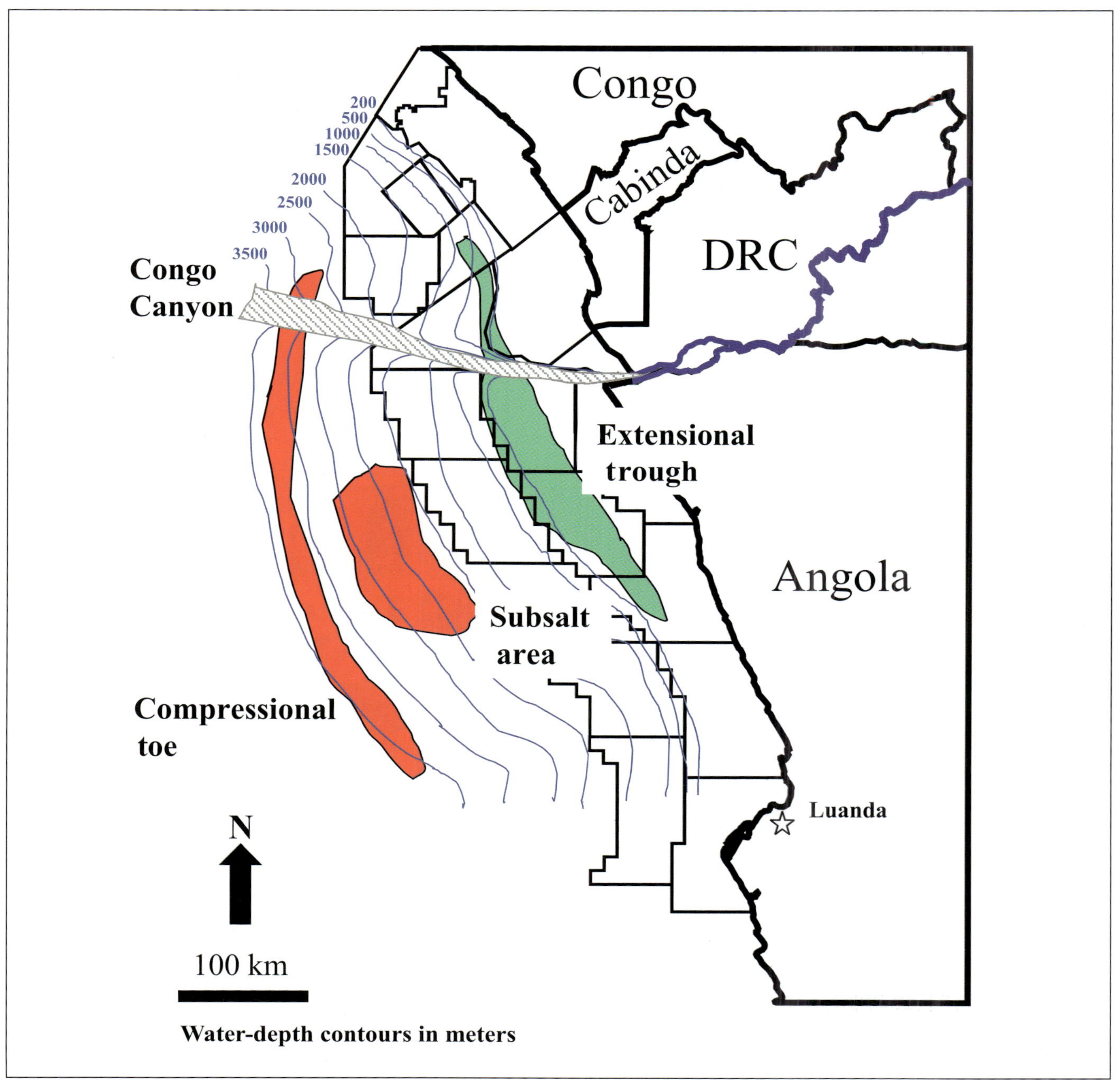

Figure 7. Lower Congo Basin with position of Congo Canyon, Tertiary extensional trough, compressional toe, and subsalt area shown. Water-depth contours and exploration-block outlines are shown.

part of the basin. A compressional wedge, very similar in geometry to the Sigsbee Escarpment in the Gulf of Mexico, exists along the western side of the basin and denotes the leading edge of deformation. A significant area of salt-cored compressional structures, thrust faults, and folds, as well as mobilized and deformed salt diapirs, occurs in the ultradeep water, directly outboard of the greatest amount of extension (Figure 7).

Extension and compression continued throughout the Tertiary and strongly influenced deposition of turbidites. Long-lived depocenters were sites of turbidite deposition through time, resulting in stacking of channel systems. Some growth occurred after deposition, as indicated by channels crossing structure with little or no deflection and channels pierced by salt domes. This relationship of concurrent deposition and structural growth across the basin has enhanced the potential for numerous trap types.

The pervasive extensional and compressional tectonics created a large number of traps in the basin. These traps are associated with rollover into extensional faults; channel truncation against updip faults; compaction closures over deeper Cretaceous-cored structures; and traps over and around salt domes, salt-cored thrusted folds, and turtle structures. In general, structural traps with a component of stratigraphic trapping dominate in the basin. Structural deformation continues today, as indicated by examples of sea-bottom expression of salt domes and surface scarps of active faults. The timing of traps must be studied carefully to avoid the potential for trap breaching caused by active faulting.

The existence of organic-rich source rocks vertically stacked throughout the stratigraphic section in the Cretaceous (Bucomazi and Iabe Formations) and the Oligocene-Miocene (Malembo Formation) provides the world-class petroleum system that has generated the large volume of hydrocarbons discovered and produced in the basin (14 billion bbl oil and gas equivalent produced and proved). The network of faults that occurs in the basin facilitates migration of hydrocarbons up through the section from deeper mature source rocks. Migration from Bucomazi source rocks (presalt) is enabled by windows in the regional Aptian Salt resulting from thinning and faulting along the regional detachment. Within the Malembo section, direct migration from source-rock shales to reservoir sands can occur.

Data gathered on discoveries made in the deep-water trend indicate that oil quality varies within the basin (Fig-

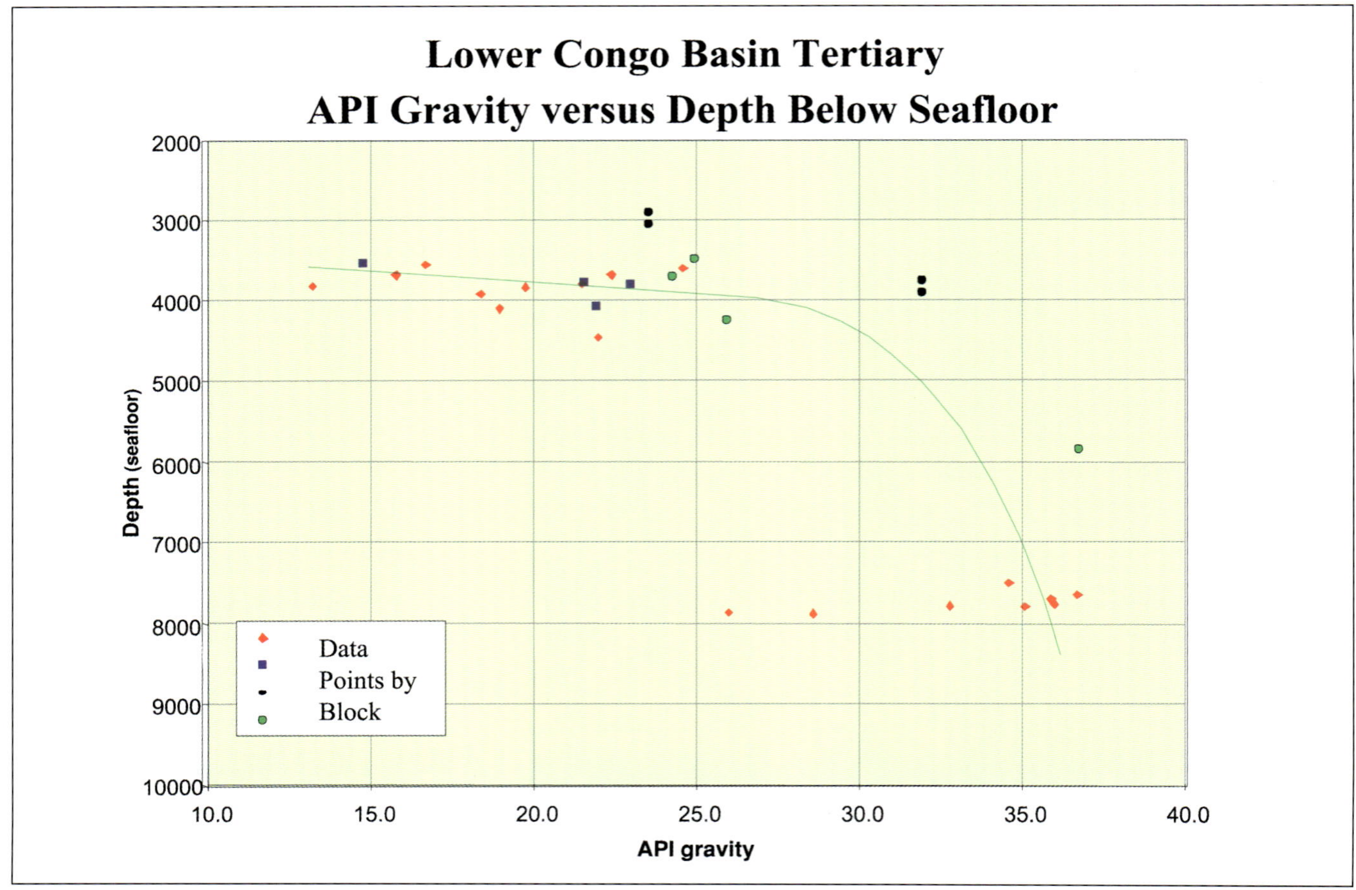

Figure 8. The relationship of oil API gravity versus depth below seabed for Lower Congo Basin Tertiary wells. Note differentiation of oil quality with depth, with deeper reservoirs containing higher-gravity oils. Low-gravity oil in shallow reservoirs may be caused by low-maturity source rocks as well as by biodegradation.

ure 8). Generally, discoveries made in sand systems buried deeper below the mudline contain higher-quality oil (higher API gravity) than shallow traps. This relationship may result from original maturity of the migrated oils, but it is also related to the effects of temperature on biodegradation of shallow trapped oils. It is also possible that some fields have received multiple charges of hydrocarbons and have oils from more than one source rock.

If we look at the history of the basin, we can understand how this occurs. The vertical distribution of source rocks through the rock column means that different source intervals generated hydrocarbons through time as burial progressed. For example, the deep Tertiary extensional basins will have more Malembo source rocks in the oil window, forming local generating systems with the potential for higher-quality oils, but the deeper Iabe and Bucomazi source rocks may be pushed through the oil window into gas or condensate generation. The middle and upper Malembo may not be buried deeply enough in some portions of the basin to put these source rocks in the oil window. Therefore, any traps in these intervals must be charged by more complex migration systems along faults from deeper source rocks. The relationship of burial history, source-rock maturity, and migration is critical because of the impact that oil quality has on the value of the crude oil produced. With most of the Lower Congo Basin occurring in technically challenging water depths, the value of the crude oil will have a significant impact on the economics of fields.

EXPLORATION HISTORY

A large number of oil and gas fields occur in the Congo Basin. Two main productive trends have been discovered (Figure 9). The first, the Cretaceous presalt and postsalt trend, stretches 350 km from the Republic of Congo to central Angola along the coastline in water depths less than 200 m. The second occurs in the Lower Congo Basin deep-water Tertiary turbidite trend and stretches 300 km from the Haute Mer block in the Republic of Congo to Block 18 in central Angola in water depths greater than 200 m. Although the producing trend from Cretaceous reservoirs has been explored for more than 30 years, additional exploratory potential exists with acquisition of new 3-D seismic data. The Tertiary deep-water turbidite trend is not mature, with three-quarters of the prospective area of the trend still undrilled. Further drilling will probably expand the Tertiary deep-water turbidite trend significantly.

Exploration began in the Congo Basin offshore province when Gulf Oil Corporation, operating the Cabinda concession under the name Cabinda Gulf Oil Company (Cabgoc), drilled an exploratory well in 1966 based on marine geophysical data. The well was a discovery (Limba field) in the Cenomanian Vermelha sandstone (postsalt section). The first presalt discovery was Malongo North field in 1967. Cabgoc exploration for large presalt fields continued through the late 1960s and early 1970s, when the giant Malongo West field (1970) and the giant Takula field (1971) were discovered (Dale et al., 1992). In 1979, drilling in Takula field resulted in the discovery of the Pinda (postsalt) accumulation. Exploration has proceeded on the trend through the 1980s and 1990s, with 3-D seismic data driving activity. Presently, the presalt and postsalt Cretaceous reservoirs account for 500,000 bbl oil/day (BOPD) in the Block 0 Cabinda Concession, with additional production along trend in the Republic of Congo to the northwest and the Democratic Republic of Congo and Angola to the southeast.

Interest in the Lower Congo Basin deep-water trend simmered during the early 1990s as companies picked up blocks in water depths of 200–1500 m (Blocks 14–18 in Angola and Haute Mer in the Republic of Congo). Expectations were high because the basin was already a proven petroleum province in the shallow-water area. Whether this proven petroleum system extended from shallow water to the deep-water area and whether the Congo River had delivered enough sand to the submarine fan to produce significant reservoir potential were the key questions waiting to be answered by drilling.

The discovery of Girassol field in Block 17 by Elf and partners in April 1996 brought interest in the trend to a high level. The Girassol accumulation was reported to be a giant accumulation of 32° API oil. Activity increased in 1997 with the discovery of Kuito and Landana fields in Block 14 by Chevron and partners, and Dahlia and Rosa fields in Block 17. An expansion of exploration in the basin in 1998 saw four new discoveries in Block 15 (Kissanje, Marimba, Hungo, and Dikanza fields) by Exxon and partners, as well as the discovery of Benguela and Belize fields in Block 14. Exploration drilling has continued at a rapid pace, with 29 additional exploration wells drilled from 1999 to 2001 (Figure 2).

Since 1994, 62 exploration wells have been drilled in the Lower Congo Basin Tertiary deep-water trend in water depths between 200 and 1500 m. Technical risk in the basin appears to be very low. Of these wells, 42 may be geologic successes (they encountered testable hydrocarbons), and as many as 31 are commercially developable accumulations. Therefore, geologic risk for exploratory wells (the risk to encounter testable hydrocarbons) has been 1:1.5, and commercial risk (the risk that a field will contain enough hydrocarbons to be developed) is 1:2.0. Several fields announced to date have been characterized as giant accumulations with recoverable reserves of 500 million bbl of oil or greater (Girassol and Dahlia fields in Block 17 and Kuito field in Block 14). The mean size of fields discovered to date is at least 200 million bbl of recoverable reserves.

The first field to produce from the turbidite trend is Kuito field in Block 14, Cabinda, Angola, discovered in April 1997 by Chevron (operator) and partners Sonangol, Agip, TotalFinaElf, and Petrogal. Oil gravity ranges from 19° to 24° API. The field lies in about 350 m of water. Production is through subsea wells to a central floating production storage offloading unit (FPSO), with export directly from offshore. The field began production in December 1999. Benguela and Belize fields in Block 14 lie only 3–5 km south of Kuito field. Benguela and Belize fields are being studied for possible development in 2002.

Girassol field in Block 17 lies in about 1400 m of water. Development plans apparently are focusing on subsea wells tied back to an FPSO, with production building to a plateau rate of 200,000 BOPD.

As the larger fields are discovered and developed in the Lower Congo Basin deep-water trend, the growing infrastructure will enable exploitation of midsize and smaller accumulations that cannot economically carry an initial exploratory well in these water depths. Much will be learned about the long-term performance of these turbidite sands as different fields come onstream across the basin. The geographic grouping of channels through the stratigraphic section around major trapping structures creates a full distribution of potential field sizes, and the existence of infrastructure will allow companies to tap into a broad inventory of prospects to keep facilities running at full capacity.

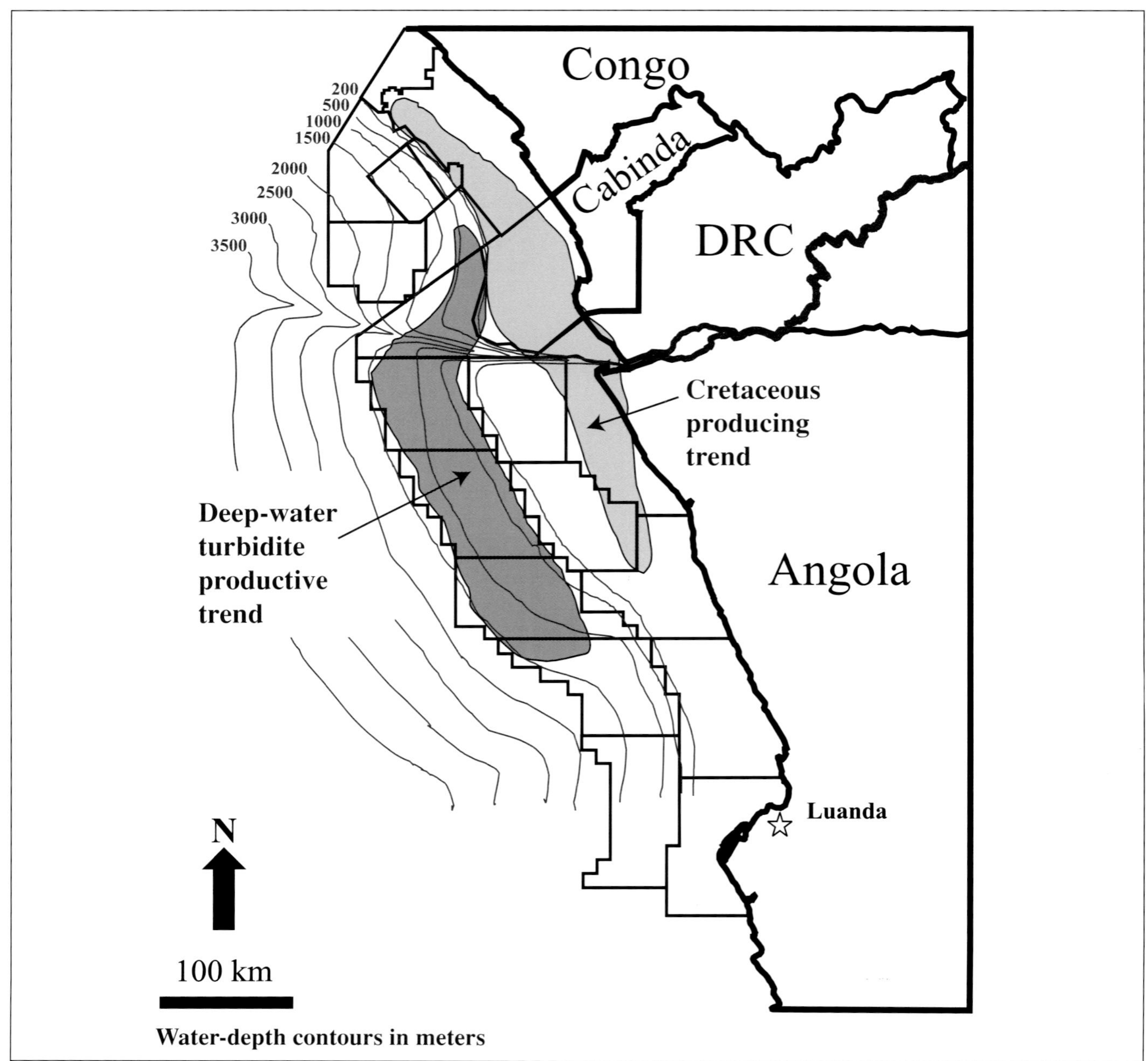

Figure 9. Position of Cretaceous (postsalt and presalt) producing trend and present area of Tertiary turbidite discoveries in the Lower Congo Basin. Water-depth contours and exploration-block outlines are also shown.

FUTURE POTENTIAL

In the primary deep-water trend explored in the last eight years. 62 exploratory wells occur in an area of approximately 30,000 km^2, with a well density of 1 well/484 km^2. Three-dimensional seismic data are the key to mapping the complex turbidite channel prospects in the Lower Congo Basin. Because this part of the trend is essentially covered by 3-D surveys, it is likely that continued drilling within this area will discover new fields.

Large areas of this new province, with numerous channels and trap features, remain undrilled (Figure 10). Approximately 60,000 km^2 of highly prospective acreage with play elements consistent with the successful Tertiary prospects in the present trend are completely unex-

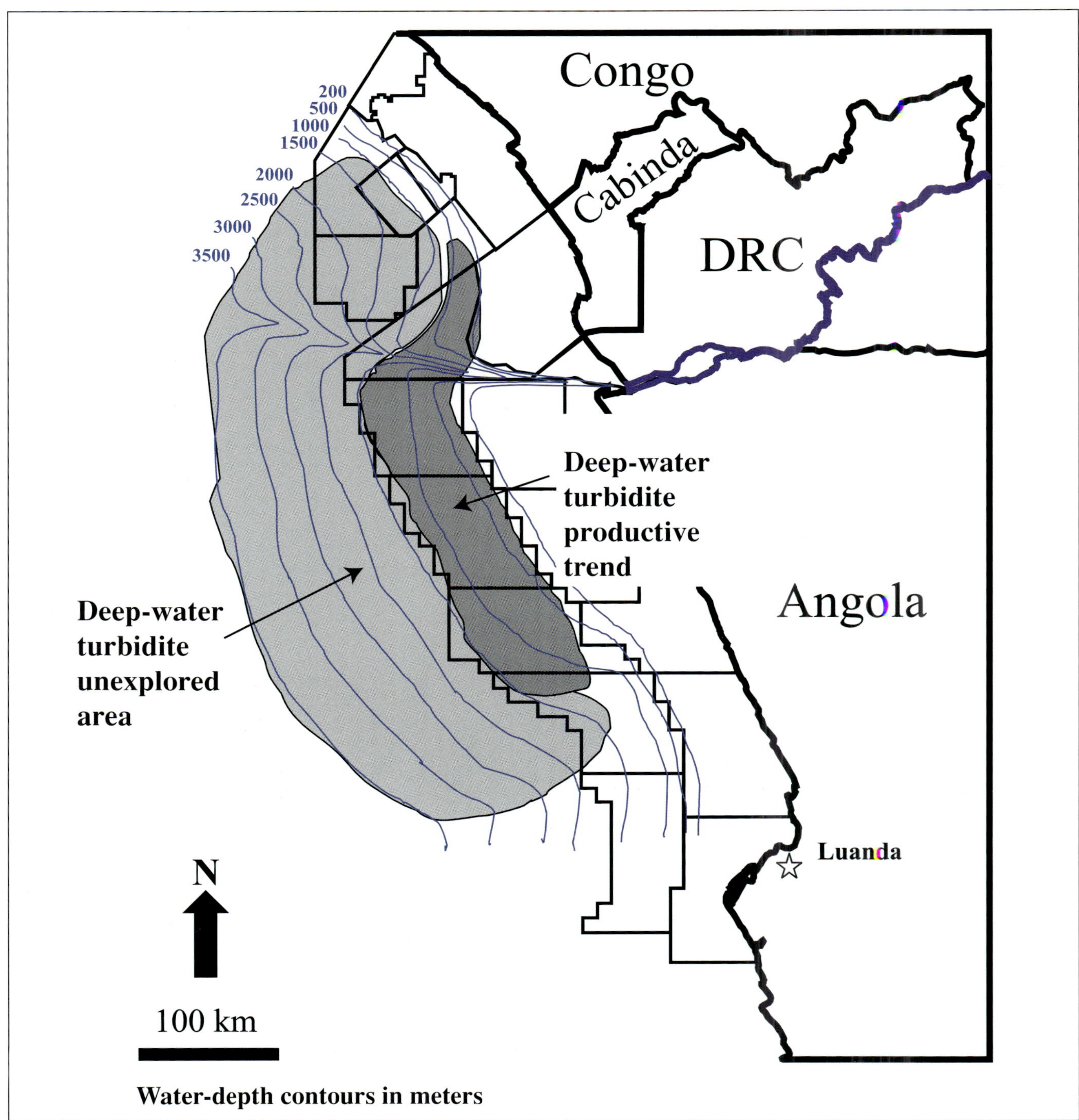

Figure 10. Location of present area of Tertiary turbidite discoveries and remaining unexplored area in the Lower Congo Basin. Within the area of present deep-water discoveries, well density is only 1 well/484 km^2, which indicates that a significant area remains to be explored. Water-depth contours and exploration-block outlines are also shown.

plored. This unexplored area lies outboard of Blocks 14–18 in Angola and Haute Mer in the Republic of Congo in water depths of 1500 to 3500 m.

It is apparent that present exploration efforts in the Tertiary turbidite trend are finding oil fields with combined structural and stratigraphic trap components. Large traps associated with drape over deeper structures, rollover associated with extensional faults, fault truncation of channels, and salt-related structures are the most obvious features. Faults appear to be an important trapping mechanism and may cause compartmentalization of fields. Stratigraphic trap components are commonly caused by lateral pinch-out of sand facies at the margin of channel deposition. Because the bulk of the rock column in the Congo submarine fan is mud/shale, with sands deposited mostly in channel systems, the lateral seals to the system should be good. In areas where channel density is high, the lateral seals may be compromised because of leakage through overbank sands and crosscutting channel systems. The lowest risks for lateral seal are within isolated channel features cut into shale.

An assessment of turbidite discoveries by Pettingill (1998a, b) shows that stratigraphic trapping usually contributes significantly to a basin's ultimate reserves, although this potential may not be seen early in a basin's exploration history. Given the Lower Congo Basin success rate to date in the combined structural/stratigraphic traps and the large area still unexplored with the same characteristics, it is likely that near-term exploration will focus on these types of features. Sometime in the future, the subtle and pure stratigraphic traps will likely be understood and will contribute to the prospectivity of the basin.

The existence of significant areas of subsalt will create a challenge to those companies prospecting for turbidite plays. The decrease in quality of the seismic imaging and difficulty in proper spatial positioning of prospects beneath the salt overhangs will raise the geologic risk of wells. However, the existence of the petroleum system and turbidite channels bordering these areas to the east means the subsalt area is still highly prospective. Companies with experience in the Gulf of Mexico subsalt play can leverage that experience to better handle the seismic-imaging challenges, to more accurately locate wells, and to evaluate predrill risk appropriately.

The Lower Congo Basin is a world-class petroleum province with giant fields discovered and significant areas still unexplored. It is probable that additional giant fields await discovery and that the basin has a full distribution of fields of various sizes.

CONCLUSIONS

The Lower Congo Basin deep-water turbidite trend, offshore Angola and the Republic of Congo, is one of the most successful and competitive new exploration plays in the world. Several giant discoveries have been announced in the last several years, and the success rate of exploratory wells is very high, indicating a basin with low geologic risk. Significant areas of the basin remain unexplored in water depths as great as 3500 m. The challenge to companies working the trend is to evaluate the distribution of prospects and build up enough of a reserve base to drive technology advances to produce oil at a profit from these water depths.

ACKNOWLEDGMENTS

This overview of the Lower Congo Basin deep-water turbidite play is based on data, interpretations, and concepts generated by Chevron Overseas Petroleum, Inc., and Sonangol staff in the last four years. The authors thank Sonangol and Chevron for permitting publication.

The authors also thank the Block 14 partners (Sonangol, Agip, TotalFinaElf, and Petrogal) for permission to include examples from Block 14 in this paper.

REFERENCES CITED

Barrett, M., A. Ruiter, T. Schirmer, and K. Pedro, 1998, Deepwater turbidite channel exploration plays, Block 14, Cabinda, Angola (abs.): AAPG Bulletin, v. 82, p. 1889.

Clifford, A. C., 1986, African oil—Past, present, and future, *in* M. T. Halbouty, ed., Future petroleum provinces of the world: AAPG Memoir 40, p. 339–372.

Cole, G. A., A. Yu, and J. Smith, 1998, Predicting oil charge types and quality in the deepwater offshore Lower Congo Basin, Angola (abs.): AAPG Bulletin, v. 82, 10, p. 1902.

Dale, C. T., J. R. Lopes, and S. Abilio, 1992, Takula oil field and the greater Takula area, Cabinda, Angola, *in* M. T. Halbouty, ed., Giant oil fields of the decade 1978–1988: AAPG Memoir 54, p. 197–215.

David, J., 1998, Evolving upstream opportunities in Angola (abs.): AAPG Bulletin, v. 82, p. 1907.

Dominey, J. R., and S. White, 1998, Salt tectonics and sedimentation—An integrated interpretation ultra deep water area, Lower Congo Basin, Offshore, Angola (abs.): AAPG Bulletin, v. 82, p. 1910.

Ewins, N., and R. Minck, 1998, Tertiary regional study: Chevron internal report.

Hartman, D. A., W. A. Swanson, P. R. Smith, F. J. Goulding, and C. A. Kelly, 1998, Structural development of the continental margin of Congo and northern Angola (abs.): AAPG Bulletin, v. 82, p. 1923.

Heezen, B. C., R. J. Menzies, E. D. Schneider, W. M. Ewing, and N. C. L. Granelli, 1964, Congo submarine canyon: AAPG Bulletin, v. 48, p. 1126–1149.

Lehner, P., and P. A. C. De Ruiter, 1977, Structural history of the Atlantic margin of Africa: AAPG Bulletin, v. 61, p. 961–981.

Marton, G., and G. Tari, 1998, Evolution of salt-related structures and their impact on the post-salt petroleum systems of the Lower Congo Basin, offshore Angola (abs.): AAPG Bulletin, v. 82, p. 1939.

Pettingill, H. S., 1998a, Lessons learned from 43 turbidite giant fields: Oil & Gas Journal, October 12, p. 93–95.

Pettingill, H. S., 1998b, Turbidite plays' immaturity means big potential remains: Oil & Gas Journal, October 5, p. 106–112.

Raillard, S., J. J. Biteau, P. Alix, and C. Chevalier, 1998, Lower Congo Tertiary Basin—Offshore West Africa structural zonation and evolution (abs.): AAPG Bulletin, v. 82, p. 1954–1955.

Raposo, A., and M. Inkollu, 1998, Tertiary reservoirs in Congo-Kwanza-Namibe Basins (abs.): AAPG Bulletin, v. 82, p. 1956.

Cochran, M. D., and L. E. Petersen, 2001, Hydrocarbon exploration in the Berkine Basin, Grand Erg Oriental, Algeria, *in* M. W. Downey, J. C. Threet, and W. A. Morgan, eds., Petroleum provinces of the twenty-first century: AAPG Memoir 74, p. 531–557.

Chapter 26

Hydrocarbon Exploration in the Berkine Basin, Grand Erg Oriental, Algeria

Michael D. Cochran and Lee E. Petersen
Anadarko Petroleum Corporation, Houston, Texas, U.S.A.

ABSTRACT

Hydrocarbon exploration during the 1990s in the Berkine Basin of eastern Algeria resulted in discovery of more than two billion barrels (bbl) of recoverable oil. Although exploration focused on Triassic quartz sandstones, flow rates of more than 5000 bbl condensate per day (BCPD) and 50 million cubic feet of gas per day (MMCFGD) have been recorded from Lower Devonian through Carboniferous reservoirs. Uppermost Devonian and Carboniferous hydrocarbon-charged reservoirs first were discovered in the basin during the early 1990s phase of exploration.

Understanding the evolution of Anadarko's exploration model is instructive in defining remaining hydrocarbon potential of the Berkine Basin. The exploration model evolved from a single Silurian hydrocarbon source filling a single primary Lower Devonian reservoir to a dual Lower Silurian and Upper Devonian hydrocarbon source with a complex maturation and trap-filling history of multiple reservoirs. Migration-path modeling shifted the primary area of exploration focus from western updip margins, where successively older Paleozoic strata are truncated by the Hercynian unconformity, to a more downdip position relative to the present-day structural axis. This shift takes advantage of a more complete Paleozoic section and use of regional fault systems as sites of vertical migration and entrapment. Employment of geophysical data as the primary tool for definition of prospects evolved from reprocessing of existing 2-D seismic data to acquisition of new 120-fold and 240-fold 2-D seismic grids. Three-D seismic surveying is now employed as the primary tool for field development.

Existence of high-quality source rocks and reservoirs, a Liassic evaporite "superseal," favorable tectonic history with respect to hydrocarbon generation and expulsion, and high-quality hydrocarbon migration carrier beds lead us to conclude that the primary exploration risk for additional reserves involves seismic imaging of reservoirs and traps.

Comparison of generated hydrocarbons to discovered volumes of hydrocarbons suggests that the Berkine Basin is immature with respect to hydrocarbon exploration, and significant undiscovered fields are feasible. Subcrop traps in Paleozoic reservoirs below the Hercynian unconformity should exist. Lochkovian, Strunian, Viséan, and Carnian siliciclastic reservoirs will continue to be important exploration reservoir objectives.

INTRODUCTION

Discoveries of giant oil and gas fields are significant events to the petroleum geoscientist. Successful drilling of a giant hydrocarbon accumulation validates the petroleum geoscientist's work and ensures that the basin or province will become one of the world's major producing areas. Prediction of such areas in advance of the drill bit, however, is full of risk. In 1977, the *Oil & Gas Journal* published a special issue, "PETROLEUM 2000," to commemorate its 75th anniversary. The issue presented a series of papers which attempted to visualize future opportunities in the petroleum industry. In a review of future onshore petroleum opportunities in that issue, Meyerhoff (1977) predicted that traps remained to be discovered in Algeria, Libya, and Egypt. Although he acknowledged the possibility of several undiscovered giant fields in northern Africa, he suggested that accumulations would be smaller than those producing in 1977 and only marginally commercial. With renewed exploratory drilling in Algeria in the 1990s, Algeria ranked first in the world in 1994, with 1.1 billion barrels (bbl) of oil discovered (*Oil & Gas Journal*, 1997). The majority of these new reserve additions was the result of exploration in the Berkine Basin portion of the greater Ghadames Basin. Exploration of the Berkine Basin in the 1990s discovered more than 2 billion bbl of recoverable oil. Although individual field reserves remain proprietary, Macgregor (1998) indicates that at least three of the Berkine Basin discoveries have reserves exceeding 250 million bbl oil equivalent (BOE), with two of them having 500 million to 1000 million BOE. The three discoveries—Hassi Berkine North, Hassi Berkine South, and El Merk—were the result of an intensive exploration effort by Anadarko Petroleum Corporation and its partners Sonatrach Oil and Gas, Lasmo Oil, and Maersk Oil and Gas. One additional field discovered by the Anadarko group in 1994, Ourhoud (the field formerly known as Qoubba), has reserves of more than 1.1 billion bbl oil (*Oil & Gas Journal*, 1999). Thus, an understanding of the Anadarko group's exploratory process and its evolution provides insight into the remaining undiscovered hydrocarbon potential of the Berkine Basin.

LOCATION

The Berkine Basin, located in extreme eastern Algeria (Figure 1), is the westernmost extension of the larger Ghadames Basin. The Ghadames Basin, a pericratonic basin with a polyphase tectonic history, covers approximately 350,000 km^2 in eastern Algeria and parts of Tunisia and Libya, and contains more than 6000 m of Paleozoic and Mesozoic siliciclastic-dominated sediments (Echikh, 1998). Located in the Grand Erg Oriental portion of the Sahara Desert, the Berkine Basin is characterized on the surface by extensive areas of eolian sand dunes more than 300 m in height and intervening areas of gravel pavement. Petroleum infrastructure, roads, and centers of population are developed only along the northern, western, and southern portions of the Grand Erg Oriental. Within the subsurface, the Berkine Basin is bounded structurally on the northeast by the Damah zone and on the northwest and west by the Hassi Messaoud Ridge. It is separated from the Illizi Basin on the south by the Mole D'Ahara. Surrounding the Berkine Basin and on its western flank is a series of giant oil and gas fields, including Rhourde El Baguel, Gassi Touil, Tin Fouye-Tabankort, and Alrar (Figure 1). The supergiant Hassi Messaoud oil field is located approximately 160 km northwest of the Berkine Basin. El Borma field, straddling the Algerian-Tunisian border, is located in the northeast portion of the Berkine Basin.

BASIN ARCHITECTURE AND DISTRIBUTION OF OIL AND GAS

With renewed interest in exploration of the Berkine Basin in the 1990s, various aspects of Berkine Basin petroleum geology were summarized in a series of papers on the basin's structural evolution (Gauthier et al., 1995; Sonatrach, 1995; Boote et al., 1998; Echikh, 1998; Guiraud, 1998) and petroleum systems (Daniels and Emme, 1995; Boote et al., 1998; Macgregor, 1998). Echikh (1998) provides a comprehensive synthesis of the basin's petroleum geology. A detailed description of the Phanerozoic sediment package, development and geographic extent of hydrocarbon-charged reservoirs, and their chronostratigraphy are incompletely known. Sonatrach (1995), Echikh (1998), and Fekirine and Abdallah (1998) provide summaries of the Phanerozoic sediment package and its reservoir potential. Details on many potential and proven hydrocarbon-bearing reservoirs remain to be established and will require additional exploratory drilling. A generalized stratigraphic column for the Berkine Basin is shown in Figure 2.

Figure 3 illustrates the distribution of oil and gas accumulations in the Berkine Basin and portions of the bounding Hassi Messaoud Ridge and Mole D'Ahara/Illizi Basin. Known hydrocarbon accumulations are presently restricted to the west and north portions of the basin and are closely related to truncation edges of various Paleozoic strata. Structurally, the Berkine Basin is a part of the Central Saharan platform, which has undergone a complex and polyphase evolution and has been influenced by Pan-African brittle basement fracturing. This portion of the Central Saharan platform is the site of a series of structural arches and continental highs (moles) which have controlled sedimentation and thus deposition of hydrocarbon source rocks, reservoirs, and seals. Tectonic features have been affected by nine major tectonic phases, ranging in age from Precambrian to Tertiary. Pre-Hercynian tectonic activity in the Early Ordovician (Taconic tectonic phase) and Late Silurian–Early Devonian (Caledonian tectonic phase) caused uplift and erosion

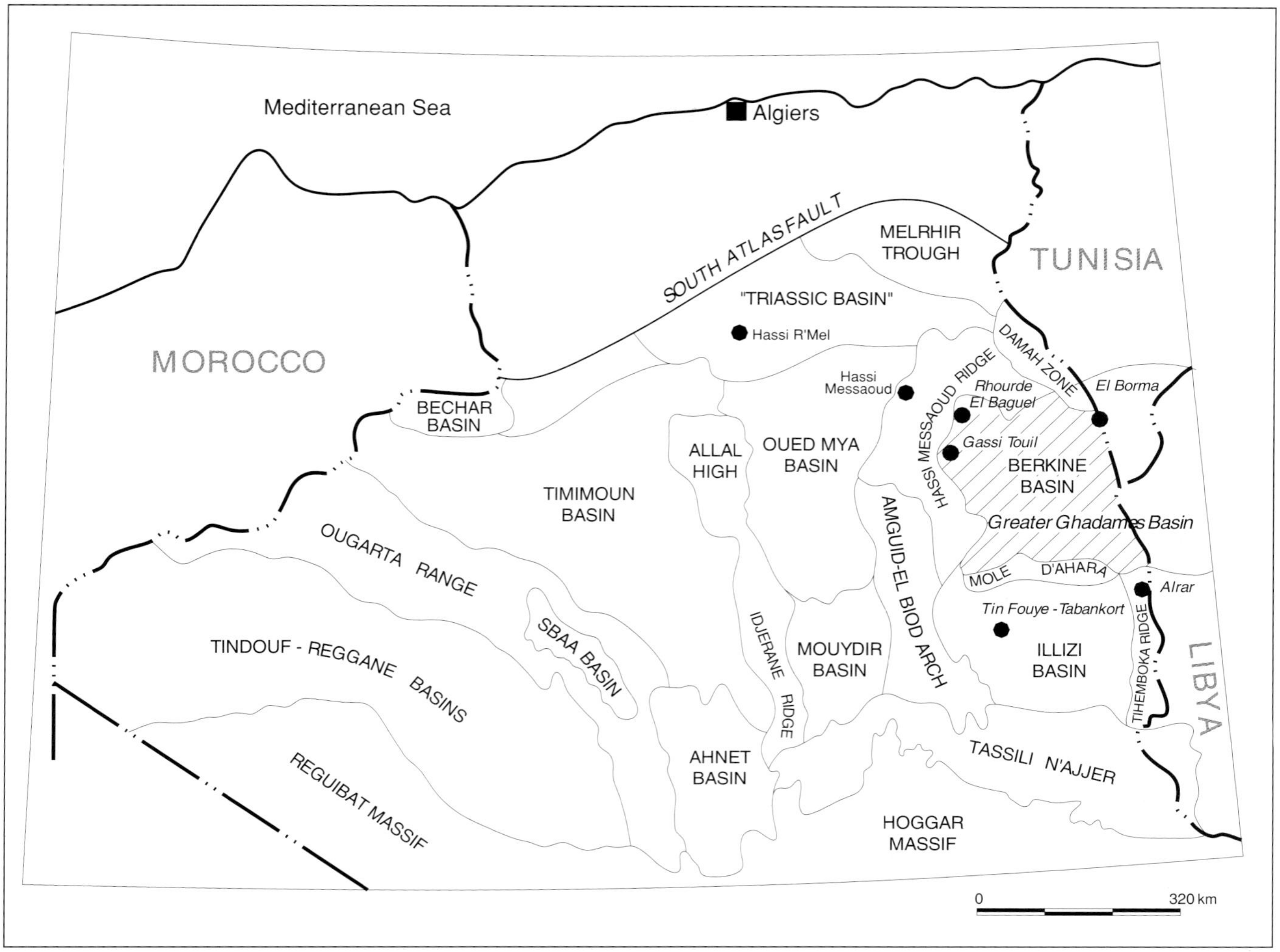

Figure 1. Tectonic-elements map of Algeria showing location of the Berkine Basin. Filled circles represent locations of giant oil and gas fields.

on the southwestern and southern flanks of the Berkine Basin (Echikh, 1998) and is responsible for the depositional geometry of lower Paleozoic reservoir intervals and maturation history of the Lower Silurian source-rock interval. Hercynian (Carboniferous through Middle Triassic) tectonic activity and accompanying erosion is responsible for the present truncated distribution of the Paleozoic along the western and northern portions of the basin (Figure 3) as well as destruction of some pre-Hercynian oil-filled traps. Topographic relief created on the Hercynian unconformity also played a major role in the deposition of Late Triassic sediments in the Berkine Basin. Mesozoic tectonic events, including Triassic-Liassic rifting, and Early Cretaceous (Austrian tectonic phase) and latest Eocene (Pyrenean tectonic phase) structural movement strongly influenced formation of hydrocarbon-charged traps, Mesozoic seal development, and maturation history of Upper Devonian source rock.

Stratigraphic nomenclature and chronostratigraphy of the Paleozoic and Late Triassic Berkine Basin sediment pile is in transition. Application of sequence-stratigraphic concepts to Paleozoic and lower Mesozoic intervals has been initiated only recently (Ford and Scott, 1997; Scott et al., 1997; Fekirine and Abdallah, 1998; O. Nykjaer, personal communication, 1998), and basinwide biostratigraphic correlation of the Paleozoic is in its infancy. Biostratigraphic correlation of the Late Triassic nonmarine section is hampered by poor to sparse recovery of diagnostic palynomorphs. A unified stratigraphic nomenclature for the Berkine Basin Paleozoic and Late Triassic is yet to be accomplished. Present nomenclature in the Paleozoic is a mixture of outcrop terminology from the Tassili N'Ajjer (Figure 1) and informal naming of hydrocarbon-producing reservoir intervals in the Illizi Basin. Correlation and extension of the Tassili N'Ajjer outcrop and Illizi Basin subsurface stratigraphic succession into the Berkine Basin is complicated by rapid thickening of the Paleozoic interval into the basin and sparse well control penetrating Lower Silurian and older sediments.

The Anadarko group's electric-log database, however, does allow reasonably reliable wireline correlation over much of the sedimentary section. Figure 4 is a generalized Paleozoic chronostratigraphic cross section along the western portion of the Berkine Basin. The influence of the Hercynian unconformity on preservation of Middle Devonian through Middle Triassic sedimentation is evident on the northern end of the section, where more than 150 million years of geologic record is missing. Carboniferous sedimentation is preserved only in the central portion of the basin. Absence of a Carboniferous rock record on the southern end of the cross section is related to the presence of the Mole D'Ahara, a continental high which persisted throughout most of the Paleozoic and well into the Mesozoic. Overall, nonmarine deposition and significant facies changes from the northern to southern portion of the Berkine Basin in the Late Triassic hamper correlation of this important hydrocarbon-bearing interval.

EVOLUTION OF THE ANADARKO GROUP'S EXPLORATION PROGRAM

The Anadarko group's exploration effort has evolved along three lines: (1) continuing development of a basin-wide petroleum system and structural model, (2) increased improvement of the seismic tool as the primary method of prospect generation and field development, and (3) continuing development of cost-saving measures associated with increased drilling efficiency.

Petroleum System and Structural Model

Beginning in 1986, Anadarko's early exploration efforts, in conjunction with block selection, focused on identification of structural traps and improvement of seismic imaging. Potential exploratory targets included the Trias Argilo Greseux Superieur (TAGS), Trias Argilo Greseux Inferieur (TAGI), and Silurian-Devonian F6 quartz sandstones (Figure 2). Although Upper Devonian

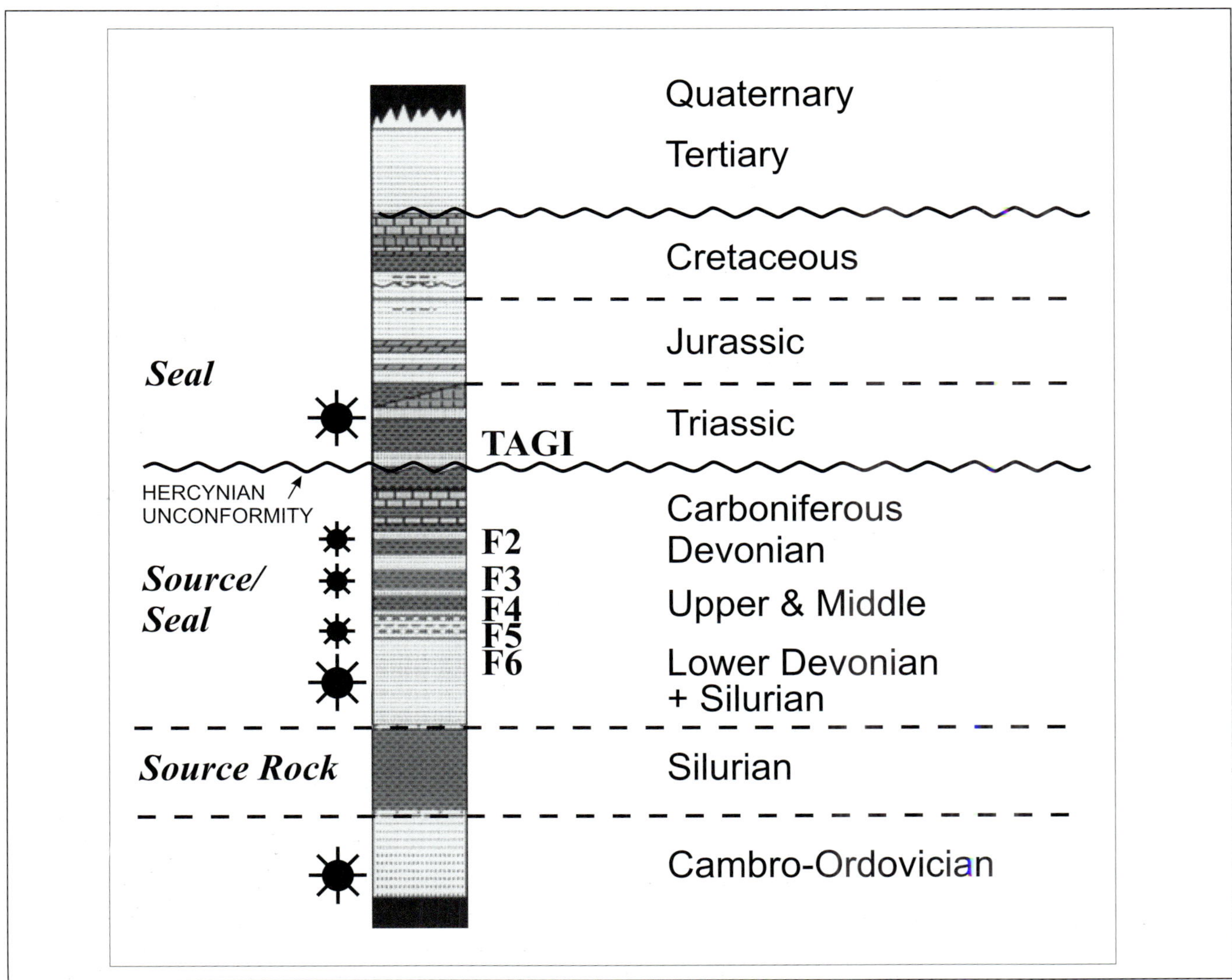

Figure 2. Generalized stratigraphic column for Berkine Basin, showing position of producing reservoirs, source rocks, regional seals, and Hercynian unconformity.

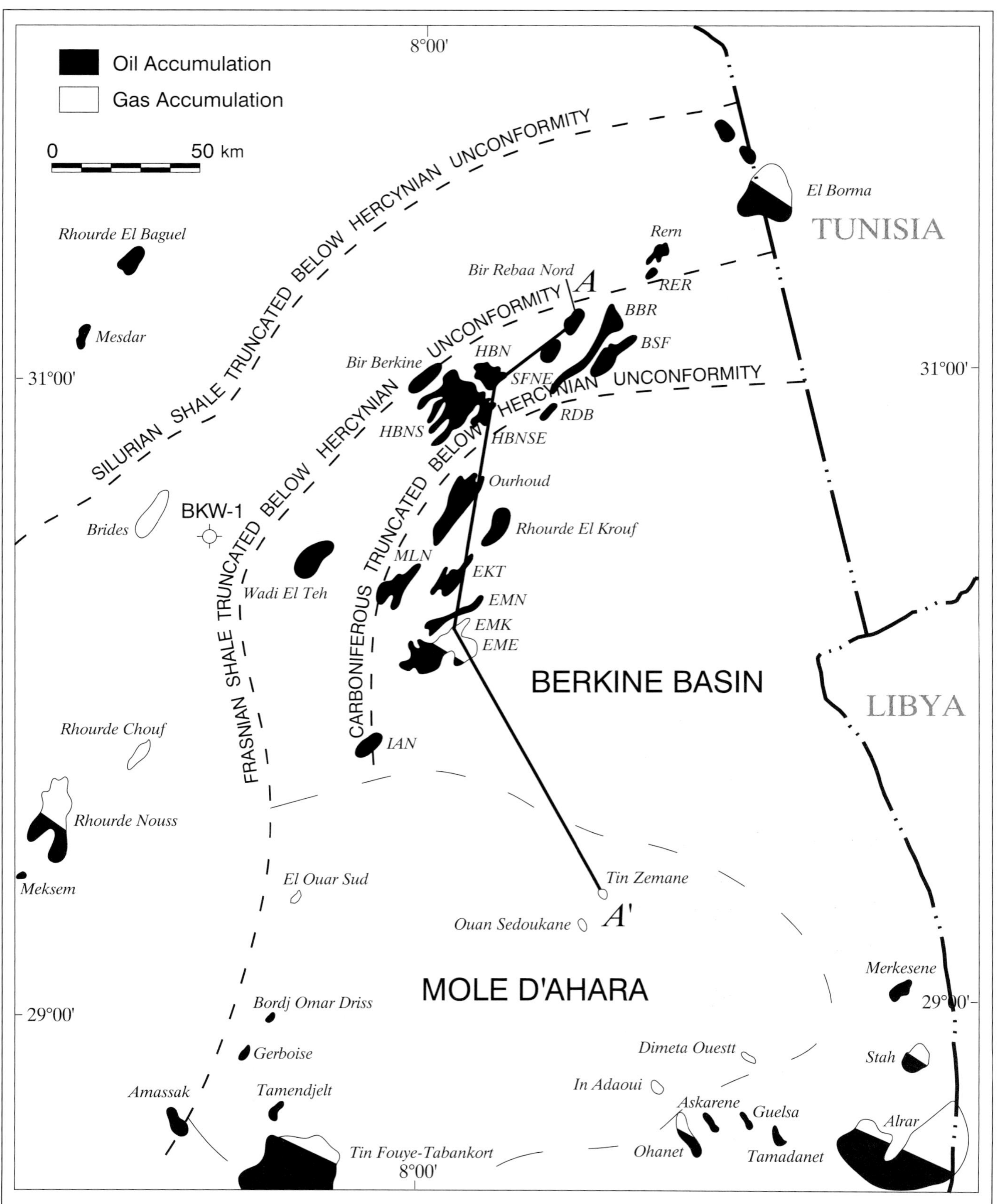

Figure 3. Geographic distribution of oil and gas accumulations in the Berkine Basin, northern portion of the Illizi Basin, and eastern portion of Amguid–El Biod Arch and Hassi Messaoud Ridge.

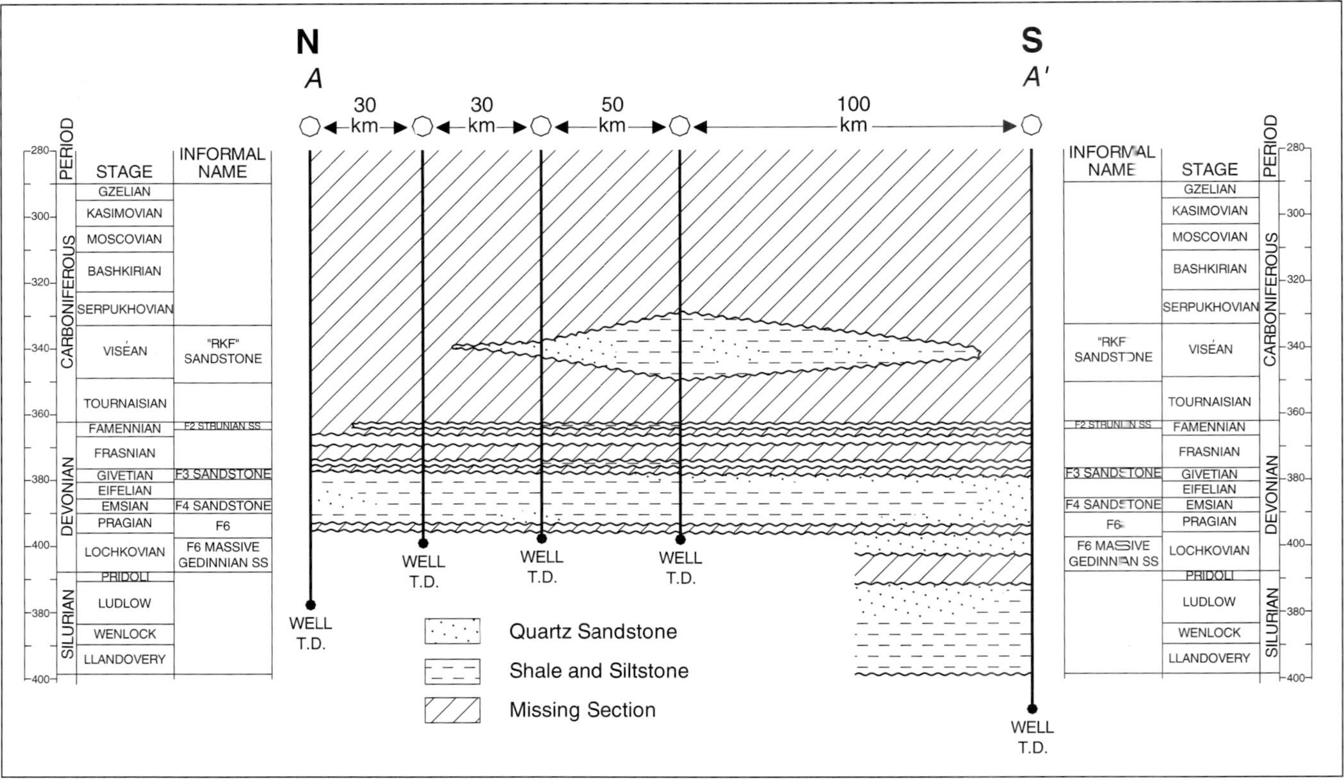

Figure 4. Paleozoic chronostratigraphic cross section across the western portion of the Berkine Basin, showing temporal and geographic distribution of Silurian, Devonian, and Carboniferous sedimentary packages. Line of section (A-A′) is illustrated in Figure 3.

(Frasnian) radioactive shales were recognized during these evaluations as a potential hydrocarbon source, Lower Silurian (Llandovery) radioactive shales were considered to be a primary hydrocarbon source. Based on the assumption that all present-day structures would be filled to spill, by 1988 Anadarko's strategy was to (1) reprocess existing seismic with improved statics corrections, (2) consider the entire basin prospective, (3) focus on the Lower Devonian as the most prospective reservoir target, and (4) decrease drilling costs. Attention focused rapidly on a potential four-way closure of 138 km² in the northwestern portion of the Berkine Basin. The structure was located 26 km east of Brides field and 27 km west of Wadi El-Teh (Figure 3), west of the Frasnian radioactive shale truncation and east of the onset of Silurian truncation. With the drilling of the BKW-1 exploratory well and its abandonment in 1991, the Anadarko group focused its attention on a series of technical evaluations to better understand source-rock quality and distribution and general migration pathways.

The BKW-1 well is noteworthy for its lack of hydrocarbon shows and reservoir-quality quartz sandstones in Triassic and Silurian-Devonian intervals. Building on Boudjema's (1987) work, it was quickly demonstrated that the BKW structure occupied a regionally structural low position within the basin at the time of peak hydrocarbon migration which precluded significant migration of hydrocarbons into the structure. Mapping of Frasnian and Llandovery radioactive shales' organic-carbon content, kerogen type, and maturity forced a reassessment of potential areas for exploration within the Berkine Basin and focused the group's attention on areas in a more optimal "fetch" position than the BKW structure.

With announcement of Cepsa's discovery at Rhourde El Krouf in 1992 and successful completion of the Anadarko group's well at EMK in 1993 (Figure 3), additional hydrocarbon-bearing reservoirs and the significance of faulting in hydrocarbon migration were recognized. Detailed studies on the TAGI siliciclastic system were undertaken in 1994 and continue today. Initial studies on the Berkine Basin Devonian and Carboniferous biostratigraphy, sequence stratigraphy, and deposition were initiated during 1995–1997. In conjunction with Institut Français du Pétrole, the Anadarko group began, in 1995, 2-D Temispack and Temiscomp basin modeling to assist in predicting hydrocarbon migration pathways and timing, and hydrocarbon composition of the reservoirs. A portion of this study was published by Rudkiewicz et al. (1997).

As the Anadarko group's understanding of the Berkine Basin evolved, the exploratory drilling program resulted in an additional 13 discoveries, 11 of which were operated by the Anadarko group. Flow rates as high as 21,395 bbl oil/day (BOPD) have been recorded from the

Triassic section. Strunian and Viséan reservoirs have recorded flow rates as high as 53.6 million cubic feet gas/day (MMCFGD) and 5,508 bbl condensate/day (BCPD).

Seismic Acquisition

During Anadarko's initial visit to Sonatrach's offices in 1986, evaluation of seismic data indicated the existence of severe statics problems associated with dynamite acquisition in the loosely packed, unconsolidated surface sand section. In addition, the size of the sand dunes covering the surface of the Berkine Basin did not allow for "straight"-line seismic acquisition. Correct processing of "crooked"-line seismic needed to be applied. Seismic misties caused by navigational errors were also common. Based on these initial observations, the Anadarko group reprocessed 36,000 km of Sonatrach 12-fold stack 2-D seismic data. Examples of unreprocessed and reprocessed Sonatrach data are shown in Figure 5. The Anadarko group's strategy was to reprocess Sonatrach data, identify

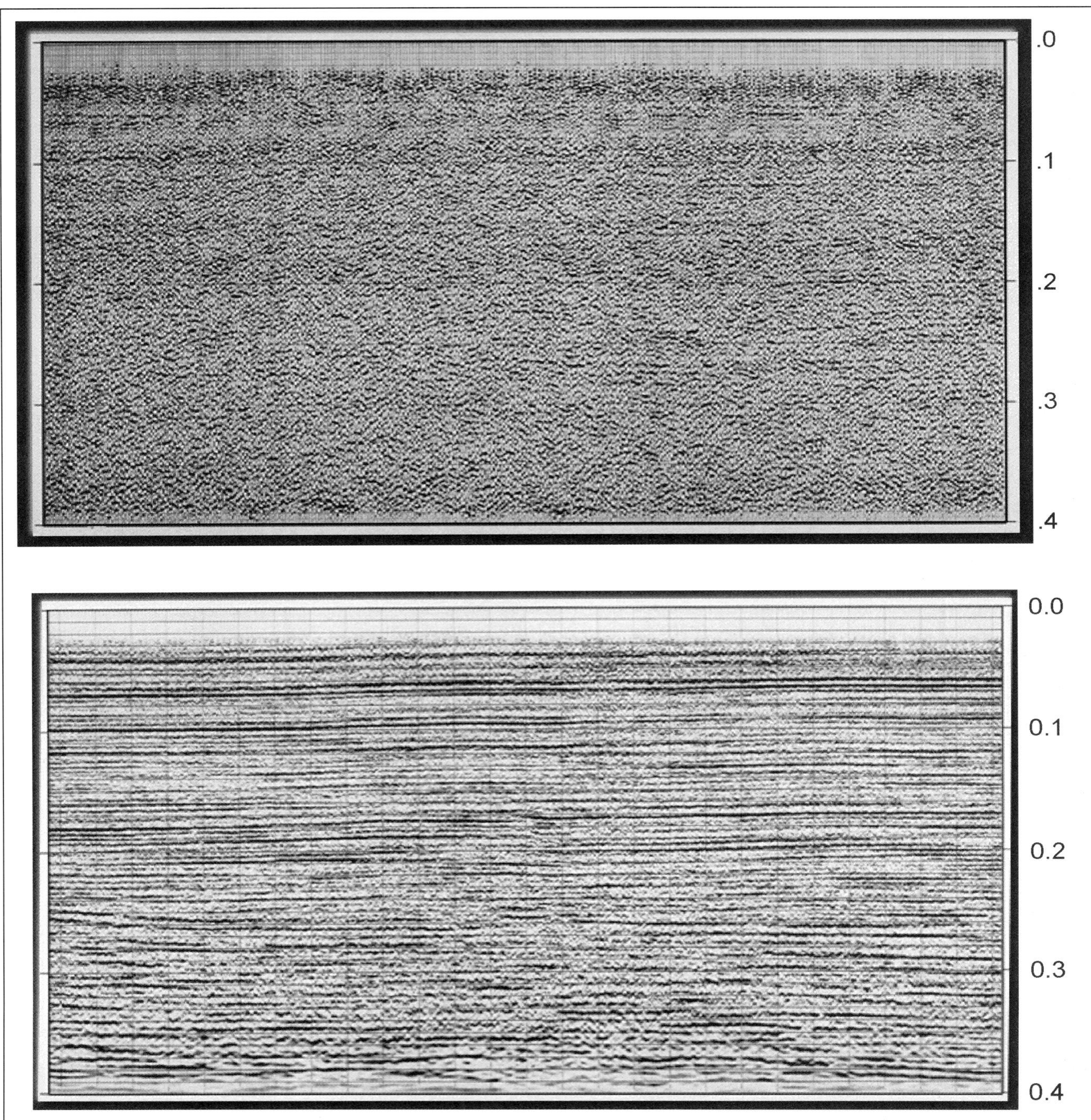

Figure 5. Comparison of 1973 Sonatrach-acquired 12-fold stack seismic line (top) and reprocessed version of the same line by the Anadarko group (bottom).

prospective areas, and supplement existing data with additional seismic acquisition.

Early exploration was directed toward late inversion structures because they were the only structures visible on the seismic data. Such traps were inherently high risk because of late structural movement. Later exploration concentrated on early-formed structures, facilitated by dramatic improvements in the resolution of the data. During the reprocessing effort, it became apparent that reprocessed data were insufficient to identify, with any degree of confidence, potential drilling locations. The decision was made in 1990 to acquire an entire new grid of 2-D seismic.

Feasibility studies indicated that alternative sources to conventional surface dynamite as an energy source for seismic acquisition were warranted. Therefore, the vibroseis technique was chosen as the energy source for new acquisition. Between 1990 and 1993, the Anadarko group acquired 6718 km of 120-fold seismic data. Two-dimensional seismic acquisition resumed in 1994, with acquisition of 2188 km of 240-fold seismic data between 1994 and 1997. Examples of 120-fold and 240-fold data are illustrated in Figure 6. Beginning in 1995, feasibility and optimization studies for acquisition of 3-D seismic were initiated. Based on these studies, 3-D seismic acquisition was started in 1996. To date, 1634 km^2 of 3-D seismic have been acquired. Additional 3-D seismic acquisition is under way. An example of 3-D seismic is shown in Figure 6.

Cost-saving Measures

Concurrent with basin studies and geophysical evaluations, substantial reductions in the cost of drilling wells were implemented. The need for costly, permanent road construction into remote locations was eliminated. Drilling-rig mobilization was accomplished directly across the sand dunes. Efficiencies in bit selection, drilling fluids, and hole size resulted in a dramatic decrease in the time necessary to drill wells (Figure 7). The Anadarko group's first well (BKW-1) took more than 175 days to drill and evaluate. By 1996, drilling time for a well of comparable depth was less than 80 days. Field-development wells at HBNS are now being completed in less than 20 days to total depths of 2900 m.

BERKINE BASIN PETROLEUM-SYSTEM COMPONENTS

Source Rocks

The most comprehensive evaluation of source rock in the Berkine Basin is by Daniels and Emme (1995). Based on organic richness and thermal-maturity analysis of more than 3900 rock samples from 84 wells and oil-oil correlation of samples from 24 fields, they concluded that Lower Silurian (Llandovery), Silurian Argileux and Argiles a Graptolites, and Middle to Upper Devonian (Givetian through Famennian) Serie de Tin Meras shales were the source for hydrocarbons trapped in the Berkine Basin. Later work by Zumberge et al. (1996), Illich et al. (1997), and Makhous et al. (1997) at least partially supported those conclusions. Recognition of Llandovery graptolitic shales as a source for reservoired hydrocarbons in the Illizi Basin was reported by Tissot et al. (1984).

Highest total-organic-carbon (TOC) content in the Lower Silurian and Upper Devonian shales is concentrated in two discrete intervals. Highest TOC content (2% to >17%) in the Lower Silurian shale section is found consistently in the Llandovery basal radioactive shale interval. Upper Devonian maximum TOC content (8%–14%) is found in Frasnian radioactive shales. Kerogen facies in Llandovery and Frasnian radioactive shales are generally oil prone with type I/II content. Present-day geographic distribution of Llandovery and Frasnian radioactive shales in the Berkine Basin is controlled by Hercynian erosion on the north and west flanks of the basin. In the central portion of the basin, Frasnian radioactive shales exceed 200 m in thickness, whereas Llandovery radioactive shales range from 25 to 35 m. Frasnian radioactive shales thin to less than 50 m across the Mole D'Ahara. TOC quality and thickness vary more regionally for Frasnian radioactive shales than for Llandovery radioactive shales.

Mapping of "equivalent" vitrinite reflectance indicates Llandovery radioactive shales are in the wet- to dry-gas window (1.75–2.0 R_o) in the central portion of the basin. In the northern portion of the basin near El Borma field, and on the Mole D'Ahara (Figure 3), Llandovery radioactive shales are within the peak to late oil-generation window (0.8–1.2+ R_o). "Equivalent" vitrinite-reflectance data for Frasnian radioactive shales demonstrate that the interval is within the late oil to early wet-gas generation window (1.1–1.4 R_o) for the central portion of the Berkine Basin. In the northern portion of the basin, south of El Borma field, and over the Mole D'Ahara, Frasnian radioactive shales are in the early to peak oil-generation window (0.5–1.1 R_o).

Based on 1-D kinetic simulations for 22 wells in the Berkine and Illizi Basins, Daniels and Emme (1995) calculated timing of peak oil generation for Llandovery basal radioactive shales and Frasnian radioactive shales. Figure 8 summarizes Daniels and Emme's (1995) interpretation of areas of peak oil-generative timing for Llandovery basal radioactive shales. The central portion of the Berkine Basin and eastern portion of the Illizi Basin reached peak oil generation from the Early Carboniferous to Late Jurassic. This is indicative of pre- and post-Hercynian oil-generative phases. Traps filled prior to the Hercynian tectonic phase have probably been destroyed in areas of significant Hercynian erosion. The northern portion of the Berkine Basin reached peak oil generation beginning in the middle Cretaceous (approximately 125 Ma) and has continued to the present. The most significant quantity of oil generated was probably during the

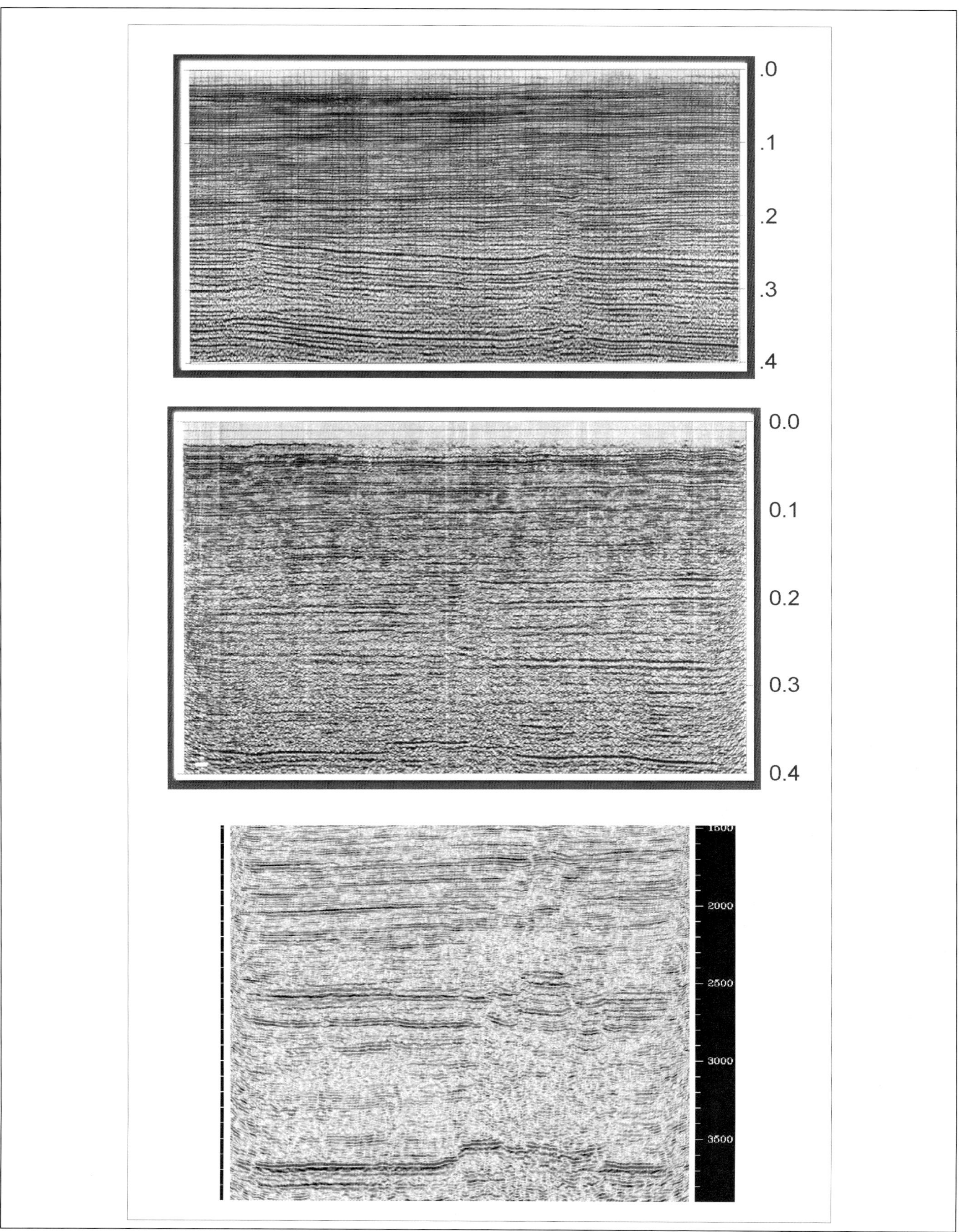

Figure 6. Comparison of 1992 acquired and processed 120-fold stack seismic line (top), 1994 acquired and processed 240-fold dip-moveout (DMO) stack seismic line (middle), and an arbitrary line from a 3-D volume (bottom). Compare with Figure 5.

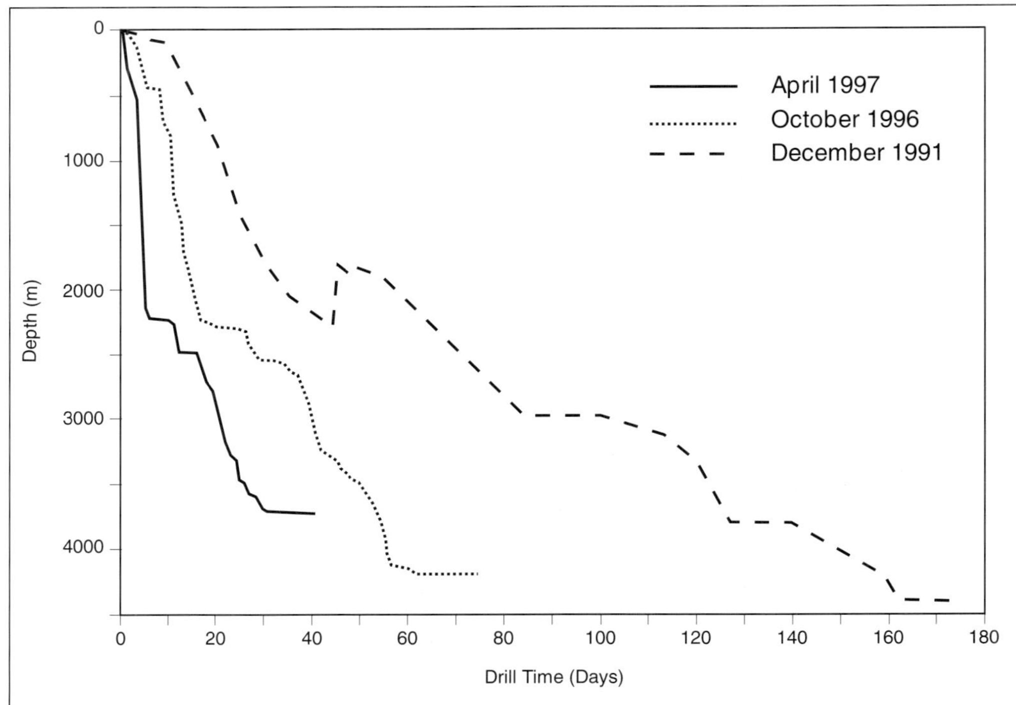

Figure 7. Plot of drill time (days) versus depth (meters) for wells drilled by the Anadarko group in 1991, 1996, and 1997. Drilling time for wells of comparable depth decreased from more than 170 days in 1991 to less than 80 days in 1996. By 1997, drilling time for wells drilled to within 500-m depth of 1996 wells decreased to less than 50 days.

last 65 million years (Daniels and Emme, 1995). The southwestern portion of the Berkine Basin and the area across the Mole D'Ahara have a broad oil-generative history which ranges from the Late Carboniferous to the present. The main generative phase appears to be middle Cretaceous (Daniels and Emme, 1995). Frasnian radioactive shales exhibit nearly the same pattern for areas of peak oil generation (Figure 9), with all timing shifted to the post-Hercynian. The northern portion of the basin reached the oil-generation window in the early Tertiary and is presently at or near peak generative capacity (Daniels and Emme, 1995). Timing of peak oil generation in the deepest portion of the basin was from the Late Triassic through Middle Jurassic.

Based on compositional analysis of oils from 24 fields, high-resolution gas chromatography, and liquid chromatographic separation, Daniels and Emme (1995) concluded that the oils are highly mature and are sourced by a similar kerogen type which is characteristic of Llandovery basal radioactive shales and Frasnian radioactive shales. The oils do not correlate with any of the Ordovician, Silurian, uppermost Devonian (Famennian), Carboniferous, or Mesozoic shale samples analyzed (Daniels and Emme, 1995). Two genetically related families of oils can be discriminated. The first family, located north and west of the Frasnian radioactive shale truncation line (Figures 3, 9), is most likely sourced from Llandovery basal radioactive shales. Reservoired oil south and east of the Frasnian radioactive shale truncation line (Figure 9) is a mixture of Llandovery and Frasnian-sourced oil.

Calculations of generated oil volumes indicate that Frasnian radioactive shales generated more than 700 billion bbl of oil in the Berkine and Illizi Basins. Llandovery basal radioactive shales generated more than 1400 billion bbl of oil (Daniels and Emme, 1995). These volumes are consistent with calculations made by Anadarko for other world-class petroleum systems (e.g., the Ellesmerian of the North Slope, Alaska).

Hydrocarbon Reservoirs

Overview

Hydrocarbon-producing reservoirs in the Berkine Basin range in age from Lochkovian (Early Devonian) through Carnian (Late Triassic) and are siliciclastic in composition. Present distribution of Lochkovian (Lower Devonian) through Viséan (Lower Carboniferous) reservoir intervals is controlled by the amount of Hercynian erosion on the north and west flanks of the basin and by paleotopographic expression on the Mole D'Ahara to the south. Sub-Lochkovian production exists along the flanks of the Berkine Basin. Ordovician siliciclastics produce gas at Brides field (Arenig-Hamra Quartzite) on the west flank of the Berkine Basin and oil and gas at Tin Fouye–Tabankort field (Ashgill through Llandovery–Unit IV) in the Illizi Basin (Figure 3). Acacus Formation siliciclastics (Silurian Wenlock through Pridoli) produce oil and gas in the Tunisian and Libyan portions of the Ghadames Basin. Equivalent quartz sandstones to the lower portion of the Acacus (lower F6 of Algeria) produce from numerous fields in the Illizi Basin. Cambrian-Ordovician (Upper Cambrian to Lower Ordovician Tremadoc-Arenig) quartz sandstones are the primary producing oil reservoirs at the supergiant Hassi Messaoud field northwest of the Berkine Basin (Figure 1). Hydrocarbon potential of the sub-Lochkovian section within the Berkine Basin is unknown

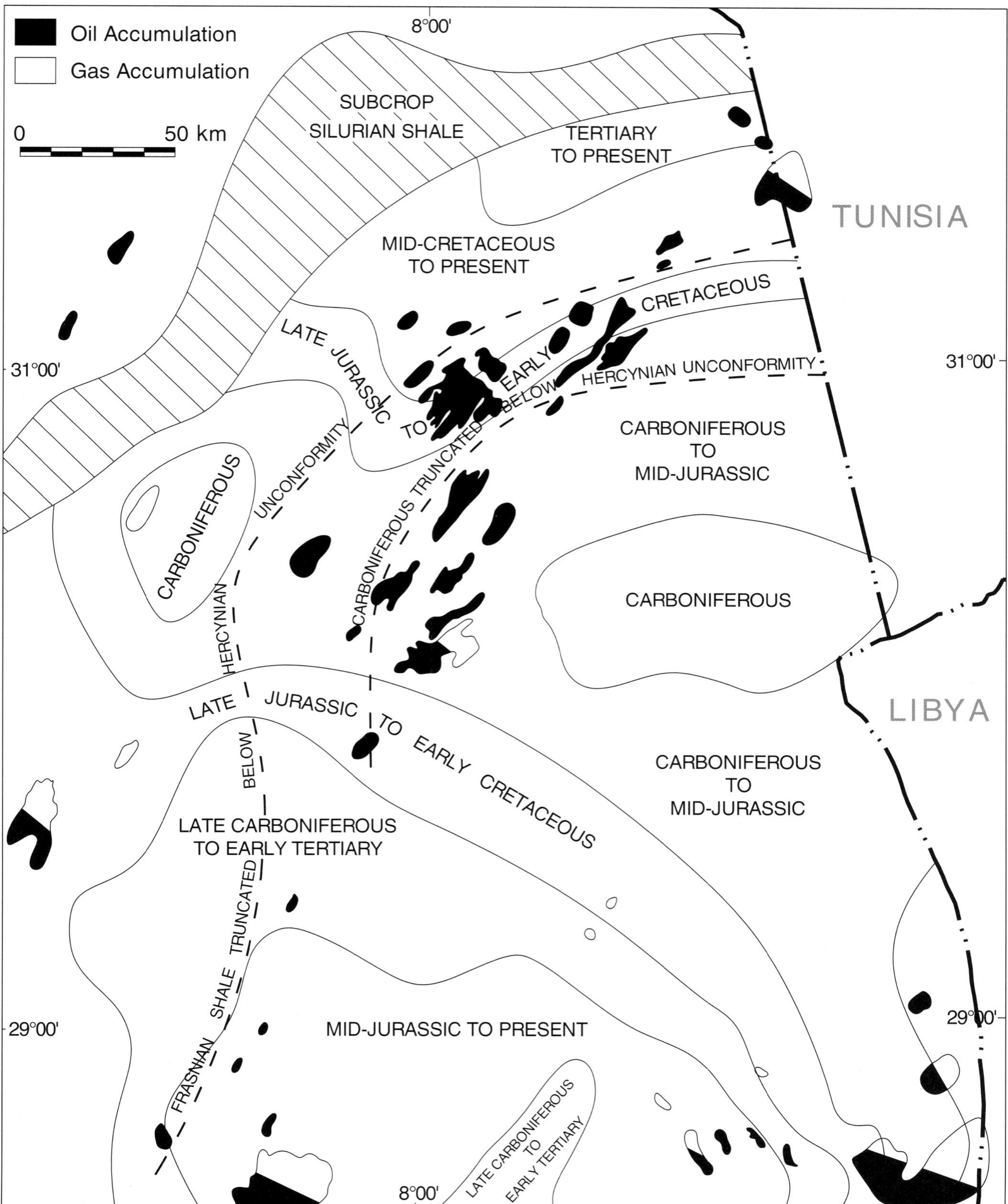

Figure 8. Peak oil-generation timing map of Early Silurian (Llandovery) radioactive shales in the Berkine Basin and northern portion of Illizi Basin. (Modified from Daniels and Emme, 1995.)

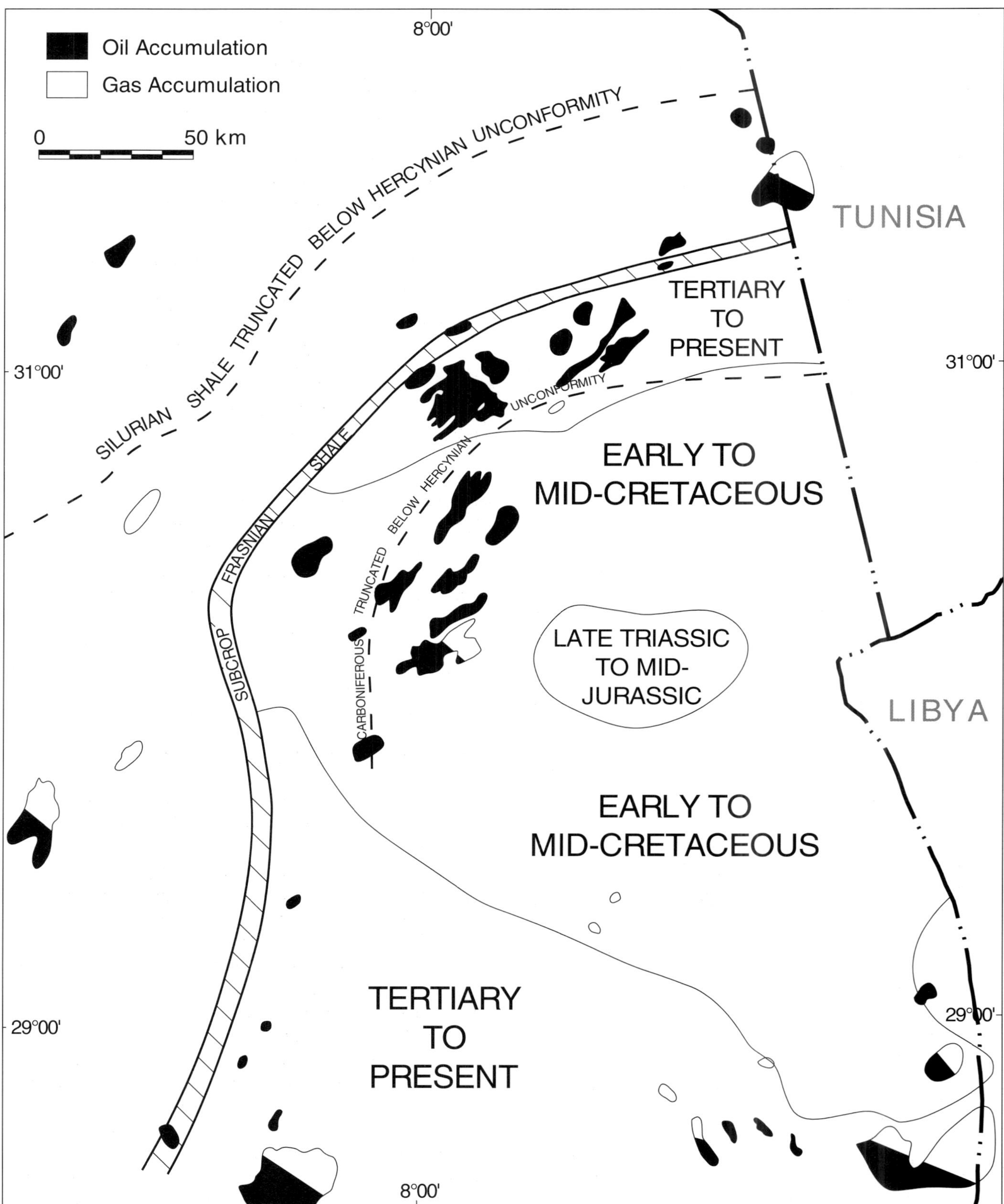

Figure 9. Peak oil-generation timing map of Late Devonian (Frasnian) radioactive shales in the Berkine Basin and the northern portion of Illizi Basin. Modified from Daniels and Emme (1995).

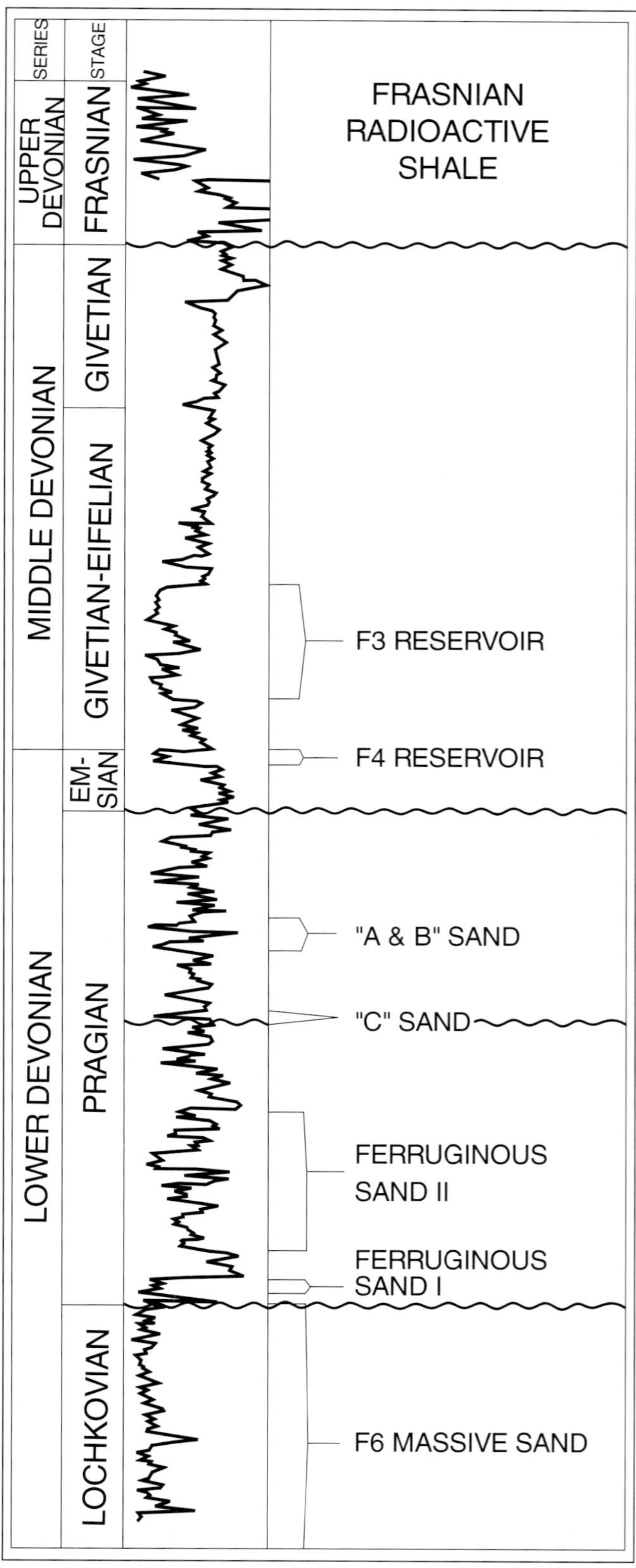

Figure 10. Gamma-ray log signature illustrating Lower and Middle Devonian quartz sandstone packages in the Berkine Basin. Data from O. Nykjaer (personal communication, 1998).

at the present time because of sparseness of penetrations.

Lower and Middle Devonian (Lochkovian through Givetian/Eifelian) siliciclastic reservoirs represent deposition in coastal-plain to marine settings. The stratigraphic position of producing reservoirs is shown in Figure 10. Stratigraphic nomenclature for these reservoirs is from the Illizi Basin. Although Berkine Basin Lower and Middle Devonian reservoirs generally correlate with the Illizi Basin section, accurate chronostratigraphic correlation requires further study.

Lower Devonian (Lochkovian-Emsian) Siliciclastics

Lochkovian F6 massive quartz sandstones (Figure 10) are present throughout the Berkine Basin. On the western and northern flanks of the basin, F6 massive sandstones are truncated by Hercynian erosion. Present-day maximum F6 quartz sandstone thickness is in the northern part of the basin (Figure 11). Porosity values of more than 20% are found in the northern portion of the basin and in the eastern portion of the Illizi Basin. In the central portion of the basin, porosity values for nonargillaceous massive F6 quartz sandstones range from 12% to 18% (Figure 11). Although there is general correlation between grain size and reservoir quality (Figure 12), the dominant limiting factor in reservoir quality is the presence of pore-lining ferriferous chlorite cement. Presence of pore-lining chlorite cement preserves high porosity (>20%) and permeability (>1000 md) at depths greater than 4700 m (Figure 13). Where pore-lining chlorite cement is not present, porosity is reduced by compaction and quartz overgrowths to less than 10% at the same depth (Figure 13). Permeability is reduced to less than 5 md. Bekkouche et al. (1993) calculate timing of formation of chlorite pore-lining cement at approximately 283 Ma based on potassium/argon dating. Formation of chlorite pore-filling cement is apparently related to the Hercynian tectonic phase.

Additional Lower Devonian (Pragian through Emsian) hydrocarbon-producing reservoirs are present in the Illizi Basin and the northwestern portion of the Berkine Basin. Fine-grained to very fine-grained laminated quartz sandstones, correlated with the F6 "C" sandstone of the Illizi Basin (Figure 10), are the primary reservoir at Sonatrach's Bir Berkine field (Figure 3). Secondary reservoirs at Bir Berkine field have been correlated with the Illizi Basin F6 "A" and "B" quartz sandstones (Figure 10). O. Nykjaer (personal communication, 1998) interprets the Bir Berkine Pragian quartz sandstones as part of a prograding deltaic complex. Reservoir properties are fair. Porosities can exceed 20%, but because of the fine-grained nature of the sandstone, permeability is limited to less than 100 md. Emsian fine-grained quartz sandstones, productive at Bir Rebaa Nord field (Figure 3), are apparently equivalent to the F4 reservoir interval of the Illizi Basin (Figure 10). They have been penetrated only in the north and northwestern portions of the Berkine

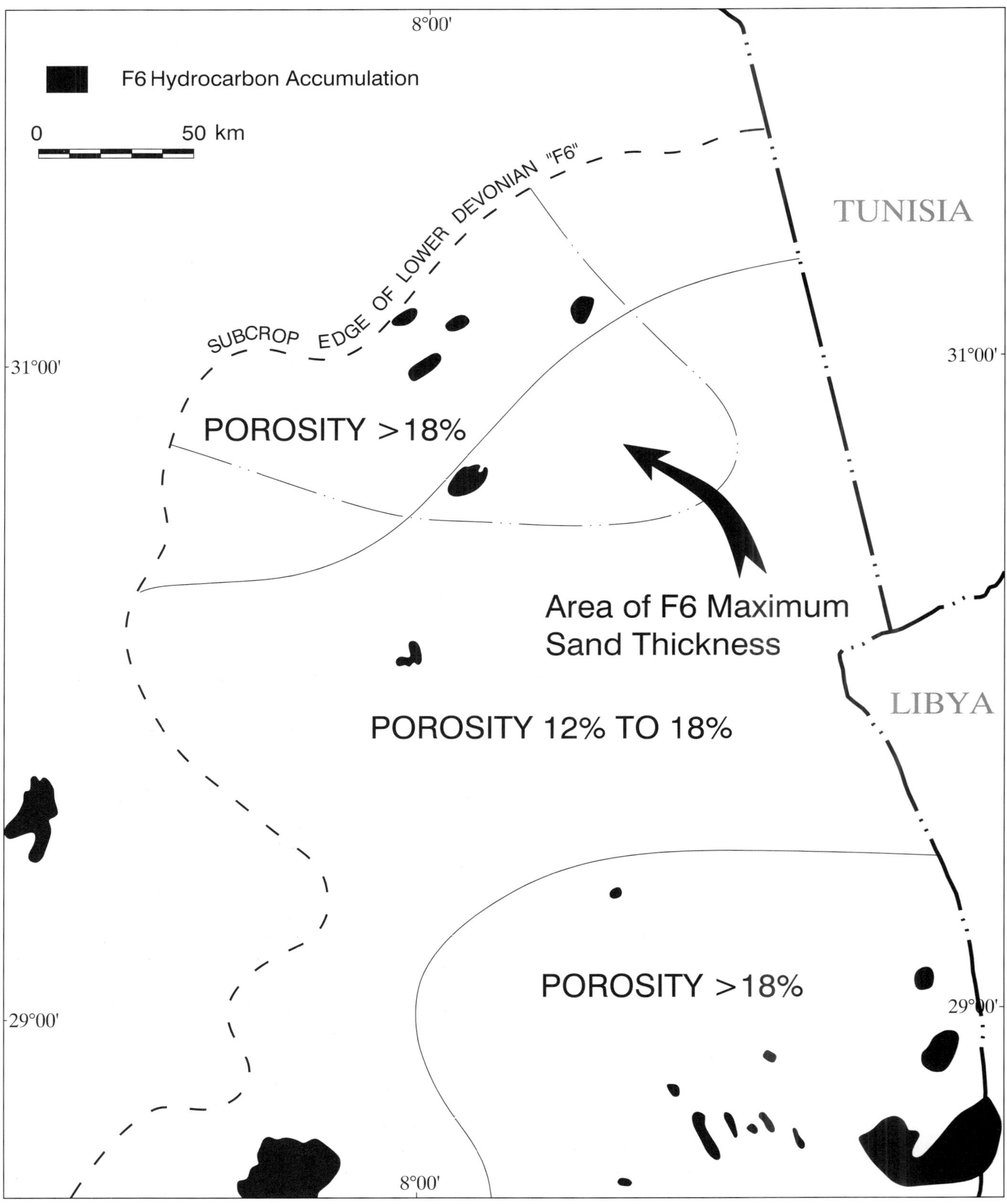

Figure 11. Generalized porosity distribution and area of maximum quartz sandstone thickness of Lower Devonian (Lochkovian-Pragian) F6 sandstone interval. Porosity distribution from Sonatrach (1995). Area of maximum F6 quartz sandstone thickness from O. Nykjaer (personal communication, 1998).

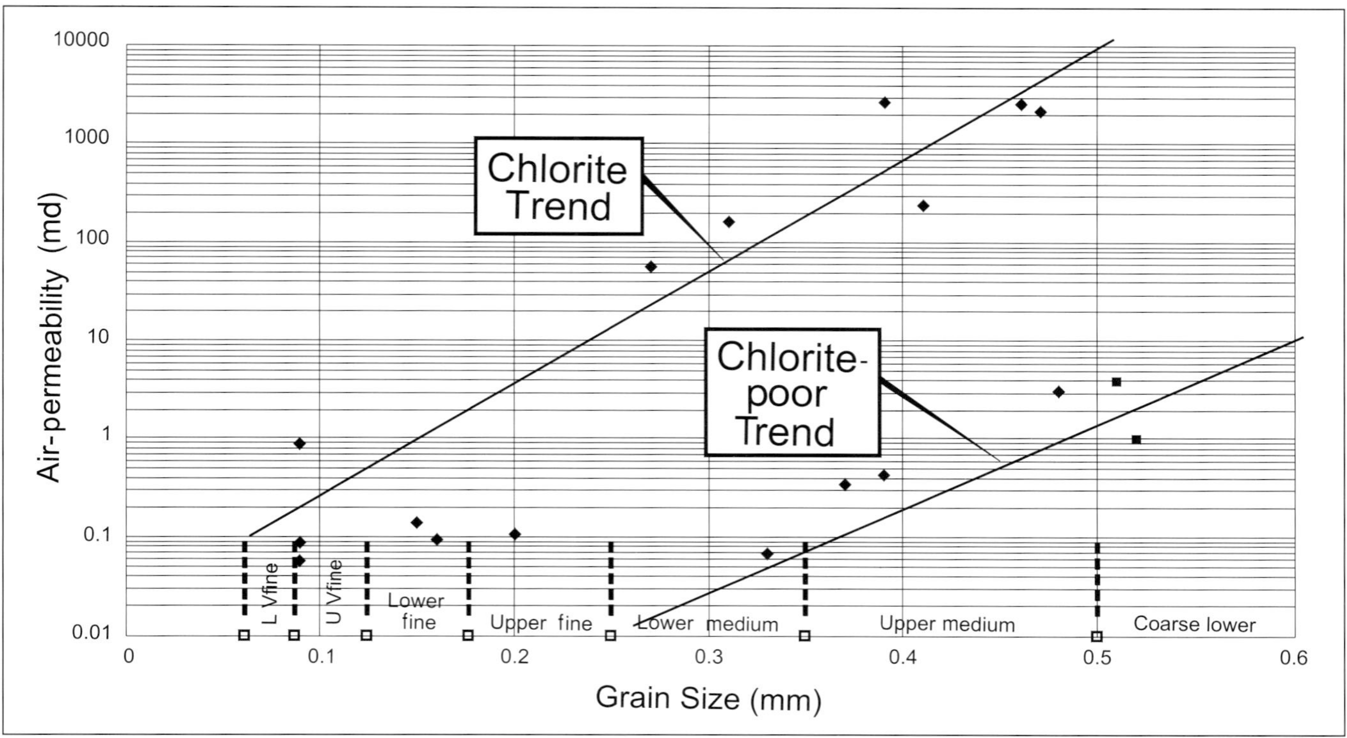

Figure 12. Grain size (millimeters) versus air permeability (millidarcys) plot for chlorite- and nonchlorite-cemented F6 (Lochkovian-Pragian) quartz sandstones. Permeability in chlorite- and nonchlorite-cemented trends typically increases with grain size. Data from O. Nykjaer (personal communication, 1998).

Basin and in the Illizi Basin. O. Nykjaer (personal communication, 1998) interprets the Bir Rebaa Nord quartz sandstones as indicative of wave-dominated shoreface deposition.

Middle Devonian (Eifelian-Givetian) Siliciclastics

Middle Devonian (Eifelian through Givetian) F3 reservoirs (Figure 10) are productive at Alrar field on the east side of the Mole D'Ahara in the Illizi Basin (Figure 3). Chaouchi et al. (1998) interpret productive siliciclastic intervals at Alrar as a series of tidally influenced barrier bars. Typical bar facies porosities range from 7% to 15%. Permeabilities range from 5 to 43 md (Chaouchi et al., 1998). F3-equivalent quartz sandstones are present along the north flank of the Mole D'Ahara. They are not present on the Mole D-Ahara, where their absence is a result of depositional thinning of the F3 interval and local erosion over depositional highs. O. Nykjaer (personal communication, 1998) suggested that longshore drift is responsible for redeposition of F3 quartz sandstones north of the Mole D'Ahara in a progradational sequence. F3-equivalent quartz sandstones pinch out northward toward the central portion of the Berkine Basin.

Upper Devonian (Strunian)–Lower Carboniferous (Viséan) Siliciclastics

Prior to Cepsa's discovery of Rhourde El Krouf field (Figure 3) in 1992, the youngest Paleozoic interval containing reservoir-quality quartz sandstones observed in the Berkine Basin was the lowermost Strunian (latest Devonian) F2 interval (Figure 14). F2 sandstones are productive in the Illizi Basin, where porosities reach 18% and permeabilities as high as 1000 md have been recorded (Sonatrach, 1995). Cepsa's discovery well, RKF-1, penetrated several thick Viséan (Lower Carboniferous) and Strunian (latest Devonian) quartz sandstone packages (Figure 14) which were informally termed the "RKF" and "Strunian beach" sandstones. The regional nature of these previously unknown sandstone packages was confirmed in 1993 by Anadarko's EMK-1 well.

The Strunian-Viséan contact is a minor angular unconformity, and the basal part of the Carboniferous (Tournaisian and earliest Viséan) is missing. Strunian and Viséan quartz sandstones are known only from the northwestern and northern portions of the Berkine Basin (Figure 15). Their present distribution is controlled on the north and west by truncation on the Hercynian unconformity. O. Nykjaer (personal communication, 1997) favors an easterly shale-out for the Strunian sandstone packages (Figure 15).

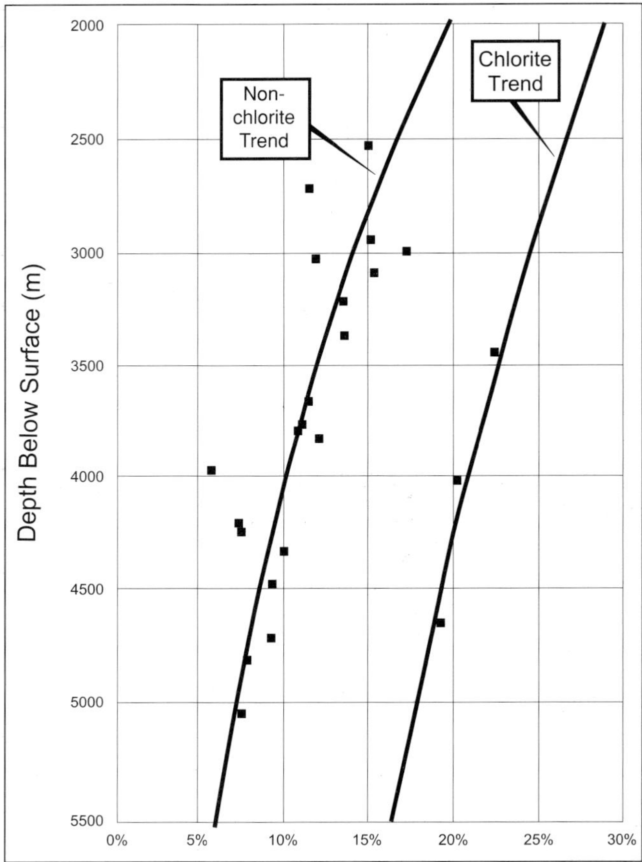

Figure 13. Porosity (percent) versus depth below surface (meters) for chlorite- and nonchlorite-cemented F6 (Lochkovian-Pragian) quartz sandstones. Chlorite-cemented F6 quartz sandstones illustrate higher porosity preservation at depth than nonchlorite-cemented F6 quartz sandstones. Data from O. Nykjaer (personal communication, 1998).

Strunian and Viséan quartz sandstone packages represent a continuum of environments ranging from shoreface through wave-dominated delta deposition.

Viséan deltaic quartz sandstone packages appear to extend along the northern flank of the basin. The quartz sandstones are most likely derived from erosion of older Paleozoic sediments present along the Hassi Messaoud Ridge and Dahar Dome to the west and north (O. Nykjaer, personal communication, 1997).

Triassic Siliciclastics

Triassic reservoirs, in particular Carnian TAGI quartz sandstones, are major producers of oil and gas along the western and northern flanks of the Berkine Basin. El Borma field (Figure 3), discovered in 1964 and extended into Algeria in 1969, was the first significant discovery in the Berkine portion of the Ghadames Basin. Production at El Borma is from TAGI quartz sandstones. Exploratory drilling in the 1990s resulted in several significant Berkine Basin discoveries. Beginning in the middle 1990s

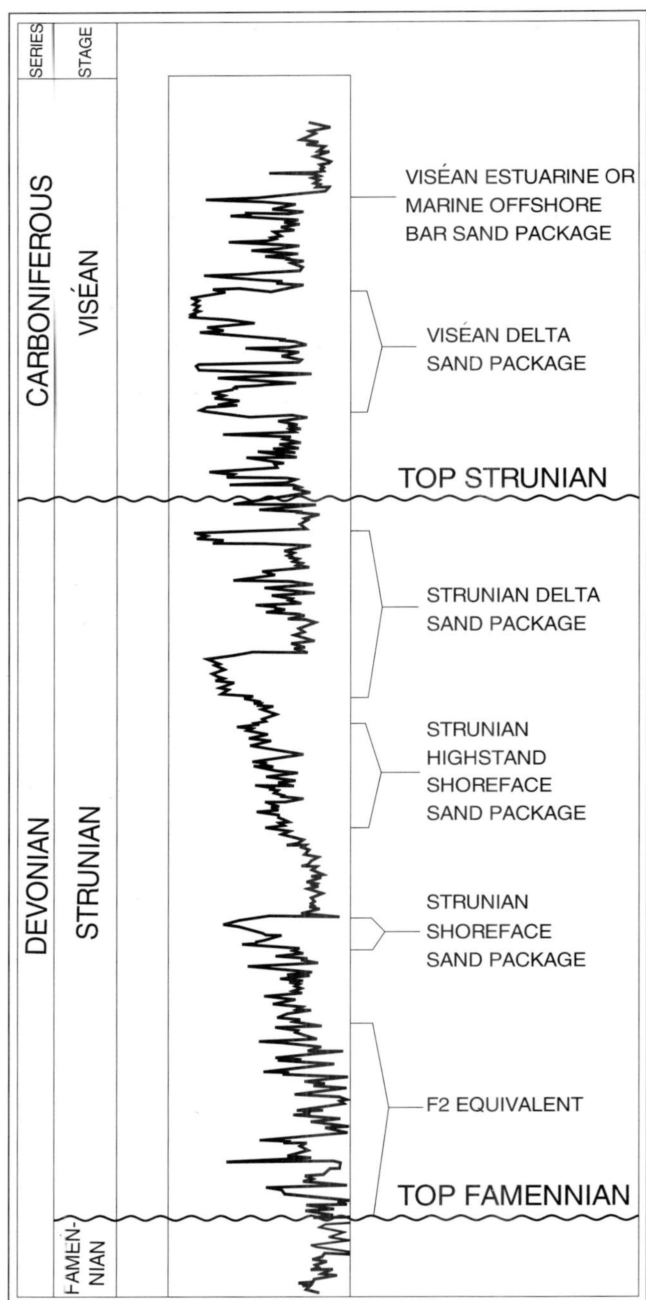

Figure 14. Gamma-ray log signature illustrating Strunian (uppermost Devonian) and Viséan (Lower Carboniferous) quartz sandstone packages in the Berkine Basin. Data from O. Nykjaer (personal communication, 1997).

(Daniels et al., 1994), these producing reservoirs have been the subject of numerous studies (Drumheller et al., 1997; Ford and Scott, 1997; Scott et al., 1997; Boote et al., 1998; Echikh, 1998; Pink et al., 1999; Scott and Wheller, 1999; Wheller et al., 1999).

The TAGI can be divided into three distinctive sandstone/mudstone cycles (Figure 16) which represent progressive filling of a pre-TAGI incised valley system (Scott

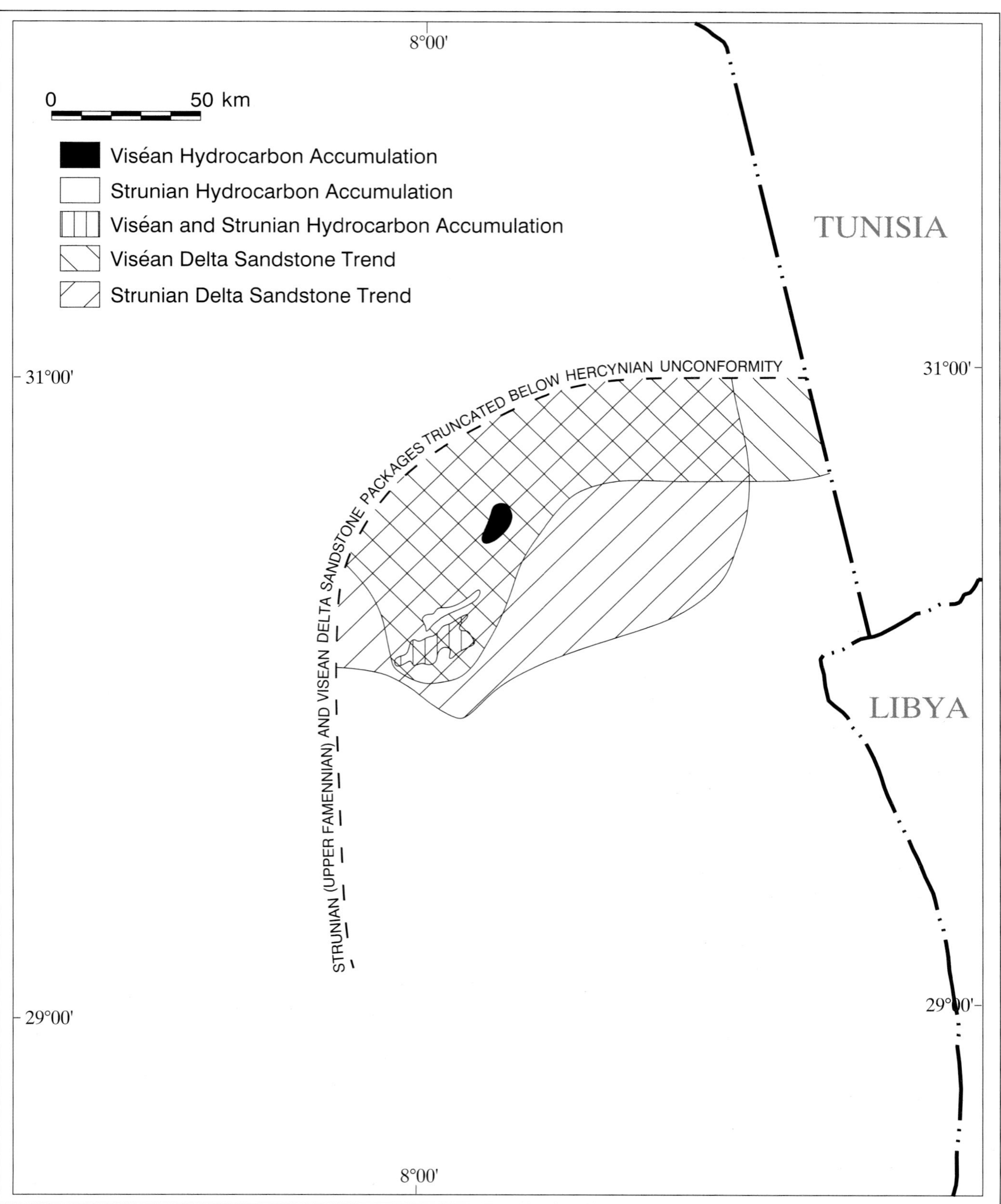

Figure 15. Generalized distribution of Strunian (uppermost Devonian) and Viséan (Lower Carboniferous) delta quartz sandstone packages and their subcrop relation to the Hercynian unconformity. Data from O. Nykjaer (personal communication, 1997).

et al., 1997). The majority of TAGI reservoirs was deposited as fluvial valley-fill and braid-plain quartz sandstones and eolian quartz sandstones. Overlying and interbedded with the quartz sandstones are pedogenic carbonate mudstones and sabkha and chott facies (A. J. Scott and G. Ford, personal communication, 1996).

Deposition of the TAGI occurred in an overall transgressive episode during the Late Triassic. TAGI siliciclastic-dominated strata were deposited on a major, low-relief, regional erosion surface, the Hercynian unconformity. Drainage systems throughout most of TAGI deposition were focused toward the east and northeast (toward the developing proto-Tethys). TAGI sediments are dominantly fluvioeolian associations, but with widespread inderbedded lacustrine/marginal sabkha deposits. Late TAGI deposition followed a trend that became predominant in the overlying Carbonaté toward the north and west, with deposition ranging from fluvioeolian to lacustrine to full sabkha conditions in the north. Few wells have penetrated the area of maximum TAGI gross sand development (Figure 17), and the depositional relationship of this area to the El Merk–El Borma producing trend is unclear.

M. Wilson (personal communication, 1996) attributes the excellent reservoir quality of the TAGI (average porosity of 17% and average permeability of 598 md) along the northwest portion of the basin to a combination of factors, including (1) sourcing of TAGI quartz sandstones from a moderately to highly quartzose source terrane, resulting in TAGI sandstones having a high quartz and low feldspathic and lithic content); (2) moderate to highly saline formation waters present in the TAGI during most of its burial history, resulting in the lack of alteration of feldspar to clays; (3) depositional environments which favored formation of detrital clay rims on framework grains; and (4) low burial temperatures, which prohibited illitization of feldspars and kaolinite.

Variation in reservoir quality is primarily associated with grain size. Medium- to coarse-grained quartz sandstones exhibit permeabilities of more than 3000 md (Figure 18). Very fine-grained quartz sandstones typically have permeabilities less than 50 md. Based on detailed petrographic work, M. Wilson (personal communication, 1996) suggested that lower TAGI quartz sandstones in the northwest portion of the basin were derived from erosion of slightly metamorphosed quartzose sandstones. Upper TAGI quartz sandstones probably were derived from granitic plutons. Lower and upper TAGI sandstones may represent a continuum of unroofing of a single source area. Provenance studies for other portions of the TAGI depositional system are under way.

The youngest producing interval within the Berkine Basin is the TAGS. Where present, TAGS quartz sandstones overlie the TAGI reservoirs (Figure 16) and are part of a complex depositional sequence referred to as the Triassic Carbonaté. TAGS productive reservoirs occur primarily west and northwest of the Berkine Basin on the

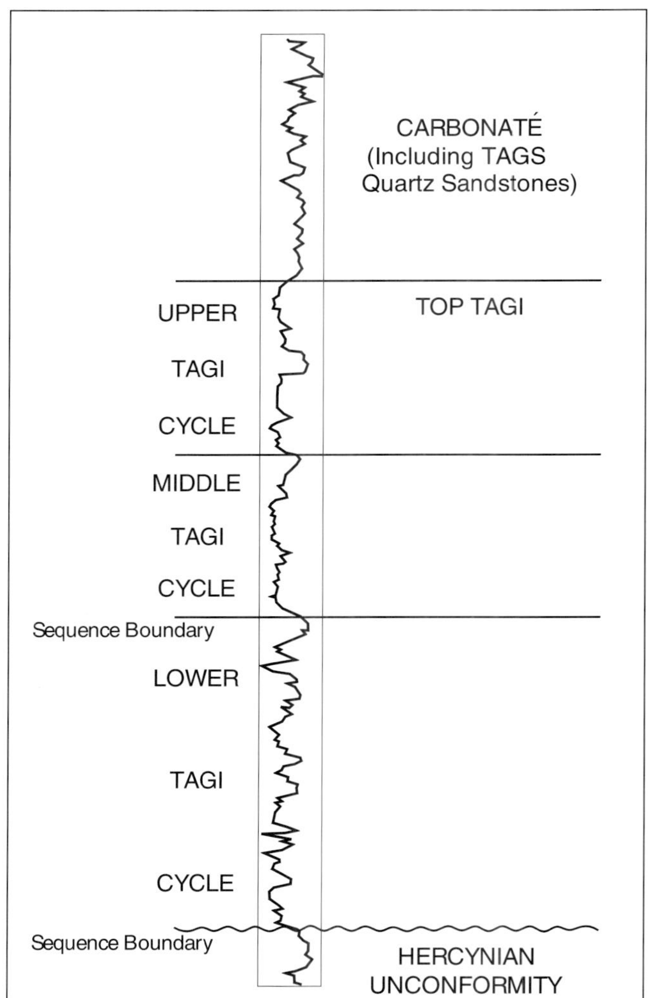

Figure 16. Gamma-ray log signature illustrating tripartite subdivision of TAGI and overlying Carbonaté/TAGS section. Data from A. J. Scott and G. Ford (personal communication, 1996).

Hassi Messaoud Ridge and in the "Triassic" Basin (Figure 1). Carbonaté quartz sandstones are productive within the western portion of the Berkine Basin at EMK and Rhourde El Krouf (Sonatrach, 1995; N. Mountford, personal communication, 1999).

Carbonaté quartz sandstones are similar in composition to those of the TAGI. Plagioclase, a very minor component of the TAGI, is present in minor to moderate amounts within the Carbonaté sandstones in the Berkine Basin. Based on amounts of feldspar and monocrystalline quartz in the Carbonaté sandstones, M. Wilson (personal communication, 1996) suggested that they were derived from a different source terrane than the TAGI, probably an adamellite or granodiorite. This may represent continued unroofing of the same source area as that from which the TAGI quartz sandstones were derived (M. Wilson, personal communication, 1996).

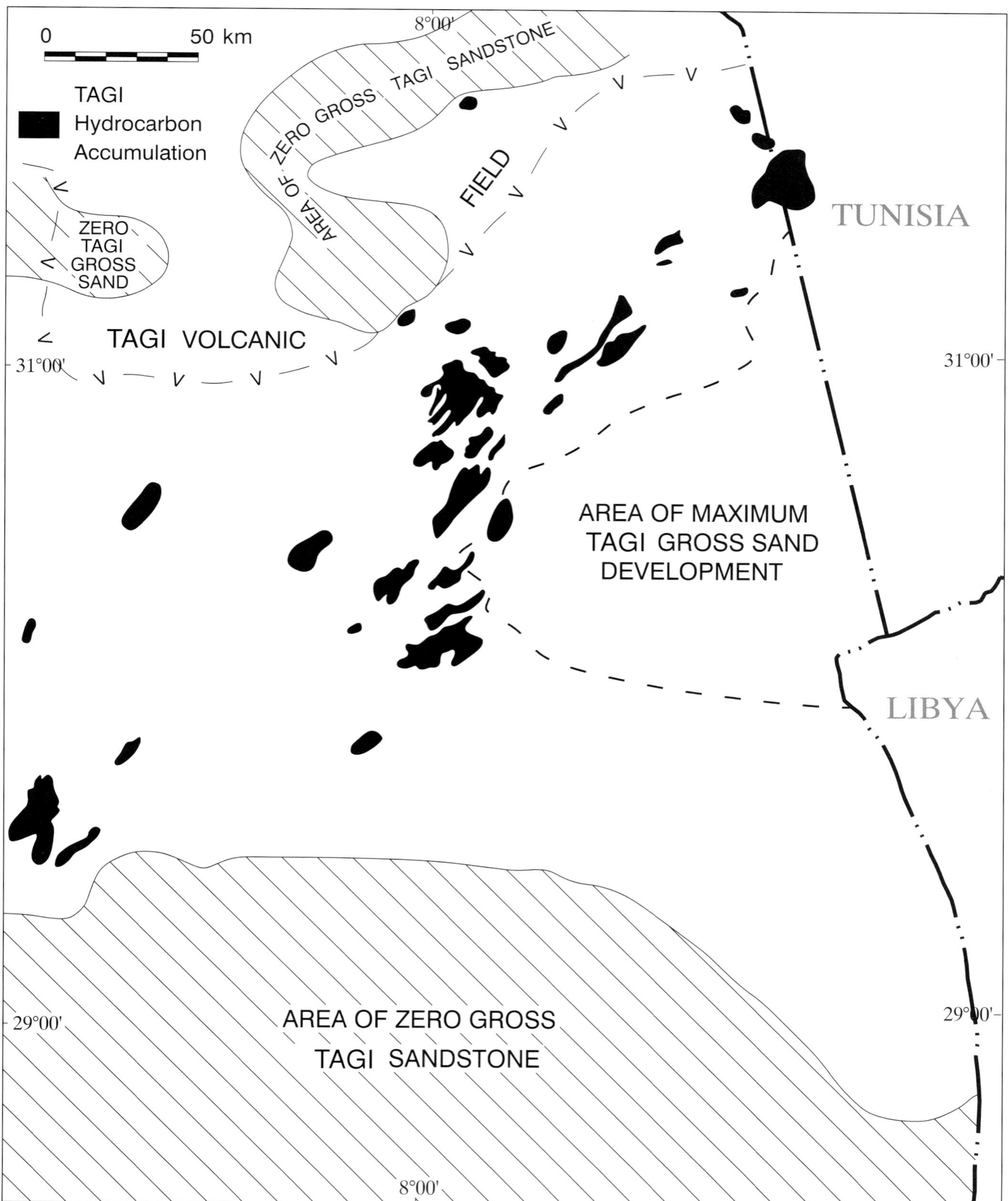

Figure 17. Generalized distribution of TAGI quartz sandstone and hydrocarbon accumulations. Data from N. Mountford (personal communication, 1999).

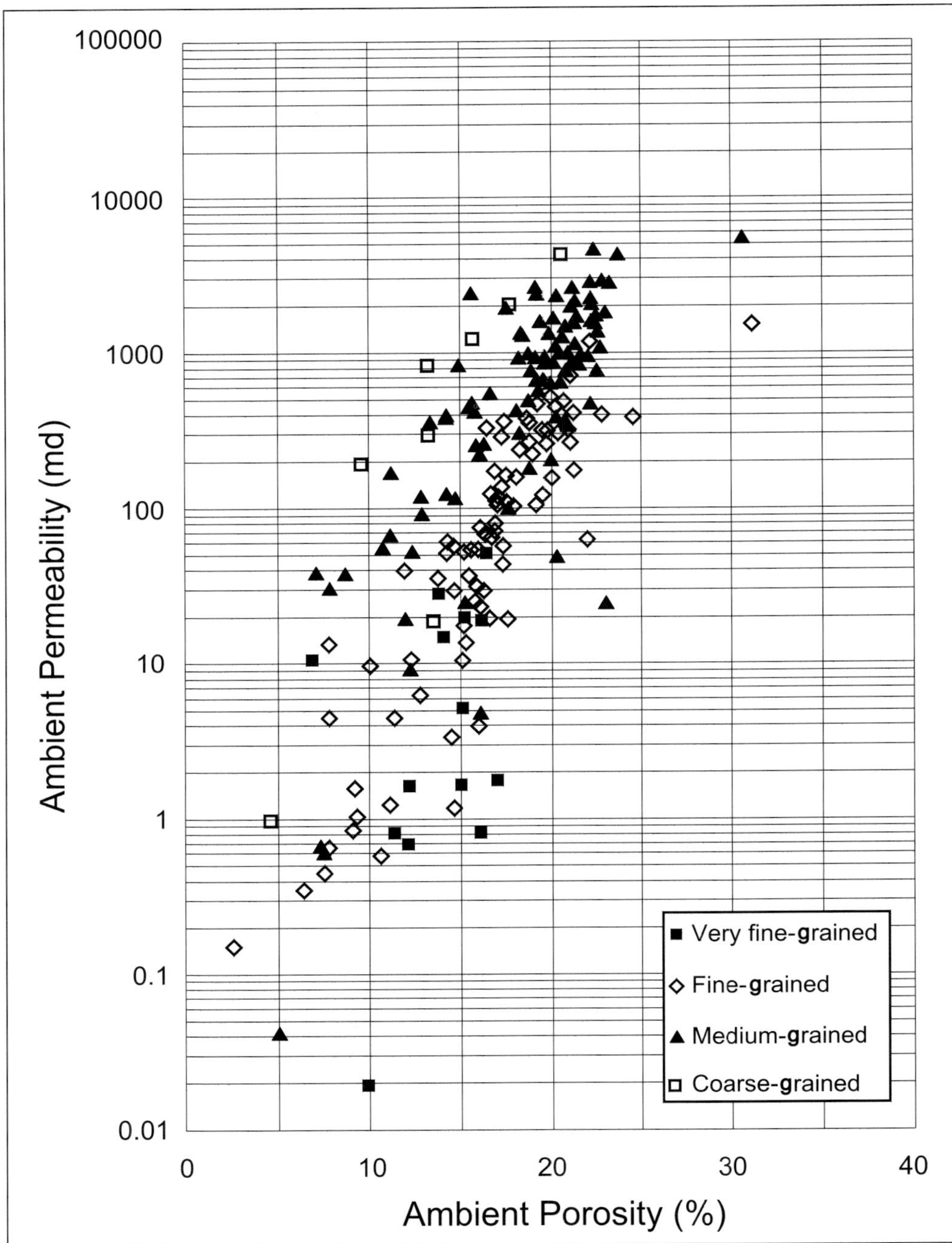

Figure 18. Porosity (percent) versus permeability (millidarcys) plot of TAGI quartz sandstones. Medium- and coarse-grained quartz sandstones typically have higher porosity and permeability than fine- and very fine-grained sandstones. Data from M. Wilson (personal communication, 1996).

Hydrocarbon Trapping Styles and Seals

With the exception of BBR field, hydrocarbon accumulations discovered in the Berkine Basin have been associated with structural traps. Although a significant stratigraphic component is associated with these structural traps, most discovered Berkine Basin accumulations appear to be associated with upthrown and downthrown faulted blocks in which primary fault movement occurred during Carbonaté deposition (Late Triassic). Many accumulations are broad, low-relief anticlines complicated by series of horsts and grabens. Many of the faults associated with these structural traps terminate in the Liassic (Figures 19, 20). In several fields, however, faulting extends into the Cretaceous, suggesting trap modification by Early Cretaceous (Austrian) and possibly latest Eocene (Pyrenean) structural movement.

Subcrop (truncation), stratigraphic onlap or updip pinch-out, and hydrodynamic hydrocarbon-bearing traps have been reported from the Illizi Basin (Chiarelli, 1978; Sonatrach, 1995; Alem et al., 1998; Echikh, 1998) but have yet to be explored for in the Berkine Basin. Considering the stratigraphic complexities of the Devonian, Carboniferous, and Triassic, subcrop and stratigraphic onlap traps should be expected to occur on the flanks of the Berkine Basin and within the TAGI depositional system. Ford and Muller (1995) suggested that the southern flank of the Berkine Basin could hold significant Silurian- and Devonian-reservoired hydrocarbon accumulations

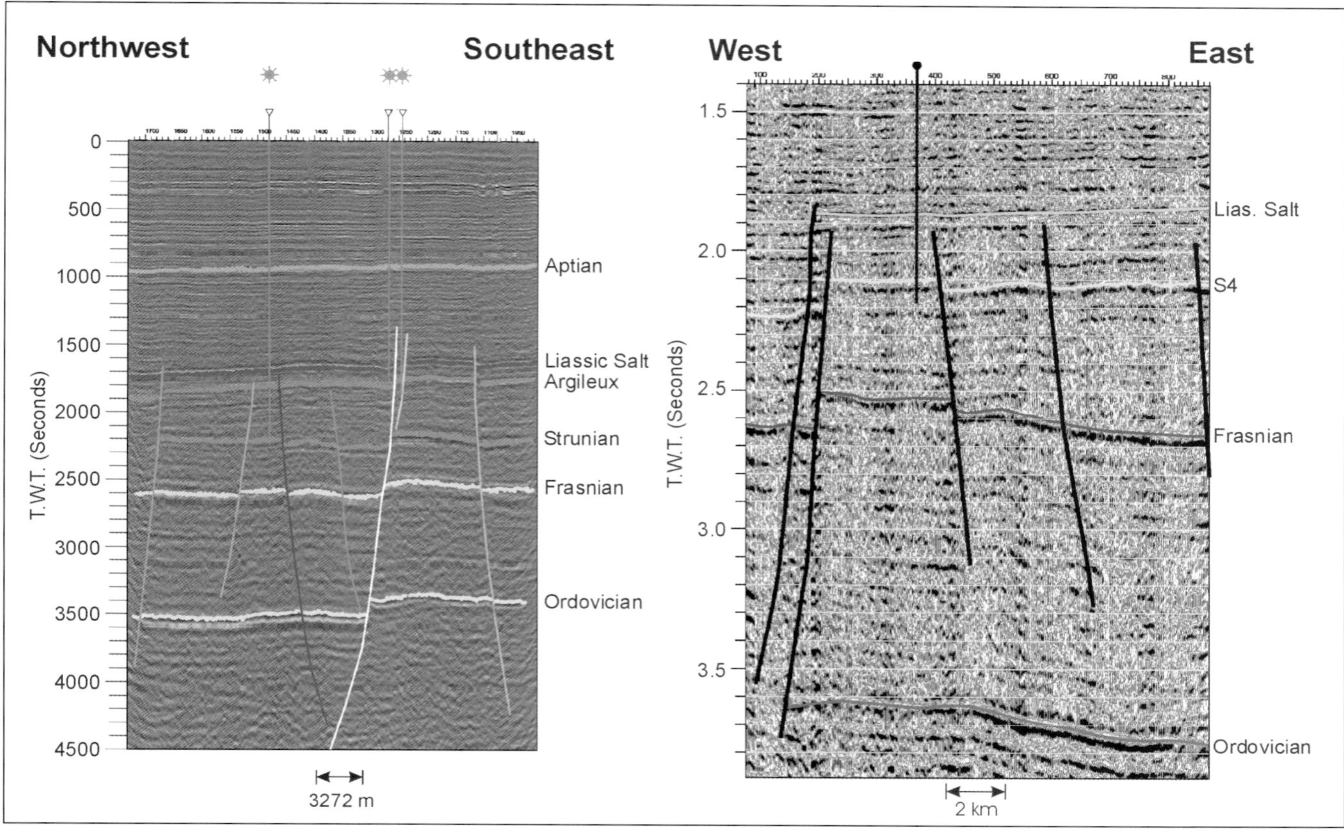

Figure 19. Seismic cross sections illustrating common types of structural traps in the Berkine Basin. Note variability of faults with respect to horizons they cut. Most faults do not cut the Liassic salt section.

associated with pre-Frasnian truncation of quartz sandstone packages across the Mole D'Ahara. Initial exploratory drilling has yet to confirm this model. Based on Chiarelli's (1978) hydrodynamic studies of southeastern Algeria, hydrodynamic trapping is not considered to be a viable trap type within the Berkine Basin. Although stratigraphic traps have not been explored for directly, the Anadarko group database allows for recognition of significant stratigraphic complexity at all levels, even within the TAGI depositional system.

The amount of fine-grained siliciclastics in the Devonian, Carboniferous, and Triassic provides for many excellent local hydrocarbon seals which play a significant role in the structural trapping of hydrocarbons. In addition, Sonatrach (1995) identified four stratigraphic intervals which probably form regional migration seals: (1) Lower Silurian (Llandovery) radioactive shales, which form the caprock for Upper Ordovician reservoirs in the Illizi Basin; (2) Upper Devonian (Frasnian) radioactive shales; (3) Carboniferous shales, which in areas of maximum development in the Berkine Basin impede hydrocarbon migration from Llandovery and Frasnian source rocks; and (4) Triassic and Liassic (Lower Jurassic) evaporites, which act as highly effective "superseals." The Triassic/Liassic evaporites act as the roof seal for the overall Berkine Basin petroleum system. Where the Triassic/Liassic evaporite section is present, no hydrocarbon accumulations have been discovered and no hydrocarbon shows have been observed in the overlying section.

Hydrocarbon Migration and Timing of Trap Fill

Daniels and Emme (1995), Rudkiewicz et al. (1997), and Echikh (1998) demonstrated the complex nature of hydrocarbon migration in the normally pressured Berkine Basin. Based on paleoreconstructions integrated with kinetic simulations and fluid-flow modeling, three primary migration paths account for distribution of oil and gas in the Berkine Basin: (1) updip migration of hydrocarbons in source-rock-adjacent reservoir couplets of Frasnian and Llandovery radioactive shales; (2) migration vertically up faults which have lower capillary entry pressures than adjacent shales, and into reservoirs; and (3) long-distance migration associated with movement of hydrocarbons from the deepest portions of the Berkine Basin west and north to a position where permeable basal Triassic siliciclastics are juxtaposed at Llandovery and Frasnian subcrop edges. Migration continues updip into Triassic reservoirs associated with the Hercynian unconformity surface. Daniels and Emme (1995) and Echikh (1998) provide diagrammatic cross sections across the Berkine Basin which illustrate the general nature of these migration pathways. Updip migration in source-reservoir

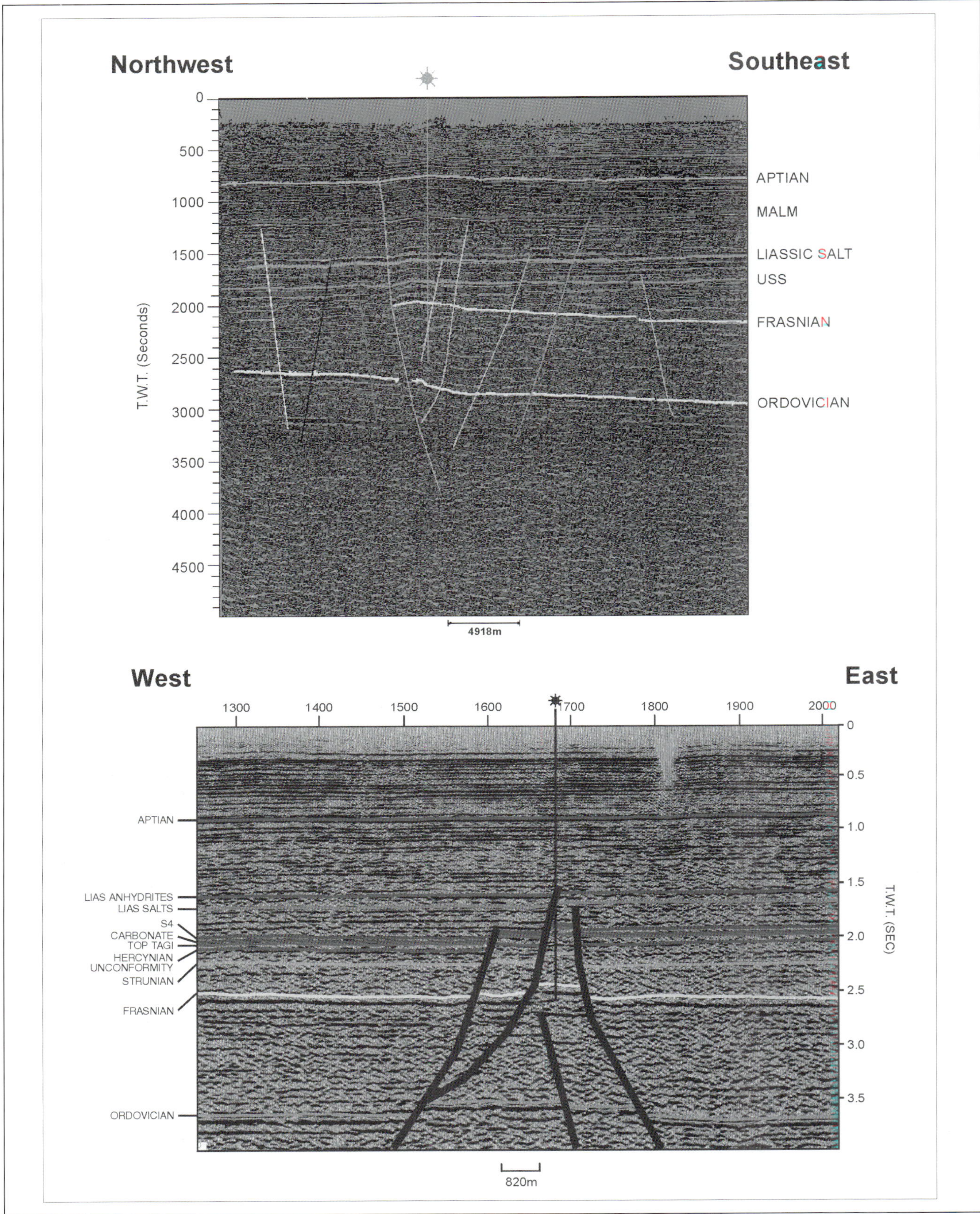

Figure 20. Seismic cross sections illustrating common types of structural traps in the Berkine Basin. Note thickening across the fault zone in the S4 to Liassic anhydrites on the upper seismic section. One fault on the upper seismic section extends up to the Aptian carbonate section.

couplets and vertical migration up faults best explain the distribution of oil and gas accumulations in the Devonian and Carboniferous of the Berkine Basin. Distribution of TAGI oil and gas accumulations fit long-distance migration pathway models.

Modeling by Daniels and Emme (1995) of the timing of peak oil generation in Llandovery radioactive shales demonstrated that many Silurian-sourced oil-filled traps generated during the Paleozoic have probably been destroyed by Hercynian erosion. Only along the northern flank and perhaps southern flank of the Berkine Basin would Paleozoic-formed oil-filled traps be preserved (Figure 8). Timing of peak oil generation for Frasnian radioactive shales (Figure 9) suggests that most present-day trapped oil is sourced from this interval. Present-day trapped oil sourced from Llandovery radioactive shales appears to be restricted to the northern portion of the basin. Modeling of several hydrocarbon-charged structural traps indicates post-Hercynian development. Mesozoic and younger structural trap development was favorable for accumulation of Frasnian-sourced oil. Because Llandovery and Frasnian radioactive shales are presently beyond the peak-oil window in the central portion of the Berkine Basin, present gas accumulations are probably a product of both Early Silurian and Late Devonian sources.

PRIMARY RISK IN EXPLORATION FOR UNDISCOVERED RESERVES

The presence of multiple organic-carbon-rich source rocks, high-porosity and high-permeability reservoirs, effective local and regional hydrocarbon seals, and favorable development of traps of a sufficient size to accumulate large quantities of oil and gas in the Berkine Basin lead us to conclude that the primary risk in exploration for additional hydrocarbon reserves is successful seismic imaging of reservoirs and traps. Although vibroseis acquisition and state-of-the-art processing have partially resolved some general imaging problems, seismic resolution of hydrocarbon-bearing reservoirs needs to be improved. Downhole multiples associated with shallow Mesozoic evaporites and carbonates (Figure 21) still provide uncertainty with respect to proper imaging of structural traps. In addition, a low-velocity zone in the shallow high-velocity interval causes “ringing,” which masks the reservoir reflections. Use of 3-D seismic techniques in association with development of discovered fields has greatly aided in resolution of fault geometries and the subtle nature of the structural traps. Accurate mapping of the Upper Triassic–Lower Jurassic (Liassic) interval is necessary with respect to growth history of faulting typically associated with structural traps. Further work with respect to proper bin size needs to be done for adequate resolution of faults.

Figure 22 illustrates the stratigraphic position of seismic horizons which can be correlated and mapped with varying degrees of confidence throughout the Berkine Basin. The Hercynian unconformity surface, an important surface in prediction of TAGI reservoir distribution and critical to identification of potential subcrop traps, is not a consistent reflector. Its seismic resolution is directly related to underlying Paleozoic lithology and fluid contents and is dependent on area. The primary oil-charged reservoir (TAGI) is presently not resolvable seismically. Imaging of Devonian and Carboniferous reservoir intervals awaits further developments in optimum fold, receiver array, and offset studies. If exploration for non-structural traps in the Berkine Basin is to become efficient, imaging at the reservoir scale needs to be resolved.

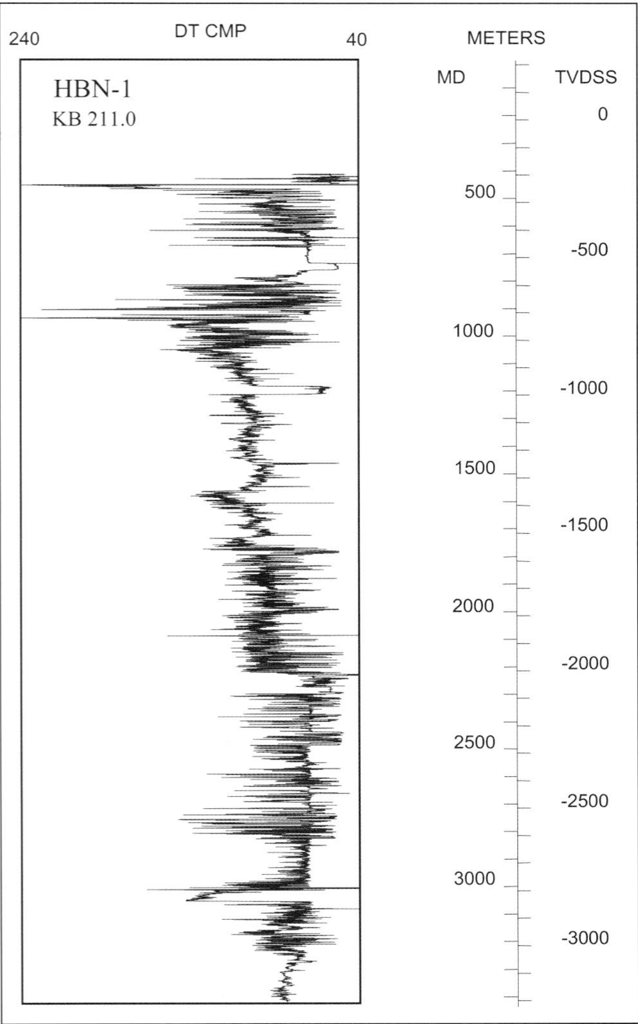

Figure 21. Sonic log of HBN-1 well illustrating shallow intervals (500–1000 m) which produce interbed multiples at reservoir intervals.

BERKINE BASIN UNDISCOVERED HYDROCARBON RESOURCE

Based on Daniels and Emme’s (1995) calculations of generated hydrocarbons from Llandovery and Frasnian

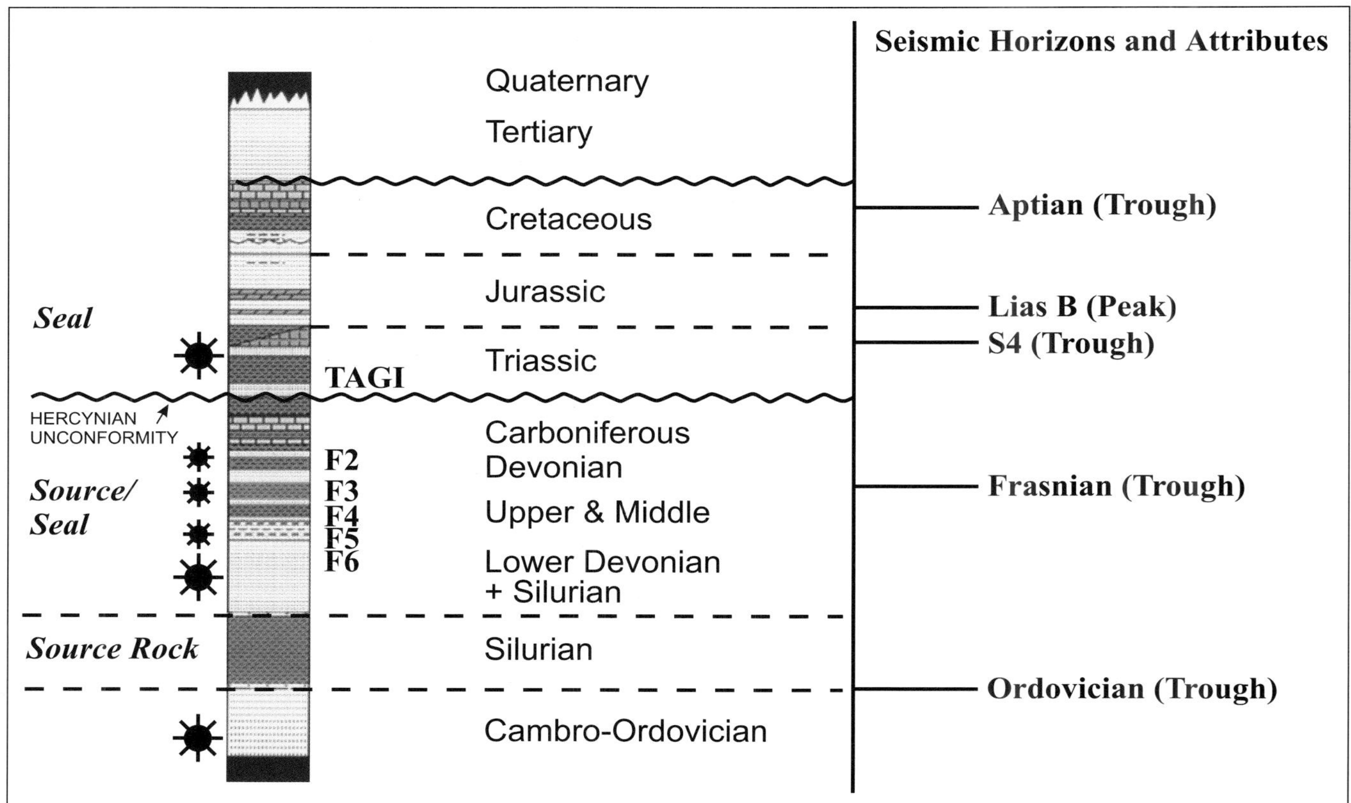

Figure 22. Stratigraphic position of seismic horizons which can be correlated and mapped with varying degrees of confidence throughout the Berkine Basin.

radioactive shales, Macgregor (1998) estimated that generated petroleum in the Berkine and Illizi Basins is more than two orders of magnitude greater than the original oil in place of discovered traps. With most of the proven reserves being concentrated in the Illizi Basin and considering the quality and thickness of Llandovery and Frasnian source rocks in the Berkine Basin, Daniels and Emme (1995) and Macgregor (1998) concluded that the Berkine Basin is immature with respect to discovered hydrocarbon reserves, and suggested that significant yet-to-be-discovered fields in the range of 250–800 million BOE are feasible. The life-cycle curve for the Berkine Basin (Figure 23) supports this conclusion. Analysis of the Berkine Basin discovery process profiles a basin in its peak exploration history.

Considering the maturation histories for Llandovery and Frasnian source rocks and timing of peak oil generation, we expect that a significant portion of the undiscovered hydrocarbon resource in the Berkine Basin will be in the form of gas and condensate, particularly with respect to exploration for Paleozoic reservoirs. Successful exploration for Upper Triassic (TAGI, TAGS, and Carbonaté) accumulations will continue to be a primary contributor to future oil reserves.

Exploration in the Berkine Basin is still in the structural-trap phase of its life cycle, and additional structural traps will continue to be discovered. Various Paleozoic intervals truncated by the Hercynian unconformity along the western and northern portions of the basin should yield significant subcrop traps, if seismic-imaging problems associated with the Hercynian surface and resolution of shallow high-velocity-induced multiples and ringing can be resolved. Distribution of Viséan and Strunian reservoir trends in the northern portion of the basin suggests that stratigraphic onlap/updip pinch-out traps will be present. The fluvial nature of the El Merk–El Borma TAGI-producing trend should also yield numerous stratigraphic traps.

Important stratigraphic intervals for future exploration of the Berkine Basin include the Lochkovian-Pragian F6 massive sandstone interval, Strunian shoreface and delta sandstone packages, Viséan delta sandstone packages, and Carnian TAGI fluvial sandstone packages. Of these intervals, the F6 massive sandstone interval is the most widespread (Figure 11), deepest, and of highest risk because of porosity-versus-depth relationships. Viséan and Strunian exploration will be the most restricted geographically (Figure 15). The northeast-trending TAGI depositional system (Figure 16) is the best-understood stratigraphic interval in the Berkine Basin and will continue to yield future discoveries. The relationship of this depositional system to additional TAGI depositional systems on the western flank of the basin and in the area of maximum sand thickness needs to be clari-

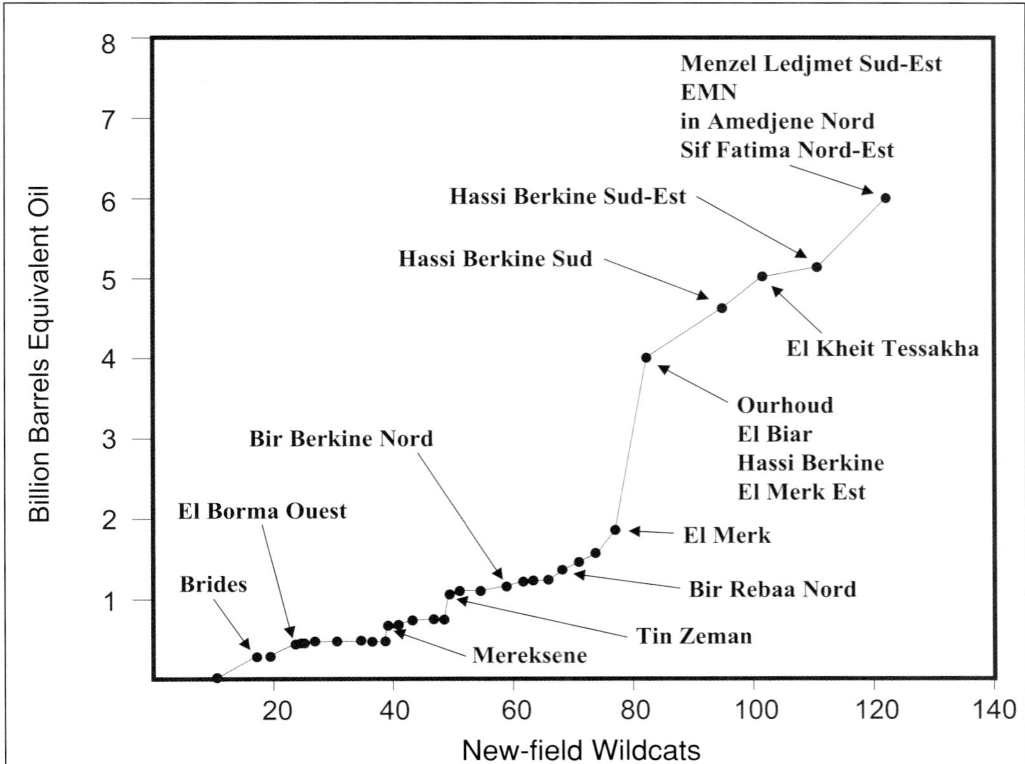

Figure 23. Life-cycle curve for the Berkine Basin portion of the greater Ghadames Basin.

fied. Pre-Lochkovian hydrocarbon potential of the Berkine Basin is unknown. Insufficient penetrations of Lower Silurian, Ordovician, and Cambrian intervals preclude any discussion of their hydrocarbon potential. Based on maturity mapping of known source-rock intervals, hydrocarbon-charged reservoirs in these stratigraphic intervals will be dry gas.

CONCLUSIONS

The Berkine Basin petroleum system compares favorably to other world-class petroleum systems. Multiple major hydrocarbon source rocks of Early Silurian and Late Devonian age are rich in total-organic-carbon content (2% to >17%) and have a generation and expulsion history which coincided with development of large structural traps resulting in giant accumulations of oil and gas. Lower Devonian through Upper Triassic reservoirs are high-porosity (as much as 20%) and permeable (in excess of 1000 md) quartz sandstones. Although migration of hydrocarbons is complex, a series of excellent regional shale and evaporite seals effectively contains the hydrocarbons.

The Anadarko group's exploration history demonstrates the value of understanding the petroleum system. Early exploration was directed toward updip portions of the present structural basin and failed as a result of inadequate mapping of source-rock "kitchens" and understanding of the hydrocarbon migration history. Reappraisal of the group's exploration model shifted attention to a position closer to the hydrocarbon "kitchen," where a more complete Paleozoic section was preserved and regional fault systems provided sites for vertical migration and entrapment. Employment of geophysical data as the primary tool for definition of prospects has evolved from reprocessing of 1970s 12-fold data to acquisition of extensive 120-fold and 240-fold 2-D seismic grids and employment of 3-D seismic techniques as an aid in development of discovered accumulations. Continued refinement of seismic acquisition and processing is needed to effectively image hydrocarbon-bearing reservoirs and to efficiently explore for nonstructural traps.

The undiscovered Berkine Basin hydrocarbon resource is probably significant. Although large oil accumulations will continue to be discovered in the Upper Triassic interval, Paleozoic accumulations will probably be gas and condensate. Future field discoveries of as much as 1000 million BOE are feasible.

ACKNOWLEDGMENTS

We thank the management of Anadarko Petroleum Corporation and its partners Sonatrach Oil and Gas, Lasmo Oil, and Maersk Oil and Gas for permission to publish this paper. Literally dozens of the Anadarko group's geoscientists and petroleum engineers have contributed to the understanding of the Berkine Basin petroleum system since 1986. We especially acknowledge the work of Gary Ford (Anadarko Petroleum Corporation), Neil Mountford (Anadarko Algeria Corporation), and

Alan J. Scott (Alan J. Scott and Associates and Anadarko Algeria Corporation) for their pioneering work on the Late Triassic depositional systems. Olav Nykjaer's (Maersk Oil and Gas) work on the Devonian and Carboniferous stratigraphy and petroleum geology was critical to understanding the Paleozoic potential of the Berkine Basin. Bob Daniels (Anadarko Algeria Corporation) and Jim Emme (Anadarko Petroleum Corporation) initiated Anadarko's first work on petroleum systems in the Berkine Basin. Bob Lunn and Brian Sunderland (Anadarko Petroleum Corporation) played a major role in developing many of the innovative seismic techniques applied to the Anadarko group's Algerian exploration. A. Attar and the late Dr. A. Boudjema of Sonatrach Oil and Gas provided critical technical expertise and guidance during the exploration phases of the Anadarko group's contract. We acknowledge John Seitz and Bruce Stover (Anadarko Petroleum Corporation) for their skillful management of the Algerian project during the early years of exploration. Mark Pease and Bill Sullivan (Anadarko Algeria Corporation) have carried this management tradition into the field-development and production stage. Finally, we acknowledge Bob Allison (CEO of Anadarko Petroleum Corporation). Without his vision and persistence, the group's discoveries in Algeria never would have happened.

REFERENCES CITED

Alem, N., S. Assassi, S. Benhebouche, and B. Kadi, 1998, Controls on hydrocarbon occurrence and productivity in the F6 reservoir, Tin Fouye–Tabankort area, NW Illizi basin, *in* D. S. Macgregor, R. T. J. Moody, and D. D. Lowes, eds., Petroleum geology of North Africa: Geological Society, London, Special Publication 132, p. 175–186.

Bekkouche, D., M. G. Bonhomme, J. Perriaux, H. Abdallah, A. M. Dokka, and A. Ghomari, 1993, Diagenetic events, porosity development and K/Ar dating of clay minerals in Lower Devonian reservoirs of Ghadames basin (northeast Algerian Sahara) (abs.): AAPG Bulletin, v. 77, p. 74.

Boote, R. D., D. D. Clark-Lowes, and M. W. Traut, 1998, Paleozoic petroleum systems of North Africa, *in* D. S. Macgregor, R. T. J. Moody, and D. D. Clark-Lowes, eds., Petroleum geology of North Africa: Geological Society, London, Special Publication 132, p. 7–68.

Boudjema, A., 1987, Evolution structurale du Bassin Petrolier Triassique du Sahara Nord-Oriental (Algerie): Ph.D. thesis, Universite Paris–Sud, 290 p.

Chaouchi, R., M. S. Malla, and F. Kechou, 1998, Sedimentological evolution of the Givetian-Eifelian (F3) sand bar of the West Alrar field, Illizi basin, Algeria, *in* D. S. Macgregor, R. T. J. Moody, and D. D. Clark-Lowes, eds., Petroleum geology of North Africa: Geological Society, London, Special Publication 132, p. 187–200.

Chiarelli, A., 1978, Hydrodynamic framework of eastern Algerian Sahara—Influence on hydrocarbon occurrence: AAPG Bulletin, v. 62, p. 667–685.

Daniels, R .P., and J. J. Emme, 1995, Petroleum system model, eastern Algeria, from source rock to accumulation: When, where and how?, *in* Proceedings of the Seminar on Source Rocks and Hydrocarbon Habitat in Tunisia: Enterprise Tunisienne D'Activités Pétrolières Memoir 9, p. 101–124.

Daniels, R. P., R. C. Hook, P. R. Sorenson, and J. J. Emme, 1994, Triassic depositional system and reservoir development, Ghadames basin, Algeria (abs.): AAPG Bulletin, v. 78, p. 131.

Drumheller, R. E., R. P. Daniels, and J. M. Yarus, 1997, Reservoir characterization of the fluvial T.A.G.I. sandstones, central Ghadames basin, Algeria (abs.): AAPG Bulletin, v. 81, p. 30.

Echikh, K., 1998, Geology and hydrocarbon occurrences in the Ghadames Basin, Algeria, Tunisia, Libya, *in* D. S. Macgregor, R. T. J. Moody, and D. D. Clark-Lowes, eds., Petroleum geology of North Africa: Geological Society, London, Special Publication 132, p. 109–129.

Fekirine, B., and H. Abdallah, 1998, Paleozoic lithofacies correlatives and sequence stratigraphy of the Saharan Platform, Algeria, *in* D. S. Macgregor, R. T. J. Moody, and D. D. Clark-Lowes, eds., Petroleum geology of North Africa: Geological Society, London, Special Publication 132, p. 97–108.

Ford, G. W., and A. J. Scott, 1997, A depositional model and sequence stratigraphy of the Trias Argilo–Greseux Inferieur (T.A.G.I.) in the Ghadames basin, Algeria (abs.): AAPG Bulletin, v. 81, p. 36.

Ford, G. W., and W. J. Muller, 1995, Potential Silurian and Devonian truncation traps across the Ahara Arch, southwest Ghadames Basin, Algeria, *in* D. Clark-Lowes et al., convenors, First Symposium on the Hydrocarbon Geology of North Africa—Abstracts, p. 24.

Gauthier, F. J., A. Boudjema, and R. Lounis, 1995, The structural evolution of the Ghadames and Illizi basins during the Paleozoic, Mesozoic and Cenozoic: Petroleum implications (abs.): AAPG Bulletin, v. 79, p. 1214–1215.

Guiraud, R., 1998, Mesozoic rifting and basin inversion along the northern African Tethyan margin: An overview, *in* D. S. Macgregor, R. T. J. Moody, and D. D. Clark-Lowes, eds., Petroleum geology of North Africa: Geological Society, London, Special Publication 132, p. 217–229.

Illich, H., J. Zumberge, C. Schiefelbein, and S. Brown, 1997, Petroleum systems of the Ghadames basin and Illizi platform (abs.): AAPG Bulletin, v. 81, p. 54.

Macgregor, D. S., 1998, Giant fields, petroleum systems and exploration maturity of Algeria, *in* D. S. Macgregor, R. T. J. Moody, and D. D. Clark-Lowes, eds., Petroleum geology of North Africa: Geological Society, London, Special Publication 132, p. 79–96.

Makhous, M., Y. Galushkin, and N. Lopatin, 1997, Burial history and kinetic modeling for hydrocarbon generation, Part II: Applying the GALO model to Saharan basins: AAPG Bulletin, v. 81, p. 1679–1699.

Meyerhoff, A. A., 1977, Best chances onshore are in China and Russia: Oil & Gas Journal, v. 75, no. 35, p. 132–138.

Oil & Gas Journal, 1997, Legal improvements brighten North African production outlook: Oil & Gas Journal, v. 95, no. 19, p. 48–54.

Oil & Gas Journal, 1999, Industry briefs: Oil & Gas Journal, v. 97, no. 11, p. 30.

Pink, A. T., S. R. Carney, R. E. Drumheller, and L. Okbi, 1999, The structural evolution and reservoir architecture of the HBNS Field from 3-D seismic, Berkine basin, Algeria: AAPG International Conference and Exhibition, Extended Abstracts, p. 398.

Rudkiewicz, J., R. Daniels, and A. Chaouche, 1997, Hydrocarbon migration through faults and successive reservoir infilling in the Ghadames basin, Algeria (abs.): AAPG Bulletin, v. 81, p. 1408.

Scott, A. J. and D. A. Wheller, 1999, Semi-arid lacustrine cycles, a controlling mechanism for Triassic reservoir geometries and characteristics, Berkine basin, Algeria (abs.): AAPG International Conference and Exhibition, Extended Abstracts, p. 444.

Scott, A. J., G. W. Ford, and H. Tourqui, 1997, High-resolution sequence stratigraphy and sedimentologic models of the Algerian Ghadames basin, Trias Argilo–Greseux Inferieur (T.A.G.I.) (abs.): AAPG Bulletin, v. 81, p. 1411.

Sonatrach, 1995, Geology of Algeria, *in* Conference Sur L'Evaluation des Puits, Algerie, p. I-1–I-93.

Tissot, B., J. Espitalie, G. Deroo, C. Tempere, and D. Jonathan, 1984, Origin and migration of hydrocarbons in the eastern Sahara (Algeria), *in* G. J. Demaison and R. J. Murris, eds., Petroleum geochemistry and basin valuation: AAPG Memoir 35, p. 315–324.

Wheller, D. A., S. R. Carney, R. E. Drumheller, and S. I. Winstanley, 1999, Sedimentology and reservoir characterization of Hassi Berkine South (HBNS) field Block 404, Berkine basin, Algeria: AAPG International Conference and Exhibition, Extended Abstracts, p. 504.

Zumberge, J. E., S. Macko, M. Engel, F. Johansson, C. Schiefelbein, and S. Brown, 1996, Silurian shale origin for light oil, condensate, and gas in Algeria and the Middle East (abs.): AAPG Bulletin, v. 80, p. 160.

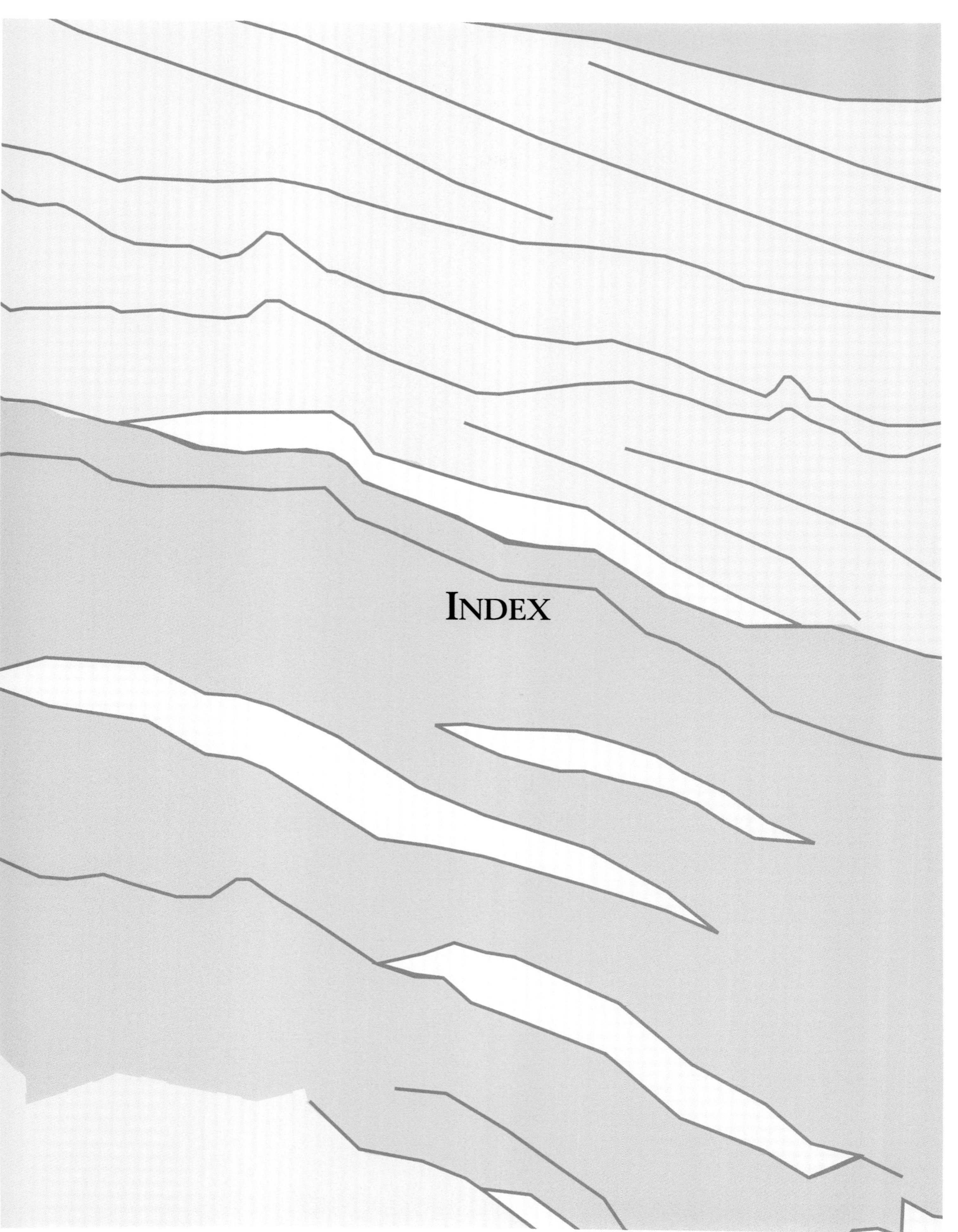

INDEX

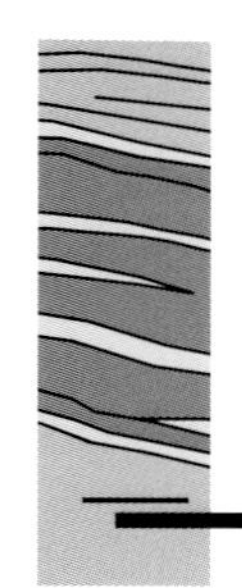

INDEX

A

Abkatun field, 345
Aborted rift systems, of southern South America, 386
Abu Zenima Formation, 462
Acacus Formation
  isopach map for, 445
  structure map on top of, 446
Active margins, in Venezuela, 354–356
Africa, *see also* Egypt; North Africa
  offshore west, 517–529
  oil types of Atlantic margin, 407
  plate reconstruction of north, 464
  subduction of plate, 462
Ajdabiya trough
  map of, 432, 438
  petroleum systems of, 433–434, 438
  stratigraphy of, 433
  structural cross section of, 438
Al Ghubar salt dome, structure of, 512
Alam el-Bueib (AEB) Formation, 459
Alaska
  discoveries of, 142–144
  exploration of, 141–142
  map of, 138
  northern, 138–161; *see also* Northern Alaska
  oil production in, 142–144
  petroleum potential of, 144–161
  petroleum systems of northern, 143, 144–145
  resources of, 139
  tectonic setting of northern, 138–139, 140
Algeria, *see also* Berkine Basin
  exploration in Berkine Basin, 532–555
  giant fields of, 533
  tectonic elements of, 533
Alliance Pipeline, 114–115
Alluvial fans, of Karamay field, China, 331
ALNG (Australian LNG), 293
Amguid–El Biod Arch, 535
Anaco trend south subthrust play, 363
Anadarko group, exploration in Berkine Basin by, 532, 534–538, 540
Ancestral Rockies orogeny, 205
Andean compression, 366
Andean-type subduction basins, of southern South America, 389–398, 401
Andes, plays in Venezuelan, 365, 366, 367
Angola
  exploration history of, 526–527
  geologic history of, 520–526
  stratigraphy of, 521
Antarctica, future exploration of, 18
Anticlines, in Venezuelan play, 366
Antler orogeny, 204
ANWR, *see* Alaska National Wildlife Refuge
API gravity, versus depth for Lower Congo Basin, 525
Arabian Plate
  basement depth map of, 486
  Cambrian paleogeography of, 489, 492, 493
  Carboniferous paleogeography of, 495–496, 497
  Devonian paleogeography of, 494–495, 496
  exploration history of, 484–485
  Hercynian orogeny in, 499–501
  hydrocarbon availability in, 502–504
  Ordovician paleogeography of, 489, 491, 494
  paleopositions of, 486, 487
  Paleozoic basin evolution of, 489–499
  Paleozoic stratigraphy of, 487–489
  Permian paleogeography of, 497–499
  petroleum systems of, 503
  plays of, 504–512
  reservoirs of, 502–504
  Silurian paleogeography of, 492, 494, 495
  tectonic setting of, 484, 485–487
Arabian shelf, stratigraphy of source rocks of, 425
Archie rock types, porosity/permeability of, 69
Arctic
  future of basins of, 17
  petroleum basins of, 125–127
  provinces of Alaska, 138–161
Arctic National Wildlife Refuge (ANWR), 140
Argentina, *see also* South America (southern)
  foreland basins of, 391–394
  petroleum history of, 376
  pipelines in, 377
Asia-Australasia region
  EUR of, 282–284
  petroleum systems of, 282–284
Athabasca, oil sands of, 118–119
Atlantic continental margin, 387, 388, 389, 390
  basins of southern South America, 387–389
  gas hydrates of, 87–90
  of United Kingdom, 186–195
Atlantic Ocean, *see* South Atlantic Ocean
ATROR, for property values, 83
Auk field
  redevelopment of, 81, 82
  production history of, 82
Aulacogens, of southern South America, 398
Austral 1, 2, 3 divisions, 298
Austral-Magallanes Basin, 396–398
Australasia region
  EUR of, 282–284
  petroleum systems of, 282–284
Australia
  eastern margin of, 301, 312, 313
  exploration history of, 288–291
  petroleum basins of, 289, 290–291
  petroleum resources of, 292

petroleum supersystems of, 290–292, 295, 296–297
promise of Paleozoic, 311–315
reserves of, 290–296
southern margin of, 296–301, 308
Australian Antarctic Territory (AAT), 287, 288
Australian Exclusive Economic Zones (AEEZ), 287, 288

B

Bahloul source rocks, 450
Baltim trend, gas pays of, 478
Barik dome, structural diagram of, 512
Barik Sandstone, 512
Barinas-Apure Stratigraphic Traps play, 367
Barrow arch, 155
Basins, *see* Petroleum basins; Sedimentary basins
Bayu-Undan field, 294, 306
Beaufort Sea, discoveries of, 125–126
Beaufortian Sequence
map of, 152
petroleum potential of, 151–153
reservoir characteristics of, 153
Beda Formation
isopach map for, 439
petroleum geology of, 438–439
Beni Suef Basin
new-field discoveries in, 468
structural section of, 471
Benioff-type subduction basins, of southern South America, 398
Berkine Basin
petroleum distribution in, 532–534, 535
Anadarko group's exploration of, 534–538
architecture of, 532–534
chronostratigraphy of Paleozoic of, 536
exploration risk in, 553
hydrocarbon migration and trap fill in, 551–553
life-cycle curve for, 555
location map of, 533
peak oil-generation timing maps for, 541, 542
petroleum systems of, 538–553
quartz sandstone reservoirs of, 543–549
radioactive shales of, 538, 541, 553
reliable seismic horizons of, 554
reservoirs of, 540–549
seismic lines of, 537, 539, 551, 552
source rocks of, 538–540
stratigraphic column for, 534
structural traps in, 550, 551, 552
undiscovered resources of, 553–555
Bight Basin
petroleum potential of, 298
regional sections of, 309
seismic lines for, 310
Biogenic gases, from coal, 211, 213
Bitumen, in Canadian oil sands, 118–119
Blake Ridge
gas hydrates of, 87–90
seismic profile of, 89
Body checking, 509, 511
Bohaiwan Basin, map of, 320
Bolivia
foreland basins of, 391–394
petroleum history of, 376
Bolsones Basin, 394–395
Boom-and-bust cycles, 35
Bottom-simulating reflectors
at gas-hydrate zone, 87–89
from Lord Howe Rise, 313
Brazil, petroleum history of, 376, 378
Brecknock field, 293–294, 304
Bremer Basin, petroleum potential of, 298–299
British Columbia, petroleum systems of, 127
Brookian fold belt, 157–159
Brookian Sequence
clinoform facies of, 155–157, 158
map of, 157, 158
petroleum potential of, 153
seismic expression of, 156
topset facies of, 153–154, 156, 157
Brooks Range, thrust belt of, 159
Burgos Basin
map of plays of, 339
petroleum geology of, 339
production profile of, 340

C

Cabinda Gulf Oil, 526
Caciporé Formation, 404
Caernarfon Bay Basin, Irish Sea, 185
Campos Basin, 520
Canada
East Coast basins of, 121–125
extra-heavy oil in, 42–43
future of East Coast, 124–125
gas hydrates of, 92–96
heavy oil of, 117
oil sands of, 113, 115, 117–121
petroleum production in, 112–113, 128–130
petroleum resources of, 111–112
petroleum systems of, 115–127
projected resources of, 128–131
sedimentary basins of, 110, 115–127
Cantarell field, 345
structure map of, 346
Cañón de Veracruz, 348
Capricorn Basin, 311
Carbonates
of Berkine Basin, 548
of Lisburne Group, 149
of Timan-Pechora Basin, 273, 275, 276
Carboniferous, of North Sea gas province, 180–182
Cardigan Bay Basin, 186, 187
Cascadia continental margin
gas hydrates of, 90–91
physiographic map of, 91
Caspian Sea, petroleum potential of, 244, 245–251
Cave Gulch gas field, 227–228
Cedar Hills oil field, 225–226
Ceduna Subbasin, seismic lines for, 310
Central Graben (North Sea)
petroleum geology of, 172–175
redevelopment of, 81, 82

Central Pampas Basin, 380
Central Patagonian basins, 383
Chaco-Formosa Platform, 380
Chaco-Paraná Basin
  petroleum geology of, 379–382
  petroleum systems of, 382
Channel systems, of Lower Congo Basin, 523
Cheshire Basin, 184
Chiapas-Tabasco province, 343, 344
Chile, petroleum history of, 376
China
  history of production in, 319–321
  reservoir potential of, 330–331
  sedimentary basins of, 320
Chloride concentration profiles, Blake Ridge, 90
Chlorite cements
  and permeability, 545
  and porosity, 546
Chronostratigraphy
  of Berkine Basin Paleozoic, 536
  of Miocene of Lower Congo Basin, 522
Chrysaor field, 294, 307
Chukchi shelf, exploration in, 140, 142
Chu-Sarysu Basin, reserves of, 255–256
Cinturón Plegado de Perdido province
  petroleum geology of, 348
  seismic section of, 349
Clinoform facies, of Brookian Sequence, 155–157, 158
Clusters, of fields, 38
Coal, future as fuel, 31
Coal-bed methane
  of United Kingdom onshore basins, 185
  of Greater Rocky Mountain Region, 210–212, 213, 217–218, 222, 228–230, 234, 235
Cold Lake, oil sands of, 119
Colombia field, 367, 368
Colorado, petroleum potential of, 201–235; *see also* Greater Rocky Mountain Region
Colorado Basin (South America)
  failed arms of, 388
  map of, 389
Comalcalco province, 344
Congo Basin, *see* Lower Congo Basin
Congo Canyon, map of, 524
Continental margins
  of Atlantic, 387, 388, 389, 390
  gas hydrates of, 87–91
  of Pacific, 90–91
  of southern Australia, 296–301, 308
  of southern South America, 387–389
  of United Kingdom, 186–195
Continuous-type accumulations, of Greater Rocky Mountain Region, 213–214, 217, 222, 233
Conventional traps, of Greater Rocky Mountain Region, 213, 214
Cordilleras Mexicanas province
  petroleum geology of, 348
  seismic section of, 350
Creaming curves
  for central North Sea, 175
  for Timan-Pechora Basin, 266, 268
Cretaceous, petroleum systems of Greater Rocky Mountain Region, 214–215, 216, 217, 218
Cross sections, of Greater Rocky Mountain Region, 206–207, 216
Crude oil
  of Canada, 111–114, 129
  estimated ultimate recovery of, 28
  future production of, 52
  resource versus production, 47
  of United States, 24
  world consumption of, 22, 23
  world production of, 22, 24
  world reserves of, 24
Cuenca de Burgos, 339
Cuenca de Tampico-Misantla, 339–341
Cuenca de Veracruz, 341–342
Cuyo Basin, 394–395
Cycle steam stimulation (CSS), for oil-sand recovery, 120

D

Daqing field (China), 320
Décollement, in Venezuelan play, 361
Deep-basin accumulation, in Greater Rocky Mountain Region, 219, 221
Deep water
  exploration of Lower Congo Basin, 518, 519
  as primary target, 36
  provinces of Mexican Gulf of Mexico, 345–349
  turbidite plays, 518–520, 522–526, 527
Defa Formation, 438
Delta del Río Bravo province, cross section of, 348
Delta Platform play, 359
Delta quartz sandstone, of Berkine Basin, 547
Deseado Massif, 383, 386
Diamictites, of southern South America, 381
Diapir Belt play, 360
Diasterane biomarkers, from Foz do Amazonas Basin, 410
Dionysis field, 294, 307
Discoveries, *see also* Exploration
  of 1990s, 35–43
  of Alaska, 142–144
  giant of world, 36–37
  recent in Greater Rocky Mountain Region, 225–229
  significant of world, 39
  in Timan-Pechora Basin, 269, 270
Discrete-type accumulations, of Greater Rocky Mountain Region, 213, 214
Drilling history, *see* Exploration; History
Drunkard's Wash coal-bed methane field, 228–229
Dry holes, 58
Dushanzi field, 330, 331

E

E&P hot spots, 38, 40–41, 42, 59
East Coast basins, of Canada, 121–125
East Texas basin, traps in, 16, 17
Eastern Desert, exploration potential of, 468–469
Eastern margin, of Australia, 301, 312, 313
Eastern Patagonia Basin, 389
Eastern Venezuela Basin, petroleum systems of, 358
Economics

of 1990s, 35–36
of Canada's production, 128
of gas-hydrate production, 105
of Greater Rocky Mountain Region petroleum potential, 223–225
of reserve additions, 83
Egypt, *see also* Upper Egypt
cumulative discovery volume of, 466
drilling history of, 464–465, 466
exploratory tests in, 456–457
giant fields of, 455
map of, 454
paleogeographic reconstructions of, 463, 465
petroleum history of, 453–458
petroleum potential of, 465–474
petroleum systems of, 460
reserves of, 460, 466
rifting in Jurassic of, 459
tectono-stratigraphy of, 458–464, 461
Egyptian General Petroleum Corporation, 470
Eifelian-Givetian quartz sandstones, of Berkine Basin, 545–546, 547
El Soldado field, 360
Ellesmerian Sequence, petroleum systems of, 144, 145–151
Endicott Group
map of, 146
petroleum potential of, 145–146
Energy consumption
of United States, 22, 46
of world, 22, 51
Energy pyramid, 48, 53
Energy sources, renewable nonpolluting, 31–32
Energy supplies, future of, 29, 30
Espino Graben play, 364
Estimated ultimate recovery (EUR)
of Asia-Australasia region, 282–284
definition of, 260
Etel Formation
isopach map for, 435
stratigraphy of, 433, 434
Eurasian petroleum systems, of Indonesia and Papua New Guinea, 283
Exploration, *see also* E&P; History
Australia's history of, 288–291
Berkine Basin's risk of, 553
Egypt's history of, 453–458
Greater Rocky Mountain Region potential, 217–220
future of world, 12–15, 17–19, 57–60
history of world, 11–12
Iran's potential for, 426
Junggar Basin's potential for, 332–333
of Lower Congo Basin, 518, 519, 526–527
of northern Alaska, 141–142
of South America (southern), 378
Tarim Basin's potential for, 331–332
of Timan-Pechora Basin, 260–262, 266–278
United Kingdom's potential for, 170–195
Venezuela's future of, 358–370
Extra-heavy oil
in Canada, 42–43
in Venezuela, 42–43

F

Fahdene-Sidi Kralif source rocks, 450
Failed arms of triple junctions, in southern South America, 386–388
Fan complex, of Faroe-Shetland Basin, 189
Faregh oil field, cross section of, 438
Faroe-Shetland Basin, 187–191
fan complex of, 189
plays of, 190, 191
Fars province (Iran)
reservoirs and seals of, 424
source rocks of, 423–424
Field-size distribution (FSD)
definition of, 260
for Egypt, 467
for Timan-Pechora Basin, 268, 271
Fluid pressure regimes, of Greater Rocky Mountain Region, 215–217
Foinaven field, 190
Fold belt, of Iran, 418–420
Forearc basins, of southern South America, 398, 401
Foredeep
idealized scheme of, 356
of Venezuela, 356, 357
Foreland basins
of Bolivia and Argentina, 391–394
of Iran, 418–420
of southern South America, 392, 389–398, 400
transgression in Egypt, 462
Fossil fuel, future of, 29–31; *see also* Petroleum potential; Future
Foz do Amazonas Basin
map of, 404
petroleum geology of, 404–407
petroleum systems of, 407, 409–412
reservoir structure of, 408
sections of, 406
stratigraphic chart of, 405
tectonic setting of, 407–409
Fractured reservoirs, of Greater Rocky Mountain Region, 210, 219, 223
Franja de Sal Aloctona province
cross section of, 349
petroleum geology of, 348
Franja Distensiva province
petroleum geology of, 346
seismic section of, 347
Frontier; *see also* Future
of Canada, 113
of South Atlantic, 403–413
of United Kingdom, 171, 188
Fuel, fossil, 29–31
Furrial trend, 360
Future
of Antarctic exploration, 18
of Arctic basins, 17
of Beaufortian Sequence, 153
of Canada's East Coast, 124–125
of crude-oil production, 52
of Greater Rocky Mountain Region, 225–229
of Egypt, 465–474

of fossil fuels, 29–31
of gas production, 51
hot spots of, 41, 42
of Indonesia, 284–286
of Kazakhstan, 243–257
of Lower Congo Basin, 517–520, 528–529
of northern Alaska, 161
of Papua New Guinea, 284–286
of petroleum supplies, 17–19
of production levels, 41–43
renewable nonpolluting energy sources of, 31–32
of United Kingdom, 194
of United States, 30
of Venezuela, 358–370
of world energy supplies, 17–19, 29
of world exploration, 12–15, 17–19, 57–60

G

"G" field, redevelopment of, 69–75
Gabes-Sabratha Basin, 442
Gamma-ray logs, for Berkine Basin, 543, 546, 548
Gas, *see also* Methane; Natural gas
of Canada, 112
in coal beds of Greater Rocky Mountain Region, 210–212, 213, 217–218, 222, 228–230
demand for, 43, 106
future production of, 43, 51
in gas hydrates, 101–103
giant discoveries of, 37, 38
of Greater Rocky Mountain Region, 232
of Green River Basin, 226–227
of Indonesia and Papua New Guinea, 284, 285
of Kazakhstan, 244–256
in Mesaverde Group, 221
of Nile Delta, 478
of northern Alaska, 145
of North Sea, 178, 179–182
of southern South America, 376–378
of Wind River Basin, 227–228
Gas condensates, from Karachaganak field, 249
Gas hydrates
of continental margins, 87–91
depth-temperature stability of, 87, 100
natural occurrence of, 96–101
of North Slope of Alaska, 91–92
in permafrost, 92–96, 97, 99
production technology of, 103–105
reservoir models for, 98–99
resource potential of, 96–106
structure of, 86–87, 97
well logs of, 92, 95
worldwide locations of, 87, 88, 97, 103
Gas-potential reserves, undiscovered, 40
Gas-to-liquids (GTL), 49
Gebel Duwi Formation, 476
Generation potential
of Junggar Basin, 326
of Tarim Basin, 324
Geochemistry, of onshore China source rocks, 327–330
Georges Banks Basin, petroleum systems of, 127
Ghaba salt basin, geologic cross section of, 501
Ghadamis Basin
exploration in, 532–555
isopach map for, 445
location map of, 441
petroleum geology of, 440–441
stratigraphic chart of, 444
structure of, 442, 443, 446
tectonic setting of, 431
Gharif Formation
play of, 507–508
structure of, 512
Ghawar area
Paleozoic section of, 504
structural cross section of, 507
Giant discoveries, of 1990s, 36–37
Giant fields
of Algeria, 533
of Egypt, 455
of Libya, 430
of Venezuela, 370
Girassol field, 526, 527
Glaciation, Ordovician of Gondwana, 491–492
Golden Lane, plays of, 340
Gondwana
on Arabian Plate, 485
megasequences of, 381–382
Ordovician glaciation of, 491–492
in southern South America, 381–382, 387
Gorgon field, 292, 300
Grabens
of Egypt, 459
of North Sea, 172–176
of Venezuela, 364
Grand Banks, petroleum systems of, 123
Grand Erg Oriental, exploration in, 532–555
Gravity flow sand play, in Oman, 509, 510, 511
Great Australian Bight, 298, 309
Greater Rocky Mountain Region (GRMR)
coal-bed methane reservoirs of, 210–212, 213, 217–218, 222, 228–230
cross sections of, 206–207
geologic history of, 204–209
geologic map of, 203
geologic provinces of, 231–232
new technology in, 220–223
petroleum basins of, 212
petroleum systems of, 209–218
recent discoveries in, 225–229
reserves of, 232
resource estimates of, 202–204, 233–234
stratigraphic chart for, 211
structure of, 205–209
traps of, 212–214
Green River Basin
pressures in, 216, 220
recent discoveries in, 226–227
Groundwater-dominated fluid systems, of Greater Rocky Mountain Region, 215, 218
Growth faults, of Foz do Amazonas Basin, 408
GTL, *see* Gas-to-liquids
Guaporé shield, 393

Guarico Basin, 362, 364
Gulf of Mexico Basin (United States), *see also* Mexican portion of Gulf of Mexico (MGOM)
  secondary recovery in, 69–75
  stratigraphic section of, 72
  structure map of, 74
  value of fields in, 83
  well logs from, 73
Gulf of Suez
  exploration potential of, 468–469, 473
  field-size distribution for, 467
  petroleum system of, 468
  rifting of, 462–463
  sequence stratigraphy of, 474
  step-out exploration of, 475
  structural map of, 473
  synrift petroleum systems of, 474
  tectonic setting of, 462–463, 464

H

Haima deep gas play, 510–511, 512
Half graben, in Venezuelan play, 364
Halokinesis, in trapping, 175
Hammerhead supersequence, 310
Hanna trough, exploration in, 139, 140, 145
Hasirah gravity-flow sand play, 509, 510, 511
Hassi Messaoud Ridge, 535
Hatton Basin, 193–195
Hawtah oil field, perspective diagram of, 508
Heavy oil, of Canada, 117
Hercynian orogeny
  tectonic history of, 495, 496, 499–501
  unconformity of, 547
Hibernia field, of Canada, 114
History
  of 1990s, 35–43
  of Australia exploration, 288–291
  of China's production, 319–321
  of Egypt's petroleum exploration, 453–458, 464–465, 466
  of Iranian petroleum exploration, 417–418
  of Kazakhstan's production, 244–245
  of Lower Congo Basin exploration, 526–527
  of Libyan petroleum exploration, 429–430
  of oil shortages, 46
  of petroleum industry, 12
  of southern South America's petroleum industry, 373–378, 398–401
  of Timan-Pechora Basin exploration, 260–262
  of United Kingdom production, 168–170
  of Venzuela's petroleum industry, 353–354
Hopanes
  from Foz do Amazonas Basin, 410, 411
  from Niger Delta Basin, 410, 411
Hormuz salt plug, satellite image of, 423
"Hot" shale
  as hydrocarbon source, 502, 503
  tectonic history of, 488, 492
Hot spots, E&P, 38, 40–41, 42, 59
Hubbert's model, of production versus consumption, 50
Hydrocarbon charge, in onshore China basins, 321
Hydrocarbon energy pyramid, 48
Hydrocarbon potential, *see* Petroleum potential
Hydrocarbon-dominated fluid systems, of Greater Rocky Mountain Region, 215–217, 219–221
Hydrocarbons, *see* Gas; Crude oil; Oil; Petroleum
Hydrodynamics, of Greater Rocky Mountain Region, 215–217

I

Iabe Formation, 520, 525
Ice, gas hydrates in, 92–96, 97, 99
Illizi Basin
  peak oil-generation timing maps for, 541, 542
  petroleum distribution in, 535
  petroleum systems of, 538–553
  radioactive shales of, 538, 541
  reservoirs of, 540–549
Imbricated thrusts, in Venezuelan play, 365
Incipient rift systems, of southern South America, 386
Indonesia
  petroleum potential of, 284–286
  petroleum systems of, 281–284
Infrastructure, *see* Pipelines
Intraarc basins, of southern South America, 398, 401
Intracratonic basins, of southern South America, 378–385
Iran
  exploration history of, 484–485
  Khuff reservoirs of, 505
  microplates of, 485, 489
  oil fields of, 419
  petroleum history of, 417
  petroleum potential of, 424–426
  plate-tectonic reconstructions of, 420, 421
  source rocks of, 423–424, 425
  stratigraphy of, 420–423, 425, 490, 505
  tectonic setting of, 418–420, 484, 485–487, 493
Iraq, *see also* Arabian Plate
  stratigraphy of source rocks of, 425
Irish Sea, basins of, 184, 185–186
Isopach maps, for Sirt Basin, 434, 435, 436, 439
Itarare Group, 381
Ivishak Sandstone, petroleum potential of, 149–151

J

Jauf Sandstone, gas in, 508
Jeanne d'Arc Basin, petroleum systems of, 123
Jonas gas field, 226–227
Jordan, *see also* Arabian Plate
  geologic cross section of, 500
  stratigraphy of, 490, 491
Junggar Basin
  map of, 320, 322
  petroleum potential of, 326, 332–333
  postmature Permian rocks of, 330
  reservoir potential of, 331
  sandstones in, 333
  source-rock potential of, 323–324, 326

K

Karachaganak field, 244, 249
Karamay field, 331
Karazhanbas oil field, 255

Kazakhstan
- history of production in, 244–245
- petroleum potential of, 243–257
- petroleum resources of, 246
- undiscovered reserves of, 245–256

Kekiktuk Conglomerate, 145–146
Kenkiyak oil field, 251
Khoreyver depression, 263–265, 273, 275
Khuff Formation reservoirs
- development of, 505
- gas fields of, 485, 506
- plays of, 504–507
- seismic impedance contrast in, 506
- structural cross section of, 507
- tectonic history of, 498

Khuzestan province (Iran)
- foreland fold-belt stratigraphy of, 420–423
- reservoirs and seals of, 424
- source rocks of, 423

Kimmeridge Clay, 170
Kingak Shale, depositional style of, 152
Kolva swell, 263–265, 274, 276, 277
Komi Republic, petroleum potential of, 266–277
Komombo Basin
- exploration in, 468
- seismic line and stratigraphy of, 472

Kuito field, 523, 526
Ku-Maloob-Zaap complex, 345
Kumkol oil and gas field, 254
Kuwait, *see* Arabian Plate
Kwanza Basin, 520

L

La Batalla del Petróleo, 376
La Luna Formation, 357, 358, 365
Laboulaye Basin, 380
Labrador, offshore basins of, 124
Lacustrine deposits, of China's basins, 323, 330
Laisky swell, 276
Lake Maracaibo
- redevelopment of fields in, 75–81
- sealing faults of, 77, 78
- structure map of, 76

Landsat image, of Egypt, 454
Laramide orogeny, 205, 216
Laurestan province (Iran), source rocks of, 423
LCU, *see* Lower Cretaceous unconformity
Leman Sandstone, 179
Libya
- offshore stratigraphy of, 447
- petroleum history of, 429–430
- petroleum potential of, 467
- petroleum systems of, 431–448
- source rocks of, 441, 442, 444
- tectonic setting of, 430–431
- western, 440–448

Light oil, from Western Canada Sedimentary Basin, 115–116
Limoeiro Formation, 406
Liquid natural gas (LNG), future production of, 43
Liquid petroleum, *see* Oil
Lisburne Group, 146–149
Litoral de Tabasco province, 345, 346
Llandovery radioactive shales, of Berkine Basin, 538, 541, 553
LNG, *see* Liquid natural gas
Lochkovian-Emsian quartz sandstones, of Berkine Basin, 543–545
Loeme Salt Formation, development of, 520
Lord Howe Rise
- bottom-simulating reflectors from, 313
- map of, 312, 313, 314
- petroleum potential of, 301, 308–309
- structural elements of, 314

Los Monos Formation, 391
Louisiana, secondary recovery in, 69–75
Lower Congo Basin
- API gravity versus depth, 525
- deep-water exploration of, 518, 519
- deep-water turbidite producing trends of, 527, 528
- exploration history of, 526–527
- future potential of, 517–520, 528–529
- geologic history of, 520–526
- map of, 519, 524
- Miocene chronostratigraphy of, 522
- petroleum systems of, 522–526
- source rocks of, 520, 523, 525, 526
- stratigraphy of, 521

Lower Cretaceous unconformity (LCU), 140, 143, 146
Lucaogou Formation, thermal maturation history of, 332

M

Macachin Basin, 380
Mackenzie Delta
- gas hydrates of, 92–96
- petroleum systems of, 125–126
- well log of, 95

Macuspana province, 344–345
Magallanes Basin, 387
Magoon and Dow methodology, 259–260
Malembo Formation, as source rock, 523, 525, 526
Malvinas Basin, 387
Mangyshlak province, reserves of, 252–253
Maracaibo Basin, petroleum systems of, 358
Maradah graben
- map of, 432
- petroleum systems of, 435–436
- stratigraphy of, 433

Maryborough Basin, 311
Mass chromatograms, of oils from Niger Delta Basin, 409–411
Mature fields
- redevelopment of, 64–83
- reserve additions from, 82–83
- of Texas, 64

Meade Basin, development of, 147
Mediterranean region
- cross section of, 477
- drilling history of, 465
- exploration potential of, 471–474
- field-size distribution for, 467
- Messinian crisis in, 463–464
- plate reconstruction of, 464

Megasequences, of southern South America, 381–382
Merida Andes, plays in, 365, 366, 367

Mesaverde Group, pressures in, 215, 220, 221
Messinian Crisis, 463–464
Methane, *see also* Gas; Natural gas
  in coal beds of Greater Rocky Mountain Region, 210–213, 217–218, 222, 228–230, 234, 235
  hydrates of, 49
Methodology, of Magoon and Dow, 259–260
Metlaoui group, structure map of top of, 448
Mexican Cordillera, 348, 350
Mexican portion of Gulf of Mexico Basin (MGOM)
  deep-water provinces of, 345–349
  maps of, 338
  petroleum potential of, 337–349
  petroleum provinces of, 338–349
  seismic sections of, 347–350
Mexico, offshore petroleum provinces of, 337–351
MGOM, *see* Mexican portion of Gulf of Mexico Basin
Mid-Atlantic Ridge, rifting on, 520
Middle Caspian Basin, reserves of, 252–253
Middle East, 42, *see also* Arabian Plate
Midlands microcraton, 184
Migration pathways, of Timan-Pechora Basin, 265–266
Mole D'Ahara, 534, 535
Montana, *see also* Greater Rocky Mountain Region
  recent discoveries in, 230
  petroleum potential of, 201-235
Moratoriums, Canadian areas under, 127–129
Moray Firth, 172–176
Mud diapir, in Diapir Belt play, 360
Murzuq Basin
  isopach map for, 445
  location map of, 441
  petroleum geology of, 441–442
  stratigraphic chart of, 444
  structure of, 442, 443, 446
  tectonic setting of, 431

N

National Petroleum Reserve of Alaska (NPRA), 141–142
Natural gas, *see also* Gas
  of Canada, 112, 114, 115–116
  hydrates of, *see* Gas hydrates
  future as fuel, 31
  future production of, 43, 51
Natural-gas liquids, of Greater Rocky Mountain Region, 232
Nenets Okrug (Russia), petroleum potential of, 266–277
Neuquen Basin, 395, 397
Newfoundland, basins offshore, 123
Niger Delta Basin
  oil types from, 409, 411
  relationship to Foz do Amazonas Basin, 403
  reservoir structure of, 408
Nile Delta
  cross section of, 477
  exploration potential of, 471–474
  field-size distribution for, 467
  gas pays of, 478
  Messinian crisis in, 463–464
  progradation of, 464
  seismic profiles of, 477, 478, 479
  tectonic setting of, 462, 464
Niobara Formation, 210
Nonpolluting energy sources, 31–32
North Africa, plate reconstruction of, 464; *see also* Africa; Egypt
North Andean Mountain Front play, 365
North Caspian Basin
  reserves of, 245–251
  stratigraphy of, 247
  structure of, 247
North Dakota, recent discoveries in, 226
North Sea
  gas province of, 178, 179–182
  gas reservoir potential of, 181
  lithofacies of, 172
  oil province of, 170–179
  redevelopment in Central Graben, 81, 82
  reservoirs of, 170–179
  source rocks of, 173
  underexplored plays of, 174
North Slope of Alaska, gas hydrates of, 91–92
North Ustyurt Basin, reserves of, 254–255
North West Shelf
  tectonic elements of, 299
  undeveloped reserves of, 292–294
  undiscovered reserves of, 294–296
Northern Alaska
  geologic setting of, 138–139, 140
  petroleum potential of, 144–151, 160–161
  petroleum province of, 138–161
  petroleum systems of, 143, 144–145
  stratigraphy of, 139–141
  thrust belt of, 157–160
Northern Offshore Venezuelan Basin play, 369
Northwest Territories, hydrocarbon potential of, 121
Nova Scotia, basins offshore, 121–123
NPRA, *see* National Petroleum Reserve of Alaska
Nubian Formation
  isopach map of, 436
  stratigraphy of, 433, 435–436
  structure map on top of, 437
Nuclear energy, 53–54
Nukhul Formation, 462

O

Ocean Drilling Program, search for gas hydrates by, 87–91
Oceanic gas hydrates, 87–91, 103
Offshore exploration, world's future of, 14–15
Oficina trend, 360, 363
Oil, *see also* Crude oil
  future as fuel, 29–30
  giant discoveries of, 36–38
  of northern Alaska, 144–145
  province of North Sea, 170–179
  ultimate production of United States, 25–26
  unconventional resources of, 48–49
  undiscovered potential reserves of, 39–40
  world consumption of, 22, 23
Oil exploration, *see* Petroleum exploration
Oil sands
  of Canada, 113, 115, 117–121
  production of, 115, 119

Oil seeps
in Foz do Amazonas Basin, 412–413
remote-sensing detection of, 412–413
Oil shale, of world, 49
Oil shortages, history of, 46
Oil types
in Foz do Amazonas Basin, 410–411
of Iran, 424
in Niger Delta Basin, 409–411
Oleananes, from South Atlantic, 410, 411
Oman
basin evolution in, 489
geologic cross section of, 501
gravity-flow sand in, 509, 510, 511
plays of, 509–510
source rocks of, 425, 502
stratigraphy of, 425, 491
Oriental Cordillera, 393
Orinoco Delta, 359
Orinoco Oil Belt, 364, 367
Overpressure, in Greater Rocky Mountain Region fluid systems, 215, 219, 220

P

Pacific continental margin, gas hydrates of, 90–91
Pacific petroleum systems, in Indonesia and Papua New Guinea, 283
Paleocañon de Chicontepec, depositional model of, 340–341
Paleozoic
Arabian Plate stratigraphy of, 487–489
Berkine Basin chronostratigraphy of, 536
Palermo sandstones, 382
Pampas Basin, central, 380
Pampas Plains, 380, 385
Pando-Madidi Arch, 393
Papua New Guinea
petroleum potential of, 284–286
petroleum systems of, 281–284
reserves of, 281, 285
Paraná Basin, petroleum geology of, 379–382
Passive margins
in Cretaceous of Egypt, 459–462
in southern South America, 387
in Venezuela, 354–356
Patagonian basins, petroleum geology of, 382–383, 385, 399
Pay-Khov Ridge, 261, 272
Peace River, oil sands of, 119
Peak, of world crude-oil production, 26, 28
Pechora Sea
exploration in, 260–261
proposed offshore terminals in, 263
Pedernales field, 360
People's Republic of China, *see* China
Peri-Caspian Basin, *see* North Caspian Basin
Permafrost, gas hydrates of, 92–96, 97, 99
Permeability
of Archie rock types, 69
chlorite cements and, 545
versus porosity in TAGI sandstones, 550
Permian, of North Sea gas province, 179
Permian Basin
secondary recovery in, 66–69
waterfloods in, 66–71, 83
Petroleum basins, *see also* Sedimentary basins
of Arctic, 17, 125–127
of Australia, 289, 290–291
of China, 320
of Greater Rocky Mountain Region, 212
of United Kingdom, 171, 182–185, 188
Petroleum consumption
of United States, 27–28
of world, 22, 23
Petroleum exploration, *see* Exploration
Petroleum history, *see* History
Petroleum industry, *see* Exploration; Production; History
Petroleum potential
of Arabian Plate, 502–504
of Egypt, 465–474
of Greater Rocky Mountain Region, 204, 223–225, 233–234
of Iran, 424–426
of Junggar Basin, China, 332–333
of Kazakhstan, 243–257
of Komi Republic, Russia, 266–277
of Lord Howe Rise, 301, 308–309
of Lower Congo Basin, 517–520, 528–529
of Mexican portion of Gulf of Mexico Basin, 337–349
of northern Alaska, 144–161, 160–161
of Northwest Territories, 121
of Russia, 259–278
of Tarim Basin, China, 331–332
of Timan-Pechora Basin, 266–278
of United Kingdom, 170–195, 194
of Yukon, 121
Petroleum production, *see* Production
Petroleum reserves, *see* Reserves
Petroleum resources, *see* Resources
Petroleum systems
of Arabian Plate, 503
of Asia-Australasia region, 282–284
of Australia, 290–292, 295, 296–297
of Berkine Basin, 538–553
of Canada, 115–127, 121–127
of Chaco-Paraná Basin, 382
of Egypt, 460
of Foz do Amazonas Basin, 407
of Greater Rocky Mountain Region, 209–215, 212–215
of Illizi Basin, 538–553
of Indonesia and Papua New Guinea, 281–284
of Lower Congo Basin, 522–526
of Libya, 431–448
of northern Alaska, 143, 144–145
of Venezuela, 356–358
Pinda Formation, development of, 520
Pipelines
Alliance (Canada), 114–115
in Argentina, 377
in southern South America, 376–378
for Timan-Pechora Basin, 262
Planicie Abisal, 349
Plataforma de Yucatán, 345
Plate-margin basins, of southern South America, 389–398

Plate-tectonic reconstructions, of Iran, 420, 421
Platform-incised rifts, of southern South America, 389–391
Play analysis, of Timan-Pechora Basin, 266–278
Plays, classification system of, 267
Population growth, and oil consumption of world, 22
Pore pressure, and gas hydrates, 100
Porosity
  of Archie rock types, 69
  chlorite cements and, 546
  versus permeability in TAGI sandstones, 550
Postrift, of North Sea oil province, 177–179
Powder River Basin, coal-bed play of, 229–230
Pratt, Wallace E., contributions of, 11–12
Pre-Cretaceous Weathered Zone play (Venezuela), 368
Prerift, of North Sea oil province, 170–175
Pridorozhnoye gas field, 256
Prirazlomnoye field, 278
Production; *see also* History
  of 1990s, 41
  Auk field's history of, 82
  of Canada, 112–113, 118, 128–130
  China's history of, 319–321
  future levels of, 41–43, 52; *see also* Future
  of future natural gas, 51
  Hubbert's model of, 50
  Kazakhstan's history of, 244–245
  of Venezuela, 354
  of Western Canada Sedimentary Basin, 115–121
  world basins' status of, 14
  world capacity of, 35
  of world crude oil, 22, 24, 52
Production-decline curves, 65
Prudhoe Bay
  exploration in, 142
  gas hydrates of, 93–94
Punta del Este Basin, 389
Pyramids
  for energy, 48, 53
  for hydrocarbon energy, 48
  for resources for Greater Rocky Mountain Region, 223–225

Q

Qatar
  geologic cross section of, 501
  stratigraphy of, 490
Qigu field, 330
Quartz sandstone reservoirs, of Berkine Basin, 543–549
Qunkuqiake field, 329

R

Rachmat Formation, isopach map of, 434
RADARSAT-1 imagery, of Foz do Amazones Basin, 412, 413
Radioactive shales, of Berkine Basin, 538, 541, 553
Recovery rates, increase of, 43
Red Sea
  breakup of, 463
  exploration potential of, 469–470
  regional gravity interpretation of, 476
  seismic line of, 476
Redevelopment, *see also* Secondary recovery
  in Central Graben, 81, 82
  in Gulf of Mexico Basin, 69–75
  in United States, 63–83
  in Venezuela, 75–81
Remote sensing, oil-seep detection by, 412–413
Renewable energy sources, 31–32
Republic of Congo, *see also* Lower Congo Basin
  exploration history of, 526–527
  geologic history of, 520–526
Reserve additions
  economics of, 83
  from secondary recovery, 66, 71, 75, 80, 82–83
Reserves
  of Australia, 290–296
  of Canada, 113–114
  of Egypt, 460
  of Greater Rocky Mountain Region, 232
  of Indonesia and Papua New Guinea, 281, 285
  of Kazakhstan, 244–245, 245–256
  of Timan-Pechora Basin, 270
  undiscovered potential, 39–40
  of United Kingdom, 169, 195
  of United States, 28, 47
  of Venezuela, 354
Reservoir models, for gas hydrates, 98–99
Reservoir potential, *see* Petroleum potential
Reservoirs
  of Arabian Plate, 502–504
  of Berkine Basin, 540–549
  of Foz do Amazonas Basin, 408
  of Greater Rocky Mountain Region, 209–212, 217–219
  of Gulf of Mexico sands, 73
  of Illizi Basin, 540–549
  of Junggar Basin, 331
  of North Sea oil province, 170–179
  of Tarim Basin, 330–331
Resource pyramid
  for Greater Rocky Mountain Region, 223–225
  for world, 224
Resources
  of Alaska, 139
  of Australia, 292
  of Berkine Basin, 553–555
  of Canada, 111–112, 128–131
  of gas hydrates, 96–106
  of Greater Rocky Mountain Region, 204, 233–234
  of Kazakhstan, 246
  of unconventional oil, 48–49
  of United Kingdom, 169
  of United States, 47
  of world, 47
Río Irati black shales, 382
Rifting
  in Congo Basin, 520
  in Gulf of Suez, 462–463
  in Jurassic of Egypt, 459
  in offshore Venezuela basin, 369
  in southern South America, 389–391
Río Mayo Basin, 396–398
Rockall Plateau, 193–195
Rockall Trough, exploration in, 191–193

Rosario Basin, 380
Russia, petroleum potential of, 259–278

S

Sable Offshore Energy Project, 114, 122
Sahara Desert, exploration in, 532–555
Salado Basin
 failed arms of, 388
 map of, 389
Salina del Golfo Profundo
 petroleum geology of, 348
 seismic section of, 350
Salina del Istmo province, 343–344
Salt plug, in Iran, 423
San Andres reservoir (Permian Basin), 67–71
San Jorge Basin, petroleum geology of, 385
San Juan Basin, underpressures in, 216, 221
Santa Rosa Formation, 391
Saudi Arabia, *see also* Arabian Plate
 geologic cross section of, 501
 Hawtah field in, 508
 Khuff reservoirs of, 505
 Paleozoic section of, 504
 source rocks of, 425
 stratigraphy of, 425, 490, 491, 505
Scarab field, discovery of, 479
Scarborough field, 294, 305
Scotian Shelf, petroleum systems of, 121–122
Scott Reef field, 293, 302
Sea-level curves, for Lower Congo Basin Miocene, 522
Sealing faults, in Lake Maracaibo reservoir, 77, 78
Secondary recovery, *see also* Redevelopment
 in Gulf of Mexico Basin, 69–75
 in Louisiana, 69–75
 in Permian Basin, 66–69
 reserve additions from, 66, 71, 75, 80, 82–83
 by WAG injection, 80
Sedimentary basins, *see also* Petroleum basins
 of Arabian Plate, 489–499
 of Canada, 110, 115–127
 of China, 320
 of Lower Congo Basin, 520–526
 of northern Alaska, 138
 production status worldwide of, 14
 of southern South America, 373, 374
Seismic technology, *see* Technology
Serra Geral basalts, 380
Serrania del Interior, 362
Serravalian paleogeography, of Egypt, 465
Shapkino-Yuraga swell, 276, 277
Shortages, *see* Oil shortages
Sierra Madre Oriental thrust belt, 341
Siliciclastic plays, in Timan-Pechora Basin, 274
Sinai, exploration potential of, 468–469, 471–474
Sirt Basin
 exploration of, 429–430
 isopach maps for, 434, 435, 436
 petroleum systems of, 432–440
 structural map of, 432, 437
 tectonic setting of, 431
Sirt Formation, isopach map of, 434
Sirt-Tibesti arch, 431
Smith Bank Graben, 176
Solan/Strathmore discoveries, 189
Sonda de Campeche province, 345
Songliao Basin, map of, 320
Sorell Basin, petroleum potential of, 301
Sorokin swell, 273, 277
Source rocks
 of Berkine Basin, 538–540
 of Foz do Amazonas Basin, 409–412
 of Greater Rocky Mountain Region, 209–212, 217, 219–220
 of Indonesia and Papua New Guinea, 284
 of Iran, 423–424, 425
 of Junggar Basin, 323–324, 326
 of Lower Congo Basin, 520, 523, 525, 526
 of Libya, 433, 441, 442, 444, 450
 of North Sea, 173
 of northern Alaska, 145–146, 148, 150, 151
 of Oman, 502
 of Tarim Basin, 321–323, 324, 325
 of Timan-Pechora Basin, 265, 267
 of Venezuela, 358
South America, tectonic history of, 354–355
South America (southern)
 basins of, 373, 374
 exploration in, 378
 foreland basins of, 389–398, 392
 intracratonic basins of, 378–385
 megasequences of, 381–382
 plate-margin basins of, 389–398
 rift basins of, 385–389
 petroleum history of, 373–378, 398–401
 relief map of, 374
 tectonic setting of, 375
South Andean Mountain Front play, 366
South Atlantic Ocean
 oil-type distribution of, 407
 petroleum frontier of, 403–413
South Tasman Rise
 map of, 312
 petroleum potential of, 301
South Turgay Basin, reserves of, 253–254
Southern Central Mountain Range play (Venezuela), 362
Southern Eastern Mountain Range play (Venezuela), 361
Southern Patagonian basins, 399
Southern South America, *see* South America (southern)
Springhill Platform, 398
Stable margin, of Beaufortian sequence, 154
Steam-assisted gravity drainage (SAGD), for oil-sand recovery, 120
Step-out exploration, of Gulf of Suez, 475
Strunian-Viséan quartz sandstones, of Berkine Basin, 545–546, 547
Subduction, of African Plate, 462
Subduction basins, of southern South America, 389–398, 401
Submarine fan, of Lower Congo Basin, 524
Subtle traps, search for, 15–16
Suncor, production of oil sands, 115, 119
Sunrise field, 293, 303
Supersystems, of Australia, 290–292, 295, 296–297
Sureste Basins, petroleum provinces of, 342–345

Sustainable energy supply, 32
Sverdrup Basin, petroleum systems of, 126–127
Sweet spots, in Greater Rocky Mountain Region, 214, 226
Syncrude, production of oil sands, 115, 119
Synrift
  of North Sea oil province, 175–177
  of Gulf of Suez, 474
Syria, *see also* Arabian Plate
  geologic cross section of, 500
  stratigraphy of, 490
Syrian Arc
  deformation of, 462, 465
  structural reactivation of, 465

T

TAGI quartz sandstone
  cycles of, 548
  deposition of, 546–548
  exploration of, 534, 536
  hydrocarbons in, 549
  porosity versus permeability in, 550
TAGS quartz sandstone, exploration of, 534
Tampico-Misantla Basin, 339–341
TAPS, *see* Trans-Alaska Pipeline System
Tar sands, 48–49
Tarim Basin
  exploration potential of, 331–332
  generation potential of, 324
  map of, 320, 321
  petroleum potential of, 331–332
  reservoir potential of, 330–331
  source-rock potential of, 321–323, 324, 325
Tasmania, offshore potential of, 301
Technology
  of gas-hydrate production, 103–105
  failure of, 58–59
  new in Greater Rocky Mountain Region, 220–223
Tectonic setting
  of Egypt, 458–464, 461
  of Libya, 430–431
  of northern Alaska, 138–139, 140
  of South America, 354–355, 375
  of South Atlantic Ocean, 407–409
  of Venezuela, 355
Temsah field, seismic profile of, 477
Tengiz oil field, 244, 248
Terra Nova field, of Canada, 114
Terrestrial gas hydrates, 92–96, 103
Texas, mature fields of, 64
Thermal maturity, of source rocks in China, 324–325, 332
Thrust belt, of Brooks Range, 157–160
Thrust faults, of Foz do Amazonas Basin, 408
Tierra del Fuego, 398
Tight-gas reservoirs, of Greater Rocky Mountain Region, 209–210, 217
Timan Ridge, 261
Timan-Pechora Basin
  annual production from, 269
  exploration history of, 260–262
  field size in, 268, 271
  geologic setting of, 262–265
  maps of, 261, 272
  offshore petroleum potential of, 277–278
  onshore petroleum potential of, 266–277
  play analysis of, 266–278
  source rocks of, 265, 267
  stratigraphic sections of, 264, 265, 266
Timing generation, of source rocks in China, 325–326
Tinat field, structural cross section of, 507
Topset facies, of Brookian Sequence, 153–154, 156, 157
Townsville Basin, 311
Trans-Alaska Pipeline System (TAPS), 138, 142
Traps
  of Greater Rocky Mountain Region, 212–214
  of northern Alaska, 146, 149, 151
  subtle, 15–16
  of Venezuela, 359–369
Trias Argilo Greseux Inferieur, *see* TAGI quartz sandstone
Trias Argilo Greseux Superieur, *see* TAGS quartz sandstone
Triassic quartz sandstones, of Berkine Basin, 546
Trinidad field, 360
Triple junctions, of southern South America, 386–388
Tripolitania Basin, 431
  cross section of, 449
  petroleum geology of, 442, 448
  structure map of, 448, 450
Troubadour field, 293, 303
Tumayam trough, petroleum systems of, 437–440
Tunisia, offshore stratigraphy of, 447
Turbidite channel, on seismic profile, 479
Turbidite plays, of Lower Congo Basin, 520, 522–526, 527
Turgay Basin, South, 253–254
Turkey
  geologic cross section of, 500
  stratigraphy of, 490
Twentieth century, exploration in, 57–59
Twenty-first century, challenges of, 59–60

U

UAE, *see* Arabian Plate
Ukhta pipeline, 262
Ultimate United States liquid petroleum, 24, 25
Ultimate world liquid petroleum, 24, 25, 28
Ultradeep-water prospects, 38
Unayzah Formation, play of, 507–508
Unconventional accumulations, of Greater Rocky Mountain Region, 213–214
Unconventional oil resources, 48–49
Underpressure, in Greater Rocky Mountain Region fluid systems, 215–216, 219, 221
Undiscovered potential reserves of world, 39–40
United Kingdom, *see also* North Sea
  coal-bed methane of, 185
  history of production in, 168–170
  onshore basins of, 182–185
  petroleum potential of, 170–195
United States, *see also* Alaska; Greater Rocky Mountain Region (GRMR)
  crude-oil production in, 52
  energy consumption of, 22, 27–28, 46
  exploration in Rocky Mountain region of, 201–235
  future energy supplies of, 30

gas hydrates of, 85–106
mature field redevelopment in, 64–83
oil production in, 12
petroleum in 1990s, 24
petroleum reserves of, 28
redevelopment of mature fields in, 63–83
reserves versus production in, 47
ultimate liquid petroleum of, 25–26
Unsupervised Semivariogram Textural Classifier (USTC), 413
Upper Egypt, *see also* Upper Egypt
exploration potential of, 468
seismic line and stratigraphy of, 472
structural section of, 471
Ural Mountains, 261, 263
Uranium, as energy source, 53–54
Usa field, 268
USGS provinces, of Greater Rocky Mountain Region, 231–232
Usinsk-Ukhta pipeline, 262
Ustyurt Basin, North, 254–255
Utah, petroleum potential of, 201-235; *see also* Greater Rocky Mountain Region
Uzen field, 244
Cretaceous plays in, 253
Jurassic plays in, 252

V

Van Krevelen–type diagram
for Junggar Basin source rocks, 327
for Tarim Basin source rocks, 325
Venezuela
exploration future of, 358–370
extra-heavy oil in, 42–43
foredeep of, 356, 357
Lake Maracaibo fields, 75–81
offshore exploration in, 369
petroleum systems of, 356–358
production of, 354
redevelopment of fields in, 75–81
reserves of, 354
source rocks of, 358
tectonosequences of, 355
traps in, 359–369
WAG injection in, 80
Venezuelan Basin, northern offshore, 369
Veracruz Basin
cross section of, 342
petroleum geology of, 341–342
seismic section of, 343
Viking Graben, 172–175
Vitrinite reflectance, for Junggar Basin, 328
VLC-363, Block III, Venezuela, 75–81
Volcanic fields, of Greater Rocky Mountain Region, 220

W

WAG injection, in Venezuela, 80
Wall Street, influence on industry, 58
Waqr field, structural cross section of, 507
Water alternating with gas, *see* WAG injection
Waterfloods
injection patterns of, 70, 71
in Permian Basin, 66–71, 83
Weald Basin, 183, 184
Weathered zone, in Venezuelan play, 368
Well log
of Beaufortian sandstones, 155
of gas hydrates, 92, 95
from Gulf of Mexico Basin, 73
from Mackenzie Delta, 95
from VLC-363 Block III, Venezuela, 79
Wessex Basin, 183, 184
West Africa, offshore exploration of, 517–529
Western Canada Sedimentary Basin, *see also* Canada
oil sands from, 117–121
petroleum systems of, 115–121
Western Desert
drilling history of, 465
exploration potential of, 465–468, 470
field-size distribution for, 467
geochemical data for, 470
structural map of, 469
tectonic setting of, 459, 462
Western Libya, petroleum geology of, 440–448
Westphalian A-C coal measures, of United Kingdom, 179–181
White Rose field, of Canada, 114
Wildcats, of Berkine Basin, 555
Williston Basin, recent discoveries in, 225–226
Wind River Basin, recent discoveries in, 227–228
World
discoveries of 1990s of, 36–38
energy consumption of, 22, 51
exploratory status of basins of, 13
future energy supplies of, 29
gas from hydrates of, 103
oil and gas undiscovered potential of, 39–40
oil resource versus production in, 47
peak liquids production of, 26, 28
petroleum in 1990s of, 23–24
petroleum production capacity of, 35
production status of basins of, 14
Wyoming
recent discoveries in, 227–228, 230
petroleum potential of, 201-235; *see also* Greater Rocky Mountain Region

Y

Yemen, *see* Arabian Plate
Yucatán platform, 345
Yukon, hydrocarbon potential of, 121

Z

Zagros Foreland Basin, 418–420
Zagros Mountains
exploration potential of, 426
fault zone of, 493
fold belt of, 485
tectonic development of, 420, 421
Zallah-Tumayam trough
Cretaceous correlation in, 440
isopach map for, 439
petroleum systems of, 437–440
stratigraphy of, 433
Zhanazhol gas condensate and oil field, 250